The Evolutionary Biology of Hearing

Douglas B. Webster   Richard R. Fay   Arthur N. Popper
Editors

The Evolutionary Biology of Hearing

With 355 Illustrations, 2 in Full Color

Springer-Verlag
New York Berlin Heidelberg London Paris
Tokyo Hong Kong Barcelona Budapest

Douglas B. Webster
Department of Otorhinolaryngology
Louisiana State University Medical Center
New Orleans, LA 70112, USA

Arthur N. Popper
Department of Zoology
University of Maryland
College Park, MD 20742, USA

Richard R. Fay
Parmly Hearing Institute
and
Loyola University of Chicago
Chicago, IL 60626, USA

Cover Illustration: William N. Tavolga

Library of Congress Cataloging-in-Publication Data
The evolutionary biology of hearing / Douglas B. Webster, Richard R.
   Fay, Arthur N. Popper, editors.
        p. cm.
     Based on a conference held at the Mote Marine Laboratory in
   Sarasota, Fla., May 20–24, 1990.
      Includes bibliographical references and index.
      ISBN 0-387-97588-8 (alk. paper). — ISBN 3-540-97588-8 (alk.
   paper)
      1. Ear—Evolution—Congresses.   2. Hearing—Congresses.
   3. Physiology, Comparative—Congresses.   I. Webster, Douglas B.
   II. Fay, Richard R.   III. Popper, Arthur N.
      [DNLM: 1. Ear—congresses.   2. Evolution—congresses.   3. Hearing—
   congresses.   4. Histology, Comparative—congresses.
   5. Invertebrates—congresses   6. Physiology, Comparative—
   congresses.   7. Vertebrates—congresses. WV 270 E92 1990]
   QP460.EP96   1992
   591.1′825—dc20
   DNLM/DLC
   for Library of Congress   91-4805
                                                                     CIP

Printed on acid-free paper.

Production managed by Terry Kornak; manufacturing supervised by Rhea Talbert.
Typeset by Publishers Service of Montana Inc., Bozeman, MT.
Printed and bound by Edwards Brothers Inc., Ann Arbor, MI.
Printed in the United States of America.

9 8 7 6 5 4 3 2 1

ISBN 0-387-97588-8 Springer-Verlag New York Berlin Heidelberg
ISBN 3-540-97588-8 Springer-Verlag Berlin Heidelberg New York

This volume is dedicated to the memory of Professor Ernest Glen Wever (1902–1991), a pioneer in auditory research during much of the 20th century, and one of the few people who kept the flame of evolutionary biology alive in the hearing sciences. Without doubt, his ideas and work have had a major influence on all the chapters in this volume.

Meeting Attendees

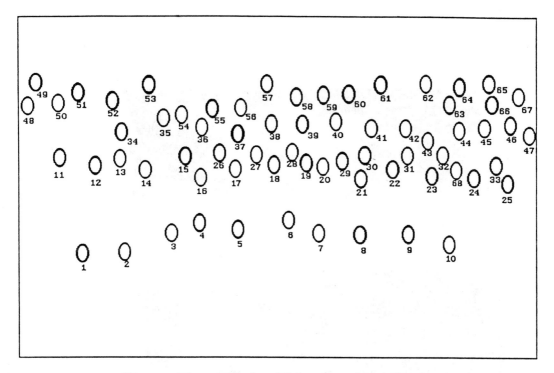

Key to Photograph of Meeting Attendees

1. W. Tavolga
2. A. Popper
3. D. Webster
4. R. Fay
5. R. Northcutt
6. C. Platt
7. R. Hoy
8. N. Fuzessery
9. J. Hall
10. R. Masterton
11. M. Lenhardt
12. J. Lu
13. J. Crawford
14. C. Köppl
15. K. Keller
16. C. Carr
17. T. Bullock
18. M. Miller
19. K. Brändle
20. A. Mason
21. S. Yack
22. L. Gunn
23. Z. Wollberg
24. J. Clack
25. K. Mahadevan
26. W. Yost
27. J. Hopson
28. C. Schreiner
29. H. Heffner
30. G. Pollak
31. J. Christensen-Dalsgaard
32. J. Bolt
33. E. Lewis
34. G. Manley
35. T. Lewis
36. N. Schellart
37. A. Musicant
38. C. Gans
39. T. Hetherington
40. E. Lombard
41. H. Römer
42. B. Lewis
43. R. Heffner
44. G. Meredith
45. W. Plassmann
46. M.G. Sneary
47. M. Weiderhold
48. C. McCormick
49. R. Dooling
50. S. Coombs
51. C. von Bartheld
52. W. Stebbins
53. E. Allin
54. J. Rosowski
55. J. Fullard
56. P. Narins
57. A. Michelsen
58. B. Budelmann
59. R. Williamson
60. J. Montgomery
61. B.L. Roberts
62. G. Patton
63. D. Ketten
64. S. Frost
65. J. Jannsen
66. P. Edds
67. K. Cortopassi
68. B. Fritzsch

Preface

To develop a science of hearing that is intellectually satisfying we must first integrate the diverse, extensive body of comparative research into an evolutionary context. The need for this integration, and a conceptual framework in which it could be structured, were demonstrated in landmark papers by van Bergeijk in 1967 and Wever in 1974. However, not since 1965, when the American Society of Zoologists sponsored an evolutionary conference entitled "The Vertebrate Ear," has there been a group effort to assemble and organize our current knowledge on the *evolutionary*—as opposed to comparative—biology of hearing.

In the quarter century since that conference there have been major changes in evolutionary concepts (e.g., punctuated equilibrium), in systematics (e.g., cladistics), and in our understanding of hearing (e.g., hair cell mechanisms). Moreover, the study of hearing and the ear has matured to the point where peripheral and central investigations are recognized as of equal importance and are often coordinated.

With these considerations in mind, we realized that the time was ripe for an international conference on the Evolutionary Biology of Hearing, which would have a threefold purpose: to focus on the evolutionary implications of comparative studies; to integrate central and peripheral auditory studies; and to include both vertebrates and invertebrates.

The five-day conference was held at the Mote Marine Laboratory in Sarasota, Florida, May 20–24, 1990. The invited participants came from the fields of comparative anatomy, physiology, biophysics, animal behavior, psychophysics, evolutionary biology, ontogeny, and paleontology. Before the conference, preliminary manuscripts of the invited papers were distributed to all participants. This facilitated—even encouraged—discussions throughout the conference which could be called, among other things, "lively." The preview of papers, along with the free exchange of information and opinion, also helped improve the quality and consistency of the final manuscripts included in this volume.

In addition to the invited papers, several studies were presented as posters during evening sessions. The poster abstracts appear at the end of each appropriate group of papers.

The final half-day of the conference was devoted to discussion, in order to allow time for topics not previously heard, to attempt consensus on controversial topics, and to suggest some fruitful avenues of future research. This final session was videotaped and a summary is presented in the final chapter of this volume.

Douglas B. Webster
Richard R. Fay
Arthur N. Popper
January, 1991

Acknowledgments

We are grateful to the National Institute of Deafness and Other Communicative Disorders (Grant NS 1-R13-DC00667), the National Science Foundation (Grant BNS-8912389) and the Office of Naval Research (Grant N00014-90-J1799) for providing funds to support the conference that led to this volume.

Without doubt the conference could not have taken place without the guidance, creativity, expertise, and support of Dr. William N. Tavolga, Senior Scientist at Mote Marine Laboratory. Dr. Tavolga has become an expert on how to make conferences work. He devoted enormous amounts of time and effort to assure that every detail was taken care of so that the conference attendees, including the editors, had no cares except scientific interactions.

We are also grateful to Dr. Selvakumaran Mahadevan, Director of the Mote Marine Laboratory. He provided outstanding facilities and the cooperation of his staff, and was always available to help us with problems. Without him, and them, the conference could not have been as successful.

While everyone at Mote has our thanks, a few deserve special mention: Dr. Robert Heuter for extraordinary help in many facets of the organization of the meeting; Mr. Daniel Bebak, Curator of the Mote Aquarium; lab photographer, Ms. Carmen Harris; Ms. Linda Franklin and Ms. Pamela James, Administrative Assistants; and Mr. Paul Shrader, Business Manager. We are most grateful to Ms. Sandy Hingtgen of the Holiday Inn/Lido Beach for doing everything possible to make our stay in Sarasota a pleasant one, and to the Sarasota Outboard Club for use of their facilities during part of the meeting.

Finally, all the participants thank Chef Alain Mons of the French Affair Delicatessen in Sarasota for once again turning a meeting into a sensory experience. By delighting our senses of taste and smell at mealtimes, he made the long intervening sessions devoted to hearing much easier.

Contents

SECTION III   ASPECTS OF HEARING AMONG VERTEBRATES

Section IV   Anamniotes

SECTION VI   MAMMALS

SECTION VII   EPILOGUE

Detailed Chapter Contents

SECTION IV ANAMNIOTES

Chapter 36 Adaptations of Basic Structures and Mechanisms
            in the Cochlea and Central Auditory Pathway
            of the Mustache Bat

SECTION VII. EPILOGUE

Chapter 37   Epilogue to the Conference on the Evolutionary
             Biology of Hearing

Contributors

Edgar F. Allin
Department of Anatomy, Chicago College of Osteopathic Medicine, Chicago, IL 60615, USA

H. Bleckman
Universität Bielefeld, Fakultät für Biologie, Lehrstuhl für Neurophysiologie, 4800 Bielefeld, FRG

John Bolt
Department of Geology, Field Museum of Natural History, Chicago, IL 60605, USA

G.S. Boylan
Molecular Neurobiology Group, Research School of Biological Sciences, Australian National University, Canberra City, ACT 2601, Australia

Kurt Brändle
Zoologisches Institüt, J.W. Goethe Universität, D-6000 Frankfurt/Main, FRG

Bernd U. Budelmann
The Marine Biomedical Institute, The University of Texas Medical Branch at Galveston, Galveston, TX 77550-2772, USA

Theodore H. Bullock
Department of Neurosciences A-001, University of California at San Diego, La Jolla, CA 92093-0201, USA

Catherine E. Carr
Department of Zoology, University of Maryland, College Park, MD 20742, USA

J.C.K. Chan
Department of Neurophysiology, University of Wisconsin Medical School, Madison, WI 53706, USA

Jacob Christensen-Dalsgaard
Institute of Biology, University of Odense, DK-5230 Odense M, Denmark

Jennifer A. Clack
University of Cambridge, Museum of Zoology, Cambridge, CB2 3EJ, England, UK

Alex M. Clark
Department of Otolaryngology and Division of Biomedical Engineering, Medical College of Virginia, Virginia Commonwealth University, Richmond, VA 23298, USA

Sheryl Coombs
Parmly Hearing Institute, Loyola University of Chicago, Chicago, IL 60626, USA

John D. Crawford
Parmly Hearing Institute, Chicago, IL 60626, USA

Robert J. Dooling
Department of Psychology, University of Maryland, College Park, MD 20742, USA

Peggy L. Edds
Department of Zoology, University of Maryland, College Park, MD 20742, USA

Andreas Elapfandt
Fakultät für Biologie, Universität Konstan D-7750 Konstanz, FRG

Richard R. Fay
Parmly Hearing Institute and Loyola University of Chicago, Chicago, IL 60626, USA

Bernd Fritzsch
Creighton University, Department of Biomedical Sciences, Division of Anatomy, Omaha, NE 68178, USA

Shawn B. Frost
Department of Psychology, Florida State University, Tallahassee, FL 32306, USA

James H. Fullard
Department of Zoology, Erindale College, University of Toronto, Mississauga, Ontario, Canada

Z.M. Fuzessery
Department of Zoology, University of Wyoming, Laramie, WY 82071, USA

Carl Gans
Department of Biology, The University of Michigan, Ann Arbor, MI 48109, USA

Edmund Gerstein
Mote Marine Laboratory, Sarasota, FL 33577, USA

Otto Gleich
Institüt für Zoologie, Technische Universität Munchen, 8046 Garching, FRG

Steven Greenberg
Department of Neurophysiology, University of Wisconsin, Wisconsin Medical School, Madison, WI 53706, USA

Henry E. Heffner
Department of Psychology, University of Toledo, Toledo, OH 43606, USA

Rickye S. Heffner
Department of Psychology, University of Toledo, Toledo, OH 43606, USA

Thomas E. Hetherington
Department of Zoology, Ohio State University, Columbus, OH 43210-1293, USA

D.E. Hind
Department of Neurophysiology, University of Wisconsin Medical School, Madison, WI 53706, USA

James A. Hopson
Department of Anatomy, University of Chicago, Chicago, IL 60637, USA

Ronald R. Hoy
Section of Neurobiology and Behavior, Cornell University, Ithaca, NY 14853, USA

John Janssen
Department of Biology, Loyola University, Chicago, IL 60626, USA

Darlene R. Ketten
Department of Otology and Laryngology, Harvard Medical School, Massachusetts Eye and Ear Infirmary, Boston, MA 02114, USA

Christine Köppl
Institüt für Zoologie, Technische Universität Munchen, 8046 Garching, FRG

Martin L. Lenhardt
Medical College of Virginia, Richmond, VA 23298-0168, USA

Edwin R. Lewis
Department of Electrical Engineering and Computer Science, University of California, Berkeley, CA 94720, USA

Brian Lewis
Faculty of Life and Environmental Sciences, City of London Polytechnic, London E1 7NT, England, UK

Eric Lombard
Department of Anatomy, University of Chicago, Chicago, IL 60637, USA

Selvakumaran Mahadaven
Mote Marine Laboratory, Sarasota, FL 33577, USA

Geoffrey A. Manley
Institut für Zoologie, Technische Universität Munchen, 8046 Garching, FRG

Andrew C. Mason
Department of Biology, Erindale College, University of Toronto, Mississauga, Ontario, Canada, L5L 1C6

R. Bruce Masterton
Department of Psychology, Florida State University, Tallahassee, FL 32306, USA

Catherine A. McCormick
Department of Biology, Oberlin College, Oberlin, OH 44074, USA

Gloria E. Meredith
Vrije Universiteit, Faculteit Der Geneeskunde, Laboratorium Voor Anatomie en Embryologie, 1007 MC Amsterdam, The Netherlands

Michael M. Merzenich
Coleman Memorial Laboratory, University of California, San Francisco, CA 94143, USA

Axel Michelsen
Institute of Biology, Odense University, DK 5230 Odense M, Denmark

Malcolm R. Miller
Department of Anatomy, University of California, San Francisco, CA 94143, USA

John Montgomery
Zoology Department, University of Auckland, Auckland, New Zealand

H.M. Müller
Universität Bielefeld, Fakultät für Biologie, Lehrstuhl für Neurophysiologie, 4800 Bielefeld, FRG

Alan D. Musicant
Department of Neurophysiology, University of Wisconsin Medical School, Madison, WI 53706, USA

Peter M. Narins
Department of Biology, University of California, Los Angeles, CA 90024, USA

U. Niemann
Universität Bielefeld, Fakultät für Biologie, Lehrstuhl für Neurophysiologie, 4800 Bielefeld, FRG

R. Glenn Northcutt
Department of Neurosciences, A-001, University of California, San Diego, La Jolla, CA 92093, USA

Geoffrey W. Patton
Mote Marine Laboratory, Sarasota, FL 33577, USA

Wolfgang Plassmann
Fachbereich Biologie, J.W. Goethe - Universitat Zoologie, 6000 Frankfurt am Main, FRG

Christopher Platt
Program Director of Sensory Systems, National Science Foundation, Washington, D.C. 20550, USA

George D. Pollak
Department of Zoology, University of Texas at Austin, Austin, TX 78712, USA

Arthur N. Popper
Department of Zoology, University of Maryland, College Park, MD 20742, USA

William S. Rhode
Department of Neurophysiology, University of Wisconsin, Madison, WI 53706, USA

Barry L. Roberts
Department of Experimental Zoology, University of Amsterdam, 1098 SM Amsterdam, The Netherlands

Heiner Römer
Ruhr-Universitat Bochum, Fakultat fur Biologie, Allg. Zoologie und Neurobiologie, D-4630 Bochum 1, FRG

John J. Rosowski
Eaton-Peabody Laboratory, Massachusetts Eye and Ear Infirmary, Boston, MA 02114, USA

Edwin W. Rubel
Hearing Development Laboratories, University of Washington RL-30, Seattle, WA 98195, USA

Nico Schellart
A.M., Laboratory of Medical Physics, University of Amsterdam, 1105 AZ Amsterdam, The Netherlands

Klaus Schildberger
Department of Biology, Erindale College, University of Toronto, Mississauga, Ontario, Canada L5L 1C6

Christoff E. Schreiner
Coleman Memorial Laboratory, University of California, San Francisco, CA 94143, USA

Michael G. Sneary
Department of Electrical Engineering and Computer Science, University of California, Berkeley, CA 94720, USA

Mitchell S. Sommers
Department of Psychology and Kresge Hearing Research Institute, University of Michigan, Ann Arbor, MI 48109-0506, USA

William C. Stebbins
Department of Psychology and Kresge Hearing Research Institute, University of Michigan, Ann Arbor, MI 48109-0506, USA

William N. Tavolga
Mote Marine Laboratory, Sarasota, FL 33577, USA

Christopher von Bartheld
Hearing Development Laboratories, Department of Otolaryngology, RL-30, University of Washington, Seattle, WA 98195, USA

Douglas B. Webster
School of Medicine in New Orleans, Louisiana State University Medical Center, New Orleans, LA 70112-2234, USA

Molly Webster
School of Medicine in New Orleans, Louisiana State University Medical Center, New Orleans, LA 70112-2234, USA

T.L.D. Williams
Max-Planck Institüt für Verhalttenphysiologie, D-8130 Seewiesen, FRG

Roddy Williamson
The Marine Laboratory, Citadel Hill, Plymouth PL1 2PB, England, UK

Jayne E. Yack
Department of Zoology, Erindale College, University of Toronto, Mississauga, Ontario, Canada L5L 1C6

William A. Yost
Parmly Hearing Institute, Loyola University of Chicago, Chicago, IL 60626, USA

Ernest Glen Wever: A Brief Biography and Bibliography

Richard R. Fay

1. Introduction

The international symposium "Evolutionary Biology of Hearing" (May 20–24, 1990, Sarasota, FL), and this volume are dedicated to Ernest Glen Wever. This brief biographical sketch and bibliography were compiled for the participants of the meeting to introduce Glen Wever to younger scientists, and to document his contributions to the field of the evolutionary biology of hearing. Scholars wishing to delve deeper into the history of Wever's scientific contributions should contact The Director, Archives of the History of American Psychology, University of Akron, Ohio, where many of Wever's notes, books, and papers are collected. Scholars wishing to study Wever's vast and well-organized histological slide collection, primarily on the heads and ear regions of reptiles and amphibians, should contact the Division of Reptiles and Amphibians, U.S. National Museum of Natural History, the Smithsonian Institution, Washington, DC. Some carcasses are also conserved for species identification. Additional tissue from Wever's research collection can be found at the Carnegie Museum, Division of Amphibians and Reptiles.

2. Biographical Sketch

Ernest Glen Wever was born October 16, 1902 in Benton, IL, the son of Ernest Sylvester and Mary Schurtz Wever. He received an A.B. degree from Illinois College in 1922, and the MA and PhD degrees from Harvard University in 1924 and 1926. In 1928 he was married to Suzanne Rinehart.

Wever taught Science at Roanoke, IL High School (1922–1923), and at Gunnery Preparatory School in Washington, CT (1924–1925). After one year as an instructor in Psychology at U.C. Berkeley (1926–1927), Wever was hired as an instructor of Psychology at Princeton University in 1927. He was promoted to Assistant Professor in 1929, and to Associate Professor in 1931. During 1936 to 1937, Wever worked on a fellowship at the Otological Research Laboratory at Johns Hopkins University with Stacy Guild and S.J. Crowe. In 1941, he was promoted to Professor, and named the Dorman T. Warren Professor (1946–1950), and the Eugene Higgins Professor (1950–1971). Wever was a consultant to the National Defense Research Committee on anti-submarine warfare during World War II. He was a Research Associate at the Lempert Institute of Otology in New York (1947–1957) and served as Psychology Department Chair from 1955 to 1958.

Wever was elected to the National Academy of Sciences and the National Academy of Arts and Sciences, and was a fellow of the Acoustical Society of America and the Society of Experimental Psychologists. He was a member of the American Psychological Association, the Association for Research in Otolaryngology, and the American Otological Society.

Wever had received the following honors: Howard Crosby Warren Medal from the Society of Experimental Psychologists (1931), the George Shambaugh Prize in Otology (1957), The Gold Medal and Certificate of Merit from the American

Otological Society (1959), Honors of the Association from the American Speech and Hearing Association (1967), The Beltone Institute for Hearing Research Award (1969), and was the Guest of Honor at the 104th meeting of the American Otological Society (1971). In 1981 Wever received the Silver Medal in Psychological and Physiological Acoustics from the Acoustical Society of America, and in 1983 received the Award of Merit from the Association for Research in Otolaryngology.

Wever's first experiments were in vision where the figure and ground distinctions (and other Gestalt concepts) in perception were influential at the time. This led to an interest in auditory masking, and to a valuable relationship with Wegel and others at Bell Labs.

During the late 1920s, Charles W. Bray returned to Princeton after a summer spent learning neurophysiology with Forbes at Harvard Medical School. Both Wever and Bray were interested in the problem of pitch perception and frequency representation in the nervous system, and their experiments resulted in the discovery of the cochlear potential. With an electrode placed near the auditory nerve of the anesthetized cat and the potentials monitored using an audio system, Bray heard Wever's voice transduced by the hair cells of the cat's ear. This resulted in the paper "Auditory nerve impulses" (*Science*, 1930, 71, p. 215, 191 words long), and the "Wever-Bray Effect." This was the beginning of cochlear electrophysiology. The general scientific response was strongly critical, and Wever and Bray spent the next few years eliminating possible artifacts. Wever had commented that Adrian visited the lab during this period and ". . . approved of everything he saw. . . ."

I recall Wever telling me that the result he got (faithful temporal encoding of the sound waveform) was unexpected—his initial motivation for the experiment was to demonstrate that the evidence would not support the improbable "telephone theory" of Rutherford. He expected to rule out a temporal code for frequency. (I have always found it interesting that von Békésy related to me a similar story about his experiments leading to the description of the traveling wave. He also got unexpected results—his initial motivation for the experiments was to demonstrate that a place-principal based on resonance could not operate in the ear. He expected to rule out a Helmholtz-type

spatial code for frequency.) Wever and von Békésy became good friends during the '50s following a period during which Wever was rather critical of von Békésy's work. Von Békésy cherished the new friendship but regretted losing his most valuable critic when relations turned personally warm. Wever translated from the German and edited all von Békésy's papers published up to 1948 to produce the book, *Experiments in Hearing*. This scholarship on Wever's part probably played a role in von Békésy's winning the 1961 Nobel Prize in Physiology and Medicine. Wever also edited von Békésy's last book, *Sensory Inhibition*.

Wever had three or more careers in hearing research: 1) With Charles Bray and Merle Lawrence, Wever discovered and further investigated of the cochlear potential, and used it as a tool for investigating the biomechanical function of the outer, middle, and inner ears. This work culminated in *Physiological Acoustics* (E.G. Wever and Merle Lawrence, Princeton University Press, 1954). This book was a "bible" in its time, and nothing of its scope has been attempted since. 2) Wever reviewed, evaluated, and developed contemporary theories of hearing, including his well-known "volley principle" which he combined with a place principle to account for frequency analysis by the auditory system. This scholarly and theoretical work was published in his *Theory of Hearing* in 1949. This book remains today the most ambitious, complete, and well-written book on hearing. 3) Wever founded, defined, and for several decades maintained the modern field that is represented in this volume, the evolutionary biology of hearing. It is clear from his bibliography that he was interested in evolutionary issues from the start (early '30s), and began studies on both vertebrates and invertebrates. These interests in comparative and evolutionary issues in hearing brought forth *The Reptile Ear* in 1978 and *The Amphibian Ear* in 1985, both from Princeton University Press. These books are unparalleled treatises on the comparative study of the ear. Wever's comparative work on hearing is responsible, directly or indirectly, for the careers of many of the contributors to this volume, and probably for the fact that the National Institutes of Health, the National Science Foundation, and the Office of Naval Research (ONR) funds this sort of research. Wever had one of the first ONR grants.

During the 1960s, Wever's comparative interests came to dominate his work and thinking, and it is during this period many new investigators entered the field of hearing research and applied new approaches and techniques. Wever didn't rush to the most modern of techniques (e.g. single fiber recording, electron microscopy), and remained truly interested in evolutionary questions when many others studying diverse species were simply developing "animal models" to study particular processes or structures. During this time, Wever seemed, to some, to be out of the "main stream" of auditory research which he dominated during the previous decades. With his help, the contributors to this volume, and many others, are carrying on a vigorous new stream of comparative and evolutionary studies on hearing. Glen Wever clearly takes his place as founder of the modern study of the evolutionary biology of hearing, and as an inspiration for all of us.

Some additional personal and professional information about E.G. Wever, including photographs, can be found in *Hearing and Other Senses: Presentations in Honor of E.G. Weaver*, R. Fay and G. Gourevitch (eds.), Amphora Press, Groton, CT, particularly in chapters by Frank Geldard, Lawry Gulick, Jim McCormick, Merle Lawrence, Jack Vernon, and Bob Ruben. This volume presents papers given at a conference in honor of Glen Wever that George Courevitch and I organized in 1982 (May 16–18). The contributors were many of Wever's students and closest scientific friends. For that occasion, a video tape was made of Frank Geldard interviewing Glen about his earliest years in research. This tape was shown at the meeting from which this volume arose.

In the 1950s, Wever delivered a brief lecture on the stimulation of the ear caused by various surgical procedures used at that time that was tape-recorded. The first few sentences of this lecture follow:

"The following recording was made in the Auditory Research Laboratories of Princeton University. Participating in the work were Dr. L.E. Wolfson of Boston Massachusetts, Dr. George von Békésy of Harvard University, and Drs. E.G. Wever, W.F. Strother, and W.E. Rahm of Princeton University. The sounds that you will hear from now on were recorded through the ear of an anesthetized cat."

The lecture was delivered by Wever in one of the surgical rooms at his laboratory, using the cochlear potentials of the cat's ear as an input to the tape recorder, and various drills and chisels were used to illustrate the possibly damaging effects of the noises they produced. This recording was kindly made available to me by Dr. William F. Strother, and has been transferred to cassette at Loyola University. Copies may be obtained from me.

During the 1970s, Wever delivered a series of Langfeld Lectures at Princeton on the reptile ear which were video-recorded by Joseph Pylka of the Department of Psychology.

3. Bibliography

This bibliography is not a piece of historical scholarship, and may contain errors of omission. In many cases, I included items from Wever's curriculum vitae such as book reviews and notices of presentations that were apparently later published. I included these because Wever himself considered them important, possibly for establishing a detailed chronology of ideas.

1927 Wever EG: Figure and ground and the visual perception of form. Am J Psychol 38:194–226.

1928 Wever EG: Attention and clearness in the perception of figure and ground. Am J Psychol 40:51–74.

1928 Wever EG: The effect of a secondary tone upon hearing. Science 67:612–613.

1928 Wever EG, Truman SR: The course of the auditory threshold in the presence of a tonal background. J Exp Psychol 11:98–112.

1928 Truman SR, Wever EG: The judgement of pitch as a function of the series. Univ Calif Publ Psychol 3:215–223.

1928 Wever EG, Zener C: The method of absolute judgement in psychophysics. Psychol Rev 35:466–493.

1929 Wever EG: Beats and related phenomena resulting from the simultaneous sounding of two tones. Psychol Rev 36:402–418.

1929 Wever EG: High-speed lamps for tachistoscopic work. J Gen Psychol 2:553–556.

1929 Zener KE, Wever EG: A multiple-choice apparatus. Am J Psychol 51:647–648.

1930 Wever EG: The upper limit of hearing in the cat. J Comp Psychol 10:221–233.

1930 Wever EG, Bray CW: Auditory nerve impulses. Science 71:215.

1930 Wever EG, Bray CW: Action currents in the auditory nerve in response to acoustical stimulation. Proc Natl Acad Sci USA 16: 344–350.

1930 Wever EG, Bray CW: The nature of acoustic response: The relation between sound frequency and frequency of impulses in the auditory nerve. J Exp Psychol 13:373–387.

1930 Wever EG, Bray CW: Present possibilities for auditory theory. Psychol Rev 37:365–380.

1930 Wever EG, Bray CW: Action currents in the auditory nerve in response to acoustical stimulation; experiments demonstrating the correspondence between sound and nerve impulse. Paper read before the National Academy of Sciences, April 28, 1930. Abstracted in Science, 1930 71:515–516.

1930 Robinson EW, Wever EG: Visual distance perception in the rat. Univ Calif Publ Psychol 4:233–239.

1931 Wever EG: Impulses from the acoustic nerve of the guinea pig rabbit and rat. Am J Psychol 43:457–462.

1931 Wever EG: Auditory nerve experiments in animals and their relation to hearing. Laryngoscope 41:387–391.

1931 Wever EG, Bray CW: Auditory nerve responses in the reptile. Acta Otolaryngol 16:154–159.

1932 Wever EG: Water temperature as an incentive to swimming activity in the rat. J Comp Psychol 14:219–224.

1932 Wever EG, Bray CW: Kreezer and Darge on auditory action currents. Science 75:267.

1932 Wever EG, Bray CW: A note on "A neglected possibility in frequency theories of hearing." Am J Psychol 44:192–193.

1932 Wever EG, Bray CW: Auditory nerve responses and auditory theory. Paper read at meeting of the Acoust Soc Am, May 3, 1932. J Acoust Soc Amer 4:10.

1933 Wever EG, Bray CW: Auditory sensitivity of katydids and crickets. Paper read at NY Branch APA, Yale University, April 1, 1933. Abstracted in Psychol Bull 30:548.

1933 Wever EG: The physiology of hearing: The nature of response in the cochlea. Physiol Rev 13:400–425.

1933 Wever EG, Bray CW: A new method for the study of hearing in insects. J Cell Comp Physiol 4:79–93.

1934 Wever EG, Bray CW, Horton GP: The problem of stimulation deafness as studied by auditory nerve techniques. Science 80: 18–19.

1935 Wever EG: A study of hearing in the sulphur-winged grasshopper (*Arphia sulphurea*). J Comp Psychol 20:17–20.

1935 Wever EG: Book review: Freeman GL *Introduction to Physiological Psychology.* Psychol Bull 32:310–313.

1935 Wever EG: Book review: Stewart GW *Introductory Acoustics.* Psychol Bull 32: 102–104.

1935 Wever EG: Audition. In EG Boring, HS Langfeld and HP Weld *Psychology, A Factual Textbook.* Wiley, New York, pp 102–139 (Chapter 5).

1935 Wever EG, Bray CW, Horton GP: Tone localization in the cochlea. Ann Otol Rhinol Laryngol 44:772–776.

1936 Wever EG, Bray CW: The nature of acoustic response: the relation between sound intensity and the magnitude of responses in the cochlea. J Exp Psychol 19:129–143.

1936 Wever EG, Bray CW: Hearing in the pigeon as studied by the electrical responses of the inner ear. J Comp Psychol 22:353–363.

1936 Wever EG, Bray CW: The nature of bone conduction as shown in the electrical response of the cochlea. Ann Otol Rhinol Laryngol 45:822–530.

1937 Wever EG, Bray CW, Willey CF: The response of the cochlea to tones of low frequency. J Exp Psychol 20:336–349.

1937 Wever EG, Bray CW: The tensor tympani muscle and its relation to sound conduction. Ann Otol Rhinol Laryngol 46:947–961.

1937 Wever EG, Bray CW: A comparative study of the electrical response of the ear. Proc Am Phil Soc 78:407–410.

1937 Wever EG, Bray CW: The perception of low tones and the resonance-volley theory. J Psychol 3:101–114.

1937 Wever EG, Bray CW: A discussion of Ruckmick's critical review of audition. Psychol Bull 34:39–43.

1937 Wever EG, Bray CW: The effects of chemical substances upon the electrical response of the cochlea I The application of sodium chloride to the round window membrane. Ann Otol Rhinol Laryngol 46:291–302.

1937 McCrady E Jr, Wever EG, Bray CW: The development of hearing in the opossum. J Exp Zool 75:503–517.

1938 Wever EG: The width of the basilar membrane in man. Ann Otol Rhinol Laryngol 47:37–47.

1938 Wever EG, Bray CW: Distortion in the ear as shown by the electrical responses of the cochlea. J Acoust Soc Am 9:227–233.

1938 Wever EG, Bray CW: The nature of acoustic response: the relation between stimulus intensity and the magnitude of cochlear responses in the cat. J Exp Psychol 22:1–16.

1939 Wever EG: The electrical responses of the ear. Psychol Bull 36:143–187.

1940 Wever EG, Bray CW, Lawrence M: A quantitative study of combination tones. J Exp Psychol 27:469–496.

1940 Wever EG, Bray CW, Lawrence M: The origin of combination tones. J Exp Psychol 27:217–226.

1940 Wever EG, Bray CW, Lawrence M: The locus of distortion in the ear. J Acoust Soc Am 11:427–433.

1940 Wever EG, Bray CW, Lawrence M: The interference of tones in the cochlea. J Acoust Soc Am 12:268–280.

1940 McCrady E, Wever EG, Bray CW: A further investigation of the development of hearing in the opossum. J Comp Psychol 30:17–21.

1941 Wever EG: The designation of combination tones Psychol Rev 48:93–104.

1941 Wever EG, Bray CW, Lawrence M: The effect of middle ear pressure upon distortion. J Acoust Soc Am 13:1892–187.

1941 Wever EG, Bray CW, Lawrence M: The nature of cochlear activity after death. Ann Otol Rhinol Laryngol 50:317–329.

1941 Wever EG, Lawrence M: Tonal interference in relation to cochlear injury. J Exp Psychol 29:283–295.

1941 Wever EG, Wedell CH: Pitch discrimination at high frequencies. Psychol Bull 38:727.

1942 Wever EG: The problem of the tonal dip. Laryngoscope 52:169–187.

1942 Wever EG, Bray CW, Lawrence M: The effects of pressure in the middle ear. J Exp Psychol 30:40–52.

1942 Wever EG, Bray CW: The stapedius muscle in relation to sound conduction. J Exp Psychol 31:35–43.

1944 Wever EG, Smith KR: The problem of stimulation deafness; I cochlear impairments as a function of tonal frequency. J Exp Psychol 34:239–245.

1946 Wever EG: Audition: In JM Luck and VE Hall (eds) Annual Review of Physiology 8:447–450.

1946 Wever EG: The acoustic characteristics of the ear In Harriman L et al Twentieth Century Psychology: Recent Developments in Psychology New York Philosophical Library Inc pp 371–386.

1947 Wever EG, Neff WD: A further study of the effects of partial section of the auditory nerve. J Comp Physiol Psychol 40:217–226.

1947 Lempert J, Wever EG, Lawrence M: The cochleogram and its clinical applications; a preliminary report. Arch Otolaryngol 45:61–67.

1948 Wever EG, Lawrence M, Smith KR: The effects of negative air pressure in the middle ear. Ann Otol Rhinol Laryngol 57:418–528.

1948 Wever EG, Lawrence M: The functions of the round window. Ann Otol Rhinol Laryngol 57:579–589.

1948 Wever EG, Lawrence M, Smith KR: The middle ear in sound conduction. Arch Otolaryngol 48:19–35.

1949 Wever EG: Theory of Hearing. New York: Wiley.

1949 Wever EG, Lawrence M: The patterns of response in the cochlea. J Acoust Soc Am 21:127–134.

1949 Wever EG, Lawrence M, Hemphill RW, Straut CB: Effects of oxygen deprivation upon the cochlear potentials. Am J Physiol 159:199–208.

1949 Lempert J, Wever EG, Lawrence M, Meltzer PE: Perilymph: Its relation to the improvement of hearing which follows fenestration of the vestibular labyrinth in clinical otosclerosis. Arch Otolaryngol 50:377–387.

1949 Smith KR, Wever EG: The problems of stimulation deafness III. The functional and histological effects of a high-frequency stimulus. J Exp Psychol 39:238–241.

1950 Wever EG: Recent investigations of sound conduction II. The ear with conductive impairment. Paper presented to meeting of Amer Otol Soc May 21, 1950. Ann Otol Rhinol and Otolaryngol 59:1037–1061.

1950 Wever EG: [A review of progress during the period June 1949–June 1950] In CP Stone and DW Taylor (eds) *Annual Review of Psychology* 2:65–75.

1950 Wever EG: Review: Meyer MF *How We Hear; How Tones Make Music*. Amer J Psychol 64:625–626.

1950 Wever EG, Lawrence M: The acoustic pathways to the cochlea. J Acoust Soc Am 22:460–467.

1950 Wever EG, Lawrence M: The transmission properties of the middle ear. Ann Otol Rhinol Laryngol 59:5–18.

1950 Wever EG, Lawrence M: The transmission properties of the stapes. Ann Otol Rhinol Laryngol 59:322–330.

1950 Lawrence M, Wever EG: Recent investigations of sound conduction Part I: The normal ear; Part 2: The ear with conductive impairment. Ann Otol Rhinol Laryngol 59:1020–1062.

1950 Lempert J, Meltzer PE, Wever EG, Lawrence M: The cochleogram and its clinical applications; concluding observations. Arch Otolaryngol 51:307–311.

1951 Wever EG: Some remarks on the modern status of auditory theory. J Acoust Soc Am 23:287–289.

1952 Wever EG: Chapters 25, 26, 27, 28 in Stevens SS *Handbook of Experimental Psychology*. Am J Psychol 65:130–132.

1952 Wever EG, Lawrence M: The place principle in auditory theory. Proc Natl Acad Sci USA 38:133–138.

1952 Wever EG, Lawrence M: Sound conduction in the cochlea. Ann Otol Rhinol Laryngol 61:824.

1952 Lawrence M, Wever EG: Effects of oxygen deprivation upon the structure of the organ of corti. Arch Otolaryngol 55:31–37.

1952 Lempert J, Meltzer PE, Wever EG, Lawrence M, Rambo JHT: Structure and function of the cochlear aqueduct. Arch Otolaryngol 55:134–145.

1952 Lempert J, Wolff D, Rambo J, Wever EG, Lawrence M: New theory for the correlation of the pathology and the symptomatology of Meniere's disease. Ann Otol Rhinol Otolaryngol 61:717–737.

1953 Wever EG, Lawrence M: Auditory theory: An experimental study of the place principle. Bull NY Acad Med 29:159–163.

1954 Wever EG, Lawrence M, Rahm WE Jr: The phase characteristics of the ear. Proc Natl Acad Sci USA 40:209–218.

1954 Wever EG, Lawrence M: *Physiological Acoustics*. Princeton, New Jersey: Princeton University Press.

1954 Wever EG, Lawrence M, von Békésy G: A note on recent developments in auditory theory. Proc Natl Acad Sci USA 40:508–512.

1954 Lempert J, Wever EG, Lawrence M: Are the membranous walls of the endolymphatic labyrinth permeable? Trans Am Acad Opthal Otolaryngol 58:460–465 (Also in Acta Otolaryngol Suppl 116:182–188).

1955 Wever EG: Sound conduction in the ear. Paper presented at ONR Symposium Pensacola, FL March 10–11, 1955, Published in proceedings: Symposium on Physiological Psychology *ONR Symposium Report* ACR-1:202–214.

1955 Wever EG: Book review: Helmholtz H *On the Sensations of Tone*. New York: Dover (1954 reprinting). Scientific Monthly 81:256.

1955 Wever EG: Book review: SS Stevens, JGC Loring and D Cohen (eds) *Bibliography on Hearing* Cambridge: Harvard University Press. Science 122:380.

1955 Wever EG: Book Review: *On the Sensations of Tone as a Physiological Basis for the Theory of Music*. The Scientific Monthly 81.

1955 Wever EG, Lawrence M: Patterns of injury produced by overstimulation of the ear. J Acoust Soc Am 27:853–858.

1955 Wever EG, Vernon JA: The effects of the tympanic muscle reflexes upon sound transmission. Acta Otolaryngol 45:433–439.

1955 Wever EG, Vernon JA, Lawrence M: The maximum strength of the tympanic muscles. Ann Otol Rhinol Otolaryngol 64:383–391.

1955 Wever EG, Vernon JA: The threshold sensitivity of the tympanic muscle reflexes. Arch Otolaryngol 62:204–213.

1956 Wever EG, Lempert J, Meltzer PE, Rambo JHT: The effects of injury to the lateral semicircular canal. Trans Amer Acad Ophthalmol Otolaryngol Sept–Oct 718.

1956 Wever EG, Vernon JA: The Sensitivity of the turtle's ear as shown by its electrical potentials. Proc Natl Acad Sci USA 42:213–220.

1956 Wever EG, Vernon JA: Sound transmission in the turtle's ear. Proc Natl Acad Sci USA 42:292–299.

1956 Wever EG, Vernon JA: The control of sound transmission by the middle ear muscles. Ann Otol Rhinol Laryngol 65:5.

1956 Wever EG, Vernon JA: Auditory responses in the common box turtle. Proc Natl Acad Sci USA 42:962–965.

1957 Wever EG, Vernon JA: Auditory responses in the spectacled caiman. J Cell Comp Physiol 50:333–339.

1957 Wever EG, Vernon JA: The auditory sensitivity of the atlantic grasshopper. Proc Natl Acad Sci USA 43:346–348.

1958 Wever EG, Vernon JA: Auditory responses in reptiles. Symposium Proceedings, Office of Naval Research Pensacola, Florida, March 191–196.

1958 Wever EG, Vernon JA, Lawrence M: The nature of the cochlear potentials in the monkey. Acta otolaryngol 49:38–49.

1958 Wever EG, Vernon JA, Rahm WE, Strother WF: Cochlear potentials in the cat in response to high-frequency sounds. Proc Natl Acad Sci USA 44:1087–1090.

1959 Wever EG: The cochlear potentials and their relation to hearing. Ann Otol Rhinol Laryngol 68:975–989.

1959 Wever EG, Rahm WE Jr, Strother WF: The lower range of the cochlear potential. Proc Natl Acad Sci USA 45:1447–1449.

1959 Wever EG, Vernon JA: The auditory sensitivity of orthoptera. Proc Natl Acad Sci USA 45:413–419.

1960 Wever EG: Translator and editor, Bekesy G von Experiments in Hearing New York: McGraw-Hill.

1960 Wever EG, Vernon JA: The problem of hearing in snakes. J Aud Res 1:77–83.

1961 Wever EG: The physiological basis of musical hearing. International Musical Society Report I 133–138.

1961 Wever EG, Vernon JA: The protective mechanisms of the bat's ear. Ann Otol Rhinol Laryngol 70:5–18.

1961 Wever EG, Vernon JA: Hearing in the bat Myotis lucifugus as shown by the cochlear potentials. J Aud Res 2:158–175.

1961 Wever EG, Vernon JA: Cochlear potentials in the marmoset. Proc Natl Acad Sci USA 47:739–741.

1961 Békésy G von, Wever EG, Rahm WE Jr, Rambo JHT: A new method of perfusion for the fixation of tissues. Laryngoscope 71:1534–1547.

1962 Wever EG: The transmission of sound in the ear. Henry Ford Hospital International Symposium: Otosclerosis. Boston: Little Brown and Company, pp 139–152.

1962 Wever EG: Development of traveling-wave theories. J Acoust Soc Am 34:1319–1324.

1962 Wever EG: Hearing. In: Farnsworth PR (ed) Annu Rev Psychol 13:225–150.

1963 Wever EG, Crowley DE, Peterson EA: Auditory sensitivity in four species of lizards. J Aud Res 3:151–157.

1963 Wever EG, Peterson EA: Auditory sensitivity in three iguanid lizards. J Aud Res 3:205–212.

1963 Wever EG, Vernon JA, Peterson EA: The high-frequency sensitivity of the guinea pig ear. Proc Natl Acad Sci USA 49:319–322.

1963 Wever EG, Vernon JA, Peterson EA, Crowley DE: Auditory responses in the tokay gecko. Proc Natl Acad Sci USA 50:806–811.

1964 Wever EG: The physiology of the peripheral hearing mechanism. In: Fields WS and Alford BR (eds) Neurological Aspects of Auditory and Vestibular Disorders Springfield, IL: Charles C. Thomas. Chapter 2, pp 24–50.

1964 Wever EG, Peterson EA, Crowley DE, Vernon JA: Further studies of hearing in the gekkonid lizards. Proc Natl Acad Sci USA 51:561–567.

1965 Wever EG: The degenerative processes in the ear of the shaker mouse. Ann Otol Rhinol Laryngol 74:5–21.

1965 Wever EG: Structure and function of the lizard ear. J Aud Res 5:331–371.

1965 Wever EG, Vernon JA, Crowley DE, Peterson EA: The electrical output of the lizard ear. Relation to hair-cell population. Science 150:1172–1174.

1966 Wever EG: Electrical potentials of the cochlea. Physiological Rev 46:102–127.

1966 Wever EG, Hepp-Reymond MC, Vernon JA: Vocalization and hearing in the leopard lizard. Proc Natl Acad Sci USA 55:98–106.

1966 Vernon JA, Dalland JI, Wever EG: Further studies of hearing in the bat *Myotis lucifugus* by means of cochlear potentials. J Aud Res 6:153–163.

1967 Wever EG: The tectorial membrane of the lizard ear: Types of structure. J Morphol 122:307–320.

1967 Wever EG: The tectorial membrane of the lizard ear: species variations. J Morphol 123:355–372.

1967 Wever EG: Tonal differentiation in the lizard ear. Laryngoscope 77:1962–1973.

1967 Wever EG: Editor: Bekesy G von *Sensory Inhibition*. Princeton, NJ: Princeton Univ Press.

1967 Wever EG, Hepp-Reymond MC: Auditory sensitivity in the fan-toed gecko *Ptyodactylus hasselquistii puiseuxi* Boutan. Proc Natl Acad Sci USA 57:681–687.

1968 Wever EG: Phonoreception. In: *Encyclopedia of Science* 3rd ed. New York: McGraw-Hill, pp. 141–145.

1968 Wever EG: The lacertid ear: *Eremias argus*. Proc Natl Acad Sci USA 61:1292–1299.

1968 Wever EG: The ear of the chameleon: *Chamaeleo senegalensis* and *Chamaeleo quilensis*. J Exp Zool 168:423–436.

1968 Wever EG, Herman PH: Stridulation and hearing in the tenrec *Hemicentetes semispinosus*. J Aud Res 8:39–42.

1969 Wever EG: The ear of the chameleon: The round window problem. J Exp Zool 171:1–6.

1969 The ear of the chameleon: *Chamaeleo hohnelii* and *Chamaeleo jacksoni*. J Exp Zool 171:305–312.

1969 Wever EG: Cochlear stimulation and lempert's mobilization theory. Arch Otolaryngol 90:68–73.

1969 Wever EG, Herman PN, Simmons JA, Hertzler DR: Hearing in the blackfooted penguin *Spheniscus demersus* as represented by the cochlear potentials. Proc Natl Acad Sci USA 63:676–680.

1969 Ridgway SH, Wever EG, McCormick JG, Palin J, Anderson JH: Hearing in the giant sea turtle *Chelonia mydas*. Proc Natl Acad Sci USA 64:884–890.

1970 Wever EG: The Lizard Ear: *Cordylus platysaurus* and *gerrhosaurus*. J Morph 130:37–56.

1970 Wever EG: The lizard ear: Scincidae. J Morph 132:277–292.

1970 Wever EG, Werner YL: The function of the middle ear in lizards: *Crotaphytus collaris* (Iguanidae). J Exp Zool 175:327–342.

1970 McCormick JG, Wever EG, Palin J, Ridgway SH: Sound conduction in the dolphin ear. J Acoust Soc Am 48:1418–1428.

1971 Wever EG: Hearing in the crocodilia. Proc Nat Acad Sci 68:1498–1500.

1971 Wever EG: The lizard ear: Anguidae. J Aud Res 11:160–172.

1971 Wever EG: The ear of *Basiliscus* (Sauria: Iguanidae); Its structure and function. Copeia 1:139–144.

1971 Wever EG: The mechanics of hair-cell stimulation. Ann Otol Rhinol Laryngol 80:786–804.

1971 Wever EG: Modes of stimulation of hair cells. In: Sachs MB (ed) *Physiology of the Auditory System*. Baltimore: National Educational Consultants, pp 55–56.

1971 Wever EG, McCormick JG, Palin J, Ridgway SH: The cochlea of the dolphin *Tursiops truncatus*: General morphology. Proc Natl Acad Sci USA 68:2381–2385.

1971 Wever EG, McCormick JG, Palin J, Ridgway SH: Cochlea of the dolphin *Tursiops truncatus*: The basilar membrane. Proc Natl Acad Sci USA 68:2708–2711.

1971 Wever EG, McCormick JG, Palin J, Ridgway SH: The cochlea of the dolphin *Tursiops truncatus*: Hair cells and ganglion cells. Proc Natl Acad Sci USA 68:2908–2912.

1971 Simmons JA, Wever EG, Pylka JM: Periodical cicada: sound production and hearing. Science 171:212–213.

1971 Pylka JM, Simmons JA, Wever EG: Sound production and hearing in the rattlesnake. Paper presented at meeting of AAAS Philadelphia, PA Dec 26–31, 1971. Abstracted in Herpetol Rev 1972 3:107.

1972 Wever EG, Gans C: The ear and hearing in *Bipes biporus* (Amphisbaenia: Reptilia). Proc Natl Acad Sci USA 69:2714–2716.

1972 Wever EG, McCormick JG, Palin J, Ridgway SH: Cochlear structure in the dolphin *Lagenorhynchus obliquidens*. Proc Natl Sci Acad USA 69:657–661.

1972 Gans C, Wever EG: The ear and hearing in amphisbaenia (Reptilia). J Exp Zool 179: 17–34.

1972 Werner YL, Wever EG: The function of the middle ear in lizards: *Gekko gecko* and *Eublepharis macularius* (Gekkonoidea). J Exp Zool 179:1–16.

1973 Wever EG: The labyrinthine sense organs of the frog. Proc Natl Acad Sci USA 70: 498–502.

1973 Wever EG: Closure muscles of the external auditory meatus in gekkonidae. J Herpetol 7:323–329.

1973 Wever EG: The function of the middle ear in lizards: *Eumeces* and *Mabuya* (Scincidae). J Exp Zool 183:225–240.

1973 Wever EG: The function of the middle ear in lizards: Divergent types. J Exp Zool 184:97–126.

1973 Wever EG: Tectorial reticulum of the labyrinthine endings of vertebrates. Ann Otol Rhinol Laryngol 82:277–289.

1973 Wever EG, Gans C: The ear in amphisbaenia (Reptilia); further anatomical observations. J Zool 171:189–206.

1973 Wever EG: The ear and hearing in the frog *Rana pipiens*. J Morph 141:461–478.

1974 Wever EG: The Ear of *Lialis burtonis* (Sauria: Pygopodidae), its structure and function. Copeia 2:297–305.

1974 Wever EG: The lizards ear: Gekkonidae. J Morph 143:121–166.

1974 Wever EG: The evolution of vertebrate hearing. In: Keidel WD and Neff WD (eds) *Handbook of Sensory Physiology Vol V-1 Auditory System* New York: Springer-Verlag, pp 423–454.

1974 Wever EG: Sound reception. In *Encyclopedia Britannica*, 15th edition 39–51.

1974 Gans C, Wever EG: Temperature effects on hearing in two species of Amphisbaenia (Squamata Reptilia). Nature 250:79–80.

1974 Ridgway SH, McCormick JG, Wever EG: Surgical approach to the dolphin's ear. J Exp Zool 188:265–276.

1975 Wever EG: The caecilian ear. J Exp Zool 191:63–72.

1975 Gans C, Wever EG: The amphisbaenian ear: *Blanus cinerus* and *Diplometopon zarudnyi*. Proc Natl Acad Sci USA 72:1487–1490.

1976 Wever EG: Origin and evolution of the ear of vertebrates. In: Masterson RB, Bitterman ME, Campbell CBG and Hotton N (eds) *Evolution of Brain and Behavior in Vertebrates* Hillsdale, NJ: Lawrence Erlbaum Assoc.

1976 Wever EG, Gans C: The caecilian ear. Further observations. Proc Natl Acad Sci USA 73:3744–3746.

1976 Gans C, Wever EG: Ear and hearing in *Sphenodon punctatus*. Proc Natl Acad Sci USA 73:4244–4246.

1978 Wever EG: Sound transmission in the salamander ear. Proc Natl Acad Sci USA 75:529–530.

1978 Wever EG: *The Reptile Ear*. Princeton, NJ: Princeton University Press.

1979 Wever EG: Middle ear muscles of the frog. Proc Natl Acad Sci USA 76:3031–3033.

1980 McCormick JG, Wever EG, Ridgeway SH, Palin J: Sound reception in the porpoise as it relates to echolocation. In: Busnel RG (ed) *Animal Sonar Systems Symposium* New York: Plenum Press, pp 449–467.

1981 Wever EG: The role of the amphibians in the evolution of the vertebrate ear. Am J Otolaryngol 2:145–152.

1985 Wever EG: *The Amphibian Ear* Princeton, NJ: Princeton University Press.

Section I
Evolutionary Perspectives

The first four chapters in this volume deal with general principles of evolution.

In the first chapter, Gans argues for multiple parallel but independent trends in the evolution of vertebrate hearing. He discusses some major evolutionary processes and how they apply to studies of hearing.

The theme of hearing systems is continued by Bullock (Chapter 2), who uses the auditory system as an example of a sensory system in which different grades of complexity have evolved. He argues that numerous adaptive differences have evolved laterally within each grade. He is baffled that scientists have yet to discern clear-cut functional differences between these grades of complexity.

In Chapter 3, Northcutt discusses the relationship of ontogeny to phylogeny, based upon Garstang's view that phylogeny reflects changes in ancestral ontogeny to phylogeny, based upon Garstang's view that phylogeny reflects changes in ancestral ontogeny rather than a succession of adult patterns. In this context, Northcutt analyzes the octavolateralis system and provides a model for studying the evolution of the auditory system, or other systems, for which there are sufficient ontogenetic data.

In Chapter 4, the final chapter of this section, Popper, Platt, and Edds review major historic ideas about the evolution of the vertebrate ear. They discuss the history of the acousticolateralis hypothesis and present divergent views on its validity. They emphasize germinal papers by Baird, van Bergeijk, and Wever. This chapter serves as a background for those which follow, many of which refer to the historical ideas discussed.

1
An Overview of the Evolutionary Biology of Hearing

Carl Gans

1. Introduction

The external linkages of vertebrate ears have been the kind of material that brings joy to the comparative morphologist. The homology patterns of the ear ossicles, first described by Reichert (1837), document that structures that currently share neither appearance nor function may still be homologous. Not only this, but the amplification of Reichert's work by other morphologists represents an outstanding and nearly unique demonstration of major structural and functional shifts and multiple radiations that can be explained only by developmental and comparative evidence.

The vertebrate hearing system thus is characterized by complex structures undergoing functional changes. In addition, we have a better historical record of its multiple branches than for the multiple auditory systems of any group of invertebrates. Hence, I here use it as a model for the evolutionary processes involved. The structures involved in vertebrate audition reflect parallel shifts of various biological roles, such as ventilation, ingestion, and the perception and production of sounds. Understanding of the shifts requires a parallel consideration of the physical principles and functional morphology of the systems, as well as of the ecology and behavior of the organisms. Most important, understanding of these shifts requires an understanding of evolutionary mechanisms, the processes generating the patterns of evolutionary change.

It seems useful to note the aim of an overview, such as the present. Clearly, it cannot serve to provide the readers of a volume on hearing with details of the auditory system, after all, my background in this is less than that of almost any member of the audience. Instead, I have tried to develop a generalist overview and to note some major evolutionary processes and to see how these may be applied to detailed studies of the hearing system. This should not be interpreted as a plea for detailed phyletic analyses of components and processes, although these would obviously be useful. Instead, I hope to provide the evolutionary (process) and phyletic (product) framework that may permit the many elegant and differently specialized studies to be viewed in an evolutionary perspective.

2. Evolutionary Patterns

2.1 Phenotype and Environment

Animals display structures and associated mechanisms; these are the phenotypes that they use to perform their daily activities. This observation implies that the individual phenotypes, representing some interaction of genotypes and environment, will be at least adequate for performing all of the biological roles demanded of the individual during its life span. The (often untested) assumption is that a substantial fraction of the phenotypic variability is genotypically based.

The environment encountered by organisms is patchy in that environments differ in space and time. Also animals of any species show phenotypic variability, implying that some individuals will be better at any given task than will others. Such variability reflects genetic differences and environmental effects during development. Generally, natural selection acts as a sieve, with its mesh varying with a time course established by the short-term and long-term fluctuations of environmental circumstances. Hence, the selection each individual is likely to incur changes with time and is rarely at a mean level. Some individual organisms that are inadequate under mean environmental circumstances will prove sufficient under the situations that the organisms actually encounter (Gans 1991).

Ontogeny produces further complications. Each intermediate state needs both to be adequate for current circumstances and to represent a precursor condition for the subsequent states, up to and even beyond sexual maturity. This dual need for matching current circumstances and still proceeding with ontogeny represents what is sometimes referred to as a set of developmental constraints that act as a complex selective filter (Gans 1989). However, it is important to note that this "filtration" only involves functional tolerance of levels of environmental conditions. Furthermore, this process indicates the difficulty of finding non-lethal mutations that will change the state of the phenotype at any particular age; to be tolerated such mutations must affect the desired condition, without further and likely deleterious effects on other ontogenetic stages. In short, they must leave the organism adequate to the environments encountered throughout the remainder of its life cycle.

A number of described genetic mechanisms will buffer the implications of these constraints. Among them are repression and inhibition of gene actions which may restrict their expression to particular time periods and similarly the enhancement of portions of genetic actions which again may occur for limited intervals. Other organisms have the innate (genetically determined) capacity to monitor the environment during their development. They use the information thus obtained to generate phenotypes that are better able to match the environment they likely will encounter during the remainder of their life span; this process is referred to by various special terms, such as physiological adaptation. However, the present account need not address the detailed characterization of the multiple genetic mechanisms. We only need to note the existence of such mechanisms and to consider why they are intrinsically important in producing a substrate for phenotypic and potentially phylogenetic (= evolutionary) change.

Any modification of the genetic or developmental instructions must at the least allow survival (viability) of subsequent stages. This makes it theoretically best to modify phenotypes by the addition of terminal stages and such addition has often been touted as being the only or the major mechanism of evolutionary change (cf. Gould 1977). However, the history of life is replete with examples of changes during the course of development and it is likely that early changes and intercalations have occurred and proven viable. For instance, the development of the mammalian morula is unlikely to have been such a terminal addition. Instead of arguing that only terminal additions occur it is less risky to consider the ubiquity of pleiotropic effects of genes and their actions within developmental cascades; the occurrence of viable mutations had best be considered probabilistically (Gans 1987). This suggests that (all other factors being equal) the probability of a viable mutation of a pleiotropic gene relates to the position at which the gene products act in the several genetic/ontogenetic pathways in which they participate. Viable early minor genetic changes with far-reaching effects remain possible, although less likely than later ones. Also there is the potential switching of entire developmental subroutines (Maderson 1975).

2.2 Physiology, Behavior, and Environmental Demands

The patchiness of the environment demonstrates a range of environmental factors. This permits organisms some freedom to select suitable circumstances. Such selection may proceed on several levels. In the simplest case, seeds or larval stages may be broadcast; only those encountering suitable patches of soil or rock will germinate or settle and develop into adults. The chance of achieving a viable descendant will be increased if the seed (larva) can differentiate among the several kinds of substrate encountered and can use the resultant information to decide on germination. Such germination (settling) only at a suitable site will increase the chance of producing offspring in turn; hence, the characters conferring an ability to do so represent the potential for a substantial advantage. Obviously, the decision process need not be conscious; as obviously, the capacity to make decisions based upon environmental cues is likely to be genetically determined and hence subject to selection.

The conditions observed in higher animals achieve further complexity beyond that seen in such simple cases, but the basic rules do not change. Whereas a kind of behavior restricts seed germination to sites and states at which survival would have a greater chance of success, higher organisms often display behaviors that let them select among aspects of the environment (i.e., Huey and Slatkin 1976). In short, they may avoid those environments that exceed mechanical or physiological capacity.

Examples are multiple and often elegant in their seeming simplicity; they generally permit one to discern cost–benefit relations. Fishes may incorporate oxygen receptors that let some species concentrate in the most oxygenated layers of an otherwise limiting low-oxygen environment (Gans 1971). Tropical reptiles routinely occupy thermal environments, the mean conditions of which far exceed their tolerance; they escape the lethal aspects by selectively occupying (nocturnal, subterranean, and shady) temporal and spatial patches and use the thermal capacity of their body to buffer brief exposures to zones of higher temperature (Huey 1982). Some vertebrates, including certain mammals, can temporarily lower their metabolic rates and thus reduce the "cost" of "waiting out" intervals of inappropriate environmental circumstances (Feaver 1977).

What is most critical in attempting to understand these examples and important in the study of environmental change is that the organisms routinely circumvent the physiological limitations of their cells and tissues. This important capacity is seen even among very simple creatures, although the increased size and capacity of nervous systems may increase the potential for integrating several levels of stereotypic and more flexible decisions.

2.3 Sequences and Experiments

It is almost a truism that natural selection looks at the totality of each organism. However, at any one time the organism is likely to encounter a scattering of limiting characteristics for which the mean of an important phenotypic aspect happens to be close to the environmental demand. Any

modification that affects either the phenotype or the frequency at which the environmental limitation is imposed is hence likely to incur a selective advantage. Whenever the limitation restricts access of members of a group to a major (and continuing) resource, different organisms may show diverse "solutions" to the problem. This process may be called an "experimental" radiation (Gans 1990). However, the aspects of the phenotype so involved will not be homologous; indeed some of the different solutions may involve changes of morphology, whereas others involve physiology and/or behavior.

A completely different success story is often referred to as an adaptive radiation. This involves a line of organisms that may share some aspect of their phenotype, enabling them to invade a newly available environment and there radiate (Gans 1990). It is often assumed that the capacity for such invasion of a new biotope reflects the same set of characteristics as that which facilitates the subsequent radiation; indeed this is sometimes the case. One can then note that the diversification involves homologous phenotypic states. However, the aspect permitting the initial invasions and that leading to speciation may be quite different; indeed, the characters enabling the entry into the situation that leads to the adaptive radiation may remain relatively constant.

2.4 Uncertainty

What is perhaps most important is that our reconstruction of history will always incorporate several levels of uncertainty. First of all, we have only part of the evidence, and the further back we go the less likely we are to understand the genetics and phenotypes of the organisms involved and the nature of the environments they faced. (Actually, of course, we lack this information even for most cases in the Recent; but for these, there remains the possibility of obtaining such information.) With this we retain the risk of misinterpreting the selective effects that influenced the organisms at the time of their formation. Then we proceed with estimates of the way that the organisms exploited the environment and compensated for its (and their own) limitations. However, these are probability statements, and misinterpretation, even of a single factor, may deflect highly logical analyses to erroneous conclusions.

One obvious example may be taken from the application of various cladistic methods to phylogenetic reconstruction. It is often claimed that Occam's razor, recently referred to as parsimony, must be used to decide which of the many alternative phylogenies shall be accepted. This is generally appropriate in operational terms; however, the operational solution only generates a logical minimum state. We know that the mutations that generate the shared-derived states (synapomorphies) are not isolated or unique events but regularly repeated ones and may often be reversible. Consequently, the branching points are much less clear cut than sometimes assumed. Perhaps we can define a degree of uncertainty, but we are unlikely to achieve a definitive solution. This suggests the merit of caution in accepting and defending the results of conceptual reconstructions of history.

3. Hearing

3.1 Aspects

Hearing represents the detection of cues involving propagated vibratory energy from the environment and the establishment of their magnitude, direction, and significance. In vertebrates and many invertebrates, detection of vibration involves the use of "ciliated" cells that show slender ciliary hairs projecting from their free surface (de Burlet 1934; Capranica 1976). In some invertebrates the detectors are membranous, but even some of these involve ciliary processes (Bullock and Horridge 1965). Deformation of the projecting hairs produces effects within the cell membrane. These effects can be used as signals by monitoring and transmitting them to other cells, i.e., these sensory cells are transducers of motion that transform mechanical into electrical energy.

The flow of liquid over the gelatinous cupulae of the hair cells generates drag, inducing forces on the projections and bending the hairs (Capranica 1976). As important for the detection of forces is the addition of calcium or other dense mineral crystals to the cupulae. Hair cells with cupulae that are calcified, rather than gelatinous, may be sensitive to gravitational or muscle-induced acceleration of the system; the effect is due to differential inserts reflecting differential density. The hairs can monitor this motion, providing the basis for the several simple detectors of the gravitational axis and of acceleration. As the effect results from a vector action, the placement of the cells and the direction of the fluid movement are critical.

Many aquatic invertebrates show such systems, their multiplicity presumably representing parallelisms or convergences (Barnes 1963). Coelenterates and urochordates show simple detectors of the gravitational axis; cephalochordates have various categories of hair cells that presumably detect flow. Various terrestrial arthropods have developed small mechanical detector systems that do not rely on hair cells, perhaps because of the constraints imposed by their exoskeletons.

Paired, external distance receptors, deriving embryologically from placodes and the neural crest, seem to be a synapomorphy of vertebrates (Northcutt and Gans 1983). The placodes underlie the several paired receptors (which are generally ciliated) and initially form superficially, which explains the oft-noted inward migration of the various components of the octavolateralis system during ontogeny. The earliest vertebrates appear to have formed these as saccular structures, a variable number of semicircular canals, and a (possibly cephalic) lateral line system, each with its patches of hair cells (Starck 1982). The sacculus clearly detected gravitational field of the earth, the semicircular canals direction of acceleration and thus turns. The lateral line system could also detect the effects of water motion near the head and body; comparison of the different signals from adjacent sensors and the orientation of hair cells within neuromasts permits detection of the position of prey near the mouth.

Whereas these chordate senses serve for the detection of cues close to the animal, where nearfield effects may cause physical motions (currents) of the system, they would also respond to the vibrations referred to as sounds, including those arising from more distant sources. Detection of such sounds is hearing. We can then state that the development of hair cells and paired external organs represented the evolutionary novelty that permitted the adaptive radiation which allowed the development of multiple gravity, motion, and sound monitoring systems.

3.2 Cues and Discrimination

What sounds might have been of interest to early fishes and what would have been the problems of their detection? As noted above, the oceans now are noisy places, filled with biologically produced signals. However, many of these sound sources likely postdate the origin of audition, i.e., they achieved their roles only after there was a system for detecting them. There may have been some signals produced by invertebrates; however, the major sources of sound would have derived from abiotic phenomena, the surface interactions of waves and wind, and waves breaking on reefs and shores. It is unclear whether these sounds would initially have been interesting to these relatively small animals as information about such environmental events is unlikely to have permitted them to take simple useful (evasive?) action.

Consequently, it seems most likely that the initial role of hearing was prey detection. An example might be the kind of auditory detection seen in sharks which can be attracted by the sounds of struggle or similar irregular sequences. Perhaps such biologically significant sound sources are emphasized by an inhibition of the background noises generated at surface and shore. As the several potentially sound-detecting organs are already bilaterally symmetrical, there would be a potential for orientation by the use of such cues.

The aquatic environment transmits a wide range of sounds with variable attenuation. Many natural processes, such as waves, breakers on shores and reefs, the actions of currents along the bottom and of wind and rain onto the surface, will produce substantial noise that generally has little biological importance to the organism. The sound system then represents a channel containing substantial noise. This establishes the need for filtering out the unimportant by processing the signal, a task that establishes a major requirement for central processing. Such processing and filtering out the signal of interest presumably is a hallmark of the hearing system of many animals. It remains to be seen how many animals possess ears that can detect sounds over the entire spectrum rather than in specific spectral regions. Whereas frequency filtering is peripheral, and habituation may gradually reduce responses due to common background noise, recognition of patterned signals would seem to require central processing.

As all of the cells of the several detecting systems could have become refined to the reception of high vibration cues, there remains the question of the homologies of the components of the inner ear. At this moment, it appears that all of the mechanoreceptive cells of the octavolateralis system may be considered to be homologous, as detecting cells. This is suggested both by their structural patterns and by their universal embryological derivation from ectodermal placodes and by the adjacent placement of their parallel central projections in the dorsal medulla oblongata (Northcutt 1979). Their placement in one or another system will then represent modulation of a common capacity. One remaining question concerns the number of morphologically and physiologically distinct mechanoreceptors, such as the lateral-line derived paratympanic organ of birds, which may serve as a baroreceptor (von Bartheld 1990). However, the recent explosive increase in our capacity to determine the genetic mechanisms underlying the construction of phenotypes suggests the merit of using these tools to address the question of detector cell homologies.

The next stage for such a potential evolutionary sequence would have involved improvement in the discrimination ability for particular sounds, refined directionality of detection, and ultimately the potential for selective recognition of sounds generated by conspecifics.

3.3 Liquids and Gases

We owe to Willem van Bergeijk the characterization of the dichotomy of near and far field effects (1966, 1967). Specifically the displacement and amplitude is not simply proportional to distance, but first decreases sharply ($1/distance^3$) and then more gradually ($1/distance$); this indicates that very close signals are easier to detect. In general, the near-field situation assures greater excursion of the liquid particles rather than the low-amplitude vibrational cues seen in the passage of sound waves. Presumably the capacity for detection of the former is evolutionarily earlier, as they involved high amplitude, low-frequency signals. The potential utility of far field effects would seem gradually to have required the detection of signals with lower amplitudes presented at higher frequencies.

Aquatic vertebrates have two potential strategies for improving the detection of far-field signals (van Bergeijk 1967). The first is the improvement of the vibration-detecting ability of particular hair cells or neurons. The second is modification of the signal, specifically the development of mechanisms for increasing its amplitude. Both have clearly occurred; I shall stress the latter as providing an excellent example of an experimental radiation.

The easiest way of achieving amplitude multiplication is by placing a gas-filled space into the sound path. Pressure and volume are inversely related in ideal gases; however, in relatively incompressible liquids, they are effectively uncoupled. Hence, gas-filled spaces undergo major volumetric changes during the passage of pressure waves. Transmission of the resultant movements to the liquid within which the ciliated portion of the detector cells rests will induce increased ciliary deflection and potentially greater magnitude of the signals, providing a greater substrate for physiological processing and behavioral decision.

Naturally, the movement of the fluid incorporates a vector component. Hence, the more peripheral apparatus also tends to orient the fluid pulsation along an axis. This suggests tubular arrangements within which the movement is constrained. Also the requirement of differential motion remains, thus the merit of fixing the cell body and base of the hairs relative to the liquid stream or as in auditory organs, fixing the cupulla and hair cells and permitting movement of the cell body.

Teleost fishes appear to have solved the air bladder problem in several ways, suggesting that we are seeing several distinct synapomorphies (Gans 1971). All appear to derive from the basis of an air sac serving as a gas exchanger or a swim bladder, and thus involving a shift in its biological role. The association suggests that this may by itself have generated some increase in the amplitude of the movement.

The differences among these organs lie in the several kinds of transmission of the vibration. Two obvious cases are the development of Weberian ossicles and direct connection of an extension of the air sac to the oval window (Starck 1982). Clearly these are not unitary nor multiple independent developments. Critical here is the develop-

ment in various fishes of new sound-generating systems that clearly support intraspecific communication (Lythgoe 1978). Certainly fishes show remarkable diversity in air sac structure.

3.4 Transition from Water to Land

The emergence onto a terrestrial situation by the animals now classified as amphibians, represented a revolutionary sequence of events that has provided the basis for literally dozens of cartoons in the *New Yorker* magazine, as well as of "Far Side" comics produced by Gary Larson. What is less often appreciated is that the Recent amphibians are hardly remnants of the transitional animals (Gans 1970). Also, the characteristics that permitted the initial invasion of the biotopes, often referred to as "the land," presumably preceded the origin of more detailed terrestrial adaptations. Thus the adaptive radiations into terrestrial environments depended on aspects independent of those allowing their initial occupation.

For instance, the development of aspiration breathing and tetrapody in sarcopterygian fishes had advantages to these organisms, and also represented protoadaptations that allowed the emergent amphibians to discharge their CO_2 and walk on land (Gans 1970; Goin, Goin, and Zug 1978). Looking at other physiological aspects we note that the discharge of wastes continued to require water and the reproductive pattern continued to demand an aquatic larval stage. This suggests that the shift was by stages, involving a gradual refinement of capacity. Finally, we see what might be termed "behavioral" (although perhaps innate) components to the shift. The newly terrestrial tetrapods apparently encountered a conflict between ventilation and locomotion (Carrier 1989). Aspiration breathing involved parallel activation of the muscles of the two sides of the trunk whereas undulant locomotion involved alternate activity of the muscles of the two sides. Evidence now indicates that the transitional animals (and many of their descendants) resolved the conflict by an adaptive compromise; they alternately breathe and walk (Carrier 1989)!

How does the hearing system fit into this pattern? Indeed, what changes are necessary once the bearer of a fish ear invades the land? First of all, we may note that all putative preterrestrial fishes were air-breathers having some sort of gas-filled chambers (Schmalhausen 1968; Gans 1971). Consequently, their ears may have shown some initial sound-detecting specializations; on the other hand, the transitional animals may have been essentially deaf. The difficulty comes in dealing with sounds that are primarily aerial, those that are generated in and transmitted by this medium.

The problem in dealing with aerial sounds is that of impedance matching. Any object has particular "natural" frequencies at which it will vibrate with minimal energy input. These depend on its shape, material, and intrinsic flexibility, as well as the dampening effects of surrounding tissues. Thus, the solid portions of an animal's body that occupy positions along the path between the source of the sound and the inner ear are critical to the capacity of the ear to detect aerial vibrations. (This leaves open for the moment the concept of detecting substrate vibrations by solid linkages.) Consequently, fish ears would encounter one additional obstacle in the sound path in that the aerial signal must either pass through the air–water interface or must be able to vibrate the portion of the body wall that defines any air chamber.

Tetrapods have apparently dealt with these problems in a series of ways. They have retained some ancestral aspects, but also show adaptive variations on a common theme. The ancestral aspect is the retention of a basically liquid-filled inner ear system in which hair cells are variously placed along a fluid path and excited by its vibration. A second ancestral aspect is the introduction of a gas-filled space adjacent to the liquid-filled ear, placed so that its volumetric changes can be transmitted. These aspects may be seen as another level of protoadaptations, those for aerial hearing. The new variation is the restriction of the aerial chamber to the side of the head, forming a middle ear system, and the modification of its external surface into a site of relative flexibility, matching the incident vibrational spectrum. With this there is a series of parallel developments or multiple inventions in each of which the external membrane becomes linked to the oval window by a direct connection and the air-filled space is vented, so that it no longer incurs major (and signal damping) pressure changes with vibration of the external membrane. The connection is universally derived from part of the hyoid chain, again involving functional shifts from an element supporting the filtration system,

to one involved in gas exchange as a gill support, to a jaw linkage, to one serving audition (Romer and Parsons 1976). The impedance of the entire linkage system must be matched to achieve and improve sound detection.

3.5 The Uses of Sounds – Benefits and Costs

Thus far we have ignored the question of the use of the hearing ability except for the initial comments on the detection of unusual signals. However, the fact that an individual has the capacity to detect vibrations in the environment, their magnitude, and the pattern of their changes allows this energy channel to be used to transmit information (Wiley and Richards 1982). This generates interest in producing the sounds for detection by conspecific organisms. However, other species may then eavesdrop on the primary signal in a variety of ways, to modify it, and to subvert it. The absorptive and reflective nature of the environment may be used to map it. Throughout, one sees a complex interplay of costs and benefits modifying a usage pattern that seems to change opportunistically with its several possible roles complexly intertwined.

The use of audition in prey detection was suggested to have been the original role of sound detection in vertebrates. However, detection has been repeatedly modified, often by enhancement of the detector ability. Examples are the insect and frog-catching ability of bats (Griffin 1958; Ryan 1985) and the rodent hearing ability of some owls (Suthers 1977a,b). Another level of modification has been in the behavioral capacity to respond to a particular pattern of sounds, as seen in the response of crocodiles to splashing water. Another presumably early improvement in the system, critical for the cases adduced, was in the ability to locate the source of sounds in three-dimensional space. Such prey-locating capacity is also used by prey to detect predators. Examples are the ability of desert rodents to jump vertically in response to the sound of a striking snake and the wind rustle of owls (Webster 1962) and of some lizards to respond specifically to the calls of predatory birds.

Another set of hearing-related roles is in the apparently independent acquisition of the ability to echolocate, seen in bats, oilbirds, and cetaceans (Suthers 1977a,b). In general, these systems combine a directional sound source, which emits regular, constant level stereotypically oriented pulses so that the echoes incorporate correct information about the sites at which they originate. With this they provide a sonar image of the space surrounding the animal. The critical aspects are the detail of its resolution and presumably the three-dimensional depth of the image.

In some ways the most interesting set of sound systems are those involving intraspecific communication. Here again we see multiple, presumably independently developed systems in teleosts, frogs, perhaps salamanders, many reptiles (lizards, rhynchocephalians, crocodilians), and birds and mammals. It is interesting to note the parallels with the organization of the system of fishes utilizing weak electroreception (Bass 1986). When speaking of independent development, we are of course making assumptions based on the diversity of sound-generating systems; it remains to be seen whether the receptors are as diversified.

The key to success of a communication system will be the specific tuning of the receptors to the source levels and signal patterns generated by the emitters. This is critical because emission systems will incur costs and demand compromises on several levels, namely in terms of the energy that has to be expended in radiating the signal and because any intraspecific signal also has a risk of informing predators of one's existence and location.

3.6 Social Vocalization in Frogs as an Example of Complexity

The vocalization patterns of frogs (and presumably their auditory systems) nicely show these influences and complications of the problem of communication (Duellman and Trueb 1986). First of all they show diversity, so that (1) communication within a species differs sufficiently from that of other forms to avoid interference (Bogert 1960; Littlejohn 1977), and (2) they may avoid entrainment of predators to a general characteristic that will lead them to any "frog."

Next, these animals appear to match their vocal output fairly specifically to the auditory capacity of their partners (Ryan et al. 1990). Also, the signals prove to be beamed on wavelengths and in burst patterns that avoid the frequency regions and differ

in other ways from common background sounds that might have a high intensity in the shared biotope. Examples are the short pulse clicks heard in frogs that occupy streams cascading down rocky slopes (a situation similar to that seen in birds occupying similar habitats; Dubois and Martens 1984; Martens and Geduldig 1988). Other calls avoid zones of high absorption or scattering by the environment.

Finally, there are diverse systems that superimpose specific messages and incorporate components that protect the emitter. The vocalization of frogs generally has a size-associated component. It is unclear whether the size-frequency association is secondary, an indirect result of the physics of larger emitters, or whether the property was selected for; however, the receiver may judge this aspect. Information about the mass of the caller is apparently utilized by the females of several species, as it may have a bearing on the caller's suitability as a prospective parent (Wells 1977).

The protection pattern opens the discussion of a spectrum of curious structures and behaviors, the meaning of which is unclear. First of all, many frogs beam their mating calls; they distribute most of the sound energy along a beam extending anterior to the animal. It is unclear whether such frogs send their beams in particular directions and how this affects success in attraction. It certainly does tend to confuse some predators; thus, I remember spending 15 minutes trying to locate a beamer that was calling from 15 cm up a 25 cm high bank at the edge of a pond. From the pond the sound appeared to emanate from atop the bank, some meter or so back of the edge. From the top of the bank the sound seemed to emanate from the middle of the pond (apparently reflected by some bushes).

There is also the curious stereotype of frog calling sites, which I have observed while collecting them on four continents. Each species calls from a characteristic place and position. The places are sometimes complex; thus there are Central American and African forms that sit and call from one branch at a place at which a second (unconnected) branch passes diagonally. Other species call while sitting vertically and others horizontally, on stems, leaves, axillae, the ground, amid short grass, and at various aquatic sites in the center of bodies of water (of various sizes), at the edge, horizontally or vertically, exposed wholly, partially, or not at all.

Clearly, the selection of any one of these sites will affect the direction, magnitude, and nature of the calls produced. In some cases it will change the partitioning of sound between water and air. However, we mostly do not know whether, why, and how these factors are important. Even more curious are certain paddy frogs that call into semicircular depressions from which the sound seems to radiate widely, making direct localization difficult (personal observation). Are such calls designed to avoid predators? What auditory tactics do the females use to locate such males?

The frog communication system also lets these animals communicate other messages. Release calls tell males that the clasped "partner" is also male, rather than female (Duellman and Trueb 1985). Territorial spacing of males, and indeed the location of temporary breeding ponds in the spring, demands that the ears be developed in males as well as females. It might be argued that attraction of other males cannot be one of the roles of the call, as the emitters would only be generating competition; however, it remains to be tested whether individual survival would be greater if the animals are distributed sparsely in many separated ponds, or when concentrated into a few. The deafening quality of concentrated choruses suggests benefits to concentration. The alarm calls of frightened or captured frogs transmit a message; however, the message may be addressed to the predators, rather than being addressed to conspecifics as it is in groups of mammals and birds.

The preceding account should document not only the diversity that occurs in the vocalization of frogs, but also the parallel diversity that must occur in the signal-detecting systems. Use of the auditory system in social communication allows tuning or parallel manipulation of signal and detector and thus differs theoretically from systems for prey detection in which the predator always tracks the cues involuntarily provided by the prey. However, the examples should indicate the complex interaction of factors that shape the two sides of this communication channel; also the similarity seen in many aspects of the vocalizations (and sound reception) of amphibians and birds shows that their communication systems often (and independently) matched similar functional problems with similar solutions.

4.Overview: Trends and Questions

The first finding that should be clear is that variants of the "ear" have evolved multiple times in phyletically distinct groups, using the hair cell detector as a symplesiomorphy or protoadaptation. We see and should describe sequences only in very broad terms, but not in detail. The use of the mechanoreceptive cells for vibration detection, the development of gas-filled amplifiers, and the capacity for central tuning may represent a sequence. However, I have tried to note that each of the uses of the ear arose multiple times, often slightly differently. Here we probably see the effect of local adaptation of the earliest members of a group and a radiation that represents variation upon a largely genetic theme.

This immediately causes me to make a plea for limiting generalization. We all like to assume that the phenomena we discover apply to organisms generally; however, the evidence suggests that very few aspects do. The results of diversity are always with us and it is often most interesting to note why results prove to be limited. Safety requires that we, at the very least, always check that the phenomenon described pertains to more than the particular grouping.

The apocryphal literature equates evolution with improvement. It confounds the short-term kind of improvement often driving natural selection with long-term improvement leading to major phyletic change. The concept is wrong for two important and some less trenchant reasons. First of all, fitness or adaptiveness is a function of the environmental circumstances, and these will not be constant throughout a sequence. Next fitness is to be evaluated for the whole organism and the contribution thereto by any particular organ system is buffered, for instance by the pleiotropic effects of the genes comprising it and by simultaneous selection for other ontogenetic stages. Finally most aspects of the phenotype simultaneously subserve multiple roles. Hence, it is unlikely that one would encounter prolonged trends for particular characteristics. More important it shows that discussions of "maladaptive intermediates" in evolutionary trends are likely to represent a look at only a single aspect. Passage of such zones may involve a reduced fitness of a particular phenotype from the viewpoint of audition, but need not imply reduced fitness, either for all aspects of the structure, or even more important for the organism.

What is also important is that the evolutionary process does not lead to complexity; rather it generally leads to simplification of structures. Each aspect of the phenotype will regularly be tested, as its genetic instructions change by random mutation. If "less" will still permit equivalent performance, it will not be acted against, thus leading to a reduced or simpler phenotype.

Most important for the comparisons seen is that there appears to be a profound capacity to achieve equivalent adaptive solutions from quite disparate precursor states. This teaches us that organisms are not constrained by the properties of the cells and organs that comprise them; rather, selection continuously modifies these properties toward the state demanded by the organism. This poses the task of mapping functional characteristics on predetermined phylogenetic diagrams. In this way one can see how often the represented states have developed and can follow the sequence of functional change as one examines the multiple terminals of the lines of an embranchment.

5. Summary

The evolution of any complex system may proceed by a series of sequential steps or by a more opportunistic approach in which the system repeatedly becomes modified to meet local environmental circumstances. The hearing system is here reviewed and shows multiple parallel but independent trends. Whereas the construction of the chain of ossicles and their homologies represents a famous success story in the study of evolution, the more detailed structural and functional changes provide the material for more complex, but as interesting discoveries.

Acknowledgments. I thank Doug Webster, Dick Fay, and Art Popper for organizing the symposium and inviting me to participate. S.L.L. Gaunt, Bruce Masterton, Kiisa Nishikawa, Ron Nussbaum, and Douglas Webster provided discussions and comments on the manuscript, the preparation of which was supported by a grant from the Leo Leeser Foundation.

References

Barnes RD (1963) Invertebrate Zoology. Philadelphia: W.B. Saunders.

von Bartheld CS (1990) Development and innervation of the paratympanic organ (Vitali organ) in chick embryos. Brain Behav Evol 35:1–15.

Bass AH (1986) Species differences in electric organs of mormyrids: Substrates for species-typical electric organ discharge waveforms. J Comp Neurol 244: 313–330.

Bogert CM (1960) The influence of sound on the behavior of amphibians and reptiles. In: Lanyon WE, Tavolga WN (eds) Animal Sounds and Communication. Washington, D.C.: American Institute of Biological Sciences, pp. 137–320.

Bullock TH, Horridge GA (1965) Structure and Function in the Nervous System of Invertebrates. San Francisco: Freeman.

Capranica RR (1976) Morphology and physiology of the auditory system. In: Llinas R, Precht W (eds) Frog Neurobiology. A Handbook. Heidelberg: Springer-Verlag, pp. 551–575.

Carrier DR (1989) Ventilatory action of the hypaxial muscles of the lizard *Iguana iguana*: a function of slow muscle. J Exper Biol 143:435–457.

de Burlet HM (1934) Vergleichende Anatomie des statoakustischen Organs. a) Die innere Ohrsphäre. In: Bolk L, Göppert E, Kallius E, Lubosch W (eds) Handbuch der vergleichenden Anatomie der Wirbeltiere, Vol. II(2), Pt. III, 2. Berlin and Wien: Urban und Schwarzenberg, pp. 1293–1380.

Dubois A, Martens J (1984) A case of possible vocal convergence between frogs and a bird in Himalayan torrents. J Ornithol 125:455–463.

Duellman WE, Trueb L (1986) Biology of Amphibians. New York: McGraw-Hill.

Feaver PA (1977) The Demography of the Michigan Population of *Natrix sipedon* with Discussions of Ophidian Growth and Reproduction. PhD Thesis, University of Michigan, Ann Arbor.

Gans C (1970) Respiration in early tetrapods — the frog is a red herring. Evolution 24:740–751.

Gans C (1971) Strategy and sequence in the evolution of the external gas exchangers of ectothermal vertebrates. Forma et Functio 3:66–104.

Gans C (1987) The neural crest, a spectacular invention. In: Maderson PFA (ed) Developmental and Evolutionary Aspects of the Neural Crest. New York: Wiley, pp. 361–379.

Gans C (1989) On phylogenetic constraints. Acta Morph Neerl-Scand 27(1–2):133–138.

Gans C (1990) Adaptations and conflicts. Evolutionary Biology: Theory and Principles. Proc Intern Symp Plzen 1988. Czech Acad Sci Praha pp. 23–31.

Gans C (1991) Efficiency, effectiveness, perfection, optimization. Their use in understanding vertebrate evolution. In: Blake RW (ed) Efficiency, Economy and Related Concepts in Comparative Animal Physiology. Oxford University Press (in press).

Goin CJ, Goin OB, Zug GR (1978) Introduction to Herpetology, 3rd Ed. San Francisco: Freeman.

Gould SJ (1977) Ontogeny and Phylogeny. Cambridge: Harvard University Press.

Griffin DR (1958) Listening in the Dark. New Haven: Yale University Press.

Huey RB (1982) Temperature, physiology, and the ecology of reptiles. In: Gans C, Pough HF (eds) Biology of the Reptilia. London: Academic Press, 12:25–91.

Huey RB, Slatkin M (1976) Costs and benefits of lizard thermoregulation. Q Rev Biol 51:363–384.

Littlejohn MJ (1977) Long range acoustic communication in anurans: an integrated and evolutionary approach. In: Taylor DH, Guttman SI. The Reproductive Biology of Amphibians. New York: Plenum Press, pp. 263–294.

Lythgoe JN (1978) Fishes: Vision in dim light and surrogate senses. In: Ali MA (ed) Sensory Ecology. Review and Perspectives. New York: Plenum Press, pp. 155–168.

Maderson, PFA (1975) Embryonic induction and evolution. Amer Zool 15:315–327.

Martens J, Geduling G (1988) Acoustic adaptations of birds living close to Himalayan torrents. Proc Int 100 DO-G, Current Topics in Avian Biol. Bonn.

Northcutt RG (1979) The comparative anatomy of the nervous system and the sense organs. In Hyman's Comparative Vertebrate Anatomy. 3rd Ed. Wake MH (ed) Chicago: University of Chicago Press, pp. 615–759.

Northcutt RG, Gans C (1983) The genesis of neural crest and epidermal placodes: A reinterpretation of vertebrate origins. Q Rev Biol 58:1–28.

Reichert C (1837) belthiere im Allg bei den Vögeln u Phys wiss Med.

Romer AS, Parson Ed. Philadelphi

Ryan MJ (1985) ' Selection and (of Chicago Pre

Ryan MJ, Fox JH selection for se *mus pustulosu*

Schmalhausen II tebrates. New

Starck D (1982 tiere auf evol Heidelberg: S

Suthers RA (1 MA (ed) Sen New York: F

Suthers RA (1 Ali MA (ed) tives. New '

Van Bergeijk, ing in verte

Van Bergeijk hearing. In Physiology

Webster, DB ear cavitie Zool 35:2

Wells KD amphibiar

Wiley RH, F communi nal detec Acoustic demic Pr

2
Comparisons of Major and Minor Taxa Reveal Two Kinds of Differences: "Lateral" Adaptations and "Vertical" Changes in Grade

Theodore H. Bullock

1. Introduction

Hearing provides a special case of a general problem in comparative neurology, indeed in evolutionary biology broadly: what has evolution wrought? What varieties of differences between taxa have evolved—from species to phyla?

The proposition here asserted is that two kinds of differences can be distinguished. (1) If we stand back and compare phyla, classes, and sometimes orders, on the one hand, we can sometimes discern advancement through major levels of complexity. I shall call these vertical grades—in the dictionary sense, not the special meaning in the literature of cladistics, which is compatible but does not necessarily imply complexity or advancement. (2) If we compare species within the same order, family or genus, on the other hand, differences appear more in the nature of adaptive radiation within a grade. I shall call this lateral radiation.

To avoid misunderstanding, it may be well to emphasize that advances in grade are not steady or inevitable and there are many exceptions. Since the regressive exceptions are frequently parasitic or sedentary groups and otherwise advances do generally occur in more recently derived major taxa, the occasional steps in vertical grade can be called progress. This does not imply a goal or predictable future, or a moral value scale. Many progressive grades, as well as some regressive ones, were obvious long before humans appeared, or primates, or mammals, or vertebrates. It is not anthropocentric to recognize the advances of the nervous system in arthropods over coelen-

terates or in cephalopods over older molluscs or in reptiles over elasmobranchs.

2. Evolution of Hearing

One of the outstanding achievements of evolution has been the development of sensory systems—specialized sense organs and central pathways particularly related to one modality and its integration with other central functions. Hearing is an elegant example. Many invertebrate phyla, including coelenterates as well as higher groups, have receptors sensitive to water movement but appear to lack specialization for acoustic stimuli. (The concluding chapter by Webster provides a definition of hearing.) Several chapters document the relatively advanced state of hearing in some insects. Relevant to the present argument are a certain degree of frequency discrimination, temporal pattern discrimination, species recognition, and call recognition—signs of substantial complexity. The chapter by Budelmann points out intermediate groups with some indication of acoustic reception less advanced than that in the best insects.

Within the vertebrates, as detailed in a number of chapters, hearing and its peripheral and central organs have evolved greatly. Differences between major taxa, such as agnathans, elasmobranchs, teleosts, amphibians, reptiles, birds, and mammals are generally more than niche-specific features such as range of hearing which might be regarded as ecological adaptations within the same grade of complexity, that is, lateral radiations. Instead, we may notice at least four vertical grades. Whereas the agnathans have not been shown to hear, the auditory system of some elasmobranchs is quite appreciable. At best, however, it appears to be simpler than that of some teleosts, in anatomy and physiology as well as behavior. Birds and mammals are more advanced than fish.

For the present argument we need not address more difficult questions. For example, even after selecting the best, that is the most advanced species of each group with respect to hearing, it is not so easy to rank the relative advancement of teleosts, amphibians, and reptiles or of birds vis-a-vis mammals, or of cetaceans compared to bats, or of either one compared to other orders of mammals. Such questions may be difficult even after we take a closer look at the estimators of complexity—upon which the grades, as used here, are based.

3. Estimation of Complexity

Complexity is not necessarily vague or subjective. In principle, it could be estimated by objective measures. There is an extensive literature that is beyond the present scope. Proposals for measures of complexity include the number of bits of information required to describe the system and its operation or the number of statements in a program to simulate it (excluding "and not" statements, since they could be indefinite in number). They extend as far as the fractal dimensionality of chaos theory (Livi et al. 1988) and the computation of mutual information or information gain (Mars and Lopes da Silva 1987). It is sufficient here to indicate that a useful approximation would be a list of the distinguishably different components of the systems under consideration: the anatomical parts and subdivisions, the physiological reactions and interactions, the discriminable stimuli, available motor actions, and items in the complete behavioral repertoire. Any one of these can be used as a first approximation, if the two or more taxa being compared are equally well described by equivalent standards.

The list-of-components measure becomes more usable when we narrow the system and speak of the grade of complexity of one organ or modality, rather than of the taxon. For example, we may compare the acoustic sense organ and its adnexa, or the acoustically excited cortex, or the number of

discriminable sounds. An experienced comparative histologist is familiar with making judgments about relatively more and less differentiated regions of the brain. One of the goals of the ethologist is to discern the whole ethogram or the complete repertoire for the species in defined domains of behavior such as courtship or reproduction or acoustic signals intrinsic and extrinsic to the species. Although our data base is seriously inadequate, such measures of comparable fragments of the whole form the basis of the judgment given above about four vertebrate grades, at least. They justify the terms simpler and more advanced, lower and higher with respect to complexity—with that degree of confidence the available data deserve.

The two kinds of evolutionary differences, vertical grades and lateral adaptations, are not symmetrical or equivalent; nor are they mutually exclusive alternatives. One important contrast is that, whereas complexity can be estimated fairly objectively by observation, adaptation is basically a subjective guess. Only in exceptional instances is it feasible to do experiments to test the adaptive value of a trait. Such experiments ideally require, in addition to selective manipulation of the trait, controls that no other hidden difference is involved, also prolonged exposure to selective pressures under natural field conditions, with adequate quantitation of the consequences over at least a generation.

Evidence of complexity of the type manifested in the bat's auditory cortex should not be used to claim higher advancement relative to other mammalian orders since one can suspect that the cat or monkey is equally complex in ways that have not been physiologically dissected. Pending new information of this kind and the application of suitable criteria of complexity that allow for equally complex but qualitatively different solutions in taxa of disparate eco-ethological types, we can usefully avoid small steps in grade and pay attention to the few obviously different levels, such as the series: agnathan, carcharhinid, anuran, microchiropteran. These are, of course, not necessarily a phylogenetic or a time series, but simply representatives of four grades of complexity. It suffices to rest the case for grades on large differences such as those between classes. Once this is granted, the agenda spelled out below becomes clear.

4. Admission of Vertical Grades

If we can agree that there are grades of complexity (Bonner 1988), at least large ones such as those represented by coelenterates and arthropods, or elasmobranchs and mammals, it will already represent significant progress, since a popular recent trend among authors of books and essays has been to question the reality of evolutionary advance apart from adaptive radiation into ecological niches (Nitecki 1989). The result has been neglect of one of the great consequences of biological evolution, the development of more and more complex nervous systems and behavior—not inevitably, not steadily, but occasionally, irregularly, and to some extent independently for such subsystems as the auditory (Demski 1984). It is not a priori clear, self-evident, or testable whether the changes in grade are adaptive, or, if they are, what the nature of the advantage may be (Northcutt 1985). The only criterion, normally, for acceptance or rejection of hypotheses in these areas is plausibility, which is not a criterion of high reliability.

5. Agenda for Future Research

Recognition of the fact of advance is enormously heuristic. It suggests the work needed to determine just what is different between taxa of distinct grades. A sketch of such an agenda with respect, first to physiology, then behavior and anatomy includes the following.

I cannot find an authority for any statement on what exactly is different with respect to the physiology of audition, at its best, in fishes, amphibians, and mammals, apart from the upper frequency limit. It has even been suggested that there may be, in fact, rather small, primarily quantitative differences, when they occur at all! One may be excused for wondering if this is a premature conclusion from inadequate evidence, at least with respect to higher level processing. Compared to other branches of comparative physiology, we are sadly lacking in just that area—*neural* physiology, where the greatest span exists between simpler and more complex taxa. We are lacking in that we don't know what we want to know; we have no list of measurable variables to be filled in as they do in respiratory, in cardiovascular, and other branches of comparative physiology.

Behavior research is similarly short of conclusions comparing major taxa (Brown 1975; Roitbatt, Bever, and Terrace 1984; Weiskrantz 1985; Schusterman, Thomas, and Wood 1986; Hoage and Goldman 1986). What exactly are the differences between the behaviorally exhibited auditory abilities of the best endowed elasmobranchs, teleosts, amphibians, reptiles, birds, and mammals? There are only fragments such as the upper frequency limit or the intensity threshold—nothing approaching the number of discriminably different sounds, including mixtures of frequencies with time structure. Humans can detect the voices of man, woman, or child; a Southern drawl; Bach from Beethoven; and the Contra Costa White Crowned Sparrow from the Santa Cruz. Can a frog? Can a mocking bird?

Anatomy is different. Thanks to a better understood set of desiderata, anatomists have provided a lot of data with which to compare taxa. Most experts are reluctant, however, to do the comparing beyond the literature on their own order or class—and are naturally unhappy with the result when a nonexpert summarizes! Like other scientists, they are also technique-bound, each tending to see the brain as stained by his favorite method. Nevertheless, one can hope they will offer tentative generalizations as to the differences between classes, over and beyond discussion of homologies.

Special attention should be drawn to the more difficult levels of characterization of audition, since most discussion and literature is exclusively concerned with more manageable and measurable properties. An adequate assessment of the complexity of hearing in any taxon should include the domain of the so-called higher neural, or cognitive functions, including emotional, motivational, and teachable capacities which give hearing, like other modalities, its meaning (Adam, Meszaros, and Banyai 1981; Evarts, Shinoda, and Wise 1984; Kesner and Olton 1990). Assessing complexity in these spheres, without which our comprehension of biology would be severely limited, calls for substantial effort in each of the disciplines. For example, the behavioral endpoints will require innovative experimental as well as observational research using multiple measures of cognitive capacity each with appropriate quantitation, such as the agenda for comparative cognition spelled out by Bullock (1986).

Physiological endpoints offer a parallel opportunity. Are there cognitive event-related potentials to "oddball" stimuli, using a variety of ethologically important dimensions – in frog, lizard, pigeon, or opossum? If so, in what respects are they different from those in humans and monkeys (not only how are they similar)? How many physiological types of cells can be found with equal search in the hippocampus of rats and in the equivalent medial cortex of lizards, defined by the behavioral contexts that elicit firing increments?

The demands of such an agenda appear dauntingly great. They may deter especially those who are the best hope – the very experts we need to do the research. Two things should be said. First, we are overlooking one of THE great features of biological evolution as long as we neglect to study differences in vertical grade, and confine ourselves to lateral adaptations. These overlap, of course, and each helps the other. Nevertheless, there is a great gap in contemporary recognition of goals of biology, in just this domain of evolutionary advancement in complexity. Neurobiology should be leading the way since the brain and behavior are by far the principal arenas for advance.

The second thing to be said concerns tactics. The dauntingly difficult agenda is much more tractable if we do not start out with the hardest questions, such as comparing cognition in dolphin vs chimpanzee, but choose species pairs much farther apart in complexity – perhaps rays vs mormyrids or goldfish vs iguanas or frogs vs geckos. The challenge appears more approachable when we narrow the problem to the domain of hearing but we cannot forget the familiar caveat: traits can change independently and we should not narrow too much. At the least we should embrace the central nervous system, the forebrain, and the cognitive levels of hearing before pronouncing too narrow a conclusion.

6. Summary

The evolution of hearing and the structures and functions underlying it may illustrate some features of the story of behavior and the nervous system, especially if we can extend our present factual base. One such feature is the important distinction between adaptive differences among species of the same grade – usually the same genus and family, and the changes in grade of complexity, which are sufficiently large to be clear only when we compare major taxa: classes or phyla.

At least three grades are evident in this sensory modality. Many invertebrate groups have water movement receptors but apparently no specialized acoustic sense organs; crustaceans and possibly a few other invertebrates appear to have modest capacity for acoustic reception; some insects represent a markedly higher grade in having special sense organs, central analyzing systems, and behavior devoted to producing and responding to acoustic signals. Whether some fish and amphibians are at the same or a higher grade, certainly birds and mammals have achieved a still more complex grade.

The advances in these grades of complexity are great, not only in the sensory apparatus but especially in the brain centers for processing acoustic input, and in the resulting behavior, including the advance from discriminating a few dozens of sounds in some species to many orders of magnitude more in the human, and controlling vocal output of similar numbers of signals. This great fact of biological evolution is often overlooked in the zealous routing out of anthropocentrism, orthogenesis, and other heterodoxies.

In principle, complexity can be estimated objectively and advancement in grade therefore justified, case by case, especially comparing phyla and classes. Small differences in grade are more difficult to establish, particularly if attention is confined to one modality or a few groups of traits. Grades of complexity are distinguishable from lateral radiations manifested by specializations adapting species, genera, and families to their habits of life. The two categories are not mutually exclusive and can overlap, but for the most part trying to assign adaptive significance to the vertical grades is pure guesswork, with only plausibility for a criterion.

Large problems, which have hardly been taken up as yet and stand as major challenges to new research, are (1) to document such differences in sufficient detail, especially in physiology and behavior, (2) to improve the objective assessment of level of complexity, and (3) to apply such knowledge, case by case, to understand better the distinction between these two classes of species differences.

References

Adam G, Meszaros I, Banyai EI (1981) Brain and Behaviour. Oxford: Pergamon. (also in Advances in Physiological Science, Volume 17. Budapest: Akademia Kiado.

Bonner JT (1988) The Evolution of Complexity by Means of Natural Selection. Princeton: Princeton University Press.

Brown JL (1975) Evolution of Behavior. New York: Norton.

Bullock TH (1986) Suggestions for research of ethological and comparative cognition. In: Schusterman RJ, Thomas JA, Wood FG (eds) Dolphin Cognition and Behavior: A Comparative Approach. Hillsdale, NJ: Lawrence Erlbaum, pp. 207–219.

Demski LS (1984) Evolution of neural systems in the vertebrates: functional-anatomical approaches. Am Zoologist 24:689–833.

Evarts EV, Shinoda Y, Wise SP (1984) Neurophysiological Approaches to Higher Brain Functions. New York: John Wiley & Sons.

Hoage RJ, Goldman L (1986) Animal Intelligence. Washington, DC: Smithsonian.

Kesner RP, Olton PS (1990) Neurobiology of Comparative Cognition. Hillsdale, NJ: Lawrence Erlbaum.

Livi R, Ruffo S, Ciliberto S, Buiatti M (1988) Chaos and Complexity. Singapore and Teaneck, NJ: World Scientific.

Mars NJI, Lopes da Silva FH (1987) EEG analysis methods based on information theory. In: Gevins AS, Rémond A (eds) Methods of Analysis of Brain Electrical and Magnetic Signals. EEG Handbook (Revised Series, Vol 1). Amsterdam: Elsevier, pp 297–307.

Nitecki MH (1989) Evolutionary Progress? Chicago: University of Chicago Press.

Northcutt RG (1985) Brain phylogeny: speculations on pattern and cause. In: Cohen MJ, Strumwasser F (eds) Comparative Neurobiology. New York: John Wiley & Sons, pp 351–378.

Roitbatt HL, Bever TG, Terrace HS (1984) Animal Cognition. Hillsdale, NJ: Lawrence Erlbaum.

Schusterman RJ, Thomas JA, Wood FG (1986) Dolphin Cognition and Behavior: A Comparative Approach. Hillsdale, NJ: Lawrence Erlbaum.

Weiskrantz L (1985) Animal Intelligence. Oxford: Clarendon Press.

3

The Phylogeny of Octavolateralis Ontogenies: A Reaffirmation of Garstang's Phylogenetic Hypothesis

R. Glenn Northcutt

"Through the whole course of Evolution, every adult Metazoan has been the climax of a separate ontogeny or life-cycle, which has always intervened between adult and adult in that succession of forms which Haeckel terms "Phylogenesis." The real Phylogeny of Metazoa has never been a direct succession of adult forms, but a succession of ontogenies or life cycles.— —Ontogeny does not recapitulate Phylogeny: it creates it."

Garstang (1922:82)

1. Introduction

The relationship of ontogeny—the development or life history of an individual—to phylogeny—the history of successive biological populations—has been a focus of biological research for almost 200 years, and a number of solutions have been proposed (see Russell 1916; Holmes 1944; de Beer 1958; Gould 1977; Kluge and Strauss 1985, and Northcutt 1990a for reviews). Garstang's proposal (1922), however, appears to be the most insightful. Garstang realized that phylogeny is usually perceived as a succession of adults, when, in fact, it is the result of changes, through time, in an ancestral life history (those stages and processes that span the development of one zygote to another zygote). Thus in Garstang's view, subsequent changes in an ancestral ontogeny create phylogeny.

Although Garstang's insight into the relationship of ontogeny to phylogeny has been widely noted (Schindewolf 1946; Danser 1950; de Beer 1958; Gould 1977; Rosen 1982; de Queiroz 1985; Kluge 1988; Roth 1988), most researchers have not applied his conclusion to their own phylogenetic

analysis but, rather, have restricted such analyses to adult stages. Initially, the implications of Garstang's hypothesis were probably neglected due to a change of emphasis in embryological research: rather than continuing to describe ontogenies and histories, researchers began to search for mechanisms underlying developmental events. Equally important, there was no sophisticated methodology at that time for comparing ontogenetic sequences and determining the polarity of character transformations, i.e., which stages are primitive and which derived. The emergence of cladistics (Hennig 1966; Eldredge and Cracraft 1980; Wiley 1981) has since provided such a methodology, and phylogenetic analyses can now be expedited in the manner originally proposed by Garstang (Northcutt and Gans 1983; Mabee 1987; Langille and Hall 1989; Northcutt 1990a). In the following section, a method for the cladistic analysis of ontogenies is briefly outlined, and the ways in which ontogenies can change are summarized. In subsequent sections, this method is utilized to reconstruct a morphotype (i.e., the features of an ancestor as based on the shared primitive features of its descendants) of the development of an octavolateralis placode (Section 3), the number and distribution of octavolateralis placodes (Section 4), and, finally the phylogeny of ontogenetic changes of these placodes (Section 5). The primary focus will be on the phylogeny of ontogenetic changes of the placodes that give rise to the lateral line system, rather than changes in the octaval placode that gives rise to the receptors and nerve of the inner ear. The related lateral line system was chosen because comparable data do not exist for the inner ear; it is hoped, however, that the present analysis will stimulate such studies on this organ.

2. A Cladistic Method for the Analysis of Ontogenies

A cladistic analysis of ontogenies involves the following basic steps: (1) recognizing and describing the stages in the development of the characters of interest; (2) comparing these stages in the relevant taxa and formulating hypotheses regarding homologous stages; (3) determining the polarity of the suspected homologous stages utilizing the outgroup criterion; (4) generating scenarios regarding the biological significance of phyletic changes in the reconstructed ancestral ontogeny and recognizing the processes that may have produced these changes; (5) testing of the scenarios and their corollaries by continued studies.

2.1 Description and Recognition of Ontogenetic Stages

Although development of a character (any definable attribute of an organism) is essentially a continuous process, it is possible to recognize a sequence of stages based on changing morphology and associated processes. For example, four major stages can be recognized in the formation of the neural tube (neurulation) in most craniates: (1) an initial stage in which induction and apicobasal thickening of the cuboidal ectodermal cells of the dorsal midline results in the formation of the neural plate; (2) continued apicobasal thickening and pseudostratification of the columnar neuroectoderm, accompanied by transverse narrowing and longitudinal lengthening of the neural plate (neural plate shaping), which leads to (3) bending of the neural plate to form a neural groove stage and (4) closure of the neural groove due to neural fold apposition and fusion (Schoenwolf and Smith 1990). As noted by Løvtrup (1978) and others (Campbell and Richie 1983; Alberch 1985), however, describing the development of a character as a sequence of stages may be meaningful only if the stages are causally related (i.e., if stage A is required for the development of stage B and so on). In order to recognize a

causally related sequence of stages, numerous experimental data are needed to demonstrate that the sequence is the product of a series of inductive events. Fortunately, experimental data exist for many vertebrate characters whose developmental stages have already been described. For example, there is an extensive literature regarding the inductive processes underlying the development and differentiation of the placodes that give rise to the inner ears (see Noden and Van De Water, 1986 for a recent review), and a similar but less extensive literature exists for the placodes that give rise to the lateral line system (Harrison 1904; Stone 1922, 1928, 1929, 1933, 1935; Winklbauer and Hausen 1983, 1985; Smith, Lannoo, and Armstrong 1988).

2.2 Formulation of Hypotheses of Homologous Ontogenetic Stages

Hypotheses regarding homology of developmental stages, like those regarding homology of adult characters, should be based on the existence of similarities that are greater than chance (Remane 1956) and the stages should also exhibit a continuous phylogenetic history with linear rather than parallel derivation (Wiley 1981). Comparison of the developmental stages of characters in different taxa frequently reveals similarities that are suspected to be due to the historical continuity of these stages from their origin in a common ancestor. For example, four similar stages can be recognized in the formation of the neural tube in members of most craniate radiations (Huettner 1949; Nelsen 1953). The degree of overall physical and temporal similarity in these stages suggests that they probably arose in the common ancestor of these radiations rather than having arisen multiple times independently.

However, neural tube formation in ray-finned fishes, as well as in the caudal trunk region of birds and some mammals, involves a stage (neural keel) different from that seen in other craniates (Nelsen 1953). In these taxa the neural plate does not form a neural groove but rather a ventrally directed midline thickening (keel) that is subsequently remodeled to form the neural tube. The occurrence of this stage in ray-finned fishes and birds does not mean we should reject a hypothesis that the other stages in neurulation are homologous or a hypothesis that the neural tubes are themselves homologous. However, these differences do raise questions regarding whether neural groove and neural keel stages are homologous and which is primitive and which is derived in craniates. It is possible that either of these conditions, or even a different condition, is primitive for craniates. Such questions can only be answered by an out-group analysis that determines the polarity of these stages.

2.3 The Polarity of Suspected Homologous Ontogenetic Stages

Ideally the polarity and the phylogeny of ontogenetic stages would be inferred from a well-corroborated genealogical hypothesis generated independently on a large number of other characters. Such is frequently the case when a cladistic analysis is undertaken for ontogenetic stages or adult characters among members of higher taxonomic categories. For example the genealogical relationships of amphibians, reptiles, birds, and mammals are well corroborated by large numbers of characters, and it is very unlikely that additional data will force a reassessment of the relationships of these taxa. When characters are analyzed in members of lower taxonomic categories, however, there may be far more uncertainty regarding the genealogical relationships of the taxa, and several competing hypotheses frequently exist. In this case a cladistic analysis becomes more complicated but the basic methodology does not change. As we will subsequently see, inferences regarding character polarity and phylogeny are strongly affected by hypotheses of genealogy.

In the case of a well-corroborated hypothesis of genealogy, which is assumed to be the case in Figure 3.1A, the ontogenetic stages (A–E) in the development of a character must be described for at least three taxa (1–4 in this case) before a cladistic analysis can be undertaken. Such an analysis is based on an out-group comparison in which the sequence of stages is compared in an increasing hierarchial manner: initially, stages and sequences in taxa 3 and 4 are compared; then those features held in common by taxa 3 and 4 are sought in taxon 2; subsequently features held in common by taxa 2, 3, and 4 are sought in taxon 1. Those features held in common by two or more taxa are inferred, on the basis of parsimony, to have been present in an ancestor rather than having evolved independently.

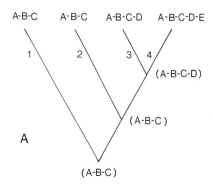

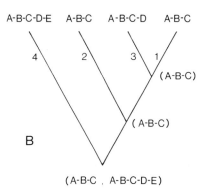

FIGURE 3.1. Stages (A–E) in the development of a character in four hypothetical taxa (1–4) and the ancestral ontogenetic stages (in parentheses) inferred from an out-group analysis. The effect that different genealogical hypotheses have on such analysis is illustrated if one first assumes that taxon 1 is the sister group of taxa 2–4 (A), then alternately assumes that taxon 4 is the sister group of taxa 1–3 (B).

Thus in the case illustrated in Figure 3.1A, the sequence of stages A–B–C–D are shared by taxa 3 and 4 and are therefore inferred to have been present in the ancestral population that gave rise to taxa 3 and 4. Comparison of the stages in taxa 2, 3, and 4 allows us to infer that only stages A–B–C were present in the ancestral population that gave rise to taxa 2, 3, and 4. This inference is supported further by the existence of the same sequence of stages in an additional outgroup (taxon 1). In this example the most parsimonious hypothesis (i.e., the one requiring the fewest number of transformations) is that the ancestral or primitive sequence of stages was A–B–C and that this sequence has been retained in taxa 1 and 2, whereas the sequence of stages in taxa 3 and 4 is derived, due to the terminal addition of a new stage (stage D) in the common ancestor of taxa 3 and 4, and the further addition of stage E in taxon 4 following the divergence of taxa 3 and 4. However it should be noted that the hypothesis that the embryonic sequence A–B–C–D–E occurred in the common ancestor of taxa 3 and 4, with the subsequent loss of stage E in taxon 3 is equally probable.

Furthermore stages A–B–C–D would be considered to be homologous in taxa 3 and 4 as would stages A–B–C in all four taxa, based on the degree of similarity and a continuous phylogenetic history with only linear transformations of these stages. For the same reasons, the adult characters (C, D, and E) exhibited in the four taxa would also be considered to be cladistically and patristically homologous (Wiley 1981; Northcutt 1984). For example if the theoretical ontogenetic sequence in Figure 3.1A actually represented the development of the proximal middle ear ossicle in craniates, stage C of taxon 2 might represent the hyomandibula of bony fishes, whereas stage D of taxon 3 and stage E of taxon 4 might represent the columella and stapes, respectively, in tetrapods. In this specific case, these three terminal stages are interpreted to be cladistically and patristically homologous due to their linear transformation.

Let us assume, however, that additional analysis utilizing a number of other characters indicated that taxon 1 in Figure 3.1A was not the out-group of taxa 2 through 4 but that taxon 4 was the out-group of taxa 1 through 3 (Fig. 3.1B). In this case, the inferences reached from an out-group analysis of the ontogenetic sequence A–E would be considerably different. Assuming that addition or deletion of ontogenetic stages is equally probable, and that such alterations need not occur in a step-wise fashion, the two sequences listed within parenthesis in Figure 3.1B have equal probability of representing the ancestral ontogenetic sequence for taxa 1 through 4, as each potential ancestral sequence would require a minimal number of two transformations to exhibit the variation seen in taxa 1 through 4 (the sequence A–B–C–D is not considered a possible ancestral sequence, as it requires at least one more event). If A–B–C represents the ancestral sequence, then stages D and E have been added in taxon 4 (1 event), and stage D has been added independently in taxon 3 (1 event).

If this were the case, stages A–B–C would be cladistically and patristically homologous in all taxa, but stage D in taxa 3 and 4 would be a case of parallel homoplasy, and stage E of taxon 4 would be a uniquely derived stage also homoplasous to the adult characters of taxa 1, 2, and 3.

If A–B–C–D–E represents the ancestral sequence, then deletion of stages D and E in the common ancestor of taxa 1, 2, and 3 would represent one event, as would the re-evolution of stage D in taxon 3. In this case, stages A–B–C would be cladistically and patristically homologous in all taxa, and the re-evolved stage D in taxon 3 would be homoplasous to stage D in taxon 4. In addition, stage D in taxon 3 would also be a homoplasous adult character to stages C and E of the other taxa, but stages C and E in taxa 1, 2, and 4 would be interpreted as cladistically homologous.

These two examples (Fig. 3.1) demonstrate the striking affects that different hypotheses of genealogy have on the results of a cladistic analysis of the same characters. The inferred ancestral ontogenetic sequence of stages and the recognition of homologues among living taxa are both altered by the particular hypothesis of genealogy that is adopted. For these reasons, an analysis should consider all relevant hypotheses of genealogy and test these hypotheses cladistically. However, this may not be feasible unless a large number of ontogenetic characters are being examined (it is not uncommon for fifty or more characters to be used in current cladistic tests of rival hypotheses of genealogies). The present analysis assumes that the craniate relationships proposed by Janvier (1981) and Young (1986) are valid.

2.4 Generation of Phylogenetic Scenarios

The determination of the polarity of ontogenetic stages and the inference of ancestral sequences isolates the points in phylogeny where biologically significant events have occurred. For example, in the theoretical phylogeny illustrated in Figure 3.1, important transformations occurred in the populations between the origin of taxon 2 and the origin of taxon 3 and in the populations following the separation of taxa 3 and 4. The analysis also suggests that both of these events occurred by terminal addition to a sequence. Such terminal alterations may include metamorphic or other re-organizational events that are correlated with major changes in life-history strategies. Thus the results of this cladistic analysis indicate that the life histories of taxa 3 and 4 should be compared to those of the out-groups in order to isolate significant differences in these histories and determine whether or not the differences appear to be correlated with the derived stages. For example, if taxa 2, 3, and 4 in the theoretical phylogeny illustrated in Figure 3.1A represented hemichordates, cephalochordates, and craniates, respectively, the ontogenetic sequences might represent the development of the cephalic neural tube, in which case the development of a complex brain (stage E) would be correlated with the shift from filter-feeding to active predation (Northcutt and Gans 1983; Gans 1989; Langille and Hall 1989). In this case, a scenario was postulated regarding the ontogenetic stages and possible underlying processes that would have transformed a simple neural groove or tube into a more complex brain. A scenario is thus a series of hypotheses regarding the biological processes and biological significance of phylogenetic transformations. The major advantage of such scenarios, beyond outlining how historical events may have occurred, is that their generation focuses our attention on the ontogenetic events and processes that are critical to such histories and frequently reveals our ignorance regarding these events and processes. In the case of craniate cephalization, a picture is emerging in which neural crest, paraxial mesoderm, and neurogenic placodes were the critical embryonic tissues, but little information currently exists regarding the morphogens and ontogenetic processes underlying the origin of these tissues.

One final advantage of generating scenarios is that their formulation may indicate that similar, if not identical, processes are associated with recurring events such as the elaboration of complex brains or the occurrence of electroreception, endothermy, flight, etc. The generation of scenarios thereby provides a powerful tool for analyzing convergent evolution and detecting the rules and constraints that govern phylogenetic change.

2.5 Testing Phylogenetic Scenarios

If phylogeny is the result of changes in an ancestral ontogeny, many of the processes postulated to underlie such changes can be identified by examining the processes currently operating in development. Some of these processes may have changed through time, even though their associated ontogenetic stages are conserved. For example all craniate embryos exhibit reorganization of the germ layers (gastrulation), but this occurs by a number of different mechanisms (Balinsky 1975). The number of mechanisms or processes associated with a particular ontogenetic stage appears limited, however, and the mechanisms themselves can be treated as characters in a cladistic framework, which means that their polarity and transformations can be determined. Thus the history of morphological stages and their processes can also be determined in many cases. Such analyses do not identify possible stages or processes that no longer exist among living taxa, but in some cases these stages and processes can be inferred from the adult or larval anatomy of extinct taxa and the range of processes exhibited by extant taxa. In other cases we may fail to detect stages and processes that no longer exist, but this is an unavoidable limitation of all historical sciences and its possible disruptive affect can be assessed only following an attempt to reconstruct such events.

2.6 Ontogenetic Changes

A number of developmental processes underlie changes in ontogeny (de Beer 1958; Gould 1977; Alberch et al. 1979; Fink 1982; Roth 1988; Thomson 1988). Changes can occur in the embryonic source or precursor of a character, or changes can occur in developmental processes (patterning, sequence, and/or timing). For example the dorsal spinal nerves of cephalochordates arise from cells within the neural tube, whereas those of craniates arise from neural crest; the epithelium of the gut may arise from the roof (as in teleosts) or the floor (as in salamanders) of the archenteron. Similarly, the patterning mechanisms that underlie the development of a character may change: different patterning mechanisms are responsible for gastrulation and neurulation among different craniate radiations.

The sequence of ontogenetic stages that culminates in a particular adult character can change in phylogeny in several ways (Fink 1982; Kluge and Strauss 1985; Wake and Larson 1987; Northcutt

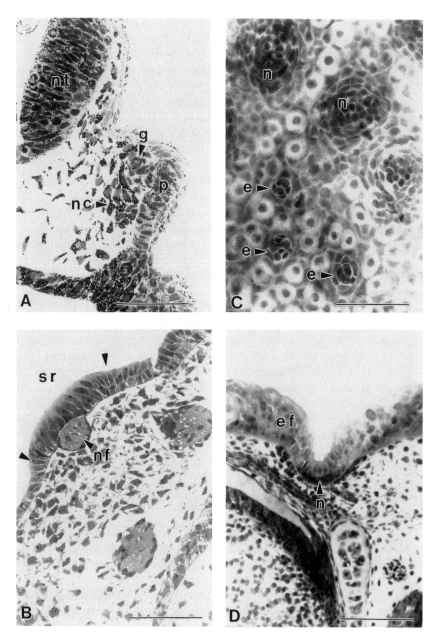

FIGURE 3.2. Photomicrographs of various stages in the development of lateral line placodes. (A) Transverse section through the head of a stage-35 axolotl embryo, illustrating the relationship of cephalic neural crest (nc) to the neural tube (nt) and the anterodorsal lateral line placode (p) and its developing sensory ganglion (g). (B) Elongated sensory ridge (sr) of a cephalic lateral line placode and its associated lateral line nerve (nf) in a stage-26 *Squalus* embryo. (C) Surface view of developing neuromasts (n) and electroreceptors (e) in the sensory ridge stage of a lateral line placode in a stage-43 axolotl embryo. (D) Transverse section through the head of a stage-49 channel catfish embryo, illustrating the development of epidermal folds (ef) adjacent to a groove neuromast (n). Bar scales equal 200 μm (A) and 100 μm (B, C, D).

1990a). An ancestral ontogenetic sequence, such as A–B–C, could change in at least seven ways: terminal addition, A–B–C–D; terminal deletion, A–B; nonterminal addition, A–D–B–C; nonterminal deletion, A–C; or terminal (A–B–X), nonterminal (A–X–C) or even complete (X–Y–Z) substitution.

Cases of terminal addition or deletion can be interpreted as changes in timing events (heterochrony), so that truncation or prolongation of an ontogenetic sequence results in deletion or addition of ontogenetic stages (de Beer 1958; Gould 1977; Alberch et al. 1979). However, substitutions and nonterminal additions or deletions of stages are most likely due to changes in embryonic source or patterning. Because multiple changes can occur in any sequence and because alterations in stages do not always occur in stepwise fashion (i.e., two or more stages can be altered simultaneously), the phylogenetic history of a given character will rarely exhibit a simple recapitulatory pattern (Alberch et al. 1979; Mabee 1989a,b; Northcutt 1990a).

3. Developmental Stages of a Primitive Octavolateralis Placode

Most craniates that have been examined exhibit a variable number of localized thickenings, termed placodes (Fig. 3.2A), in the inner layer of cephalic ectoderm shortly after neurulation (von Kupffer 1895; Platt 1896; Landacre 1912; Landacre and Conger 1913; Coghill 1916; Stone 1922; Knouff, 1935; Holmgren, 1940). These cephalic placodes are grouped into a dorsolateral, or octavolateralis, series and a ventrolateral, or epibranchial, series, based on their topography and subsequent embryonic fate (von Kupffer 1895; Platt 1896; Landacre 1910; Stone 1922). The dorsolateral placodes give rise to the hair-cell based sensory organs of the inner ear and lateral lines, as well as to the sensory cranial nerves that innervate these organs. The ventrolateral placodes give rise to the sensory components of the facial, glossopharyngeal, and vagal nerves that innervate chemoreceptive taste buds. However the embryonic origin of taste buds is still unknown. It is not clear whether taste buds arise from the ventrolateral placodes or from other embryonic tissues.

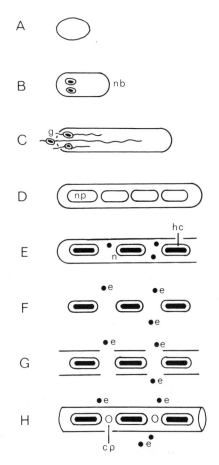

FIGURE 3.3. Stages (A–H) in the development of a primitive lateral line placode. cp, canal pore; e, electroreceptor; g, sensory ganglion of a lateral line nerve; n, neuromast; hc, hair cells of a neuromast; nb, neuroblasts of a sensory ganglion of a lateral line nerve; np, neuromast primordium.

Although the placode that gives rise to the inner ear—traditionally termed the otic placode but here termed the octaval placode to avoid confusion with an otic placode that gives rise to the receptors of the otic lateral line and nerve—arises earlier in development than the placodes that give rise to the lateral lines, the octaval and lateral line placodes are believed to constitute a single series, as they all give rise to sensory organs whose receptors consist of hair cells and to the sensory ganglia that innervate these receptors. This interpretation is further reinforced by similarities in the early stages of the development of placodes that give rise to lateral lines (Fig. 3.3) and those that give rise to the inner ear (see Noden and Van De Water 1986 for a recent review). Both the octaval and lateral line placodes can initially be recognized as localized thickenings of the ectoderm due to the apicobasal elongation of the initially cuboidal cells of the inner layer of the ectoderm (p, Fig. 3.2A). Shortly after the formation of the placodes (stage A, Fig. 3.3), cells (neuroblasts) located within the placodes proximal to the developing neural tube begin to migrate out of the placodes (g, Fig. 3.2A) and form the primordial sensory ganglionic cells (stages B and C, Fig. 3.3) that will innervate the receptors that develop subsequently from the placodes (Platt 1896; Landacre 1910; Coghill 1916; Stone 1922).

Further development of the octaval and lateral line placodes diverges at this stage. The octaval placode invaginates to form the octaval vesicle, and multiple receptors (hair cell maculae) develop at various positions within the epithelial octaval sac, whereas the lateral line placodes elongate, forming sensory ridges on the head (Fig. 3.2B; stage C, Fig. 3.3), or actually migrate along the trunk, leaving sensory primordia in their wake (Platt 1896; Johnson 1917; Ruud 1920; Stone 1922, 1933; Winklbauer and Hausen 1983; Metcalfe, Kimmel, and Schabtach 1985; Vischer 1989a).

Unfortunately, details regarding the subsequent development of the sensory maculae of the inner ear in each of the major craniate radiations are insufficient to allow a cladistic analysis of the ontogenies of the inner ear. For this reason the present analysis will be limited to the fate of the better understood lateral line placodes.

The cells of lateral line placodes remain morphologically undifferentiated at the sensory ridge stage (sr, Fig. 3.2B, Fig. 3.3C), but the sensory ganglionic cells are much further developed, and distinct bundles of peripheral neurites, already associated with developing Schwann cells, are closely aligned with the sensory ridge (Fig. 3.2B) along its entire rostrocaudal length. Similarly, the axons of the sensory ganglionic neurons have already reached the medulla, and the cranial nerve roots formed by these fibers enter the future octavolateralis area of the medulla (unpublished observations).

The next major morphological event that characterizes the development of lateral line placodes is the formation of neuromast primordia (stage D, Fig. 3.3). The morphologically undifferentiated epithelial cells of the placode begin to aggregate into oval clusters of 20 to 30 cells (Stone 1922, 1933; Winklbauer and Hausen 1983; Fritzsch and Bolz 1986; Northcutt 1986a). Each of the neuromasts is characterized by major and minor axes, with the major axis usually oriented parallel to the long axis of the sensory ridge. Darker staining and more superficially located sensory hair cells can be distinguished from the paler staining and more deeply located support and mantle cells within primordial neuromasts (Fig. 3.2C, Fig. 3.3E) within hours of the initial formation of these sensory primordia.

Primitively, the lateral line system of gnathostomes (Northcutt 1986b), as well as the lateral line system of lampreys (Bodznick and Northcutt 1981; Ronan 1986), consists of electroreceptors in addition to mechanoreceptive neuromasts. Craniate electroreceptors, however, were discovered subsequent to the initial studies of the development of lateral line placodes, and their embryonic origin remained unknown until recently. In salamanders (Northcutt 1986a; Northcutt, Fritzsch, and Brändle 1990) and elasmobranchs (unpublished observations) the primordia of electroreceptors appear after the initial formation of primordial neuromasts. The primordia of electroreceptors can be recognized initially as small clusters of three or four cells, each comprising a single hair cell surrounded by two or more sickle-shaped support cells which differentiate primarily from the lateral portions of the sensory ridges (Fig. 3.2C, 3.3E). At this stage (stage E), the sensory ridges and their developing receptors are still covered by the outer layer of ectoderm, and subsequent development involves movements of the overlying ectoderm as well as the receptors themselves. Apical cavities develop between the overlying ectodermal cells and the sensory macular regions of the neuromasts, followed by retraction of the ectodermal cells lying above the neuromasts, resulting in exposure of the neuromasts on the surface (Allis 1889; Sato 1956, 1976; Srivastava and Srivastava 1969; Vischer 1989a) and their frequent occurrence in shallow depressions (stage F, Fig. 3.3). Superficially located neuromasts at this stage have traditionally been termed pit organs, and lines of such neuromasts occur widely in all anamniotic vertebrate radiations (Northcutt 1989).

Apical cavities also form over the sensory region of developing electroreceptors and are followed by retraction of the overlying ectoderm (Ruud 1920; Sato 1956; Fritzsch and Bolz 1986; Northcutt

1986a, 1987; Vischer 1989b), although the sensory maculae of these receptors do not move to the surface but remain stationary in salamanders and actually invaginate into the underlying dermis in most other gnathostomes. These movements result in the electroreceptors forming ampullary organs, in which the sensory maculae are housed in blind sacs (ampullae) located within the dermis and connected to surface pores by elongated tubes lined with cuboidal epithelium. Although there are marked similarities in the morphogenesis of electroreceptors and neuromasts, all stages of neuromast development temporally precede similar developmental stages of electroreceptors, and these two receptor classes are easily distinguished from one another
in all stages.

Following their initial formation, electroreceptor primordia appear to migrate out of the sensory ridges and may undergo several divisions, creating large fields of receptors, prior to the final formation of their tubes and pores. In adults, these organs and their pores may lie adjacent to lines of superficial neuromasts, or they may be widely scattered over the surface of the body (Northcutt 1986b).

Formation of the canals that enclose many lines of neuromasts represents the final stages of placodal differentiation in most craniates (stages G and H, Fig. 3.3). The beginning of canal formation is marked by the creation of ectodermal ridges paralleling the major axes of the superficially located neuromasts (Figs. 3.2D, 3.3G). Subsequently, the apical surfaces of a pair of ectodermal ridges arch over each neuromast and fuse, creating an epithelial tube with the neuromast located at the bottom of the tube (Allis 1889; Ruud 1920; Lekander 1949; Northcutt 1987; Vischer 1989a; Webb 1989). The ectodermal ridges do not completely fuse between neuromasts, but leave gaps that subsequently become canal pores that remain open to the surface (stage H, Fig. 3.3). The source and range of tissue movements involved in canal formation are poorly known. It is not known whether the ectodermal ridges are of placodal or general ectodermal origin, and the epithelial canals are usually associated with a secondary bony canal whose embryonic origin is also unknown. Finally, the canals and their enclosed neuromasts are frequently located well within the dermis, which suggests that ectodermal ridge formation may occur initially by an evagination of general or placodal

ectoderm adjacent to the neuromast, followed by active invagination of these tissues.

Although details of the development of lateral line placodes do not exist for all vertebrate radiations, the data are sufficient to indicate that the stages outlined in Figure 3.3 represent the primitive pattern of development for vertebrates. Lateral line placodes have been described for lampreys (von Kupffer 1895; Damas 1944, 1951) and for all extant anamniotic radiations (Platt 1896; Landacre 1910; Landacre and Conger 1913; Ruud 1920; Stone 1922; Knouff 1935; Holmgren 1940; Lekander 1949; Pehrson 1949). The details of neuromast development are essentially the same for all gnathostome radiations (Johnson 1917; Ruud 1920; Stone 1933; Lekander 1949; Vischer 1989a; Webb 1989). Given the homology of electroreceptors in lampreys and all nonteleost anamniotic gnathostomes (McCormick 1982; Bullock, Bodznick, and Northcutt 1983; McCormick and Braford 1988), there is reason to believe that electroreceptors developed primitively from lateral line placodes, as they do in modern elasmobranchs and salamanders. Finally, the occurrence of lateral line grooves and/or canals in most members of all extinct, as well as extant, craniate radiations (Northcutt 1989) suggests that the lateral line placodes of the earliest vertebrates, if not the earliest craniates, exhibited stages A through G or H. As will be seen in the subsequent two sections; however, the number and developmental fate of individual placodes varies widely in the different craniate radiations.

4. The Ancestral Number and Distribution of Octavolateralis Placodes

Sufficient details regarding the development of the octavolateralis placodes are known for species of the six genera illustrated in Figures 3.4–3.6 to allow a cladistic determination of the number and distribution of placodes in the earliest gnathostomes. Comparable stages have been selected at a point in development which corresponds to sensory ridge formation (stage D, Fig. 3.3) or to early neuromast formation (stage E, Fig. 3): *Squalus*, 22 mm total body length (Holmgren 1940); *Protopterus*, 12.4 mm total body length (Pehrson 1949); *Lepisosteus*,

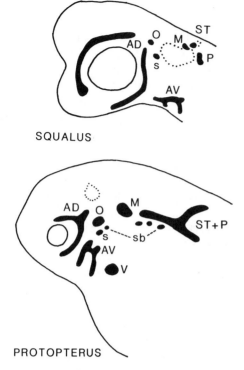

FIGURE 3.4. Number and distribution of the octavolateralis placodes at comparable developmental stages in the spiny dogfish, *Squalus acanthias*, and the African lungfish, *Protopterus annectens*, based on the studies of Holmgren (1940) and Pehrson (1949), respectively. The lateral line placodes are indicated in black; the octaval placodes are indicated by dotted ovals. AD, anterodorsal placode; AV, anteroventral placode; M, middle lateral line placode; O, otic placode; P, posterior placode; s, spiracular primordium; sb, suprabranchial neuromast primordia; ST, supratemporal placode; V, ventral trunk line placode.

8.5 mm total body length (Landacre and Conger 1913); *Leuciscus*, 141 hours (Lekander 1949); *Ambystoma*, stage 35 (Stone 1922); and *Rana*, stage 7 (Knouff 1935). This point in placodal development was chosen because the total number of placodes that have been described are present at this stage and have given rise to the sensory ganglia of the cranial nerves that will innervate placodally derived receptors, but the placodes themselves are only beginning to form sensory ridges. At later stages, the sensory ridges will vary greatly in terms of their topographical extent and number of subdivisions, which increases the likelihood that homologous placodes will be identified erroneously. In a new nomenclature introduced here, each placode bears the same name as the lateral line nerve to whose sensory ganglion it gives rise. Previously, these placodes have generally been identified according to the lateral lines to which they give rise, but as the names of these lines change from species to species, whereas the names of the cranial nerves that innervate these lines do not change, it is preferable to identify the placodes according to the cranial nerves (see Northcutt 1989 and Song and Northcutt 1991 for recent reviews of the lateral line cranial nerves).

An out-group analysis of placodal variation among these six genera (Figs. 3.4–3.6) suggests that the embryos of the earliest gnathostomes possessed at least six lateral line placodes in addition to an octaval placode. At least three lateral line placodes (anterodorsal, anteroventral, and otic) appear to have existed anterior to the octaval placode in the embryos of the earliest gnathostomes, as anterodorsal (AD) and anteroventral (AV) placodes occur in all six extant ontogenies, and an otic placode (0) occurs in five of the six ontogenies. The AD placode gives rise to the anterodorsal lateral line nerve (the superficial ophthalmic and buccal rami of the facial nerve of the older literature), which innervates the receptors of the supra- and infraorbital lines as well as the receptors of the anterior pit line in all six taxa. In three of the illustrated taxa (*Squalus*, *Lepisosteus*, and *Leuciscus*) separate placodes appear to give rise to the supra- and infraorbital lines (Figs. 3.4, 3.5). However reexamination of a *Squalus* embryo of 15 mm total body length reveals that only a single placode exists at this stage; its dorsal and ventral arms give rise to the supra- and infraorbital lines, respectively. It has not been possible to reexamine comparable stages in ray-finned fishes; it may be that their AD placodes have been reconstructed incorrectly, or they may exhibit a derived pattern of development.

An anterior pit line develops in only four of the six taxa being examined (*Protopterus*, *Lepisosteus*, *Leuciscus*, and *Ambystoma*), but the widespread occurrence of this pit line among the extinct placoderms and acanthodians (see Northcutt 1989 for a recent review) suggests that ancestrally the AD placode gave rise to an anterior pit line, as it does in modern bony fishes and salamanders.

The AV placode gives rise to the anteroventral lateral line nerve (the external mandibular ramus

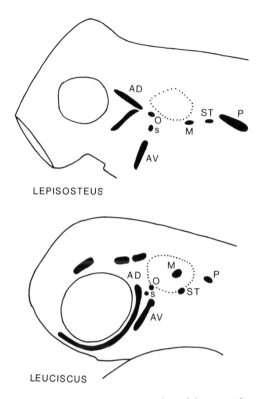

 is described above with labels LEPISOSTEUS and LEUCISCUS.

FIGURE 3.5. Number and distribution of the octavolateralis placodes at comparable developmental stages in the longnose gar, *Lepisosteus osseus*, and in a cyprinid fish, *Leuciscus rutilus*, based on the studies of Landacre and Conger (1913) and Lekander (1949), respectively. Abbreviations are the same as in Figure 3.4.

(s, Figs. 3.4 and 3.5), it is presently unclear whether the spiracular primordium arises as a subdivision of the otic placode or from an entirely separate placode (Hammarberg 1937). If the spiracular placode is a separate primary placode, it is the only member of the octavolateralis series that does not also give rise to a distinct sensory ganglion. For this reason, it is tentatively interpreted as a subdivision of the otic placode, pending additional studies of its development. Interestingly, although a spiracular organ does not exist in teleost fishes, Lekander (1949) described a transient spiracular placode in *Leuciscus* (Fig. 3.5), and Pehrson (1949) described a series of transient placodes or neuromasts in lungfishes (sb, Fig. 3.4), in which the spiracular primordium may be the most rostral member of a suprabranchial series. Suprabranchial neuromasts do not occur in other gnathostomes, but a comparable pit line does occur in lampreys (Holmgren 1942). Unfortunately, the embryonic origin of the suprabranchial neuromasts of lampreys as well as

of the facial nerve of the older literature); in all six taxa this nerve innervates the receptors of the oral and preoperculo-mandibular lines, as well as a variable number of additional cheek lines whose topography and character states (canals, grooves, or pit lines) vary greatly among living gnathostomes (Northcutt 1989).

The otic placode gives rise to the otic nerve (otic ramus of the facial nerve of the older literature) and the receptors of the otic line. Although an otic line and otic lateral line nerve do not exist in *Ambystoma* or possibly other salamanders, they do appear to exist in *Rana* (Knouff 1935) and in members of all other anamniotic gnathostome radiations (Northcutt 1989). The otic nerve also innervates the receptors of the spiracular organ. Although this organ appears to develop from a separate placode

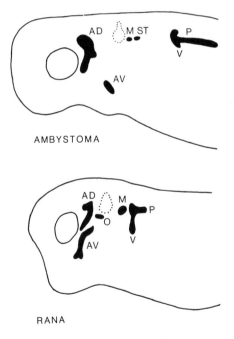

FIGURE 3.6. Number and distribution of the octavolateralis placodes at comparable developmental stages in the spotted salamander, *Ambystoma maculatum*, and the leopard frog, *Rana pipiens*, based on the studies of Stone (1922) and Knouff (1935), respectively. Abbreviations are the same as in Figure 3.4.

the innervation of these neuromasts is unknown. Therefore there is neither direct (embryonic) nor indirect (neural) evidence in lampreys for a suprabranchial placode. Although there is insufficient evidence to demonstrate that an additional preoctaval lateral line placode existed in the earliest gnathostomes, it is possible that the spiracular organ represents the vestige of a suprabranchial line that arose from an additional lateral line placode.

An out-group analysis of the six taxa for which there is sufficient ontogenetic data (Figs. 3.4–3.6) indicates that at least three lateral line placodes (middle, supratemporal, and posterior) occurred posterior to the octaval placode in the embryos of the earliest gnathostomes, as these three placodes occur in representatives of all anamniotic gnathostome radiations. A middle (M) lateral line placode gives rise to the middle lateral line nerve (lateral line ramus of the glossopharyngeal nerve of the older literature) and to the receptors of the temporal line and the middle pit line. The temporal line is present in the adults of all anamniotic gnathostome radiations but a middle pit line in elasmobranch fishes has been reported only in adult *Etmopterus* (Allis 1923), although it may exist as a transitory pit line in *Squalus* (Landacre 1916). Knouff (1935) claimed that a post-octaval placode in *Rana*, interpreted here as the middle lateral line placode, gives rise to ganglionic cells but no receptors. Strong (1895), however, claimed that the ramus currently identified as the middle lateral line nerve does innervate neuromasts, but it is not clear whether these neuromasts are homologous to the neuromasts of temporal and/or middle pit lines of other gnathostomes. Unfortunately a recent description of the development of the lateral line system of *Rana pipiens* (Smith, Lannoo, and Armstrong 1988) does not clarify the issue.

A supratemporal (ST) lateral line placode gives rise to the supratemporal lateral line nerve (supratemporal ramus of the vagal nerve of the older literature) and to the receptors of the post-temporal line, supratemporal commissure, and the posterior pit line in members of all four extant anamniotic gnathostome radiations. Anuran amphibians, however, apparently fail to develop a supratemporal placode, as only two post-octaval placodes form ganglionic cells, and neuromasts clearly do not form in positions comparable to those of the lateral lines in other gnathostomes.

A posterior (P) lateral line placode gives rise to the posterior lateral line nerve (the lateralis rami of the vagal nerve of the older literature) and to the receptors of the dorsal and lateral trunk lines in all six taxa whose ontogenies have been described. The dorsal trunk line exists as a pit line (stage F, Fig. 3.3) in all six taxa, whereas the lateral trunk line usually occurs as a closed canal (stage H, Fig. 3.3), which indicates that ontogenetic trajectories of the derivatives of a single placode can be uncoupled.

A ventral trunk line of superficial neuromasts exists embryonically in some elasmobranch fishes (Johnson 1917; Disler 1977), in lungfishes, and in amphibians, but not in living nonteleost ray-finned fishes (Northcutt 1989). Although the placodal origin of the ventral or yolk trunk line of elasmobranch fishes is unknown, Pehrson (1949) claimed that the ventral trunk line of lungfishes arises from a placode (V) distinctly separate from the placode (P) that gives rise to the dorsal and lateral trunk lines (Fig. 3.4). The condition in amphibians is less clear. Smith, Lannoo, and Armstrong (1988) have claimed that experiments in *Ambystoma* demonstrate that the ventral trunk line arises from a separate placode, although the detailed observations of Stone (1922) and Knouff (1935) would indicate that the ventral trunk line initially arises as a subdivision of the posterior placode (Fig. 3.6). The experiments performed by Smith, Lannoo, and Armstrong (1988) involved the transplanting of large sections of pigmented ectoderm from wild-type embryos into albino embryos. The orientation of these transplants was such that placodal tissue could have been recognized migrating both caudally and ventrally, but without detailed histology it could not have been determined whether one or two placodes originally existed. Therefore, although there is some evidence that the ventral trunk line of gnathostomes arises from a separate placode, the innervation of the receptors of all three trunk lines by a single nerve (Northcutt 1990b) suggests that only a single posterior lateral line placode existed primitively and that this placode initially gave rise to three separate secondary placodes that migrated onto the trunk, forming the dorsal, lateral, and ventral trunk lines, respectively.

At present there is no firm basis for reconstructing the number and fate of the lateral line placodes in the earliest craniates. Although lateral line placodes have been described in lampreys (von Kupffer 1895; Damas 1944, 1951), and it is clear that these placodes give rise to the lateral line receptors as well as the cranial nerves that innervate these receptors, a comprehensive account of the development of the lateral line system in lampreys does not exist. Based on the number of cranial nerves that innervate lateral line receptors, it is probable that anterodorsal, anteroventral, otic, middle, supratemporal, and posterior lateral line placodes exist in lampreys, but the distribution of the lateral lines is distinctly different from that in gnathostomes. Given the even larger number and unique distribution of the lateral lines of the extinct heterostracans (reviewed in Northcutt 1989), it is probable that the pattern of placodal differentiation in these taxa differed greatly from that in gnathostomes and that additional lateral line placodes existed in heterostracans.

Not only does the number of lateral line placodes vary among living gnathostomes (Figs. 3.4–3.6), the receptor cell lineages of single placodes also vary within a species. For example, in elasmobranch fishes, such as *Squalus*, the anterodorsal and anteroventral placodes give rise to both electroreceptors and neuromasts, whereas the remaining placodes give rise only to neuromasts (R.G. Northcutt and E. Gilland unpublished observations). In *Ambystoma*, however, all of the lateral line placodes give rise to both electroreceptors and neuromasts, except for the posterior placode, which gives rise only to neuromasts (unpublished observations). Surprisingly, the electroreceptors that arise from the middle and supratemporal placodes in *Ambystoma* are not innervated by the cranial nerves that arise from these placodes; rather they are innervated by a caudally directed ramus of the anteroventral lateral line nerve (Northcutt 1990b). Thus the ganglionic cells of one placode (AV) appear to be able to induce the cells of postoctaval placodes to develop electroreceptors. Unfortunately, it is presently impossible to determine whether all lateral line placodes gave rise to both types of receptors primitively, as comparable developmental studies do not exist for other craniates. Circumstantial evidence regarding the distribution and/or innervation of electroreceptors in extinct and extant vertebrates, however, suggests that, primitively, all lateral line placodes may have given rise to both electroreceptors and neuromasts. Living holocephalans possess a rudimentary field of electroreceptors that may be innervated by the otic lateral line nerve (Fields 1982), and both lampreys and lungfishes possess trunk electroreceptors that are innervated by a caudally directed ramus of the anteroventral lateral line nerve (Northcutt 1986b; Ronan and Northcutt 1987). The innervation of the electroreceptors by the otic nerve in holocephalans suggests that an otic placode may give rise to these receptors, and the innervation of trunk electroreceptors by the anteroventral lateral line nerve presents a pattern that is markedly similar to that of the development and innervation of postoctaval electroreceptors in *Ambystoma*. Given their absence in living cartilaginous fishes, it is possible that the development of postoctaval electroreceptors has occurred independently in lampreys and bony fishes. The ptyctodontid placoderms, however, the suspected sister group of all other gnathostomes, exhibit distinct fields of dermal pores, believed to have housed electroreceptors, which are located caudally on the head in association with the rostral terminus of the dorsal trunk lateral line (Ørvig 1971). This pattern of electroreceptor distribution in placoderms indicates that the postoctaval placodes in some members of all radiations of vertebrates, with the exception of cartilaginous fishes, may develop electroreceptors and that this was a primitive ontogenetic feature of the earliest vertebrates.

A number of placodal features vary extensively among craniate taxa: the number of placodes, the final topographical distribution of the developed lines and fields of receptors, the receptor cell lineages of the placodes, and the innervation of these electroreceptors. Given this variation, the phylogeny of lateral line ontogenies becomes quite complex, as will be seen in the subsequent section.

5. The Phylogeny of Ontogenetic Changes in Lateral Line Placodes

The development of the lateral line placodes has not been described for members of all sixteen radiations of craniates illustrated in Figure 3.7. Ontogenies cannot be examined in five of the sixteen craniate radiations, as they are extinct, and there are few if any developmental data for three

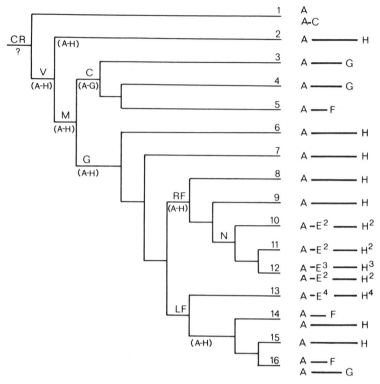

FIGURE 3.7. A cladistic analysis of lateral line placode development in anamniotic craniates. Developmental stages for the placodes in each radiation are indicated to the right of the dendrogram, and inferred ancestral stages for each clade are indicated in parentheses. C, cephalaspidomorphs; CR, craniates; G, gnathostomes; LF, lobe-finned bony fishes; RF, ray-finned fishes; V, vertebrates; (1) hagfishes; (2) heterostracans; (3) osteostracans; (4) anaspids; (5) lampreys; (6) placoderms; (7) cartilaginous fishes; (8) polypteriform fishes; (9) chondrostean fishes; (10) gars; (11) *Amia*; (12) teleost fishes; (13) *Latimeria*; (14) lungfushes; (15) rhipidistian fishes; (16) tetrapods.

of the extant radiations (hagfishes, polypteriform fishes, and coelacanth fishes). Nevertheless, there are sufficient data for eight of the eleven extant radiations to allow a cladistic analysis, and, even where direct ontogenetic information is absent, a number of developmental events can be inferred from the adult anatomy of the lateral line system and its innervation. For example, only a single late embryonic stage is known for the living coelacanth *Latimeria*; even so, the presence of well developed lateral line canals and the innervation of the lateral line receptors indicates that the embryos of this fish probably retain the primitive number of lateral line placodes for gnathostomes and that these placodes must develop through stage H (Fig. 3.3). The first inference is based on the fact that adult *Latimeria* exhibit all lateral lines and all of the lateral line nerves that are known to arise from lateral line

placodes in any gnathostome (R.G. Northcutt and W.E. Bemis, unpublished observations). The second inference—that some of the placodes in embryonic *Latimeria* must develop through stage H—is based on the fact that many of the lateral lines of *Latimeria* exist as canals. The cladistic analysis of the craniate radiations illustrated in Figure 3.7 is, therefore, based on both direct observations and inferences regarding lateral line ontogenies. In each case the most complete ontogenetic sequence for the development of any placode has been listed. Where more than one sequence is listed for a single radiation, either there is doubt which sequence actually exists (hagfishes, radiation 1), or more than one sequence is known to exist (teleosts, lungfishes, and amphibians, radiations 12, 14, and 16, respectively), and both the primitive and derived sequences for that radiation are listed.

TABLE 3.1. Changes in placodal ontogenies

Radiation	Stages	Position	Process
1	Loss of receptors, A–G(H) to A(A–C)	TD	T
3–5	Loss of canals, A–H to A–G	TD	T
5	Loss of grooves, A–G to A–F	TD	T
10–12	Loss of electroreceptors, E to E^2	ND	P
10–12	Secondary ridges, D to D^2	NS	P
12	Teleost electroreceptors, E^2 to E^3	NA	P + S
12	Accessory lines, F to F^2	NA	P
13	Rostral organ, E^1 to E^4	NA	P
14	Loss of grooves, A–H to A–F	TD	T
16	Loss of grooves, A–H to A–F	TD	T
16	Loss of electroreceptors, E to E^2	ND	P
16	Loss of neuromasts, D to D^2	ND	P
16	Accessory neuromasts, F^2	NA	P
16	Degeneration of system, A–H to A–I	TA	P
16	Amniotic loss of placodes	ND	P

NA, nonterminal addition; ND, nonterminal deletion; NS, nonterminal substitution; P, patterning; S, embryonic source; T, timing; TA, terminal addition; TD, terminal deletion.

An examination of the variation in lateral line ontogenies indicates that the following categories of change occur: (1) the number of placodes can change, resulting in the loss of lateral lines and their associated cranial nerves; (2) the ontogenetic stages of a placode can change, resulting in alteration of the adult character state of a line (i.e., a line can exist as a pit line, groove or canal); (3) the cell lineages of a placode can change, resulting in the presence or absence of electroreceptors and/or neuromasts; (4) elongation or migration of placodes can change, resulting in alterations of the adult topography of individual lateral lines and fields of electroreceptors. An analysis of all these types of change is beyond the scope of the present chapter, and only changes in ontogenetic stages and cell lineages will be emphasized herein. However, some idea of the changes in number and movement of placodes in the various craniate radiations can be realized from Section 4, and in an earlier study of the phylogenetic distribution of the lateral lines of craniates (Northcutt 1989).

5.1 Changes in Ontogenetic Stages of Placodes

In the case of lateral line placodes, changes in cell lineages primarily involve the loss and re-evolution of electroreceptors, which are nonterminal

changes in the ontogenetic sequence and account for approximately 65% of all changes (Table 3.1). All other changes, which primarily involve the alteration of the character states (canals, grooves and pit lines) of the lateral lines, are terminal changes (35%). An out-group analysis of the placodal ontogenies illustrated in Figure 3.7 suggests that the ontogenetic sequence that characterized the earliest vertebrates, if not the earliest craniates, was the sequence A to H illustrated in Figure 3.3. Although modern amphibians (radiation 16) either exhibit pit lines (stage F) or lose their lateral line system at metamorphosis (stage I), the earliest amphibians (Moodie 1908) exhibited groove lateral lines (stage G), and their sister group, the rhipidistian (radiation 15) fishes (Jarvik 1980) exhibited canal lateral lines (stage H) as does the modern Australian lungfish, Neoceratodus (radiation 14), and the only living coelacanth, Latimeria (radiation 13). Thus three of the four radiations of lobe-finned fishes exhibit canal lateral lines, suggesting that stages A to H represent the ancestral ontogenetic sequence for these fishes. If so, then terminal deletion of stages G and H has occurred independently in the South American lungfish, Lepidosiren, and in modern amphibians. In this context, it should be noted that the embryos of modern amniotic vertebrates fail to develop lateral line placodes (Noden and De Lahunta 1985), which should be considered a nonterminal deletion if loss of larval lateral line receptors at metamorphosis in amphibians is interpreted as the addition of a terminal stage.

The ancestral ontogenetic sequence for the development of lateral lines in ray-finned fishes also appears to have been the sequence A to H, as this sequence occurs in all five radiations. Similarly, the ancestral ontogenetic sequence for these placodes in gnathostomes appears to have been the sequence A to H, as this is the case for cartilaginous fishes (radiation 7) (Johnson 1917; Rudd 1920), and, apparently, the extinct placoderms (radiation 6) (Jarvik 1980).

The ontogenetic sequence for the development of the lateral lines in the earliest cephalaspidomorph fishes appears to have been the sequence A to G, as groove lateral lines (stage G) occurred in osteostracans and anaspids (radiations 3 and 4, respectively) (reviewed in Northcutt, 1989). Modern lampreys (radiation 5), however, exhibit only pit lines (stage F), which suggests that the condition in lampreys represents truncation of the ancestral cephalaspidomorph sequence. Although the ancestral cephalaspidomorph sequence appears to have been characterized by stages A to G, the ancestral sequence for myopterygians, as well as the vertebrates, was probably the sequence A to H, as well developed canal lateral lines (stage H) occurred in heterostracans (radiation 2) (Denison 1964), which are believed to be the sister group of the myopterygians (Janvier 1978, 1981). If these inferences are correct, the phylogeny of the lateral lines in cephalaspidomorph fishes was characterized by truncation of the ancestral myopteryian sequence (stages A to H) and a further truncation in modern lampreys (stages A to F).

Unfortunately, it is presently impossible to infer with any confidence what the ancestral sequence of lateral line development was in the earliest craniates, as hagfishes are the only sister group to myopterygians, and the nature of their lateral line system is disputed. The living hagfishes, or myxinoids, comprise two radiations: the Atlantic myxinids and the Pacific eptatretids. Peripheral lateral lines or superficial receptors have never been reported in the myxinids, although short lateral line grooves reportedly exist immediately rostral and caudal to the eye in eptatretids (Ayers and Worthington 1907) and are innervated by anterior and posterior lateral line nerves (Kishida et al.

1987). Although Kishida et al. (1987) labeled these nerves with horseradish peroxidase and traced their central projections into the brain, they did not present evidence that the peripheral processes of these nerves do, in fact, innervate receptors in the lateral line grooves. Furthermore, Fernholm (1985) has demonstrated by scanning and transmission electron microscopy that these grooves consist only of small mucous cells in *Eptatretus*. Thus there is little evidence that hagfishes possess a lateral line system. It is possible that the cranial nerves described by Kishida et al. (1987) do represent lateral line nerves that have arisen from placodes and that the lateral line system of eptatretids is arrested at stage C (Fig. 3.3). It is also possible, however, that these nerves have been erroneously interpreted as lateral line nerves. If so, the "lateral line grooves" may represent an arrested stage A or may even have nothing to do with the lateral line system of vertebrates. Clearly, these possibilities must be examined before further inferences can be made regarding the ontogenetic sequence of the lateral line placodes of the earliest craniates.

In addition to the numerous terminal alterations that have been described for craniates, two distinct nonterminal alterations in the ontogenies of lateral line placodes occur in ray-finned fishes. Unlike most craniates, the central nervous system of most, if not all, ray-finned fishes is formed by a ventrally directed thickening of the neural plate that produces a neural keel stage. An epithelial thickening termed the lateral mass (Beckwith 1907; Landacre 1910) occurs on either side of the neural keel, so that in embryos examined at this stage in the transverse plane the neural keel forms the body of a T-shaped structure with each lateral mass forming one arm of the T. Subsequently, the neural keel forms the neural tube by a process of delamination, whereas the lateral masses exhibit a more complicated developmental history. Initially, an octaval vesicle forms within each lateral mass, dividing it into rostral and caudal segments. Subsequently, each segment is subdivided further into ventromedial and dorsolateral components (Landacre 1910). The ventromedial component appears to be homologous to the neural crest of other vertebrates, based on the migratory behavior and further differentiation of its cells. Some cells of the dorsolateral component appear to give rise to the sensory ganglionic cells of the lateral line nerves

(Landacre 1910), but the fate of the remaining cells is uncertain. Landacre (1910) claimed that cells of the dorsolateral component transform into mesenchyme and that a few hours later individual neuromasts differentiate directly from the deeper layers of the epidermis. A number of workers (Clapp 1898; Beckwith 1907; Landacre and Conger 1913; Lekander 1949), however, have described neuromasts in ginglymodes (gars, radiation 10, Fig. 3.7), halecomorphs (*Amia*, radiation 11, Fig. 3.7), and teleosts (radiation 12, Fig. 3.7) as differentiating from sensory ridges (stage D, Fig. 3.3). It is therefore possible that the dorsolateral component of the lateral masses of some, if not all, ray-finned fishes represents a modification of stages B and C: individual lateral line placodes do not form but are replaced by a continuous placode-like lateral mass that initially gives rise to the ganglionic cells of the lateral line nerve and then directly to lateral line receptors. It is also possible, however, that following neurogenesis of the lateral line nerves, the dorsolateral component is reorganized into sensory ridges comparable to those of other vertebrates (stage D, Fig. 3.3), as cephalic sensory ridges have been described in many of these taxa and a distinct posterior lateral line placode is known to migrate onto the trunk in a number of teleosts (Metcalfe, Kimmel, and Schabtach 1985; Vischer 1989a). Clearly, additional studies on the development of the lateral line system of ray-finned fishes are needed, but it appears that distinct ontogenetic stages, corresponding to stages A through C or even D in other vertebrates, may not exist in ray-finned fishes.

The early formation of the octaval vesicles within the lateral masses of teleosts has given rise to the erroneous interpretation that both the octaval and lateral line organs arise from a single placode (Beard 1884; Wilson 1889; Wilson and Mattocks 1897). Because formation of the octaval vesicles divides the lateral masses into rostral and caudal components, termed pre- and postauditory placodes, which are known to be related, albeit unclearly, to the subsequent development of the lateral line system, it has been claimed that both the inner ear and the lateral line system arise from a single placode (i.e., the lateral masses). Whatever the actual relationship of the lateral masses to lateral line formation, out-group analysis clearly indicates that the ontogenetic pattern in teleosts is derived and that primitively separate placodes gave rise to the inner ear and the lateral line organs, along with their respective nerves.

A second nonterminal alteration occurs in the ontogeny of the lateral line system in some teleosts: neuromast formation is modified prior to the occurrence of stage G (Fig. 3.3). In all nonteleost ray-finned fishes, as well as in most groups of teleosts, neuromast differentiation occurs as outlined in Figure 3.3, and if superficial lines of neuromasts occur, they do so by truncation of the ontogenetic sequence at stage F. Some ostariophysine teleosts, however, exhibit extensive numbers of superficial neuromasts adjacent to canal lines or even scattered over the entire body (Lekander 1949; Puzdrowski 1989). To date, the lines of superficial neuromasts (the accessory lines of Coombs, Janssen, and Webb 1988) that occur adjacent to those neuromast lines that form canals appear to arise directly from the neuromasts of the canal lines. Thus formation of accessory neuromasts would appear to be temporally constrained by the formation of canals, after which the migration of accessory neuromasts could not physically occur. Further studies are needed to determine the ontogenetic origin of more scattered neuromast fields on the trunk, but it appears that an ancestral nonterminal stage (D, E, or F) was modified. A similar nonterminal alteration occurs in some salamanders: both the primary and secondary numbers of neuromasts are increased in pit lines (Lannoo 1985, 1987) that are clearly homologous to the canal lines of other craniates (Northcutt 1989).

5.2 Changes in Placodal Cell Lineages

An out-group analysis of variation in the two major groups of receptors that develop from lateral line placodes indicates that the loss and re-evolution of electroreceptors accounts for most of the changes in placodal cell lineages. Some members of all extant gnathostome radiations possess electroreceptors (see Northcutt 1986b and McCormick and Braford 1988 for recent reviews); however, two different classes of electroreceptors can be identified: cathodally activated ampullary receptors, which are innervated by pre-octaval lateral line nerves that terminate in the dorsal octavolateralis nucleus of the brain stem; and anodally activated electroreceptors, which are innervated by pre- and/or postoctaval lateral line nerves that terminate in an electrosensory lateral line lobe of the brain stem. An analysis of the distribution of these two classes of electroreceptors indicates that cathodally activated ampullary organs occur in most members of all anamniotic gnathostome radiations except for the members of the neopterygian ray-finned fishes (radiations 10–12, Fig. 3.7). Gars (ginglymodes, radiation 10) and *Amia* (halecomorphs, radiation 11) do not possess either class of electroreceptors, nor do most groups of teleost fishes (radiation 12), although anodally activated electroreceptors do occur in some osteoglossomorph and ostariophysine teleosts. An out-group analysis indicates that cathodally activated ampullary electroreceptors probably occurred in the common ancestor of gnathostomes and were lost with the origin of neopterygian ray-finned fishes. This inference is supported further by the occurrence of cephalic fields of flask-shaped ducts and pores, believed to have housed ampullary organs, in the extinct placoderms (Ørvig 1971; Jarvik 1980) which represent the sister group of all extant gnathostomes.

Both descriptive (Northcutt 1986a) and experimental (Northcutt, Fritzsch, and Brändle 1990) data indicate that the primitive cathodally activated electroreceptors of gnathostomes arise from lateral line placodes at stage E (Fig. 3), but there is little information concerning the origin of the anodally activated electroreceptors of teleosts. Several researchers have described electroreceptor primordia immediately adjacent to differentiating sensory ridges in teleosts (Sato 1956; Northcutt 1987; Vischer 1989b) which suggests that teleost electroreceptors also arise from placodes, but their origin has not been directly observed. Given the widespread occurrence of cathodal electroreceptors and the similarities in their placodal origin in elasmobranch fishes and amphibians, the ontogenetic sequence of placodal development outlined in Figure 3.3 is inferred to have been the ancestral sequence for primitive gnathostomes. If so, then alteration of stage E and all subsequent stages (E^2–H^2) occurred in the common ancestor of all neopterygians and resulted in the loss of cathodal receptors. Further alterations in stage E^2 must have occurred independently, in some osteoglossomorph and ostariophysine teleosts (stage E^3–H^3), assuming that the anodal electroreceptors of these teleosts also arise from lateral line placodes.

The distribution and innervation of teleost electroreceptors allows one to make several additional inferences regarding their ontogeny. Although electroreceptors are distributed on both the head and trunk in osteoglossomorph and ostariophysine teleosts, they are characterized by different patterns of innervation. Both cephalic and trunk electroreceptors are innervated by the anterodorsal, anteroventral, and posterior lateral line nerves in osteoglossomorph and silurid ostariophysine teleosts (Finger, Bell, and Carr 1986), whereas both cephalic and trunk electroreceptors are innervated by only the anteroventral lateral line nerve (Northcutt and Vischer 1988) in gymnotiform ostariophysine teleosts. If homologous lateral line placodes exist in teleosts and other gnathostomes, there must have been changes in the embryonic sources of both the electroreceptors and their innervation. The absence of trunk electroreceptors in ancestral ray-finned fishes, inferred from the absence of such receptors in living polypteriform and chondrostean fishes (radiations 8 and 9), suggests that the posterior lateral line placode did not give rise to electroreceptors in ancestral ray-finned fishes. If so, then independent changes must have occurred in the developmental sequence of this placode at least three times (once in osteoglossomorph teleosts and at least twice in ostariophysine teleosts). Furthermore, other placodes, such as the AV placode, must give rise to sensory neurons that innervate and probably induce electroreceptors from their own placode of origin and in additional placodes as well.

A number of changes in placodal cell lineages also characterize the phylogeny of lobe-finned fishes (radiations 13–16, Fig. 3.7). There is evidence that three of the four radiations (the condition in the extinct rhipidistians is uncertain) possess cathodal ampullary receptors, but the distribution and innervation of these receptors differs in each radiation. Many members of two of the three amphibian orders (apodans and urodels) possess cathodal electroreceptors that are restricted to the head and innervated by the anterodorsal and anteroventral lateral line nerves (Fritzsch 1989; Northcutt 1990b). Lungfishes (radiation 14) also possess cathodal electroreceptors, but theirs are located on both the head and trunk; the cephalic electroreceptors are known to be innervated by at least the anterodorsal and anteroventral lateral line nerves, whereas the trunk electroreceptors appear to be innervated by only the anteroventral lateral line nerve (Northcutt 1986b). The living coelacanth, *Latimeria* (radiation 13), does not possess cathodal electroreceptors as ampullary organs scattered in the skin; rather these receptors appear to be housed in three pairs of epithelial diverticula that form the cephalic rostral organ (Bemis and Hetherington 1982). The rostral organ consists bilaterally of an anterior diverticulum, an inferior posterior diverticulum, and a superior posterior diverticulum. The receptors of the anterior diverticulum are innervated by fibers of the buccal ramus of the anterodorsal lateral line nerve, whereas the receptors of the two posterior diverticula are innervated by fibers of the superficial ophthalmic ramus of the anterodorsal lateral line nerve (R.G. Northcutt and W.E. Bemis, unpublished observations).

An out-group analysis of the distribution and innervation of electroreceptors in lobe-finned fishes and amphibians suggests that the ancestral lobe-finned fishes, like the ancestral ray-finned fishes, were characterized by cathodal electroreceptors that were restricted to the head and innervated by at least the anterodorsal and anteroventral lateral line nerves. Given extensive information on the development of lateral line receptors in amphibians, it appears that the sequence for placode development outlined in Figure 3.3 also characterized the earliest lobe-finned fishes. If so, this must have been subsequently altered in coelacanths (radiation 13), with electroreceptor development being restricted to the anterodorsal placode. Furthermore, changes in the development of this placode must have occurred prior to stage F, resulting in generation of electroreceptors by only the distal portions of the dorsal and ventral sensory ridges of the AD placode (Fig. 3.4) and extensive invagination of these ridges to form the diverticula of the rostral organ (stage E^4). Lungfishes have apparently retained the ancestral sequence of placodal development. Peripheral processes of some of the sensory neurons that arise from the AD placode, however, may have induced electroreceptors to arise from the posterior placode, as lungfishes have trunk electroreceptors that are innervated by a caudally directed ramus of the anteroventral lateral line nerve. Two of the three orders of amphibians, apodans and urodels, have also apparently retained the ancestral sequence of placodal development, but anurans would appear to have altered stage E of this sequence, as no larval anurans are known to develop electroreceptors. Finally, some apodans have apparently altered one or more of the early stages (stage D and/or E), as taxa such as *Typhlonectes* possess cathodal electroreceptors but no neuromasts (Fritzsch and Wake 1986; McCormick and Braford 1988).

Changes in placodal cell lineages for craniate radiations other than gnathostomes can be inferred only with great difficulty, as three (radiations 2–4) of the five radiations are extinct, and hagfishes (radiation 1) possess neither electroreceptors nor neuromasts. Although lampreys do possess cathodal electroreceptors, these receptors do not occur as ampullary organs extending into the underlying dermis, as in most gnathostomes, but exist as "terminal buds" located intraepidermally (Ronan and Bodznick 1986). These nondescript electroreceptors are widely distributed on the head and trunk, and those on the head are innervated by at least the anterodorsal and anteroventral lateral line nerves; those on the trunk are innervated by a caudally directed ramus of the anteroventral lateral line nerve (Ronan and Northcutt 1987) as in gymnotiform teleosts and lungfishes. The morphological and physiological similarities in the cathodal electroreceptors and their innervation in lampreys and gnathostomes suggest that similar ontogenetic sequences characterize placodal development in vertebrates, but this possibility must be evaluated in the context of further descriptive studies of the development of the octavolateralis system in lampreys.

There are only two known morphological criteria for inferring the occurrence of electroreceptors in extinct taxa: (1) the presence in the dermal armor of flask-like depressions and surface pores which may have housed ampullary organs, a n d
(2) the presence within ossified neurocrania of periotic canals which may have housed a caudally directed ramus of an anteroventral lateral line nerve innervating trunk electroreceptors. The first criterion is applicable only if electroreceptors occur as well developed ampullary organs, and the second is applicable only if electroreceptors occur on the trunk and are innervated by the anteroventral lateral line nerve rather than the posterior lateral line nerve.

Although well developed fields of flask-shaped cavities opening to the surface of cephalic dermal armor have been described in fossil placoderms

not been reported in cephalaspidomorphs or heterostracans. The dermal armor of many of these taxa,
however, is characterized by pores of smaller
diameter and cavities, collectively termed the pore-
canal system which has been interpreted as housing
electroreceptors (Thomson 1977; Northcutt and
Gans 1983; Gans 1989). Unfortunately, the pore-
canal system in these fossil taxa is several orders of
magnitude more dense than any known sensory system in living craniates. Moreover, the pore-canal
system occurs in addition to larger diameter pores in
fossil lungfishes, where the distribution of the larger
pores matches that of electroreceptors in living
lungfishes (Jarvik 1980; Bemis and Northcutt
1987). Finally the snout of the Australian lungfish
Neoceratodus is characterized by a complex series
of vascular loops extending into the epidermis
whose morphology is remarkably similar to that of
cosmine and the pore-canal system in fossil lungfishes (Bemis and Northcutt 1987).

These data are consistent with the interpretation
that electroreceptors as ampullary organs probably
arose with the origin of gnathostomes, as there is
clear evidence that both fossil placoderms and
bony fishes possessed ampullary organs. If electroreceptors existed in the earliest cephalaspidomorphs or the earliest vertebrates, they must have
been similar to the intra-epidermal electroreceptors of lampreys, as the dermal armor of the extinct
cephalaspidomorphs (radiations 3 and 4) and heterostracans (radiation 2) is not characterized by
pores comparable to ampullary organs. There is
some evidence that at least one group of extinct
cephalaspidomorphs did possess electroreceptors.
In some of the osteostracans (radiation 3) the
neurocranium possesses a periotic canal which
may have housed a caudally directed ramus of
a pre-octaval cranial nerve (Janvier 1974). It is
therefore possible that the earliest myopterygians
possessed cathodal electroreceptors whose placodal development was comparable to the stages outlined in Figure 3.3. If so, the electroreceptor
primordia of stage F would have matured as intra-
epidermal receptors, comparable to those of lampreys, and differentiated dermally associated
ampullary organs would have occurred with the
origin of gnathostomes.

If the earliest vertebrates were electroreceptive,
their receptors must have occurred as intra-epidermal organs, as the dermal armor of heterostracans shows no convincing evidence of ampullary
pores. Unfortunately, nothing is known about the
neurocrania of these fossil fishes, so it is impossible to reach any conclusions regarding their cranial
nerves. Similarly, the absence of electroreceptors
and neuromasts in hagfishes and the uncertain
status of the morphological features interpreted as
"lateral line" grooves and nerves requires that two
hypotheses be given equal consideration at this
time: (1) the sequence of placodal development
outlined in Figure 3.3 is the ancestral one for craniates and has been severely truncated in hagfishes;
(2) the sequence of placodal development outlined
in Figure 3.3 is the ancestral one, at least for myopterygians and perhaps for vertebrates. The
presence of placodes or sensory ridges in ancestral
adult craniates (A, A–C, Figs. 3.3, 3.6) is not considered a viable hypothesis, as possible functions
cannot be envisioned for these character stages.

5.3 Frequency of Different Types of Ontogenetic Change

No general topic in morphological studies is more
fraught with intuitive assumptions and lack of
empirical data than the nature and frequency of
ontogenetic changes that produce phylogenies.
Some evolutionary biologists (Nelson 1973, 1978,
1985; Nelson and Platnick 1981; Rosen 1982) have
assumed that character phylogeny occurs solely by
the addition of new states at the end of character
ontogenies, whereas others (Mayr 1960; Liem
1973; Gans 1974; Northcutt and Gans 1983; Gans
1989) have noted that early ontogenetic changes
can be critical in underlying novel phylogenetic
innovations, although they assumed such events
are rare. Generally, deletions whether terminal or
nonterminal are both viewed as occurring more
frequently than additions, based on theoretical
assumptions regarding the nature of genetic and
developmental processes regulating phenotypes.
Similarly, changes in the timing of developmental
events have been assumed to be the major process
underlying phylogenetic change (de Beer 1958;
Gould 1977; Alberch et al., 1979). Although these
assumptions exist, few studies have actually
attempted to assess the nature and frequency of
ontogenetic changes. In one recent study of the
ontogenies of 58 characters in 29 species, Mabee
(1989a) argues that additions and deletions (36%

and 40%, respectively) occur with approximately equal frequency. In a similar out-group analysis of changes in pharyngeal pouch ontogenies (Northcutt 1990a), additions (62%) appeared to be more frequent than deletions (38%); data from the present study on changes in the ontogenies of lateral line placodes (Table 3.1) suggest that additions (41%) and deletions (53%) are approximately equal, whereas substitutions (6%) occur infrequently. Mabee (1989a) reported a higher frequency of ontogenetic substitution (21%) than is suggested by the data on pharyngeal pouches and lateral line placodes, but all three studies indicate that additions occur at least as frequently as deletions. The three studies suggest, however, that the frequency of nonterminal and terminal changes can vary greatly in different characters. Terminal (78%) were far more frequent than either nonterminal (5%) or substitutional (15%) changes in Mabee's characters, whereas nonterminal (81%), as opposed to terminal (19%), changes appear to account for the majority of changes in the ontogenies of pharyngeal pouches. The present analysis of changes in the ontogenies of lateral line placodes also suggests that nonterminal changes (65%) are more frequent than terminal changes (35%). At present it is difficult to account for these reported differences. Although variations in interpretation—i.e., does loss of a cell lineage in a particular stage represent deletion, alteration, or substitution of a stage—would influence collective data on the frequencies of additions, deletions, and substitutions, interpretational differences would not affect the reported frequencies of terminal versus nonterminal changes. It is possible that different developmental and historical constraints affect the probability of ontogenetic change for different types of characters. If so, Mabee's analysis of the ontogenies of a large number of characters may be a more accurate reflection of the frequency of terminal versus nonterminal changes. At any rate, it is clear in all three studies that terminal additions account for only a fraction of ontogenetic changes, and these studies strongly support the contention that out-group analysis is the only method for determining the polarity of both ontogenetic and adult characters (Brooks and Wiley 1985; de Queiroz 1985; Kluge 1985; Mabee 1989a).

Although the processes underlying ontogenetic changes in lateral line placodes (Table 3.1) can only be inferred at present, it is possible that changes in patterning (i.e., changes in inductive interactions) account for 53% of ontogenetic changes, whereas changes in timing (i.e., heterochrony) and changes in patterning and source are less frequent, 29% and 18%, respectively. The loss of lateral line receptors and, possibly, even lateral line nerves in hagfishes is interpreted as truncation of an ancestral ontogenetic sequence (A to G or H), based on the assumption that the cephalic grooves of eptatretid hagfishes are developmentally arrested lateral line placodes. Similarly, the absence of lateral line canals or grooves in cephalaspidomorphs, lepidosirenid lungfishes, and amphibians is interpreted as truncation of an ancestral myopterygian ontogenetic sequence, based on the out-group analysis of these characters and on the widespread occurrence in these taxa of other characters interpreted as paedomorphic (Hardisty 1979; Bemis 1984; Duellman and Trueb 1986).

The remaining ontogenetic changes involving the lateral line placodes—which primarily concern the loss and re-evolution of electroreceptors—are interpreted as changes in patterning or source, as these changes do not involve the loss of a placodal stage; rather, the cell lineages (neuromasts and electroreceptors) that are developed at a given stage are altered. Such changes likely comprise inductive interactions involving the neural tube, the sensory ganglionic cells of the lateral line nerves, and the placodal cells, but the details of these interactions are still uncertain. The re-evolution of electroreceptors in teleost fishes has been interpreted as changes in both patterning and source, as it appears that electroreceptors and their innervation develop from different placodes in the various electroreceptive taxa.

5.4 Future Directions

Although the phylogeny of octavolateralis ontogenies can be broadly outlined at the present time, many details of these ontogenies and their underlying processes remain to be elucidated. The single largest gap in our knowledge of the octavolateralis system is our lack of information on the development of the inner ear in all major craniate radiations. Without such information, it is essentially impossible to determine the phylogeny of the complex and highly variable sensory maculae of the inner ear.

Additional information on the organization and development of the lateral line system is also needed in order to provide a more complete understanding of its phylogeny. Inferences regarding the ontogeny of lateral line placodes of the earliest craniates will remain tenuous until far more information exists on the "lateral line" nerves and grooves of eptatretid hagfishes. Much of the needed information can be derived from new tracing studies involving the functional components of those cranial nerves that have been identified as lateral line nerves, but developmental studies focusing on the origin of the "lateral line" grooves are also needed.

Considerable information can also be gained by reexamining the distribution and character states of the lateral lines in heterostracans, osteostracans, anaspids, placoderms, acanthodians, and rhipidistian fishes, as well as in early amphibians. Much of our information on the organization of the lateral lines in members of these extinct radiations is based on studies that predate the discovery of electroreceptors, as well as predating our grasp of the nature of lateral line innervation and our awareness of the significance of pit lines, grooves, and canals and the importance of accurate descriptions in this regard. It is likely, therefore, that reexamination of these fossil taxa will provide considerable new information that is critical to inferences regarding the ontogenies of these extinct radiations, and, because of their strategic phylogenetic positions, inferences regarding the ancestral ontogenies of at least three of the major craniate clades (myopterygians, cephalaspidomorphs, and vertebrates).

Additional descriptive ontogenetic studies of members of extant radiations, particularly the ray-finned fishes, are also needed, as these fishes may have extensively modified a number of early stages in the development of the lateral line system, in addition to the fact that they exhibit the loss and re-evolution of electroreceptors. At present, few details regarding any of these ontogenetic events are known. Continued experimental studies are also needed to provide an understanding of the nature of tissue interactions and the biochemical inductors that link stages of placodal development. A clearer view of these inductors and their sources offers our best chance to understand the phylogenetic loss of electroreceptors and whether this loss was directly or indirectly linked to environmental pressures.

Finally, far more information is needed on the physiology and biological roles of lateral line receptors and their spatial distribution, before meaningful scenarios can be constructed and tested. In this context, the interpretation of superficial lines of neuromasts in lampreys, lepidosirenid lungfishes, and modern amphibians as paedomorphic character states suggests that many of the ontogenetic changes observed in specific characters may not be adaptive, per se, but the consequence of selective pressures acting on life history strategies at a more global level—a reminder of the hierarchical nature of biological systems and the absurdity of extreme reductionist approaches.

6. Summary

Garstang's realization—that changes in an ancestral ontogeny create phylogeny—suggests that phylogenetic analyses should involve comparisons of descendent ontogenies and their underlying processes in order to allow one to reconstruct an ancestral ontogeny and to isolate the points in phylogeny where significant transformations have occurred. This involves a cladistic analysis of the ontogenies, which may involve five steps: (1) recognizing and describing the developmental stages of one or more characters; (2) comparing these stages in the relevant taxa and formulating hypotheses regarding homologous stages; (3) determining the polarity of suspected homologous states by using the outgroup criterion; (4) generating scenarios regarding the biological significance of phyletic changes; and (5) testing these scenarios by continued studies. An out-group analysis of the development of octavolateralis placodes in several groups of vertebrates suggests that the placodes that give rise to the inner ear and the lateral line system exhibit comparable early stages before placodal elongation, at which point the developmental trajectories of octaval and lateral line placodes diverge. Unfortunately, a cladistic analysis of the ontogenies of the octaval placode cannot be undertaken at present, as detailed descriptive information on the development of its sensory maculae in different craniates is lacking.

There is sufficient information to allow an out-group analysis of the development of lateral line placodes, and such analysis suggests that at least seven, perhaps eight, stages existed in the development of a lateral line placode in the earliest vertebrates. Each placode appears initially to give rise to sensory neurons whose cell bodies and processes form the sensory ganglion and peripheral processes of a lateral line nerve. Subsequently, mechanoreceptive neuromasts, followed by electroreceptive ampullary organs, arise from each placode. The primordia of these receptors initially develop within the placode, which is a specialized portion of the inner layer of the epidermis, but the neuromasts subsequently erupt to the surface, where they pass through pit line and open groove stages prior to becoming enclosed in canals.

An out-group analysis of the variation in lateral line ontogenies indicates that three types of changes occur: (1) there can be a change in the number of placodes that form, which results in the loss of lateral lines and their associated cranial nerves; (2) ontogenetic sequences can become truncated, resulting in alteration of the adult character state of a line; and (3) the cell lineages of a placode can undergo changes, resulting in the presence or absence of neuromasts and/or electroreceptors.

Analysis indicates that additions and deletions of ontogenetic stages occur with approximately equal frequency as regards lateral line placodes, and that substitutions are infrequent. Nonterminal changes in placodal sequences, however, are approximately four times more frequent than terminal changes. Although the processes underlying such changes can only be inferred at present, changes in patterning (inductive interactions) appear to account for approximately 50% of all changes, whereas changes in timing and source appear to account for the remaining changes with approximately equal frequency.

Acknowledgments. This research was supported in part by NIH research grants (NS24669 and NS24869). William E. Bemis, Georg Striedter, and Helmut Wicht offered many helpful comments on the manuscript, as did Wolfgang Plassmann, who also assisted with the illustrations. I am grateful to Susan M. Commerford for word processing and library retrieval, and to Mary Sue Northcutt for her assistance in numerous phases of my research. Finally, I thank the editors of the volume for inviting me to participate in the symposium and to contribute to its proceedings.

References

Alberch P (1985) Problems with the interpretation of developmental sequences. Syst Zool 34:46–58.

Alberch P, Gould SJ, Oster GF, Wake DB (1979) Size and shape in ontogeny and phylogeny. Paleobiology 5: 296–317.

Allis EP (1889) The anatomy and development of the lateral line system in *Amia calva*. J Morphol 2:463–566.

Allis EP (1923) The cranial anatomy of *Chlamydoselachus anguineus*. Acta Zool 4:123–221.

Ayers H, Worthington J (1907) The skin end-organs of the trigeminus and lateralis nerves of *Bdellostoma dombeyi*. Am J Anat 7:327–336.

Balinsky BI (1975) An Introduction to Embryology, 4th Ed. Philadelphia: W. B. Saunders.

Beard J (1884) On the segmental sense organs of the lateral line, and on the morphology of the vertebrate auditory organ. Zoologisch Anzeig 7:123–126.

Beckwith CJ (1907) The early development of the lateral line system of *Amia calva*. Biol Bull 14:23–28.

Bemis WE (1984) Paedomorphosis and the evolution of the Dipnoi. Paleobiology 10:293–307.

Bemis WE, Hetherington TE (1982) The rostral organ of *Latimeria chalumnae*: morphological evidence of an electroreceptive function. Copeia 1982:467–471.

Bemis WE, Northcutt RG (1987) Pore canals of devonian lungfishes did not house electroreceptors. Am Zool 27:168A.

Bodznick D, Northcutt RG (1981) Electroreception in lampreys: evidence that the earliest vertebrates were electroreceptive. Science 212:465–467.

Brooks DR, Wiley EO (1985) Theories and methods in different approaches to phylogenetic systematics. Cladistics 1:1–11.

Bullock TH, Bodznick DA, Northcutt RG (1983) The phylogenetic distribution of electroreception: evidence for convergent evolution of a primitive vertebrate sense modality. Brain Res Rev 6:25–46.

Campbell RL, Richie DM (1983) Problems in the theory of developmental sequences: prerequisites and precursors. Hum Dev 26:156–172.

Clapp CM (1898) The lateral line system of *Batrachus*. J Morphol 15:223–265.

Coghill GE (1916) Correlated anatomical and physiological studies of the growth of the nervous system of Amphibia. J Comp Neurol 26:247–340.

Coombs S, Janssen J, Webb JF (1988) Diversity of lateral line systems: evolutionary and functional considerations. In: Atema J, Fay RR, Popper AN, Tavolga WN (eds) Sensory Biology of Aquatic Animals. New York: Springer-Verlag, pp 553–594.

Damas H (1944) Recherches sur le développement de *Lampetra fluviatilis* (L.), contribution a l'étude de la cephalogenèse. Arch Biol 55:1–284.

Damas H (1951) Observations sur le développement des ganglions crâniens chez *Lampetra fluviatilis* (L.). Arch Biol 62:65–95.

Danser BH (1950) A theory of systematics. Bibl Biotheoret 4:117–180.

de Beer G (1958) Embryos and Ancestors, 3rd Ed. London: Oxford University Press.

de Queiroz K (1985) The ontogenetic method for determining character polarity and its relevance to phylogenetic systematics. Syst Zool 34:280–299.

Denison RH (1964) The Cyathaspididae. Fieldiana. Geology 13:307–473.

Disler NN (1977) The lateral line system sense organs or sharks (Elasmobranchii). Moscow: Science Publications.

Duellman WE, Trueb L (1986) Biology of Amphibians. New York: McGraw-Hill.

Eldredge N, Cracraft J (1980) Phylogenetic Patterns and the Evolutionary Process. New York: Columbia University Press.

Fernholm B (1985) The lateral line system of cyclostomes. In: Foreman RE, Gorbman A, Dodd JM, Olsson R (eds) Evolutionary Biology of Primitive Fishes. New York: Plenum, pp 113–122.

Fields RD (1982) Electroreception in the ratfish (subclass Holocephali): anatomical, behavioral, and physiological studies. PhD Thesis, Department of Biology, San Jose State University, San Jose, CA.

Finger TE, Bell CC, Carr CE (1986) Comparisons among electroreceptive teleosts: why are electrosensory systems so similar? In: Bullock TH, Heiligenberg W (eds) Electroreception. New York: John Wiley & Sons, pp 465–481.

Fink WL (1982) The conceptual relationship between ontogeny and phylogeny. Paleobiology 8:254–264.

Fritzsch B (1989) Diversity and regression in the amphibian lateral line and electrosensory system. In: Coombs S, Görner P, Münz P (eds) The Mechanosensory Lateral line. New York: Springer-Verlag, pp 99–114.

Fritzsch B, Bolz D (1986) On the development of electroreceptive ampullary organs of *Triturus alpestris* (Amphibia: Urodela). Amphibia-Reptilia 7:1–9.

Fritzsch B, Wake MH (1986) A note on the distribution of ampullary organs in gymnophions. J Herpetol 20:90–93.

Gans C (1974) Biomechanics. Philadelphia: Lippincott.

Gans C (1989) Stages in the origin of vertebrates: analysis by means of scenarios. Biol Rev 64:221–268.

Garstang W (1922) The theory of recapitulation: a critical re-statement of the biogenic law. Zool J Linnean Soc Lond 35:81–101.

Gould SJ (1977) Ontogeny and Phylogeny. Cambridge: Harvard University Press.

Hammarberg F (1937) Zur Kenntnis der ontogenetischen Entwicklung des Schädels von *Lepidosteus platystomus*. Acta Zool 18:209–337.

Hardisty MW (1979) Biology of the Cyclostomes. London: Chapman Hall.

Harrison RG (1904) Experimentelle Untersuchungen über die Entwicklung der Sinnesorgane der Seitenlinie bei den Amphibien. Arch Mikroscop Anat Entwicklungsgesch 63:35–149.

Hennig W (1966) Phylogenetic Systematics. Urbana: University of Illinois Press.

Holmes SJ (1944) Recapitulation and its supposed causes. Quarterly Rev Biol 19:319–331.

Holmgren N (1940) Studies on the head in fishes; embryological, morphological, and phylogenetical researches. Acta Zool 21:51–267.

Holmgren N (1942) General morphology of the lateral sensory line system of the head in fish. Kungl Sven Handling 20:1–46.

Huettner AF (1949) Fundamentals of Comparative Embryology of the Vertebrates. New York: Macmillan.

Janvier P (1974) The sensory line system and its innervation in the Osteostraci (Agnatha, Cephalaspidomorphi). Zool Scripta 3:91–99.

Janvier P (1978) Les negeoires paires des Ostéostracés et la position systématique des Céphalaspidomorphes. Ann Paléontol (Vertébres) 64:113–142.

Janvier P (1981) The phylogeny of the craniata, with particular reference to the significance of fossil "agnathans". J Vert Paleontol 1:121–159.

Jarvik E (1980) Basic Structure and Evolution of Vertebrates, 2 Vols. London: Academic Press.

Johnson SE (1917) Structure and development of the sense organs of the lateral canal system of selachians (*Mustelus canis* and *Squalus acanthias*). J Comp Neurol 28:1–74.

Kishida R, Goris RC, Nishizawa H, Koyama H, Katoda T, Amemiya F (1987) Primary neurons of the lateral line nerve and their central projections in hagfishes. J Comp Neurol 264:303–310.

Kluge AG (1985) Ontogeny and phylogenetic systematics. Cladistics 1:13–27.

Kluge AG (1988) The characterization of ontogeny. In: Humpries CJ (ed) Ontogeny and Systematics. New York: Columbia University Press, pp 57–81.

Kluge AG, Strauss RE (1985) Ontogeny and systematics. Ann Rev Ecol Syst 16:247–268.

Knouff RA (1935) The developmental pattern of ectodermal placodes in *Rana pipiens*. J Comp Neurol 62: 17–65.

Kupffer C von (1895) Studien zur vergeichenden Entwicklungsgeschichte des Kopfes der Kranioten. Heft. 3, Die Entwicklung der Kopfnerven von Ammocoetes Planeri. München, J.F. Lehmann: pp. I–80.

Landacre FL (1910) The origin of the cranial ganglia in *Ameiurus*. J Comp Neurol 20:309–411.

Landacre FL (1912) The epibranchial placodes of *Lepisosteus osseus* and their relation to the cerebral ganglia. J Comp Neurol 22:1–69.

Landacre FL (1916) The cerebral ganglia and early nerves of *Squalus acanthias*. J Comp Neurol 27: 19–67.

Landacre FL, Conger AC (1913) The origin of the lateral line primordia in *Lepidosteus osseus*. J Comp Neurol 23:575–633.

Langille RM, Hall BK (1989) Developmental processes, developmental sequences and early vertebrate phylogeny. Biol Rev 64:73–91.

Lannoo MJ (1985) Neuromast topography in *Ambystoma* larvae. Copeia 535–539.

Lannoo MJ (1987) Neuromast topography in urodele amphibians. J Morphol 191:247–263.

Lekander B (1949) The sensory line system and the canal bones in the head of some *Ostariophysi*. Acta Zool 30:1–131.

Liem KF (1973) Evolutionary strategies and morphological innovations: cichlid pharyngeal jaws. Syst Zool 22:424–441.

Løvtrup, S (1978) On von Baerian and Haeckelian recapitulation. Syst Zool 27:348–352.

Mabee PM (1987) Phylogenetic change and ontogenetic interpretation in the family Centrarchidae (Pisces: Perciformes). PhD Thesis, Duke University, Durham, North Carolina.

Mabee PM (1989a) An empirical rejection of the ontogenetic polarity criterion. Cladistics 5:409–416.

Mabee PM (1989b) Assumptions underlying the use of ontogenetic sequences for determining character state order. Trans Am Fish Soc 118:151–158.

Mayr E (1960) The emergence of evolutionary novelties. In: Tax S (ed) Evolution after Darwin, Vol. 2. Chicago: University of Chicago Press, pp 349–380.

McCormick CA (1982) The organization of the octavolateralis area in actinopterygian fishes: a new interpretation. J Morphol 171:159–181.

McCormick CA, Braford MR, Jr (1988) Central connections of the octavolateralis system: evolutionary considerations. In: Atema J, Fay RR, Popper AN, Tavolga WN (eds) Sensory Biology of Aquatic Animals. New York: Springer-Verlag, pp 733–756.

Metcalfe WK (1985) Sensory neuron growth cones comigrate with posterior lateral line primordial cells in zebrafish. J Comp Neurol 238:218–224.

Metcalfe WK, Kimmel CB, Schabtach E (1985) Anatomy of the posterior lateral line system in young larvae of the zebrafish. J Comp Neurol 233:377–389.

Moodie RL (1908) The lateral line system in extinct Amphibia. J Morphol 19:511–541.

Nelsen OE (1953) Comparative Embryology of the Vertebrates. New York: The Blakiston Co.

Nelson G (1973) The higher-level phylogeny of the vertebrates. Syst Zool 22:87–91.

Nelson G (1978) Ontogeny, phylogeny, paleontology, and the biogenetic law. Syst Zool 27:324–345.

Nelson G (1985) Outgroups and ontogeny. Cladistics 1:29–45.

Nelson G, Platnick NI (1981) Systematics and Biogeography: cladistics and vicariance. New York: Columbia University Press.

Noden DM, de Lahunta A (1985) The Embryology of Domestic Animals. Baltimore: Williams & Wilkins.

Noden DM, Van De Water TR (1986) The developing ear: tissue origins and interactions. In: Ruben RJ, Van De Water TR, Rubel EW (eds) The Biology of Change in Otolaryngology. Amsterdam, New York and Oxford: Excerpta Medica, pp. 15–46.

Northcutt RG (1984) Evolution of the vertebrate central nervous system: patterns and processes. Am Zool 24:701–716.

Northcutt RG (1986a) Embryonic origin of amphibian electroreceptors. Soc Neurosci Abstr 12:103.

Northcutt RG (1986b) Electroreception in non-teleost bony fishes. In: Bullock TH, Heiligenberg W (eds) Electroreception. New York: Wiley, pp 257–285.

Northcutt RG (1987) Development of the lateral line system of the channel catfish. Soc Neurosci Abstr 13:133.

Northcutt RG (1989) The phylogenetic distribution and innervation of craniate mechanoreceptive lateral lines. In: Coombs S, Görner P, Münz H (eds) Neurobiology and Evolution of the Lateral Line System. New York: Springer-Verlag, pp 17–78.

Northcutt RG (1990a) Ontogeny and phylogeny: a reevaluation of conceptual relationships and some applications. Brain Behav Evol 36:116–140.

Northcutt RG (1990b) The lateral line system of the axolotl. Axolotl Newsletter 19:5–14.

Northcutt RG, Fritzsch B, Brändle K (1990) Experimental evidence that ampullary organs of salamanders derive from placodal material. Soc Neurosci Abstr 16:129.

Northcutt RG, Gans C (1983) The genesis of neural crest and epidermal placodes: a reinterpretation of vertebrate origins. Q Rev Biol 58:1–28.

Northcutt RG, Vischer HA (1988) *Eigenmannia* possesses autapomorphic rami of the anterior lateral line nerves. Soc Neurosci Abstr 14:54.

Ørvig T (1971) Comments on the lateral line system of some brachythoracid and ptyctodontid arthrodires. Zool Scripta 1:5–35.

Pehrson T (1949) The ontogeny of the lateral line system in the head of dipnoans. Acta Zoolog 30:153–182.

Platt JB (1896) Ontogenetic differentiations of the ectoderm in *Necturus*. Q J Microscop Sci 38:485–547.

Puzdrowski RL (1989) The peripheral distribution and central projections of the lateral line nerves in goldfish, *Carassius auratus*. Brain Behav Evol 34:110–131.

Remane A (1956) Die Grundlagen des natürlichen Systems der vergleichenden Anatomie und Phylogenetik. Leipzig: Geest und Portig, K. G.

Ronan M (1986) Electroreception in cyclostomes. In: Bullock TH, Heiligenberg W (eds) Electroreception. New York: Wiley, pp 209–224.

Ronan MC, Bodznick D (1986) End buds: non-ampullary electroreceptors in adult lampreys. J Comp Physiol 158:9–16.

Ronan MC, Northcutt RG (1987) Primary projections of the lateral line nerves in adult lampreys. Brain Behav Evol 30:62–81.

Rosen DE (1982) Do current theories of evolution satisfy the basic requirements of explanation? Syst Zool 31:76–85.

Roth VL (1988) The biological basis of homology. In: Humphries CJ (ed) Ontogeny and Systematics, New York: Columbia University Press, pp. 1–26.

Russell ES (1916) Form and Function. London: John Murray (Reprinted, 1982, Chicago: University of Chicago Press)

Ruud G (1920) Über Hautsinnesorgane bei *Spinax niger* Bon. II. Die embryologische Entwicklung. Zool Jahrbüher Anat Ontogen 41:459–546.

Sato M (1956) Studies on the pit organs of fishes. IV. The distribution, histological structure and development of the small pit organs. Annotat Zool Jpn 29:207–212s.

Sato M (1976) Electron microscopic study of the developing lateral line organs in the embryo of *Triturus pyrrhogaster*. Anat Rec 233:377–389.

Schindewolf OH (1946) Zur Kritiq des "Biogenetischen Grundgesetzes." Naturwissenschaften 33:244–249.

Schoenwolf GC, Smith JL (1990) Mechanisms of neurulation: traditional viewpoint and recent advances. Development 109:243–270.

Smith SC, Lannoo MJ, Armstrong JB (1988) Lateral-line neuromast development in *Ambystoma mexicanum* and a comparison with *Rana pipiens*. J Morphol 198:367–379.

Song JS, Northcutt RG (1991) Morphology, distribution, and innervation of the lateral line receptors of the Florida gar, *Lepisosteus platyrhincus*. Brain Behav Evol 37:10–37.

Srivastava MDL, Srivastava CBL (1969) The development of neuromasts in *Cirrhina mrigala* Ham. Buch. (Cyprinidae) and *Ophicephalus* (*Channa*) *punctatus* Block (Channidae). J Morphol 122:321–344.

Stone LS (1922) Experiments on the development of the cranial ganglia and the lateral line sense organs in *Amblystoma punctatum*. J Exp Zool 35:421–496.

Stone LS (1928) Experiments on the transplantation of placodes of the cranial ganglia in the amphibian embryo. III. Preauditory and postauditory placodal materials interchanged. J Comp Neurol 47:117–154.

Stone LS (1929) Experiments on the transplantation of placodes of the cranial ganglia in the amphibian embryo. IV. Heterotopic transplantations of the postauditory placodal material upon the head and body of *Amblystoma punctatum*. J Comp Neurol 48:311–330.

Stone LS (1933) Development of the lateral-line sense organs in amphibians observed in living and vitally stained preparations. J Comp Neurol 57:507–540.

Stone LS (1935) Experimental formation of accessory organs in midbody lateral-line of amphibians. Proceedings of the Society for Exp Biol Med 33:80–82.

Strong OS (1895) The cranial nerves of Amphibia. J Morphol 10:101–231.

Thomson KS (1977) On the individual history of cosmine and possible electroreceptive function of the pore-canal system in fossil fishes. In: Mahala Andrews S, Miles RS, and Walker AD (eds) Problems in Vertebrate Evolution. Linnean Soc Symp Ser 4, pp 247–270.

Thomson KS (1988) Morphogenesis and Evolution. New York: Oxford University Press.

Vischer HA (1989a) The development of lateral-line receptors in *Eigenmannia* (Teleostei, Gymnotiformes). I. The mechanoreceptive lateral-line system. Brain Behav Evol 33:205–222.

Vischer HA (1989b) The development of lateral-line receptors in *Eigenmannia* (Teleostei, Gymnotiformes). II. The electroreceptive lateral-line system. Brain Behav Evol 33:223–236.

Wake DB, Larson A (1987) Multidimensional analysis of an evolving lineage. Science 238:42–48.

Webb JF (1989) Gross morphology and evolution of the mechanoreceptive lateral-line system in teleost fishes. Brain Behav Evol 33:34–53.

Wiley EO (1981) Phylogenetics. Theory and Practice of Phylogenetic Systematics. New York: John Wiley & Sons.

Wilson HV (1889) The embryology of the sea bass (*Serranus atrarius*). Bull US Fish Comm 9:209–277.

Wilson HV, Mattocks JE (1897) The lateral sensory anlage in the salmon. Anat Anzeig 13:658–660.

Winklbauer R, Hausen P (1983) Development of the lateral line system in *Xenopus laevis*. I. Normal development and cell movement in the supraorbital system. J Embryol Exp Morphol 76:265–281.

Winklbauer R, Hausen P (1985) Development of the lateral line system in *Xenopus laevis*. IV. Pattern formation in the supraorbital system. J Embryol Exp Morphol 88:193–207.

Young, GC (1986) The relationships of placoderm fishes. Zool J Linnean Soc 88:1–57.

4
Evolution of the Vertebrate Inner Ear: An Overview of Ideas

Arthur N. Popper, Christopher Platt, and Peggy L. Edds

1. Introduction

When considering the evolution of hearing it is useful to have an historical perspective of ideas about the origin of the vertebrate inner ear. Literature explicitly addressing this question is not particularly extensive, but it does go back more than a century. As a consequence, a small number of papers have had a large influence on how we think about the evolution of the inner ear. The purpose of our chapter is to provide a limited overview of some of these papers and to consider questions about the evolution of inner ear structures in the context of hearing. We see our contribution as a synopsis, rather than a new synthesis, of some key ideas and questions about the vertebrate inner ear and hearing. We hope that our chapter will be useful in framing questions for future research.

In this chapter, we first briefly review theories on the origin of the vertebrate ear and the major ideas in three important papers, concentrating on issues that are germane to evolution. (Readers are urged to look at the original papers for the extensive valuable details about morphology and function of the ear.) Then, we discuss some controversies that are still worthy of consideration.

1.1 A Caution About Terminology

The literature on the "ear" of vertebrates is complicated by the structural variety, multiple functions, and diverse neural connectivity of the portions of the ear implicated in hearing, or that are considered in some way comparable or related to the mammalian cochlea (see Lewis, Leverenz, and Bialek 1985). Many terms describing the ear, parts of the ear, and its functions, including "hearing" itself, have been used in various ways by different authors, thereby further complicating the literature, and its interpretation.

For example, the terms basilar papilla, cochlear duct, and cochlea are not uniformly applied; all three have been used for the same part of the reptilian ear. Both amphibians and reptiles have a "basilar papilla," but this term has been used for what may be two different structures (Wever 1974; Lombard 1980; Fritzsch Chapter 18). In order not to add to the confusion in this chapter, we will use "basilar papilla" only for the amphibian structure, and refer to the auditory endorgan of reptiles as the "cochlear duct," that of birds as the "avian cochlear duct," and of mammals, the "cochlea."

There is also disagreement in defining "hearing." Some definitions invoke behavior while others are purely in terms of physiology or acoustics. Placing hearing in the context of evolution makes it particularly important to consider the physical parameters of the medium (water, air, substrate) that carries the sound. See Webster (Chapter 37), Michelsen (Chapter 5), E. Lewis (Chapter 11), Schellart and Popper (Chapter 16), and Stebbins (1980; Chapter 13) for additional considerations of these problems.

Here are some of the diverse historical definitions of hearing, as examples:

"An animal hears when it behaves as if it has located a moving object (a sound source) not in contact with it." (Pumphrey 1950, p. 3)

"...ascribe the ability of hearing only to those animals which *first* are shown to be sensitive to air- or water-borne sound, i.e., to a succession of pressure waves propagating with a characteristic velocity throughout the medium involved, and which *secondly* detect these stimuli with special receptors primarily used for this purpose. If the latter condition is not fulfilled we merely speak of sound reception. Responsiveness to sound or vibrations reaching the animal through the solid substrate will be termed vibration reception." (Dijkgraaf 1960, p. 51)

"The stimulation of macular hair cells by sounds no doubt occurs in all animals that possess these organs, but the effects are interpreted as auditory only when the proper neural arrangement is provided." (Wever 1974, p. 98)

"Hearing is the response of an animal to sound vibrations by means of a special organ for which such vibrations are the most effective stimulus. [That special organ] ... (which we shall call an ear) is more sensitive to sound than it is to any other form of energy." (Wever 1976, p. 425)

2. Germinal Papers

Hearing is a very active topic of research, but papers specifically dealing with evolution have been infrequent. Discussions by van Bergeijk, Baird, and Wever have been especially influential (see below). Among these and other papers some of the central issues include origins of the inner ear and structural changes in the ear as vertebrates emerged and radiated into terrestrial habitats.

2.1 Origin of the Vertebrate Inner Ear

It is difficult to relate vertebrate ears evolutionarily to any invertebrate precursors. No fossil evidence has been found to establish the presence or absence of inner ear structures, such as semicircular canals or otolithic organs, in the earliest chordates (see Baird 1974; Wever 1974). The most ancient examples of a vertebrate inner ear occur as labyrinths in the skulls of Silurian ostracoderms (Stensiö 1927). Some of these early labyrinths had two semicircular canals, much as in some modern agnathans, while others had as many as five or six canals and undivided ampullae (Stensiö 1927; Jarvik 1980). Among invertebrates, even coelenterates have

mechanoreceptive hair cells in gravitationally sensitive statolith organs, which are presumed to be the most ancient sensory organs of all (see Bullock and Horridge 1966). Statocyst organs apparently have arisen independently in a variety of forms in many invertebrate phyla (see Budelmann, Chapters 9 and 10). Thus, it is also reasonable that an organ for postural equilibrium or for hearing could have arisen independently in the chordates, without postulating an invertebrate receptor organ as a precursor to the vertebrate inner ear.

2.2 The Acousticolateralis Hypothesis

The most prevalent idea regarding the origin of the vertebrate inner ear has been the acousticolateralis hypothesis (see Lowenstein 1967 and Wever 1974 for detailed historical reviews). The acousticolateralis hypothesis proposed that the inner ear was evolutionarily derived from the lateral line, based on evidence from structure, ontogeny, and function. The hypothesis, as discussed by Wever, includes the presumption of several investigators that "the ear is considered to be derived from the lateral line sensory system" (Wever, 1974), and of Ayers (1892), who "concluded that all of these structures had a single function, that of hearing." (Wever 1974).

An early paper linking the ear to the lateral line by structural similarities was by Mayser (1882), who traced central neural projections in cyprinid fishes. Mayser considered the lateral line an accessory hearing organ and argued that both the lateral line and inner ear nerves terminate at the same area in the medulla, the "acoustic tubercle." As pointed out by Northcutt (1981), the "acoustic tubercle" became known as the "acousticolateral area," so "the inner ear's proposed origin from a portion of the lateral line system is known as the acousticolateralis hypothesis." Note, however, that the acousticolateralis hypothesis is not directly related to the current acceptance of the anatomical term "octavolateralis system" (McCormick, 1978, 1982; see Section 3.1).

Wilson and Mattocks (1897) linked the inner ear and lateral line ontogenetically by observing embryonic development in salmon, suggesting that the two systems are derived from a common embryonic placode. Wilson and Mattocks postulated that the inner ear actually originated as a part of the lateral line, but sank beneath the surface of the skin in the region of the head (see van Bergeijk 1967, and Section 2.3 below). Ayers also proposed that the labyrinth developed from the lateral line (see Wever 1974).

Several workers suggested that the lateral line functions as a primitive version of the ear as a hearing organ (see Lowenstein 1967). Ayers' (1892) influential paper stated that "every sense organ in the ear executes an auditory function, as regards essentials, in the same manner as its prototype, the lateral line sense organ; there exist between the auditory sense organs differences only in degree, not in kind, of the function observed."

The acousticolateralis hypothesis was strongly supported by van Bergeijk (1966, 1967) and Baird (1974). However, Wever (1974, 1976) concluded that the evidence for the hypothesis was weak, and that it was more plausible that the lateral line and inner ear evolved concurrently. The arguments for and against the acousticolateralis hypothesis will be discussed in the following sections.

Subsequent evolution of the inner ear itself, once formed, appears less controversial. Most later authors follow Ayers' idea that the earliest form of the inner ear was a labyrinthine complex functioning primarily as an equilibrium receptor, and that the cochlea was later derived from the labyrinth. Other special endorgans for hearing also are considered to be derived from parts of the basic labyrinthine plan by structural modifications for acoustic reception (see below).

2.3 Van Bergeijk (1967): "Evolution of Vertebrate Hearing"

Van Bergeijk published two papers (1966, 1967) that have been important not only for their impact on ideas about the evolution of hearing, but also for their impact on questions about mechanisms of hearing by aquatic and terrestrial vertebrates. In his 1967 paper, van Bergeijk dealt specifically with the evolution of hearing in fishes and amphibians.

Van Bergeijk agreed with the view that the inner ear and the lateral line are closely related because: (1) they develop from the same embryonic mesodermal anlage, (2) they have the same kind of ciliated sensory hair cells, and (3) their covering and innervation are very similar. He further argued

that the acousticolateralis system can be considered as a "unitary system" for hearing, which he believed to be a secondarily acquired capability of a generalized mechanoreceptor system.

In his analysis supporting the derivation of the ear from the lateral line, van Bergeijk suggested that a superficial structure with sensory hair cells, the lateral line, would be sensitive to hydrodynamic stimuli. If part of the lateral line had submerged and become enclosed it could have become a "vehicle-oriented inertial accelerometer" with equilibrium function. Linear acceleration sensitivity would have been improved by loading with increased inertial mass in the otolithic organs. Van Bergeijk asserted that this structure of fluid ducts and otolithic organs in the vertebrate inner ear was essentially unchanged during evolution from jawed fishes on, except for the gradual emergence of a special hearing organ. He reasoned that for an otolithic organ to become auditory, there had to be other changes that would allow the otolithic organ to detect pressure waves. He suggested that one of the major changes in fishes to facilitate this detection was the involvement of an accessory gas-filled chamber as a pressure-to-displacement transducer.

Van Bergeijk pointed out that anurans were the first vertebrates to develop a terrestrial middle ear, and they evolved a variety of structural specializations to match the impedance difference from the air to the fluids of the inner ear. These structures carry the acoustic signal to various specialized auditory endorgans which, in amphibians, include the amphibian papilla and the basilar papilla.

2.4 Baird (1974): "Anatomical Features of the Inner Ear in Submammalian Vertebrates"

Baird presented a review comparing the diverse structures of the inner ear in a phylogenetic framework. The high impact of this paper results from its very comprehensive scope and detailed morphological analysis of the ear in diverse vertebrates. (Unlike evolutionary systematists' particular usage, Baird used the term "submammalian" for all nonmammalian vertebrates.)

Baird argued that the early chordates had lateral lines, and that simple inner ear structures developed from enclosed lateral line canal organs. He agreed with van Bergeijk (1967) that the evidence from differentiation of lateral placodes during development provides a "sound ontogenetic basis" for the derivation of the inner ear from neuromast structures of the lateral line system. He often used the term "otic labyrinth" to describe this internal membranous sensory organ, which in more advanced vertebrates became more completely encased in the bony labyrinth at the base of the skull. Such inner ears might first have functioned for equilibrium and possibly vibration detection, since Lowenstein (1970) demonstrated these functions in lamprey inner ears. The evolution of parts of the inner ear toward acoustic sensitivity then could follow with specialization of particular structures. For example, differences in the gross structure of the pouches, ducts, and windows allow different mechanical access to the fluids surrounding the sensory endorgans.

Baird believed that functional segregation may have been linked to anatomical separation and/or specialization. From observations on extant agnathans and elasmobranchs, he surmised that the lateral semicircular duct may be the least ancient of the three canals. He also noted that in some bony fishes (such as the mormyrids), the semicircular ducts and one otolithic organ, the utricle, are almost completely separated from the two more ventral otolithic organs, the saccule and lagena. The semicircular canal apparatus and utricle are usually considered sensors for equilibrium function, while in many fishes the saccule has been implicated in auditory function. So the structural separation may reflect a functional division during evolution, as the lower parts of the membranous labyrinth became increasingly specialized for audition.

Baird pointed out that the macula neglecta (or papilla neglecta) of fishes is quite diverse in whether it is present or not, its size, and its location among different species (also see Retzius 1881; Corwin 1978). He discussed how this epithelium, or parts of it, might have evolved into the amphibian papilla or basilar papilla of amphibians (but see Fritzsch 1987 and Chapter 18). Baird wrote that "it is probable that the macula neglecta shifted ventrally" to lie closer to the saccular than to the utricular macula. While Baird admitted that the ontogenetic evidence does not support this shift, he believed the arching of the nerve and its branching patterns in different species supports the con-

cept. Similarly, Baird argued that the cochlear duct (he used the term "basilar papilla") of reptiles also could have formed by migration of the basilar papilla from the saccular wall in amphibians toward the lagenar area in reptiles. This argument implies homology of the reptilian cochlear duct with the basilar papilla of amphibians.

A central aspect to the evolution of the structure of the inner ear is that the epithelium must be located in a place that can receive appropriate fluid displacements, driven by acoustic pressure waves in the environmental medium. Baird summarized the specializations for hearing in aquatic animals that exploited the compressibility of gaseous spaces to produce a large displacement component in response to an acoustic pressure stimulus. These displacements are transmitted to the inner ear directly from hollow ducts extending from the abdominal swimbladder rostrally to the inner ear (e.g., clupeids, the herrings), or from gas-filled bullae near the inner ear (e.g., mormyrids such as the elephant-nosed fish, and anabantids, the bubble-nest builders), or indirectly by bony mechanical linkages extending from the swimbladder to the inner ear (e.g., otophysans such as goldfish and catfish). In terrestrial vertebrates, a series of bony linkages evolved from jaw bones to become the middle ear, which transmits air-driven vibrations of the tympanum to the endolymphatic system of the inner ear (see also Allin and Hopson Chapter 28; Bolt and Lombard Chapter 19; Clack Chapter 20).

The cochlear duct in reptiles exhibits a range of forms that Baird noted can be put into a sequence showing an elongation of the cochlear duct, that becomes increasingly separated from the saccule (see also Miller, Chapter 23). The lagena remains at the distal end of the cochlear duct in all reptiles and birds. The development of cellular specializations in the receptors with an overlying tectorial membrane reaches the most advanced reptilian form in the crocodilians, where the structure is comparable to that of birds.

Baird pointed out that this development essentially completes the nonmammalian story, with the exception of comparative data on the inner ear of birds (see Manley and Gleich, Chapter 27). Baird suggested that further refinements of the elongated cochlea are primarily histological and ultrastructural specializations.

2.5 Wever (1974): "The Evolution of Vertebrate Hearing"

Wever presented his thoughts about the evolution of vertebrate hearing in several papers (e.g., Wever, 1971, 1976, 1981) and two important books (Wever, 1978, 1985). His 1974 paper probably remains the most comprehensive and influential of this group, at least with regard to overall trends in the evolution of the ear. Wever pointed out that studying the evolution of the vertebrate inner ear is not easy because none of the animal groups thought to be ancestral to vertebrates have an inner ear as we know it. Wever did not consider the agnathans to have a true inner ear that acts as an auditory receptor; he considered the bony fishes to have the first "real ear" that could serve as a hearing organ, although he indicated that sharks may also hear.

Wever concluded that the inner ear was not derived from the lateral line. He asserted that much of the evidence for the origin of the labyrinth from the lateral line is based upon "assumptions" and he suggested abandoning the acousticolateralis hypothesis. He agreed that the two systems share similarities that suggest a relationship between them, but he argued that their evolution from a common mechanosensory system using hair cells was more likely than the inner ear evolving from the lateral line itself.

Wever agreed with the generally accepted view that the auditory portion of the inner ear arose from the labyrinth. He felt that only slight changes of an equilibrium receptor could provide an endorgan that could detect energy of rapid oscillations, such as a vibration or sounds, in addition to (or in place of) equilibrium signals. Associated with such mechanical changes would be changes in neural organization that would filter out signals pertaining to equilibrium, thus leaving a hearing endorgan. According to Wever, this change would start with detection of low frequencies with poor sensitivity, followed by improvements in sensitivity and in the ability to discriminate frequencies and/or sound source direction. He contended that this sequence may be reflected, to some degree, in differences in the auditory system among various extant vertebrates, with fishes having hearing abilities closest to the original. But Wever also cautioned that extant animals may not accurately represent what

occurred in the past because hearing systems have had time to evolve beyond ancestral forms.

Wever proposed that the transformation from equilibrium receptor to auditory receptor occurred four or possibly five discrete times among the vertebrates. This idea implies that although the auditory portions of the inner ears in vertebrates evolved from the labyrinthine organs, changes did not occur in a sequential pattern in the vertebrate lineage. For example, the auditory portion of an amphibian inner ear might not be derived from the auditory portion of a fish inner ear, and reptilian auditory structures, in turn, might not be derived from those of amphibians and fishes.

Wever's (1974) evidence for two of these discrete derivatives for auditory receptors came from otolithic endorgan function in teleost fishes (also see Schellart and Popper, Chapter 16). First, he suggested that the saccule, presumably an equilibrium organ in its initial stages of evolution, became a hearing organ through reorganization of central projections of the saccular nerve in many fish groups. He inferred a second auditory derivation in the clupeids where the utricle, an equilibrium organ in most vertebrates, became a hearing organ. The third and fourth derivations could have evolved in the amphibians by conversion of some unknown organs into the amphibian papilla and the basilar papilla (see also Wever 1981; Fritzsch, Chapter 18). A possible fifth type of derived hearing organ is the cochlear duct of reptiles, which, according to Wever, is not directly related to these other four organs of fishes and amphibians. The source for the evolution of the reptilian cochlear duct was not clear to Wever, but he argued that neither the basilar papilla of amphibians nor the lagena (as proposed by other investigators) was the source. Instead, he proposed that the cochlear duct may have been derived from "its own specific part of the embryonic mass, a part that earlier grew into the labyrinth, and not as a secondary development out of cells that in the embryo first produced the lagenar macula or any other labyrinthine endings." (Wever 1974)

Although the avian cochlear duct and the mammalian cochlea were considered by Wever to be derived from the reptilian cochlear duct, he pointed out that their derivation began very early in the evolution of the reptilian inner ear. Both avian and mammalian cochlear structures diverged substantially from the reptilian cochlear duct. He again emphasized that inner ears are highly derived structures in all extant vertebrate groups, so present structures may not be reliable for inferring ancestral lineage.

In summary, Wever argued against the acousticolateralis theory and against the idea of a linear evolution of the auditory portion of the inner ear. Instead of a phylogenetic and ontogenetic sequence from lateral line to inner ear, he felt the evidence allows for concurrent independent evolutionary and developmental elaboration of the inner ear and lateral line organs and their innervation, exploiting a common class of mechanosensory receptor cell. Instead of a progressive evolutionary sequence leading to the mammalian cochlea, he saw the vertebrate labyrinth as the potential source for multiple, independently derived auditory endorgans. He argued that the endorgans for hearing in some different vertebrate groups are so distinct structurally that it is unlikely that one gave rise to another.

3. Issues from the Reviewed Papers

These three papers have been widely cited but represent two fundamentally different views about the evolution of the vertebrate inner ear. Van Bergeijk and Baird elaborated on a set of ideas that were widely accepted for a good part of this century. Wever's opposing view was based upon modern embryology and anatomy and seems much more tenable than the earlier suggestions. In the following sections, we will discuss two questions that are of particular interest.

3.1 Acousticolateralis Hypothesis

The acousticolateralis hypothesis argued for the lateral line being the antecedent of the labyrinth (Mayser 1882; Ayers 1892; Wilson and Mattox 1897; van Bergeijk 1967). Baird (1974) made the blanket statement, "It is generally agreed that the inner ear of vertebrates represents the evolutionary result of an inward migration and specialization of elements of the lateral line system." In contrast, in the same volume as the Baird paper, Wever (1974) argued that the inner ear and lateral line may be related to one another, but that one is not the antecedent of the other.

Much of the rationale for the acousticolateralis hypothesis now seems poorly founded. One of the basic assumptions, of a common "acoustic tubercle" in the medulla, has been shown to be oversimplified (reviewed in Wever 1974, 1976; Northcutt 1981). A second assumption, that the two systems come from a common embryonic placode, is now considerably modified, as separate otic and lateral line placodal regions can be identified within this original placode (Wever 1974, 1976; Northcutt 1980, Chapter 4). Third, in contrast to Ayers (1892) claim, there apparently are not any neuromasts in the lateral line of hagfish, an agnathan (Fernholm, 1985; Northcutt, 1989); while this lack might be a derived character, it may instead reflect a primitive condition of an early chordate with an inner ear but no lateral line (Fritzsch, personal communication).

Another basis for suggesting the inner ear arose from the lateral line is the similarity of the mechanoreceptive hair cells common to both systems. However, the similarities between the hair cells of the inner ear and lateral line need not imply that one system was derived from the other, because many similar cells are found in other vertebrate epithelia and sensory organs (Wever 1974, 1976). In the vertebrates, the hair cells of the inner ear and mechanosensory lateral line, receptor cells of the electrosensory lateral line, ciliated chemosensory cells, and photoreceptor cells of the eye all are believed to have come from a stem form of ciliated epithelial cell. Thus the similarity between hair cells in the inner ear and lateral line may simply reflect use of a common epithelial sensory cell class. Generally similar mechanosensory cells are also found among diverse invertebrate phyla (see Bullock and Horridge 1966; Budelmann, Chapter 10). The origin of the vertebrate mechanoreceptive hair cell is not clear, although the invertebrate data suggest that similar cells could have arisen independently more than once during the course of animal evolution.

Usage of the anatomical term "octavolateralis" has become common and may provide a good way to close the issue. This term is applied to the mechano- and electro-sensory systems utilizing hair cells as receptors and branches of the eighth (octaval) and the lateral line nerves (McCormick 1978, 1982; Northcutt 1980, 1981; Platt, Popper, and Fay 1989; and see Coombs, Görner, and Münz 1989; McCormick, Chapter 17). It acknowledges a basic morphological and functional similarity among the various mechanosensory endorgans involved, including those found within the inner ear (maculae, cristae, and papillae) and those of the lateral line (superficial and canal neuromasts, of the head and trunk). However, octavolateralis does not inherently suggest acoustic function, nor imply evolutionary or ontogenetic origins when considering these sensory systems.

The term octavolateralis appeared in 1918 in the title of a dissertation by Schepman (cited in Äriens Kappers, Huber, and Crosby 1936). However, Schepman used the term in a very broad sense and not to specifically refer to the central nervous system (CNS). The term was first used as a specific reference to the CNS by Nieuwenhuys, who, in 1967, stated that " . . . the rhombencephalic part of [the dorsal column of the medulla oblongata] gives rise to two sensory areas, *viz.*, the end station of the trigeminal nerve (area trigemini), and the nuclear complex of the 8th cranial and lateral line nerves (area octavo-lateralis)" (p. 2). Broad acceptance of the term began in the early 1980's with papers by McCormick (1978, 1982) and Northcutt (1980, 1981).

3.2 Auditory Organ Homologies

It is difficult to clarify the degree of homology among the variety of present vertebrate auditory organs, yet this issue needs to be addressed for both comparative and evolutionary interpretations of structure and function. There apparently is a consensus that the earliest inner ear structures acted as equilibrium receptors and that the auditory function arose one or more times. The traditional view, as discussed by van Bergeijk (1967) and Baird (1974), is that the auditory structures in terrestrial vertebrates arose separately from those in fishes. But these and other authors have argued that within the terrestrial radiations, there was a linear derivation from a papilla in amphibians to the reptilian cochlear duct, then to the avian cochlear duct and the mammalian cochlea. Wever (1974) challenged this view. He agreed that the inner ear has separate parts for audition and equilibrium, but suggested that there were multiple derivations for the auditory parts in different vertebrate groups. He argued that the basic structure and mechanics of the auditory endorgans differ so substantially

among the different groups that one would parsimoniously expect separate derivations.

This argument is supported, at least in part, by Lombard (1980), who argued that while the amphibian and amniote inner ears share common ancestry with one another, each tetrapod group develops "an auditory periphery de novo (no common morphology or function) and in which a cline of 'improvement' on the *scala natura* of tetrapods is clearly not appropriate." Lombard supported his argument by demonstrating that major components of the amphibian inner ear and more peripheral structures, including the amphibian and basilar papillae, the tympanum, the tympanum-stapes articulation, and the opercular muscle, are unique to amphibians, making the amphibian peripheral auditory system distinct from that of other tetrapods.

In this context, the lagena is particularly interesting. In function, it has been implicated in both hearing and equilibrium in fishes (see Lowenstein 1967); in structure, it shows marked interspecific structural variation in bony fishes (see Platt and Popper 1981). These properties raise speculation that the lagena might have multiple origins. The picture is complicated further in terrestrial vertebrates where the lagena is present in amphibians, reptiles, birds, and monotremes, but not in placental mammals. Where present, the lagena lies close to the cochlear duct (if present), and this relation has led to the suggestion that the lagena may have given rise to the mammalian cochlea (Baird 1974).

The other enigmatic endorgan is the macula neglecta. Its presence and its structure have been documented in many diverse vertebrate species (Retzius 1881; Igarashi 1965; Jørgensen 1972; Corwin 1978).There is substantial variation in the location of the macula neglecta within the labyrinth. Moreover, in some species the macula neglecta has a single sensory patch, while other species have more than one; the sizes of these patches are also quite variable (Corwin 1978). Thus the term "macula neglecta" may be in use for different endorgans. The macula neglecta also has been implicated in the origin of several different auditory endorgans including the basilar papilla (Fritzsch 1987), but the evidence in each case is circumstantial. Until we know more about the function of the "macula neglecta" in different species, it will be difficult to resolve these issues.

The possibility that the term "macula neglecta" may represent multiple endorgans highlights again the serious problem of assigning the same name to what may be different inner ear structures in different species. Homology was assumed when assigning the names "utricle," "saccule," and "lagena" for regions of the single elongate sensory macula in petromyzontids (Lowenstein, Osborne, and Thornhill 1968). The rationale was based largely on spatial relations, without further compelling reasons to propose homology between parts of this single macula and the separate otolithic endorgans found in gnathostomes (Popper and Hoxter 1987). Similarly, Weber (1820) inappropriately used the terms "malleus," "incus," and "stapes" for the three bones that connect the swimbladder to the inner ear in otophysan fishes; these bones, now called the Weberian ossicles, are not related to those named bones in tetrapods. It is clear that more comparative data are needed to resolve cases of putative homology, analogy, and adaptive evolutionary changes in the various kinds of vertebrate ears.

4. Summary

In this chapter, we have emphasized the papers that we consider to be of fundamental importance in the history of ideas regarding the evolution of the vertebrate inner ear. Many of the questions raised in those papers are addressed in various chapters in this volume; however, others remain, and we now have a plethora of exciting new questions that have been generated.

Acknowledgments. We want to express our thanks to our colleagues Drs. D.B. Webster and R.R. Fay for enriching discussions and perspectives that certainly affected the development of our ideas. We also thank Drs. B. Fritzsch, W. Plassmann, and W.M. Saidel for critically reading the MS; Dr. J. Clack for helping us understand the uses of the word "submammalian" by different groups of biologists; and Drs. C.A. McCormick, G. Meredith, R. Nieuwenhuys, and R.G. Northcutt for discussions about the origin of the term "octavolateralis." Preparation of this paper was supported by NIH Grant DC-00140 and ONR Contract N-00014-87-K-0604.

References

Äriens Kappers CU, Huber GC, Crosby EC (1936) The Comparative Anatomy of the Nervous System of Vertebrates, Including Man. Reprinted 1967. New York: Haffner.

Ayers H (1892) Vertebrate cephalogenesis. J Morph 6:1–360.

Baird IL (1974) Anatomical features of the inner ear in submammalian vertebrates. In: Keidel WD, Neff WD (eds) Handbook of Sensory Physiology, Vol. V/1, Auditory System. Berlin: Springer-Verlag, pp. 159–212.

Bergeijk WA van (1966) Evolution and the sense of hearing in vertebrates. Am Zool 6:371–377.

Bergeijk WA van (1967) The evolution of vertebrate hearing. In: Neff WD (ed) Contributions to Sensory Physiology. New York: Academic Press, pp. 1–49.

Bullock TH, Horridge GA (1966) Structure and Function of the Nervous Systems of Invertebrates. San Francisco: Freeman.

Coombs S, Görner P, Münz H (1989) The Mechanosensory Lateral line, Neurobiology and Evolution. New York: Springer-Verlag.

Corwin JT (1978) The relation of inner ear structure to the feeding behavior in sharks and rays. In: Johari OM (ed) Scanning Electron Microscopy/1978, Vol. II. Chicago: SEM. Inc.: pp. 1105–1112.

Dijkgraaf S (1960) Hearing in bony fishes. Proc Roy Soc B 152:51–54.

Fernholm B (1985) The lateral line system of cyclostomes. In: Forman RE, Gorbman A, Dodd JM, Olsson R (eds) Evolutionary Biology of Primitive Fishes. New York: Springer-Verlag, pp. 113–122.

Fritzsch B (1987) Inner ear of the coelacanth fish *Latimeria* has tetrapod affinities. Nature 327:153–154.

Igarashi M (1965) Redefinition of the macula neglecta in mammals. J Morphol 125:287–294.

Jarvik E (1980) Basic Structure and Evolution of Vertebrates, Vol. 1. London: Academic Press.

Jørgensen JM (1972) The avian Neglecta Retzii. Acta Zoologica 53:155–163.

Lewis ER, Leverenz EL, Bialek WS (1985) The Vertebrate Inner Ear. Boca Raton, Fl: CRC Press.

Lombard E (1980) The structure of the amphibian auditory periphery: a unique experiment in terrestrial hearing. In: Popper AN, Fay RR (eds) Comparative Studies of Hearing in Vertebrates. New York: Springer-Verlag, pp. 121–138.

Lowenstein O (1967) The concept of the acoustico-lateral system. In: Cahn PH (ed) Lateral Line Detectors. Bloomington: Indiana University Press, pp. 3–11.

Lowenstein O (1970) The electrophysiological study of the responses of the isolate labyrinth of the lamprey (*Lampetra fluviatilis*) to angular acceleration, tilting and mechanical vibration. Proc Roy Soc Lond B 174:419–434.

Lowenstein O, Osborne MP, Thornhill RA (1968) The anatomy and ultrastructure of the labyrinth of the lamprey (*Lampetra fluviatilis* L.). Proc Roy Soc Lond B 170:113–134.

Mayser P (1882) Vergleichend anatomische Studien über das Gehirn der Knochenfisch besonderer Berücksichtigung der Cyprinoiden. Z wiss Zool 36:259–264.

McCormick CA (1978) Central projections of the lateralis and eighth nerves in the bowfin, *Amia calva*. PhD Thesis, University of Michigan, Ann Arbor.

McCormick CA (1982) The organization of the octavolateralis area in actinopterygian fishes: A new interpretation. J Morphol 171:159–181.

Nieuwenhuys R (1967) Comparative anatomy of the cerebellum. In: Fox CA, Snyder RS (eds) Progress in Brain Research, Vol. 25. New York: Elsevier, pp. 1–93.

Northcutt RG (1980) Central auditory pathways in anamniotic vertebrates. In: Popper AN, Fay RR (eds) Comparative Studies of Hearing in Vertebrates. New York: Springer-Verlag, pp. 79–118.

Northcutt RG (1981) Audition and the central nervous system of fishes. In: Tavolga WN, Popper AN, Fay RR (eds) Hearing and Sound Communication in Fishes. New York: Springer-Verlag, pp. 331–355.

Northcutt RG (1989) The phylogenetic distribution and innervation of craniate mechanoreceptive lateral lines. In: Coombs S, Görner P, Münz (eds) The Mechanosensory Lateral Line: Neurobiology and Evolution. New York: Springer-Verlag, pp. 17–78.

Platt C, Popper AN (1981) Structure and function in the ear. In: Tavolga WN, Popper AN, Fay RR (eds) Hearing and Sound Communication in Fishes. New York: Springer-Verlag, pp. 3–38.

Platt C, Popper AN, Fay RR (1989) The ear as part of the octavolateralis system. In: Coombs S, Görner P, Münz H (eds) The Mechanosensory Lateral Line: Neurobiology and Evolution. New York: Springer-Verlag, pp. 633–651.

Popper AN, Hoxter B (1987) Sensory and non-sensory ciliated cells in the ear of the sea lamprey, *Petromyzon marinus*. Brain Behav Evol 30:43–61.

Pumphrey RJ (1950) Hearing, in Physiological Mechanisms in Animal Behavior. Symp Soc Exp Biol 4:3–18.

Retzius G (1881) Das Gehörorgan der Wirbelthiere, Vol. I. Stockholm: Samson and Wallin.

Schepman AMH (1918) De Octavo-laterale zintuigen en hun verbindingen in de hersenen der vertebraten. Amsterdam: Dïsserti.

Stebbins WC (1980) The evolution of hearing in mammals. In: Popper AN, Fay RR (eds) Comparative Studies of Hearing in Vertebrates. New York: Springer-Verlag, pp. 421–436.

Stensiö EA (1927) The Downtownian and Devonian vertebrates of Spitzbergen. I. Family Cephalaspidae. Skr om Svalbard og Nordishavet, No. 12 (2 Vols.).

Weber EH (1820) De Aure et Auditu Hominis et Animalium. Pars I. De Aure Animalium Aquatilium. Leipzig: Gerhard Fleischer, 134 pp.

Wever EG (1971) The mechanics of hair-cell stimulation. Ann Otol Rhinol Laryngol 80:1–19.

Wever EG (1974) The evolution of vertebrate hearing. In: Keidel WD, Neff WD (eds) Handbook of Sensory Physiology, Vol. V/1, Auditory System. Berlin: Springer-Verlag, pp. 423–454.

Wever EG (1976) Origin and evolution of the ear of vertebrates. In: Masterton RB, Bitterman ME, Campbell CBG, Hotton N (eds) Evolution of Brain and Behavior in Vertebrates. Hillsdale, NJ: Lawrence Erlbaum, pp. 89–105.

Wever EG (1978) The Reptile Ear. Princeton: Princeton University Press.

Wever EG (1981) The role of the amphibians in the evolution of the vertebrate ear. Amer J Otolaryngol 2:145–152.

Wever EG (1985) The Amphibian Ear. Princeton: Princeton University Press.

Wilson HV, Mattocks JE (1897) The lateral sensory anlage in the salmon. Anat Anz 13:658–660.

Section II
Invertebrates

The extensive diversity of invertebrate auditory systems has received little attention in most previous works dealing with the evolution of hearing. As demonstrated in this section, there is a wealth of data on invertebrate auditory systems, which are immensely diverse in both structure and function.

In Chapter 5, Michelsen presents general ideas that are germane to all hearing animals. He points out that the sense of hearing, including both peripheral mechanisms and central processing, evolved independently many times, in various groups of insects and invertebrates. He also argues that "... some features [of the auditory systems] are consequences of the physical nature of sound waves, and we therefore find functionally analogous mechanisms [of hearing] in rather unrelated animals." While he concentrates on the special problems of hearing and social communication in small animals, he makes the significant point that a number of the physical constraints on small animals are also important in the evolution of hearing in large animals, including vertebrates.

In Chapter 6, Römer takes to the field for an examination of insect sound production and hearing. These findings are placed in the context of the selective pressures that have shaped the character of the signals and the types of acoustic communication systems evolved in insects.

Römer's analysis of the auditory periphery is followed by Brian Lewis' Chapter 7, in which he provides a detailed analysis of the central processing of acoustic communication signals in perhaps the best-known acoustic insects, the orthopterans.

As part of a discussion on how selective pressures have affected the evolution of insect auditory systems, Hoy (Chapter 8) gives a brief overview of insect paleontology and argues that insects from the Permian Period, 225 millions years ago, are likely to have used song to mediate reproduction. However, he points out that bats, arriving on the scene perhaps 50 million years ago, placed intense predation pressure on insects, which may have encouraged the evolution of ultrasound detection in widely diverse insect species.

Budelmann discusses invertebrates other than insects in two chapter. In Chapter 9 he documents sound production in aquatic and semiterrestrial crustaceans, and their receptor systems that are potentially used as acoustic receptors. In Chapter 10 he argues that although "typical" hearing organs do not exist in nonarthropod invertebrates, many of them are able to detect particle motion in the aquatic environment. He discusses whether this ability should be called hearing, and describes the structures which are likely used for detection of particle displacement.

5
Hearing and Sound Communication in Small Animals: Evolutionary Adaptations to the Laws of Physics

Axel Michelsen

1. Introduction

A sense of hearing (including the hearing organs and central processing of auditory information) obviously evolved many times and independently in various groups of insects and vertebrates. Many of the features of the auditory systems reflect the evolutionary prehistory of the animals and thus follow their systematic positions. However, some features are consequences of the physical nature of sound waves, and we therefore find functionally analogous mechanisms in rather unrelated animals.

Unfortunately, only little is known about the evolution of hearing in insects. It is obvious that the hearing organs evolved from already existing mechanoreceptors (see Hoy, Chapter 8), and that the present-day hearing organs are adapted to serve social communication or the detection of predators. Some evidence suggests that hearing evolved originally for the detection of predators, and that some hearing organs were later adapted for social communication (see Lewis, Chapter 7), but other evidence supports the opposite scenario (see Hoy, Chapter 8).

In order to understand *how* hearing evolved, we have to consider the physical laws governing the emission, propagation and reception of sounds. An understanding of *why* hearing evolved requires a consideration of the strategies available to the animals for the detection of predators and prey and for social communication. To disclose whether hearing (and sound signaling) are reasonable strategies for these purposes, we must consider the advantages and disadvantages of sounds as cues and social signals. In other words, which alterna-

tive sensory capacities could be used? In order to understand the reasons for the complexity of the hearing organs and of the auditory processing in the central nervous system, we have to separate features that are adaptations to the laws of physics from those which reflect the evolutionary origins.

Obviously, this is an enormous task that cannot be undertaken in a single chapter. Here, we shall concentrate on the special problems of social communication in small animals (with body lengths of a few millimeters to a few centimeters). In exploring their situation we shall, however, discuss a number of physical mechanisms, which have also been important in the evolution of hearing in large animals.

In brief, small animals trying to exploit sounds for social communication are facing two problems: It is difficult (and metabolically expensive) to emit proper sounds, and it is difficult when listening to locate other sound sources. Furthermore, the frequency band available for communication is limited by a high-pass filter (the efficiency of sound emission increases with frequency) and a low-pass filter (sound penetration of the environment generally decreases with frequency). The efficiency of sound emission tends to decrease as the animals become smaller, and communication with airborne sounds is thus impossible in very small animals.

2. The Transmission Channel

Small animals, such as insects can be severely restricted by the laws of physics when trying to emit sounds (see Section 3). Sound communication is further complicated by the fact that not all

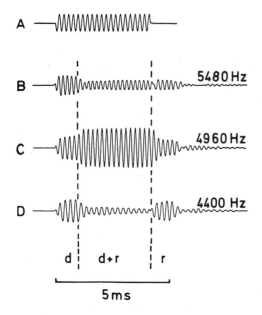

FIGURE 5.1. Interference patterns established by the addition of a direct sound wave (d) and a sound wave reflected from the ground (r). A pure tone with the duration shown in (A) is travelling 4.6 m and in a height of 80 cm parallel to ground. In (C) the direct and reflected waves were in phase when they reached the microphone, in (B) and (D) almost out of phase. A single reflection from the ground thus causes the spectrum of the received sound to show a number of maxima and minima (spectral distortion). (From Michelsen and Larsen 1983.)

types of sounds travel equally well from the emitter to the receivers. Furthermore, it is not sufficient for the sound signals to reach the receivers. The sounds should also be recognizable, detectable, and locatable.

The properties of the habitats as transmission channels for sounds were most certainly one of the major factors determining the evolution of sound communication, including the evolution of sound emission and hearing. In addition, distortion of the signals by processes in the transmission channel is the main reason for the existence of the elaborate information processing in the auditory parts of the central nervous system (see Römer, Chapter 6 and Lewis, Chapter 7).

The transmission channel should therefore be considered an integral part of the sound com-

munication system. It is important to know how the sound signals are adapted to the animals' present habitats. However, in order to understand the evolution of sound communication we have to know whether the species in question may have changed its habitat during evolution. Some animals have only recently changed their habitat and are thus unlikely to have adapted much to their present environment (for example, some eager investigators, who shall remain anonymous here, recently described the acoustical environment of the cat by recording the sounds of various farm animals).

Sound signals propagating through natural habitats are subjected to geometric spreading, reflection, absorption, refraction and diffraction. These physical processes have been reviewed elsewhere (Michelsen 1978; Michelsen and Larsen 1983) and are further discussed by Römer in Chapter 6. Here, we shall concentrate on the effects on the sound signals: attenuation, frequency filtering, temporal distortion, and decreasing directionality.

Most studies on acoustic transmission channels have concentrated on attenuation, which may provide a rough idea of the maximum distance of communication. Unfortunately, some pitfalls in the performance of the measurements make the results of many studies less useful (see Michelsen 1978). The general finding in habitats dominated by plants is that the attenuation increases with the frequency of sound. Sonic signals travel reasonably well in such habitats, but ultrasound is generally heavily attenuated (this probably explains why only few terrestrial vertebrates can produce or hear ultrasound). Some investigators have reported the existence of "sound windows," that is, frequency bands penetrating certain environments particularly well, but the experimental evidence seems insufficient in most cases.

Data on attenuation are interesting, but it should be realized that the signals may have degraded to become useless long before they are no longer audible. Reflections and diffraction (scattering) are the processes mainly responsible for signal degradation. The frequency spectrum may become heavily distorted after just a single reflection (Fig. 5.1). Single frequency components in the signal may suffer attenuations of 30 to 40 dB or more.

Pure tone components may therefore almost disappear, but they can be heard if the receiving animal moves around when listening (the frequency filter is critically dependent on the positions of the sender and receiver in the habitat). Multiple reflections and scattering causes the signal to be followed by a long tail of reflected components, thus smearing the temporal structure of the signal (Fig. 5.2). The same processes also change the sound propagation from unidirectional (free field) to more omnidirectional (diffuse field); see Römer, Chapter 6.

It should be noted that these degradations of the signal occur without the addition of external noise to the signal. Added noise further smears the signal. The recognition of the signals after the degradation is the main task of the auditory nervous system, and improvements of the information processing to handle this task has probably been the central event in the evolution of hearing during the millions of years after the appearance of the hearing organs. It is a pity that so few students of auditory processing have related their findings to the signal degradation occurring in the animals' habitats.

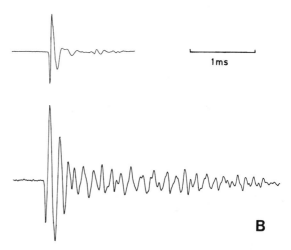

FIGURE 5.2. Temporal distortion. A sound impulse recorded in a free sound field (above) and after having travelled 2 m through a vegetation of herbs (below). (From Michelsen and Larsen 1983.)

3. Sound Emission

Sound waves are normally generated by the action of a vibrating structure on the surrounding medium. The sound emission depends on the vibration amplitude, the properties of the medium, and the size of the vibrator relative to the wavelength of sound. Maximum efficiency requires that the diameter of the sound source is of the same order of magnitude as the wavelength of sound—or larger (for simple and technical explanations of the laws of sound emission see Michelsen 1983 and Beranek 1954, respectively). Frequency and wavelength are inversely related, and the efficiency of a certain source thus increases with frequency up to a certain value point (indicated by a star in Fig. 5.3) beyond which it stays almost constant. (For a parallel discussion of underwater acoustics see Schellart and Popper, Chapter 16).

To the left of the star in Figure 5.3, the efficiency depends on the nature of the source. A pulsating sphere and a disc surrounded by an "infinite" wall (baffle) are examples of *monopoles* (solid curve in Fig. 5.3). Here, all parts of the surface "agree" on either producing a compression or a rarefaction. The exact magnitude of the sound power may be found with a little further calculation (see Michelsen 1983). For a constant displacement amplitude, the emitted sound power depends on the fourth power of frequency (f) and on the fourth power of the radius (r) of the disc. The sound pressure depends on the second power of f and r.

These relations hold for both air and water (Schellart and Popper, Chapter 16), but the emitted power depends on the so-called characteristic impedance of the medium (ρc, where ρ is the density of the medium, and c is the propagation velocity of sound). ρc is 3,500 timer larger for water than for air, and the emitted sound power is correspondingly larger (at a constant f, r, and displacement amplitude). Although it requires more energy to cause the source to vibrate with a certain amplitude in water than in air, it is still easier to emit sounds at low frequencies in water than in air. It can also be surprisingly easy to generate some vibrational waves travelling through solid

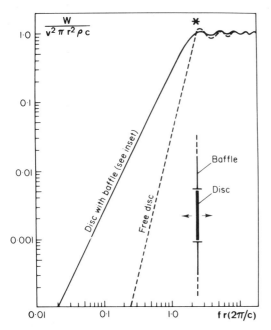

FIGURE 5.3. The emitted sound power (W), measured in watts, at a constant vibrational displacement of a free disc (a dipole) and of a disc surrounded by an "infinite" baffle (a monopole), as a function of the frequency (f) and radius (r) of the disc. (v) is the vibration velocity. Other symbols are explained in the text. (From Michelsen 1983.)

substrates. We shall return to the biological significance of this.

If a sphere (or a free disc) oscillates to and fro, one side will produce a compression, while the opposite side produces a rarefaction—and vice versa when it moves in the opposite direction a little later. The compression and rarefaction are separated by a small distance if the sphere or disc is small relative to the wavelength (which is the distance that the disturbance can spread during one oscillation period), and they may therefore cancel. This phenomenon is called "acoustic short-circuiting," and it causes the sound output from *dipoles* like small oscillating spheres to be much smaller than that from a pulsating sphere of the same *f*, *r*, and vibration amplitude. We avoid this problem when we place a loudspeaker in a cabinet, thus separating the front and back surfaces. The sound power emitted by a free disc is shown in the dotted curve in Figure 5.3. At low frequencies (to the left of the star) the power output from a vibrating free

disc (or an oscillating sphere) is proportional to the sixth power of *f* and *r* (and the sound pressure produced at some distance from the disc is proportional to the third power of *f* and *r*, again provided that the displacement amplitude is kept constant).

3.1 Animal Monopoles and Dipoles

Most terrestrial vertebrates emitting sound are probably not far from acting as monopoles, and their efficiency is thus determined by the size of the emitter and the solid curve in Figure 5.3. The same is true for the cicadas, where the timbals are a part of the body wall and thus situated in a "cabinet." Most insects, however, emit sounds by vibrating their wings or other appendages, and their efficiency is determined either by the dotted curve or it may be found somewhere between the two curves.

A circular baffle with a radius of one quarter of the wavelength acts approximately as an "infinite" wall. In the European field cricket *Gryllus campestris*, for example, sound is emitted mainly by a central part of the wing (the harp) which is about 7 mm from the distal border of the wing. The wavelength of the 5 kHz calling song is 7 cm, so the radius of the baffle is about half of that required for placing the cricket on the solid curve in Figure 5.3 (Nocke 1971). The sound emission is improved by the position of the femora and lateral parts of the wings (Fig. 5.4A). Some South African field crickets (*Oecanthus*) cut a hole in the middle of a large leaf and place themselves in the hole when singing (Fig. 5.4C), thus moving to the solid curve and producing a 10 dB higher sound pressure (Prozesky-Schulze et al. 1975). An alternative strategy followed by mole crickets is to dig a hornshaped hole in the ground and to place themselves at the throat of the horn (Fig. 5.4B), improving the output by at least 10 dB (Bennet-Clark 1970).

Although such strategies may be quite common (Forrest 1982), they only marginally improve the sound output from an animal. Insects with a body length of, for example, 5 mm cannot avoid having a very low efficiency at sonic frequencies. Some insects are specialized in using ultrasound for communication, but these insects are normally so large that they would also have been reasonably efficient sound emitters at sonic frequencies. The main reason why only few of the small insects communicate with ultrasound seems to be that ultrasound is heavily attenuated and distorted by most habitats. In addition, small ears are less sensitive than large ears. So, moving the emitted frequencies to the ultrasonic range does not improve the situation very much for the small animals.

3.2 Communicating Through a Solid Substrate

Many groups of small insects produce very quiet sounds at sonic frequencies and yet communicate over appreciable distances (up to meters). The solution to this apparent mystery became clear when several investigators found evidence that the animals communicate through a solid substrate (review: Markl 1983). Vibration signals with displacement amplitude of 100 to 1,000 Å may travel (for example through a host plant) to reach other animals which detect the vibrations with suitable sense organs (mainly the subgenual organs). Many insects (and some vertebrates such as frogs) have vibration thresholds around 1 Å. This is three to four decades below the threshold of human vibration receptors (in the skin), and the animals' signals are therefore outside the range of our sensory capacity.

Several different kinds of substrate-borne vibrational waves exist, and careful examinations of the physical properties of the substrate are needed in order to disclose the kind of wave motion responsible for carrying the animals' signals. In plants, insect vibrational songs appear to be carried mainly as bending waves (Michelsen et al. 1982). The physical properties of plants vary widely, but their properties as transmission paths for vibration signals are surprisingly uniform.

3.3 The Metabolic Costs of Signaling

The power input needed in order to set up a vibration signal in a plant stem is mainly proportional to the first power of the mass (per unit length) and to the second power of the vibration amplitude. A power of 10^{-8} to 10^{-7} W may be sufficient for producing 1,000 Å amplitude at 500 Hz which may travel to all parts of the herbs with only little attenuation. This corresponds to the power output of one µg of muscle (assuming only little loss in the transfer of the vibration to the plant). Communication by means of substrate-borne vibration signals should thus be realistic for herb-

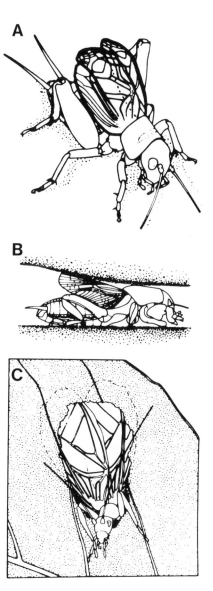

FIGURE 5.4. Strategies for improving the sound emission from dipole sources. (A) A singing cricket in which the lateral fields of the wings and the large hind femora form part of a "speaker cabinet." (B) Calling mole cricket in a subsurface tunnel. (C) A cricket looking up through and calling in a hole it has made in a leaf. (Drawings from Forrest 1982.)

associated insects down to 1 to 2 mm body length (Michelsen et al. 1982).

In theory, the situation should be less favorable for insects living on or in tree stems, but such signals are common also in bark beetles and other wood-borers. However, the vibration signals may be heavily attenuated in solid wood. The rapping signals of carpenter ants (main energy at 4 to 5 kHz) are attenuated by 2 dB/cm (Markl and Fuchs 1972). Vibrations may also travel at the surface of the soil or through ground, but here the vibrations suffer even larger attenuation with distance (for example, Markl (1968) found an attenuation of 6 dB/cm in soil).

Communication over meters by means of air-borne sound is more costly. For example, a power output of 6×10^{-5} W has been estimated for a singing field cricket (Bennet-Clark 1971). In addition, the efficiency of converting muscle power into acoustic power is only 5%, so the power output of the muscles is 20 times larger. Measurements of the oxygen consumption show that the energetic costs of calling by male crickets may exceed 10 times that of resting (Prestwich and Walker 1981).

Typical values for the sound pressures measured close to large singing insects and for the threshold of hearing are 90 to 100 dB SPL and 40 to 60 dB SPL (relative to 20 μPa), respectively. In free space, the difference between these figures would allow communication to take place over tens of meters. In dense vegetation and at ultrasonic frequencies, however, the distance may be less than one meter (Michelsen and Larsen 1983).

3.4 Long-Range Signaling: A Summary

In summary, whether insects may communicate by means of airborne sound depends on their size, habitat and population density. Large insects can emit sounds above ca. 1 kHz. Sonic frequencies penetrate vegetation reasonably well (although with much distortion; see Römer, chapter 6), whereas ultrasound is useful only in free space or at short range. Insects with body lengths below 1 cm are generally restricted to ultrasounds, which are less useful in most habitats. However, many small insects (body size 1 to 10 mm) may communicate by means of substrate-borne vibrations through their host plants and other suitable, solid media; these signals may travel up to some meters (for example, from one end of a reed to the other end), and they are "cheap" to produce in terms of muscle power.

Sound is thus not an obvious choice for small animals which communicate over distances many times their own body length. At short range, how-ever, sound may be useful, either as ultrasound or as near-field oscillations of the air. Both components of the sound may be received with suitable sense organs. The pressure component requires an ear, and the air oscillations may be received by means of lightly articulated appendages (hair, antennae). It is a matter of taste (and terminology) whether the latter type of receivers should also be called hearing organs.

3.5 Close-Range Signals

A male fruitfly (*Drosophila*) communicates with the female by means of wing vibrations. The wings have an area of 1.2 mm² and are vibrated with a displacement amplitude of 1 mm (peak–peak) at 170 Hz (Bennet-Clark 1971). The female is 5 mm away from the edge of the (free-disc) wing and receives a sound pressure of about 40 dB SPL and an air oscillation of 3 mm/s. The latter components causes a part of the antennae (the aristae) to vibrate. The power needed for this communication is about 3×10^{-16} W.

The fruitfly has a body length of 2 to 3 mm. The much larger honeybee (body length 13 mm) emits similar wing vibration signals as a part of the dance language (signaling the direction and distance to food sources). Here, the wings are vibrated with a displacement amplitude (peak–peak) of 0.5 mm at 280 Hz, causing a sound pressure of about 1 Pa (94 dB SPL) at each wing surface (the wings are kept together over the abdomen and vibrated as a unit). Close to the edge of the wings the air oscillates with a velocity of 0.5 to 1 m/s (Michelsen et al. 1987).

The dancing honeybees are surrounded by bees following the dance. Some of these bees later leave the hive and fly to the position indicated by the dance. Several receptors are available at their surface for receiving the air oscillations, the ampli-tude of which decreases with the third power of the distance to the wings (Fig. 5.5). At a distance of 1 to 2 cm the air oscillations can probably be received by sensitive hairs on the heads of the bees. At close range, the air oscillations are so intense that they activate the antennae, which are otherwise used for monitoring the flight of the bee. In passing, we may note that the air oscillations seem to play an important role in the dance lan-guage. It is possible to tell the bees where to fly, if a mechanical model of a dancing bee is made

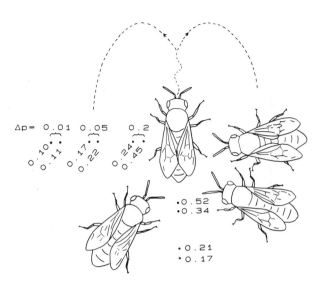

FIGURE 5.5. The acoustic near field around a dancing honeybee. Sound pressures (average peak values in Pa) measured in different experiments with a pair of minute microphones. Note that the pressure difference (Δ p) decreases more rapidly with distance than does the pressure. A part of the dance path is indicated (dashed line). Three follower bees are also shown. (From Michelsen et al. 1987.)

to include air oscillations as found around live dancers (Michelsen et al. 1989a).

Why do relatively large animals like honeybees make use of air oscillations for communication instead of the pressure component of sound? A probable answer may be found by considering the situation in the hive. About 50,000 individuals live together in a space of about 100 liters. Under these conditions it may be practical to be able to have a "private talk" with the neighbors without telling everybody. The bees may dance just a few centimeters from other dancers, and it is important for the follower bees to be able to pick up the message from one dancer at a time. By "listening" to the air oscillations rather than to the pressure fluctuations, the bees may take advantage of the fact that the pressure gradients and air oscillations decrease with the third power of distance to the source. In contrast, the sound pressure decreases only with the second power of distance (Fig. 5.5).

Only a few investigations have been performed on close-range and substrate-borne sound communication, and one can only speculate how common such signals may turn out to be. It should be noted that close range mechanical communication does not necessarily have to involve sounds. For example, the nonstridulating cricket *Phaeophilacris* generates travelling vortex rings (similar to those made by Dick Fay and other passionate cigar smokers). The rings are made by wing flicks and have an initial diameter of about 2 cm; they may grow to up to 5 cm while they travel a maximal distance of 15 cm (Heinzel and Dambach 1987).

3.6 Physical Restraints on Signal Coding

So far, we have not discussed how the small animals manage to produce the high vibration frequencies that can be emitted as airborne sounds. At the low frequencies of many substrate-borne signals the muscle contractions are sufficiently fast (for example, honeybees use their flight muscles for producing various 200 to 500 Hz vibration signals for communication within the hive, Michelsen et al. 1986b). Although some insect muscles can contract up to 1,000 times per second, they are too slow for generating vibrations in the kHz range.

The solution to this problem is to let the muscles act on a *frequency multiplier*. This is a device which converts each muscle twitch into a vibration of higher frequency. Several types of frequency multipliers exist (Michelsen and Nocke 1974). Here we shall examine the stridulatory organs which occur in many insects, spiders and fishes. Despite their different evolutionary origin, they are built in a similar way. A "scraper" located on one part of the body is moved over a "file" on another part of the body. The scraper may have the shape of a ridge, and the file may consist of a row of "teeth." During each muscular contraction the scraper may hit several hundred teeth, thus producing a vibration of high frequency.

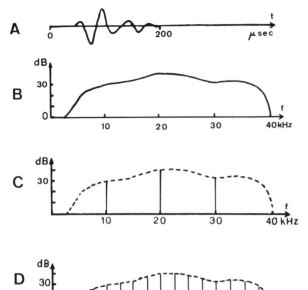

FIGURE 5.6. Spectral properties of stridulatory sound (the song of a grasshopper). (A) and (B) An impulse (caused by a single tooth-impact) and its spectrum. (C) and (D) Line spectra caused by a regular repetition of the impulse (at a rate of 10 kHz and 2.5 kHz in (C) and (D), respectively). [Redrawn (altered) from Skovmand and Pedersen 1978.]

The vibration spectrum is determined by the physical properties of the stridulatory apparatus and is generally broad. The impact of a single tooth causes a short impulse (of some hundred microseconds duration in many insects). If the teeth are hit at regular time intervals, a line spectrum may be approached (Fig. 5.6). Note that the amplitude of each of the lines is determined by the spectrum of the individual impulses, and that there are fewer lines at high repetition frequencies than at low ones because of the harmonic nature of the series. Frequency sweeps may be produced by varying the distance between the teeth or the velocity of the scraper (many insects make use of one of these possibilities).

In some insects, such as shorthorned grasshoppers, the sound radiating structures do not appear to cause much distortion of the primary vibration, and here the individual tooth impacts may be observed in the recorded sound (Elsner and Popov 1978). Other insects use one of the harmonic spectral components of the primary vibration to drive a resonance in a lightly damped sound emitter, thus boosting up the intensity of the emitted sound (many crickets provide examples of this).

The strategies used for sound production have consequences for the coding of the social signals and for the reception of the sounds (including the performance of the hearing organs). Boosting up the sound output by means of resonances may limit the animals to the resonance frequency(ies) of the sound radiator, which in insects are generally cuticular structures. Birds are able to affect the resonance frequency of their vocal membranes, but such a control of the sound emitter has not evolved in insects.

Another disadvantage of resonant systems is that they are "slow" and thus less suited for transmitting rapid amplitude modulations of the sound. This shortcoming appears to be an integral part of the behavior of resonant systems, and the animals cannot do much to affect the time resolution. Small animals may therefore have to choose between either having fast sound radiating systems (with the additional possibility of emitting frequency sweeps) or being able to produce high sound intensities.

4. The Hearing Organs

Several insect ears have adapted to the reception of the ultrasonic cries of hunting bats (see Hoy, Chapter 8), and some of these ears are also used for social sound signals (e.g., the ears of some moths, Spangler 1987). One may speculate that other ears evolved as detectors of social sounds and of predators other than bats. Although many insect ears are now primarily used for receiving social sound signals, the ears and signals generally do not match very well. For example, although the calling song of the European field cricket is carried by a very pure sinusoid around 5 kHz, the shape of the threshold curve of the ear does not reflect this (but a relatively large number of receptor calls and auditory neurons are tuned to 5 kHz; review: Michelsen and Larsen 1985).

One may further speculate that the ears, like the sound emitters, might be specialized either in exploiting resonances (obtaining a high sensitivity and a slow amplitude response around the resonance frequencies) or being critically damped (and thus having a lower sensitivity, but a faster response within a broad band of frequencies). Again, only few examples support such a notion.

4.1 Directional Hearing

4.1.1 Diffraction and Temporal Cues

Large animals (and man) can use two mechanisms for detecting the direction of sound waves: the interaural difference in sound pressure caused by diffraction of sound by the body (head in vertebrates), and the difference in time of arrival of the sound at the two ears. Differences in sound pressure arise when the dimensions of the body (head) are above one-tenth of the wavelength of the sound. At the lowest frequencies where diffraction occurs, a surplus pressure is found at the ear facing the sound source. At higher frequencies the ears may experience a number of maxima and minima of sound pressure. The frequency spectrum at each ear varies with the angle of incidence, and the spectra at the two ears are different for most directions. It is thus possible for the brain to estimate the direction of the sound source by comparing the sound spectra at the two ears. Obviously, this task is easier with broad-band sounds than with pure tones or narrow-band sounds.

In many hearing animals (e.g., insects, frogs, birds) the part of the body carrying the ears is about 10 to 50 times smaller than the human head. A difference in the sound pressure at the two ears is available only at very high frequencies. For example, the wavelength of the calling song of many crickets and frogs is about 10 times larger than the

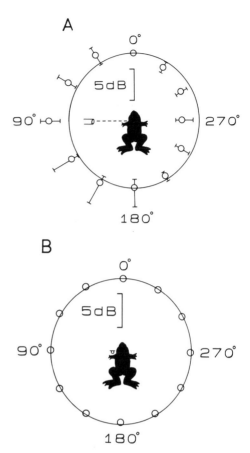

FIGURE 5.7. Directional hearing without diffractional cues in a green treefrog. (A) The average vibration velocity of the left ear drum at 1–3 kHz measured with laser vibrometry in an awake, unrestrained frog at 12 different directions of sound incidence. (B) The average sound pressure at the ear drum. (From Michelsen et al. 1986.)

animal's body, and hardly any difference in pressure is found at the ears during the calling song (Fig. 5.7B). It should be noted, however, that the possible use of diffraction is not determined by the size of the animal, but by its size relative to the wavelength of the sounds to be heard. For example, in moths exposed to the ultrasonic cries of bats, the two ears may experience a pressure difference of 10 to 20 dB (and even 30 to 40 dB in very large moths), which is more than sufficient for the purpose of determining the direction of the bat (Payne, Roeder, and Wallman 1966).

Sound travels in air with a velocity of about 340 m/s. In humans (head diameter about 17 cm), the maximum difference in the time of arrival at the

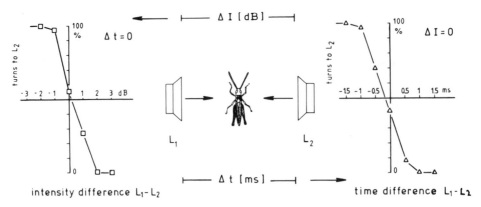

FIGURE 5.8. The frequency of turns in a male grasshopper (*Chorthippus biguttulus*) towards one of two loudspeakers which simultaneously emitted female song (as a response to the male's own singing). The songs emitted by the speakers differed in intensity (left graph) or time (right graph). (From von Helversen and von Helversen 1983.)

ears is about 0.5 ms. We can localize long pure tones below about 1,400 Hz by means of this mechanism, since there is a cycle-by-cycle following of the sound wave by the nerve impulses (phase detection). At higher frequencies, the time of arrival can be detected if the sound is sufficiently modulated in amplitude or frequency content.

In principle, the time and diffraction mechanisms are quite different, but the latency of the nervous response depends upon the magnitude of the force acting on the ear. The minute difference in the time of arrival of the sound is seen as a difference in the timing of the nerve impulses from the two ears, only if the force (sound pressure) is the same on both ears. The nervous system has to compare the information about time and intensity at the two ears, however, if this is not the case. Obviously, a complicated central processing is necessary in order to make use of the time cue for directional hearing.

Some insects and other arthropods (spiders, scorpions) are able to process information about differences in the time of excitation of their sense organs. Scorpions and spiders receive vibration signals through their legs, and they will turn to the side from which they first receive a stimulus (Brownell and Farley 1979; Hergenröder and Barth 1983). Time differences of 0.2 to 1 ms are necessary, and these values correspond to those expected from the size of the animals and the propagation velocity of the signals. A male grasshopper listening to female song transmitted by two loudspeakers placed to its left and right side (Fig. 5.8) will likewise turn to the side of the earlier signal when the songs are time shifted by 0.5 to 1 ms (von Helversen and von Helversen 1983). However, the maximum time difference expected when the grasshoppers are locating each other is only 20 to 30 µs, and there is no evidence that the insect central nervous system can handle time differences of this order of magnitude.

Although differences in the time of arrival of sound to the ears does not seem to be utilized in insects, these observations are nevertheless very interesting when speculating about the evolution of hearing. Central auditory neurons in grasshoppers make use of both the number of nerve impulses arriving from the two ears and of the difference in the time delays (Mörchen, Rheinlaender, and Schwartzkopff 1978). Furthermore, insects and other arthropods make use of both cues when processing the neural activity of other pairs of sense organs such as vibration receivers. These observations suggest that the neural machinery necessary for making use of minute time cues may have evolved before a sense of hearing appeared.

4.1.2 Movement and Pressure Difference Receivers

Most small animals are obviously unable to determine the direction of sound by means of these two mechanisms. The solutions to this problem were first realized by Autrum (1940), who pointed out that the direction of sound waves is estimated by insects either by means of pressure difference receivers or by means of movement receivers.

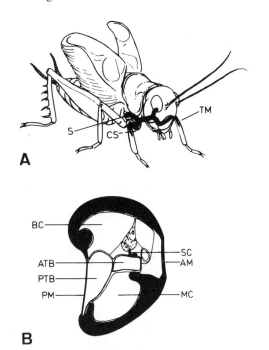

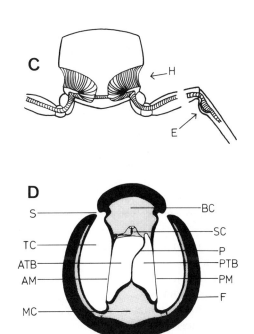

FIGURE 5.9. Schematic diagrams of the hearing organs in crickets (A) and (B) and bush crickets (C) and (D). (A) The "H" shaped tracheal tubes connecting the back surface of the tympanal membranes of one ear with the other ear and with ipsi- and contralateral spiracles (S). CS: the cross-section shown in (B) (symbols as in D). (C) In bush crickets the horn-shaped trachea, H, from each ear, E, are connected by a thin tube in the thorax. (D) Cross-section through the ear in the region of the leg just below the "knee." AM: anterior tympanic membrane; ATB: anterior tracheal branch; BC: blood channel; MC: muscle channel; P: partition; PM: posterior tympanic membrane; PTB: posterior tracheal branch; SC: sense cells; TC: tympanal cavity. From Michelsen and Larsen (1978) and Larsen and Michelsen (1978). A: Redrawn from Michel (1974).

One obvious solution is to measure the direction of the oscillatory movements of the air particles. Insects do this by means of long, lightly articulated sensory hairs protruding from their body surface. Such movement receivers are inherently directional. They are insensitive to sound propagating in the direction of the hair. In addition, their shape, their articulation, and the attachment of the sensory cell(s) may all contribute to the directionality (Tautz 1979). The main disadvantage of such hairs is their modest sensitivity in acoustic far fields (where the air particle vibration velocity is proportional to the sound pressure; see Schellart and Popper, Chapter 16). However, the movement component of sound increases more than does the pressure component when one approaches a sound source, and at close range to the sound source the hairs may be as sensitive as pressure or pressure difference receivers.

In a pressure difference (pressure gradient) receiver, the sound waves are able to reach both surfaces of a (tympanal) membrane. This is possible in many animals, since the ears are often connected by an air-filled passage. Alternatively, the sound waves may enter the body and reach the inner surface of the tympanal membrane through some other route (e.g., a tracheal tube in crickets, Fig. 5.9). For several years, the pressure difference receiver was regarded as a speciality of insect hearing. We now know that the directional hearing of the majority of small terrestrial animals is based upon this kind of sound receiver mechanism. It has been found in several, systematically distant insect groups and in some frogs, reptiles, birds, and even mammals (Lewis 1983). It has obviously evolved independently on several occasions.

The directionality of pressure difference receivers can easily be understood, at least in principle (consider a piece of paper, which is rotated in a free sound field). Quantitative descriptions of the physical operation of, for example, band microphones also exist. The biological pressure difference receivers are generally more complicated, however, and have so far not been subjected to a proper physical analysis. It is therefore not possible to pinpoint the crucial steps in their evolution.

The existence of an anatomical air space leading to the back surface of the tympanum is a necessary prerequisite, but it does not automatically create a pressure difference receiver with proper directionality. The sound has to arrive at the back surface of the membrane with a proper amplitude and phase. In addition, the phase of the sound inside the animal should be affected in a suitable manner by the direction from which the sound reaches the outer surface of the animal. One complicating factor in biological pressure difference receivers is that the sound arriving at the back surface may have entered the body through several inputs (for example, through the other ear and through two spiracles in crickets; through the other ear and through the lung in frogs).

The transmission of the sound to the back surface of the tympanum is often a function of frequency. In locusts (Michelsen 1971) and some birds (Hill et al. 1980) low-frequency sound is transmitted from one ear to the other with little attenuation. High-frequency sounds are attenuated, however, and the ear becomes a pressure receiver. In most animals the ears are at the surface of the head or body, and at high frequencies diffraction causes a difference in the sound pressure at the ears that is sufficient to provide the animal with a reliable directional cue.

This is not so in crickets (*Gryllidae*) and bush crickets (katydids, *Tettigoniidae*), however. The ears are in the thin front legs far away from the body surface (Fig. 5.9). Measurements with a thin microphone probe demonstrate that body diffraction does not (even at very high frequencies) provide a sufficient difference in pressure at the ear (Michelsen, Heller, and Stumpner 1991). Two different solutions to this problem are found in crickets and bush crickets. In crickets, the pressure difference receiver mechanism appears to operate at 2 to 18 kHz (Larsen, Kleindienst, and Michelsen 1989), that is also at high frequencies where the

body diffraction is significant. In contrast, in many bush crickets the horn-shaped tracheal tube ("hearing trumpet") has a pressure gain well above 1 at frequencies above a few kHz. At the frequency of the calling song (typically above 10 kHz), the sound carried by the acoustic trachea dominates the ear (i.e., the ear drum is driven by the sound pressure at its inner surface, and the sound at the outer surface does not play a significant role). The acoustic trachea faithfully transmits (an "amplified" version of) the pressure at the body surface to the inner surface of the tympanum. The ears thus obtain the same (diffraction-based) directionality as if they had been situated at the body surface (Michelsen, Heller and Stumpner 1991). The ears of crickets and bush crickets obviously have a common evolutionary origin, and yet they have exploited two widely different mechanisms for directional hearing.

The air space leading to the back surface of the tympanum is often a part of (or connected to) the respiratory pathways. This may have undesirable consequences, since the large pressure fluctuations during respiration may affect the ears. In grasshoppers the tympana may be displaced outside their linear range (so that Hooke's law is no longer obeyed). This may affect the threshold for hearing and seriously distort the frequency analysis (Michelsen, Hedwig, and Elsner 1990). Large displacements coupled with the respiration can also be observed in frogs (in which the middle ear cavity and the mouth are connected through a wide Eustachian tube). Obviously, a reduction of such effects may have been an important factor in the evolution of pressure difference systems. The air-filled spongy bone connecting the middle ears in birds, moles and some reptiles would appear to be ideal in this respect.

4.1.3 The Ideal Solution and the Need to Compromise

Before leaving the subject of directionality, it may be interesting to compare ideal methods with the biological solutions to the problem of estimating direction. In order to perform an accurate determination of the direction of sound in three dimensional space one needs four sound receivers, widely spaced, but not placed in the same plane (Møhl and Miller 1976). A computation of the differences in the time of arrival of sound at the receivers would provide the necessary information. Preferably, the receivers should be small and placed far away from large (sound reflecting) bodies. The receivers would then also be available for making exact records of the temporal and spectral characteristics of the sounds.

Most animals have only two ears, and they are thus inherently not well equipped for solving a three-dimensional problem (we therefore have trouble in localizing sound sources in the median plane of the head). Pressure is a scalar quantity, and pressure-sensitive ears are thus inherently omnidirectional. The ears may obtain some directionality by being parts of (or neighboring) a large body (head) and exploiting its diffraction of the sound. Further directionality may be obtained by means of, for example, a horn (the meatus and outer ear). Diffractional effects are very frequency dependent, and the frequency spectrum of the sounds arriving at the ear drum thus depends very much on the direction of sound. The spectral composition of the sound can therefore be used as a cue for the direction to the sound source.

Despite our having two ears and not the ideally required four, we have established a quite satisfactory directional hearing. There is a price to be paid, however, since we have "sold" some of our ability to determine the exact frequency spectrum of the sounds. One could say that the direction and spectrum are complementary in a system based on diffraction, and that one cannot determine both with maximum accuracy when doing a few measurements. In passing, it should be noted that diffraction changes not only the sound pressure, but also the phase. Attempts to make the ear more directional are thus blurring the time information.

The pressure receiver is inherently omnidirectional. In contrast, pressure difference receivers are inherently directional. But again, the price for this useful property is that the force driving the membrane depends on the phase angle between the sounds acting on the two surfaces of the membrane, that is, on the frequency and the angle of incidence of the sound. An optimum performance with respect to directionality requires that the sounds arriving at the two surfaces do not differ much with respect to their amplitude and phase. The driving force will then vary very much with frequency. At certain combinations of frequency and direction the force may drop to almost zero, because the sounds acting on the two sides have almost the same amplitude and phase.

So far, evolution has not produced an ideal solution to the problem of performing a perfect, simultaneous determination of the temporal, spectral, and directional properties of the sounds (for example, an animal carrying ears at the tip of four tentacles). Arthropods carrying many mechanoreceptive hairs on their body may perhaps be close to the ideal as far as the reception of close range oscillations of the medium are concerned. Most present-day animals have only two ears, and their performance has to be a compromise. In theory, one should expect the different animals to have arrived at different compromises depending on their needs. Future research may perhaps locate animal specialists in temporal and/or spectral analysis, but in most animals evolution appears to have emphasized a good directional hearing at the expense of temporal and (especially) spectral analysis.

4.2 Frequency Analysis

A sense of hearing evolved in insects for the detection of predators and social sounds. Some nocturnal insects can only hear ultrasound. They respond to ultrasonic pulses with a characteristic avoidance behavior, and their hearing seems to be used exclusively for the detection of hunting bats (see Hoy, Chapter 8). The ability to perform a frequency analysis has not been found in this group. This is not surprising, since the survival value of a frequency analysis would seem to be limited (it is not very interesting for the insect to know which kind of bat is looking for its dinner). In contrast, most (all ?) insects using hearing for social communication have frequency analyzers.

Auditory receptors may signal the frequency of sound in two different ways. The telephone (volley) principle is based upon the tendency for phase locking at low frequencies in the neural activity travelling in the auditory nerve. Although used up to some kHz by the central nervous system of vertebrates, this ability has not been proven in insects. Behavioral observations suggest, however, that it may be used in some of the few insects that listen to sounds of low frequency. The flight sounds of mosquitoes are around 400 Hz, and the neural response in the auditory (antennal) nerve is phase

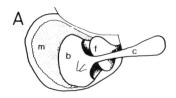
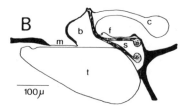

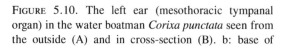

FIGURE 5.10. The left ear (mesothoracic tympanal organ) in the water boatman *Corixa punctata* seen from the outside (A) and in cross-section (B). b: base of the clubbed process; c: clubbed process; f: bottlelike body; m: tympanal membrane; s: sensory cells; t: tracheal air sac. (From Prager and Larsen 1981.)

locked to the sound wave (Tischner 1953). Male mosquitoes respond to a new frequency when they have become adapted to one frequency (Roth 1948). The flight sounds of mosquitoes are of much lower frequency than most insect sound signals, and the telephone principle would therefore be less useful in most insects. Furthermore, in insect ears activated by mechanical vibration the phase locking disappears at frequencies above 200 to 400 Hz (Michelsen 1973).

Frequency analysis in insects is generally carried out by means of the place principle: Receptor cells with different anatomical locations differ as to characteristic frequency. Three different mechanisms are known. The tympanum may be used both for receiving the sound and for analyzing its frequency content (locusts, mole crickets). In most insects, however, the tympanum is only used for picking up the sound energy, and the frequency analysis takes place in an "inner ear" (crickets, bush crickets). A third mechanism is based upon frequency-dependent, oscillatory rocking movements of heavy structures in the ear [water boatmen (*Corixa*), locusts].

These mechanisms and the neural processing of frequency information have been reviewed (Michelsen and Larsen 1985). The receptor cells are able to respond to a broad range of frequencies (from some Hz to some hundred kHz) when subjected to mechanical vibration. The tympanal membrane acts as a band-pass filter, however, and the response of the ear to sounds is therefore limited (for example, in locusts the ear responds within the frequency range 1 to 40 kHz). The receptor cells appear to derive their different frequency sensitivities from their position on a structure with complex and frequency-dependent vibrations. In contrast to the situation in (at least some)

vertebrates, frequency analysis in insects thus seems to be a purely mechanical phenomenon (this notion has been questioned for some years, but attempts to disprove it have failed).

The number of frequency bands in insect sound frequency analyzers vary from two (water boatmen), over three (mole crickets) and four (grasshoppers) to 24 (bush crickets). It is not at all obvious that these varying capacities are correlated with the needs of the animals for performing a frequency analysis (see also Römer, Lewis, and Hoy, Chapters 6, 7, and 8).

Very little is known about the evolution of the capacity for frequency analysis, but the situation within the *Ensifera* (bush crickets, crickets, and mole crickets) suggests that the capacity for frequency analysis evolved later than the hearing organs. The ears are in the front legs in these animals, and apparently the ears had a common origin. In mole crickets the frequency analysis is carried out in the tympanum (where groups of receptor cells attach to parts of the tympanum with different frequency responses). In contrast, the tympana vibrate as rather homogeneous and broadly tuned units in crickets and bush crickets, and the frequency analysis takes place in a separate structure inside the leg and at some distance from the tympanum.

The situation in water boatmen is particularly amusing and interesting from an evolutionary point of view. The frequency analysis is not a property of the individual ears, but the left ear has a higher characteristic frequency and sensitivity than the right ear (Prager 1976). In *Corixa punctata* the threshold is a minimum (38 dB) at 2.35 kHz in the left ear and at 1.73 kHz (45 dB) in the right ear. The difference is due to different vibration characteristics in the two ears of a large club which attaches to the tympanum (Fig. 5.10) and vibrates in two planes (Prager and Larsen 1981). Water boatmen carry an air bubble (a physical gill) when submerged. The bubble gradually decreases in size and is replenished when the animal comes to the water surface. In the submerged animal, the bubble is caused to oscillate by sound, and the different frequency sensitivities of the two ears appear to be an adaptation to the variation in resonance frequency in air bubbles of various sizes (Prager and Streng 1982). The "frequency analysis" and its function in water boatmen is thus very untypical — but an interesting example of an adaptation to a special habitat and way of life.

5. Summary and Conclusions

Hearing in insects evolved for two purposes: social communication and detection of predators (mainly bats). The songs of many large insects are audible to man, but most insects are too small to be efficient sound emitters at sonic frequencies. In theory, the small animals could use ultrasound for their social communication, but ultrasound is heavily attenuated and distorted by most habitats. Consequently, sound communication is not an obvious strategy for most insects or other small animals. In trying to understand the evolution of hearing one has, therefore, to consider the alternative communication strategies available to small animals. Substrate-borne vibrations may carry social signals over distances of up to some meters and yet require much less muscular power than airborne sounds. At shorter distances, various sorts of air currents and tactile contacts may be exploited.

Many insect ears possess elaborate mechanisms for frequency analysis and directional hearing. The central auditory neurons often show selective preferences for certain temporal patterns. Obviously, these abilities evolved independently many times. Which factors were responsible for their evolution? It is argued that the use of temporal coding is limited by the physics of the sound emitters and by the temporal distortion of the sound signals in the habitat. Furthermore, the directionality and frequency spectrum of the sound signals are also being degraded by the habitat. The various mechanisms of frequency analysis (now found in most insects that use sounds as social signals) probably evolved to cope with this problem. The recognition of the signals after the degradation in the transmission channel is the main task of the auditory nervous system, and improvements of the information processing to handle this task has probably been the central event in the evolution of hearing after the appearance of the hearing organs.

Acknowledgments. Original research for this chapter was supported by grants from the Danish Natural Science Research Council. The author is most grateful to Ole Næsbye Larsen for his comments on the manuscript.

References

Autrum H (1940) Das Richtungshören von *Locusta* und Versuch einer Hörtheorie für Tympanalorgane der Locustidentyp. Z Vergl Physiol 28:326–352.

Bennet-Clark HC (1970) The mechanism and efficiency of sound production in mole crickets. J Exp Biol 52: 619–652.

Bennet-Clark HC (1971) Acoustics of insect song. Nature 234:255–259.

Beranek LL (1954) Acoustics. New York: McGraw-Hill.

Brownell P, Farley RD (1979) Orientations to vibrations in sand by the nocturnal scorpion *Paruroctonus mesaensis*: mechanisms of target localization. J Comp Physiol 131:31–38.

Elsner N, Popov AV (1978) Neuroethology of acoustic communication. Adv Insect Physiol 13:229–355.

Forrest TG (1982) Acoustic communication and baffling behaviors of crickets. Fl Entomol 65:33–44.

Heinzel H-G, Dambach M (1987) Travelling air vortex rings as potential communicative signals in a cricket. J Comp Physiol 160:79–88.

Helversen D von, Helversen O von (1983) Species recognition and acoustic localization in acridid grasshoppers: a behavioral approach. In: Huber F, Markl H (eds) Neuroethology and Behavioral Physiology. Berlin Heidelberg: Springer-Verlag, pp. 95–107.

Hergenröder R, Barth FG (1983) Vibratory signals and spider behavior: How do the sensory inputs from the eight legs interact in orientation? J Comp Physiol 152: 361–371.

Hill KG, Lewis B, Hutchings ME, Coles RB (1980) Directional hearing in the Japanese quail. I. Acoustic properties of the auditory system. J Exp Biol 86:135–151.

Larsen ON, Michelsen A (1978) Biophysics of the ensiferan ear. III. The cricket ear as a four input system. J Comp Physiol 123:217–227.

Larsen ON, Kleindienst H-U, Michelsen A (1989) Biophysical aspects of sound reception. In: Huber F, Moore TE, Loher W (eds) Cricket behavior and neurobiology. Cornell University Press, pp. 364–390.

Lewis B (1983) Directional cues for auditory localization. In: Lewis B (ed) Bioacoustics, a comparative approach. London: Academic Press, pp. 233–257.

Markl H (1968) Die Verständigung durch Stridulationssignale bei Blattschneiderameisen. II. Erzeugung und Eigenschaften der Signale. Z Vergl Physiol 60:103–150.

Markl H (1983) Vibrational communication. In: Huber F, Markl H (eds) Neuroethology and Behavioral Physiology. Berlin Heidelberg: Springer-Verlag, pp. 332–353.

Markl H, Fuchs S (1972) Klopfsignale mit Alarmfunktion bei Rossameisen (Camponotus, Formicidae, Hymenoptera). Z Vergl Physiol 76:204–255.

Michelsen A (1971) The physiology of the locust ear. Z Vergl Physiol 71:49–128.

Michelsen A (1973) The mechanics of the locust ear: an invertebrate frequency analyzer. In: Møller Å (ed) Mechanisms in hearing. London: Academic Press, pp. 911–934.

Michelsen A (1978) Sound reception in different environments. In: Ali MA (ed) Sensory ecology, review and perspectives. New York: Plenum Press, pp. 345–373.

Michelsen A (1983) Biophysical basis of sound communication. In: Lewis B (ed) Bioacoustics, a comparative approach. London: Academic Press, pp. 3–38.

Michelsen A, Nocke H (1974) Biophysical aspects of sound communication in insects. Adv Insect Physiol 10:247–296.

Michelsen A, Larsen ON (1978) Biophysics of the ensiferan ear. I. Tympanal vibrations in bushcrickets (Tettigoniidae) studied with laser vibrometry. J Comp Physiol 123:193–203.

Michelsen A, Larsen ON (1983) Strategies for acoustic communication in complex environments. In: Huber F, Markl H (eds) Neuroethology and Behavioral Physiology. Berlin Heidelberg: Springer-Verlag, pp. 323–331.

Michelsen A, Larsen ON (1985) Hearing and sound. In: Kerkut GA, Gilbert LI (eds) Comprehensive insect physiology, biochemistry and pharmacology. Oxford: Pergamon Press, Vol. 6, pp. 495–556.

Michelsen A, Fink F, Gogala M, Traue D (1982) Plants as transmission channels for insect vibrational songs. Behav Ecol Sociobiol 11:269–281.

Michelsen A, Jørgensen M, Christensen-Dalsgaard J, Capranica RR (1986a) Directional hearing of awake unrestrained treefrogs. Naturwissenschaften 73:682.

Michelsen A, Kirchner WH, Andersen BB, Lindauer M (1986b) The tooting and quacking vibration signals of honeybee queens: a quantitative analysis. J Comp Physiol 158:605–611.

Michelsen A, Towne WF, Kirchner WH, Kryger P (1987) The acoustic near field of a dancing honeybee. J Comp Physiol 161:633–643.

Michelsen A, Andersen BB, Kirchner WH, Lindauer M (1989a) Honeybees can be recruited by a mechanical model of a dancing bee. Naturwissenschaften 76:277–280.

Michelsen A, Hedwig B, Elsner N (1990) Biophysical and neurophysiological effects of respiration on sound reception in the migratory locust, Locusta migratoria. In: Gribakin FG, Wiese K, Popov AV (eds.) Sensory systems and communication in Arthropods. Basel: Birkhäuser Verlag, pp. 199–203.

Michelsen A, Heller K-G, Stumpner A (1991) Biophysics of directional hearing in bush crickets (in preparation).

Michel K (1974) Das Tympanalorgan von Gryllus bimaculatus Degeer (Saltatoria, Gryllidae). Z Morphol Tiere 77:285–315.

Møhl B, Miller LA (1976) Ultrasonic clicks produced by the peacock butterfly: A possible bat-repellent mechanism. J Exp Biol 64:639–644.

Mörchen A, Rheinlaender J, Schwartzkopff J (1978) Latency shift in insect auditory nerve fibers. Naturwissenschaften 65:656.

Nocke H (1971) Biophysik der Schallerzeugung durch die Vorderflügel der Grillen. Z Vergl Physiol 74:272–314.

Payne R, Roeder KD, Wallman J (1966) Directional sensitivity of the ears of noctuid moths. J Exp Biol 44:17–31.

Prager J (1976) Das mesothorakale Tympanalorgan von Corixa punctata Ill. (Heteroptera, Corixidae). J Comp Physiol 110:33–50.

Prager J, Larsen ON (1981) Asymmetrical hearing in the water bug Corixa punctata observed with laser vibrometry. Naturwissenschaften 68:579.

Prager J, Streng R (1982) The resonance properties of the physical gill of Corixa punctata and their significance in sound reception. J Comp Physiol 148:323–335.

Prestwich KN, Walker TJ (1981) Energetics of singing in crickets: effect of temperature in three trilling species (Orthoptera: Gryllidae). J Comp Physiol 143:199–212.

Prozesky-Schulze L, Prozesky OPM, Anderson F, van der Merwe GJ (1975) Use of a selfmade sound baffle by a tree cricket. Nature 255:142–143.

Roth LM (1948) A study of mosquito behavior. An experimental laboratory study of the sexual behavior of *Aedes aegypti* (Linnaeus). Am Mdld Natural 40: 265–352.

Skovmand O, Pedersen SB (1978) Tooth impact rate in the song of a shorthorned grasshopper: a parameter carrying specific behavioral information. J Comp Physiol 124:27–36.

Spangler HG (1987) Ultrasonic communication in *Corcyra cephalonica* (Stainton) (*Lepidoptera: Pyralidae*). J Stored Products Res 23:203–211.

Tautz J (1979) Reception of particle oscillation in a medium—an unorthodox sensory capacity. Naturwissenschaften 66:452–461.

Tischner H (1953) Über den Gehöhrsinn von Stechmücken. Acoustica 3:335–343.

6
Ecological Constraints for the Evolution of Hearing and Sound Communication in Insects

Heiner Römer

1. Introduction

Many insects, in particular those belonging to the orthoptera, use acoustic signals both for the attraction of mates and for aggressive interactions with competitors (for review see Cade 1985). There are a number of reasons for the great diversity of acoustic signals used in social behavior, and these may include a response to (1) changes in the physical properties of the environment, (2) the presence of other acoustically communicating species competing for the same transmission channel, (3) the presence of predators, or (4) sexual selection. No more than 10 years ago, the function of (and the evolution of) hearing in insects was almost reduced to a mechanism for avoiding mismating. Evidence that predator avoidance could be part of the selection pressure for the evolution of hearing was largely ignored (with the exception of the obvious interaction between bat echolocation and moth hearing; Roeder 1965; but see Hoy, Chapter 8. Only recently, the role of sexual selection in shaping acoustic communication systems through female choice is receiving increased attention from behavioral biologists (West-Eberhard 1984), but unfortunately not from physiologists interested in insect acoustics. Therefore, although this chapter mainly deals with the environmental constraints on acoustic signaling and hearing, and the way in which the animal and its nervous system can cope with these constraints, I incorporate other obvious constraints as well, evident from the field work of sociobiologists.

2. Range of Communication

The size of insects limits the distance over which communication may take place. Most small animals can produce sound with reasonable loudness only in the high sonic or ultrasonic frequency range (for the physics of sound production see Bennet-Clark 1970; Michelsen and Nocke 1974). One should consider that the lower end of the frequency scale of insect sounds represents almost the upper end used in the songs of most vertebrates. Since the mechanisms resulting in the attenuation of sound over distance are highly frequency dependent, we would expect the effects to be far more pronounced for the transmission of insect sound compared to sound of birds or other vertebrates (Wiley and Richards 1982).

2.1 Signal Attenuation

The distance over which two individuals may communicate depends on the intensity of the emitted sound, the hearing threshold of the receiver, and the attenuation of sound. Given a sound output of 90 dB SPL at a distance of 1 m, a hearing threshold of about 30 dB SPL (for the bushcricket *Tettigonia viridissima*), and the 6 dB decrease for each doubling of distance under ideal conditions, one could calculate a maximum communication distance of about 1,000 m! The attenuation properties of the actual habitat, acting as the transmission channel between sender and receiver, make this estimate far from realistic. In the example given in Figure 6.1 the intensity of a song of a male *T. viridissima*

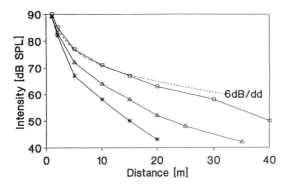

FIGURE 6.1. Calling song intensity of a male *Tettigonia viridissima* recorded at various distances in open grassland and bushland. The height of the calling male and the height of the microphone in the open grassland was 2.0 m (squares), in the bushland habitat 1.5 m (triangles) and 0.75 m, respectively. The decrease in intensity due to spherical spread (6 dB/dd) is indicated by the broken line.

was measured in two habitats (open grassland and bushland) at varying distances. Attenuation in excess of the 6 dB per doubling of distance relationship can amount to 25 dB at a distance of 20 m, depending on the density of vegetation and the heights of sender and receiver.

The contribution of the possible mechanisms (atmospheric absorption, scattering of sound, and boundary effects) remains unclear from this figure. Scattering refers to the reflection, diffraction, and refraction of sound waves by objects in the transmission channel. Scattering is a strongly frequency dependent mechanism, since it depends on the relationship between the wavelength of the sound and the size of the intervening object. For example, if the size of the foliage is about 5 cm and the wavelength of a 40 kHz sound less than 1 cm, most of the sound energy will be reflected in all directions. We also have to consider that insects are far from ideal sound sources that radiate sound energy equally well into all directions; instead, the sound output from an insect is often highly directional (Keuper and Kühne 1983; Bailey 1985), particularly for high frequencies. For example, in the bushcricket *Metrioptera sphagnorum*, a high-frequency component of the song at 33 kHz is up to 15 dB more intense from the dorsal than that from the ventral surface (Römer and Morris unpublished). Scattering will reduce the intensity along this beam by redirecting the sound energy out-

ward. This effect is not linear with distance and, therefore, requires measurements at various distances from the source.

Scattering of sound by atmospheric turbulences is difficult to measure experimentally, but may be worth considering as an important source for excess attenuation of insect sound, since it is frequency dependent as well. Theory predicts that it can amount to 3.8 dB/m at 10 kHz (Lighthill 1953). Yet another source of sound loss is through the dissipation of energy into heat (Griffin 1971; Piercy, Embleton, and Sutherland 1977). A 10 kHz signal attenuates by 0.1 dB/m at 20°C and a relative humidity of 70%. Hence, the loss due to absorption is probably not the most important limiting factor for long-range acoustic communication for insects. Even in the near ideal condition when both sender and receiver are placed on the top of the vegetation, it is hard to separate the contribution of absorption and scattering effects for the total amount of excess attenuation. The data presented in Figure 6.1 as well as the signaling and receiving positions of many species that have been investigated indicate that the effects due to scattering are by far the most important in limiting the broadcast range of insect social signals.

Boundary interferences may cause excess attenuation of sound as well, which in theory should be a serious problem for individuals communicating acoustically at some height from the ground. As a consequence of sound waves being reflected from the ground, sound intensities at the site of the receiver may either be close to zero or may even be doubled, depending on the path length, the wavelength of the sound, and the surface impedance of the ground. However, measurements of sound attenuation in natural environments show only little evidence for boundary interference above 2 kHz (Embleton, Piercy, and Olsen 1976; Marten and Marler 1977) or 10 kHz (Michelsen and Larsen 1983), most probably because the natural boundary (the ground) represents a poor "mirror" of those frequencies used by insects (see scattering effects mentioned above). In fact, many insects make use of the sound propagation path parallel to the ground, often referred to as "the forbidden mode of propagation," most probably to avoid strong attenuation effects due to scattering in the vegetation (see below).

There are two questions arising from these theoretical considerations. One concerns the relevance of these habitat limitations to the evolution of insect sound communication behaviour. The other concerns the scientist investigating insect sound communication and hearing outdoors. The directionality of speakers and microphones may strongly affect measurements of sound attenuation and degradation in natural habitats. Other properties of technical emitters and receivers differ from those of insect stridulatory and hearing organs as well. For example, bushcricket ears may be sensitive to frequencies from about 3 kHz up to 100 kHz with an absolute sensitivity of about 20 to 30 dB SPL (Autrum 1960; Rheinlaender 1975; Kalmring, Lewis, and Eichendorf 1978). Such a combination of features can hardly be equalled by technical receivers. Rheinlaender and Römer (1986) tried to bypass these problems by using a portable recording unit to monitor the activity of single, identified auditory interneurons in the field, whose characteristics (tuning, directionality, etc.) were known in great detail from laboratory experiments. Thus the insect became a "biological microphone," and it was possible to study the hearing capacity in the natural situation of a receiver as well as the transmission properties of the habitat.

2.2 Frequency Filtering

Figure 6.2 gives an example for the use of such a preparation to study the frequency-dependent attenuation properties of the habitat quantitatively. A speaker was placed at a height of 1 m in a bushland habitat broadcasting pure tones at variable frequencies and intensities. The threshold of the auditory neuron (the so-called omega neuron) was determined at a distance of 1 m as a reference, and then at increasing distances as indicated. The apparent increase in threshold for any given frequency is the result of the overall attenuation over distance. Whereas a 5 kHz sound attenuates approximately according to the spherical spread of sound, there is increasing excess attenuation with increasing frequency, so that a 40 kHz tone attenuates at a distance of 10 m by 30 dB in excess of the spherical spread. These measurements also confirm the existence of two kinds of excess attenuation, as predicted previously (Michelsen 1985). Whereas for low frequencies a dB/m relationship

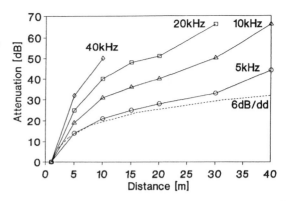

FIGURE 6.2. Attenuation of pure tones in a bushland habitat, at frequencies as indicated. Threshold responses of an auditory interneuron (omega-neuron) in the ventral nerve cord of a bushcricket were determined at a distance of 1 m as a reference, and then at increasing distances. The apparent increase in threshold of the neuron corresponds to the attenuation of the sound stimulus. The broken line indicates the theoretical attenuation due to spherical spread of sound. (After Römer and Lewald, unpublished.)

was found (0.3 dB/m for 5 kHz), for higher frequencies and for distances up to 20 m a nonlinear relationship was demonstrated (Römer and Lewald, unpublished).

Figure 6.3 demonstrates how the excess attenuation affects the frequency content of a broadband song, a situation typical for many bushcrickets (Keuper and Kühne 1983). At a distance of 2 m the signal contains the full frequency spectrum up to 50 kHz, whereas at 25 m all higher components appear filtered out. In this figure the tuning curves of three (out of a total of 34) different auditory receptors are superimposed on the spectra to show the effect of frequency filtering for the activation of single receptor units in the ear. It is obvious that a signal received at 25 m can only activate the receptor with a characteristic frequency of 10 kHz. At an intermediate distance of 15 m, the spectrum is broader and the overall intensity has increased, so that now the receptors tuned to 10 kHz and 20 kHz respond. Finally, at a close distance of 2 m, the signal energy is high enough to stimulate all receptors. This simple example demonstrates that a receiver with an ability for frequency discrimination could make use of the frequency filtering properties of the transmission channel for estimating acoustic distances (Römer 1987).

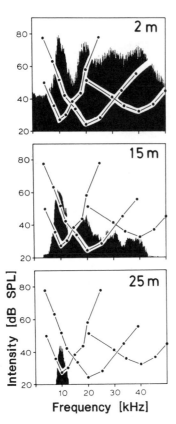

FIGURE 6.3. Frequency spectrum of the song of the bush-cricket *Tettigonia viridissima* measured at distances of 2, 15, and 25 m in a bushland habitat. Note the decrease of high frequency components with increasing distance from the signaller. For comparison, the threshold curves of three auditory receptors (CF 10, 20, and 40 kHz) are superimposed on the spectra.

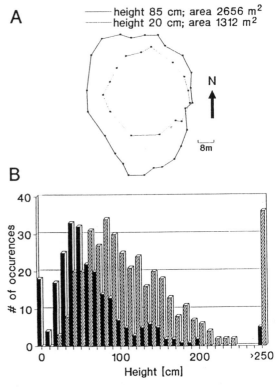

FIGURE 6.4. (A) Polar plots of the broadcasting area of the song of *Mygalopsis marki* at two singing heights, as determined by the threshold activity of the omega-preparation in various directions around a conspecific singing male. Threshold was evaluated at each position for a period of at least 30 seconds. The two experiments were superimposed so that the place of the singing male overlaps. (After Römer and Bailey 1986.) (B) Singing height of the bog katydid *Metrioptera sphagnorum* (black bars) and corresponding available height of the trees from which the males were singing. (Römer and Morris, unpublished.)

2.3 Consequences for the Evolution of Signaling

Given that size restricts the generation of the signal, forcing most insects to use higher frequencies, and given the degree of excess attenuation shown above, the consequence for the evolution of acoustic signaling should be that the animal has to adopt behavioral strategies and make use of optimal positions in its environment for signaling. Since scattering of sound within the vegetation seems to be the major source for attenuation, insects should call on top of the vegetation or at least close to it. It would appear that many insects sing from elevated perches rather than from the ground (Paul and Walker 1979; Doolan and MacNally 1981;

Dadour and Bailey 1985). Although experimentally shown only in rare cases, the advantage from such an elevated singing position is a greatly increased broadcast area (Paul and Walker 1979; Römer and Bailey 1986; Fig. 6.4A). Selection on calling strategies appears to be a solution to optimize signaling strategies commensurate with our physical measurements. Males of the bog katydid *M. sphagnorum*, for example, prefer to sing from isolated spruce trees at an average height of about 60 cm, although the available trees were more than twice that high (Fig. 6.4B). However, compared to males singing from ground level, males at 60 cm increase their broadcast range threefold, whereas there is only a minute effect when singing further up the tree (Römer and Morris unpublished).

It has been argued from studies of sound transmission in terrestrial habitats that maximum range of detection probably is not the primary selection pressure on animal vocalizations (Wiley and Richards 1978). However, virtually all studies performed on different species of acoustic insects support the idea that there is strong selection on the signals or the signaling behavior to maximize broadcast range. Fruitflies (Partridge, Hoffmann, and Jones 1987), mole-crickets (Forrest 1983), crickets (Shuvalov and Popov 1973; Thorsen, Weber, and Huber 1982), and bushcrickets (Latimer and Sippel 1987; Bailey and Yeoh 1988) preferentially approach the louder of two conspecific signals of different intensity. A female bushcricket *Requina verticalis* in the field will always move to the male producing the more intense sound at the position of the female (Bailey et al. in press). To achieve an increased sound output, selection has favored the use of resonators, amplifying burrows, and baffles. The mole cricket *Scapteriscus acletus*, for example, uses his underground burrow to increase sound output. By matching the dimensions of the burrow exactly to the carrier frequency of his call, a male can increase the call intensity by 18 dB and significantly increase its broadcast range (Bennet-Clark 1987).

However, there are a number of species using acoustic signals that seem not to be optimized for long-range communication at all. For example, Morris, Klimas, and Nickle (1989) characterized the calling songs of some katydids from the tropical rain forest as being very sharply tuned to frequencies up to more than 100 kHz, which certainly will not be detected by conspecifics over more than a few meters. It is very likely that in all these cases, as shown for the Australian tettigoniid species *Zaprochillus sp.*, the calling song does not serve to attract females from a distance. Instead, individuals of both sexes gather around food resources and any acoustic interaction can then occur at close range using relatively faint, high-frequency sound (Gwynne and Bailey 1988; Simmons and Bailey in press).

The attraction to the most intense signal and the sound amplification mechanisms found in many acoustic insects do not, however, rule out strong selection pressures that act against maximizing broadcast range. An increased song intensity can be achieved with an increased force of coupling the wings. However, a greater frictional force must then be overcome and consequently the energetic costs of wing closure are higher (Bennet-Clark 1970). Calling will probably attract parasites as well (Cade 1975) or a preferred singing position will make the signaler more conspicuous to predators guided by visual cues. Furthermore, an increased broadcast range will also elicit increased intraspecific competition with the increased risk of damage in physical contacts or fights with rivals. It will therefore be the net profit from different, counteracting selection pressures that determines in each situation whether or not a greater broadcast range will be selected for.

3. Signal Degradation

The attenuation of sound by whatever mechanism primarily affects the receivers' ability to detect a broadcast signal. For social communication to occur, however, the receiver must be able to discriminate among different signals as well. The most crucial sound parameter for signal discrimination in insects is the temporal patterning of the song (Elsner and Popov 1978). Consequently, any degradation of temporal parameters should impose severe limitations for long range social communication.

3.1 Degradation of Temporal Cues

A constant tone broadcast from a speaker in the field shows irregular amplitude fluctuations after propagating through a turbulent atmosphere or in the presence of moving objects on the transmission channel (Wiley and Richards 1978). Amplitude fluctuations of up to 36 dB under various conditions have been reported. Theory predicts that these fluctuations increase in direct proportion to the square of frequency; thus, stronger effects should be expected with transmission of high-frequency insect sound. Michelsen and Larsen (1983) measured amplitude fluctuations for sonic and ultrasonic carriers in insect habitats and confirmed the trend reported by Richards and Wiley (1980), with the exception that they also found substantial fluctuations up to 100 Hz.

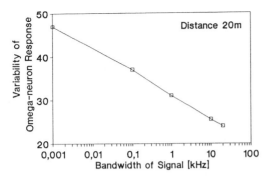

FIGURE 6.5. Effect of signal bandwidth on the amount of irregular amplitude fluctuations in the received signal. The same signal was broadcast 500 times, and the variability of the responses of the omega-preparation was taken as a measure for amplitude fluctuations on the transmission channel (distance 20 m; after Römer and Lewald, unpublished). For further explanations see text.

However, since most insects do not use pure, high-frequency tones for long-range communication, the same experiment was repeated varying the bandwidth of the carrier from a pure tone to almost white noise. Again, by using the insects' nervous system as a "biological microphone" we were able to obtain direct information on how amplitude fluctuations affect the representation of temporal patterns in the afferent auditory pathway of an insect receiver (Fig. 6.5). The results clearly indicate that amplitude fluctuations increase with decreasing bandwidth under these broadcast conditions. They suggest that insects using amplitude-modulated song patterns as a means to broadcast the signaler's identity (and probably individual quality as well, see below) should avoid high-frequency, pure tones for long-range communication. Indeed, many orthopteran insects use broadband signals, which are merely a consequence of the sound production mechanism during stridulation (Bennet-Clark 1970; Michelsen and Nocke 1974). On the other hand, broadband rather than pure tone signals seem to be less vulnerable to amplitude fluctuations in the habitat. Thus, they may represent some kind of adaptation for reliable information transfer in a communication system that almost exclusively uses amplitude-modulated patterns, thereby avoiding masking by unpredictable amplitude fluctuations on the transmission channel.

As a result of reverberations, a temporally distinct sound pulse at the source will be received at some distance with a long tail of scattered sound (Michelsen and Larsen 1983). Since reverberations decay by more than 3 dB/50 ms, they primarily mask rapid amplitude modulations in a signal (Wiley and Richards 1982). Many insect songs exhibit finer temporal details as well (short gaps between syllables, tooth impact rates, etc.) that may provide information for the receiver other than just the signaler's identity. However, very little is known about the effect of reverberations on insect sound in the habitat, nor do we know much about the mechanical or physiological time resolution of insect ears (Schiolten, Larsen, and Michelsen 1981; Ronacher and Römer 1985; Surlykke, Larsen, and Michelsen 1988). For the field cricket *Gryllus bimaculatus*, Simmons (1988) observed a reduction in the interval between pulses and chirps due to reverberations with increasing distance from a singing male. In playback experiments with degraded and nondegraded songs (controlled for intensity) males never searched in response to degraded songs and no male responded to songs that were both attenuated and degraded by reverberations. The most likely explanation for this result is that the temporal pattern of the degraded signal no longer fits with the species-specific song pattern, in which a certain intersyllable interval is crucial for eliciting phonotaxis (Popov and Shuvalov 1977; Thorsen, Weber, and Huber 1982). Clearly more studies are needed in other acoustic insects as well to establish signal degradation as a mechanism for distance estimation.

3.1.1 Consequences for the Evolution of Signals

It should be kept in mind that, despite the various sources for temporal degradation of a signal, the gross temporal pattern of insect songs can be received by conspecifics over quite large distances, when examined using the insect's own hearing system (see Fig. 6.6; Rheinlaender and Römer 1986; Römer and Bailey 1986; Römer and Lewald 1989). Given the environmental constraints for long-range communication it is not surprising that selection has favored temporal cues rather than others to carry information about species identity. However, most of the reported studies have been carried out with species using highly redundant signals, in which simple phonatomes are repeated at

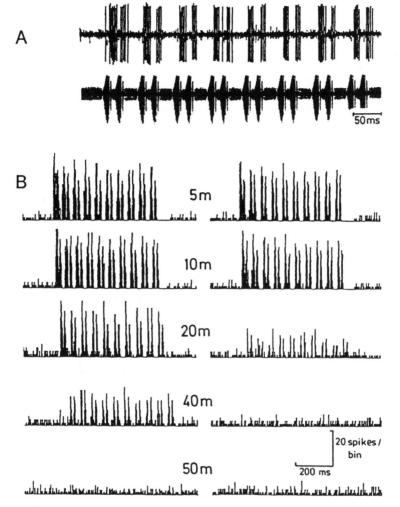

FIGURE 6.6. (A) Response of the omega-neuron of the bushcricket *Tettigonia viridissima* (upper trace) as monitored with a portable recording unit in the field at a distance of 2 m. The sound stimulus (lower trace) was the conspecific stridulatory signal. (B) Post-stimulus-time histograms of the omega-cell responses in two different habitats at various distances. Sound stimulus as in A, repeated 10 times to obtain each histogram. The difference in the maximum hearing distance in the two habitats is about 20 m. Note that the temporal pattern of chirps and interchirp intervals of the song is well reflected in the activity of the neuron up to distances of 40 m and 20 m, respectively. (After Rheinlaender and Römer 1986.)

a high rate, often for many hours per day or night. These highly stereotyped repetitions allow the receiving insect (and the investigating physiologist in the field) to predict the entire signal when part of it is lost in the noise of signal degradation on the transmission channel. However, although redundancy is quite typical for acoustic communication in insects, we nevertheless find large differences in the degree of redundancy among species. Unfortunately, no study exists on the range of communi-

cation for species having low and high redundant song, although nonsystematic observations suggest that discontinuously singing species of bushcricket occur in higher density populations than continuously singing species (Greenfield 1990).

Of course, highly redundant and intense signals such as those found in many bushcrickets are disadvantageous with respect to predation. Thus, when selection pressure due to predation is high, then calling should become more cryptic: redundancy

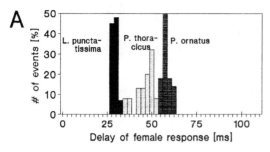

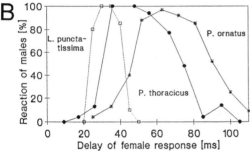

FIGURE 6.7. (A) Distribution of delay times of the female responses to the songs of conspecific males in three Phaneropterid species. The delay time was measured from the onset of the male song to the onset of the female reply. (B) Reaction curve of males of the same three species to a female signal presented at different delay times. The reaction was either a turn to the speaker (*P. ornatus*) or successful approaches to the speaker (*P. thoracicus* and *L. punctatissima*). (Modified from Heller and v. Helversen 1986; data for *L. punctatissima* after Robinson, Rheinlaender and Hartley 1986.)

and song duration should be reduced (Belwood and Morris 1987). Furthermore, different types of communication between male and female may have evolved, such as in phaneropterine bush-crickets, that may cope with these selection pressures in quite a different way (Zhantiev and Dubrovin 1977; Robinson, Rheinlaender, and Hartley 1986; Heller and von Helversen 1986). Here, pair formation is achieved by duetting: the male elicits an acoustic reply in the female to which the male then responds by phonotaxis. The female reply is extremely short, sometimes less than 1 ms, leaving the male with the problem of identifying and localizing such a short signal. Identification is a problem since a click of 0.5 ms duration could hardly contain species-specific amplitude modulations. However, the female responds to the call of the male after a very short delay time, which is species-specific and remarkably constant for each individual female (Robinson, Rheinlaender, and Hartley 1986; Heller and von Helversen, 1986; Fig. 6.7A). This delay time of the female, and not the amplitude-modulation of the response, could be used by the male as a temporal feature for recognition. Indeed, experimental variation of the delay time revealed that the female response must occur within a certain time window in order to elicit phonotaxis by the male. Moreover, this time window is species-specific (Fig. 6.7B) and matches the species-specific female delay time (Robinson, Rheinlaender, and Hartley 1986; Heller and von Helversen 1986). These authors point out that there are advantages in using extremely brief, low information signals, combined with the narrow time window of the male and the corresponding delay time of the female. By listening only for a short time period for a rather unspecific female signal, the chance of confusion due to random events on the transmission channel is greatly reduced. Species identification can be achieved with a signal that would otherwise offer little chance for identification.

3.2 Degradation of Directional Cues

Numerous reports about directional hearing in insects have been published that describe behavioral, physical, and neurophysiological mechanisms by which both sexes may find each other for mating (see Lewis, Chapter 7). Such studies should be performed under acoustic conditions that may approach an ideal free field, in which no redirected components are present and the sound wave is moving in only one direction. It is clear, however, from the findings described in the previous section that the acoustic conditions of many insect habitats may be close to the other extreme of a diffuse sound field, in which sound waves arrive with equal intensity from all directions. This is particularly true for scattering environments with intervening vegetation, where at some distance from the source most of the received sound energy will be scattered sound. Since this is a frequency-dependent pathway (see above), the directionality of high-frequency sound will be degraded more than that of low-frequency sound.

Furthermore, the frequency-dependent attenuation of sound over distance contributes to the quality of directional hearing in the habitat as well. Since most insects seem to rely on the diffraction mechanism at high frequencies to create interaural intensity differences (for review see Lewis 1983), and on the other hand high frequencies suffer from strong excess attenuation (see Fig. 2), a receiver at some distance from the signaler probably has to rely on very poor directional cues. For example, the interaural intensity difference (IID) for different frequencies in the conspecific song of the bushcricket *T. viridissima* may vary from less than 4 dB at 5 kHz to 17 dB at 20 kHz (Rheinlaender and Römer 1980). Therefore, given the frequency filtering properties of the environment, the directionality of the received signal should be expected to improve with decreasing distance, or, in more general terms, with decreasing number of redirecting and sound-attenuating objects between the sender and receiver.

Again, probably the most straightforward quantitative approach for studying directional hearing of insects outdoors is to use the insect's own nervous system as a measure for the directional properties of the habitat. A pair of directionally sensitive interneurons proved to be useful for these kinds of studies (Rheinlaender and Römer 1986). As an example, Figure 6.8 shows the directional characteristics of this pair of interneurons recorded in dense bushland, at two different heights from the ground, and at a distance of 10 m from the speaker. At each of the two positions the preparation was turned in the horizontal plane, from 90° right to 90° left, in order to examine the responses of the two neurons at different source angles. With the preparation on the ground, the directionality of the sound field is lost since there is no systematic change in the neuronal response as a function of source azimuth. However, since each sound pulse elicits a response in the neurons, we can conclude that although the animal receives the sound it is unable to localize it. This is a clear-cut example demonstrating the different effects of signal attenuation and signal degradation: in this particular receiving position, far from unusual for insects, directional cues are completely degraded, but the signal is not attenuated below the threshold of detection. This situation changes when the preparation is on top of the vegetation at a height of 1.5 m. The neuron on the stimulated side clearly responds with the greatest number of spikes and

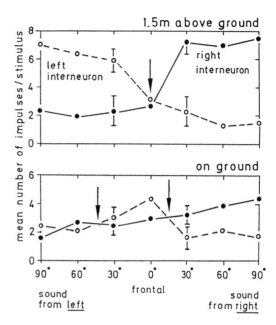

FIGURE 6.8. Directional characteristics of a pair of auditory interneurons of the bushcricket *Tettigonia viridissima*, recorded in dense bushland at a distance of 10 m from the sound source, at two different heights within the vegetation (carrier frequency 20 kHz; intensity 90 dB SPL at a distance of 1 m). Note the loss of directionality when the preparation is on the ground. (Modified from Rheinlaender and Römer 1986.)

the crossing point of the curves coincides exactly with frontal stimulation. This and similar results at different distances demonstrate that directional sensitivity strongly depends on the configuration of the pathway between sender and receiver, and is not determined simply by the inherent properties of the auditory system.

3.2.1 Consequences for the Evolution of Directional Hearing

There are two consequences arising from these studies: the female, which usually localizes and approaches the calling male, is left with the problem of receiving reliable directional cues only at somewhat exposed positions in the vegetation, thereby increasing the risk of predation. The male has a similar problem, since singing from exposed perches increases broadcast range at the expense of increased susceptibility to predators (see above). However, apart from a study of Bailey et al. (1990) virtually no quantitative information is available

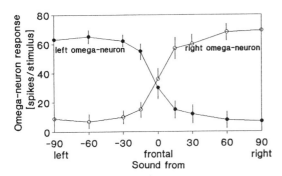

FIGURE 6.9. Directional characteristics of the right and left omega-neuron in the prothoracic ganglion of the bushcricket *Gampsocleis gratiosa*. Both neurons were recorded simultaneously in the anechoic room, with the conspecific song as the sound stimulus. Note that significant right–left differences in the discharge of the neurons are established when the sound source is only 5 to 10 degrees lateral, corresponding to 1 to 2 dB interaural intensity differences. (Courtesy of R. Plewka, unpublished.)

about phonotactic approaches of insects in the field, which could then be compared with the ideal situation in the anechoic room. This will certainly be an important area for future research.

The other consequence of selection pressures for directional hearing may have been for the evolution of a neuronal network, capable of processing directional information. Due to the degradation of directional cues on the transmission channel, there may have been a demand for shaping the nervous system to achieve high directionality using poor directional cues. In all orthopteran insects studied so far, networks using lateral inhibition provide the neuronal basis for an accurate side-to-side discrimination, although sometimes at the expense of angular discrimination abilities (Rheinlaender and Römer 1990). The pair of omega neurons in crickets and bushcrickets (Wohlers and Huber 1982; Römer, Marquardt, and Hardt 1988) is a well-documented example for such a system. Each omega cell provides reciprocal inhibition onto its counterpart on the opposite auditory side (Selversten, Kleindienst, and Huber 1985). The result is an extreme sensitivity for small IIDs in the frontal zone of the animal, as documented in a simultaneous recording of the right and left omega neuron in the bushcricket *Gampsocleis gratiosa* (Fig. 6.9). A change of the speaker position from 30° right to 30° left results in a robust change in the activity

of both neurons, even though the corresponding IIDs associated with these angular changes are less than 3 dB. An IID of only 1 dB is sufficient to induce a significant right–left difference in the auditory pathway. Using a recently developed technique for independent ear stimulation in the unrestrained animal, it was shown that IIDs of 1 to 2 dB were sufficient to elicit reliable turns to either side in the phonotactically orienting female (Rheinlaender, Römer, and Shen unpublished).

4. Noise and Hearing in the Field

The problem of noise on the transmission channel has often been neglected in studies on animal sound communication, although it is the nature and amplitude of background noise—and thus the signal-to-noise ratio—that ultimately determines the range of communication using any sensory modality. Several species may communicate acoustically at the same time in the same habitat and each is presented with the problem of detecting and discriminating its own signals in the noise produced by the other species.

4.1 Interspecific Interference

Song interference is particularly severe for those species that use broadband songs and therefore do not transmit information on different frequency channels. *Hemisaga denticulata* and *Mygalopsis marki* are two such sympatrically occurring species of bushcricket which have almost complete overlap in their song frequencies. In these species, the temporal pattern of their songs differs considerably; males of *M. marki* sing continuously compared with short bursts of song produced by *H. denticulata* (Fig. 6.10A). A comparative study between two populations of *H. denticulata* showed that in the presence of singing males of *M. marki*, song production in *H. denticulata* was suppressed (Fig. 6.10B; Römer, Bailey, and Dadour 1989; Greenfield 1988).

There are two possible explanations that may account for such inhibitory interspecific interactions, both not mutually exclusive. One is that the inhibition represents a by-product of temporal interactions that usually occur between individuals of one species (for a detailed discussion of this

FIGURE 6.10. (A) Oscillograms (left panel) and spectra (right panel) of the songs of *M. marki* and *H. denticulata*. (B) Number of calling males of both species and mean background noise level between 16.00 and 22.00 in the habitat. Note the inhibition of singing activity of *H. denticulata* with the onset of singing activity of *M. marki* (black bars). (After Römer, Bailey, and Dadour 1989.) For further explanation see text.

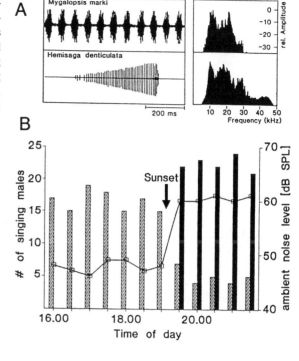

hypothesis see Greenfield 1990). The other explanation refers to the singing inhibition as a masking effect for intraspecific communication, in that critical song elements, necessary for the receiver to respond in an appropriate way to the signal, can no longer be detected in the noise produced by the inhibiting species. Neurophysiological experiments conducted both in the laboratory and the field indeed confirm a strong masking effect on the conspecific signal when *M. marki* (the constant singer) was providing the high background noise level. In assuming that singing in *H. denticulata* mainly serves to attract mates from a distance, the inhibited species therefore responded adequately; it avoided producing essentially useless signals which could hardly be detected by prospective mates.

4.2 Intraspecific Interference

Another source of "noise" that is usually overlooked in studies of hearing in insects is simply a result of insects calling in more or less dense choruses (sensu Greenfield and Shaw 1983). In its most simple form, these choruses may comprise the unsynchronized singing activity of many individual signalers, each with the species-specific

amplitude-modulated song. Due to random superposition of nonsynchronized amplitude-modulated patterns in time and space, the resultant pattern received at any position in the chorus should be largely obscured. However, insects within such a chorus must attend to and analyze acoustic signals from individual signalers despite the presence of other conspecific signals. The spacing of a focal male *M. marki*, for example, is the result of responding to the nearest conspecific neighbors, although the hearing range of this insect allows it to detect many more calling conspecifics (Römer and Bailey 1986). It is remarkable, therefore, that in recordings of the activity of auditory neurons in such choruses, the species-specific amplitude pattern may be quite well preserved at the position of a focal male. An independent laboratory study (Pollack 1986, 1988) has provided evidence for a similar effect in the field cricket *Teleogryllus commodus*. In an analogy with the "cocktail party phenomenon" familiar to humans (Cherry 1966), Pollack called this effect selective attention and presented a neuronal mechanism to account for it as well. Pollack showed that when a low- and high-intensity stimulus was presented simultaneously from the ipsilateral side, an auditory interneuron (the omega neuron) responded selectively to the high-intensity stimulus, even though the low-intensity stimulus was effective when presented alone. However, in Pollack's study the intensity difference between the two stimuli was 20 dB, quite high and unusual for the natural situation in a chorus where intensities between the signals of two or more conspecifics on one auditory side may differ by only a few decibels. Future research may reveal a high degree of intensity "selectivity" in the insects' auditory pathway, which is necessary for segregating individual sources in choruses.

5. Reliable and Nonreliable Acoustic Cues

The message arising from the previous sections clearly is that the gross temporal pattern of insect signals is the only cue that withstands the strong degradation effects in the natural habitat. However, in contrast to the importance of fixed, stereotyped signal patterns for species identification, increasing attention has been paid to the intraspecific variability of these acoustic patterns with reference to the obvious role that sexual selection plays for the evolution of social signals. Thus, a calling male may provide the receiver with information referring both to his identity and his individual "quality." A female receiving the signals of two males could then make a choice based on this information. Again, when it comes to the possible acoustic cues carrying such detailed information, the only reliable one is the temporal cue. In the cricket *Gryllus integer* females are able to base their choice on the length of song elements (Hedrick 1986), and in *G. bimaculatus* females prefer males calling with more song elements (Simmons 1988). In contrast, both the intensity or frequency content of the received signal are rather ambiguous and potentially misleading. They depend on a number of variables, unpredictable for the receiver: height, position and aspect of the signaler, density of the vegetation, macro- and microclimatic conditions, etc. Similar arguments hold for the frequency content of the perceived signal, which strongly depends on the density of the vegetation between sender and the receiver and, even more profoundly, on the unknown posture of the sender. Experiments with females of crickets and bush-

crickets demonstrate a preference for stimuli containing higher frequency components (Latimer and Sippel 1987; Gwynne and Bailey 1988; Bailey et al. 1990). Since these high-frequency stimuli provide the female with better directional cues than low-frequency ones and increase the phonotactic acuity (Latimer and Lewis 1986; Bailey and Yeoh 1988), the observed preference may be just passive attraction, with the female approaching the sound that is easier to localize (Bailey et al. in press).

This does not, however, rule out the important role these signal parameters have for a female approaching one of several signalers: simply moving toward the most intense signal seems to be a common strategy in many examined cases, both in the laboratory and in the field (for a detailed discussion about arguments on active choice or passive attraction in acoustic insects see Bailey 1990). Animals passively attracted to the most intense stimulus show a high degree of intensity discrimination acuity (in the order of 1 to 2 dB, similar to that described in directional hearing experiments by von Helversen and Rheinlaender 1988). The song amplitude of nearby conspecific individuals plays an important role in intermale spacing as well. This evidence is based on field measurements showing that the variability in signal amplitude received from nearest neighbors was less than the variability in distances between neighbors (Römer and Bailey 1986). The insects thus maintain acoustic rather than absolute distances.

Acousticians should be aware of the fact that insect sound signals may undergo strong degradation effects of the environment and still serve to atract a mate over surprisingly great distances. The final mating decision close to the signaler (the "real" choice), however, may then be a matter of close-range, nonacoustic stimuli such as odor, substrate-borne vibrations, contact-mechanical stimulation, or others. This stimulation "cocktail" probably provides the choosing insect with more reliable information about the quality of the signaler.

6. Summary

Insect calling and hearing is examined under field conditions and placed in the context of the possible selection pressures that may have shaped both the character of the signals and the different types of acoustic communication systems used by insects.

The high sonic or ultrasonic frequencies of the signals result in strong frequency filtering and attenuation of the signal in excess to the spherical spread of sound, especially when insects call within stands of vegetation. Behavioral mechanisms to compensate for the signal loss are presented. Distortions of acoustic signals in the time domain may be severe, but using broadband, highly stereotyped and redundant signals the species-specific temporal pattern can be received over remarkable distances. Frequency-dependent scattering and redirection of sound waves limits the locatability of insect songs in the field. It is suggested that central mechanisms, which sharpen weakly peripheral directional components, represent evolutionary adaptations to the necessity of localizing a mate under these conditions. Masking of the conspecific temporal pattern as a result of communicating in aggregations of several conspecific and heterospecific individuals is described as well as behavioral and neurophysiological mechanisms to avoid this masking. The possible roles of the various sound parameters in carrying reliable information about the identity or quality of the signaler are discussed.

Acknowledgments. This article was largely inspired by my recent collaboration with W.J. Bailey. I benefited in particular from discussions with him about the evolutionary aspects of acoustic signaling and hearing, but also from numerous discussions with J. Rheinlaender, I. Dadour, G.K. Morris, and D. Gwynne during our field work. The research for this article has been supported by grants of the Deutsche Forschungsgemeinschaft (Ro 728/2-1 and Heisenberg-Programm).

References

Autrum H (1960) Phasische und tonische Antworten vom Tympanalorgan von *Tettigonia viridissima*. Acoustica 10:339–348.

Bailey WJ (1985) Acoustic cues for female choice in bushcrickets (Tettigoniidae). In Kalmring K and Elsner N (eds) Acoustic and vibrational communication in insects. Berlin: Paul Parey, pp. 107–111.

Bailey WJ, Yeoh PB (1988) Female phonotaxis and frequency discrimination in the bushcricket *Requena verticalis*. Physiol Entomol 13:363–372.

Bailey WJ, Cunningham RC, Lebel L (1990) Song power, spektral distribution and female phonotaxis in the bushcricket *Requena verticalis* (Tettigoniidae,

Orthoptera): active female choice or passive attraction. Anim Behav 40:33–42.

Bennet-Clark HC (1970) The mechanism and efficiency of sound production in mole crickets. J Exp Biol 52:619–652.

Bennet-Clark HC (1987) The tuned singing burrow of mole crickets. J Exp Biol 128:383–409.

Belwood J, Morris GK (1987) Bat predation and its influence on calling behavior in neotropical katydids. Science 238:64–67.

Cade WH (1975) Acoustically orienting parasitoids: fly phontaxis to cricket song. Science 190:1312–1313.

Cade WH (1985) Insect mating and courtship. In Kerkut GA, Gilbert LI (eds) Comprehensive Insect Physiology, Biochemistry and Pharmacology Oxford: Pergamon Press, pp. 591–619.

Cherry C (1966) On human communication. Cambridge: MIT Press.

Dadour IR, Bailey WJ (1985) Male agonistic behaviour of the bushcricket *Mygalopsis marki* Bailey in response to conspecific song (Orthoptera; Tettigoniidae). Z Tierpsychol 70:320–330.

Doolan JM, MacNally RC (1981) Spatial dynamics and breeding ecology in the cicada *Cystosoma saundersii*: The interaction between distributions of resources and intraspecific behaviour. J Anim Ecol 50:925–940.

Elsner N, Popov AV (1978) Neuroethology of acoustic communication. Adv Insect Physiol 13:229–355.

Embleton TFW, Piercy JE, Olson N (1976) Outdoor sound propagation over ground of finite impedance. J Acoust Soc Am 59:267–277.

Forrest TG (1983) Phonotaxis and calling in Puerto Rican mole crickets (Orthoptera: Gryllotalpidae). Ann Entomol Soc Am 76:797–799.

Greenfield MD (1988) Interspecific acoustic interactions among katydids (Neoconocephalus): inhibition-induced shifts in diel periodicity. Anim Behav 36:684–695.

Greenfield MD (1990) Evolution of acoustic communication in the genus Neoconocephalus: Discontinuous song, synchrony, and interspecific interactions. In Bailey WJ, Rentz DCF (eds) The Tettigoniidae: Biology, Systematics and Evolution. Bathurst: Crawford House Press.

Greenfield MD, Shaw KC (1983) Adaptive significance of chorusing with special reference to the Orthoptera. In: Gwynne DT, Morris GK (eds) Orthoperan Mating Systems: sexual competition in a diverse group of insects. Boulder Colorado: Westview Press, pp. 1–27.

Griffin DR (1971) The importance of atmospheric attenuation for the echolocation of bats (Chiroptera). Anim Behav 19:55–61.

Gwynne DT, Bailey WJ (1988) Mating system, mate choice and ultrasonic calling in a zaprochiline katydid (Orthoptera: Tettigoniidae). Behaviour 105:202–223.

Hedrick AV (1986) Female preferences for calling bout duration in a field cricket. Behav Ecol Sociobiol 19:73–77.

Heller K-G, Helversen D von (1986) Acoustic communication in phaneropterid bushcrickets: species-specific delay of female stridulatory response and matching male sensory time window. Behav Ecol Sociobiol 18:189–198.

Helversen D von, Rheinlaender J (1988) Interaural intensity and time discrimination in an unrestrained grasshopper: a tentative behavioural approach. J Comp Physiol 162:333–340.

Kalmring K, Lewis B, Eichendorf A (1978) The physiological characteristics of the primary sensory neurons of the complex tibial organ of *Decticus verrucivorus L*. J Comp Physiol 127:109–121.

Keuper A (1989) Sound production and sound emission in seven species of european Tettigoniids. III. Determination of the mechanism of sound production using cepstrum analysis. Bioacoustics 1:287–306.

Keuper A, Kühne R (1983) The acoustic behaviour of the bushcricket *Tettigonia cantans*. II. Transmission of airborne sound and vibration signals in the biotope. Behav Processes 8:125–145.

Latimer W, Lewis DB (1986) Song harmonic content as a parameter determining acoustic orientation behaviour in the cricket *Teleogryllus oceanicus* (Le Guillous). J Comp Physiol 158:583–591.

Latimer W, Sippel M (1987) Acoustic cues for female choice and male competition in *Tettigonia cantans*. Anim Behav 35:887–910.

Lewis B (1983) Directional cues for auditory localization In: Lewis B (ed) Bioacoustics: A Comparative Approach. London: Academic Press, pp. 233–260.

Lighthill MJ (1953) On the energy scattered from the interaction of turbulence with sound or shock wave. Proc Cambridge Soc 49:531–551.

Marten K, Marler P (1977) Sound transmission and its significance for animal vocalizations. I. Temperate habitats. Behav Ecol Sociobiol 2:271–290.

Michelsen A (1985) Environmental aspects of sound communication in insects. In Kalmring K and Elsner N (eds) Acoustic and vibrational communication in insects. Berlin: Paul Parey, pp. 1–9.

Michelsen A, Larsen O (1983) Strategies for acoustic communication in complex environments. In Huber F, Markl H (eds) Neuroethology and behavioural physiology. Berlin, Heidelberg, New York: Springer, pp. 321–332.

Michelsen A, Nocke H (1974) Biophysical aspects of sound communication in insects. Advanc Insect Physiol 10:247–296.

Morris GK, Klimas DE, Nickle DA (1989) Acoustic signals and systematics of false-leaf katydids from ecuador (Orthoptera, Tettigoniidae, Pseudophyllinae). Trans Am Entomol Soc 114:215–264.

Parker GA (1983) Mate quality and mating decisions. In: Bateson P (ed) Mate Choice. Cambridge: Cambridge University Press, pp. 141–164.

Partridge L, Hoffmann A, Jones JS (1987) Male size and mating success in *Drosophila melanogaster* and *D. pseudoobscura* under field conditions. Anim Behav 35:468–476.

Paul RC, Walker TJ (1979) Arboreal singing in a burrowing cricket, *Anurogryllus arboreus*. J Comp Physiol 132:217–223.

Piercy JE, Embleton TFW, Sutherland, LC (1977) Review of noise propagation in the atmosphere. J Acoust Soc Am 61:1402–1418.

Pollack GS (1986) Discrimination of calling song models by the cricket, *Teleogryllus oceanicus*: the influence of sound direction on neural encoding of the stimulus temporal pattern and on phonotactic behavior. J Comp Physiol 158:549–561.

Pollack GS (1988) Selective attention in an insect auditory neuron. J Neurosc 8:2635–2639.

Popov AV, Shuvalov VF (1977) Phonotactic behaviour of crickets. J Comp Physiol 119:111–126.

Rheinlaender J (1975) Transmission of acoustic information at three neuronal levels in the auditory system of *Decticus verrucivorus* (Tettigoniidae: Orthoptera). J Comp Physiol 97:1–53.

Rheinlaender J, Römer H (1980) Bilateral coding of sound direction in the CNS of the bushcricket *Tettigonia viridissima* (Orthoptera, Tettigoniidae). J Comp Physiol 140:101–111.

Rheinlaender J, Römer H (1986) Insect hearing in the field. I. The use of identified nerve cells as 'biological microphones'. J Comp Physiol 158:647–651.

Rheinlaender J, Römer H (1990) The neuroethology of sound reception in bushcrickets. In Bailey WJ, Rentz DCF (eds) The Tettigoniidae: Behaviour, Systematics, Evolution. Bathurst: Crawford House Press.

Richards DG, Wiley RH (1980) Reverberations and amplitude fluctuations in the propagation of sound in a forest: Implications for animal communication. Am Nat 115:381–399.

Robinson D, Rheinlaender J, Hartley JC (1986) Temporal parameters of male-female sound communication in *Leptophyes punctatissima*. Physiol Entomol 11:317–323.

Roeder KD (1965) Moths and ultrasound. Sci Am 212:94–102.

Römer H (1987) Representation of auditory distance within a central neuropil of the bushcricket *Mygalopsis marki*. J Comp Physiol 161:33–42.

Römer H, Bailey WJ (1986) Insect hearing in the field. II. Male spacing behaviour and correlated acoustic cues in the bushcricket *Mygalopsis marki*. J Comp Physiol 159:627–638.

Römer H, Bailey WJ, Dadour IR (1989) Insect hearing in the field. III. Masking by noise. J Comp Physiol 164:609–620.

Römer H, Lewald J (1989) Degradation and filtering of insect acoustic signals in the habitat: Biophysical and neurophysiological studies outdoors. In: Elsner N, Singer W (eds) Proc. 17th Göttingen Neurobiology Conference. Stuttgart, New York: Thieme Verlag p. 143.

Römer H, Marquart V, Hardt M (1988) The organization of a sensory neuropil in the auditory pathway of grasshoppers and bushcrickets. J Comp Neurol 275:201–215.

Ronacher B, Römer H (1985) Spike synchronization of tympanic receptor fibres in a grasshopper (*Chorthippis biguttulus L.*, Acrididae). J Comp Physiol 157:631–642.

Schiolten P, Larsen ON, Michelsen A (1981) Mechanical time resolution in some insect ears. I. Impulse responses and time constants. J Comp Physiol 143:289–295.

Selverston AI, Kleindienst H-U, Huber F (1985). Synaptic connectivity between auditory interneurons as studied by photoinactivation. J Neurosci 5:1283–1292.

Shuvalov VF, Popv AV (1973) Significance of some of the parameters of the calling songs of male crickets *Gryllus bimaculatus* for phonotaxis of females (in Russian). J Evol Biochem Physiol 9:177–182.

Simmons LW (1988) The calling song of the field cricket, *Gryllus bimaculatus* (De Geer): constraints on transmission and its role in intermale competition and female choice. Anim Behav 36:380–394.

Simmons LW, Bailey WJ (1990) Resource influenced sex roles of Zaprochiline tettigoniids (Orthoptera: Tettigoniidae). Anim Behav (in press).

Surlykke A, Larsen ON, Michelsen A (1988) Temporal coding in the auditory receptor of the moth ear. J Comp Physiol 162:367–374.

Thorson J, Weber T, Huber F (1982). Auditory behaviour of the cricket. II. Simplicity of calling song recognition in Gryllus, and anomalous phonotaxis at abnormal carrier frequencies. J Comp Physiol 146:361–378.

West-Eberhard MJ (1984) Sexual selection, competitive communication and species-specific signals in insects. In Lewis T (ed) Insect Communication. London: Academic Press, pp. 283–324.

Wiley RH, Richards DB (1978) Physical constraints on acoustic communication in the atmosphere: Implications for the evolution of animal vocalizations. Behav Ecol Sociobiol 3:69–94.

Wiley RH, Richards DB (1982) Adaptations for acoustic communication in birds: Sound transmission and signal detection. In: Kroodsma DE, Miller EH, Quellet H (eds) Acoustic Communication in Birds. New York: Academic Press pp. 131–181.

Wohlers DW, Huber F (1982). Processing of sound signals by six types of neurons in the prothoracic ganglion of the cricket, *Gryllus campestris* L. J Comp Physiol 146:161–173.

Zhantiev RD, Dubrovin NN (1977) Sound communication in the genus *Isophya* (Orthoptera, Tettigoniidae) (in Russian). Zool Zurnal 56:40–51.

7

The Processing of Auditory Signals in the CNS of Orthoptera

Brian Lewis

1. Introduction

1.1 Signal Structure and Behavior

Sound production in the Orthoptera is based on a mechanism known as *stridulation* where one cuticular surface is repeatedly applied to another. In the katydids (Tettigoniidae) and crickets (Gryllidae) the effective appendages are the forewings (elytra) whereas in the locusts and grasshoppers (Acrididae) sound is most commonly produced by the movement of both hind legs against the elytra. The periodicity of elytral or leg movement is rate multiplied by means of a series of teeth on one or other of the structures over which a hardened edge (plectrum) is passed during one stroke. Each tooth impact results in a pulse of energy that is passed to a radiating surface.

In the Ensifera (katydids and crickets) sound is produced during the closing stroke of the elytra, opening being silent: periods of sound (*syllables*) therefore alternate with periods of silence. The number of *syllables* grouped together to form *chirps* or *trills*, the syllable repetition rate, chirp rate, chirp duration etc., have all evolved as features that are species specific and the *temporal characteristics* have all been tested behaviorally. In the Acrididae, the situation is rather more complex: both hind legs are used and may be moved together or out of phase with one another; the phase difference may change and even reverse during a singing bout, with the result that the chirp structure of the song may be more or less obliterated. As a result, the extent to which time cues are important for species recognition in grasshoppers is an intriguing problem. The frequency spectra of the sounds produced depend on the relationship between the tooth impact rate and the frequency of natural vibration of the sound radiator. Close correspondence results in a narrow band, resonant sound with a slow onset and decay; wide separation of rate and frequency produces a highly damped, broad band sound. The majority of cricket species produce a highly tuned calling song with a carrier frequency around 3 to 5 kHz and, sometimes, with harmonics extending into the ultrasound range. Most katydids and the grasshoppers produce a broad band song often with considerable energy in the ultrasound region. In the katydids and grasshoppers, the tooth impact rate is detectable in the frequency spectrum as a frequency-modulated energy band. Skovmand and Pedersen (1978) showed that grasshoppers preferred signals with normal tooth-impact rates to signals with half the rate and that the amplitude function seem to play a minor role in the recognition process. Stiedl, Bickmeyer, and Kalmring (1990) have recently shown that female katydids of the genus *Ephippiger* can also distinguish songs that differ only in the tooth impact rate. Because the file teeth are lost or worn away with age, songs of young adult males are therefore preferred to those of older males.

In addition to the calling song, the crickets, almost exclusively, also produce an aggression song during male/male interactions and a courtship song during male/female close encounters. The aggression song is usually an extended calling chirp composed of a greater number of syllables and is accompanied by additional aggressive behavior such as opened mandibles, biting, and lashing antennae. The courtship song is clearly distinct from both calling and aggression either on the basis of frequency and time (European *Gryllus* species, for example, where the calling frequency is 5 kHz and the courtship frequency is around 16 kHz) or on the basis of amplitude modulation pattern alone (e.g., the Australian *Teleogryllus* species). Very little behavioral or physiological work has been carried out on the recognition of courtship song patterns.

Apart from the katydid *Amblycorypha uhleri*, which produces the most complex of insect songs and which includes both amplitude and carrier frequency modulation (Walker and Dew 1972), insect calling songs are relatively simple especially when compared to bird songs, for example. Their very simplicity and their repetitive nature, however, have allowed detailed investigation of the relative importance of the frequency and time characteristics of the calling songs in eliciting phonotactic behavior. Natural songs and simulations of natural songs as well as pulsed pure tones have been used as test stimuli in the investigation of the roles of identified neurons in acoustic behavior.

1.2 The Input to the CNS

In the Ensifera (crickets and katydids) the tympanal organs are on the first pair of legs and consist of a linear array of differently tuned receptors; in the Acrididae (locusts and grasshoppers) the ear is on the first abdominal segment and contains four groups of receptors.

Insect auditory receptors consist of an array of bipolar neurons whose cell bodies are located peripherally (Gray 1960; Hutchings and Lewis 1983). The whole organ, as judged from whole tympanic nerve recordings, is preferentially sensitive to the carrier frequency of the species-specific song and the maximum tympanic membrane velocity. Investigations of single auditory neurons have shown that the vast majority respond in a tonic fashion to pulsed sound. Single units are more or less sharply tuned to frequencies over the range from about 1 kHz to at least 100 kHz (Michelsen 1971; Kalmring, Lewis, and Eichendorf 1978; Hutchings and Lewis 1981; Oldfield 1982). The auditory receptor therefore codes the frequency of the signal and, as a pulse train, the syllable/chirp repetition rate. Tonotopic organization of the

receptors has now been confirmed in locusts (Michelsen 1971), katydids (Oldfield 1983), and crickets (Oldfield, Kleindienst, and Hüber 1986). In the locust, the tuning of the four groups of receptor cells in the Müller's organ is mainly the result of changing patterns of tympanic membrane vibration (Michelsen 1971; Stephen and Bennett-Clark 1982). The mechanical basis of receptor tuning in katydids and crickets, however, is not known at this time. Neither is there substantive information available about any "second filter" mechanism related to the electrical characteristics of the receptor membrane.

The net result of the transduction and peripheral coding mechanisms is the projection to the Central Nervous System (CNS) of:

(a) The frequency content of the signal based on the place principle of line coding;
(b) The intensity of the signal based on an intensity response characteristic for each neuron that extends over about 30 dB and a range fractionation of the intensity scale that extends the range to about 90 dB;
(c) The sound stimulus repetition rate based on the production of a tonic spike train.

In addition, the CNS receives information about the position of the sound source in space. The biophysical characteristics of the accessory auditory structures produce differences in the magnitude of the neural responses from the two ears, even at frequencies whose wavelengths are far greater than the dimensions of the body (Lewis 1983 for review).

It is possible to attack a review of the central auditory pathways of insects in a number of ways: one could, for example, replace an interest in stamp collecting with one of cataloging all the "identified" neurons (with their various aliases) that have been described for the different orthopteran groups; alternatively, and more interestingly, one can take a more functional approach and consider the extent to which phonotactic behavior and life-style can be understood on the basis of known neuronal mechanisms. I shall describe the levels of response complexity in the central auditory pathway and focus on the second of these approaches. However, not all the permutations of neuronal activity in the three major groups can be described. In many cases the principles are known,

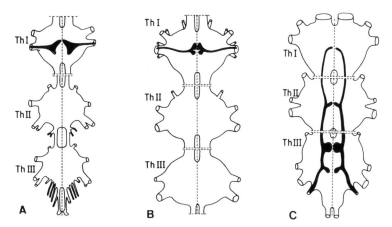

FIGURE 7.1. Central projections of tympanal receptors in: (A) Crickets; (B) Katydids; (C) Locusts. THI–III: pro- meso- and metathoracic ganglia. (From Michelsen and Larsen 1984.)

or suspected, to be similar and these will be taken as read; in other cases there are significant differences that relate to life-style and these will be highlighted.

2. The Central Auditory Pathway

2.1 Primary Sensory Neuropil

The primary auditory fibers of the Ensifera project to a single auditory neuropil in the prothoracic ganglion while those of the acridids project to the metathoracic ganglion. In the acridids they form two distinct acoustic neuropils, the frontal and caudal metathoracic neuropils. About 30% of fibers proceed forward to the mesothoracic ganglion and about 5% of these pass to the prothoracic auditory neuropil (Rehbein 1973) (Fig. 7.1). Despite these differences, the tympanal receptor fibers in both groups of insects terminate in apparently homologous areas of neuropil (Kalmring, Lewis, and Eichendorf 1978; Oldfield 1983; Romer 1983; Wohlers and Hüber 1985; Römer, Marguart, and Hardt 1988) known as the *anterior intermediate sensory neuropil* (aISN) (Römer et al. 1988). This area is almost exclusively occupied by the terminals of the tympanal receptors and branches of the first and second order interneurons with which they synapse. In katydids the receptors terminate in an organization corresponding to their characteristic frequencies which, in turn, correlates with the position of the sensory cells in the receptor array. This tonotopicity runs in both the anteroposterior and dorsoventral axes of the neuropil. In the locust the neuropil appears to be less structured; the only tonotopic order found so far being related to the different termination areas of the group of high-frequency receptors compared to the groups of low-frequency receptors.

In katydids, the whole aISN contains terminations of auditory afferents and both local and intersegmental interneurons receive direct inputs from afferents. This contrasts with the organization in the locust where almost no overlap occurs between primary afferents and intersegmental neurons: the primary afferents occupy a distinct zone in the posterior and ventral aISN where they connect with the local interneurons; these local segmental interneurons project to the secondary auditory area on the contralateral side to make connections with the ascending interneurons.

2.2 Local and Intersegmental Neurons

The first-order auditory interneurons processing the information from receptors are located in the prothoracic ganglion of crickets and katydids but may be in the meta-, meso-, and/or prothoracic ganglion of grasshoppers and locusts.

The first-order auditory interneurons in the prothoracic ganglion of crickets and katydids show many common morphological and physiological characteristics. The best-studied of the neurons that are local to the ganglion link the auditory

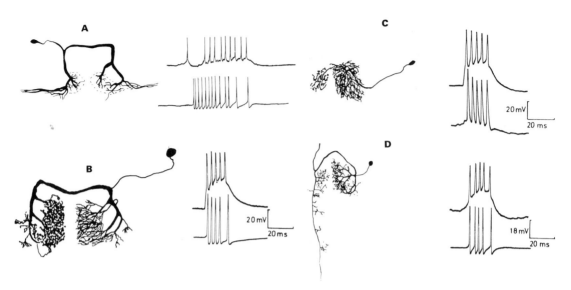

FIGURE 7.2. (Left). Branching patterns of local auditory interneurons in the prothoracic ganglion of: (A) the cricket *Teleogryllus oceanicus*; (B) the katydid, *Tettigonia cantans*; (C)(D) the locust, *Locusta migratoria*, together with intracellular recordings made in their ipsilateral (upper traces) and contralateral (lower traces) branches. The responses indicate a clear separation of input and output areas. (A, from Harrison and Lewis, unpublished; B–D from Römer et al. 1988.)

neuropil of each hemi-ganglion (Fig. 7.2). The branches on the cell body (ipsilateral) side appear to be postsynpaptic (i.e., input) in character while those on the contralateral side are presynaptic (i.e., output) in character. The most intensively studied of these interneurons are the omega neurons in the cricket (Wohlers and Hüber 1978, 1982; Harrison, Horseman, and Lewis 1988). Recordings from the input branches show excitatory postsynaptic potentials (EPSPs) and spike activity in response to spikes in receptor axons. Latencies as short at 0.5 to 1.0 ms have been recorded, indicating monosynaptic connections (Wohlers and Hüber 1982; Römer, Marquart, and Hardt 1988).

As well as the local interneurons, all groups of orthopterans have intersegmental acoustic interneurons with axons ascending to the brain, both ascending and descending (T-fibers) or only descending to more posterior ganglia. In the Ensifera, at least, much more is known about the first two types that is known about the descending fibers whose roles and functions in acoustic behavior are rather speculative. The response characteristics of these intersegmental neurons are more complex than those of the local neurons considered above. Examples of this complexity are shown in Figs. 7.3, 7.4, 7.5, and 7.6.

2.3 Brain Neurons

A detailed analysis of brain neurons has been carried out by Schildberger (1984) in the cricket (Fig. 7.4). The neurons ascending from the prothoracic ganglion project to an area in the anterodorsal region of the diffuse neuropil lateral to the alpha lobe of the mushroom body. This field overlaps one of the projection fields of the so-called BNC1 type neurons; their other projection fields are frequently found in the posterior ventral region, at the boundary between the proto- and deutocerebrum. Here, they overlap the projection fields of the BNC2 type cells. All three classes include cells with low-frequency (5 kHz) and broadband (2 to 20 kHz) tuning. Only AN2 and some BNC1 cells showed high-frequency (10 to 20 kHz) tuning. Sensitivity decreased and response latency increased in the order AN, BNC1, BNC2. There are, therefore, indications of a stepwise transfer of auditory information in the brain from input to output.

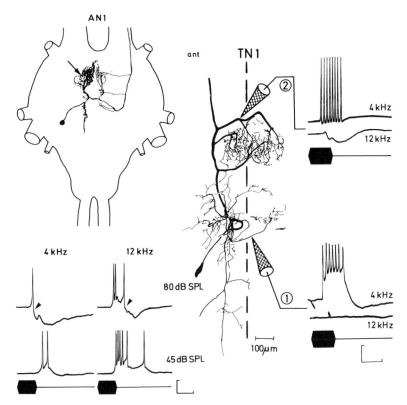

FIGURE 7.3. Morphology and physiology of AN1 (left) and TN1 (right) in the metathoracic ganglion of the locust (*Locusta migratoria*). In AN1, note the onset of IPSP (arrowhead) which is delayed relative to the EPSP. Recordings from different parts of TN1 provide evidence for different sites of input and output synapses: low-frequency stimulation (4 kHz, 75 dB SPL) elicits only spike responses in the enlarged anterior crossing segment compared to EPSP and spikes in its posterior fine branches. In contrast, the IPSP response resulting from high-frequency (12 kHz, 75 dB SPL) stimulation can only be recorded in the large anterior crossing segment near to the frontal auditory neuropil. Stimulation from the axon side; calibration: 15 mV, 20 ms. (From Römer and Marquart 1984.)

2.4 Descending Brain Neurons

Interneurons descending from the brain have been shown to respond to auditory stimuli at biologically relevant frequencies (Zhantiev and Korsunovskaya 1977). More recently Boyan and Williams (1981) have described two descending neurons responding to high intensity auditory stimuli in the 13 to 16 kHz range (i.e., the range of the courtship song). The ipsilateral descending brain neuron (IDBN) has extensive branching in the central and dorsal regions of the deutocerebrum and the axon exits the brain in the ipsilateral connective. The contralateral descending brain neuron (CDBN) arborizes extensively near the dorsal surface of the neuropil in both the proto- and deutocerebrum. The axon crosses the midline, gives off several branches to the dorsal deutocerebrum, and exits via the contralateral connective. The IDBN responds with a maintained discharge of EPSPs and a variable number of spikes, the latency from stimulus onset being 24 ms at 80 dB SPL and independent of stimulus intensity. The response of CDBN consists of long-lasting compound EPSP with variable spike threshold and spike latencies of up to 120 ms at 90 dB. Stimulation with a light source directed at the ipsilateral eye also evokes EPSPs in the IDBN to both "on" and "off," suggesting that the neuron may be bimodal. The output fields of IDBN and CDBN are not

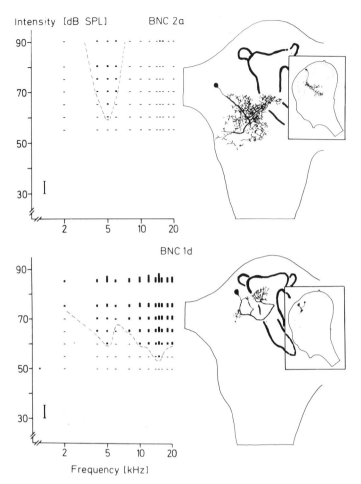

FIGURE 7.4. Response maps of a BNC1 type and a BNC2 type neuron in the brain of the cricket *Gryllus bimaculatus*. The arborization regions of the BNC1 neurons overlap the projection fields of ascending neurons, while those for BNC2 overlap the fields of BNC1 but not ascending cells. Inset, schematic reconstruction (not to scale). In the response maps the responses to synthesized four-syllable chirps varying in intensity and carrier frequency are plotted; response magnitude (total number of spikes) is represented by bar height and dashes indicate no response. Calibration: 50 spikes. (From Schildberger 1984.)

known but they may project to posterior thoracic ganglia and, together with auditory neurons descending from the prothorax, either initiate or modulate the activity of thoracic motor neurons involved in phonotaxis.

3. Central Processing

The most detailed investigations of the neuronal mechanisms underlying phonotaxis have been carried out in the Gryllidae. However, despite the fact that the auditory organs are located on the forelegs in Ensifera and on the abdomen in Acrididae, at the present time there is no reason to believe that the basic principles are any different in the different orthopteran groups. Although the weightings of the different inputs from the peripheral sense organs or local interneurons may be different and

related to the particular ecological niche occupied by the species, the central systems may actually predate the separation of the two major groups from their common ancestor.

Directional hearing in Orthoptera depends initially on biophysical mechanisms at the periphery which establish a difference in the effective net pressure acting on the two ears. In the katydids the dimensions of the body approximate the wavelengths of the sounds used in communication and the two ears therefore act effectively as independent pressure receivers; intensity differences are established initially because of the +6 dB excess pressure ipsilaterally and, at higher frequencies, contralateral shadowing. In the crickets, body dimensions are significantly smaller than the wavelength of the communication sound and ipsi/contra differences are established by means of a pressure difference system whereby sound is

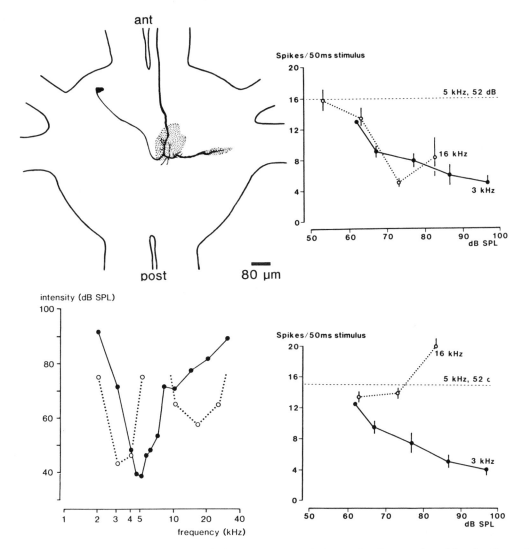

FIGURE 7.5. Reconstruction of the AN3 neuron in the prothoracic ganglion of the cricket *Gryllus campestris* (top left) and (below left) the threshold curve (solid line) flanked by the thresholds of inhibitory effects on a 5 kHz, 52 dB SPL tone (dotted lines). The right column shows two-tone suppression in an AN3 neuron in the intact state (top) and after cutting the contralateral fore-leg (bottom). The curves show the reduction of the response to a constant control tone (5 kHz, 52 dB SPL) by the simultaneous presentation of a 3 kHz and a 16 kHz test tone of increasing intensity. Note that the 16 kHz inhibition originates contralaterally. (From Boyd et al. 1984.)

transmitted both across its body through the tracheal system and around the body without significant loss at the carrier frequency (Lewis 1983 for review, Michelsen Chapter 5). In the grasshoppers and locusts the peripheral structures act as mixed pressure (low frequencies)/pressure difference (high frequencies) systems. Differences of up to 25 dB can be established between the two ears with a 90° angle of incidence by these mechanisms, with

the consequence that the spike input to the CNS differs significantly on the two sides, both in terms of spike number and in latency.

In the cricket *Gryllus campestris*, for example, the phonotactic course of the female to the singing male in the field, on a treadmill, or in an open arena, is characterized by a directed path during which the female meanders about 30° to 60° to either side. The threshold for phonotactic response is around 40 to 50 dB and the female's course deviations decrease with increasing sound intensity up to about 70 dB (Weber, Thorson, and Hüber 1981).

Boyan (1979) recorded the directional characteristics of two ascending auditory neurons in the cricket and showed that the directional information provided by these interneurons does not increase in a linear manner with target angle. Rather, the gain in directional information is scaled so that an angular deviation of the sound source from the midline of only 30° produces 80% of the maximum directionality established at an incident angle of 90°. At 60°, up to 99% of that available at 90° is established, compared to the 75% of maximum which would be obtained if the directionality were of purely biophysical origin. The evidence therefore suggests that a central neuronal sharpening occurs with a focusing of acuity to the frontal field, the physiological acuity correlating well with that reported from behavioral experiments.

In crickets, this neuronal sharpening is believed to occur as a result of the interaction of the pair of local prothoracic neurons termed omega neuron 1 (ON1) (Fig. 7.2). Each member of the omega pair receives excitatory input only from the ear ipsilateral to its cell body and is inhibited by sound presented to the contralateral ear. IPSPS recorded in an omega cell's ipsilateral dendritic field are the result of spike activity generated in the partner cell's ipsilateral dendritic field and propagated via the omega-shaped axon to the mirror-image partner. Thus, if the animal's right ear receives the highest sound intensity, the right omega neuron will show the highest spike activity consequently inhibiting its left partner. Simultaneous intra- and extracellular recordings of the two omega cells provides evidence that this reciprocal inhibition is mediated directly by the partner cell, without an intervening interneuron (Wohlers and Hüber 1982; Harrison and Lewis, unpublished observations).

The fact that Boyan (1979) recorded more acute directional responses in ascending interneurons than could be explained on the basis of biophysical data alone, and the fact that female crickets cannot perform phonotaxis in the absence of head ganglia (Hüber 1983), suggests connections between the ON1 cells and plurisegmental ascending auditory neurons that carry directional and probably frequency and pattern information to the head ganglia (but see below).

Two types of mirror-image plurisegmental ascending neurons whose axons terminate in the brain have been intensively studied in the cricket. They have been termed AN1 and AN2 in *G. campestris*.

AN1 shows a small and rather dense field of dendritic arborizations entirely restricted to the ganglion half contralateral to the cell body. Its dendrites overlap completely with the terminations of primary auditory fibers from the ipsilateral ear and with the arborizations of ON1 dendrites which receive excitatory input from the same ear and with ON1 output of the mirror-image partner. AN1 copies the temporal pattern of the calling song at all sound intensities and does not respond to the courtship song. According to Hüber (1983), AN1 receives only monaural excitatory input from the ear ipsilateral to its dense dendritic field and an inhibitory influence could not be detected. Further, mutual interactions between AN1 neurons of the two sides seem to be absent. AN1 has been said to be the only ascending prothoracic neuron found so far which is preferentially tuned to the calling song carrier frequency of 4 to 5 kHz. However, Boyd et al. (1984) and Silver, Kühne, and Lewis (1984) described a neuron in *G. campestris* (AN3) with similar tuning characteristics to AN1 but which differed morphologically and physiologically from AN1: AN3 has a more dorsal dendritic field and has a main process passing out towards the leg nerve; it is sharply tuned to 5 kHz but is about 15 dB more sensitive than AN1. Further, Boyd et al. (1984) showed that, in contrast to AN1, more complex stimuli identified inhibitory inputs to this neuron (Fig. 7.5). They showed that two pure tone stimuli of differing frequencies presented simultaneously (the classic two-tone paradigm) resulted in the inhibition of the normal spike response to one of these tones presented alone (see also Boyan 1981; Hutchings and Lewis 1984). AN3 is most sensitive to the carrier frequency of the calling song but the highly sensitive characteristic frequency is bordered by inhibitory sidebands on both the high- and low-frequency sides. The inhibition peaks center on 3 kHz and 20 kHz. The source of this inhibition is at present unknown although Wiese (1981b) has presented some evidence that the ON1 output inhibits not only its partner cell but also the AN1 and AN2 neurons (see below). However, the ON1 cannot be the source of the side-band inhibition of

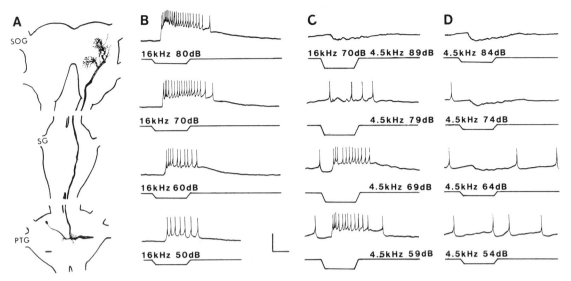

FIGURE 7.6. (A) Camera lucida drawing of the projections of a neuron ANA (= AN2) in the prothoracic (PTG), sub-oesophageal (SG) and supra-oesophageal (SOG) ganglia of the cricket, *Teleogryllus oceanicus*. Scale bar 0.1 mm, B, C, D. Intracellular recordings from one ANA in response to 50 ms tones presented at the intensities and frequencies shown. Scale bars: ordinate 25mV; abscissa 25ms. (From Harrison et al. 1988.)

AN3 because the characteristic frequency of ON1 corresponds to that of AN3 and not to the inhibitory sideband peaks.

The AN2 of *Gryllus* differs from AN1 in its less dense dendritic arborizations in the medial part of the prothoracic ganglion (Fig. 7.6). Also, the main dendrite of AN2 courses further toward the entrance of the auditory nerve. AN2 is a broadband unit, most sensitive to the frequencies contained in the courtship song. It responds to the calling song only at high intensities. Silver, Kühne, and Lewis (1984) state that the presentation of low-frequency stimuli results in inhibition of the spike activity in response to high frequencies, and Popov and Markovich (1982) and Wohlers and Hüber (1982) state that it also receives subthreshold excitatory and inhibitory input from the contralateral ear: binaural stimulation reduces the spike response by an amount depending on the strength of the inhibition. Harrison, Horseman, and Lewis' (1988) intracellular recordings of the response of ANA in *Teleogryllus oceanicus* (a neuron homologous to AN2 and Int 1 (Cassaday and Hoy 1977, Hoy, Chapter 8) showed (Fig. 7.6) IPSPs in response to high intensities of low frequencies; however, there was an additional underlying excitatory response at low frequencies which was revealed when the preparation was treated with picrotoxin. A broadly tuned inhibition of contralateral origin, apparent as a suppression of spontaneous discharge in Int 1 when the ipsilateral ear is removed and the contralateral ear is stimulated has been shown in *T. oceanicus* (Moiseff and Hoy 1983) and intracellular recordings have shown that this suppression is a result of postsynaptic inhibition (Nolen and Hoy 1987). Although it has not been demonstrated, it may well be that this inhibition is produced by the input contralateral ON1 as has been shown in the homologous neurons in *G. bimaculatus* (Selverston, Kleindienst, and Hüber 1985). However experiments in which two-tone inhibition of ANA has been tested before and after removal of the input contralateral ear indicate that the source of inhibition is primarily ipsilateral in origin (Moiseff and Hoy 1983; Hutchings and Lewis 1984). Harrison, Horseman, and Lewis' (1988) series of double electrode experiments in *T. oceanicus* designed to investigate the source of the neurally mediated inhibition of ANA provided no evidence for the existence of effective synaptic connections between the paired ON1s and ANA (Fig. 7.7). Thus, the result of the complex processing of directional information by the paired, mutually inhibitory ON1 neurons is to enhance the left/right intensity differences; however, the output

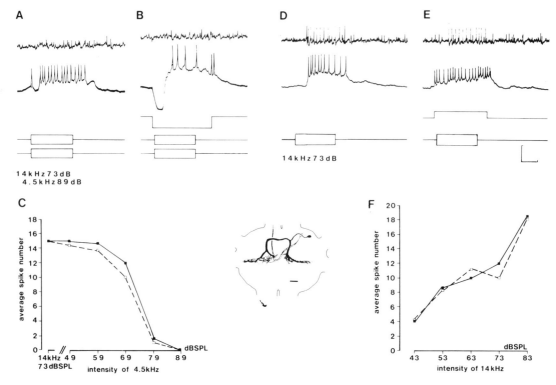

FIGURE 7.7. Tests for inhibitory connections between ANA and the 'input contralateral' ON1. In A, B, D, E, the upper trace shows ANA activity recorded via a suction electrode on the cut neck connective; the second trace shows intracellular recordings from the input contralateral ON1; lower traces are monitors of one or two auditory stimuli and also in B and E current injection into ON1. (A) Responses of ANA and the contralateral ON1 to tones of 14 kHz 73 dB SPL together with 4.5 kHz 89 dB SPL: ON1 is excited and ANA's response to the 14 kHz tone is suppressed. (B) Reduction of contralateral ON1 activity by inward current injection (−4 nA) does not reduce the suppression of ANA. (C) Response of ANA to 14 kHz 73 dB SPL alone and together with various intensities of 4.5 kHz before current manipulation of the response of the contralateral ON1 (solid line) and when the contralateral ON1 activity is suppressed by −4nA of injected current (dashed line): suppression of ON1 activity has no effect on the inhibition of ANA by low-frequency tones over the intensity range tested. (D) ANA and input contralateral ON1 responses to 14 kHz and 73 dB SPL. (E) Increasing the number of spikes produced in ON1 by current injection (+2 nA) during the excitatory tone does not reduce the activity in ANA. (F) Intensity/response functions for ANA to 14 kHz before current injection of the contralateral ON1 (solid line) and with ON1 activity increased by current injection (+2 nA: dashed line). ANA responses are essentially unaffected by contralateral ON1 activity over the intensity range tested. Scale bars: ordinate 5 nA and 30 mV; abscissa 25 ms. (Similar results were obtained for the 'input ipsilateral' ON1 and ANA). (From Harrison et al. 1988.)

connections of these neurons is still unclear. If, as Harrison, Horseman, and Lewis (1988) suggest, the ipsilateral inhibition is not mediated by ON1, then the inhibition must be mediated either by some other unknown interneuron type or directly via primary afferents. The short minimum latency of the IPSP, almost identical to that of the excitatory input (Nolen and Hoy 1987), supports the latter hypothesis. Perhaps two separate groups of afferents tuned to 4 to 5 kHz but with different absolute sensitivities and magnitudes of suprathreshold

responses, as has been shown for *G. bimaculatus* (Oldfield, Kliendienst, and Hüber 1986) could be responsible for the excitatory and inhibitory inputs at this frequency. If this eventually proves to be the case, the function of ON1 remains equivocal.

In the locust, Römer and Dronse (1982) have shown that only a minority of units (3 out of 43) could be activated monaurally and all three units were excitatory. Binaural units showed more complex responses; in some cases, synaptic inputs from both sides were excitatory; others showed ipsilateral excitation but contralateral inhibition; in still others both excitatory and inhibitory potentials are generated by either the ipsi- or the contralateral ear (Fig. 7.8). As a result, the intensity characteristics of interneurons (and therefore directional responses) can differ greatly from those of receptors. The greater complexity of the responses of the central neurons in the locust probably results from the intercalation of local neurons between the primary afferent and the intersegmental neurons as discussed above.

3.1 Temporal Selectivity of Central Neurons

Behavioral experiments have shown that in choice situations female crickets clearly prefer (and can therefore recognize) the pattern of the calling song produced by the conspecific male. It is perhaps for this reason that the vast majority of the work on temporal pattern recognition has been carried out in crickets.

3.1.1 The Calling Song

The mechanisms mediating stimulus pattern recognition (as opposed to stimulus copying) have long been under discussion. A song recognition mechanism in the female cricket was postulated (Hoy, Hahn, and Paul 1977) which was the genetic reciprocal of the male song pattern generator (Bentley and Hoy 1972). However, further behavioral experiments with crickets (Pollack and Hoy 1979) and genetically based phonotaxis experiments with acridids (von Helversen and von Helversen 1975) have called this concept into question. Even the location of this putative signal analyzer is still in question although the brain is the most likely site. In *Gryllus campestris* the most critical temporal parameter for calling song recognition and phonotaxis is the syllable rate within the chirp. At 21 °C phonotaxis is initiated and maintained with syllable intervals ranging from 25 to 55 ms (Thorson, Weber, and Hüber 1982). However, at the level of the prothoracic ganglion, neither the ascending nor the descending neurons are specifically tuned to phonotactically effective temporal patterns. These neurons copy, reasonably faithfully, both phonotactically effective and ineffective patterns. Brain neurons are less well synchronized

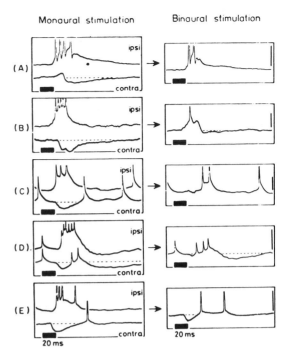

FIGURE 7.8. Intracellular responses of five different interneurons (A–E) evoked by monaural (left) and binaural (right) stimulation. Stimulus intensity was equal at both ears and in all cases 20 dB SPL above ipsilateral threshold. Stimulus duration 20 ms, rise and fall time 1 ms. Note the different latencies of the ipsi- relative to the contralaterally generated synaptic potentials varying in the five units and the resulting different temporal interaction with binaural stimulation. (From Römer and Dronse 1982.)

with the stimulus pattern but here again there is no increase in synchronization when the stimulus repetition rate is that of the natural song. However, the magnitude of the BNC2-type neurons' response (the number of action potentials discharged per chirp) does depend on the stimulus repetition rate and is often maximal for intervals that are phonotactically effective (Schildberger 1984). That is, these BNC2 neurons are band-pass filters of syllable rate with characteristics very close to the band-pass characteristics of the behavioral response. Schildberger (1984) has also described brain neurons that could act as high-pass and low-pass temporal pattern filters and suggests that a BNC2 type band-pass filter response could arise via the ANDing of the responses of these neuron types (Fig. 7.9) as in the toad brain (Rose and Capranica 1984).

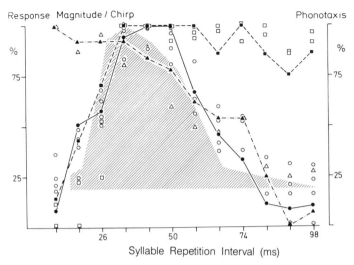

FIGURE 7.9. The relative magnitudes of the responses of various brain auditory neurons in the cricket *Gryllus bimaculatus* to chirps varying in syllable repetition rates. Each of the three curves (maximum response = 100%) shows the averaged responses (12 presentations) of single identified neurons with particular temporal filter properties. ■, BNC1d; ▲, BNC2b; ●, BNC2a. Open symbols show the extent of the variability. Hatched area shows the relative effectiveness of the syllable repetition intervals in eliciting phonotactic tracking. Frequency, 5 kHz; intensity, 80 dB SPL. (From Schildberger 1984.)

3.1.2 The Courtship Song

Courtship behavior, because it occurs during close proximity of the male and female, must involve several modalities. On meeting, individuals antennate each other (Loher and Rence 1978) and chemical cues detected by antennal contact can produce information on species and sex of individuals (Otte and Cade 1976). The significance of the courtship song during this behavioral sequence has been a matter of debate but evidence in favor of courtship song recognition has been provided by the behavioral experiments of Crankshaw (1979) and Burk (1982). They showed that production of the courtship song is a prerequisite for successful mating in *A. domesticus* and *T. oceanicus*. Burk (1982) also showed that in *T. oceanicus* the courtship song acts as a cue for the female to choose a dominant male since subordinate males tend not to sing.

In the *Gryllus* species the courtship song differs from the calling and aggression songs, not only in temporal pattern but also in carrier frequency (Nocke 1972). However, the sound levels of all the spectral components of the calling song and the dominant frequency of the courtship song within one meter of a singing male in the field (Popov

et al. 1974) are suprathreshold to identified neurons in the brain of *G. bimaculatus* (Boyan 1980). Ambiguous spectral information may therefore be present since the same neurons could respond to the courtship song and the higher harmonics of the calling song. Boyan (1981) therefore suggested a two-network system where courtship song recognition could be based on brain neurons which are excited by the high carrier frequency of the courtship song (e.g., some BNC1 cells, see Fig. 7.4) but which are suppressed by the lower frequency contained within the calling and aggression songs. Calling song on the other hand would be recognized by a different neural network, excited by low frequencies and acting as temporal filters tuned to the calling song pattern as discussed above.

In *T. oceanicus*, however, all three songs have similar spectral compositions, with carrier frequencies of 4.5 to 5 kHz and harmonics from 10 up to 55 kHz (Hutchings and Lewis 1984; Latimer and Lewis 1986; Nolen and Hoy 1986a,b) but their temporal patterns are very different. Thus, recognition may depend more on the temporal patterns of the songs than on their frequency spectra. Hutchings and Lewis (1984) showed that at certain intensities, the high frequency ascending neuron

ANA (= AN2 = Int 1) produced a coded response to the temporal patterns of all three song types. Harrison, Horseman, and Lewis (1988) extended these experiments and showed that at higher intensities, similar to those obtained when crickets are in close proximity during courtship behavior, the response of ANA maintains the fidelity of the temporal pattern of the courtship song but not the calling or aggression songs (Fig. 7.10). ANA's response to the trill phrase of the courtship song is based on the integration of strongly excitatory high frequencies and weakly excitatory low frequencies with strongly inhibitory low frequencies. The syllable rate, intensity, and frequency content can all affect the extent of this integration. However, other neurons must also be involved in song recognition and a low frequency ascending neuron tuned to the carrier frequency of the songs and which codes all three songs, is also present in this species; it is probably homologous with AN1 (Wohlers and Hüber 1982) and/or AN3 (Boyd et al. 1984). Courtship may therefore be distinguished from the other songs because of the different weighting of the low frequency inhibitory input to these different neurons.

3.1.3 Courtship Success and Predator Avoidance

Substantial evidence has now accumulated that Int 1 (ANA) is involved in predator avoidance behavior. During flight, crickets turn away from sound sources containing ultrasonic frequencies (the strongly excitatory region of the neuron), a strategy which may serve to avoid hunting bats (Popov and Shuvalov 1977); negative phonotaxis has also been demonstrated in response to ultrasound in *T. oceanicus* under conditions of tethered flight (Moiseff, Pollack, and Hoy 1978; Pollack, Hüber, and Weber 1984; Nolen and Hoy 1986a). Electrical stimulation of Int 1 to produce a response rate above 180 to 220 spikes/s results in negative phonotaxis and therefore suggest that Int 1 is both necessary and sufficient to elicit this behavior (Nolen and Hoy 1986b). Nolen and Hoy (1986b) also investigated the phonotactic response to model calling songs (30-ms pulses repeated at a rate of 15 to 16 pulses/s) when these pulses were composed of 10 kHz. The majority of the animals responding to this stimulus showed negative

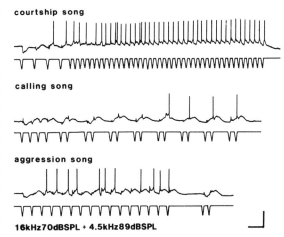

courtship song

calling song

aggression song

16kHz70dBSPL + 4.5kHz89dBSPL

FIGURE 7.10. Responses of an ANA in *Teleogryllus oceanicus* to simulations of the courtship, calling and aggression songs presented at 16 kHz 70 dB SPL plus 4.5 kHz 89 dB SPL. Scale bars: ordinate 25 mV; abscissa 100 ms. (From Harrison et al. 1988.)

phonotactic steering. However, when a high intensity 5 kHz tone (the natural carrier frequency of the species songs) was added in a two-tone paradigm, all the crickets showed positive phonotactic steering. The authors suggest that the addition of 5 kHz to a high-frequency harmonic of the calling song "masked" the aversive nature of this stimulus.

The question then arises as to how this neuron can be involved in both courtship and avoidance behaviors?

Despite the obvious differences in the behavioral strategies necessary for predator avoidance and courtship it is possible to hypothesize that ANA/Int 1 is involved in both behaviors if the switch between avoidance and courtship is based on, or involves, a difference of spike rate in the neuron. Predator avoidance does not occur if activity in Int 1 is below 180 spikes/s; spike activity in response to the courtship song is of the order of 35 spikes/s. In general, the inhibition elicited by the carrier frequency prevents ANA's response to the songs from exceeding this level of activity. It is possible therefore that ANA provides an input to separate neural networks in the brain concerned with negative phonotaxis and song recognition. In any case, the flight steering pathway would need to include a gating mechanism so that the avoidance behavior would not occur outside the context of flight and

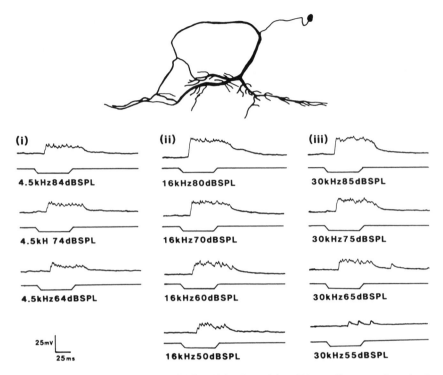

(i)

4.5kHz84dBSPL

4.5kH 74dBSPL

4.5kHz64dBSPL

25mV

25ms

(ii)

16kHz80dBSPL

16kHz70dBSPL

16kHz60dBSPL

16kHz50dBSPL

(iii)

30kHz85dBSPL

30kHz75dBSPL

30kHz65dBSPL

30kHz55dBSPL

FIGURE 7.11. Camera lucida drawing of the ON2 cell in the cricket *Teleogryllus oceanicus* (top) and (below) intracellular recordings of the responses of this cell to the frequencies and intensities shown (Harrison and Lewis unpublished.)

have a high pass filter, possibly a high-frequency interneuron with high threshold and facilitating synaptic inputs so that the postsynaptic cells would not be excited sufficiently to elicit avoidance behavior except at high input impulse rates.

ANA/Int 1 could provide input to ipsilateral song recognition networks, particularly to a courtship song system of the type implicated in calling song recognition in *G. bimaculatus* (Schildberger 1984). However, negative phonotaxis involves the activation of contralateral motor activity. Are there any known neurons, at any level, that might be implicated in this behavioral activity? This problem has not yet been investigated in any detail but a second local prothoracic auditory neuron, known as ON2, may be a possible candidate (Fig. 7.11). ON2 is similar to ON1 except that its axon is very much smaller in diameter and it has a relatively large process which originates from the dendritic field ipsilateral to the cell body (input side) and crosses the midline to arborize in the contralateral auditory neuropile (Wohlers and Hüber 1982). ON2 is a difficult cell to characterize since

its spikes are often greatly attenuated but on the basis of the amplitude of its EPSPs it is clearly a preferentially high-frequency neuron. Its output target is not known but its morphology suggests an involvement in high-frequency processing in the contralateral auditory pathway and it may therefore have some part to play in the modulation or activation of contralateral motor networks.

3.2 The Integration of Sound and Vibration

Most central vibratory neurons receive excitatory inputs from all six legs but the inputs from one leg pair dominate and are responsible for the characteristic sensitivity peak of the neuron. The discrimination of the low and high vibration frequencies is thought to result from the need to distinguish between signals of different biological significance. For example, in katydids, low-frequency vibrations may signal the presence of predators, whereas high frequencies are prevalent in the vibration signals generated by stridulating conspecific males. This difference is less likely to be of importance for the ground-living crickets, as high frequencies will be more quickly damped by this substrate. However, discrimination of low and high vibration frequencies has been demonstrated in crickets by v. Dolen (1981): low-frequency substrate vibration causes the cessation of singing and evasive behavior in stridulating males whereas high-frequency vibration produces only a slight pause in the normal song rhythm. Also, central discrimination of vibration frequencies appears to be more pronounced in crickets than in either katydids or locusts (Fig. 7.12). Excitatory and inhibitory inputs can act in parallel to influence the responses of a vibratory interneuron and in some high-frequency vibratory neurons, inhibitory sidebands occur that enhance the frequency selectivity (Kühne, Silver, and Lewis 1985). In other vibratory interneurons, the occurrence of contralateral inhibition suggests the presence of central nervous directionality within the vibration sense of the cricket, which may be based, at least partly, on the activity of ON1 (Wiese 1981) since this unit's responses to sound are inhibited when vibration stimuli are applied simultaneously to the forelegs.

In both locusts and katydids, most of the auditory and vibratory interneurons receive inputs from both receptors i.e., they are bifunctional (Kalmring, Kühne, and Lewis 1983). In the katydid *Tettigonia cantans* this convergence allows the combination of acoustic and vibratory signals produced by a stridulating male to be used during phonotaxis (Latimer and Schatral 1983). Although the auditory receptor systems of these two groups differ in both anatomical position and structural organization, the physiological responses of the auditory/vibratory interneurons are very similar. In the cricket, the only neuron types in which a convergence of vibratory and tympanal receptors have been found are ON1 described above (Wiese 1981a) and the so-called TN1 and AN2 (Kühne, Silver, and Lewis 1984). TN1 (Fig. 7.13) produces phasic responses to both airborne sound and vibration stimuli. The auditory threshold curve is broadbanded with a best sensitivity between 10 and 20 kHz; it is as much as 20 to 30 dB less sensitive than AN2. In terms of its responses to vibration, TN1 can be classified as a high-frequency, relatively insensitive unit. A characteristic feature of TN1 neurons is their habituation to repetitive stimulation. However, when TN1 is subjected to a

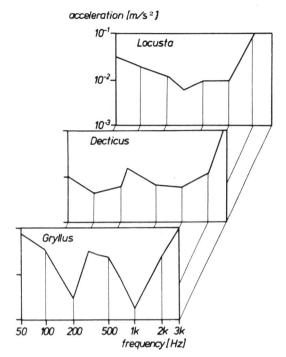

FIGURE 7.12. Discrimination of vibration frequencies at the ventral cord level in *Gryllus*, *Decticus*, and *Locusta*. Each diagram was constructed by superimposing threshold curves of 10–15 individual interneurons to show the difference in overall vibration sensitivity. (From Kühne et al. 1985.)

combination of sound and vibration stimuli, the responses are enhanced and habituation is almost completely abolished. This effect is particularly noticeable for the courtship song with accompanying vibration, a condition that will occur during the close encounter of the male and female during courtship and copulation (Fig. 7.14). The responses of AN2-type neurons to auditory stimuli have already been described (Fig. 7.6). Simultaneous presentation of vibratory stimuli results in inhibition of the auditory response. This effect occurs at all frequencies of airborne sound although the effect is greater with sound signals of around 5 kHz (the calling song frequency) than at high frequencies. Since AN2 may also be inhibited by low-frequency sound this may be part of the neuronal "switch" mechanism for distinguishing between the calling and courtship songs in this group.

Although different song types do not occur in other orthopteran groups the inhibition of high-frequency auditory neurons by simultaneous

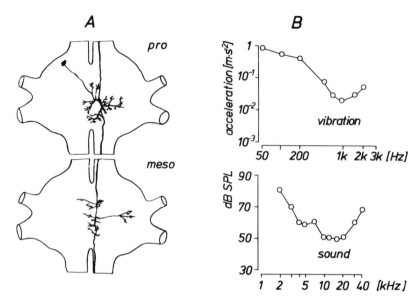

FIGURE 7.13. The TN1 neuron in the cricket *Gryllus campestris*. (A) Reconstruction TN1 in the prothoracic and mesothoracic ganglia. (B) Vibration and airborne sound thresholds of this neuron. (From Kühne et al. 1985.)

vibration is a common occurrence especially in those neurons with the largest axon diameters (AN2 in *Gryllus* (Wohlers and Hüber 1982), S1 in *Decticus* (Kalmring, Rehbein, and Kühne 1979) and B (alias S1) in *Locusta* (Čokl, Kalmring, and Wittig 1977). It is reasonable to suppose, therefore, that these neurons may also have an arousal or warning function.

4. Conclusions

The physiologists' "simple" approach usually means that the various characteristics of the sound signal (frequency, intensity, temporal pattern, and direction) have been considered separately. However, the nervous system is much more holistic in its approach: for example, a neuron's response can depend on the signal's frequency as well as its direction or on its intensity as well as its frequency *spectrum*. The central neurons know only of the number and pattern of the spike trains or the amplitude and pattern of synaptic potentials received from the primary or lower-order interneurons. Of course, biophysical and environmental constraints on both the production and the reception of the species signal have tuned the communication system to a particular frequency or frequency band. In this connection it is interesting to note that the majority (and most easily recorded) of

the *primary* neurons respond to the carrier frequency of the species song (Oldfield 1982; Hutchings and Lewis 1984; Oldfield, Kleindienst, and Hüber 1986); however, the largest and most fast-conducting of the *central* neurons in the katydids and crickets respond preferentially to ultrasound. Further, it is now clear that the so-called "auditory" neurons in the CNS are not simple relay pathways ascending from the primary auditory neuropile to the brain in a sequentially organized hierarchical fashion. Many neurons in crickets (Kühne, Silver, and Lewis 1984), katydids, and locusts (Kalmring et al. 1985) also respond to leg vibration and many show complex integration of these two modalities. Some of the large intersegmental auditory neurons in acridids (locusts and grasshoppers), such as the G and B neurons (Rehbein 1976), have collaterals that project to motor neuropiles in the meso- and metathoracic ganglia where their signals combine with the neurons descending from the brain such as the contralateral movement detector neurons (DCMD) to prime leg motor neurons for the escape jump (Heitler and Burrows 1977; Pearson, Heitler, and Steeves 1980). In the locust, sound can also initiate sequences of rhythmic activity in indirect flight muscles and in flight inter- and motorneurons (Boyan 1985). Finally, Hedwig and Elsner (1985), in an elegant study that allowed intracellular recording from central neurons

while stridulation was in progress showed that
some so-called "auditory" neurons do not respond
to external sound while stridulation is in progress;
neuronal activity is, however, phase coupled to the
stridulatory movements even when this movement
produces no sound.

Römer, Marquart, and Hardt (1988) have speculated on the reasons for the different organizations
of the auditory neuropile in crickets, katydids, and
acridids. Most crickets and katydids sing in the
evening or early morning and both are heavily
preyed upon by bats. Crickets have been shown to
display negative phonotaxis to ultrasound (Popov
and Shuvalov 1977; Moiseff, Pollack, and Hoy
1978). Römer, Marquart, and Hardt (1988) therefore suggest that the relatively simple wiring diagram of the katydid aISN was evolved as part of a
neural subsystem mediating rapid escape behavior
using giant T-shaped fibers as fast conducting
channels. Later, the system might have been incorporated into, and modified for, conspecific acoustic communication. The acridids, on the other
hand, are less sensitive to ultrasound and are probably more highly preyed upon by birds and lizards,
which could be detected by lower frequency sound,
vibration, and visual cues. The acridids may therefore need to combine signals from more than one
modality, so there is no premium for a direct input
to the second order neurons in the aISN.

Smart (1963) presents evidence that the tympanal organs of orthopterans were present before
their sound-generating mechanisms evolved and
therefore before the stridulatory form of conspecific communication started. If this is the case,
there would be considerable pressure on:

(1) The peripheral receptor structures to become
    modified to emphasize frequencies consistent
    with those that could be produced by the evolv-
    ing generating mechanisms;
(2) Central processing systems capable of trans-
    mitting and recognizing the species-specific
    patterns of the generating movements;
(3) However, neither of these evolutionary changes
    could be allowed to occur at the expense of a
    predator-avoidance system. (Avoid the predator
    and copulate tomorrow is a good rule to follow.)

The "reductionist" approach to investigating the
responses of identified neurons to changes in one
of the parameters of a species song has greatly

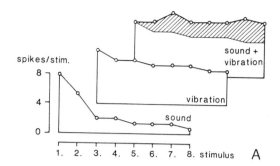

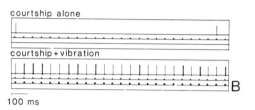

FIGURE 7.14. (A) Response magnitude of a TN1 neuron
in *Gryllus campestris* during repetitive stimulation (1/s)
with airborne sound (5 kHz, 70 dB SPL), with vibration
(500 Hz, 0.56 m.s.$^{-2}$) and with a combination of both
stimuli. The shaded area shows the enhancement with
respect to vibratory stimulation alone. (B) Responses of
a TN1 unit to repeated courtship syllables (rate 3/s, 16
kHz, 70 dB SPL) alone and in combination with modulated 200 Hz, 0.56 m.s.$^{-2}$ vibration stimuli. (From Silver et al. 1984.)

increased our understanding of the neuronal basis
of insect acoustic behavior. And, changing a single
parameter at a time is the standard—and laud-
able—"scientific" approach, providing, that is, that
the various parameters do not interact with each
other in unknown fashions in the CNS. There is
now evidence that they do and future work must
move into a more "holistic" phase by (1) investigating more complex stimulus changes, and (2) utilizing "parallel processing" approaches to these "simple" nervous systems.

Acknowledgments. Thanks to James Fullard and
David Yager for criticizing the manuscript; remaining faults are the result of my stubborness. My
special thanks are due to Jackie Ford, Jeanette
Jones, and Irene Rawlins for repeated corrections
of the manuscript to my very difficult deadlines.
Thanks also to Patrick Foster for the production of
the figures.

References

Bentley DR, Hoy RR (1972) Genetic control of the neuronal network generating cricket (*Teleogryllus gryllus*) song patterns. Anim Behav 20:478–492.

Boyan GS (1979) Directional responses to sound in the central nervous system of the cricket (*Teleogryllus commodus* (Orthoptera, Gryllidae), I. Ascending interneurones. J Comp Physiol 130:137–150.

Boyan GS (1980) Auditory neurons in the brain of the cricket *Gryllus bimaculatus* (De Geer). J Comp Physiol 140:81–93. .

Boyan G (1981) Two-tone suppression of an auditory neuron in the brain of the cricket: proposed role in phonotactic behaviour. J Comp Physiol 144:171–126.

Boyan GS (1985) Auditory input to the flight system of the locust. J Comp Physiol 156:79–91.

Boyan GS, Williams JLD (1981) Descending interneurons in the brain of the cricket. Naturwiss 67:486.

Boyd P, Kühne, Silver S, Lewis B (1984) Two-tone suppression and song coding by ascending neurons in the cricket *Gryllus campestris* L. J Comp Physiol 154:423–430.

Burk T (1982) Male aggression and female choice in a field cricket *Teleogryllus oceanicus*: the importance of courtship song. In: Gwynne DT, Morris GK (eds) Orthopteran Mating Systems: Sexual Competition in a Diverse Group of Insects. Boulder, Colorado: Westview Press, pp. 97–119.

Cassaday GB, Hoy RR (1977) Auditory interneurons in the cricket *Teleogryllus oceanicus*: Physiological and anatomical properties. J Comp Physiol 121:1–13.

Čokl A, Kalmring K, Wittig H (1977) The responses of auditory ventral cord neurons of *Locusta migratoria* to vibration stimuli. J Comp Physiol 120:161–172.

Crankshaw OS (1979) Female choice in relation to calling and courtship songs in *Acheta domesticus*. Anim Behav 24:1274–1275.

Dolen Pv (1981) Üntersuchüngen züm Vibrationsinn der Feldgrille. PhD Thesis, Koln.

Gray EG (1960) The fine structure of the insect ear. Phil Trans Roy Soc B 243:75–94.

Harrison L, Horseman G, Lewis B (1988) The coding of the courtship song by an identified auditory neuron in the cricket *Teleogryllus oceanicus* (Le Guillou). J Comp Physiol 163:215–225.

Hedwig B, Elsner N (1985) Sound production and sound detection in a stridulating acridid grasshopper (*Omocestus viridulus*). In: Kalmring K, Elsner N (eds) Acoustic and Vibrational Communication in Insects. Berlin and Hamburg: Paul Parey, pp. 61–72.

Heitler WJ, Burrows M (1977) The locust Jump: II. Neural circuits of the motor programme. J Exp Biol 66:221–241.

Helversen Dv, Helversen Ov (1975) Verhaltensgentische Üntersuchüngen am aküstischen Kommünicationssystem der Feldheuschrecken (Orthoptera: Acrididae) I, II. J Comp Physiol 104:273–323.

Hoy RR, Hahn J, Paul RC (1977) Hybrid cricket auditory behaviour: evidence for genetic coupling in animal communication. Science: 195:82–84.

Hüber F (1983) Neural correllates of orthopteran and cicada phonotaxis. In: Hüber F, Markl H (eds) Neuroethology and Behavioural Physiology. Berlin and Heidelberg: Springer-Verlag, pp. 108–131.

Hutchings M, Lewis B (1981) Response properties of primary auditory fibres in the cricket *Teleogryllus oceanicus* (Le Guillou). J Comp Physiol 143:129–134.

Hutchings M, Lewis B (1983) Insect Sound and vibration receptors. In: Lewis B (ed) Bioacoustics. A Comparative Approach. London: Academic Press, pp. 181–206.

Hutchings M, Lewis B (1984) The role of two-tone suppression in song coding by ventral cord neurons in the cricket *Teleogryllus oceanicus* (Le Guillou). J Comp Physiol 154:103–112.

Kalmring K, Lewis B, Eichendorf A (1978) The physiological characteristics of the primary sensory neurons of the complex tibial organ of *Decticus verrucivorus* L. (Orthoptera, Tettigoniidae). J Comp Physiol 127:109–121.

Kalmring K, Rehbein, H-G, Kühne R (1979) An auditory giant neuron in the ventral cord of *Decticus verrucivorus* (Tettigoniidae). J Comp Physiol 123:225–234.

Kalmring K, Kühne R, Lewis B (1983) The acoustic behaviour of the katydid *Tettigonia cantans*. III. Responses of ventral cord neurons. Behav Proc 8:213–228.

Kalmring K, Kaiser WOC, Kühne R, (1985) Co-processing of vibratory and auditory information in the CNS of different tettigoniids and locusts. In: Kalmring K, Elsner N (eds) Acoustic and Vibrational Communication in Insects. Berlin and Hamburg: Paul Parey, pp. 193–202.

Kühne R, Silver S, Lewis B (1984) Processing of vibratory and acoustic signals by ventral cord neurons in the cricket *Gryllus campestris*. J Insect Physiol 30:575–585.

Kühne R, Silver S, Lewis B (1985) Processing of vibratory signals in the central nervous system of the cricket. In: Kalmring K, Elsner N (eds) Acoustic and Vibrational Communication in Insects. Berlin and Hamburg: Paul Parey, pp. 183–192.

Latimer W, Lewis B (1986) Song harmonic content as a parameter determining acoustic orientation behaviour in the cricket *Teleogryllus oceanicus* (Le Guillou). J Comp Physiol 158:583–591.

Latimer W, Schatral A (1983) The acoustic behaviour of the katydid *Tettigonia cantans*. I. Behavioural responses to sound and vibration. Behav Proc 8:113–124.

Lewis B (1983) Directional cues for auditory localisation. In: Lewis B (ed) Bioacoustics. A Comparative Approach. London: Academic Press, pp. 233–260.

Loher W, Rence B (1978) The mating behaviour of *Teleogryllus commodus* (Walker) and its central and peripheral control. Z Tierpsychol 46:225–259.

Michelsen A (1971) The physiology of the locust ear. Z vergl Physiol 71:49–128.

Michelsen A, Larsen ON (1984) Hearing and Sound. In: Kerkut GA, Gilbert LI (eds) Comprehensive Insect Physiology, Biochemistry and Pharmacology. Oxford, New York: Pergamon Press, pp. 496–556.

Moiseff A, Hoy R (1983) Sensitivity to ultrasound in an identified auditory interneuron in the cricket: a possible neural link to phonotactic behaviour. J Comp Physiol 152:155–167.

Moiseff A, Pollack GS, Hoy R (1978). Steering responses of flying crickets to sound and ultrasound: mate attraction and predator avoidance. Proc Natl Acad Sci USA 75:4052–4056.

Nocke H (1972) Physiological aspects of sound communication in crickets (*Gryllus campestris* L.) J Comp Physiol 80:141–162.

Nolen TG, Hoy RR (1986a) Phonotaxis in flying crickets. I. Attraction to the calling song and avoidance of bat-like ultrasound are discrete behaviours. J Comp Physiol 159:423–439.

Nolen TG, Hoy RR (1986b) Phonotaxis in flying crickets. II. Physiological mechanisms of two-tone suppression of the high frequency avoidance steering behaviour by the calling song. J Comp Physiol 159: 441–456.

Nolen TG, Hoy RR (1987) Postsynaptic inhibition mediates high frequency selectivity in the cricket *Teleogryllus oceanicus*; implications for flight phonotaxis behaviour. J Neurosci 7:2081–2096.

Oldfield BP (1982) Tonotopic organisation of auditory receptors in Tettigoniidae (Orthoptera: Ensifera). J Comp Physiol 147:461–469.

Oldfield BP (1983) Central projections of primary auditory fibres in Tettigoniidae (Orthoptera: Ensifera). J Comp Physiol 151:389–395.

Oldfield BP, Kleindienst H-Ü, Hüber F (1986) Physiology and tonotopic organisation of auditory receptors in the cricket *Gryllus bimaculatus* De Geer. J Comp Physiol 159:457–464.

Otte D, Cade W (1976) On the role of olfaction in sexual and interspecies recognition in crickets (*Acheta* and *Gryllus*). Anim Behav 24:1–6.

Pearson KG, Heitler WJ, Steeves JD (1980) Triggering of the locust jump by multimodal inhibitory interneurons. J Neurophysiol 43:257–278.

Pollack GS, Hoy RR (1979) Temporal pattern as a cue for species-specific calling song recognition in crickets. Science 204:429–432.

Pollack GS, Hüber F, Weber T (1984) Frequency and temporal pattern dependent phonotaxis of crickets *Teleogryllus oceanicus* during tethered flight and compensated walking. J Comp Physiol 154:13–26.

Popov AV, Markovich AM (1982) Auditory interneurons in the prothoracic ganglion of the cricket *Gryllus bimaculatus*. II. A high frequency ascending neuron (HF1AN). J Comp Physiol 146:351–359.

Popov AV, Shuvalov VF (1977) Phonotactic behaviour of crickets. J Comp Physiol 119:119–126.

Popov AV, Shuvalov VF, Svetlogorskaya ID, Markovich AM (1974) Acoustic behaviour and auditory system in insects. In: Schwartzkopf J (ed) Mechanoreception. Rhein-Westf Acad Wiss Abh 53:281–306.

Rehbein HG (1976) Auditory neurons in the ventral cord of the locust; morphological and functional properties. J Comp Physiol 110:233–250.

Rehbein HG (1973) Experimentelle-anatomische Üntersuchüngen über den Verlauf der Tympanalnervenfasern im Baüchmark von Feldheuschrecken, Laubheuschrecken ünd Grillen. Verh Dtsch Zool Ges 66: 184–189.

Römer H (1983) Tonotopic organisation of the auditory neuropile in the katydid, *Tettigonia virridissima*. Nature 306:60–62.

Römer H, Dronse R (1982) Synaptic mechanisms of monaural and binaural processing in the locust. J Insect Physiol 28:365–370.

Römer H, Marquart V, Hardt M (1988) Organisation of a sensory neuropile in the auditory pathway of two groups of Orthoptera. J Comp Neurol 275:201–215.

Rose G, Capranica RR (1984) Processing amplitude modulated sounds by the auditory midbrain of two species of toads: matched temporal filters. J Comp Physiol 154:211–219.

Schildberger K (1984) Temporal selectivity of identified auditory neurons in the cricket brain. J Comp Physiol 155:171–185.

Selverston AI, Kleindienst H, Hüber F (1985) Synaptic connectivity between cricket auditory interneurons as studied by selective photoinactivation. J Neurosci 5:1283–1292.

Silver SO, Kühne R, Lewis DB (1984) Two-tone interactions and song coding in identified ventral cord neurons in the cricket. Acoustics Letters 7:135–140.

Skovmand O, Pedersen SB (1978) Tooth impact rate in the song of a short-horned grasshopper: a parameter carrying specific behavioural information. J Comp Physiol 124:27–36.

Smart J (1963) Explosive evolution and the phylogeny of insects. Proc Linn Soc Lond 1974:125–126.

Steidl O, Bickmeyer Ü, Kalmring K (1990) Tooth impact rate alteration in the song of males of *Ephippiger ephippiger* Fiebig (Orthoptera, Tettigoniidae) and its consequences for phonotactic behaviour of females. Bioacoustics (In press).

Stephen RO, Bennet-Clark HC (1982) The anatomical and mechanical basis of stridulation and frequency analysis in the locust ear. J Exp Biol 99:279–414.

Thorson J, Weber T, Hüber F (1982) Auditory behaviour of the cricket. II. Simplicity of calling song recognition in *Gryllus* and anomalous phonotaxis at abnormal carrier frequencies. J Comp Physiol 146: 361–378.

Walker TJ, Dew D (1972) Wing movements of calling katydids: fiddling finesse. Science 178:174–176.

Weber T, Thorson J, Hüber F (1981) Auditory behaviour of the cricket. I. Dynamics of compensated walking and discrimination paradigms on the Kramer treadmill. J Comp Physiol 141:215–232.

Wiese K (1981a) Influence of vibration on cricket hearing: interaction of low frequency vibration and acoustic stimuli in the Omega neuron (*Gryllus bimaculatus*). J Comp Physiol 143:135–142.

Wiese K (1981b) Aküstische, vibratorische ünd efferente Eingange am Omega-Neuron der Grillen-hörbahn. Verh Dtsch Zool Ges 168.

Wohlers DW, Hüber F (1978) Intracellular recording and staining of cricket auditory interneurons (*Gryllus campestris* L., *Gryllus bimaculatus* De Geer). J Comp Physiol 127:11–28.

Wohlers DW, Hüber F (1982) Processing of sound signals by six types of neurons in the prothoracic ganglion of the cricket *Gryllus campestris* L. J Comp Physiol 146:161–173.

Wohlers DW, Hüber F (1985) Topographic organisation of the auditory pathway within the prothoracic ganglion of the cricket *Gryllus campestris* L. Cell Tissue Res 239:555–565.

Zhantiev RD, Kursunouskaya OS (1977) Reaction to sound of descending neurons in cervical connections of the cricket *Gryllus bimaculatus* DeGeer (Orthoptera, Gryllidae) (in Russian). Ent. Obozr. 54:248–257.

8
The Evolution of Hearing in Insects as an Adaptation to Predation from Bats

Ronald R. Hoy

1. Introduction

In a volume devoted to the evolutionary biology of hearing it is worth emphasizing that among terrestrial animals, only the vertebrates and insects have evolved specialized receptor systems for hearing. Like other sensory modalities hearing subserves survival behavior of which two stand out: reproductive behavior and predator detection. Hearing mediates both behaviors in a wide variety of vertebrates and insects, and especially among nocturnally active species in both. I will briefly touch on the well-known role of hearing in the reproductive hearing of insects since it is the subject of other chapters in this volume (Michelsen, Chapter 5; Romer, Chapter 6; Lewis, Chapter 7), and I will devote most of this chapter to the evolution of insect hearing in relation to predators.

2. Old Problems, Sound Answers

The use of ears to hear potential mates and predators is evolutionarily ancient in both vertebrates and insects. The evolution of lateral-line and cochlear-based hearing organs is a strong theme in this volume. In insects, air-borne acoustic signals are detected both by tympanal organs and particle-displacement-sensitive receptor hairs (Michelsen and Larsen 1985). Because this chapter will emphasize the detection of high-frequency acoustic signals, and since particle-displacement receptors are low-frequency detectors, we shall be primarily concerned with tympanal organs.

2.1 Sexual Signals and Ears

The use of vibrational signals to mediate sexual and reproductive signals is likely to be an ancient mode of communication in insects. Sensory mechanisms to detect substrate-borne vibrations are widespread among present-day insects and in some instances are similar to sensory organs used to detect sound (review: Kalmring and Elsner 1985). For example, in crickets and katydids, the auditory tympanal organ (TO) shows structural and embryological similarities with the vibration-sensitive subgenual organs (SGO) (Meier and Reichert in press). Evolutionarily, the SGO is undoubtedly older than the TO, since SGOs occur in all six legs of these insects, and moreover SGOs occur in insects that lack TOs. It has been shown that TOs and SGOs even share common embryological origins (review: Dethier 1963; Meier and Reichert 1989; Meier and Reichert in press). Tympanal receptor organs are used in intraspecific communication in many insects, as distantly related as cicadas, moths, and crickets. Clearly, tympanal organs have evolved independently several times in several orders, in the evolution of insects (Michelsen and Larsen 1985). In all known examples of loud and conspicuous (to the human ear) insect acoustic reproductive calls or signals, the insects themselves hear with tympanal organs.

2.2 Predator Detection and Ears

Substrate vibration detectors are ubiquitous among insects, and many of them detect their predators by using a variety of vibrational detectors, including cercal detectors, various sensory hairs, and SGOs. Among the best-known receptors are the cercal hair detectors of orthopteroid insects that are sensitive to low-frequency (infrasound to 500 Hz) airborne stimuli, and are the "front-end" of a giant-fiber mediated escape response (Roeder 1967; Camhi 1984). Selection pressure from potential predators was undoubtedly a driving force in the evolution of vibration detectors in insects, although there are many examples of substrate-borne intraspecific communication signals in a variety of taxa (Henry 1980). This subject has a large literature and is beyond the scope of this chapter. I mention it to point out that vibrational signals can be divided into those that are transmitted through the substrate and those transmitted through the air; both modes are significant channels for conspecific signaling and predator detection. Moreover, from the point of view of physiological mechanisms they are often functionally related. That said, I now turn to the subject of acoustic signals in predator detection.

Thanks to pioneering work of the late K.D. Roeder, the best-documented acoustic interaction between identified predator and prey species is that which occurs between insectivorous microchiropteran bats, which use biosonar to detect flying prey, and moths, which use hearing to avoid bats (Roeder 1967). Roeder showed that several families of moths have independently evolved tympanal organs that are located in different parts of the body and that vary in complexity of the sensory apparatus. In all cases, the sensory apparatus has been shown to be a chordotonal organ associated with an identifiable tympanal membrane. Subsequent work on moths by workers in Canada (e.g., Fullard 1987) and Denmark (Surlykke 1984) have extended Roeder's findings.

It is now abundantly apparent that the moth-bat "story" is not an isolated oddity of animal behavior. Indeed, contemporaneous with Roeder's studies were those of Lee Miller, who demonstrated that green lacewings (*Chrysopa carnea*) also have a bat avoidance behavior based on hearing with a chordotonal organ (Miller 1970, 1971, 1975). In the past decade, further investigations of ultrasonic hearing have revealed that bat predation may have influenced the evolution of hearing in numerous other nocturnally active, flying insects, including crickets (Popov and Shuvalov 1977; Moiseff,

Pollack, and Hoy 1978), praying mantises (Yager and Hoy 1986a), tiger beetles (Spangler 1988), migratory locusts (Robert 1989), and katydids (Libersat and Hoy 1989). One suspects that this is but a glimpse at a much more widespread auditory behavior than previously realized. Bats utilize ultrasonic frequencies in their biosonar, ranging from 20 kHz to 200 kHz, depending on species. In all cases studied to date, insects detect the high frequencies of bat biosonar with tympanal hearing organs. These organs are highly variable in placement on the body (wings, abdomen, thorax, legs, head) and sensory complexity (ranging from a single auditory receptor to over a hundred). Clearly then, bats have exerted significant predation potential on insects, which have in turn evolved ears to detect bats. Thus, ears capable of detecting ultrasound have evolved independently several times within insects, resulting in a morphological diversity. However, the auditory response properties of the various hearing organs in insects exhibit a suite of common characters that reflect convergence resulting from commonalities in the biosonar emissions of microchiropteran bats.

2.3 Which Came First: Bats or Ears?

Speaking only of tympanal hearing organs, it is virtually certain that some acoustically active insects used ears to hear other conspecific insects before they were employed to detect bats. After all, even an admittedly spotty fossil record provides evidence that katydids and crickets, among other sonorous insects, sang their songs by the time of the Jurassic period (Alexander 1962). Sharov (1971) has proposed that crickets, katydids, and grasshoppers descended from a common ancestral type that lived in the Carboniferous period, about 300 million years ago. The fossil evidence that crickets stridulated relies on key characters such as specialized forewings (tegmina), which are associated with sound production. Fossil evidence for ears is harder to come by, and given their small size and delicate structure, fossil specimens would be a find indeed. If one assumes that the specialized tegmina of fossilized crickets reflect their use for sound production as in modern species, it is reasonable to presume that those calls were heard by conspecific listeners, as they are now (Alexander 1962). Like Alexander earlier (1962),

I believe that from their earliest arrival in the fossil record, ancestral crickets and katydids were producing sounds, and that hearing evolved in a conspecific context, for communication of social signals. As will be discussed later, other insects (e.g., moths) may have evolved their ears in response to bats.

What about acoustically active predators? The earliest fossil Microchiropteran bats did not make their appearance until 50 million years ago, in the Eocene (Novacek 1985). These creatures are *Paleochiropteryx tupaiodon* (45 Myr BP, found in Western Germany), and *Icaronycteris index* (50 Myr BP, found in Wyoming). Did these early bats have the sophisticated biosonar system that is the hallmark of modern microchiroptera? Michael Novacek (1985) presents morphological evidence that they did. In particular, he points out that the early Eocene bats and living microchiropterans both have a relatively large cochlear apparatus; this is significant because modern microchiropterans have a large cochlea in relation to the rest of the skull, presumably to accommodate the basal turn region that encodes high frequencies in echolocation. Thus, Novacek argues that the ability to echolocate was present already with the first appearance of bats in the fossil record, and was derived within the group. And possibly, this ability was an adaptation that allowed bats to exploit the rich food niche of nocturnally flying insects.

While it is not possible to discount the possibility that the small nocturnal mammals of the mesozoic preyed upon the stridulating orthoptera (they probably did, but using direct hearing as they do today, rather than echolocation), only bats among mammals took to the wing, and no modern-day birds hunt insects at night with biosonar. Thus, parsimony dictates that when insects took to the wing in the Carboniferous period, those that adopted a nocturnal habit probably had no significant predatory threats on the wing until Eocene times, with the coming of the Microchiroptera. By the Eocene, all the major insectan orders had been long established, and in all likelihood, conspecific acoustic signalling systems abounded as well. As is the case in many modern insects, it is likely that high frequency, ultrasound components were part of acoustic signals of mesozoic/cenozoic insects, and that their hearing organs were tuned to ultrasounds. It is but a short, speculative step to propose that the ability to hear bats was an exaptation (*sensu* Gould and Vrba 1982) from the ability to hear conspecific mating signals.

In the rest of my review, I will briefly sketch out the kinds of ultrasound-sensitive, tympanal ears found so far among insects and describe behavioral responses to ultrasound. I will argue that whatever its origin, the ability to hear ultrasound and react to it is a surprisingly common capacity among nocturnally active, flying insects.

3. Hearing Organs in Insects

3.1 Cytological Aspects

High-frequency-sensitive ears in insects (sensitivity to 1 kHz and above) are of the chordotonal type and it is worthwhile briefly describing their morphology (Dethier 1963; Moulins 1976; Michelsen and Larsen 1985). Arthropod (here, insect and crustacean) chordotonal organs are characterized by specialized sensory sensilla called *scolopidia*. Chorodotonal organs are used to subserve a variety of mechanoreceptive functions, including hearing, but all show a common morphological plan. In fact, the first ultrastructural study of a scolopidium was made in the tympanal hearing organ of the migratory locust, *Locust migratoria* (Gray 1960), whose acoustic behavior will be described later in this chapter. A scolopidium is a complex sensillum consisting of one or more sensory cells, and a pair of accessory cells, a scolopale cell and an attachment cell; these form a morphological unity (Fig. 8.1). Scolopidia are recognized by a special structural relationship between the sensory cell and its accessory cells. The scolopale cell ensheaths the distal portion of the sensory cell, including its dendrite, as well as its ciliary root which is derived from the centriole. The attachment cell (also known as the "cap" cell) ensheaths the distal portions of the scolopale cell, as well as the distal-most, ciliary, portion of the sensory cell. Precisely how these receptors cells transduce vibratory stimuli is still not known, but recently microelectrode recordings have been made from receptors in the tympanic organs of katydids and crickets (Oldfield 1985; Oldfield, Kleindienst, and Huber 1986).

The sensilla of the tympanic organ of katydids and crickets form a compact, linear array that

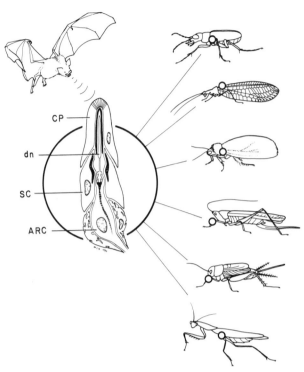

FIGURE 8.1. Predation by insectivorous microchiropteran bats, which hunt for and locate their prey by means of biosonar signals, is presumed to have influenced the evolution of hearing in six different insects, as depicted here. The cellular morphology of the hearing organ in these insects has a complex organization and is referred to as a chordotonal organ. A "typical" chordotonal sensillum is drawn here, and is described in the text. Abbreviations: ARC, auditory receptor cell; dn, dendrite of the ARC; SC, scolopale cell; CP, cap cell, also called an attachment cell. A typical chordotonal hearing organ contains from 1 (in some moths) to upwards of 70 (in field crickets) chordotonal sensilla. This depiction shows the kinds of insects in which ultrasound sensitive hearing has been demonstrated as well as behavioral evidence of an ultrasound startle or escape behavior. From top to bottom: tiger beetle, *Cicindela lemniscata*; green lacewing, *Chrysopa carnea*; moth (many species); katydid, *Neoconocephalus ensiger*; cricket (many species); mantis (several species). The drawings are not meant to be taxonomically accurate, and are not drawn to a common scale. The chordotonal sensillum is drawn after Gray (1960) and Moulins (1976). In the case of the tiger beetle, the histological and cellular structure of the tympanic hearing organ is not yet known (Spangler 1988).

reflects a tonotopic map of frequency encoding (Oldfield 1985), and is depicted in Figure 8.2. In these orthopterous insects, sensilla of the tympanic organ are not directly inserted on the tympanic membrane but instead on a basement membrane associated with tracheae. The tracheae are in turn closely apposed to the tympanic membrane. This association of the tympanic organ with the leg tracheae adds to the complexity of the functional "ear" of crickets, for there are acoustic contributions from excitation of the thoracic tracheal system as well as that from the legs (Michelsen and Larsen 1985). The complexities of tympanic organs in the orthoptera are in contrast to the much simpler arrangement in moths, where one to four auditory receptor sensilla insert directly upon the tympanic membrane (summarized in Roeder 1967).

3.2 Development and Evolution

The similarity of specialized tympanal hearing organs to more generalized vibration receptive chordotonal organs has been noted from the earliest investigations, and has given rise to speculations about homologous origins. Recent investigations by Meier and Reichert (1988, 1990) have provided some definitive findings about the ontogeny of hearing organs in both of the major orthopteroids that are acoustically active: acridids (abdominal ears) and ensiferans (leg ears).

Receptor cells destined to be the scolopales of the hearing organ originate by epithelial invagination during embryonic development. This was shown earlier by Slifer (1935) for acridids. Meier and Reichert (1989, 1990) confirm this not only for acridids, but also for the tibial ears of katydids. The receptor organ (crista acustica) begins its development in the embryo, but is completed in postembryonic free-living instars. In the case of the locust, the protochordotonal organs migrate as a cellular aggregate, to take their places in the pleurae, whereas the proto-crista acustica develops in situ in the prothoracic legs. It is interesting to note that in the ensifera, the cells of the ear are added to a linear array in a systematic way (Ball and Young 1974; Meier and Reichert, 1990) and that may provide a structural basis for the functional tonotopic organization of the acoustically competent adult ear.

Of evolutionary interest is the important finding of Meier and Reichert that in both locust and katydid, the specialized tympanal organs of hearing are probably segmentally homologous to similar chordotonal receptor organs. They present strong evidence that, in locusts, the pleural chordotonal organs — including wing hinge stretch receptors — share embryological origins with the tympanal organ, and that some of these are even sensitive to acoustic stimulation, but not to the same degree as tympanal hearing organs. The same is true of the tibial subgenual organs in the walking legs of katydids. The similarity of the structural organization of the SGOs of orthoptera to that of the specialized prothoracic tympanal organ has long been noted (review: Dethier 1963), and it is known that the SGO is sensitive to vibrational stimulation (review: Kalmring and Elsner 1985).

A fascinating example of the relation of SGOs to hearing organs has been reported in an abstract by Nelson (1980 and in preparation) in the hissing cockroach, *Gromphadorhina portentosa*. That this species uses its hisses as intraspecific communication signals in male–male aggression and in courtship has been amply demonstrated (Nelson and Fraser 1979). In a follow-up study, the hearing organs of *G. portentosa* were found to reside in the tibia of the walking legs, particularly the meso-

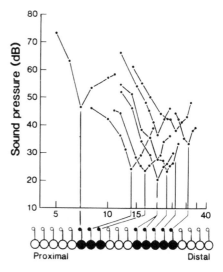

FIGURE 8.2. Tonotopic organization of the katydid ear. Schematic representation of the linear array of chordotonal sensilla in the hearing organ of katydids and crickets. Data shown here are from *Mygalopsis marki* (Orthoptera: Tettigoniidae), from Oldfield (1985). The graph shows a tuning curve of seven identified auditory receptors, and the row of large circles at the bottom represent auditory receptors. Low-frequency sounds are coded by receptors at the proximal portion of the organ, whereas higher frequencies are coded by more distally located receptors. (From Oldfield 1985.)

thoracic pair (Nelson 1980 and personal communication). Here, the number of scolopales were increased from the usual 14 to 20, to over 40. Destroying these SGOs abolished neural activity recorded from central connectives that was evoked by acoustic stimulation. It would be satisfying to report that in adult Madagascar cockroaches a conspicuous tympanal membrane can be found in the tibial segment analogous to that which occurs in adult crickets and katydids; however, none has been found (Nelson, personal communication). However, Nelson (personal communication) has found that this enlarged SGO is associated on the one hand with a transtibial membrane and on the other with an enlarged trachea, bringing it into parallel with more specialized ears.

Nonetheless, one can argue that the tympanal hearing organs of acridids and ensiferans are exaptations of body and limb chordotonal organs sensitive to vibrational stimuli. This notion is reinforced by investigations in insecta orders that have representatives that are acoustically active as well as species that are deaf. Examples of such studies include moths (Yack and Fullard in press), cockroaches (Nelson 1980 and personal communication), locusts (Meier and Reichert in press), and praying mantis (Yager in preparation).

4. Ultrasound Startle/Escape Reactions in Insects

To date, ultrasound-sensitive ears have been reported in the five diverse orders of insects: Lepidoptera, Neuroptera, Orthoptera, Dictyoptera, and Coleoptera; moreover in some of these orders ears have evolved independently several times. In each of these orders some evidence has been presented that such ears evolved in response to predation by insectivorous bats that hunt their prey by echolocation. The behavior elicited by ultrasound in these cases takes the form of an acoustic startle response (ASR). Acoustic startle responses are phylogenetically widespread behavioral acts that are displayed in virtually all animals that have a sense of hearing (reviewed in Eaton 1984). Acoustic startle—a rapid, coordinated series of muscle contractions elicited by an intense, sudden stimulus (often containing high frequencies) undoubtedly has survival value. ASRs vary from a "freezing" response, in which the animal does not move, to an escape movement, and it may involve jumping, running away, or flying away. In particular, ultrasound-mediated startle responses and/or evasive behavior are elicited from stimulation by bat-like ultrasound. Much of this material has been reviewed elsewhere (e.g., Roeder 1967; Miller 1971, 1975; Michelsen and Larsen 1985; Spangler 1988; Hoy 1989; Hoy, Nolen, and Brodfuehrer 1989). However, a brief, selective summary will be given for each group below, with an emphasis on evolutionary themes.

4.1 Hearing and Startle in Moths

Moths have the simplest of all scolopophorous hearing organs, ranging from a single auditory receptor in each ear of notodontid moths (Surlykke 1984), to two receptors in noctuids to four in geometids (Roeder 1967). In general, moth tympanal organs are tuned to ultrasonic frequencies ranging from 20 to 50 kHz, although some species are sensitive to over 200 kHz (Fenton and Fullard 1983; Fullard 1987).

While moth hearing appears to be mostly directed at the biosonar signals of insectivorous bats, a few moth species use hearing to detect conspecific reproductive signals. Most available evidence indicates that these signals are used to mediate short-range communication, but Gwynne and Edwards (1986) cite an example of a moth, *Syntonarchia iriastris*, that produces very loud emissions of ultrasound for the putative purpose of long-range mate-calling, such as occurs in the stridulating orthoptera. Presumably, these moths have anti-bat evasion responses as well, and sorting out prospective mates from predatory bats may be done on the basis of other cues that are dependent on the behavioral context of interaction, as well as the acoustic properties of the signals.

The typical evasive response exhibited by flying moths that are stimulated by actual or simulated trains of bat biosonar signals is a directional steering response away from the ultrasound source (Roeder 1967), which would suffice to take the flying moth out of detection range of the bat; moths hear bats before bats hear echoes from moths. However, when a flying moth is stimulated by a very loud, sudden burst of ultrasound, as would occur from a close encounter between predator and prey, the moth responds with a startle reaction in an unpredictable direction (Roeder 1967). This may be an unsteered power dive or looping flight, or simply the cessation of flight altogether and dropping to the ground. These evasive responses are characterized by short latencies of reaction (40 to 100 ms). In an important study, Dunning and Roeder (1965) demonstrated that under natural field conditions, the ability to hear and steer away from an approaching bat confers considerable survival value to moths compared to those that do not exhibit evasive behavior. The neural analysis of acoustic startle was set on firm foundations by Roeder's pioneering work, and has been considerably advanced in recent years (Fullard 1987).

Surlykke (1984) has studied notodontid moths that have hearing organs consisting of but a single auditory receptor, yet these insects are capable of the full complement of ultrasound-escape maneuvers. Fullard has expanded the story of coevolution between moths and bats. In Hawaiian moths, he has discovered that a state of heightened auditory sensitivity to lower frequencies to permit them to detect the social signals of the Hawaiian hoary bat (1984). Progressing from the receptor level, the central nervous processing of ultrasound has been the subject of recent work (Boyan and Fullard 1986; Boyan, Williams, and Fullard in press) in which numbers of auditory interneurons have been identified.

4.2 Hearing and Ultrasound Startle in Crickets

This has recently been reviewed (Hoy 1989; Hoy, Nolen, and Brodfuehrer 1989) and will be simply summarized now. Hearing in crickets is obviously related to reproductive behavior. The conspicuously noisy chirping of field crickets has been known since antiquity to mediate social interactions among conspecifics. Male crickets produce rivalry songs when they fight and courtship songs when they court females; solo males produce calling songs to attract females to their territories. Depending on the singing species, these songs vary in their spectral content; in general, calling and aggression songs have dominant frequencies ranging from 3 to 6 kHz, but courtship songs may have higher frequencies. In all cases, however, higher-order harmonics are produced that may extend into the ultrasound (reviews: Huber et al. 1989; Ewing 1989). It has also been well documented by the agricultural community that field crickets can undergo mass dispersal on the wing, and that this occurs at night (Ulagaraj and Walker 1973; Walker and Masaki 1989).

Thus, it should not be surprising that crickets are a potential food source for insectivorous bats (Cranbrook 1965). The question is: do crickets attempt to evade hunting bats? The first investigator to suggest this was A.V. Popov, on the basis of field observations (Popov and Shuvalov 1977). Concurrently but independently, they and our laboratory were also conducting laboratory studies of phonotaxis in tethered, flying crickets. We reported that short-latency aversive steering occurs to directional ultrasound (Moiseff, Pollack, and Hoy 1978). The results of this work suggested that crickets, like moths, use their sense of hearing to detect the ultrasonic biosonar of bats, and steer directionally away from the sound source. The

argument was based on similarity between moth and cricket in their short-latency responses and in their behavioral and neural tuning curves in the ultrasonic range (Moiseff et al. 1978; Moiseff and Hoy 1983). A similar argument has been advanced for evasive behavior in flying locusts (Robert 1989).

4.3 Hearing Organs

Crickets have a scolopophorous tympanal organ near the "knee" of the tibia of the foreleg. In *Teleogryllus oceanicus*, this ear is sensitive over a wide range of frequencies ranging from about 3 to over 70 kHz (Moiseff et al. 1978). Brian Oldfield has shown that the approximately 70 auditory receptors in the ear of *Gryllus bimaculatus* are arrayed linearly, forming an orderly tonotopic map of frequency (Zhantiev and Korsunoskaya 1978; Oldfield 1985). Thus, while the mammalian cochlea is renowned for its orderly, linear tonotopy, it appears that orthopteroid insects might have "invented" this ear-map first. The same hearing organ is sensitive to the acoustic signals that mediate both social (other crickets) and predatory (bats) behavior.

4.4 Central Nervous Mechanism

In *Teleogryllus oceanicus* an identified, second-order, auditory interneuron, T.O.-501-T-1 (formerly Int 1, Casaday and Hoy 1977, and which will be referred to herein as "501"), is likely to be a key element in ultrasound-startle (Fig. 8.3). Neural activity in this neuron has been shown to be both necessary and sufficient to *trigger* the aversive steering response to ultrasound in tethered, flying crickets (Nolen and Hoy 1984). It is important to observe that while neuron 501's activity is sufficient to initiate the steering response, it is but one member of an extensive neural system that coordinates the steering response. As we have found, the ultrasound startle-escape response involves coordinated movements of the antennae, head, legs, wings, and abdomen. This is a large number of motor neurons and the upstream set of interneurons that drive the motor neurons remains to be identified. However, it seems likely that the steering interneurons are driven by 501 in the brain, and we have thus far identified 20 brain neurons that are sensitive to ultrasound, some of which may be

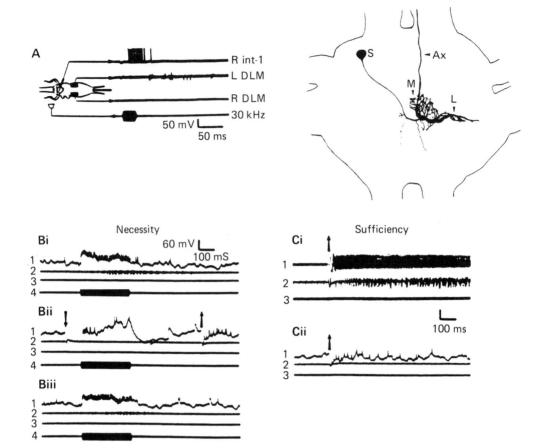

FIGURE 8.3. Ultrasound escape response and its physiological correlates. Simultaneous recordings made from the right (R) 501 (intracellular record) and both left (L) and right (R) abdominal dorsal longitudinal steering muscles (DLMs) (electromyographic records). (A) A 30 kHz, 85 dB, 30 ms duration sound pulse was played from the loudspeaker at 90 degrees to the cricket's right. The identity of cell 501 was established by lucifer yellow injection [shown at the top right of the figure, drawn from a stain in the prothoracic ganglion; abbreviations: S, cell body; Ax, axon; M, medial dendrites; L, lateral dendrites]. Stimulation by ultrasound elicited a burst of action potentials in 501, followed 50 ms later by a burst of muscle action potentials in the contralateral DLMs. (B) Activity in 501 is necessary to initiate escape during flight. Recordings as in (A), except that the microelectrode is inserted nearer the integrating segment of the cell (recording synaptic potentials) instead of nearer the axon (recording action potentials). Acoustic parameters: 30

kHz, 82 dB, 300 ms ultrasound pulse presented to the right ear. (i) Control: normal 501 excitation, as in (A) above. (ii) Depressing 501's excitability to ultrasound by injection of a -15 nA hyperpolarizing current just preceding the sound stimulus prevents activation of the DLMs. (iii) Control trial conducted immediately after the hyperpolarization trial in (ii) shows that restoring 501's excitability to ultrasound restores the steering response in the abdomen. (C) Activity in 501 is sufficient to elicit escape in the absence of ultrasound. Recording arrangements as above. (i) The right 501 was electrically excited to discharge at a spike rate exceeding 400 spike/sec, following prolonged hyperpolarizing current injection (anode break excitation). The contralateral—but not the ipsilateral—abdominal steering muscle was activated about 60 ms later. (ii) If the anode break excitation produced spike rates in 501 below about 170 spikes/sec, the steering muscles were not activated. (After Nolen and Hoy 1984.)

postsynaptic to 501 (Brodfuehrer and Hoy 1990). It appears that the circuitry for ultrasound startle-escape will be complex and involve scores of neurons. However, the fact that the motor act itself is highly stereotyped and coordinated from segment to segment, and is completed on the order of 50 to 100 ms after acoustic stimulation by a pulse of ultrasound gives us hope that we will arrive at an understanding of this behavior by the conventional techniques of single unit analysis.

4.5 Hearing and Startle in the Praying Mantis

For sheer bizarreness, no example of hearing in insects tops the story of the praying mantis. In the first place, it is "common knowledge" that the praying mantis, with its large "predatory" eyes, hunts by light of day, and is therefore, according to conventional wisdom, diurnal by habit. However, field workers who study the acoustically active nocturnal insects (crickets and katydids) often have noted a mantis nearby the loudly singing insect. In the early 1970s, two of my graduate students, Bob Paul and Bill O'Neill, together with Jon Copeland who was doing his thesis, the neuromuscular basis of the predatory strike behavior of the mantis, were convinced that the Chinese mantis (*Tenodera sinensis*) hunted crickets and katydids by "ear." To test this hypothesis, they put mantises on a specially designed Y-maze to measure the phonotactic attractiveness of cricket and katydid songs, and tested tethered, walking mantises. Unfortunately, the tethered mantises were not attracted to recorded orthopteran songs, by this measure. Ten years later, my Cornell colleague, Paul Sherman (personal communication) and his undergraduate field class, while making a census of singing katydids over a several acre field site, repeatedly found katydids and mantises in close proximity. They also speculated that mantises were hunting katydids by hearing them.

On a hot September night in 1983, I observed a cricket singing, with three mantises (*Mantis religiosa*) in close proximity, and also wondered—can mantises hear? David Yager joined the project and our study of mantis hearing began in earnest. Again, as in 1973, behavioral choice experiments in the laboratory failed to show that mantises were attracted to cricket songs. This time, however, we

followed Roeder's example. We turned to the nervous system to record its response to sound, and this turned out to be the key strategy. Recordings from the connectives of the nervous system showed no activity in response to stimuli in the 3 to 6 kHz frequencies, so dominant in cricket calls—the nervous system was "deaf" to crickets, as far as we could tell. However, an auditory response was recorded from single units to sounds of higher frequencies: particularly in the range 20 to 60 kHz, the universal bat biosonar band. This suggested that mantises (we looked at *Mantis religiosa*), which definitely fly, may be preyed upon by bats (Yager and Hoy, 1986a,b). Unfortunately, tethered, flying *M. religiosa* do not respond behaviorally at all to acoustic stimulation with ultrasound. However, another mantis that we tested (*Creobroter sp.*) also flies on a tether, and it responded to ultrasound with a stereotyped startle response that in normal flight would certainly alter its flight path (Yager and Hoy 1986b). Ultrasound evasion was later confirmed convincingly by Yager and May in still another species, *Parasphendale agrionina*, in free-flight as well as in tethered-flight, and in the field (in reaction to hunting bats) (Yager et al. 1990). For the record, while our behavioral data indicate a role for hearing in detecting predaceous bats, physiological recordings from the auditory system of some species reveal a sensitivity to lower frequencies, such as 5 kHz, which is prominent in the calls of many orthopterans, including crickets and katydids (Yager and Hoy 1986b). Thus, the original hypothesis that mantis hearing serves for prey detection remains a viable one.

4.6 The Mantis Hearing Organ

The search for the location of the mantis hearing organ was frustrating until Yager found that it lies recessed in a deep groove, in the animal's midline, right between its hind (metathoracic) legs (Fig. 8.4). While like ears of other animals in that it consists of a right–left pair, the hearing apparatus of *M. religiosa* lies deep in a groove. The two tympana are closely apposed, separated by less than 150 microns. Functionally, this hearing apparatus acts as a single organ in the midline—in effect, a "cyclopean ear" (Yager and Hoy 1986a, 1987). This interpretation predicts that the ear is not sensitive to the direction of ultrasound, and this was confirmed in physiological studies of *M. religiosa* (Yager and Hoy 1989).

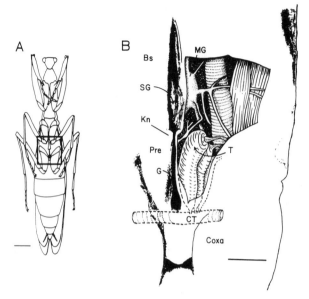

FIGURE 8.4. Location of the hearing organ in an adult female praying mantis, *Mantis religiosa*. (A) Ventral view of mantis; boxed area highlights the portion of the metathorax that contains the ear and is enlarged in the drawing in B; size bar: 5 mm. (B) Drawing of the gross anatomy of the auditory organ, from a dissected specimen in which the cuticle had been removed on the left side. A deep longitudinal groove (G) lies in the midline, between the metathoracic coxae, and is demarcated anteriorly by two sclerotized cuticular knobs (Kn). A shallow groove (SG) continues anteri- orly from G. The anterior portions of the tracheal sacs (TS) were reflected slightly laterally to expose the tympanal organ. The tracheal sacs arise from a large first abdominal commissural trachea (CT). The "Y- shaped" structure just anterior to the tympanal organ is the pair of ligamentous processes that normally attach to the ventral cuticle. In this dissection, two branches of the tympanal nerve near the metathoracic ganglion (MG) are *not* shown. Other abbreviations: B1, basister- num; Pre, preepisternum. Size bar: 1 mm. (From Yager and Hoy 1987.)

4.7 Central Auditory Pathways

The sensory organ of the mantis ear appears to be unique in its terminal projection within the CNS. Anatomical investigation of the central auditory projection in several orthopteran (Woh- lers and Huber 1982: crickets; Romer, Marquart, and Hardt 1988: locusts and katydids) and lepi- dopteran (Boyan, Williams, and Fullard 1990: moths) insects indicates that the auditory nerve terminates in the anterior and lateral portions of the ring tract (after the terminology of Gregory 1974), forming a discrete auditory neuropil. The auditory axons of *M. religiosa*, on the other hand, project not within the ring tract, but out- side it, making a less discrete, more fan-like ter- minal projection, prominently near the third

dorsal commissure, DCIII (Yager and Hoy 1987). Investigation of the neural basis of hearing in mantises has just begun, but in *M. religiosa*, a pair of large interneurons (MR-501-T3) has been identified, that have response properties match- ing those of the auditory nerve, with prominent sensitivity in the 20 kHz to 40 kHz frequency band (Yager and Hoy 1989). Some of our find- ings are summarized in Figure 8.5. The response properties of MR-501-T3 to ultrasound are very similar to those of TO-501-T-1, and suggest a possible analogy in function. It is important to note that the nomenclatural designation "501" does not imply homology between these neurons; it merely signifies an interneuron with an ascend- ing axon and a contralateral cell body (after the inventors of this terminology, Robertson and

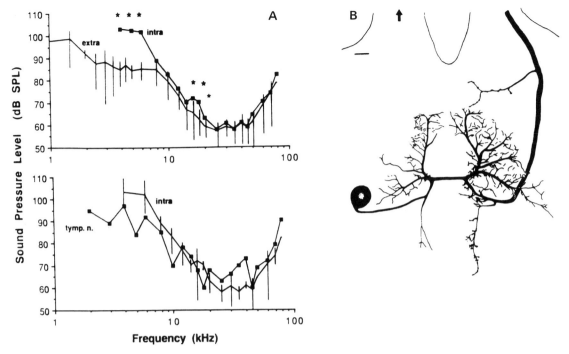

FIGURE 8.5. (A) Auditory tuning curves of MR 501. Top: Data from intracellular (n = 6) and extracellular (n = 21) recordings of MR 501. Bars indicate standard deviations and asterisks denote statistical significance. Bottom: Tuning curve of a single tympanal nerve (contain-ing the axons of auditory receptors) is plotted, and for comparison, the responses of the intracellular curve, shown above, from recordings of MR 501 (n = 6). (B) Camera lucida drawing of MR 501 in the metathoracic ganglion. (From Yager and Hoy 1989.)

Pearson 1983). In fact, these neurons are clearly not homologous, in the sense of sharing common ances-try by the criterion of developmental cell lineage. Since the question of homology has been raised, interneurons closely resembling MR-501 have been found in other orthoptera, namely locusts (where it is called the G-cell; Pearson et al. 1985) and cock-roaches (Yager and Read unpublished; Ritzmann personal communication). In both locusts and roaches, the G-cell appears to respond to sound, although not necessarily to ultrasounds. The proper-ties of the G-cell in locusts are particularly worth investigating since the recent discovery by Robert (1989) that *Locusta migratoria* has an ultrasound-startle/escape response that closely resembles that of crickets. Naturally, this leads us to wonder if there is anything "special" about interneurons with G-cell morphologies and auditory function. It should not be long before we find out. Finally, a cyclopean ear should not be directionally responsive to sound, and recordings from MR-501-T3 show

that it is also insensitive to the direction of an ultra-sound source (Yager and Hoy 1989).

5. Summary and Conclusions

I began with a general discussion of insect ears and the problem of bat predation. I then presented brief summaries of ultrasound hearing and startle behavior in three different insects, moths, crickets, and mantises. Each species has "solved" its bat problem by evolving different kinds of ears and different neural systems in the CNS. Nonetheless, certain themes recur in the solution to the bat problem.

For example, what is special about the anterior ring tract in moths, katydids, crickets, and locusts, such that it tends to be where the auditory afferents terminate to form a discrete neuropil? Are the G-neurons of locusts and mantises like 501 in *Teleogryllus oceanicus*, in that they are triggers of the ultrasound startle response?

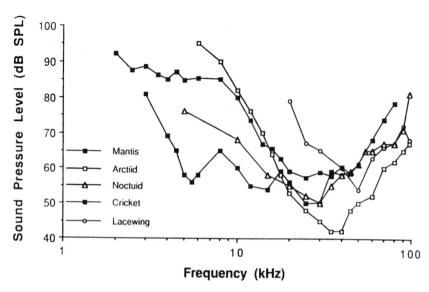

FIGURE 8.6. The tuning curves of five different nocturnal, flying insects, known to have ultrasound-sensitive hearing. Identifications: mantis: *Mantis religiosa*; arctiid moth: *Halysidota tessellaris*; noctuid moth: *Agrotis segetum*; cricket: *Teleogryllus oceanicus*; lacewing: *Chrysopa carnea*. (From Yager and Hoy 1989.)

In textbooks of animal behavior and neuroethology, the story of moth and bat is frequently cited as an example of the coevolutionary process that occurs between a predator and its prey; implicitly, this is also presented as one of those rare, "special" cases of animal behavior that when uncovered presents a fortuitous, "gem-like" insight about adaptation. I hope it is clear from this essay that there is nothing rare about bats preying upon insects, and that many insects have evolved a sense of hearing and a startle/escape response in response to these predators. To review, examples of moths, green lacewings, beetles, crickets, katydids, locusts, and praying mantises have been found to possess ultrasound-sensitive ears and have ASRs (Fig. 8.6). Why has it taken so long to be aware of the widespread occurrence of this behavior? First, in the field, bat–insect interactions take place at night, and usually high above our heads. Moreover, ultrasonic signals dominate the interaction. Thus, the interaction occurs out of both sight and sound in the human range. Secondly, even if we want to study the interaction, detecting ultrasound presents special problems, for our electronic instruments are designed for ourselves—the audio range is defined by our aural sensitivities. Specialized (and expensive) apparatus must be employed to study bat–insect interactions. However, as it

becomes ever more clear that any nocturnal flying insect is potentially prey to insectivorous bats, we can expect more investigations of this fascinating behavior, both in the field and in the laboratory. And correspondingly, we can expect an increase in the number of insects discovered to have evolved a sense of hearing to detect the biosonar signals of hunting bats.

Finally, we return to some evolutionary issues. I have presented evidence that hearing with a tympanal ear is in part an adaptation to bat predation, yet most ethologists are familiar with insect hearing mainly in the context of communication during reproductive behavior. Which came first, hearing as an adaptation for reproductive calling or for predator detection? Both Lewis (Chapter 7) and Romer (Chapter 6) in this volume argue the latter. Although my entire chapter is about ultrasound hearing as an adaptation to bat predation, I believe that many insects, notably crickets, katydids, and locusts, evolved their sense of hearing as an adaptation for conspecific communication, and that when bats appeared on the evolutionary scene, ultrasound sensitivity was built upon the prior function. As for the ultrasound hearing of moths, lacewings, beetles, and mantises, it is possible that their specialized ears evolved in response to bat predation. Lewis and Romer make their arguments (about crickets and katydids) primarily from the neuroethological performance of present-day insects, whereas my arguments (about crickets and katydids) are based on a combination of historical evidence and present-day performance. Paleoentomologists argue that crickets and katydids have been around for a long, long time, over 255 million years. Their identification is based on fossil remains, such as forewings and hindwings, which were already specialized (by comparison with modern-day descendants); it must be admitted that there is an unfortunate dearth of insect fossils in the paleological record to date. However, the specialization of forewings in modern crickets and katydids is related to their adaptation for song production. Thus, it is not unreasonable to assume that then (the Permian Period—225 million years ago) as now, songs were used to mediate reproductive functions. Bats would not appear on the evolutionary stage for another 175 million years, based on fossil evidence. However, a caveat is appropriate since the available fossil evidence indicates that the Eocene bat is already a fully recognizable microchiropteran bat; a more primitive ancestral form would presumably be found considerably earlier (Novacek 1985). Given the prominence of acoustic signals throughout the order Orthoptera, it seems likely that insects of this order modified an already sophisticated sense of hearing to accommodate the ultrasonic vocalizations of bats. However, arguments can be made for "uncoupling" song production from reception. In an abstract presented to these meetings, Andrew Mason (Abstract C) has shown that in a primitive orthopteran insect, a haglid—an evolutionary "relict" species—the best frequency of the auditory system (12 kHz) seems to be mismatched to the spectral peak of its calls (2 kHz). Mason argues that the primary hearing function in this species is predator detection. Here, the predator is not likely to be a bat, because Haglids are flightless species, and 12 kHz is below the typical biosonar range of most hunting bats. We must await further results from this fascinating study to provide interesting insights into the relation between conspecific communication and predator detection. However, let us end on the question of bats and bugs.

Bats exploited the food niche consisting of nocturnal flying insects so successfully that they became the most speciose of mammals, as well as one of the most numerous, until recent times when the coming of mankind has threatened their very existence. In their 50 million-plus years on earth, bats have exerted a formidably strong predation pressure upon insects such that hearing in the ultrasound is not unusual among nocturnal flying insects. The coevolutionary struggle between insects and the bats that prey upon them has fascinated neuroethologists for over three decades, and has become even more interesting in the past five years. It is truly a topic worthy of study in evolutionary biology.

Acknowledgments. I thank the organizers of this superb meeting in Sarasota—Art, Dick, and Doug—for a job well done. I also thank James Fullard, Jayne Yack, Andrew Mason, and Margy Nelson for sharing unpublished data and for stimulating discussions, and Dr. Fullard for his thoughtful review of the manuscript. Dan Otte generously helped by steering me to Sharov's work. This work was funded by NS-11630 and a Jacob Javits Neuroscience Investigator Award from the NIH to RRH; the mantis work was funded in part by a Hatch Award No. NYC 191-6403, and an award from the Eppley Foundation.

References

Alexander RD (1962) Evolutionary change in cricket acoustical communication. Evolution 16:443–467.

Ball EE, Young D (1974) Structure and development of the auditory system in the prothoracic leg of the cricket *Teleogryllus commodus* (Walker). II. Postembryonic development. Z Zellforsch Mikrosk Anat 147:313–324.

Belwood JJ, Morris GK (1987) Bat predation and its influence on calling behavior in neotropical katydids. Science 238:64–67.

Boyan GS, Fullard JH (1986) Interneurons responding to sound in the tobacco budworm moth *Heliothis virescens* (Noctuidae): morphological and physiological characteristics. J Comp Physiol 158:391–404.

Boyan G, Williams L, Fullard J (1990) Organization of the auditory pathway in the thoracic ganglia of noctuid moths. J Comp Neurol 295:248–267.

Brodfuehrer PD, Hoy RR (1990) Ultrasound sensitive neurons in the cricket brain. J Comp Physiol 166:651–662.

Camhi J (1984) Neuroethology. Sunderland: Sinauer Associates.

Casaday GB, Hoy RR (1977) Auditory interneurons in the cricket *Teleogryllus oceanicus*: Physiological and anatomical properties. J Comp Physiol 121:1–13.

Cranbrook, the Earl of, Barrett HF (1965) Observations on noctule bats (*Nyctalus noctula*) captured while feeding. Proc Zool Soc Lond 144:1–24.

Dethier VG (1963) The Physiology of Insect Senses. New York: John Wiley & Sons.

Dunning DC, Roeder KD (1965) Moth sounds and the insect catching behavior of bats. Science 147:173–174.

Eaton RC (1984) Neural Mechanisms of Startle Behavior. New York: Plenum Press.

Ewing AW (1989) Anthropod Bioacoustics. Ithaca, NY: Comstock Publ. Assoc.

Fenton MB, Fullard JH (1983) Moth hearing and the feeding strategies of bats. Amer Sci 69:266–275.

Fullard JH (1984) Acoustic relationships between tympanate moths and the Hawaiian hoary bat (*Lasiurus cinereus somotus*). J Comp Physiol 155:795–801.

Fullard JH (1987) Sensory ecology and neuroethology of moths and bats: interactions in a global perspective. In: Fenton MB, Racey P, Rayner JMV (eds) Recent Advances in the study of bats. Cambridge: Cambridge U. Press, pp. 244–272.

Gould SJ, Vrba ES (1982) Exaptation – a missing term in the science of form. Paleobiology 8:4–15.

Grey EG (1960) The fine structure of the insect ear. Phil Trans R Soc Lond B 243:75–94.

Gregory GE (1974) Neuroanatomy of the mesothoracic ganglion of the cockroach *Periplaneta americana* (L.). I. The roots of the peripheral nerves. Phil Trans R Soc Lond B 267:421–465.

Gwynne DF, Edwards ED (1986) Ultrasound production by genital stridulation in *Syntonarca iriastris* (Lepidoptera: Pyralidae): long-distance signalling by male moths? Zool J Linn Soc 88:363–376.

Henry CS (1980) The importance of low-frequency, substrate-borne sounds in lacewing communication (Neuroptera: Chrysopidae). Ann Ent Soc Amer 73:617–621.

Hoy RR (1989) Startle, categorical response, and attention in acoustic behavior of insects. Ann Rev Neurosci 12:355–375.

Hoy RR, Nolen TG, Brodfuehrer PD (1989) The neuroethology of acoustic startle and escape in flying insects. J Exp Biol 146:287–306.

Huber R, Markl H (1983) Neuroethology and Behavioral Physiology. Heidelberg: Springer-Verlag.

Huber F, Moore TE, Loher W (1989) Cricket Behavior and Neurobiology. Ithaca: Cornell University Press.

Kalmring K, Elsner N (1985) Acoustic and Vibrational Communication in Insects. Berlin: Springer-Verlag.

Libersat F, Hoy RR (1989) Soc Neurosci Abst 15:348.

Meier T, Reichert H (1989) Development and evolution of segmentally homologous sensory systems in the locust. Soc Neurosci Abst 15:1286.

Meier T, Reichert H Embryonic development and evolutionary origins of the orthopteran auditory organs. J Neurobio 21:592–610.

Michelsen A, Larsen ON (1985) Hearing and Sound. In: Kerkut GA, Gilbert LI (eds) Comprehensive Insect Physiology Biochemistry and Pharmacology. New York: Pergamon, pp. 495–555.

Miller LA (1970) Structure of the green lacewing tympanal organ (Chrysopa carnea). J Morphol 131:359–382.

Miller LA (1971) Physiological responses of green lacewings (Chrysopa, Neuroptera) to ultrasound. J Insect Physiol 17:491–506.

Miller LA (1975) The behavior of flying green lacewings. Chrysopa carnea, in the presence of ultrasound. J Insect Physiol 21:205–219.

Moiseff A, Hoy RR (1983) Sensitivity to ultrasound in an identified auditory interneuron in the cricket: a possible neural link to phonotactic behavior. J Comp Physiol 152:155–167.

Moiseff A, Pollack GS, Hoy RR (1978) Steering responses of flying crickets to sound and ultrasound: mate attraction and predator avoidance. Proc Natl Acad Sci USA 75:4052–4056.

Moulins M (1976) Ultrastructure of chordotonal organs. In: Mill PJ (ed) Structure and Function of Proprioceptors in the Invertebrates. London: Chapman & Hall.

Nelson MC (1980) Are subgenual organs "ears" for hissing cockroaches? Soc Neurosci Abst 6:198.5.

Nelson M, Fraser J (1979) Sound production in the cockroach, *Gromphadorhina portentosa*: evidence for communication by hissing. Behav Ecol and Sociobiol 6:305–314.

Nolen TG, Hoy RR (1984) Initiation of behavior by single neurons: the role of behavioral context. Science 226:992–994.

Novacek MJ (1985) Evidence for echolocation in the oldest known bats. Nature 315:140–141.

Oldfield BP (1985) The tuning of auditory receptors in bushcrickets. Hearing Research 17:27–35.

Oldfield BP, Kleindienst H-U, Huber F (1986) Physiology and tonotopic organization of auditory receptors in the cricket *Gryllus bimaculatus* DeGeer. J Comp Physiol 159:457–464.

Pearson KG, Boyan GS, Bastiani MJ, Goodman CS (1985). Heterogeneous properties of segmentally homologous interneurons in the ventral nerve cord of locusts. J Comp Neurol 233:133–145.

Popov AV, Shuvalov VF (1977) Phonotactic behavior of crickets. J Comp Physiol 119:111–126.

Rehbein HG (1976) Auditory neurons in the ventral cord of the locust: morphological and functional properties. J Comp Physiol 110:233–250.

Robert D (1989) The auditory behavior of flying locusts. J Exp Biol 147:279–301.

Robertson RM, Pearson KG (1983) Interneurons in the flight system of the locust: distribution, connections, and resetting properties. J Comp Neurol 215:33–50.

Roeder KD (1967) Nerve Cells and Insect Behavior. Cambridge: Harvard University Press.

Romer H (1983) Tonotopic organisation of the auditory neuropile in the bushcricket, *Tettigonia viridissima*. Nature 306:60–62.

Romer H, Marquart V, Hardt M (1988) Organization of a sensory neuropile in the auditory pathway of two groups of orthoptera. J Comp Neurol 275:201–215.

Roeder KD, Treat AE (1957) Ultrasonic reception by the tympanic organs of noctuid moths. J Exper Zool 134:127–158.

Sharov A (1971) Phylogeny of the Orthopteroidea. Jerusalem: Israel Program for Sciencitif Publications.

Slifer EH (1935) Morphology and development of the femoral chordotonal organs of *Melanoplus differentialis* (Orthoptera, Acrididae). J Morph 58:615–637.

Spangler HG (1988) Hearing in tiger beetles (Cicindelidae). Physiol Entomol 13:447–452.

Surlykke A (1984) Hearing in notodontid moths. A hearing organ with only a single auditory neurone. J Exp Biol 113:323–335.

Ulagaraj SM, Walker TJ (1973). Phonotaxis of crickets in flight: attraction of male and female crickets to male calling songs. Science 182:1278–1279.

Walker TJ, Masaki S (1989) Natural History. In: Huber F, Moore TE, Loher W (eds) Cricket Behavior and Neurobiology. Ithaca: Cornell University Press.

Wohlers DW, Huber F (1982) Processing of sound signals by six types of neurons in the prothoracic ganglion of the cricket, *Gryllus campestris* L. J Comp Physiol 146:161–173.

Yack JE, Fullard JH The mechanoreceptive origin of insect tympanal organs: a comparative study of similar nerves in tympanate and atympante moths. J Comp Neurol (in press).

Yager DD, Hoy RR (1986a) The cyclopean ear: a new sense for the praying mantis. Science 231:727–729.

Yager DD, Hoy RR (1986b) Neuroethology of audition in the praying mantis, *Creobroter gemmatus*. Soc Neurosci Abst 12:202.

Yager DD, Hoy RR (1987) The midline metathoracic ear of the praying mantis, *Mantis religiosa*. Cell Tissue Res 250:531–541.

Yager DD, Hoy RR (1989) Audition in the praying mantis, *Mantis religiosa* L.: identification of an interneuron mediating ultrasonic hearing. J Comp Physiol 165:471–493.

Yager DD, May ML, Fenton MB (1990) Ultrasound-triggered, flight-gated evasive maneuvers in the praying mantis *Parasphendale agrionina*. J Exp Biol 152:17–39.

Zhantiev RD, Korsunovskaya OS (1978) Morphological organization of tympanal organs in *Tettigonia cantans* (Orthoptera, Tettigoniidae). Zool J 57:1012–1016 (in Russian).

9
Hearing in Crustacea

Bernd U. Budelmann

1. Introduction

Crustaceans are the only invertebrates besides insects and spiders in which communication via acoustic signals is well known (e.g., Horch and Salmon 1969; Altevogt 1970; Salmon and Horch 1972; Markl 1983, for general aspects of vibrational communication; Salmon 1983; Aicher and Tautz 1990; Römer and Tautz 1991, for a recent review). Acoustic communication necessarily includes both effector structures to produce sound and sensory structures to receive sound. In the different crustacean species, these structures show a variety of morphological expressions, depending on whether the animals are aquatic (such as lobsters and shrimps) or semiterrestrial (such as ghost and fiddler crabs.)

In aquatic crustacea, the production of sound has been described only in a few species but the detection of sound is widespread and well documented (see below for more details). However, intraspecific acoustic interactions are not known for any of these species. This is in contrast to some semiterrestrial crabs which use sound as a species-specific calling signal during the breeding season or in courtship behavior (Salmon 1983).

This chapter will briefly describe the different ways crustacea produce sound in water and on land, and it will also summarize the mechanisms of sound detection, with particular reference to the various types of receptor systems involved.

2. Production of Sound

2.1 Production of Sound in Water

Only two groups of crustacea are known to make sound in water: barnacles (*Cirripedia*) and decapods. Barnacles produce a rhythmic crackling sound that consists of 1 to 3 ms pulses with peak amplitudes of 70 dB (measured at a distance of 50 cm). This sound is produced incidentally, however, presumably when the chitinous, fringed body appendages scrape on the calcareous shells during feeding (Busnel and Dziedzic 1962; Fish 1967). In decapods, stridulatory movements during which hard and often fringed body parts are scratched against each other (e.g., the antennal coxa against tubercles in front of the eye stalks on the cephalothorax) produce creaky sounds. These are known in different species of spiny lobsters (e.g., Dijkgraaf 1955; Hazlett and Winn 1962; Takemura 1971, also for references), in crayfish (Sandeman and Wilkens 1982), and in many shrimps and crabs (Balss 1921; Johnson, Everest, and Young 1947). Particularly well known are the short powerful sound pulses of the snapping shrimp *Alpheus* which are produced by a forceful closing of the chela (in addition to a strong jet of water) (Johnson, Everest, and Young 1947; Knowlton and Moulton 1963; Fish 1967; Takemura and Mizue 1968; Ritzmann 1974).

The biological significance of these sounds is unclear. Since they are often produced when the animals are disturbed or seized by a predator, they may serve to scare off potential attackers (Frings 1964; Takemura 1971).

This paper is dedicated to Professor Hermann Schöne, Seewiesen, on the occasion of his 70th birthday.

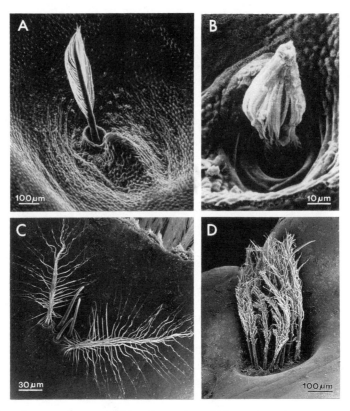

FIGURE 9.1. Scanning electron micrographs of superficial water movement receptors in crustacea. (A) One of the feathered hairs on the dorsal and lateral parts of the exoskeleton of the rock lobster *Palinurus vulgaris*. (From Vedel and Clarac 1976; published with the permission of Gordon and Breach Science Publishers, S.A.) (B) Hair-fan organ on the carapace of the lobster *Homarus gammarus*. (Courtesy of M.S. Laverack.) (C, D) Hair-pit organs on the chelae of young (C) and adult (D) crayfish *Cherax destructor*. (From Tautz and Sandeman 1980; reprinted with the permission of The Company of Biologists Ltd.)

2.2 Production of Sound on Land

Male fiddler and ghost crabs are the only crustaceans that produce sound on land. This acoustic behavior is prevalent in a semilunar rhythm and is used to attract females when males are in possession of a mating burrow. In the fiddler crab *Uca*, a substrate vibration signal is produced outside the entrance of the mating burrow by beating the ground with the large chela. Typical pulses are 30 to 50 ms at a rate of 10 to 20 per second (e.g., Salmon 1965; Salmon and Atsaides 1968; Horch and Salmon 1969; Altevogt 1970; Salmon and Horch 1972; von Hagen 1985; Tautz 1990; Aicher and Tautz 1990). In contrast, the ghost crab *Ocypode* calls from deep inside the burrow by using a stridulatory organ on the propodus of their major claw. This organ produces a species-specific temporal pattern of pulses (e.g., Horch and Salmon 1969; Horch 1971, 1975; Salmon and Horch 1972; Salmon 1983).

3. Reception of Sound

As discussed in the chapter for nonarthropod invertebrates (Budelmann, Chapter 10), definitions of sound reception, i.e., hearing, especially in an underwater habitat, are controversial. When using a narrow definition, hearing almost certainly is absent in most crustacea (perhaps with the exception of the fiddler and the ghost crabs). Conversely, when using a broad definition, then almost all crustaceans are able to hear. Because of this

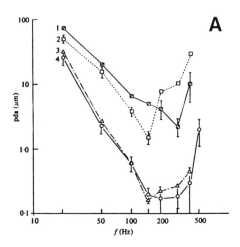
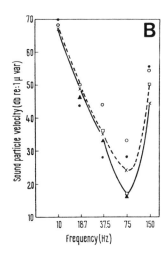

FIGURE 9.2. (A) Threshold curves from hair-pit organs on the chelae of the crayfish *Cherax destructor* (shown in Fig. 1 C,D) to water vibrations. The ordinate shows the stimulus amplitude (given in half peak-to-peak water particle displacement), and the abscissa shows stimulus frequency. Threshold curves are given for two single units of one pit organ (upper curves 1 and 2), for the summed activity from several units from one hair-pit (lower curve 3), and for the summed activity from units of several hair-pit organs (lower curve 4) (n = 15). (From Tautz and Sandeman 1980; reprinted with the permission of The Company of Biologists Ltd.) (B) Behaviorally obtained threshold curves of probably hair-fan organs of the American lobster *Homarus americanus* to water-borne vibrations. Ordinate shows particle velocity and abscissa frequency of vibration stimulus. (From Offutt 1970; reprinted with the permission of Birkhäuser Verlag.)

predicament, this chapter will briefly describe all crustacean receptor systems that may possibly be involved in some kind of sound reception.

3.1 Reception of Sound in Water

The crustacean receptor systems involved in the reception of underwater sound (i.e., of mechanical disturbance of water) can be classified into three groups: superficial receptor systems on the body surface, internal statocyst receptor systems, and chordotonal organs.

3.1.1 Superficial Receptor Systems

Superficial receptor systems for the detection of water disturbances are well known as hydrodynamic receptor systems. They are found all over the external body surface of many crustacea (for references see Laverack 1968, 1976; Bush and Laverack 1982; Budelmann 1989). Each receptor system has either a single cuticular hair ("sensillum") (e.g., a feathered hair; Fig. 9.1A), or a group of hairs (e.g., hair-fan and hair-pit organs; Fig. 9.1B–D). The hair(s) has a flexible basal joint and

is mechanically coupled to between one and four sensory cells (e.g., Schöne and Steinbrecht 1968; Ball and Cowan 1977; Vedel 1985). The hairs are between 50 and 2,000 μm in length and of varying form (Fig. 9.1; Bush and Laverack 1982; Derby 1982; Vedel 1985). The hairs can easily be bent by water movements, thereby mechanically stimulating the sensory cell(s).

The cuticular hairs are well described in decapod crustacea, and particularly in lobsters and crayfish. They occur on their carapace (Laverack 1963; Mellon 1963) and all over the body surface (Laverack 1962a,b; Offutt 1970; Vedel and Clarac 1976; Derby 1982), but they are especially abundant and well studied on the two large and small antennae (Tazaki and Ohnishi 1974; Tazaki 1977; Vedel 1985; Phillips and Macmillan 1987), on the chelae (Laverack 1963; Solon and Cobb 1980; Tautz and Sandeman 1980), and on the telson (Wiese 1976). Only few data are available so far for nondecapod crustacea (Bush and Laverack 1982, for most references; Crouau 1985, 1986).

The cuticular hairs are very sensitive to water displacements of between 0.05 and 300 Hz, depending

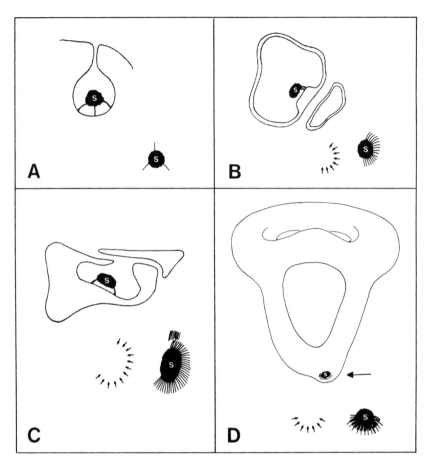

FIGURE 9.3. Morphological diversity of gravity receptor systems in crustacea. Diagrammatic transverse sections (A–C) and lateral view (D) of the statocysts (not to scale). *Small insets* show dorsal views of the statoliths (S), with the arrangement of the cuticular hairs and their direction of polarization (*small arrows*). (A) Statocyst in the telson of the isopod *Cyathura polita*. (From Budelmann 1988; based on Rose and Stokes 1981.) (B) Statocyst in the right uropod of the mysid shrimp *Prau-* *nus flexuosus*. (From Budelmann 1988; based on Neil 1975, Espeel 1985.) (C) Statocyst in the basal segment of the right antennule of the crayfish and lobster. (From Budelmann 1988; based on Schöne 1954, Takahata and Hisada 1979.) (D) Statolith organ (*large arrow*) in the vertical canal of the crab statocyst. (From Budelmann 1988; based on Sandeman and Okajima 1972). Note that in (C) and (D) the sensory hairs of the angular acceleration receptor systems of the statocyst are not shown.

upon the species. Their minimum receptor amplitude threshold is at least 0.2 μm water displacement (Fig. 9.2; Offutt 1970; Wiese 1976; Tautz and Sandeman 1980; Plummer, Tautz, and Wine 1986; Heinisch and Wiese 1987; Breithaupt and Tautz 1990; Goodall, Chapman, and Neil 1990; Wiese and Marschall 1990).

3.1.2 Statocyst Receptor Systems

Statocysts of invertebrates are primarily equilibrium receptor systems and most are gravity receptor systems only. However, statocysts of cephalopods and decapod crustacea include angular acceleration receptor systems as well (Budelmann 1988, for a review). Because of their gross morphology as linear accelerometers one should not automatically exclude, without testing, that the statocyst gravity receptor systems can additionally function as acoustic particle detectors and are thus involved in underwater hearing as well (compare Hawkins and Myrberg 1983).

In crustacea, the statocysts basically are similar in design among all species, but they vary in location between the basal segment of the antennule (in decapods) and the uropods or telson of the tail

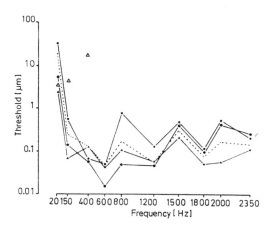

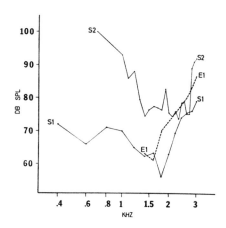

FIGURE 9.4. Three examples of threshold curves (solid lines) representing the extreme values and mean threshold curve (n = 26; dashed line) of gravity receptors in the statocyst of the crayfish *Orconectes limosus* to sinusoidal oscillations of the statocyst. Ordinate shows peak-to-peak displacement of the preparation at threshold, abscissa shows frequency of oscillation. (From Breithaupt and Tautz 1988.)

FIGURE 9.5. Electrophysiologically obtained hearing threshold curves of the ghost crab *Ocypode quadrata* to airborne sound, presented by loudspeaker (S) to whole animal, or to individual walking legs with a small earphone (E) (S1 and E1 are from the same animal). Ordinate shows sound pressure and abscissa frequency of stimulus. (From Horch 1971.)

(mysids and isopods) (Fig. 9.3). The flexible receptor structures are cuticular sensory hairs with an asymmetric basal joint that polarizes the hair in one particular direction. Shearing of the cuticular hair by the overlying statolith structure stimulates three sensory cells that are mechanically connected to the cuticular hair (Schöne and Steinbrecht 1968; Stein 1975; Takahata and Hisada 1979; Janse 1980). The number of cuticular hairs per statocyst varies from three (in the isopod *Cyathura*; Fig. 9.3A) to about 400 (in the decapod *Homarus*); they are generally arranged in two to four rows in a crescent pattern, with the direction of polarization always towards the center of the crescent (Fig. 9.3; Cohen 1955, 1960; Sandeman and Okajima 1972; Neil 1975; Rose and Stokes 1981). For the function of these organs as gravity receptor systems, see Schöne 1954, 1971; Sandeman 1976, 1983; Jansen 1980).

There is clear evidence that statocysts of decapod crustacea are sensitive to vibrations as well. In the lobster, some units of the statocyst nerve are sensitive to substrate-conducted vibrations (Cohen, Katsuki, and Bullock 1953; Cohen 1955), and in crayfish statolith hair receptors clearly respond to both sound and vibration stimuli below 200 Hz (Sugaware 1965). Breithaupt and Tautz (1988) used an isolated statocyst/brain preparation of the

crayfish *Orconectes* to measure under controlled stimulation conditions the threshold curve of statocyst units to substrate-conducted vibrations. They found the threshold of the units to be about 0.1 μm peak-to-peak displacement of the preparation, with a broad maximum from 150 to 2,350 Hz (Fig. 9.4).

3.1.3 Chordotonal Organs

Chordotonal organs are widespread among crustacea and are generally associated with joints of flexible body appendages (Bush and Laverack 1982). In water, these appendages easily follow an oscillation of the water column surrounding it, whereby they stimulate the chordotonal sensory cells. A chordotonal organ with two sets of sensory cells has been described in the basal segment of the antennal flagellum of the hermit crab *Petrochirus* (Taylor 1967) and comparable organs exist on the large and small antenna of the rock and the spiny lobster (Laverack 1964; Hartman and Austin 1972; Rossi-Durand and Vedel 1982). An extremely sensitive system that is associated with intersegmental joints of the flagellum of the first and second antenna has been described for the crayfish *Astacus* (Tautz et al. 1981; Masters et al. 1982; Bender, Gnatzy, and Tautz 1984).

3.2 Reception of Sound on Land

The acoustic signals that male fiddler and ghost crabs produce are detected by their females with Barth's myochordotonal organ located in the merus of each walking leg (Barth 1934; Horch 1971; Salmon, Horch, and Hyatt 1977). Hearing curves for two species of *Ocypode* range from 800 to 3,000 Hz, with maximum sensitivities at between 1,000 and 2,000 Hz (Fig. 9.5); the substrate vibration curve extends to 5,000 Hz (Horch 1971; Salmon 1983 for further details).

4. Concluding Remarks

Besides insects, crustacea reach the highest level in sound production and sound reception among all invertebrates. Although "hearing" is difficult to define, especially for an aquatic environment, crustacea might well be considered as being able to "hear," even when a more narrow definition for hearing is used. Since crustacea do not have gas-filled cavities associated with sensory structures, a reception of the pressure component of sound is highly unlikely, however.

Except for insects, crustacea are the only invertebrates that not only produce and receive sound, but also interact via acoustic signals. It has been assumed that the species-specific parts of the calling sounds of fiddler and ghost crabs are heritable and therefore are influenced by natural selection (Salmon 1983 for further details).

It will be of particular interest to learn from future anatomical tracing experiments to which degree the central projections of the statocyst, vibration, hydrodynamic, and chordotonal receptors are identical, overlap, or are diverse. Such data would help more clearly to separate between the various mechanical senses. To date, in crustacea only few data are available on central projections of identified receptors or neurons (Sandeman and Okajima 1973; Tautz and Tautz 1983; Yoshino, Kondoh, and Hisada 1983; Hall 1985; Roye 1986).

5. Summary

This chapter very briefly summarizes the different ways crustacea produce sound on land and in water, and describes the receptor systems by which sound can be received: by superficial receptor systems on the animal's body surface, by statocyst receptor systems, and by chordotonal organs.

Acknowledgments. The author thanks Dr. Jürgen Tautz, Würzburg, for critically reading the manuscript.

References

Aicher B, Tautz J (1990) Vibrational communication in the fiddler crab, *Uca pugilator*, I. Signal transmission through the substratum. J Comp Physiol A 166:345–353.

Altevogt R (1970) Form und Funktion der vibratorischen Signale von *Uca tangeri* und *Uca inaequalis* (Crustacea, Ocypodidae). Forma Functio 2:178–187.

Ball EE, Cowan AN (1977) Ultrastructure of the antennal sensilla of *Acetes* (Crustacea, Decapoda, Natantia, Sergestidae). Phil Trans R Soc Lond B 277:429–456.

Balss H (1921) Über Stridulationsorgane bei dekapoden Crustaceen. Naturw Wochenschr 36:697–701.

Barth G (1934) Untersuchungen über Myochordotonalorgane bei dekapoden Krebsen. Z Wiss Zool 145:576–624.

Bender M, Gnatzy W, Tautz J (1984) The antennal feathered hairs in the crayfish: a non-innervated stimulus transmitting system. J Comp Physiol A 154:45–47.

Breithaupt T, Tautz J (1988) Vibration sensitivity of the crayfish statocyst. Naturwissenschaften 75:310–312.

Breithaupt T, Tautz J (1990) The sensitivity of crayfish mechanoreceptors to hydrodynamic and acoustic stimuli. In: Wiese K, Krenz WD, Tautz J, Reichert H, Mulloney B (eds) Frontiers in Crustacean Neurobiology. Basel, Boston, Berlin: Birkhäuser Verlag, pp. 114–120.

Budelmann BU (1988) Morphological diversity of equilibrium receptor systems in aquatic invertebrates. In: Atema J, Fay RR, Popper AN, Tavolga WN (eds) Sensory Biology of Aquatic Animals. New York: Springer-Verlag, pp. 757–782.

Budelmann BU (1989) Hydrodynamic receptor systems in invertebrates. In: Coombs S, Görner P, Münz H (eds), The Mechanosensory Lateral Line: Neurobiology and Evolution. New York: Springer-Verlag, pp. 607–631.

Bush BMH, Laverack MS (1982) Mechanoreception. In: Atwood HL, Sandeman DC (eds). The Biology of Crustacea, Vol. 3. Neurobiology: Structure and Function. New York: Academic Press, pp. 399–468.

Busnel RG, Dziedzic A (1962) Rythme du bruit de fond de la mer a proximite des côtes et relations avec l'activite acoustique des populations d'un cirripede fixe immerge. Cahiers Ocean, XIVᵉ annee 5:293–322.

Cohen MJ (1955) The function of receptors in the statocyst of the lobster *Homarus americanus*. J Physiol 130:9–34.

Cohen MJ (1960) The response pattern of single receptors in the crustacean statocyst. Proc R Soc Lond B 152:30–49.

Cohen MJ, Katsuki Y, Bullock TH (1953) Oscillographic analysis of equilibrium receptors in Crustacea. Experientia 9:434–435.

Crouau Y (1985) Étude du comportement rhéotaxique d'un mysidacé cavernicole. Crustaceana 50:7–10.

Crouau Y (1986) Antennular mechanosensitivity in a cavernicolous mysid crustacean. J Crustac Biol 6: 158–165.

Derby CD (1982) Structure and function of the cuticular sensilla of the lobster *Homarus americanus*. J Crustac Biol 2:1–21.

Dijkgraaf S (1955) Lauterzeugung und Schallwahrnehmung bei der Languste (*Palinurus vulgaris*). Experientia 11:330–331.

Espeel M (1985) Fine structure of the statocyst sensilla of the mysid shrimp *Neomysis integer* (Leach, 1814) (Crustacea, Mysidacea). J Morphol 186:149–165.

Fish MP (1967) Biological source of sustained ambient sea noise. In: Tavolga WN (ed) Marine Bio-Acoustics, Vol. 2. Oxford, U.K.: Pergamon Press, pp. 175–194.

Frings H (1964) Problems and prospects in research on marine invertebrate sound production and reception. In: Tavolga WN (ed) Marine Bio-Acoustics, Oxford, U.K.: Pergamon Press, pp. 155–173.

Goodall C, Chapman C, Neil D (1990) The acoustic response threshold of the Norway lobster, *Nephrops norvegicus* (L.) in a free sound field. In: Wiese K, Krenz WD, Tautz J, Reichert H, Mulloney B (eds) Frontiers in Crustacean Neurobiology. Basel, Boston, Berlin: Birkhäuser Verlag, pp. 106–113.

Hall J (1985) Neuroanatomical and neurophysiological aspects of vibrational processing in the central nervous system of semi-terrestrial crabs. J Comp Physiol A 157:91–104.

Hartman HB, Austin WD (1972) Proprioceptor organs in the antenna of decapod crustacea. J Comp Physiol 81:187–202.

Hawkins AD, Myrberg AA (1983) Hearing and sound communication under water, In: Lewis B (ed) Bio-Acoustics, A Comparative Approach, London: Academic Press, pp. 347–405.

Hazlett BA, Winn HE (1962) Characteristics of a sound produced by the lobster *Justitia longimanus*. Ecology 43:741–742.

Heinisch P, Wiese K (1987) Sensitivity to movement and vibration of water in the north sea shrimp *Crangon crangon* L. J Crustac Biol 7:401–413.

Horch K (1971) An organ for hearing and vibration sense in the ghost crab *Ocypode*. Z Vergl Physiol 73:1–21.

Horch K (1975) Acoustic behavior of the ghost crab *Ocypode cordimana* Latreille, 1818 (Decapoda, Brachyura). Crustaceana 29:193–205.

Horch K, Salmon M (1969) Production, perception and reception of acoustic stimuli by semiterrestrial crabs (Genus *Ocypode* and *Uca*, Family Ocypodidae). Forma Functio 1:1–25.

Janse C (1980) The function of the statolith-hair and free-hook-hair receptors in the statocyst of the crab, *Scylla serrata*. J Comp Physiol 137:51–62.

Johnson MW, Everest FA, Young RW (1947) The role of snapping shrimp (*Crangon* and *Synalpheus*) in the production of underwater noise in the sea. Biol Bull 93:122–138.

Knowlton RE, Moulton JM (1963) Sound production in the snapping shrimps *Alpheus (Crangon)* and *Synalpheus*. Biol Bull 125:311–331.

Laverack MS (1962a) Response of cuticular sense organs of the lobster, *Homarus vulgaris* (Crustacea), I. Hair-peg organs as water current receptors. Comp Biochem Physiol 5:319–325.

Laverack MS (1962b) Response of cuticular sense organs of the lobster, *Homarus vulgaris* (Crustacea), II. Hair-fan organs as pressure receptors. Comp Biochem Physiol 6:137–145.

Laverack MS (1963) Response of cuticular sense organs of the lobster, *Homarus vulgaris* (Crustacea), III. Activity invoked in sense organs of the carapace. Comp Biochem Physiol 10:261–272.

Laverack MS (1964) The antennular sense organs of *Panulirus argus*. Comp Biochem Physiol 13:301–321.

Laverack MS (1968) On the receptors of marine invertebrates. Oceanogr Mar Biol Annu Rev 6:249–324.

Laverack MS (1976) External proprioceptors. In: Mill PJ (ed) Structure and Function of Proprioceptors in the Invertebrates, London: Chapman and Hall, pp. 1–63.

Markl H (1983) Vibrational communication. In: Huber F, Markl H (eds) Neuroethology and Behavioral Physiology. Berlin, Heidelberg: Springer-Verlag, pp. 332–353.

Masters WM, Aicher B, Tautz J, Markl H (1982) A new type of water vibration receptor on the crayfish antenna. J Comp Physiol A 149:409–422.

Mellon D (1963) Electrical responses from dually innervated tactile receptors on the thorax of the crayfish. J Exp Biol 40:137–148.

Neil DM (1975) The mechanism of statocyst operation in the mysid shrimp *Praunus flexuosus*. J Exp Biol 62:685–700.

Offutt GC (1970) Acoustic stimulus perception by the American Lobster *Homarus americanus* (Decapoda). Experientia 26:1276–1278.

Plummer MR, Tautz J, Wine JJ (1986) Frequency coding of waterborne vibrations by abdominal mechanosensory interneurons in the crayfish, *Procambarus clarkii*. J Comp Physiol A 158:751–764.

Phillips BF, Macmillan DL (1987) Antennal receptors in puerulus and postpuerulus stages of the rock lobster *Panulirus cygnus* (Decapoda: Palinuridae) and their potential role in puerulus navigation. J Crustac Biol 7:122–135.

Ritzmann RE (1974) Mechanisms for the snapping behavior of two alpheid shrimp, *Alpheus californiensis* and *Alpheus heterochelis*. J Comp Physiol 95:217–236.

Römer H, Tautz J (1991) Invertebrate auditory receptors. In: Ito F (ed) Comparative Aspects of Mechanoreceptor Systems, New York: Springer-Verlag (in press).

Rose RD, Stokes DR (1981) A crustacean statocyst with only three hairs: light and scanning microscopy. J Morphol 169:21–28.

Rossi-Durand C, Vedel JP (1982) Antennal proprioception in the rock lobster *Palinurus vulgaris*: Anatomy and physiology of a bi-articular chordotonal organ. J Comp Physiol A 145:505–516.

Roye DB (1986) The central distribution of movement sensitive afferent fibers from the antennular short hair sensilla of *Callinectes sapidus*. Mar Behav Physiol 12:181–196.

Salmon M (1965) Waving display and sound production in *Uca pugilator*, with comparison to *U. minax* and *U. pugnax*. Zoologica 50:123–150.

Salmon M (1983) Acoustic 'calling' by fiddler and ghost crabs. Rec Me Aust Mus 18:63–76.

Salmon M, Atsaides SP (1968) Visual and acoustical signalling during courtship by fiddler crabs (Genus *Uca*). Am Zool 8:623–639.

Salmon M, Horch K (1972) Sound production and acoustic detection by Ocypodid crabs. In: Winn HE, Olla B (eds) Recent Advances in the Behavior of Marine Organisms, Vol. 1. New York: Plenum Press, pp. 60–96.

Salmon M, Horch K, Hyatt GW (1977) Barth's myochordotonal organ as a receptor for auditory and vibrational stimuli in fiddler crabs (*Uca pugilator* and *U. minax*). Mar Behav Physiol 4:187–194.

Sandeman DC (1976) Spatial equilibrium in the arthropods. In: Mill PJ (ed) Structure and Function of Proprioceptors in the Invertebrates. London: Chapman and Hall, pp. 485–527.

Sandeman DC (1983) The balance and visual systems of the swimming crab: their morphology and interaction. Fortschr Zool 28:213–229.

Sandeman DC, Okajima A (1972) Statocyst-induced eye movements in the crab *Scylla serrata*. I. The sensory input from the statocyst. J Exp Biol 57:187–204.

Sandeman DC, Okajima A (1973) Statocyst-induced eye movements in the crab *Scylla serrata*. III The anatomical projections of sensory and motor neurons and the responses of the motor neurons. J Exp Biol 59:17–38.

Sandeman DC, Wilkens LA (1982) Sound production by abdominal stridulation in the Australian Murray-river crayfish, *Euastacus armatus*. J Exp Biol 99:469–472.

Schöne H (1954) Statocystenfunktion und statische Lageorientierung bei dekapoden Krebsen. Z Vergl Physiol 36:241–260.

Schöne H (1971) Gravity receptors and gravity orientation in Crustacea. In: Gordon SA, Cohen MJ (eds) Gravity and the Organism. Chicago: University of Chicago Press, pp. 223–235.

Schöne H, Steinbrecht RA (1968) Fine structure of statocyst receptor of *Astacus fluviatilis*, Nature 220: 184–186.

Solon MH, Cobb JS (1980) The external morphology and distribution of cuticular hair organs on the claws of the American lobster, *Homarus americanus* (Milne-Edwards). J Exp Mar Biol Ecol 48:205–215.

Stein A (1975) Attainment of positional information in the crayfish statocyst. Fortschr Zool 23:109–119.

Sugawara K (1965) Electrical responses of the statocysts and the central transmission of impulses in the crayfish. Zool Magazine 74:295–304.

Takahata M, Hisada M (1979) Functional polarization of statocyst receptors in the crayfish *Procambarus clarkii* Girard. J Comp Physiol 130:201–207.

Takemura A (1971) Studies on underwater sounds, III. On the mechanism of sound production and the underwater sounds produced by *Linuparus trigonus*. Marine Biol 9:87–91.

Takemura A, Mizue K (1968) Studies on the underwater sound, I. On the underwater sound of the genus *Alpheus fabricus* in the costal waters of Japan. Bull Fac Fish Nagasaki Univ 26:37–48.

Tautz J (1990) Coding of mechanical stimuli in crustaceana—what and why? In: Wiese K, Krenz WD, Tautz J, Reichert H, Mulloney B. (eds) Frontiers in Crustacean Neurobiology. Basel, Boston, Berlin: Birkhäuser Verlag, pp. 200–206.

Tautz J, Sandeman DC (1980) The detection of waterborne vibration by sensory hairs on the chelae of the crayfish. J Exp Biol 88:351–356.

Tautz J, Tautz JM (1983) Antennal neuropile in the brain of the crayfish: morphology of neurons. J Comp Neurol 218:415–425.

Tautz J, Masters WM, Aicher B, Markl H (1981) A new type of water vibration receptor on the crayfish antenna. I. Sensory physiology, J Comp Physiol A 144:533–541.

Taylor RC (1967) The anatomy and adequate stimulation of a chordotonal organ in the antennae of a hermit crab. Comp Biochem Physiol 20:709–717.

Tazaki K (1977) Nervous responses from mechano-sensory hairs on the antennal flagellum in the lobster, *Homarus americanus* (L.). Mar Behav Physiol 5:1–18.

Tazaki K, Ohnishi M (1974) Responses from the tactile receptors in the antenna of the spiny lobster *Panulirus japonicus*. Comp Biochem Physiol 47A: 1323–1327.

Vedel JP (1985) Cuticular mechanoreception in the antennal flagellum of the rock lobster *Palinurus vulgaris*. Comp Biochem Physiol 80A:151–158.

Vedel JP, Clarac F (1976) Hydrodynamic sensitivity by cuticular organs in the rock lobster *Palinurus vulgaris*. Morphological and physiological aspects. Mar Behav Physiol 3:235–251.

von Hagen HO (1985) Visual and acoustic display in *Uca mordox* and *U. burgersi*, sibling species of neotropical fiddler crabs, II. Vibration signals. Behaviour 85: 204–228.

Wiese K (1976) Mechanoreceptors for near-field water displacements in the crayfish. J Neurophysiol 39: 816–833.

Wiese K, Marschall HP (1990) Sensitivity to vibration and turbulence of water in context with schooling in antarctic krill *Euphausia superba*. In: Wiese K, Krenz WD, Tautz J, Reichert H, Mulloney B (eds). Frontiers in Crustacean Neurobiology, Basel, Boston, Berlin: Birkhäuser Verlag, pp. 121–130.

Yoshino M, Kondoh Y, Hisada M (1983) Projection of statocyst sensory neurons associated with crescent hairs in the crayfish *Procambarus clarkii* Girard. Cell Tissue Res 230:37–48.

10
Hearing in Nonarthropod Invertebrates

Bernd U. Budelmann

1. Introduction

Hearing, generally defined as the process of perceiving sound by means of sensory organs specialized for sound reception, is best known in terrestrial, i.e., air-exposed, vertebrates. But it is also well known in aquatic vertebrates (Schellart and Popper, Chapter 16), as well as in many arthropods (e.g., Römer and Tautz 1991; Michelsen, Chapter 5; Hoy, Chapter 8; Lewis, Chapter 7; Römer, Chapter 6). Whether hearing is present in nonarthropod invertebrates, however, is still controversial and difficult to decide. Basically the problem lies with the definition of *underwater sound* and of *underwater hearing* (compare Markl 1973, 1978, 1983; Hawkins and Myrberg 1983; Kalmijn 1988; Webster, Chapter 37).

If using a narrow definition of hearing, such as "hearing is the detection of far-field sound, or pressure waves (van Bergeijk 1964)," or "hearing is the response of an animal to sound vibrations by means of a special organ for which such vibrations are the most effective stimulus (Wever 1974)," or "hearing is the sense modality concerned with the perception of sound entering the ears at low and moderate intensities, as distinct from the perception of touch, vibration, and pain (Webster, Chapter 37)," then the ability to hear almost certainly is absent in all nonarthoropod invertebrates, or at least hearing is not developed in an "ear-like" sensory organ that is specialized for hearing. Consequently, in a number of reviews there are no data at all on hearing in invertebrate species other than arthropods (e.g., Autrum 1963; Markl 1972; Schwartzkopff 1977; Myrberg 1978). Conversely,

when using a broad definition of hearing, such as "hearing is the reception of vibratory stimuli of any kind and nature, provided that the sound source is not in contact with the animal's body (Pumphrey 1950)" or "an animal hears when it behaves as if it has located a sound source (= mechanical disturbance not in contact with the animal) (Pumphrey 1950)," then hearing is widespread among nonarthropod invertebrates and includes a variety of receptor systems for the detection of local water movements (compare Frings 1964; van Bergeijk 1967; Frings and Frings 1967; Hawkins and Myrberg 1983).

In any discussion of hearing of nonarthropod invertebrates three facts are important to realize: (1) the vast majority of these animals are aquatic, i.e., are rarely exposed to air, (2) they have no air-filled cavities associated with sensory structures that are capable of detecting the pressure component of sound, and (3) with few exceptions (e.g., some coelenterates and ctenophores), nonarthropod invertebrates are permanently, or at least sporadically, in contact with a substrate that does not follow exactly, or at all, the movement of the surrounding water. Therefore, during underwater sound exposure there is a relative movement between the surface of these animals and the oscillating water column; this situation is different from free-swimming neutrally buoyant animals, such as most fishes. Furthermore, in the aquatic medium there is no sharp borderline between sound, vibration, and water flow (Hawkins and Myrberg 1983). Consequently, in nonarthropod invertebrates it is almost impossible to distinguish between behavioral reactions that are based on the

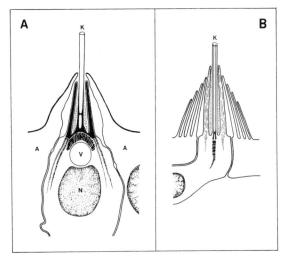

FIGURE 10.1. Presumed hydrodynamic receptor cells in coelenterates. (A) Receptor cell on the tentacle of the hydroid polyp *Coryne pintneri*. (Redrawn from Tardent and Schmid 1972.) (B) Distal part ("ciliary cone") of the receptor cell on the tentacle of anthozoan coelenterates. (From Budelmann 1989; based on data from Lyons 1973; Peteya 1975.) A, accessory cell; K (modified) kinocilium; N, nucleus; V, vacuole.

reception of sound, on the reception of waterborne and substrate-borne vibrations, and on the reception of local water movements.

Because of the problem to define "hearing" and because of a basic lack of precise physiological and behavioral data, especially in combination with morphology, this chapter cannot offer an overall concept of hearing of nonarthropod invertebrates. Focusing on those systems for which at least some morphological as well as some physiological or behavioral data are known, it will, instead, review those receptor systems that may possibly be involved in some kind of sound reception. Future physiological and behavioral experiments, together with a generally accepted definition of hearing, have to be used to decide which of the receptor systems may be regarded as "true" sound receptor systems.

2. Possible Receptor Systems for Hearing

There are at least two types of possible receptors systems for hearing in nonarthropod invertebrates: (1) superficial receptor systems on the animal's

body surface, and (2) statocyst receptor systems, most of which are internal (see also Frings and Frings 1967). This chapter will briefly describe these two types of receptor systems in the various nonarthropod invertebrate groups; more details are given in two recent reviews (Budelmann 1988, 1989).

2.1 Superficial Receptor Systems

The superficial receptor systems, in most cases, are epidermal detector systems for vibration and other local water movements. They are known as "hydrodynamic receptor systems" and can be compared in structure and function with the fish and amphibian lateral lines (compare Coombs, Janssen, and Montgomery, Chapter 15). Their receptor cells are epidermal sensory cells that carry between 1 and 7, or up to 100 kinocilia. These kinocilia can be mechanically deflected by local water movements that occur relative to the animal's body surface and thus relative to the receptor cells. If only one or few kinocilia are present in a single cell, they are often surrounded by a collar of between 6 and 10 stereovilli. In some cases the cilia are embedded in an accessory cupula structure (Budelmann 1989, for further details).

Protozoans

Possible superficial receptors for some kind of hearing are not yet known in protozoans, although at least some species are able to detect water disturbances and respond to vibrations (Schaeffer 1916). A positive vibrotaxis has been described in *Amoeba*, with the most effective stimulus frequency at 50 Hz. This frequency is identical to the frequency of the ciliary beat of their prey *Glaucoma* (Kolle-Kralik and Ruff 1967).

Coelenterates

Sensitivity to low-frequency water oscillations is well known in some coelenterates. For example, the hydroid polyp *Syncoryne* bends toward an object oscillating at frequencies below 5 Hz in its vicinity (Josephson 1961). The hydromedusa *Eutonia* is also sensitive to low-frequency oscillations (Horridge 1966) and the sea anemone *Sagartia* reacts differently to slow and strong water currents (Frings and Frings 1967). The sensory structures

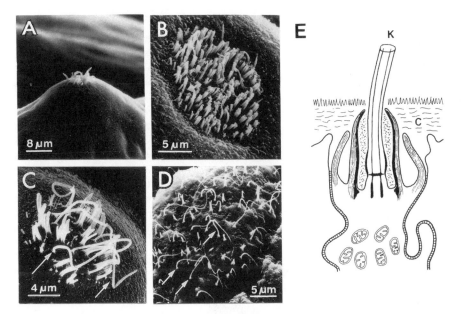

FIGURE 10.2. Presumed hydrodynamic receptor systems of leeches. (A–D) Scanning electron micrographs of the sensory buds ("segmental sensilla") on the body surface. The long kinocilia (arrows in C and D) belong to the presumed hydrodynamic receptor cells, the group of short kinocilia to presumed chemoreceptor cells. (A–C) *Erpobdella montezuma*. (From Blinn, Wagner, and Grim 1986; reprinted with the permission of the American Microscopical Society.) (D) *Hirudo medicinalis*. (From Derosa and Friesen 1981.) (E) Ciliary region of the receptor cell shown in (D) of *Hirudo medicinalis*. (From Budelmann 1989; redrawn from Phillips and Friesen 1982.) C, cuticle; K, kinocilium.

involved in these reactions are still unknown. Presumably they are hair cells similar to those described in the tentacle ectoderm of other coelenterates, with a single kinocilium surrounded by a collar of several shorter stereovilli of the same or of neighboring cells (Fig. 10.1; Budelmann 1989, for references).

Ctenophores

Sensitivity to water oscillations is known in two ctenophores, the carnivorous comb jelly *Leucothea* and the sea walnut *Pleurobrachia*. The former shoots out finger-like projections toward a nearby source of low-frequency (10 Hz) oscillation, such as a copepod (Horridge 1965). In contrast, *Pleurobrachia* contracts its tentacles when water disturbances come near (Horridge 1966). The receptor cells in *Leucothea* are hair cells with a single kinocilium and a specialized basal body (Horridge 1965; Hernandez-Nicaise 1974).

Flatworms

Many sensory cells, each with a single kinocilium, have been described in flatworms that may sense local water movements (Budelmann 1989, for references). However, no physiological data exist on the function of these cells. The only behavioral evidence is for the trematode *Nicolla*, where their cercariae are sensitive to small water currents (Pariselle and Matricon-Gondran 1985).

Annelids

Sensitivity to water movements and vibration is well documented in annelids. Leeches, for example, exhibit a positive taxis when stimulated by water currents (Friesen 1981). The most likely receptor organs are the "segmental sensilla" (Fig. 10.2) which are disk-like sensory buds that occur all over the body surface. These organs contain three types of ciliated epidermal cells, one of which ("S hair") projects at least 12 μm beyond the cuticle and is the most likely candidate for water movement detection (Derosa and Friesen 1981; Phillips and Friesen 1982; Blinn, Wagner, and Grim 1986). Other aquatic annelids are able to sense water movements as well. *Harmothoe* locates a prey via vibration receptors in its tentacular cirri and palps

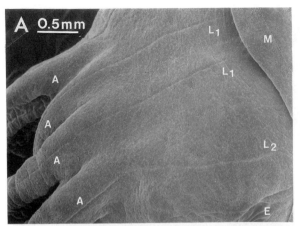

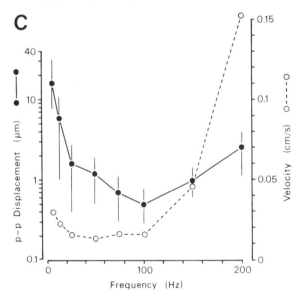

FIGURE 10.3. Lateral line analogue system of the cuttlefish *Sepia officinalis*. (A, B) Scanning electron micrographs of (A) the lines on the head of a young animal, two weeks after hatching, and (B) the ciliary groups of the receptor cells of line L4 running below the eye. (C) Threshold of epidermal lines to local sinusoidal water movements. Peak-to-peak water displacement (closed circles, solid line; left ordinate) and velocity (open circles, dashed line; right ordinate), as a function of stimulus frequency (abscissa). Average of six recordings of five animals. [(A) and (C) from Budelmann and Bleckmann 1988.] A, base of arm; E, eye; L1, L2, two of the four epidermal lines that run on each side of the head; M, anterior edge of mantle.

(Daly 1973) and tube worms withdraw rapidly into their tubes when their crown of tentacles is stimulated with low-frequency vibrations (Laverack 1968, for references). In annelids, there are several other descriptions of epidermal hair cells and of possible water movement receptors (Knapp and Mill 1971; Budelmann 1989, for further references). However, they are without physiological or behavioral proof.

Molluscs

In molluscs other than cephalopods there are few morphological data on superficial receptor systems that, on the basis of analogy, might be involved in hearing. This is surprising since these animals are commonly known to respond to various kinds of water movements (Frings 1964;

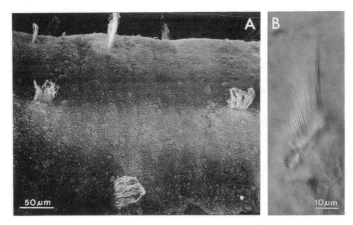

FIGURE 10.4. Hydrodynamic receptor cells of the chaetognath arrowworm *Spadella cephaloptera*. (A) Scanning electron micrograph showing the orientation of the ciliary bundles ("ciliary fences") on the body surface. (B) Oblique view of a single "ciliary fence" (Nomarski interference contrast.) (From Bone and Pulsford 1978; reprinted with the permission of Cambridge University Press.)

Frings and Frings 1967; Creutzberg 1975; Budelmann 1989, for references).

In cephalopods, the ability to sense local water movements has been known for a long time (Baglioni 1910). Whether cephalopods actually can hear, however, has recently become a subject of intense discussion (Moynihan 1985; Reid, Eckert, and Muma 1986; Taylor 1986; Hanlon and Budelmann 1987; Packard, Karlsen, and Sand 1990; Bullock and Budelmann 1991). Parts of cephalopod statocysts are likely receptor candidates for hearing. However, cephalopods also have superficial receptor systems that are analogous to the amphibian and fish lateral lines and that are able to sense local water movements. These receptor systems are formed of eight (cuttlefish) to 10 (squid) "epidermal lines" of ciliated sensory cells that run parallel in longitudinal direction over the head and onto the arms (Fig. 10.3A). Each sensory cell in the line carries a group of kinocilia up to 12 μm in length (Sundermann-Meister 1978; Arnold and Williams-Arnold 1980; Sundermann 1983; Fig. 10.3B). It has been recently shown that these lines are highly sensitive to local water oscillations in the range of 0.5 to 400 Hz, with a threshold at 100 Hz as low as 0.06 μm peak-to-peak water displacement at the receptor cells (Fig. 10.3C). The lines are especially sensitive to frequency and/or amplitude modulations of a given stimulus, with a frequency modulation of 1% being sufficient to elicit a neural response (Budelmann and Bleckmann 1988; Bleckmann, Budel-

mann, and Bullock 1991). In addition to the ciliated cells of the epidermal head and arm lines, there are others with shorter cilia that occur scattered all over the animal's body surface. These may also be involved in the detection of water movements (Sundermann-Meister 1978; Arnold and Williams-Arnold 1980; Hulet 1982).

Chaetognathes

Sensitivity to water oscillations is well known in chaetognathe arrowworms. These are predators of marine plankton that wait motionless in the water until prey, such as a copepod or a small fish, or another source of vibration comes near. Their attack occurs best in response to vibration frequencies of 12, 30, or 150 Hz, depending on the chaetognathe species. It seems possible that chaetognathes even select specific prey by the vibration rate the prey produce (Horridge and Boulton 1967; Newbury 1972; Feigenbaum and Reeve 1977). The receptor systems responsible for this behavior are "ciliary fences" on the animal's body surface (Fig. 10.4). Each fence consists of 75 to 100 stiff kinocilia of up to 100 μm in length. Although each kinocilium of a fence belongs to a separate cell, they are all polarized in the same direction, i.e., at right angles to the long axis of the fence. Since all fences are arranged either parallel or across the long axis of the animal (Fig. 10.4), it seems possible that together they are able to detect the

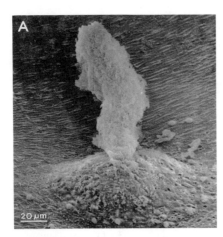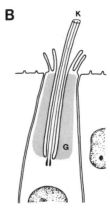

FIGURE 10.5. Hydrodynamic receptor system of the urochordate *Ciona intestinalis*. (A) Cupula organ in the exhalent siphon. (From Bone and Ryan 1978; reprinted with the permission of the Zoological Society of London.) (B) Distal part of the ciliated receptor cell of the cupula organ. (From Budelmann 1989; redrawn from Bone and Ryan 1978.) K, kinocilium; G, granular region.

direction of water movements (Bone and Pulsford 1978; Feigenbaum 1978).

Urochordates

In this group of lower chordates, the sessile ascidians are very sensitive to water movements, even after removal of their brains. *Ciona*, for example, closes its siphons and retracts its body when stimulated by oscillating water movements between 25 and 400 Hz. The most likely organs that sense these water movements are the "cupular organs" that occur in abundance in the exhalent siphon of the animal (Fig. 10.5; Hecht 1918; Bone and Ryan 1978). Another receptor organ that may sense local water movements is the "Langerhans Receptor" on either side of the trunk of the larvacean tunicate *Oikopleura*. Its sensory cell carries a modified cilium but has no cupular structure attached to it (Bone and Ryan 1979).

Cephalochordates

The lancelet *Branchiostoma* (Amphioxus) has two different types of ciliated sensory cells on its anterior body region. One type occurs on the buccal cirri and carries a normal kinocilium of up to 15 μm in length. The other type occurs on the velar tentacles and bears a short modified cilium, twice as thick as that of type one. There is experimental evidence that both types of cells

are mechanoreceptors that may respond to local water movements (Bone 1961; Guthrie 1975; Bone and Best 1978).

2.2 Statocyst Receptor Systems

All the systems described so far are superficial receptor systems on the animal's body surface. The other and even more likely candidates for possible sound receptors are the statocyst receptor systems. With a few exceptions (e.g., coelenterate medusae and ctenophores) they are all internal and show many features in common with the vertebrate inner ear, especially the otolith organs. In fact, at the time when their function as equilibrium receptor organs was not yet established, they generally were called "otocysts," in analogy to the vertebrate systems.

With the exception of cephalopods, no data exist for nonarthropod invertebrates on sensitivity of their statocyst receptor systems to vibration or underwater sound. Statocysts, of course, are first of all equilibrium receptor organs. Based on gross morphology alone, however, it is possible to conclude that statocysts, as linear accelerometers, can also detect acoustic particle motion (since the whole animal vibrates together with the water column). Therefore statocysts might be involved in some kind of underwater hearing (compare Hawkins and Myrberg 1983).

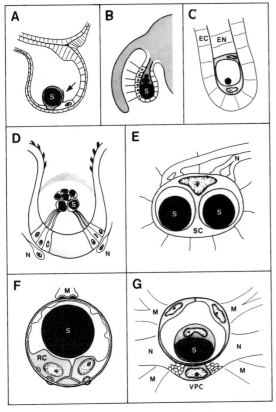

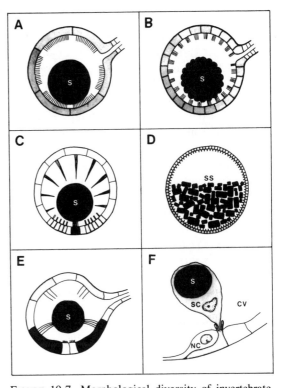

FIGURE 10.6. Morphological diversity of invertebrate statocysts. (A) Coelenterate medusa *Tiaropsis*. Arrow points to the kinocilium of the single receptor cell and the statolith carrying cell ("lithocyte"). (From Budelmann 1989; based on Singla 1975.) (B) Coelenterate medusa *Nausithoe albida*. (From Budelmann 1989; based on Hertwig and Hertwig 1878.) (C) Coelenterate polyp *Corymorpha palma*. The statocyst is formed by the two terminal entoderm cells (stippled). (From Budelmann 1989; based on Campbell 1972.) (D) "Apical organ" of the comb jelly *Pleurobrachia*. Kinocilia of four groups of hair cells carry the statoliths. (From Budelmann 1989; based on Krisch 1973; Aronova 1974.) (E) Nemertine worm *Ototyphlonemertes pallida*. The statocyst cell carries 2–6 statoliths, each enclosed in a separate cell cavity. (From Budelmann 1989; based on Brüggemann and Ehlers 1981.) (F) Turbellarian flatworm *Retronectes*. (Redrawn from Ehlers 1985.) (G) Turbellarian flatworm *Convoluta psammophila*. (From Budelmann 1989; based on Ferrero 1973.) EC, ectoderm cell; EN, entoderm cell; M, muscle, N, neuron, nerve fibre(s); RC, possible receptor cell; S, statolith; SC, statolith cell; VPC, ventral polar cell.

FIGURE 10.7. Morphological diversity of invertebrate statocysts. The diagrams show cross-sections of the statocysts in the (wherever possible) normal "upright" position of the animal, i.e., the direction of gravity is from top to bottom of page. (A) Statocysts in pulmonate and opisthobranch gastropods, with 13 large hair cells. (From Budelmann 1989; data given in Budelmann 1976.) (B) Right (lower) statocyst of the scallop bivalve *Pecten* and statocysts of other bivalves. (From Budelmann 1989; based on data from Barber and Dilly 1969; Tsirulis 1981.) (C) Prosobranch gastropod *Pterotrachea*. Receptor cell in lower part of the statocyst is strippled. (From Budelmann 1989; based on Tschachotin 1908.) (D) Prosobranch gastropod *Pomacea paludosa*. Statocyst with up to 3,000 hair cells. (From Budelmann 1989; based on data from Schmidt 1912; McClary 1968; Stahlschmidt and Wolff 1972.) (E) Larvae of opistobranch gastropod *Rostranga pulchra*. (From Budelmann 1989; based on Chia, Koss, and Bickell 1981.) (F) Tadpole larvae of ascidian tunicate *Ciona intestinalis*. (From Budelmann 1989; based on Dilly 1961; Eakin and Kuda 1971.) CV, cerebral vesicle; NC, nerve cell; S, statolith; SC, statolith cell, SS, mass of statoconia.

The statocysts of nonarthropod invertebrates range from simple gravity receptor systems (Figs. 10.6 and 10.7) to highly complex organs in cephalopods, with one or several receptor systems for linear and angular accelerations (Figs. 10.8 and 10.9). All their receptor systems, however, are composed of only two basic structural elements. The first is a mass, the statolith or statoconia, whose instantaneous position depends on the forces applied. The second are sensory elements that are mechanically affected by the position of the mass. Most of the sensory elements are hair cells that carry between 1 and 700 kinocilia which are in contact with the heavy mass. Sensory cells lacking kinocilia also occur, but in those cases the heavy mass is surrounded by, or included in, the sensory cell.

In a taxonomical order, Figures 10.6 and 10.7 summarize the morphological diversity of statocysts of nonarthropod invertebrates, except cephalopods. The statocysts show a variety of morphological expressions and levels of complexity. In coelenterates, statocysts vary in number from 1 to 300 and can best be described as external, or internal, pendulum-like projections that contain the heavy mass. The position of the pendulum is monitored by one (Fig. 10.6A) or by several hair cells (Fig. 10.6B) (Horridge 1969; Markl 1974; Singla 1975; Budelmann 1988, for further references). The pelagic ctenophores have only a single statocyst ("apical organ") of unique structure. In an ectodermal groove bundles of kinocilia of four groups of hair cells balance the statolith; the whole organ is roofed by a transparent dome (Fig. 10.6D; Krisch 1973; Aronova 1974). Statocysts lacking hair cells occur in the holdfasts of the coelenterate polyp *Corymorpha* (Fig. 10.6C; Campbell 1972), in the nemertine worm *Ototyphlonemertes* (Fig. 10.6E; Brüggemann and Ehlers 1981), and in the turbellarian flatworms *Retronectes* and *Convoluta* (Figs. 10.6F, G; Ehlers 1985; Ferrero 1973). The process of stimulus detection in these statocysts is still unclear. The differential membrane contact of the statolith and the surrounding sensory cell(s), or membrane distortions, might be important (see Budelmann 1988). The "typical" invertebrates statocyst is present in most gastropod, bivalve, and scaphopod molluscs (Fig. 10.7A–E). It is a sphere that is lined by between 10 and 3,000 hair cells, each bearing between 5 and 700 kinocilia. The cyst

contains either a single statolith or a mass of statoconia (Budelmann 1988, for references). A rather unique statocyst is present in the free-swimming larvae, but not in the sessile adult, of the ascidian tunicate *Ciona* (Fig. 10.7F). The organ is a single cell that carries a large pendulum-like projection but no ciliary structures. The neck and collar of that cell are presumably the transducer region (Dilly 1961, 1962).

For all the statocysts mentioned so far, no data exist on sensitivity of their receptors to vibration or to underwater sound. However, in octopod and decapod cephalopods sensitivity to vibration is well documented and parts of their sophisticated statocysts (with linear and angular acceleration receptor systems) are known to be involved. In octopods, the statocyst is a sphere-like sac (Fig. 10.8). It contains a single gravity receptor system, the macula plate of which is vertically oriented and has a compact attached statolith. The angular acceleration receptor system is a ridge of cells that runs along the inside of the statocyst sac in three almost orthogonal planes. It is subdivided into nine crista segments. Either a large or a small cupula is attached to each segment (Fig. 10.8B) (e.g., Young 1960; Budelmann, Sachse, and Staudigl 1987; Budelmann 1988, for further references). In decapods, such as cuttlefish and squid, the statocysts are even more complex (Fig. 10.9). Their cyst cavity is quite irregular because of many cartilaginous protrusions. Its angular acceleration receptor system is subdivided into only four segments, but again arranged in the three main planes. Its gravity receptor system is subdivided into three systems that are arranged at almost right angles to each other and thus resemble the utricle, saccule, and lagena of fish. Each system has a unique pattern of morphological polarization of its hair cells (Budelmann 1979). One of these three systems is covered by a large calcareous statolith, whereas the others are covered by statoconial layers (Stephens and Young 1982; Budelmann 1990, for references). In both the octopod and decapod statocysts, the cellular organization of the maculae and cristae are highly complex, with the maculae being composed of secondary sensory hair cells and first-order afferent neurons and the cristae being composed additionally of primary sensory hair cells (Fig. 10.8C). Both systems receive a high degree of efferent innervation (Colmers 1981; Budelmann, Sachse, and Staudigl 1987).

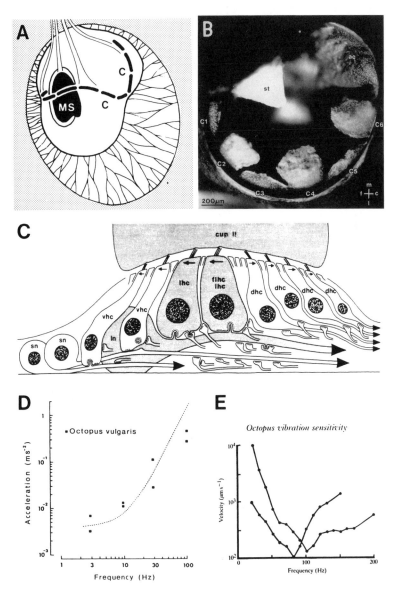

FIGURE 10.8. Statocyst of the octopod cephalopods *Octopus vulgaris* (A–C) and *Eledone cirrosa* (D). (A) Lateral view of left statocyst, with gravity receptor system (macula-statolith, MS) and nine segments of the angular acceleration receptor system (crista-cupula; C.) (From Budelmann 1989; based on Young 1960.) (B) Ventral view of cut-open statocyst sac, showing the statolith (st) of gravity receptor system and the six horizontally arranged crista segments (C1–C6) with alternating small and large cupulae attached to them. Crossbar indicates caudal (c), frontal (f), lateral (l), and medial (m). (From Budelmann, Sachse, and Staudigl 1987; reprinted with permission of The Royal Society.) (C) Cross section through a crista segment. (From Budelmann, Sachse, and Staudigl 1987; reprinted with the permission of The Royal Society.) (D) Acceleration threshold to low frequency sound in *Octopus vulgaris*, analyzed by classical conditioning (from Packard, Karlsen, and Sand 1990). (E) Threshold response curves from two afferent units of the crista to substrate vibration (from Williamson 1988; reprinted with the permission of The Company of Biologists Ltd.); ac, anticrista lobe; C, nine crista-cupula segments (= angular acceleration receptor system); C1–C6, six crista-supula segments with alternating small and large cupulae; cup, cupula; dhc, dorsal primary sensory hair cell; flhc, lhc, large secondary sensory hair cells; ln, large first-order afferent neuron; MS, macula-statolith (= gravity receptor system); sn, small first-order afferent neuron; st, statolith; vhc, ventral secondary sensory hair cells.

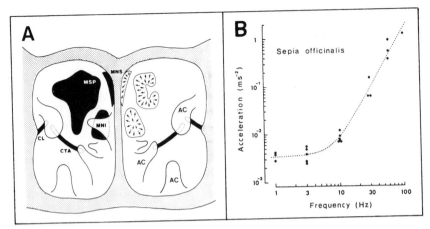

FIGURE 10.9. Statocyst of the decapod cephalopod *Sepia officinalis*. (A) Anterior parts (forward view) of the two statocysts. *Arrows* show polarization pattern of the different macula epithelia. (From Budelmann 1988.) (B) Acceleration threshold to low-frequency sound in *Sepia officinalis*, analyzed by classical conditioning. (From Packard, Karlsen, and Sand 1990.) AC, anticrista crista lobe; CL, CTA, two of the four crista segments of the angular acceleration receptor system; MNI, MNS, MSP, gravity receptor systems.

Although in octopods and decapods the gravity receptor systems are known to be responsible for compensatory equilibrium reactions (e.g., Budelmann 1970, 1990), it is tempting to speculate that, in analogy to the fish systems, parts of the three decapod statolith systems are involved in some kind of hearing as well. In contrast to *Nautilus* with its hard external shell, cuttlefishes are the only nonarthropod invertebrates with gas-filled cavities that can be compressed by the pressure waves of underwater sound. To date, however, no sensory structures are known to be associated with these cavities, nor have any structures been described that could transmit a compression of the gas cavities onto the statolith or statoconia systems of the statocysts. Thus, there is no evidence that the gas-filled cavities—besides their function as flotation device—are used for hearing, in a way similar to the swim bladder of fishes.

In cephalopods, behavioral evidence for some kind of underwater sound reception has been available for many years (Baglioni 1909, 1910). For example, blind *Octopus* are able to locate the direction of a source of vibration, caused by tapping the side of the tank, from a distance of 0.5 m (Wells and Wells 1956). Squids also show clear reactions to this kind of vibrational stimuli (Hanlon and Budelmann 1987). Cuttlefish respond to sound stimuli of 180 Hz given at a distance of 15 to 20 cm (Dijkgraaf 1963), and the squid *Todarodes* might be attracted by pure tones of 600 Hz (Maniwa 1976). Further experiments are necessary to decide whether these data imply selective responses.

Best evidence for low-frequency sound detection is for octopus, cuttlefish, and squid (Karlsen, Packard, and Sand 1989; Packard, Karlsen, and Sand (1990). Classical conditioning in a standing-wave acoustic tube showed that cephalopods respond to particle motion rather than to the pressure of sound and that they can be trained to stimuli below 100 Hz, with best results in the range of 1 to 3 Hz (Figs. 10.8D, 10.9B).

In cephalopods, it is still unclear whether the behavioral sensitivity to vibration is a property of the statocyst receptor systems themselves, or of the lateral line analogue system (or even of another, so far unknown, receptor system). Young (1960) reported that *Octopus* clearly responds to a tap on the laboratory tank even after removal of the statocysts, and Hubbard's (1960) negative behavioral data on hearing were equivocal, presumably because of an inappropriate training paradigm. Unfortunately Karlsen, Packard, and Sand (1989) and Packard, Karlsen, and Sand (1990) did not experimentally destroy the statocysts and so did not clarify this point.

In electrophysiological experiments, sensitivity to substrate vibration (directly applied to the preparation) could clearly be attributed to the crista/cupula system of the octopus statocyst (Maturana and Sperling 1963). When applying vibrations of different frequencies, the crista hair cells respond in synchrony with vibrations in the range of 20 to 100 Hz at 60 dB (see Budelmann 1976), and it has recently been shown that the highest sensitivity occurs at frequencies of 75 to 100 Hz (Fig. 10.8E; Williamson 1988). Preliminary data show that the macula hair cells are sensitive to vibration as well (see Budelmann 1976).

A new preparation for brain physiology in cuttlefish has recently been developed by Bullock and Budelmann (1991). Recordings of evoked potentials from mostly supraesophageal areas of the brain of unanesthetized and unrestrained cuttlefish to 5 ms clicks and high-intensity 300 Hz tone bursts were negative so far. This is not necessarily a contradiction to all the behavioral results, since, for technical reasons, the authors were unable to deliver well-controlled sound stimuli below 200 Hz and did not yet record from the more likely afferent subesophageal "sound" projection areas of the brain.

3. Evolutionary Aspects and Conclusions

It has been the intent of this chapter to show that in nonarthropod invertebrates "typical" hearing organs, i.e. receptor system for which sound is the most effective stimulus, do not exist. This does not exclude, however, that many animals of this group are able to detect some kind of particle motion in their aquatic environment. Whether this ability should be called "hearing" is basically a matter of definition. It will be interesting to learn from future experiments whether distinct brain pathways or projection areas exist for acceleration and water movements in those invertebrates that are able to sense both stimuli. If so, this would be a strong argument to carefully distinguish between them in any further discussion of hearing in (nonarthropod) invertebrates or of any related mechanoreception (see also Webster, Chapter 37).

There is no doubt that (nonarthropod) invertebrates have evolved independently several times during the course of evolution (Markl 1974). Consequently, any attempt to draw a direct evolution-

ary line of sound receptors or of hearing from any of the invertebrate groups, or from protozoans via hemi- and urochordates, to vertebrates must certainly fail. Also, we should not expect to find a direct *homolog* precursor of the vertebrate hearing organ, or any other sense organ, in any of the invertebrate groups. With respect to hearing organs, invertebrate statocysts or hydrodynamic receptor systems are certainly not homologous to the vertebrate, or even arthropod, ear. At best, we may find *analog* receptor systems that serve similar function in both vertebrates and invertebrates, but have evolved independently during the course of evolution.

In this context, it is interesting to note that in urochordates, which are close but specialized relatives of the vertebrate species, the statocyst design (Fig. 10.7F) is very different from the design of the vertebrate inner ears, as well as different from all other invertebrate statocyst designs (Figs. 10.6–10.9). Most importantly, their statocysts lack hair cells as sensory receptors. Conversely, the statocysts that show the highest degree of analogy with the vertebrate inner ear are those of the cephalopod molluscs (Figs. 10.8 and 10.9), an animal group that, from the evolutionary point of view, is much further away from vertebrates than are urochordates. Since statocysts might well be involved in some kind of underwater vibration reception, future physiological experiments on statocyst function should—whenever possible—include experiments not only on gravity reception but on vibration sensitivity as well.

Clearly, morphological diversity in (nonarthropod) invertebrates, including the variety of expressions and level of complexity of their sense organs, must be understood as an evolutionary response of this heterogeneous polyphyletic animal group to the different demands of their life-styles and the environments in which they live, rather than as an expression of their evolutionary relationship.

It is also important to realize that in nonarthropod invertebrates the production of sound is rare. It is known in only a few mollusc species, the bivalve *Mytilus*, which makes some kind of crackling sound with its shells (Fish 1964), and in some squids, which sometimes emit a popping sound with peak intensities of approximately 1,500 Hz during jet propulsion (Nishimura 1961; Iversen, Perkins, and Dionne 1963). In both cases the sound is obviously produced incidentally and is not used in an intra-specific communication system. This fact might be used as an argument that nonarthropod invertebrates do not necessarily need to hear. However, the most basic function of the ear is not to analyze but to localize sound, an often vital function in predator avoidance and prey detection.

4. Summary

This chapter briefly summarizes those mechanosensory organs in nonarthropod invertebrates that might possibly be involved in some kind of hearing: (1) superficial receptor systems on the animal's body surface, and (2) internal statocyst receptor systems. Because of problems of definition of *underwater* sound and *underwater* hearing, this chapter does not offer a "concept of hearing" of nonarthropod invertebrates. Since none of them have gas filled cavities associated with sensory structures, the reception of at least the pressure component of sound is highly unlikely.

Acknowledgments. Part of the author's work mentioned in the text was supported by grants from the Deutsche Forschungsgemeinschaft (Bu 404/1-3 and SFB 4/A2) and the National Institutes of Health (EY08312-01). The author thanks Dr. Jürgen Tautz, Würzburg, for critically reading the manuscript.

References

Arnold JM, Williams-Arnold LD (1980) Development of the ciliature pattern on the embryo of the squid *Loligo pealei*: A scanning electron microscope study. Biol Bull 159:102–116.

Aronova M (1974) Electron microscopic observations on the aboral organ of ctenophora. Z Mikrosk Anat Forsch 88:401–412.

Autrum H (1963) Anatomy and physiology of sound receptors in invertebrates. In: Busnel RG (ed) Acoustic Behaviour of Animals. Amsterdam, London, New York: Elsevier, pp. 412–433.

Baglioni S (1909) Zur Physiologie des Geruchsinnes und des Tastsinnes der Seetiere. Versuche an *Octopus* und einigen Fischen. Zentralbl Physiol 22:719–723.

Baglioni S (1910) Zur Kenntnis der Leistungen einiger Sinnesorgane (Gesichtssinn, Tastsinn und Geruchs-

sinn) und des Zentralnervensystems der Zephalopoden und Fische. Z Biol 53:255–286.

Barber VC, Dilly PN (1969) Some aspects of the fine structure of the statocysts of the molluscs *Pecten* and *Pterotrachea*. Z Zellforsch Mikrosk Anat 94:462–478.

Bleckmann H, Budelmann BU, Bullock TH (1991) Peripheral and central nervous responses evoked by small water movements in a cephalopod. J Comp Physiol A 168:247–257.

Blinn DW, Wagner VT, Grim JN (1986) Surface sensilla on the predaceous fresh-water leech *Erpobdella montezuma*: Possible importance in feeding. Trans Am Microsc Soc 105:21–30.

Bone Q (1961) The organization of the atrial nervous system of amphioxus (*Branchiostoma lanceolatum* (Pallas). Phil Trans R Soc Lond B 243:241–269.

Bone Q, Best ACG (1978) Ciliated sensory cells in amphioxus (*Branchiostoma*). J Mar Biol Assoc U.K. 58:479–486.

Bone Q, Pulsford A (1978) The arrangement of the ciliated sensory cells in *Spadella* (Chaetognatha). J Mar Biol Assoc U.K. 58:565–570.

Bone Q, Ryan KP (1978) Cupular sense organs in *Ciona* (Tunicata: Ascidiacea). J Zool Lond 186:417–429.

Bone Q, Ryan KP (1979) The Langerhans receptor of *Oikopleura* (Tunicata, Larvacea). J Mar Biol Assoc U.K. 59:69–75.

Brüggemann J, Ehlers U (1981) Ultrastruktur der Statocyste von *Ototyphlonemertes pallida* (Keferstein 1862) (Nemertini). Zoomorphology 97:75–87.

Budelmann BU (1970) Die Arbeitsweise der Statolithenorgane von *Octopus vulgaris*. Z Vergl Physiol 70:278–312.

Budelmann BU (1975) Gravity receptor function in cephalopods with particular reference to *Sepia officinalis*. Fortschr Zool 23:84–96.

Budelmann BU (1976) Equilibrium receptor systems in molluscs. In: Mill PJ (ed) Structure and Function of Proprioceptors in the Invertebrates. London: Chapman and Hall, pp. 529–566.

Budelmann BU (1979) Hair cell polarization in the gravity receptor systems of the statocysts of the cephalopods *Sepia officinalis* and *Loligo vulgaris*. Brain Res 160:261–270.

Budelmann BU (1988) Morphological diversity of equilibrium receptor systems in aquatic invertebrates. In: Atema J, Fay RR, Popper AN, Tavolga WN (eds) Sensory Biology of Aquatic Animals. New York: Springer-Verlag, pp. 757–782.

Budelmann BU (1989) Hydrodynamic receptor systems in invertebrates. In: Coombs S, Görner P, Münz H (eds) The Mechanosensory Lateral Line: Neurobiology and Evolution. New York: Springer-Verlag, pp. 607–631.

Budelmann BU (1990) The statocyst of squid. In: Gilbert DL, Adelman WJ, Arnold JM (eds) Squid as Experimental Animals. New York and London: Plenum Press, pp. 421–439.

Budelmann BU, Bleckmann H (1988) A lateral line analogue in cephalopods: Water waves generate microphonic potentials in the epidermal head lines of *Sepia officinalis* and *Lolliguncula brevis*. J Comp Physiol A 164:1–5.

Budelmann BU, Young JZ (1984) The statocyst-oculomotor system of *Octopus vulgaris*: extraocular eye muscles, eye muscle nerves, statocyst nerves and the oculomotor centre in the central nervous system. Phil Trans R Soc Lond B 306:159–189.

Budelmann BU, Sachse M, Staudigl M (1987) The angular acceleration receptor system of the statocyst of *Octopus vulgaris*: morphometry, ultrastructure, and neuronal and synaptic organization. Phil Trans R Soc Lond B 315:305–343.

Bullock TH, Budelmann BU (1991) Sensory evoked potentials in unanesthetized unrestrained cuttlefish: a new preparation for brain physiology in cephalopods. J Comp Physiol A 168:141–150.

Campbell RD (1972) Statocyst lacking cilia in the coelenterate *Corymorpha palma*. Nature 238:49–51.

Chia FS, Koss R, Bickell LR (1981) Fine structural study of the statocysts in the veliger larva of the nudibranch, *Rostanga pulchra*. Cell Tissue Res 214:67–80.

Colmers WF (1981) Afferent synaptic connections between hair cells and the somata of intramacular neurons in the gravity receptor system of the statocyst of *Octopus vulgaris*. J Comp Neurol 197:385–394.

Creutzberg F (1975) Orientation in space: Animals. Invertebrates. In: Kinne O (ed) Marine Ecology, Vol. 2. Physiological Mechanisms, Part 2. London: John Wiley, pp. 555–656.

Daly JM (1973) The ability to localize a source of vibrations as a prey-capture mechanism in *Harmothoe imbricata* (Annelida, Polychaeta). Mar Behav Physiol 1:305–322.

Derosa YS, Friesen WO (1981) Morphology of leech sensilla: Observations with the scanning electron microscope. Biol Bull 160:383–393.

Dijkgraaf S (1963) Versuche über Schallwahrnehmung bei Tintenfischen. Naturwissenschaften 50:50.

Dilly N (1961) Electron microscope observations of the receptors in the sensory vesicle of the ascidian tadpole. Nature 191:786–787.

Dilly N (1962) Studies on the receptors in the cerebral vesicle of the ascidian tadpole. I. The otolith. Q J Microsc Sci 103:393–398.

Eakin RM, Kuda A (1971) Ultrastructure of sensory receptors in ascidian tadpoles. Z Zellforsch Mikrosk Anat 112:287–312.

Ehlers U (1985) Organisation der Statocyste von *Retronectes* (Catenulida, Plathelminthes). Microfauna Marina 2:7–22.

Feigenbaum DL (1978) Hair-fan patterns in the Chaetognatha. Can J Zool 56:536–546.

Feigenbaum D, Reeve MR (1977) Prey detection in the Chaetognatha: Response to a vibrating probe and experimental determination of attack distance in large aquaria. Limnol Oceanogr 22:1052–1058.

Ferrero E (1973) A fine structural analysis of the statocyst in turbellaria acoeola. Zool Scr 2:5–16.

Fish MP (1964) Biological sources of sustained ambient sea noise. In: Tavolga WN (ed) Marine Bio-Acoustics. Oxford, U.K.: Pergamon Press, pp. 175–194.

Friesen WO (1981) Physiology of water movement detection in the medicinal leech. J Exp Biol 92:255–275.

Frings H (1964) Problems and prospects in research on marine invertebrate sound production and reception. In: Tavolga WN (ed) Marine Bio-Acoustics. Oxford, U.K.: Pergamon Press, pp. 155–173.

Frings H, Frings M (1967) Underwater sound fields and behavior of marine invertebrates. In: Tavolga WN (ed) Marine Bio-Acoustics. Oxford, U.K.: Pergamon Press, pp. 261–282.

Guthrie DM (1975) The physiology and structure of the nervous system of amphioxus (the lancelet) *Branchiostoma lanceolatum* (Pallas). Symp Zool Soc Lond 36:43–80.

Hanlon RT, Budelmann BU (1987) Why cephalopods are probably not "deaf." Am Nat 129:312–317.

Hawkins AD, Myrberg AA (1983) Hearing and sound communication under water. In: Lewis B (ed) Bioacoustics, A Comparative Approach. London: Academic Press, pp. 347–405.

Hecht S (1918) The physiology of *Ascidia atra* Lesuer. II. Sensory Physiology. J Exp Zool 25:261–299.

Hernadez-Nicaise ML (1974) Ultrastructural evidence for a sensory-motor neuron in Ctenophora. Tissue & Cell 6:43–47.

Hertwig O, Hertwig R (1878) Das Nervensystem und die Sinnesorgane der Medusen. Leipzig: Vogel.

Horridge GA (1965) Non-motile sensory cilia and neuromuscular junctions in a ctenophore independent effector organ. Proc R Soc Lond B 162:333–350.

Horridge GA (1966) Some recently discovered underwater vibration receptors in invertebrates. In: Barnes H (ed) Some Contemporary Studies in Marine Science. London: Allen and Unwin, pp. 395–405.

Horridge BA (1969) Statocyst of medusae and evolution of stereocilia. Tissue & Cell 1:341–353.

Horridge GA, Boulton PS (1967) Prey detection by Chaetognatha via a vibration sense. Proc R Soc Lond B 168:413–419

Hubbard SJ (1960) Hearing and the octopus statocyst. J Exp Biol 37:845–853.

Hulet WH (1982) Commentary on the international symposium on functional morphology of cephalopods. Malacologia 23:203–208.

Iversen RTB, Perkins PJ, Dionne RD (1963) An indication of underwater sound production by squid. Nature 199:250–251.

Josephson RK (1961) The response of a hydroid to weak water-borne disturbances. J Exp Biol 38:17–27.

Kalmijn AJ (1988) Hydrodynamic and acoustic field detection. In: Atema J, Fay RR, Popper AN, Tavolga WN (eds) Sensory Biology of Aquatic Animals. New York: Springer-Verlag, pp. 83–130.

Karlsen HE, Packard A, Sand O (1989) Cephalopods are definitely not deaf. J Physiol 415:75P.

Knapp MF, Mill PJ (1971) The fine structure of ciliated sensory cells in the epidermis of the earthworm Lumbricus terrestris. Tissue & Cell 3:623–636.

Kolle-Kralik U, Ruff PW (1967) Vibrotaxis von Amoeba proteus (Pallas) im Vergleich mit der Zilienschlagfrequenz der Beutetiere. Protistologica 3:319–323.

Krisch B (1973) Über das Apikalorgan (Statocyste) der Ctenophore Pleurobrachia pileus. Z Zellforsch Mikrosk Anat 142:241–262.

Laverack ML (1968) On the receptors of marine invertebrates. Oceanogr Mar Biol Annu Rev 6:249–324.

Lyons KM (1973) Collar cells in planula and adult tentacle ectoderm of the solitary coral Balanophyllia regia (Anthozoa, Eupsammiidae). Z Zellforsch Mikrosk Anat 145:57–74.

Maniwa Y (1976) Attraction of bony fish, squid and crab by sound. In: Schuijf A, Hawkins AD (eds) Sound Reception in Fish. Amsterdam: Elsevier, pp. 271–283.

Markl H (1972) Neue Entwicklungen in der Bioakustik der wirbellosen Tiere. J Ornithol 113:91–104.

Markl H (1973) Leistungen des Vibrationssinnes bei wirbellosen Tieren. Fortschr Zool 21:100–120.

Markl H (1974) The perception of gravity and of angular acceleration in invertebrates. In: Kornhuber HH (ed) Handbook of Sensory Physiology, Vol. 6. Vestibular System, Part 1. Basic Mechanisms. Berlin: Springer-Verlag, pp. 17–74.

Markl H (1978) Adaptive radiation of mechanoreceptors. In: Ali MA (ed) Sensory Ecology. Review and Perspectives. New York and Amsterdam: Plenum Press, pp. 319–341.

Markl H (1983) Vibrational communication. In: Huber F, Markl H (eds) Neuroethology and Behavioral Physiology. Berlin, Heidelberg: Springer-Verlag, pp. 332–353.

Maturana HM, Sperling S (1963) Unidirectional response to angular acceleration recorded from the middle cristal nerve in the statocyst of Octopus vulgaris. Nature 197:815–816.

McClarey A (1968) Statoliths of the gastropod Pomacea paludosa. Trans Am Microsc Soc 87:322–328.

Moynihan M (1985) Why are caphalopods deaf? Am Nat 125:465–469.

Myrbert AA (1978) Ocean noise and the behavior of marine animals: Relationships and implications. In: Diemer FP, Vernberg FJ, Mirkes DZ (eds) Advanced Concepts in Ocean Measurements for Marine Biology. University of South Carolina Press, pp. 461–491.

Newbury TK (1972) Vibration perception by Chaetognaths. Nature 236:459–460.

Nishimura M (1961) Frequency characteristics of sea noise and fish sound. Tech Rep Fish Boat Lab Min Agr For Japan 15:111–118.

Packard A, Karlsen HE, Sand O (1990) Low frequency hearing in cephalopods. J Comp Physiol A 166:501–505.

Pariselle A, Matricon-Gondran M (1985) A new type of ciliated receptor in the cercariae of Nicolla gallica (Trematoda). Z Parasitenkd 71:353–364.

Peteya DJ (1975) The ciliary-cone sensory cell of anemones and cerianthids. Tissue & Cell 7:243–252.

Phillips CE, Friesen WO (1982) Ultrastructure of the water-movement sensilla in the medicinal leech. J Neurobiol 13:473–486.

Pumphrey RJ (1950) Hearing. Symp Soc Exp Biol 4:3–18.

Reid ML, Eckers CG, Muma KE (1986) Booming odontocetes and deaf cephalopods: Putting the cart before the horse. Am Nat 128:438–439.

Römer H, Tautz J (1991) Invertebrate auditory receptors. In: Ito F (ed) Comparative Aspects of Mechanoreceptor Systems. New York: Springer-Verlag (in press).

Schaeffer AA (1916) On the feeding habits of amoeba. J Exp Zool 20:529–584.

Schmidt W (1912) Untersuchungen über die Statocysten unserer einheimischen Schnecken. Z Med Naturwiss 48:515–562.

Schwartzkopff J (1977) Auditory communication in lower animals: Role of auditory physiology. Ann Rev Psychol 28:61–84.

Singla CL (1975) Statocysts of hydromedusae. Cell Tissue Res 158:391–407.

Stahlschmidt V, Wolff HG (1972) The fine structure of the statocyst of the prosobranch mollusc Pomacea paludosa. Z Zellforsch Mikrosk Anat 133:529–537.

Stephens PR, Young JZ (1982) The statocyst of the squid Loligo. J Zool Lond 197:241–266.

Sundermann G (1983) The fine structure of epidermal lines on arms and head of postembryonic *Sepia officinalis* and *Loligo vulgaris* (Mollusca, Cephalopoda). Cell Tissue Res 232:669–677.

Sundermann-Meister G (1978) Ein neuer Typ von Cilienzellen in der Haut von spätembryonalen und juvenilen *Loligo vulgaris* (Mollusca, Cephalopoda). Zool Jahrb Abt Anat Ontog Tiere 99:493–499.

Tardent P, Schmid V (1972) Ultrastructure of mechanoreceptors of the polyp *Coryne pintneri* (Hydrozoa, Athecata). Exp Cell Res 72:265–275.

Taylor MA (1986) Stunning whales and deaf squids. Nature 323:298–299.

Tschachotin S (1908) Die Statocysten der Heteropoden. Z Wiss Zool 90:343–422.

Tsirulis TP (1981) The ultrastructural organization of statocysts of some bivalve molluscs (*Ostrea edulis, Mytilus edulis, Anodonta cygnea*). Tsitologiya 23:631–637.

van Bergeijk WA (1964) Directional and nondirectional hearing in fish. In: Tavolga WN (ed) Marine Bio-Acoustics. Oxford, U.K.: Pergamon Press, pp. 281–299.

van Bergeijk WA (1967) The evolution of vertebrate hearing. In: Neff WD (ed) Contributions to Sensory Physiology. New York: Springer-Verlag, pp. 1–49.

Wells MJ, Wells J (1956) Tactile discrimination and the behaviour of blind *Octopus*. Pubbl Stn Zool Napoli 28:94–126.

Wever EG (1974) The evolution of vertebrate hearing. In: Keidel WD, Neff WD (eds) Handbook of Sensory Physiology, Vol 5/1. Auditory System. Berlin: Springer-Verlag, pp. 423–454.

Williamson R (1988) Vibration sensitivity in the statocyst of the northern octopus, *Eledone cirrosa*. J Exp Biol 134:451–454.

Young JZ (1960) The statocysts of *Octopus vulgaris*. Proc R Soc Lond B 152:3–29.

Abstract A

The Mechanoreceptive Origin of Insect Tympanal Organs: A Comparative Study of Homologous Nerves in Tympanate and Atympanate Moths

Jayne E. Yack and James H. Fullard

Insect tympanal organs (ears) have evolved independently in several different taxa. They occur on many different regions of the body wall and range in complexity from the simple organ of the Notodontid moth, which has only one auditory receptor to that of the cicada with more than one thousand (see Michelson and Larsen, 1985 for review). Despite such variations in distribution and complexity, all tympanal organs are alike: they are characterized by the presence of a thinned region of exoskeleton (the tympanum), exposed on one side to the external air, and having associated with its inner surface, an air sac and a chordotonal organ (CO).

Several authors have speculated that the primitive, undifferentiated form of the insect tympanal organ was a chordotonal proprioceptor, which detected the displacement of one part of the exoskeleton with respect to another. This theory is feasible for two reasons. First, insect nervous systems are conservative in their evolution: it appears that changes in neuronal architecture are few compared to changes that take place in peripheral structures, and that preexisting neural elements are modified to serve new functions. Second, both tympanal receptors and proprioceptors are mechanotransducers responding to deformations of the cuticle. It is believed that the progressive evolution of the tympanal organ may have involved the reduction of an area of cuticle to which a proprioceptor was attached, to the extreme thinness of a tympanic membrane.

In this study we have combined anatomical and electrophysiological techniques to examine the characteristics of the IIIN1b nerve (homologue to the tympanal nerve of the nuctuoid moth) in a moth representing the ancestral atympanate form (*Actias luna*, Bombicoidea: Saturniidae). We discovered a CO which appears to be homologous to the noctuoid tympanal organ. Peripherally, the organ attaches to the undifferentiated membranous region underlying the hind-wing, the same region in the noctuoid which has become specialized as the tympanal membrane. Extracellular recordings from the IIIN1b nerve indicate the presence of single units (presumed to originate in the CO) which respond phasically to the movements of the hind-wing and also to low frequency (2 kHz) sounds played at high intensities (83 to 96 dB, SPL). Also present in the recording is a large, spontaneously active cell which resembles the nonauditory "B" cell of the noctuoid tympanum. We suggest that the CO represents the evolutionary precursor to the noctuoid tympanal CO, and that it acts as a proprioceptor monitoring hind-wing movements. This simple system, consisting of only a few sensillae, is a promising one for studying the changes to the nervous system (both peripheral and central) which accompanied the evolutionary development of the insect tympanum.

Reference

Michelsen A and Larsen OH (1985) Hearing and Sound. In: Kerkut GA and Gilbert LI (eds) *Comprehensive Insect Physiology, Biochemistry, and Pharmacology.* New York: Pergamon Press, pp. 495–552.

Abstract B

Organization of the Auditory Pathway in Noctuoid Moths: Homologous Auditory Evolution in Insects

G.S. Boyan, James H. Fullard, and J.L.D. Williams

The auditory pathways of insects at both the peripheral and central nervous system (CNS) levels have been described as conservative, with homologous tracts used for information transmission. The Orthoptera (crickets, katydids, and grasshoppers) has yielded considerable information regarding the neuronal circuitry of auditory systems and these data support the idea that certain CNS pathways appear pre-adapted for auditory processing. When using the comparative method for testing evolutionary hypotheses it is necessary to examine taxa that are phylogenetically disparate. Understanding the mechanics of hearing within one taxon (e.g., Orthoptera) may tell us something about the evolutionary pathways of that specific group but will not provide insights into the general principles of auditory evolution. To demonstrate homologous auditory evolution in all tympanate insects the neural centers of phylogenetically distant taxa must be examined. Toward this end, the auditory pathway within the pterothoracic ganglion of the noctuoid moth (Lepidoptera) is described.

Noctuoids possess only two auditory receptors (A1, A2), in each ear. The axon of the A1 cell projects initially to a glomerulus located ventrally and medially in the metathoracic ganglion, where it bifurcates. One axon ascends in the ventral intermediate tract (VIT) to the brain, the other descends in VIT into abdominal neuromeres of the metathoracic ganglion. The A1 arborizes in the median ventral tract (MVT) and intermediate neuropile (Ring tract, anterior intermediate sensory neuropile) in each neuromere. The central projections of the A2 cell remain largely within the metathoracic ganglion. The axon bifurcates at the midline, and directs arborizations dorsally to the dorsal intermediate and median dorsal tracts, and ventrally, in the Ring tract where the arborizations overlap those of the A1 afferent. The afferent projections remain ipsilateral to the ear of origin.

We describe a posterior auditory association area (PAA) in the metathoracic ganglion in which the major arborizations of several previously identified interneurones overlap those of the A1 afferent, and make monosynaptic connections with it. We have also identified the projections of the A1 afferent, interneurones, and motor neurons in the segmentally equivalent anterior auditory association area (AAA) of the mesothoracic ganglion. The arborizations of higher-order interneurones lie mainly in dorsal tracts along with those of flight motor neurones. One of these interneurones (501) responds to afferent input with very short excitatory postsynaptic potentials and may function as a "noise-filter" allowing the moth to discriminate sounds in its environment.

A taxonomic comparison of the organization of the auditory pathway in various tympanate insects reveals that the VIT, MVT, dVCLII and parts of the Ring tract contain the projections of auditory afferents and first-order interneurons in moths (Lepidoptera), grasshoppers, katydids and crickets (Orthoptera), mantids (Dictyoptera) and possibly cicadas (Hemiptera). At the gross level, auditory neuropiles clearly occupy homologous regions of the CNS in different insect groups even when the ears are located on different parts of the body. Audition and phonotactic behavior therefore derive from an anatomical organization that is conserved in insects with very different evolutionary histories and present-day lifestyles.

Hearing in the Primitive Ensiferan *Cyphoderris monstrosa* (Orthoptera: Haglidae)

Andrew C. Mason and Klaus Schildberger

Anatomical and physiological studies of the auditory system of *Cyphoderris monstrosa* (Orthoptera: Haglidae) reveal important differences between these primitive animals and the modern acoustic Ensifera, despite an apparent similarity in acoustic behavior.

In *C. monstrosa*, two identical tympana are borne naked on the surface of each fore-tibia. The anatomy of the tibial organ of the foreleg is similar to the tettigoniid arrangement. The organ comprises a subgenual organ, intermediate organ, and crista acustica, with the latter two components attached to one of two nearly symmetrical tracheal vesicles underlying the tympanal membranes. Surprisingly, these three components of the tibial organ are well developed in all three legs, although only the forelegs bear tympana. Central projections of the tympanal nerve and its homologues in the meso- and metathorax are similar.

Auditory responses in *C. monstrosa* show a distinct mismatch between auditory best frequency and the carrier frequency of the calling song of the species. Male *C. monstrosa* produce a very narrow band (high Q) calling song at a frequency of 12 kHz. Auditory best frequency (BF), measured in whole-nerve responses, is 2 kHz. Tibial organs of the mesothoracic legs also respond to acoustic stimulation at low frequencies (BF = 1 kHz). Threshold intensities are similar in pro- and mesothoracic organs at their respective BFs. However, at the calling song frequency, mesothoracic thresholds are 40 to 60 dB SPL higher than prothoracic (tympanal) thresholds.

Directionality of tympanal organ responses is limited. The maximum side-to-side sensitivity difference is about 10 dB SPL at the calling song frequency. The ear is essentially nondirectional at auditory BF.

Despite a well-developed foreleg trachea, tracheal sound conduction does not appear to be important for *C. monstrosa* hearing. Directionality and tuning are unaffected by either blockage of the prothoracic spiracle, or removal of the spiracle covers.

Lucifer yellow fills of prothoracic auditory interneurons have revealed morphologies corresponding to many of the identified types from gryllid auditory systems, as well as some novel cell types. Interneuron types corresponding to ON-1, ON-2, AN-1, DN-1, and TN-1 have been identified. No AN-2 cell type has yet been found. However, a second ascending cell type, of a novel morphology has been filled, as well as two previously unidentified descending cells, and two through neurons.

Prothoracic auditory interneurons, in general, show similar tuning to the whole auditory nerve, with BFs of 2 to 4 kHz. Descending and through neurons show little response at the calling song frequency (thresholds above 70 dB SPL). Omega and ascending neurons, although still low-frequency biased, are 10 to 20 dB more sensitive at 12 kHz.

These results suggest that intraspecific communication was not the primitive function of hearing in the Ensifera, and that the neural mechanisms of auditory processing in this group evolved in another context, such as predator detection.

Abstract D
Hair Cell Sensitivity in the Cephalopod Statocyst

Roddy Williamson

The cephalopod statocyst is analogous to the vertebrate vestibular system in that it has receptor systems for the detection of linear and angular accelerations. Recent intracellular recordings from the hair cells in the latter system, the crista/cupula system, have given a measure of the sensitivity of the hair cells and how this might be varied by the animal.

The crista in colleoid cephalopods is similar to that found in the vertebrate semicircular canals in that it comprises a ridge of hair cells underlying a cupula that is deflected by endolymph flow. However, the cephalopod crista is made up of both primary hair cells, and secondary hair cells and their afferent neurones. Recordings from these cells have shown that they are physiologically polarised in opposite directions, i.e., a cupula displacement in one direction will depolarise the primary hair cells but hyperpolarise the secondary hair cells, and vice versa. The displacement/response curve of the second-order hair cells has the same sigmoidal shape found for vertebrate hair cells, but the cephalopod cells show a pronounced adaptation. The sensitivity of the squid hair cells can be measured as 0.5 mV depolarisation per degree angle of cupula displacement. This compares with figures of 3 mV per degree cilia displacement for frog saccular hair cells and 10 mV per degree for turtle basilar papillar hair cells.

Cephalopods, however, seem to have developed several ways of changing the overall sensitivity of the system. First, they can have different sizes of cupula. The octopus has nine crista segments and these have, in alternation, either a large or a small cupula. It has been proposed that this is a method of fractionating the response range in order to cope with the two different modes of locomotion, i.e. slow crawling and jet propulsion. Secondly, at least some of the second order hair cells in the crista are electrically coupled. This may increase the sensitivity of the system by improving its signal to noise ratio. If the degree of coupling was under efferent control, as for example in the vertebrate retina, this would clearly be a means of directly varying the hair cell sensitivity. Thirdly, the crista receives a massive efferent innervation. This is a dual system that can either depress or enhance the afferent output by directly depolarising or hyperpolarising the hair cells and afferent neurones.

Thus, the cephalopod statocyst is a sophisticated sense organ, with a sensitivity similar to that found in analogous vertebrate systems, but with a variety of mechanisms for altering the level of afferent output.

Section III
Aspects of Hearing Among Vertebrates

The chapters in this section, which discuss specific parameters of vertebrate hearing, deal with multiple groups of vertebrates. They point out how different vertebrate groups have evolved similar ends in signal detection and analysis, despite widely varying receptor systems.

Ted Lewis (Chapter 11) describes diverse and non-homologous sound receptors in amphibians, reptiles, and mammals. He argues that although the elements from which auditory filters are constructed differ among these groups, the receptors have converged on the same sort of frequency selectivity derived from high dynamic order filtering rather than a high degree of undamping of resonance. Because of this convergence, all the receptors retain good temporal resolution.

Efferent systems are found in almost all vertebrate hearing systems. Roberts and Meredith (Chapter 12) review these systems. Roberts and Meredith (Chapter 12) review these systems, and stress that efferent systems to the ear are just one component of a larger and more diverse *Octavolateralis Efferent System* (OES), which is a basic feature of hair cell systems. They suggest that the OES may function similarly to the efferent innervation of the retina and muscle spindles, in that it matches sense organ sensitivity to the animal's behavioral requirements.

In Chapter 13, Stebbins and Sommers argue that the perception of and response to complex, biologically relevant sounds are adaptive behaviors that, like all systems, have evolved to maintain reproductive success. They describe a research program, including field observations and laboratory experiments, designed to achieve a better understanding of an animal's perception of conspecific communication sounds.

Fay (Chapter 14) presents behavioral data from many species of vertebrates that have converged on similar sound discrimination abilities within their respective hearing ranges. He argues that mammals and birds have similar mechanisms for analyzing frequencies, and that these have probably had a stable evolutionary history. He develops a method to estimate frequency acuity in extant and extinct species based on estimates of the frequency range of hearing and on basilar membrane length.

11
Convergence of Design in Vertebrate Acoustic Sensors

Edwin R. Lewis

1. Introduction

The senses of the vertebrate inner ear are divided into two categories: senses of balance, which convey information about orientation and motion of the head; and acoustic senses, which convey information about vibrations propagated to the ear from remote sources. In the mammalian inner ear, acoustic senses are concentrated in the cochlea, and senses of balance in the vestibule (that part of the inner ear that serves as the entrance to the cochlea). This division led to lumping of vertebrate inner-ear senses into *vestibular senses* and *auditory senses*—a mixture of structural and functional nomenclature that is inappropriate for inner ears of all vertebrates other than therian mammals. In monotremes, birds, and many reptiles, for example, the lagena resides at the distal end of the cochlea or cochlear homolog—not in the vestibule (Baird 1974); and we do not know yet whether the sensitivities of the lagenae in these animals are auditory, nonauditory, or both. In fish and amphibians, there evidently is no cochlear homolog with which to identify an entrance; and exquisite acoustic sensitivity has been found in organs seemingly homologous to—and bearing the same names as— two of those in the mammalian vestibule: the sacculus and the utriculus.

Physiological distinctions between the orientation and motion senses and acoustic senses usually are based on stimulus frequency—stimuli at frequencies below approximately 10 Hz being orientational or motional, and those at frequencies greater than approximately 10 Hz being acoustic. Thus eighth-nerve axons (along with their periph-

eral associations) often are classified as being specialized for orientation or motion sensing or for acoustic sensing, depending on their relative responsiveness at frequencies above and below 10 Hz. Axons found to be comparably responsive to frequencies above and below 10 Hz are classified as having combined (e.g., gravitational and vibrational) specialization (Lewis et al. 1982; Platt 1983). Morphological evidence for acoustic specialization includes peripheral structures (e.g., those associated with the periotic system) that appear well suited to channeling acoustic energy to a particular inner-ear organ (Lombard 1980), and neural projections from a particular inner-ear organ to known or suspected acoustic centers in the brain (e.g., Boord and Rasmussen, 1963; Hamilton 1963).

Although there is disagreement regarding the homologies among them, at least eight catalogued organs of the inner ear in one vertebrate group or another are putative acoustic sensors: the common maculae of some cyclostomes (Lowenstein 1970); the sacculi of some elasmobranchs (Lowenstein and Roberts 1951; Corwin 1981), teleosts (Fay and Popper 1980), and amphibians (Ashcroft and Hallpike 1934; Cazin and Lannou 1975); the utriculi of some elasmobranchs (Lowenstein and Roberts 1951) and teleosts (Denton, Gray, and Blaxter 1979); the lagenae of some teleosts (Enger 1963; Furukawa and Ishii, 1967), amphibians (Caston, Precht, and Blanks 1977), and possibly reptiles (Hamilton, DW 1963) and birds (Boord and Rasmussen 1963); the papillae neglectae of some elasmobranchs (Fay et al. 1974; Corwin 1981); the amphibian papilla and amphibian basilar papilla

164

1963; Feng, Narins, and
(Fris... d, 1980); and the mam-
Ca... ind its commonly recog-
m... in and avian basilar papil-
... , four of these organs, the
... sacculus, the utriculus, and
... e been identified as being
... on sensors. In some cases, the
... dently has a region of acoustic
... separate region of orientation
... ation (Lowenstein and Roberts
... nstein 1970; Budelli and Maca-
... , Precht, and Blanks 1977). In
... entire organ evidently is special-
... acoustic sensor (e.g., Furukawa
... ; Fay and Popper 1980; Koyama et
an orientation or motion sensor (e.g.,
... in and Thornhill 1970; Lowenstein and
... 1949; Baird and Lewis 1986). The image
... emerges is that of an inner ear remarkably
... astic in the face of selective forces. What is not
clear, however, is how many times the acoustic specialization has arisen independently among all of these organs. It is possible, for example, that acoustic sensitivity in various otoconial or otolithic organs (sacculus, utriculus, and lagena) was inherited directly or indirectly from that of a common ancestor—possibly homologous to the common macula of the cyclostomes. Nevertheless, the presence of so many separate organs with acoustic specialization in the vertebrate ear provides us an opportunity to use convergences as clues to the phenotypic consequences of selective pressures—to deepen our understanding of the ear. Among other things, we are challenged to define more sharply the physiological distinctions between orientation or motion sensors and acoustic sensors, and to relate those distinctions to evolutionary contexts.

In this chapter I approach the problem in reverse, arguing from a hypothetical evolutionary context that certain properties would be compellingly advantageous in organs specialized for acoustic sensing. Then I present evidence of convergence on those properties in three of the eight inner-ear organ types listed in the previous paragraph—the sacculus, the anuran amphibian papilla, and the organ of Corti and homologs. The properties are—peripheral spectral filters each with a relatively broad pass band, bounded by very

steep band edges that extend over very wide dynamic ranges. Such filters are capable of providing excellent resolution in both time and frequency, the selective advantages of which I shall argue in terms of the processing of acoustic signals and the generation of acoustic images.

2. Theoretical and Conjectural Preamble

Although signal theory and network theory may seem to many biologists to be too abstract or too concerned with man-made systems to be of value in natural science, they both embrace concepts and logical tools that have proved valuable for biologists. The arguments in this chapter are based on a few elementary examples of those concepts and tools.

2.1 Some Technical Formalities

Network-theory terms are commonly used in the hearing-research literature; but their definitions appear to vary. In this section, in an attempt to avoid misinterpretation, I define various network-theory terms that will be used in biological discussions later in the chapter.

2.1.1 Signals and Noise

For dealing with transducers, it is convenient to define three categories of energy: (1) Energy that carries specific, relevant information is a *signal*. (2) Energy that can sum with signal energy, but carries irrelevant information or no information is *noise*. (3) The third category is energy that is neither signal nor noise (e.g., energy held in reserve—as in a battery, that can be converted to signal energy or noise energy). The acoustic energy impinging upon the tympanum of an animal can arise from the thermal motion of air molecules a few micrometers away, from storms thousands of miles away, from wind blowing through nearby bushes and trees, from footfalls of nearby predators, from advertisement calls of nearby conspecifics, and so forth. Depending on his hypotheses concerning selective advantage, one would place some of these energy components in the signal category and some in the noise category.

2.1.2 Passive and Active

A passive element or process is one that transfers, stores, or dissipates signal energy, but does not convert nonsignal energy into signal energy. An active element or process is one that converts nonsignal energy into signal energy. Thus the assignment of the labels *passive* and *active* to elements or processes depends entirely on what has been defined to be signal. All of the theorems from network theory related to passivity and activity operate under this definition (e.g., see Guillemin 1957).

2.1.3 Linearity and Natural Frequencies

Taylor's theorem from elementary analysis implies that when they are carried through very small dynamic ranges, processes should behave in an affine manner: $y = a_0 + a_1 x$. Most physical processes of concern to biophysicists fall into the category of nonequilibrium process in which a potential gradient is linked to a directed motion or flow (e.g., see Yourgrau, van der Merwe, and Raw 1966). The potential difference between two places or states is defined to be the free-energy change per unit of stuff that moves or flows between them. The second law of thermodynamics implies that when the free-energy change is zero, the directed flow or motion must be zero. Therefore, for small dynamic range, one expects the relationships between potentials and flows or motions to be linear rather than affine: $y = a_1 x$. This expectation is consistent with experience.

If a system comprising linearly operating processes is stimulated briefly and then allowed to return to rest, the dynamic behavior of the system beginning immediately after the end of the stimulus can be described as a sum of two kinds of functions: simple exponentials — $A_j \exp[a_j t]$, and exponentially modulated sinusoids — $A_k \exp[a_k t] \cos[\omega_k t + b_k]$. Using the conventional definition of exponentiation of complex numbers, one can use the same notation for both kinds of functions: $A_n \exp[z_n t]$, where A_n and z_n are either real or complex numbers: e.g., $z_j = a_j$ (a real number) or $z_k = a_k + i\omega_k$ (a complex number). The members of the set $\{z_j\}$ of all real and complex exponential coefficients in the description of its response are the *natural frequencies* of the linearly operating system (Guillemin 1957).

The natural frequencies are prop[...] tem itself, and are absolutely inde[...] stimulus that excited it. Furtherm[...] energy in the system can flow from e[...] or process either directly or indirect[...] other element or process, then one will o[...] cisely the same set of natural frequencies [...] of where in the system the stimulus is ap[...] regardless of where in the system the res[...] observed. In that respect the natural frequen[...] robust. In another respect, they are ephe[...] when the system is connected in a way that allo[...] to share signal energy in both directions [...] another system, then all of the natural frequenc[...] in both systems will change, and the combined sys[...] tem will have both sets of altered natural frequen-cies. If a system comprises only passive elements, then the real parts of all of its natural frequencies will be negative (the first law of thermodynamics prevents them from being positive, the second law prevents them from being zero).

2.1.4 Dynamic Order

The dynamic order of a system comprising linearly operating processes is equal to the order of the differential equation describing its behavior in response to residual signal energy, after all external stimuli have ceased. Generally, the dynamic order is equal to the number of natural frequencies and to the number of locations or processes in which signal energy can be stored independently in the system.

2.1.5 Transduction

In general, transduction is defined to be a process of transferring energy from one system to another (Guralnik 1970). For information-processing systems, the definition requires modification: Transduction is the process of transferring information from energy in one physical realm (e.g., acoustic energy) to energy in another realm (e.g., electrical energy). When all of the signal energy in the second realm is derived from signal energy in the first, the transducer is passive. When some of the signal energy in the second realm is derived from nonsignal energy, the transducer is active. Strain-sensitive channels are active transducers: information carried by mechanical energy is transferred to electrical energy, but the electrical signal energy is derived largely from nonsignal energy that drive ions through

166 ...search no further to find

(the f...ar.

the (...

ac...ncerning

...es

...culate about the ecological

...: rise to each inner-ear acous-

...ulpted its signal-processing

...mple, an acoustic sensor may

...ponse to its adaptive value in

...tors or other dangers (e.g., as an

...tem), its adaptive value in detec-

...her resources, its adaptive value in

...conspecific communication (e.g.,

...and locating offspring or prospective

...rospective rivals), or some combination

...A relevant signal would comprise the var-

...acoustic cues to the presence of the event or

...mal to be detected.

2.2.1 Evolution in the Presence of Noise and Interference

Regardless of the ecological interaction involved, one can be certain about at least one aspect of the environmental context of each evolving acoustic sensor: every relevant signal always was accompanied by noise and often was accompanied by other, relevant but interfering signals. At any point in space the acoustic signals, interference, and noise all were combined into one single-valued variable (e.g., sound pressure as a function of time); and, under (common) adverse circumstances, only a tiny fraction of the amplitude of that variable was determined by a particular relevant signal. Operating far from 0 deg K, the evolving acoustic sensory system certainly also faced the problem of noise added by the thermal motion of its own elements — e.g., the Brownian motion of its mechanical elements, the random opening and closing of its various ion channels, and the random emissions of its synaptic transmitters (e.g., DeFelice 1981; Bialek 1983; Bialek and Schweitzer 1985; Holton and Hudspeth 1986).

No matter which components of acoustic energy one hypothesizes to be signal and which to be noise, it is clear that the animal would derive conspicuous selective advantage from being able to separate those components — e.g., the predator's footfall from the sound of the wind. Whether it is noise or interference, the sound of the wind tends to cover the predator's footfall, and the system that uncovered it would be selectively advantageous indeed. The selective advantage of the ability to separate acoustic signals from one another and from noise seems so compellingly obvious that I take it to be axiomatic.

2.2.2 Acoustic Features and Objects

Human psychophysical literature on segregation and integration of auditory stimuli into discrete percepts provides numerous examples of features that an evolving nervous system might use to separate signals. For examples, when several tones are presented together, human listeners tend to integrate them (hearing them as a single object) as long as their frequencies are harmonically related, their amplitudes conform to a smooth spectral envelope, and they share common onset times and common modulation (Hartmann 1988). These are spectral and temporal attributes that one would expect in the signal from a single, localized acoustic source. The human auditory system evidently is highly tuned to these attributes; very slight deviation in the pattern (e.g., slight mistuning of one or more components, sight asynchrony of onset, slight lack of commonality of modulation) allows the listener to segregate the collection of tones into more than one object. Integration and segregation can be achieved with monaural stimulus presentation, suggesting that the auditory system subjects the single-valued acoustic variable (e.g., sound pressure as a function of time) impinging on each ear to the following processes: (1) decomposition into spectral components; (2) analysis of the components for harmonic relationships, smoothness of spectral envelope, synchrony of onset, and commonality of modulation; and (3) segregation of the components into subsets, each exhibiting spectral and temporal attributes (shared among its members) that are appropriate for an acoustic signal from a single physical source. Clearly, spectral and temporal features both are valuable for this sort of signal separation — the auditory system that used both would have conspicuous advantage over the system that used just one or the other.

2.3 Appropriate Signal Processing Schemes

In terms of discriminability of similar, interfering signals, evolving acoustic sensory systems would have derived considerable selective advantage from being able to achieve spectral decomposition with high resolution in frequency, while retaining high temporal resolution in the resolved spectral components. Note that spectral decomposition in this context is distinct from perception of pitch by human listeners. Pitch evidently is an identifier attached to each acoustic object segregated (by the auditory CNS) in step 3 of the previous paragraph (Hartmann 1988). Spectral decomposition is step 1, and it is accomplished by filters (tuned structures) at the auditory periphery. It is those filters that would be expected to achieve high resolution in frequency while retaining high temporal resolution.

2.3.1 Desirable Frequency Responses of Spectral Filters

In terms of frequency response, an ideal spectral filter with one pass band would allow all spectral components within the pass band to pass freely and would completely reject all components outside the pass band. On the conventional log–log plot of gain (response amplitude divided by stimulus amplitude) vs frequency, this ideal filter would have band edges with infinite slope and infinite extent (corresponding to infinite dynamic range). A real spectral filter can approach this ideal by having very steep band edges that extend indefinitely. In a real filter constructed with linear elements, the frequency response properties can approach the ideal as the filter's dynamic order increases. Alternatively, a low-order filter with a resonance can approach the ideal by becoming increasingly underdamped.

2.3.2 Properties of Spectral Filters that Preserve Time Resolution

The uncertainty relation of Fourier transform theory states that the product of the minimum resolvable time (Δt) and minimum resolvable frequency (Δf) for any signal is equal to or greater than 0.08 Hz s (Bracewell 1978). Thus any frequency–time presentation of a signal (such as that which one might hypothesize to account for the segregation of acoustic objects evident in human psychophysics)

will have finite grains of resolution no matter how perfect the spectral filters are. For example, a presentation in which time is resolved to 0.1 ms can have frequency resolved to no better than 800 Hz. It very likely was advantageous for evolving acoustic sensory systems to establish multiple presentations, some with high time resolution and some with high frequency resolution. The temporal response of a band-pass filter is a simple frequency–time presentation, with just one frequency – resolved to an accuracy of approximately one bandwidth, Δf. The corresponding temporal resolution would be no better than $0.08/\Delta f$.

The alternative routes to approaching ideal frequency characteristics in linearly operating filters are: (1) increasing dynamic order, and (2) employing resonance and reducing damping. In either case, the goal is to increase the steepness and range of each edge of the pass band. When this is accomplished by increase of dynamic order, bandwidth need not change; when it is done by reduction of damping, bandwidth necessarily must decrease and temporal resolution necessarily must suffer. In terms of physical explanation, a filter based on underdamped resonance operates by accumulation of signal energy over time. Thus, when such a filter is stimulated by a tone of steady amplitude and frequency within its pass band, the signal energy entering the filter during the present stimulus cycle is only a fraction of the total energy accumulated from previous stimulus cycles. As the damping decreases, so does that fraction; the signal becomes increasingly contaminated with residual energy from long-past events. Considering this fact, one reasonably would expect evolving acoustic sensory systems to have found much greater selective advantage in alternative (1). Consequently, one would expect to find that the filters associated with inner-ear acoustic sensors exhibit steep band edges derived from high dynamic order rather than a high degree of undamping.

2.3.3 Singularities

An analytic function of time is one for which the function itself and all of its time derivatives are continuous. Taylor's theorem tells us that the trajectory of an analytic function of time is determined, for all time, by the set of values of the function and all of its time derivatives at any point in time. Any signal that changes its trajectory cannot be analytic; it must exhibit at least one singular point (singularity) at the moment of the change—a point in time at which the signal itself or one or more of its time derivatives are discontinuous. Since this event occurs, in principle, in infinitesimal time, it provides the most precise of temporal markers—it is the quintessential temporal event. It is not an impulse. Formally, an impulse is a stimulus of infinitesimal duration that delivers finite energy. Practically, any stimulus of finite energy whose duration is very short in comparison to the reciprocal of the magnitude of the highest natural frequency of a system is equivalent to an impulse. The response to an impulse comprises the natural frequencies of the system, each excited with a characteristic relative amplitude. An impulse or its practical equivalent may include one or many singularities. In that sense, the singularity is a more primitive descriptor of the stimulus waveform. The singularity itself possesses no energy.

Corresponding to every singularity, a linearly operating filter will produce a transient response comprising its own natural frequencies and timed precisely to the instant of the singularity. The energy for the response is accumulated from portions of the stimulus signal immediately following the singularity. The shape of the response is determined by the change in the stimulus trajectory at the singularity and, in general, is not the same as the shape of the impulse response. By definition, the singularity is the most precisely timed event to which any filter is capable of responding. The filter also will produce a continuing response to the ongoing stimulus waveform; and in the overall response of the filter the transient response to the singularity ordinarily will be absorbed into the continuing response and thus lost as a temporal cue. If the dynamics of the filter are of sufficiently high order, however, providing one or more sufficiently steep band edges (with unlimited dynamic range), then it can selectively attenuate the continuing response and allow the transient response to stand alone as a precise temporal marker (Lewis and Henry 1989a).

Singularities are ubiquitous in natural acoustic signals, including those produced by animals. In individual mammalian cochlear afferents and in compound action potentials from the entire mammalian auditory nerve, responses to singularities do stand alone, indicating that the acoustic filters of the cochlea are of sufficiently high dynamic order to make that happen (Lewis and Henry 1989b). The same thing should be true in other acoustic sensors with steep band edges derived from high-order dynamics. This is another selective advantage of high-order filters at the periphery of acoustic sensors.

2.3.4 Contrast with Filters for Orientation and Motion Senses

According to the evidence published so far, inner-ear orientation and motion sensors exhibit low dynamic order—e.g., order two or three (Fernandez and Goldberg 1976; Blanks and Precht 1976; Dickman and Correia 1989). Such low-order dynamics are sufficient to translate jerk, acceleration, or velocity into strain, or to translate the equivalent acceleration of gravity into strain. Thus it seems reasonable to propose that a fundamental distinction between acoustic sensors and orientation or motion sensors is peripheral filters of high dynamic order in the former, low dynamic order in the latter. Test cases for this hypothesis should be provided by the various inner-ear organs already identified as having one region specialized for acoustic sensing and a separate region specialized for orientation sensing.

2.3.5 Electrical Resonances

During the past decade, second-order electrical resonances were identified in the hair cells of putative homologs of the organ of Corti—the basilar papillae of reptiles (Crawford and Fettiplace 1981) and birds (Fuchs 1988), as well as in the amphibian papillae and sacculi of anurans (Ashmore 1983; Pitchford and Ashmore 1987; Lewis and Hudspeth 1983). Consequently, it has become increasingly common for hearing researchers to associate those resonances with tuning in the peripheral filters of those organs. According to the arguments in Section 1.3.3, however, a peripheral filter based on high-order dynamics would be considerably more effective than one based on a second-order resonance. Thus, if the observed electrical resonances were involved in the peripheral filters, one would expect them not to be acting alone, but to be immersed in peripheral filters whose dynamic order is much greater than two. In the following section, this expectation is tested in the three (non-homologous) inner-ear organs in which electrical resonances were first discovered.

3. High-Order Dynamics in Peripheral Acoustic Filters

Observing a response (e.g., spike rate in a single acoustic axon) to a one-dimensional stimulus (e.g., sound pressure at the tympanum), the physiologist is treating the ear as a single-input single-output (SI/SO) system. There are two conventional ways to characterize the dynamic properties of a linearly operating SI/SO system: the impulse response (response amplitude as a function of time) and the sinusoidal steady-state response (gain and phase shifts as functions of frequency). Both characterizations bear clues to the order of the system.

If there is a frequency beyond which the sinusoidal response of a linearly operating SI/SO system decreases monotonically with increasing frequency, then in the conventional log–log plot of gain versus frequency the dynamic order of the system is equal to or greater than the magnitude of the asymptotic slope (given in decades of gain per decade of frequency) as the frequency approaches infinity. Because noise prevents one from tracking the gain to indefinitely low levels, the best one can do is follow the gain to the noise floor (usually established by the time available for observation and by the maximum practical stimulus amplitude). If there also is a frequency below which the gain of the system declines monotonically as the frequency decreases further, then the order of the system is equal to or greater than the sum of the magnitudes of the slopes as frequency approaches zero and infinity. The order of a linearly operating system also is greater than or equal to the range of phase shift (of the response sinusoid relative to the stimulus sinusoid) measured in quarter cycles.

The impulse response of an underdamped second-order resonance is a monotonically damped oscillation, maximally asymmetric, with the largest amplitude occurring in the first cycle of oscillation (Fig. 11.1, top). An impulse response that is not maximally asymmetric implies a system of order greater than two. When plotted on the conventional log–log scale, the sinusoidal steady-state gain versus frequency tuning curve of an under-

damped second-order resonance exhibits conspicuously concave flanks and asymptotic slopes (at high and low frequencies) whose magnitudes sum to two (Fig. 11.1, center). Only in the highest 3 dB range is the log–log gain tuning curve of the underdamped resonance convex. The phase shift of the second-order resonance is strictly limited to one-half cycle (Fig. 11.1 bottom).

3.1 SI/SO Functions from Reverse Correlation

The reverse-correlation (REVCOR) method of obtaining impulse responses and gain and phase-shift tuning curves has been well documented in the auditory physiology literature (de Boer 1968; de Boer and de Jongh 1978; Eggermont, Johannesma, and Aertsen 1983), and has been used very effectively recently by Evans (1989) and others. The stimulus is band-limited white noise and the measured response is the occurrence of a spike. For estimates of peripheral filter dynamics, the spikes are monitored in a primary afferent axon. The noise waveform occurring in a fixed time interval immediately before each spike is averaged with those for all of the other spikes in a sample. As the sampling progresses, a time-inverted estimate of the impulse response emerges. Fourier transformation of the impulse response yields estimates of the gain and phase-shift tuning curves. REVCOR has the advantage of operating with relatively low power densities in time and frequency. Being based on noise stimulus, it has the disadvantage of relatively slow convergence. A sampling of data over approximately 4,000 spikes typically is required to reveal the upper 20 dB range of gain in a tuning curve – i.e., to move the noise floor to a position approximately 20 dB below the response amplitude at the center frequency (e.g., see Evans 1989). Thereafter, the noise floor recedes at a rate of 10 dB for every tenfold increase in the sample size. Thus, tuning curves obtained with the REVCOR method are limited, practically, to a dynamic range of less than 40 dB. REVCOR also has the disadvantage of introducing (as an artifact) a spectral zero. Depending on its position, this zero will add +1, 0, or −1 to the order of the system as estimated by gain and phase curves.

The SI/SO functions obtained by REVCOR reflect the dynamics of the signal path from the

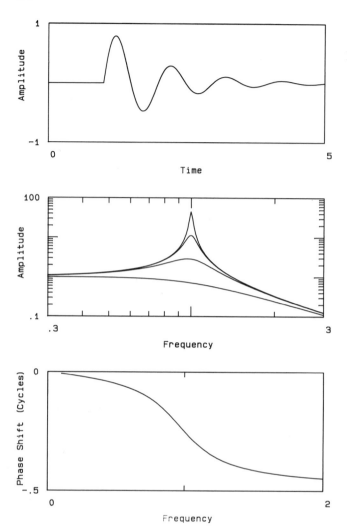

FIGURE 11.1. Response functions of a spectral filter based on second-order resonance. (Top) Impulse response. (Center and bottom) Sinusoidal steady-state gain and phase curves.

point at which the noise stimulus is monitored to the point at which the spikes are monitored. This path will include linear time delay processes (a short acoustic wave propagation path and a short spike propagation path), which have indefinitely high dynamic order and thus produce indefinitely large phase shifts (one cycle of phase lag for each $1/T$ Hz, where T is the propagation time in seconds). If the propagation times can be estimated, then the phase tuning curve can be used to estimate the dynamic order of the rest of the signal path (which presumably comprises simply the peripheral filter). The axon and its synaptic signal path from the hair cell also will affect the gain versus frequency tuning curve (Weiss and Rose 1988a). In the alligator-lizard basilar papilla (Weiss and Rose

1988b), the processes between receptor potential generation and spike generation evidently introduces three negative real natural frequencies, which produce a third order, low pass filter with corner frequency between 300 and 500 Hz. A similar effect occurs in anuran auditory papillae (Narins and Hillery 1983; Hillery and Narins 1987) and in mammalian organ of Corti (Johnson 1980; Palmer and Russell 1986), with corner frequencies comparable to those of the lizard in the anuran and higher (between 1 and 3 kHz) in the mammals (Weiss and Rose 1988b). Above the corner frequencies, these processes add approximately three decades of gain per decade of frequency to the slope of the tuning curve band edge. Below the corner frequencies, they have little

FIGURE 11.2. REVCOR-derived response functions from a low-frequency afferent axon from the basilar papilla of the red eared turtle (*P. scripta*). The noise floor in the center panel occurs at a relative amplitude of approximately 60. (Courtesy of M. Sneary.)

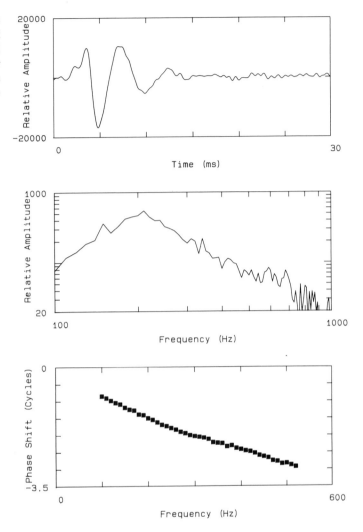

effect on the gain tuning curve; but they can be expected to contribute approximately 3/8 cycle of phase shift over the frequency decade below the corner, and approximately another 3/8 cycle over the decade above the corner.

3.1.1 Basilar Papilla of the Red-Eared Turtle

Being the first inner-ear organ in which electrical resonances were observed in hair cells (Crawford and Fettiplace 1981), the basilar papilla of the red-eared turtle (*Pseudemys scripta*) is an appropriate place to look for high-order filters. Crawford and Fettiplace (1981) found that the amplitude tuning curves in individual hair cells implied order five — two evidently contributed by the electrical

resonance, two evidently contributed by middle-ear dynamics (Moffat and Capranica 1978), and one contributed by an unknown process. The center of the gain tuning curve, however, was dominated by the second-order resonance, with its characteristic concave flanks. SI/SO functions derived from REVCOR data from the red-eared turtle basilar papilla suggest a minimum dynamic order of approximately five to eight (Figs. 11.2 and 11.3). Units with characteristic frequencies in the lower range for this organ — i.e., below approximately 400 Hz, exhibit relatively asymmetric impulse responses, similar to those of second-order resonances. Units with higher characteristic frequencies are found to have more symmetric impulse responses (Fig. 11.3, top). The gain versus

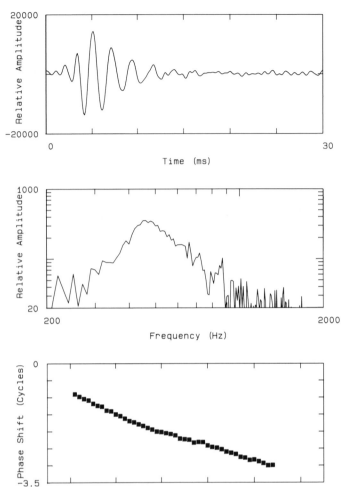

FIGURE 11.3. REVCOR-derived response functions from a mid-frequency afferent axon from the basilar papilla of *P. scripta*. Noise floor in center panel approximately 60. (Courtesy of M. Sneary.)

frequency tuning curves of Figures 11.2 and 11.3 do not show the concave flanks typical of resonances. Thus the electrical resonances of the hair cells appear to be absorbed into the higher order dynamics of the whole tuning system.

3.1.2 Sacculus of the American Bullfrog

The sacculus of the American bullfrog (*Rana catesbeiana*) is extraordinarily sensitive to seismic stimuli in the frequency range between approximately 10 Hz and 150 Hz (Koyama et al. 1982). Second-order electrical resonances were found in the hair cells of this organ and subsequently attributed to voltage-sensitive calcium channels and calcium-sensitive potassium channels, coupled through the electrical properties of the cell mem—

brane and intracellular diffusion of calcium (Hudspeth and Lewis 1988). SI/SO responses from the bullfrog sacculus typically imply tuning dynamics of order nine or more (Lewis 1988). In the majority of axons penetrated, impulse responses and gain and phase tuning curves show no indications of resonances (Fig. 11.4); the impulse responses show almost no ringing, and the gain tuning curves are relatively broad, convex, with very steep band edges. In the REVCOR produced gain tuning curve of Figure 11.4, only the beginning of the steep high-frequency band edge is visible above the noise floor. Since bullfrog saccular axons show no adaptation to sustained, single-frequency sinusoids, it is easy to continue the tuning curves one frequency at a time to levels well below the REVCOR noise floor. This has been done in Figure 11.5, which

FIGURE 11.4. REVCOR-derived response functions from a mid-frequency afferent axon from the sacculus of the bullfrog (*R. catesbeiana*). Noise floor in center panel approximately 50. (Courtesy of X. Yu.)

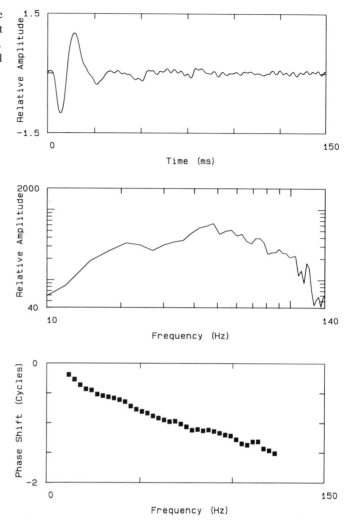

shows the high-frequency band edges for two units. These slopes imply dynamic order six or more. When the (comparably steep) low-frequency slopes are added, the lower bound on the dynamic order is considerably higher. The high-frequency band edges typically fall within an octave of those shown in Figure 11.5. The position of the low-frequency band edge is more variable; in that respect, the tuning curve of Figure 11.4 is unusually broad.

3.1.3 Amphibian Papilla of the American Bullfrog

The anuran amphibian papilla was the third inner-ear organ in which electrical resonances were found in hair cells (Pitchford and Ashmore 1987).

Figures 11.6 and 11.7 show representative SI/SO functions derived from REVCOR data from the amphibian papilla of *R. catesbeiana*. The impulse responses are not those of simple resonances; in fact they occasionally are very nearly symmetric rather than maximally asymmetric (Lewis 1990). Gain versus frequency curves show none of the concavity expected from resonances. Before they disappear into the noise floor, the band edges in Figures 11.6 and 11.7 indicate dynamic orders of at least eight and twelve, respectively. Phase shift in the bullfrog amphibian papilla has been followed over as many as five full cycles (Lewis and Lombard 1988). Independent measure of the combined acoustic and axonal delay in the same units yielded corrections of between one and two cycles, leaving three or more cycles of phase shift

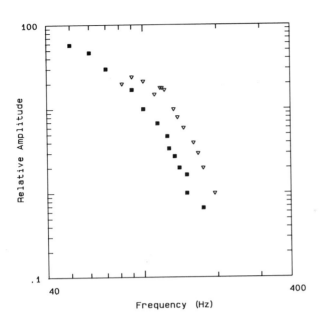

FIGURE 11.5. High-frequency band edges of gain tuning curves derived from single-frequency measurements on two saccular afferents in *R. catesbeiana*. Vertical axis corresponds to the peak–peak stimulus amplitude. (See Lewis, 1986.)

attributable to the filter dynamics, implying a dynamic order of twelve or more.

3.1.4 Cochlea of the Mongolian Gerbil

Although electrical resonances have been sought in mammalian hair cells, none has been found so far (Russell and Richardson 1987; Ashmore 1988). Numerous REVCOR derived cochlear SI/SO functions have been published (e.g., Evans 1989); those in Figure 11.8 were taken from an afferent axon of the gerbil (*Meriones unquiculatus*) and are representative for mammalian axons with characteristic frequencies below 4 kHz. Before they disappear into the noise floor, the band edges of the gain versus frequency tuning curve in this figure indicate a dynamic order of at least twelve.

3.2 How High-Order Dynamics Might Be Achieved

In the last half of the nineteenth century and the first third of the twentieth century, it was popular to attribute tuning in the mammalian cochlea to second-order resonance (Helmholtz 1885). With his traveling-wave theory, Békésy replaced this low-order dynamic model of the cochlea with a model of very high order (see Békésy 1960). In fact, most models of cochlear tuning dynamics subsequent to Békésy's work treat the cochlea as a

system with infinite dynamic order (Ranke 1950; Zwislocki 1950; Siebert 1974; Taber and Steele 1981). Therefore, finding evidence of at least order twelve in REVCOR-derived, cochlear SI/SO functions is not surprising. It may be surprising to many in the hearing research community, however, to find evidence of comparable dynamic order in the SI/SO functions of lower vertebrates, particularly those in which electrical resonances have been found. On the other hand, one might argue that SI/SO functions of all systems of finite spatial extent, when pressed to sufficiently high frequency, will reflect indefinitely high order. Therefore, the surprise is not high dynamic order per se, but the fact that over and over again in these various acoustic sensors, the individual peripheral filter is sculpted by a very large number of natural frequencies clustered in such a way as to provide very steep band edges with indefinitely large dynamic range. In contrast, the band edges of the underdamped second-order resonance remain steep only through the resonance peak, eventually giving way to weakly sloping skirts (whose slopes sum to two decades of gain per decade of frequency).

3.2.1 Low-Pass Filtering After the Receptor Potential

Weiss and Rose (1988a) list a sequence of nine dynamic processes that are presumed to occur

FIGURE 11.6. REVCOR-derived response functions from a low-frequency afferent axon of the amphibian papilla of *R. catesbeiana*. Noise floor in center panel approximately 80. (Courtesy of X. Yu.)

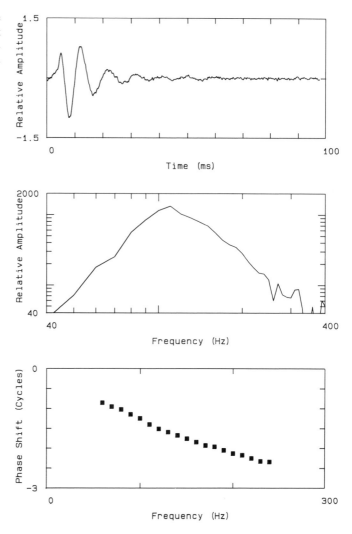

between the generation of the receptor potential in the hair cell and the initiation of a spike in the primary afferent axon. Each of these processes should add at least one natural frequency to the system. Some of the nine are diffusion processes, which could add indefinitely high dynamic order. From the differences between the SI/SO responses at the hair cell and that observed in the spike train of the afferent axon, Weiss and Rose concluded that the effective increase in dynamic order imposed by all of these processes was three. In other words, in the frequency range over which they observed SI/SO responses, only three of the natural frequencies were effective. The other natural frequencies, although they must have been present, had values that made them ineffective in the observed frequency range. The frequency range itself was not arbitrarily selected. In each case it was extended to the values at which the SI/SO response had been driven below the noise floor by the three effective natural frequencies.

It is possible that natural frequencies contributed by processes following generator potential production form part of the spectral filter sculpted by natural selection. On the other hand, acoustic sensors typically have been found to operate with considerable even-order (rectifying) nonlinearity in the production of the hair-cell generator potential. In response to an amplitude-modulated sinusoidal stimulus, such a nonlinearity would transfer a large portion of the signal energy to frequency components that reflect the modulation envelope rather

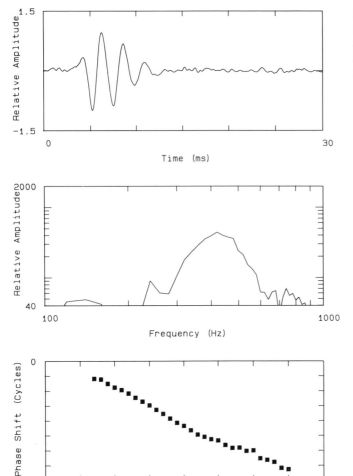

FIGURE 11.7. REVCOR-derived response functions from a mid-frequency afferent axon of the amphibian papilla of *R. catesbeiana*. Noise floor in center panel approximately 60. (Courtesy of X. Yu.)

than the frequency of the sinusoid itself. Applying appropriate filters to modulation envelopes would be a potentially valuable tool for segregating signals; but it would not eliminate the value of spectral decomposition of the modulated waveforms (carriers) themselves. If rectification plays a major role in generator-potential production, then processes central to the hair cell cannot participate effectively in that sort of decomposition. It must be left to a more peripheral filter.

Among the four inner-ear organs for which tuning curves were presented in the previous subsection of this chapter, the bullfrog sacculus is the only one whose afferent spike trains typically show no sign of rectification during receptor potential generation. Thus, in the usual terminology of the auditory physiology literature, they show an ac (or phase locking) component, but no dc component (Lewis 1986). Therefore, for those fibers, the low-pass filtering that occurs between receptor potential generation and spike initiation evidently could be an effective part of the peripheral spectral filter. In order to be so, however, the corner frequencies would have to range over more than an octave and to have been shifted approximately two to three octaves below those inferred by Weiss and Rose (1988b) in the lizard and tree-frog auditory papillae.

3.2.2 Micromechanics

The micromechanics associated with the mammalian cochlea are the subject of considerable,

FIGURE 11.8. REVCOR-derived response functions from an afferent axon of the Mongolian gerbil (*M. unguiculatus*). Noise floor in center panel approximately 150.

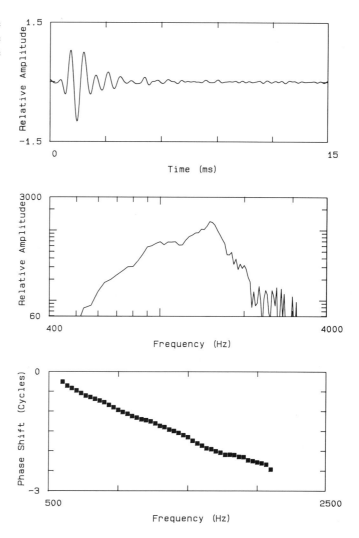

continuing discussion. Many of the current models include local mass and compliance of the basilar membrane which, if isolated, would form a local resonance. The dynamics of these local basilar membrane regions are coupled through the distributed inertia of the fluids of the cochlea duct, evidently forming a structure capable of supporting a slow, travelling wave. Thus, would-be local resonances are prevented from operating independently and participate instead in a filter system with very high dynamic order.

For the turtle basilar papilla, the frog amphibian papilla, and the frog sacculus, there have been no direct observations of travelling waves. Based on phase-shift data, the presence of slow, travelling waves has been suggested for the frog amphibian papilla (Hillery and Narins 1984; Lewis 1984). Because that organ is embedded in a thick wall of limbic tissue rather than a thin, compliant wall analogous to the basilar membrane, it has been suggested that the tectorium might be a travelling-wave structure. Other suggestions include local resonances involving the hair bundles and the tectorial mass (Shofner and Feng 1983, 1984; Lewis and Leverenz 1983), and a travelling-wave structure in which the dynamics of the electrical resonances of individual hair cells are coupled to one another (bidirectionally) through the ciliary bundles and the tectorial mass (Lewis 1986; Lewis and Lombard 1988). All of these theories are preliminary and speculative, however. None of them, for example, takes into account the elaborate and highly organized system of channels in the tectorium of the frog amphibian papilla (Lewis 1976; Wever 1985). The micromechanics of the anuran sacculus also is obscure. The sensory macula is a kidney-shaped patch of epithelium residing toward the center of a large, taut membrane (McNally and Tate 1925). On one side of the membrane is the massive otoconial structure (approximately 0.01% of the mass of the total animal), in which individual otoconial crystals are suspended in an organic matrix of unknown structure. On the other side of the membrane is a fluid space. The saccular twig of the eighth nerve passes through the fluid on its way from the thick outer wall of dense limbic tissue to the macula. Several thin, nonneural fibers also pass through the fluid, between the macula and the dense wall. The participations of these various elements in the micromechanics of the sacculus have not been investigated.

Presently, the predominant theories of tuning in the frog sacculus and turtle basilar papilla include major roles for electrical resonances. For the lizard basilar papilla, on the other hand, the theories have focused on the micromechanics of hair bundles (Weiss et al. 1976). With appropriate coupling (e.g., through endolymph or tectorium), the mechanics of many individual hair bundles could interact to form a filter with high-order dynamics. Electrical resonances in hair cells could do the same thing if transduction between the mechanical and electrical processes of the cell were bidirectional (Weiss 1982).

3.2.3 Electrical Tuning

Given the mismatch between the observed SI/SO response curves from the intact frog ear (Figs. 11.4 to 11.8) and the dynamics of resonances (Fig. 11.1), as well as the elaborate micromechanical structures in those organs, one might question whether or not the electrical resonances observed in frog saccular and amphibian-papillar hair cells actually have anything to do with tuning. The significance of the question is intensified by the fact that when the resonances were observed, the hair cell milieu had been altered either by removal of the tectorial or otoconial membrane or by complete isolation of the hair cell itself, and also by the fact that similar resonances were found in the isolated hair cells of the frog semicircular canals — nontuned, nonacoustic sensors (Housley, Norris, and Guth 1989). Probably the strongest evidence for the actual involvement of electrical elements in the high-order tuning of the intact frog ear is found in the temperature dependence of tuning curves. Two of the three elements (mass and stiffness) involved in purely mechanical tuning typically are only weakly dependent on temperature. Channel dynamics, on the other hand, typically are strongly dependent on temperature. Therefore, one expects tuning curves involving hair-cell ion channels to be strongly temperature dependent and tuning curves involving only mechanical elements to be weakly temperature dependent. Moffat and Capranica (1976) reported strong temperature dependence in the threshold tuning curves of the intact amphibian papilla of the toad *Bufo americanus*, and weak temperature dependence of those of the basilar papilla from the same species. Recently, van Dijk, Lewis, and Wit (1990) and Stiebler and Narins (1990) extended these studies by examining the temperature dependence of other SI/SO response functions, including linear gain and phase-shift tuning curves from the frog amphibian papilla. They found strong temperature dependence in lower-frequency (100 to 500 Hz) axons of the amphibian papilla (the range of frequencies over which Pitchford and Ashmore observed electrical resonances), weak temperature dependence in basilar papillar axons, and intermediate temperature dependence in higher-frequency (> 500 Hz) axons of the amphibian papilla. Of two frog saccular axons observed by van Dijk, one exhibited strong temperature dependence and the other intermediate.

An example of the results from van Dijk's study are shown in Figure 11.9. For this bullfrog amphibian papillar axon, the peak of the gain tuning curve shifted by approximately 40% in frequency as the temperature was varied by 7°C. Comparably strong temperature dependence has been observed in threshold tuning curves from the basilar papillae of gecko (Eatock and Manley 1981), caiman (Smolders and Klinke 1984), and pigeon (Schermuly and Klinke 1985), suggesting the involvement of electrical elements in each case. Direct observations of electrical resonances recently have been extended to isolated hair cells of the basilar papillae of alligator and chick (Fuchs 1988; Fuchs and Evans 1988). Thus it appears that electrical elements (ion channels) are involved in the filters of acoustic sensors

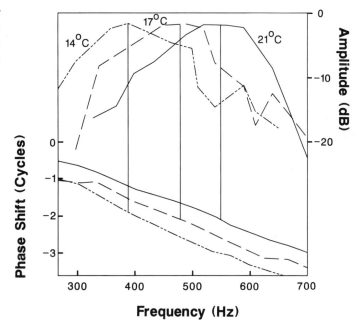

FIGURE 11.9. Temperature dependence of REVCOR derived tuning curves from an afferent axon of *R. catesbeiana*. Vertical lines connect approximate gain tuning peaks to the corresponding phase curves. (See van Dijk, Lewis, and Wit, 1990.)

in amphibians, reptiles, and birds. In the guinea pig cochlea, on the other hand, the temperature dependence of tuning is very weak (Gummer and Klinke 1983), suggesting that electrical elements are not conspicuously involved in tuning in the mammalian cochlea. Electrical resonances have been sought directly in mammalian hair cells, but none has been found so far (Russell and Richardson 1987; Ashmore 1988).

The strong temperature dependence of the gain tuning curve in some acoustic sensors and its absence in the mammalian cochlea strongly suggests that tuning is accomplished by electrical elements in the former, mechanical elements in the latter. From the SI/SO response functions in Figures 11.2 through 11.8, one would conclude that the tuning in both cases is accomplished with systems of high dynamic order, with none of the usual signs of resonance (sharp peaks with concave flanks, associated with half-cycle phase shifts and maximally asymmetric ringing in the impulse response). For the anuran amphibian papilla, the data of van Dijk, Lewis, and Wit clarify the significance of the phase tuning curves. Recently, several investigators have suggested that, in nonmammalian acoustic sensors, phase tuning curves of the types shown in Figures 11.2 through 11.7 should be attributed to second-order

dynamics in the filter itself (contributing one-half cycle) in combination with time delays extraneous to the filter. If that were the case, then only the half-cycle phase shift attributed to the filter should shift concomitantly with the peak of the gain tuning curve as temperature is changed. What is observed, however, is a concomitant shift of the entire phase tuning curve (Fig. 11.9). This implies that extraneous time delays do not contribute conspicuously to the phase tuning curve, and that the electrical tuning of the filter itself is of high dynamic order (e.g., at least order eight in the case of Fig. 11.9).

3.2.4 High Dynamic Order with Electrical Tuning

There are two fundamentally different ways that the second-order electrical tuning observed in individual hair cells could be combined to produce the high-order, evidently electrical tuning reflected in individual afferent axons: parallel connection and cascade connection. Implications of parallel connection are discussed in Lewis (1990). In the space remaining in this chapter, I shall focus the discussion on cascade connection.

In cascade connection, signal energy could be transferred from the electrical elements of one hair

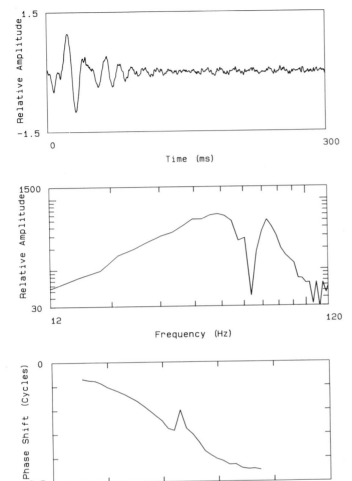

FIGURE 11.10. REVCOR-derived response functions from an afferent axon from the sacculus of *R. catesbeiana,* showing an antiresonant notch. (Courtesy of X. Yu.)

cell to those of another, allowing many such elements to participate in a single peripheral filter channel. This sort of exchange of signal energy would be possible if electromechanical transduction at the hair cell were bidirectional; and there now is considerable evidence for bidirectional transduction in turtle basilar papilla, frog sacculus, and frog amphibian papilla. Transduction from hair-bundle motion to membrane voltage already is well established in hair cells in general. In turtle basilar papilla, Crawford and Fettiplace (1985) found that electrical oscillations of the hair-cell membrane voltage, elicited by electrical stimuli, were accompanied by vibrations of the hair bundle, implying transduction in the other direction as well, from membrane voltage to hair-bundle motion. Comparing the spontaneous fluctuations of membrane voltage with motion of the hair bundle in bullfrog saccular hair cells, Denk and Webb (1989) concluded that reverse transduction takes place there as well. For the amphibian papilla, the best evidence is the presence of spontaneous otoacoustic emissions (Palmer and Wilson 1982; van Dijk and Wit 1987).

Since the hair bundles in each case are connected to a single tectorium or otoconial membrane, and thus can interact mechanically, bidirectional coupling implies that the electrical elements of the various hair cells will interact with each other (through the mechanical linkage). Thus one would expect the natural frequencies observed in an individual hair cell to be determined by the dynamics in neighboring hair cells as well as those in the hair cell under observation. Thus, when the tectorium or gelatinous membrane is in place, the calcium and potassium channels of the individual hair cell are part of a higher-order dynamic system.

Figure 11.10 shows what may be an anomalous consequence of bidirectional transduction in the bullfrog sacculus. In the gain tuning curves of approximately 20% of the recorded axons from this organ, one finds deep antiresonances, identifiable as such by positive phase jumps that accompany them. These antiresonances occur over a wide range of frequencies, normally have high Q's, and are accompanied by impulse responses with considerable ringing. Since antiresonances do not, by themselves, produce ringing, the simplest interpretation of these observations is the presence of a high-Q resonance in a stable feedback loop. The electrical elements of a single hair cell could provide the resonance, bidirectional transduction could place that resonance in a feedback loop. In order to produce the notch, however, the resonance cannot be in the direct signal path for the SI/SO function. Therefore, if the notch is produced by a hair cell resonance, it would not be in a hair cell contacted by the axon being recorded; it would be in a neighboring hair cell. From the slopes of the gain tuning curve and the range of the phase shift tuning curve before it begins to flatten out at each end, presumably to residual slopes determined by time delays extraneous to the filter, one would place the dynamic order of the complete filter of Figure 11.10 between six and nine.

4. Summary and Conclusions

We have examined the three, nonhomologous inner-ear acoustic sensors in the hair cells of which electrical resonances were first observed. In each case, REVCOR-derived response functions of the peripheral filter associated with individual afferent axons failed to show the usual signatures of underdamped resonance. Nevertheless, strong temperature dependencies of those response functions (examined in two of the three organs) imply participation of the hair cell's electrical elements in tuning. Bidirectional transduction in three organs, for which there now is abundant evidence, evidently couples the electrical elements of each hair cell,

through mechanical elements, to the electrical elements of its neighbors. In that way, the natural frequencies observed in isolated hair cells are altered and combined with those of the micromechanical system and those of other hair cells to yield a peripheral filter of high dynamic order (see Section 1.1.3).

The sharp, labile tuning peaks of the mammalian cochlea evidently do not occur in the lower 80% of the octaves of human hearing (including the entire frequency range required for intelligible speech). For cochlear tuning in that range, the response functions of Figure 11.8 are typical; tuning is accomplished by filters with steep band edges derived from high dynamic order. Tuning in the turtle basilar papilla, the frog amphibian papilla and the frog sacculus also is accomplished by filters whose steep band edges are derived from high dynamic order. Thus, although the elements from which the filters are constructed may differ, being largely mechanical in the mammalian cochlea and having a major electrical contingent in the other organs, the filters have converged on the same sort of dynamics, deriving frequency selectivity from high dynamic order rather than from a high degree of undamping of resonance. Doing so enabled all of the filters to retain good temporal resolution.

Acknowledgments. Original research for this chapter was supported by Grant DC 00112 from the National Institute on Deafness and other Communicative Disorders. I thank Eva Hecht-Poinar for help with the figures, and Xiaolong Yu, Dr. Michael Sneary, David Feld, and Dr. Kenneth R. Henry for contributions to the physiological studies.

References

Ashcroft DW, Hallpike CS (1934) On the function of the saccule. J Laryngol 49:450–460.

Ashmore JF (1983) Frequency tuning in a frog vestibular organ. Nature 304:536–538.

Ashmore JF (1988) Ionic mechanisms in hair cells of the mammalian cochlea. Progr Brain Res 74:1–7.

Baird IL (1974) Anatomical features of the inner ear in submammalian vertebrates. In: Keidel WD, Neff WD (eds) Handbook of Sensory Physiology, Vol. V/1, Berlin: Springer-Verlag, pp. 159–212.

Baird RA, Lewis ER (1986) Correspondences between afferent innervation patterns and response dynamics in the bullfrog utricle and lagena. Brain Res 369: 48–64.

Bekesy G von (1960) Experiments in Hearing. New York: McGraw-Hill.

Bialek WS (1983) Thermal and quantum noise in the ear. In: de Boer E, Viergever MA (eds) Mechanics of Hearing. Delft: Delft University Press, pp. 185–192.

Bialek WS, Schweitzer A (1985) Quantum noise and the threshold of hearing. Phys Rev Lett 54:725–728.

Blanks RHI, Precht W (1976) Functional characterization of primary vestibular afferents in the frog. Exp Brain Res 25:369–390.

Boord RL, Rasmussen GL (1963) Projection of the cochlear and lagenar nerves on the cochlear nuclei of the pigeon. J Comp Neurol 120:463–475.

Bracewell RN (1978) The Fourier Transform and Its Application. New York: McGraw-Hill.

Budelli R, Macadar O (1979) Statoacoustic properties of utricular afferents. J Neurophysiol 42:1479–1493.

Caston J, Precht W, Blanks RHI (1977) Response characteristics of frog's lagena afferents to natural stimulation. J Comp Physiol 118:273–289.

Cazin L, Lannou L (1975) Response du saccule a la stimulation vibratoire directe de la macule, chez la grenouille. CR Seance Soc Biol Rouen 169:1067–1071.

Corwin JT (1981) Peripheral auditory physiology in the lemon shark: evidence of parallel otolithic and nonotolithic sound detection. J Comp Physiol 142: 379–390.

Crawford AC, Fettiplace R (1981) An electrical tuning mechanism in turtle cochlear hair cells. J Physiol 312:377–412.

Crawford AC, Fettiplace R (1985) The mechanical properties of ciliary bundles of turtle cochlear hair cells. J Physiol 364:359–379.

de Boer E (1968) Reverse correlation. I. A heuristic introduction to the technique of triggered correlation with application to the analysis of compound systems. Proc K Ned Akad Wet C71:472–486.

de Boer E, de Jongh HR (1978) On cochlear encoding: potentialities and limitations of the reverse-correlation technique. J Acoust Soc Am 63:115–135.

DeFelice LJ (1981) Introduction to Membrane Noise. New York: Plenum.

Denk W, Webb WW (1989) Simultaneous recording of fluctuations of hair-bundle deflection and intracellular voltage in saccular hair cells. In: Wilson JP, Kemp DT (eds) Cochlear Mechanisms. New York: Plenum. pp. 125–133.

Denton EJ, Gray JAB, Blaxter JHS (1979) The mechanics of the clupeid acousticolateralis system: frequency responses. J Mar Biol Assoc UK 59:27–47.

Dickman JD, Correia MJ (1989) Responses of pigeon horizontal semicircular canal afferent fibers II: high frequency mechanical stimulation. J Neurophysiol 62:1102–1112.

Eatock RA, Manley GA (1981) Auditory nerve fiber activity in the Tokay Gecko II. temperature effect on tuning. J Comp Physiol 142:219–226.

Eggermont JJ, Johannesma PIM, Aertsen AMHJ (1983) Reverse correlation methods in auditory research. Q Rev Biophys 16:341–414.

Enger PS (1963) Single unit activity in the peripheral auditory system of a teleost fish. Acta Physiol Scand 59(Suppl. 3):9–48.

Evans EF (1989) Cochlear filtering: a view seen through the temporal discharge patterns of single cochlear nerve fibers. In: Wilson JP, Kemp DT (eds) Cochlear Mechanisms. Plenum, New York: pp. 241–250.

Fay RR, Kendall JI, Popper AN, Tester AL (1974) Vibration detection by the macula neglecta of sharks. Comp Biochem Physiol 47A:1235–1240.

Fay RR, Popper AN (1980) Structure and function in teleost auditory systems. In: Popper AN, Fay RR (eds) Comparative Studies of Hearing in Vertebrates. Berlin: Springer-Verlag, pp. 3–42.

Feng AS, Narins PM, Capranica RR (1975) Three populations of primary auditory fibers in the bullfrog (R. catesbeiana): their peripheral origins and frequency sensitivities. J Comp Physiol 100:221–229.

Fernandez C, Goldberg JM (1976) Physiology of peripheral neurons innervating otolith organs of the squirrel monkey. III. Response dynamics. J Neurophysiol 39:996–1008.

Frischkopf LS, Goldstein MH (1963) Responses to acoustic stimuli from single units in the eighth nerve of the bullfrog. J Acoust Soc Am 35:1219–1228.

Fuchs PA (1988) Electrical tuning in hair cells isolated from the chick cochlea. J Neurosci 8:2460–2467.

Fuchs PA, Evans MG (1988) Evidence for electrical tuning in hair cells isolated from the alligator cochlea. ARO Abstr 11:16.

Furukawa T, Ishii Y (1967) Neurophysiological studies on hearing in goldfish. J Neurophysiol 30:1377–1403.

Guillemin EA (1957) Synthesis of Passive Networks. New York: John Wiley, pp. 1–67, 186–210.

Gummer AW, Klinke R (1983) Influence of temperature on tuning of primary-like units in the guinea pig cochlear nucleus. Hear Res 12:367–380.

Guralnik DB (1970) Webster's New World Dictionary of the American Language. New York: World Publishing Co., p. 1509.

Hamilton DW (1963) Posterior division of the eighth cranial nerve in Lacerta vivipara. Nature (London) 200:705–706.

Hartmann WM (1988) Pitch perception and the segregation and integration of auditory entities. In: Edelman GM, Gall WE, Cowan WM (eds) Auditory Function: Neurological Bases of Hearing. New York: Wiley, pp. 623–645.

Helmholtz HLF (1885) On The Sensations of Tone. New York: Dover. (1954 Republication of the 1885 English-language edition.)

Hillery CM, Narins PM (1984) Neurophysiological evidence for a traveling wave in the amphibian inner ear. Science 225:1037–1039.

Hillery CM, Narins PM (1987) Frequency and time domain comparison of low-frequency auditory fiber responses in two anuran amphibians. Hear Res 25:233–248.

Hillman DE (1969) New ultrastructural findings regarding a vestibular ciliary apparatus and its possible significance. Brain Res 13:407–412.

Holton TL, Hudspeth AJ (1986) The transduction channels of hair cells from the bullfrog characterized by noise analysis. J Physiol (London) 375:195–227.

Houseley GD, Norris CH, Guth PS (1989) Electrophysiological properties and morphology of hair cells isolated from the semicircular canal of the frog. Hear Res 38:259–276.

Hudspeth AJ, Lewis RS (1988) Kinetic analysis of voltage and ion dependent conductances in saccular hair cells of the bullfrog, *Rana catesbeiana*. J Physiol 400:275–297.

Johnson DH (1980) The relationship between spike rate and synchrony in responses of auditory nerve fibers to single tones. J Acoust Soc Am 68:1115–1122.

Koyama H, Lewis ER, Leverenz EL, Baird RA (1982) Acute seismic sensitivity in the bullfrog ear. Brain Res 250:168–172.

Lewis ER (1976) Surface morphology of the bullfrog amphibian papilla. Brain Behav Evol 13:196–215.

Lewis ER (1984) On the frog amphibian papilla. Scan Electr Microsc 1984:1899–1913.

Lewis ER (1986) Adaptation, suppression and tuning in amphibian acoustical fibers. In: Moore, BJC and Patterson RD (eds) Auditory Frequency Selectivity. New York: Plenum, pp. 129–136.

Lewis ER (1988) Tuning in the bullfrog ear. Biophys J 53:441–447.

Lewis ER (1990) Electrical tuning in the ear. Comm Theoret Biol 1:253–273.

Lewis ER, Baird RA, Leverenz EL, Koyama H (1982) Inner ear: dye injection reveals peripheral origins of specific sensitivities. Science 215:1641–1643.

Lewis ER, Henry KR (1989a) Transient responses to tone bursts, Hear Res 37:219–240.

Lewis ER, Henry KR (1989b) Cochlear nerve responses to waveform singularities and envelope corners. Hear Res 39:209–224.

Lewis ER, Leverenz EL (1983) Morphological basis for tonotopy in the anuran amphibian papilla. Scan Electr Microsc 1983:189–200.

Lewis ER, Lombard RE (1988) In: Fritsch B (ed) The Evolution of the Amphibian Auditory System. New York: Wiley, pp. 93–123.

Lewis RS, Hudspeth AJ (1983) Frequency tuning and ionic conductances in hair cells of the bullfrog sacculus. In: Klinke R, Hartmann R (eds) Hearing – Physiological Bases and Psychophysics. Berlin: Springer-Verlag, pp. 17–24.

Lombard RE (1980) The structure of the amphibian auditory periphery: a unique experiment in terrestrial hearing. In: Popper AN, Fay RR (eds) Comparative Studies of Hearing in Vertebrates. New York: Springer-Verlag, pp. 121–138.

Lowenstein O (1970) The electrophysiological study of the responses of the isolated labyrinth of the lamprey (*Lampreta fluviatilis*) to angular acceleration, tilting and mechanical vibration. Proc R Soc London Ser B 174:419–434.

Lowenstein O, Roberts TDM (1949) The equilibrium function of the otolith organs of the thornback ray (*Raja clavata*). J Physiol (London) 110:392–415.

Lowenstein O, Roberts TDM (1951) The localization and analysis of the response to vibration from the isolated elasmobranch labyrinth. A contribution to the problem of the evolution of hearing in vertebrates. J Physiol (London) 114:471–489.

Lowenstein O, Thornhill RA (1970) The labyrinth of *Myxine*: anatomy, ultrastructure and electrophysiology. Proc Roy Soc London Ser B 176:21–42.

McNally WJ, Tait J (1925) Ablation experiments in the labyrinth of the frog. Am J Physiol 75:155–179.

Moffat AJM, Capranica RR (1976) Effects of temperature on the response of the auditory nerve in the American toad (*Bufo americanus*). J Acoust Soc Am 60 (suppl. 1):S80.

Moffat AJM, Capranica RR (1978) Middle ear sensitivity in anurans and reptiles as measured by light scattering spectroscopy, J Comp Physiol 127:97–107.

Narins PM, Hillery CM (1983) Frequency coding in the inner ear of anuran amphibians. In: Klinke R, Hartmann R (eds) Hearing – Physiological Bases and Psychophysics. Heidelberg: Springer-Verlag, pp. 70–76.

Palmer AR, Russell IJ (1986) Phase-locking in the cochlear nerve of the guinea pig and its relation to the receptor potential of inner hair cells. Hear Res 24:1–15.

Palmer AR, Wilson JP (1982) Spontaneous and evoked acoustic emissions in the frog, *Rana esculenta*. J Physiol 324:66P.

Pitchford S, Ashmore JF (1987) An electrical resonance in hair cells of the amphibian papilla of the frog *Rana temporaria*. Hear Res 27:75–84.

Platt C (1983) The peripheral vestibular system of fishes. In: Northcutt TG, Davis RE (eds) Fish Neurobiology, Vol. 1. Ann Arbor: University of Michigan Press, pp. 89–123.

Ranke OF (1950) Theory of operation of the cochlea: a contribution to the hydrodynamics of the cochlea. J Acoust Soc Am 22:772–777.

Russell IJ, Richardson GP (1987) The morphology and physiology of hair cells in organotypic cultures of the mouse cochlea. Hear Res 31:9–24.

Schermuly L, Klinke R (1985) Change of characteristic frequency of pigeon primary auditory afferents with temperature. J Comp Physiol 156:209–211.

Shofner WP, Feng AS (1983) A quantitative light microscopic study of the bullfrog amphibian papilla tectorium: correlation with tonotopic organization. Hear Res 11:103–116.

Shofner WP, Feng AS (1984) Quantitative light and scanning electron microscopic study of the developing auditory organs in the bullfrog: implications on their functional characteristics. J Comp Neurol 224:141–154.

Siebert WM (1974) Ranke revisited—a simple shortwave cochlear model. J Acoust Soc Am 56:594–600.

Smolders JWT, Klinke R (1984) Effects of temperature on the properties of primary auditory fibers of the spectacled caiman, *Caiman crocodilus*. J Comp Physiol 155:19–30.

Sneary M, Lewis ER (1989) Response properties of turtle auditory afferent nerve fibers: evidence for a high-order tuning mechanism. In: Wilson JP, Kemp DT (eds) Cochlear Mechanisms. New York: Plenum, pp. 235–240.

Steibler I, Narins PM (1990) Temperature dependence of auditory nerve response properties in the frog. Hear Res 46:63–81.

Taber LA, Steele CR (1981) Cochlear model including three-dimensional fluid and four modes of partition flexibility. J Acoust Soc Am 70:426–436.

van Dijk P, Lewis ER, Wit HP (1990) Temperature effects on auditory nerve fiber response in the American bullfrog. Hear Res 44:231–240.

van Dijk P, Wit HP (1987) Temperature dependence of frog spontaneous otoacoustic emissions. J Acoust Soc Am 82:2147–2150.

Weiss TF (1982) Bidirectional transduction in vertebrate hair cells: a mechanism for coupling mechanical and electrical processes. Hear Res 7:353–360.

Weiss TF, Mulroy M, Turner R, Pike R (1976) Tuning of single fibers in the cochlear nerve of the alligator lizard: relation to receptor morphology. Brain Res 115:71–90.

Weiss TF, Rose C (1988a) Stages of degradation of timing information in the cochlea: a comparison of hair-cell and nerve-fiber responses in the alligator lizard. Hear Res 33:167–174.

Weiss TF, Rose C (1988b) A comparison of synchronization filters in different auditory receptor organs. Hear Res 33:175–180.

Wever EG (1985) The Amphibian Ear, Princeton: Princeton University Press.

Yourgrau W, van der Merwe A, Raw G (1966) Treatise on Irreversible and Statistical Thermophysics. New York: Macmillan.

Zwislocki JJ (1950) Theory of the acoustical action of the cochlea. J Acoust Soc Am 22:778–784.

12
The Efferent Innervation of the Ear: Variations on an Enigma

Barry L. Roberts and Gloria E. Meredith

1. Introduction

For more than forty years, the efferent supply to the mammalian ear provided by the olivocochlea (OC) bundle (Rasmussen 1946) has been an enigma. Rasmussen (1953) recognized that because of this pathway "the receptors would possess a double innervation which would be most interesting . . . particularly from the point of view of neural mechanisms of hearing" and initial studies soon implicated the OC bundle in auditory function. When the OC axons were transected, nerve endings near hair cells degenerated (Smith and Rasmussen 1963) and when they were stimulated, the total sensory activity of the auditory nerve was reduced (Galambos 1956). In the two decades following these fundamental observations, little insight into the significance of the OC system was obtained and some authors concluded that it served no auditory function (see Klinke and Galley 1974). Indeed, it was difficult to see how the efferent system could operate, since in mammals the majority of the auditory sensory nerve fibers connect with inner hair cells yet it is the outer hair cells which are the main target of the efferent innervation. Considerably more information has accumulated in the last few years, particularly with the advent of intracellular recordings from hair cells and with the application of modern neuroanatomical techniques; yet much of this detail has only added to the mystery of the efferent system. Its existence and importance are no longer in doubt but its biological significance remains enigmatic. In particular, we have little idea of how and when the efferent supply is brought into action under natural conditions.

We shall argue here that the OC neurons form just one component of a larger, much more diverse octavolateralis efferent system (OES) that is found in all classes of vertebrates, and even in some invertebrates. The OES is a basic feature of hair cell sense organs that involves several sensory modalities, emphasizing its adaptive significance and, in common with the efferent supply to the retina and the muscle spindle, provides an essential mechanism whereby an animal can match sense organ sensitivity to its behavioral requirements.

Our goal in the present chapter is to provide a comparative overview of the organization of the OES and to delineate fundamental features and the extent of morphological and functional diversity. By recognizing phyletic trends we shall consider how the OES has changed in the course of evolution but we are severely limited in our ability to generate evolutionary hypotheses because of the incompleteness of the morphological, ontogenetic, and functional data.

2. Targets for the Efferent Innervation

Octavolateral efferent neurons can terminate, peripherally, on the hair cells and sensory nerve fibers and, centrally, in hindbrain sensory nuclei. The peripheral innervation has been extensively investigated, both morphologically and physiologically, in a variety of vertebrates, whereas the central projections of the efferent axons have, as yet, been little studied (see Section 4).

The basic features of the efferent innervation at the periphery are illustrated in Figure 12.1 which

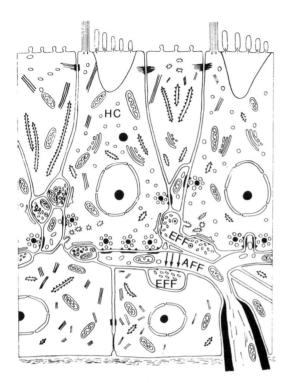

FIGURE 12.1. Diagrammatic representation of the sensory epithelium of the goldfish saccular macula. Note the efferent and afferent terminals at the base of each hair cell. Arrows indicate the synapse on an afferent nerve fiber. AFF, afferent fiber; EFF, efferent fiber or terminal; HC, hair cell. (Taken from Nakajima and Wang 1974.)

is taken from Nakajima and Wang's (1974) study of the sacculus of the goldfish. The hair cells make synaptic connection with the dendrites of sensory neurons that project to the brainstem and which have their perikarya in a sensory ganglion close to the brain. The hair cells receive on their basal portions synaptic input from axons that originate outside the sense organ. These efferent axons also form axodendritic synapses on the sensory fibers. Such axodendritic and axosomatic efferent synapses are characteristic of the efferent supply to hair cell sense organs of many vertebrates.

The richness of the axosomatic innervation varies considerably—for example, there are from 1 to 3 efferent endings per hair cell in the fish *Colisa* (Wegner 1982); from 1 to 40 in various species of lizards (Miller and Beck 1988) and from 6 to 13 in the guinea pig (Altschuler and Fex 1986)—and

some hair cells have no axosomatic contacts (see Section 3). The ultrastructural features of the axosomatic synapses seen in the mammalian organ of Corti, the most studied system (reviewed in Kimura 1975), have been observed, by and large, in other hair cell sense organs in other classes of vertebrates. In mammals, these efferent endings, which are usually larger than the afferent terminals, are located at the base of the hair cells and contain numerous spherical vesicles (30 to 40 nm diameter) and some larger vesicles (60 to 80 nm) with dense cores. The synaptic contact may be as long as 5 μm and the postsynaptic membrane is underlaid by a subsynaptic cistern.

The physiological impact of the axosomatic efferent innervation on the target cells appears to be threefold: a change in membrane potential, an alteration of the sharpness of tuning, and a mechanical response. Efferent nerve stimulation hyperpolarizes the hair cells by 5 to 25 mV [*Lota* lateral line (Flock and Russell 1976), frog sacculus (Ashmore and Russell 1982), and turtle cochlea (Art et al. 1982)] and results in a reduction in the evoked and spontaneous synaptic potentials that can be recorded from the underlying sensory fibers (Furukawa 1981). These effects are eliminated when cholinergic blockers are applied to the sense organs. Figure 12.2A shows the hyperpolarization of a turtle hair cell induced by efferent nerve stimulation, the impact of which depends strongly on the frequency of sound stimulation and is most marked around a hair cell's characteristic frequency (Art et al. 1982). Consequently, the iso-intensity tuning curve becomes much flatter as a result of efferent nerve stimulation (Fig. 12.2C). The mechanical response of hair cells that might involve the efferent system has been observed in isolated hair cells maintained in vitro. Such hair cells can expand and contract in response to current injection or to iontophoretic application of acetylcholine (Brownell et al. 1985; Ashmore 1987), which are procedures that mimic the action of the axosomatic efferent terminals.

Axodendritic synapses have only been studied in detail in the organ of Corti. They are up to 1.5 μm in diameter and are made with the type I radial afferent fibers (Warr, Guinan, and White 1986) although a few synapses are formed with type II spiral fibers (Ginsberg and Morest 1984). The synapses contain two types of vesicles: clear (20 to 50 nm in diameter) and dense cored (70 to 120 nm).

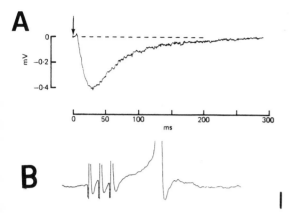

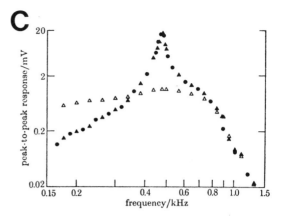

FIGURE 12.2. The impact of efferent nerve stimulation on sense organ output. (A) average intracellular recording (999 presentations) from turtle cochlear hair cell in response to single shock given to efferent nerve fibers, indicated at arrow. On ordinate, 0 = resting potential, 47 mV. (From Art et al. 1984.) (B) Intracellular recording from sensory nerve fiber under semicircular canal crista in *Opsanus* during efferent nerve stimulation. An action potential takes off from an excitatory postsynaptic potential evoked by three efferent shocks. Vertical scale: 8 mV (from Highstein and Baker, 1985). (C) isointensity tuning curve for hair cell from turtle cochlea before (▲), during (△) and after (●) efferent nerve stimulation at 50 stimuli/s. The ordinate shows the peak to peak amplitude of the membrane response to a continuous tone at various frequencies. (From Art et al. 1982.)

The action of the axodendritic efferent innervation on the afferent fibers remains uncertain. When recording from afferent fibers close to semicircular canal hair cells of *Opsanus*, Highstein and Baker (1985) observed excitatory postsynaptic potentials during efferent nerve stimulation (Fig. 12.2B). However, Lin and Faber (1988) found only inhibitory postsynaptic potentials during efferent nerve activation when recording from saccular fibers of the goldfish.

The most commonly reported effect of efferent nerve stimulation has been a decrease or inhibition of sensory nerve activity, but excitatory efferent effects have also been observed in vestibular sense organs of fish, frog and monkey (Rossi et al. 1980; Hartmann and Klinke 1980; Goldberg and Fernández 1980; Highstein and Baker 1985). These differences may involve different sets of neurons, perhaps with different transmitters (see Section 5) or perhaps terminating as axosomatic or as axodendritic synapses. Much evidence suggests that the axosomatic efferent innervation exerts solely an inhibitory impact on the hair cells but in their recordings from turtle cochlear hair cells, Art, Fettiplace, and Fuchs (1984) showed the response to efferent nerve stimulation to consist of a slow hyperpolarization that followed an initial depolarization, which could indicate that the efferent input is not entirely inhibitory. Similarly, Williamson (1989b) has reported that depolarizing, hyperpolarizing, and mixed responses can all be recorded from single hair cells of cephalopods during efferent nerve stimulation. How the excitatory and inhibitory actions of the OES are mediated remains, therefore, an important question.

3. Patterns of Efferent Innervation of Auditory Endorgans

There is considerable diversity among the various vertebrate classes in the relationship between the efferent axons and the auditory end organs. Differences between species are substantial and there is much variation in the number of synapses per hair cell and in the proportions of axosomatic and axodendritic endings.

We do not know whether cyclostomes can hear (see Popper and Hoxter 1987), but lampreys certainly respond to labyrinthine vibration (Currie and Carlsen 1987). The cyclostome labyrinth has only a single macula, the macula communis, the hair cells of which receive an efferent axosomatic innervation in *Lampetra* (Lowenstein, Osborne, and Thornhill 1968; Hoshino 1975) and *Petromyzon* (Popper and Hoxter 1987) but probably not in *Myxine* (Lowenstein and Thornhill 1970).

In fishes, the sense organs of the pars inferior of the labyrinth (sacculus, lagena, and macula neglecta) have been suggested as being involved in acoustic behavior (Schellart and Popper, Chapter 16). Most attention has been focused on the sacculus which is now known to be inhomogeneous in organization. The rostral part of its macula contains "short" hair cells that are innervated by large-diameter afferent nerve fibers, whereas the more caudal hair cells, which are tall, are innervated by smaller nerve fibers (Sento and Furukawa 1987). Different neural activity patterns have been recorded from the fibers and the hair cells have different membrane properties—the short hair cells show low-grade electrical tuning whereas the tall hair cells respond to injected depolarizing current by producing action potentials (Sugihara and Furukawa 1989).

The hair cells of the endorgans of the pars inferior receive a rich axosomatic efferent innervation (elasmobranchs: *Raja*, Lowenstein, Osborne, and Wersäll 1964; *Carcharhinus*, Corwin 1977; teleosts: *Anguilla*, Mathiesen 1985; *Gymnothorax*, Popper 1979; *Salmo*, Drescher, Drescher, and Hatfield 1987; *Opsanus*, Sokolowski and Popper 1988; *Carassius*, Hama 1969; Nakajima and Wang 1974). Efferent endings have been observed to contact afferent fibers in the goldfish sacculus and then only "on rare occasions" (Nakajima and Wang 1974), and they have also been reported in the semicircular canal cristae of *Opsanus* (Sans and Highstein 1984).

Furukawa (1981) recorded from afferent fibers of the goldfish sacculus during sound presentation and showed that the intracellular response consisted of excitatory postsynaptic potentials and spikes which were reduced when the sound was preceded by a train of electrical stimuli to the saccular nerve. This suggests that the efferent innervation exerts purely an inhibitory effect on the auditory endorgan. The effect on vestibular endorgans is, however, more complex since excitatory as well as inhibitory effects of efferent nerve stimulation have been reported (Hartmann and Klinke 1980; Highstein and Baker 1985).

The auditory sensory epithelia of amphibia are in the basilar and amphibian papillae, but the sacculus also responds to airborne sound (Moffat and Capranica 1976; Lewis et al. 1982). There is an efferent innervation of the papillae and sacculus in most species (Flock and Flock 1966; White 1986) although the basilar papilla lacks an efferent supply in the few ranids that have been examined so far (Robbins, Bauknight, and Honrubia 1967; Frischkopf and Flock 1974). Axosomatic, and less commonly, axodendritic contacts have been found in the urodelian basilar papilla (White 1986).

The hearing organ of the reptilian ear is the basilar papilla, the sensory epithelium of which differs considerably in form and function in the different groups (Miller, Chapter 23; Köppl and Manley, Chapter 24). The hair cells of the turtle are electrically tuned and are distributed along the papilla in a tonotopic fashion (Art and Fettiplace 1987). These hair cells receive axosomatic efferent terminals (Jørgensen 1974; Sneary 1988), as do the inner and outer hair cells in *Caiman* (von Düring 1974) and hair cells in the apical region of the papilla of some lizards (Miller and Beck 1988). Efferent endings on afferent fibers have been reported in the apical and middle regions of the papilla of turtles (Sneary 1988) and in *Caiman* (von Düring 1974) but are absent from the majority of lizards although they have been observed in *Ameiva* (Miller and Beck 1988).

Intracellular recording studies from hair cells of the basilar papilla of the turtle, *Pseudemys*, where both axosomatic and axodendritic efferent synapses are found (Sneary 1988), have shown that efferent nerve stimulation generates hyperpolarizing synaptic potentials and a concomitant loss of frequency selectivity (Art et al. 1982).

The basilar papilla of the cochlear duct of the avian ear contains two morphological types of hair cells, tall and short, which are thought to function like the inner and outer hair cells, respectively, of mammals (see Smith 1985, Manley and Gleich, Chapter 27 for review). Tall hair cells isolated from different portions of the chick cochlea differ in their electrical properties, the most apically situated being able to generate action potentials (Fuchs, Nagai, and Evans 1988), and these properties could play a role in frequency selectivity. The efferent innervation arises from fibers that branch extensively and end exclusively as axosomatic synapses that form large calyces on the short hair cells and small boutons on the tall hair cells (*Columba*, Takasaka and Smith 1971). Axodendritic synapses are apparently present below the tall hair cells during late development of the chick cochlea (Whitehead and Morest 1985). The efferent innervation has been shown in the pigeon to exert an inhibitory effect on the inner ear (Desmedt and Delwaide 1965).

The efferent innervation of the mammalian organ of Corti originates from axons of the olivocochlear bundle and ends as terminals both on sensory dendrites and hair cells. Thick (>0.7 μm) axons terminate primarily on the outer hair cells, whereas thin (<0.7 μm) axons end on the radial sensory dendrites under the inner hair cells (Liberman 1980; Ginsberg and Morest 1984; Guinan, Warr, and Norris 1983; Liberman and Brown 1986; Brown 1987; Brown et al. 1988).

Efferent action at the mammalian cochlea is not yet fully understood. Galambos (1956) was the first to show that electrical stimulation of the efferent axons leads to a reduction in the auditory nerve response to acoustic stimuli. This general finding has been replicated in several studies (see Klinke and Galley 1974, for review) and is the basis for the idea that the olivocochlear bundle exerts an inhibitory action on cochlear hair cell function. Recordings from single afferent fibers have confirmed that olivocochlear stimulation suppresses firing rate (Fex 1962) and have shown that the suppression is greatest when the acoustic stimulus is at the fiber's characteristic frequency (Wiederhold and Kiang 1970). Efferent nerve stimulation results, therefore, in a broadening of the tuning curve.

As more information becomes available about the arrangement of hair cell types it is becoming clear that there are, within many auditory end organs, hair cells that completely lack an efferent axosomatic supply. There is, for example, no efferent supply to the ear of *Myxine* (Lowenstein and Thornhill 1970) nor to the hair cells of the basilar papilla of ranid frogs (Robbins, Bauknight, and Honrubia 1967; Frischkopf and Flock 1974). In teleost fishes, some portions of the lagenar (*Colisa*, Wegner 1982) and saccular (*Astronotus*, Popper, personal communication) epithelia lack an efferent supply. In lizards the situation is complex: an efferent supply is present on hair cells that code for low frequencies and are located in that part of the papilla where hair cell orientation is unidirectional, whereas hair cells located in parts of the epithelium showing bidirectional orientation and

that code for high frequencies appear to lack an efferent innervation entirely (Miller and Beck 1988; Miller, Chapter 23; Köppl and Manley, Chapter 24). Bagger-Sjöbäck (1976) was unable to find any efferent endings in the ear of the lizard, *Calotes*. In birds, the tall hair cells on the neural side of the epithelium generally lack an efferent innervation, whereas the abneurally situated short hair cells are generously supplied with efferent endings, particularly at the apical end of the epithelium where the lowest frequencies are coded (Manley and Gleich, Chapter 27). In mammals, the inner hair cells seem to lack an efferent supply, although Saito (1980) reports an occasional axosomatic contact, and even outer hair cells in certain types of bat lack or have a reduced efferent innervation (Bruns and Schmieszek 1980; Bishop 1987; Pollak, Chapter 36).

There is no evidence that efferent nerve stimulation exerts a direct action on the inner hair cells of mammals, the majority of which lack an axosomatic supply. Rather it appears that efferent action on the outer hair cells influences the vibration pattern of the basilar membrane and this change leads to a reduced stimulation of the inner hair cells during sound stimulation (see Siegel and Kim 1982; Nuttall 1986; and Patuzzi and Robertson 1988, for a discussion of the efferent role in tuning). This raises two questions: what is the significance of the axodendritic innervation that underlies the inner hair cells; and are there similar mechanical interactions in other hair-cell sense organs between regions with and without efferent supplies?

4. Central Organization of the Auditory Efferent System

In all vertebrates examined so far, the efferent supply to the ear arises from cell somata that are located in the rhombencephalic tegmentum in one or more nuclei. In all classes except mammals, these neurons comprise a single nucleus, the octavolateralis efferent nucleus (OEN), which also houses the efferent neurons that supply the lateral line, should it be present (Roberts and Meredith 1989). A tabulation of what is known about the distribution, numbers and sizes of efferent neurons in various vertebrate genera is provided in Table 12.1.

TABLE 12.1. Number, laterality, and size of efferent neurons.

Animal	Total to one labyrinth[1]	Total to auditory endorgan[1]	Percentage contralateral	Size range[2]	References
Fish					
Epatretus	0		0		Amemiya et al. (1985)
Lampetra	22		0	12–20 μm	Fritzsch et al. (1989)
Lampetra	18		0	15–50 μm	Koyama et al. (1989)
Scyliorhinus	24		31	2,500 μm^2	Meredith and Roberts (1986a)
Amia			0	50–80 μm	McCormick and Braford 1979)
Opsanus		131 (saccule)	19	65 × 25 μm	Highstein and Baker 1986)
Gnathonemus		31 (saccule)	38		Bell (1981a)
Anguilla		40 (saccule)	8	418 μm^2	Meredith and Roberts (1987)
Amphibia					
Ichthyophis	8		0	18–24 μm	Fritzsch and
Boulengerula	13		0	18–24 μm	Crapon de Caprona (1984)
Hyla	15		0	25 μm	Strutz et al. (1982)
Xenopus	16		0	20–34 μm	Will (1982)
Xenopus	17		0		González (1986)
Hymenochirus	10		0	20–27 μm	Will (1982)
Rana	38		0	18–30 μm	Will (1982)
Salamandra	22		41	25–30 μm	Fritzsch (1981)
Bufo	25		0	28 × 16 μm	Pellegrini et al. (1985)
Pleurodeles	20		small %		González (1986)
Discoglosus	5		0		González (1986)
Reptilia					
Terrapene	63		33		Strutz (1982b)
Varanus	62		15		Barbas-Henry and Lohman (1988)
Caiman		263	63		Strutz (1981)
Aves					
Gallus		203	60		Whitehead and Morest (1981)
Mammalia					
Guinea pig		1,375	42	20–38 μm	Robertson (1985)
Guinea pig		2,107	31		Robertson et al. (1987)
Guinea pig		2,540	30		Aschoff and Ostwald (1987)
Rat		1,441	24	14 × 9 (sm^3)	Aschoff and Ostwald (1987)
				26 × 17 (lrg^3)	Aschoff and Ostwald (1987)
Rat		578	36	9 × 6	White and Warr (1983)
Rhinolophus		983	0		Aschoff and Ostwald (1987)

TABLE 12.1. (Continued)

Animal	Total to one labryrinth[1]	Total to auditory endorgan[1]	Percentage contralateral	Size range[2]	References
Mammalia (Continued)					
Cat		2,166	59-75		Warr (1975)
Saimiri		880	27 (sm[3])	165–324 μm[2]	Thompson and
			58 (lrg[3])		Thompson (1986)

[1]Expressed as the maximum number in the best case
[2]Expressed as major diameter, major × minor diameters, or as square area
[3]lrg, large OC efferent neurons; sm, small OC efferent neurons.

4.1. Organization in Anamniotes

No efferent somata are retrogradely labeled in the brainstem when horseradish peroxidase (HRP) is applied to the eighth nerve of the hagfish, *Epatretus* (Amemiya et al. 1985), a result that is in line with the ultrastructural studies indicating that myxinoid labyrinthine hair cells totally lack an efferent supply. However, efferent neurons have been found in the brainstem of lampreys (*Lampetra*, Fritzsch et al. 1989; Koyama et al. 1989). The labeled cells are large and multipolar and have widely spread, unilateral dendrites that extend ventrally into the reticular formation, although some course close to entering eighth nerve afferent fibers. The cells are located only ipsilaterally and are closely associated with the facial motor nucleus. They do not project to lateral line endorgans (Yamada 1973; Fritzsch et al. 1984).

The efferent somata that supply the ear have now been located in several species of cartilaginous and actinopterygian fishes. In the elasmobranch *Scyliorhinus*, eighth nerve efferent axons originate from a few, large multipolar neurons that lie in OEN at the rostral end of the motor column just rostral to and overlapping with the facial motor nucleus (Meredith and Roberts 1986a). As in the lamprey, in *Scyliorhinus* the dendrites extend ventrally into the reticular formation and laterally towards the octaval nuclei but, in contrast to the lamprey, many somata lie contralateral to the sense organs they innervate. In *Scyliorhinus*, some efferent axons branch and innervate sense organs of both the inner ear and lateral line (Meredith and Roberts 1986a). From the available, very limited data, it seems that the arrangement of OEN in other elasmobranchs

(*Rhinobatos*, Dunn and Koester 1987; *Raja*, Barry 1987) and in chondrostean (New and Northcutt 1984) and holostean (McCormick and Braford 1979) fishes resembles that seen in *Scyliorhinus*. Unfortunately, the data are insufficient for us to say whether OEN can be subdivided with respect to the different labyrinthine targets.

The arrangement of the efferent supply for teleosts differs in several respects from that of other fishes. In particular, the nuclei of each side unite across the midline rostrally, and in some species, the contralateral component is markedly smaller (Table 12.1) than that in elasmobranchs (Meredith 1988). However, as in *Scyliorhinus*, single efferent neurons can innervate sense organs of different modalities (e.g., *Poecilia*, Claas, Fritzsch, and Münz 1981). The centrifugal saccular and lagenar projections are known for three teleost species: *Gnathonemus* (Bell 1981a), *Opsanus* (Highstein and Baker 1986), and *Anguilla* (Meredith and Roberts 1987). In *Opsanus* most of the saccular neurons are located dorsally in the nucleus; in *Anguilla* they lie rostrally, closely associated with, but caudal to, the presumed superior olivary nucleus (Meredith and Roberts 1986b).

Labyrinthine efferent neurons have been identified in several amphibian species (reviewed in Will and Fritzsch 1987). They are multipolar, have extensive dendrites that spread laterally and ventrally, and are closely associated with the facial motor nucleus. In urodeles they are bilaterally arranged (Fritzsch 1981; González 1986), whereas in anurans (Strutz, Bielenberg, and Spatz 1982; Will 1982; González 1986) and apodans (Fritzsch and Crapon de Caprona 1984) they are only ipsilateral to the labeled nerve (Table 12.1). Some

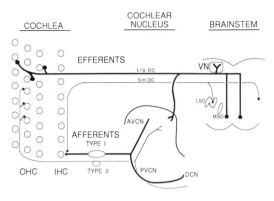

FIGURE 12.3. Schematic diagram of the efferent path-ways from the large (lrg) and small (sm) mammalian OC neurons in relation to the type I and type II afferent projections. Thick lines indicate thick axons and thin lines, thin axons. AVCN, anteroventral cochlear nuc-leus; DCN, dorsal cochlear nucleus; IHC, inner hair cells; OHC, outer hair cells; PVCN, posteroventral cochlear nucleus; LSO, lateral superior olive; MSO, medial superior olive; VN, vestibular nucleus. (Modi-fied from Brown et al. 1988.)

neurons innervate hair cells of both the inner ear and the lateral line (Claas, Fritzsch, and Münz 1981; Fritzsch and Wahnschaffe 1987).

4.2. Organization in Amniotes

In view of the significant role played by reptiles in the evolution of the ear, it is unfortunate that we have only very limited information about the central organization of efferent neurons in various reptilian radiations. In the turtle, *Terrapene* (Strutz 1982b), crocodilian, *Caiman* (Strutz 1981), and lizard, *Varanus* (Barbas-Henry and Lohman 1988), ves-tibular and auditory efferent neurons constitute a diffuse group that lies medial and just rostral to the facial motor nucleus and extends from the abducens nucleus to the superior olive. These neurons inter-mingle in the turtle, whereas according to Strutz (1981), they are separated in *Caiman*, with the most ventral neurons innervating only the cochlea.

Injections of HRP into the inner ear of birds have shown that the efferent neurons lie in a position similar to that in reptiles (Whitehead and Mor-est 1981; Strutz and Schmidt 1982). Strutz and Schmidt (1982) consider that the vestibular and auditory efferent neurons in the chicken lie in sep-

arate groups, with the vestibular component being the more dorsal. However, Whitehead and Morest (1981) in a more complete study, showed that there is, in fact, an extensive overlap, although the ves-tibular neurons tend to lie more dorsally than the auditory efferent neurons. Schwarz et al. (1981) report that vestibular and cochlear efferent neurons in the pigeon overlap extensively within the nucleus, although again the vestibular efferent neurons tend to lie more dorsally. In mammals, the axons of the olivocochlear (OC) bundle originate bilaterally from neurons in the superior olivary complex of the medulla (Fig. 12.3). Most data are available for the cat (Warr 1975), rat (White and Warr 1983), and guinea pig (Strutz and Bielenberg 1983; Robert-son 1985) but there have also been studies on old and new world monkeys (Strominger et al. 1981; Thompson and Thompson 1986), mouse (Taniguchi 1980), gerbil (Aschoff, Müller, and Ott 1985), and various bats (Aschoff and Ostwald 1987). The OC neurons can be divided into two groups on the basis of their morphology and location. One group projects primarily to the ipsilateral cochlea and contains small neurons that lie in or around the lateral superior olive (LSO). These cells have small, unmyelinated axons that predominantly end as axodendritic synapses under the inner hair cells (Fig. 12.3). These are referred to as lateral OC neu-rons by Warr and Guinan (1979) but, because they differ in location in different species, Adams (1983) designated them as small OC neurons. The other group projects predominantly to the contralateral cochlea and contains larger neurons that are asso-ciated with the medial superior olive (MSO). Their myelinated axons mostly form axosomatic synapses on the outer hair cells (Fig. 12.3). These neurons have been designated as medial OC neurons by Warr and Guinan (1979) and as large OC neurons by Adams (1983). These large neurons are absent from rhinolophid bats (Aschoff and Ostwald 1987). The small and large OC neurons differ in the neuroac-tive substances they contain (summarized in Table 12.2 and discussed further in Section 5) and also in their relative amounts of cytoplasm, Nissl sub-stance, and mitochondria, features that may reflect differences in their activities (Warr, Guinan, and White 1986).

The axons of the large OC neurons give off col-lateral branches before exiting from the brainstem (Brown et al. 1988). As well as terminating in the ventral cochlear nucleus (Fig. 12.3), where they apparently end among granule cells, some project to vestibular nuclei (interstitial nucleus of the vestibular nerve in cat (Fig. 12.3) and inferior vestibular nucleus in rat). In the mouse, fibers of large OC cells form terminal and en passant boutons that make asymmetric synapses with various neuronal elements in the granule cell regions of the cochlear nucleus (Benson and Brown 1990).

The efferent supply to the vestibular organs has been less well investigated than that to the cochlea. The central organization has been described in the cat (Gacek and Lyon 1974; Warr 1975; Ito et al. 1983; Dechesne et al. 1984), rat (Schwarz et al. 1986; Aschoff and Ostwald 1987), guinea pig (Strutz 1982a), gerbil (Perachio and Kevetter 1989), and squirrel monkey (Goldberg and Fernández 1980). Although there are differences in the results of these studies, the general picture of the supply to the vestibular organs is the same. In contrast to the efferent supply to the cochlea, that to the vestibular system arises from fewer neurons, most of which are ipsilateral to their targets. In the gerbil, however, the contralateral efferent innervation of saccule and utricle is greater than the ipsilateral supply (Perachio and Kevetter 1989). The vestibular efferent neurons are contained within two nuclei: a dorsal vestibular efferent nucleus (Goldberg and Fernánde 1980; Gacek 1981; Ito et al. 1983; Schwarz et al. 1986), split in two by the genu of the facial nerve tract and a ventral nucleus, which is a part of the caudal pontine reticular formation in some species (Strutz 1982a; Schwarz et al. 1986).

5. Putative Neurotransmitters of the Auditory Efferent System

For many years it seemed that the transmitter manufactured by the efferent neurons had to be acetylcholine (ACh). Acetylcholine and its synthetic [choline acetyltransferase (ChAT)] and degradative [acetylcholinesterase (AChE)] enzymes have been demonstrated in both axons and perikarya of efferent neurons, not only for the auditory (Klinke and Galley 1974; Altschuler and Fex 1986), but also for vestibular (Flock and Lam 1974; Goldberg and Fernández 1980; Schwarz et al. 1986) and lateral line (reviewed in Roberts and Meredith 1989) systems.

Bobbin and Konishi (1971) first established that ACh could be a neurotransmitter in the efferent system by showing that it can mimic the effects of OC bundle stimulation. The effect of electrical stimulation of the OC bundle could also be enhanced by the cholinergic agonist, carbachol, and blocked by muscarinic and nicotinic antagonists and by the inhibition of choline uptake at the terminal (reviewed in Klinke and Galley 1974; Klinke 1981). Although ACh is now widely accepted as being a major neurotransmitter, its action may not be as straightforward as previously thought. For example, it inhibits stimulus-induced hair cell and afferent nerve activity but seems to have no effect on the spontaneous activity of afferent fibers. Moreover, nicotinic antagonists appear to be more effective than muscarinic antagonists in blocking the impact of olivocochlear bundle stimulation (Klinke 1986).

As well as the uncertainties about the action of ACh, substantial evidence has come in the last decade or so from immunohistochemical, biochemical and pharmacological studies that indicates that other neurochemicals could also be playing a role in efferent neurotransmission. Table 12.2 is an inventory of the types of neurochemicals reportedly present in efferent neurons for different species.

5.1. Neurotransmitter Distribution in Animals Other Than Mammals

Acetylcholine (Auerbach and Budelmann 1986), noradrenaline, and dopamine (Budelmann and Bonn 1982) have been identified histochemically in the efferent nerve fibers that supply the statocyst of *Octopus* and electrophysiological studies have shown that ACh probably has an inhibitory, and catecholamines an excitatory, effect in the cephalopod statocyst (Williamson 1989a).

Octavolateralis efferent neurons of the teleost fish, *Anguilla*, stain positively for AChE (Meredith and Roberts 1986b) and immunoreact for ChAT (van der Jagt, Maslam, and Roberts 1990, Fig. 12.4A). Choline acetyltransferase-immunoreactive efferent neurons have also been identified in the fish *Carassius* (Danielson et al. 1988) and *Porichthys* (Brantley and Bass 1988) and in the reptile, *Gekko* (Fig. 12.4B). Further, the presence of ChAT has been demonstrated radiochemically in the labyrinths of the frog, *Rana* (López and Meza 1988), and the chick (Meza and Hinojosa 1987).

TABLE 12.2. Neuroactive Substances Associated with the Octavolateralis Efferent Neurons.

Neurochemical Animal	Large OC neurons	Small OC neurons	Vestibular Efferent neurons	CPR[1] (ventral) Efferent neurons	OEN neurons	References
AChE						
Anguilla					+	Meredith and Roberts (1986b)
Cat	+	+	+			Warr (1975)
Rat	+	+	+	−		Schwarz et al. (1986)
Guinea pig	+	+				Altschuler et al. (1983)
Squirrel monkey			+			Goldberg and Fernández (1980)
Rhinolophus	+ (nucleus olivocochlearis)					Aschoff and Ostwald (1987)
Gerbil			+	−		Perachio and Kevetter (1989)
ChAT						
Porichthys					+	Brantley and Bass (1988)
Carassius					+	Danielson et al. (1988)
Anguilla					+	van der Jagt et al. (1990)
Rana					+	López and Meza (1988)
Gekko					+	Fig. 12.4B
Rat	+	−	+	−		Schwarz et al. (1986)
Rat	+	+				Abou-Madi et al. (1987)
Guinea pig	+	+				Altschuler et al. (1984)
Cat	+	+				Jones and Beaudet (1987)
Gerbil			+	−		Perachio and Kevetter (1989)
GABA						
Guinea pig	+	+				Peyret et al. (1986); Fex et al. (1986)
Gerbil	+	+				Schwartz and Ryan (1986)
Gerbil			−	−		Perachio and Kevetter (1989)
GAD						
Anguilla					+	van der Jagt et al. (1990)
Guinea pig	+	+				Fex and Altschuler (1984)
D-Aspartic Acid						
Gerbil	−	+	−			Ryan et al. (1987)
Aspartate amino transferase and glutaminase						
Guinea pig	+	−				Altschuler et al. (1981); Altschuler and Fex (1986)

TABLE 12.2. (Continued)

Neurochemical Animal	Large OC neurons	Small OC neurons	Vestibular Efferent neurons	CPR[1] (ventral) Efferent neurons	OEN neurons	References
TH						
Guinea pig	−	+				Altschuler and Fex (1986)
Peptides						
ENK						
Rat	−	+				Abou-Madi et al. (1987)
Guinea Pig	−	+				Hoffman et al. (1984); Altschuler et al. (1985)
Gerbil			+	−		Perachio and Kevetter (1989)
DYN						
Guinea pig	−	+				Altschuler et al. (1985)
Rat	−	+				Abou-Madi et al. (1987)
CGRP						
Xenopus					+	Adams et al. (1987)
Cat	−	+				Schweitzer et al. (1985)
Rat	−	+				Schweitzer et al. (1985)
Gerbil			+	−		Perachio and Kevetter (1989)
Colocalizations						
AChE + ENK						
Guinea pig	−	+				Altschuler et al. (1983)
ChAT + ENK						
Rat	−	+				Abou-Madi et al. (1987)
Guinea pig	−	+				Abou-Madi et al. (1987)
ChAT + DYN						
Rat	−	+				Abou-Madi et al. (1987)
Guinea pig	−	+				Abou-Madi et al. (1987)
ENK + DYN						
Rat	−	+				Abou-Madi et al. (1987)
Guinea pig	−	+				Abou-Madi et al. (1987)

[1]CPR, the caudal pontine reticular formation contains a few neurons that are efferent to vestibular endorgans

Abbreviations: AChE, acetylcholinesterase; CGRP, calcitonin gene-related peptide; ChAT, choline acetyltransferase; DYN, dynorphin; ENK, enkephalin; GABA, gamma aminobutyric acid; GAD, glutamate decarboxylase; TH, tyrosine hydroxylase.

Little is known about the distribution of neuro-transmitters other than ACh, in nonmammalian vertebrates. There is new evidence (van der Jagt, Maslam, and Roberts 1990), however, that some OEN neurons in the teleost *Anguilla* immuno-react positively with antibodies directed against glutamate decarboxylase (GAD), the synthetic enzyme for the production of gamma aminobu-tyric acid (GABA). Calcitonin gene-related pep-tide (CGRP) has been localized to small fibers in the peripheral lateral line system of *Xenopus* and has been shown to exert an excitatory action on afferent nerve activity (Adams, Mroz, and Sewell 1987).

5.2. Neurotransmitter Distribution in Mammals

On the basis of AChE histochemistry and ChAT immunocytochemistry, it seems that ACh is present in both divisions of the olivocochlear system (see Fig. 12.4C–D and Table 12.2). It is apparently the major efferent neurotransmitter and is probably responsible for the inhibitory action of the olivoco-chlear system.

It has been suggested that transmitters other than ACh might be present in efferent neurons ever since studies in the vestibular system showed that efferent stimulation could exert an excitatory as well as an inhibitory effect (Goldberg and Fernán-dez 1980). It is now known that neurochemicals such as γ-aminobutyric acid (GABA) and various peptides are present in both small and large OC neu-rons (Table 12.2; see Altschuler and Fex 1986, for review) and several peptides have been colocalized with ACh in small OC neurons (Altschuler et al. 1984; Abou-Madi et al. 1987) suggesting a coaction of these substances. Using in situ hybridization, Harlan et al. (1987) identified, in the superior olive, neurons that contain the messenger RNA that codes for preproenkephalin; these neurons seem to be similar in location to OC neurons.

Choline acetyltransferase immunoreactivity has been identified in neurons of the vestibular effer-ent nucleus of the rat (Schwarz et al. 1986) as has immunoreactivity for the peptide, calcitonin gene-related peptide (CGRP) (Tanaka et al. 1989). Ter-minals labeled for CGRP have recently been identi-fied as ending on both inner and outer hair cells as well as on afferent fibers in the inner spiral bundle

of the rat (Kuriyama et al. 1990). The ventromedial group of vestibular efferent neurons in the caudal pontine reticular formation is negative both for ChAT and CGRP (Schwarz et al. 1986).

6. Circuitry Underlying the Activation of the Auditory Efferent System

Herrick (1948) was the first to recognize that the olivocochlear pathway would provide a system whereby "the excitation of a peripheral sense organ may be followed by an efferent discharge back to the receptor." In such a circuit, the efferent neurons would be driven by sensory input and pro-vide a feedback control of sense organ sensitivity. Although subsequent studies have confirmed that hair cell stimulation can activate efferent neurons, they have also shown other stimuli to be equally or even more effective. This responsiveness to a range of sensory modalities raises the question as to whether the efferent neurons are being driven by the primary sensory stimulus or by some central program that is initiated by sensory stimulation. This essential difference in the activation of the OES can be expressed in feedback and feedforward models of efferent neuron function (Fig. 12.5; see Roberts and Meredith 1989).

6.1. Sensory Activation of Efferent Neurons

There is now a lot of evidence to show that the mammalian OC neurons can be activated by sound stimulation to the ear (Fex 1962; 1965; Liberman and Bell 1979; Rupert et al. 1968; Cody and John-stone 1982; Robertson 1984; Robertson and Gum-mer 1985) and this would seem to support a feed-back model. At first it was thought that the sounds had to be loud but more modern work has shown that efferent neurons will respond to sound levels that are similar to those that drive the primary audi-tory fibers (Robertson and Gummer 1988). Efferent neurons are seldom spontaneously active in the absence of sound and are sharply tuned to restricted frequencies (Cody and Johnstone 1982). However, the latencies of activation are long (up to 40 ms) and

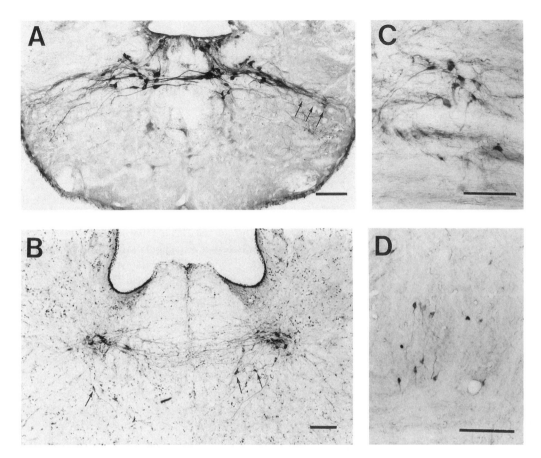

FIGURE 12.4. Photomicrographs of ChAT-immunore-active neurons that are presumed to be efferent neurons, seen in transverse sections of the brainstem (A) OEN of *Anguilla*; arrows indicate the lateral extent of the dendrites; (B) OEN of *Gekko*; arrows indicate scattered ChAT-immunoreactive neurons ventral to the main group (preparation kindly loaned by Dr. P. Hoogland); (C) Large OC neurons of rat; (D) Small OC neurons of rat; note the difference in size with the large OC cells in (C). Scale bars (A–D) = 100 μm.

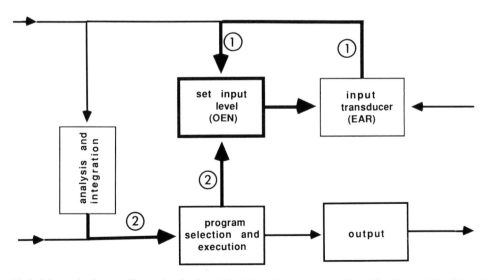

FIGURE 12.5. Schematic diagram illustrating feedback (1 – driven by sensory input) and feedforward (2 – driven by central circuits) models of OEN activation.

197

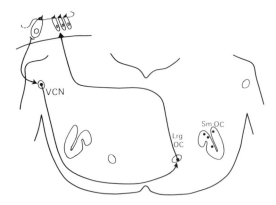

FIGURE 12.6. Probable circuit involved in the acoustic activation (inner hair cells) of contralateral large (Lrg) OC neurons in the guinea pig. (Taken from Robertson and Winter 1988.) Sm OC, small OC neurons; VCN, ventral cochlear nucleus.

this suggests that an indirect, multisynaptic pathway is involved, although shorter latencies have been obtained with electrical stimulation of the auditory nerve (Fex 1962).

The basis for the feedback action initiated by sound stimulation in mammals could be the connections that have been demonstrated between the cochlear nuclei and the superior olivary nuclei (Fig. 12.6). Axons originating from cells in the dorsal cochlear nucleus pass directly to the contralateral inferior colliculus, whereas those from the ventral cochlear nuclei end ipsilaterally in the LSO, and bilaterally in the MSO and surrounding periolivary regions (guinea pig, Robertson and Winter 1988). It is clear from dual-labeled material that the axons of the cochlear nuclei are terminating among contralateral medial OC neurons (Robertson and Winter 1988); consequently, acoustic stimulation presented monaurally will activate contralateral large OC neurons which then project back to the stimulated ear (see Wiederhold 1986, for review).

Further support for a feedback efferent model has come from studies on other vertebrates that have shown hair cell sense organ stimulation to evoke efferent nerve activity. In *Scyliorhinus* (Roberts and Russell 1972) and in *Lota* (Flock and Russell 1973), vibration excites lateral line efferent neurons and airborne sound, vibration, and head rotation are effective stimuli in activating labyrinthine efferent neurons in amphibians

(Schmidt 1963; Schmidt 1965; Gleisner and Hendricksson 1964). An anatomical basis for this reflex activation could be provided by the mono- and disynaptic projections of the primary sensory input onto efferent neurons that have been reported for several species. Some eighth nerve afferent fibers from the semicircular canals pass through the octavolateralis sensory region and enter OEN directly, where they may terminate (*Raja*, Barry 1987; *Anguilla*, Meredith, Roberts, and Maslam 1987; *Varanus*, Barbas-Henry and Lohman 1988). Highstein and Baker (1986) provide physiological evidence for such monosynaptic efferent activation in *Opsanus*, following stimulation of semicircular canal efferent nerves. These findings suggest that there could be a monosynaptic connection between primary afferent (vestibular?) and efferent neurons. However, another reflex pathway could be provided by the neurons of the octavolateralis sensory nuclei which receive the primary afferent input (reviewed in Meredith 1988 and McCormick, Chapter 17). The axons of secondary cells then ascend via the lateral lemniscus to the torus semicircularis (Bell 1981b; Will 1982; de Wolf, Schellart, and Hoogland 1983; Meredith and Roberts 1986b; Barry 1987; González and Muñoz 1987), the presumed homologue of the mammalian inferior colliculus, and as they do so, they pass through OEN en route, where they presumably could make synaptic contact with the efferent neurons. Such disynaptic connections, which have recently been documented for *Anguilla* (Mos, van der Jagt, and Roberts 1989), provide a pathway that is similar to the cochlear nucleus-superior olivary connection reported in mammals (Fig. 12.6).

6.2. Central Activation of Efferent Neurons

We turn now to the evidence underpinning the feedforward model of efferent neuron activation. Support for this idea comes not only from the finding that the latency to hair cell stimulation is longer than would be expected of a short circuit but also from the reports of studies in anamniotes that have shown stimulation of other sensory receptors, in addition to those in the ear and lateral line, to activate efferent neurons. Efferent neuron activity is seen, for example, in response to visual (Späth and Schweickert 1975; Hartmann and Klinke 1980; Münz and Claas 1986; Tricas and Highstein 1990) and tactile (Russell 1971; Roberts and Russell 1972) stimuli. However, the best evidence for the feedforward model is provided by the good correlation that has been observed in certain anamniotes between efferent nerve activity and motor performance—the efferent neurons become active when an animal is moving or is about to move (evidence reviewed in Roberts and Meredith 1989). The actual movement, however, is not essential because efferent nerve activity is still evoked by tactile (Roberts and Russell 1972), reticulospinal tract (Flock and Russell 1973), and optokinetic (Klinke and Schmidt 1970) stimulation in paralyzed animals. It seems, therefore, that efferent neurons are activated not by the sensory input or by the motor output but by the initiation of certain central circuits.

A good example of the involvement of the efferent system in a complete piece of behavior, triggered by labyrinthine stimulation, is provided by the C-start rapid escape that is displayed by many fishes. This movement, which can be evoked by eighth nerve stimulation, sudden vibrations, and by certain visual stimuli, is controlled by a network of brainstem and spinal cord neurons, of which the best known is the Mauthner cell. Efferent neurons supplying the lateral line and the ear are active during the C-start (Russell 1974) and saccular hair cells (Furukawa 1966) and nerve fibers (Lin and Faber 1988) are inhibited. The peak of the inhibition occurs just after the Mauthner cell has been activated and shortly before the tail flip begins. It is likely that the efferent population discharges throughout the course of the escape movement because other reticulospinal neurons participate in the escape movement and also drive the efferent neurons (Russell 1974).

It remains unknown at what level in the central nervous system the OES is activated but it now seems to involve those circuits that underlie arousal and attention. It is well known that when one sensory system dominates a piece of behavior, then other sensory inputs become less effective. For example, in a cat concentrating on a visual target (mouse in a jar) the auditory input, as measured electrophysiologically in the cochlear nucleus, is reduced (Hernández-Peón, Scherrer, and Jouvet 1956). This reduction in responsiveness could, of

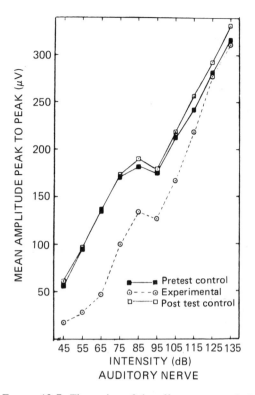

FIGURE 12.7. The action of the efferent system during selective attention. Summed response (N1) of auditory nerve fibers to sound presentations of increasing intensity under control conditions before and after (solid lines) the performance (dotted line) of a visual discrimination task. (Taken from Oatman 1976.)

course, be entirely determined centrally but there is evidence to indicate that such selective attention involves inhibition at the periphery, mediated via the OC bundle (Fig. 12.7; Oatman and Anderson 1977) and perhaps regulated by the descending auditory pathway (Glenn and Oatman 1980). Although there are also data for other vertebrates indicating that central processes and the behavioral state can influence the input from the ear to the brainstem (Piddington 1971a, b; Laming and Brooks 1985), it is presently unclear whether these adaptive processes are occurring centrally or peripherally at the sense organ. A recent series of elegant experiments, reported by Tricas and Highstein (1990), has implicated the OES in selective attention by showing that lateral line afferent nerve activity recorded from a toadfish (*Opsanus*) is reduced when the fish attends to a visual target.

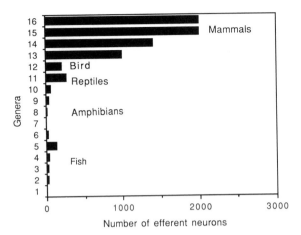

FIGURE 12.8. Bar chart of the number of efferent neurons reported for selected genera. 1. *Epatretus*; 2. *Lampetra*; 3. *Scyliorhinus*; 4. *Anguilla*; 5. *Opsanus*; 6. *Salamandra*; 7. *Discoglosus*; 8. *Xenopus*; 9. *Rana*; 10. *Terrapene*; 11. *Caiman*; 12. *Gallus*; 13. *Rhinolophus*; 14. *Rattus*; 15. *Cavia*; 16. *Felis*. References given in Table 12.1.

Very little is known for any vertebrate about the neural pathways involved in the central activation of the efferent neurons. We would expect from the feedforward model that the central control of efferent activity would lie in those parts of the brain responsible for behavioral (locomotor) initiation. However, as far as we are aware, the central connections of the OC neurons have been investigated only in relationship to the descending auditory pathways. The output of the auditory cortex in rats is directed ipsilaterally to the medial geniculate body and bilaterally to the inferior colliculus (IC). Fibers from the external cortex and central nucleus of the IC descend ipsilaterally to terminate among large OC neurons (Faye-Lund 1986; Syka et al. 1988). No projections to the small OC neurons have yet been reported. In line with these anatomical findings, Rajan (1990) has recently demonstrated in the guinea pig that electrical stimulation of the inferior colliculus has a protective effect on the cochlea, reducing the neural desensitization that results from loud sound exposure, and apparently involving the large OC neurons.

It seems likely that homologous pathways exist in other vertebrates but the data are too scanty to confirm this. However, in bony fishes (Echteler 1984; Carr and Maler 1985) and in *Rana* (Jacoby and Rubinson 1984) descending fibers from the torus semicircularis are known to pass into the medial reticular formation, where they could contact the extensive dendritic arborizations of the efferent neurons.

7. Trends in the Evolution of OEN

Several evolutionary trends are evident: an increase in the number of efferent neurons (Fig. 12.8), a decrease in their size (Table 12.1) and dendritic arborization, and a subdivision and migration of cell masses in relation to the loss of a lateral line system and the acquisition of a cochlea.

In our survey of the distribution of the efferent neurons in the brainstem, we can recognize three patterns: piscine, reptilian, and mammalian. The piscine pattern, which occurs in fishes and amphibia, consists of a single nucleus, the OEN. These neurons innervate auditory, vestibular, and lateral line sense organs and are intermingled, bearing little or no topographical relationship to the peripheral endorgans (Meredith and Roberts 1987). Indeed, some cells have axons that branch and innervate sense organs of different modalities (Claas, Fritzsch, and Münz 1981).

The reptilian pattern, seen in both reptiles and birds, also consists of a single nucleus, the neurons of which exhibit some topography in their organization because the neurons that supply auditory endorgans tend to lie more ventrally in the nucleus than do those that innervate the vestibular organs. However, there is considerable overlap of these neurons and some have collaterals that supply both otolithic and semicircular canal organs (Schwarz et al. 1981).

The auditory and vestibular efferent neurons are completely separated in the mammalian pattern, which is based therefore on more than one cell group. The auditory (olivocochlear) cells lie ventrally and can be subdivided into two groups that differ in their location, morphology, and content of neuroactive substances. The vestibular neurons lie dorsally, close to the fourth ventricle and are also organized into more than one group.

Phyletic comparison of these three patterns suggests several common features and leads to ideas about the evolution of the efferent system, some of which are illustrated in Figure 12.9. A striking common feature is that the perikarya of efferent

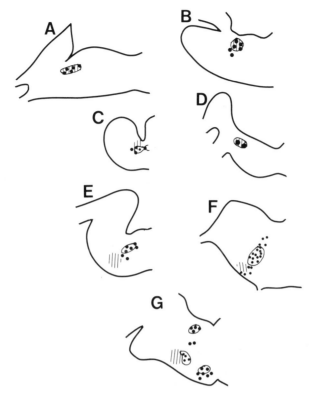

FIGURE 12.9. Schematic diagrams of the rhomben-cephalic tegmentum of a variety of vertebrates illustrating the relationship between the facial motor nucleus (cross-hatched) and the efferent neurons (solid dots) in the octavolateralis efferent nuclei (shown in outline). In (A) *Lampetra*, the overlap between efferent neurons and facial motoneurons is complete at a rostral level of the facial motor nucleus. In (B) *Scyliorhinus*, the facial motor nucleus is contiguous caudally with OEN. In (C) *Anguilla*, the overlap occurs at all levels of both nuclei. In (D) *Pleurodeles*, the overlap is complete at all levels for both nuclei. In (E) *Varanus*, the facial motor nucleus appears at the same level as the caudalmost cells of OEN. In (F) *Gallus*, the facial motor nucleus appears at the same level as OEN. In (G) the albino rat, the facial motor nucleus appears at the caudal pole of auditory (more ventral) and vestibular (more dorsal) efferent neurons. Drawings (A–G) are based on data in Koyama et al. (1989); Meredith and Roberts (1986a); Meredith and Roberts (1987); González (1986); Barbas-Henry and Lohman (1988); Whitehead and Morest (1981); White and Warr (1983), respectively.

neurons are closely associated with the rhomben-cephalic motor column, overlapping with facial motoneurons in the piscine pattern and with both facial and abducens motoneurons in the reptilian and mammalian patterns. Because of this striking overlap, we have suggested (Meredith and Roberts 1986a) that the efferent neurons are phylogenetically derived from the motor column even though they innervate neuroepithelium rather than muscle. Efferent neurons, in common with motoneurons, synthesize ACh as neurotransmitter and have axons that follow the same pathway as those of the

facial motoneurons. Indeed, Bischoff (1899), who was the first to recognize efferent fibers in the vestibular root in mammals, considered them to be "aberrant" facial nerve fibers.

Unlike most rhombencephalic motor neurons, some efferent neurons lie contralateral to their target(s). This decussating feature of the efferent system, which was emphasized by Rasmussen in his first report (1946), varies considerably between species: the percentage of cells contained in the nucleus that have contralateral targets ranges from 0 to 60% (see Table 12.1). It is presently a matter of opinion as to whether the purely ipisilateral condition, as seen in the lamprey, is primitive.

Within the motor column of fishes, the facial motor nucleus lies just under the floor of the fourth ventricle. In amphibia, it migrates slightly into the reticular formation but is located further ventrally in reptiles, birds, and mammals and in mammals, the ventrolateral position is known to come about during development as the facial neuroblasts migrate ventrally and laterally (Windle 1933). The efferent neurons, which in fishes are closely associated with the facial motoneurons, in other vertebrates can be seen (Fig. 12.9) to retain a close juxtaposition during the phylogenetic migrations of the motor column. In mammals it is the auditory efferent neurons that lie close (rostral) to the now ventrally located facial nucleus whereas the vestibular neurons retain a dorsal and, presumably therefore, primitive position.

Auditory efferent neurons in both the reptilian and mammalian patterns are associated with the superior olivary complex. However, the overlap between these two neuronal groups is not seen in fishes where the presumed superior olive lies rostral to the efferent neurons and can be distinguished from them (Meredith and Roberts 1986b). It could be argued, therefore, that the close relationship of certain efferent neurons with the superior olive is a derived mammalian condition that is brought about by the ventral migration of the motor nuclei.

The similarities between efferent neurons in a wide range of species, in terms of cell somata location, transmitter synthesis, peripheral innervation patterns, and central connections, make it difficult to believe that the three patterns arose independently. Consequently, we suggest that the single efferent nucleus that is seen in the piscine pattern represents the primitive condition and is homologous to the vestibular efferent nucleus of mammals. We would argue that the separation in vestibular and auditory groupings seen in mammals (the large OC neurons) results from a splitting and migration from the single nucleus. The small OC neurons, however, in view of their distinctive morphological and neurochemical features (see Section 4 and Table 12.2), may not be homologous to OEN neurons and may be independently derived.

8. Other Hair Cell Systems with an Efferent Innervation

An efferent supply to hair cell sense organs is found not only in the inner ear but also in other sense organs in vertebrates and in invertebrates. The best known example is the mechanoreceptive lateral line of fishes and amphibians, the efferent innervation of which we have reviewed elsewhere (Roberts and Meredith 1989). Both axodendritic and axosomatic synapses have been observed in lateral line neuromasts, given off by axons that arise from neurons in the OEN. The action of the efferent supply to the lateral line is inhibitory.

The vesicles of Savi of electric rays have an efferent supply (Nickel and Fuchs 1974) but the proprioceptive spiracular organ, found in several groups of fishes, apparently lacks efferent synapses (Barry and Bennett 1989). The electroreceptive lateral line system is also composed of modified hair cells and appears to have a close evolutionary relationship to the mechanoreceptive lateral line yet it completely lacks an efferent innervation (see Bodznick 1989, for discussion). A final example is provided by the paratympanic organ of birds, a sense organ in the middle ear that may serve as a pressure receptor, the hair cells of which receive an efferent innervation (Jørgensen 1984) that arises from somata that lie adjacent to the motoneurons of the facial motor nucleus (von Bartheld 1990). Some invertebrate hair cell sense organs also have an efferent supply. For example, the statocyst of cephalopods contains secondary hair cells that are densely innervated by efferent endings (Fig. 12.10; see Budelmann 1988, for review). Indeed, more than 70% of the axons that supply the statocyst of *Octopus* are efferent and when stimulated can reduce or enhance the sensory discharge (Williamson 1985; 1989b). Efferent synapses on the primary hair cells of gastropod molluscs also exert inhibitory (Detwiler and Alkon 1973) and excitatory (Janse et al. 1988) effects.

9. The Origins of the OES

The evolutionary origins of the OES are unknown but we assume that the cephalopod and vertebrate systems represent parallel developments that arose independently. In both groups the primary function of the OES has to be sought in its action at the periphery because tasks such as the reduction of incoming information, or changes in the content of that information, could all be adequately performed centrally at the first or subsequent synapses within the central nervous system. It is important, therefore, to focus on the organization of the peripheral efferent system; in particular, it might be most revealing to understand how some hair cells function without any efferent supply. In our subsequent discussion we shall assume that the noninnervated hair cell reflects the primary, primitive condition because such hair cells are found in protochordates and hemichordates (see Jørgensen 1989 for review). Beneath these cells in the amphioxus, *Branchiostoma*, vesicle-filled nerve fibers have been observed (Bone and Best 1978, their Plate 1H) that might represent a simple form of efferent system should they release neuroactive substances to modulate sensory cell activity, as has been reported for a simple crustacean mechanoreceptor (Pasztor and Bush 1987). In this way, the first efferent system would have been able to modulate, nonsynaptically, the performance of both hair cells and sensory nerve fibers, an action that could be better specified by the formation of axosomatic and axodendritic contacts.

Ontogenetic evidence suggests that the efferent supply arose secondarily to the afferent innervation because in the development of the mammalian and avian ear, afferent fibers are in place and forming mature synapses before the efferent fibers grow into the sense organ from the medulla (Pujol and Lenoir 1986; Whitehead 1986).

We have seen (Section 3) that the OES is absent from myxinoids but present in cyclostomes, where it innervates labyrinthine but not lateral line hair cells. Should these modern genera represent primitive conditions, it is likely that the OES originated very early in vertebrate evolution, most probably in association with the vestibular system. Perhaps the significance of the OES can be seen in the role that is played by the vestibular system in the coordination of movement and the matching of vestibular endorgan performance to the range of stimuli reaching the ear as a result of head and body movement. This matching involves the excitatory OES which enables semicircular canal afferent fibers

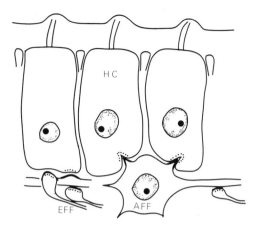

FIGURE 12.10. Schematic diagram illustrating the pattern of afferent (AFF) and efferent (EFF) terminals in the sensory epithelium of a statocyst of cephalopods. Note the axosomatic and axodendritic contacts of the EFF endings. HC, hair cell. (Modified from Budelmann 1988.)

to discharge bidirectionally throughout entire stimulus cycles (Boyle and Highstein 1990) and the inhibitory OES which reduces hair cell sensitivity, so as to protect the sensory synapse during rapid movement. We suggest that the important role played by the OES in audition is a phylogenetically later development that came about with the adoption of terrestrial life and the evolution of a cochlea.

10. Summary and Conclusions

In the course of vertebrate evolution, auditory endorgans have become more specialized and the efferent system has developed in parallel with these peripheral changes. This coevolution might be taken to suggest that the OES is an adaptation for hearing. However, the centrifugal supply to the auditory endorgans is just one component of an extensive efferent system (OES) that has been recognized in hair cell sense organs subserving other modalities. Within OES, a number of features are ubiquitous: the neurons comprise one group, OEN, in anamniotes, which subsequently proliferates and migrates; these cholinergic cells are associated with the brainstem motor column and are relatively few in relation to the sensory cells they innervate; their patterns of central and peripheral pathways are remarkably similar throughout vertebrate phylogeny and the fundamental role of the system is to be understood in terms of its yet undefined action in controlling sense organ sensitivity in the aroused animal.

Several features are still puzzling: for example, why in some species there is a crossed component yet in others the projection is entirely uncrossed; why an efferent innervation is widespread yet not totally universal so that some hair cells operate without any efferent control; why ACh is used as the transmitter but other neuroactive substances are also synthesized; and why both an inhibitory and an excitatory impact on sensory activity can follow efferent nerve stimulation. Until these questions are answered, the efferent system will remain an enigma.

Acknowledgments. We are grateful to Drs. W. Mos, B. van der Jagt, and R. Williamson for their comments on the manuscript and to Drs. P. Hoogland and A. González for the loan of histological material prepared from *Gekko* and *Pleurodeles*, respectively. Some of our own experimental work was supported by a National Research Service Award to Dr. G. Meredith from the NINCDS, NRSA, 5F32NS07286.

References

Abou-Madi L, Pontarotti P, Tramu G, Cupo A, Eybalin M (1987) Coexistence of putative neuroactive substances in lateral olivocochlear neurons of rat and guinea pig. Hearing Res 30:135–146.

Adams JC (1983) Cytology of periolivary cells and the organization of their projections in the cat. J Comp Neurol 215:275–289.

Adams JC, Mroz EA, Sewell WF (1987) A possible neuro-transmitter role for CGRP in a hair-cell sensory organ. Brain Res 419:347–351.

Altschuler RA, Fex J (1986) Efferent neurotransmitters. In: Altschuler RA, Hoffman DW, Bobbin RP (eds) Neurobiology of Hearing: The Cochlea, Raven Press, New York, pp. 383–396.

Altschuler RA, Neisses GR, Harmison GC, Wenthold RJ, Fex J (1981) Immunocytochemical localization of aspartate aminotransferase immunoreactivity in cochlear nucleus of the guinea pig. Proc Natl Acad Sci (USA) 78:6553–6557.

Altschuler RA, Parakkal MH, Fex J (1983) Localization of enkephalin-like immunoreactivity in acetylcholin-

esterase-positive cells in the guinea-pig lateral superior olivary complex that project to the cochlea. Neuroscience 9:621–630.

Altschuler RA, Fex J, Parakkal MH, Eckenstein F (1984) Colocalization of enkephalin-like and choline acetyltransferase-like immunoreactivities in olivocochlear neurons of the guinea pig. J Histochem Cytochem 32:839–843.

Altschuler RA, Hoffman DW, Reeks KA, Fex J (1985) Localization of dynorphin B-like and alpha-neoendorphin-like immunoreactivities in the guinea pig organ of Corti. Hearing Res 17:249–258.

Amemiya F, Kishida R, Goris RC, Onishi H, Kusunoki T (1985) Primary vestibular projections in the hagfish. *Eptatretus burgeri.* Brain Res 337:73–79.

Art JJ, Fettiplace R (1987) Variation of membrane properties in hair cells isolated from the turtle cochlea. J Physiol 38:5 207–242.

Art JJ, Crawford AC, Fettiplace R, Fuchs PA (1982) Efferent regulation of hair cells in the turtle cochlea. Proc R Soc Lond B 216:377–384.

Art JJ, Fettiplace R, Fuchs PA (1984) Synaptic hyperpolarization and inhibition of turtle cochlear hair cells. J Physiol 356:525–550.

Aschoff A, Ostwald J (1987) Different origins of cochlear efferents in some bat species, rats, and guinea pigs. J Comp Neurol 264:56–72.

Aschoff A, Müller M, Ott H (1985) Efferent olivocochlear neurons in gerbils. Neurosci Lett (Suppl) 22:431.

Ashmore JF (1987) A fast motile response in guinea-pig outer hair cells: the cellular basis of the cochlear amplifier. J Physiol 388:323–349.

Ashmore JF, Russell IJ (1982) Effect of efferent nerve stimulation on hair cells of the frog sacculus. J Physiol (Lond) 329:25–26.

Auerbach B, Budelmann BU (1986) Evidence for acetylcholine as a neurotransmitter in the statocyst of *Octopus vulgaris.* Cell Tissue Res 243:429–436.

Bagger-Sjöbäck D (1976) The cellular organization and nervous supply of the basilar papilla in the lizard, *Calotes versicolor.* Cell Tissue Res 165:141–156.

Barbas-Henry HA, Lohman AHM (1988) Primary projections and efferent cells of the VIIIth cranial nerve in the monitor lizard, *Varanus exanthematicus.* J Comp Neurol 277:234–249.

Barry M (1987) Afferent and efferent connections of the primary octaval nuclei in the clearnose skate *Raja eglanteria.* J Comp Neurol 266:457–477.

Barry MA, Bennett MVL (1989) Specialized lateral line receptor systems in elasmobranchs: The spiracular organs and vesicles of Savi. In: Coombs S, Görner P, Münz H (eds) The Mechanosensory Lateral Line. New York: Springer-Verlag, pp. 591–606.

Bell CC (1981a) Central distribution of octavolateral afferents and efferents in a teleost (*Mormyridae*). J Comp Neurol 195:391–414.

Bell CC (1981b) Some central connections of medullary octavolateral centers in a mormyrid fish. In: Tavolga WN, Popper AN, Fay RR (eds) Hearing and Sound Communication in Fishes. New York: Springer-Verlag, pp. 383–392.

Benson TE, Brown MC (1990) Synapses formed by olivocochlear axon branches in the mouse cochlear nucleus. J Comp Neurol 295:52–70.

Bischoff E (1899) Uber den intramedullaren Verlauf des Facialis. Neur Centralbl 18:1014–1016.

Bishop A (1987) The efferent auditory system in Doppler shift compensating bats. In: Nachtigal P (ed) Animal Sonar Systems. New York: Plenum Press.

Bobbin RP, Konishi T (1971) Acetylcholine mimics crossed olivocochlear bundle stimulation. Nature 231:222–223.

Bodznick D (1989) Comparisons between electrosensory and mechanosensory lateral line systems. In: Coombs S, Görner P, Münz H (eds) The Mechanosensory Lateral Line. New York: Springer-Verlag, pp. 653–678.

Bone Q, Best ACG (1978) Ciliated sensory cells in *Amphioxus* (Branchiostoma). J Mar Biol Assoc UK 58:479–486.

Boyle R, Highstein SM (1990) Efferent vestibular system in the toadfish: action upon horizontal semicircular canal afferents. J Neurosci 10:1570–1582.

Brantley RK, Bass AH (1988) Cholinergic neurons in the brain of a teleost fish (*Porichthys notatus*) located with a monoclonal antibody to choline acetyltransferase. J Comp Neurol 27:87–105.

Brown MC (1987) Morphology of labelled efferent fibers in the guinea pig cochlea. J Comp Neurol 260:591–604.

Brown MC, Liberman MC, Benson TE, Ryugo DK (1988) Brainstem branches from olivocochlear axons in cats and rodents. J Comp Neurol 278:591–603.

Brownell WE, Bader CR, Bertrand D, De Ribaupierre Y (1985) Evoked mechanical responses of isolated cochlear outer hair cells. Science 227:194–196.

Bruns V, Schmieszek E (1980) Cochlear innervation in the greater horseshoe bat: Demonstration of an acoustic fovea. Hearing Res 3:27–43.

Budelmann BU (1988) Morphological diversity of equilibrium receptor systems in aquatic invertebrates. In: Atema J, Fay RR, Popper AN, Tavolga WN (eds) Sensory Biology of Aquatic Animals. New York: Springer-Verlag, pp. 757–782.

Budelmann BU, Bonn U (1982) Histochemical evidence for catecholamines as neurotransmitter in the statocyst of *Octopus vulgaris*. Cell Tissue Res 227:475–483.

Carr CE, Maler L (1985) A Golgi study of the cell types of the dorsal torus semicircularis of the electric fish *Eigenmania*: Functional and morphological diversity in the midbrain. J Comp Neurol 235:207–240.

Claas B, Fritzsch B, Münz (1981) Common efferents to lateral line and labyrinthine hair cells in aquatic vertebrates. Neurosci Lett 27:231–235.

Cody AR, Johnstone BM (1982) Acoustically evoked activity of single efferent neurons in the guinea pig cochlea. J Acoust Soc Am 72:280–282.

Corwin JT (1977) Morphology of the macula neglecta in sharks of the genus *Carcharhinus*. J Morphol 152:341–362.

Currie SN, Carlsen RC (1987) Modulated vibration-sensitivity of lamprey Mauthner neurones. J Exp Biol 129:41–51.

Danielson PD, Zottoli SJ, Corrodi JG, Rhodes KJ, Mufson EJ (1988) Localization of choline acetyltransferase to somata of posterior lateral line efferents in the goldfish. Brain Res 448:158–161.

Dechesne C, Raymond J, Sans A (1984) The efferent vestibular system in the cat: a horseradish peroxidase and fluorescent retrograde tracers study. Neuroscience 11:893–902.

Desmedt JE, Delwaide PJ (1965) Functional properties of the efferent cochlear bundle of the pigeon revealed by stereotaxic stimulation. Exp Neurol 11:1–26.

Detwiler PB, Alkon DL (1973) Hair cell interactions in the statocyst of *Hermissenda*. J Gen Physiol 62:618–642.

De Wolf FA, Schellart NAM, Hoogland PV (1983) Octavolateral projections to the torus semicircularis of the trout, *Salmo gairdneri*. Neurosci Lett 38:209–214.

Drescher MJ, Drescher DG, Hatfield JD (1987) Potassium-evoked release of endogenous primary amine-containing compounds from the trout saccular macula and saccular nerve *in vitro*. Brain Res 417:39–50.

Dunn RF, Koester DM (1987) Primary afferent projections to the central octavus nuclei in the elasmobranch, *Rhinobatos* sp., as demonstrated by nerve degeneration. J Comp Neurol 260:564–572.

Echteler SM (1984) Connections of the auditory midbrain in a teleost fish, *Cyprinus carpio*. J Comp Neurol 230:536–551.

Fay-Lund H (1986) Projection from the inferior colliculus to the superior olivary complex in the albino rat. Anat Embryol 175:35–52.

Fex J (1962) Auditory activity in centrifugal and centripetal cochlear fibres in cat. A study of a feedback system. Acta Physiol Scand 55 (Suppl) 189:1–68.

Fex J (1965) Auditory activity in uncrossed centrifugal cochlear fibres in cat. A study of a feedback system II. Acta Physiol Scand 64:43–57.

Fex J, Altschuler RA (1984) Glutamic acid decarboxylase immunoreactivity of olivocochlear neurons in the organ of Corti of guinea pig and rat. Hearing Res 15:123–131.

Fex J, Altschuler RA, Kachar B, Wenthold RJ, Zempel JM (1986) GABA visualized by immunocytochemistry in the guinea pig cochlea in axons and endings of efferent neurons. Brain Res 366:106–117.

Flock Å, Flock B (1966) Ultrastructure of the amphibian papilla in the bullfrog. J Acoust Soc Am 40:1262.

Flock Å, Lam DMK (1974) Neurotransmitter synthesis in inner ear and lateral line sense organs. Nature 249:142–144.

Flock Å, Russell IJ (1973) The postsynaptic action of afferent fibres in the lateral line organ of the burbot *Lota lota*. J Physiol 235:591–605.

Flock Å, Russell IJ (1976) Inhibition by efferent nerve fibres: action on hair cells and afferent synaptic transmission in the lateral line canal organ of the burbot, *Lota lota*. J Physiol 257:45–62.

Frishkopf LS, Flock Å (1974) Ultrastructure of the basilar papilla, an auditory organ in the bullfrog. Acta Otolaryngol 77:176–194.

Fritzsch B (1981) Efferent neurons to the labyrinth of *Salamandra salamandra* as revealed by retrograde transport of horseradish peroxidase. Neurosci Lett 26:191–196.

Fritzsch B, Crapon de Caprona D (1984) The origin of centrifugal inner ear fibers of gymnophions (amphibia). A horseradish peroxidase study. Neurosci Lett 46: 131–136.

Fritzsch B, Wahnschaffe V (1987) Electronmicroscopical evidence for common inner ear and lateral line efferents in urodeles. Neurosci Lett 81:48–52.

Fritzsch B, Crapon de Caprona M-D, Wächtler K, Körtje K-H (1984) Neuroanatomical evidence for electroreception in lampreys. Z Naturforsch 39C:856–858.

Fritzsch B, Dubuc R, Ohta Y, Grillner S (1989) Efferents to the labyrinth of the river lamprey (*Lampetra fluviatilis*) as revealed with retrograde tracing techniques. Neurosci Lett 96:241–246.

Fuchs PA, Nagai T, Evans MG (1988) Electrical tuning in hair cells isolated from the chick cochlea. J Neurosci 8:2460–2467.

Furukawa T (1966) Synaptic interaction at the Mauthner cell of goldfish. Prog Brain Res 21A:46–70.

Furukawa T (1981) Effects of efferent stimulation on the saccule of goldfish. J Physiol 315:203–215.

Gacek RR (1981) The afferent and efferent vestibular pathways: Morphologic aspects. In: Gualtierotti T (ed) The Vestibular System: Function and Morphology. New York: Springer-Verlag, pp. 38–63.

Gacek RR, Lyon M (1974) The localization of vestibular efferent neurons in the kitten with horseradish peroxidase. Acta Oto-laryngol 77:92–103.

Galambos R (1956) Suppression of auditory nerve activity by stimulation of efferent fibres to the cochlea. J Neurophysiol 19:424–437.

Ginsberg RD, Morest DK (1984) Fine structure of cochlear innervation in the cat. Hearing Res 14:109–127.

Gleisner L, Henriksson NG (1964) Efferent and afferent activity pattern in the vestibular nerve of the frog. Acta Oto-laryngol (Suppl) 192:90–103.

Glenn JF, Oatman LC (1980) Stimulation studies in the descending auditory pathway. Brain Res 196:258–261.

Goldberg JM, Fernández C (1980) Efferent vestibular system in the squirrel monkey: Anatomical location and influence on afferent activity. J Neurophysiol 43:986–1025.

González A (1986) Estudio comparado de las proyecciones nerviosas de los sistemas de la linea lateral y acustico-vestibular de anfibios. Organizacion del area octavo-lateral rombencefalica. Doctoral thesis, Univ. Complutense, Facultad de Biologia, Madrid.

González A, Muñoz M (1987) Some connections of the area octavolateralis of *Pleurodeles waltlii*. A study with horseradish peroxidase under in vitro conditions. Brain Res 423:338–342.

Guinan JJ Jr, Warr WB, Norris BE (1983) Differential olivocochlear projections from lateral versus medial zones of the superior olivary complex. J Comp Neurol 221:358–370.

Hama K (1969) A study of the fine structure of the saccular macula of the goldfish. Z Zellforsch 94:155–171 (1969).

Harlan RE, Shivers BD, Romano GJ, Howells RD, Pfaff DW (1987) Localization of preproenkephalin mRNA in the rat brain and spinal cord by *in situ* hybridization. J Comp Neurol 258:159–184.

Hartmann R, Klinke R (1980) Efferent activity in the goldfish vestibular nerve and its influence on afferent activity. Pflügers Arch 388:123–128.

Hernández-Peón R, Scherrer H, Jouvet M (1956) Modification of electrical activity in cochlear nucleus during "attention" in unanesthetized cats. Science 123:331–332.

Herrick CJ (1948) The Brain of the Tiger Salamander (*Ambystoma trigrinum*). Chicago: Chicago University Press.

Highstein SM, Baker R (1985) Action of the efferent vestibular system on primary afferents in the toadfish, *Opsanus tau*. J Neurophysiol 54:370–384.

Highstein SM, Baker R (1986) Organization of the efferent vestibular nuclei and nerves of the toadfish, *Opsanus tau*. J Comp Neurol 243:309–325.

Hoffman DW, Rubio JA, Altschuler RA, Fex J (1984) Several distinct receptor binding enkephalins in olivocochlear fibers and terminals in the organ of corti. Brain Res 322:59–65.

Hoshino T (1975) An electron microscopic study of the otolithic maculae of the lamprey (*Entosphenus japonicus*) Acta Oto-laryngol 80:43–53.

Ito J, Takahashi H, Matsuoka I, Takatani T, Masashi S, Takaori S (1983) Vestibular efferent fibers to ampulla of anterior, lateral and posterior semicircular canals in cats. Brain Res 259:293–297.

Jacoby J, Rubinson K (1984) Efferent projections of the torus semicircularis to the medulla of the tadpole, *Rana catesbeiana*. Brain Res 292:378–381.

Janse C, van der Wilt GJ, van der Roest M, Pieneman AW (1988) Intracellularly recorded responses to tilt and efferent input of statocyst sensory cells in the pulmonate snail, *Lymnaea stagnalis*. Comp Biochem Physiol 90A:269–278.

Jones BE, Beaudet A (1987) Distribution of acetylcholine and catecholamine neurons in the cat brainstem: A choline acetyltransferase and tyrosine hydroxylase immunohistochemical study. J Comp Neurol 261: 15–32.

Jørgensen JM (1974) The sensory epithelia of the inner ear of two turtles, *Testudo graeca* L. and *Pseudemys scripta* (Schoepff). Acta Zool 55:289–298.

Jørgensen JM (1984) Fine structure of the paratympanic organ in the avian middle ear. Acta Zool 65:89–94.

Jørgensen JM (1989) Evolution of octavolateralis sensory cells. In: Coombs S, Görner P, Münz H (eds) The Mechanosensory Lateral Line. New York: Springer-Verlag, pp. 115–145.

Kimura RS (1975) The ultrastructure of the organ of Corti. Int Rev Cytol 42:173–222.

Klinke R (1981) Neurotransmitters in the cochlear and the cochlear nucleus. Acta Oto-laryngol 94:541–554.

Klinke R (1986) Neurotransmission in the inner ear. Hearing Res 22:235–243.

Klinke R, Galley N (1974) Efferent innervation of vestibular and auditory receptors. Physiol Rev 54:316–357.

Klinke R, Schmidt CL (1970) Efferent influence on the vestibular organ during active movements of the body. Pflügers Arch 318:325–332.

Koyama H, Kishida R, Goris RC, Kusunoki T (1989) Afferent and efferent projections of the VIIIth cranial nerve in the lamprey *Lampetra japonica*. J Comp Neurol 280:663–671.

Kuriyama H, Shiosaka S, Sekitani M, Tohyama Y, Kitajiri M, Yamashita, Kumazawa T, Tohyama M (1990) Electron microscopic observation of calcitonin gene-related peptide-like immunoreactivity in the organ of Corti of the rat. Brain Res 517:76–80.

Laming PR, Brooks M (1985) Effects of visual, chemical and tactile stimuli on the averaged auditory response of the teleost *Rutilus rutilus*. Comp Biochem Physiol 82A:667–673.

Lewis ER, Baird RA, Leverenz EL, Koyama H (1982) Inner ear: Dye injection reveals peripheral origins of specific sensitivities. Science 215:1641–1642.

Liberman MC (1980) Efferent synapses in the inner hair cell area of the cat cochlea: An electron microscopic study of serial sections. Hearing Res 3:189–204.

Liberman MC, Bell DG (1979) Hair cell condition and auditory nerve response in normal and noise-damaged cochleas. Acta Oto-laryngol 88:161–176.

Liberman MC, Brown MC (1986) Physiology and anatomy of single olivocochlear neurons in the cat. Hearing Res 24:17–36.

Lin JW, Faber DS (1988) An efferent inhibition of auditory afferents mediated by the goldfish Mauthner cell. Neuroscience 24:829–836.

López I, Meza G (1988) Neurochemical evidence for afferent GABAergic and efferent cholinergic neurotransmission in the frog vestibule. Neuroscience 25:13–18.

Lowenstein O, Thornhill RA (1970) The labyrinth of *Myxine*: Anatomy, ultrastructure and electrophysiology. Proc Roy Soc Lond B 176:21–42.

Lowenstein O, Osborne MP, Thornhill RA (1968) The anatomy and ultrastructure of the labyrinth of the lamprey (*Lampetra fluviatilis* L) Proc Roy Soc B 170: 113–134.

Lowenstein O, Osborne MP, Wersäll J (1964) Structure and innervation of the sensory epithelia of the labyrinth in the thornback ray (*Raja clavata*). Proc Roy Soc B 160:1–12.

Mathiesen C (1985) Structure and innervation of inner ear sensory epithelia in the European eel (*Anguilla anguilla L.*) Acta Zoologica 65:189–207.

McCormick CA, Braford MR (1979) Identification of eighth nerve efferent cells in the bowfin, *Amia calva*. Soc Neurosci Abstr 5:144.

Meredith GE (1988) Comparative view of the central organization of afferent and efferent circuitry for the inner ear. Acta Biol Hung 39:229–249.

Meredith GE, Roberts BL (1986a) Central organization of the efferent supply to the labyrinthine and lateral line receptors of the dogfish. Neuroscience 17:225–233.

Meredith GE, Roberts BL (1986b) The relationship of saccular efferent neurons to the superior olive in the eel, *Anguilla anguilla*. Neurosci Lett 68:69–72.

Meredith GE, Roberts BL (1987) Distribution and morphological characteristics of efferent neurons innervating end organs in the ear and lateral line of the European eel. J Comp Neurol 265:494–506.

Meredith GE, Roberts BL, Suharti Maslam (1987) Distribution of afferent fibers in the brainstem from end organs in the ear and lateral line in the European eel. J Comp Neurol 265:507–520.

Meza G, Hinojosa R (1987) Ontogenetic approach to cellular localization of neurotransmitters in the chick vestibule. Hearing Res 28:73–78.

Miller MR, Beck J (1988) Auditory hair cell innervational patterns in lizards. J Comp Neurol 271:604–628.

Moffat AJ, Capranica RR (1976) Auditory sensitivity of the saccule in the American Toad (*Bufo americanus*). J Comp Physiol 105:1–8.

Mos W, van der Jagt B, Roberts BL (1989) Reflex activity of hair cell efferent neurons of the European eel. Eur J Neurosci (Suppl) 2:38.

Münz H, Claas B (1986) Activity of efferent neurons in the lateral line system. Neurosci Lett (Suppl) 26:S375.

Nakajima Y, Wang DW (1974) Morphology of afferent and efferent synapses in hearing organ of the goldfish. J Comp Neurol 156:403–416.

New JG, Northcutt RG (1984) Central projections of lateral line nerves in the shovelnose sturgeon. J Comp Neurol 225:129–140.

Nickel E, Fuchs S (1974) Organization and ultrastructure of mechanoreceptors (Savi vesicles) in the elasmobranch *Torpedo*. J Neurocytol 3:161–177.

Nuttall AL (1986) Physiology of hair cells. In: Altschuler RA, Hoffman DW, Bobbin RP (eds) Neurobiology of Hearing: the Cochlea. New York: Raven Press, pp. 47–75.

Oatman LC (1976) Effects of visual attention on the intensity of auditory evoked potentials. Exp Neurol 51:41–53.

Oatman LC, Anderson BW (1977) Effects of visual attention on tone burst evoked auditory potentials. Exp Neurol 57:200–211.

Pasztor VM, Bush BMH (1987) Peripheral modulation of mechanosensitivity in primary afferent neurons. Nature 326:793–795.

Patuzzi R, Robertson D (1988) Tuning in the mammalian cochlea. In: Giebisch GH, Boron WF (eds) Physiological Reviews. Am Physiol Soc 68:1010–1067.

Pelligrini M, Ceccotti F, Magherini P (1985) The efferent vestibular neurons in the toad (Bufo bufo L.): Their location and morphology. A horseradish peroxidase study. Brain Res 344:1–8.

Perachio AA, Kevetter GA (1989) Identification of vestibular efferent neurons in the gerbil: histochemical and retrograde labelling. Exp Brain Res 78:315–326.

Peyret D, Geffard M, Aran J-M (1986) GABA immunoreactivity in the primary nuclei of the auditory central nervous system. Hearing Res 23:115–121.

Piddington RW (1971a) Central control of auditory input in the goldfish. I. Effect of shocks to the midbrain. J Exp Biol 55:569–584.

Piddington RW (1971b) Central control auditory input in the goldfish. II. Evidence of action in the free-swimming animal. J Exp Biol 55:585–610.

Popper AN (1979) Ultrastructure of the sacculus and lagena in a moray eel. J Morphol 161:241–256.

Popper AN, Hoxter B (1987) Sensory and nonsensory ciliated cells in the ear of the sea lamprey, *Petromyzon marinus*. Brain Behav Evol 30:43–61.

Pujol R, Lenoir M (1986) The four types of synapses in the organ of Corti. In: Altschuler RA, Hoffman DW, Bobbin RP (eds) Neurobiology of Hearing: The Cochlea. New York: Raven Press, pp. 161–172.

Rajan R (1990) Electrical stimulation of the inferior colliculus at low rates protects the cochlea from auditory desensitization. Brain Res 506:192–204.

Rasmussen GL (1946) The olivary peduncle and other fiber connections of the superior olivary complex. J Comp Neurol 84:141–219.

Rasmussen GL (1953) Further observations of the efferent cochlear bundle. J Comp Neurol 99:61–74.

Robbins R, Bauknight RS, Honrubia V (1967) Anatomical distribution of efferent fibers in the VIIIth cranial nerve of the bullfrog, *Rana catesbeiana*. Acta Otolaryngol 64:436–448.

Roberts BL, Meredith GE (1989) The efferent system. In: Coombs S, Görner P, Münz H (eds) The Mechanosensory Lateral Line. New York: Springer-Verlag, pp. 445–460.

Roberts BL, Russell IJ (1972) The activity of lateral-line efferent neurones in stationary and swimming dogfish. J Exp Biol 57:435–448.

Robertson D (1984) Horseradish peroxidase injection of physiologically characterized afferent and efferent neurones in the guinea pig spiral ganglion. Hearing Res 15:113–121.

Robertson D (1985) Brainstem location of the efferent neurones projecting to the guinea pig cochlea. Hearing Res 20:79–84.

Robertson D, Gummer M (1985) Physiological and morphological characterization of efferent neurones in the guinea pig cochlea. Hearing Res 20:63–77.

Robertson D, Gummer M (1988) Physiology of cochlear efferents in the mammal. In: Syka J, Masterton RB (eds) Auditory Pathway. New York: Plenum Press, pp. 269–278.

Robertson D, Winter IM (1988) Cochlear nucleus inputs to olivocochlear neurones revealed by combined anterograde and retrograde labelling in the guinea pig. Brain Res 462:47–55.

Robertson D, Cole KS, Harvey AR (1987) Brainstem organization of efferent projections to the guinea pig cochlea studied using the fluorescent tracers fast blue and diamidino yellow. Exp Brain Res 66:449–457.

Rossi ML, Prigioni I, Valli P, Casella C (1980) Activation of the efferent system in the isolated frog labyrinth: Effects on the afferent epsps and spike discharge recorded from single fibers of the posterior nerve. Brain Res 185:125–137.

Rupert AL, Moushegian G, Whitcomb MA (1968) Olivocochlear neuronal response in medulla of cat. Exp Neurol 20:575–584.

Russell IJ (1971) The role of the lateral-line efferent system in *Xenopus laevis*. J Exp Biol 54:621–641.

Russell IJ (1974) Central and peripheral inhibition of lateral line input during the startle response in goldfish. Brain Res 80:517–522.

Ryan AF, Schwartz IR, Helfert RH, Keithley E, Wang Z-X (1987) Selective retrograde labeling of lateral olivocochlear neurons in the brainstem based on preferential uptake of ³H-D-Aspartic acid in the cochlea. J Comp Neurol 255:606–616.

Saito K (1980) Fine structure of the sensory epithelium of the guinea pig organ of corti: afferent and efferent synapses of hair cells. J Ultrastruct Res 71:222–232.

Sans A, Highstein SM (1984) New ultrastructural features in the vestibular labyrinth of the toadfish, *Opsanus tau*. Brain Res 308:191–195.

Schmidt RS (1963) Frog labyrinthine efferent impulses. Acta Oto-laryngol 56:51–64.

Schmidt RS (1965) Amphibian acoustico-lateralis efferents. J Cell Comp Physiol 65:155–162.

Schwartz IR, Ryan AF (1986) Amino acid labeling patterns in the efferent innervation of the cochlea: An electron microscopic autoradiographic study. J Comp Neurol 246:500–512.

Schwarz IE, Schwarz DWF, Fredrickson JM, Landolt JP (1981) Efferent vestibular neurons: A study employing retrograde tracer methods in the pigeon (*Columba livia*). J Comp Neurol 196:1–12.

Schwarz DWF, Satoh K, Schwarz IE, Hu K, Fibiger HC (1986) Cholinergic innervation of the rat's labyrinth. Exp Brain Res 64:19–26.

Schweitzer LF, Lu SM, Dawbarn D, Cant NB (1985) Calcitonin gene-related peptide in the superior olivary complex of cat and rat: A specific label for the lateral olivocochlear system. Soc Neurosci Abstr 11:1051.

Sento S, Furukawa T (1987) Intra-axonal labeling of saccular afferents in goldfish, *Carassius auratus*: correlations between morphological and physiological characteristics. J Comp Neurol 258:352–367.

Siegel JH, Kim DO (1982) Efferent neural control of cochlear mechanics? Olivocochlear bundle stimulation affects cochlear biomechanical nonlinearity. Hearing Res 6:171–182.

Smith CA (1985) Inner ear. In: King AS, McLelland J (eds) Form and Function in Birds. London: Academic Press, pp. 273–310.

Smith CA, Rasmussen GL (1963) Recent observations on the olivocochlear bundle. Ann Oto-rhinolaryngol 72:489–507.

Sneary M (1988) Auditory receptor of the red-eared turtle: I. Afferent and efferent synapses and innervation patterns. J Comp Neurol 276:588–606.

Sokolowski BHA, Popper AN (1988) Transmission electron microscopic study of the saccule in the embryonic, larval and adult toadfish *Opsanus tau*. J Morph 198:49–69.

Späth M, Schweickert W (1975) Lateral-line efferents to mechanical and visual stimuli. Naturwissenschaften 62:S579–S580.

Strominger NL, Silver SM, Truscott TC, Goldstein JC (1981) The cells of origin of the olivocochlear bundle in New and Old World monkeys. Anat Rec 199:246.

Strutz J (1981) The origin of centrifugal fibers to the inner ear in *Caiman crocodilus*. A horseradish peroxidase study. Neurosci Lett 27:95–100.

Strutz J (1982a) The origin of efferent vestibular fibers in the guinea pig. Acta Oto-laryngol 94:299–305.

Strutz J (1982b) The origin of efferent fibers to the inner ear in a turtle (*Terrapene ornata*). A horseradish peroxidase study. Brain Res 244:165–168.

Strutz J, Bielenberg K (1983) Efferent acoustic neurons within the lateral superior nucleus of the guinea pig. Brain Res 299:174–177.

Strutz J, Schmidt CL (1982) Acoustic and vestibular efferent neurons in the chicken (*Gallus domesticus*). Acta Oto-laryngol 94:45–51.

Strutz J, Bielenberg K, Spatz WB (1982) Location of efferent neurons to the labyrinth of the green tree frog (*Hyla cinerea*). Arch Oto-rhinolaryngol 234:245–251.

Sugihara I, Furukawa T (1989) Morphological and functional aspects of two different types of hair cells in the goldfish sacculus. J Neurophysiol 62:1330–1343.

Syka J, Popelár J, Druga R, Vlková A (1988) Descending central auditory pathway – structure and function. In: Syka J, Masterton RB (eds) Auditory Pathway: Structure and Function. New York: Plenum Press, pp. 279–292.

Takasaka T, Smith CA (1971) The structure and innervation of the pigeon's basil papilla. J Ultrastruct Res 35:20–65.

Tanaka M, Takeda N, Senba E, Tohyama M, Kubo T, Matsunaga T (1989) Localization, origin and fine structure of calcitonin gene-related peptide-containing fibers in the vestibular end-organs of the rat. Brain Res 504:31–35.

Taniguchi I (1980) Auditory afferent and efferent neurons in the mouse brain stem studied by axonal transport of HRP. Nagoya J Med Sci 42:75–77.

Thompson GC, Thompson AM 1986) Olivocochlear neurons in the squirrel monkey brainstem. J Comp Neurol 254:246–258.

Tricas TC, Highstein SM (1990) Visually mediated inhibition of lateral line primary afferent activity by the octavolateralis efferent system during predation in the free-swimming toadfish. *Opsanus tau*. Exp Brain 83:233–236.

van der Jagt B, Maslam S, Roberts BL (1990) Immuno-histochemical investigations of efferent neurons innervating the ear and lateral line in the eel. Europ J Neurosci (Suppl) 3 157.

Von Bartheld CS (1990) Development and innervation of the paratympanic organ (Vitali organ) in chick embryos. Brain Behav Evol 35:1–15.

Von Düring M (1974) The fine structure of the inner ear in *Caiman crocodilus*. Z Anat Entwickl-Gesch 145:41–65.

Warr WB (1975) Olivocochlear and vestibular efferent neurons of the feline brain stem: their location, morphology and number determined by retrograde axonal transport and acetylcholinesterase histochemistry. J Comp Neurol 161:159–182.

Warr WB, Guinan JJ Jr (1979) Efferent innervation of the organ of Corti: Two separate systems. Brain Res 173:152–155.

Warr WB, Guinan JJ Jr, White JS (1986) Organization of the efferent fibers: The lateral and medial olivo-chochlear systems. In: Altschuler RA, Hoffman DW, Bobbin RP (eds) Neurobiology of Hearing: The Cochlea. New York: Raven Press, pp. 333–348.

Wegner N (1982) A qualitative and quantitative study of a sensory epithelium in the inner ear of a fish (*Colisa labiosa*; Anabantidae). Acta Zoologica (Stockh) 63:133–146.

White JS (1986) Differences in the ultrastructure of labyrinthine efferent neurons in the albino rat. Assoc Res Oto-laryngol Abstr 9:34–35.

White JS, Warr WB (1983) The dual origins of the olivocochlear bundle in the albino rat. J Comp Neurol 219:203–214.

Whitehead MC (1986) Development of the cochlea. In: Altschuler RA, Hoffman DW, Bobbin RP (eds) Neurobiology of Hearing: The Cochlea. New York: Raven Press, pp. 191–211.

Whitehead MC, Morest DK (1981) Dual populations of efferent and afferent cochlear axons in the chicken. Neuroscience 6:2351–2365.

Whitehead MC, Morest DK (1985) The development of innervation patterns in the avian cochlea. Neuroscience 14:255–276.

Wiederhold ML (1986) Physiology of the olivocochlear system. In: Altschuler RA, Hoffman DW, Bobbin RP (eds) Neurobiology of Hearing: The Cochlea. New York: Raven Press, pp. 349–370.

Wiederhold ML, Kiang NYS (1970) Effects of electric stimulation of the crossed olivocochlear bundle on single auditory-nerve fibers in the cat. J Acoust Soc Am 48:950–965.

Will U (1982) Efferent neurons of the lateral-line system and the VIII cranial nerve in the brainstem of anurans. Cell Tissue Res 225:673–685.

Will U, Fritzsch B (1987) The eighth nerve of amphibians. Peripheral and central distribution. In: Evolution of the Amphibian Auditory System. New York: Wiley and Sons.

Williamson R (1985) Efferent influences on the afferent activity from the octopus angular acceleration receptor system. J Exp Biol 119:251–264.

Williamson R (1989a) Electrophysiological evidence for cholinergic and catecholaminergic efferent transmitters in the statocyst of octopus. Comp Biochem Physiol 93:23–27.

13
Evolution, Perception, and the Comparative Method

William C. Stebbins and Mitchell S. Sommers

1. Introduction

Arguably, it is premature to consider the question of the evolution of perception. Until recently (Berkley and Stebbins 1990; Stebbins and Berkley 1990) any substantial discussion of comparative perception has been rare in the literature of animal behavior (for a noteworthy exception see von Uexkull 1934, in his discussion of umwelt), yet animal behaviorists in both field and laboratory have frequently engaged far more diffuse and unsettled concepts such as intelligence (Weiskrantz 1985), mental faculties [Darwin 1981 (1871)], awareness (Griffin 1981), and memory (Roitblat 1984). Notwithstanding, animals have achieved reproductive success in the perception of their environment—in recognizing mates, in eluding predators, in locating prey, and finally in discriminating and responding to the myriad fine details in the communication signals of conspecifics, which is the subject matter of this chapter. Adaptation and natural selection are plainly evident in all of these behaviors, but it is the knowledge of the evolutionary process, particularly in these more complex forms of auditory perception, that evades us and will continue to do so for good reason.

Unfortunately, perception defies a precise and universally agreed upon definition. Even the revered Oxford English Dictionary spends the better part of a page displaying the many definitions of those in the past who have bravely (or foolishly) attempted to define it unambiguously. "Perception is used here in one of its original meanings from the Latin *percipere*—to catch, get hold of, gather in; but also in the more generic and metaphorical sense of awareness or cognizance of the environment." (Stebbins 1990, p. 1) It is in this latter sense that cognition plays a role in perception and may include, for example, such phenomena as abstraction, categorization, imaging, memory, symbol manipulation, and so on. For our purposes perception must be thought of not only as the relatively simple relationship between sensory input and an animal's response to that input (i.e., threshold of audibility), but also as the far more complex relationship between sensory events, their subsequent processing by the nervous system, and the animal's behavior. Perception's recondite quality is undoubtedly due to the fact that we must infer it from the behavior that we observe and measure, and from the stimuli, and from the animal's experience and endowment, all of which we hold accountable for the control that they exercise over that behavior. The rigor lies in the way that we use it rather than in how we define it.

Bullock (1984) is quite right that our understanding of comparative behavior lags that of comparative anatomy. But the fault lies not in the "tools" of the behaviorist as Bullock has suggested, but in their complexity and the time required for their effective application. It is one thing to measure brain volume in a dead animal or the number of cortical laminations in inert tissue (Bullock's examples) but quite another to evaluate, with the necessary precision, of course, an awake behaving animal's perception of the subtly changing acoustic signals that are characteristic of its environment. At least the behavioral methods for the study of animal perception are in place and constantly improving, but it is the time necessary for their implementation that is a major bottleneck. Bullock is also correct that findings on a single species are "innocent of comparison" (ibid., 1984, p. 695) and certainly limited with regard to our understanding of the evolutionary process. Here we would counsel patience. Achieving insight into the vocal communication or auditory perceptual system of even a single species is not entered into lightly. These are hard-won data that often combine field and laboratory approaches (see Beecher and Stoddard 1990, for example, and below) and usually require extensive research on a small number of animals. Large groups for purposes of statistical comparisons are not feasible and usually superfluous because the findings from individual animals are often robust. Smash and grab techniques (the use of the pinna reflex or evoked response audiometry, for example) will not provide the kind of answer that I think Bullock wants and that is necessary in order to make species comparisons that would permit us to establish relations between anatomical structure and behavioral function and reasoned conjecture concerning the evolutionary process.

With regard to the morphological evolution of the auditory system, considerable progress has been made in understanding the phylogenesis of the mammalian middle ear (chapters by Rosowski; Bolt and Lombard; Hetherington; Allin and Hopson; and Clack), and because soft tissue leaves no trace in the fossil record, there has been far less progress in our understanding of the evolving inner ear (chapter by Popper, Platt, and Edds). The evolution of auditory function is even more uncertain although there is substantial speculation on the more fundamental properties of auditory perception such as detection, discrimination, and sound localization, the first two of which are related, to a major extent, to peripheral receptor characteristics, for example, frequency discrimination and frequency coding (Masterton, H. Heffner, and Ravizza 1969; Stebbins 1980). These basic processes of hearing are perhaps less influenced by cortical control mechanisms than the more judgmental form of acoustic perception in animals that will be the dominant focus of this chapter. The concern will be less with an animal's mere detection of stimuli and its limits of resolution and more with its perceptual evaluation and categorization of communication sounds and their critical features. Although the results are not inconsequential, they

are limited in their application to a few species. For this reason, and because we are dealing with complex behavior which is not yet easily or precisely related to structure, it is too early to contemplate the evolutionary sequence, ancestral form, or even underlying mechanisms. Alternatively, we might dismiss this developing field of study as not yet ripe for the picking. That would be unfortunate, for the findings are promising and may even allow a word or two about selective pressures. This is also a research program that, at this stage, should be given for scrutiny to the evolutionist for consideration and comment but also as a reminder that the structure with which she is so often preoccupied has a behavioral outcome. Furthermore, this is a book entitled *The Evolutionary Biology of Hearing*, and yet its major thrust is on the evolution of structure. Given the present state of our knowledge, this is not unreasonable. But hearing is not structure; it is perception, even behavior, and it would be in error to omit in a book so titled any discussion, no matter how preliminary, of behavioral function to which anatomical structure is so clearly dedicated.

2. Precursors

When comparing the mental powers of man and other animals, Charles Darwin, in his *Descent of Man and Selection in Relation to Sex* [1981 (1871)], remarked on the six distinctive calls of the Cebus monkey which "excite in other monkeys similar emotions" (ibid., p. 54). He reminded us also that dogs bark "in distinct tones" (ibid.) which may denote different affective states or even convey a request for some form of action. His illustration of geographical variation in the bird song of conspecifics was presented as a model for human dialects (ibid., pp. 55–56). While he assured us that human language is a distinctive and unique form of communication, he left no doubt that its origins can be found in the "voices" (ibid.) of other animals. The difference is in complexity, not in kind but in degree, and these forms of communication represent different stages in the evolutionary process. Darwin's observations were amplified and extended and given a more secure experimental basis later in this century with the research of Marler and Tamura (1962) on geographical dialects in the white crown sparrow, of Struhsaker (1967) and Seyfarth et al. (1980) on the external referents for the different alarm calls of vervet monkeys, and of S. Gouzoules et al. (1984) on representational signaling in rhesus monkeys. The evidence (see also H. Gouzoules, S. Gouzoules, and Marler 1985) supports semantic content (or meaning) in animal communication, and, in so doing, provides a closer link to human language and suggests a form of cognitive processing in other animals. This is in sharp contrast to earlier conceptions that viewed animal signals, at least in the acoustic mode, as reflecting little more than the affective or motivational state of the signaler (Rowell and Hinde 1962; see H. Gouzoules, S. Gouzoules, and Marler 1985, for a brief review). The earlier view was decidedly centered on the signaler, suggested a discontinuity between animal and human vocal communication, and provided an impoverished image of the recipient of the signal as a perceiver relative to its human counterpart. In comparison, the more recent conception offers a much tighter evolutionary link to human speech and language and the prospects for a productive analysis of some of the significant parallels between man in his perception of speech and other animals in their perception of the communication sounds of conspecifics.

Darwin's view of the mental powers of other animals was in the sensationalist tradition of philosophers such as Locke and Hume and also of his own grandfather, Erasmus Darwin, that ideas were developed directly from sensory information and that thoughts sprang from the consequent sensory images (Boakes 1984; Richards 1987). For Charles Darwin, reason or cognition was based upon the processing and comparison of sensory images from the environment and was evident in most animals that could profit from experience and demonstrate behavioral flexibility in the face of significant environmental change (Richards 1987). As Richards has suggested, in Darwin's day "intelligence was not opposed to instinct, but rather grew out of it" (ibid., p. 109), and there was far less contention about reasoning in other animals in nineteenth century England than there is today. Then, "the more significant question was: Can animals make moral judgments as we do?" (ibid., p. 109). Animals, man included, have the ability to make use of either acquired, developmentally based,

or innate strategies in processing environmental information that go beyond the fundamental detection and discrimination of stimuli that is demonstrable even in the simplest of organisms. From a bottom-up perspective there is built in selection or filtering at the peripheral receptor and, more centrally, a top-down selection or influence that determines an animal's perception of the world and guides its behavior in important ways that are adaptive, given its evolutionary history. Both forms of processing are an endowment from reproductively successful ancestors; in many animals top-down processing is a composite of endowment and experience. The top-down processing is cognitive, if only in the sense that the animal's behavior is not strictly determined by the stimulus but is heavily influenced by other variables, many of which are based on its experience. Consequently, in the course of evolution, perception became a major adaptive strategy by which animals, either in the course of development or through learning, selectively process information from the environment in order to react successfully to the world and to the other animals that occupy it.

For a process of such importance it is unfortunate that perception is two steps removed from the evolutionary record. For perception is inferred from an animal's behavior, and behavior, in its turn, must be inferred from the material left to us in the evolutionary record. This is not dissimilar to what Simpson (1958) described as second order behavior, which, unlike locomotion or food gathering and consumption, can never be inferred directly from the fossil record. "The evolution of second-order behavior is even more important and more interesting, and this is quite properly the principal occupation of the evolutionary psychologist" (ibid., p. 11). This of course makes the whole matter highly speculative but well worth the risk if we are to assign perceptual function to accompany the evolution of structure with which paleontologists and morphologists have presented us. For without function the discovery of structure is a hollow achievement (Stebbins 1980). That Darwin understood this is evident in his lengthy discourse on animal behavior, particularly in his separate treatments of bodily structure and mental faculties, in *The Descent of Man* [1981 (1871)].

3. The Comparative Method

The comparative method provides the essential means of unravelling the evolution of behavior. The approach and the significant issues have been reviewed and thoroughly discussed (see Masterton, Hodos, and Jerison 1976; and earlier, Roe and Simpson 1958). Our objective is to describe the method more specifically as it relates to the study of the evolution of perception. At the outset it is understood that the issues are complex (for some this would be considered an understatement), and predictions based on findings, no matter how convincing and well substantiated, are never certain, but only more or less probable. Given that caveat there is much that can be done, and we are beginning to see some interesting results. Laboratory studies of sensory capacity (chapter by Fay; Stebbins 1970; 1990) are an important part of perception and have given us considerable insight into the relationship of stimulus detection and discrimination to underlying physiological mechanisms (Stebbins, Brown, and Peterson 1984) and to the evolution of, for example, high frequency hearing in mammals (Masterton, H. Heffner, and Ravizza 1969; R. Heffner and Masterton 1990).

Perception in general, and specifically as it is treated here as in the perception of complex acoustic signals, is not merely selective but contiguously both synthetic and analytic. It is argued that perceptual analysis and synthesis represent successful adaptations particularly when they are combined in the same nervous system. At one level auditory systems are selective for sounds that are utilized in such biologically essential activities as mating, prey capture, and predator evasion. At another level, the central nervous system can direct attention to certain immediately compelling acoustic events at the expense of others, which for the moment are less compelling. Humans, and now our evidence includes other animals as well, demonstrate a synthetic perceptual constancy in responding similarly to widely varying acoustic waveforms in the communication sounds of conspecifics (you all vs y'all, for example), and yet when circumstances require it, they can with little difficulty analyze and discriminate between these different waveforms, as in individual recognition. Such perceptual categorization seems efficacious, particularly in noisy environments, since consider-

able variance in the acoustic structure of the signal will not render the message ambiguous to the receiver. Yet the ability to dismember the category (analysis) when the context changes is also advantageous. It is a case of having it both ways. In either condition there are certain features or components in the signal itself that engage and direct perception. These features in vocal signals may have evolved as a consequence of ecological pressures related to characteristics of the environment, specifically the properties of sound transmission in that environment. Examples of these features include the frequency modulation in the cry of the echoranging microchiropteran bat or in the stop consonants used in human speech. Perceptual constancy, categorical perception, and the critical nature of significant features of biologically relevant stimuli are then the focal points of this chapter. In addition, the apparently asymmetrical processing of such stimuli by the brain's left cerebral hemisphere, long recognized in man, has been extended to include other animals (Petersen et al. 1978; H. Heffner and R. Heffner 1986; 1990). The advantages of such asymmetry for the processing of species signals and its evolution have been the object of much discussion (see Bradshaw and Nettleton 1983, for a review) but the matter remains unresolved. It may serve to "avoid redundant duplication of functions in a limited processing space" (ibid., p. 187). It may have evolved from an important separation of function between the two hands or paws in animals able to use one for holding and the other for manipulating objects. Obviously the close relationship between limb or hand (paw) gesture, facial gesture, and vocal communication is an important ingredient, but we can only surmise the selective advantage or evolutionary sequence.

In its most general sense the comparative method is integral to the study of evolution and the evolutionary process. Comparison for its own sake is difficult to justify except in a very limited sense and will do little to further the progress of biological or behavioral science. Comparison for the purpose of affirming (or perhaps, in the minds of some, reaffirming) a natural scale with man on top, or as the final objective or achievement of the evolutionary process, dominated the earliest efforts in comparative psychology in this country and abroad, particularly with regard to intelligence (cognition). Although it has been effectively dismissed as legitimate scientific inquiry (Hodos and Campbell 1969; 1989), it continues to find favor and expression not only in the popular culture, but even in the scientific literature [when animals are referred to as lower (sub or infra) or higher relative to man, or for example, as the endpoint of the evolutionary process; Antinucci 1989]. The comparative method has been amply discussed both in relation to structure and to behavior (Hailman 1976; Hodos 1976; Webster 1976), but for the most part the method has been applied to structure.

If we agree with Hodos (1976) among others that the recognition of homologous behaviors must depend upon an adequate knowledge of the underlying homologous structures, then for the most part our hands are tied in the study of evolutionary process in complex perception. The discovery of hemispheric laterality in other animals (Petersen et al. 1978; H. Heffner and R. Heffner 1986) is still too preliminary to permit conjecture regarding the responsible neural mechanisms, although it is a beginning. Comparative studies of perception in both closely and distantly related species can give us insights into the ecological pressures that have guided certain forms of perception and the use of particular communicative signals. In the perception of simpler signals, the relation between head size and high-frequency hearing in mammals provides a good model of what can be done given a sufficient sample of diverse species (R. Heffner and Masterton 1990). Perceptual similarities and differences and exceptions to frequently observed perceptual capabilities uncover the presence or absence of certain ecological demands on perception as in above-ground versus underground hearing (R. Heffner and Masterton 1990; chapter by H. Heffner and R. Heffner). Developmental studies can reveal certain life history variables in individuals (Hailman 1976), although these have proven difficult to engage in the auditory domain because of the formidable task of removing all auditory stimulation during the course of development.

Of course, the close interdependent relationship between the perception of species signals and their vocal utterance requires that we study the process in both sender and receiver. This is one example where it is necessary to combine both field and laboratory study. Extensive observation of a social group (i.e., primate troop) under natural or quasi-

natural conditions permits determination of the relationship of social and environmental context to message form and content (Green 1975, for example). With a vastly improved acoustic recording technology we are achieving a far better understanding of the complexity and diversity of the species signals used in animal communication. In addition the judicious use of vocal playback experiments in the field has revealed the distinct possibility that communication in nonhumans is not solely under the control of affect or drive state, but, in fact, is also cognitively based (Seyfarth, Cheney, and Marler 1980; S. Gouzoules, H. Gouzoules, and Marler 1984). By direct implication the evidence for cognition in the vocal message requires cognitive processing by the receiver. Playback experiments permit some separation of the signal from its usual context, thus preventing any visible or perhaps olfactory indication of the emotional state of the signaler (H. Gouzoules, S. Gouzoules, and Marler 1985). In addition, some elementary studies of perception are feasible under field conditions (D. Nelson and Marler 1989).

Understanding the perception of these signals must ultimately depend on the methodology of the laboratory where the animal can be completely detached from its rich environmental context, thus enabling the analysis of acoustic perception per se unaffected by countless other modes of stimulation. At the same time the laboratory scientist is dependent upon her colleague from the field for the field recordings and a complete description of the social and environmental context of the signal. The laboratory methodology is based upon a combination of operant conditioning procedures devised to overcome the language barrier between subject and experimenter and ensure that the animal becomes a reliable observer of sensory events, and classical psychophysics with its protocols for stimulus presentation and data treatment (chapter by Fay; Stebbins, Brown, and Petersen 1984).

A major criticism of laboratory experimentation is directed at the removal of the animal from its natural setting, which may distort the phenomenon under study—a biological version of the principle of indeterminacy. It is likely a criticism that arises from a misunderstanding of the nature of the experimental question that is being asked. Our knowledge of auditory perception and its basic underlying physiological mechanisms requires the animal's isolation from its natural environment in order to determine the limits of resolution of the auditory system and the animal's judgment of various auditory events uninfluenced by nonauditory stimulation or even other auditory signals. It is perfectly true that under natural conditions the auditory system acts in concert with other systems, and for that reason it is appropriate in certain circumstances to evaluate the animal in that context. But if one is to understand the unique contribution that the auditory system makes to acoustic communication, perception, and behavior, then it is imperative to determine its specific participation exclusive of the participation of the visual system, the olfactory system, and so on. The information from these experiments is valuable to field researchers both because it establishes sensory limits for different species in the detection and discrimination (or localization) of acoustic stimuli, and because like the field playback experiment it focuses on the attention given to the auditory events themselves by the animal and thereby allows the experimenter to gauge the importance of these events to the animal.

4. The Research Program

In the following section we outline the details of a research program on the perception of complex species signals in a species of nonhuman primate. It includes a brief description of the field research that provided the essential background including the signals themselves and a description of their context (Green 1975). This is followed by a more detailed discussion of the subsequent laboratory experiments on the acoustic perception of these signals and their putative physiological mechanisms. Finally, we will examine the feature analysis of one of these signals, the importance of an exemplary information-bearing element, the presence of that particular element in the communication systems of other animals, and its possible adaptive significance for communication and perception.

The Japanese monkey's (*Macaca fuscata*) communication system and, in particular, its contact or "coo" call affords a good model for the study of the perception of complex species signals and of their physiological basis. Green's (1975) original field

work provided a very thorough analysis of this animal's complete vocal repertoire and the context for the various signals that it employs in communicating with conspecifics. It was an exceptional study because of the extended period of time over which it occurred and the number of observations that were carried out on the species signals and their context (6×10^4 vocalizations and 2×10^3 observation hours) and because of the excellent technology that was used in recording the signals themselves. Green's understanding of both the literature of animal communication and of human speech perception gave the study a theoretical richness it might not have had otherwise. Perhaps its most important contribution was the finding of the close relationship between variation in the acoustic structure of the call and the circumstances in which the call occurred. The coo call proved a good choice because of its relatively simple acoustic morphology (Fig. 1 in Green 1975, p. 11). This made it readily adaptable for laboratory experimentation, since it would be comparatively easy to measure in a perceptual experiment and susceptible to being copied and even synthesized for further experiments on those features that might be critical for its detection and discrimination.

The Japanese monkey is a macaque found only in Japan that lives in relatively large provisioned troops, often in selected areas such as national parks. It has been studied intensively with careful record keeping maintained on individual identity, kinship, matrilineal relationships, dominance, and life history. Its communication system is similar, though not identical, to that of other macaques that have been studied. Green has identified ten different sound classes, including the coo call, in a system of communication that is considered graded rather than discrete (see, for example, Marler 1976); this refers to the almost continuous variation in call structure and presumably communicative significance with context. According to Green (1975), the coo call is affinitive—calling for some form of contact, nonagonistic, and commonly displayed under conditions of low arousal or affect. Of all this species' calls the coo is probably the least complicated acoustically (see Fig. 13.1). The seven different variants that Green described are characterized by tonality, by seldom more than two significant harmonics, by a high signal-to-noise ratio, by a duration of about 0.5 second or slightly

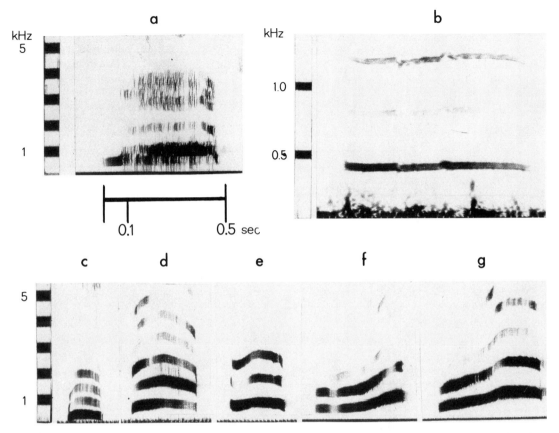

FIGURE 13.1. Seven patterns of coos differing in shape and used differently: (A) double; (B) long low; (C) short low; (D) smooth early high; (E) dip early high; (F) dip late high; (G) smooth late high. (From Green 1975.)

less, and by a frequency modulation and peak within the call itself. The context for the call often includes a solitary animal sometimes out of visual contact with the other group members, and the call is often answered in kind. Figure 13.2 provides a frequency analysis of the occurrence of the different coo variants according to the situation in which they occurred (Green 1975, p. 15). Although there is some overlap, the calls are specific, for the most part, to one or two well-defined contexts.

Two of the coo calls were selected for several experiments on their perception in the laboratory (Beecher et al. 1979). These were the smooth early high (SEH) and the smooth late high (SLH) (Fig. 13.1). The SEH is often employed by a young animal sitting alone or by a group member out of sight of the main body of the troop, while the SLH is usually given by an estrus female soliciting a male consort. The calls were chosen because they were

employed in very different circumstances. They were also selected because they were similar acoustically in many respects but differed on one dimension that was observable by the experimenters—the point in the signal when the frequency inflection or modulation occurred, near the beginning in the first two-thirds of the call (SEH) or nearer the end in the last third (SLH). The protypical SEH and SLH calls represented approximate ends of a continuum of time of inflection or modulation and thus might be thought of as an animal communication analogy or model for the phonemic continuums in human speech (i.e., *ba* vs *da*). By employing the speech perception model we could investigate in an animal communication system the same kind of phenomena found in studies of human speech perception and in similar experiments on the perception of speech sounds by other animals. These phenomena include perceptual constancy,

Type of COO Vocalization

| Situation | LOW | | | HIGH | | | |
| | | | | Early | | Late | |
	Double	Long	Short	Smooth	Dip	Dip	Smooth
Separated Male	XXXX XXXXXX XXXXXX		XX	XX		XX	XX
Female Minus Infant	XXXX	XXX					
Nonconsorting Female	XX	XXXXXX	X				
Female at Young		XXX XXXXXXX	X				
Dominant at Subordinate				XXX XXXXXXX		XX	
Young Alone				X XXXXXXX XXXXXX	XXXXXX	X	
Dispersal				X XXXXXXX	XX	XXXXX	XX
Young to Mother				XXXX	XXX XXXXXXX	XXXX	X XXXXXXX
Subordinate to Dominant				XXX	XXXX XXXXXXX	XXX XXXXXXX XXXXXXX XXXXXX	XXXXX
Estrus Female					X	XXXX XXXXXXX XXXXXXX	XXXXX XXXXXXX XXXXXXX XXXXXXX

FIGURE 13.2. Occurrence of coo sounds in different situations. A sample of 226 utterances of coo sounds were separated into the listed types by spectrographic analysis. Each occurrence of a type is scored in a behavioral situation as indicated by the marks. (From Green 1975.)

categorical perception, and neural lateralization. The other important objective was the determination of the acoustic features in the calls that might be especially salient in their perception. The risk was in using a model developed strictly for human use on other animals; the experiment was approached with that caveat in mind and not fully resolved. A significant question was whether the subjects would be able to perceive the differences among these vocal species signals when both animal and signal were removed from their rich natural environment. The finding would have important implications for the feasibility of studying these phenomena in the laboratory and of determining the responsible auditory mechanisms without input from the other sensory systems.

The laboratory setting included a sound proof chamber and a small computer for controlling and recording the conditions of the experiment. The Japanese monkeys were employed in all of the experiments; in addition other macaques (*M. nemestrina*) were used as controls for the perception of the Japanese monkey vocalizations. The animals were trained and tested in the sound chamber while wearing earphones so that each ear could be tested separately. Subjects were trained by operant conditioning procedures to touch a small metal tube in response to auditory stimulation and then release it when the auditory stimulation, which began as soon as the tube was touched, changed from one form to another, for example, from a SEH to a SLH. Correct responses (to the change in signal type from SEH to SLH or vice versa) were reinforced with food, while incorrect responses were followed by a brief time out from the experiment. In the earliest study the animals were required to discriminate between exemplars of the two call types (SEH and SLH). If they were able to do this to a criterion of at least 90 % correct, new exemplars of each type were added until the animals were discriminating between as many as eight calls of each type. In spite of considerable variation in many parameters of the calls within each class such as level, duration, and starting frequency, the Japanese monkeys acquired the discrimination very quickly with few errors. (Beecher et al. 1979) On the other hand, the control animals (*M. nemestrina*) had great difficulty even with the simplest discrimination between only two signals. As an additional control, in a second experiment with different subjects, the discrimination was changed to one in which the calls were grouped according to starting frequency. This time the Japanese monkeys found the discrimination formidable, while the control animals learned it relatively quickly. On the basis of these findings we argued that the Japanese monkeys, in showing clear evidence of perceptual constancy within the two call types and in acquiring the discrimination with such facility (as opposed to the congeneric controls), were showing evidence of perceptual specialization for their own species signals.

In these same experiments, the calls were presented randomly to right or left ear and responses tabulated accordingly. Our assumption, supported in human studies of cerebral laterality, was that if the animal performed with greater accuracy when the stimuli were delivered to one ear as opposed to the other, the cerebral hemisphere opposite to that ear would be implicated in the processing of the call. In fact, that is what happened; all of the five Japanese monkeys showed a right ear, thus left hemisphere, advantage for their species calls. The three other macaques, used as nonspecies controls, showed no ear advantage for these same calls, although the one vervet monkey (*Cercopithecus aethiops*) in the experiment also showed a right ear advantage. The evidence thus far supported a left hemisphere specialization in the Japanese monkey for this particular call (Petersen et al. 1978). These results later received strong support from investigators in another laboratory who took the process a step further and showed that left temporal lobe lesions produced a severe although temporary deficit in the discrimination between the two call types, SEH and SLH (H. Heffner and R. Heffner 1986; 1990). Moreover, the Heffners found that lesions in the right temporal lobe had no effect on the discrimination, while lesions in both lobes abolished the discrimination which could not then be relearned.

The same methodology that had been employed in demonstrating both perceptual constancy and cerebral dominance in the Japanese monkey's coo call could be used with appropriate modification to determine whether this animal with its graded communication system would categorically perceive the coo call on a continuum from the SEH, with the frequency peak or inflection near the beginning of the call, to the SLH, with the peak near the end. The continuum was based, then, on the temporal location of the frequency inflection within the call. The experimental paradigm is analogous to that which is used with the phonemic continua in the perception of human speech. *Ba* and *da*, for example, lie at opposite ends of a continuum that is characterized by a fast frequency change in the first 50 ms of the signal before the steady-state vowel. *Ba* is distinguished by a sharp rise in the frequency of the second format; *da* by an equally sharp fall in the frequency of the same formant. Intermediate positions on the continuum are characterized by intermediate transitions in the starting frequency of the second format. Both human and animal subjects perceive the phonemes categorically—that is, they label the signals in one half of the continuum as *ba* and in the other half as *da*. Subjects discriminate between pairs of these phonemes more accurately if members of the pair are in different categories than if they belong to the same category (Kuhl 1986).

The question was whether the Japanese monkeys would perceive their own call in a manner similar to the way in which human speech is perceived. It was necessary in these experiments to use synthetic signals since natural coo calls intermediate between the SEH and SLH were not always equally spaced. The initial question that had to be answered was whether the animals would accept the synthetic signals as examples of their own natural calls. In fact, when synthesized versions of the SEH and SLH were first inserted as probes into the experiment along with the natural calls, the animals labeled them appropriately. Only then could the intermediate versions of these stimuli be inserted into the procedure. This was a particularly difficult experiment to carry out with animals because we were quite sure, based on

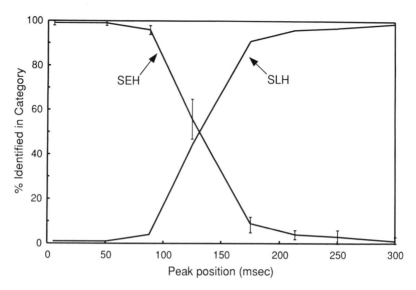

FIGURE 13.3. Identification of synthetic "coo" calls. Data are presented as the mean and the standard error of the mean for four subjects (Japanese monkeys). A category boundary, as indicated by the sharp transition in the behavioral response, was observed near the midpoint of the function. Although most of the synthetic stimuli were placed in only one of the two vocal categories, responses to the stimulus with a 125 ms peak position were more equivocal. (From May et al. 1989.)

previous evidence of the acuity of the macaque auditory system, that the animals could discriminate among all of the signals. But in this experiment we were not interested in what the animal could do at the limits of its auditory resolution but what it would do, "naturally," given this particular set of stimuli. Thus the manner of asking the question had important implications for the design of the experiment. We continued the discrimination procedure with the natural calls that the animals had acquired previously with food reinforcement (see above). We then probed infrequently with the synthetic stimuli but with no consequences provided for responding to these stimuli. Thus we examined the perception of these stimuli without direction from the experimenter—that is, without explicit training. We found that the subjects did categorize the stimuli as shown in Figure 13.3 and similarly to the way in which speech sounds are categorized. They responded in like manner to those stimuli in the SEH class with the frequency peak early in the call and to those in the SLH class with a late-occurring peak, but they differentiated between signals from the two classes. The category boundary separating the two classes is shown by the sharp transition in the behavioral response near

the midpoint of the function. While most of the signals were placed in only one category most of the time, the placement of the signal with the frequency inflection at 125 ms was equivocal (May, Moody, and Stebbins 1989).

The evidence suggests that the animals are able to perceive and distinguish at least one of their vocalizations outside their natural setting, and when the auditory system alone is engaged under the highly controlled and artificial conditions of the laboratory. In this context they respond to synthetic signals and the natural calls similarly, they demonstrate perceptual constancy and categorical perception, and provide evidence for left hemisphere processing and laterality for this vocalization. At this point in the experimental program it seemed important to examine the coo call more closely in order to determine if there was a certain information-bearing feature (or features) that was responsible for its perception. The original hypothesis from the field research and from the laboratory studies indicated that it was the frequency inflection that was important, but there were other candidates that could not yet be ruled out. The following experiments were undertaken with that in mind. For these experiments synthetic stimuli

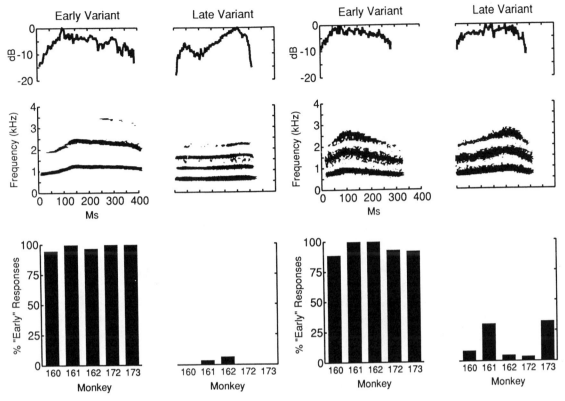

FIGURE 13.4. Sonogram for natural early (upper left) and late (upper right) inflected coo calls. The contours at the top of the panel represent the amplitude changes while those just below represent the frequency changes for the fundamental and harmonics. Percent "early" responses to the early inflected call (lower left) and percent "early" responses to the late inflected call (lower right). All animals responded to the early variant as an early inflected call but not similarly to the late variant. (From May et al. 1988.)

FIGURE 13.5. Sonogram for synthetic early (upper left) and late (upper right) inflected coo calls. The contours at the top of the panel represent the amplitude changes while those just below represent the frequency changes for the fundamental and harmonics. Responses to the calls in lower panels as described for Figure 13.4. Animals responded to the early variant as an early inflected call, but not similarly to the late variant. (From May et al. 1988.)

were essential since they could be easily modified or degraded so that those features that were salient in the perception of the signal could be isolated and identified. The method was similar to that used in the experiment on categorical perception just described. The natural calls were continued and the synthetic calls were introduced as probes infrequently without consequences or reinforcement. A more complete description of the procedure and findings is available elsewhere (May, Moody, and Stebbins 1988).

The findings for these experiments are illustrated in Figure 13.4. Sonograms for two new natural stimuli, not previously used in the experiment, are presented in the upper panel of the figure. The highest contour on the figure gives a relative measure of signal amplitude while the contours immediately below indicate the changes in frequency with time. These two signals were initially presented as probes; the animals' response to them is shown in the histogram in the lower panel of the figure. The signal in the upper left entitled "early variant" is almost invariably identified as a SEH, while the signal labeled "late variant" is almost never selected as an exemplar of the SEH category. The use of natural stimuli in this instance served to validate the probe procedure which could then be used with the synthetic

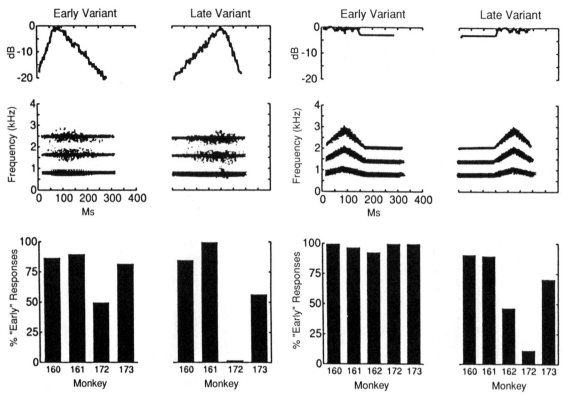

FIGURE 13.6. Sonogram for synthetic early (upper left) and late (upper right) altered coo calls. The contours at the top of the panel represent the amplitude changes while those just below represent the frequency changes for the fundamental and harmonics. The frequency inflection has been eliminated although the amplitude change has been retained. Responses to the calls in lower panels as described for Figure 13.4. For the most part the animals responded similarly to both early and late variants. (From May et al. 1988.)

FIGURE 13.7. Sonogram for synthetic early (upper left) and late (upper right) altered coo calls. The contours at the top of the panel represent the amplitude changes while those just below represent the frequency changes for the fundamental and harmonics. The frequency inflection has been shortened. Responses to the calls in lower panels as described for Figure 13.4. For the most part the animals responded similarly to both early and late variants. (From May et al. 1988.)

signals. The synthetic prototypes designed to replace the SEH and SLH are shown in the upper panel of Figure 13.5. From the behavioral evidence in the histogram below it is apparent that these signals are accepted by the animals as legitimate substitutes for their natural counterparts.

With these results it was now possible to alter the parameters, one at a time, in the prototypes. In the first such experiment we removed the frequency inflections from both smooth-early and late-high prototypes, leaving the position of the amplitude peak (early or late in the signal) as a dominant feature of both signals. With one partial exception, the animals did not differentiate between the two synthetic and altered signals as shown in Figure 13.6.

It seemed unlikely that amplitude was a significant information-bearing element of the coo call. We then shortened the frequency inflection at the beginning and end of each prototype call, on the assumption that we were isolating and perhaps therefore emphasizing the frequency inflections at each end of the call. To our surprise, with one exception (the same animal referred to in Fig. 13.6 above) the animals again did not distinguish between the two altered calls (see Fig. 13.7), casting some doubt on the importance of frequency inflection per se. We therefore removed the frequency peaks completely so that all that remained were two tone glides, a downward frequency modulation for the early variant and an upward modulation for the late variant as shown in the upper part of Figure 13.8. The findings were robust for all subjects who clearly discriminated between the two synthetic calls much as they had for the two natural calls (Fig. 13.4) and for the two original prototype copies (Fig. 13.5), leading us to argue for the importance of simple frequency modulation as a significant informational component of the Japanese monkeys' coo call. The specific parameters of the modulation remain to be identified. Additional research on artificial, laboratory-generated stimuli (Moody and Stebbins 1989) has confirmed the salience of frequency modulation for nonhuman primate perception. We have also provided evidence that, at least near threshold for the detection of frequency modulation, these animals may, in the course of processing the signal, recode it in some manner into a discrete frequency percept. It is then the discrete frequency change upon which their discrimination of FM may ultimately be based. This processing and coding strategy may reflect an adaptation of the auditory system to deal with these commonly used vocal signals that are simultaneously near modulation threshold and rapidly modulating in time.

The ubiquity of frequency modulation in a variety of animal vocalizations in many ecologically diverse species argue for its consideration here as a feature of complex biologically relevant signals that is adaptive generally without regard to specific habitat characteristics. It is a feature of the calls of many amphibian and avian species (Falls 1963; Capranica 1978; Ryan 1983) as well as aerial, aquatic, and terrestrial mammals (Green 1975; Simmons and Stein 1980; Popov et al. 1986). For example, it is a principal component of male sexual solicitation calls in some anurans; its functional significance has been revealed in a series of playback experiments in which females demonstrated a preference for calls containing FM over those from which all FM components had been removed (Rand and Ryan 1981). Ryan (1983) showed that the female vocalizations also contained an FM component. Synthetic renditions of these calls in which the FM sweep was in the same direction as in the natural version of the call led the species males to approach a loudspeaker more often and with a shorter latency than when the sweep was in the reverse direction.

Several species of microchiropteran bats use FM in echoranging; the reflection of the various frequencies in the call provide a far more complete

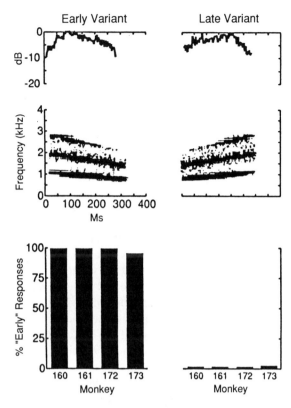

FIGURE 13.8. Sonogram for synthetic early (upper left) and late (upper right) altered coo calls. The contours at the top of the panel represent the amplitude changes while those just below represent the frequency changes for the fundamental and harmonics. The inflection has been removed from the frequency change. Responses to the calls in lower panels as described for Figure 13.4. Animals responded to the early variant as an early inflected call, but not similarly to the late variant. (From May et al. 1988.)

image of the target than single frequencies or narrow bands (Simmons and Stein 1980). The requisite physiological mechanisms are in place (chapters by Merzenich and Schreiner, and by Pollak). Suga (1965; 1969) has confirmed the existence of single neurons in the bat's auditory system sensitive to FM with fully ten percent of those sensitive to sweep direction. Similarly, the cat's auditory system is equipped with FM-sensitive neurons and a subset sensitive to the direction of the FM sweep (Whitfield and Evans 1965; P. Nelson, Erulkar, and Bryan 1966; Erulkar, Butler, and Gerstein 1968). In one of the earliest studies on the significance of FM in human speech, Licklider and Miller (1951) report that removing all of the amplitude-modulated segments from speech sounds had little effect on intelligibility as long as the frequency-modulated components were retained. Removal of the FM components, however, resulted in severe losses in speech comprehension. In consonant–vowel stimuli the first 50 to 100 ms particularly in the second and third speech formants are defined by an FM sweep whose characteristics provide important perceptual distinctions for stop consonants (*ba*, *da*, and *ga*, for example) (Liberman et al. 1967). Schwartz and Tallal (1980) have argued convincingly, on the basis of their findings, that it is not language per se that is lateralized to the brain's left hemisphere but, instead, the initial frequency transitions in consonant–vowel combinations. Psychophysical experiments with human subjects using an adaptation paradigm have demonstrated the sensitivity of the human listener to FM and to FM-sweep direction outside the context of speech and have suggested the existence of FM channels (Kay and Matthews 1972; Gardner and Wilson 1979; Regan and Tansley 1979; Tansley and Regan 1979) although there is some question concerning the permanence of the adaptation effect (Moody et al. 1984).

On the basis of the importance of FM in a variety of vocal communication systems we argue here for its general adaptive value irrespective of the specialized requirements or selective pressures of particular ecologies. It is likely, therefore, that it is an effective component in signal transmission and in certain aspects of signal identification and discrimination. The use of a frequency-modulated signal may increase the effective signal-to-noise ratio and facilitate signal detection in noisy environments. Support for this hypothesis comes from the work of Collins and Cullen (1978) who found that detection thresholds for human subjects for frequency-modulated signals in broad-band noise were significantly lower than for any pure tone within the frequency range of the modulation presented alone. Dolphins are also more sensitive to FM than to pure tones in background noise (Saprykin et al. 1980). In some instances the use of FM may permit the signal to escape the frequency limits of environmental ambient noise and thus increase transmission distance and intelligibility. At the same time FM has an added advantage over equivalent broad-band noise for added distance in that the signaler is able to put more energy per cycle into the call when the different frequencies are temporally separated. There is

increasing evidence that FM has been incorporated into vocal communication systems because of its relatively low susceptibility to distortion and degradation over distance (Marler 1969; 1972; 1973; Waser and Waser 1977). In fact, the temporal structure and amplitude modulation in primate calls appear to be more vulnerable to environmental degradation than the FM components (Wiley and Richards 1978). Finally, the use of FM in species communication signals may enhance their locatability. Several species of macaques in the laboratory were able to localize macaque vocalizations that were frequency modulated far more accurately than unmodulated vocalizations (Brown et al. 1979).

5. Summary

It has been our objective in this chapter to discuss the issues and the problems regarding the study of the evolution of the perception of complex, biologically significant, species vocal communication signals. The difficulty of such an endeavor quickly becomes obvious but should in no way detract from its importance. The likelihood of substantial progress in the near future is problematic, however, not only because of the time-intensive nature of behavioral research, but also because of the current prevailing view first of the auditory system as an engineering problem to be solved rather than a biological one, and second of auditory perception and psychophysics as disciplines predominantly or even exclusively for the study of human behavior. We have outlined a research program that includes both field and laboratory investigation and argue that both are essential if we are to make any substantial progress toward a better understanding of both the ultimate and proximate mechanisms in complex perception. Although the paleontological evidence is unobtainable and our knowledge of the morphological basis of perception is preliminary this should not slow the behavioral research, for it is behavior from which we infer perception. Only by measuring behavior in ecologically similar and diverse species in response to naturally occurring sounds including the complex signals of conspecifics in the field and in the laboratory can we begin to put together a systematic and complete account of the perceptual process. Therefore, it is quite clear that more comparative data are required, since many of our current findings are indeed "innocent of comparisons." With these data in hand we can begin to create more sensible hypotheses with regard to the selective pressures that have imposed themselves in the course of evolution, about the forms of adaptation that have resulted from these ecological conditions, and perhaps even about the evolutionary sequence and process of those two interdependent, but not identical phenomena, perception and communication including human speech and language. It should be possible to gain additional insight, for example, into such perceptual characteristics as constancy, neural laterality, and categorization. It is the nature of these behavioral findings that can and should guide further research into some of the basic physiological and anatomical mechanisms underlying complex perception.

References

Antinucci F (1989) (ed) Cognitive Structure and Development in Nonhuman Primates. Hillsdale: Erlbaum.

Beecher MD, Petersen MR, Zoloth SR, Moody DB, Stebbins WC (1979) Perception of conspecific vocalizations by Japanese macaques: Evidence for selective attention and neural lateralization. Brain, Behavior, and Evolution 16:433–460.

Beecher MD, Stoddard PK (1990) The role of bird song and calls in individual recognition: Contrasting field and lab perspectives. In: Stebbins WC, Berkley MA (eds) Comparative Perception, Vol 2: Complex Signals. New York: Wiley.

Berkley MA, Stebbins WC (eds) (1990) Comparative Perception, Vol 1: Basic Mechanisms. New York: Wiley.

Boakes R (1984) From Darwin to Behaviorism. Cambridge: Cambridge University Press.

Bradshaw JL, Nettleton NC (1983) Human Cerebral Asymmetry. Englewood Cliffs: Prentice-Hall.

Brown CH, Beecher MD, Moody DB, Stebbins WC (1979) Locatability of vocal signals in Old World monkeys: Design features for the communication of position. Journal of Comparative and Physiological Psychology 93(5):806–819.

Bullock TH (1984) The future of comparative neurology. American Zoologist 24:693–700.

Capranica RR (1978) Auditory processing and animal sound communication. Federation Proceedings, Federation of American Societies for Experimental Biology 37:2315–2359.

Collins MJ, Cullen JK (1978) Temporal integration of tone glides. Journal of the Acoustical Society of America 63(2):469–473.

Darwin C (1981) The Descent of Man and Selection in Relation to Sex. Princeton: Princeton University Press, (Photoreproduction of 1871 edition published by J. Murray, London).

Erulkar SD, Butler RA, Gerstein GL (1968) Excitation and inhibition in cochlear nucleus. II. Frequency-modulated tones. Journal of Neurophysiology 31:537–548.

Falls JB (1963) Properties of bird song eliciting response from territorial males. Proceedings of the XIIIth International Ornithological Congress, pp. 259–271.

Gardner RB, Wilson JP (1979) Evidence for direction-specific channels in the processing of frequency modulation. Journal of the Acoustical Society of America 66(3):704–709.

Gouzoules H, Gouzoules S, Marler P (1985) External reference and affective signaling in mammalian vocal communication. In: Zivin G (ed) The Development of Expressive Behavior, Biology-Environment Interactions. Orlando: Academic Press.

Gouzoules S, Gouzoules H, Marler P (1984) Rhesus monkey (Macaca mulatta) screams: Representational signaling in the recruitment of agonistic aid. Animal Behaviour 32:182–193.

Green S (1975) Variation of vocal pattern with social situation in the Japanese monkey (Macaca fuscata): A field study. In: Rosenblum L (ed) Primate Behavior. New York: Academic Press, pp. 1–102.

Griffin DR (1981) The Question of Animal Awareness. New York: Rockefeller.

Hailman JP (1976) Uses of the comparative study of behavior. In: Masterton RB, Hodos W, Jerison H (eds) Evolution, Brain, and Behavior: Persistent Problems. Hillsdale: Erlbaum.

Heffner HE, Heffner RS (1986) Effect of unilateral and bilateral auditory cortex lesions on the discrimination of vocalizations by Japanese macaques. Journal of Neurophysiology 56:683–701.

Heffner HE, Heffner RS (1990) Role of primate auditory cortex in hearing. In: Stebbins WC, Berkley MA (eds) Comparative Perception Vol 2: Complex Signals. New York: Wiley.

Heffner RS, Masterton RB (1990) Sound localization in mammals: Brainstem mechanisms. In: Berkley MA, Stebbins WC (eds) Comparative Perception Vol 1: Basic Mechanisms. New York: Wiley.

Hodos W (1976) The concept of homology and the evolution of behavior. In: Masterton RB, Hodos W, Jerison H (eds) Evolution, Brain, and Behavior: Persistent Problems. Hillsdale: Erlbaum.

Hodos W, Campbell CBG (1969) Scala naturae: Why there is no theory in comparative psychology. Psychological Review 76:337–350.

Hodos W, Campbell CBG (1989) Evolutionary scales and comparative studies of animal cognition. In: Kesner R, Olton D (eds) The Neurobiology of Animal Cognition. Hillsdale: Erlbaum.

Kay RH, Matthews DR (1972) On the existence in human auditory pathways of channels selectively tuned to the modulation present in frequency-modulated tones. Journal of Physiology 255:657–677.

Kuhl PK (1986) Theoretical contributions of tests on animals to the special mechanisms debate in speech. Experimental Biology 45:233–265.

Liberman AM, Cooper FS, Shankweiler DP, Studdert-Kennedy M (1967) Perception of the speech code. Psychological Review 74:431–461.

Licklider JCR, Miller GA (1951) The perception of speech. In: Stevens SS (ed) Handbook of Experimental Psychology. New York: Wiley.

Marler P (1969) Colobus guereza: territoriality and group composition. Science 179:1083–1090.

Marler P (1972) Vocalizations of East African monkeys. II. Black and white Colobus. Behaviour 42:175–197.

Marler P (1973) A comparison of vocalizations of red-tailed and blue monkeys. Cercopithecus ascanius and C. mitis in Uganda. Zeitschrift für Tierpsychologie 33:223–247.

Marler P (1976) Social organization, communication and graded signals: the chimpanzee and the gorilla. In: Bateson PPG, Hinde RA (eds) Growing Points in Ethology. Cambridge: Cambridge University Press.

Marler P, Tamura M (1962) Song "dialects" in three populations of white-crowned sparrows. Condor 64:368–377.

Masterton RB, Heffner HE, Ravizza RJ (1969) The evolution of human hearing. Journal of the Acoustical Society of America 45:966–985.

Masterton RB, Hodos W, Jerison H (1976) Evolution, Brain, and Behavior: Persistent Problems. Hillsdale: Erlbaum.

May BJ, Moody DB, Stebbins WC (1988) The significant features of Japanese monkey coo sounds: a psychophysical study. Animal Behaviour 36:1432–1444.

May BJ, Moody DB, Stebbins WC (1989) Categorical perception of conspecific communication sounds by Japanese macaques, Macaca fuscata. Journal of the Acoustical Society of America 85:837–847.

Moody DB, Cole D, Davidson LM, Stebbins WC (1984) Evidence for a reappraisal of the psychophysical selective adaptation paradigm. Journal of the Acoustical Society of America 76:1076–1079.

Moody DB, Stebbins WC (1989) Salience of frequency modulation in primate communication. In: Dooling RJ, Hulse SJ (eds) The Comparative Psychology of Audition: Perceiving Complex Sounds. Hillsdale: Erlbaum.

Nelson DA, Marler P (1989) Categorical perception of a natural stimulus continuum: Birdsong. Science 244:976–978.

Nelson PG, Erulkar SD, Bryan JS (1966) Responses of units of the inferior colliculus to time-varying acoustic stimuli. Journal of Neurophysiology 29:834–860.

Petersen MR, Beecher MD, Zoloth SR, Moody DB, Stebbins WC (1978) Neural lateralization of species-specific vocalizations by Japanese macaques (Macaca fuscata). Science 202:324–327.

Popov VV, Ladygina TF, Supin AY (1986) Evoked potentials of the auditory cortex of the porpoise, Phocoena phocoena. Journal of Comparative Physiology 158:705–711.

Rand AS, Ryan JK (1981) The adaptive significance of a complex vocal repertoire in a neotropical frog. Zeitschrift für Tierpsychologie 57:209–214.

Regan D, Tansley BW (1979) Selective adaptation to frequency-modulated tones: Evidence for an information-processing channel selectively sensitive to frequency changes. Journal of the Acoustical Society of America 65(5):1249–1257.

Richards RJ (1987) Darwin and the Emergence of Evolutionary Theories of Mind and Behavior. Chicago: University of Chicago Press.

Roe A, Simpson GG (1958) Behavior and evolution. New Haven: Yale University Press.

Roitblat HL (1984) Representations in pigeon working memory. In: Roitblat HL, Bever TG, Terrace HS (eds) Animal Cognition. Hillsdale: Erlbaum.

Rowell TE, Hinde RA (1962) Vocal communication in the Rhesus monkey (Macaca mulatta). Proceedings of the Zoological Society of London 138:279–294.

Ryan MJ (1983) Frequency modulated calls and species recognition in a neotropical frog. Journal of Comparative Physiology 150:217–221.

Saprykin VA, Korolev VI, Belov BI, Garanin VM, Serebrennikov VA (1980) Discrimination of frequency-modulated signals against a noise background by the dolphin. Neuroscience and Behavioral Physiology 10(2):178–179.

Schwartz J, Tallal P (1980) Rate of acoustic change may underlie hemispheric specialization for speech perception. Science 207:1380–1381.

Seyfarth RM, Cheney DL, Marler P (1980) Vervet monkey alarm calls: Semantic communication in a free-ranging primate. Animal Behaviour 28:1070–1094.

Simmons JA, Stein RA (1980) Acoustic imaging in bat sonar: Echolocation signals and the evolution of echolocation. Journal of Comparative Physiology 135:61–84.

Simpson GG (1958) The study of evolution: Methods and present status of theory. In: Roe A, Simpson GG (eds) Behavior and Evolution. New Haven: Yale University Press.

Stebbins WC (1970) Animal Psychophysics: The Design and Conduct of Sensory Experiments. New York: Appleton-Century-Crofts.

Stebbins WC (1980) The evolution of hearing in the mammals. In Popper AN, Fay RR (eds) Comparative Studies of Hearing in Vertebrates. New York: Springer-Verlag.

Stebbins WC (1990) Perception in animal behavior. In: Berkley MA, Stebbins WC (eds) Comparative Perception, Vol 1. New York: Wiley.

Stebbins WC, Berkley MA (eds) (1990) Comparative Perception, Vol 2: Complex Signals. New York: Wiley.

Stebbins WC, Brown CH, Petersen MR (1984) Sensory processes in animals. In: Darian-Smith I, Brookhart J, Mountcastle VB (eds) Handbook of Physiology, Sensory Functions, Vol 1. Bethesda: American Physiological Society.

Struhsaker TT (1967) Auditory communication among vervet monkeys (Cercopithecus aethiops). In: Altmann SA (ed) Social Communication among Primates. Chicago: University of Chicago Press.

Suga N (1965) Analysis of frequency modulated sounds by auditory neurones of echo-locating bats. Journal of Physiology (London) 179:26-53.

Suga N (1969) Classification of inferior collicular neurones of bats in terms of responses to pure tones, FM sounds and noise bursts. Journal of Physiology 200: 555-574.

Tansley BW, Regan D (1979) Separate auditory channels for unidirectional frequency modulation and unidirectional amplitude modulation. Sensory Processes 3:132-140.

Uexkull J von (1934) Strafzuge durch die Umwelten von Tieren und Menschen. Berlin: Springer.

Waser PM, Waser MS (1977) Experimental studies of primate vocalization: Specializations for long-distance propagation. Zeitschrift für tierpsychologie 43: 239-263.

Webster DB (1976) On the comparative method of investigation. In: Masterton RB, Hodos W, Jerison H (eds) Evolution, Brain, and Behavior: Persistent Problems. Hillsdale: Erlbaum.

Weiskrantz L (1985) Animal Intelligence. Oxford: Clarendon.

Whitfield IC, Evans EF (1965) Responses of auditory cortical neurons to stimuli of changing frequency. Journal of Neurophysiology 28:655-672.

Wiley RH, Richards DG (1978) Physical constraints on acoustic communication in the atmosphere: Implications for the evolution of animal vocalizations. Behavioral Ecology and Sociobiology 3:69-94.

14
Structure and Function in Sound Discrimination Among Vertebrates

Richard R. Fay

1. Introduction

A sense of hearing can only be demonstrated and described through an analysis of sound's effects on behavior. Since we can observe only the behavior of living animals, the study of the historical evolution of the sense of hearing seems impossible.

The study of the evolution of the ear and other structures of the auditory system is possible, as many of the chapters in this volume illustrate. If the auditory structures of a living species are similar to those of its ancestors, we could suggest that those aspects of hearing that are primarily determined by such structures are similar as well. In order to pursue this sort of investigation, the relations between auditory system structures and behavioral hearing functions must be determined.

1.1 What Are the Functions of Hearing?

Imagine being at a picnic at a park. As you talk with your friends about choosing teams for softball, an airplane cruises overhead, a crow calls in the woods nearby, a radio plays behind you, and traffic moves on the highway behind a stand of trees. Each of these sources of sound is located in a different direction and at a different distance. Each source, too, is made up of different materials (such as biological tissue, wood, etc.) and vibrates according to a different pattern. Some of the energy from each source reaches your two ears as a complex mixture of waves.

Now imagine what you hear. You are able to hear each sound source independently, and you are able to identify and perceive the location of each. Your perception of this auditory scene is a spatial model of the local world. The "problems" that the human auditory system solves so well are to analyze the complex mixture into component frequencies and temporal events, group those components that "belong" to each source, and then reconstruct a representation of the auditory scene (Bregman 1990) in which each source is perceived as an object, appropriately located in space.

In the popular essay *A Brief History of Time*, the physicist Stephen Hawking (1988) has used a simple interpretation of natural selection to argue why it is at least possible that we humans will be able to understand correctly the fundamental organization of the physical world. Hawking says, "in any population of self-reproducing organisms, there will be variations in the genetic material and upbringing that different individuals have. These differences will mean that some individuals are better able than others to draw the right conclusions about the world around them, and to act accordingly. These individuals will be more likely to survive and reproduce and so their pattern of behavior and thought will come to dominate" (p. 12).

In the most general sense, "drawing the right conclusions about the world" would seem to be of primary survival value to individuals of all species, and the auditory system has probably had an important, general role to play in informing individuals about the world of sound sources (and scatterers) (see Römer, Chapter 6; Michelsen, Chapter 5; Lewis, Chapter 11; Schellart and Popper, Chapter 16). The physical characteristics of sound sources are the same now as ever, including size, material, distance, direction, frequency components present, and their respective temporal envelopes. This means that we should expect to find in all species auditory mechanisms that accurately encode these features of sound sources so that perceptual objects can be formed, corresponding to those formed through other senses, and confirmed through behavioral interaction with the environment. Accurate encoding of acoustic features depends on the structures and physiological processes of the auditory periphery.

Jerison (1973) has suggested that biological intelligence may be viewed as the capacity to construct a perceptual world, or in the context of hearing, to synthesize an auditory scene. Jerison goes on to argue that only the mammals (and to some extent the birds) are able to construct the sorts of perceptual worlds we humans are familiar with; i.e., having the ability to "bind" images together in time and space long enough to form objects that remain constant in perception during the passage of time and relative movements between the perceiver and the perceived. This synthesis of information analyzed at the periphery is the function of the brain, and many or all of these functions (e.g., sound localization) probably arise from the structural organization of brain subsystems.

Hearing therefore depends upon the quality and quantity of the information encoded by the ears, and upon the organization and complexity of brain mechanisms devoted to auditory perceptual processing. A broad approach to the evolution of hearing thus includes a study of the relations between hearing (defined behaviorally) and both ear and brain structures among present species (see Chapter 34 by Heffner and Heffner, and Chapter 13 by Stebbins and Sommers). The most optimistic goal of this investigation would be to successfully predict hearing functions form ear and brain morphology, and then to apply these predictions to ancestral forms based on the fossil record. Although predictions based on fossil material cannot be properly tested, they may help to steer our thinking about the evolution of hearing in profitable directions.

In this chapter, I interpret the results of those experiments that focus on the sound discrimination capacities of vertebrate animals. I identify those measures of hearing that seem to be determined by receptor organ structure (e.g., frequency analysis), and those that seem more to reflect primitive brain processes probably common to all vertebrates (e.g., aspects of intensity and envelope processing). It is possible, I suggest, that those hearing capacities closely determined by ear structure can be estimated for ancestral forms based on the dimensions of fossil ears.

I only briefly treat the large behavioral data set we have on the sensitivity and bandwidth of hearing in vertebrates, since these topics are the primary focus of other chapters in this volume (see Chapter 26 by Dooling). Sound localization is discussed in detail in the Chapter 34 by Heffner and Heffner. Nor do I consider the possible relations between brain structure and complex aspects of auditory perception (Chapter 13 by Stebbins and Sommers). Although introspection suggests to us that we are able to form invariant auditory objects into internal representations of the auditory scene, very few experiments have been carried out in any species to quantitatively investigate these aspects of complex auditory perception and the brain mechanisms responsible for them.

2. Sound Discrimination

Studies on sound discrimination essentially answer questions about the finest distinctions between simple sounds an animal is capable of making under rather unusual laboratory conditions. The results from discrimination studies do not necessarily suggest the form of the animal's *Umwelt*, and do not tell us how sound is used, or what distinctions or decisions the animal normally makes about sound in its usual environment. Rather, discrimination thresholds provide a rather neutral, quantitative way to evaluate and compare the limiting capabilities of different auditory systems. Using the analogy of an electroacoustic system, I assume that such specifications as the signal-to-noise ratio, dynamic range, bandwidth, time constants for summation and resolution, and linearity (see, for example, Schwartz 1970) are not accidentally determined. Rather, they have been designed "in," possibly by an evolutionary process in which strict specifications and their costs rise only to the point of ensuring the desired success within their market niche. Similarly, it makes sense that comparable design characteristics for auditory systems arise from structural features that have evolved to maximize success at acceptable costs. Some sound discrimination capabilities appear to arise from ear structure and others from brain mechanisms, as discussed below.

2.1 How Is Sound Discrimination Measured?

Psychophysics is a subfield of experimental psychology which has been developed to investigate objectively the quantitative relations between well-defined behavioral responses indicating a judgement or decision, and the physical dimensions of stimuli upon which these decisions are based (see Green and Swets 1966).

Psychoacoustics is the behavioral study of hearing using psychophysical methods. During the past several decades, psychoacoustic research has revealed and defined many fundamental aspects of human hearing that seem closely tied to the structures and physiological functions of the human ear. For example, the lowest detectable sound pressure of a simple tone (the absolute threshold) can be understood with reference to the acoustic properties of the outer and middle ears, the biomechanical responses of the cochlea, the transduction characteristics of hair cell receptors, the innervation patterns of auditory nerve fibers, and probable detection schemes used by the brain. Psychoacoustic studies have been generally criticized since they seem to reveal *only* the roles that the ear and lower levels of the brain play in determining sound perception. In other words, many of the fundamental psychoacoustic phenomena can be understood without postulating complex processes of the brain other than statistical decision-making. In accepting this criticism, it is recognized that experiments designed to measure detection and discrimination thresholds cannot provide a complete description of or explanation for all the many and complex phenomena of hearing. At the same time, psychoacoustics is quite well suited to the task of defining hearing in a way that might be closely related to ear structure.

Animal psychoacoustics is the study of hearing in nonhuman animals using psychophysical methods. Ideally, the primary difference between animal and human psychophysics is simply that animals must be trained and motivated to make the responses that cooperative human observers readily give after instruction. Over 100 years of research on animal conditioning and learning has developed a powerful technology for behavior analysis and control, and hundreds of animal psychoacoustical studies of sound discrimination have been carried out (reviewed by Fay 1988).

Psychoacoustic thresholds depend not only on the characteristics of the auditory system investigated, but also on the methods and assumptions used in measurement. Space does not allow a detailed treatment of the various methods that have been used in animal psychoacoustics; the following is a brief outline of the general features of the methods most often used (see Stebbins 1970).

2.2 Operant Methods with Reward

In operant reward conditioning, a food- or water-deprived animal is trained to make a simple set of motor responses in order to obtain a small quantity of food or water (reward or reinforcer). If a selected response occurs in the presence of the acoustic signal of interest, the animal is rewarded and the probability of subsequent responses under these conditions is increased. Once this pattern of behavior is established, response probability is measured as a function of stimulus magnitude, and threshold is defined as the stimulus magnitude required to produce a criterion response probability. In discrimination experiments, the animal is usually exposed to a constant stimulus (e.g., tone bursts of constant frequency), and trained to respond when the stimulus changes (e.g., in frequency). The smallest stimulus difference required to produce a criterion response probability is defined as the discrimination threshold (e.g., the frequency discrimination threshold, or limen). Even in a well designed experiment, the discrimination threshold could depend on the level of motivation, the animal's capacities for memory, and decision efficiency (Patterson and Moore 1986), as well as the animal's sensory acuity. Experiments are usually designed to optimize motivation and minimize memory requirements to the extent possible. Operant methods with positive reinforcement have been useful in psychophysical studies on birds (e.g., Hienz, Sinnott, and Sachs 1980), and mammals (e.g., Sinnott, Petersen, and Hopp 1985), but have not been successful for amphibians and reptiles.

2.3 Avoidance Methods

In avoidance conditioning, an animal is trained to make a simple response in the presence of an auditory signal in order to avoid a noxious stimulus (e.g., foot shock). Although motivation is assumed to be maximized in operant and avoidance methods, it is theoretically difficult to equate the motivation established in these two methods. In general, however, many replications of psychophysical thresholds determined using both methods on the same species have revealed no systematic differences in measured acuity (see Figs. 14.1, 14.2). Avoidance methods have been usefully applied to fishes (e.g., Popper and Tavolga 1981), one reptile (the turtle; Patterson 1966), birds (e.g., Dooling and Searcy 1981), and mammals (e.g., Elliot and McGee 1965).

2.4 Classical Conditioning

In some forms of classical conditioning, an unconditioned stimulus (US) such as mild electric shock is used to elicit a reflex behavior (e.g., brief respiratory suppression, or the suppression of an ongoing behavior stream), termed the unconditioned response (UR). After several pairings, an acoustic signal presented just prior to the US tends to evoke a response similar to the UR, termed the conditioned response (CR). The probability of a criterion CR is measured as a function of signal magnitude, and a threshold defined as above. The CR is usually a graded rather than a discrete response, and it is difficult to equate CR magnitude, or the probability of a criterion CR magnitude, with the probabilities of discrete responses.

Since it is not clear how CR probability is adaptive for the individual, questions have arisen regarding the validity of thresholds determined using classical conditioning. However, replications of experiments using operant, avoidance, and classical conditioning methods on the same species have revealed that comparable threshold estimates are often obtained (Fay 1988). Classical conditioning has been successfully used with fishes (e.g., Fay and Coombs 1983), birds (Kuhn, Leppelsack, and Schwartzkopff 1980), and mammals (e.g., Heffner, Heffner, and Masterton 1971).

Megela-Simmons and her colleagues (e.g., Megela-Simmons, Moss, and Daniel 1985; Moss and Megela-Simmons 1986; Megela-Simmons 1988) have used the method of reflex inhibition (Yerkes 1904) to study sound discrimination in anuran amphibians. In this case, a UR normally evoked by a US is suppressed in magnitude or probability by the presence of an acoustic signal just preceding the US.

In general, thresholds determined using each of these behavioral methods have been defined in various ways, some of which are rather arbitrary and subject to response bias (tendencies to respond that are independent of the acoustic signal or its sensory effect). Methods consistent with the theory of signal detection (Green and Swets 1966) are specifically designed to obtain sensitivity estimates which are independent of response bias (e.g., Costalupes 1988). These experimental designs are clearly preferred when it is possible to apply them to any given method of behavioral control.

3. Hearing and Sound Discrimination in Vertebrates

In the following sections, I review the results of animal psychoacoustical experiments and their possible relations to ear structure. Tone detection thresholds measured in quiet (audiograms) are presented primarily to indicate the quantity of behavioral hearing data in the literature, and to generally describe the variation in hearing range among species within and between vertebrate classes. I next treat sound level ("intensity") discrimination, and then the various aspects of temporal envelope processing. I suggest that level and time processing are generally determined by primitive neural subsystems operating on the time structure of peripherally encoded representations of the sound stimulus, and that these functions are not clearly related to gross structures of the ear. Rather, they are probably determined by neural decision strategies common to most nervous systems.

I discuss last the questions of frequency analysis. Here, there are reasons to believe that various measures of spatial resolution over the basilar membranes of birds and mammals can be predicted from measures of cochlear length and the bandwidth of hearing (Greenwood 1961, 1962, 1990). Since the dimensions of fossil middle and inner

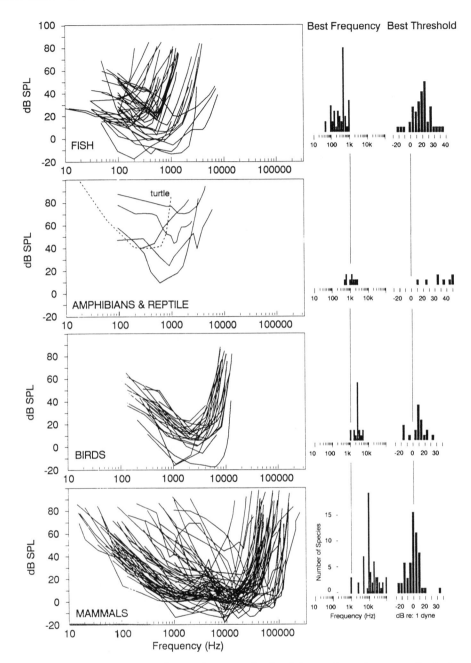

FIGURE 14.1. All behavioral audiograms for vertebrates. Thresholds are expressed in dB re: .0002 dynes cm⁻². Thresholds for aquatic animals tested underwater are the sound pressure levels in air corresponding to the sound intensity levels underwater calculated from underwater sound pressure thresholds assuming far field conditions. On the right are shown frequency distributions for each class of "best frequency" (frequency with lowest threshold) and "best threshold" (lowest threshold) derived from the audiograms on the left. [Modified after Fay (1988) with permission. The references and tabled data for all audiograms can be found in Fay (1988).]

ears might allow estimates of these parameters (see Chapter 29 by Rosowski), I suggest that there may be a basis for estimating capacities for frequency resolution in ancestral species.

3.1 Hearing Sensitivity and Bandwidth (Audiograms)

Figure 14.1 is intended to give an idea of the amount of audiogram data that exists in the literature. Each line represents a different species.

3.1.1 Fishes

The audiograms for 49 fish species are plotted to take into consideration the impedance difference between air and water. Thresholds are the sound pressure levels in air (dB re: .0002 dyne cm^{-2}) corresponding to the sound intensity levels (in Watts cm^{-2}) existing underwater in the far-field at sound pressure threshold.

Those species having best sensitivity and the widest bandwidth of hearing are generally the "hearing specialists" (see Chapter 16 by Schellart and Popper) having a specialized mechanical pathway between the swimbladder or other bubble of gas in the abdomen or head, and the ear. The movements of the bladder wall in a pressure field are transmitted to the otolithic ear by specially adapted mechanical links. These adaptations increase sound pressure sensitivity and extend bandwidth (see Chapter 15 by Coombs, Janssen, and Montgomery). There is some evidence that the specialists have additional receptor and neural elements tuned in different frequency ranges to "cover" this extended bandwidth (Fay and Ream 1986; McCormick and Braford 1988; McCormick, Chapter 17).

Fishes with a swimbladder but lacking an efficient mechanical pathway to the ear may respond to sound pressure but with reduced sensitivity compared with the specialists. Fishes lacking a swimbladder altogether probably respond directly to acoustic particle motion in both the near and far fields (their otolith ears act like accelerometers). Sound pressure thresholds are inappropriate for animals that primarily detect particle motion.

Figure 14.1 makes three major points: (1) Best sensitivity falls between 100 and 1,000 Hz for most species. (2) There is wide species variation in the high-frequency limit of hearing and in best sen-

sitivity. (3) The most sensitive fishes have equivalent thresholds at about 0 dB (+ and − 10 dB).

3.1.2 Amphibians and Reptiles

The second panel shows that there are very few data at all for amphibians and reptiles (eight anuran species and the turtle, *Pseudemys scripta*). This results from the difficulty in controlling the behavior of amphibians and reptiles. The two lowest curves were obtained for the bullfrog and green tree frog (Megela-Simmons, Moss, and Daniel 1985) using relatively sensitive methods (reflex inhibition). The others are based on unconditioned, or naturally occurring responses to sound (see Chapter 22 by Narins), and are included here only to suggest the frequency range of best hearing.

The one audiogram for a turtle (dashed line) shows rather poor sensitivity and a fish-like hearing bandwidth. Clearly there is room for more data here, particularly for the reptiles which show such wide variation in the anatomical organization of the ear (see Chapter 23 by Miller and Chapter 24 by Köppl and Manley). My impression of these data is that best sensitivity probably approaches 0 dB and that the bandwidth of hearing for anurans is only slightly wider than for the hearing specialists among the fishes.

3.1.3 Birds

The audiograms for 20 bird species have been determined using a variety for different methods. The birds are remarkably homogeneous in sensitivity and bandwidth, with best sensitivity in the range between −15 and 25 dB and best frequency between 1,000 and 6,000 Hz (see Chapter 26 by Dooling). The exceptional animals are owls with particularly sensitive hearing in the range between 2 and 9 kHz. Some of the birds represented here are songbirds with widely different types of vocalizations. Yet, we see little hint of specialization in the audiograms. As is the case for fishes and amphibians, the most sensitive birds tend to have lowest thresholds in the region of 0 dB SPL.

3.1.4 Mammals

The audiograms for 57 mammalian species are remarkable in this comparative context for the inter-species variation in the frequency range of hearing. The limit of low frequency hearing (defined as the lowest frequency at which the threshold is at or below 60 dB) varies from 20 Hz to 10 kHz — about nine octaves. The high frequency upper limit of hearing varies from 11 kHz to about 150 kHz — about 4 octaves. No single species covers this range. Marine mammals and bats define the upper range, and specialized rodents, the elephant, and primates define the lower. Most mammals have best thresholds in the region of 0 dB SPL.

The frequency distributions of best frequency and best sensitivity (Fig. 14.1, right) show: (1) The distributions of best threshold show relatively small variation among animal classes. Some species among all classes of vertebrates are capable of a sensitivity to sounds in the region of −20 to 0 dB SPL. (2) Best frequency of hearing clearly tends to rise in the following sequence: fishes, amphibians, birds, mammals. (3) Fishes, amphibians, reptiles, and birds hear best in the region from 100 to about 5,000 Hz. This is the frequency region in which auditory nerve fibers are capable of phase-locking to the sound waveform. (4) Only the mammals have extended their hearing into the frequency range where phase-locking cannot carry information about the structure of the sound waveform.

3.1.5 Audiogram Replications Using Different Methods

Figure 14.2 shows 13 audiograms determined for the chinchilla (*Chinchilla laniger*) and 10 audiograms for the domestic cat (*Felis catus*) determined using operant conditioning methods (solid lines) and avoidance conditioning methods (dashed lines). The purpose of this figure is to illustrate the sort of variability that occurs across experiments on the same species in different laboratories. At least for these species, the ordering of measured sensitivity cannot be predicted from the conditioning methods used. Although there are differences of up to 30 dB among these studies in particular frequency ranges, keep in mind that for each curve, there were differences in the number of individuals tested, the training and psychophysical methods used, the definition of threshold, and probably ambient noise levels. The results of any one study has to be viewed with caution, and accepted only tentatively. We do not yet know how to determine which of these estimates are "best" or most valid.

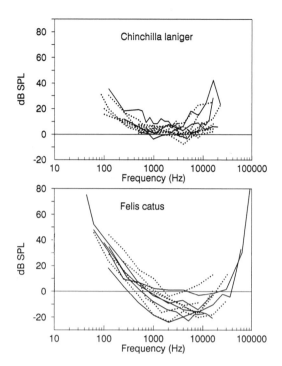

FIGURE 14.2. All published audiograms for the *Chinchilla laniger* (chinchilla) and *Felis catus* (domestic cat) reviewed in Fay (1988). The dashed lines indicate data obtained using avoidance conditioning, and the solid lines indicate operant conditioning for reward.

Chinchilla: Miller (1970); Saunders, Mills, and Miller (1977); Clark et al. (1974); Davis and Ferraro (1984); Ades et al. (1974); Salvi, Hamernik, and Henderson (1983); Halperin and Dallos (1986); Dallos et al. (1978); McGee, Ryan, and Dallos (1976); Henderson et al. (1983); Blakeslee et al. (1978); Henderson et al. (1969); Clark and Bohne (1986).

Cat: Costalupes (1983a); Elliot, Stein, and Harrison (1960); Heffner and Heffner (1985); Neff and Hind (1955); Gerken and Sandlin (1977); Gerken et al. (1985); Trahiotis and Elliot (1970); McGill (1959); Miller, Watson, and Covell (1963); Sokolovski (1973).

3.2 Level Discrimination

The ability to detect a change in the overall level of a sound is a simple yet important hearing function for all animals. Not only does this ability seem to have survival value (e.g., its role in the perception of source distance and changes in distance), but it is also most important in binaural sound localization (e.g., interaural intensity difference cues), monaural sound localization (see Chapter 34 by Heffner and Heffner), and the identification of sources through their characteristic spectral shapes (e.g., perceiving the relative amplitudes of multiple frequency components). Level discrimination also plays an important role in the detection of sound in noisy backgrounds, since signal detection in noise can be a decision about an increment in level at one or several peripheral detection filters (see Chapter 11 by E. Lewis).

Figure 14.3 summarizes most of what we know about the abilities of vertebrates (one fish, one anuran, four birds, and six mammals) to discriminate between simple sounds differing only in level. For most of these experiments, animals were exposed to tone bursts of constant level and were trained to respond when the level changed. The top panel shows the general effect of tone frequency, and the bottom panel shows the effect of overall level on the level discrimination threshold (LDT).

The effect of frequency is slight with a subtle overall trend for the LDT to rise at the higher frequencies. The generalization that the LDT is independent of frequency is not far wrong. There are large species differences in level discrimination that are difficult to summarize. The four most sensitive vertebrates appear to be the human, the parakeet, the rat, and the goldfish. However, the parakeet and rat data are not strictly comparable because in these cases, the task was to detect an increment in the level of a continuous pure tone. This may be an essentially different task from the detection of a level change between successive short tone bursts (Lamming 1986).

The effect of overall level shows a more clear and consistent effect; the LDT declines with increasing level. This decline has been termed the "near miss" to Weber's Law (Viemeister 1972). Weber's Law states that the "just-detectable" change in stimulus magnitude is proportional to the overall stimulus magnitude. If the LDT is presented in decibels (as in Fig. 14.3), Weber's Law predicts that the LDT is independent of overall level (giving horizontal functions). Obviously, Weber's Law doesn't quite hold for all vertebrates tested, with the possible exception of the goldfish. On the other hand, the approximate truth of Weber's law indicates a conspicuous convergence of function among these vertebrates. It is not clear whether sound level processing is limited or determined at the periphery or

by the brain. Some theorists focus on changes in the shapes of cochlear excitation patterns as the basis for level discrimination (Florentine and Buus 1984; Ehret 1975b), while others focus on the brain's processing of spike count statistics arriving along individual auditory nerve fibers (e.g., Viemeister 1983). In a recent experiment, Fay, Shofner, and Dye (1989) showed that spike count statistics could account for level discrimination thresholds (and the effect of sound duration on these thresholds) in humans and in goldfish. This suggests that the LDT is probably independent of cochlear structure, reflecting more the ways that all nerve fibers encode stimulus level and the ways that nervous systems integrate neural activity. Thus, at present there are no indications of specially adapted mechanisms for level discrimination, and no reasons to believe that this hearing capacity has changed significantly during vertebrate evolution.

New experiments on level discrimination have been reported by Green and his colleagues (Green 1988) that define the phenomenon of "profile analysis." In these experiments, the LDT for one frequency component within a multitone complex has been shown to depend on the spectral complexity and bandwidth of the complex. This task could be termed a timber discrimination since it is, in effect, a measure of spectral shape discrimination independent of pitch. This seems a promising experimental approach to take (there are no animal experiments on profile analysis yet reported) since it takes the question of LDT processing to a level of complexity that more accurately simulates the sorts of discrimination problems that might be encountered in the real world. To the extent that different species may represent the sound spectrum through different coding schemes that depend on ear structure, comparative studies of profile analysis might shed some light on the mechanisms for and the evolution of timber perception.

3.3 Temporal Analysis in Hearing

Most sounds have time-varying characteristics such as overall envelope, the envelopes of individual frequency components, patterns of level and frequency fluctuation, and waveform structure represented in the time domain. The time patterns of envelope fluctuation are probably important for

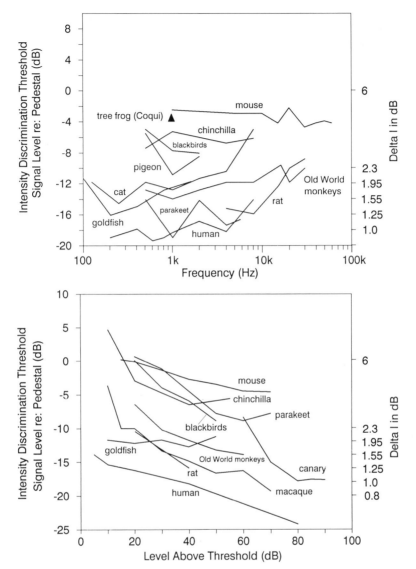

FIGURE 14.3. Sound level (intensity) discrimination thresholds as functions of frequency (top) and as functions of level (bottom) for vertebrates. The level discrimination threshold is indicated as a sound pressure level increment (Delta I in dB) on the right ordinate. The left ordinate indicates the discrimination threshold as the level of a pure tone (signal) which is just detectable when added to another pure tone of identical frequency and phase (pedestal). Note that when the two tones are added at 0 dB S/N (at equal levels), the level increment ("Delta I") is 6 dB.

Carassius auratus (goldfish) (Fay 1985, 1989); *Eleutherodactylus coqui* (neotropical tree frog) (Zelick and Narins 1983); *Columbia livia* (pigeon) and *Agelaius phoeniceus* and *Molothrus ater* (blackbirds) (Hienz, Sinnott, and Sachs 1980); *Serinus canarius* (canary) (Okanoya and Dooling 1985); *Melopsittacus undulatus* (parakeet) (Dooling and Saunders 1975a; Dooling and Searcy 1979); *Mus musculus* (mouse) (Ehret 1975b); *Rattus norvegicus* (rat) (Hack 1971); *Chinchilla laniger* (chinchilla) (Saunders, Shivapuja, and Salvi 1987); *Felis catus* (domestic cat) (Elliot and McGee 1965); *Macaca fuscata* and *Cercopithecus aethiops* (monkeys) (Sinnott, Petersen, and Hopp 1985); *Macaca mulatta* (rhesus monkey) (Clopton 1972); human (Jesteadt, Weir, and Green 1977).

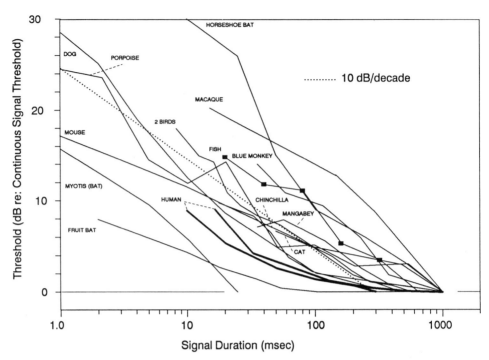

FIGURE 14.4. Thresholds for pure tones as functions of tone duration. Thresholds given in dB re: threshold for a long-duration or continuous pure tone. These functions are medians across tone frequency for the species indicated, taken from the review by Fay (1988).

Symbols are *Carassius auratus* (goldfish) (Fay and Coombs 1983). Birds are *Melopsittacus undulatus* (parakeet) and *Spizella pusilla* (field sparrow) (Dooling and Searcy 1975; Dooling 1979). *Mus musculus* (mouse) (Ehret 1976a); *Chinchilla laniger* (chinchilla) (Wall,

Ferraro, and Dunn 1981); *Rousettus aegyptiacus* (fruit bat) (Suthers and Summers 1980); *Myotis oxygnathus* (bat) and *Rhinolophus ferrumequinum* (horseshoe bat) (Ayrapet'yants and Konstantinov 1974); *Tursiops truncatus* (porpoise) (Johnson 1968a); *Felis catus* (domestic cat) (Costalupes 1983b); *Canis canis* (dog) (Baru 1971); *Macaca mulatta* (rhesus monkey) (Clack 1966); *Cercopithecus mitis* (blue monkey) and *Cercocebus torquatus* (mangabey) (Brown and Maloney 1986); human (Watson and Gengel 1969).

the detection, identification, and classification of sound sources and play important roles in the perception of species-specific vocalizations (e.g., human speech) (e.g., Moore 1982). The fine structure of sound waveforms probably play a role in spectral analysis at low frequencies (below several kilohertz), particularly for those systems in which frequency-domain spectral analysis is crude or nonexistent (e.g., in the octavolateralis system of fishes and the basilar papilla of amphibians) (Fay 1982).

Issues of temporal processing include: (1) temporal summation in detection and discrimination, (2) detection of temporal envelope fluctuations, (3) sound duration discrimination, (4) silent interval ("gap") detection, and (5) acuity of temporal pattern discrimination.

3.3.1 Temporal Summation in Sound Detection

The pressure threshold for detecting a brief sound generally declines as sound duration increases in all vertebrate species studied. This is generally true for all sensory systems as well. There are two ways that this effect can be understood. First, to the extent that sensory systems detect stimulus energy (the product of duration and intensity), this effect follows. Second, stimuli of longer duration offer the opportunity for a greater number of "looks" or independent decisions about their presence. Various hypotheses about how these decisions may be combined optimally predict that detectability generally improves with duration (Viemeister and Wakefield 1989).

Figure 14.4 summarizes the effect of tone duration on detection threshold for several vertebrates. To a first approximation, these functions are not strongly dependent on frequency and tend to be linear functions in the coordinates of log sound pressure and log duration. Gerken, Bhat, and Hutchinson-Clutter (1990) have analyzed temporal summation data from a number of species and concluded that a "power function model" (a linear function in the coordinates of threshold in dB and $\log_{10}$ duration, as plotted in Figure 14.4) provides the single best description of the data. The approximately linear portions of these functions are limited at short durations due to the spread of spectral energy that may be an additional detection cue (Watson and Gengel 1969). At long durations, these functions are limited by the time constant of a hypothetical neural integrator that combines neural activity, or the decisions based on this activity, over time.

Plomp and Bouman (1959) have modeled data such as these as exponential functions of time with an hypothetical time constant as a parameter. The best-fitting time constants for humans range from 400 to 100 ms, generally declining with increasing tone frequency. In a comparative study, Brown and Maloney (1986) showed a comparable range among mammals and one bird, with a few exceptionally high estimates at some frequencies. We are unable, as yet, to reject the hypothesis that vertebrates are homogeneous with respect to the time constant of temporal summation.

These data may also be viewed as estimates of the trading relations between sound intensity and duration (i.e., the slopes of the functions). If the auditory system were an energy detector within the range of these functions, a slope of 10 dB per decade of duration would be predicted. The dashed line in Figure 14.4 shows this slope and suggests this as a good "rule of thumb" (but see Gerken, Bhat, and Hutchinson-Clutter 1990). The echolocating mammals tested at high frequencies may be exceptions, with generally lower slopes.

Zwislocki (1969) has pointed out that if the brain is the integrator, neural activity and not sound intensity is what is integrated. This means that the trading relation between intensity and time revealed in the behavioral studies is determined by the relation between the number of spikes evoked and sound intensity (the slope of the rate-level

function). Fay and Coombs (1983) tested this idea in the goldfish by comparing the slopes of rate-level functions for auditory nerve fibers with behavioral temporal summation functions. They concluded that the effect of sound duration on threshold could be predicted from the rate-level functions of peripheral fibers. This suggests that temporal summation is determined at two levels of the auditory system; by the transfer functions for sound level at the receptor cells and innervating nerve fibers, and by the rather linear way the brain seems to accumulate neural activity in decision making. Thus, there are no compelling reasons to believe that temporal summation is a special or recently adapted hearing function, or one determined by observable structure. As is the case for level processing, temporal summation seems to be determined by rather primitive functions of receptors, nerve cells, and simple neural processing.

3.3.2 Temporal Resolution

Green (1973) has termed the time constant for temporal summation as the "maximum integration time," and has pointed out that this function can be used to advantage in efficiently detecting long duration sounds. However, when organisms are motivated to detect brief, broad band sounds, the time constant of the integrator used in these tasks seems to be at least two orders of magnitude shorter. Green (1973) has termed these "minimum integration times."

Minimum integration time has been estimated in essentially two ways for several vertebrate species: Measurements of the shortest detectable silent "gap," and of the "temporal modulation transfer function" (TMTF). Gap detection is measured by asking listeners to discriminate between sounds with unvarying envelopes and those containing a brief period of silence. The shortest silent interval detectable is the gap detection threshold. Figure 14.5 shows noise gap thresholds for the human, parakeet (*Melopsittacus undulatus*), starling (*Sturnus vulgaris*), chinchilla, rat (*Rattus norvegicus*), and goldfish (*Carassius auratus*) plotted as a function of noise bandwidth. The single points for the goldfish, parakeet and rat were determined in noise about equal in bandwidth to the entire hearing range. For the human and chinchilla, the gap threshold clearly depends on noise bandwidth,

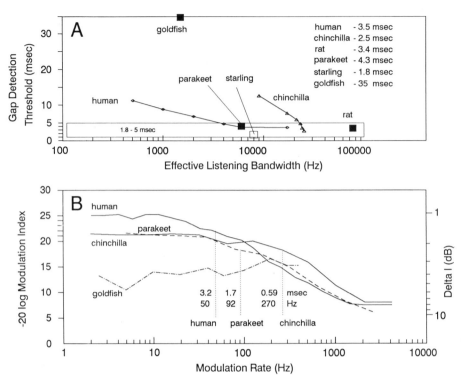

FIGURE 14.5. (A) Minimum detectable gap in noise as a function of noise bandwidth for those vertebrates tested. The box encloses gap thresholds between 1.8 and 5 ms.

Carassius auratus (goldfish) (Fay 1985); *Melopsitta-cus undulatus* (parakeet) (Dooling and Searcy 1981); *Sturnus vulgaris* (starling) (Klump and Maier 1989); *Rattus norvegicus* (rat) (Ison 1982); *Chinchilla laniger* (chinchilla) (Salvi and Arehold 1985); human (Fitzgibbons 1983).

(B) Temporal modulation transfer functions for those vertebrates tested. The vertical dashed lines indicate the frequency at which the low-pass filter function falls to half-power (at the 3 dB-down point). The frequencies and their respective time constants, $(2\,\mathrm{pi}\,f)^{-1}$, are given for three species.

Carassius auratus (goldfish) (Fay 1985); *Melopsitta-cus undulatus* (parakeet) (Dooling and Searcy 1981); *Chinchilla laniger* (chinchilla) (Salvi et al. 1982; Henderson et al. 1984); human (Viemeister 1979).

with wider bands leading to the lowest thresholds (in the 2–4 ms range for both species). The effect of bandwidth seems to be a frequency effect; the gap threshold falls as noise components are added at higher frequencies (Fitzgibbons 1983). Salvi and Arehole (1985) showed in the chinchilla that experimentally induced hearing loss above 1 kHz could increase the gap threshold to as much as 23 ms. The very large gap threshold for the goldfish (about 35 ms) could reflect its relatively narrow, low-frequency sharing range compared with the other species in Figure 14.5A.

It is not entirely clear what structures or mechanisms determine the gap detection threshold. It could well reflect the shortest time over which

nervous systems are capable of resolving envelope events. At the same time, the gap threshold's dependence on bandwidth (or frequency) would be consistent with the notion that sensitivity depends on the bandwidth of peripheral filters (primary fiber tuning curves). The statistical variability of the envelope of a random noise increases as the noise bandwidth narrows. If the detection of gaps (envelope fluctuation) is limited by the variability with which the gap is encoded within single channels, we would expect auditory systems with very narrow tuning or particularly poor high frequency hearing to have poor gap thresholds. Human listeners with some types of hearing impairment that tend to widen or detune auditory detection filters perform more poorly in gap detection than normal listeners. This suggests that the bandwidth of peripheral channels is not the only factor determining the gap threshold. Although the data are few and our understanding of gap detection incomplete, I suggest the conclusion that the equivalence of mammals and birds in gap detection reflects an essential limitation of neural time processing in the brain, but that species with particularly poor high-frequency hearing (i.e., the fishes and amphibians) will be limited in gap detection acuity due to the variability inherent in peripheral envelope coding.

The temporal modulation transfer function (TMTF) is another way to estimate the temporal resolution with which sound envelopes are detected and processed. In this type of experiment, broad band noise is sinusoidally amplitude-modulated at a given rate and depth. Modulation depth is measured as the index m, defined as $(P - T)/(P + T)$, usually scaled in dB ($-20 \log (m)$). P and T are the sound pressures at the envelope peaks and troughs, respectively. Listeners are trained to discriminate between modulated and unmodulated noise. The smallest modulation depth required for discrimination is measured as a function of modulation rate. This TMTF function describes the frequency response characteristic of an hypothetical envelope-detection filter. The assumptions of this experiment are that there are no spectral cues available for the discrimination, and that envelope fluctuations per se are the only cues. The upper frequency (f) at which sensitivity declines by one half (the "3 dB-down" point) defines the integrator's bandwidth, and $(2 \pi f)^{-1}$ is an estimate of its time-constant.

Figure 14.5B shows the TMTFs for the human, chinchilla, parakeet, and goldfish. The two mammals and the bird show similar functions resembling low-pass filters with -3 dB per octave upper slopes. The cut-off frequencies and their time-constant estimates are given in this figure. The time-constant for humans (3.2 ms) is quite similar to the gap detection threshold. The time constants for the parakeet and chinchilla are about 2 and 5 times shorter, respectively. The same variables thought to influence the gap-detection threshold probably play some role in the form of the TMTF as well. The TMTF for the chinchilla is significantly reduced in bandwidth following experimentally induced hearing loss at 8 kHz and above (Henderson, Salvi, Pavek, and Hamernik 1984). This indirectly suggests that the time constant of the TMTF varies inversely with hearing bandwidth. To this extent, the TMTF might reflect peripheral structures that determine hearing range.

Comparisons of TMTFs across species are complicated since the two ends of the function could represent somewhat different hearing functions. At the lowest modulation rates (e.g., 3 Hz), sensitivity is limited by capacities for level discrimination. The thresholds at 3 Hz in Figure 14.5B are thus comparable to the level discrimination thresholds for the same species in Figure 14.3. At the highest modulation rates (above several hundred Hz), sensitivity may be limited by temporal processing mechanisms in the brain. The 3-dB cut-off frequencies determined from these functions may thus reflect an interaction of level and time processing capacities, and may not be comparable across species having different level discrimination thresholds. The TMTF for the goldfish seems qualitatively as well as quantitatively unlike those for the other vertebrates tested. These differences remain to be understood.

These limited data are roughly consistent with gap thresholds in estimating the minimum integration time for birds and mammals to be in the range between 0.6 and 4 ms. The chinchilla seems superior to human observers in both tasks, while the parakeet and human seem to be essentially similar. The goldfish data from both tasks suggest, at least, that animals with poor high frequency hearing will be limited in the temporal resolution of envelope.

3.3.3 Duration Discrimination

These experiments ask listeners to discriminate the difference between sounds differing only in duration. This task can be viewed differently depending on the range of sound durations studied. At durations shorter than several milliseconds, changes in duration may produce detectable changes in the signal's bandwidth. At durations up to several hundred milliseconds, changes in duration will change the effective energy of the signal, and thus may affect its level above threshold and its loudness. At durations longer than this (where detection threshold is independent of energy), the task may be a test of short-term memory. In the present context, duration discrimination at the longer durations is most interesting.

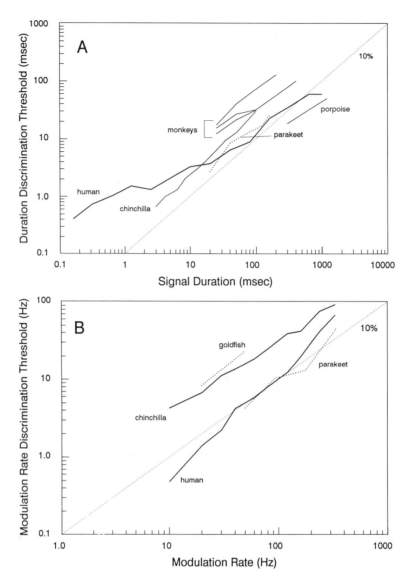

FIGURE 14.6. (A) Sound duration discrimination thresh-olds for all vertebrates tested. *Melopsittacus undulatus* (parakeet) (Dooling and Haskell 1987); *Chinchilla laniger* (chinchilla) (Long and Clark 1984); *Tursiops truncatus* (porpoise) (Yunker and Herman 1974); *Macaca mulatta*, *Cercopithecus aethiops*, *Cercopithecus neglectus* (monkeys) (Sinnott, Owren, and Petersen 1987); human (Abel 1972).

(B) Modulation or repetition rate discrimination thresholds for all vertebrates tested. *Carassius auratus* (goldfish) (Fay 1982); *Melopsittacus undulatus* (parakeet) (Dooling and Searcy 1981); *Chinchilla laniger* (chinchilla) and human (Long and Clark 1984).

Figure 14.6A shows sound duration discrimina-tion thresholds for four primates, a porpoise (*Tur-siops truncatus*), and the parakeet. These functions are remarkable in having slopes approximating unity in log–log coordinates, indicating that Weber's Law seems to hold pretty well for duration discrimi-nation. The porpoise is most sensitive, followed by the human, parakeet, and the other primates. The accuracy of auditory short-term memory for simple sounds seems to vary among species, but the order-ing of species does not suggest an obvious interpre-tation. It would be interesting to have results for a wider range of vertebrate species, perhaps extended to longer sound durations.

3.3.4 Repetition Rate Discrimination

These experiments ask listeners to discriminate between sounds differing in the rate at which a series of identical events occur (or in the time interval between successive events). Data are shown for the human, chinchilla, parakeet, and goldfish in Figure 14.6B. As is the case for duration discrimination, the functions for each species tend to have slopes roughly consistent with Weber's Law. Here again, the human and the parakeet are virtually identical in performance, with the chinchilla and goldfish somewhat less sensitive. It is possible that within the range of repetition rates studied here, this task is essentially similar to the duration discrimination task in requiring the coding of and memory for a time interval. If the duration discrimination thresholds for the human and parakeet are recalculated as rate discrimination thresholds, they are somewhat higher than those of Figure 14.6B. This may occur because the repetition rate discrimination experiments exposed the listeners to multiple samples of the intervals to be discriminated. I assume that duration and time interval discrimination are independent of ear structure and peripheral encoding and are purely a function of central processing. It may be significant that these tasks do not show a clear advantage for humans who are usually credited with superior cognitive capacities. These data give us no reason to expect that ancestral species were particularly poor at these tasks, and suggest either a long history for Weber's Law or its adaptive value in temporal processing.

4. Frequency Analysis

All vertebrate (and some invertebrate) auditory systems investigated employ the strategy of decomposing complex sounds into their constituent frequency components. This process aids in optimizing signal-to-noise ratio for sound detection, in a spectral analysis that is fundamentally important in sound localization (see Chapter 34 by Heffner and Heffner), and in source identification. Psychophysical measures of frequency analysis

inform us about the resolution with which sound spectra can be represented in the nervous system, and the resolution limits of the auditory images formed in perception.

4.1 Four Behavioral Measures of Frequency Resolution

I now concentrate on behavioral experiments that have estimated four aspects of frequency analysis in birds and mammals: frequency discrimination thresholds, critical masking ratio bandwidth (derived from the signal-to-noise ratio at masked threshold), critical bandwidths, and psychophysical tuning curves. Each of these estimates a frequency range or bandwidth that can be thought of as a perceptual unit in the frequency domain, and each can, in principle, be translated to a position and distance along the basilar membrane. Based on biomechanical and human psychophysical data, von Békésy (1960) and Greenwood (1961, 1962, 1990) have argued that, to a first approximation, the cochleas of mammals and perhaps birds behave essentially similarly as spatial processors except for scale.

4.1.1 Frequency Discrimination Thresholds (FDT)

Frequency discrimination is measured by asking listeners to discriminate between pure tone sounds differing only in frequency. The smallest difference in frequency (Δf) required for reliable discrimination is the frequency discrimination threshold (FDT) which is typically measured throughout an animal's hearing range. The FDT generally declines as sound level is raised, but tends to remain constant at levels above 40 dB above absolute threshold.

A frequency discrimination task could possibly be solved using two, quite different types of neurally coded information. Wever's (1949) physiological volley principle, since repeatedly confirmed by experiment (e.g., Rose et al. 1967), states that evidence about sound frequency is represented within auditory nerve fibers in terms of distributions of inter-spike times. Although it has never been clearly demonstrated that this evidence is used by the brain in discriminating sound frequency, many investigators assume that it is at least a possibility in the frequency range within which neural phase locking occurs (up to several thousand Hertz). Von Békésy's (1960) biomechanical place principle, since confirmed in several species (e.g., Rhode 1971), states that information about sound frequency is represented by the spatial pattern of excitation along the length of the basilar membrane, and thus in the auditory nerve as an across-fiber distribution of activity. Most investigators believe that this evidence is used by the brain in discriminating sound frequency above several thousand Hertz, and may be used in addition to temporal evidence (e.g., Young and Sachs 1979), at much lower frequencies as well. In this view, frequency discrimination is accomplished by analyzing a shift in the spatial pattern of excitation along the basilar membrane. For the purposes of this discussion, I assume that this is the evidence used in frequency discrimination by birds and mammals, at least above 1 kHz. The outcome of the present analysis may help evaluate this assumption.

4.1.2 Critical Masking Ratio Bandwidth (CRB)

Critical masking ratios (CR) are measured by asking listeners to detect the presence of a pure tone against a background of broad band noise. The CR is a signal-to-noise ratio, or the level of the signal at masked threshold (in dB) minus the spectrum level (level per Hz, in dB) of the masking noise. The CR is typically measured for signal frequencies throughout the hearing range. CRs decline somewhat as the masking noise level is raised.

The CR is can be interpreted as a measure of cochlear filtering using the following assumptions (Fletcher 1940): (1) The signal tone is detected at threshold by monitoring the output of the hypothetical cochlear filter centered on the signal frequency. Thus, only the masking noise falling within this filter is effective in interfering with signal detection. (2) At threshold, the power of the tone signal is equal to the power of the noise within the filter. Thus, the width of the filter can be estimated by determining how wide a symmetrical, rectangular band of noise (at the spectrum level used in the masking experiment) would have to be to equal the power of the signal at threshold. Since the power within a band B Hz wide is equal to its spectrum level in dB plus 10 log (B), then

$B_{Hz} = 10^{(CR/10)}$. This bandwidth estimate is termed the critical ratio bandwidth (CRB), and is assumed to map onto the basilar membrane of birds and mammals as a distance.

The CRB is an indirect measure of the bandwidth of psychophysical detection filters, and depends on the assumptions and interpretations above. The assumption that signal power equals noise power at threshold is arbitrary. If this were true, it would mean that the level at the filter's output (analogous to voltage) would be incremented by 3 dB when the signal was added to the noise. We know that the level discrimination threshold can be significantly different from this in many species (Fig. 14.3), and therefore that the CRB may underestimate or overestimate the filter's true bandwidth depending on the size of the LDT. For a variety of reasons, the CRB multiplied by 2.5 has been used as an estimate of the bandwidth of detection filters for humans (Zwicker, Flottorp, and Stevens 1957).

Note that while the CRB is an indirect estimate of hypothetical detection filters, and may vary across species due to differences in level discrimination acuity and the efficiency of decision making (Patterson and Moore 1986), it is valid and clearly useful for predicting the detectability of narrow band signals in noisy environments for any given species.

4.1.3 Critical Bandwidth (CB)

The critical bandwidth (CB) is a more direct measure of the bandwidth of hypothetical detection filters. In the classical critical band experiment (e.g., Greenwood 1961), listeners detect the presence of a pure tone in the presence of masking noise bands of different width centered on the signal frequency. Signal levels at threshold are measured and plotted as a function of noise bandwidth. For very narrow bands (presumably more narrow than the detection filter, or critical band), incrementing the noise bandwidth results in more noise power falling within the filter, and signal threshold rises. For very wide noise bands (wider than the filter, or the critical band), further increments in noise bandwidth result in no additional noise power within the filter, and signal threshold becomes independent of noise bandwidth. The bandwidth at which signal threshold just becomes independent of noise bandwidth is defined as the critical band.

CBs are laborious to measure since each estimate requires the determination of many signal thresholds. On the other hand, CB estimates are generally independent of the listener's efficiency and performance in signal detection (Patterson and Moore 1986), and do not depend systematically on level discrimination acuity. CB estimates tend to grow slightly when determined at higher noise spectrum levels. For humans, CRB values multiplied by 2.5 are approximately equal to CB values. As with the CRB, the CR can be viewed as a position and distance on the basilar membrane in mammals and birds.

4.1.4 Psychophysical Tuning Curve (PTC)

Psychophysical tuning curves (PTC) are measures of frequency selectivity in some ways analogous to neurophysiological tuning curves for single auditory nerve fibers. A brief, narrow band signal is fixed in frequency and level just above absolute threshold. Listeners must detect the signal in the presence of narrow band maskers. Threshold is defined as the masker level required to just mask the signal. Masker level at signal threshold plotted as a function of masker frequency defines the PTC (see Fig. 14.15). These functions resemble neural tuning curves in shape; threshold masker level generally increases as the separation between signal and masker frequency widens. The measure of PTC bandwidth most often used is the width (in Hz) of the function for a masker level 10 dB above that required to mask the signal at the "tip" (where masker and signal frequency are usually equal). In principle, the 10-dB bandwidth can be viewed as a distance on the basilar membrane.

4.2 Structure and Function in Frequency Analysis

The goal of the treatment to follow is to evaluate several questions suggested by the notion that vertebrate cochleas are scale models of one another: (1) Can psychophysical estimates of frequency resolution be identified with particular places and distances along the basilar membrane? (2) What are the structural correlates of the frequency intervals which are behaviorally equivalent? (3) Are these structural correlates invariant across bird and mammal species? (4) Can psychophysical performance in

frequency analysis be predicted from gross cochlear structure? (5) Can we reasonably guess performance in frequency analysis from the fossil record?

4.2.1 Position-Frequency Maps on the Basilar Membrane

The first step in relating cochlear structure to behavioral performance in frequency analysis is to describe how sound frequency is mapped onto the receptor organ surface. Von Békésy (1960) observed the traveling wave envelopes and measured basilar membrane elasticity in the elephant, cow, human, guinea pig, rat, mouse, and chicken (von Békésy did not give the generic names for these animals). These measurements resulted in empirical position-frequency maps, and demonstrated that the nature of mechanical frequency analysis on the basilar membrane seems to be based on the same principles for the species studied. For present purposes, I tentatively assume that this is true for most extant species of birds and mammals.

In his psychophysical studies of critical bandwidths, Greenwood (1961) investigated the notion that critical bands represent equal distances on the basilar membrane regardless of center frequency (within limits) (Zwicker, Flottorp, and Stevens 1957). Greenwood's strategy was to derive a position-frequency map for the human cochlea under the assumption that critical bands represent equal distances, and then evaluate the assumption by the correspondence between his derived map and von Békésy's (1960) empirical map. After measuring the critical bandwidths (CB) at several center frequencies, Greenwood simply integrated the CB function of frequency, assuming that each CB represents a unit distance along the basilar membrane. The map function derived is

$$f = A(10^{ax} - 1) \qquad (1)$$

where x is distance, f is frequency, a is slope, and A is a parameter determining the frequency range of the map (the "minus 1" term defines the position at zero Hz as zero millimeters). For humans, Greenwood estimated a as 0.06, and A as 165.4. (This map is compared to von Békésy's data in Figure 14.7) Greenwood pointed out that it is significant that this function derived from psychophysical data has precisely the same form and slope as the function relating basilar membrane

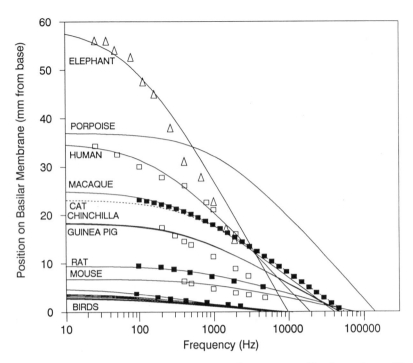

FIGURE 14.7. Position-frequency maps for the mammal species indicated and for several birds. The lines plot the function:

$$P_{mm} = (1 - (\log_{10}[f_{Hz}/(0.008\ F_{max} + 1)/2.1)]L_{mm}$$

where P_{mm} is the position on the basilar membrane corresponding to the frequency f_{Hz} for a species which can hear at frequencies up to F_{max} Hz, and having a basilar membrane or papilla Lmm long. The symbols and dashed line for cat plot the function determined by Liberman (1982) to best fit his data based on horseradish peroxidase tracing. The other symbols plot the maps determined in direct observation by von Békésy (1960) for elephant, human, guinea pig, rat, mouse, and chicken (graphically derived from figures from von Békésy 1960).

elasticity to distance from the apex determined by von Békésy (1960). This striking correspondence between biomechanical and psychophysical data provides support for Greenwood's assumption that in humans critical bands correspond to equal distances on the basilar membrane.

Greenwood then used the equivalence of the position-compliance function and the derived position-frequency function (Equation 1) to predict position-frequency maps for the cat, elephant, guinea pig, rat, mouse, and chicken based on von Békésy's compliance measurements. These predictions corresponded reasonably well with von Békésy's position-frequency measurements (Fig. 14.7) and led Greenwood to the conclusion that the cochleas of mammals (and possibly birds) may be scale models of one another. One prediction from this conclusion is that, as for humans, the critical bandwidths represent equal distances on the basilar membrane for all mammal (and bird) species.

Liberman (1982) carefully determined the position-frequency map for the cat cochlea by comparing auditory nerve fiber characteristic frequencies with their position of cochlear innervation using horseradish peroxidase tracing. Liberman's detailed map for the cat can be described exceptionally well with Greenwood's (1961) position-frequency function (Fig. 14.7).

4.2.2 Cochlear Position-Frequency Maps Can Be Derived Using Two Species-Specific Parameters: Cochlear Length and the High Frequency Hearing Limit

Recently, Greenwood (1990) has pointed out that a single slope parameter (a) can be used to model all

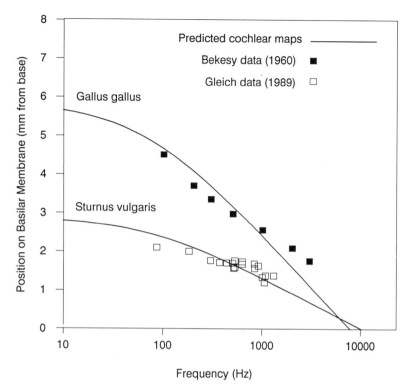

FIGURE 14.8. Position-frequency maps for chicken and *Sturnus vulgaris* (starling) (thick, solid lines) defined as in Figure 14.7. Solid symbols are data from von Békésy (1960). Open symbols are "unambiguous" data points determined using horseradish peroxidase tracing by Gleich (1989). See new empirical map data on the chicken and the starling in Chapter 27 by Manley and Gleich.

species investigated if cochlear length is normalized to unity. This means that only one additional parameter, the species-specific constant A, is required to estimate the normalized cochlear position-frequency map for any species:

$$f_{Hz} = A(10^{(2.1p)} - 1) \qquad (2)$$

where p is a proportion of cochlear length (from the apex).

Greenwood's (1961; 1990) strategy has been to estimate A by finding the value producing the best fit between equation (1) and empirical map data (e.g., Liberman's 1982 data for the cat). However, in the absence of data of this kind, A can also be estimated by finding the value that forces equation (1) to equal the highest audible frequency for a given species (F_{max}) when $p = 1$. I define F_{max} as the highest frequency still audible at 60 dB SPL. I have found that $A = .008\,F_{max}$ for all maps of this kind. Thus,

$$f_{Hz} = .008\,F_{max}(10^{(2.1p)} - 1) \qquad (3)$$

Solving for p,

$$p = \log_{10}(f_{Hz}/(0.008F_{max}) + 1)/2.1 \qquad (4)$$

Multiplying $1 - p$ by cochlear length, L_{mm}, gives the absolute position, P_{mm} (in mm from the base), of the peak of the excitation pattern at fHz for a species having an upper frequency hearing limit of F_{max}.

$$P_{mm} = (1 - (\log_{10}(f_{Hz}/(0.008F_{max}) + 1)/2.1))L_{mm} \qquad (5)$$

The map functions (Eq. 5) for several mammals and birds are plotted in Figures 14.7 and 14.8 along with some of the empirical map data for these species. Note that for each species, the maps were constructed using the two species-specific parameters F_{max} and L_{mm} (see also Table 14.1). I suggest that if the fossil record allows estimates of cochlear length (e.g., Graybeal et al. 1989), and if the frequency range of hearing can be estimated from middle ear morphology (see Chapter 29 by Rosowski), then we can, in principle, estimate absolute position-frequency maps for ancestral species.

The fits between the lines derived from Equation (5) and the points observed by von Békésy (1960), Liberman (1982), Gleich (1989), and Manley and Gleich (Chapter 27) are reasonably good. Greenwood (1990) has reviewed additional data on cochlear maps for the human, chinchilla, guinea pig, macaque, and gerbil that are generally well fit by the map equation. Greenwood's (1990) strategy was to adjust the parameter A for the best fit rather than to set A based on an estimate of the frequency range of hearing, as I have done here. In addition, Greenwood set the integration constant at -0.8 or -0.85 for nonhumans, while I have set this constant to -1 for all species. The only important differences between the Greenwood (1990) maps and the present maps for the same species occurred for the chinchilla for which the frequency range of the cochlea was estimated to be about 20 kHz on the basis of data on hearing loss following cochlear lesions from Eldredge, Miller, and Bohne (1981). Behavioral audiograms for the chinchilla (Fig. 14.2) suggest that the upper frequency limit is actually higher than this (between 30 and 50 kHz).

It should be remembered, in addition, that von Békésy's observations were made on dead animals. Experiments now show that the mechanical response of the basilar membrane is highly vulnerable to disturbances of normal metabolic integrity (e.g., Khanna and Leonard 1982), and that the von Békésy's maps could be in error for this reason (but see discussion by Greenwood 1990). In general, the fits between the predicted maps and the few data that exist are not at all perfect, and many contexts for speculation exist regarding the correspondence between the prediction and the data for any one species. I leave this sort of speculation to others. I simply point out that the function derived by Greenwood (1961) and applied here predicts the cochlear frequency-position map for many bird and mammal species reasonably well with no free parameters. I am not aware of a more accurate map function that is also as simple, general, and elegant as the Greenwood function, and of no better way to estimate a cochlear map based on fossil evidence.

4.3 Can Cochlear Maps Predict Behavioral Performance in Frequency Analysis?

Greenwood (1961) derived the cochlear map for humans by integrating the critical band function of frequency under the assumption that critical bands represent equal distances on the basilar membrane. Now that we have a general way to estimate the cochlear map for mammals and birds using two parameters (F_{max} and L_{mm}), is it possible to predict critical bands (and other perceptually equivalent units of frequency) for mammals and birds generally? This sort of prediction requires the equal-distance assumption (for critical bands within a species), and one additional parameter, the critical cochlear distance (CD) corresponding to a critical band. Greenwood (1961) determined that the CD is about 1 mm for humans, but did not estimate this value for other species.

The equal-distance assumption can be tested for other species by evaluating the fit between behavioral data on frequency analysis and the predictions from the map function, with CD as a free parameter. If the map predictions can be made to fit the data well by adjusting CD, then we have reason to accept the equal-distance hypothesis. Further, the CD value leading to the best fit between the data and the predictions can be determined and compared across species. If there is an intelligible relationship between CD and ear structure, then we may be able to use ear structure to estimate absolute behavioral performance in frequency analysis for any species, including fossil ones.

To begin this investigation, the position-frequency map function (Eq. 5), expressed in terms of distance from the apex in order to give positive slopes, is differentiated to give dP_{mm}/df, or an expression for "millimeters per Hertz." The inverse gives df/dP_{mm}, or an expression for "Hertz per mm,"

$$df/dP_{mm} = (f + (.008\ F_{max}))\ 2.1\ \mathrm{Ln}10\ L_{mm}^{-1}$$

$$(6)$$

This is a linear equation with slope proportional to L_{mm}^{-1} ($2.1\ \mathrm{Ln}10\ L_{mm}^{-1}$) and y-intercept jointly determined by L_{mm}^{-1} and F_{max} ($.008\ F_{max}\ 2.1\ \mathrm{Ln}10\ L_{mm}^{-1}$). Equation (6) describes the resolution with which frequency is spatially mapped onto the basilar membrane (Hz/mm) as a function of frequency for a species with cochlea L_{mm} long and which hears within the frequency range bounded by F_{max}. Multiplying Equation (6) by the parameter CD (the distance covered by a critical band) estimates the critical bandwidth in Hz at center frequency f.

In order to evaluate how well Equation (6) describes frequency analysis performance, it is first used as a model for human psychophysical

TABLE 14.1. Best-Fitting Critical Cochlear Distances (CD in mm) and the Standard Error of Estimate (SEE, log units) Corresponding to Behaviorally Equivalent Frequency Intervals (FTD, CRB, CB, PTC).

Species	F_{max} (kHz)	L_{mm} (mm)	FDT CD (SEE) [range kHz]	CRB × 2.5 CD (SEE) [range kHz]	CB CD (SEE) [range kHz]	PTC (10dB) × 2.5 CD [range kHz]
Human	20	31.5	.017 (.232) [0.4–8]	1.1 (.079) [0.4–9]	1.15 (.120) [0.35–13]	1.01 [1–6]
Cat	60	22.5	0.44 (.205) [0.5–32]	1.36 (.239) [0.5–16]	.89 (.118) [1–1 6]	.96 [1–2]
Macaque	42	25	.046 (.238) [0.5–30]	.88 (.231) [0.5–4]	1.07 (.137) [0.6– 16]	1.1 [0.13–8]
Chinchilla	40	18.3	.07 (.253) [0.5–8]	2.6 (.250) [0.5–21]	1.54 (.145) [0.5–4]	.89 [0.5–12]
Mouse	90	6.8	.013 (.063) [1–40]	1.48 (.191) [5–60]	.76 (.246) [2–60]	
Porpoise	150	37	.035 (.269) [2–130]	1.2 (.217) [5–100]		
White rat	80	8.6	.09 (.174) [1–42]	2.2 (.084) [1–80]		
Horseshoe bat	100	16		.26 (.026) [40–80]	0.2 (.085) [35–75]	
Guinea pig	50	17.6	.087 (.310) [0.125–42]			
Elephant	12	60	.061 (.304) [0.25–4]			
Parakeet	8	3.7	.007 (.273) [0.5–5.7]	.22 (.453) [0.63–5.7]	0.22 (0.11) [0.63–5.7]	.29 [0.32–6]
Redwing blackbird	10	3.65	.009 (.272) [0.5–80]	.58 (.362) [0.5–60]		
Cowbird	10	3.65	.007 (.204) [0.5–80]	.58 (.362) [0.5–6]		
Pigeon	7	3.65	.012 (.072) [0.25–4]	.30 (.147) [0.5–4]		
Starling	9	3.5	.005 (.310) [0.8–6.4]	.38 (.156) [0.5–4]		
Barn owl	12	9	.011 (.119) [3.5–10]			
Chicken	6	6	.038 (.158) [0.5–2.8]			
Cockatiel	6	3.5		.45 (.097) [0.5–4]		
Canary	9	2.4		.63 (.037) [1–5.7]		
Song sparrow	12	2.4		.28 (.112) [0.5–8]		
Swamp sparrow	12	2.4		.30 (.165) [0.5–8]		
Zebra finch	7	2.4		.35 (.149) [1–5.7]		

Note: The genus, species, and references for the data used in this table are given in the captions for Figures 14.10 and 14.11.

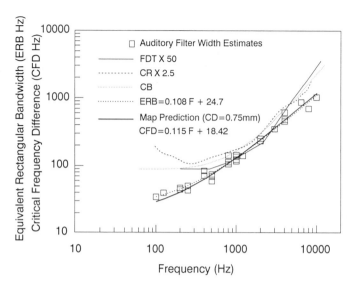

FIGURE 14.9. Comparison of human psychophysical data on frequency analysis with predictions from the cochlear map (with a CD of 0.75 mm), and the function best fitting "equivalent rectangular bandwidth" (ERB) psychophysical estimates of the auditory filter shape (symbols) compiled and reviewed by Glasberg and Moore (1990). The equation for ERB is the best linear fit to the data points as given by Glasberg and Moore (1990), with frequency in hertz. The equation for "Critical Frequency Difference" (CFD) is Eq. 6 with $F_{max} = 20$ kHz, $L_{mm} = 31.5$ mm, and CD = 0.75 mm. See text for references for other psychophysical data shown.

data in Figure 14.9 (heavy solid line). The three thin lines plotted are critical bandwidths (CB) (Zwicker, Flottorp, and Stevens 1957), critical ratio bandwidths (CRB) (Hawkins and Stevens 1951) multiplied by 2.5, and frequency discrimination thresholds (FDT) (Wier, Jesteadt, and Green 1976) multiplied by 50. These multiplication factors have been shown to roughly equate these three measures of frequency acuity.

Plotted as symbols in Figure 14.9 are estimates of the equivalent rectangular bandwidths (ERB) for "auditory filters" compiled and reviewed by Glasberg and Moore (1990). The thick, dotted line labeled ERB plots the linear function which Glasberg and Moore found to best fit the auditory filter width estimates. The psychophysical auditory filters reviewed by Glasberg and Moore were obtained from experiments critically designed to overcome many of the problems of interpretation that arise in psychophysical studies of auditory filters based on critical ratio bandwidth, the critical band, and psychophysical tuning curve experiments. Space does not permit a treatment of these experiments (only once applied to nonhumans; Halperin and Dallos 1986). Note, however, that

these auditory filter width estimates correspond closely to the FDT, CRB, and CB measures between 1 and 3 kHz, but are significantly smaller than these measures at the lowest and highest frequencies. The important point is that the critical frequency differences predicted using the human cochlear map (Eq. 6) and a critical cochlear distance (CD) of 0.75 mm corresponds very well with the best available psychophysical estimates of auditory filter width (ERB function of Fig. 14.9).

The implications of Figure 14.9 are the following: The performance of human listeners in frequency analysis (ERB function and the data used to derive it) is successfully predicted using the cochlear map function (Eq. 5) and the assumption that the ERBs occupy equal distances along the basilar membrane regardless of center frequency (Eqs. 5 and 6). Thus, the methods used and reviewed by Glasberg and Moore (1990) to estimate the bandwidths of auditory filters are probably more valid than the "classical" methods (CB, CRB, and FDT) at frequencies below 1 kHz and above 3 kHz. If we can generalize from human to nonhuman listeners, we would expect that the fit between predicted performance in frequency analysis (using Eqs. 5 and

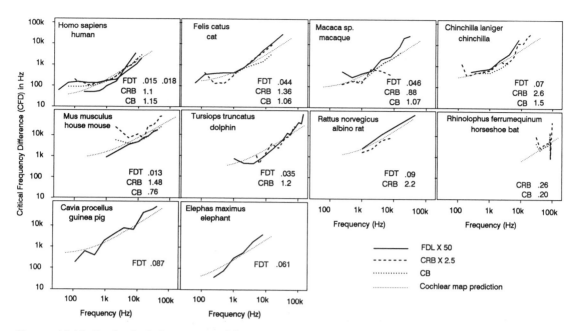

FIGURE 14.10. Psychophysical measures of frequency analysis (thick lines) for 10 mammal species compared with predictions (thin, dotted lines) from the cochlear maps of Figure 14.7 made under the assumption that for each species, behaviorally defined critical frequency differences correspond to a constant critical cochlear distance (CD) on the basilar membrane.

Solid lines are frequency discrimination thresholds (FDT) multiplied by 50. Dashed lines are critical ratio bandwidths (CRB) multiplied by 2.5. Dotted lines are critical bandwidths (CB). The numbers appearing in each panel are the CDs (in mm) which resulted in the best fit between the predictions and the psychophysical data at the mid frequencies. See also Table 14.1.

Human (Shower and Biddulph 1931; Wier, Jesteadt, and Green 1976; Hawkins and Stevens 1950; Zwicker, Flottorp, and Stevens 1957); *Felis catus* (domestic cat) (Elliot, Stein, and Harrison 1960; Watson 1963; Nienhuys and Clark 1979); *Macaca sp.* (macaques) (Stebbins 1970; Clack 1966; Gourevitch 1970); *Chinchilla laniger* (chinchilla) (Nelson and Kiester 1978; Miller 1964; Seaton and Trahiotis 1975); *Mus musculus* (mouse) (Ehret 1975a, 1975b, 1976b); *Tursiops truncatus* (porpoise) (Herman and Arbeit 1972; Thompson and Herman 1975; Johnson 1968b); *Rattus norvegicus* (rat) (Kelly 1970; Gourevitch 1965); *Rhinolophus ferrumequinum* (horseshoe bat) (Long 1977, 1980); *Cavia procellus* (guinea pig) (Heffner, Heffner, and Masterton 1971); *Elephas maximus* (elephant) (Heffner and Heffner 1982).

6) and the data from CB, CRB, and FDT experiments would be best in the mid-frequency range, and would deviate, as it does for humans, at higher and lower frequencies.

Figures 14.10 and 14.11 show much of the published data on CB, CRB, and FDT for 10 mammal and 12 avian species compiled in an extensive review of this literature by Fay (1988). Again, the CRBs are multiplied by 2.5 and the FDT multiplied by 50 to roughly equate these measures to CBs. (In general, while two or more data sets may have been published for a given measure and species, I have chosen to use the single set with the most data and the lowest thresholds.) Clearly, the

relations among these measures are similar for most mammals and birds investigated so far. For each species in Figures 14.10 and 14.11, Equation (6) was evaluated and plotted as a function of frequency using F_{max} and L_{mm} values as listed in Table 14.1. For plotting the map predictions, a CD value was chosen for each species which best summarized the various behavioral data as plotted. Also for each species, CD values are given that provided the best fit to the behavioral functions (FDTs and CRBs not multiplied by the factors used in plotting them here). The best fit is defined as the CD value which minimized the standard error of estimate (given in

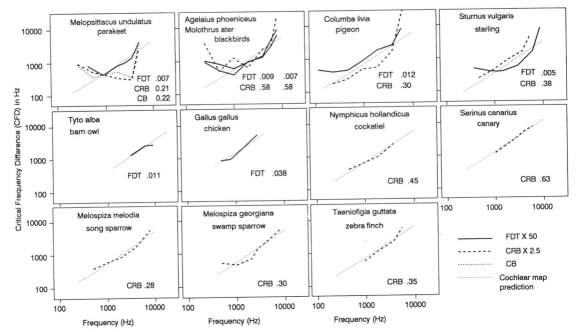

FIGURE 14.11. Psychophysical measure of frequency analysis (thick lines) for 12 avian species compared with predictions (thin, dotted lines) from the cochlear maps of Figure 14.8 made under the assumption that for each species, behaviorally defined critical frequency differences correspond to a constant critical distance (CD) on the basilar papilla.

Solid lines are frequency discrimination thresholds (FDT) multiplied by 50. Dashed lines are critical ratio bandwidths (CRB) multiplied by 2.5. Dotted lines are critical bandwidths (CB). The numbers appearing in each panel are the CDs (in mm) which resulted in the best fit between the predictions and the psychophysical data at the mid frequencies.

Melopsittacus undulatus (parakeet) (Dooling and Saunders 1975b; Saunders, Denny, and Bock 1978); *Molothrus ater* and *Agelaius phoeniceus* (blackbirds) (Sinnott, Sachs, and Hienz 1980; Hienz and Sachs 1987); *Columba livia* (pigeon) (Sinnott, Sachs, and Hienz 1980; Hienz and Sachs 1987); *Sturnus vulgaris* (starling) (Kuhn, Leppelsack, and Schwartzkopf 1980; Okanoya and Dooling 1987); *Tyto alba* (barn owl) (Quine and Konishi 1974); *Gallus gallus* (chicken) (Gray and Rubel 1987); *Nymphicus hollandicus* (cockatiel), *Serinus canarius* (canary), *Melospiza melodia* (song sparrow), *Melospiza georgiana* (swamp sparrow), *Taeniofigia guttata* (zebra finch) (Okanoya and Dooling 1987).

Table 14.1), comparing the logarithms of the predicted values with the logarithms of the data values. (The logarithmic transformations prevented the large values at high frequencies from dominating the fitting procedure).

The following generalizations can be made from these data:

(1) For most species, the cochlear map predictions (Eq. 6) summarize the psychophysical data rather well. The predictions tend to be best at mid frequencies and tend to fail at the lowest (and sometimes highest) frequencies. As suggested by the human data in Figure 14.9, these deviations from the predictions could be due to underestimations of the acuity of frequency analysis that tend

to occur at the lowest and highest frequencies. The FDT functions are sometimes (but not always) steeper than the predictions. At frequencies between 100 and 1,000 Hz, particularly low values for the FDT could be due to the use of cues in addition to the spatial pattern of excitation on the basilar membrane; spike timing cues, for example.

(2) The psychophysical data for two species (*Melopsittacus undulatus* and *Rhinolophus ferrumequinum*) deviate significantly from the predictions. In the case of *Rhinolophus* (horseshoe bat), we know that the predictions are in error since they do not take into consideration the striking inhomogeneity in cochlear organization (e.g., Bruns 1976; Pollak, Chapter 36) which is most

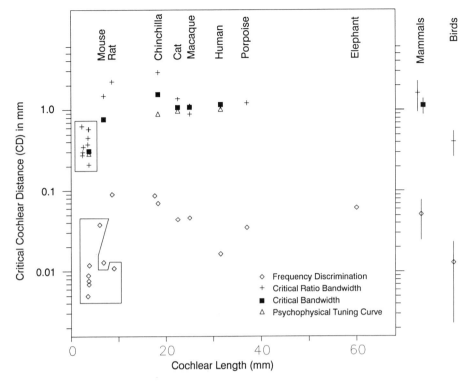

FIGURE 14.12. (A) Best-fitting basilar membrane critical distances (CD) for the species of Figs. 10 and 11 plotted as a function of basilar membrane length for frequency discrimination thresholds (FDT), the critical masking ratios (CRB) multiplied by 2.5, the critical bandwidths (CB), and the widths of psychophysical tuning curves 10 dB above the tip, multiplied by 2.5 (PTC from Fig. 14.15). Data for birds enclosed in polygons. Means plus and minus one standard deviation are shown at the right.

likely a specialization for processing Doppler-shifted biosonar echoes. While we are not yet aware of a similar structural specialization in the ear of *Melopsittacus* (Gleich and Manley 1988), the psychophysical data suggest we should continue to look for one.

(3) Although data are limited and variable, there are no clear indications that birds and mammals essentially differ in these measures of frequency analysis or in the spatial processing underlying them. This suggests that for all the similarities and differences in cochlear structure between the birds and mammals (see Manley and Gleich, Chapter 27), the cochlea is apparently adapted in both classes to function similarly in hearing.

(4) There are no clear indications that the three functions (FDT, CRB, and CB) differ systematically in their correspondence with the predictions, at least above 1 kHz. This means that one function can be used to estimate the others, and that all three

may be reasonably well predicted from the cochlear map function and an estimate of CD.

In summary, the function describing predicted acuity in frequency analysis (equation 6) provides a rather good qualitative fit to most of the psychophysical data for several mammals and birds. This correspondence suggests that, as for humans, behaviorally defined "units" of frequency analysis for a given species represent equal distances on the basilar membrane.

Figure 14.12 shows the best-fitting CDs for each of the behavioral measures plotted versus cochlear length for each species (also listed in Table 14.1). For this plot, CRBs have been multiplied by 2.5, and data are included for the widths of psychophysical tuning curves (PTCs—see also Fig. 14.15). The data for birds are enclosed in polygons. For mammals, the critical cochlear distances for CBs, CRBs, and PTCs average about 1.3 mm, and there is no clear tendency for CDs to

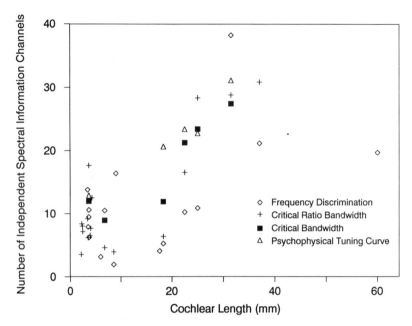

FIGURE 14.13. The number of independent spectral information channels plotted as a function of basilar membrane length derived from CBs, CRBs multiplied by 2.5, FDTs multiplied by 50, and PTC widths 10 dB above the tip multiplied by 2.5. Estimates were made by dividing basilar membrane length by the best-fitting CD determined for each species and psychophysical data set.

vary as a function of cochlear length. We are therefore unable to reject the hypothesis that the mammals investigated so far have the same CD. An apparent trend for the CDs for FDTs to decline for longer cochleas is not supported by FTD data for the elephant. The largest deviations from the mean mammal CD occurs in the CRB data for chinchilla and rat. The CDs for birds are clearly smaller than for mammals by a factor of about 3.3. This is true for all four behavioral measures. This suggests that absolute spatial resolution within the cochlea is significantly greater in birds than in mammals.

If we wish to estimate the behavioral frequency resolving abilities of extant mammals that have not yet been studied, our best estimate would be to use Equation (6) multiplied by the mean mammal CD of 1.3 mm. Similarly, our best estimate for a bird's behavioral performance would be to use Equation (6) multiplied by the mean avian CD of 0.4 mm. If we wish to estimate the most probable behavioral frequency-resolving ability of an extinct species, fossil evidence would have to supply estimates of cochlear length and the upper frequency limit for hearing. Then Equation (6) would be multiplied by

the mean mammal or bird CD depending on whether the ear appears bird-like or mammal-like.

The approximate equivalence of mammal CDs despite large differences in cochlear length means that species with longer cochleas have a greater number of independent spectral "information channels" within their hearing bandwidth. Figure 14.13 shows estimates of the number of these channels plotted as a function of cochlear length. The number of channels is defined here as cochlear length divided by CD (CD determined using CRB times 2.5, FDT times 50, and PTC widths at 10 dB above the tip times 2.5). Among the mammals, the number of information channels declines toward zero for shorter cochleas. In this context, the birds stand out as somehow avoiding an unacceptably small number of information channels by increasing the absolute spatial resolution on the basilar membrane.

The greatest absolute frequency acuity will occur in animals having poor high frequency hearing and long cochleas (e.g., human and elephant). Poorest absolute frequency acuity is associated with short cochleas and high upper frequency hearing limits

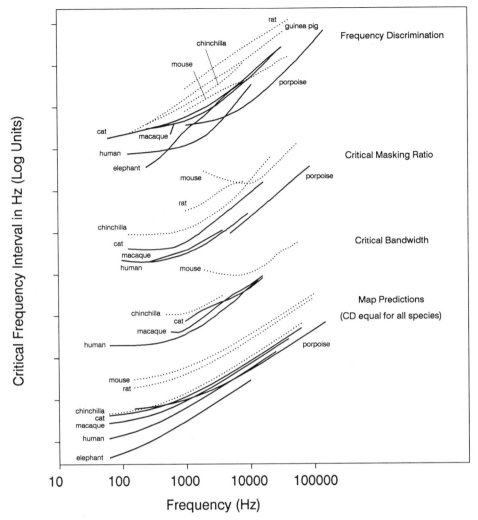

FIGURE 14.14. Predictions of performance in frequency analysis compared with data on frequency discrimination, critical bandwidth, and critical ratio bandwidth for several mammal species. Predictions of species relations were made under the assumption that all species shown have identical CDs. The ordinate is in logarithmic units and the four sets of curves are positioned vertically to separate them visually.

(e.g., rodents—dashed lines). Figure 14.14 illustrates this point for the mammals. Here, smoothed functions representing FDTs, CBs, and CRBs for the several mammals represented earlier are compared with cochlear map predictions assuming the same CD for all species. Each group of functions is arbitrarily placed on the ordinate to facilitate comparisons.

4.4 Psychophysical Tuning Curves

The comparisons and conclusions above were formed partially on the basis of data from psycho-

physical tuning curves (PTC). Figure 14.15 shows the PTCs used in the above analysis and illustrates the methods used to measure the bandwidth 10 dB above threshold, in millimeters along the basilar membrane. Using the cochlear map functions for each species (macaque, human, chinchilla, and parakeet), frequency was simply transformed to distance in millimeters. The plots of all PTCs were then aligned at the tips, and the distance corresponding to the bandwidth 10 dB above threshold was determined graphically on a high resolution graphics screen. The major features of the

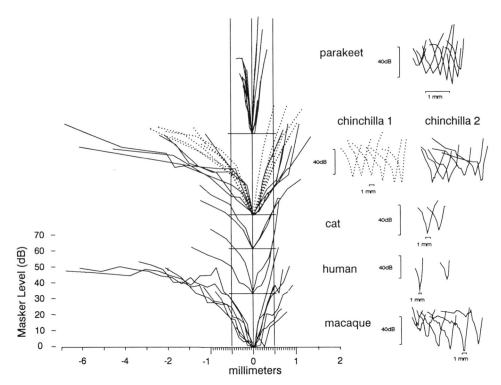

FIGURE 14.15. Psychophysical tuning curves (PTC) in simultaneous masking. On the right, PTCs are plotted as a function of the cochlear positions corresponding to the masker frequencies, as derived from the cochlear map for each species. In the center, all PTCs are plotted on the same cochlear distance scale with tips normalized at the same arbitrary point.

Melopsittacus undulatus (parakeet) (Saunders, Rintelmann, and Bock 1979); *Chinchilla laniger* (chinchilla) (McGee, Ryan, and Dallos 1976; Salvi et al. 1982); *Felis catus* (domestic cat) (Pickles 1979); *Macaca nemestrina* (pig-tailed macaque) (Serafin, Moody, and Stebbins 1982); human (Moore 1978).

PTCs plotted as functions of cochlear distance are: (1) The millimeter "bandwidths" are not systematically dependent on frequency. (2) PTCs for the four mammals investigated are quite similar in form and width. (3) The parakeet PTCs are clearly significantly more sharply tuned than the PTCs for mammals. Spatial processing acuity on the basilar membrane as defined by PTCs is greater in the parakeet (and perhaps other birds) than in mammals (including the human).

4.5 Fishes May Be Similar to Mammals and Birds in Patterns of Frequency Analysis

The goldfish (*Carassius auratus*) is the only anamniotic vertebrate for which we have a full range of quantitative behavioral data on frequency analysis, including frequency discrimination thresholds (Jacobs and Tavolga 1968; Fay 1970), critical masking ratios (Fay 1974), critical bands (Tavolga 1974) and psychophysical tuning curves (Fay, Ahroon and Orawski 1978). Although fibers of the saccular nerve (the saccule is the primary hearing organ in the goldfish) have been shown to be frequency selective to some degree (Fay and Ream 1984), there is at present no clear evidence that the saccular macula is tonotopically organized to perform a spatial analysis of frequency like that observed in birds and mammals. Nevertheless, it is interesting that the form of the functions relating acuity in frequency analysis (from FDT, CRB, CB, and PTC measurements) to frequency is essentially the same as those found for mammals and birds (Fig. 14.10 and 14.11). Figure 14.16 shows these data for the goldfish and the best-fitting

FIGURE 14.16. Psychophysical data
for frequency analysis in *Carassius
auratus* fit with Eq. 6. FDT (frequency discrimination thresholds)
from Fay (1970); CRB (critical
masking ratios) from Fay (1973);
CB (critical bands) from Tavolga
(1974); PTC (psychophysical tuning
curves) from Fay, Ahroon, and
Orawski (1978).

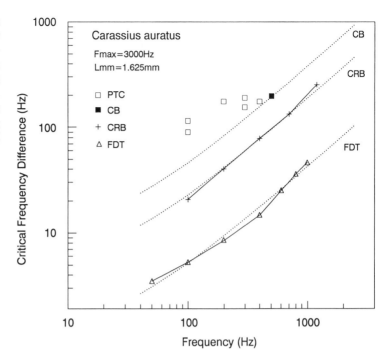

linear functions (Eq. 6), using F_{max} = 3,000 Hz
(Fay 1988) and L_{mm} = 1.625 mm (Platt 1977).
The CD value resulting in the best fit to the data are
0.29 mm for CB, 0.144 mm for the CRB, and
0.032 mm for the FDT. Not only are the fits
between Eq. (6) and the data at least as good as for
humans (Fig. 14.9), but the quantitative relations
among the four measures of frequency analysis in
fishes are similar to those characteristic of mammals and birds. Moreover, the CD that best fits the
CB data (and 2.5 times the CRB data) is about 0.3
mm, the same as found for birds. What is the
meaning of this?

One possible explanation for this correspondence is that, in fact, the goldfish saccule is a spatial frequency analyzer that functions similarly to the basilar membrane of the birds and
mammals. If this were the case, the goldfish would
represent a remarkable example of the parallel
evolution of peripheral frequency analysis among
vertebrates. However, it remains to be shown
that the goldfish saccule functions as a spatial
frequency analyzer, and we know already that
the characteristic frequencies of saccular fibers
are not uniformly distributed across the animal's
hearing range (Fay and Ream 1984) as they are in
most mammals and birds.

Another view of the correspondence in behavioral frequency acuity is that the goldfish has
arrived at the same solution as mammals and
birds for performance in frequency analysis
based on quite different mechanisms. In general,
the solution is that frequency resolution is an
approximately constant proportion of center frequency, or that Weber's Law holds for frequency
analysis as well as for quantitative (intensive)
analysis in hearing. Whether or not this solution
is based on the same principles operating among
the mammals and birds, the goldfish's capacities for frequency resolution remain a remarkable example of convergence of function in vertebrate hearing.

5. Summary and Conclusions

A review of the psychophysical data on hearing
and sound discrimination reveals that vertebrates
are qualitatively and quantitatively similar in
many essential hearing functions, including absolute sensitivity, sound level processing, temporal
summation at threshold, temporal resolution,
duration and repetition rate discrimination, and
in several aspects of frequency analysis. Capacities

for sound level processing and temporal processing do not seem to be predictable from cochlear structure, and are probably determined by rather primitive properties of receptor and nerve cells, and their interactions. There are, at present, no indications of special adaptations for these hearing functions among vertebrates. We have no basis for predicting changes in these hearing functions since the time of the common ancestor of the modern birds and mammals.

Absolute position-frequency maps of the cochleas of birds and mammals can be estimated using only two parameters (the upper frequency limit of hearing—F_{max}, and cochlear length—L_{mm}). Both parameters may be, in principle, derived from the fossil record. Thus, cochlear position-frequency maps may be drawn, in principle, for ancestral species.

Capacities for frequency analysis among mammals and birds seem to be closely related to gross cochlear structure. Specifically, several behaviorally defined units of frequency analysis tend to occupy equal distances (independent of frequency or position) along the basilar membrane in living mammals and birds. This means that frequency analysis acuity can be well predicted for mammals and birds using the position-frequency map of the cochlea and one additional parameter, CD or the critical cochlear distance corresponding to a critical bandwidth. Among extant species, the CD is approximately constant across species at about 1.35 mm for mammals, and about 0.4 mm for birds. These "rules of thumb" may be used to predict frequency analysis acuity from measures of the upper frequency range of hearing and cochlear length. Further animal psychoacoustical studies will help evaluate and refine these predictions.

Von Békésy (1960) and Greenwood (1961; 1962; 1990) argued that the cochleas of mammals (and possibly birds) are scale models of one another and operate under the same principles as spatial frequency analyzers. The data from behavioral studies of hearing in a number of mammalian and avian species are generally in accord with this hypothesis. With the exceptions of the horseshoe bat and the parakeet, the quantitative details of spatial processing of frequency on the basilar membrane, and their consequences for sound perception, seem to recur widely among extant birds and mammals.

These hearing functions are thus quite conservative and likely have had a long and stable history. It seems likely that evolution has selected the consequences of frequency analytic mechanisms for perception rather than the mechanisms themselves. In this way, fishes, birds, mammals, and perhaps all vertebrates that hear, have evolved the same or quite similar solutions to common problems of perception based on quite different peripheral and central mechanisms.

Acknowledgments. This research and the writing of this chapter were supported by a Center Grant to the Parmly Hearing Institute from the NIH, NIDCD. Thanks to Jud Crawford, Bill Yost, Sheryl Coombs, Bill Shofner, Toby Dye, Stan Sheft, Mark Stellmack, and Andre Gauri for discussions that have contributed to some of the ideas in this chapter.

References

Abel S (1972) Duration discrimination of noise and tone bursts. J Acoust Soc Am 51:1219–1223.

Ades HW, Trahiotis C, Kokko-Cunningham A, Averbuch A (1974) Comparison of hearing thresholds and morphological changes in the chinchilla after exposure to 4 kHz tones. Acta Otolaryngol 78:192–206.

Ayrapet'yants E Sh, Konstantinov AI (1974) Echolocation in Nature. An English translation of the National Technical Information Service, Springfield, VA, JPRS 63328-1 and 2.

Baru AV (1971) Behavioral thresholds and frequency difference limen as a function of sound duration in dogs deprived of the auditory cortex. In: Gersuni GV (ed) Sensory Processes at the Neuronal and Behavioural Levels. New York: Academic Press.

Békésy G von (1960) Experiments in Hearing. Wever EG (ed) New York: McGraw-Hill.

Blakeslee EA, Hynson K, Hamernik RP, Henderson D (1978) Asymptotic threshold shift in chinchillas exposed to impulse noise. J Acoust Soc Am 63:876–882.

Bregman AS (1990) Auditory Scene Analysis: The Perceptual Organization of Sound. Cambridge, MA: MIT Press.

Brown CH, Maloney CG (1986) Temporal integration in two species of Old World monkeys: Blue monkeys (*Cercopithecus mitis*) and grey-cheeked mangabeys (*Cercocebus albigena*). J Acoust Soc Am 79:1058–1064.

Bruns V (1976) Peripheral auditory tuning for fine frequency analysis by the CF-FM bat *Rhinolophus ferrumequinum*. I. Mechanical specialization of the cochlea. J Comp Physiol 106:77–86.

Clack TD (1966) Effect of signal duration on the auditory sensitivity of humans and monkeys (*Macaca mulatta*). J Acoust Soc Am 40:1140–1146.

Clark WW, Bohne BA (1986) Cochlear damage: Audiometric correlates. In: Collins MJ, Glattke TJ, Harker LA (eds) Sensorineural Hearing Loss: Mechanisms, Diagnosis, and Treatment. Iowa City: University of Iowa Press.

Clark WW, Clark CS, Moody DB, Stebbins WC (1974) Noise-induced hearing loss in the chinchilla, as determined by a positive-reinforcement technique. J Acoust Soc Am 56:1202–1209.

Clopton BM (1972) Detection of increments in noise intensity by monkeys. J Exp Anal Behav 17:473–481.

Costalupes JA (1983a) Broad band masking noise and behavioral pure tone thresholds in cats. J Acoust Soc Amer 74:758–764.

Costalupes JA (1983b) Temporal integration of pure tones in the cat. Hear Res 9:43–54.

Costalupes JA (1988) Frequency difference thresholds for tone and noise combinations in the cat. In: Duifhuis H, Wit H, Horst J (eds) Basic Issues in Hearing. London: Academic Press.

Dallos P, Harris D, Ozdamar O, Ryan A (1978) Behavioral, compound action potential, and single unit thresholds: relationship in normal and abnormal ears. J Acoust Soc Am 64:151–157.

Davis R, Ferraro J (1984) Comparison between AER and behavioral thresholds in normally and abnormally hearing chinchillas. Ear Hearing 5:153–159.

Dooling RJ (1979) Temporal summation of pure tones in birds. J Acoust Soc Am 65:1058–1060.

Dooling RJ, Haskell RJ (1987) Auditory duration discrimination in the parakeet (*Melopsittacus undulatus*. J Acoust Soc Am 63:1640–1642.

Dooling RJ, Okanoya K, Downing J, Hulse S (1986) Hearing in the starling (*Sturnus vulgaris*): Absolute thresholds and critical ratios. Bull Psychon Soc 24:462–464.

Dooling RJ, Saunders JC (1975a) Auditory intensity discrimination in the parakeet (*Melopsittacus undulatus*). J Acoust Soc Am 58:1308–1310.

Dooling RJ, Saunders JC (1975b) Hearing in the parakeet (*Melopsittacus undulatus*): Absolute thresholds, critical ratios, frequency difference limens, and vocalizations. J Comp Physiol Psychol 88:1–20.

Dooling RJ, Searcy MH (1979) The relation among critical ratios, critical bands, and intensity difference limens in the parakeet (*Melopsittacus undulatus*). Bull Psychonom Soc 13:300–302.

Dooling RJ, Searcy MH (1981) Amplitude modulation thresholds for the parakeet (*Melopsittacus undulatus*). J Comp Physiol 143:383–388.

Dooling RJ, Searcy MH (1985) Temporal integration of acoustic signals by the budgerigar. J Acoust Soc Am 77:1917–1920.

Ehret G (1975a) Masked auditory thresholds, critical ratios, and scales of the basilar membrane of the housemouse (*Mus musculus*). J Comp Physiol 103: 329–341.

Ehret G (1975b) Frequency and intensity difference limens and nonlinearities in the ear of the housemouse (*Mus musculus*). J Comp Physiol 102:321–336.

Ehret G (1976a) Temporal auditory summation for pure tones and white noise in the house mouse (*Mus musculus*). J Acoust Soc Am 59:1421–1427.

Ehret G (1976b) Critical bands and filter characteristics of the ear of the housemouse (*Mus musculus*). Biol Cybernetics 24:35–42.

Eldredge DH, Miller JD, Bohne BA (1981) A frequency-position map for the chinchilla cochlea. J Acoust Soc Am 69(4):1091–1095.

Elliot DN, McGee TM (1965) Effect of cochlear lesions upon audiograms and intensity discrimination in cats. Ann Otol Rhinol Laryngol 74:386–408.

Elliot D, Stein L, Harrison M (1960) Determination of absolute intensity thresholds and frequency difference thresholds in cats. J Acoust Soc Am 32: 380–384.

Fay RR (1970) Auditory frequency discrimination in the goldfish (*Carassius auratus*). J Comp Physiol Psychol 73:175–180.

Fay RR (1974) Masking of tones by noise for the goldfish (*Carassius auratus*) J Comp Physiol Psychol 87:708–716.

Fay RR (1982) Neural mechanisms of an auditory temporal discrimination by the goldfish. J Comp Physiol 147:201–216.

Fay RR (1985) Sound intensity processing by the goldfish. J Acoust Soc Am 78:1296–1309.

Fay RR (1988) Hearing in Vertebrates: A Psychophysics Databook. Winnetka, IL: Hill-Fay Associates.

Fay RR (1989) Intensity discrimination of pulsed tones by the goldfish (*Carassius auratus*). J Acoust Soc Am 85:500–502.

Fay RR Ahroon W, Orawski A (1978) Auditory masking patterns in the goldfish (*Carassius auratus*): Psychophysical tuning curves. J Exp Biol 74:83–100.

Fay RR, Coombs S (1983) Neural mechanisms in sound detection and temporal summation. Hear Res 10: 69–92.

Fay RR, Ream TJ (1986) Acoustic responses and tuning in saccular nerve fibers of the goldfish (*Carassius auratus*). J Acoust Soc Am 79:1883–1895.

Fay RR, Shofner W, Dye R (1989) Effects of duration on intensity discrimination: Psychophysical data and predictions from single cell response. J Acoust Soc Am 85 (1):S13.

Fitzgibbons PJ (1983) Temporal gap detection in noise as a function of frequency, bandwidth, and level. J Acoust Soc Am 74:67–72.

Fletcher H (1940) Auditory patterns. Rev Mod Physics 12:47–65.

Florentine M, Buus S (1984) Intensity DL as a function of stimulus duration. J Acoust Soc Am 76 (1):S12.

Gerken GM, Bhat VKH, Hutchinson-Clutter M (1990) Auditory temporal integration and the power function model. J Acoust Soc Am 88:767–778.

Gerken GM, Sandlin D (1977) Auditory reaction time and absolute threshold in cat. J Acoust Soc Am 61: 602–607.

Gerken GM, Saunders S, Simhadri-Sumithra R, Bhat KHV (1985) Behavioral thresholds for electrical stimulation applied to auditory brainstem nuclei in cat are altered by injurious and noninjurious sound. Hear Res 20:221–231.

Glasberg BR, Moore BCJ (1990) Derivation of auditory filter shapes from notched-noise data. Hear Res 47: 103–138.

Gleich O (1989) Auditory primary afferents in the starling: Correlation of function and morphology. Hear Res 37:255–268.

Gleich O, Manley G (1988) Quantitative morphological analysis of the sensory epithelium of the starling and pigeon basilar papilla. Hear Res 34:69–86.

Gourevitch G (1965) Auditory masking in the rat. J Acoust Soc Am 37:439–443.

Gourevitch G (1970) Detectability of tones in quiet and in noise by rats and monkeys. In: Stebbins WC (ed) Animal Psychophysics: The Design and Conduct of Sensory Experiments. New York: Appleton-Century-Crofts, pp. 67–97.

Gray L, Rubel EW (1987) Development of auditory thresholds and frequency difference limens in chickens. In: Gottlieb G, Krasnegor N (eds) Measurement of Audition and Vision in the First Year of Postnatal Life: A Methodological Overview. Norwood, NJ: Ablex, pp. 145–165.

Graybeal A, Rosowski J, Ketten D, Crompton A (1989) Inner ear structure of in *Morganucodon*, an early Jurassic mammal. Zool J Linnean Soc 96:107–117.

Green DM (1973) Minimum integration time. In: Moller A (ed) Basic Mechanisms in Hearing. London: Academic Press.

Green DM (1988) Profile Analysis. Oxford: Oxford University Press.

Green DM, Swets JA (1966) Signal Detection Theory and Psychophysics. New York: Wiley.

Greenwood DD (1961) Critical bandwidth and the frequency coordinates of the basilar membrane. J Acoust Soc Am 33:1344–1356.

Greenwood DD (1962) Approximate calculation of the dimensions of traveling-wave envelopes in four species. J Acoust Soc Am 34:1364–1369.

Greenwood DD (1990) A cochlear frequency-position function for several species — 29 years later. J Acoust Soc Am 87(6) 2592–2605.

Hack MH (1971) Auditory intensity discrimination in the rat. J Comp Physiol Psychol 74:315–318.

Halperin L, Dallos P (1986) Auditory filter shapes in the chinchilla. J Acoust Soc Am 80:765–775.

Hawking S (1988) A Brief History of Time. New York: Bantam.

Hawkins JH, Stevens SS (1950) The masking of pure tones and of speech by white noise. J Acoust Soc Am 22:6–13.

Heffner RS, Heffner HE (1982) Hearing in the elephant (*Elephas maximus*): Absolute sensitivity, frequency discrimination, and sound localization. J Comp Psychol 96:926–944.

Heffner RS, Heffner HE (1985) Hearing range of the domestic cat. Hear Res 19:85–88.

Heffner R, Heffner H, Masterton RB (1971) Behavioral measurement of absolute and frequency-difference thresholds in guinea pig. J Acoust Soc Am 49:1888–1895.

Henderson D, Hamernik RP, Salvi RJ, Ahroon WA (1983) Comparison of auditory-evoked potentials and behavioral thresholds in the normal and noise-exposed chinchilla. Audiology 22:172–180.

Henderson D, Onishi S, Eldredge D, Davis H (1969) A comparison of chinchilla auditory evoked response and behavioral response thresholds. Percept Psychophys 5:41–45.

Henderson D, Salvi R, Pavek G, Hamernik RP (1984) Amplitude modulation thresholds in chinchillas with high-frequency hearing loss. J Acoust Soc Am 75: 1177–1183.

Herman LM, Arbeit WR (1972) Frequency difference limens in the bottlenose dolphin: 1-70 kc/s. J Aud Res 2:109–120.

Hienz RD, Sachs MB (1987) Effects of noise on pure-tone thresholds in blackbirds (*Agelaius phoeniceus* and *Molothrus ater*) and pigeons (*Columba livia*). J Comp Psych 101:16–24.

Hienz RD, Sinnott JM, Sachs MB (1980) Auditory intensity discrimination in blackbirds and pigeons. J Comp Physiol Psych 94:993–1002.

Ison JR (1982) Temporal acuity in auditory function in the rat. J Comp Physiol Psychol 96:945–954.

Jacobs DW, Tavolga WN (1968) Acoustic frequency discrimination in the goldfish. Anim Behav 16:67–71.

Jerison HJ (1973) Evolution of the Brain and Intelligence. New York: Academic Press.

Jesteadt W, Wier CC, Green DM (1977) Intensity discrimination as a function of frequency and level. J Acoust Soc Am 61:169–177.

Johnson CS (1968a) Relation between absolute threshold and duration-of-tone pulses in the bottlenosed porpoise. J Acoust Soc Am 43:757–763.

Johnson CS (1968b) Masked tonal thresholds in the bottle-nosed porpoise. J Acoust Soc Am 44:965–967.

Kelly JB (1970) The effects of lateral lemniscal and neocortical lesions on auditory absolute thresholds and frequency difference thresholds in the rat. PhD Thesis, Vanderbilt University, University Microfilms 70-16:429.

Khanna S, Leonard D (1982) Basilar membrane tuning in the cat cochlea. Science 215:305–306.

Klump CM, Maier EH (1989) Gap detection in the starling (Sturnus vulgaris). J Comp Physiol 164:531–538.

Kuhn A, Leppelsack HJ, Schwartzkopff J (1980) Measurement of frequency discrimination in the starling (Sturnus vulgaris) by conditioning of heart rate. Naturwiss 67:102.

Lamming D (1986) Sensory Analysis. London: Academic Press.

Liberman MC (1982) The cochlear frequency map for the cat: Labeling auditory nerve fibers of known characteristic frequency. J Acoust Am 72:1441–1449.

Long GR (1977) Masked auditory thresholds from the bat, Rhinolophus ferrumequinum. J Comp Physiol 116:247–255.

Long G (1980) Some psychophysical measurements of frequency processing in the greater horseshoe bat. In: van der Brink G, Bilsen FH (eds) Psychophysical, Physiological, and Behavioral Studies in Hearing. Delft: Delft University Press, pp. 132–135.

Long G, Clark W (1984) Detection of frequency and rate modulation by the chinchilla. J Acoust Soc Am 75:1184–1190.

McCormick CA, Braford MR (1988) Central connections of the octavolateralis system: Evolutionary considerations. In: Atema J, Fay RR, Popper AN, Tavolga WN (eds) Sensory Biology of Aquatic Animals. New York: Springer-Verlag, pp. 733–755.

McGee T, Ryan A, Dallos P (1976) Psychophysical tuning curves of chinchillas. J Acoust Soc Am 60:1146–1150.

McGill TE (1959) Auditory sensitivity and the magnitude of cochlear potentials. Ann Otol Rhinol Laryngol 68:193–207.

Megela-Simmons A (1988) Masking patterns in the bullfrog (Rana catesbeiana), I: Behavioral effects. J Acoust Soc Am 83:1087–1092.

Megela-Simmons A, Moss CF, Daniel KM (1985) Behavioral audiograms of the bullfrog (Rana catesbeiana) and the green tree frog (Hyla cinerea). J Acoust Soc Am 78:1236–1244.

Miller JD (1964) Auditory sensitivity of the chinchilla in quiet and in noise. J Acoust Soc Am 36:2010 (abstract).

Miller JD (1970) Audibility curve of the chinchilla. J Acoust Soc Am 48:513–523.

Miller JD, Watson CS, Covell WP (1963) Deafening effects of noise on the cat. Acta Oto-laryngol Suppl 176:1–81.

Moore BCJ (1982) An Introduction to the Psychology of Hearing. London: Academic Press.

Moore BCJ (1978) Psychophysical tuning curves measured in simultaneous and forward masking. J Acoust Soc Am 63:524–532.

Moss CF, Megela-Simmons A (1986) Frequency selectivity of hearing in the green treefrog, Hyla cinerea. J Comp Physiol 159:257–266.

Neff W, Hind J (1955) Auditory thresholds of the cat. J Acoust Soc Amer 27:480–483.

Nelson DA, Kiester TE (1978) Frequency discrimination in the chinchilla. J Acoust Soc Am 64:114–126.

Nienhuys TW, Clark GM (1979) Critical bands following the selective destruction of cochlear inner and outer hair cells. Acta Oto-Laryngol 88:350–358.

Okanoya K, Dooling RJ (1985) Colony differences in auditory thresholds in the canary. J Acoust Soc Am 78:1170–1176.

Okanoya K, Dooling RJ (1987) Hearing in passerine and psittacine birds: A comparative study of absolute and masked auditory thresholds. J Comp Psychol 101:7–15.

Patterson RD, Moore BCJ (1986) Auditory filters and excitation patterns as representations of frequency resolution. In: Moore BCJ (ed) Frequency Selectivity in Hearing. London: Academic Press.

Patterson WC (1966) Hearing in the turtle. J Aud Res 6:453–464.

Pickles JO (1975) Normal critical bands in the cat. Acta Oto-laryngol 80:245–254.

Pickles JO (1979) Psychophysical frequency resolution in the cat as determined by simultaneous masking and its relation to auditory nerve resolution. J Acoust Soc Am 66:1725–1732.

Platt C (1977) Hair cell distribution and orientation in goldfish otolith organs. J Comp Neurol 172:283–298.

Plomp R, Bouman A (1959) Relation between hearing threshold and duration for tone pulses. J Acoust Soc Am 31:749–758.

Popper AN, Tavolga WN (1981) Structure and function in the ear of the marine catfish *Arius felis*. J Comp Physiol 144:27–34.

Quine DB, Konishi M (1974) Absolute frequency discrimination in the barn owl. J Comp Physiol 93: 347–360.

Rhode W (1971) Observations on the vibrations of the basilar membrane in squirrel monkeys using the Mossbauer technique. J Acoust Soc Am 49: 1218–1231.

Rose J, Brugge J, Anderson D, Hind J (1967) Phase locked response to low frequency tones in single auditory nerve fibers of the squirrel monkey. J Neurophysiol 30:769–793.

Salvi RJ, Ahroon WA, Perry JW, Gunnarson AD, Henderson D (1982) Comparison of psychophysical and evoked-potential tuning curves in the chinchilla. Am J Otolaryngol 3:408–416.

Salvi RJ, Arehole S (1985) Gap detection in chinchillas with temporary high-frequency hearing loss. J Acoust Soc Am 77:1173–1177.

Salvi RJ, Giraudi DM, Henderson D, Hamernik RP (1982) Detection of sinusoidal amplitude modulated noise by the chinchilla. J Acoust Soc Am 71: 424–429.

Salvi RJ, Hamernik RP, Henderson D (1983) Response patterns of auditory nerve fibers during temporary threshold shift. Hear Res 10:37–67.

Saunders SS, Shivapuja BG, Salvi RJ (1987) Auditory intensity discrimination in the chinchilla. J Acoust Soc Am 82:1604–1607.

Saunders JC, Denny RM, Bock GR (1978) Critical bands in the parakeet (*Melopsittacus undulatus*). J Comp Physiol 125:359–365.

Saunders JC, Mills JH, Miller JD (1977) Threshold shift in the chinchilla from daily exposure to noise for six hours. J Acoust Soc Am 61:558–570.

Saunders JC, Rintelmann W, Bock G (1979) Frequency selectivity in bird and man: A comparison among critical ratios, critical bands and psychophysical tuning curves. Hear Res 1:303–323.

Schwartz M (1970) Information Transmission, Modulation, and Noise. New York: McGraw-Hill.

Schwartzkopff J (1949) Über Sitz und Leistung von Gehör und Vibrationsinn bei Vögeln. Z vergl Physiol 31:527–603.

Seaton WH, Trahiotis C (1975) Comparison of critical ratios and critical bands in the monaural chinchilla. J Acoust Soc Am 57:193–199.

Serafin SV, Moody DB, Stebbins WC (1982) Frequency selectivity of the monkey's auditory system: Psychophysical tuning curves. J Acoust Soc Am 71: 1513–1518.

Shannon RV (1983) Multichannel electrical stimulation of the auditory nerve in man I Basic psychophysics. Hear Res 11:157–189.

Shower EG, Buddulph R (1931) Differential pitch sensitivity of the ear. J Acoust Soc Am 3:275–287.

Sinnott JM, Sachs MB, Hienz RD (1980) Aspects of frequency discrimination in passerine birds and pigeons. J Comp Physiol Psychol 94:401–415.

Sinnott JM, Owren MJ, Petersen MR (1987b) Auditory duration discrimination in Old World Monkeys (*Macaca, Cercopithecus*) and humans. J Acoust Soc Am 82:465–470.

Sinnott JM, Petersen MR, Hopp SL (1985) Frequency and intensity discrimination in humans and monkeys. J Acoust Soc Am 78:1977–1985.

Sokolovski A (1973) Normal threshold of hearing for cat for free-field listening. Arch Klin Exp Ohr- Nas- u Kehlk Heilk 203:232–240.

Stebbins WC (1970) Studies of hearing and hearing loss in the monkey. In: Stebbins WC (ed) Animal Psychophysics: The Design and Conduct of Sensory Experiments. New York: Appleton-Century-Crofts.

Suthers RA, Summers CA (1980) Behavioral audiogram and masked thresholds of the megachiropteran echolocating bat, *Rousettus*. J Comp Physiol 136:227–233.

Tavolga WN (1974) Signal/noise ratio and the critical band in fishes. J Acoust Soc Am 55:1323–1333.

Thompson RK, Herman LM (1975) Underwater frequency discrimination in the bottlenose dolphin (1-140 kHz) and the human (1-8 kHz). J Acoust Soc Am 57:943–948.

Trahiotis C, Elliot DN (1970) Behavioral investigation of some possible effects of sectioning the crossed olivocochlear bundle. J Acoust Soc Am 47:592–596.

Viemeister NF (1972) Intensity discrimination of pulsed sinusoids: the effects of filtered noise. J Acoust Soc Am 51:1265–1269.

Viemeister NF (1979) Temporal modulation transfer functions based upon modulation thresholds. J Acoust Soc Am 66:1364–1380.

Viemeister NF (1983) Auditory intensity discrimination at high frequencies in the presence of noise. Science 221:1206–1208.

Viemeister NF and Wakefield GH (1989) Multiple looks and temporal integration. J Acoust Soc Am 86(1):S23.

Wall LG, Ferraro JA, Dunn DE (1981) Temporal integration in the chinchilla. J Aud Res 21:29–37.

Watson CS (1963) Masking of tones by noise for the cat. J Acoust Soc Amer 35:167–172.

Watson CS, Gengel RW (1969) Signal duration and signal frequency in relation to auditory sensitivity. J Acoust Soc Am 46:989–997.

Wever EG (1949) Theory of Hearing. New York: Wiley.

Weir C, Jesteadt W, Green D (1976) Frequency discrimination as a function of frequency and sensation level. J Acoust Soc Am 61:178–184.

Yerkes RM (1904) Inhibition and reinforcement of reaction in the frog *Rana clamitans*. J Comp Neurol Psychol 14:124–137.

Young ED, Sachs MB (1979) Representation of steady-state vowels in the temporal aspects of the discharge patterns of populations of auditory nerve fibers. J Acoust Soc Am 66:1381–1404.

Yunker MP, Herman LM (1974) Discrimination of auditory temporal differences by the bottlenose dolphin and by the human. J Acoust Soc Am 56: 1870–1875.

Zelick RD, Narins PM (1983) Intensity discrimination and the precision of call timing in two species of neotropical treefrogs. J Comp Physiol 153:403–412.

Zwicker E, Flottorp G, Stevens S (1957) Critical bandwidths in loudness summation. J Acoust Soc Am 29:548–557.

Zwislocki J (1969) Temporal summation of loudness. J Acoust Soc Am 46:431–441.

Section IV
Anamniotes

Hearing first arose in aquatic vertebrates and the inner ear evolved as an adaptive response to the special characteristics of sound in the aquatic environment. When vertebrates came onto land there had to be major modifications in sound reception in order for the inner ear to receive sufficient stimulation. In this section, various chapters describe hearing in water and on land, and the overlap between aquatic and terrestrial hearing.

As described in Chapter 3, there is a phylogenetic relationship between the embryonic development of the lateral line and the inner ear. In Chapter 15, the first in this section on anamniotes, Coombs, Janssen, and Montgomery describe the structural and physiological diversity of the lateral line, as well as behavioral measures of lateral line capabilities. Various types of evidence point to functional diversity among the lateral line systems of fishes. The authors argue that there are clear functional distinctions between the lateral line and the auditory system of fishes.

Schellart and Popper (Chapter 16) discuss the evolution of the auditory system of fishes in the context of the environmental constraints placed on acoustic propagation underwater. They describe basic structures and mechanisms of hearing in diverse groups, and present factors that may have affected the evolution of fish ears.

In Chapter 17, McCormick describes the central auditory circuitry of fishes and amphibians in the context of evolution. She discusses the relationship between the diversity of inner ear receptors (as described in Chapter 16) and the organization of auditory pathways.

Fritzsch (Chapter 18), using broad comparative anatomical data, makes a persuasive argument that the basilar papilla, a perilymphatic sense organ, first appeared in sarcopterygian fishes and has since been modified in various extant tetrapods.

After reviewing 470 paleontological manuscripts, Bolt and Lombard (Chapter 19) have determined primitive and derived states of four key otic characters during the evolution of the ear of early tetrapods. They have successfully mapped out a cladogram of these characters in 16 Paleozoic tetrapod groups. From these cladograms, one can make evolutionary hypotheses about the structure/function relationships.

Clack (Chapter 20) presents paleontological data supporting the hypothesis that the earliest tetrapod stapes was involved with mastication rather than audition. She argues that it was not until the Temnospondyls, the group from which extant amphibians arose, that the stapes became solely auditory in function.

Hetherington (Chapter 21) presents abundant data consistent with the hypothesis that, in amphibians, body size has been significant in determining mechanisms of acoustic reception. This may explain the adaptations of non-tympanic mechanisms of auditory reception in small tetrapods.

Narins (Chapter 22) gives ethological data on frog auditory function, determined in the natural habitat. His emphasis is on understanding the factors affecting the evolution of acoustic communication systems — a theme set forth by Michelsen in Chapter 5.

15
Functional and Evolutionary Implications of Peripheral Diversity in Lateral Line Systems

Sheryl Coombs, John Janssen, and John Montgomery

1. Introduction

Any treatise on the evolution of hearing inevitably raises the possibility that the vertebrate auditory system has evolved from the mechanosensory lateral line system known to exist in the earliest vertebrates. The arguments for (van Bergeijk 1967; Jørgensen 1989) and against (Wever 1976; Northcutt 1981) this "octavolateralis" hypothesis, first proposed by Ayers (1892), have depended primarily on anatomical and developmental comparisons between the two systems. It is questionable whether this issue can or ever will be resolved, but the frequency with which it has been addressed in the past and is currently being addressed in this volume (Popper, Platt, and Edds, Chapter 4; Fritzsch, Chapter 18) testifies to the fascination it holds for anyone interested in the evolution of ears and hearing.

Although theories on the evolutionary origin of vertebrate ears must rely on anatomical evidence, ideas about selective pressures in the evolution of ears and lateral line come primarily from evidence of function. In this chapter, we will focus on what is known about lateral line function in teleost fish (and to some extent on aquatic amphibians), drawing most heavily from psychophysical evidence of sensory capabilities, neurophysiological evidence of the response properties of peripheral endorgans, and behavioral evidence of the biological relevance of the system to the animal. Not only do these various lines of evidence point to functional diversity within the lateral line system of fishes, but also to some clear functional distinctions between lateral line and auditory system of fish.

Because of these differences, it is probably no longer useful to think of the lateral line system simply as an "accessory" auditory structure (van Bergeijk 1967), as Dijkgraaf (1963) has argued all along. Nonetheless, when thinking about the evolution of hearing, it may be useful to think about the major functions of the lateral line system as part of a range of functions performed by octavolateralis systems.

In the first half of this chapter we will (1) describe the basic anatomy of the lateral line system in fishes (Section 2) and summarize what is known about its phylogenetic distribution and origin (Section 3); (2) identify anatomical dimensions of variability and common trends in the evolution of lateral line systems across phylogenetically diverse taxa (Sections 4 to 4.2) and (3) identify some of the developmental mechanisms (Section 5) and selective pressures (Section 6) that may have led to the differentiation of lateral line systems. In the second half of the chapter we will compare and contrast the lateral line and auditory system of teleost fishes in terms of (1) modes of stimulation and principles of operation (Section 7), (2) the functional consequences of these operational differences (Section 7 and 8), and (3) how the two systems can be thought of as part of a functional continuum on the one hand, and where they appear to overlap in function (Section 9), on the other hand. In an effort to make direct comparisons between subsurface detection abilities of lateral line and ear, we have largely neglected a significant body of information on surface-wave detection by the lateral line system of fishes and amphibians. For further information on this

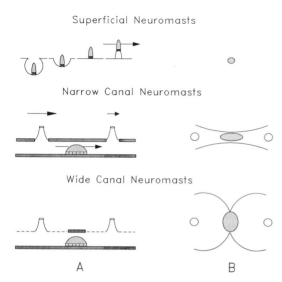

FIGURE 15.1. A schematic representation of major variations in the lateral line system of fish. (A) Sagittal sections through the skin and underlying canal showing differences in the elevation of superficial neuromasts (top panel) and differences between narrow (middle panel) and wide (bottom panel) canal systems. A single arrow depicts the flow of water past the cupula (stippled area) on a papillate superficial neuromast (top panel) and on a narrow canal neuromast (middle panel). Arrows above the canal pores (middle panel) indicate the size and direction of the pressure gradient between the two pores that induces water flow inside the canal. The heavily stippled bars above and below each canal indicate regions where the canal is covered by hard tissue, such as bone or scale, whereas the dashed lines indicate flexible, soft tissue. Water flow inside wide canals with compliant walls (bottom panel) may be more complicated than depicted for narrow canals and is therefore, not shown in this figure. (B) Schematic coronal sections at the level of the cupular base to show the relative size and shape of the neuromast and the surrounding canal.

specialized function of the lateral line system, the reader is referred to Bleckmann, Tittel, and Blubaum-Gronau (1899), Elepfandt (1989), and Görner and Mohr (1989).

2. Peripheral Anatomy and Innervation of the Lateral Line System

The basic endorgan in the lateral line system of fishes and amphibians is the neuromast, a patch of

specialized epithelial cells that sits above the basement membrane in the epidermis. Hair cells in the center of the neuromast are surrounded by two types of nonsensory cells: support cells dispersed between the hair cells and mantle cells around the periphery. The ciliary bundle on the apical surface of each hair cell is anatomically polarized with stereocilia increasing in length in a stepwise fashion leading up to a single, eccentrically placed kinocilium (Flock 1967). Adjacent hair cells typically have their kinocilia on opposite sides of the cell such that for any given neuromast, hair cells will respond maximally to one of two directions (180° out of phase) along a single axis of ciliary bundle deflection (e.g., Janssen et al. 1987).

Overlying the ciliary bundle is a gelatinous cupula. Because the cupula is such a fragile structure, not much is known about its architecture or infrastructure in most fishes. Yet this information is critical to the biomechanical properties of the system. However, the cupula does appear to have an inner and outer core, the former presumably secreted by support cells and the latter by mantle cells (Münz 1979). Moreover, recent optical scanning of the cupula from a head canal neuromast in the clown knifefish (*Notopterus chitala*) revealed a very ordered linkage between the underlying hair cells and evenly sized compartments of the central region of the cupula (Kelly and van Netten 1991).

In aquatic amphibians, neuromasts are found only on the skin surface (Russell 1976; Lanoo 1987a,b), whereas in fishes, neuromasts can be additionally found enclosed in canals under the skin (Fig. 15.1). Pores intermittently spaced along the canal connect the fluid-filled interior with the environment. The distribution of neuromasts on the head and trunk of amphibians and fishes is similar in that there are generally lines of neuromasts found above and below the eye, along the side of the head (cheek), along the preopercle and lower jaw, across the top of the head, and down the trunk. Figure 15.2 shows the distribution of neuromasts found in an anuran amphibian, *Xenopus*, a teleost fish, *Cottus bairdi*, and the proposed ancestral distribution of neuromasts in gnathostome fishes (Northcutt 1989).

The lateral line system of all vertebrates is innervated by a distinct set of lateral line nerves that are no longer regarded as components of the trigeminal (V), facial (VII), glossopharyngeal (IX), and vagal (X) cranial nerves (see Northcutt 1989, for most recent review). At least three separate lateral line nerves, each with its own ganglion, are now generally recognized in teleosts. The dorsal and ventral anterior lateral line nerves innervate superficial and canal neuromasts on the head and the posterior lateral line nerve innervates neuromasts predominantly on the trunk. A middle lateral line nerve, innervating superficial and canal neuromasts just behind the eye, has been reported for several extant noneuteleostean (Allis 1889; 1895; Norris 1925; Pehrson 1947) and teleostean species (Herrick 1901; Pollard 1892; Pehrson 1945; Puzdrowski 1989; Wulliman and Senn 1981). Thus, it would appear that the earliest craniates possessed at least four and perhaps as many as seven lateral line nerves (Northcutt 1989).

3. Phyletic Distribution and Origin of the Lateral Line System

According to a rigorous out-group analysis of the phylogenetic distribution of lateral line organs (Northcutt 1989), there is no evidence for reinvention of the mechanosensory lateral line system in different phyla, as there seems to be for the closely related electrosensory system (Bullock 1974; McCormick 1982; Bullock, Northcutt, and Bodznick 1982; Bullock, Bodznick, and Northcutt 1983) and for the auditory portion of the inner ear (Wever 1976; Popper, Platt, and Edds, Chapter 4). The mechanosensory lateral line system is present in nearly all anamniotic vertebrates (Northcutt 1989), being found in petromyzontid agnathans, in all extant cartilaginous and bony fishes, and in most larval and many postmetamorphic amphibians (Fritzsch 1989). It has also been found in the earliest vertebrates (heterostracans) to appear in the fossil record (Moy-Thomas and Miles 1971). Unfortunately, conclusions about the existence of the lateral line system in the presumed sister group to vertebrates (myxinoid hagfish) cannot be made from the existing data (Northcutt 1989). One of the extant hagfish genera in this group has apparently retained some vestigate of the lateral line system (canal-like grooves without neuromasts), whereas a second has retained no outward appearance of a lateral line system at all (Fernholm 1985).

As Northcutt (1989) has pointed out, without further information about myxinoid hagfish, the earliest origin and morphotype of the lateral line

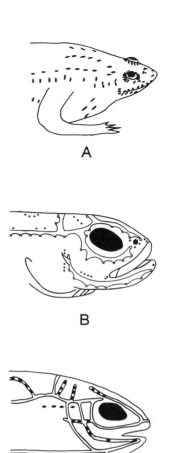

A

B

C

FIGURE 15.2. Distribution pattern of superficial neuromasts on an amphibian, *Xenopus laevis* (after Shelton 1970) (A); and of superficial neuromasts and canals on a teleost, *Cottus bairdi* (after Janssen, Coombs, Hoekstra, Platt 1987) (B); and the proposed ancestral distribution of canals on the earliest gnathostomes (after Northcutt 1989) (C). Individual neuromasts within canals (indicated by double lines in B and C) are depicted as black circles in C and represent the location of canal neuromasts thought to give rise to certain populations of superficial neuromasts in teleosts, as shown by black dots for the mottled sculpin in B. Individual canal neuromasts are not depicted in B, but occur between every two canal pores, indicated by small open circles. Black ovals in A represent stitches, or rows of 6 to 12 superficial neuromasts.

system cannot be determined. However, it is most probable that the lateral line system of the earliest craniates existed as a series of canals (Fig. 15.2c) rather than superficial neuromasts and that the reduced state of the lateral line system in hagfish represents a degenerate, rather than primitive state (Northcutt 1989). The first lines of superficial neuromasts (pit lines, see Fig. 15.2) found in most primitive fish probably arose as a result of incomplete canal formation during development (see Section 5). In contrast, additional lines of superficial neuromasts are thought to have arisen secondarily in many teleost taxa (Northcutt 1989). In many postmetamorphic amphibians, the lateral line system regresses or disappears after metamorphosis from the aquatic, larval stage to the terrestrial phase and in some species, even reappears during aquatic, reproductive phases in the animal's life cycle (Fritszch 1989).

4. Anatomical Dimensions of Variability and Functional Consequences

Although there is no evidence for a polyphyletic origin of the lateral line system, a polyfunctional interpretation of the lateral line system, based largely on the inter and intraspecific diversity of structures within the lateral line system of bony fishes (Dijkgraaf 1963; Coombs, Janssen, and Webb 1988; Denton and Gray 1988) is frequently assumed. Among cartilaginous and bony fishes, at least four kinds of endorgans, distinguished primarily by surrounding support structures, have been classified as mechanosensory lateral line organs: superficial neuromasts, canal neuromasts, spiracular organs, and vesicles of Savi. Spiracular organs are housed in diverticula of the hyoid pouch in cartilaginous and most nonteleost bony fishes and are thought to function as proprioceptors of hyomandibular motion (Barry and Bennett 1989). It has been hypothesized that the middle ear of tetrapods has evolved from the spiracular pouch (see Fritzch, Chapter 18) and that the paratympanic organ of the middle ear in birds is homologous to the spiracular organ of fishes (Jørgensen1989; von Bartheld 1990). Vesicles of Savi are found on the ventral surface of certain cartilaginous fishes, most notably those such as the skates and rays that are dorsoventrally flattened. These vesicles, which have no pores connecting them to the environment,

contain a cluster of about three neuromasts and may function in substrate vibration detection (Barry and Bennett 1989). Spiracular organs and vesicles of Savi are both effectively isolated from surrounding water motions and represent very specialized cases for which the usual mechanisms of lateral line stimulation (see Section 7) do not apply. As such, these two classes of endorgans merely underscore the diversity of functions possible within the lateral line system of fishes.

The remaining two types of endorgans, superficial and canal neuromasts, represent the most common endorgans found in the lateral line system of fishes. Even these two types of endorgans, however, may differ significantly in their shapes, sizes, and orientation and especially in the structures surrounding them (see Coombs, Janssen, and Webb 1988 for review). Dimensions of variability of surrounding structures include (1) elaborations of epidermal structures surrounding or underlying free neuromasts; (2) the relative width of canals and the compliance of the canal covering; (3) variation in the size and number of canal pores, in interpore distances, and in the degree of branching in tubules leading to canal pores; and (4) differences in the size and shape of the cupula and degree to which it fills the canal (Blaxter, Gray, and Best 1983; Denton and Gray 1988, 1989). Furthermore, the distribution pattern and number of neuromasts on the body of fish may vary (Webb 1989b).

As one might expect, the degree of morphological variation known to exist in the lateral line system of teleost fishes is quite extensive, but the full extent of this variation in all species of fish (estimated at between 25,000 and 30,000) is unknown. Although a few major taxa can be defined by characteristics of the lateral line system (e.g., Notopteroidei and Clupeomorpha), it is more generally the case that they cannot. Rather than present a comprehensive catalog of the variation known to exist, as has been done earlier by Dijkgraaf (1963), Marshall (1971), and more recently by Coombs, Janssen, and Webb (1988) and Webb (1989b), we will focus on two major categories of anatomical differentiation (superficial vs canal neuromasts and canal width), for which some data on functional properties or behavioral significance exist.

4.1 Superficial Neuromasts vs Canal Neuromasts

Neuromasts that appear superficially on the skin surface and those that are enclosed in canals constitute the biggest and most obvious anatomical differentiation of the lateral line system (Fig. 15.1). Most fish that have been looked at carefully have both superficial and canal neuromasts, but many fish, particularly certain teleosts, show a reduction in canals and a concomitant increase in the number of superficial neuromast lines (Fig. 15.3; Coombs, Janssen, and Webb 1988 for review). Although there is now a general belief that developmental mechanisms play a key role in how the lateral line system gets from the more primitive canal state to the more derived superficial neuromast state (see Section 5), questions about why this has happened repeatedly remain largely unanswered. By describing some of the anatomical and functional differences between these two classes of endorgans, we hope to provide some insight into the selective forces that may have been involved in this evolution.

Before considering the differences between superficial and canal neuromasts, however, it is important to emphasize that neuromasts within each class are not homogeneous with respect to neuromast and surrounding support structures (see Section 4.2) and that superficial neuromasts may differ with respect to developmental origins (see Section 5). Without knowing whether between class differences are greater than within class differences, the following generalizations, based to some extent on comparisons between superficial and canal neuromasts in the same species (see Münz 1989 for review) can be made. Superficial neuromasts are generally smaller (<100 μm in diameter) and have fewer hair cells and fewer fibers innervating them than canal neuromasts. As a group, superficials tend to be more symmetrical and more homogeneous in shape (Janssen, Coombs, and Hoekstra 1987). Their cupulae also tend to be less heterogeneous in shape than canal neuromasts (Blaxter, Gray, and Best 1983), with those from free neuromasts in fish (Best and Gray 1982; Blaxter, Gray, and Best 1983) being very similar in shape to what has been reported in *Xenopus* (Dijkgraaf 1963) and the mudpuppy, *Necturus maculosus* (Liff and Shamres 1972; Oman 1972).

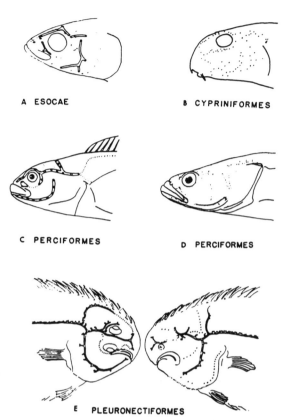

A ESOCAE   B CYPRINIFORMES

C PERCIFORMES   D PERCIFORMES

E PLEURONECTIFORMES

FIGURE 15.3. Examples of cephalic lateral line canal loss and replacement by superficial neuromasts in (A) *Umbra limi* (Umbridae) (from Nelson 1972); (B) *Cobitis taenia* (Cobitae) (from Lekander 1949); (C) *Opisthocentrus dybowskii* (Stichaeidae) (From Makushok 1961); (D) *Eleotris pisonis* (Gobiidae) (from Miller and Wongrat 1979); and (E) *Solea vulgaris* (Solidae). (From Applebaum and Schemmel 1983.) (Composite figure from Coombs, Janssen, and Webb 1988.)

Most importantly, afferent fibers innervating canal and superficial neuromasts have different response properties. If response properties are characterized by the frequency causing maximum displacement sensitivity or responsiveness, the frequency range of maximum sensitivities is much lower (between 10 and 60 Hz) for fibers innervating superficial neuromasts than that for fibers innervating canal neuromasts (50 to 200 Hz) (Münz 1989). If, on the other hand, response properties are characterized in terms of whether fibers are responding predominantly to the velocity or acceleration of external water motion, the general finding is that superficial neuromast fibers

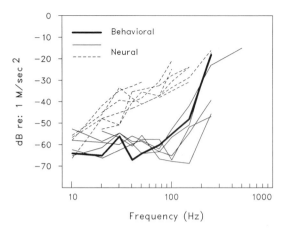

FIGURE 15.4. Neural and behavioral measures of threshold detection as a function of frequency for the mottled sculpin, *Cottus bairdi*. Behavioral thresholds were obtained with the stimulus (a vibrating sphere) positioned 1.5 cm away from the middle of the trunk lateral line canal. Neural thresholds were obtained from posterior lateral line nerve fibers innervating organs on the trunk, with stimulus position the same as for behavioral measures. Dashed lines are from velocity-sensitive fibers presumed to innervate superficial neuromasts, whereas solid lines are from acceleration-sensitive fibers presumed to innervate canal organs. (From Coombs and Janssen, in press.)

respond to the velocity and canal neuromast fibers to the acceleration component of a periodic stimulus (Denton and Gray 1983; Münz 1985; Kroese and Schellart 1987; Coombs and Janssen 1989). Although canal and superficial neuromasts appear to be innervated by completely different afferent fiber populations in cichlid fish (Münz 1985), data on whether these different fiber populations project to different areas in the brain are largely lacking. Thus, we have no clear idea in any fish, let alone most fish, as to whether information from these two subsets of lateral line organs are processed separately in the central nervous system (but see Song and Northcutt 1991).

Unfortunately, generalizations about how superficial and canal neuromasts might subserve different behavioral functions are currently impossible to make across species. Feeding responses of the mottled sculpin, *Cottus bairdi*, for example, appear to be driven primarily by acceleration-sensitive endorgans, as the threshold level of detection for evoking the feeding response is independent of

frequency when stimulus levels are plotted in acceleration units (Fig. 15.4, Coombs and Janssen, in press). Although direct evidence is lacking, it is presumed that acceleration-sensitive canal organs underly the feeding behavior in this species, as tuning curves from acceleration-sensitive fibers do a remarkably good job of predicting behavioral threshold curves, both in terms of bandwidth and absolute sensitivity, whereas tuning curves from velocity-sensitive fibers presumed to innervate superficial neuromasts do not (Fig. 15.4).

In contrast, both larval sculpin (Jones and Janssen 1990) and adult *Xenopus* (Görner and Mohr 1989; Elepfandt 1989) use their lateral line systems in feeding, yet neither have canal neuromasts. Moreover, the feeding response of the blind cavefish, *Astyanax jordani* (= *Anoptichthys jordani*), appears to be driven primarily by superficial neuromasts, as destruction of canal organs results in no reductions in the feeding response (biting at a vibrating sphere), but in significant reductions in the animal's ability to detect nonmoving obstacles in the environment (Abdel-Ledif, Hassan, and von Campenhausen 1990).

Many questions remain as to how information from superficial and canal neuromasts is used by the animal, whether it is integrated or segregated in the CNS and what happens to the wiring and function of the system as it changes from one of entirely superficial neuromasts in the earliest stages of development in all species to one of combined superficial and canal neuromasts in most adult fishes.

4.2 Canal Width

Although there are no published data on canal width in the earliest vertebrates, it is generally assumed that head canals in primitive fishes were relatively narrow and that widened head canals (e.g., covering up to over half the surface of the head) represent a derived state in fishes (Coombs, Janssen, and Webb 1988; Northcutt 1989; Webb 1989b). Among bony fishes, head canal diameters vary from approximately 100 μm to 7 mm (Denton and Gray 1988). The phyletic distribution of fish taxa that have widened head canals clearly shows that this trait has evolved independently a number of different times (Fig. 15.5, Coombs, Janssen, and Webb 1988). A number of other structural

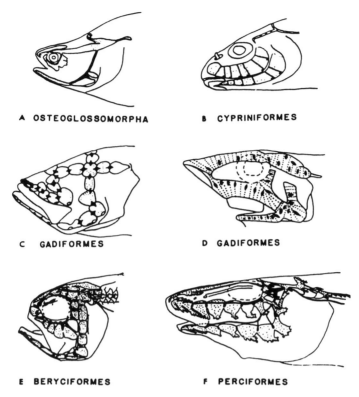

A OSTEOGLOSSOMORPHA   **B** CYPRINIFORMES

C GADIFORMES   **D** GADIFORMES

E BERYCIFORMES   **F** PERCIFORMES

FIGURE 15.5. Examples of widened cephalic canals in various fish taxa: (A) *Notopterus chitala* (Notopteridae) (from Kapoor 1964); (B) *Ericymba buccata* (Cyprinidae) (from Hoyt 1972); (C) *Typhliasina pearsei* (Brotulidae) (from Schemmel 1977); (D) *Coelorhynchus flabellispinis* (Macrouridae) (from Marshall 1965); (E) *Hoplosthethus mediterraneus* (Trachichthyidae) (from Jakubowski 1974); and (F) *Aspro zingle* (Percidae) (from Jakubowski 1967). Darkened triangular structures represent neuromast positions in C, D, E and F. Canal pores are absent in A, not depicted in E and F, and indicated by open circles in B and C. (Composite figure from Coombs, Janssen, and Webb 1988.)

modifications are associated with some, but not all, widened canals, including (1) a reduction in the bony covering of canals and the replacement of bone with soft tissue as the canal covering; (2) a change in the shape of canal neuromasts such that their major axis is oriented across the canal rather than along the canal, as typically occurs in narrow canals; (3) a compartmentalization of the canal by narrow bony constrictions or membranous lamellae in the vicinity of the neuromast; and (4) cupulae which tend to completely, or almost completely, block the canal (e.g., *Lota* (Flock 1967), *Coryphenoides* (Fänge, Larsson, and Lidman 1972)).

The functional consequences of canal width and these other structures have been examined in an elegant series of theoretical predictions and empirical measurements by Denton and Gray (1988, 1989). By measuring the amplitude of water displacements inside and outside capillary-tube models of canals and real canals in bony fish, they have determined the filtering properties of canals with different diameters and different compliances. Since the response of the neuromast is proportional to inside velocity which in turn is proportional to outside acceleration (for rigid canals), we have used their mathematical model to generate filter functions showing the effects of canal diameter on the ratio between outside acceleration and inside velocity levels (top panel, Fig. 15.6) and between outside velocity and inside velocity levels (bottom panel, Fig. 15.6). With respect to outside acceleration levels, rigid canals have the effect of decreasing sensitivity to lower frequencies as canal diameter decreases and

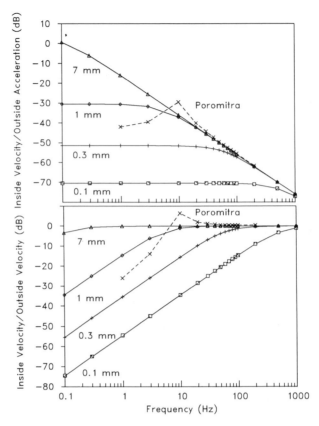

FIGURE 15.6. Filtering properties of canals of different diameters, based on the equations of Denton and Gray (1988, 1989) for rigid-walled tubes (solid lines). Dashed-line functions for *Poromitra* based on both empirical and theoretical data from Denton and Gray (1989) for a canal, 7 mm in diameter with a membranous covering. Bottom graph depicting the ratio of inside velocity to outside velocity as a function of frequency shows that for a canal of 7 mm, the widest canal found in fish, this ratio is nearly 1 (or 0 dB, 20 × Log(ratio) for all frequencies. For canals of smaller diameters, this ratio changes at the rate of about 6 dB/frequency doubling over a large portion of the frequency range. This rate of change is equivalent to saying that the ratio of inside velocity to outside acceleration is constant, as shown in the top graph. The "view" of filtering properties shown in the top graph is more consistent with our understanding of canal neuromasts as acceleration-sensitive, showing that acceleration-dependent responsiveness is independent of frequency for much of the frequency range. As the canal diameter increases, the responsiveness (or sensitivity) to lower frequencies increases and the range over which acceleration sensitivity remains constant shrinks.

increasing the bandwidth over which the response of the system is proportional to acceleration. At a canal diameter of 7 mm (the widest canal diameter reported in fish), the canal has essentially no effect, with the ratio of outside-to-inside velocity being 1 (0 dB, Fig. 15.6b) over most of the frequency range.

Given that the widest canals in bony fish are frequently covered by soft tissues that are much less rigid than bone, it has been suggested that this more compliant tissue might behave as a tympanum with low-frequency resonant characteristics (Denton and Gray 1988). In fact, the response of such a system, modeled after 7 mm wide head canals in the deep water fish, *Poromitra*, indicate a resonance around 10 Hz (Fig. 15.6, dashed lines). Based on this measured and predicted resonance, Denton and Gray (1988) have estimated that *Poromitra* would be about 8 times more sensitive for a given signal than a fish of the same size with smaller bony canals (the sprat) in the frequency range of 5 to 15 Hz. Add to this resonance the enhanced sensitivity expected from a larger neuro-mast with increased hair cell/nerve fiber ratios, steeper velocity gradients along the canal wall, and a cupula with a low profile (all features associated with the widened canal in *Poromitra*), and it is predicted that *Poromitra* could be up to 100 times more sensitive than the sprat in this narrow fre-quency range (Denton and Gray 1988). Although increased sensitivity at low frequencies appears to be an advantage of systems like the one in *Poro-mitra*, there are two major drawbacks. One is that enhanced sensitivity is relatively useless in envi-ronments with high levels of low-frequency noise, since signals would be masked. Another is that enhanced sensitivity through resonance results in poorer time resolving capabilities (see E. Lewis, Chapter 11).

In terms of frequency response, an increase in canal width can be viewed as approaching the fre-quency response expected and measured for super-ficial neuromasts (e.g., see response of 7 mm canal in Fig. 15.6). From a functional point of view, then, widening of canals and the complete reduction (loss) of canals (and hence transition to superficial neuromasts) could be regarded as converging on the same functional endpoint, resulting in a system whose response is proportional to velocity over a bandwidth extending down to lower frequencies. Adding resonance to the system via a membrane-like covering on the canal further enhances sensitiv-ity over a narrow band of low frequencies. In lieu of what appears to be functional similarities, it is interesting to note that proliferation of superfi-cial neuromasts is found in close association with widened head canals in at least some species of fish where this has been looked at (Moore and Burris 1956; Schemmel 1977). Furthermore, fish having a preponderance of superficial neuromasts tend to be sedentary and/or to live in relatively still (lentic) waters (Dijkgraaf 1963; Marshall 1971; Merri-lees and Crossman 1973a), as do percid fish with widened canals (Jakubowski 1967).

Very little information exists on the behavioral relevance of widened canals to *Poromitra* or to any other species that have them, although it has been suggested that the head canal system in *Poromitra* could be used for invertebrate prey detection (Den-ton and Gray 1988). In general, there appears to be a high incidence of widened canals in deep-water species or in species living in other habitats with low light levels and relatively still water conditions (Marshall 1971). These correlations suggest that ambient noise characteristics (i.e., the flow charac-teristics of the surrounding water) and light levels may play significant roles in the evolution of both superficial neuromasts and widened canal systems. The mechanisms underlying such changes and the validity of using correlations like these to come to conclusions about selective pressures in the evolu-tion of the lateral line system will be addressed in the following two sections.

5. Developmental Mechanisms in the Evolution of the Lateral Line System

It is likely that all, if not most variation in lateral line endorgans, can be accounted for, mechanisti-cally at least, by heterochrony, or changes in the timing of developmental events (Northcutt 1989, Chapter 3; Webb 1989a). In all developing lateral line systems studied so far, the first neuromasts to be laid down are differentiated from a migrating primordium originating from epidermal thickenings called placodes in the developing embryo (e.g., Stone 1933; Metcalfe 1989; Northcutt 1989). As such, these first or "primary" neuromasts are super-ficially located in the epidermis. Only in fish, do some of these primary neuromasts become enclosed in canals later in development. Canal formation involves at least three steps, including invagination of the epidermal neuromast into the dermal layer, ossicle (bone or scale) formation in the vicinity of the neuromast, and ossicle fusion between neuro-masts (Webb 1989a). Through an alteration or trun-cation in the process of invagination and canal for-mation, many lines of superficial neuromasts (e.g., the anterior, middle and posterior pit lines on the head, Fig. 15.2) may very well have had their evolu-tionary origin as neuromasts in canals or grooves (Northcutt 1989, Fig. 15.2).

In teleost fishes, but apparently not in amphibians or more primitive fishes (Northcutt 1989), "second-ary" superficial neuromasts may appear later during larval and juvenile stages, either by budding off from primary neuromasts or by arising *in situ* from the basal layer of the epidermis, possibly under neu-ral induction (Allis 1889; Lekander1949; Disler

1960; O'Connell 1981). According to Lekander (1949), these secondary neuromasts never become canal neuromasts. It is quite likely that superficial neuromasts occurring just above or below a canal and appearing to follow the same course as the canal (the so-called "accessory lines" of Lekander 1949; and Coombs, Janssen, and Webb 1988) have arisen as secondary neuromasts during development, whereas those that appear as a continuation of an existing canal or in positions normally occupied by canals in closely related species (the so-called "replacement lines" of Nelson 1972; and Coombs, Janssen, and Webb 1988) have arisen as primary neuromasts. Since the orientation of hair cells within the sensory strip of canal neuromasts is always (so far as we know) aligned with the canal axis, but those of adjacent superficial neuromasts in accessory lines commonly have orthogonal orientations (Münz 1979; Marshall 1986; Janssen, Coombs, and Hoekstra 1987; Vischer 1989; Hama 1978), the orientation of superficial neuromasts may provide an important clue as to whether any given superficial line arose as primary canal neuromasts or as secondary superficial neuromasts.

The mechanism of secondary neuromast addition, apparently newly evolved in teleosts (Northcutt 1989), may account for the widely different patterns and numbers of superficial neuromasts found in teleost fishes (Coombs, Janssen, and Webb 1988; Münz 1989; Northcutt 1989; Webb 1989b). It may also account for the fact that superficial neuromasts, but not canal neuromasts, appear to be capable of postembryonic proliferation (Peters 1973; Münz 1986). It is quite likely that the proximal mechanisms for some or nearly all of these phenomena will be more clearly understood as we begin to gather more information on cellular mechanisms of neuromast development, as has been done with *Xenopus* (Winklbauer and Hausen 1983a,b; 1985a,b).

Although it is relatively easy to see how alterations in the timing of canal formation could account for canal reductions (see Webb 1989a for further detail) and how a postembryonic mechanism for the addition of superficial neuromasts could account for superficial neuromast proliferation, a similarly simple hypothesis for mechanisms underlying the evolution of widened canals has not been articulated. Such a mechanism must account for the primary growth of neuromasts across the width of the canal, rather than along its length and for various other structural modifications of the canal, such as incomplete coverage by bone. It is quite likely that the order, rate, and onset and offset times of canal formation is involved in the development of some or all of these features (Webb 1989a). One can imagine, for example, that since neuromasts reportedly have morphogenetic or at least trophic influences on bone and scale formation (Kapoor 1970; Merilees and Crossman 1973b; Graham-Smith 1978), a simple truncation of the canal-forming process after bone formation in the vicinity of the neuromast could easily account for the incomplete coverage by bone common to widened canals.

The cellular mechanisms determining the axis (orientation) and rate of growth of neuromasts in teleost fish are presently unknown, although there is some information on growth mechanisms operating in *Xenopus* (Winklbauer and Hausen 1983a,b; 1985a,b). Whatever mechanisms determine neuromast orientation and growth in fish, however, may not determine the orientation of hair cells on the neuromast. In the few cases where hair cell orientation has been examined, it is always parallel to the long axis of the canal, regardless of the orientation of the neuromast within the canal (Coombs, Janssen, and Webb 1988; Vischer 1989). Clearly, much more information is needed on the cellular mechanisms underlying the development of widened canals before the full role of heterochrony in the evolution of widened canals can be evaluated.

6. Selective Pressures in the Evolution of Lateral Line Systems

In preceding sections, we have tried to describe some of the functional consequences of anatomical diversity in the lateral line system in fishes. Trying to infer convergence of function from anatomical similarities in diverse taxa has several pitfalls, however. One is that structures that look similar do not necessarily perform similar functions. It is important to have quantitative and systematic descriptions of the response properties of the system and furthermore, information on the biological relevance of these response properties to the animal. We have a fairly good handle on the response properties of superficial neuromasts and neuromasts in narrow canals and a modest idea of the response properties of neuromasts in widened canals, but rather sketchy information on how these various "subsystems" are actually used by the animal. Secondly, the fact that certain anatomical trends tend to occur over and over again in phylogenetically unrelated taxa does not necessarily mean that the driving force for such trends is active selection on the trait in question (Northcutt 1988). Morphogenetic interactions and developmental mechanisms must be taken into consideration (Gould and Lewontin 1979) and there are a number of examples of how selection for nonlateral-line traits (e.g., small body size through retention of larval characteristics) linked to lateral-line traits (e.g., reduction in canals), might be equally or more compelling than adaptive explanations of structural diversity in the lateral line system (Northcutt 1989; and Webb 1989a). Finally the role that direct environmental influences, such as water temperature, pH, etc., play in determining the diversity of lateral line structures is largely unknown (but see Münz 1986).

With these caveats in mind, we give two examples of where it might be profitable to look for functional adaptation as the primary driving force in the evolution of lateral line diversity. Our criteria for choosing these examples are that (1) we have established, at least to some degree, the functional properties of the lateral line trait in question, (2) we have some information on the ecology and behavior of the animals having different expressions of that trait, and (3) diversity in the trait is expressed in closely related species for which a phylogeny, based on a recent cladistic analysis, has been proposed.

6.1 The Evolution of Widened Head Canals in Percid Fishes

Within the teleost family Percidae, Jakubowski (1967) has identified several genera of fish that have widened canals with neuromasts oriented across the canal and several that have narrow canals with neuromasts oriented along the canal. A recent cladistic analysis of the Percidae tells us that species in the most primitive percid genus had narrow canals and parallel neuromasts, as do many of the putative sister groups (e.g., Centrarchidae) (Fig. 15.7, Wiley, in press). Most importantly, it shows us that the distribution of genera with widened canals (e.g., *Gymnocephalus*, *Zingel*, and *Percerina*) does not show any phylogenetic trends within the Percidae.

Although there are no data on the functional properties of neuromasts in the head canal system in *Percerina* or *Zingel*, there are considerable data from *Gymnocephalus* (alias the ruff *Acerina*), including information on summed extracellular (microphonic) responses (Kuiper 1956), on micromechanical properties of the cupula (Jielof, Spoor, and De Vries 1952; van Netten and Kroese 1987, 1989), on intracellular hair cell responses (Kroese and van Netten 1989), on response properties of afferent fibers (Denton and Gray 1989; Wubbels 1989) and on behavioral sensitivity (Kuiper 1967). Kuiper's data show best acceleration sensitivity at 10 Hz and Denton and Gray (1989) have shown a low frequency resonance (in terms of displacement sensitivity) around 10 to 15 Hz for afferent fibers that is similar to that shown for *Poromitra* (Fig. 15.6), suggesting once again that enhanced sensitivity over a narrow range of low frequencies may be one of the major functional consequences of widened head canal systems.

Although there is a good deal of information about the ecology and behavior of these percid fishes, there is little information about how these fish actually use their lateral line system. Jakubowski (1967) and Disler and Smirnov (1977) make the observation that percid fish with widened canals tend to be bottom dwellers living in slow currents or under conditions of low light levels (or with nocturnal feeding behavior), whereas fish with narrow canals tend to live in swiftly flowing waters or to be diurnal, visual predators. Although exceptions to these correlations can be found, they nevertheless point to low light levels and low-level, low-frequency noise characteristics of the environment as possible candidates for selection pressures in the evolution of widened canal systems. Clearly we need more specific information on how the lateral line system is actually used by these fishes before drawing any firm conclusions about precise selection pressures, but such a functional analysis within a phylogenetic context affords real possibilities for this endeavor.

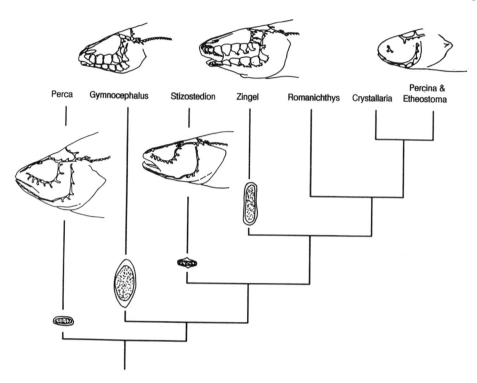

FIGURE 15.7. Proposed phylogeny of Wiley (in press) for percid fish. The shape and width of head canals and of neuromasts within canals are shown for each genus for which data exists (from Jakubowski 1967). The stippled area represents the extent of the hair cell epithelium and the outer line, the outer boundary of the neuromast (i.e., surrounding mantle cells). Neuromasts are oriented according to how they would actually sit in the major portion of the infraorbital and mandibular canals shown above for each fish.

6.2 The Evolution of Supernumerary Free Neuromasts in Amblyopsid Blind Cavefish

Probably no other family holds as much promise for understanding the role of selection in the evolution of the lateral line system as do the amblyopsids. The family Amblyopsidae consists of six known species in five genera: a noncave (epigean) species (*Chologaster cornua*) with eyes and pigment, a facultative cave (troglophilic) species (*Forbesichthys agassizi*) with reduced eyes and pigment, and several obligate (troglobitic) cave species (*Typhlichthys subterraneus*, *Amblyopsis spelaea*, *Amblyopsis rosae*, and *Speoplatyrhinus poulsoni*) with degenerate eyes and no pigment (Cooper and Kuehne 1974). This family has been studied extensively with respect to a wide range of factors, including genetic variability and biochemical relationships (Swofford, Branson, and Sievert 1980) and ecology, life history, metabolic rate, growth, morphology, and behavior (Poulson 1963). From quantitative and systematic measures of various indicators in the above categories, Poulson (1963, 1985) has identified two basic trends in the epigean-to-troglobite continuum which cannot be explained on the basis of phylogenetic affinities (Swofford 1976). One is a reducing trend in which certain traits (e.g., eye size, melanophore dispersion) were simply reduced or lost (Fig. 15.8). The second was what Poulson (1963) termed a troglomorphic trend in which certain traits appeared to vary systematically as a function of whether species were epigean, troglophilic, or troglobitic. Troglomorphic trends included an increase in head size, decrease in the absolute growth rate, an increase in longevity, and an increase in the age at first reproduction. With respect to the lateral line system, these trends include an increase in the number of superficial neuromast lines, an increase

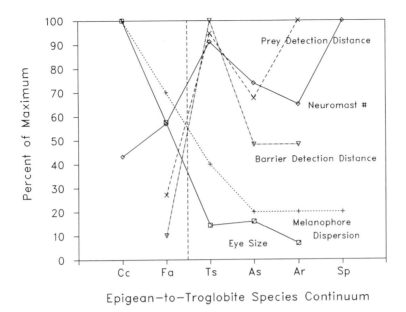

FIGURE 15.8. Reducing and increasing (troglomorphic) trends in 6 amblyopsid species: *Chologaster cornuta* (Cc), *Forbesichthys agassizi* (Fa), *Typhlichthys subterraneus* (Ts), *Amblyopsis spelaea* (As), *Amblyopsis rosae* (Ar) and *Speoplatyrhinus poulsoni* (Sp). Raw data on eye size, melanophore dispersion (measured on a scale from 1 to 5), neuromast # (from three superficial lines found in the nares region of all species), ability to avoid barriers (distance at which fish turned away from a novel barrier), and ability to detect prey (distance at which fish detected *Daphnia magna*) have been taken from Poulson (1963, 1985) and converted to percentage of maximum values. Dashed line indicates the transition from facultative to obligate cave dwelling species.

in the number and elevation of superficial neuromasts (i.e., neuromasts on papillae) within a line and an increase in the ability of fish to avoid obstacles and capture invertebrate prey (Fig. 15.8).

Based on an analysis of the reducing and troglomorphic patterns across the epigean-to-troglobitic continuum, Poulson (1985) concluded that reducing trends were not pleiotropically linked to troglomorphic trends and that reducing trends could be regarded as selectively neutral. He further concluded that many troglomorphic trends in amblyopsid fish could best be explained as adaptations in response to limited food supplies in the cave environment. Whether or not these conclusions are correct, the complex nature of the analysis and the comprehensive data base on which it is based illustrates the difficulty in determining how or if selection is operating. Even assuming that Poulson's analysis is correct, we are still left with the problem of whether superficial neuromast proliferation was directly selected for or simply linked to some other troglomorphic trend (e.g., an increase in

head size). Until we have more information on the developmental and cellular mechanisms underlying such changes and the genetic linkage between traits, such questions are unanswerable. Clearly, however, the increased ability of cave-adapted species to detect invertebrate prey (a lateral line ability documented for many different species; see Montgomery 1989 for review) and to avoid obstacles speaks directly to the functional advantages that would be accrued by fish with an hypertrophied lateral line system.

7. Lateral Line and Inner Ear Stimuli—From Incompressible Water Flow to Propagated Pressure Waves

In an historical review on the sensory function of the lateral line system, Sand (1981) cites a description by Parker (1904) on the views held by

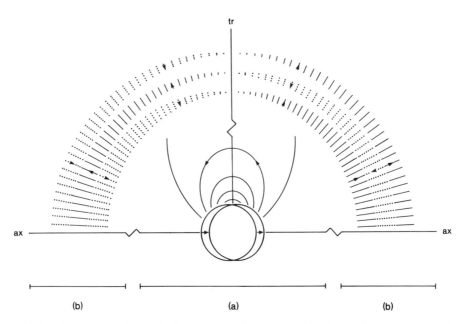

FIGURE 15.9. Flow field surrounding a dipolar source. (a) Local flow region near the source where the incompressible flow dynamics predominate. Arrows along the axis (ax) of sphere vibration show the displacement of the sphere from left to right and the resulting direction of flow around the sphere during this phase of sphere vibra-tion. (b) Far field flow region where propagated pressure waves predominate. The amplitude of pressure fluctuations (indicated by the extent of solid (compression) and dashed (rarefaction) lines) are maximum along the axis of vibration and zero along the transverse (tr) axis. (From Kalmijn 1988.)

scientists at the turn of the century: "the majority of investigators disagree, some maintaining that the lateral line organs are simply organs of touch (Merkel 1880; de Sede 1884), others that they are organs belonging to an independent class, probably intermediate between touch and hearing (Leydig 1850, 1851; Schulze 1870; Dercum 1880) and lastly, those that believe them to be accessory auditory organs (Mayser 1881; Bodenstein 1882)." Sand then proceeded to observe how amazingly well that statement, then written 76 years ago, described the situation at the time of his review.

Fortunately, recent theoretical (Kalmijn 1988, 1989) and empirical (Denton and Gray 1982, 1988; Gray 1984) work has significantly clarified our understanding of the stimuli to which the lateral line responds. Based on this understanding, most investigators today would probably agree that lateral line perception is intermediate between touch and hearing, as was first proposed by Knox (1825) and later championed as "distance-touch" by Dijkgraaf (1963). Given that both the lateral line and otolithic ears of fishes can respond to low frequency, nearby

"sounds" produced by an underwater loudspeaker, however, it is important to describe the way in which both systems can respond to the same hydrodynamic stimulus. It is the way in which the stimulus is transduced that leads to a clearer understanding of where lateral line perception falls on the touch–hearing continuum.

Many of the biologically relevant stimuli likely to be transduced by either the lateral line system or the otolithic ear of fishes can be approximated by a dipolar source like a vibrating sphere (Kalmijn 1988, 1989). As a vibrating sphere is displaced from its starting position, pressure is elevated in front of the source and reduced in the rear, resulting in an incompressible flow that predominates close to the source and compressions and rarefactions that predominate farther away. The path of water as it is pushed out in front of the advancing sphere and as it returns to the back of the sphere along the pressure gradient is shown by the local flow lines in Figures 15.9. The incompressible flow or "bulk" movement of water represented by these flow lines can be characterized in terms of there vector components: displacement, velocity or acceleration, with acceleration being proportional to the pressure gradient (Denton and Gray 1988). Farther from the source, there is very little bulk movement of water, but rather a propagated pressure wave that consists of radially directed water oscillations. The local water movements or oscillations in the pressure wave can also be measured in terms of displacement, velocity or acceleration and additionally, the pressure wave can be measured by a pure scalar quantity, pressure.

One of the consequences of fluids, such as air and water, that are neither completely nor incompletely compressible is that the hydrodynamic principles governing particle motion change as a function of distance. This is because totally incompressible flow generated by a dipolar source attenuates at the rate of $1/distance^3$, whereas the amplitude of particle motion associated with pressure changes attenuates as $1/distance$. There is a third hydrodynamic principle associated with dipolar sources, called intermediate flow. Although this phenomenon is difficult to conceptualize as a separate entity, it is readily described mathematically as attenuating at the rate of $1/distance^2$ (Kalmijn 1988, 1989). The different rates of attenuation with distance result in a picture (Fig. 15.10) for which incompressible flow dynamics essentially govern up to distances of $\lambda/2\pi$ ($\lambda = 1$ wavelength, defined as c/f, where $f =$ frequency of sinusoidal sphere vibration and c = speed of sound in water), intermediate flow in the region from $\lambda/2\pi$ to λ/π, and propagated pressure changes at distances beyond λ/π (the so called far-field, as defined by van Bergeijk (1964) and Siler (1969)) (Fig. 15.2).

At this stage, one might be tempted to distinguish the lateral line system from the fish ear as a detector of incompressible flow, rather than propagated pressure waves. But this is not entirely correct, because both the otolithic ear and the lateral line of fish are capable of responding in the near field where incompressible flow dynamics predominate (Kalmijn 1989). Furthermore, all otolithic ears of fishes can presumably operate in a motion-transducing mode (e.g., Myrberg and Spires 1980; Chapman and Hawkins 1973), but only fish with special pressure-transducing air cavities linked with the ear are capable of detecting pressure changes per se (see Schellart and Popper, Chapter 16; Popper and Coombs 1980a,b for overview).

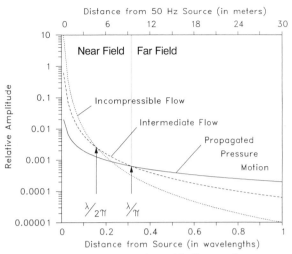

FIGURE 15.10. Flow attenuation with distance from a dipolar source for incompressible flow, intermediate flow, and propagated pressure flow. Distances at which intermediate and propagated pressure flows begin to predominate are marked with arrows.

For fish, this may not be such a big problem, since their "acoustic" world is predominantly confined to the near field. This is because the actual distance at which pressure dynamics start to govern depends on both frequency and the compressibility of the medium: the lower the frequency and the more incompressible the medium, the larger the extent of the near field. The hypothetical near field/far field boundary for 10 Hz and 100 Hz sources, for example, is 47.8 and 4.8 m respectively in water, but 11 and 1.1 m respectively in air. Since many underwater sounds of biological origin are relatively low in frequency (see Schellart and Popper, Chapter 16 for review), there may have been few selective pressures to evolve far field detection capabilities in fish.

If the detection of propagated pressure waves vs the detection of incompressible flow cannot distinguish auditory from lateral line function, what can? The answer is that the response of the lateral line depends upon differential movement between the fish and the surrounding water, whereas the otolithic inner ear does not. This difference is illustrated in Figure 15.11, which shows the relative motion of the fish and the surrounding water when the fish is close to (e.g., inner near field) or far away from (e.g., far field) a vibrating source. The middle panel (B) shows that in both the near and

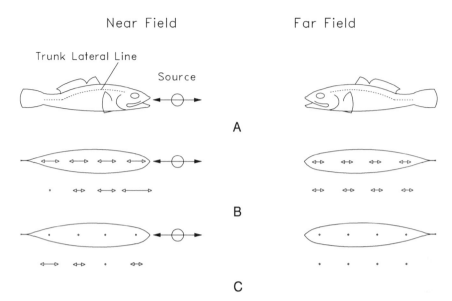

FIGURE 15.11. Diagram showing differential movement between fish and surrounding water in the near and far field. (A) Fish facing a dipolar source of vibration in the near (left) and far (right) field. (B) Dorsal view of the fish showing the amplitude of vibration (lines with double, open arrowheads) of the fish and the adjacent water. In both near and far fields, the amplitude of fish vibration is constant along the length of the fish. The amplitude of adjacent water vibration falls off as 1/distance$_3$ in the near field and 1/r (shown here as negligible along the length of the fish) in the far field. Surrounding water motion in (C) illustrates the magnitude of relative motion arising in (B) with respect to a motionless fish. In the near field, large net movements, of opposite polarity, will be seen at the head and tail and no net movement somewhere near the center of the body. This figure illustrates two extremes of a continuum—the inner field, where net movements can be substantial, depending on the length of the fish, amplitude of the source etc., and the far field, where there are essentially no net movements. Net movements may be negligible and undetectable in the outer near field as well. (From Coombs and Janssen 1990, adapted from Denton and Gray 1988.)

far field, the fish moves as a rigid body (Denton and Gray 1982), meaning that the amplitude of fish movement (indicated by the length of double-headed arrows on the body of the fish) is equal and in phase at every point along the fish's body. The amplitude of surrounding water motion, however, attenuates steeply (i.e., at the rate of 1/distance^3) in the near field, but much more gradually (at the rate of 1/distance) in the far field (shown here as negligible along the length of the fish). The bottom panel of Figure 15.11 shows the movement of the surrounding water relative to that of the fish. This figure illustrates that in the far field, where attenuation of water motion with distance is gradual and the distance of the source is great relative to the length of the fish, there is little if any differential movement between the fish and the surrounding water. In the near field, however,

where the attenuation of water motion is steep, differential movement between the fish and the surrounding water is substantial and may result in complicated amplitude and phase patterns along the length of the fish (Denton and Gray 1988).

The outcome of near and far field differences in relative motion between fish and surrounding water is that it is simply not possible for the typical lateral line system to respond in the far field. Otolithic ears, on the other hand, can respond in both near and far fields because all that is required is that the fish (and hair cell epithelium) be accelerated. The inertia of the denser otolith, tied to the ciliary tips by the otolithic membrane, causes the differential movement between otolith and hair cells necessary to cause bending of the cilia and generation of receptor potentials. A second, pressure-transducing pathway, exists in the far field for some fish (see Schellart and Popper, Chapter 16). In this pathway, pressure changes cause the swimbladder to expand and contract, owing to the greater compressibility of the gas in the bladder. This motion is then transmitted to the fluids of the inner ear, either indirectly through intervening tissue or by a direct mechanical coupling, like the Weberian ossicles in otophysans (e.g., goldfish), a series of modified vertebrae considered to be analogous to the middle ear bones of mammals.

In the lateral line system, however, the hair cell cilia project into a gelatinous cupula, presumed to be of approximately the same density as the surrounding water. Since the cupula and surrounding water are coupled by viscous forces, the cupula and cilia will not move relative to the fish unless there is relative movement between the fish and surrounding water. Moreover, since the amplitude of relative or net motion attenuates at a rate close to $1/distance^4$ (Denton and Gray 1988; Kalmijn 1988, 1989), this means that the effective lateral line stimulus diminishes even more steeply with distance than does the incompressible flow of water alone. For this reason, Kalmijn (1988, 1989) has concluded that lateral line function should be restricted to the inner most regions of the near field, whereas auditory function could be extended to greater distances — much farther out in the near field and possibly into the far field, especially for systems with auxillary structures capable of transducing pressure changes.

8. Operational and Functional Trends in the Evolution of the Lateral Line and Inner Ear

Based on our current understanding of how octavolateralis systems of fish are stimulated, it is possible to think of their principles of operation as ranging from being motion (displacement–velocity–acceleration) detectors to absolute pressure detectors. Whereas hair cells operate as displacement transducers, the lateral line system, as a whole, behaves as either a velocity transducer, in the case of superficial neuromasts, or as an acceleration transducer, in the case of neuromasts in narrow rigid canals. A similar dichotomy exists for the vestibular portion of the fish ear for which both

velocity- and acceleration-sensitive fibers have been reported to innervate the horizontal semicircular canal (Boyle and Highstein 1990). The otolithic inner ear of fish also responds to acceleration, but of the whole animal with respect to a single point, rather than to the difference between the acceleration of the animal and the surrounding water, as does the lateral line canal system (Kalmijn 1988, 1989). Put in a slightly different way, the ear, organized as a single cluster of hair cells, can operate in a spatially uniform stimulus field (as well as a spatially nonuniform field; Kalmijn 1988, 1989), whereas the response of the lateral line system, organized as numerous dispersed patches of hair cells, is limited to the spatial nonuniformities in the stimulus field (Peters and Buwalda 1986). Whereas the lateral line canal system operates essentially as a pressure-gradient detector, as do many invertebrate and small vertebrate auditory systems (Michelson, Chapter 5), the inner ear of fish can additionally operate as a pressure transducer in fishes which have compressible air cavities near the ear. As the principle of operation changes from velocity- to acceleration- to pressure-transducer, there is a corresponding increase in the sensitivity of the system to higher frequencies (see also Schellart and Popper, Chapter 16), leading to an increase in the overall bandwidth of detection (Fig. 15.12). Tuning curves (expressed in terms of acceleration) from fibers innervating the lateral line system can be thought of as relatively simple, low pass filters with those innervating superficial neuromasts having a high-frequency cut-off no higher than around 30 Hz and those from canal neuromasts with a higher cut-off of around 100 Hz. Saccular fibers from ears with no specialized pathway to a pressure transducing cavity, such as those in the pumpkinseed sunfish, *Lepomis gibossus* may respond up to 200 to 300 Hz (Coombs, unpublished data) and those innervating a pressure-sensitive ear, like in the goldfish, *Carassius auratus* up to around 1000 Hz (Fay and Ream 1986).

Behavioral measurements of threshold-detection curves show similar trends in terms of increasing bandwidth. Best threshold detection is only as high as 100 Hz for surface-wave detection by the lateral line system of the topminnow, *Aplocheilus lineatus* (Bleckmann 1980) and of the mottled sculpin, *Cottus bairdi* in response to a subsurface vibrating

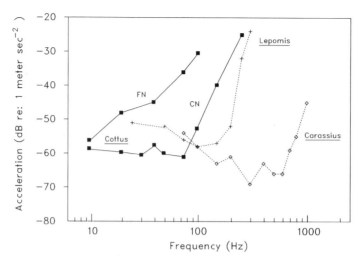

FIGURE 15.12. Tuning curves from afferent fibers innervating different octavolateralis organs in fish. Dashed lines
represent tuning curves from saccular fibers in two different species of fish—one with pressure-sensitivity (the
goldfish, *Carassius auratus*, Fay (1981) and one without
(the pumpkinseed sunfish, *Lepomis gibbosus*, Coombs
and Fay, unpublished data). In both cases, the otolithic ear
was stimulated directly by vibrating the whole fish, thus

bypassing the normal pressure-transducing step of the
swimbladder in these fish. Solid lines represent tuning
curves from posterior lateral line fibers innervating both
free (FN) and canal (CN) neuromasts on the trunk of the
mottled sculpin, *Cottus bairdi* (from Coombs and Janssen
1989). Tuning curves were obtained in response to the
vibration of a small sphere located 1.5 cm away from the
lateral line canal on the trunk.

sphere (Coombs and Janssen 1989), but up to
1,000 Hz for acoustic pressure thresholds in the
goldfish (Fay 1969; Popper 1971). Moreover,
although the lateral line system has at best only two
differently tuned populations of fibers, behavioral
detection can be accounted for, both in terms of
bandwidth and absolute sensitivity, on the basis of
information from a single population of fibers (Fig.
15.4, Coombs and Janssen 1989; Bleckmann and
Topp 1981; Topp 1983). Although there are saccular fibers from the goldfish ear as broadly tuned as
the behavioral audiogram, there are a total of
between 2 and 4 differently tuned fiber populations
innervating this pressure sensitive endorgan (Fay
and Ream 1986). This compares to between 10 and
30 frequency-tuned channels likely to exist in the
peripheral nervous system of pressure-receptive
tetrapod ears. In these auditory systems, information must be integrated across a wide range of differently tuned fibers in order to account for the
bandwidth of behavioral detection (Fig. 15.13).

A second outcome of the operational differences
between the lateral line and ear of fishes is that the
distance at which signals can be detected increases.

The lateral line system detects water motion in the
inner regions of the near field, the unaided otolithic
ear detects motion farther out in the near field, and
the pressure-aided otolithic ear detects motion well
into the far field (Kalmijn 1988, 1989).

Surprisingly, there are very few data that actually establish the relative working distances of different hair cell systems. One of the obvious problems with comparisons of this type is that the
working distance will depend on source amplitude,
frequency and the sensitivity of the system. In the
real world, the working distance is also affected by
background noise levels and the sound transmission characteristics of the environment. In a controlled lab situation in which signal losses due to
interfering noise levels, absorption and scattering
are minimized, there are two ways to get around
the problem of source amplitude and frequency.
One is to measure the rate at which stimulus amplitude must be incremented as a function of distance
in order for the stimulus to be detected. These rates
can then be compared to attenuation rates for the
amplitudes of incompressible flow, intermediate
flow and propagated pressure changes. Another is

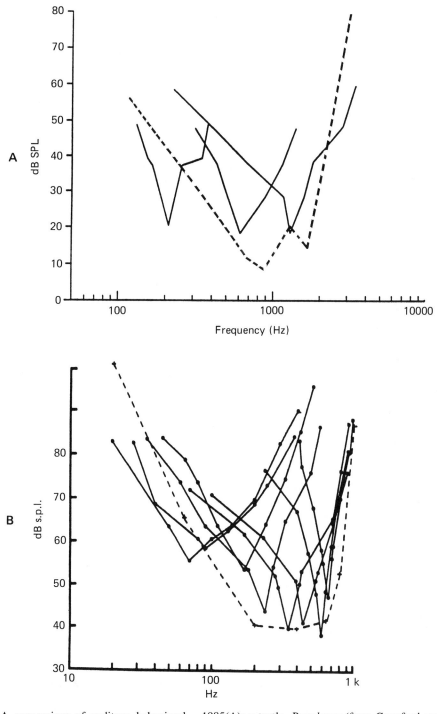

FIGURE 15.13. A comparison of auditory behavioral threshold curves (dashed lines) and neural tuning curves from different fibers (solid lines) in three species: the bullfrog, *Rana catesbeiana* (based on data from Megala 1984 and Megala–Simmons, Moss and Daniel 1985(A); a turtle, *Pseudemys* (from Crawford and Fettiplace 1980) (B); and the cat, *Felis catus* (from Coombs and Janssen 1989, based on data from Kiang 1965 and from Heffner and Heffner 1985).

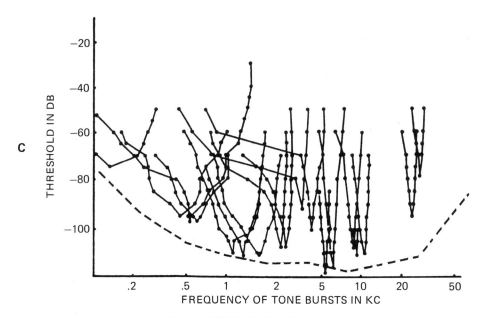

FIGURE 15.13. Continued

to measure the average distance or "threshold" distance at which an animal can detect a biologically relevant stimulus that is reasonably fixed in amplitude and spectral properties.

Both approaches have been used in estimating the working distance of the lateral line system in the mottled sculpin, *Cottus bairdi*. In making these estimates, the unconditioned and naturally occurring feeding behavior of the mottled sculpin was used as an indicator of detection by the lateral line system (Coombs and Janssen 1989). Preliminary results from the first approach indicate that the vibration amplitude of a sphere (6 mm in diameter) must be increased at the rate of between 12 and 18 dB/distance doubling for a wide range of stimulus frequencies in order for the mottled sculpin to report a just detectable stimulus in the near field (Coombs and Janssen 1990). This is equivalent to signal attenuation between the rates of 1/distance^2 and 1/distance^3. The interpretation of these results is made somewhat difficult by the fact that this benthic fish, being denser than water, may be more coupled to the substrate than to the surrounding water. If the fish were completely locked to the substrate, then only the surrounding water, and not the fish, would move in response to the vibrating sphere. This would mean that the effective stimulus for the lateral line system in this case

would attenuate at the rate of incompressible flow alone, or at 1/distance^3, and not at 1/distance^4, the predicted rate if both fish and water moved in response to the incompressible flow. Regardless of how these results are interpreted, however, they clearly show that the lateral line of the mottled sculpin is not following pressure changes (6 dB/distance doubling or 1/distance) and that the effective stimulus declines rather steeply with distance.

For the second approach, the distance at which the mottled sculpin could detect live *Daphnia* (2 mm in length) 50% of the time (Coombs and Janssen 1989) was measured to be 7 mm near the head, the most sensitive region of the animal's body. This is very similar to the distance of *Daphnia* detection by noncave amblyopsid fish (10 mm), but somewhat less than that reported for cave species (20 to 40 mm) (Poulson 1963). Detection of much larger, vertebrate (fish) prey by the bluegill, *Lepomis machrochirus* has been reported out to 20 to 30 mm by Enger, Kalmijn, and Sand (1989). When expressed in terms of body lengths, however, these distances are all less than one body length of the fish. Even the greatest distance at which a 7 to 10 cm mottled sculpin can detect a 6 mm vibrating sphere (at its maximum vibration amplitude) is still less than one body length.

Unfortunately, comparisons of working distances between the auditory and lateral line system of fish are difficult to make, primarily because monopolar, rather than dipolar sources have been used to measure acoustic thresholds in terms of sound pressure levels rather than water motion levels. Nevertheless, changes in detection thresholds with distance indicate that auditory sensitivity is proportional to either water motion (attenuating at the rate of $1/distance^2$ for a monopolar source) or pressure ($1/distance$). Motion sensitivity usually occurs in the near field at low frequencies (< 50 Hz) and pressure sensitivity typically at higher frequencies in fish with pressure-transducing cavities (Enger 1966; Chapman and Sand 1974; Chapman and Hawkins 1973). Moreover, acoustic thresholds have been measured out to several meters (corresponding to 3 to 10 $\times$ to body length of the fish) from the source for fishes lacking pressure-sensitive structures (Enger and Anderson 1967; Chapman and Sand 1974) and as far away as 50 m (100$\times$ the body length) for fish that are believed to be pressure sensitive (Chapman and Hawkins 1973). Despite the paucity of existing data and problems of interpretation, the current data seem to be consistent with what Kalmijn (1988, 1989) has put forward as a general rule of thumb: the lateral line system does not extend much further than one or two body lengths of the fish. We might add to this that the unaided otolithic ear seems to increase this capability by one order of magnitude and the pressure-sensitive ear by another order.

Although the occurrence of pressure-sensitive ears in diverse fish taxa has been recognized for some time (see Popper and Coombs 1982; Schellart and Popper, Chapter 16 for review), it is not widely known that pressure-sensitive lateral lines may also exist (see Coombs, Janssen, and Webb 1988 for review and Bleckmann et al. this volume). Of the possibilities that exist, the best known and most thoroughly described is that existing in clupeomorph fishes, where a thin flexible membrane (recessus lateralis) separates the perilymphatic fluids of the inner ear in the cranial cavity and the sea water outside the cavity in lateral line head canals (see Blaxter, Denton, and Gray 1981 for review). Pressure-induced changes in a gas-filled bulla associated with each ear are transmitted through the perilymphatic fluid and to the lateral recess membrane, making it possible for neuromasts in the canal to respond to pressure changes from the bulla and/or to water acceleration outside the canal.

Thus, the ability to detect both particle motion and pressure may have evolved independently in a number of taxonomically unrelated groups for both the ear and the lateral line system of fish. Gray (1984) has shown that significant changes in the relative excitation of different lateral line canal neuromasts can result from mixing the pressure-induced and motion-induced signals inside the infraorbital canal of the sprat. It has been suggested that these changes could encode information about the distance of a source, about sudden changes in the acceleration of the source, and about small changes in its relative position (Gray 1984). Similar suggestions have been made for the comparison of pressure- and motion-induced responses by the otolithic ear of fishes (see Schellart and Popper, Chapter 16, for review).

9. Functional Overlap Between the Lateral Line and Auditory System

Since the lateral line and auditory system of fish have been traditionally described as sharing common central pathways, it is worth pointing out here that more recent anatomical and physiological studies generally show a segregation of pathways, at least below the level of the midbrain (McCormick 1989) where integration of visual, as well as lateral line and auditory input takes place (Schellart and Kroese 1989). For further information on central processing and possible cites of overlap, the reader is referred to the excellent reviews of McCormick (1989) and Schellart and Kroese (1989).

Although there are very few psychophysical data on the sensory capabilities of the lateral line system relative to a much larger data base on the capabilities of the fish auditory system (Fay 1988), the following generalizations can be made:
(1) Contrary to earlier beliefs, the lateral line system is as sensitive as the fish ear and most other vertebrate auditory systems, with displacement sensitivity at best frequency in the nanometer range (Kroese and van Netten 1989).
(2) Although the upper limit of the detection range differs between ear and lateral line, there is no

evidence for a difference in the lower frequency limit. In fact, detection by the ear of the cod, *Gadus morhua*, extends down to 0.1 Hz (Sand and Karlsen 1986) and lateral line nerve fiber responses in the thornback ray, *Platyrhinoidis triseriata*, have been measured down to 1 Hz (Görner and Kalmijn 1989).

(3) Frequency discrimination by the lateral line system (based on surface wave detection by *Xenopus* (Elepfandt, Seiler, and Aicher 1985) and *Aplocheilus* (Bleckmann, Waldner, and Schwartz 1981) ranges from 4 to 15% of the base frequency, compared to between 3 and 20% for fish auditory systems (Fay 1988). Fish lateral line and auditory abilities in this task fall on the upper margins of the distributions for birds and mammals (varying between 0.1% and 10%) (Fay 1988).

10. Summary and Conclusions

Unlike the auditory portion of the ear and the closely related electrosensory system, the mechanosensory lateral line is believed to have arisen from a single common ancestor being present in nearly all anamniotic vertebrates. Although the proximal mechanisms in the evolution of diverse lateral line morphologies are likely to have involved changes in the timing of developmental events, the primary driving forces for these changes remain unknown. In order to begin to understand whether some of this diversity may have evolved as functional adaptations in direct response to selection pressures, we have examined three different morphological components of the lateral line system (superficial neuromasts, neuromasts enclosed in narrow rigid canals and neuromasts enclosed in wide canals with compliant walls) for which the functional properties have been measured. This examination revealed some clear differences in the functional properties of these subcomponents, but led to no generalizations of how these differences are used by different species in behaviorally relevant contexts. We then identified two teleost families, Amblyopsidae and Percidae, for which it was unlikely that variations between species in the relative development of or presence or absence of these components could be accounted for on the basis of phylogenetic relationships and for which it

might be possible to identify selective pressures. Of these two groups, the amblyopsid cavefish seemed to provide the strongest, but far from conclusive, evidence for selection of lateral line characteristics. The addition of new lines of superficial neuromasts and an increase in the number of neuromasts per line found in different obligate cave-dwelling species seems to have evolved as a functional adaptation for increased abilities to avoid obstacles and detect small invertebrate prey in an environment without light and with limited prey availability. Widened head canals in various percid species might also have evolved for enhanced prey detection in response to environments with low light levels or low-level, low-frequency noise characteristics.

Although there is no evidence presented here to suggest that the inner ear of fish evolved from the lateral line, the operation and functional properties of the two systems suggest a "functional" evolution of sorts, as suggested most recently by Kalmijn (1988, 1989) and summarized in Table 15.1. In general, the superficial and canal neuromasts of the lateral line system can be regarded as velocity/acceleration transducers, whereas the motion-sensitive ear (unaided otolithic ear) and the pressure-sensitive ear (otolithic ear aided by compressible air cavity) of fishes can be regarded as acceleration and pressure transducers, respectively. Pressure-sensitive systems have the advantage of being able to detect signals over longer distances and over a broader bandwidth. The peripheral lateral line system appears to be comprised of few frequency-tuned channels, but many different spatially tuned channels for the detection of spatially nonuniform stimulus fields very close to the source. Since the lateral line system responds only to low-frequency stimuli, information about the spectrum of signals can be obtained by temporal coding of the time waveform. As the bandwidth of detection increases and temporal coding of higher frequencies becomes limited by the refractory properties of the nervous system, hair cell systems must employ additional mechanisms, such as differently tuned filters, for extracting spectral information. Thus, the peripheral auditory system of most tetrapod vertebrates is comprised of many frequency-tuned channels for the detection of spectrally complex, but spatially uniform stimulus fields.

TABLE 15.1.

	Lateral Line System		Auditory System	
	Superficial	Canal	Otolithic ear	Otolithic ear + air cavity
Stimulus	Net velocity/acceleration		Whole animal acceleration	Pressure
	Spatially nonuniform field		Spatially uniform field	
Receptor distribution	Dispersed		Clustered	
Working distance	1 body length		10 Body lengths	100 Body lengths
Upper frequency limit	100 Hz		500 Hz	2,000 Hz
No. of frequency-tuned channels	1	1	1–2	2–4

Acknowledgments. Support for mottled sculpin research reported here came from the National Institutes of Health and from the Max Goldenberg Foundation. We would like to thank Dr.'s Jacqueline Webb and Tom Poulson for their comments and suggestions for improvement on an earlier draft of this chapter.

References

Abdel-Latif H, Hassan ES and C von Campenhausen (1990) Sensory performance of blind mexican cave fish after destruction of the canal neuromasts. Naturwissenschaften 77:237–239.

Allis EP (1889) The anatomy and development of the lateral line system in *Amia calva*. J Morphol 2 (3):463–542.

Allis EP (1895) The cranial muscle and first spinal nerves in *Amia calva*. J Morphol 12:489–808.

Applebaum S, Schemmel C (1983) Dermal sense organs and their significance in the feeding behavior of the common sole *Solea vulgaris*. Mar Ecol Prog Ser 13:29–36.

Ayers H (1892) Vertebrate cephalogenesis. J Morp 6:1–360.

Barry MA, Bennett MVL (1989) Specialized lateral line receptor systems in elasmobranchs: The spiracular organs and vesicles of savi. In: Coombs S, Görner P, Münz H (eds) The Mechanosensory Lateral Line: Neurobiology and Evolution. New York: Springer-Verlag, pp. 591–606.

von Bartheld CS, (1990) Development and innervation of the Paratympanic Organ (Vitali organ) in chick embryos. Brain Behav Evol 35:1–15.

van Bergeijk WA (1964) Directional and non-directional hearing in fish. In: Tavolga WN (ed) Marine Bioacoustics. Oxford: Pergamon Press, pp. 199–281.

van Bergeijk WA (1967) The evolution of vertebrate hearing. In: Neff WD (ed) Contributions to Sensory Physiology. New York: Academic Press, pp. 1–49.

Best A, Gray J (1982) Nerve fibre and receptor counts in the sprat utriculus and lateral line. J Mar Biol Assoc UK 62:201–213.

Blaxter J, Denton E, Gray J (1981) Acoustico-lateralis system in clupeoid fishes. In: Tavolga W, Popper A, Fay R (eds) Hearing and Sound Communication in Fishes. New York: Springer-Verlag, pp. 39–60.

Blaxter JHS, Gray JAB, Best ACG (1983) Structure and development of the free neuromasts and lateral line system of the herring. J Mar Biol Assoc UK 63:247–260.

Bleckmann H (1980) Reaction time and stimulus frequency in prey localization in the surface-feeding fish *Aplocheilus lineatus*. J Comp Phys 140:163–172.

Bleckmann H, Topp G (1981) Surface wave sensitivity of the lateral line organs of the topminnow *Aplocheilus lineatus*. Naturwissenschaften 68:624–625.

Bleckmann H, Waldner I, Schwartz E (1981) Frequency discrimination of the surface-feeding fish *Aplocheilus lineatus*—A prerequisite for prey localization. J Comp Phys 143:485–490.

Bleckmann H, Tittel G, Blubaum-Gronau E (1989) The lateral line system of surface feeding fish: Anatomy, physiology and behavior. In: Coombs S, Görner P, Münz H (eds) The Mechanosensory Lateral Line: Neurobiology and Evolution. New York: Springer-Verlag, pp. 501–526.

Bodenstein E (1882) Der Seitenkanal von *Cottus gobio*. Z Wiss Zool 37:121–145.

Boyle R, Highstein SM (1990) Resting discharge and response dynamics of horizontal semicircular canal afferents of the toadfish, *Opsanus tau*. J Neurosci 10:1557–1569.

Bullock TH (1974) An essay on the discovery of sensory receptors and the assignment of their functions together with an introduction to electroreceptors. In: Fessard A (ed) Handbook of Sensory Physiology, Vol. III/3. New York: Springer-Verlag, pp. 1–12.

Bullock TH, Northcutt RG, Bodznick D (1982) Evolution of electroreception. Trends Neurosci 5:50–53.

Bullock TH, Bodznick D, Northcutt RG (1983) The phylogenetic distribution of electroreception: Evidence for convergent evolution of a primitive vertebrate sense modality. Brain Res Rev 6:25–46.

Chapman CJ, Hawkins AD (1973) A field study of hearing in the cod, *Gadus morhua* L. J Comp Physiol 85:147–167.

Chapman CJ, Sand O (1974) Field studies of hearing in two species of flatfish, *Pleuronectes platessa* (L.) and *Limanda limanda* (L.) (Family Pleuronectidae). Comparative Biochemistry and Physiology 47A:371–385.

Coombs S, Janssen J, Webb J (1988) Diversity of lateral line systems: Evolutionary and functional considerations. In: Atema J, Fay RR, Popper AN, Tavolga WN (eds) Sensory Biology of Aquatic Animals. New York: Springer-Verlag, pp. 553–593.

Coombs S, Janssen J (1989) Peripheral processing by the lateral line system of the mottled sculpin (*Cottus bairdi*). In: Coombs S, Görner P, Münz H (eds) The Mechanosensory Lateral Line: Neurobiology and Evolution. New York: Springer-Verlag, pp. 299–319.

Coombs S, Janssen J (1990) Water flow detection by the mechanosensory lateral line. In: Stebbins WE, Berkley MA (eds) Comparative Perception—Vol. II: Complex Signals. pp. 89–123.

Coombs S, Janssen J. Behavioral and neurophysiological assessment of lateral line sensitivity in the mottled sculpin, *Cottus bairdi*. J Comp Physiol A (in press).

Cooper JE, Kuehne RA (1974) *Speoplatyrhinus poulsoni*, a new genus and species of subterranean fish from Alabama. Copeia 486–493.

Crawford AC, Fettiplace R (1980) The frequency selectivity of auditory nerve fibres and hair cells in the cochlea of the turtle. J Physiol 306:79–125.

Denton EJ, Gray JAB (1982) The rigidity of fish and patterns of lateral line stimulation. Nature 297:679–681.

Denton EJ, Gray JAB (1983) Mechanical factors in the excitation of clupeid lateral lines. Proc R Soc Lond B 218:1–26.

Denton EJ, Gray JAB (1988) Mechanical factors in the excitation of the lateral lines of fishes. In: Atema J, Fay RR, Popper AN, Tavolga WN (eds) Sensory Biology of Aquatic Animals. New York: Springer-Verlag, pp. 595–617.

Denton EJ, Gray JAB (1989) Some observations on the forces acting on neuromasts in fish lateral line canals. In: Coombs S, Görner P, Münz H (eds) The Mechanosensory Lateral Line: Neurobiology and Evolution. New York: Springer-Verlag, pp. 229–246.

Dercum F (1880) The lateral sensory apparatus of fishes. Proc Acad Nat Sci. Philadelphia, 152–154.

Dijkgraaf S (1963) The functioning and significance of the lateral line organs. Biological Reviews 38:51–105.

Disler N (1960) Lateral line sense organs and their importance in fish behavior. Acad. Sci. USSR Severtsov Institute of Animal Morphology (translated from Russian: Israel Program for Scientific Translations, Jerusalem, 1971. U.S. Department of Commerce, National Technical Information Service, Springfield, Va. 22151).

Disler NN, Smirnov SA (1977) Sensory organs of the lateral-line canal system in two percids and their importance in behavior. J Fish Res Board Can 34:1492–1503.

Elepfandt A (1989) Wave analysis by amphibians. In: Coombs S, Görner P, Münz H (eds) The Mechanosensory Lateral Line: Neurobiology and Evolution. New York: Springer-Verlag, pp. 527–541.

Elepfandt A, Seiler B, Aicher B (1985) Water wave frequency discrimination in the clawed frog, *Xenopus laevis*. J Comp Physiol A 157:255–261.

Enger P (1966) Acoustic threshold in goldfish and its relation to the sound source distance. Comp Biochem Physiol 18:859–868.

Enger P, Andersen R (1967) An electrophysiological field study of hearing in fish. Comp Biochem Physiol 1967(22):517–525.

Enger P, Kalmijn AJ, Sand O (1989) Behavioral identification of lateral line and inner ear function. In: Coombs S, Görner P, Münz H (eds) The Mechanosensory Lateral Line: Neurobiology and Evolution. New York: Springer-Verlag, pp. 575–587.

Fänge R, Larsson A, Lidman U (1972) Fluids and jellies of the acousticolateralis system in relation to body fluids in *Corphaenoides rupestris* and other fishes. Mar Biol 17:180–185.

Fay RR (1969b) Behavioral audiogram for the goldfish. J Aud Res 9:112–121.

Fay RR (1978) Coding of information in single auditory-nerve fibers of the goldfish. J Acoust Soc Am 63(1):136–146.

Fay RR (1981) Coding of acoustic information in the eighth nerve. In: Tavolga WN, Popper AN, Fay RR (eds) Hearing and Sound Communication in Fishes. Berlin: Springer-Verlag, pp. 189–222.

Fay RR (1988) Hearing in Vertebrates: A Psychophysics Databook. Winnetka, IL: Hill-Fay Associates.

Fay RR, Ream TJ (1986) Acoustic response and tuning in saccular nerve fibers of the goldfish (*Carassius auratus*). J Acoust Soc Am 79 (6):1883–1895.

Fernholm B (1985) The lateral line system of cyclostomes. In: Forman RE, Gorbman A, Dodd JM, Olsson R (eds) Evolutionary Biology of Primitive Fishes. New York: Plenum Press, pp. 113–122.

Flock Å (1967) Ultrastructure and function in the lateral line organs. In: Cahn PH (ed) Lateral Line Detectors. Bloomington: Indiana University Press, pp. 163–197.

Fritzch B (1989) Diversity and regression in the amphibian lateral line and electrosensory system. In: Coombs S, Görner P, Münz H (eds) The Mechanosensory Lateral Line: Neurobiology and Evolution. New York: Springer-Verlag, pp. 99–114.

Görner P, Kalmijn AJ (1989) Frequency response of lateral-line neuromasts in the thornback ray (*Platyrhinoidis triseriata*). In: Erber J, Menzel R, Pfluger H, Todt D (eds) Neural Mechanisms of Behavior. Stuttgart/NewYork: Georg Thieme/Verlag, p. 82.

Görner P, Mohr C (1989) Stimulus localization in *Xenopus*: The role of directional sensitivity of lateral line stitches. In: Coombs S, Görner P, Münz H (eds) The Mechanosensory Lateral Line: Neurobiology and Evolution. New York: Springer-Verlag, pp. 543–560.

Gould SJ, Lewontin RC (1979) The spandrels of San Marco and Panglossian paradigm: A critique of the adaptationist programme. Proc R Soc Lond B 205: 581–598.

Graham-Smith W (1978) On the lateral lines and dermal bones in the parietal region of some crossopterygian and dipnoan fishes. R Soc Lond Philos Trans Biol Sci 282:41–105.

Gray JAB (1984) Interaction of sound pressure and particle acceleration in the excitation of the later-line neuromasts of sprats. Proceedings of the Royal Society of London B 220:299–325.

Hama K (1978) A study of the fine structure of the pit organ of the common japanese sea eel *Conger myriaster*. Cell Tissue Res 189:375–388.

Hassan ES (1985) Mathematical analysis of the stimulus for the lateral line organ. Biological Cybernetics 52:23–36.

Hassan ES (1986) On the discrimination of spatial intervals by the blind cave fish (*Anoptichthys jordani*). J Comp Physiol A 159:701–710.

Heffner RS, Heffner HE (1985) Hearing range of the domestic cat. Hear Res 19:85–88.

Herrick CJ (1901) The cranial nerves and cutaneous sense organs of the North American siluroid fishes. J Comp Neurol 11:177–249.

Hoyt RD (1972) Anatomy and osteology of the cephalic lateral line system of the silverjaw minnow, *Ericymba buccata*. Copeia 1972:812–816.

Jakubowski M (1967) Cutaneous sense organs of fishes. VIII. The structure of the system of lateral-line canal organs in the *Percidae*. Acta Biol Cracov Ser Zool 10:69–81.

Jakubowski M (1974) Structure of the lateral line canal system and related bones in the berycoid fish *Hoplostethus mediterraneus* Cuv. et Val. (Trachichthyidae, Pisces). Acta Anat 87:261–274.

Janssen J, Coombs S, Hoekstra D, Platt C (1987) Anatomy and differential growth of the lateral line system of the mottled sculpin, *Cottus bairdi* (Scorpaeniformes: Cottidae). Brain Behavior and Evolution 30: 210–229.

Jielof R, Spoor A, De Vries HL (1952) The microphonic activity of the lateral line. J Physiol 116:137–157.

Jones WR, Janssen J (1990) Lateral line development and feeding behavior in the mottled sculpin (*Cottus bairdi*). Abstracts of the Thirteenth Midwinter Research Meeting, Association for Research in Otolaryngology.

Jørgensen J (1989) Evolution of octavolateralis sensory cells. In: Coombs S, Görner P, Münz H (eds) The Mechanosensory Lateral Line: Neurobiology and Evolution. New York: Springer-Verlag, pp. 115–145.

Kalmijn AJ (1988) Hydrodynamic and acoustic field detection. In: Atema J, Fay RR, Popper AN, Tavolga WN (eds) Sensory Biology of Aquatic Animals. New York: Springer-Verlag, pp. 83–130.

Kalmijn AJ (1989) Functional evolution of lateral line and inner-ear sensory systems. In: Coombs S, Görner P, Münz H (eds) The Mechanosensory Lateral Line: Neurobiology and Evolution. New York: Springer-Verlag, pp. 187–215.

Kapoor AS (1964) Functional morphology of laterosensory canals in the Notopteridae (Pisces). Acta Zool Stockh 65:77–91.

Kapoor AS (1970) Development of dermal bones related to sensory canals of the head in the fishes *Ophicephalus punctatus* Bloc (Ophicephalidae) and *Wallago attu* Bl. and Schn. (Siluridae). Zool J. Linn Soc 49:69–97.

Kelly J, van Netten S (1991) Topography and mechanics of the cupula in the fish lateral line. I. Variation of Cupulan Structure and composition in three dimensions. J Morph 207:23–36.

Kiang N (1965) Discharge Patterns of Single Fibers in the Cat's Auditory Nerve. Cambridge, MA: MIT Press, 154 pp.

Kirk K (1984) Water flows produced by *Daphnia* and *Diaptomus*: Implications for prey selection by mechanosensory predators. Limnol Oceanogr 30(3): 679–686.

Knox R (1925) On the theory of the 6th sense in fishes. Edinburgh J Sci 2:12.

Kroese ABA, van Netten SM (1989) Sensory transduction in lateral line hair cells. In: Coombs S, Görner P, Münz H (eds) The Mechanosensory Lateral Line: Neurobiology and Evolution. New York: Springer-Verlag, pp. 265–284.

Kroese ABA, Schellart NAM (1987) Evidence for velocity-and acceleration-sensitive units in the trunk lateral line of the trout. J Physiol 394:13.

Kroese ABA, Van der Zalm JM, Van den Bercken J (1978) Frequency response of the lateral-line organ of *Xenopus laevis*. Pflugers Archives 375:167–175.

Kuiper JW (1956) The microphonic effect of the lateral line organ. PhD Thesis, University of Groningen, The Netherlands.

Kuiper JW (1967) Frequency characteristics and functional significance of the lateral line organ. In: Cahn PH (ed) Lateral Line Detectors. Bloomington: Indiana University Press, pp. 105–122.

Lekander B (1949) The sensory line system and the canal bones in the head of some Ostariophysi. Acta Zool (Stockh) 30:1–131.

Liff H, Shamres S (1972) Structure and motion of cupulae of lateral line organs in *Necturus maculosus*. III. A technique for measuring the motion of free-standing lateral line cupulae. Quarterly Progress Report. Research Laboratory of Electronics, Massachusetts Institute of Technology 204:331–336.

Lannoo M (1987a) Neuromast topography in urodele amphibians. J Morph 191:247–263.

Lannoo M (1987b) Neuromast topography in anuran amphibians. J Morph 191:115–129.

Leydig F (1850) Uber die schleimkanale der knochenfische. Mull Arch Anat Physiol 170–181.

Leydig F (1851) Uber die nervenknopfe in den schleimkanalen von *Lepidoleprus*, *Umbrina* und *Corvina*. Mull Arch Anat Physiol 235–240.

Makushok VM (1961) Some peculiarities in the structure of the seismosensory system of the northern Blenniids (Stichaeoidae, Blennoideii, Pisces). Tr Inst Okeanol Akad Nauk USSR 43:225–269.

Marshall NB (1965) Systematic and biological studies of the macrourid fishes (Anacanthini-teleostei). Deep-Sea Res 12:299–322.

Marshall NB (1971) Explorations in the Life of Fishes. Cambridge: Harvard University Press, p. 204.

Marshall NJ (1986) Structure and general distribution of free neuromasts in the black goby, *Gobius niger*. J Mar Biol Assoc UK 66:323–333.

Mayser P (1881) Vergleichend anatomische Studien uber das Gehirn der Knochenfische mit besonderer Berucksichtigung der Cyprinoiden. Z Wiss Zool 36: 259–364.

McCormick C (1982) The organization of the octavolateralis area in actinopterygian fishes: A new interpretation. J Morphol 171:159–181.

McCormick C (1989) Central lateral line mechanosensory pathways in bony fish. In: Coombs S, Görner P, Münz H (eds) The Mechanosensory Lateral Line: Neurobiology and Evolution. New York: Springer-Verlag, pp. 341–364.

Megela AL (1984) Diversity of adaptation patterns in responses of eighth nerve fibers in the bullfrog, *Rana catesbeiana*. J Acoust Soc Am 75(4):1155–1162.

Megela Simmons A, Moss CF, Daniel KM (1985) Behavioral audiograms of the bullfrog (*Rana catesbeiana*) and the green tree frog (*Hyla cinerea*). J Acoust Soc Amer 78:1236–1244.

Merkel F (1880) Uber die endigungen der sensiblen nerven in der haut der wirbeltiere. Rostock.

Merrilees M, Crossman E (1973a) Surface pits in the family Esociadae. I. Structure and Types. J Morphol 141 (3):321–343.

Merrilees M, Crossman E (1973b) Surface pits in the family Esociadae. II. Epidermal-dermal interaction and evidence for aplasia of the lateral line sensory system. J Morphol 141 (3):321–343.

Metcalfe W (1989) Organization and development of the zebrafish posterior lateral line. In: Coombs S, Görner P, Münz H (eds) The Mechanosensory Lateral Line: Neurobiology and Evolution. New York: Springer-Verlag, pp. 147–159.

Miller PJ, Wongrat P (1979) A new goby (Teleostei: Gobiidae) from the South China Sea and its significance for gobioid classification. Zool J Linn Soc 67:239–257.

Montgomery JC, Macdonald JA (1987) Sensory tuning of lateral line receptors in Antarctic fish to the movements of planktonic prey. Science 235:195–196.

Montgomery JC, Macdonald JA, Housley GD (1988) Lateral line function in an antarctic fish related to the signals produced by plantonic prey. J Comp Physiol A 163:827–833.

Montgomery JC (1989) Lateral line detection of planktonic prey. In: Coombs S, Görner P, Münz H (eds) The Mechanosensory Lateral Line: Neurobiology and Evolution. New York: Springer-Verlag, pp. 561–574.

Moore F, Burris W (1956) Description of the lateral line system of the pirate perch, *Aphredoderus Sayanus*. Copeia 1956:18–20.

Moy-Thomas JA, Miles RS (1971) Palaeozoic Fishes. Philadelphia: Saunders.

Münz H (1979) Morphology and innervation of the lateral line system in *Sarotherodeon niloticus* L. J Comp Physiol A 157:555–568.

Münz H (1985) Single unit activity in the peripheral lateral line system of the cichlid fish *Sarotherodon niloticus* L. J Comp Physiol A 157:555–568.

Münz H (1986) What influences the development of canal and superficial neuromasts? Ann Kon Mus Mid Afr Zool Wetensch 251:85–89.

Münz H (1989) Functional organization of the lateral line periphery. In: Coombs S, Görner P, Münz H (eds) The Mechanosensory Lateral Line: Neurobiology and Evolution. New York: Springer-Verlag, pp. 285–297.

Myrberg AA, Jr, Spires JY (1980) Hearing in damselfishes: an analysis of signal detection among closely related species. J Comp Physiol 140:135–144.

Nelson GJ (1972) Cephalic sensory canals, pitlines, and the classification of Esocoid fishes, with notes on Glaxiid and other teleosts. Am Mus Novit 2492: 1–49.

van Netten SM, Kroese ABA (1987) Laser interferometric measurements on the dynamic behavior of the cupula of the fish lateral line. Hear Res 29:55–61.

van Netten SM, Kroese ABA (1989) Dynamic Behavior and Micromechanical Properties of the Cupula. In: Coombs S, Görner P, Münz H (eds) The Mechanosensory Lateral Line: Neurobiology and Evolution. New York: Springer-Verlag, pp. 247–263.

Norris HW (1925) Observation upon the peripheral distribution of the cranial nerves of certain ganoid fishes (*Amia*, *Lepidosteus*, *Polyodon*, *Scaphirhynchus*, and *Acipenser*). J Comp Neurol 39:345–432.

Northcutt RC (1981) Audition and the central nervous system of fishes. In: Tavolga WN, Popper AN, Fay RR (eds) Hearing and Sound Communication in Fishes. New York: Springer-Verlag, pp. 331–355.

Northcutt RG (1988) Sensory and other neural traits and the adaptationist program: Mackerals of San Marco? In: Atema J, Fay RR, Popper AN, Tavolga WN (eds) Sensory Biology of Aquatic Animals. New York: Springer-Verlag, pp. 869–883.

Northcutt RG (1989) The phylogenetic distribution and innervation of craniate mechanoreceptive lateral lines. In: Coombs S, Görner P, Münz (eds) The Mechanosensory Lateral Line: Neurobiology and Evolution. New York: Springer-Verlag, pp. 17–78.

O'Connell CP (1981) Development of organ systems in the northern anchovy, *Engrauliis mordax* and other teleosts. Am Zool 21:429–446.

Oman C (1972) Structure and motion of cupuale of lateral line organs in *Necturus maculosus*. IV. Preliminary model for the dynamic response of the free-standing lateral line cupula, based on measurements of cupula stiffness. Quarterly Progress Report. Research Laboratory of Electronics, Massachusetts Institute of Technology 104:336–342.

Parker GH (1904) The function of the lateral line organs in fishes. Bull US Bur Fish 24:185–207.

Pehrson T (1945) The system of pit organ lines in *Bymnarchus niloticus*. Acta Zool 26:1–8.

Pehrson T (1947) Some new interpretations of the skull in *Polypterus*. Acta Zool 28:400–454.

Peters HM (1973) Anatomie und entwicklungsgeschichte des lateralissystems von Tilapia (Pisces, Cichlidae). Z Morphol Tiere 74:89–161.

Peters RC, Buwalda JA (1986) The octavo-lateralis perception in lower aquatic vertebrates: Clustered versus dispersed sensor configurations. Netherlands Journal of Zoology 36 (3):381–392.

Pollard HB (1892) The lateral line system in siluroids. Zool Jahrb [Anat] 5:525–551.

Popper AN (1971) The effects of size on auditory capacities of the goldfish. J Aud Res 11:239–249.

Popper AN, Coombs S (1980a) Acoustic Detection by Fishes. In: Ali MA (ed) Environmental Physiology of Fishes. New York: Plenum Press, pp. 403–430.

Popper AN, Coombs S (1980b) Auditory mechanisms in teleost fishes: Significant variations in both hearing capabilities and auditory structures are found among species of bony fishes. American Scientist 68:429–440.

Popper AN, Coombs S (1982) The morphology and evolution of the ear in Actinopterygian fishes. American Zoologist 22:311–328.

Poulson T (1963) Cave adaption in amblyopsid fishes. Am Midl Nat 70:257–290.

Poulson T (1985) Evolutionary reduction by neutral mutations: Plausibility arguments and data from albyopsid fishes and linyphid spiders. NSS Bulletin 47(2):109–117.

Puzdrowski RL (1989) Peripheral distribution and central projections of the lateral-line nerves in goldfish, *Carassius auratus*. Brain, Behavior and Evolution 34:110–131.

Russell IJ (1976) Amphibian lateral line receptors. In: Llinas R, Precht W (eds) Frog Neurobiology. New York: Springer-Verlag, pp. 513–550.

Sand O (1981) The Lateral Line and Sound Reception. In: Tavolga WN, Popper AN, Fay RR (eds) Hearing and Sound Communication in Fishes. New York: Springer-Verlag, pp. 459–580.

Sand O, Karlsen HE (1986) Detection of infrasound by the Atlantic cod. J Exp Biol 125:197–204.

Schellart NAM, Kroese ABA (1989) Interrelationship of acousticolateral and visual systems in the teleost midbrain. In: Coombs S, Görner P, Münz H (eds) The Mechanosensory Lateral Line: Neurobiology and Evolution. New York: Springer-Verlag, pp. 421–443.

Schemmel C (1967) Bergiechende untersuchungen an den Hautsinnesorganen oberund unterirdisch lebender *Astyanax* Formen. Ein beitrag zur evolution der cavernicolen. Z Morph Tiere 61:255–316.

Schemmel C (1977) Zur morphologie und funktion der sinnesorgane von *Typhliasina pearsei* (Hubbs) (Ophidioidea, Teleostei). Zoomorphologie, 87:191–202.

Schulze FE (1870) Uber die sinnesorgane der seitenlinie bei fischen und amphibien. Arch Mikr Anat 6:62–88.

de Sede P (1884) La Ligne laterale des poissons osseux. Rev Scient (Serie 3).

Shelton PMJ (1970) The lateral line system at metamorphosis in *Xenopus laevis* (Daudin). J Embryol Exp. Morph 24(3):511–524.

Siler W (1969) Near- and farfields in a marine environment. J Acoust Soc Am 16 (3):483–484.

Song J, and RG Northcutt (1991) Morphology, distribution and innervation of the lateral-line receptors of the Florida gar, *Lepisosteus platyrhincus*. Brain Behav Evol 37:10–37.

Stone L (1933) The development of lateral line sense organs in amphibians observed in living and vital stained preparations. J Comp Neurol 5:507–540.

Swofford DL (1976) Genetic variability, population differentiation, and biochemical relationships in the family Amblyopsidae. Masters' Thesis, Eastern Kentucky University (submitted 1982).

Swofford DL, Branson BA, Sievert GA (1980) Genetic differentiation of cavefish populations (Amblyopsidae). Isozyme Bulletin 13:109–110.

Topp G (1983) Primary lateral line response to water surface waves in the topminnow *Aplocheilus Lineatus* (Pisces, Cyprinodontidae). Pflugers Arch 397: 62–67.

Vischer H (1989) The development of lateral-line receptors in *Eigenmannia* teleostei, gymnotiformes). I. The mechanoreceptive lateral-line system. Brain Behav Evol 33:205–222.

Vischer H (1989) The development of lateral-line receptors in *Eigenmannia* (Teleostei, Gymnotiformes). II. The electroreceptive lateral-line system. Brain Behav Evol 33:223–236.

Webb JF (1989a) Developmental constraints and evolution of the lateral line system in teleost fishes. In: Coombs S, Görner P, Münz H (eds) The Mechanosensory Lateral Line: Neurobiology and Evolution. New York: Springer-Verlag, pp. 79–97.

Webb JF (1989b) Gross morphology and evolution of the mechanoreceptive lateral-line system in teleost fishes. Brain Behav Evol 33:34–53.

Wever EG (1976) Origin and evolution of the ear of vertebrates. In: Masterton RB, Bitterman ME, Campbell CBG, Hotton N (eds) Evolution of Brain and Behavior in Vertebrates. New Jersey: Lawrence Erlbaum Assoc, pp. 89–105.

Wiley EO. Phylogenetic Relationships of the Percidae (Teleostei: Perciformes): A Preliminary Hypothesis. In: Mayden, RL (ed) Systematics, Historical Ecology and North American Freshwater Fishes. Stanford University Press (in press).

Winklbauer R, Hausen P (1983a) Development of the lateral line system in *Xenopus laevis*. I. Normal development and cell movement in the supraorbital system. J Embryol Exp Morph 76:265–281.

Winklbauer R, Hausen P (1983b) Development of the lateral line system in *Xenopus laevis*. II. Cell multiplication and organ formation in the supraorbital system. J Embryol Exp Morph 76:283–296.

Winklbauer R, Hausen P (1985a) Development of the lateral line system in *Xenopus laevis*. III. Development of the supraorbital system in triploid embryos and larvae. J Embryol Exp Morph 88:183–192.

Winklbauer R, Hausen P (1985b) Development of the lateral line system in *Xenopus laevis*. IV. Pattern formation in the supraorbital system. J Embryol Exp Morph 88:193–207.

Wubbels R (1989) Afferent activity in the supra-orbital canal of the ruff lateral line. PhD Thesis, University of Groningen, The Netherlands.

Wullimann MF, Senn DG (1981) Zur Morphologie der Lateralis-Innervation bei Mormyriden-Fischen (*Brienomyrus* spec. Taverne 1971, Mormyridae, Teleostei). Verh Naturforsch Ges Basl 92:63–72.

16
Functional Aspects of the Evolution of the Auditory System of Actinopterygian Fish

Nico A.M. Schellart and Arthur N. Popper

1. Introduction

There are many interspecific differences in the structure of the auditory system among the more than 25,000 extant species of actinopterygian fishes. These differences include: (a) structural variation in the regions of the inner ear associated with sound detection (based on studies of species in 65 genera) (e.g., Platt and Popper 1981; Popper and Coombs 1982; Popper 1983); (b) differences in hearing capabilities (behavioral data from species representing 45 genera) (Fay 1988a); and (c) variation in physiological response properties of the ear (albeit, with data from species representing only 10 genera) (e.g., Fay and Popper 1983). Beyond these differences, there is significant variation in structures more peripheral to the ear (e.g., swimbladder, Weberian ossicles) (Popper and Coombs 1982), and possibly in central pathways although data on the central auditory pathways of fishes are limited in scope (McCormick 1981, Chapter 17).

Despite knowing a good deal about the morphology, behavior, and physiology associated with fish hearing, we still do not know whether any or all of the interspecific differences are correlated with differences in the ways various species use sounds or process acoustic signals. In other words, we do not know whether interspecific variations reflect ways to derive different types of information, or whether there are diverse strategies to deal with the same acoustic problems (e.g., parallel evolution). It is also possible that fishes with similar peripheral structures have different central mechanisms for sound processing, thereby leading to other variations in sound analysis mechanisms.

The goal of this chapter is to explore some of these questions. More specifically, we will present a number of hypotheses regarding the relationships between structure and function of the fish auditory periphery from an evolutionary point of view. Since the number of species used for neurophysiological and neuroanatomical studies is too limited for relevant interspecific comparison, the chapter is mainly restricted to data derived using morphological and psychophysical (behavioral) studies. In order to keep the chapter at a reasonable length, we have chosen not to discuss, for the most part, fishes other than extant actinopterygians (but see Section 8) or data from the fossil record and elasmobranchs.

We want to point out that we often cite review articles rather than primary references. Much of the earlier material is cited in several volumes including Schuijf and Hawkins (1976); Tavolga, Popper, and Fay (1981); Atema, Fay, Popper, and Tavolga (1988); Fay (1988a); and Coombs, Görner, and Münz (1989). The reader is directed to these sources for the bulk of the literature.

2. The Diversity of Underwater Acoustic Environments

Before considering the auditory system of fishes, we will discuss certain aspects of the physics of underwater sound that are likely to have had a dominant effect on the evolution of acoustic behavior, hearing, and hearing structures in fishes. A more detailed discussion of underwater acoustics is found in Kalmijn (1988) and Rogers and Cox

(1988). Michelsen (Chapter 5) also discusses similar acoustical concepts although primarily in terms of terrestrial animals.

2.1 Underwater Acoustics

In any medium, an acoustic wave can be described both as a pressure variation within the medium and as a to-and-fro oscillation of the component particles of the medium. Both waves are longitudinal for a propagating acoustic field. For a free propagating wave with a plane wave front, the sound pressure (the scaler p) and the particle velocity (the vector $\mathbf{v}$) are mutually related such that $p = \rho c \mathbf{v}$, with ρ being the (characteristic) density of the medium and c the sound velocity in the medium. The acoustic impedance is ρc, $\mathbf{v}$ is the acoustic analogue of electric current, and p the analogue of voltage. Thus, $\mathbf{v}p$ gives the intensity or power (in W/cm^2) of the sound.

The easiest sound source to deal with mathematically is the monopole, a pulsating sphere that produces spherical waves. At distances from the source that are large with respect to the wavelength (λ), the amplitudes of $\mathbf{v}$ and p decrease linearly with distance (R), as does the time integral and derivative of $\mathbf{v}$ (i.e., the particle displacement, and the acceleration, respectively). It is clear that intensity (i.e., $\mathbf{v}p$) of the sound decreases with R^2. For distances from the source that are much less than a wavelength, pressure decreases linearly but displacement decreases with the square of distance (also see van Bergeijk 1964).

The phase difference between p and $\mathbf{v}$ depends upon R. When R is much greater than λ, p and $\mathbf{v}$ are in phase with one another. However, for short distances from the source, p leads $\mathbf{v}$ up to a maximum of 90°. For distances less than $\lambda/2\pi$, phase differences are more than 45°. These shorter distances have been called the "near-field," while larger distances have been called the "far-field" (see also van Bergeijk 1964). For distances approximately $\lambda/2\pi$, there is a gradual transition from the near- to the far-field (Siler 1969).

In addition to the near- and far-field effects of λ, there is also an effect of the frequency, f (where $f = c/\lambda$). Under the condition that the particle velocity at the interface of the pulsating source is the same for all frequencies, p is proportional to f, irrespective of R. In the near-field, displacement is proportional to the reciprocal of f, but in the far-field displacement is independent of f. The dependency upon f is important since the inner ears of all fishes are directly stimulated by a displacement stimulus.

Many biological sound sources are dipoles (vibrating spheres) rather than monopoles. Dipole sources produce complicated sound fields that are composed of radial and tangential components. The tangential component, in contrast to the radial, is very small in the far-field. All frequency effects for the radial component are a factor of f (= frequency in Hz) times stronger in a dipole field than in monopole field.

In the far-field of a dipole source, distance effects are the same as for a monopole source, aside from the effect of the angle between the direction of oscillation of the dipole and the direction of observation (multiplication by a cosine function). The near-field of a dipole is very complicated. A more complete discussion of the fields of monopole and dipole sources can be found in Kalmijn (1988).

2.2 Constraints of Underwater Sound Characteristics on the Evolution of Fish Hearing

The characteristics of underwater sound must have affected the evolution of fish hearing and sound production systems. The acoustic laws are the same in air and in water. Still, there are sufficient differences in magnitude of the system parameters so that one cannot readily extrapolate from what is known about hearing in air to hearing in water. The differences in acoustic environment include a near-field that is biologically much more important in water than in air (Section 2.1). Differences also include other characteristics of sound such as its speed and the reflective and scattering properties of underwater objects (including parts of the fish's body itself). All of these will impose constraints on the way fishes can use sounds that differ from the constraints imposed on terrestrial vertebrates. It is likely that differences in the structures of the auditory system reflect various strategies to deal with underwater sound.

2.2.1 Speed of Sound

The velocity of sound in water, and thus wavelength of the sound, is about 4.5 times larger than in air. The wavelength of a 100 Hz sound in water is 15 m, as compared to 3.3 m in air. Wavelengths of audible sounds are generally larger than the size of fish, and often larger than the distance of the fish to a relevant sound source. As a direct consequence of this high speed, the difference in time of a sound's arrival between the two inner ears of fishes (which are at most a few centimeters apart) is extremely small. Thus, the interaural time and intensity cues used by many terrestrial species for determination of sound source direction are probably not available to fishes (van Bergeijk 1964). At the same time, since frequencies used in acoustic behavior have wavelengths of meters, fishes are often in the near-field of sound sources. Therefore, the complicated structure of near-field dipole fields will affect the sound localization strategies of fish.

Water is approximately 830 times more dense than air and its acoustic impedance is about 3,600 times greater. Consequently v is 3,600 times smaller in water than in air for the same p. Although it might be expected that pressure would be the most suitable stimulus for sound detection by agnathans and fishes, as it is for terrestrial animals, the stimulus is, in fact, the particle motion of the sound wave (Fay and Popper 1974). As will be discussed below (Section 4), particle motion provides the "direct" stimulus to the ear.

Particle motion is the only possible acoustic stimulus for fish without a swimbladder (or other air cavity). However, fish with a swimbladder are potentially able to detect a secondary "indirect stimulus" that emanates from the swimbladder which acts as a pressure-to-displacement transducer. The swimbladder thus serves to enhance hearing sensitivity in many (though not all) species (e.g., Popper and Coombs 1982) and potentially provides the fish with the ability to detect sounds in the far-field using this indirect stimulus. This may be called pressure hearing. In contrast, near-field hearing in these species is primarily mediated by the direct stimulus and may be called displacement hearing.

2.2.2 Reflection, Refraction, Scatter, and Depth

The propagation of underwater sound is affected by reflection, refraction, scatter, and water depth. Ideally, the air–water interface is a perfect reflec-

tor. However, the reflected unattenuated wave is phase shifted 180°, thereby canceling the impinging wave. Under natural conditions, and close to the surface of the water, the remaining signal is so small that it is difficult to detect even by a nearby receiver. However, this only holds (to a first approximation) for a surface that does not have much rippling, as on small lakes. In contrast to the surface, the bottom is generally a poor reflector, although the actual amount of reflection depends upon the composition and the texture of the bottom (Urick 1975). The reflectance also depends upon the angle of incidence of the sound to the bottom. Angle of incidence is much more important for the bottom than for the water surface.

Sound scattering occurs at objects with an acoustic impedance that is significantly different from that of the propagating medium. Underwater objects that cause scattering include the swimbladder, gas-filled cavities, and air bubbles in the water. Nonbiological bubbles may be caused by heavy rain, fast surface currents (white water), and waterfalls. Since such bubbles have a strong resonance, they significantly increase background pressure noise by providing reflective surfaces for any ambient sounds.

2.3 Background Noise

Background noises are all sounds that are not paid attention to by the listener at a particular moment. The level of nonbiological background noise in the sea is mainly determined by the ripple and wave effects of wind on the water surface. Less important components of nonbiological noises are rain (which produces moderate and high frequencies) and seismic events in the sea bottom (producing low frequencies). Background noise is much higher in the sea than on land owing to the extremely low absorption of sound in water. Figure 16.1 gives the combined data of Urick (1975) and Hawkins and Myrberg (1983) for some wind levels and sea states.

Background noise potentially has some behavioral relevance since it may provide the fish with information about sea state, and the presence of beaches, rocky coasts, rapids, or other types of environments. At the same time, background noise can potentially mask, or prevent, detection of biologically important sounds. Thus, it is likely that if

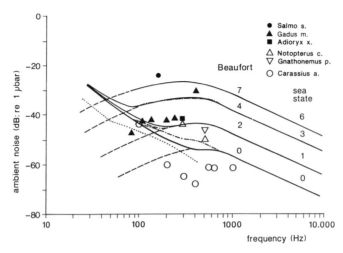

FIGURE 16.1. Psychophysical sensitivity of some teleosts for noise in relation to ambient noise in marine and fresh water bodies. The solid and dashed lines represent ambient noise levels (= spectrum level) for the open sea (Urick 1975), the dashed-stippled lines for a 15 m deep Scottish loch (Hawkins and Myrberg 1983), and the stippled line for a small, shallow lake with a slow current (recalculated data from Abbott 1973). The latter curve has a level that is probably 4 to 10 dB too high resulting from artifacts of the set up. On the low frequency side each of the curves approach a single asymptote with a slope of −8 to −10 dB/octave and at the high frequency side the curves approach a 1/f attenuation (Wenz 1971). Theoretical sensitivities for ambient noise have been calculated from data of the nonspecialists *Adioryx*, *Gadus* and *Salmo*, the otophysan *Carassius* and the nonotophysan specialists *Notopterus* and *Gnathonemus* (see text).

marine fishes hear background signals they are also likely to have some mechanism(s) to enable them to extract the biologically important signals from the background. In fact, the ability to extract signals from noise is well known among several freshwater species, and there is also evidence that certain marine species can perform the same task (reviewed in Fay 1988a).

It is not known if many species of fish can hear background noise in their natural environments. However, it is theoretically possible that the ability to detect noise can be estimated from a fish's pure tone threshold, S_f, and the critical bandwidth B_{3dB}. Since behavioral hearing data are primarily derived from pressure thresholds for fishes that use the swimbladder in hearing, we will primarily consider this approach for sound pressure. The threshold for hearing ambient noise (i.e., background noise per 1-Hz band) centered around a given frequency (f) can theoretically be approximated by $S_f - 20 \log B_{3dB,f}^{0.5}$ with S_f in dB re: 1 µbar and $B_{3dB,f}$ the bandwidth centered around f, in Hz. (It is assumed that neighboring 1-Hz noise bands are uncorrelated, resulting in an amplitude

of the filtered noise that is proportional to the square root of the bandwidth. For correlated noise bands, the threshold is presumably reached for lower background levels.)

Our calculations to estimate noise sensitivity have been made using data in the literature for S_f and $B_{3dB,f}$ for three species that do not hear very well, *Salmo salar* (Atlantic salmon), *Gadus morhua* (cod), and *Adioryx xantherythrus* (Hawaiian squirrelfish), and three species with good hearing, *Carassius auratus* (goldfish), *Notopterus chitala* (a knifefish) and *Gnathonemus petersii* (elephant nose fish). Since most of the time the wind force is greater than two on the Beaufort scale, *Gadus morhua* will generally hear background noise up to 300 Hz (calculated from data of Chapman and Hawkins 1973; Offutt 1973; Hawkins and Chapman 1975), and the same is probably true of *Adioryx xantherythrus* (calculated from data of Coombs and Popper 1979 and 1981). As illustrated in Figure 16.1, *Salmo salar* (calculated from Hawkins and Johnstone 1978) should hear background noises only when the wind force is greater than seven. In contrast, the goldfish (data points calculated from Jacobs and Tavolga 1967; Popper 1971; Fay, Ahroon and Orawski 1978) will hear even the lowest levels of background noise. *Notopterus* (calculated from Coombs and Popper 1981; Coombs and Popper 1982) and *Gnathonemus* (calculated from McCormick and Popper 1984) will have intermediate capabilities to hear the background noise.

The actual level of background noise in the sea varies considerably in different environments and under different conditions. For example, in the 50 to 500 Hz bandwidth, the background noise of coastal waters is 5 to 15 dB greater than that in the open sea (Urick 1975). Close to the beach, the noise of heavy surf will increase the background noise of coastal waters even more (10–15 dB) (calculated from Wilson, Wolf, and Ingenito 1985), and this effect will be even stronger near a rocky coast. Heavy rain will increase background levels by 5 to 10 dB (Urick 1975). The noise levels close to the bottom will be greater in deep water than in shallow water (Arase and Arase 1967).

Bodies of fresh water are generally quieter than the sea due to the smaller influence of winds. Figure 16.1 gives the ambient noise level – which is close to sea state 3 – of a small lake that has some water currents (Abbott 1973). The generally lower noise levels of biological sound sources (see Section 3), and especially the very strong high pass filtering in very shallow waters (Rogers and Cox 1988), are additional reasons that ambient noise levels are lower in fresh water as compared to marine environments. The effect of high pass filtering is stronger for soft than for hard bottoms, since the latter have lower cutoff frequencies. This filtering effect is of less importance in marine waters, since the depth, even very close to the coast, is often many meters and the sea bottom at the continental shelf is generally hard.

Because the fresh water environment is generally much quieter than the marine environment, fresh water fish with poor hearing sensitivity, such as *Perca* (perch), *Gymnocephalus* (ruff) and *Lucioperca* (pike-perch) (Table 16.1, column 1) probably do not hear background noises under normal weather conditions according to our theoretical extrapolation. In contrast, hearing specialists (fishes that have special adaptations to enhance hearing) such as the otophysans are likely to hear such noise. Moreover, the typical high frequency sensitivity of otophysan fishes (species with Weberian ossicles connecting the swimbladder to the inner ear, see Section 5) might be a way for these species to compensate for the substantial low-frequency attenuation in the shallow waters with soft bottoms that these fish often inhabit.

2.4 Conclusions

From the above discussion, some general rules and predictions can be formulated regarding the effect

TABLE 16.1. Basic anatomical and psychophysical data of the auditory system and environmental and behavioral characteristics.

Order Family Species	0 OC	 HP	1 f_b (Hz)	2 S_{fb} (dB)	3 Q_{10dB}	4 HI	5 Habitat		6 fb	7 sp
A. Bony fish with a swimbladder										
Osteoglossiformes										
Notopteridae										
Notopterus chitala[1]	ss	O	500	−33	1.1	950	f		Pe	pc	
Mormyridae		V									
Gnathonemus petersii[2]	sa		500	−33	0.4	1600	f		B	pr	+
Elopiformes										
Albulidae										
Albula vulpes[3]	sl		300	−26	0.7	410	m	s	Pe	pr	180
Anguilliformes										
Anguillidae										
Anguilla anguilla[4]			40	−19	0.4	89	b	d	B	pr+	+$_g$

TABLE 16.1. (*Continued*)

Order Family Species	0 OC	HP	1 f_b (Hz)	2 S_{fb} (dB)	3 Q_{10dB}	4 HI	5 Habitat			6 fb	7 sp
Cypriniformes											+
Cyprinidae											
Carassius auratus[5]	w	V	500	−48	0.5	7900	f	s	Pe	pr	
Carassius carassius[6]	w	V	1000	−33	0.4	2200	f	ds	Pe	o*	
Cyprinus carpio[7]	w		500	−42	0.43	4300	f	ds	Pe	o	
Leucaspius delineatus[8]	w		1000	−17	0.60	290	f	s	Pe	pl	
Semotilus atromaculatus[9]	w		300	−21	0.35	500	f	s	Pe	o	
Characiformes											
Characidae		V									+
Astyanax jordani[10]	w		1000[b]	−48	0.8	8900	f	s	Pe	o	
Astyanax mexicanus[10]	w		1000[b]	−41	0.4	5600	f	s	Pe	h	
Siluriformes											
Ictaluridae											
Ictalurus nebulosus[11]	w	V	600	−34	0.35	2600	f	sd	B	pr	+
Ariidae											
Arius felis[12]	w	V	200	−55	0.7	9500	m	s	B	pr[+]	+
Salmoniformes											
Salmonidae											
Salmo gairdneri[13]		S	150	−6	0.6	32	fbm	s	Pe	pr[+]	
Salmo salar[14]		S	160	−5	0.77	26	fbm	s	Pe	pc	
Gadiformes											
Gadidae											
Gadus morhua[15]		D	150	−36	1.0[l]	760	m	sd	BPe	pc	30
Melanogrammus aeglefinus[16]			200	−20	0.57	190	m	sd	Pe	pc	50
Molva molva[16]			200	−19	0.53	170	m	d	B	pc	
Pollachius pollachius[16]			120	−19	0.50	140	m	sd	BPe	pr[+]	+g
Batrachoidiformes											
Batrachoididae											
Opsanus tau[17]		D	<38	−2	1.3[l]	<6.7	m	s	B	pr[+]	70
Beryciformes											
Holocentridae											
Adioryx xantherythrus[18]		S	500	−29	1.1[l]	600	m	s	Pe	pr	+
Holocentrus vexillarius[19]			600	−12	1.9	71	m	s	Pe	pr	90
Holocentrus ascensionis[20]	se		800	−24	1.0	450	m	s	Pe	pr	75
Myripristis kuntee[18]	se	O	1000[b]	−50	0.5	14000	m	s	Pe	pr[+]	+
Scorpaeniformes											
Triglidae											
Prionotus scitulus[19]			400	4	<0.7	>15	m	sd	B	pr	+g
Perciformes											
Serranidae											
Epinephelus guttatus[19]			200	−11	0.54	68	m	s	Pe	pl	50
Percidae											
Gymnocephalus cernuus[21]			100	12	0.56	3.4	f	d	BPe	pr[+*]	
Stizostedion lucioperca[21]			100	0	0.4	16	f	sd	Pe	pc	
Perca fluviatilis[22]		S	100	−14	3.2	6.3	f	d	Pe	pc	
Lutjanidae											
Lutjanus apodus[19]			350[b]	5	0.92	11	m	s	Pe	pr[+]	100
Haemulidae											
Haemulon sciurus[23]			150	−23	0.81[l]	192	m	s	Pe	pr	50
Sparidae											
Lagodon rhomboides[24]	se	S	300	−21	0.6	250	m	s	Pe	h,pr	50
Sargus annularis[25]	se		800	17	0.91[l]	4.2	m	s	Pe		
Sciaenidae											
Equetus acuminatus[19]			600	−39	0.75	2500	m	s	B	pr[+]	60

TABLE 16.1. (*Continued*)

Order / Family / Species	OC	HP	1 f_b (Hz)	2 S_{fb} (dB)	3 Q_{10dB}	4 HI	5 Habitat			6 fb	7 sp
Cichlidae		S									
Tilapia macrocephala[24]		S	310	−9	1.1[l]	48	b	s	Pe	pl	+$_g$
Pomacentridae		S									
Eupomacentrus dorsopunicans[26]			500	−19	1.43	170	m	s	Pe	o	180
Labridae											
Tautoga onitis[27]			150	−20	1.0[l]	120	m			pr	70
Thalassoma bifasciatum[19]			500	5	0.86	14	m	s	Pe	pr	
Gobiidae		D									
Gobius niger[25]			<100	3	1.0[l]	<14	m	sd	B	pr*	+$_g$
Scombridae											
Thunnus albacares[28]			500	−11	0.83	88	m	s	Pe	pc	
B. Bony fish without a swimbladder											
Perciformes											
Scombroidae											
Euthynnus affinus[32]		S	500	7	.91	11	m	sd	Pe	pc	
Pleuronectiformes											
Pleuronectidae											
Pleuronectus platessa[33]		S	110	−3	.79[l]	17	m	s	B	pr	
Limanda limanda[33]		S	110	−12	.79[l]	47	m	s	B	pr	
C. Cartilagenous fishes											
Heterodontiformes											
Heterodontidae											
Heterodontus francisci[29]		B	40	17	2.1	0.61	m		Pe	pc	
Lamniformes											
Carcharhinidae											
Negaprion brevirostris[30]		B	320	−12	0.81[l]	79	m		B	pc	
Carcharhinus leucas[31]		B	450	0	0.96	22	m		Pe	pc	

The common names of the fishes are found in Fay (1988a) and in this text. [1]Coombs and Popper (1982), [2]McCormick and Popper (1984), [3]Tavolga (1974b), [4]Jerkø et al. (1989), [5]Popper (1971), [6]Siegmund and Wolff (1973), [7]Popper (1972), [8]van Schade (1971), [9]Kleerekoper and Chagnon (1954), [10]Popper (1970), [11]Poggendorf (1952) and Weiss et al. (1969), [12]Popper and Tavolga (1981), [13]Abbott (1973), [14]Hawkins and Johnstone (1978), [15]Offutt (1973), [16]Chapman (1973), [17]Fish and Offutt (1972), [18]Coombs and Popper (1979), [19]Tavolga and Wodinsky (1963), [20]Wodinsky and Tavolga (1964), [21]Wolff (1968), [22]Wolff (1967), [23]Tavolga and Wodinsky (1965), [24]Tavolga (1974a), [25]Dijkgraaf (1952), [26]Myrberg and Spires (1980), [27]Offutt (1971), [28]Iversen (1967), [29]Kelly and Nelson (1975), [30]Banner (1967), [31]Kritzler and Wood (1961), [32]Iversen (1969), [33]Chapman and Sand (1974). (*Gymnocephalus cernuus*, *Stizostedion lucioperca* and *Sarotherodon melanotheron* are the current names for *Acerina cernua*, *Lucioperca sandra* and *Tilapia macrocephala*, respectively).

Column 0: OC, type of otophysic connection; sa, separate anterior air bubble close to the ear; se, swimbladder makes external contact with auditory region of the skull; sl, swimbladder invades auditory region; ss, swimbladder invades inner ear, terminating on or near saccule; w, Weberian ossicles; HP, hair cell patterns (Fig. 16.4); A, alternating; D, dual; O, opposing; S, standard; V, vertical.

Column 1: f_b, best frequency (or in case of low pass characteristic the cutoff frequency) of the pressure audiogram. [b], bimodal audiogram. (All data are reviewed in Fay (1988a), except that of *Anguilla* and *Salmo gairdneri*.)

Column 2: S_{fb}, threshold of f_b in dB re: 1 µbar (1 µbar = 0.1 Pa). In case of a low pass audiogram, the highest sensitivity has been taken as the value of S_{fb}.

Column 3: Q_{10dB}, quality factor of audiogram measured at 10 dB attenuation, [l]low pass audiogram. (For low pass audiograms the bandwidth was set at the high frequency side by the frequency with 10 dB attenuation – with respect to the horizontal asymptote – and at the lower frequency side arbitrarily at 25 Hz. Accordingly Q_{10dB} has been calculated.)

Column 4: HI – hearing index (for sound pressure). $HI = (f_b/Q_{10dB})^{0.5} x \ 10\{exp(-S_{tb}/20)\}$.

Column 5: left: hab, habitat; b, brackish; f, fresh water; m, marine. center: d, deep water (for fresh water > 5 m, brackish > 10 m and marine > 20 m); s, shallow water. right: Pe, pelagic; B, benthic.

Column 6: fb, feeding behavior; h, herbivorous; o, omnivorous, no fish food; pr, predator of invertebrates; pr$^+$, predator of invertebrates and small fish; pc, piscivorous; pl, planktonivorous. *, is itself prey for larger fish.

Column 7: sp, produces sounds with internal mechanism (+), with frequency of fundamental in Hz, $_g$, known to occur in other species of the same genus. The listed fundamental frequencies are not always for the same type of vocalization.

of underwater sounds and underwater acoustics on fish hearing, and perhaps on its evolution.

Due to the relatively high noise levels (mostly due to wind) in the oceans and seas, detection of low intensity biological sounds is likely to be impossible because of extensive masking. Such sounds will not be detected by Gadidae (codfishes) and Scrombridae (mackerels and tunas). These fish live in open waters where biological sounds are generally weak. Poor and moderate (absolute) sensitivity is also found in marine fishes living near exposed coasts and fish living in turbulent estuaries. In contrast, we would predict that fishes living in quiet environments (e.g., lagoons, mangroves) would have high sensitivity and the ability to hear a wide sound bandwidth. The reverse seems to hold for fresh water fish living in waters with strong currents (e.g., *Salmo*). Sounds in very shallow water with soft bottoms, as often found in lower courses of rivers, will be so low that they demand special hearing adaptations in order to be detected. Although the number of species studied is limited, the data lead us to suggest that these rules are valid and may have served as constraints on the evolution of fish auditory mechanisms. However, there are a number of notable exceptions to these rules. In order to understand these exceptions we need a better understanding of fish hearing.

3. Use of Sounds by Fishes

Before discussing how fishes hear and the evolutionary trends in hearing, we will consider acoustic behavior of fishes. Sounds are produced by many fishes, invertebrates, and marine mammals (reviewed in Moulton 1963; Fish and Mowbray 1970; Tavolga 1971; Demski, Gerald, and Popper 1973). As a consequence, the aquatic environment is a biologically "noisy" place.

As discussed by Tavolga (1971), sound is more appropriate for underwater communication than other signals in a number of ways. Since the wavelengths used in biological signals are generally long, rocks and other barriers probably only minimally impede communications. Although there are some limitations as to when biological sounds would be most useful to fishes, it is likely that selective pressures resulted in the use of swim-

bladder for vocalizations. It is reasonable to suggest that improvement of hearing and refinement of vocalizations co-evolved.

Acoustic theory leads us to predict that fish are likely to communicate with higher frequencies in very shallow waters than in deep waters (Section 2.3). Even in such environments, however, frequencies would still be well below those used by terrestrial vertebrates and with fundamental frequencies well within the constraints of the muscle systems used to produce the sounds.

Biological sounds in the aquatic environment are many and variable (e.g., Moulton 1963). Such sounds can be louder than the sounds produced by wind (Tavolga 1971). Fish vocalizations are often generated using the swimbladder either as an amplifier for sounds produced with other body parts or as an actual sound generator using muscles that cause direct vibration of the swimbladder. The swimbladder is close to being a monopole sound source. These sounds generally have an upper frequency of 800 Hz, with fundamental frequencies between 25 and 250 Hz (Fish and Mowbray 1970; Tavolga 1971; Demski, Gerald and Popper 1973; see Table 16.1, column 7). Swimming sounds, produced as a result of hydrodynamic interactions between parts of the fishes body and water, may contain energy up to 1 kHz (Moulton 1960). Invertebrates, and particularly crustaceans and mollusks, produce loud (nonmonopole) sounds that range from several hundred Hz to many kHz (Moulton 1963; Fish and Mowbray 1970; Tavolga 1971; Hawkins and Myrberg 1983).

Many swimbladder sounds are used for intraspecific communication, including courtship and territorial behavior (reviewed in Tavolga 1971; Demski, Gerald, and Popper 1973; Myrberg 1981; Hawkins and Myrberg 1983). There is a moderate but significant concurrence of vocalization bandwidth and hearing bandwidth for 15 species for which data are available on both hearing and sound production (Fig. 16.2). The geometric mean of the bandwidth of the sonograms increases nearly proportionally with that of the audiogram, suggesting co-evolution of vocalization and frequency sensitivity. Further analysis of the data also indicates that larger species generally produce sounds with lower frequencies than small species.

Fishes, as other vertebrates, don't just respond to the presence or absence of sound (e.g., Dooling,

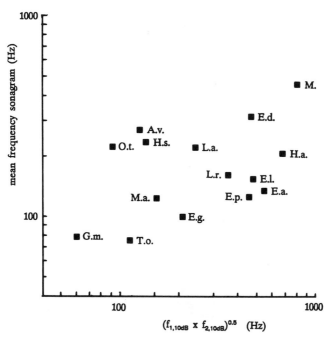

FIGURE 16.2. Graph showing the relationship between the swimbladder sounds produced by selected species and their hearing ability. The vertical axis shows the geometric mean of the (nearly) whole range of the bandwidth of the swimbladder produced sounds (estimates from sonagrams in Fish and Mowbray 1970; Horch and Salmon 1973 and Hawkins and Myrberg 1983). The horizontal axis shows the geometric mean of the frequencies f_1 and f_2 limiting the 10 dB bandwidth of the hearing range of the species, determined from the psychophysical data in Fay (1988a) (also see Table 16.1). The correlation between the two parameters is 0.56 ($p = 0.01$, $n = 15$). A.v., *Albula vulpes*; E.a., *Equetus acuminatus*; E.d., *Eupomacentrus dorsopunicans*; E.g., *Epinephelus guttatus*; E.l., *Eupomacentrus leucostictus* (not in Table 16.1); E.p., *Eupomacentrus partitus* (not in Table 16.1); G.m., *Gnathonemus petersii*; H.a., *Holocentrus ascenscionis*; L.a., *Lutjanus apodus*; L.r., *Lagodon rhomboides*; M, *Myripristis*; M.a., *Melanogrammus aeglefinus*; O.t., *Opsanus tau*; T.o., *Tautoga onitis*. In all cases but *Myripristis*, the sound production and hearing data are for the same species.

Chapter 26). Instead, many species respond behaviorally to the temporal pattern or to specific frequency components of sounds (e.g., Winn, Marshall, and Hazlett 1964: Winn 1972; Spanier 1979; Myrberg 1981; Hawkins and Myrberg 1983). In this way fishes can discriminate between behaviorally relevant and behaviorally irrelevant sounds. As an example, various species of pomacentrids (damselfishes) produce sounds that differ from one another in temporal parameters and not in frequency range. The various pomacentrid species, discriminating between sounds of conspecifics and those of other species, do this on the basis of these temporal patterns (Spanier 1979).

4. Structure and Function of Teleost Ears

The structure and function of fish ears has been reviewed several times (e.g., Platt and Popper 1981; Popper and Coombs 1982; Popper 1983; Popper and Fay 1984). Consequently, we will only present a short overview of the subject.

4.1 Ear Structure

Fish ears have three semicircular canals and three otolith organs—the saccule, lagena, and utricle

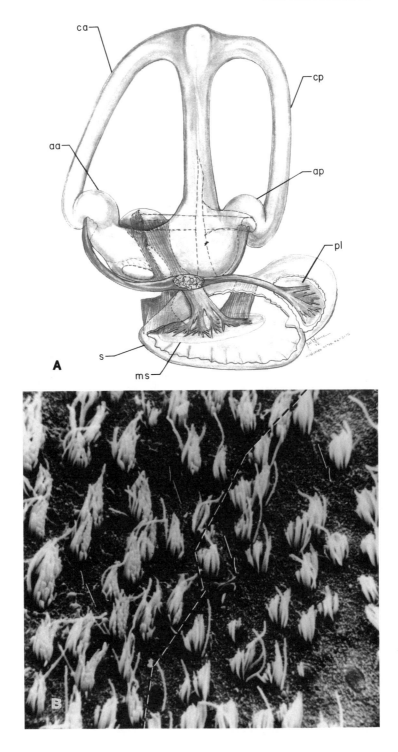

A

B

(Fig. 16.3A). Some species have an additional endorgan, the macula neglecta, the function of which is not understood in most fishes. Each otolith organ in teleost fishes has a single, calcareous otolith lying close to a sensory epithelium (or macula) which has numerous sensory hair cells and supporting cells. A thin otolith membrane connects to the otolith surface and to the microvilli on the supporting cells of the macula (Dale 1976; Popper 1977; Popper and Hoxter 1981) and mechanically couples the otolith to the epithelium (Popper 1978, 1983; Fay and Popper 1980).

The apical surface of each sensory hair cell has a ciliary bundle that projects into the narrow space between the macula and otolith (Fig. 16.3B). The ciliary bundle has many stereocilia and one eccentrically positioned kinocilium (or true cilium). Each epithelium is organized so that all the hair cells in a particular region have their kinocilia in the same orientation (see Fig. 16.3B). As a consequence, each macula may be divided, morphologically, into "orientation" groups (Fig. 16.4).

4.2 Ear Stimulation

Since the density of a fish's body is similar to that of water, the motion of the body in response to a sound follows the displacements of the medium. However, the otoliths lag the rest of the body because of their weak binding with surrounding tissues and their nearly three times higher density than water. The actual sensory stimulus is the shearing that occurs when the dense otolith and the far less dense macula move at different amplitudes and phases relative to each other during acoustic stimulation. This relative motion forces the ciliary bundle to bend, thereby providing the mechanical stimulus for hair cell stimulation (Flock 1971; Hudspeth and Corey 1977).

Depolarization of the hair cell occurs when the ciliary bundle is bent, with maximum depolarization resulting from bending towards the kinocilium. This direction defines the functional orientation of the hair cell. Bending in other directions results in depolarization or hyperpolarization of the hair cell, with the level of response being a cosine function of the direction of bending (Flock 1971; Hudspeth and Corey 1977).

The vibrating otoliths behave like damped harmonic oscillators. According to measurements by de Vries (1950), the mechanical cutoff frequency is approximately 50 Hz in a fish several decimeters in length. The displacements of the fish's body is (ideally) a direct measure of the displacement of the hair cells with respect to the immobile otoliths for frequencies above 50 Hz. Thus, the ultimate stimulus of the otolith organs is a displacement stimulus.

Since the gross structure of the three otolith organs are similar to one another, at least to a first approximation, they have the same response potential with respect to low- and high-frequency hearing, directional hearing, and vestibular functions. The roles of the individual otolith organs are certainly affected by their specific connections to other structures in the auditory pathway (for example, the swimbladder), morphological characteristics of the otoliths, orientation and position of the sensory epithelia, hair cell orientation pattern and density, and ciliary bundle length. As a consequence of these and other characteristics, one or more of the otolith organs may have multiple functions (see Platt and Popper 1981; Popper, Platt, and Saidel 1983). Moreover, interspecific differences

◁————————————————————————

FIGURE 16.3. (A) Medial view of the right ear (total length about 5 mm) of the Atlantic herring *Clupea harengus* (anterior to the left, dorsal to the top). This ear is typical of teleost fishes other than the otophysans (where the lagena is larger than the saccule). The otoliths are shown in each of the otolith organs, and the innervation is typical of that for other species. aa, crista of anterior semicircular canal; ap, crista of posterior semicircular canal; ca, anterior semicircular canal; cp, posterior semicircular canal; ms, macula of saccule; pl, macula of lagena; s, saccule. (Redrawn from Retzius, 1881). (B) Scanning electron micrograph of the saccule epithelium of the Hawaiian lizardfish (*Saurida gracilis*). The cells shown are divided into two groups that differ in their orientation (separated by dashed line). All the cells in each group are oriented in the same direction and have their kinocilia pointed in the same way (arrows). (From Popper 1983, reprinted with permission of University of Michigan Press.)

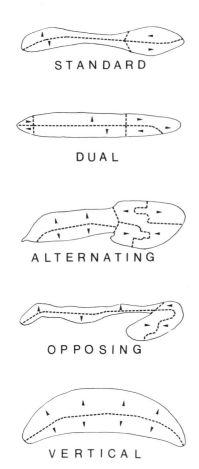

FIGURE 16.4. Schematic drawings of saccular sensory epithelia from five species showing the different general class of hair cell orientation patterns found among teleost fishes. The dashed lines separate groups of hair cells oriented in different directions while the arrows point in the direction of the kinocilia in each cell group (see Fig. 16.3B). Anterior is to the right and dorsal to the top in all drawings. (A) "Standard" pattern from the salmoniform, *Coregonus clupeaformis*. (B) "Dual" pattern found in the gadiform *Lota lota*. Fishes with this pattern have horizontally oriented cell groups at the rostral and caudal ends of the saccular epithelium. (C) "Alternating" pattern as found in the moray eel, *Gymnothorax* ssp. As compared to the standard pattern (A), the caudally oriented cells divide the rostrally oriented cells into two distinct groups. (D) "Opposing" pattern as found in the anabantid *Trichogaster trichopterus*. The two groups of horizontally oriented cells oppose one another rather than lie parallel to one another as in the standard pattern. (E) "Vertical" pattern in the mormyrid fish *Gnathonemus*. Animals with this orientation pattern (including all of the otophysans) do not have horizontally oriented saccular hair cells. (From Popper and Coombs 1982.)

in the morphology and ultrastructure of the ear and peripheral structures may result in significant differences in hearing capabilities (e.g., Popper and Coombs 1982).

5. Hearing Abilities, Specializations and Habitat

Data on behavioral hearing capabilities have been extensively reviewed by Fay (1988a). These data mainly concern species with a swimbladder and were obtained using sound pressure as the relevant stimulus. (Pressure signals are generally preferred experimentally to a displacement signal since a pressure stimulus is easier to generate and calibrate under laboratory conditions.) Instead of presenting the threshold data (audiograms) again (but see Fay, 1988a), it is more interesting to look at the relationship between the frequency of greatest sensitivity, f_b, and the pure tone pressure sensitivity at this frequency (S_{f_b}) for 40 species that have a swimbladder (Fig. 16.5a). Analyzed in this way, the data show that a higher f_b is related to greater sensitivity. A regression analysis yielded a regression coefficient deviating from 0 ($P < 0.005$). On average, a 10-fold higher f_b results in a 19.8 dB lower threshold.

A distinction has been made among three groups of teleost fish in Fig. 16.5a. One group (I) includes most fishes of the Otophysi (i.e., the Cypriniformes, Characiformes, and Siluriformes), all of which have a series of bones, the Weberian ossicles, that are thought to enhance hearing by improving the acoustic coupling between the swimbladder and the inner ear (e.g., Platt and Popper 1981). A second group (II) includes species from several unrelated taxonomic orders. These species have some specialization thought to enhance hearing by bringing the pressure to displacement transducer closer to the ear than in species without such specializations (Table 16.1, column 0). These specializations include anterior projections of the swimbladder or special air bubbles in or near the ear cavity. The third group (III) of fish also represents diverse taxa. None of these species have structures known to enhance hearing.

For the sake of convenience, fishes in the first two groups have often been referred to as hearing

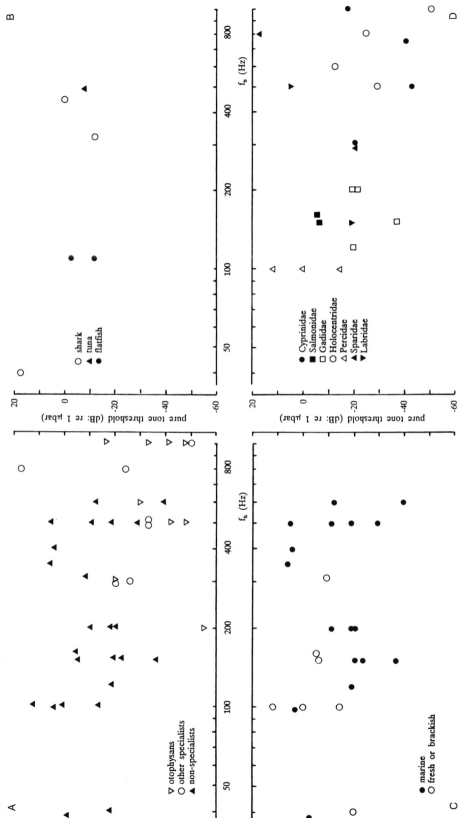

FIGURE 16.5. Relationships between psychophysical pure tone pressure thresholds and best frequency (f_b). The data of this figure are obtained from Fay (1988a), Abbott (1973) (*Salmo gairdneri*), and Jerkø et al. (1989) (*Anguilla*) and are listed in Table 16.1. For experiments done with particle motion, equivalent pressure thresholds have been calculated for the far field condition. (A) Relation between the best frequency f_b and the corresponding pressure threshold S_{fb} for swimbladder-bearing teleosts. (B) As a., for six actinopterygian species that do not have a swimbladder. They are the sharks *Heterodontus francisci*, *Negaprion brevirostris* and *Carcharhinus leucas*, the tuna *Euthynnus affinus*, and the flatfishes *Limanda limanda* and *Pleuronectes platessa* (Fay 1988a). (C) As a., for nonspecialists, distinguished by habitat. The data points of this panel are a subset of those of a. (D) As a., for genera of the same family. (*Salmo gairdneri* probably belongs to the genus *Oncorhynchus*.)

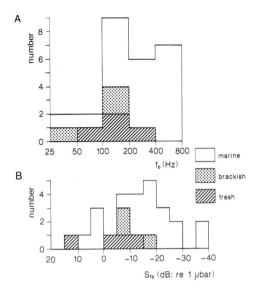

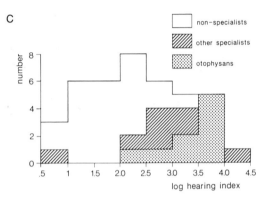

FIGURE 16.6. Occurrence of best frequency f_b (A) and sensitivity at best frequency, S_{fb} (B) of nonspecialized fish, distinguished to habitat. (C) Hearing index (HI = $(f_b/Q_{10dB})^{0.5} \times 10\{\exp(-S_{fb}/20)\})$ of otophysans, hearing specialists, and nonspecialized fish.

"specialists." Fishes without peripheral specializations for hearing will be referred to as hearing "nonspecialists." Hearing specialists generally have higher best frequencies and greater sensitivities at f_b than nonspecialists (Fig. 16.5A). Moreover, hearing specialists often have a greater sensitivity over the whole audiogram than nonspecialists. Psychophysical experiments show that fishes without a swimbladder (shown in Fig. 16.5B, also see Table 16.1), such as the Pleuronectiformes (flatfishes) and elasmobranchs, have poor pressure sensitivity (Fay 1988a). Theoretically, pressure sensitivity of these animals should even be absent (but see van

den Berg and Schuijf 1983). Their best frequencies f_b are similar to those of the nonspecialized fish.

The maximum f_b of fish (1,000 Hz) is low compared to the f_b of many terrestrial tetrapods and particularly to that of birds and mammals (see Fay 1988a; Fay Chapter 14, Dooling, Chapter 26, for relevant data). One possible reason for the low f_b is that extant fish descended from ancestors without a swimbladder, and thus without pressure sensitivity and therefore sensitivity for high frequencies. An alternative explanation is that fishes can only produce low frequency sounds (see Section 3) due to constraints on the sound producing mechanisms (Tavolga 1971; Demski, Gerald, and Popper 1973) and so development of high frequency hearing would not provide any advantage to fish.

A distinction is made between habitats (e.g., fresh water, brackish, or marine) for the nonspecialists in Fig. 16.5C. Although the number of species for which data are available is small, marine nonspecialists often have a higher f_b (Fig. 16.6A) and a higher sensitivity (Fig. 16.6B) than fresh water nonspecialists. The difference is may result from the noisier environments of the marine fishes, particularly if they live near the coast of shallow tropical waters and/or on coral reefs. Interestingly, *Salmo* and *Anguilla*, genera that invade fresh, brackish, and sea water depending on the period of their life cycle, have an intermediate position.

Looking more closely at the habitat, there is a tendency for bottom dwelling nonspecialists to have lower best frequencies than pelagic species. Table 16.1 demonstrates that perciform fishes have a wide range of f_b and maximal sensitivity values (S_{fb}). Genera belonging to the same family of perciforms and families of other orders show less variation in their combined f_b–S_{fb} values than would be expected by chance.

The occurrence of a smaller variation among family members was examined statistically by taking the distances in the f_b–S_{fb} plane between all possible pairs of the 21 data points (depicted in Fig. 16.5D) and by taking the distances between pairs of data points of the same family. A t-test was applied to the both resulting distributions ($P = 0.001$). The frequency selectivity of the audiograms (i.e., the quality factor Q_{10dB}) does not show a difference among the three groups. However, an audiogram with low pass character has never been observed for otophysans, but has been found for 29% of the nonspecialized genera. Among all taxa studied, Cypriniformes appear to have the broadest bandwidth (although selected nonotophysans have equally wide bandwidths).

Due to the considerable interspecific variability in the pressure sensitivity of species that have a swimbladder, we introduce a dimensionless index, the hearing index (HI), which gives a numerical indication of the quality of the audiogram. The limited number of species with a known audiogram does not justify a multidimensional index. Moreover, this approach makes interspecific comparison easier to handle. Pressure sensitivity, and not displacement sensitivity, has been chosen since (a) pressure is more important for detection of higher frequencies and (b) far-field hearing, favoring the biologically interesting high frequencies, is mainly pressure-mediated.

Although HI is a single index, it is comprised of the most characteristic parameters of the audiogram, f_b, Q_{10dB}, and S_{fb}. HI is defined as $(f_b/Q_{10dB})^{0.5} \times 10\{\exp(-S_{fb}/20)\}$. (Note that $f_b/Q_{10dB} = f_2 - f_1$. f_1 and f_2 are the limiting frequencies in calculating Q_{10dB}). The square root of f_b seems reasonable to use since above 500 Hz the signal-to-noise ratio of biological sounds versus ambient noise improves roughly with this factor up to about 2 kHz. Moreover, it is probably more effective to include a frequency range at higher frequencies instead of having much overlap between the auditory and lateral line systems. The square root of Q_{10dB}^{-1} (i.e., the relative bandwidth) is chosen on the basis of signal to noise considerations and because it is considered less important than the linearly weighted sensitivity at f_b.

The occurrence of the HIs listed in Table 16.1 are plotted in Fig. 16.6C. Since the audiogram depends upon the stimulus conditions (and may vary with different experimental paradigms — Popper and Fay 1973; Fay 1988a), only the order of magnitude of HI is relevant. It is remarkable that the range varies by a factor of 3,000. This considerable range is mainly caused by the large interspecific variability of S_{fb}. Species with a low HI (e.g., < 10) are generally not known to use sound (the one known exception is *Opsanus*) as extensively as species with a higher HI. The otophysans and nonspecialized fish form two rather homogeneous groups, but the specialized fish other than

the otophysans appear to be more scattered in their HI. The HI values of genera belonging to the same family are, on average, closer together than those of all species given in Table 16.1 (statistically examined by an analysis of variance, $P = 0.001$). The one known exception to this is the Holocentridae, a family of fishes that has specialized and nonspecialized genera (Table 16.1 column 0) (although there is now some question as to whether these fishes are in the same taxonomic family). The consistency within a family is more pronounced for fresh water than for marine fish.

The rather primitive *Notopterus*, *Gnathonemus*, and *Albula* (bonefish) show moderate to high indices. *Notopterus* and *Gnathonemus* hear well, and *Gnathonemus* is known to use sound for communication (Crawford Abstract F, this volume). In contrast, *Anguilla* (eel) has poor hearing.

There is a large scatter among the Perciformes, an order with thousands of species having substantial variability in behavior and ecology. Among perciform fishes, the pisciforous fish and fresh water predators have relatively low HIs. Auditory specializations have evolved only occasionally among perciform fishes, and such species generally live in less quiet waters than many cypriniforms.

Figure 16.7 gives the relationship between the number of species and HI, distinguished by food type: (1) plankton, plants (and its debris) and/or small invertebrates, (2) invertebrates, and (3) invertebrates with fish or exclusively fish. The data suggest that predators generally do not hunt by the sounds produced by their prey. The otophysans are primarily in the first group, whereas most Gadidae and freshwater Percidae belong to the third group.

Displacement sensitivity has been directly determined for only a few of the more than 25,000 teleosts and elasmobranch species (Fay 1988a). The thresholds and best frequencies of *Gadus morhua*, a nonspecialized fish, and two flatfish without a swimbladder, *Limanda limanda* (dab) and *Pleuronectus platessa* (plaice), are similar to one another. However, displacement sensitivity is about 40 dB lower for the goldfish, *Carassius auratus*, a hearing specialist (Fay 1988a). When the pressure sensitivity of *Carcharhinus leucas* (bull shark) is recalculated to give sensitivity in terms of displacement rather than pressure, sensitivity is similar to that of the flatfish, with a best frequency

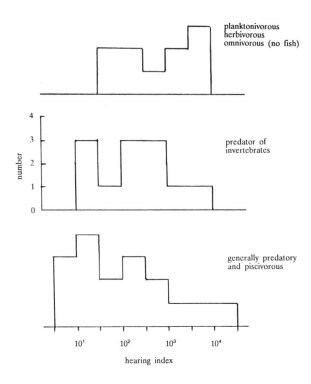

FIGURE 16.7. Occurrence of hearing index values for 39 species with swimbladders as related to three types of food preferences. The figure is a graphical representation of data from Table 16.1.

of 500 Hz (Fay 1988a). The displacement sensitivity of *Heterodontus francisci* (horn shark) and *Negaprion brevirostris* (lemon shark) is much lower, with *Heterodontus* having a best frequency of 80 Hz (Fay 1988a).

The data reviewed by Fay (1988a) for frequency discrimination, pure tone masking and critical bandwidth suggest that otophysans are probably superior to other taxa in these aspects of hearing, as they are with frequency sensitivity. However, this suggestion is only provisional since data are available for very few species, and even these are subject to significant uncertainties due to differences in the ways in which the data were obtained.

6. The Swimbladder

The swimbladder evolved tens of millions of years ago. Although its initial function was probably for buoyancy (Blaxter 1981), its position in the abdominal cavity and its significantly different density from water has resulted in its becoming involved with sound detection and sound production in many species. The swimbladder provides the potential for an increase in the workable hearing bandwidth of a fish by lowering thresholds. This is particularly true in species where the swimbladder comes near the ear (as in hearing specialists). The otophysan Weberian ossicles have further enhanced this effect by effectively bringing the movements of the swimbladder into direct contact with the ear. The role of the swimbladder in hearing is a significant adaptation, particularly in the far-field where the pressure to displacement ratio increases with frequency. As a consequence, the swimbladder (and ossicles) extended the auditory "scope" in the space and frequency domains.

Since behaviorally determined bandwidths are basically similar to those measured electrophysiologically (Fay and Popper 1974, 1975), it is likely that the shape of the audiogram is predominantly a feature of the otolith organs. The shift of the best frequency by one to two octaves to the higher frequencies may have evolved together with the ossicles or other adaptations that bring the swimbladder very close to the inner ear.

6.1 Swimbladder Mechanics

Although the swimbladder behaves like a harmonic oscillator, it is damped so heavily by the surrounding tissues that, for frequencies near the resonance frequency, f_r, the indirect displacement wave (due to the pressure dependent bladder pulsations) does not increase much compared to the indirect displacement wave at low frequencies. The latter effect, however, depends upon various system parameters of the swimbladder (e.g., Schuijf 1976a, b; De Munck and Schellart 1987).

Theoretically, f_r is roughly proportional to the inverse of the (linear) size of the swimbladder. The f_r of the swimbladder of a fish less than 20 cm long will be too high to affect the shape of the audiogram. For larger fish, and fish with high-frequency sensitivity, the resonance frequency may play a role in hearing (Blaxter 1981). Since f_r is proportional to the hydrostatic pressure, resonance is not important for fish living at many tens of meters of depth, such as the Gadidae.

The particle displacement amplitude produced by the swimbladder is, to a first approximation, proportional to its linear size (Poggendorf 1952). The average swimbladder volume of pelagic and benthic nonotophysans is 5.5% and 2.3% of body volume, respectively (calculated from data in Alexander 1959). These differences are directly related to density differences of the fish's body (Alexander 1959). For pelagic fish, larger swimbladders give a theoretical sensitivity enhancement of 2.5 dB. Some species such as *Gymnocephalus cernuus* (Alexander 1959) and *Salmo gairdneri* (rainbow trout – (Schellart, unpublished data) have larger swimbladders (8.7% and 8%, respectively). It is suggested that the increase in size of the swimbladder of *Salmo gairdneri* is probably related to the need for an increase of buoyancy in white water as well as an increase of leaping force by a sudden stretching of the swimbladder.

6.2 The Swimbladder and Hearing

In otophysans and other hearing specialists, hearing mediated by pressure (via the swimbladder) will dominate over hearing by direct particle displacement, particularly at frequencies above 100 or 200 Hz (Fay and Popper 1974, 1975). Psychophysical data of the nonspecialized *Salmo* (Abbott 1973; Hawkins and Johnstone 1978) indicate that a mixture of displacement and pressure information is used by that species.

Theories of directional hearing of nonspecialized fish indicate that both pressure and displacement information are needed (Buwalda, Schuijf, and Hawkins 1983; Schellart and De Munck 1987; Popper et al. 1988; Rogers et al. 1988). It has been shown that *Salmo gairdneri* can use the vectorial combination of two inputs for threshold detection (Schellart and Buwalda 1990). The answer to the general question of how extensively the swimbladder is used appears to depend on stimulus conditions (distance and frequency) and the behavioral paradigm used. In any case, the evolution of swimbladder involvement in hearing resulted in development of spatial hearing mechanisms. It is supposed that later adaptive changes in the swimbladder-otolith coupling will have further improved spatial hearing. Since data about directional hearing are available for only a couple of teleosts, this point cannot be evaluated further.

6.3 Swimbladder Shape and Hearing

The shape of the swimbladder varies among different species. Shapes may be of importance for hearing in some fishes (e.g., Poggendorf 1952; Tavolga 1977; Popper 1983). It has been demonstrated theoretically (Schuijf 1976a,b; De Munck and Schellart 1987) and experimentally (Poggendorf 1952; Tavolga 1977) that the displacement of the wall at the apex (rostral end) of the swimbladder is larger than of the side walls, at least for the few species that have been tested. For swimbladders of equal volume (Fig. 16.8), the displacement of the apex increases strongly with eccentricity (Fig. 16.8B, curve 0.0). An axis ratio of 1:2 increases the amplitude of the scattered displacement wave by a factor of 1.6 with respect to that of the sphere, and an axis ratio of 1:16 gives an improvement of 6.4. For the few species for which we know axis ratios, a relationship between the ratio and the audiogram cannot be simply established since another important factor is the distance between the apex and the otolith organs. For a sphere, the scattered particle displacement wave decreases with the square of the distance (measured from the center). However, for an elongate swimbladder, the particle displacement in front of the apex diminishes more strongly as the curvature of the apex becomes sharper. The effect is such that for larger distances from the apex, slender bladders have smaller displacements. The displacements as a function of the axis ratio for some distances from the apex are illustrated in Fig. 16.8B. When the volume of the swimbladders

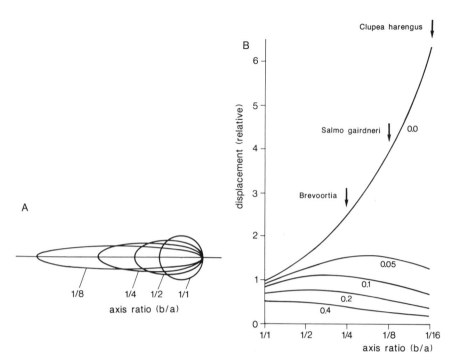

FIGURE 16.8. Increase of displacement amplitude as a function of shape and of the distance from the apex for gas filled bladders with the same volume. When the swimbladder is approximated by an ellipse of revolution the displacement amplitude at the apex is proportional to $(b/a)^{-2/3}$, where a is the semimajor axis and b the semiminor axis of the long shaped bladder. The displacement at the extended part of the major axis is proportional to $(b/a)^{4/3}\{x^2-1 + (b/a)^2\}^{-1}$, with x the distance to the center of the swimbladder, expressed in terms of a ($x >$ 1). (A) Sketch of elliptical shaped swimbladders with various axis ratio's but equal volumes, with the diameter of the equivalent sphere as reference for distance. (B) Displacement as a function of the axis ratio (b/a) for some shapes. The various curves give the distance from the coinciding apices of (a), expressed as a fraction of the radius of the sphere.

decreases, such as during a change in depth, it would be favorable not to increase the distance between the apex and the inner ears. In *Salmo gairdneri*, changes in volume mainly appear to affect the width of the bladder and not the distance between the apex and the otolith (unpublished data of Schellart). The principle of improving pressure sensitivity by making the distance of the bladder (or another air cavity) to the inner ear as short as possible has clearly been implemented by the nonotophysine hearing specialists. To what extent this also applied for nonspecialists is hard to conclude, since systematic data about the distance of the bladder to the inner ear are not available. However, at least one exception is known.

In *Anguilla anguilla*, the distance between the slender swimbladder (axis ratio 9) and the inner ear is very large (10 cm in a 50 cm long eel). Consequently, the indirect displacement is extremely small. Despite this, *Anguilla* has a reasonable pressure sensitivity for higher frequencies (Jerkø et al. 1989). Thus, the amplitude of the scattered displacement wave from the swimbladder is less attenuated than would be predicted for tissue that behaves as water (except for damping). Since specialized anatomical structures to improve the coupling between bladder and inner ear are lacking in the eel (Jerkø et al. 1989), the backbone, which is located very close to the swimbladder, may be involved. We expect a compression wave through the bar-like backbone to be less attenuated than the wave through the soft tissues.

The angle between the major axis of the swimbladder and the horizontal plane varies from 7° to 18° in different species (Blaxter 1981). This angle is such that the axis points more or less to the inner ears. It is also possible that the bicornual shape of the (rostral) apex of the swimbladder found in a number of species, including two unrelated hearing specialists, *Myripristis* (soldierfish) and *Notopterus* (Popper 1983), further enhances hearing by bring the swimbladder closer to the inner ear when compared to species without this shape.

Despite the apparent adaptive advantage of some swimbladder shapes, the variability in shape is remarkable. Many specific shapes may be of no significance with regard to hearing abilities of fishes. However, the final swimbladder anatomy is no doubt related to a mixture of anatomical and physiological constraints of the various functions of the swimbladder (Blaxter 1981), as well as constraints imposed by morphologic factors such as the size, shape, and location of the internal organs.

6.4 Conclusions

The acoustic role of the swimbladder of hearing specialists is still not well understood. However, as illustrated with the data from *Anguilla anguilla*, this holds for nonspecialized species (e.g. Tavolga 1977; Jerkø et al. 1989; Schellart and Buwalda 1990). Extension of the swimbladder model to include the contributions of the backbone and skull, as well as a better understanding of the mechanics of the otoliths and micromechanics of the hair cells, seems to be a prerequisite to further development of comparative theories on fish hearing.

7. Functional Morphology of the Inner Ear

7.1 Hair Cell Orientation Patterns

One of the striking variants in the structure of the fish ear is in the orientation patterns of the sensory cells on the sensory epithelia (or maculae) of the otolith organs. Such variability is most frequently associated with the saccule, the endorgan thought to be primarily involved with audition in most species (e.g., Popper and Fay 1973; Platt and Popper 1981). While morphologic variation is not as great in the lagena and in the utricle as it is in the saccule, hair cell orientation patterns in species using the utricle or lagena for hearing are strikingly different from other species (Popper and Coombs 1982). This is most apparent in the herrings (Clupeiformes), where the utricle is highly specialized for hearing and has a hair cell orientation pattern that is very different from utricles of most other vertebrates (Popper and Platt 1979). Similarly, the lagena of the otophysans may have a different auditory role than in other species. The much larger size of the lagena in otophysans than in other species seems to be related to this different role.

The majority of teleost species have saccular hair cells oriented in four directions (Figs. 16.4A–D; Popper and Coombs 1982). Two of the orientation "groups" are along the rostral–caudal axis of the fish, while the other two groups are oriented on the fish's vertical axis. Several other fish groups only have saccular hair cells oriented vertically (Fig. 16.4E).

The four-quadrant "standard" pattern is the most common among most bony fishes (Table 16.1, column 0). The "standard" pattern has two horizontally oriented groups at the rostral end of the saccular epithelium and two vertical groups at the caudal end of the epithelium (Fig. 16.4). Although each of the four quadrants overlaps slightly with its neighbors, the differences in orientation between cells from the various groups is distinct—there is no gradation from one orientation to the other.

There are three additional four-quadrant patterns (Popper and Coombs 1982). In each instance (Fig. 16.4), the major differences are associated with the hair cells oriented along the animals horizontal axis. In some species, such as the moray eel (*Gymnothorax*), the horizontal groups alternate in position (Popper 1979). In other species, such as *Gadus* and its relatives, there are horizontally oriented cells on the caudal as well as rostral end of the epithelium (Dale 1976).

Perhaps the most interesting is the "opposing" pattern, where horizontal cell groups abut one another. This pattern is found in such taxonomically diverse groups as *Myripristis* (Holocentridae, Popper 1977), *Notopterus* (Notopteridae, Coombs, and Popper 1982), and *Trichogaster* (three-spot gourami, Anabantidae, Popper, and Hoxter 1981). Although these species are not related to one another, and even though it is likely that the "opposing" pattern was derived independently from the "standard" pattern in each group, all the species share one common characteristic—the anterior end of the swimbladder abuts the saccular chamber and, presumably, enhances audition (Section 6.3). In fact, behavioral data for a number of species with the "opposing" pattern show that they have a hearing bandwidth and sensitivity as broad as that of the otophysans (Fig. 16.5A) (Coombs and Popper 1979, 1982). Although the functional significance of the "opposing" pattern is not clear, the correlation between the pattern and hearing specialization is high.

Several deep-sea fishes of the family Myctophidae have the same pattern (Popper 1978). Although we know nothing about hearing in the myctophids, several of these species possibly communicate acoustically (Marshall 1967). Based upon these observations, we would predict that myctophids are hearing specialists.

The bidirectional "vertical" pattern is also of considerable interest with regard to the evolution of the ear. The "vertical" pattern in the otophysans is clearly derived, and evolved from a four-quadrant pattern in ancestral species (Popper and Platt 1983). In fact, *Chanos chanos* (milkfish), an otophysan thought to be typical of primitive members of this group (Rosen and Greenwood 1970), has an ear structure that has characteristics of both modern otophysans and nonotophysans. The saccule is elongate, the lagena is large (Popper and Platt 1983), and there are structures that could be primitive Weberian ossicles (Rosen and Greenwood 1970). However, the saccule has a four-quadrant orientation pattern.

The vertical pattern did not only arise in the otophysans, although it is in that group that the pattern is most ubiquitous (Popper and Platt 1983). The pattern evolved separately in the mormyrids, a group of sound-producing osteoglossomorph fish unrelated to the Otophysi. Other osteoglossomorphs have various saccular orientation patterns, but all are four-quadrant (Popper 1981).

It is perhaps significant that the vertical pattern evolved at least twice among the teleost fishes,

and in each case the ancestral ear structure had four quadrants. Moreover, the overall structure of the ear in the mormyrids and otophysans have other characteristics in common, including the relative sizes of the lagena and saccule. Although the mormyrids do not have Weberian ossicles, they have a small bubble of air intimately tied to the inner ear. The mormyrids have hearing capabilities similar to otophysans (McCormick and Popper 1984).

Although teleost fishes generally have a four-quadrant orientation pattern in the saccule, the vertical pattern is very likely to be the more "primitive" ancestral pattern among fishes. The basic vertical pattern has been identified in primitive fishes [e.g., *Amia* (bowfin), *Polypterus* (reedfish), *Scaphyrynchus* (sturgeon)] (Popper 1978; Popper and Northcutt 1983), at least two species of lungfish (Popper, unpublished), and all of the elasmobranchs that have been studied to date (see Section 9) (Popper and Fay 1977; Barber and Emerson 1980; Corwin 1981, 1989).

Despite the pervasiveness of the vertical pattern, there is some variation on the pattern that is of interest, and of potential significance for sound localization (see below). The saccular macula in primitive fish is often observed to twist in such a way that a significant portion of the hair cells are oriented perpendicular to the rest of the cells (this also happens in otophysans). This results in a pattern that bears some resemblance to the four-quadrant pattern found in teleosts. The difference between this modified vertical pattern and the four-quadrant pattern of teleosts is the gradual shift in orientation of the hair cells as the sensory epithelium changes direction, as opposed to no such change in the teleosts. It is reasonable to argue that among the nonteleost fishes having the vertical pattern, the "horizontal" cells are just curved onto a different plane (Popper and Northcutt 1983).

We conclude that there is a global consistency between the occurrence of the type of hair cell pattern and the division in the three functional groups of teleosts. The vertical pattern mainly occurs in the otophysans and the mormyrids, the alternating and opposing pattern in the nonotophysan hearing specialists, and the standard and dual type is especially found among the nonspecialists.

7.2 Otolith Structure and Relationship to the Epithelium

The functional significance of otolith shape has not yet been considered experimentally, but it may have some relevance to the function of the hearing system in fishes and in telling us about the evolution of the fish ear. Since the otoliths are extremely dense, they survive well in the fossil record. Consequently, there is a body of data indicating that the otoliths of many extant species resemble those found in many extinct species. This is significant since the shape of the otolith, and particularly of the sulcus or groove in which the sensory epithelium lies, is characteristic of particular species and may provide insight into the history of particular epithelial structures and even hair cell orientations.

While the shapes of the lagenar and utricular otoliths are very stable among different teleosts, there is wide variability in the saccular otolith shape, particularly in fishes other than the otophysans. The saccular otolith may be somewhat ellipsoid with a simple medial groove (sulcus) in which the saccular epithelium sits. Alternatively, the saccular otolith may be highly complex with intricate sculpting along the edges and striking changes in thickness from rostral to caudal or dorsal to ventral ends. As indicated earlier, the otolith (in the species studied) does not move above 50 Hz (de Vries 1950), and when it moves above 50 Hz motion is relatively simple. However, it is hard to imagine that the motion of the otolith relative to the sensory epithelium at frequencies near and below 50 Hz is simply a motion of rotation, especially considering the otoliths visco-elastic suspension and complex shape in some species. While there are no data to support this hypothesis, it has been suggested that the sculpting, together with the suspension of the otolith, results in differential stimulation of the parts of the saccular epithelium in various species (Popper 1983).

7.3 Functional Specialization of Otolith Systems: Multifunctionality and Directional Hearing

The biological significance of spatial hearing, its monaural and binaural cues, its physical (sound velocity) and anatomical (distance between ears) constraints, and the role of the swimbladder have been reviewed recently (Fay and Feng 1987; Fay 1988b). Here we will discuss the problem in terms of how various types of peripheral hearing structures are adapted to enhance spatial hearing.

Since the fish otolith system works as inertial detectors of linear motion (see Section 4) its function may be for detection of: (1) the kinetic component of sound waves; (2) linear acceleration of the fish body for very low frequencies; and (3) tilt of the head or of the body for frequencies approaching zero. The efficiency of the detectors depends on the amount of coincidence of its directional sensitivity with the direction of the stimulus.

A hearing system comprised of three orientation groups of hair cells with their axes of maximum sensitivity perpendicular to one another is potentially able to obtain sufficient information to detect the direction of the sound source, albeit with an ambiguity of 180° (see Schuijf 1981 for a discussion of this problem). Moreover, additional information provided by the displacement from the swimbladder [or, potentially, surface or bottom reflection in fish without a swimbladder (Schuijf 1981)] enables unique localization (direction and distance) of a monopole source, thereby solving the 180° ambiguity. Thus, the orientation of the three macular planes of both ears, together with their hair cell orientation patterns, provides more than sufficient information to suggest several possible mechanisms for localizing a monopole sound source (Buwalda, Schuijf, and Hawkins 1983; Schellart and De Munck 1987; Popper et al. 1988; Rogers et al. 1988).

Information for directional hearing in the median plane is provided for in nonotophysans by the four-quadrant saccular hair cell orientation pattern (Section 7.1). Each paired quadrant is innervated by a saccular branchlet of the eighth nerve (Saidel and Popper 1983), potentially enabling coding of the direction of the particle displacement. Directional hearing in the transverse plane is also possible using responses from the lagena.

The situation for hearing in the horizontal plane is somewhat different and more complex. Since, in many species, neither the saccule nor the lagena have hair cells with sensitivity along the mediolateral axis (Popper and Coombs 1982), directional hearing in the horizontal plane will be relatively inefficient using the saccule and/or lagena. The presence of a large angle between the two saccules (seen from the swimbladder apex) in fishes other than otophysans, may help these species localize in the horizontal plane. As a result of this angle, there is bilaterally a group of saccular hair cells with sensitivity in the horizontal plane. The angle between the directional sensitivity of these two groups of hair cells is the same as the above-mentioned anatomical angle between both saccules. For various nonspecialized fish, this angle may exceed 45°, which makes the saccule a competitor of the utricle for hearing in the horizontal plane (see below). When the direction of the swimbladder mediated-displacement stimulus deviates from the medial plane, as proposed by Schellart and Buwalda (1990), such a saccular pair is well-equipped for directional hearing in the horizontal plane.

In addition to using the saccules for horizontal localization, it is also potentially possible to involve the lagena. At the same time, most models for localization proposed to date do not involve the utricle in localization, and it is generally only implicated in vestibular functions (except in the clupeids) (Blaxter, Denton and Gray 1981). However, the fan-like hair cell orientation pattern of the utricle found in all fishes (and other vertebrates) is clearly suited for encoding the direction of the compound displacement stimulus in the horizontal plane according to a spatial "place" principle as shown experimentally (Fay 1984) and modelled theoretically (Schellart and De Munck 1987).

It is unlikely that the saccule in the Otophysi is able to obtain vectorial information for localization in the horizontal plane since the saccule does not have hair cells oriented in that plane. Moreover, it is likely that the otophysan saccule provides little or no vectorial information in any plane since it is only poorly driven by the particle displacement wave directly emitted by the sound source. In contrast, the otophysan saccule is likely to provide the pressure information needed to solve the 180° ambiguity.

A second, but major role of the otophysan saccule appears to be for higher frequency detection and better sensitivity when compared to the saccule in most nonotophysans (exceptions may be the specialized saccules found in some nonotophysans). Low-frequency detection in the otophysans

has become a role of the lagena, and the utricle, together with the lagena, should be the organs for providing vectorial information for spatial hearing (Popper 1983). This hypothesis is in accordance with the best direction vector of the utricular and lagenar fibers in goldfish (Fay 1984).

From the above considerations, it seems likely that bifunctionality of the utricle should not be excluded in teleost fish. Various models, the one more conceptual (Buwalda, Schuijf, and Hawkins 1983), the other more in the direction of a hardware (neuronal network) description (Popper et al. 1988; Rogers et al. 1988; Schellart 1989), have been proposed to describe the underlying mechanisms of directional hearing. A serious problem for the models using phase-locking of spike firing as a basic tool, is the finding that fibers from the same otolith organ show considerable variation in the phase of their spiking (Fay 1988b).

For localization over short distances, fish have to deal with the complex fields of multipole sources. If fishes are able to localize such sources – something that is not known – then the underlying peripheral and central mechanisms would have to be significantly different from those described to this point. Although we have no suggestion as to how such localization might occur, it is possible that the hair cell orientation patterns and macular orientations are crucial for this task.

In addition to the hair cell patterns, the location of the otolith organs with respect to each other and the distance between the inner ears is of importance for localization. The models of directional hearing directly show that a larger distance between the inner ears (Schellart and de Munck 1987) and a larger distance between the otolith systems of the inner ear (favorable for the model of Rogers et al. 1988) will improve directional hearing. Since comparative anatomical knowledge about this point is limited, and comparative behavioral data on localization are lacking, the influence of otolith position cannot be determined with any reliability.

It may be argued that directional hearing is less important for fishes that swim constantly, such as tunas, than for fish that stay in the same locale such as on coral reefs. Continuously moving fishes could use intensity gradients to swim towards a sound source (something that may be done by elasmobranchs and other species not having a swimbladder). However, multiple surface and bottom reflections in shallow water could hamper directional hearing in this manner.

8. Evolution of the Ear in Agnathans and Elasmobranchs

8.1 Agnathans

The earliest known vertebrate ear is found in the fossil record of primitive jawless cephalaspids, ancestors of the extant petromyzontids (lampreys) (Stensiö 1927; reviewed in Jarvick 1980). Although some of these ears resembled those in extant petromyzontids, the fossil record leads to the suggestion that there was considerable variation in structure, and particularly in the number of semicircular canals (e.g., Stensiö 1927; Jarvick 1980). There is no indication of the functional significance of having multiple canals, but these findings may mean that there was considerable "experimentation" before the vertebrates finally settled on the three canal system found in gnathostomes (jawed fishes). At the same time, it is likely that basic features of the ear, such as the semicircular canals, sensory hair cells, and otoconia had evolved before the gnathostomes separated from the jawless fishes sometime in the early Silurian Period (Hardisty 1979; Jarvick 1980).

Despite some basic similarities in the ears of modern petromyzontids and gnathostomes, the differences between the modern jawless fishes and jawed fishes is striking. At the same time, it is not clear how much modern petromyzontids differ from their earliest ancestors in terms of ear structure—or if any jawless fish has ever been able to detect sound.

The ear in the petromyzontids includes only two (vertical) semicircular canals and their associated ampullae as well as a single sensory epithelium, the macula communis (Lowenstein, Osborne, and Thornhill 1968; Popper and Hoxter 1987). The macula communis contains several sensory hair cell regions that various investigators have thought to be homologous to the saccule, lagena and utricle of gnathostomes. However, evidence for such homology is, at best, weak (reviewed in Popper and Hoxter 1987) and it is possible that any simi-

larities in organization between the parts of the macula communis and those of the otolith organs reflects functional rather than evolutionary similarities (see discussion of such similarities in Popper, Platt, and Edds, Chapter 4).

Even less is known about the structure and function of the ear in Myxiniformes (hagfishes). The ears in these animals have a single ring-shaped semicircular canal, although several ampullae are present (Lowenstein and Thornhill 1970). Very little is known about the function of the ear and hearing (if it occurs) in hagfishes, and it is likely that the hagfish represent a divergent line of ear structure that has little bearing on that of extant fishes.

8.2 Elasmobranchs

The elasmobranch ear and hearing has been reviewed extensively (e.g., Popper and Fay 1977; Corwin 1981, 1989) and will not be dealt with here in great detail. However, several points regarding the ear are worthy of note, particularly with regard to evolution of the auditory system.

The elasmobranch ear consists of three semicircular canals and associated ampullae as well as three otolith organs—the saccule, lagena, and utricle. In addition, many elasmobranchs have an additional sensory epithelium, the macula neglecta. While the macula neglecta is present in some teleost fishes as well as in many other vertebrates (reviewed in Retzius, 1881; Popper, Platt, and Edds Chapter 4), it has become a major sensory epithelium in a number of elasmobranchs (Corwin 1978, 1981). In these species, the macula neglecta is in a position that could receive direct acoustic stimulation (Fay et al. 1974; Corwin 1981), and there is a good correlation between the size of the macula neglecta and feeding strategies of various species (Corwin 1978). This is an excellent example where behavior and ecology of a group of animals is correlated with inner ear structure.

Due to a paucity of data, it is not clear how the inner ear is involved with hearing in elasmobranchs. It is also not clear how elasmobranchs localize sounds, but behavioral experiments show that sharks not only detect particle motion in a sound field, but moreover show sensitivity to sound pressure (van den Berg and Schuijf 1983). To what extent other parts of the ear, in addition to the macula neglecta, are involved with sound detection is not clear.

9. Summary

Although the basic peripheral structures involved with hearing are similar among virtually all fishes, variants on the basic themes are rampant, and many of these are involved with improving hearing capabilities among different species. Such variation is certainly to be expected when we consider that the acoustic environment of fish is rich with sound, and highly variable in space and time. After analysis of what we know about structure, function, and hearing behavior, as well as what we know of acoustics and fish behavior, it is possible to derive several general principles, or hypotheses, regarding fish hearing. At the same time, it is also clear that among the 25,000 extant species, there are numerous exceptions to each rule.

The following hypotheses are very general and they do not account for all of what we know about fish behavior and fish hearing. Despite their limitations, they are presented as a way of summarizing the paper, and also to provide some suggestions for future work.

- It is likely that differences in the structure and mechanism of the fish auditory system reflect various strategies to deal with constraints imposed by characteristics of underwater acoustics such as sound speed and the reflecting and scattering properties of sound in water.
- Fishes with the best hearing (as determined using behavioral paradigms) are generally found in shallow, fresh water habitats with soft bottoms. They are planktonivorous, herbivorous and/or predators of small invertebrates.
- Fishes living in noisy environments (open seas, oceans, and upper courses of rivers) tend to have poor to moderate hearing, which probably prevents masking by response saturation due to the high background noises in the environment. Their life style is predatory or even purely piscivorous.
- The frequencies of swimbladder-produced sound are higher for fishes with sensitivity at higher frequencies. The swimbladder is generally a prerequisite for long-distance and spatial hearing.

Fishes without a swimbladder, or not having another air cavity in some intimate relationship to the inner ear, do not hear sounds from distances much greater than outside the near-field. Such species are not able to detect sound source direction, unless special conditions are fulfilled (e.g., surface reflections are present and usable).

- The best hearing, in terms of sensitivity and bandwidth, is only found in species with auxiliary structures to improve the coupling of the air cavity to the ear.
- Auxiliary structures that improve the swimbladder-to-ear coupling have always co-evolved with specialized hair cell patterns of the sensory epithelia of the otolith organs.
- The precise biophysical role of the swimbladder-ear-coupling of hearing specialists as well as non-specialized species is not well understood, but it is likely to be of great importance for directional hearing. The differences in shape, size, and position of the swimbladder are still enigmatic.
- Specialized hair cell orientation patterns are likely to be adaptations for spatial hearing, a crucial feature in a three-dimensional environment.
- The saccule and lagena are necessary for spatial hearing. It is also likely that the utricle is involved in spatial hearing in the horizontal plane in a number of species (e.g., otophysans).

Although we have considered adaptations and evolution of the ear and other peripheral hearing structures, we have not considered adaptive changes in the central nervous system. Though there is interspecific variation in the structure of auditory nuclei in the fish brain, far too little is known about the function of these regions, and their interconnections, for us to even begin to speculate on this interesting area (also see McCormick Chapter 17).

Acknowledgments. Preparation of the manuscript and some of the work reported were supported in part by grants from the National Institutes of Health (DC-00140) and Office of Naval Research (N-00014-87-K-0604). The authors thank Drs. R.R. Fay, J.D. Crawford, W.M. Saidel, and R.J. Wubbels and Ms. Peggy Edds for critically reading the manuscript and for their many suggestions to improve the text.

References

Abbott RR (1973) Acoustic sensitivity of salmonids, PhD Thesis, Univ. of Washington.

Alexander RMcN (1959) The physical properties of the swimbladders of fish other than Cypriniformes. J Exp Biol 36:347–355.

Arase EM, Arase T (1967) Ambient sea noise in the deep and shallow ocean. J Acoust Soc Amer 42:73–77.

Atema J, Fay RR, Popper AN, Tavolga WN (1988) Sensory Biology of Aquatic Animals. New York: Springer-Verlag.

Banner A (1967) Evidence of sensitivity to acoustic displacements in the lemon shark, *Negaprion brevirostris* (Poey). In: Cahn PH (ed) Lateral Line Detectors. Bloomington, Indiana: Indiana University Press, pp. 265–273.

Barber VC, Emerson CJ (1980) Scanning electron microscopic observations on the inner ear of the skate, *Raja ocellata*. Cell Tissue Res 205:199–215.

Berg AV van den, Schuijf A (1983) Discrimination of sounds on the phase difference between particle motion and acoustic pressure in the shark *Chiloscyllium griseum*. Proc R Soc Lond B 218:127.

Bergeijk WA van (1964) Directional and nondirectional hearing in fish. In: Tavolga WN (ed) Marine Bio-Acoustics. Oxford: Pergamon Press, pp. 281–299.

Bergeijk WA van, Alexander S (1962) Lateral line canal organs on the head of *Fundulus heteroclitus*. J Morphol 110:333–346.

Blaxter JHS (1981) The swimbladder and hearing. In: Tavolga WN, Popper AN, Fay RR (eds) Hearing and Sound Communication in Fishes. New York: Springer-Verlag, pp. 61–71.

Blaxter JHS, Denton EJ, Gray JAB (1981) Acoustico-lateralis systems in clupeid fishes. In: Tavolga WN, Popper AN, Fay RR (eds) Hearing and Sound Communication in Fishes. New York: Springer-Verlag, pp. 39–59.

Buwalda RJA, Schuijf A, Hawkins AD (1983) Discrimination by the cod of sounds from opposing directions. J Comp Physiol 150:175–184.

Chapman CJ (1973) Field studies of hearing in teleost fish. Helgolander wiss Meeresunters 24:371–390.

Chapman CJ, Hawkins AD (1973) A field study of hearing in the cod, *Gadus morhua* L. J Comp Physiol 85:147–167.

Chapman CJ, Sand O (1974) Field studies of hearing in two species of flatfish, *Pleuronectes platessa* (L.) and *Limanda limanda* (L.) (Family Pleuronectidae). Comp Biochem Physiol 47A:371–385.

Coombs S, Görner P, Münz H (1989) The Mechanosensory Lateral line. Neurobiology and Evolution. New York: Springer-Verlag.

Coombs S, Popper AN (1979) Hearing differences among Hawaiian squirrelfish (family Holocentridae) related to differences in the peripheral auditory system. J Comp Physiol 132A:203–207.

Coombs S, Popper AN (1981) Comparative frequency selectivity in fishes: Simultaneously and forward-masked psychophysical tuning curves. J Acoust Soc Amer 71:133–141.

Coombs S, Popper AN (1982) Structure and function of the auditory system in the clown knife fish, *Notopterus chitala*. J Exp Biol 97:225–239.

Corwin JT (1978) The relation of inner ear structure to the feeding behavior in sharks and rays. In: Johari OM (ed) Scanning Electron Microscopy/1978, Vol. II, pp. 1105–1112.

Corwin JT (1981) Audition in elasmobranchs, in Hearing and Sound Communication. In: Tavolga WN, Popper AN, Fay RR (eds) Fishes. New York: Springer-Verlag, pp. 81–105.

Corwin JT (1989) Functional anatomy of the auditory system in sharks and rays. J Exp Zool Suppl 2:62–74.

Dale T (1976) The labyrinthine mechanoreceptor organs of the cod *Gadus morhua* L. (Teleostei: Gadidae). Norw J Zool 24:85–128.

Demski L, Gerald GW, Popper AN (1973) Central and peripheral mechanisms in teleost sound production. Amer Zool 13:1141–1167.

Dijkgraaf S (1952) Bau und Funktionen der Seitenorgane und des Ohrlabyrinths bei Fischen. Experientia 8:205–216.

Fay RR (1984) The goldfish ear codes the axis of acoustic particle motion in three dimensions. Science 225:951–954.

Fay RR (1988a) Hearing in Vertebrates: a Psychophysics Databook, Hill-Fay Associates, Winnetka, IL USA.

Fay RR (1988b) Peripheral adaptations for spatial hearing. In: Atema J, Fay RR, Popper AN, Tavolga WN (eds) Sensory Biology of Aquatic Animals, New York: Springer-Verlag, pp. 711–731.

Fay RR, Ahroon W, Orawski A (1978) Auditory masking patterns in the goldfish (*Carassius auratus*): psychophysical tuning curves. J Exp Biol 74:83–100.

Fay RR, Feng AS (1987) Mechanisms for directional hearing among nonmammalian vertebrates. In: Yost WA, Gourevitch G (eds) Directional Hearing. New York Berlin: Springer Verlag, pp. 179–213.

Fay RR, Kendall JI, Popper AN, Tester AL (1974) Vibration detection by the macula neglecta of sharks. Comp Biochem Physiol 47:1235–1240.

Fay RR, Popper AN (1974) Acoustic stimulation of the ear of the goldfish (*Carassius auratus*). J Exp Biol 61:243–260.

Fay RR, Popper AN (1975) Modes of stimulation of the teleost ear. J Exp Biol 62:379–388.

Fay RR, Popper AN (1980) Structure and function in teleost auditory systems. In: Popper AN, Fay RR (eds) Comparative studies of Hearing in Vertebrates. New York: Springer-Verlag, pp. 1–42.

Fay RR, Popper AN (1983) Hearing in fishes: Comparative anatomy of the ear and the neural coding of sensory information. In: Fay RR, Gourevitch G (eds) Hearing and Other Senses: Presentations in Honor of E.G. Groton, CT: The Amorpha Press, pp. 123–148.

Fish MP, Mowbray WH (1970) Sound of Western North Atlantic Fish. Baltimore and London: The Johns Hopkins Press.

Fish JF, Offutt GC (1972) Hearing thresholds from toadfish, *Opsanus tau* measured in the laboratory and field. J Acoust Soc Amer 51:1318–1321

Hardisty MW (1979) Biology of the Cyclostomes. London: Chapman and Hall.

Harris GG, Bergeijk WA van (1962) Evidence that the lateral line organ responds to near-field displacements of sound sources in water. J Acoust Soc Amer 34:1831–1841.

Hawkins AD, Chapman CJ (1975) Masked auditory thresholds in the cod, *Gadus morhua* L. J Comp Physiol 103:209–226.

Hawkins AD, Johnstone ADF (1978) The hearing of the Atlantic salmon, *Salmo salar*. J Fish Biol 13:655–673.

Hawkins AD, Myrberg AA Jr (1983) Hearing and sound communication under water. In: Lewis B (ed) Bioacoustics: A Comparative Approach. New York: Academic Press, pp. 347–405.

Horch K, Salmon M (1973) Adaptations to the acoustic environment by the squirrelfishes *Myripristis violaceus* and *M. pralines*. Mar Behav Physiol 2:121–139.

Hudspeth AJ, Corey DP (1977) Sensitivity, polarity, and conductance change in the response of vertebrate hair cells to controlled mechanical stimuli. Proc Natl Acad Sci 74:2407–2411.

Iversen RTB (1967) Response of the yellowfish tuna (*Thunnus albacares*) to underwater sound. In: Tavolga WN (ed) Marine Bio-Acoustics II. Oxford: Pergamon Press, pp. 105–121.

Iversen RTB (1969) Auditory thresholds of the scrombrid fish *Euthynnus affinis*, with comments on the use of sound in tuna fishing. FAO Conference on Fish Behaviour in Relation to Fishing Techniques and Tactics. FAO Fisheries Rep. No. 62. 3:849–859.

Jacobs DW, Tavolga WN (1967) Acoustic intensity limens in the goldfish. Anim Behav 15:324–355.

Jarvik E (1980) Basic Structure and Evolution of Vertebrates, Vol. 1. London: Academic Press.

Jerkø H, Turunen-Rise I, Enger PS, Sand O (1989) Hearing in the eel (*Anguilla anguilla*). J Comp Physiol A 165:455–459.

Kalmijn AJ (1988) Hydrodynamic and acoustic field detection. In: Atema J, Fay RR, Popper AN, Tavolga WN (eds) Sensory Biology of Aquatic Animals. New York: Springer-Verlag, pp. 83–130.

Kelly JC, Nelson DR (1975) Hearing thresholds of the horn shark, *Heterodontus francisci*. J Acoust Soc Amer 58:905–909.

Kleerekoper H, Chagnon EC (1954) Hearing in fish with special reference to *Semotilus atromaculatus atromaculatus* (Mitchill). J Fish Res Bd Can 11:130–152.

Kritzler H, Wood L (1961) Provisional audiogram for the shark, *Carcharhinus leucas*. Science 133:1480–1482.

Lowenstein O, Osborne MP, Thornhill RA (1968) The anatomy and ultrastructure of the labyrinth of the lamprey (*Lampetra fluviatilis* L.). Proc Roy Soc Lond B 170:113–134.

Lowenstein O, Thornhill RA (1970) The labyrinth of *Myxine*: anatomy, ultrastructure and electrophysiology. Proc R Soc B 176:21–42.

Marshall NB (1967) Sound-producing mechanisms and the biology of deep-sea fishes. In: Tavolga WN (ed) Marine Bio-Acoustics II. Oxford: Pergamon Press, pp. 123–133.

McCormick CA (1981) Comparative neuroanatomy of the octavolateralis area of fishes. In: Tavolga WN, Popper AN, Fay RR (eds) Hearing and Sound Communication in Fishes, New York: Springer-Verlag, pp. 375–382.

McCormick CA, Popper AN (1984) Auditory sensitivity and psychophysical tuning curves in the elephant nose fish, *Gnathonemus petersii*. J Comp Physiol 155:753–761.

Moulton JM (1960) Swimming sounds and the schooling of fishes. Biol Bull 119:210–223.

Moulton JM (1963) Acoustic behaviour of fishes. In: Busnel RG (ed) Acoustic Behaviour of Animals. Amsterdam: Elsevier, pp. 655–693.

Munck JC de, Schellart NAM (1987) A model for the nearfield acoustics of the fish swimbladder and its relevance for directional hearing. J Acoust Soc Amer 81:556–560.

Myrberg AA Jr, Spires JY (1980) Hearing in damselfishes: an analysis of signal detection among closely related species. J Comp Physiol 140:135–144.

Myrberg AA Jr (1981) Sound communication and interception in fishes. In: Tavolga WN, Popper AN, Fay RR (eds) Hearing and Sound Communication in Fishes. New York: Springer-Verlag, pp. 395–426.

Offutt GC (1971) Response of the tautog (*Tautoga onitis* Teleost) to acoustic stimuli measured by classically conditioning the heart rate. Conditional Reflex 6:205–214.

Offutt GC (1973) Structures for the detection of acoustic stimuli in the Atlantic codfish *Gadus morhua*. J Acoust Soc Am 56:665–671.

Platt C, Popper AN (1981) Fine structure and function of the ear. In: Tavolga WN, Popper AN, Fay RR (eds) Hearing and Sound Communication in Fishes. New York: Springer-Verlag, pp. 3–38.

Poggendorf D (1952) Die absoluten Hörschwellen des Zwergwelses (*Amiurus nebulosus*) und Beiträge zur Physik des Weberschen Apparatus der Ostariophysen. Z verg Physiol 34:222–257.

Popper AN (1970) Auditory capacities of the Mexican blind cave fish (*Astyanax jordani*) and its eyed ancestor (*Astyanax mexicanus*). Anim Behav 18:552–562.

Popper AN (1971) The effects of size on auditory capacities of the goldfish. J Aud Res 11:239–247.

Popper AN (1972) Pure-tone auditory thresholds for the carp, *Cyprinus carpio*. J Acoust Soc Amer 52:1714–1717.

Popper AN (1977) A scanning electron microscopic study of the saccule and lagena in the ears of fifteen species of teleost fishes. J Morph 153:397–418.

Popper AN (1978) Scanning electron microscopic study of the otolithic organs in the bichir (*Polypterus bichir*) and shovel-nose sturgeon (*Scaphirhynchus platorynchus*). J Comp Neurol 181:117–128.

Popper AN (1979) The ultrastructure of the saccule and lagena in a moray eel (*Gymnothorax* sp.). J Morphol 161:241–256.

Popper AN (1981) Comparative scanning electron microscopic investigations of the sensory epithelia in the teleost saccule and lagena. J Comp Neurol 200:357–374.

Popper AN (1983) Organization of the inner ear and auditory processing. In: Northcutt RG, Davis RE (eds) Fish Neurobiology. Ann Arbor: University of Michigan Press, pp. 125–178.

Popper AN, Coombs S (1982) The morphology and evolution of the ear in Actinopterygian fishes. Amer Zool 22:311–328.

Popper AN, Fay RR (1973) Sound detection and processing by fish: A critical review. J Acoust Soc Amer 53:1515–1529.

Popper AN, Fay RR (1977) The structure and function of the elasmobranch auditory system. Amer Zool 17:443–452.

Popper AN, Fay RR (1984) Sound detection and processing by teleost fish: A selective review. In: Bolis L, Keynes RD, Maddrell SHP (eds) Comparative Phys-

iology of Sensory Systems. Cambridge: Cambridge University Press, pp. 67–101.

Popper AN, Hoxter B (1981) The fine structure of the saccule and lagena of a teleost fish. Hearing Res 5:245–263.

Popper AN, Hoxter B (1987) Sensory and non-sensory ciliated cells in the ear of the sea lamprey, *Petromyzon marinus*. Brain Behav Evol 30:43–61.

Popper AN, Northcutt RG (1983) Structure and innervation of the inner ear of the bowfin, *Amia calva*. J Comp Neurol 213:279–286.

Popper AN, Platt C (1979) The herring ear has a unique receptor pattern. Nature 280:832–833.

Popper AN, Platt C (1983) The ear in ostariophysan fishes: its ultrastructure and function. J Morphol 177:301–317.

Popper AN, Platt C, Saidel WM (1982) Acoustic function in the fish ear. Trends Neurosci 5:276–280.

Popper AN, Rogers PH, Saidel WM, Cox M (1988) The role of the fish ear in sound processing. In: Atema J, Fay RR, Popper AN, Tavolga WN (eds) Sensory Biology of Aquatic Animals. New York: Springer-Verlag, pp. 687–710.

Popper AN, Tavolga WN (1981) Sound detection and inner ear structure in the marine catfish *Arius felis*. J Comp Physiol 144:27–34.

Retzius G (1881) Das Gehörorgan der Wirbeltiere. Vol. I. Stockholm: Samson and Wallin.

Rogers PH, Cox M (1988) Underwater sound as a biological stimulus. In: Atema J, Fay RR, Popper AN, Tavolga WN (eds) Sensory Biology of Aquatic Animals. New York: Springer-Verlag, pp. 131–149.

Rogers PH, Popper AN, Hastings MA, Saidel WM (1988) Processing of acoustic signals in the auditory system of bony fish. J Acoust Soc Am 83:338–349.

Rosen DE, Greenwood PH (1970) Origin of the Weberian apparatus and the relationships of the Ostariophysan and Gonorynchiform fishes. Amer Mus Novitates 2428:1–25.

Saidel WM, Popper AN (1983) Spatial organization in the saccule and lagena of a teleost: Hair cell pattern and innervation. J Morph 177:301–317.

Schade R van (1971) Experimentell Untersuchungen zum Hörvermögen an *Leucaspius delineatus*. Biol Zentr 90:337–356.

Schellart NAM (1989) Strategies of teleost directional hearing. In: Erber J, Menzel R, Pflüger H-J, Todt D (eds) Neural Mechanisms of Behavior. Stuttgart: Georg Thieme Verlag, p. 83.

Schellart NAM, Buwalda RJA (1990) Directional variant and invariant hearing thresholds in the rainbow trout (*Salmo gairdneri*). J Exp Biol 149:113–131.

Schellart NAM, de Munck JC (1987) A model for directional and distance hearing in swimbladder-bearing fish based on the displacement orbits of the hair cells. J Acoust Soc Am 82:822–829.

Schuijf A (1976a) The phase model of directional hearing in fish. In: Schuijf A, Hawkins AD (eds) Sound Reception in Fish. Amsterdam: Elsevier, pp. 63–86.

Schuijf A (1976b) Time analysis and directional hearing in fish. In: Schuijf A, Hawkins AD (eds) Sound Reception in Fish. Amsterdam: Elsevier, pp. 87–112.

Schuijf A (1981) Models of acoustic localization. In: Tavolga WN, Popper AN, Fay RR (eds) Hearing and Sound Communication in Fishes. New York: Springer-Verlag, pp. 267–310.

Schuijf A, Hawkins AD (1976) Sound Reception in Fish. Amsterdam: Elsevier, 288 pp.

Siegmund R, Wolff DK (1973) Experimentelle Untersuchungen sur Bestimmung des Hörvermögens der Karausche (Carassius carassius L.) Fisch.-Forsch 11:117–124.

Siler W (1969) Near- and farfields in a marine environment. J Acoust Soc Amer 46:483–484.

Spanier E (1979) Aspects of species recognition by sound in four species of damselfish, genus Eupomacentrus (Pisces: Pomacentridae). Z Tierpsychol 51:301–316.

Stensiö EA (1927) The Downtownian and Devonian vertebrates of Spitzbergen. I Family Cephalaspidae. Skr om Svalbard og Nordishavet No. 12 (2 vols.)

Sterba G (1962) Freshwater Fishes of the World. London: Vista Books.

Tavolga WN (1971) Sound production and detection. In: Hoar WS, Randall DJ (eds) Fish Physiology. Vol. V. New York: Academic Press, pp. 135–205.

Tavolga WN (1974a) Signal/noise ratio and the critical band in fishes. J Acoust Soc Amer 55:1323–1333.

Tavolga WN (1974b) Sensory parameters in communication among coral reef fishes. Mount Sinai J Med 41:324–340.

Tavolga WN (1977) Mechanisms for directional hearing in the sea catfish (Arius felis). J Exp Biol 67:97–115.

Tavolga WN, Popper AN, Fay RR (1981) Hearing and Sound Communication in Fishes. New York: Springer-Verlag.

Tavolga WN, Wodinsky J (1963) Auditory capacities in fishes. Pure tone thresholds in nine species of marine teleosts. Bull Amer Mus Nat Hist 126:177–240.

Tavolga WN, Wodinsky J (1965) Auditory capacities in fishes: Threshold variability in the blue-striped grunt, Haemulon sciurus. Animal Behaviour 13:301–311.

Urick RJ (1975) Principles of Underwater Sound. New York: McGraw-Hill.

Vries HL de (1950) The mechanics of the labyrinth otoliths. Acta Oto-Laryngol 38:262–273.

Weiss BA, Strother WF, Hartig GM (1969) Auditory sensitivity in the bullhead catfish (Ictalurus nebulosus). Proc Nat Acad Sci 64:552–556.

Wenz GM (1971) Review of underwater acoustic research: noise. J Acoust Soc Am 51:1010–1024.

Wilson Jr OB, Wolf SN, Ingenito F (1985) Measurements of acoustic ambient noise in shallow water due to breaking surf. J Acoust Soc Amer 78:190–195.

Winn HE (1972) Acoustic discrimination by the toadfish with comments on signal systems. In: Winn HE, Olla BL (eds) Behavior of Marine Animals, Vol. 2. New York: Plenum Press, pp. 361–385.

Winn HE, Marshall JA, Hazlett BA (1964) Behavior, diel activities, and stimuli that elicit sound production and reaction to sounds in the longspine squirrelfish. Copeia pp. 413–425.

Wodinsky J, Tavolga WN (1964) Sound detection in teleost fishes. In: Tavolga WN (ed) Marine Bio-Acoustics. Oxford: Pergamon, pp. 269–280.

Wolff DL (1967) Das Hörvermögen des Flussbarsches (Perca fluviatilis L.). Biol Zentr 86:449–460.

Wolff DL (1968) Das Hörvermögen des Kaulbarsches (Acerina cernua L.) und des Zanders, (Lucioperca sandra Cuv. und Val.). Z verg Physiol 60:14–33.

17
Evolution of Central Auditory Pathways in Anamniotes

Catherine A. McCormick

1. Introduction

An inner ear containing multiple sensory endorgans arose early in vertebrate history. Among jawed vertebrates the ear includes seven endorgans believed to be primitive for jawed fishes (otolithic endorgans: utricle, saccule, lagena; macula neglecta; three semicircular canals) as well as the various derived papillar endorgans of anamniotes (amphibian and basilar papillae) and amniotes (basilar papilla or cochlear duct). What is remarkable about this collection of endorgans is that, with the exception of the semicircular canal cristae, each of the remaining endorgans has been implicated in hearing in one species or another. The classic notion that the otolithic endorgans do not contribute to hearing in land vertebrates has been recently disproven in amphibians (reviewed in Lewis et al. 1985), a discovery that challenges us to reexamine otolithic endorgan function in other vertebrates. Moreover, it has also been recently claimed that the evolution of the papillar endorgans, at least of anamniotes, predates the emergence of vertebrates onto land (Fritzsch 1987, Chapter 18), a view that requires us to reconsider the selective pressures that influenced the appearance of acoustic receptors.

It is the goal of this chapter to discuss the central auditory circuits of anamniotes in an evolutionary context. One obvious aspect of such a discussion is to ask how the diversity of acoustic inner ear receptors is reflected in the organization of brain auditory pathways. For example, was the appearance of the anamniote papillar endorgans paralleled by the evolution of a new set of auditory

lemniscal circuits? In this context, it is important to recognize that there is no a priori reason to expect brain auditory circuitry to be less complex in fishes than in amphibians, or to assume that a set of auditory pathways demonstrated in one fish species represents the "type" or "standard" pathway for all other fishes, including those that were ancestral to amphibians. Brain circuits, like other morphological features, are composites of primitive features, reflecting common ancestry, and specialized features, reflecting the independent evolution that has occurred within each vertebrate lineage. We should therefore expect some degree of diversity in auditory circuitry not only between vertebrate classes, but also within classes. Within the Class Osteichthyes (bony fishes), for example, it is possible that variations in central auditory pathways might parallel differences in inner ear endorgan function or might reflect the specialized participation of the lateral line system in sound reception. Both inter- and intraclass analyses are necessary to generate the data base which allows us to distinguish those features of auditory circuitry that are primitive from those that are derived. The recognition that auditory circuitry has not evolved in a linear fashion must underlie any attempt to describe this circuitry in a comparative sense.

2. Class Agnatha

2.1 Inner Ear Structure and Function

The inner ears of petromyzontids (lampreys) and myxinoids (hagfish) are distinctly different from those of jawed vertebrates (Lowenstein et al. 1968;

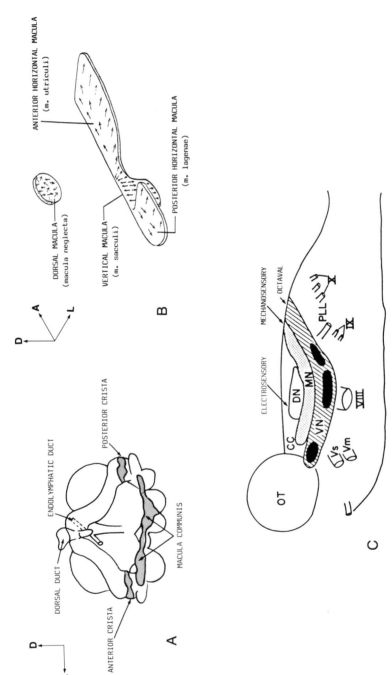

FIGURE 17.1. Inner ear and hindbrain of petromyzontid agnathans. (A) Medial view of the right inner ear of a lamprey. (After Jarvik 1980.) (B) Hair cell orientations along the common and dorsal maculae in *Lampetra fluviatilis*. (After Lowenstein et al. 1968.) (C) Lateral view of the hindbrain of *Petromyzon marinus* showing the nuclei comprising the electrosensory column (clear), mechanosensory column (stippled) and octaval column (area with diagonal lines and darkened regions). (After Northcutt 1981.) CC, cerebellar crest; DN, dorsal nucleus; IX, glossopharyngeal nerve; MN, nucleus medialis; OT, optic tectum; PLL, posterior lateral line nerve; VIII, eighth nerve; VN, ventral nucleus; Vm, motor root of the trigeminal nerve; Vs, sensory root of the trigeminal nerve; X, vagus nerve.

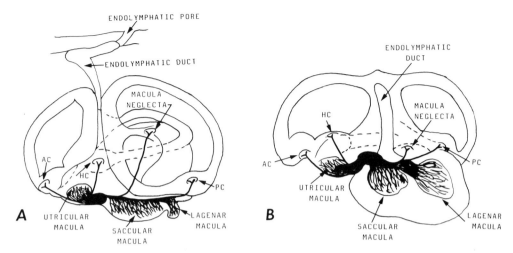

FIGURE 17.2. Medial views of the right inner ears of (A) *Raja clavata* and (B) *Amia calva*. (Modified from Retzius 1881.) AC, anterior semicircular canal crista; HC, horizontal semicircular canal crista; PC, posterior semicircular canal crista.

Lowenstein and Thornhill 1970). For example, there are only two semicircular canal ampullae, and individualized otolithic endorgans are absent (Fig. 17.1A). Lampreys and hagfish both possess a single macula communis, which is loaded by an otolith. Based on variations in hair cell orientation patterns along this common macula in the lamprey *Lampetra fluviatilis*, Lowenstein et al. (1968) recognized subdivisions that may correspond to the utricle, saccule, and lagena of jawed vertebrates (Fig. 17.1B). The subdivision corresponding to the saccule appears to be absent in the hagfish *Myxine* (Lowenstein and Thornhill 1970). However, Popper and Hoxter (1987) caution that macular hair cell orientation patterns are highly variable among gnathostomes, making either functional or evolutionary comparisons with agnathans highly speculative when based on this characteristic alone. Lampreys, but not hagfish, also possess a dorsal macula that has been interpreted as the macula neglecta.

There are no behavioral data that demonstrate hearing in any agnathan. However, eighth nerve responses to vibratory stimuli in the range of 80 Hz have been recorded in *Lampetra* but not *Myxine*. Based on circumstantial evidence, Lowenstein (1970) suggested that the presumed macula neglecta and saccule of lampreys are vibratory receptors. Clearly, more data are needed before we can conclude whether inner ear acoustic reception is present in extant agnathans. The reader is also referred to Jarvik's (1980) discussion of the inner ear of extant versus extinct petromyzontids and myxinoids.

2.2 Central Representation of the Agnathan Inner Ear

In lampreys and hagfish the eighth nerve projects largely into the octavus cell column of the medulla, a region containing first-order octaval nuclei that lies largely ventral to the termination sites of the lateral line nerve(s). There is no information about functional subdivisions within this column, although anatomical subdivisions are recognized (Fig. 17.1C). In the lamprey *Icthyomyzon unicuspis* the octavus cell column consists of a small-celled ventral nucleus that contains three large-celled octavomotor nuclei (Northcutt 1979). Koyama et al. (1989) also recognize an additional lateral octaval nucleus in *Lampetra japonica*. The ventral nucleus of the hagfish *Eptatretus* contains one large-celled subdivision (Amemiya et al. 1985). Eighth nerve fibers also course to the ipsilateral cerebellum in lampreys but not in hagfish. A limited number of fibers may also penetrate the overlying lateral line nucleus, nucleus medialis.

3. Class Chondrichthyes

3.1 Inner Ear Structure and Function

As mentioned in the Introduction, the inner ear of all jawed fishes primitively contains three distinct otolithic endorgans (the utricle, saccule, and lagena), a macula (or papilla) neglecta overlain by a tectorial mass (sometimes referred to as a cupula),

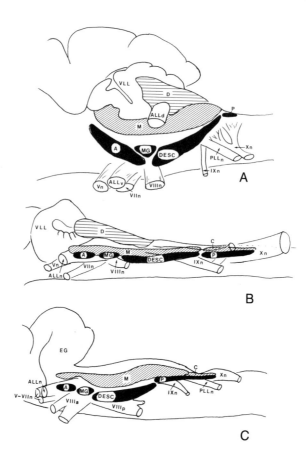

FIGURE 17.3. Lateral views of the hindbrains of (A) the skate *Raja eglanteria* (after Barry 1987), (B) the sturgeon *Scaphyrinchus platorynchus* (after New and Northcutt 1984), and (C) the bowfin *Amia calva* (after McCormick 1981) showing the nuclei of the octavolateralis area of the medulla. *Symbols*: Horizontal lines = electrosensory nucleus dorsalis; diagonal lines = mechanosensory nucleus medialis; darkened areas = octaval nuclei. A, anterior octaval nucleus; ALLn, anterior lateral line nerve; ALLd, dorsal root of anterior lateral line nerve; ALLv, ventral root of anterior lateral line nerve; C, nucleus caudalis; D, dorsal nucleus; DESC, descending octaval nucleus; EG, eminentia granularis of the cerebellum; IXn, glossopharyngeal nerve; M, nucleus medialis; MG, magnocellular octaval nucleus; P, posterior octaval nucleus; PLLn, posterior lateral line nerve; Vn, trigeminal nerve; V-VIIn, trigeminal and facial nerves; VIIn, facial nerve; VIIIn, octaval nerve; VIIIa, anterior ramus of the octaval nerve; VIIIp, posterior ramus of the octaval nerve; VLL, vestibulolateral lobe of the cerebellum; Xn, vagus nerve.

and three semicircular canal ampullae containing cupula-covered sensory cristae (Fig. 17.2A,B). There are, however, two anatomical features of the chondrichthyan inner ear that distinguish it from that of osteichthyan fishes. First, the endolymphatic duct is not a blind pouch, but instead opens as an endolymphatic pore on the dorsal surface of the head (Fig. 17.2); the functional significance of this is unknown (Tester et al. 1972). Second, there is an opening, the fenestra ovalis, in the cartilaginous wall of the otic capsule. The fenestra ovalis is physically associated to varying degrees with the macula neglecta and with the parietal fossa. The latter is a cranial depression filled with loose connective tissue which has been hypothesized to provide a sound path directed toward the macula neglecta (Tester et al. 1972; Fay et al. 1974; Corwin 1981). It is believed by some investigators that the physical relationship between the macula neglecta, fenestra ovalis, and parietal fossa accounts for the known sound/vibratory sensitivity of this endorgan (Corwin 1981).

The sensitivity of the macula neglecta as a sound detector may vary among species. Corwin (1978)

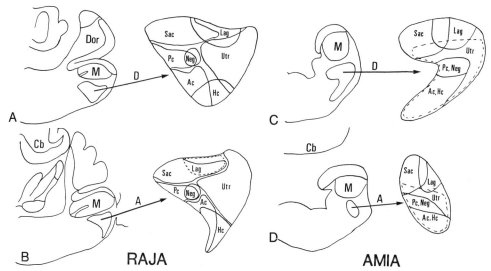

FIGURE 17.4. Line drawings of transverse hemisections through the medulla of the skate *Raja eglanteria* (after Barry 1987) and the bowfin *Amia calva* (after McCormick and Braford 1988) showing the termination sites of the inner ear endorgans in the descending octaval nucleus (A, C) and the anterior octaval nucleus (B, D). A, anterior octaval nucleus; Ac, terminals of anterior semicircular canal nerve; Cb, cerebellum; D, descending octaval nucleus; Dor, nucleus dorsalis; Hc, terminals of horizontal semicircular canal nerve; Lag, terminals of lagenar nerve; M, nucleus medialis; Neg, terminals of macula neglecta nerve; Pc, terminals of posterior semicircular canal nerve; Sac, terminals of saccular nerve; Utr, terminals of utricular nerve.

has shown that the duct containing the macula neglecta is positioned closer to the fenestra ovalis in some species (such as *Carcharinus* and *Notorhychus*) than in others (for example, *Musteleus* and *Torpedo*). Furthermore, the species in which these two structures are more closely associated tend to have a larger macula neglecta with more tightly organized hair cells. Corwin (1981) reviews other functional characteristics of the chondrichthyan macula neglecta.

Like osteichthyan fishes, chondrichthyans also utilize other otolithic endorgans in sound detection. Sound and/or vibratory sensitivity has been attributed to the lacinia of the utricle and to a portion of the saccule (Lowenstein and Roberts 1951; Budelli and Macadar 1979; Corwin 1981). It is important to realize that other portions of the saccule and utricle (as well as the entire lagena) are vestibular in function (Lowenstein and Roberts 1951; Budelli and Macadar 1979). The maculae of the saccule and utricle are therefore, in the terminology of Platt (1983), receptor mosaics, subserving both acoustic and vestibular modalities. This mosaicism may be reflected in the complex physiological properties of cells in the primary octaval nuclei (Plassmann 1983).

3.2 Central Representation of Inner Ear Endorgans

In several chondrichthyan species the eighth nerve is known to project largely to a ventral octaval cell column, to the reticular formation, and to the cerebellum (Boord and Roberts 1980; Koester 1983; Barry 1987; Dunn and Koester 1987). The ventral octaval column is subdivided into four main nuclei that are likely to be homologous to similarly named nuclei in osteichthyans: the anterior, magnocellular, descending, and posterior nuclei (Fig. 17.3A). In the skate *Raja eglanteria*, an additional periventricular nucleus has been recognized (Barry 1987). Central projections of individual inner ear endorgans to the octaval nuclei are known in *Raja* (Barry 1987), and it is likely that the two largest nuclei (the anterior and the descending) each contain subpopulations that contribute to the ascending auditory lemniscal pathway. Although the projection fields of the individual otic endorgans do overlap, the dorsal portions of the descending and anterior nuclei likely process acoustic input and provide input to higher-order auditory centers, since the neurons in this region receive input from endorgans implicated in hearing—the saccule, the utricle (Figs. 17.4A,B),

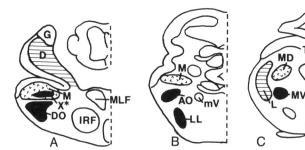

FIGURE 17.5. Line drawings of caudorostral series of hemisections through the brainstem of the skate *Raja eglanteria*. *Symbols*: Horizontal lines = electrosensory lateral line areas; floating "v's" = mechanosensory lateral line areas; darkened areas = octaval structures believed to process auditory input. Compiled from Northcutt 1978; Boord and Northcutt 1982; Corwin and Northcutt 1982. (After McCormick and Braford 1988.) AO, anterior octaval nucleus; D, nucleus dorsalis; DO, descending octaval nucleus; G, dorsal granule cell ridge; IRF, inferior reticular formation; III, oculomotor nucleus; L, lateral nucleus of lateral mesencephalic nuclear complex; LL, nucleus of lateral lemniscus; M, nucleus medialis; MD, dorsomedial nucleus of lateral mesencephalic nuclear complex; MLF, medial longitudinal fasciculus; MV, ventromedial nucleus of lateral mesencephalic nuclear complex; mV, motor nucleus of trigeminal nerve; R, nucleus ruber; T, optic tectum; X*, cell plate x.

and the macula neglecta (projection not figured). Most descending nucleus cells that project to the medial mesencephalic nucleus, a region that includes the auditory midbrain area (Corwin and Northcutt 1982), lie between the saccular and macula neglecta projection zones outlined in Figure 17.4 (Barry 1987). Barry (1987) also speculates that the periventricular octaval nucleus may have an auditory role.

It is unclear whether other first-order octaval nuclei participate in acoustic processing. The major projection of the magnocellular nucleus to the spinal cord suggests that any acoustic information received by this nucleus may be used to drive startle-type reflexes (Barry 1987). Afferent fibers from all inner ear endorgans also ramify dorsally within the boundaries of the first-order lateral line nucleus (nucleus medialis or intermedius); utricular and canal inputs reach the electrosensory nucleus dorsalis as well (Barry 1987). The functional significance of such projections is unclear and they are further considered in Section 6.

3.3 Higher Order Acoustic Areas

Acoustic centers in chondrichthyans have been demonstrated only to the level of the midbrain (Fig. 17.5), although forebrain acoustic centers are likely present. Various studies suggest that auditory information from primary octaval nuclei and higher-order medullary and isthmal nuclei is relayed to a midbrain region believed homologous to the torus semicircularis of tetrapods. This midbrain region, the lateral mesencephalic complex (LMC), varies in its degree of differentiation among chondrichthyans (Boord and Montgomery 1989). In the sharks *Squalus* and *Cephalloscyllum* clear nuclear boundaries within the LMC are not present, making it difficult to determine whether or not separate electrosensory, mechanosensory, and auditory zones are present (Boord and Northcutt 1988; Boord and Montgomery 1989).

In contrast, the LMC is well differentiated in batoids such as *Raja* (Fig. 17.5C), containing lateral (lateral line electrosensory), mediodorsal (lateral line mechanosensory), medioventral, and anterior nuclei (Boord and Northcutt 1982). An anatomical study (Barry 1987) suggests that the medioventral nucleus (MV) receives axons from cells in the dorsomedial portion of the anterior and descending octaval nuclei (bilaterally), the area that probably processes acoustic input from the saccule and macula neglecta. Projections to MV also originate from the contralateral cell plate X (Fig. 17.5A; group C1 and C2 of Smeets et al. 1983, Barry 1987) and from the nucleus of the lateral lemniscus (Fig. 17.5B) (at least contralaterally). This study did not rule out lateral line or other inputs to the MV. However, a metabolic activity/electrophysiological study in another batoid, the guitarfish *Platyrhinoides* revealed similar findings; acoustic zones were identified in the anterior octaval nucleus, cell plate X, the nucleus of the lateral lemniscus, and MV (Corwin and Northcutt 1982).

Cell plate X, which has been suggested to be homologous to part of the amniote superior olivary complex (Corwin and Northcutt 1982) lies within the first-order lateral line nucleus (nucleus medialis) and is present in both sharks and batoids. It is unknown how acoustic input reaches these cells; they may also receive mechanosensory lateral line input (Bodznick and Schmidt 1984). Similarly, the pathway(s) through which acoustic information reaches the nucleus of the lateral lemniscus, which is located in the rostral medulla, have not been demonstrated. It is clear that more information is needed on the connections of these two higher-order nuclei before we can reasonably homologize them with similarly named acoustic centers in tetrapods. On the other hand, the direct projection of auditory portions of the primary octaval nuclei to the MV supports the hypothesis that the MV is homologous to a portion of the tetrapod torus semicircularis.

4. Class Osteichthyes

4.1 Inner Ear Structure and Function

As far as is known, all bony fishes possess the seven inner ear endorgans present in chondrichthyan fishes (Fig. 17.2B). The macula neglecta is usually small and thus difficult to observe, particularly in material that is not optimally preserved. Therefore, descriptions or figures of bony fish inner ears that do not mention or include this endorgan (as in Retzius 1881) should not be interpreted as indicating with certainty its absence. The coelacanth *Latimeria chalumnae* appears to be alone among extant fish in possessing a basilar papilla (Fritzsch 1987; Chapter 18).

Osteichthyan inner ears can be grouped into two broad categories: those that are uncoupled, as opposed to those that are coupled, to gas-filled chambers (see Schellart and Popper, Chapter 16). An uncoupled ear is believed to be the primitive state among bony fishes, and the saccule and lagena are thought to be the endorgans most commonly used for hearing. Recently, the utricle has also been hypothesized to have a role in sound localization widespread among osteichthyans (Schellart and Popper, Chapter 16). It is again important to realize that, as is the case in chondrichthyans, the otolithic endorgans are

likely to be receptor mosaics containing acoustic and vestibular subpopulations (Platt 1983).

Physical coupling between the inner ear and a gas-filled chamber has occurred several times independently among fishes. Although the swimbladder is always a potential source of inner ear stimulation through reradiation of sound, physical coupling results in enhancement of acoustic abilities (reviewed in Popper 1983, Schellart and Popper, Chapter 16). In most cases, the coupling involves a contact with the *pars inferior* of the inner ear and most directly involves the saccule. In otophysans, for example, the Weberian ossicles link the swimbladder with a perilymphatic space abutting a portion of the saccule, a major acoustic endorgan in these fishes. Likewise, physical coupling between some type of gas-filled chamber and the *pars inferior* is present in certain osteoglossomorphs (Stipetić 1939; Dehadrai 1957), some Holocentridae (Nelson 1955), and in the "labyrinthine fishes" (reviewed in van Bergeijk 1967). An unusual case of coupling occurs in the Clupeidae, where it is the utricle that is contacted by a gas-filled chamber (Allen et al. 1976). The utricle has both acoustic and vesicular functions in clupeids; the functions of the saccule and lagena are unknown (Best and Gray 1980; Denton and Gray 1980).

It is important to note that the relative contributions of all three otolithic endorgans to hearing are unknown in any osteichythyan. The belief that the saccule and lagena play major acoustic roles are assumptions based largely on studies of the goldfish, an otophysan. Before a complete understanding of hearing in bony fishes can be obtained, a broader information base about the functions of all three otolithic endorgans, as well as the macula neglecta, is required.

The osteichthyan macula neglecta is not known to be closely associated with a potential sound transmission route, as is believed to be the case in chondrichthyans. Its central connections in the one species for which they are known are described below (Section 4.2.1) and a vestibular function is suggested. Fritzsch (Chapter 18) speculates on the function of another nonotolithic endorgan, the basilar papilla of the coelocanth *Latimeria*.

The largest body of data on central auditory pathways involves the medulla and midbrain; relatively little information is available on forebrain auditory connections. Species studied include those with and without physical coupling between the inner ear and a gas-filled chamber. Interestingly, variation in connectivity, at least at the level of the medulla, may in some way reflect the presence or absence of such coupling.

4.2 Inner Ear Projections in Osteichthyans: Primitive Pattern

4.2.1 Actinopterygians

Nonteleost ray-finned fishes have four first-order octaval nuclei (Figs. 17.3B,C) which are probably homologous to the four major nuclei in chondrichthyans: the anterior, magnocellular, descending, and posterior octaval nuclei (McCormick 1981; 1982; New and Northcutt 1984). These four nuclei lie largely ventral to the first-order lateral line nuclei. In the halecomorph *Amia calva* (bowfin) they receive the bulk of the afferent eighth nerve fibers that terminate in the medulla (McCormick 1981). As in chondrichthyans, the eighth nerve in *Amia* also projects to the reticular formation, the eminentia granularis of the cerebellum, and sparsely within the boundaries of the lateral line mechanosensory nucleus.

Figure 17.4 compares the central distribution of fibers from each of the inner ear endorgans within the anterior and descending octaval nuclei in *Amia* and *Raja*. As in *Raja*, acoustic neurons in *Amia* are probably present in the dorsal portion of the descending and anterior nuclei (McCormick 1983a). That these two species, members of two different vertebrate classes, are so similar in the organization of their first-order otic inputs strongly suggests that this basic organization represents the primitive pattern for jawed vertebrates.

There are some differences between inner ear inputs between *Raja* and *Amia*. For example, *Raja* has an additional periventricular nucleus (not shown) (Barry 1987). The connections of the macula neglecta are also different in some ways. In *Raja*, the bulk of the macula neglecta projections terminate in the medulla where they overlap those of the posterior semicircular canal (Fig. 17.4A,B) and, to a lesser degree, the saccule (not shown); no fibers project to the cerebellum. In contrast, the majority of macula neglecta fibers in *Amia* project to the eminentia granularis of the cerebellum; the few fibers terminating in the medulla do so along with those of the posterior semicircular canal (Fig. 17.4C,D; Saidel and McCormick 1985). These

connectional differences presumably reflect functional differences between these receptors in the two species.

4.2.2 Sarcopterygians

The octaval area of *Latimeria* is anatomically like that of *Amia* (Northcutt 1980). Four octaval nuclei—the anterior, descending, magnocellular, and posterior—lie ventral to the lateral line mechanosensory and electrosensory zones. The central connections of inner ear endorgans to these nuclei are not known.

The medulla of lungfish is like that of urodele amphibians in that the majority of neurons are positioned close to the ventricular surface. The four octaval nuclei are thus more difficult to recognize. Preliminary analysis of the South American lungfish *Lepidosiren paradoxa* indicates that, as in urodeles, the eighth nerve terminates upon cell populations located ventral to the lateral line zones (McCormick, unpublished observations). The anterior, magnocellular, and descending nuclei are probably present. Within the octaval column, saccular and lagenar fibers terminate on more dorsally positioned cells, whereas semicircular canal fibers terminate more ventrally. Though all the details of the inner ear inputs are not yet known, the termination pattern described above resembles that of *Amia* and *Raja* (Fig. 17.4).

4.3 Inner Ear Projections in Teleosts: Minor Variations on the Primitive Pattern

Teleosts have the four first-order nuclei noted above and, in addition, a nucleus tangentialis (McCormick 1982). The latter is probably vestibular since it receives input primarily from the semicircular canals and the utricle (Bell 1981a; Meredith and Butler 1983; Meredith et al. 1987). In most species studied to date, the majority of eighth nerve fibers terminate ipsilaterally in the five medullary nuclei and in the eminentia granularis of the cerebellum (Bell 1981a; Northcutt 1981; McCormick 1983b; Meredith and Butler 1983; Finger and Tong 1984; Meredith et al. 1987). There is a smaller eighth nerve projection to the reticular formation, to the region of the Mauthner cell, and in some species within the boundaries of the lateral line mechanosensory nucleus.

FIGURE 17.6. Line drawings of hemisections through the medullas of four osteoglossomorphs, (A) the mormyrid *Gnathonemus*, (B) the knifefish *Xenomystus*, (C) the arawana *Osteoglossum*, and (D) the butterfly-fish *Pantodon*. These hemisections represent levels at which the octaval column extends dorsomedially to lie medial to the lateral line column(s). Projection zones of the saccular and lagenar nerves are shown in black, while those of the utricular and semicircular canal nerves are indicated by the horizontal lines. Areas of overlap at the border of these two zones are present but are not shown. McCormick, personal observations. CC, cerebellar crest; M, nucleus medialis, PLLn, posterior lateral line nerve; VIIIn, eighth nerve.

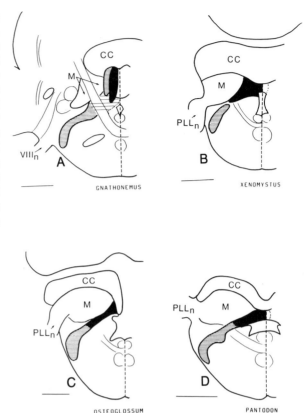

Teleosts that have contralateral eighth nerve projections are discussed in Section 4.4.

The projections of the macula neglecta are unknown in any teleost. However, the representation of other inner ear endorgans is known in a number of species. Interestingly, in two teleosts, the oscar *Astronotus* and the eel *Anguilla*, the representation is similar to that seen in *Amia* and *Raja* (Meredith and Butler 1983; Meredith et al. 1987). Minor variations include a medio-lateral reversal of the saccular and lagenar terminal fields in *Astronotus*, and a relatively heavy octaval input to the rostral pole of the lateral line nucleus (nucleus medialis) in *Anguilla*. Thus, these two teleosts appear to possess most features of the inner ear projection pattern hypothesized to be primitive for jawed fishes.

4.4 More Elaborate Variations on the Primitive Pattern in Teleosts

In certain teleosts, a portion of the octaval column extends dorsally and thereby lies medial to the first-order lateral line mechanosensory area, nucleus medialis (Figs. 17.6, 17.8, 17.9). This dorsomedial octaval zone is present along with the five standard first-order octaval nuclei and has been considered to be either a dorsal extension of one or more of these nuclei, or possibly an additional octaval nucleus (McCormick and Braford 1988). Such teleosts include members of the Osteoglossomorpha and the Otophysi. Some, but not all, of these teleosts have physical coupling between a gas-filled structure and the inner ear. The projection of the eighth nerve to the dorsomedial octaval cells gives the first impression that this is a dorsal zone of heavy overlap between lateral line and inner ear inputs (Bell 1981a; McCormick 1982). However, cytoarchitectonic analysis of a number of related species in each group, combined with inner ear projection patterns, reveals that dorsal eighth nerve projections are largely separate from those of the lateral line mechanosensory system (McCormick, personal observations).

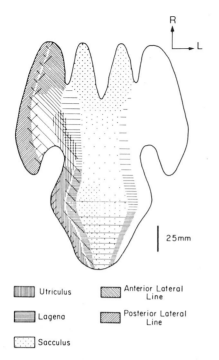

R
L

25mm

▓ Utriculus   ▨ Anterior Lateral Line

▤ Lagena   ▧ Posterior Lateral Line

░ Sacculus

FIGURE 17.7. This figure, taken from Bell (1981a), is a flattened view of the neurons that lie just ventral to the cerebellar crest in the mormyrids *Gnathonemus* and *Brienomyrus*. Note the medial, bilateral projection of the saccule and lagena as compared to the projections of the utricle and mechanosensory lateral line fibers.

4.4.1 Osteoglossomorphs

Unusual octaval projections were first demonstrated in two mormyrid species, *Gnathonemus* and *Brienomyrus* (Bell 1981a). Following the terminology of Maler (1974), a dorsally positioned anterior lateral line lobe was recognized as a site at which lateral line mechanosensory and inner ear projections overlapped extensively. However, cytoarchitectonic analysis of this zone in *Gnathonemus* and other osteoglossomorphs suggests that it is divided into a dorsomedial octaval population, which receives input largely from the eighth nerve, and a more laterally positioned mechanosensory area which is part of nucleus medialis (McCormick, personal observations; Fig. 17.6a). Bell (1981a) demonstrated the essentially nonoverlapping nature of the octaval and lateralis projections to the anterior lateral line lobe (Fig. 17.7). Figure 17.7 also illustrates the bilateral projections of the saccule, lagena, and utricle to the dorsomedial area, as well as the finding that the few lateral line fibers that extend medially towards this octaval population overlap the utricular and lagenar terminal fields, but not the saccular field.

It is possible that the dorsomedial octaval area is a displaced portion of the dorsal descending octaval nucleus. As seen in Figure 17.6, a dorsomedial octaval area is present in a number of osteoglossomorphs. The arawana *Osteoglossum* (Fig. 17.6c) and the butterfly fish *Pantodon* (Fig. 17.6d) do not have inner ears coupled to a gas-filled chamber, as do mormyrids. In these two species, cells of the descending nucleus clearly arch in a dorsomedial direction to contact the ventral border of the most medial portion of the cerebellar crest, and saccular and lagenar fibers project to these dorsomedial cells. These projections are overlapped ventrally by fibers of the utricle which, along with semicircular canal afferents, also project to the remaining (ventral) portion of the descending nucleus (McCormick, personal observations).

Essentially the same projection pattern is present in *Gnathonemus* (Bell 1981a) and the notopterid *Xenomystus* (Fig. 17.6b, Braford, personal communication). However, in these species, the continuity between the dorsomedial and ventral octaval regions is not as obvious as in *Pantodon* and *Osteoglossum* because the cells in these two regions are morphologically quite different. Despite this difference in cell morphology, the cells in these two regions are likely all part of the descending nucleus based on the total endorgan projection pattern and on the boundaries of the descending nucleus in other osteoglossomorphs.

If the hypothesis that the mormyrid descending nucleus is bipartite (with dorsomedial and ventral portions) is correct, then these osteoglossomorphs have a first-order octaval column that is more similar to that of other fishes than was previously realized (McCormick 1982). Posterior, descending, tangential, magnocellular, and anterior octaval nuclei can be recognized (personal observations, McCormick 1982). Otolithic endorgans project primarily dorsally in the descending and anterior nuclei, and semicircular canals tend to project ventrally. Bell pointed out that of all the otic endorgans in mormyrids, the projections of the saccule (likely a major auditory endorgan; Stipetić 1939) are the most discrete. In addition to its prominent bilateral dorsomedial projection (Fig. 17.7), there is another particularly dense bilateral projection zone described within the "nucleus octavius" (Bell 1981a). This saccular terminal field is likely to be located in the most medial portion of the anterior nucleus. Lagenar fibers project lateral to this field,

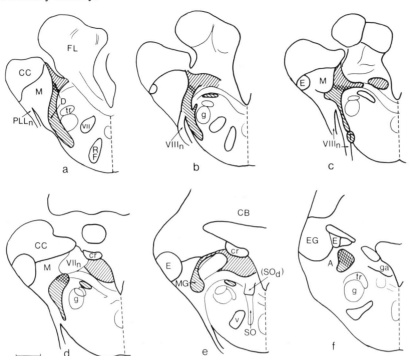

FIGURE 17.8. A series of hemisections through the medulla of the catfish *Ictalurus punctatus* showing the major projections of the saccule and lagena (▨) versus those of the utricle and semicircular canals (▨). Level a is the most caudal section. (McCormick and Braford, personal observations.) A, anterior octavus nucleus; CB, cerebellum; CC, cerebellar crest; cr, medial extension of the cerebellar crest; D, descending octaval nucleus; E, electrosensory lateral line lobe; EG, eminentia granularis of the cerebellum; FL, facial lobe; g, secondary gustatory trace; ga, auditory granule cells of Finger and Tong 1984; M, nucleus medialis; MG, nucleus magnocellularis; PLLn, posterior lateral line nerve; RF, reticular formation; SO, superior olive; (SOd), presumed dorsal division of the superior olive; T, nucleus tangentialis; tr, spinal trigeminal tract; v, trigeminal motor nucleus; VIIn, facial nerve; VIIIn, eighth nerve.

as in the anterior nucleus of other species, and utricular fibers project lateral to it as well.

In osteoglossomorphs, the presence of a dorsomedial extension of the descending nucleus is not correlated with the presence of physical coupling between the inner ear and a gasfilled chamber. However, in species which do have such coupling (*Gnathonemus* and *Xenomystus*), the dorsomedial extension is composed of cells that are morphologically different from those of the remainder of the descending nucleus, and also appears larger in volume than the corresponding region in species without coupling.

4.4.2 Otophysans

Inner ear projections are known in three otophysan species, the catfishes *Ictalurus punctatus* (Fig. 17.8; McCormick and Braford 1988 and personal observations) and *Ancistrus* (Fritzsch, Niemann, and Bleckmann 1990), and the goldfish *Carassius auratus* (Fig. 17.9; McCormick and Braford 1988

and personal observations). In *Ictalurus* and *Carassius* (Figs. 17.8, 17.9) the descending nucleus extends in a dorsomedial direction and contacts the cerebellar crest (see also Finger and Tong 1984). The saccule and lagena each project to this portion of the descending nucleus in the two species, and utricular fibers overlap the ventral portion of their terminal fields. Utricular and canal fibers project to the ventral part of the descending nucleus.

In more rostral areas of the medulla in *Ictalurus*, neurons in the dorsomedial area come to lie in progressively more medial positions (Fig. 17.8B–E). These displaced cells eventually occupy a position on the border of the fourth ventricle (Fig. 17.8D,E), where they have also been recognized in the catfish *Ancistrus* (Fritzsch, Niemann, and Bleckmann 1990). This ventricular population has previously been considered to be a second-order nucleus, the medial auditory nucleus, comparable to the superior olive of mammals (Finger and Tong

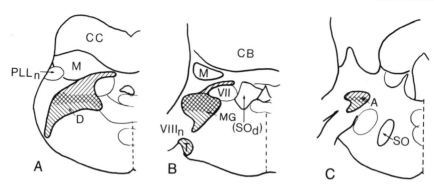

FIGURE 17.9. Hemisections through the medulla of the goldfish *Carassius auratus* showing the major projections of the saccule and lagena (▨) versus those of the utricle and canals (◨). Level a is the most caudal. McCormick and Braford, personal observations. A, anterior octavus nucleus; CB, cerebellum; CC, cere-bellar crest; D, descending octaval nucleus; M, nucleus medialis; MG, magnocellular nucleus; PLLn, posterior lateral line nucleus; SO, superior olive; (SOd), presumed dorsal division of superior olive; T, nucleus tangentialis; VII, facial nerve; VIIIn, eighth nerve.

1984). However, these neurons are clearly a first-order population (see also Fritzsch et al. 1990) which is separated from the remainder of the descending nucleus by afferent facial nerve fibers. In *Ictalurus*, saccular fibers terminate here. Similarly displaced cells which lie ventral to the arcuate tract (Fig. 17.8B–E) receive utricular projections.

The remainder of the octaval column in otophysans appears to be organized like that in *Amia*. Therefore, as appears to be the case in osteoglossomorphs, otophysans are specialized in that they possess a dorsomedial extension of the descending nucleus. Analysis of the otic projections in the related nonotophysan *Chanos chanos*, which does not have swimbladder-inner ear coupling, should allow us to conclude whether the presence of this dorsomedial extension in otophysans is strictly associated with the presence of such coupling.

4.4.3 Clupeids

The clupeid utricle, which is coupled to a gas bladder, has both auditory and vestibular functions distributed among its three maculae (Best and Gray 1980; Denton and Gray 1980). According to Meredith (1985) two of these maculae, which are entirely or partly vestibular, have some projections to the first-order octaval nuclei that overlap those of the semicircular canals. These two maculae, along with the remaining, auditory macula, also send fibers to dorsal portions of the octaval nuclei (e.g., Fig. 17.10). The above connections are simi-

lar to what is seen in other fishes; an exception is that the auditory macula has a bilateral projection to a rostral portion of the anterior octaval nucleus which is free of input from the other two maculae. Interestingly, fibers from all three utricular maculae also terminate bilaterally in a dorsomedial area (Fig. 17.10) located in a position similar to that of the dorsomedial octaval area in osteoglossomorphs. Whether the clupeid dorsomedial area is a portion of the lateralis column (Meredith 1985) or of the octavus column remains to be determined.

4.5 Higher Order Acoustic Projections

The midbrain terminus of auditory input is a subdivision of the torus semicircularis that appears to be separate, at least to some extent, from lateral line toral populations (Fig. 17.11). In some cases, this portion of the torus has been shown electrophysiologically to be auditory; in other cases, its auditory nature has been inferred from its connections. The auditory torus is usually located medial to all or part of the toral lateral line region. In *Ictalurus*, the auditory torus is composed of two rostrocaudally continuous areas, the central nucleus and the medial pretoral nucleus (Finger and Tong 1984; Streidter 1990b).

As is the case in chondrichthyans, axons terminating in the known/presumed auditory torus originate from cells in the descending and anterior nuclei, including the dorsomedial octaval zone of osteoglossomorphs and otophysans, and from two other brainstem populations (described below). All or some combination of these afferent sources to the torus are present in the halecomorph *Amia* and in diverse teleosts (Bell 1981b; Finger and Tong 1984; Echteler 1984; Murakami et al. 1986; McCormick and Braford 1988; McCormick, personal observations). In many species, rostral medullary cells associated with the reticular formation have been termed the superior olive, while clusters of perilemniscal cells have been referred to as the nucleus of the lateral lemniscus. As is the case in chondrichthyans, it is necessary to know more about the connectivity of these cells before their homologues in other vertebrates can be determined.

In otophysans, the "superior olive" appears to have dorsal and ventral portions. Echteler illustrated both of these populations in the carp *Cyprinus* but labeled only the ventral cells (Fig. 17.6 in Echteler 1984); the location of both populations (SOd and SO) is shown in the related species *Carassius* in Figure 17.9. Both populations are also present in the catfish *Ictalurus* (Fig. 17.9; see also Finger and Tong 1984; Figs. 17.5d, 17.7h,g).

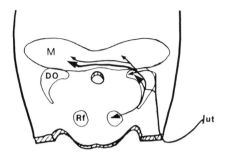

FIGURE 17.10. Line drawing of a transverse hemisection through the medulla of the clupeid *Clupea harengus* L. at the level of the descending octaval nucleus showing the projections of auditory (small arrow heads) versus vestibular (large arrow heads) maculae of the utricle. (After Meredith 1985.) DO, descending octaval nucleus; M, nucleus medialis; Rf, reticular formation; ut, utricular nerve fibers.

It is unclear whether or not the dorsal division (SOd) is present in other bony fishes.

Three main diencephalic areas have been implicated as acoustic areas: the central posterior nucleus, a portion of the preglomerular complex,

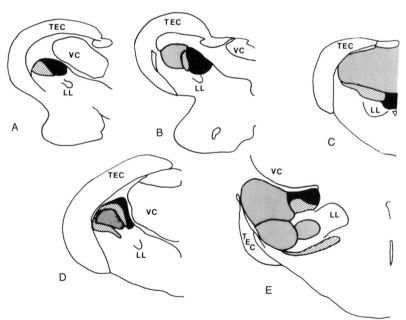

FIGURE 17.11. Line drawings of hemisections through the torus semicircularis of a number of teleosts showing the location of the known or presumed auditory zone (darkened area) as opposed to the mechanosensory zone (diagonal lines) and the electrosensory zone(s) (stippled areas). (A) *Carassius auratus*, (B) *Ictalurus punctatus*, (C) *Eigenmannia* sp., (D) *Xenomystus nigri*, (E) *Gnathonemus petersii*. This is reproduced from McCormick 1989, and represents a summary and interpretation of both published and unpublished data from a number of laboratories. LL, lateral lemniscus; TEC, optic tectum; VC, valvula cerebelli.

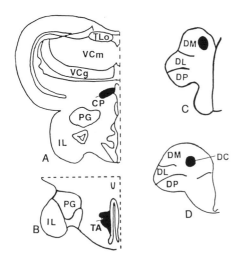

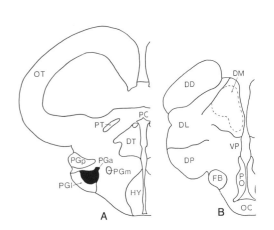

FIGURE 17.12. Forebrain hemisections showing areas implicated in auditory processing (darkened regions) in the carp, *Cyprinus carpio*. (Based on the work of Echteler 1984, 1985.)

FIGURE 17.13. Forebrain hemisections showing areas implicated in auditory processing (darkened regions) in the knifefish *Xenomystus nigri*. The auditory area within DM is a subpopulation lying within the boundaries of the dashed lines. (Unpublished data of Braford.) Abbreviations for Figures 17.12 and 17.13: CP, central posterior thalamic nucleus; DC, dorsocentral telencephalic area; DL, dorsolateral telencephalic area; DM, dorsomedial telencephalic area; DP, dorsoposterior telencephalic area; DT, dorsal thalamic area; FB, forebrain bundle; HY, hypothalamus; IL, inferior lobe of hypothalamus; OC, optic chiasm; OT, optic tectum; PC, posterior commissure; PG, preglomerular nuclear complex of diencephalon; PGa, preglomerular nuclear complex, pars anterior; PGl, preglomerular nuclear complex, pars lateralis; PGm, preglomerular nuclear complex, pars medialis; PGp, preglomerular nuclear complex, pars posterior; PO, preoptic area; PT, pretectal nucleus; TA, anterior tuberal nucleus of hypothalamus; TLo, torus longitudinalis; VCg, granule layer of valvula cerebelli; VCm, molecular layer of valvula cerebelli, VP, ventral posterior nucleus.

and, in otophysans, the anterior tuberal nucleus (Figs. 17.12, 17.13; Braford and McCormick 1979; Bell 1981b; Finger and Bullock 1982; Echteler 1984; Striedter 1990, 1991; Braford, personal communication). Acoustic input probably also reaches two other thalamic areas in otophysans, the anterior nucleus (Striedter 1991) and the ventromedial nucleus (Echteler 1984; Murakami et al. 1986).

It is difficult to determine how circuits involving these nuclei compare to the auditory forebrain circuits of amphibians and amniotes. The central posterior nucleus (CP) is a periventricular dorsal thalamic nucleus similar in position, and probably homologous, to the central nucleus of amphibians (Braford and Northcutt 1983). It is known to be reciprocally connected to the auditory torus in the carp *Cyprinus*, as is another dorsal thalamic population, the ventromedial nucleus (Echteler 1984).

The preglomerular complex (PGC) is a migrated portion of the posterior tuberculum (Braford and Northcutt 1983) that contains a number of functional subdivisions. Input to one (or more) subdivision from the auditory torus has been demonstrated in mormyrids, *Ictalurus*, and *Xenomystus* (Fig. 17.13a; Bell 1981b; Striedter 1991; Braford, personal communication). In *Xenomystus*, this portion of the PGC projects to the dorsomedial area

of the telencephalon (Fig. 17.13B). In *Ictalurus*, Striedter (1991) finds that the acoustic subdivision of the PGC (the lateral preglomerular nucleus) also receives lateral line mechanosensory and electrosensory inputs. Other portions of the PGC receive other classes of input, such as tectal (Northcutt and Butler 1980) and gustatory (Finger 1983, discussed in Striedter 1990). Because a portion of the PGC relays auditory information to the telencephalon, the PGC has been homologized with the mammalian medial geniculate body (Murakami et al. 1986). However, the probable origin of the PGC from the posterior tuberculum argues against such a comparison. Braford and Northcutt (1983) suggest that a migrated PGC complex is a unique actinopterygian feature, and that the PGC is a possible field homologue of the subthalamic region (Keyser 1972) of mammals.

An auditory projection to the anterior tuberal nucleus of the hypothalamus (TA) has been demonstrated only in otophysans and may be unique to this group of teleosts. It is reminiscent of, though not homologous to, the projection of the auditory system to the hypothalamus in anurans (see below). The reciprocal connections, present at least in *Cyprinus*, between the TA (probably Finger and Bullock's "medial thalamic zone"; see Striedter 1991) and the auditory midbrain suggest that sound may play a role in reproductive and/or aggressive behavior (Echteler 1984).

A recent study by Striedter (1991) concluded that the three acoustic diencephalic centers in *Ictalurus* – the CP, the PGC (its lateral portion), and the TA, also receive input from the electro- and mechanosensory lateral line systems by means of widespread interconnections among midbrain and diencephalic nuclei. A number of functional explanations for such convergence of sensory inputs can be suggested. For example, subpopulations of neurons in each nucleus might integrate input from the three octavolateralis modalities; such multimodal integration areas are known in the torus semicircularis (Schellart and Kroese 1989). Another possibility is that the three diencephalic nuclei, which are likely functionally distinct, each perform unique operations on the three modalities; perhaps subpopulations for each modality are present in each nucleus. Since parallel processing of auditory information occurs at various levels of the neuraxis in other vertebrates, such a possibility is not unlikely (e.g., Warr 1982; Hall and Feng 1987). Striedter's finding that at least two of the three diencephalic areas in *Ictalurus* (PGC and TA) project to largely nonoverlapping areas of the telencephalon is consistent with the hypothesis that these areas are physiologically different.

Two acoustic areas have been identified in the caudal part of the area dorsalis of the telencephalon: the dorsomedial area (DM), and the dorsocentral area (DC) (Figs. 17.12, 17.13; Echteler 1985; Murakami et al. 1986; Braford, personal communication). Auditory projections to DM have been demonstrated both anatomically and physiologically. In *Xenomystus* such inputs arise, at least in part, from a subdivision of the PGC (Fig. 17.13A). In the carp *Cyprinus*, DC has been shown electrophysiologically to be an acoustic area, though the source of this input is unknown (Echteler 1985). DC projects to the auditory midbrain in this species. Like the preglomerular complex, DM and DC are not anatomically homogeneous, and therefore probably contain functionally distinct zones (Northcutt and Braford 1980; Bass 1981; Striedter 1991).

5. Class Amphibia

5.1 Structure and Function of the Inner Ear

Amphibians possess a variable complement of inner ear receptors, some "phylogenetically old" and some "phylogenetically new" (the amphibian and basilar papillae) (Fig. 17.14; e.g., Baird 1974; Lewis et al. 1985; White 1986). The receptors common to all amphibians are the utricle, saccule, amphibian papilla, and the three semicircular canals. In addition to those receptors, all urodeles possess a lagena, and primitive urodeles have a basilar papilla (BP). The macula neglecta has traditionally been considered to be absent in urodeles, although homology with the amphibian papilla has been proposed (see below). Similarly, a macula neglecta is not recognized in anurans. All anurans have the three otolithic endorgans and both the AP and the BP. In gymnophionans all three otolithic endorgans can be present, although some species lack a lagena. All gymnophionans have an AP. The BP is generally absent; it has been identified in only one genus (White and Baird, 1982). Gymnophionans are unique among amphibians in that all species possess the macula neglecta.

The evolutionary histories of the AP and BP have long been debated. Some have suggested the AP to be a unique amphibian acoustic receptor that has no counterpart in fishes or amniotes (de Burlet 1928; Wever 1985). A contrasting hypothesis is that the AP originated from all or part of a displaced macula neglecta (Retzius 1881; Sarasin and Sarasin 1888; Baird 1974; White and Baird 1982; Fritzsch and Wake 1988). A recent

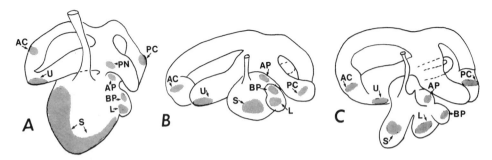

FIGURE 17.14. Medial views of the inner ears of three amphibians: (A) the gymnophionan *Siphonops*, (B) the urodele *Salamandra* and (C) the anuran *Bufo*. Stippled areas indicate the approximate positions of the sensory surfaces. (Based on a figure from Baird 1974.) AC, anterior semicircular canal crista; AP, amphibian papilla; BP, basilar papilla; L, lagenar macula; PC, posterior semicircular canal crista; PN, papilla (macula) neglecta; S, saccular macula; U, utricular macula.

study in gymnophionans provides support for this hypothesis on both morphological and embryological grounds (Fritzsch and Wake 1988). There has likewise been disagreement on the origin of the basilar papilla. The recent discovery of a basilar papilla in *Latimeria* (Fritzsch 1987) suggests that this endorgan predates the origin of amphibians. The assertion that the sarcopterygian basilar papilla is homologous to the BP in both anamniotes and amniotes (Fritzsch Chapter 18) contrasts with morphological analyses that suggest nonhomology between the basilar papillae of anurans and amniotes (Lombard and Bolt 1979; Wever 1985). Wever (1985) further postulates that the basilar papilla arose several times within the Class Amphibia.

The majority of anurans possess a tympanic ear, but other amphibians do not. Hearing among the amphibia therefore refers, as in other tetrapods, to the reception of both vibrations and airborne sound (see Webster, Chapter 37). Under this definition, various studies have shown that the papillar endorgans, the saccule, and the lagena may respond to sound in both anurans and urodeles (e.g., Ross and Smith 1980; Lewis et al. 1982; Narins 1975, reviewed in Lewis et al. 1985). The saccule and lagena also have vestibular functions in some species (reviewed in Lewis et al. 1985). No experimental studies of endorgan function are available for gymnophionans, but on morphological grounds Wever (1985) suggests that both the BP and AP (his utricular papilla) respond to sound.

5.2 Inner Ear Projections to the Medulla

Amphibians exhibit two basic patterns of organization of the primary octaval nuclei—that of anurans (Fig. 17.15E) and that of nonanurans (Fig. 17.15B). The nonanuran pattern is reminiscent of the primitive pattern in bony fishes (Fig. 17.15A,B; McCormick and Braford 1988), although in these amphibians the majority of neurons are concentrated along the ventricular surface. The primary octaval nuclei are positioned ventral to the lateral line nuclei. Three octaval subdivisions, comparable in position to the anterior, magnocellular, and descending nuclei of fishes are present. There is a general tendency for papillar, saccular, and lagenar fibers to terminate dorsal to fibers from the utricle and semicircular canals (McCormick 1988, Will 1988), which is likewise comparable to the dorsal acoustic and ventral vestibular projections in fish. Will (1988) has also identified in urodeles a group of neurons in the ventral portion of the mechanosensory lateral line nucleus that, based on dendritic orientation, might receive only saccular input. These neurons, called "nucleus saccularis," are thought also to be present in gymnophionans and may be homologous to the "nucleus saccularis" of anurans (the dorsal portion of the anuran caudal nucleus, see below). In urodeles, nucleus saccularis is known to project to the mesencephalon (Will 1988). Further experimental information on the organization of acoustic pathways is not available for urodeles and gymnophionans.

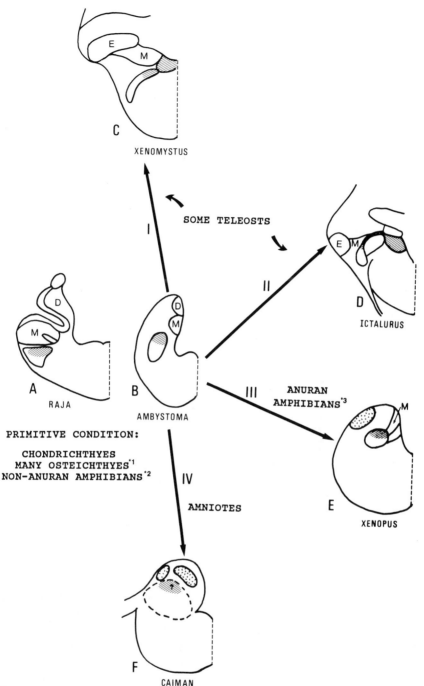

FIGURE 17.15. Schematic summary of the organization of first-order acoustic populations in anamniotes (A–E) and amniotes (F). *Symbols:* Diagonal lines = first-order populations in the primitive, ventral octavus column which receive auditory input from otolithic endorgans and, in amphibians, from papillar acoustic receptors. Areas with floating "v's" = Specialized first-order acoustic nuclei which receive a major input from papillar acoustic receptors and, at least in some taxa, a proportionately smaller input from otolithic endorgans. (*1) Many actinopterygians do not possess nucleus dorsalis (D), an electrosensory structure. (*2) Nucleus dorsalis (D) and the electric sense are absent in certain nonanuran amphibians. (*3) Most adult anurans possess neither nucleus medialis (M) nor the mechanosensory system. D, nucleus dorsalis; E, electrosensory lateral line nucleus; M, nucleus medialis.

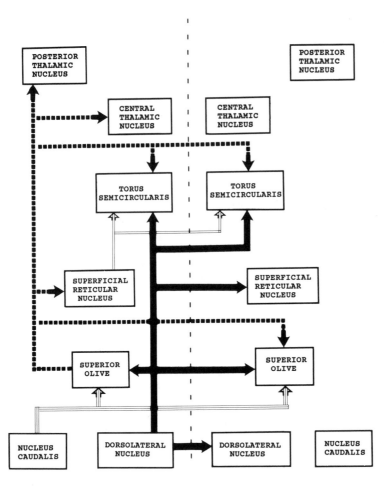

FIGURE 17.16. Summary of the ascending connections of hindbrain acoustic nuclei in ranid anurans. Four arrow widths are used to indicate the origins of these pathways from four areas: the dorsolateral nucleus, nucleus caudalis, the superior olive, and the superficial reticular nucleus.

Anurans have a more complexly organized octaval area consisting of a dorsal zone and a ventral zone (Fig. 17.15E). The dorsolateral nucleus (DLN), the only constituent of the dorsal zone, receives input (presumably entirely acoustic) from the AP, BP, and lagena (reviewed in Wilczynski, 1988) and saccule (Kuruvilla et al. 1985). The first three inputs are known to be tonotopically organized within the DLN (Wilczynski 1988). The ventral zone has been variously subdivided by different investigators (e.g., Gregory 1972; Matesz 1979; Opdam et al. 1976; Will et al. 1985; Kuruvilla et al. 1985). Will (1988) recognizes four nuclei in the ventral zone in a number of species: the anterior, lateral octaval, medial vestibular, and caudal (which includes nucleus saccularis). Whereas traditionally this ventral zone was considered to be vestibular in function, recent studies indicate that various ventral zone nuclei receive acoustic as well as vestibular input (Will et al. 1985; Will 1988). For example, the AP, BP, and

lagena project not only to the DLN but also to the dorsal portion of the lateral octaval nucleus. Likewise, the saccule projects to the dorsal lateral octaval nucleus, and has an additional projection field, near the DLN, that contacts a dorsal portion of nucleus caudalis (nucleus saccularis). Other components of the ventral zone receive vestibular projections. Thus, acoustic information appears to be processed in the dorsal portion of the ventral octavus zone of anurans. This organization is similar to that present in nonanuran amphibians and fishes, where acoustic areas are present in the dorsal portion of the single ventral octavus column (Fig. 17.15A,B). On this basis, the anuran ventral octavus zone may, as a whole, be a field homologue of the entire octavus column of fishes.

Although the DLN has been considered to be a homologue of the amniote cochlear nuclei (e.g., Ariens Kappers et al. 1967; Larsell 1934; Matesz 1979), it is likely to be an independently evolved dorsal acoustic zone (Will et al. 1985; McCormick

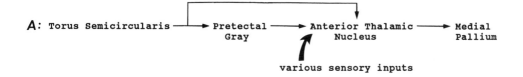

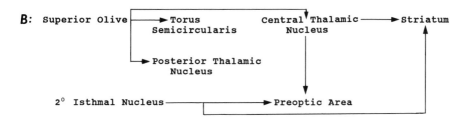

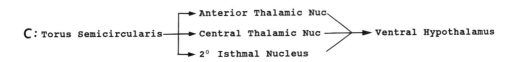

FIGURE 17.17. Summary of three pathways through the forebrain that convey acoustic information in ranid anurans.

and Braford 1988; McCormick 1988, Will 1988). The evolution of the DLN, a discrete primary acoustic nucleus, may be related to the highly specialized acoustic behavior characteristic of anurans (Will et al. 1985).

The embryonic origin of the anuran DLN is unknown (Fritzsch 1988). Jacoby and Rubinson (1983) showed that the DLN is not a transformed lateral line nucleus, as claimed by Larsell (1934). Fritzsch et al. (1984) speculated that two populations of cells, a ventral and a dorsal, potentially contribute to the formation of the DLN. They hypothesized that the ventral population also gives rise to the ventral octavus zone, and that the dorsal population is the same one that gives rise to the electrosensory dorsal lateral line nucleus in some other anamniotes. Other possibilities include (1) the derivation of the DLN entirely from the ventral octavus zone, either as a migrated subset of cells or as a duplicated part of the ventral zone, and (2) the derivation of the DLN from a different, perhaps newly evolved, embryonic field (McCormick and

Braford 1988). Further developmental studies are needed to resolve these issues.

5.3 Higher Order Auditory Connections

Auditory connections beyond the level of the first-order nuclei are essentially unstudied in nonanurans (see Section 5.2). In contrast, a great deal is known about these connections in anurans, particularly ranids. This information is reviewed thoroughly by Neary (1988), Wilczynski (1988), and Will (1988), and will only be summarized here (Fig. 17.16).

Auditory processing has been confirmed both anatomically and physiologically in three brain stem areas: the superior olivary nucleus, the superficial reticular nucleus, and a portion of the torus semicircularis. Axons of the DLN project directly to each of these areas, which are themselves interconnected in complex ways (Fig. 17.16). One other brain stem population, the secondary isthmal nucleus, derives its auditory input from the torus semicircularis (Fig. 17.17C).

The major ascending connections of medullary and isthmal auditory areas are shown schematically in Figures 17.16 and 17.17. While the DLN projects to all auditory areas through the midbrain, most of its efferents terminate bilaterally in the single superior olivary nucleus. The topographic location of the anuran superior olive is similar to that of the presumed superior olive of bony fishes. The nature of the input to the superior olive from nucleus caudalis (Fig. 17.16) is unknown; conceivably, these axons are carrying saccular information. In certain anurans, some eighth nerve fibers (of unknown endorgan origin) synapse directly in the superior olive (Matesz 1979; Will, and Fritzsch 1988). The major outflow of the superior olive is to the ipsilateral torus semicircularis. Olivary axons also supply two dorsal thalamic nuclei: the posterior thalamic nucleus (a multimodal processing area) and the caudal portion of the central thalamic nucleus (a major acoustic nucleus). The superficial reticular nucleus, similar in location to the nucleus of the lateral lemniscus in amniotes and fishes, has only one known ascending auditory connection — a bilateral projection to the torus semicircularis.

The complex descending pathways that interconnect many of the above nuclei are not shown in Figure 17.16. In short, the torus semicircularis projects to the superficial reticular and superior olivary nuclei, and the latter two nuclei project to the DLN. The torus also has other descending inputs to nonacoustic areas. Further details concerning all of the connections are provided by Wilczynski (1988).

The torus semicircularis has a complex organization in ranids. It contains a number of zones that receive various sensory and nonsensory inputs (reviewed in Wilczynski, 1988). The ventral toral zone is considered to be the main auditory region, although small numbers of fibers from lower-order acoustic nuclei also reach other toral zones. Lateral line afferents to the torus in *Xenopus* terminate in the ventral zone lateral to those from auditory structures, a pattern reminiscent of that seen in many fishes (Fig. 17.11).

Many brain areas either receive input from the torus semicircularis or are known physiologically to process acoustic information (reviewed in Neary 1988). According to Neary (1988), only two of the many nuclei that receive toral efferents may be primarily devoted (in whole or in part) to acoustic processing: the ventromedial portion of the central thalamic nucleus and the secondary isthmal nucleus. By analogy with the ascending acoustic pathways in amniotes, one might expect a forebrain pallial region to be the major terminus of the amphibian ascending acoustic pathway. However, at least in ranid anurans, this does not appear to be the case.

The three sets of pathways shown in Figure 16.17 illustrate three different features of acoustic forebrain connections in ranids. Figure 16.17A is the only known route of acoustic input to the pallium. This pathway does not appear to be dedicated to the auditory system. The anterior thalamic nucleus is a major source of various sensory inputs to the medial pallium; thus, both of these areas are polymodal rather than being concerned only with acoustic processing (Neary 1988).

Figure 17.17B summarizes acoustic projections to the striatum, preoptic area, and the posterior thalamic nucleus. Two principal auditory regions, the central thalamic nucleus, which processes the temporal components of sound (Hall and Feng 1987), and the secondary isthmal nucleus, are components of these pathways. The preoptic area is involved in the control of vocalization (Schmidt 1984). Likewise, the descending connections of the striatum may influence vocalizations and also affect acoustic processing in the central thalamic nucleus and torus semicircularis. The posterior thalamic nucleus receives significant acoustic, visual, and perhaps somatosensory inputs, and is another polymodal area.

According to Neary (1988), the ventral hypothalamus may be the major forebrain terminus of the ascending acoustic pathway, at least in *Rana catesbeiana*. The three known sources of auditory input to the hypothalamus are shown in Figure 17.17C; Neary speculates that such projections may provide the anatomical substrate by which mating calls influence hormonally driven reproductive behavior.

6. Discussion

This review has outlined what is known about central auditory circuitry in each of the four classes of anamniotes. These circuits have clearly not evolved in a linear fashion, but rather appear to

FIGURE 17.18. Summary of the organization of
ascending auditory circuitry in vertebrates. (*1)
Hypothalamic pathway in otophysans. (*2)
Hypothalamic pathway in anurans. (*3) It is
uncertain whether pallial acoustic areas are
present in jawed fishes.

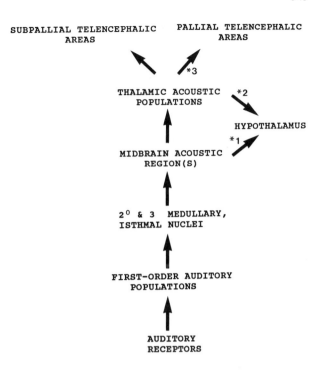

share a common organization plan – a multisynaptic, ascending lemniscal system terminating in the forebrain – upon which various specializations have been superimposed (Fig. 17.18). The auditory pathways of amniotes (Carr, Chapter 25; Frost and Masterton, Chapter 32) also share this organizational plan, which is likely a primitive, or generalized, feature at least among jawed vertebrates. However, while the route by which auditory information ascends the neuraxis is similar among vertebrates, individual auditory structures or nuclei at each level are not necessarily homologous across taxa.

The latter statement is particularly true of the auditory periphery. The inner ear acoustic receptors of vertebrates are more diverse than the receptors of any other sensory system. Among anamniotes, variation in peripheral acoustic receptors, including the evolution of dedicated acoustic endorgans such as the basilar and amphibian papillae, is not necessarily paralleled by reorganization of first order auditory neuronal populations.

Figure 17.15 illustrates variations in the organization of first-order acoustic populations among vertebrates, as well as the relationship between octaval and lateral line nuclei. It is currently believed that

the primitive pattern for the organization of first-order octavolateralis inputs is one in which lateral line electrosensory, lateral line mechanosensory, and inner ear inputs are processed in three separate regions, or cell columns, each containing one or more first-order nuclei (Figure 17.15A,B). The absence of a first-order nucleus within the octaval (inner ear) cell column devoted exclusively to acoustic processing appears to be a primitive vertebrate characteristic. Rather, in all anamniotes excepting anurans, otolithic and papillar acoustic endorgans project to dorsal subpopulations of octaval nuclei that also contain vestibular subpopulations. Therefore, the evolution of papillar endorgans is not strictly linked to the evolution of a separate first-order acoustic nucleus. These axons apparently tapped into preexisting pathways established by the phylogenetically older otolithic auditory receptors.

The presence of specialized primary acoustic nuclei in anurans (Fig. 17.15E) and amniotes (Fig. 17.15F) is considered by many recent investigators to be an example of homoplasy. The occurrence of large, dorsomedially located octaval areas in some osteoglossomorph and otophysan teleosts (Fig. 17.15C,D) appears to be another instance of the independent evolution of an expanded first-order auditory area, although these octaval areas do not appear to represent new nuclei.

It is interesting to note that both papillar and otolithic acoustic receptors in anurans retain their connections to phylogenetically old acoustic centers within the octaval column (located, in *Xenopus*, ventral to the lateral line nuclei) in addition to their connections with the new, dorsolateral nucleus. Traditionally, the dorsally located acoustic ("cochlear") nuclei of anurans and amniotes have been thought to be the domain of nonotolithic acoustic receptors, while other first-order octaval nuclei have been considered to be vestibular. Because the two dorsal acoustic nuclei of amniotes (nucleus magnocellularis and nucleus angularis of reptiles and birds; dorsal and ventral cochlear nuclei of mammals) probably evolved in parallel with the anuran dorsolateral nucleus, there is no reason to expect the first-order connections of these nuclei in anurans and amniotes to be similar. However, the traditional notion that there is a strict dichotomy in amniotes between dorsal acoustic nuclei (receiving basilar papilla, or cochlear duct, input) and ventral vestibular areas (receiving input from all other otic endorgans) may be incorrect.

Perhaps because this dichotomy is generally not questioned, little information is available that might refute it. However, there is some indication in amniotes that otolithic endorgans have dual dorsal (acoustic) and ventral (vestibular) projections, and that auditory areas may exist within the traditionally recognized first-order vestibular area.

The lagena of the pigeon and the saccule of the gerbil both project to the cochlear nuclei as well as to vestibular areas (Boord and Rasmussen 1963; Kevetter and Perachio 1989). In the crocodilian *Caiman*, Leake (1974) described a group "S," distinct from other first-order octaval nuclei, which appears to be innervated mainly by the saccule, and possibly also the utricle. In the cat and guinea pig, the saccule is known to have a discrete projection to Brodal's cell group Y in addition to its projection to the major vestibular nuclei. These saccular zones are reminiscent of the amphibian dorsal saccular projection area ("nucleus saccularis," Section 5.2). There is also a report of a projection of the cochlea to the medial vestibular nucleus in a mammal (Tickle and Schneider 1982). However, in no instance is the functional significance of the above

projections known. Furthermore, experimental evidence of auditory function in an amniote otolithic endorgan is limited to two mammalian species (guinea pig and squirrel monkey) and one endorgan (the saccule) (Young et al. 1976; Cazals et al. 1980, 1982, 1983), although others have postulated otolithic auditory functions in reptiles and birds (reviewed in Lewis et al. 1985).

In summary, the "first-order acoustic populations" of Figure 17.18 refers to neurons that may be located in discrete auditory nuclei, or may be subpopulations of primary octaval nuclei that also process vestibular information.

Before discussing acoustic processing at higher levels of the neuraxis, it is necessary to briefly summarize possible acoustic projections in anamniotes to areas that are not strictly part of the auditory system. Such areas include the reticular formation, portions of the cerebellum, and, when present, the Mauthner cell and first-order lateral line nucleus.

The strong projection of the eighth nerve, including saccular (presumably acoustic) fibers, to the Mauthner cell helps drive the startle response in many species (Eaton, Bombardieri, and Meyer 1977; Zottoli 1977). It is not clear whether first-order acoustic input per se reaches the other structures listed above, and if so, what the physiological implications of such inputs are. For example, since otolithic endorgans may have dual functions, the presence of a first-order saccular projection to the cerebellum does not necessarily mean that this input is auditory. Physiological studies are likewise needed to determine the degree and meaning of overlap between eighth nerve and lateral line input at various levels of the neuraxis. For example, a small number of eighth nerve fibers from all otic endorgans, including the semicircular canals, ramify within the boundaries of nucleus medialis, a first-order lateral line area. While some of this input may be directed to lateral line neurons, other octaval fibers might actually be terminating on long dendrites of octaval column cells that extend into nucleus medialis. Conversely, it is uncertain whether these octaval cell dendrites receive lateral line input, and if so, why. It is known that integration between auditory and lateral line input occurs at higher levels of the neuraxis (e.g., Schellart and Kroese 1989) but again the functional significance of cross talk between the two sensory systems is not understood. At present, it is generally believed that the auditory and lateral line mechanosensory systems each have specific, dedicated lemniscal pathways, at least up to the level of the midbrain, reflecting the known functional differences (Kalmijn 1988; Coombs et al., Chapter 15) between these two sensory systems.

Regarding brain stem auditory circuitry, Wilczynski (1988) has hypothesized that "evolutionary changes in the tetrapod auditory system have not involved fundamental changes in the pattern of brain stem pathways such as the addition of major new auditory centers or significant changes in long connecting pathways, but rather has involved a reorganization of nuclei within each station of a preset chain of brain stem auditory centers." This suggestion can be extended to include jawed fishes. It is clearly applicable to the organization of auditory input at the first-order level, where variable numbers of acoustic populations are present in different vertebrate taxa, and to the midbrain torus semicircularis, which is likewise variably organized both among anamniotes and amniotes. It is striking that higher order brain stem centers, similar in topographic position and connectivity to the amniote superior olivary and lateral lemniscal nuclei, have been identified in various anamniotes. The connections of these higher order centers in anurans and amniotes suggest that they are components of a set of parallel acoustic-processing channels, which are likely to have different functions (reviewed in Warr 1982, Wilczynski 1988, and Carr, Chapter 25). While these nuclei are believed to be homologous among all amniotes, it is unclear whether they are homologous between anurans and amniotes, or between jawed fishes and other vertebrates. Further studies on the distribution of these nuclei (i.e., are they present in nonanuran amphibians?) and their connections (i.e., are they in fact auditory in fishes?) are needed before they can be compared in an evolutionary context among vertebrates.

The forebrain connections of the auditory system may be the most variable, as indicated by the variety of diencephalic and telencephalic areas known or speculated to be involved in acoustic processing. In general, evolutionary analysis of these connections is hindered by lack of appropriate comparative data and various unresolved issues. For example, in jawed fishes, it is unclear whether auditory information is relayed from the thalamus to both pallial and subpallial telencephalic areas, as is the case

in anurans and amniotes. It is also uncertain in fishes whether all sources of auditory input to the telencephalon have been identified. On the other hand, it is apparent that different tetrapods have emphasized one set of telencephalic connections over the other, as evidenced by the predominantly striatal auditory representation in anurans versus the predominantly cortical auditory representation in placental mammals (also see Frost and Masterton, Chapter 32). Recall, however, that ranid anurans apparently direct the majority of forebrain acoustic fibers into the hypothalamus rather than the telencephalon. Neary (1988) points out that it is currently impossible to determine how specialized the various hypothalamic auditory pathways are in anurans without more extensive data from other vertebrates. It is possible that both anurans and catfish are unique not in having such connections, but rather in the large quantity of acoustic input they direct to the hypothalamus. In any case, the difference in the connectivity of currently identified auditory-hypothalamic pathways in catfish and anurans suggests that these pathways are nonhomologous. It can be hypothesized that the many forebrain auditory areas present in anamniotes form the anatomical substrate of multiple parallel processing channels, as occurs in the brainstem; functional data such as that presented by Hall and Feng (1987) on characteristics of the anuran thalamic nuclei are clearly needed in other taxa to evaluate this hypothesis.

7. Summary

This chapter summarizes our current understanding of the neuroanatomy of audition in the three extant classes of jawed anamniotes. Of the fourth anamniote class, the Agnatha, we can say nothing until the existence of an auditory sense is proved or disproved.

The auditory pathways of all jawed vertebrates share many organizational features (Fig. 17.18). This suggests that acoustic pathways were established early in vertebrate history, and that these pathways formed a neural substrate that was modified as changes in peripheral acoustic receptors and other relevant characteristics took place. As far as we can tell, the primitive acoustic receptors of anamniotes are some subset of the otolithic endorgans, and in the case of chondrichthyans, the macula neglecta. It can be hypothesized that the evolutionary origins of the central auditory pathways are tied in with such receptors, rather than with the lateral line system (e.g., Larsell 1934), or with the otic receptors of amphibians that evolved later (e.g., van Bergeijk 1967).

It is evident that many variations on the basic auditory circuitry of Figure 17.18 have occurred independently in various vertebrate lineages. Changes in auditory circuitry have thus not been linear. The recognition that significant variations occur within classes, as well as between classes, reinforces this point as well. Future challenges in our understanding of vertebrate hearing include investigations of the functional consequences of this variation, as well as the evolutionary and developmental mechanisms that produced them.

One might pose the following question: If different vertebrates utilize different acoustic receptors, has hearing evolved independently several times? The existence of a common organizational plan of vertebrate auditory circuitry (Fig. 17.18) might weaken such an argument, but is not in itself proof. A better understanding of the specific auditory functions of the primitive acoustic endorgans in anamniotes might provide insight. For example, if the utricle indeed plays a role in hearing in most actinopterygians (Schellart and Popper Chapter 16), the specialized auditory utricle of clupeids might be unique only in the *degree* to which it is auditory, rather than representing an independent development of the sense of hearing. Likewise, if future studies conclude that otolithic acoustic functions (be they for seismic or airborne sound) are a primitive characteristic of amphibians and amniotes, then the acoustically dedicated papillar endorgans of these vertebrates could not correctly be thought of as heralding the independent evolution of hearing in these organisms. Rather, the acoustic role of these otolithic endorgans, along with the central circuitry processing this information, would indicate that the sense of hearing is a conservative feature of jawed vertebrates. In this view, the papillar acoustic endorgans would be new contributors to hearing that augment, and in some cases supplant, more primitive acoustic receptors.

Acknowledgments. I thank Mark R. Braford for critically reading this manuscript, for our many valuable discussions about the evolution of brain circuitry, and for the use of data from studies in progress. Many thanks also to Sandra Grellinger, who provided expert technical assistance with the original studies reported here. I would also like to extend my appreciation to Douglas B. Webster for introducing me to comparative vertebrate biology as an undergraduate and for sparking my interest in the evolution of hearing. This work was supported by NSF grant BNS-8820095.

References

Allen JM, Blaxter JHS, Denton EJ (1976) The functional anatomy and development of the swimbladder-inner ear-lateral line system in herring and sprat. J Mar Biol Assoc UK 56:471–486.

Amemiya F, Kishida R, Goris RC, Onishi H, Kusonoki T (1985) Primary vestibular projections in the hagfish *Eptatretus burgei*. Brain Res 337:73–79.

Ariens Kappers CU, Huber GC, Crosby EC (1967) The Comparative Anatomy of the Nervous System of Vertebrates, Including Man. New York: Hafner Publishing Company.

Baird IL (1974) Anatomical features of the inner ear in submammalian vertebrates, In: Keidel WD, Neff WD (eds) Handbook of Sensory Physiology, Vol V/1. New York: Springer-Verlag, pp. 159–212.

Barry MA (1987) Afferent and efferent connections of the primary octaval nuclei in the clearnose skate, *Raja eglanteria*. J Comp Neurol 266:457–477.

Bass AH (1981) Organization of the telencephalon in the channel catfish, *Ictalurus punctatus*. J Morphol 169: 71–90.

Bell CC (1981a) Central distribution of octavolateral afferents and efferents in a teleost (Mormyridae). J Comp Neurol 195:391–414.

Bell CC (1981b) Some central connection of medullary octavolateral centers in a mormyrid fish. In: Tavolga WN, Popper AN, Fay RR (eds) Hearing and Sound Communication in Fishes. New York: Springer-Verlag, pp. 383–392.

Best ACG, Gray JAB (1980) Morphology of the utricular recess in the sprat. J Mar Biol Assoc UK 60:703–715.

Bodznick D, Schmidt AW (1984) Somatotopy within the medullary electrosensory nucleus of the little skate *Raja erinacea*. J Comp Neurol 225:581–590.

Board RL, Montgomery JC (1989) Central mechanosensory lateral line centers among the elasmobranchs. In: Coombs S, Gorner P, Munz H (eds) The Mechanosensory Lateral Line. New York: Springer-Verlag, pp. 323–339.

Board RL, Northcutt RG (1982) Ascending lateral line pathways to the midbrain of the clearnose skate, *Raja eglanteria*. J Comp Neurol 207:274–282.

Boord RL, Northcutt RG (1988) Medullary and mesencephalic pathways and connections of lateral line neurons of the spiny dogfish *Squalus acanthias*. Brain Behav Evol 32:76–88.

Boord RL, Rasmussen GL (1963) Projections of the cochlear and lagenar nerves on the cochlear nuclei of the pigeon. J Comp Neurol 120:463–471.

Boord RL, Roberts BL (1980) Medullary and cerebellar projections of the statoacoustic nerve of the dogfish, *Scyliorhinus canicula*. J Comp Neurol 193:57–68.

Braford MR, Jr, McCormick CA (1979) Some connections of the torus semicircularis in the bowfin, *Amia calva*: A horseradish peroxidase study. Soc Neurosci Abst 5:139.

Braford MR Jr, Northcutt RG (1983) Organization of the diencephalon and pretectum of the ray-finned fishes. In: Davis RE and Northcutt RG (eds) Fish Neurobiology, Vol. 2. Ann Arbor: University of Michigan Press, pp. 117–163.

Budelli R, Macadar O (1979) Statoacoustic properties of utricular afferents. J Neurophysiol 42:1479–1493.

Cazals Y, Aran J-M, Erre J-P (1982) Frequency sensitivity and selectivity of acoustically evoked potentials after complete cochlear hair cell destruction. Brain Res 231:197–203.

Cazals Y, Aran J-M, Erre J-P, Guilhaume A (1980) Acoustic responses after total destruction of the cochlear receptor: brainstem and auditory cortex. Science 210:83.

Cazals Y, Aran J-M, Erre J-P, Guilhaume A, Aurousseau C (1983) Vestibular acoustic function in the guinea pig: A saccular function? Acta Otolaryngol 95:211–217.

Corwin JT (1978) The relation of inner ear structure to the feeding behavior in sharks and rays, Scanning Electron Micros II:1105–1112.

Corwin JT (1981) Peripheral auditory physiology in the lemon shark: evidence of parallel otolithic and nonotolithic sound detection. J Comp Physiol 142:379–390.

Corwin JT, Northcutt RG (1982) Auditory centers in the elasmobranch brainstem: deoxyglucose autoradiography and evoked potential recording. Brain Res 236:261–273.

de Burlet HM (1928) Uber die Papilla neglecta. Anat Anz 66:199–209.

Dehadrai PV (1957) On the swimbladder and its relation with the internal ear in genus *Notopterus* (Lacepede). J Zool Soc India 9:50–61.

Denton EJ, Gray JAB (1980) Receptor activity in the utriculus of the sprat. J Mar Biol Assoc UK, 60:717–740.

Dunn RF, Koester DM (1987) Primary afferent projections to the central octavus nuclei in the elasmobranch, *Rhinobatos* sp. as demonstrated by nerve degeneration. J Comp Neurol 260:564–572.

Eaton RC, Bombardieri RA, Meyer DL (1977) The Mauthner-initiated startle response in teleost fish. J Exp Biol 66:65–81.

Echteler SE (1984) Connections of the auditory midbrain in a teleost fish, *Cyprinus carpio*. J Comp Neurol 230:536–551.

Echteler SE (1985) Organization of central auditory pathways in a teleost fish, *Cyprinus carpio*. J Comp Physiol A:156:267–280.

Fay RR, Kendall JI, Tester AL, Popper AN (1974) Vibration detection by the macula neglecta of sharks. Comp Biochem Physiol 47A:1235–1240.

Finger TE (1983) The gustatory system in teleost fish. In: Northcutt RG, Davis RE (eds) Fish Neurobiology, Vol. I. University of Michigan Press, pp. 286–309.

Finger TE, Bullock TH (1982) Thalamic center for the lateral line system in the catfish *Ictalurus nebulosus*: Evoked potential evidence. J Neurobiol 13:39–47.

Finger TE, Tong S-L (1984) Central organization of eighth nerve and mechanosensory lateral line systems in the brainstem of ictalurid catfish. J Comp Neurol 229:129–151.

Foster RE, Hall WC (1978) The organization of central auditory pathways in a reptile, *Iguana iguana*. J Comp Neurol 178:783–832.

Fritzsch B (1987) The inner ear of the coelacanth fish *Latimeria* has tetrapod affinities. Nature 327:153–154.

Fritzsch B (1988) Phylogenetic and ontogenetic origin of the dorsolateral auditory nucleus of anurans. In: Fritzsch B, Ryan MJ, Wilczynski W, Hetherington TE, Walkowiak W (eds) The Evolution of the Amphibian Auditory System. New York: John Wiley and Sons, pp. 561–585.

Fritzsch B, Wake MH (1988) The inner ear of gymnophione amphibians and its nerve supply: a comparative study of regressive events in a complex sensory system. Zoomorphology 108:210–217.

Fritzsch B, Nikundiwe AM, Will U (1984) Projection patterns of lateral line afferents in anurans: a comparative HRP study. J Comp Neurol 229:451–469.

Fritzsch B, Niemann U, Bleckmann H (1990) A discrete projection of the sacculus and lagena to a distinct brain stem nucleus in a catfish. Neurosci Lett 111:7–11.

Gregory, KM (1972) Central projections of the eighth nerve in frogs. Brain Behav Evol 5:70–88.

Hall JC, Feng AS (1987) Evidence for parallel processing in the frog's auditory thalamus. J Comp Neurol 258:407–419.

Jacoby J, Rubinson K (1983) The acoustic and lateral line nuclei are distinct in the premetamorphic frog, *Rana catesbeiana*. J Comp Neurol 216:152–161.

Jarvik E (1980) Basic structure and evolution of vertebrates. New York: Academic Press.

Kalmijn AJ (1988) Hydrodynamic and acoustic field detection. In: Sensory Biology of Aquatic Animals. Atema J, Fay RR, Popper AN, Tavolga WN (eds) New York: Springer-Verlag, pp. 83–130.

Kevetter GA, Perachio AA (1989) Projections from the sacculus to the cochlear nuclei in the mongolian gerbil. Brain Behav Evol 34:193–200.

Keyser A (1972) The development of the diencephalon of the Chinese hamster. Acta Anat 83 Suppl 59:1–181.

Knudsen EI (1977) Distinct auditory and lateral line nuclei in the midbrain of catfishes. J Comp Neurol 173:417–432.

Koester DM (1983) Central projections of the octavolateralis nerves of the clearnose skate, *Raja eglanteria*. J Comp Neurol 221:199–215.

Koyama H, Kishida R, Goris R, Kusunoki T (1989) Afferent and efferent projections of the VIIIth cranial nerve in the lamprey, *Lampetra japonica*. J Comp Neurol 280:663–671.

Kurivilla A, Sitko S, Schwartz IR, Honrubia V (1985) Central projections of primary vestibular fibers in the bullfrog: I. The vestibular nuclei. Laryngoscope 95:692–707.

Larsell O (1934) The differentiation of the peripheral and central acoustic apparatus in the frog. J Comp Neurol 60:473–527.

Leake PA (1974) Central projections of the statoacoustic nerve in *Caiman crocodilus*. Brain Behav Evol 10: 170–196.

Lewis ER, Baird RA, Leverenz EL (1982) Inner ear: Dye injection reveals peripheral origins of specific sensitivities. Science 215:1641–1643.

Lewis ER, Leverenz EL, Bialek WS (1985) The Vertebrate Inner Ear. Boca Raton: CRC Press.

Lombard RE, Bolt JR (1979) Evolution of the tetrapod ear: an analysis and reinterpretation. Biol J Linn Soc 11:19–76.

Lowenstein O (1970) The electrophysiological study of the responses of the isolated labyrinth of the lamprey (*Lampetra fluviatilis*) to angular acceleration, tilting, and mechanical vibration. Proc Roy Soc Lond Series B Biol Sci 174:419–434.

Lowenstein O, Thornhill RA (1970) The labyrinth of *Myxine*: anatomy, ultrastructure and electrophysiology. Proc Roy Soc Lond Series B Biol Sci 176:21–42.

Lowenstein O, Osborne MP, Thornhill RA (1968) The anatomy and ultrastructure of the labyrinth of the lamprey (*Lampetra fluviatilis* L.). Proc Roy Soc Lond Series B Biol Sci 170:113–134.

Lowenstein O, Roberts TDM (1951) The localization and analysis of the responses to vibration from the isolated elasmobranch labyrinth: a contribution to the problem of the evolution of hearing in vertebrates. J Physiol (Lond) 114:471–489.

Maler L (1974) The acousticolateral area of bony fishes and its cerebellar relations. Brain Behav Evol 10: 130–145.

Matesz C (1979) Central projections of the VIIIth cranial nerve in the frog. Neuroscience 4:2061–2071.

McCormick CA (1981) Central connections of the lateral line and eighth nerves in the bowfin, *Amia calva*. J Comp Neurol 197:1–15.

McCormick CA (1982) The organization of the octavolateralis area in actinopterygian fishes: a new interpretation. J Morphol 171:159–181.

McCormick CA (1983a) Central projections of inner ear endorgans in the bowfin, *Amia calva*. Am Zool 23:895.

McCormick CA (1983b) Central connections of the octavolateralis nerves in the pike cichlid, *Crenicichla lepidota*. Brain Res 265:177–185.

McCormick CA (1988) Evolution of auditory pathways in the Amphibia. In: Fritzsch B, Ryan MJ, Wilczynski W, Hetherington TE, Walkowiak W (eds) The Evolution of the Amphibian Auditory System. New York: John Wiley and Sons. pp. 587–612.

McCormick CA, Braford MR Jr (1988) Central connections of the octavolateralis system: evolutionary considerations. In: Sensory Biology of Aquatic Animals. New York: Springer-Verlag, Atema J, Fay RR, Popper WN, Tavolga WN (eds) pp. 733–756.

McCormick CA (1989) Central lateral line mechanosensory pathways in bony fish. In: Coombs S, Gorner P, Munz H (eds). The Mechanosensory Lateral Line. New York: Springer-Verlag, pp. 341–364.

Meredith GE (1985) The distinctive central utricular projections in the herring. Neurosci Lett 55:191–196.

Meredith GE, Butler AB (1983) Organization of eighth nerve afferent projections from individual endorgans of the inner ear in the teleost *Astronotus ocellatus*. J Comp Neurol 220:44–62.

Meredith GE, Roberts BL, Maslam S (1987) Distribution of afferent fibers from end organs in the ear and lateral line in the European eel. J Comp Neurol 265:507–520.

Murakami T, Fukuoka T, Ito H (1986) Telencephalic ascending acousticolateral system in a teleost (*Sebasticus marmoratus*), with special reference to the fiber connections of the nucleus preglomerulosus. J Comp Neurol 247:383–397.

Narins PM (1975) Electrophysiological determination of the function of the lagena in terrestrial amphibians. Biol Bull 149:438.

Neary TJ (1988) Forebrain auditory pathways in ranid frogs. In: Fritzsch B, Ryan MJ, Wilczynski W, Hetherington TE, Walkowiak W (eds) The Evolution of the Amphibian Auditory System. New York: John Wiley and Sons, pp. 233–252.

Nelson EM (1955) The morphology of the swimbladder and auditory bulla in the Holocentridae. Fieldiana Zool 37:121–130.

New JG, Northcutt RG (1984) Central projections of the lateral line nerves in the shovelnose sturgeon. J Comp Neurol 225:129–140.

Northcutt RG (1978) Brain organization in the cartilaginous fishes. In: Sensory Biology of Sharks, Skates and Rays, Hodgson ES, Mathewson RF (eds). Arlington, VA: Office of Naval Research, pp. 117–194.

Northcutt RG (1979) Central projections of the eighth cranial nerve in lampreys. Brain Res 167:163–167.

Northcutt RG (1980) Central auditory pathways in anamniotic vertebrates. In: Popper AN, Fay RR (eds) Comparative Studies of Hearing in Vertebrates. New York: Springer-Verlag, pp. 79–118.

Northcutt RG (1981) Audition and the central nervous system of fishes. In: Tavolga WN, Popper AN, Fay RR (eds). Hearing and Sound Communication in Fishes. New York: Springer-Verlag, pp. 331–355.

Northcutt RG, Braford MR Jr (1980) New observations on the organization and evolution of the telencephalon of actinopterygian fishes. In: Ebbesson SOE (ed). Comparative Neurology of the Telencephalon. New York: Plenum, pp. 41–98.

Opdam P, Kemali M, Nieuwenhuys R (1976) Topological analysis of the brain stem of the frogs *Rana esculenta* and *Rana catesbeiana*. J Comp Neurol 165:307–332.

Plassmann W (1983) Sensory modality interdependence in the octaval system of an elasmobranch. Exp Brain Res 50:283–292.

Platt C (1983) The peripheral vestibular system of fishes. In: Fish Neurobiology, Vol. 1. Northcutt RG, Davis RE (eds). Ann Arbor: University of Michigan Press, pp. 89–123.

Popper AN (1983) Organization of the inner ear and auditory processing. In: Northcutt RG, Davis RE (eds). Fish Neurobiology, Vol. 1. Ann Arbor: University of Michigan Press, pp. 125–178.

Popper AN, Hoxter B (1987) Sensory and nonsensory ciliated cells in the ear of the sea lamprey, *Petromyzon marinus*. Brain Behav Evol 30:43–61.

Retzius G (1881) Das Gehororgan der Wirbelthiere, Vol. 1. Stockholm: Samson and Wallin.

Ross RJ, Smith JJB (1980) Detection of substrate vibrations by salamanders: frequency sensitivity of the ear. Comp Biochem Physiol 47A:387–390.

Saidel W, McCormick CA (1985) Morphology of the macula neglecta in the bowfin, *Amia calva*. Soc Neurosci Abst 11:1312.

Sarasin P, Sarasin F (1888) Zur Entwicklungsgeschichte und Anatomie der ceylonesischen Blindwuhle Ichthyophis glutinosus. Das Gehororgan. Ergebnisse Naturwiss Forsch auf Ceylon. Wiesbaden: Kreidels Verlag.

Schellart NAM, Kroese ABA (1989) Interrelationship of acousticolateral and visual systems in the teleost midbrain. In: Coombs S, Gorner P, Munz H (eds). The Mechanosensory Lateral Line. New York: Springer-Verlag, pp. 421–443.

Schmidt RS (1984) Neural correlates of frog calling: Preoptic area trigger of "mating call." J Comp Physiol A 154:847–853.

Smeets WJAJ, Nieuwenhuys R, Roberts BL (1983) The Central Nervous System of Cartilaginous Fishes. New York: Springer-Verlag.

Stipetić E (1939) Uber das Gehororgan der Mormyriden. Z Vergl Physiol 26:740–752.

Striedter GF (1990) The diencephalon of the channel catfish, Ictalurus punctatus: II. Retinal, tectal, cerebellar, and telencephalic connections. Brain Behav Evolu 36:355–377.

Striedter GF (1991) Auditory, electrosensory, and mechanosensory lateral line pathways through the diencephalon and telencephalon of channel catfishes. Brain Behav Evol (in press).

Tester AL, Kendall JI, Milisen WB (1972) Morphology of the ear of the shark genus *Carcharhinus* with particular reference to the macula neglecta. Pacif Sci 26:264–274.

Tickle DR, Schneider GE (1982) Projection of the auditory nerve to the medial vestibular nucleus. Neurosci Lett 28:1–7.

van Bergeijk WA (1967) The evolution of vertebrate hearing. In: Neff WD (ed). Contributions to Sensory Physiology, Vol. 2. New York: Academic Press, pp. 1–49.

Warr WB (1982) Parallel ascending pathways from the cochlear nucleus: Neuroanatomical evidence of functional specialization. Contrib Sens Physiol 7:1–38.

Wever EG (1985) The Amphibian Ear. Princeton, NJ: Princeton University Press.

White JS (1986) Comparative features of surface morphology of the basilar papilla in five families of salamanders (Amphibia: Caudata). J Morphol 187:201–217.

White JS, Baird IL (1982) Comparative morphological features of the caecilian inner ear with comments on the evolution of amphibian auditory structures. Scanning Electron Microsc III:1301–1312.

Will U (1988) Organization and projections of the area octavolateralis in amphibians. In: Fritzsch B, Ryan MJ, Wilczynski W, Hetherington TE, Walkowiak W (eds) The Evolution of the Amphibian Auditory System. New York: John Wiley and Sons, pp. 185–208.

Will U, Fritzsch B (1988) The eighth nerve of amphibians: peripheral and central distribution. In: Fritzsch B, Ryan MJ, Wilczynski W, Hetherington TE, Walkowiak W (eds) The Evolution of the Amphibian Auditory System. John Wiley and Sons, Inc., New York, pp. 159–183.

Will U, Luhede G, Görner P (1985) The area octavolateralis in *Xenopus laevis*. I. The primary afferent connections. Cell Tissue Res 239:147–161.

Wilczynski W (1988) Brainstem auditory pathways in anuran amphibians. In: Fritzsch B, Ryan MJ, Wilczynski W, Hetherington TE, Walkowiak W (eds) The Evolution of the Amphibian Auditory System. New York: John Wiley and Sons, pp. 209–231.

Young ED, Fernandez C, Goldberg JM (1976) Sensitivity of vestibular nerve fibers to audiofrequency sound and head vibration in the squirrel monkey. J Acoust Soc Am 59:Suppl S1 S47.

Zottoli SJ (1977) Correlation of the startle reflex and Mauthner cell auditory responses in unrestrained goldfish. J Exp Biol 66:243–254.

18

The Water-to-Land Transition: Evolution of the Tetrapod Basilar Papilla, Middle Ear, and Auditory Nuclei

Bernd Fritzsch

1. Introduction

The evolution of the auditory system of tetrapods has been the topic of numerous investigations on the middle ear (e.g., Reichert 1837; Gaupp 1898, 1913; de Burlet 1934; Thomson 1966; Henson 1974; Lombard and Bolt 1988), the inner ear (e.g., Retzius 1881, 1884; de Burlet 1934; Werner 1960; Baird 1974; Lewis, Leverenz, and Bialek 1985), and the auditory pathways in the central nervous system (e.g., Larsell 1934; Ariens-Kappers, Huber, and Crosby 1936; Northcutt 1980). The consensus reached by many of these studies was that the water-to-land transition apparently coincided with the coevolution of a tympanic middle ear, a basilar papilla, and a periotic labyrinth in the inner ear, as well as neural pathways devoted to the processing of airborn sound in tetrapods (Fig. 18.1).

The middle ear cavity and the stapes (Gaupp 1898, 1913; de Burlet 1934; Werner 1960) have been considered to be derived from ancestral structures such as the spiracular pouch and the hyomandibular bone, respectively. Both are assumed to have changed their structure and respective function considerably during evolution. A functional transformation turned structures that were involved in breathing (spiracular pouch) and suspension of the jaws (hyomandibular bone) into structures devoted to sound conduction.

In contrast, apparent primordia of the basilar papilla had not been recognized until recently in vertebrates other than tetrapods (Fritzsch 1987). Therefore, the basilar papilla was considered to be a unique sensory adaptation devoted to airborne sound detection that developed at the time tetrapods moved onto land (Retzius 1881, 1884). More recently, it has been proposed that the basilar papilla of amphibians has evolved independently of that of amniotes and perhaps twice within amphibians (Wever 1974, 1985).

The intimate structural relationship of the basilar papilla with the sound conducting perilymphatic space recognized in tetrapods (de Burlet 1934; Baird 1974; Lombard 1977, 1980; Lombard and Bolt 1979, 1988) and subtetrapods (Werner 1960; Blaxter, Denton, and Gray 1981) has led to the argument that: "The perilymphatic space plays a critical role for the phylogeny of *pressure-sensitive* auditory organs, in particular the cochlea of mammals and man" (translated after Werner 1960, p. 127; italics are mine). Earlier de Burlet (1934) suggested that the evolution of perilymphatic spaces enables pressure changes to be perceived by "perilymphatic endorgans" which are specialized towards perception of these stimuli and this conviction was elaborated on by van Bergeijk (1966).

Besides the reorganization of existing structures to accomplish new functions in the tetrapod middle ear and the addition of the basilar papilla and the perilymphatic labyrinth in the inner ear, the water-to-land transition is associated with the loss of a prominent aquatic sensory system, the lateral line (Fritzsch 1989). The loss of one part (the lateral line), and the simultaneous gain of another part (the inner ear) of the octavolateral system, both of which project with their afferents to adjacent nuclei in the brain stem, was assumed to be accompanied by the appearance of primary auditory nuclei in the brain stem of most tetrapods (Ariens-

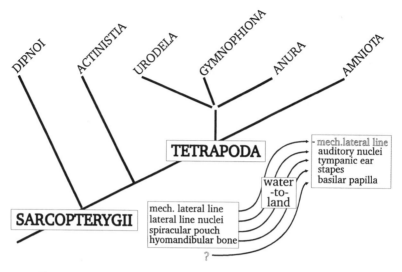

FIGURE 18.1. The prevailing view about the evolution of the tetrapod auditory system is shown as expressed in many textbooks up to 15 years ago. The transformation of the spiracular pouch into a tympanic middle ear, the transfor- mation of the hyomandibular bone into a stapes, the loss of the mechanosensory lateral line, the appearance of the basilar papilla and the auditory nuclei were all supposed to have happened when ancestral tetrapods moved on land.

Kappers, Huber, and Crosby 1936). The assumed phylogenetic coincidence of loss and gain in parts of a sensory system derived from comparable embryonic material was suggested to be causally related. The second order lateral line neurons, after having lost their afferents, had been thought to be captured by ingrowing afferents from the tetrapod basilar papilla, and functionally converted into auditory nuclei (Ariens-Kappers, Huber, and Crosby 1936). Thus the central connections of the "perilymphatic endorgans" (de Burlet 1934) were supposedly achieved through functional transformation of preexisting central pathways of a closely related sensory modality, the lateral line. In addition, it has been suggested that this phylo- genetic reorganization is recapitulated to some extent during the metamorphosis of frogs (Larsell 1934, 1967).

This review presents the evidence obtained within the last 10 years which suggests that many of the prerequisites for terrestrial hearing devel- oped and already were functional in water. The respective structures (spiracular pouch, hyoman- dibular bone, tympanic membrane, basilar pap- illa) were subsequently adapted to terrestrial hearing. It is stressed throughout this review that the phylogenetic coincidences of all these changes emphasized in earlier papers reflect occasional

examples rather than the rule and cannot be used to establish causal relationships as suggested in earlier papers.

Following the suggestion of an inside-out evo- lution of the auditory system (Lombard and Bolt 1988), I will first present the data on the evolu- tion of the inner ear of sarcopterygian fish and its adaptation to receive the pressure component of aquatic sound. I will then discuss how this system might have been adapted to terrestrial hearing, i.e., how the tetrapod inner and mid- dle ear was evolved. The last part of this chap- ter will deal with the phylogenetic loss of the electrosensory and mechanosensory lateral line in amphibians and the appearance of auditory nuclei in anurans.

2. Presumed Adaptations of the Ear of Sarcopterygian Fish to Detect the Pressure Component of Aquatic Sound

Our knowledge of the inner ear of vertebrates still has gaps: how, when, and why was the inner ear—originally developed as a sense to detect dynamic and static aspects of the position in

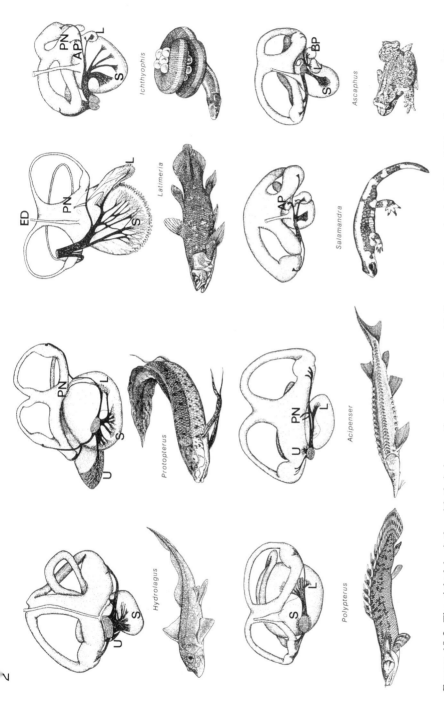

FIGURE 18.2. The right labyrinths with their innervation viewed medially are shown in several anamniotic species. The utricular macula (U) is in a separate recess only in cartilaginous fish (*Hydrolagus colliei*) and lungfish (*Protopterus annectens*), but in no other vertebrate. A lagenar macula (L) lies in a separate recess only in bony fish (not shown), Actinistia (*Latimeria chalumnae*), and most tetrapods. Except for most amphibians, gnathostome vertebrates have a papilla neglecta (PN) innervated by fibers that run with the posterior canal twig. All amphibians have an amphibian papilla (AP) together with (Gymnophionans: *Ichthyophis*), or without (frogs, salamanders), a papilla neglecta. *Latimeria* is the only subtetrapod vertebrate that has a basilar papilla (BP) in the same relative position as amphibians and with a comparable pattern of innervation. ED, endolymphatic duct; S, sacculus. Anterior is to the left, posterior to the right.

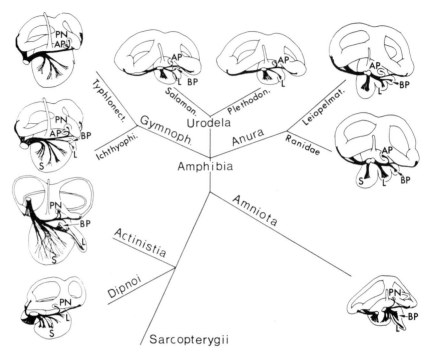

FIGURE 18.3. The form and the pattern of the innervation, in particular of the basilar papilla (BP), the amphibian papilla (AP), the lagena (L), and the papilla neglecta (PN) are shown for some sarcopterygians. Note the absence of a papilla neglecta in all but gymnophionan amphibians. A basilar papilla is absent in lungfish and lost in derived urodeles and gymnophionans. In contrast, all anurans have a basilar papilla, but there are changes in the pattern of innervation. The BP fibers run primitively with the lagenar fibers (Leiopelmatidae) but course with the posterior canal twig in derived anurans (Ranidae). All amniotes (represented by *Sphenodon*) have a basilar papilla innervated much as in *Latimeria* and all primitive amphibians. Modified after Fritzsch and Wake (1988). Lines indicate presumed phyletic relationships based on non-otic criteria.

space—transformed into an auditory system, in particular into the auditory system of land vertebrates. This problem may be subdivided into the following components:

—How and when was this change achieved in vertebrates, i.e., what is the historical context and the embryological repatterning of this change in the inner and "middle ear"?
—Why should aquatic organisms evolve from particle motion detection, to which all epithelia covered with an otolith are capable (Kalmijn 1988), to sound pressure detection, an altogether different parameter of the physical environment both in water and in air? That is, what is the driving force for the novel adaptation?

2.1 Perilymphatic, Pressure Detection Endorgans in Vertebrates

In several lineages of teleosts (Werner 1960; van Bergeijk 1967; Baird 1974) special mechanisms for the perception of sound pressure evolved independently (Popper et al. 1988). In all teleosts this is achieved with the help of a gas bubble which transforms a sound wave into near-field water motions (Van Bergeijk 1966). These motions may be received by the ear, if it is either near the gas bubble or has specialized connections to the gas bubble. Van Bergeijk (1966) suggested that the inner ear evolved specialized sensory sites that deal exclusively with fluid motions originating from these gas bubbles. In addition, the inner ear evolved a low-loss transmission pathway through changes in the perilymphatic system.

The data for the sarcopterygian lineage are scanty and almost nothing is known about the supposed tetrapod ancestors (Jarvik 1980). The morphology of the spiracular pouch and inner ear of *latimeria chalumnae* (the coelacanth) described below, is the best evidence so far for a presumed pressure-transducing system in an ear of a subtetrapod sarcopterygian. Along with the theoretical considerations of van Bergeijk (1966) the data on *Latimeria* suggest that a pressure-transducing, gas-filled spiracular pouch may have been present in at least some crossopterygian fish.

2.2 Definition and Distribution of a Basilar Papilla

The basilar papilla or cochlea of tetrapods is the most important pressure-receiving "auditory" receptor in the tetrapod ear (Retzius 1881; Baird 1974; Figs. 18.2–18.5). If one wants to understand the evolution of tetrapod hearing one must consequently explain the origin and differentiation of the basilar papilla to achieve this dominant status. The presence of this epithelium in different tetrapods has caused many controversies and a consensus still has not been reached (Baird 1974; Wever 1985). I will define criteria for the identification of the basilar papilla, describe it's distribution and try to explain the differences observed in a sensory epithelium which I consider homologous.

The term basilar papilla will be used here for any sensory epithelium of the inner ear that is (1) close to the saccular/lagenar recess, (2) receives innervation from nerve fibers that run together with the lagenar branch, (3) is not covered by an otolith but by

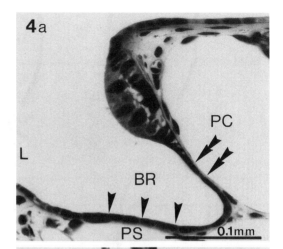

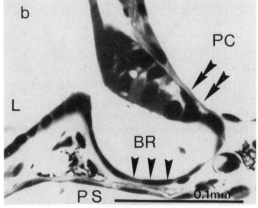

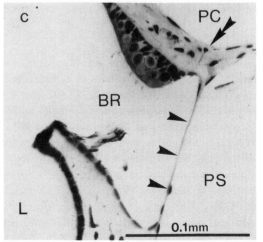

----->

Figure 18.4. Photomicrographs of parasagittally sectioned labyrinths of larvae of a gymnophionan (*Ichthyophis kohtaoensis*, a), a salamander (*Salamandra salamandra*, b) and a frog (*Ascaphus truei*, c) show the organization of the basilar papilla and its recess (BR) with respect to the periotic canal (PC). The PC may underlie the basilar papilla or may be separated by limbic tissue (double arrowheads). The BR is also in contact with the periotic sac (PS) at a thin or thick contact membrane (arrowheads). L = lagena.

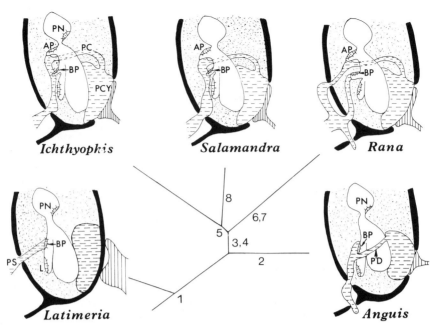

FIGURE 18.5. Schematic drawings of the cross section of a left inner ear viewed rostrally to show the presumed sequence of change of some sensory epithelia and the perilymphatic system in sarcopterygians. The presence of a basilar papilla, a perilymphatic (periotic) sac (PS) and a perilymphatic (periotic) cisternae (PCY) is an autapomorphy (uniquely derived feature) of *Latimeria* and tetrapods (1). The formation of a perilymphatic duct (PD) anteriorly around the labyrinth is an autapormorphy of amniotes (2). The formation of a periotic canal (PC) posteriorly around the labyrinth and formation of the amphibian papilla by shifting a part of the papilla neglecta into the sacculus, are autapomorphies of amphibians (3, 4). Formation of a basilar recess may have occurred twice: adjacent to the lagenar orifice but within the lagena (gymnophionans, urodeles, 5) or adjacent to the lagenar orifice but within the sacculus (anurans, 6). Loss of the papilla neglecta is considered to have been independently achieved in anurans (7) and urodeles (8). The stapes or hyomandibular bone is indicated by vertical lines. Modified after Fritzsch and Wake (1988).

gelatinous or tectorial structures and (4) rests at least partially on a basilar membrane that separates a peri- from an endolymphatic space. I argue that the detailed equality in position within a complex entity, a comparable pattern of innervation and detailed structural similarities indicate homology and are not easily compatible with suggestions of homoplasy.

Using the above criteria, a basilar papilla can be identified in *Latimeria* (Fritzsch 1987; Fritzsch and Wake 1988; Figs. 18.2, 18.3, 18.6) and all amniotes (Retzius 1884; de Burlet 1934; Werner 1960; Lewis et al. 1985). I propose that the basilar papilla evolved once within sarcopterygians and was retained in *Latimeria* and in most tetrapods, i.e., the basilar papilla defined by the criteria given above is regarded as an autapomorphy of all species who have it. These species—crossopterygian fish and tetrapods—form a monophylum defined by two autapomorphies, the intracranial joint (Jarvik 1980) and the basilar papilla. The basilar papilla was transformed independently three times in the amphibian lineages and was lost in many gymnophionans (White and Baird 1982; Fritzsch and Wake 1988) and salamanders (Lombard 1977; Fig. 18.3). This loss can proceed with or without the loss of the basilar recess (Lewis and Lombard 1988; Fritzsch and Wake 1988).

Alternative views of the above-presented scenario have been proposed in the past. Some amphibians, notably modern frogs, have an epithelium in about the same position as the basilar papilla. Nevertheless, this epithelium is different enough in both structure and innervation that it has been considered unrelated to the basilar papilla of other tetrapods (Wever 1985; Lewis and Lombard 1988). Recently, a morphocline was demonstrated

Plate 1

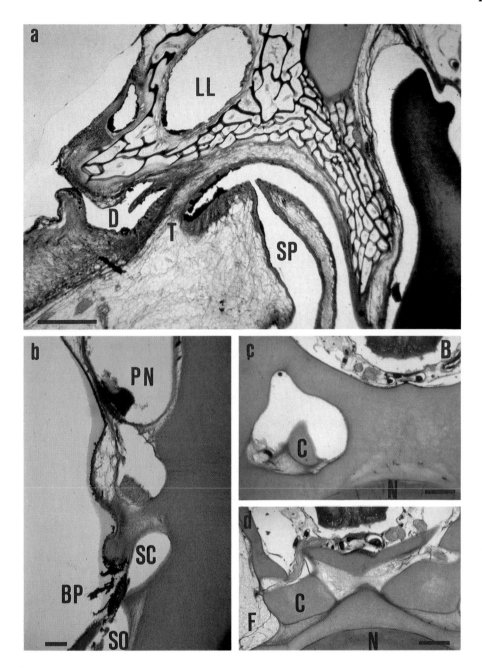

FIGURE 18.6a–d. These micrographs show the spiracular pouch and the tympanum (a), the basilar papilla and the papilla neglecta (b), and two sections through the canalis communicans (c, d) of an intraoviducal embryo of *Latimeria*. (a) The duct (D) leading from the tympanum (T) to the surface merges further rostrally with lateral line canals (LL). The spiracular pouch (SP) is in contact with the ear capsule and the hyomandibular bone. (b) The basilar papilla (BP) overlies a perilymphatic space (SC) that leads to the canalis communicans through the perilymphatic foramen. Nerve fibers run in another perilymphatic space (SO). The papilla neglecta is also visible (PN) with its gelatinous coverage. Note the tectorial-like structures covering the basilar papilla. (c, d) The canalis communicans runs through the cartilage above the notochord (N) to end both within the cranial cavity and at a complex organization of cartilaginous pieces (C). The cartilage abuts lateral on fat (F). (B) indicates the spinal cord. Bar indicates 1 mm in a, 0.2 mm in b, and 0.5 mm in c, d.

Plate 2

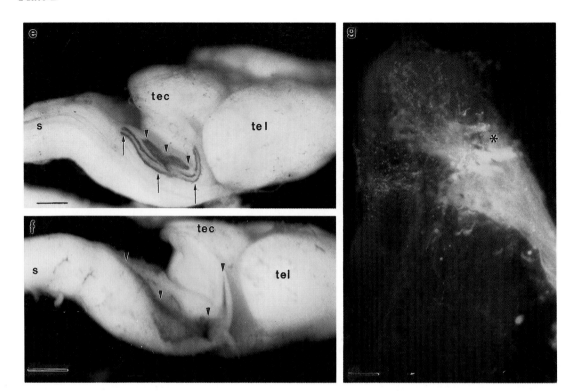

FIGURE 18.6e–g. These micrographs show the projections of the anterior lateral line (e) and the inner ear (g)
in whole mounted brains of a larval gymnophionan
(*Ichthyophis kohtaoensis*) and the inner ear (*red*) and
lateral line (*green*) projection in a transverse section
through the hindbrain of a larval frog (*Xenopus laevis*,
g). The brown HRP reaction product reveals the projection of ampullary organs to the dorsal nucleus (*arrowheads*) and the two longer ventral fascicles to the intermediate nucleus (*arrows* in e). The inner ear (*arrow-

heads*) projects without overlap ventral to the lateral line
(f). The fluorescent dyes rhodamine dextran amine (*red*)
and fluoresceine dextran amine (*green*), if applied to the
lateral line (*green*) and the inner ear (*red*), respectively,
reveal no overlap but a projection of the inner ear to a
dorsolateral auditory nucleus (*asterisk*) already in premetamorphic tadpoles (g). tel, telencephalon; tec, tectum; s, spinal cord. Bar indicates 0.5 mm in e, f and 0.1
mm in g.

in the pattern of innervation of the basilar papilla of anura (Will and Fritzsch 1988; Fritzsch and Wake 1988; Fig. 18.3). This morphocline suggests that the unique innervation of the basilar papilla in modern frogs is derived from a pattern much like that in *Latimeria*, salamanders, caecilians, and amniotes. This morphocline from primitive to modern frogs also shows that the changed pattern of innervation of the basilar papilla of modern frogs is not a valid argument to suggest a separate evolution of the anuran basilar papilla.

Likewise, the basilar papilla is situated partially on a basilar membrane in at least some species in two out of three amphibian orders (Fritzsch and Wake 1988; Fig. 18.4). This finding is in line with the suggestion that structural specializations in other species of that lineage are derived from such a structure. The alternative view that a basilar papilla has evolved within each of these lineages (Wever 1974, 1985) would require four parallel evolutionary events leading to an almost identical sensory epithelium (Fig. 18.5). For the sake of parsimony, the basilar papilla of anurans is instead interpreted here as a transformed crossopterygian basilar papilla. The unique recess and structure of the anuran basilar papilla are the only features that cannot be resolved (Fritzsch and Wake 1988).

2.3 Relationship of the Basilar Papilla to the Lagenar Macula and Recess

Assuming that the interpretation about the monophyly of the basilar papilla in tetrapods and *Latimeria* is correct, we must explain the correlation not only in position (the basilar papilla is always around the orifice of the lagenar recess) but also the coincidence with the presence of a lagenar recess.

Within the sarcopterygian lineage, the lungfish have no basilar papilla (Figs. 18.3). They have a lagenar macula but no lagenar recess (Retzius 1881; Jarvik 1980; Fig. 18.2). Lagenar and saccular maculae within a single recess are found in some chondrichthyan and primitive actinopterygian fish (Retzius 1881; de Burlet 1934; Jarvik 1980; Fig. 18.2). A somewhat segregated epithelium exists in lamprey and is called the lagenar macula (de Burlet 1934; Baird 1974). Thus, the presence of both lagenar and saccular maculae within a single recess is apparently the primitive situation whereas the location of each in a separate recess is

a derived feature of gnathostomes. The distribution suggests that a lagenar recess formed independently at least three times, once in the chondrichthyan, once in the actinopterygian and once in the sarcopterygian lineage as a posterior evagination of the saccule (Werner 1960; Lewis et al. 1985; Fig. 18.7). The data on the polarization of the hair cells in the lagenar epithelium – chaotic in elasmobranchs, away from a dividing line in tetrapods, towards or parallel to the dividing line in actinopterygian fish (Lewis, Leverenz, and Bialek 1985) – is in agreement with this suggestion. Moreover, some fossil evidence for a separation of the saccular and lagenar maculae into more or less discrete pouches exists in crossopterygian fish (Millot and Anthony 1965) and even a basilar papilla was once indicated together with a lagenar recess in the cast of an ear of an osteolepiform fish (Romer 1937). I predict that the polarity of hair cells in the lagenar macula of *Latimeria* follows the tetrapod pattern and not the pattern of bony fish or elasmobranchs.

The lagenar macula was lost several times independently (Lewis, Leverenz, and Bialek 1985; Fritzsch and Wake 1988). This loss occurred with (actinopterygians; de Burlet 1934; Werner 1960) or without (caecilians; White, and Baird 1982; Fritzsch and Wake 1988) loss of the lagenar recess (Fig. 18.3). Pending the actual systematic position of crossopterygian fish and lungfish (Forey 1988), the absence of a lagenar recess in lungfish may be interpreted as a primitive similarity to chondrichthyan fish or as secondarily derived, i.e., a similarity achieved through the unique loss of such a recess without the loss of the lagenar macula. I will reconsider the evidences in favor of both views after the structural changes in the ear and the functional implications are presented.

2.4 Ontogenetic Events Related to the Phylogenetic Diversity of the Lagena and Basilar Papilla

Given the coincidence in the appearance of a basilar papilla and separate lagenar recess in the sarcopterygian lineage (Fig. 18.3), one is tempted to speculate that both events are causally related. However, the event that has led to the formation of

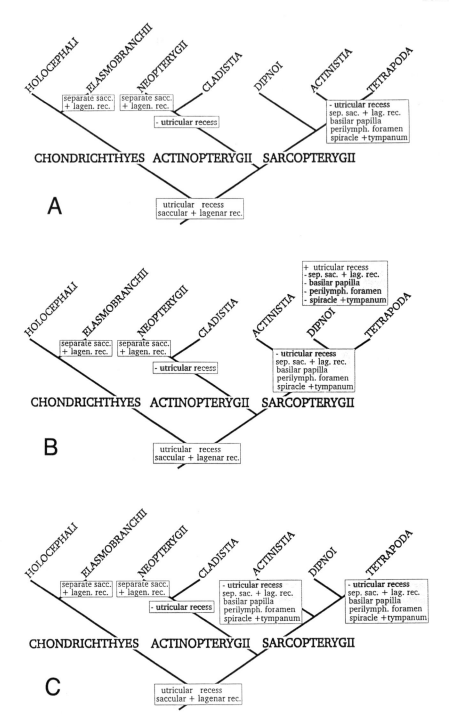

the lagenar recess has occurred independently at least three times (Figs. 18.2, 18.7). Obviously, this segregation resulted in the formation of an additional epithelium at the orifice of the lagenar recess only in the sarcopterygian lineage. In the species studied, there is evidence that the basilar papilla separates during development from the lagenar anlage after the latter has separated from the saccular anlage (de Burlet 1934; Werner 1960; Baird 1974). In contrast to the latter two epithelia, the basilar papilla was, through unknown ontogenetic changes, transformed into an otolith-free epithelium covered with a tectorial membrane. Owing to the absence of an otolith, the basilar papilla probably would not respond to direct acoustic particle motion (Popper et al. 1988). However, a relative movement of the basilar papilla with respect to the tectorial structure overlying it could stimulate the hair cells. Further research is necessary to reveal the changes in ontogeny that have led to the formation of the lagenar recess, the segregation of the basilar papilla, and the loss of an otolith in this epithelium.

Our own experiments have revealed some insights into the potential mechanism of formation of additional evaginations at the caudal end of the ear. Using lithium as a teratogene in developing clawed toads, *Xenopus laevis*, we showed that an altered gastrulation can result in the formation of multiple additional ear vesicles at the caudal aspect of the primordial ear vesicle (Gutknecht and Fritzsch 1990). The data imply that changes in the gastrulation of mesoderm may underly phylogenetic events that have led to the formation of the lagenar recess and, in addition, to the basilar papilla in ancestral sarcopterygian fish. Similar teratogenic transformations should be performed in other vertebrates to show that the results on the

clawed toad can be generalized. Lungfish and paddle fish, in particular, which have both a saccular and lagenar macula in their unpartitioned saccular recess, would be interesting in this respect.

2.5 The Evolution of the Perilymphatic Labyrinth

As introduced above, there is an apparent correlation between the phylogenetic appearance of a specialized perilymphatic labyrinth and a pressure-transducing mechanism in bony fish (Werner 1960; Van Bergeijk 1966; Blaxter, Denton, and Gray 1981) and tetrapods (de Burlet 1934; Lewis and Lombard 1988). In the latter, the perilymphatic labyrinth has been proposed to be diphyletic based on structural differences.

For example, the periotic canal of amphibians passes caudally around the inner ear and in amniotes the periotic duct passes rostrally around the inner ear (Lombard 1980; Lewis and Lombard 1988; Figs. 18.5, 18.8).

Apart from these basic differences there is substantial variability with respect to the actual course of the periotic canal, the relationship to the periotic sac, and the periotic cisterna, as well as the round window (de Burlet 1934; Lewis and Lombard 1988). More recently, the conclusion about a diphyletic origin of the tetrapod perilymphatic labyrinth was corroborated based on data on *Latimeria* (Fritzsch and Wake 1988). *Latimeria* (Figs. 18.5, 18.6) has a perilymphatic foramen that underlies the basilar papilla. The perilymphatic space extends caudally and eventually reaches into the cranial cavity. Topographically and structurally this foramen and the adjacent perilymphatic spaces resemble the perilymphatic

◁————————————————————————

FIGURE 18.7. The distribution of otic characters is shown in three different scenarios of the phyletic relationship of Actinistia and Dipnoi to tetrapods. The simplest scenario is provided when *Latimeria* is the closest living ancestor of tetrapods (A). This would require the independent loss of the primitive utricular recess twice, the formation of a separate lagenar recess three times and the formation of a basilar papilla, perilymphatic foramen and tympanic ear only once. In contrast, when

Dipnoi are the closest living tetrapod relatives, the Dipnoi have either achieved ancestral patterns by complete reversal to conditions otherwise only found in Holocephali (B), or *Latimeria* and tetrapods have achieved very similar changes of the inner and middle ear independently (C). Loss of structures are indicated by open letters, ancestral and newly evolved structures by filled letters.

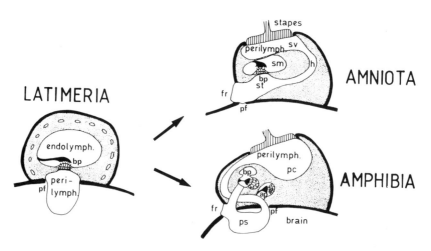

FIGURE 18.8. These horizontal sections show an ear with a perilymphatic foramen (pf) and a basilar papilla but no periotic canals in *Latimeria*. In contrast, amniotes evolved an ear with a helicotrema (h, or perilymphatic duct) leading from the perilymphatic scala vestibuli (sv) to the scala tympani (st) with the round window (fr) and the perilymphatic foramen (pf). In amphibians, a periotic canal runs posteriorly around the ear and connects the periotic cisterna (pc) with the periotic sac (ps). Note that both perilymphatic pathways obviously evolved in conjunction with the insertion of a movable element (stapes and/or operculum) into the ear capsule. Only amphibians have evolved a perilymphatic sensory epithelium, the amphibian papilla (ap) in addition to the basilar papilla (bp). Modified after Lombard (1980).

foramen of tetrapods (de Burlet 1934; Werner 1960; Lombard and Lewis 1988).

Moreover, casts of the ear of osteolepiform fish have revealed a canal leading from the ear into the brain cavity which was interpreted as a perilymphatic duct by Romer (1937). Topographically and in size this duct closely resembles the perilymphatic foramen of *Latimeria*. In addition, *Latimeria* has a large space, filled with fat, underlying the lateral wall of the otic capsule adjacent to the lateral aspect of the sacculus (Millot and Anthony 1965). This lateral space has been compared to the perilymphatic cisterna of tetrapods (Millot and Anthony 1965; Fig. 18.5). Thus, major parts of the perilymphatic labyrinth apparently are present in *Latimeria* (Fritzsch and Wake 1988).

Again we are left with the problem of the phylogenetic origin of the perilymphatic space and the perilymphatic foramen, and the changes in development that cause their formation. Clearly, the perilymphatic systems of those modern bony fish with a pressure transducing system are structurally too different to be ancestral to those of *Latimeria* and tetrapods (van Bergeijk 1967). For example in the herring, the perilymphatic system runs from a gas bladder in the ear along the utricle to the lateral line canal. Nevertheless, the features displayed by bony fish show a general principle which reflects functional needs and therefore also applies to *Latimeria* and tetrapods.

The otic capsule must be closed around the sound-conducting system so that the direction of near-field particle movements caused by sound pressure acting on the swimbladder is unequivocally determined by way(s) of low resistance for flow of the almost incompressible otic fluids. In other words, the perilymphatic labyrinth must form a low-loss transmission pathway from the gas-filled structure abutting one window, along a "perilymphatic endorgan," and leave through another window in the otic capsule into an extraotic space.

Clearly, perilymphatic tissue without specialization in terms of a perilymphatic canal is primitive for craniate vertebrates (de Burlet 1934). Lungfish have no such specializations and *Latimeria* has large, ill-defined perilymphatic spaces as well as two clearly delineated spaces, the fat-filled "perilymphatic cisterna" and the perilymphatic sac (Fig. 18.8). I propose that these differences, as compared to tetrapods, reflect the different ways pressure changes are transmitted onto the ear in *Latimeria* (see below).

2.6 The Evolution of the Perilymphatic Foramen to the Cranial Cavity

Whatever has caused the formation of a perilymphatic labyrinth, its outlet out of the ear capsule is, in tetrapods and *Latimeria*, through the perilymphatic foramen into the cranial cavity. This foramen, which gives rise to the aquaeductus cochlea and the round window in mammals (Werner 1960), is closely adjacent but clearly different from the foramen for nerves and the endolymphatic duct. No perilymphatic foramen is present in the tightly closed capsule of hagfishes, lampreys, and chondrichthyan fish (Werner 1960; Jarvik 1980). However, ratfish and primitive bony fish have virtually no wall between the cranial and the otic cavity (Baird 1974; Jarvik 1980). The latter condition has been variously interpreted as primitive (de Burlet 1934) or as an independently derived character (Werner 1960). A similar large window is present in modern lungfish (unpublished observations).

In contrast, *Latimeria*, fossil coelacanths (Bjerring 1985), and tetrapods have narrow foramina for nerves, endolymphatic duct, and the perilymphatic foramen. The distribution of the perilymphatic duct among gnathostomes indicates that it may be a derived feature of the sarcopterygian lineage (Fig. 18.7). More data on fossils are needed to solve this issue which may be crucial for an understanding of the evolution of a pressure-detecting mechanism in the sarcopterygian lineage. Given the uncertainty of the real distribution of this character, it is impossible to decide whether chance or design correlates with the known variation. Moreover, the simplicity of this character does not allow one to easily distinguish between parallelisms and homologous structures.

2.7 The Spiracular Pouch and the Possible Evolution of a Tympanic Ear in Water

Obviously, the presence of gas-filled chambers is a necessary prerequisite for the formation of a pressure-detection system (van Bergeijk 1966). Sharks have no swimbladder and consequently possess no pressure detection system. They have one epithelium, the papilla neglecta (Fig. 18.2), which is the likely candidate for sound detection (Corwin 1981). The mechanism(s) through which sound impinges on this epithelium is not yet clear (Kalmijn 1988) but must be different from otolithic organs (Rogers and Cox 1988).

All primitive actinopterygian fish have a swimbladder or a lung but no specialized connection between swimbladder and inner ear (Liem 1989). Likewise, these animals have no specialized epithelium other than those already known in chondrichthyans and have no specializations of perilymphatic sound-conducting channels around the inner ear.

Among modern bony fish, only the more primitive species have developed, obviously independently, quite different ways to connect the swimbladder either directly or through a chain of ossicles to the inner ear (Greenwood 1963; van Bergeijk 1967; Liem 1989). Many of these fishes have specialized channels for sound conduction in the perilymphatic system (Schellart and Popper, Chapter 16). In addition, otoliths of endorgans close to the perilymphatic channels show some reduction in size and mass (Werner 1960). Interestingly enough, the majority of the modern bony fish have no connection between swimbladder and the inner ear at all. Given that at least some connection appears to be primitive for this lineage, this lack is curious and must await a more detailed understanding of the functions of the swimbladder in pressure transduction (Schellart and Popper, Chapter 16) before the obvious reduction of this connection can be explained.

Among sarcopterygian fish, the proximity of a spiracular pouch to the ear could transmit pressure changes right onto the inner ear much like in anabantid fish (van Bergeijk 1967), provided the spiraculum is closed by a membrane and the pouch is filled with gas. A spiracular pouch is present in *Latimeria* (Fig. 18.6), osteolepiform fish such as *Eusthenopteron foordi*, primitive actinopterygians, and very prominent in elasmobranchs (Jarvik 1980). Many crossopterygian fish have a spiracular opening in the skull (Schultze 1986). In contrast, both modern and fossil lungfish have only a small spiracular canal (Jarvik 1980) which nevertheless has a spiracular organ (Northcutt 1986b). It has been proposed that the formation of an operculum, with its role in ventilation, may be related to the multiple parallel reduction and loss of the spiracular pouch (Werner 1960).

The spiraculum will act as a pressure transducer if filled with air, i.e., if closed by a membrane.

Latimeria has such a membrane (Fig. 18.6). If we assume that the spiracular pouch is filled with gas, then the equations calculated for the rhipidistian fish *Eusthenopteron* (van Bergeijk 1966), apply also for *Latimeria*. The displacement of the gas bubble's surface in the spiracular pouch will be determined by the quotient of the spiracular volume and the area of the pouch (van Bergeijk 1966). In essence, the spiracular pouch of *Latimeria* could work as a pressure-transducing device, provided it is filled with gas. In contrast, being closed by a membrane, it can play no role in ventilation of water as in cartilaginous fish and as suggested for some fossil tetrapods (Clack 1989). Also I interpret the presence of a membrane which closes the spiracular pouch as primitive for sarcopterygian fish, I cannot rule out that the presence of a membrane in *Latimeria* is independently derived from the tympanic membrane of tetrapods. Because membranes are unlikely to appear in the fossil record, this issue will probably remain controversial and unresolved.

2.8 The Canalis Communicans

Millot and Anthony (1965) were the first to describe bilateral canals from the perilymphatic foramen of the ear to the occipital region of *Latimeria*. These canals pass through the cartilage around the notochord and merge below the foramen magnum (Fig. 18.6). Small canals leave the canalis communicans halfway between the foramen perilymphaticum and the foramen magnum and lead to the brain case. A short canal continues further caudal, and a few pieces of cartilage form a rather complex and conspicuous articulation where it ends. These pieces of cartilage are abutting to a mass of fat. Such a canal has also been described in *Nesides schmitii*, a fossil coelacanth (Bjerring 1985), but is otherwise unknown. Given the current knowledge about its distribution, this canal must be considered as an autapomorphy of Actinistia. However, more research on other fossils is clearly needed.

The articulations at the end of this canal, as well as the presence of the canal suggest that somehow the perilymph — and not endolymph as believed by Millot and Anthony (1965) — must be moved. If there would be no movement of the perilymph, it is difficult to understand how this canal forms and is

maintained and how the cartilaginous pieces are prevented from fusing. Clearly, any pressure differences between the belly and the ear could cause movement of the perilymph inside the canalis communicans. Because the lungs are filled with fat in *Latimeria* (Millot and Anthony 1965) and most likely will not respond to pressure changes by changing their volume, it is unclear which other structure(s) may cause movement of perilymph in the canalis communicans.

Millot and Anthony (1965) have suggested that the similarity in organization of living, and in particular fossil coelacanths, and some ostariophysine fish may indicate a similar function, i.e., that pressure changes in the belly will act on the canalis communicans. The findings of some fossil coelacanth with lungs surrounded by bone (Forey 1988) are compatible with such a view. This bony capsule around the lung is reminiscent of the organization in some bottom-dwelling catfish which are known to use their swimbladder surrounded by bone only for pressure detection (Henson 1974). Direct measurement in a living or well-preserved *Latimeria* are needed to evaluate this suggestion about movement of the perilymph in the canalis communicans induced by pressure change.

Interestingly, some anuran tadpoles have a connection from the lungs to the perilymphatic foramen that resembles, in a way, the canalis communicans of coelacanth (van Bergeijk 1966; Hetherington 1988). The so-called bronchial columella of ranid tadpoles can transmit pressure changes, which act on the lungs, to the round window and may drive the perilymphatic fluid of the ear in a reverse way, i.e., from the round to the oval window (van Bergeijk 1967). A similar function is conceivable for the canalis communicans (see below).

2.9 What Could Cause Differences in Pressure on Either Side of the Basilar Papilla?

Millot and Anthony (1965) described a slit-like hole in their specimens of *Latimeria* where the basilar papilla was later found (Fritzsch 1987). The fact that this epithelium (Fig. 18.6) is obviously ruptured in all cases of adult *Latimeria* brought up from a depth of 200 or more meters

(Fricke et al. 1987) within a few minutes suggests that the basilar papilla may be moved by pressure changes between the perilymphatic and the endolymphatic spaces bordering on it. The only intact epithelia were found in intraoviducal embryos (Fritzsch 1987), which may not yet have fully developed the structures to mediate pressure differences on both sides of the basilar papilla. Pressure changes may impinge through several ways onto the basilar papilla:

a. Pressure differences between the spiracular pouch and the cranial cavity may lead to pressure differences on either side of the basilar papilla.
b. Pressure differences between the belly and the spiracular cavities may lead through the canalis communicans to pressure differences on both sides of the papilla basilaris.
c. Movement of the lower jaw may lead through the hyomandibular bone to pressure changes between the inner ear and the brain cavity.
d. Movement of the intracranial joint may change the volume of the brain cavity and lead to pressure changes at the basilar membrane.
e. Gas may bubble out of the lymph owing to the pressure difference. In the latter case, however, the basilar papilla of the embryos should have been ruptured as well.

Clearly, the second possibility is difficult to envisage for modern *Latimeria* because of the fatty degeneration of its lungs. Nevertheless, the presence of a canalis communicans which leads from the ear to the occipital region shows that some fluid motion of the perilymph likely occurs. This pathway may have been more pronounced in ancient coelacanths (Bjerring 1985). Little can be said about the other two possibilities except that they are unlikely to cause the rupture of the basilar papilla because they are self-inflicted pressure changes.

In terms of evolution the most interesting aspect is the possibility that pressure changes acting on a gas-filled spiracular pouch may cause movement of the basilar papilla in ancestral rhipidistians (van Bergeijk 1966) and modern *Latimeria*. If true, pressure detection in water would have been rather similar in these fish to the one land vertebrates perform. Moreover, the long-noticed conservation of the spiracular pouch as the tetrapod middle ear would make sense in terms of conservation of this new function. Likewise, the presence of a

tympanum-like closure of the spiracular pouch in *Latimeria* (Fig. 18.6) could be taken as an anticipation of the tetrapod tympanum.

Changes in volume in a gas-filled spiracular pouch will result in near-field water motions in any adjacent compartment, provided there are ways of low resistance flow through the ear. Water movements would be possible through the spiracular orifice into the mouth, through the tympanic membrane into the spiracular canal, and further on into lateral line canals. Water movements would also be possible through the cartilaginous wall of the ear into the inner ear and subsequently through the perilymphatic foramen into the canalis communicans and eventually into the cranial cavity and the belly. The impedance of the tissue will determine the actual particle movement, i.e., through the ear or the other two pathways. If the particle motion passes through the ear, then the ratio of the ear capsule surface that abuts the spiracular chamber and the perilymphatic foramen could result in pressure amplification as discussed for the tympanic surface to oval window ratio (Moller 1974).

From this scenario it follows that a gas-filled spiracular pouch could effectively drive fluid in the ear and stimulate otic endorgans only if (1) the pathway of the least resistance for pressure release leads through the ear capsule, (2) the pressure release from the ear into the cranial cavity is not through a large opening that minimizes amplification and not through a small foramen that increase resistance of flow and thereby impedance, and (3) a sensory epithelium is in a position to monitor minute fluid movements (Hudspeth 1989). All three criteria of middle ear function—(1) low acoustical impedance, (2) amplification, and (3) directionality of sound (Jaslow, Hetherington, and Lombard 1988)—could exist in the ear of *Latimeria* but in positions different from those of terrestrial ears. In *Latimeria* this could be provided by (1) the cartilaginous rather than osseous lateral wall of the otic capsule, (2) the ratio of the spiracular pouch to the comparatively small perilymphatic foramen, and (3) the basilar papilla overlying this foramen. Thus, the ear of *Latimeria* could be an example of a pressure-transducing middle ear and could provide a testable example of conditions which might have prevailed in other crossopterygian fish (van Bergeijk 1966) and ancestral, aquatic tetrapods.

2.10 Physical Constraints
of Underwater Hearing

This subject was presented in detail in a recent review (Kalmijn 1988) and will be dealt with in detail elsewhere (Schellart and Popper, Chapter 16). Therefore it suffices here to point out the differences in range of particle movement and sound pressure reception. It is clear that for the extension of range a sound pressure receiver, i.e., the swimbladder or other gas-filled structures, are the only physically possible ways. While this adaptation allows for hearing at larger distances it creates problems for directional hearing, largely related to the high speed of sound in water. Fish could use interaural time cues for sound localization only if they have two independent pressure transducers at a significantly larger distance from each other than on land, or a nervous system that is able to deal with the minute interaural time differences available under aquatic conditions. Instead, most fish have only one pressure receiver, the swimbladder, or two gas bubbles at a small distance, as in herrings or mormyrids (van Bergeijk 1967). Likewise, intensity differences caused by the shadowing of the head or the body will play little role in directional hearing in water given the largely similar density of the fish's body and the water.

Consequently, fish are presumed to rely on the interplay of particle motion detection through otolithic organs (i.e., direct stimulation; Popper et al. 1988) in conjunction with pressure detection (i.e., indirect stimulation; Popper et al. 1988) or on particle motion detection alone in case there is no pressure transducer (Popper et al. 1988; Schellart and Popper, Chapter 16). An organization that will submit movement caused by the direct and the indirect (Popper et al. 1988) stimulus on one epithelium has some theoretical disadvantages, given the complex pattern produced by the interaction of both signal components (Kalmijn 1988). Both components, if perceived with only one epithelium, bear the risk of uncertainty of the relative motion caused by either sound components, or both in conjunction at any hair cell. Moreover, the interplay of both components will vary at different distances and directions and with the source being a monopole, or a dipole, or mixtures of different poles (Kalmijn 1988).

Conceivably, a better solution would be a decoupling of pressure and particle motion detection by creating an epithelium with no otolith. This epithelium could become specialized for detection of near-field particle movement caused by sound pressure changes through its association with the pressure transducer, and a second epithelium could become specialized for particle motion detection without any coupling to the pressure-transducing system (Schellart and Popper, Chapter 16), i.e., two separate channels for the direct and the indirect sound could evolve. Apparently, *Latimeria* (and perhaps other crossopterygian fish as well) may have achieved such a system with the evolution of a basilar papilla with no otolith. The basilar papilla may be driven by near-field fluid motions caused by the pressure changes alone (see above), and there are other otolithic endorgans (sacculus, lagena, utriculus) which will receive little, if any, energy through the pressure-transducing channel. Thus, both components of aquatic sound could be perceived, at least in principle, at different endorgans and their information may be combined in the central nervous system (CNS) to compute the direction of the sound source. Again, appropriate tests on living *Latimeria* should be designed to test this hypothesis which could perhaps be performed in their habitat at 200 or more meters in depth.

2.11 The Ear and the Systematic
Position of *Latimeria*

The most parsimonious interpretation of the available set of data is that the inner ear of modern lungfish represents the primitive pattern of the sarcopterygian lineage and the ear in *Latimeria* and tetrapods represent a shared derived (autapomorphic) pattern (Fig. 18.7). As suggested above, the formation of a lagenar recess may have led to the segregation of a basilar papilla which was transformed into an otolith-free epithelium overlying a perilymphatic space to a perilymphatic foramen. Motion of the epithelium was perhaps achieved through pressure changes acting on a gas bubble in the spiracular pouch, which was closed by a membrane. None of these features are present in modern lungfish and are not yet described for primitive lungfish (Jarvik 1980). However, some evidence suggests that lungfish, rather than crossopterygian fish, are the closest living relatives of tetrapods (Forey 1986, 1988). If this suggestion reflects the real systematic relationship, then the distribution of the inner ear morphology among sarcopterygian fish can be resolved only with two rather complicated assumptions:

a. The pattern in modern lungfish is not primitive, but highly derived, and similarities uniquely shared with chondrichthyan fish reflect convergence that was achieved through regressive evolution of the lungfish ear (Figs. 18.2, 18.7). Data on regressive events of the lagena and the basilar papilla in tetrapods show structural differences from the pattern in lungfish (the lagenar macula but not the lagenar recess is lost in tetrapods whereas the lagenar recess rather than the macula is lost in lungfish), and therefore do not support this view. Moreover, the unique position of the utricular macula in a separate recess is a character found only in cartilaginous fish and lungfish and no other vertebrate lineage (Retzius 1881; Werner 1960). However, changes in the "middle ear" such as loss of the spiracular pouch and fusion of the hyomandibular bone with the cranium could well be interpreted as being derived owing to the autostylous fusion of the upper jaw with the brain case of modern lungfish (Jarvik 1980; Campbell and Barwick 1986).

b. The similarities in the basilar papilla of *Latimeria* and tetrapods are achieved through parallelism and represent parallel culminations towards pressure detection of an ancestral sarcopterygian predisopsition. Other pressure-transducing systems in teleosts obviously evolved parallel to those in sarcopterygians, and show such clear-cut structural differences (van Bergeijk 1967) that they cannot be taken to support this scenario. The only argument in favor of this scenario is the existence of the canalis communicans thus far apparently unique to coelacanths (Bjerring 1985). If this canal indeed plays a role in sound pressure perception of *Latimeria* it could reflect an autapomorphy of Actinistia dedicated to a unique function, the conduction of pressure differences between the belly and the ear. The bronchial columella of some tadpoles appears to be so different structurally from the canalis communicans that it is interpreted here as a parallelism.

Although the data presented here can be reconciled with different systematic relationships among sarcopterygian fish, it is clear that the interpretation of a monophylum comprising crossopterygian fish and tetrapods, but not lungfish, is the easiest way to interpret the distribution of the available data (Fig. 18.7). Whatever the systematic affinity of lungfish will ultimately turn out to be, only modern lungfish are too specialized in feeding, and the associated functional transformation of the skull, to be considered as ancestral to tetrapods in this respect. However, more recent data on fossil lungfish show that these were closer to crossopterygian fish with respect to a basicranial articulation, a movable palatoquadrate (Carroll 1988), and a large hyomandibular bone (Forey 1986; Campbell and Barwick 1986). These animals were therefore much less specialized than modern lungfish, and their otic region should be examined in more detail.

For further discussion, I will argue as if the water to land transition had started in animals with a functional tympanic ear adapted to aquatic pressure perceptions as proposed by van Bergeijk (1966), and supported to some extent by the data on *Latimeria* without denying that similar specializations might have been present, although not yet discovered, in ancestral lungfish.

3. The Water-to-Land Transition: De Novo Development of a Tympanic Ear or Transformation of an Aquatic Tympanic Ear?

This issues is a long-standing controversy (Lombard and Bolt 1979) and revolves around the question of whether or not the tympanum of tetrapods is homologous, i.e., evolved only once (Goodrich 1930), or evolved three times independently in amphibians, "reptiles," and mammals (Gaupp 1898, 1913; Lombard and Bolt 1979). Irrespective of these different standpoints, almost all investigators so far believe that a tympanic ear is a neomorph of early tetrapods, and that the formation of a tympanic membrane was somehow correlated with the insertion of the hyomandibular bone in the ear capsule (perhaps in different ways, Lombard and Bolt 1988). Most of the arguments around these ideas have been published and their evidence reviewed elsewhere and in this volume. I therefore shall concentrate here not on the de novo evolution of a tympanic ear but on the less explored suggestion that the tympanic ear of ancestral sarcopterygian fish was transformed into a tetrapod middle ear by the insertion of the hyomandibular bone.

Clearly, this could have happened only after the stapes had lost its primary function as a support of the lower jaw in sarcopterygians. This again could have happened only after these fish had developed the tetrapod type of an autostylus fusion of the upper jaw with the neurocranium and had thereby freed the hyomandibular bone from its suspensorial function. Conditions reminiscent of this hypothetical stage are described in a fossil tetrapod (Clack 1989; Chapter 20) where the hyomandibular bone is freed of its suspensorial function. Likewise, in primitive lungfish a hyomandibular bone is described which may have moved in conjunction with the operculum but not with the palatoquadrate bone (Campbell and Barwick 1986).

Autostyly is considered to be derived three times independently among vertebrates: in holocephalans, lungfish, and tetrapods (Schultze 1986). Only in the tetrapod lineage was the hyomandibular bone apparently transformed to perform novel function(s) (Gaupp 1898; Carroll 1988; Clack 1989), whereas it became a largely reduced appendix in holocephalans and modern lungfish (Jarvik 1980; Campbell and Barwick 1986). This unique fate of the hyomandibular bone may be related in tetrapods to its decoupling of ancestral functions and its early involvement in its new function(s); either ventilation (Clack 1989), bracing the braincase (Carroll 1988), or audition. The latter function was never denied, but other function(s) of the stapes in an autostylus skull are clearly possible.

3.1 Changed Physical Conditions Necessitate the Transformation of an Aquatic Tympanic Ear

The impedance problem is usually considered as the single major reason for the evolution of a terrestrial tympanic ear (van Bergeijk 1966; Moller 1974). This view seems reasonable even if we consider the many cases of hearing, especially among

amphibians, with no tympanic middle ear (Jaslow, Hetherington, and Lombard 1988), and that the impedance mismatch of the ear is much less than the theoretical values for the air/water mismatch (Lewis, Leverenz, and Bialek 1985). Among anuran amphibians and reptiles, there is no doubt that the absence of a tympanic ear is derived, but the remaining structures of the middle ear (e.g., the stapes) may nevertheless aid in aerial hearing (Jaslow, Hetherington, and Lombard 1988). It then follows that without any sort of impedance-matching device, an aquatic ear, as discussed above, was probably not well adapted to the impedance loss it would suffer when exposed to air (van Bergeijk 1966).

Moreover, given the anatomical relationship of the spiracular pouch and the lateral, cartilaginous otic wall in *Latimeria*, it appears unlikely that such an organization would, in air, result in noticeable movement in the otic fluid if pressure changes were applied. Theory argues instead (van Bergeijk 1966) that changes in volume of the spiracular gas bubble will result in displacement of the tympanum. The ratio of the area of the spiracular pouch to the area of the perilymphatic foramen, which may even have amplified the movement caused by the pulsating spiracular pouch in water, will be less effective in air because it enhances the impedance mismatch. This problem would be even more pronounced if the otic capsule were ossified as is the case in many fossil tetrapods (Carroll 1988).

This problem could not be overcome by the small area of the tympanic membrane as compared to the area of the pouch (van Bergeijk 1966) because the volume change in the spiracular pouch would result in displacement of the tympanum only as long as the inertial load on this membrane would be much smaller than that at any other contact area of the spiracular pouch.

From the reasoning above it follows that one way to overcome this problem would be the insertion of a movable element in the otic capsule. Such a movable element could represent a starting point for a low-resistance flow through the inner ear along the already existing "perilymphatic endorgan" and through this existing perilymphatic foramen. Orientation of movement towards these endorgans could be selected for, even in the absence of pronounced impedance improvement and amplification of the signal, as in some amphibians (Jaslow et al. 1988). This movable element need not be in contact with the tympanic membrane to function in that way, but any stiffening of the tympanic membrane would lessen the displacement of this membrane (van Bergeijk 1966) and could, as a result, improve the movement of the element in the otic wall. Moreover, such an ear could work both in water and on land. In a way, this tympanic ear would resemble the ear of ostariophysine fish. As in these fish, a movable bony element conducts displacement of a gas bubble to the perilymphatic labyrinth of the ear. What matters here is that such an ear would not necessarily result in a conspicuous external opening, e.g., a large otic notch. In fact, this area should be small (van Bergeijk 1966), and could be partly included into lateral line canals as is the case with the spiracular opening of *Latimeria* (Fig. 18.6).

It is conceivable that the insertion of such a movable element was not straightforward, but rather the result of a checkered history (Lombard and Bolt 1988). Given that the spiracular pouch abuts the hyomandibular bone in fossil (Jarvik 1980) and modern crossopterygian fish, it is a likely candidate for such a movable element, particularly after it had lost its function as a suspensorium of the jaws and the opercular ossicles. In fact, the existence of the stapes footplate in all tetrapods could be easily understood based on such a scenario. However, the presence of an operculum as an additional element in the oval window of most amphibians would argue for two separate events in each of these lineages. Unfortunately, no operculum has ever been described in the oval window of fossil amphibians (Lombard and Bolt 1988).

A more effective solution to this problem would be the insertion of the stapes between the tympanum and the ear capsule (van Bergeijk 1966) rather than just the stapes footplate into the ear capsule. The insertion of the stapes between the tympanum and the ear may have happened with or without an intermediate stage of an ear with a movable element only in the otic capsule as described above. Whether or not this was the case, the insertion of a stapes into the tympanum and the ear would cause a conflicting situation: in order to maximize displacement of the tympanum, the tympanum should have a small area compared to the surface area of the spiracular pouch (van Bergeijk 1966) when no stapes is present. In contrast, in order to drive effectively the stapes footplate in the oval window and to overcome the impedance of the perilymph when a stapes is present, the area of tympanum should be large compared to the area of the oval window (Moller 1974).

In this context, it is interesting to note that the otic notch of fossils, which was often hypothesized to harbor a tympanum (Lombard and Bolt 1988), varies substantially in size (Clack 1989). It is possible that this variation may be related to the different directions of change a tympanic ear adapted to aquatic sound could take to become a terrestrial middle ear. The already existing structures for sound conduction in ancestral fish may have been extremely beneficial to reduce the impedance problem by the availability of a low-resistance flow channel. This would have been completed by the formation of connections between the perilymphatic cisterna at the oval window and the perilymphatic sac at the perilymphatic foramen. The two configurations of these connections found in tetrapods suggest that the insertion of a movable element and the evolution of an oval window happened at least twice and independently (Fig. 18.8). Given the uncertainties of the systematic affinities of many fossils to different tetrapod lineages, in particular to amphibians (Carroll 1988), it is unclear what the sequence of events was that led to the formation of the different terrestrial tympanic ears. In any case, a likely explanation from our knowledge of the fossils is that the hyomandibular bone was inserted into the ear capsule at least twice (Lombard and Bolt 1988).

3.2 Insertion of the Hyomandibular Bone and Development of the Perilymphatic Labyrinth

As outlined above, the insertion of a movable element into the ear capsule must be considered as a crucial event in the evolution of terrestrial hearing. The insertion of the stapes footplate might have led to the completion of the perilymphatic system into the two different patterns currently recognized among tetrapods (Lewis and Lombard 1988; Fritzsch and Wake 1988; Figs. 18.5, 18.8). The phylogenetic evidence for this suggestion will be presented elsewhere (Bolt and Lombard, Chapter 19). I will here consider ontogenetic evidence which suggests that the stapes in the three lineages might not be completely identical.

Descriptive and experimental anatomy has long suggested that the footplate and parts of the stylus of the stapes of frogs are entirely derived from the ear capsule (Luther 1924; Hetherington 1988). In contrast to frogs, comparable extirpation experiments on salamanders (Toerien 1963) indicate that the otocyst does add little to the stapes which may be entirely of neural crest origin. Corresponding to these claims in salamanders, the entire stapes including the footplate seems to develop in mammals without contribution of the capsular mesoderm (Werner 1960; Van de Water et al. 1986). Moreover, in birds, the stapes forms from two sources, the footplate comes from the otic capsule and the stylus from neural crest cells (Noden 1987). These data suggest that either the stapes has drastically changed its embryonic source(s) during evolution (Noden 1987) or, alternatively, that only those parts of the stapes with the same source, e.g., the neural crest, can be considered as homologous. The latter view would be compatible with the suggestion that either the hyomandibular bone was inserted into a preformed opening in the otic capsule, or had fused with the otic capsule and separated a piece of the wall (Lombard and Bolt 1988). The possibility that the insertion of the hyomandibular bone into the otic capsule was different in the various tetrapod lineages may be reflected in the different contributions of the neural crest and otic capsule wall to the stapes. Together these data strengthen the view expressed above that the evolution of a terrestrial tympanic ear followed a rather tortuous path and was quite different in the various lineages. It also suggests that the stapes is only partly homologous to the hyomandibular bone and that the oval window and the perilymphatic system evolved at least twice, as discussed earlier (Lombard and Bolt 1988).

In the light of these suggestions it is possible that a perilymphatic labyrinth might have evolved independently from an ancestral condition like in *Latimeria*. The tympanum can likewise be considered as independently derived or as a transformation of the ancestral tympanic membrane which was likely present in crossopterygian fish. The rather featureless morphology of the tympanum makes it difficult to exclude the possibility of parallelism. Its suggested absence in some fossils (Clack 1989)

may indicate in fact a loss and a regain of the tympanic membrane in some tetrapods. Admittedly, the distribution of certain patterns, and the large variety of different interpretations, together with a lack of hard data on many issues discussed above, make it difficult to exclude any scenario at present.

It is also possible that during the refinement of the aquatic ear into a terrestrial ear in different tetrapod lineages some lineages may have lost the spiracular pouch, e.g., salamanders and gymnophionans. Moreover, given the rather different way the spiracular pouch and the tympanum of frogs develop as compared to amniotes (Werner 1960; Hetherington 1988), may indicate that the tympanic ear of frogs represents a newly evolved terrestrial ear completely unrelated to that of amniotes.

3.3 The Evolution of the Amphibian Papilla

This perilymphatic endorgan is present only in amphibians, hence its name (de Burlet 1934). Previous work has suggested that it may at least be partially derived from the papilla neglecta, an endorgan present in most vertebrates, but notably lacking in frogs and salamanders (Sarasin and Sarasin 1892; Baird 1974). A close examination of different developmental stages in a caecilian amphibian, the only vertebrates possessing both a papilla neglecta and an amphibian papilla, has shown that the papilla neglecta and the papilla amphibiorum are in fact more related to each other than to any other epithelium of the ear (Fritzsch and Wake 1988). Both epithelia are innervated by the same nerve branch (Fig. 18.2). Both separate together during development from the saccular macula and both separate from each other only later by the constriction of the foramen utriculosacculare. The ontogenetic argument, the pattern of innervation and the close topographical association of both epithelia in some adult caecilians are in favor of homology of the papilla neglecta and the papilla amphibiorum.

In caecilians, the papilla neglecta is supposed to be partly shifted into a new position to become the papilla amphibiorum, and that part of the papilla neglecta of anurans and salamanders that was not transformed into the papilla amphibiorum was suggested to be lost altogether (Fritzsch and Wake 1988). The different interpretations of these end-organs in Wever (1985) have been dealt with elsewhere (Fritzsch and Wake 1988) and need not be repeated here.

These amphibian-specific transformations can be related to the course of the periotic labyrinth (Figs 18.5, 18.8). It is only in amphibians that the periotic canal passes close to the papilla neglecta. This proximity of the pressure-conducting canal may lead to the transformation of those parts of the papilla neglecta into a "perilymphatic endorgan" (de Burlet 1934), which are in the pathway of low-resistance for sound-induced pressure changes (van Bergeijk 1966). In essence, it is proposed that the amphibian-specific pathway of the periotic labyrinth is causally related to the changes of parts of the papilla neglecta into the papilla amphibiorum (Fritzsch and Wake 1988). Moreover, the changes in the papilla basilaris, which rests largely on the periotic limbic tissue, may also be related to the same phenomenon, the specific pathway taken by the sound from the oval window to the perilymphatic foramen (Wever 1985). What role, if any, the unique operculum of amphibians may play for the formation of the amphibian perilymphatic labyrinth is currently unclear.

3.4 Biological Context of Hearing of Ancestral Tetrapods

Little attention has been given in the past to this problem which might be stated in the following way.

Depending on where an animal mates, where it hunts and what its major predators are, its hearing will adapt more to the air or more to the water. Unfortunately, we do not know very much about this problem, but it will seriously bias our reasoning. For example, an ear fully adapted to pressure detection under water will not work effectively in air, e.g., the hypothetical ear of *Eusthenopteron* (van Bergeijk 1966). Nevertheless, it may have represented the only ear at its time that worked, at least to some extent, in air. Thus, even a small improvement that causes only 99.8% impedance loss would help to double the sensitivity as compared to an animal with 99.9% loss of acoustic energy. If only some aspects of the life history are performed on land, then the best compromise for both terrestrial and aquatic hearing will be

selected for until the animal is completely terrestrial. This applies not only for sensitivity and range of hearing, but even more for directional hearing which makes use of different parameters in terrestrial (Eggermont 1988) and aquatic (Popper et al. 1988) vertebrates.

4. Reorganizations in the Central Nervous System During the Water-to-Land Transition

This subject has been recently reviewed extensively (Fritzsch 1988; Fritzsch 1989) and will be discussed in detail elsewhere (McCormick, Chapter 17). Most ideas have revolved around two questions: (1) What makes an endorgan that requires a novel function, like pressure reception, project into a specific area of termination? (2) What are the cell sources for primary auditory nuclei of the rhombencephalic alar plate?

The following description will focus on the correlation of regressive and progressive changes and their potential bearing for the cellular source of the anuran auditory nuclei, i.e., is there evidence for functional transformation of neuronal tissue as suggested for the evolutionary transformation of the hyomandibular bone and the spiracular pouch?

4.1 Metamorphic Loss of the Lateral Line System Does Not Correlate with the Development of Auditory Nuclei in Frogs

Amphibians display different patterns of loss, suppression, or retention of the lateral line organs during metamorphosis, or during their entire life history (Fritzsch 1989). This is true for all three orders and concerns both the ampullary electroreceptive system and the mechanosensory neuromast system. The latter is the sole component of the anuran lateral line (Fritzsch 1989). The remarkable fact is that changes in the lateral line system during metamorphosis correlate more with life-style than with systematic affinity. For example, retention of lateral line organs during metamorphosis may happen in primitive frogs like many aglossids (e.g., *Xenopus laevis*, the clawed toad) or in modern frogs like bufonids (e.g., *Lepidobatrachus laevis*, a South American bufonid). Their common denominator is not systematic affinity but life-style: they are all aquatic (Fritzsch 1990a).

Even frogs which keep most of their lateral line organs past metamorphosis have auditory nuclei as adults. Thus, the loss or the retention of lateral line organs apparently does not correlate with the formation of auditory nuclei in adult frogs, and therefore cannot be considered to be causally related to their formation as suggested previously (Larsell 1934, 1967). Apparently, the auditory nuclei form before metamorphosis in some frogs, and therefore coexist with the lateral line nuclei in premetamorphic tadpoles (Jacoby and Rubinson 1983). Moreover, many neurons of the lateral line nuclei in the alar plate were found to degenerate together with the lateral line organs in frogs which loose the lateral line (Wahnschaffe, Bartsch and Fritzsch 1987). Thus the previous attempt to correlate the loss of lateral line organs with the formation of auditory nuclei is not supported by all recent data in frogs and was consequently refuted (Fritzsch 1988, 1990a).

4.2 Loss of Electroreception Shows Phylogenetic Coincidence with the Appearance of Primary Auditory Nuclei

Auditory nuclei as discrete cytoarchitectonic entities were, until recently, only known for amniotes and frogs. However, some bony fish with an ear specialized for pressure perception have not only discrete projections of "perilymphatic endorgans" to the brain stem but also recognizable cytoarchitectonic entities which receive these projections (Fritzsch, Niemann, and Bleckmann 1990; McCormick, Chapter 17). Comparable to tetrapods, these neuronal entities may be named auditory nuclei.

Among tetrapods, the distribution of discrete primary auditory nuclei in the alar plate shows an interesting correlation: all species with such a nucleus have lost the primary sense of electroreception (Fig. 18.9). Among amphibians, only frogs have lost this ancient vertebrate sense (Northcutt 1986a), and all frogs have primary auditory nuclei. However, the reverse is not true, i.e., not all vertebrates that have lost the ancient sense

of electroreception have auditory nuclei. In fact, most teleosts probably do not perceive the pressure component of aquatic sound but only the physically rather different direct sound through otolithic endorgans (Schellart and Popper, Chapter 16).

Given the phylogenetic coincidence between the loss of one and the gain of another projection and its associated cell masses (Fig. 18.6), the functional transformation of second order electroreceptive neurons into auditory nuclei could therefore be possible, at least for amphibians (Fig. 18.9). However, there are other possible interpretations of the available data, and at this point one can only remark that the loss of electroreception may destabilize the neuronal organization of the alar plate. Subsequent reorganization under certain conditions may lead to the formation of auditory nuclei. For example, the anuran auditory nuclei apparently have two sources: the ventral zone and a dorsal area hypothesized to represent the anlage of the dorsal electroreceptive nucleus of urodeles and gymnophionans (Fritzsch 1988). How this suggestion relates to other taxa is not at all clear. In particular bony fish may have both electroreception and discrete primary auditory nuclei (Fritzsch, Niemann, and Bleckmann 1990; McCormick, Chapter 17).

4.3 What Causes the Formation of Separate Projections of "Perilymphatic Endorgans" in Vertebrates?

Largely discrete projections have been described for many endorgans of the inner ear of vertebrates (McCormick and Braford 1988; Will and Fritzsch 1988). However, these projections are typically restricted to the area of termination of the other otic endorgans and partially overlap with them. The condition found for the supposed pressure sensitive "auditory" endorgans is different. Projections from these organs in fish (McCormick and Braford 1988; Fritzsch, Niemann, and Bleckmann 1990), amphibians (Will and Fritzsch 1988), and amniotes not only are distinct with little overlap but are usually more or less remote from projections of other otic endorgans (Figs. 18.6, 18.9).

The discrete projection of an endorgan occurs independently of its previous function when it

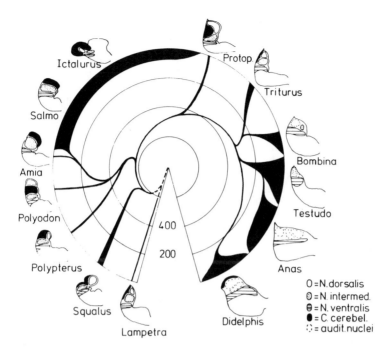

FIGURE 18.9. The variation in the left half of transversely sectioned alar plates of vertebrates and there supposed phylogenetic relationship is shown. The number of species in each vertebrate class or order is indicated by the size of the black bar, million years are indicated by circles. A reduction of the dorsal nucleus occurs in three independent lines: among teleosts (silurids, gymnotids and mormyrids have an electroreceptive nucleus that is not homologous to the dorsal nucleus in nonteleosts), among anurans, and among amniotes. Whether the auditory nuclei are homologous in anurans, reptiles and birds, and mammals is not yet clear. Note that rearrangements of the primitive pattern of nuclei occurs only after loss of the primitive sensory modality of electroreception. After Fritzsch (1988).

becomes a "perilymphatic endorgan," e.g., parts of the utricle in clupeids or the saccule in mormyrids (McCormick and Braford 1988). On the other hand, being a "perilymphatic endorgan" is apparently a necessary but not a sufficient condition in order to project separately from other endorgans. For example, the projection of the basilar papilla and the amphibian papilla in salamanders and gymnophionans are confined to the area of all other octaval projections, whereas these endorgans also reach the dorsolateral auditory nucleus in anurans (Fig. 18.6). It is unclear what has caused the doubling of the projection in anurans (Will and Fritzsch 1988). Given the similarity of the brain stem organization in *Latimeria* and salamanders, it is likely that they also have the same pattern of eighth nerve projection. This would then imply that the separate projection of the basilar papilla in sarcopterygians is correlated with the novel evolution of a dorsolateral auditory nucleus in frogs or the auditory nuclei of amniotes. The latter nuclei may or may not be homologous, i.e., derive from the homologous ancestral cell source(s) (Fritzsch 1988).

The essence is that the appearance of the basilar papilla as a unique perilymphatic endorgan of sarcopterygian fish was not necessarily accompanied by an immediate formation of a separate projection and the formation of auditory nuclei in the alar plate. Data on salamanders and gymnophionans rather suggest that an unrelated phylogenetic event has rendered the brain stem plastic to such an extent that both the separate projections and the formation of auditory nuclei were possible in amniotes and anurans (Fig. 18.6). Comparison with modern bony fish indicates that this event might have been the loss of the primitive ampullary receptor system present in all vertebrates that have retained the basic

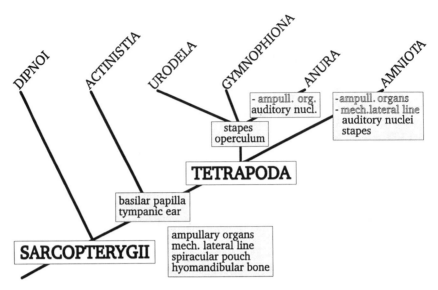

FIGURE 18.10. The distribution of ancestral and new otic and non-otic characters (filled letters) and the loss of ancestral characters (open letters) are shown as discussed in the text. A basilar papilla and a tympanic ear is considered to be an autapomorphy of actinistians and tetrapods. The insertion of the stapes and the operculum into the ear capsule is proposed to have happened at least twice, once within the amphibian and once (or twice) within the amniotic ancestors. Auditory nuclei evolved independently in anurans and amniotes. In anurans this coincides with the phylogenetic loss of ampullary organs whereas in amniotes it coincides with the phylogenetic loss of both the ampullary organs and the mechanosensory lateral line.

organization of the alar plate irrespective of the organization of the ear (Fig. 18.9).

Whether or not the anlage of the dorsal electroreceptive nucleus, freed from its input, plays any role in this reorganization, as suggested for frogs (Fritzsch 1988), must await further research. There is a chance that cells derived from this anlage could be transformed to function in a different context. In fact, ablation of the inner ear does not result in full depletion of auditory and vestibular neurons in frogs, birds and mammals (Fritzsch 1990b). These neurons, deprived of their inner ear input, will be maintained probably through other inputs which become more important and could perhaps alone be able to drive these neurons in a context in which they would not be active normally, i.e., they are transformed functionally. Selective ablation of the ampullary organs, while retaining the mechanosensory lateral line in salamanders or gymnophionans, would be necessary to test whether under such conditions a transformation of the alar plate, as occurs in frogs, takes place in these species.

5. Summary

The distribution and pattern of a sarcopterygian "perilymphatic endorgan," the basilar papilla, is described. The most parsimonious explanation for the available data is that this epithelium evolved only once within ancestral crossopterygian fish and was subsequently transformed into the amphibian and the amniotic type of basilar papilla (Fig. 18.10). This transformation coincides with a different organization of some aspects of the periotic labyrinth in amphibians and amniotes. These differences are suggested to be causally linked with each other. They are also suggested to be linked to the transformation of parts of the papilla neglecta into an amphibian specific "perilymphatic endorgan," the amphibian papilla.

A scenario is presented that proposes that the basilar papilla of *Latimeria* may work as a pressure detector in water. It is argued that this hypothesized function may have been conserved during the water-to-land transition. In order to achieve this, it is suggested that a tympanic ear evolved as a gas-filled spiracular pouch in crossopterygian fish. This organization was to some extent functional in air but must have been improved by the insertion of osseous movable elements. This is hypothesized to have occurred at least twice and differently in amphibians and amniotes. It is proposed that the oval window and the perilymphatic labyrinth of amphibians and amniotes have evolved independently.

The transformations of the central nervous system during the water-to-land transition are described. There is no evidence in amphibians that shows that mechanosensory lateral line neurons have been transformed into auditory neurons. However, there is a phylogenetic coincidence between the loss of the ancestral electroreceptive system and the novel appearance of "auditory nuclei" in bony fish, amphibians and amniotes. It is suggested that destabilizations in the nervous system with subsequent reorganization and stabilization in a novel pattern are consequences of the loss of a major aquatic sensory system, the ampullary electroreceptive sense. This may or may not coincide with the formation of a novel electroreceptive system in some bony fish. Whether or not brain stem neurons of the ancestral electroreceptive anlage play a direct role in this transformation is not yet clear.

Please see Note Added in Proof, p. 375.

Acknowledgments. Numerous colleagues have over the years helped me to understand more of the auditory system of aquatic vertebrates than I had ever anticipated. I am indebted to all, but in particular to H. Bleckmann, T. H. Bullock, R. R. Fay, A. J. Kalmijn, C. A. McCormick, H. Münz, R. G. Northcutt, A. N. Popper, and M. H. Wake. In particular, I wish to express my sincere gratitude to M-D Crapon de Caprona, who not only helped me with the drawings, but also supported my research morally. I thank S. Commerfort and R. R. Fay for linguistic advice. The continuous support of the Deutsche Forschungsgemeinschaft over the last ten years is also gratefully acknowledged.

This paper is dedicated to the memory of W. A. van Bergeijk whose flawless thinking has paved the way for most of the ideas expressed throughout this chapter.

References

Ariens-Kappers CU, Huber GC, Crosby EC (1936) The comparative anatomy of the nervous system of vertebrates, including man. New York: Hafner Publishing.

Baird IL (1974) Anatomical features of the inner ear in submammalian vertebrates. In: Keidel WD, Neff WD (eds) Handbook of Sensory Physiology, Vol. V/1. Berlin: Springer, pp. 159–212.

Blaxter JHS, Denton EJ, Gray JAB (1981) Acoustico-lateralis system in clupeid fishes. In: Tavolga WN, Popper AN, Fay RR (eds) Hearing and Sound Communication in Fishes. New York: Springer Verlag, pp. 39–59.

Bjerring HC (1985) Facts and thoughts on piscine phylogeny. In: Foreman RE, Gorbman A, Dodd JM, Olsson R (eds) Evolutionary Biology of Primitive Fishes. New York: Plenum Press, pp. 31–57.

Campbell KSW, Barwick RE (1986) Paleozoic lung-fishes—a review. J Morphol Suppl 1:93–131.

Carroll RL (1988) Vertebrate Palaeontology and Evolution. New York: Freeman and Co.

Clack JA (1989) Discovery of the earliest-known tetrapod stapes. Nature 342:425–427.

Corwin JT (1981) Audition in elasmobranchs. In: Tavolga WN, Popper AN, Fay RR (eds) Hearing and Sound Communication in Fishes. New York: Springer Verlag, pp. 81–105.

de Burlet HM (1934) Vergleichende Anatomie des stato-akustischen Organs. a) Die innere Ohrsphäre; b) Die mittlere Ohrsphäre. In: Bolk L, Göppert E, Kallius E, Lubosch W (eds) Handbuch der Vergleichenden Anatomie der Wirbeltiere, Vol. 2. Berlin: Urban and Schwarzenberg, pp. 1293–1432.

Eggermont JJ (1988) Mechanisms of sound localization in anurans. In: Fritzsch B, Ryan M, Wilczynski W, Hetherington T, Walkowiak W (eds) The Evolution of the Amphibian Auditory System. New York: Wiley and Sons, pp. 561–586.

Forey PL (1986) Relationships of lungfishes. J Morphol Suppl 1:75–91.

Forey PL (1988) Golden jubilee for the coelacanth *Latimeria chalumnae*. Nature 336:727–732.

Fricke H, Reinicke O, Hofer H, Nachtigall W (1987) Locomotion of the coelacanth *Latimeria chalumnae* in its natural environment. Nature 329:331–333.

Fritzsch B (1987) The inner ear of the coelacanth fish *Latimeria* has tetrapod affinities. Nature 327:153–154.

Fritzsch B (1988) Phylogenetic and ontogenetic origin of the dorsolateral auditory nucleus of anurans. In: Fritzsch B, Ryan M, Wilczynski W, Hetherington T, Walkowiak W (eds) The Evolution of the Amphibian Auditory System. New York: Wiley and Sons, pp. 561–586.

Fritzsch B, Wake MH (1988) The inner ear of gymnophione amphibians and its nerve supply: a comparative study of regressive events in a complex sensory system. Zoomorphology 108:210–217.

Fritzsch B (1989) Diversity and regression in the amphibian lateral line system. In: Coombs S, Görner P, Münz H (eds) The Mechanosensory Lateral Line. Neurobiology and Evolution. New York: Springer Verlag, pp. 99–115.

Fritzsch B (1990a) The evolution of metamorphosis in amphibians. J Neurobiol 21:1011–1021.

Fritzsch B (1990b) Experimental reorganization in the alar plate of the clawed toad, *Xenopus laevis*. I. Quantitative and qualitative effects of embryonic otocyst extirpation. Develop Brain Res 51:113–122.

Fritzsch B, Niemann U, Bleckmann H (1990) A discrete projection of the sacculus and lagena to a distinct brain stem nucleus in a catfish. Neurosci Lett 111:7–11.

Gaupp E (1898) Ontogenese und Phylogenese des schalleitenden Apparates bei den Wirbeltieren. Erg Anat Entwicklungsgesch 8:990–1149.

Gaupp E (1913) Die Reichertsche Theorie (Hammer-, Ambos und Kieferfrage). Arch Anat Anat Abt Suppl 1–416.

Goodrich ES (1930) Studies on the structure and development of vertebrates. London: MacMillan.

Greenwood PH (1963) The swimbladder in African notopteridae (Pisces) and its bearing on the taxonomy of the family. Bull Br Mus Nat Hist (Zool) 11:377–412.

Gutknecht D, Fritzsch B (1990) Lithium induces multiple ear vesicles in *Xenopus laevis* embryos. Naturwissenschaften 77:235–237.

Henson OW (1974) Comparative anatomy of the middle ear. In: Keidel WD, Neff WD (eds) Handbook of Sensory Physiology. V/1: Auditory System. Berlin: Springer, pp. 40–110.

Hetherington TE (1988) Metamorphic changes in the middle ear. In: Fritzsch B, Ryan M, Wilczynski W, Hetherington T, Walkowiak W (eds) The Evolution of the Amphibian Auditory System. New York: Wiley and Sons, pp. 339–357.

Hudspeth AJ (1989) How the ear's works work. Nature 341:397–404.

Jacoby J, Rubinson K (1983) The acoustic and lateral line nuclei are distinct in the premetamorphic frog. *Rana catesbeiana*. J Comp Neurol 216:152–161.

Jarvik E (1980) Basic structure and evolution of vertebrates. Vol. 1. London: Academic Press.

Jaslow AP, Hetherington TE, Lombard RE (1988) Structure and function of the amphibian middle ear. In: Fritzsch B, Ryan M, Wilczynski W, Hetherington T, Walkowiak W (eds) The Evolution of the Amphibian Auditory System. New York: Wiley and Sons, pp. 69–91.

Kalmijn AJ (1988) Hydrodynamic and acoustic field detection. In: Atema J, Fay RR, Popper AN, Tavolga WN (eds) Sensory Biology of Aquatic Animals. New York: Springer, pp. 84–130.

Larsell O (1934) The differentiation of the peripheral and central acoustic apparatus in the frog. J Comp Neurol 60:473–527.

Larsell O (1967) The Comparative Anatomy and Histology of the Cerebellum from Myxinoids through Birds. Jansen J (ed) Minneapolis: University of Minnesota Press, pp. 163–178.

Lewis ER, Leverenz EL, Bialek W (1985) The vertebrate inner ear. Boca Raton: CRC Press, pp. 256.

Lewis ER, Lombard RE (1988) The amphibian inner ear. In: Fritzsch B, Ryan M, Wilczynski W, Hetherington T, Walkowiak W (eds) The Evolution of the Amphibian Auditory System. New York: Wiley and Sons, pp. 93–123.

Liem KF (1989) Respiratory gas bladders in teleosts: Functional conservatism and morphological diversity. Amer Zool 29:333–352.

Lombard RE (1977) Comparative morphology of the inner ear in salamanders (Caudata: Amphibia). Contrib Vert Evol 2:1–143.

Lombard RE, Bolt JR (1979) Evolution of the tetrapod ear: an analysis and reinterpretation. Biol J Linn Soc 11:19–76.

Lombard RE (1980) The structure of the amphibian auditory periphery: A unique experiment in terrestrial hearing. In: Popper AN, Fay RR (eds) Comparative Studies of Hearing in Vertebrates. New York: Springer Verlag, pp. 121–138.

Lombard RE, Bolt JR (1988) Evolution of the stapes in paleozoic tetrapods. In: Fritzsch B, Ryan M, Wilczynski W, Hetherington T, Walkowiak W (eds) The Evolution of the Amphibian Auditory System. New York: Wiley and Sons, pp. 37–67.

Luther A (1924) Entwicklungsmechanische Untersuchungen am Labyrinth einiger Anuren. Soc Sc Fenn Comment Biol 2:1–48.

McCormick CA, Braford MR (1988) Central connections of the octavo-lateralis system: evolutionary considerations. In: Atema J, Fay RR, Popper AN, Tavolga WN (eds) Sensory Biology of Aquatic Animal. New York: Springer, pp. 750–767.

Millot J, Anthony J (1965) Anatomie de *Latimeria chalumnae*. Tome II. Paris: CNRS.

Moller AR (1974) Function of the middle ear. In: Keidel WD, Neff WD (eds) Handbook of Sensory Physiology, V/1: Auditory System. Berlin: Springer, pp. 492–517.

Noden DM (1987) Interactions between cephalic neural crest and mesodermal populations. In: Maderson PFA (ed) Developmental and Evolutionary Aspects of the Neural Crest. New York: Wiley, pp. 89–120.

Northcutt RG (1980) Central auditory pathways in anamniotic vertebrates. In: Popper AN, Fay RR (eds) Comparative Studies of Hearing in Vertebrates. New York: Springer Verlag, pp. 79–118.

Northcutt RG (1986a) Electroreception in nonteleost bony fishes. In: Bullock TH, Heiligenberg W (eds) Electroreception. New York: Wiley and Sons, pp. 257–287.

Northcutt RG (1986b) Lungfish neural characters and their bearing on sarcopterygian phylogeny. J Morphol Suppl 1:277–297.

Popper AN, Rogers PH, Saidel WM, Cox M (1988) Role of the fish ear in sound processing. In: Atema J, Fay RR, Popper AN, Tavolga WN (eds) Sensory Biology of Aquatic Animals. New York: Springer, pp. 687–710.

Reichert C (1847) Über die Visceralbögen der Wirbeltiere im allgemeinen und deren Metamorphose bei den Vögeln und Säugetieren. Arch Anat Physiol 120–222.

Retzius G (1881) Das Gehörorgan der Wirbeltiere: I. Das Gehörorgan der Fische und Amphibien. Stockholm: Samson and Wallin, pp. 286.

Retzius G (1884) Das Gehörorgan der Wirbeltiere: II. Das Gehörorgan der Amnioten. Stockholm: Samson und Wallin, pp. 345.

Rogers PH, Cox M (1988) Underwater sound as a biological stimulus. In: Atema J, Fay RR, Popper AN, Tavolga WN (eds) Sensory Biology of Aquatic Animals. New York: Springer, pp. 131–149.

Romer AS (1937) The braincase of the carboniferous crossopterygian *Megalichthyes nitidus*. Bull Mus Comp Zool 82:1–73.

Sarasin P, Sarasin F (1892) Über das Gehörorgan der Caeciliiden. Anat Anz 7:812–815.

Schultze H-P (1986) Dipnoans as sarcopterygians. J Morphol Suppl 1:39–74.

Thomson KS (1966) The evolution of the tetrapod middle ear in the rhipidistian-amphibian transition. Amer Zool 6:379–397.

Toerien MJ (1963) Experimental studies on the origin of the cartilage of the auditory capsule and columella in *Ambystoma*. J Embryol Exp Morphol 11:459–473.

van Bergeijk WA (1966) Evolution of the sense of hearing in vertebrates. Am Zool 6:371–377.

van Bergeijk WA (1967) The evolution of vertebrate hearing. In: Neff WD (ed) Contributions to Sensory Physiology. New York: Academic Press, pp. 1–49.

Van de Water TR, Noden DM, Maderson PFA (1986) Embryology of the ear: Outer, middle, and inner ear. In: Alberti PW, Ruben RJ (eds) Otological Medicine and Surgery. New York: Churchill-Livingstone.

Wahnschaffe U, Bartsch U, Fritzsch B (1987) Metamorphic changes within the lateral-line system of Anura. Anatomy and Embryology 175:431–442.

Werner G (1960) Das Labyrinth der Wirbeltiere. Jena: Fischer Verlag, pp. 309.

Wever EG (1974) The evolution of vertebrate hearing. In: Keidel WD, Neff WD (eds) Handbook of Sensory Physiology, V/1: Auditory System. Berlin: Springer, pp. 423–454.

Wever EG (1985) The Amphibian Ear. New Jersey: Princeton University Press, p. 405.

White JS, Baird IL (1982) Comparative morphological features of the caecilian inner ear with comments on the evolution of amphibian auditory structures. Scann Electron Micr 3:1301–1312.

Will U, Fritzsch B (1988) The octavus nerve of amphibians: Patterns of afferents and efferents. In: Fritzsch B, Ryan M, Wilczynski W, Hetherington T, Walkowiak W (eds) The Evolution of the Amphibian Auditory System. New York: Wiley and Sons, pp. 159–183.

Note Added in Proof:

The debate about the systematic position of lungfish and *Latimeria,* traditionally based on morphological data, has been extended to molecular data. Analysis of the 28S ribosomal RNA strongly suggests a close affinity of *Latimeria* and tetrapods (Hills, D.M., Dixon, M.T. and Ammerman, L.K. (1991) Env. Biol. Fish, 32: 119–130. Moreover, analysis of nerve myelin indicates that lungfish are the sister group of *Latimeria* and terapods (Waehneldt, T.V. and Malotka (1989) J. Neurochem. 52: 1941–1943). Likewise most recent data on the *B*-haemoglobin sequence suggests a tree with *Latimeria* as the sister group of terapods and lungfish as sister group of teleosts, *Latimeria* and tetrapods (Gorr, T., Kleinschmidt, T. and Fricke, H. (1991) Nature, 351: 394–397). It is particularly interesting that this tree ((((*Latimeria*- tetrapods) teleosts) lungfish) elasmobranchs) would be the most parsimonious tree for the olic characters analysed in this chapter. In contrast to these data, there is only one set of molecular data based on mitochrondrial DNA arguing for a closer lungfish-tetrapod affinity (Meyer, A., and Wilson, A.C. (1990) J. Molec. Evol., 31: 359–364).

19
Nature and Quality of the Fossil Evidence for Otic Evolution in Early Tetrapods

John R. Bolt and R. Eric Lombard

1. Introduction

A frequent, and frequently appropriate, role of paleontologists at a conference on the evolutionary biology of anything, is to provide temporal perspective via an overview of morphology and function based on fossils. The recipients of such paleontological benediction can then incorporate this testimony of the rocks into their own morphological and functional investigations. We have presented such papers on the otic region (e.g., Lombard and Bolt 1988), and remain convinced of their value, in fact of their necessity in evolutionary biology. However, the usefulness of a scientific paper is heavily dependent on the reader's ability to assess the quality of both the data and interpretations presented. In the present context, factors affecting "quality" include accuracy of observation, state of preservation of fossils, techniques used to estimate relationships among species and higher taxa, and evidence and assumptions employed in inferences of function. Paleontologists often, and nonpaleontologists nearly inevitably, overlook these points when considering otic evolution.

With this in mind, the purpose of this chapter is to assess the quality of the evidence bearing on the early evolution of the tetrapod ear. By "early" we mean here the Paleozoic Era, a time span that covers the first third of the evolution of terrestrial vertebrates and encompasses the origin and early radiation of both amphibians and amniotes. The chapter is divided into four sections sequentially dependent on one another. In the first section, the Paleozoic fossil record for terrestrial vertebrates and the major taxa presently recognized are

reviewed. In the second, present estimates of the relationships for these taxa are considered. Next, we present an overview of the morphology of the ear region in these taxa, with an emphasis on the features known for most or all. Finally, the data available on ear morphology is combined with the estimates of evolutionary relationships to present a hypothesis for the evolution of some otic features in the early land vertebrates.

The data presented in this chapter result from a review of the paleontological literature published over the past 60 years. Of the approximately 1,300 articles published in that span that mention the Paleozoic taxa considered here, about 470 present at least one original illustration of at least one fragment of a Paleozoic tetrapod. These illustrations are the morphological data easily accessible to science and this review is constructed from the illustrations in these articles, augmented by first hand examination of some of the material. The protocol used in collecting the data from the literature and a description of the resulting database may be found in the Appendix. The bibliography of papers containing original illustrations that we assembled for this review is available in either printed or electronic form to interested colleagues.

2. Paleozoic Tetrapods

2.1 Time and Space

The earliest evidence of tetrapods is a trackway in rocks about 370 million years old in Victoria, Australia. The oldest body fossils, only slightly

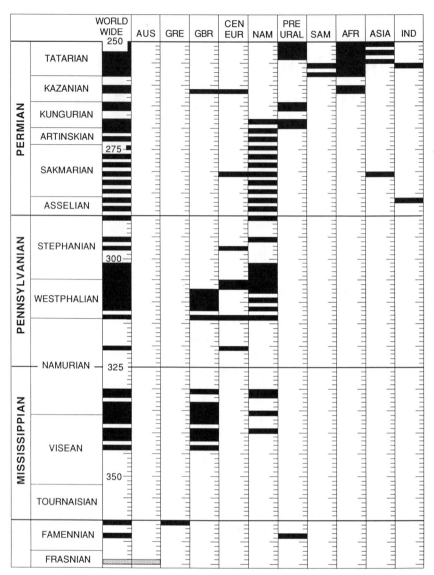

FIGURE 19.1. Geographic region and approximate age for sites where Paleozoic tetrapod fossils have been found. Except for the trackway in the Frasnian only data for body fossils are included. Data for Upper Devonian from Milner et al. (1986); for Mississippian from Carroll (1984), Smithson (1985), Bolt et al. (1988); for Pennsylvanian from Boy and Bandel (1973), Carroll (1984), Milner (1987), Mapes and Mapes (1989); and for Permian from Barbarena et al. (1985), Chatterjee and Hotton (1986),

Olson (1957, 1962, 1989). Correlation of time with the boundaries for Series (Mississippian and Pennsylvanian) and Stages (Upper Devonian and Permian) from Haq and Van Eysinga (1987). The error for any one of these correlations is about 5 to 10 million years. Abbreviations: AFR, Africa; AUS, Australia; CEN EUR, Central Europe; GBR, Great Britain; GRE, Greenland; IND, India; NAM, North America; PRE-URAL Soviet Union east of the Ural Mountains; SAM, South America.

younger than the footprints, are from East Greenland and near Tula in the Soviet Union. Fossil tetrapods have so far been found discontinuously in both time and space from these earliest occurrences in the Upper Devonian to the end of the Permian 120 million years later (Fig. 19.1). Broad continuity

of the temporal sequence is broken by three time periods in which no or no significant fossils have so far been recovered: the lower half of the Mississippian; the 15 million years between the major occurrences in the Mississippian and Pennsylvanian; and the latter third of the Pennsylvanian. Even within time spans which appear to present temporal continuity the localities producing fossils rarely extend over a significant time range. In fact, most represent an assembly of discrete events of quite short time periods on the order of a day to several thousand years and these events do not overlap in time to any significant extent. Only the last of the Permian in the Karroo of South Africa perhaps represents long term relatively continuous deposition of fossiliferous strata over several millions of years. Figure 19.1 also clearly demonstrates that the periods of time with even generally significant continuity in time are discontinuous in space. The sequential sites of the Mississippian occur in Scotland, of the Pennsylvanian first in Great Britain then in North America, of the Permian first in North Central Texas and Oklahoma, and then at the close of the Paleozoic in the Soviet Union and South Africa. The longest, though quite discontinuous, span occurs in North America from the middle of the Mississippian to the middle of the Permian. The major point here is that we do not have a continuous series of fossils from the origin of the tetrapods to the close of the Paleozoic. Some short sequences do occur but in general the record is punctuate temporally and spatially.

Not all fossil-producing areas are equal. Though several hundred discrete Paleozoic tetrapod localities are known world wide, just 10 geographically restricted areas are of major significance in that they produce well-preserved remains in any number and or taxonomic diversity for a given age (Table 19.1). Eighty percent of all Paleozoic tetrapod genera are known from these 10 areas. Upper Devonian fossils mostly come from East Greenland and Mississippian fossils from the Midland Valley of Scotland. In the Pennsylvanian most fossils come from four areas: Czechoslovakia; Ohio, and Illinois, USA; and Nova Scotia, Canada. In the U.S. Permian many genera have been found in Oklahoma and north central Texas, and in the Four Corners region (the adjacent parts of the states of Colorado, New Mexico, Arizona and Utah). The central and southern Pre-Ural region of the Soviet Union and the Karroo of South Africa produce numbers of fossils and a diversity of taxa at the close of the Permian. These latter two areas produce fossils continuously into the Triassic. The major point here is that most of the fossils we have come from restricted geographic areas. And some of these are quite small: in the extreme, the Linton, Ohio site is only a few acres in extent (Hook and

TABLE 19.1. Major producing areas for Paleozoic tetrapods

Stratigraphic Interval (SI)	Geographic Region	Number of Genera		Percent of genera for SI
		Total for SI	For each Region	
Upper Devonian		4		
	East Greenland		3	75%
	Total		3	75%
Mississippian		18		
	Scotland		14	78%
	Total		14	78%
Pennsylvanian		121		
	Czechoslovakia		30	25%
	Ohio, USA		33	27%
	Illinois, USA		15	12%
	Nova Scotia, Canada		24	20%
	Total		87	72%
Permian		232		
	Texas & Oklahoma, USA		93	40%
	Four Corners, USA		32	14%
	Pre-Urals, Soviet Union		36	16%
	Karroo, South Africa		43	19%
	Total		183	79%
Total Paleozoic		358	287	80%

Numbers of genera illustrated since 1928 for major taxa considered in this paper. See Appendix for list. Totals may be less than expected because some genera occur in more than one region or Stratigraphic Interval. Sum of Percent of Genera for SI may not equal Total for SI because some genera occur in more than one region. These are counted only once for any one SI.

Total for Paleozoic is higher than for any one SI because some genera occur in more than one stage and are counted only once to calculate Total Percentage.

Baird 1988). This means that much of the earth's surface that may have supported tetrapods in the Paleozoic is barren or otherwise unavailable for sampling.

Other factors may be at play as well. Figure 19.2 illustrates the distribution of fossil sites that have produced tetrapods, plotted on world maps on which the continents are shown in the positions they occupied in the Paleozoic. In the Late Devonian through Pennsylvanian the tetrapod-producing localities all lie quite close to the equator. Only in the Permian are fossils found which would have

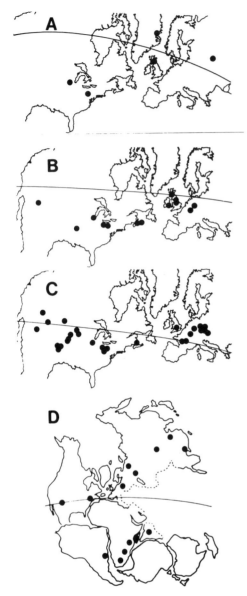

FIGURE 19.2. Abbreviated world maps indicating the Paleozoic positions of present-day locations where tetrapod fossils have been found. (A) Upper Devonian and Mississippian sites (modified from Panchen, 1977). (B) Mid Pennsylvanian sites (mostly Westphalian, modified from Milner, 1987). (C) Late Pennsylvanian and Early Permian sites (modified from Milner and Panchen 1973). (D) Mid to Late Permian sites (modified from Olson 1989). Approximate paleoequator indicated on each from Smith et al. 1981.

been at more temperate latitudes. It is possible that the data are a good representation of fact, and that during most of the Paleozoic tetrapods were restricted to equatorial climates. Continental land masses which were in those times far from the equator would thus not be expected to ever produce tetrapod fossils. On the other hand, it is also possible that through a twist of fate (and continents!) what were equatorial locales now lie in countries with a Northern European tradition of natural science and have been prospected intensively. Ninety-five percent of genera of the major taxa considered here come from North America, Europe and the Cape Province of South Africa. Land masses that were at more temperate latitudes in the Paleozoic may have indeed supported tetrapods whose remains have yet to be discovered.

The major sites that produce Paleozoic tetrapods are known to represent a restricted range of paleoenvironments, mostly low-lying flood plains, and deltas at continental margins (Milner et al. 1986). Not all types of environments are conducive to the entombment and preservation of animal remains, and this is particularly true of uplands. The major point here is that, for more than one reason, the known fossils represent relatively few environments, and there are organisms that will never be available for study.

Overall, it should be clear from this brief review that the fossils we do have cannot be placed in approximation – like a line – to represent a continuous series of organisms; there are major time gaps in the record and only a couple of sequences transgressing any significant time span. Further, it is possible and likely that these remains cannot even represent an interrupted series – like a dashed line – because we are missing faunas from nonpreserving environments or have not yet looked in the right places.

2.2 The Nature of the Fossils

A paleontological specimen may be a complete individual or a single fragment of a bone. Most, unfortunately, are closer to the latter than the former. Generally, entombment is preceded by decomposition and transport by water currents. Both usually result in disarticulation and the loss of parts, surprisingly often including the stapes. Transport also may result in breakage or abrasion as the skeletal parts are tumbled along in sand and gravel. After two or three hundred million years in the earth, specimens may still be perfectly three-dimensional or distressingly two dimensional. Most are somewhere in between, that is, crushed or sheared in a simple or complex way as a consequence of the compaction and movement of the entombing sediments. Finally, as a general rule, the older the sediments the more distorted the fossils and the harder the rock. Paleozoic tetrapods in rocks 250 to 370 hundred million years in age are therefore on average the most abused looking fossil tetrapods we have. These fragmentary and distorted remains in resistant rock present a challenge for any evolutionary morphologist.

Not surprisingly, then, no complete and undistorted remains are known for any Paleozoic tetrapod, although a very few specimens approach this ideal. Even in relatively undistorted partial remains many parts are inaccessible since the fossil would have to be partly destroyed to expose them for study. Unfortunately, this usually includes the braincase including the otic capsules and the region where the middle ear would lie, all internal parts in the primitive tetrapod skull. Furthermore, rarely is a genus known from several reasonably complete skeletons. At the other extreme a few are represented by a single broken and distorted element. Most, however, are known from one or two partial skeletons and a few to hundreds (or in a few cases, thousands) of fragments. The summary of all of the above is that the word specimen means something different to a paleontologist than it does to a neontologist. Paleontological specimens are distorted, partial and broken. We should expect that in the incomplete temporal and geographic sequence that we have to work with, many specimens will simply not have the parts of interest available for study. In those that do, the parts may be distorted or fragmentary or only partly accessible. We will be lucky if the component of interest, the stapes for example, is represented by more than one example for any given genus, if available at all.

No datum for the overall numbers of Paleozoic tetrapod specimens accessible for study in the world's collections is available. However, some recent estimates for the numbers of specimens from three localities are available. The major fossil sites at Linton, Ohio (Hook and Baird 1988) and Nyrany, Czechoslovakia (Milner 1980a,b) are very

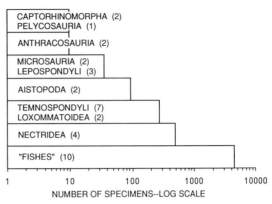

FIGURE 19.3. Numbers of vertebrate specimens in major collections from the highly productive Middle Pennsylvanian locality at Linton, Ohio, after 100 years of collecting. Described genera per taxon shown in parentheses. The "Lepospondyli" at this locality consist of a presumably monophyletic group (lysorophoids) of uncertain relationships but represented here by many more specimens than the Microsauria. Modified from Hook and Baird (1986, Figure 4).

close to being contemporary. Both represent deposits formed in swamps about 305 million years ago, and both were discovered in the course of coal mining activities. At present, about equal numbers of tetrapod genera have been found in each (Table 19.1). Figure 19.3 indicates the number of specimens of some major taxa of vertebrates available for study from the Linton site. On first inspection one is impressed by the sheer numbers of specimens in scientific collections from this locality. However, "specimen" here represents the usual wide range of preservational quality. Linton fossils at best are flat impressions in cannel coal. Some of these are high-fidelity natural molds of substantial portions of a skeleton; but often a "specimen" is a faint impression of a skeletal fragment, identifiable to genus but otherwise of little morphological significance. Milner's census of the available specimens from Nyrany is less complete in that not all known depositories were inventoried. However, his data give a good idea of what is available. Of the 21 genera of tetrapods he recognizes as present at Nyrany, 6 are represented by 1 specimen, 6 by 2–10 specimens, 5 by 11 to 50 specimens, and 2 by 50 to fewer than 100 specimens. Nyrany specimens are essentially two-dimensional impressions similar to those from Linton. A less species-rich

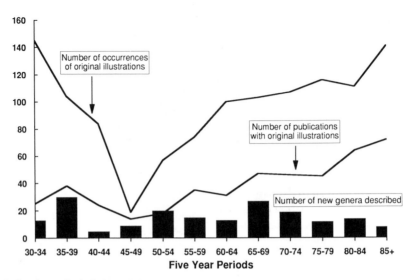

FIGURE 19.4. Indications of scholarly activity related to the discovery and study of Paleozoic tetrapods from 1930 to 1990. Number of genera described as new taxa per five year period shown as solid columns. This is an indication of field activity as well as the recognition of the uniqueness of specimens in collections as a consequence of scholarly study. Of some 1300 publications over this time period, about 470 have original illustrations. This latter subset is graphed here per five year period. An "occurrence" of an original illustration is the publication of at least one illustration of at least one skeletal remain for one species. This is an index of activity and not a count of illustrations, as an "occurrence" may include several illustrations.

locality of about 310 million years in age at Newsham, Northumberland (Great Britain) has been censused by Boyd (1984). Eight genera are recognized at Newsham each represented by 1 to 61 specimens in museums. Boyd makes an interesting assessment of the minimum numbers of individuals for each taxon represented by the specimens: the count ranges from 1 to 11. This is a cautionary note: the number of individuals represented in the world's scientific collections is certainly less than the number of specimens available.

These localities are exceptional, especially the first two. Most published localities (which total in the low hundreds for the Paleozoic) have produced partial and/or fragmentary remains of only a few genera each. These localities are important for faunal, ecological, and paleogeographic studies, but rarely important for morphological work. Thus we can determine a provisional range in time and space for a genus based on fragmentary remains of vertebrae in one place, teeth in another, and a toe bone in yet a third, but nothing, unfortunately, about the structure of its ear.

Figure 19.4 provides information useful in gauging the rates at which these Paleozoic tetrapod fossils have been discovered and studied in the past 60 years. Since 1930 new genera have been described relatively steadily at a rate of about two to five per year (average 3.1). Over that time a total of 185 genera (307 species) have been named, described and at least partially illustrated. This represents about half of the approximately 360 genera known. The other half were described from about 1840 to 1929 with the majority from 1870 on. Thus, perhaps surprisingly, new genera have been discovered at a relatively steady rate for well over 100 years. This is cause for optimism that paleontology is not yet approaching exhaustion of the fossil resources needed to understand the early evolution of land vertebrates.

We have located 470 publications published since 1928 that contain original illustrations of the taxa considered in this chapter. Figure 19.4 shows that publications in this category appeared at a rate of about five or six per year (with a dip for the Second World War) until the mid-1960s. The rate has since increased, to double that figure at the present time. On average, these papers contained illustrations of about 2.5 genera each. Overall, the data indicate that the rate of publication on

TABLE 19.2. Statistics for Publications that Describe the Otic Region of Paleozoic Tetrapods

	TOTAL	Amphibians									Amniotes								
		IC	TE	MI	NE	AI	LO	AN	SE	DD	PE	PA	ML	PR	ME	CA	DI	LE	IS
Taxon size and range																			
Number of genera illustrated since 1928	358	3	98	35	11	8	5	36	10	8	49	22	9	3	3	29	14	9	6
Approximate range within the Paleozoic (millions of years)		16	91	43	55	75	33	91	40	43	41	14	14	14	6	60	54		
Genera per million years		0.19	1.08	0.81	0.20	0.11	0.15	0.40	0.25	0.19	1.20	1.57	0.64	0.21	0.50	0.48	0.26		
General publication																			
Publications with original illustrations since 1928	470	3	175	35	20	14	8	64	21	23	59	39	8	6	8	63	26	15	12
Publications per genus		1.0	1.8	1.0	1.8	1.8	1.6	1.8	2.1	2.9	1.2	1.8	0.9	2.0	2.7	2.2	1.9	1.7	2.0
Occurrences of original illustrations since 1928	1195	8	352	89	30	28	18	100	32	25	174	134	16	6	13	99	34	24	13
Occurrences per genus		2.7	3.6	2.5	2.7	3.5	3.6	2.8	3.2	3.1	3.6	6.1	1.8	2.0	4.3	3.4	2.4	2.7	2.2
Stapes																			
Occurrences of stapes illustrations since 1928	156	1	44	24	0	4	0	6	3	6	24	0	8	1	0	30	4	1	0
Number of genera with stapes equal to: 0	273	2	71	23	11	6	5	34	8	6	38	22	3	2	3	14	11	8	6
equal to: 3	28	0	12	1	0	1	0	0	2	1	3	0	1	1	0	4	2	0	0
equal to: 2	33	1	9	3	0	1	0	1	0	0	3	0	5	0	0	8	1	1	0
equal to: 1	24	0	6	8	0	0	0	1	0	1	5	0	0	0	3	3	0	0	0
Otic capsule																			
Occurrences of otic capsule illustrations since 1928	173	0	49	23	2	5	3	14	7	9	23	5	6	0	0	20	5	2	0
Number of genera with otic capsules equal to: 0	248	3	63	22	9	5	2	26	7	5	37	18	4	3	3	16	11	8	6
equal to: 3	68	0	23	7	2	2	2	7	2	0	4	1	4	0	0	10	3	1	0
equal to: 2	33	0	10	3	0	1	1	3	1	3	5	3	1	0	0	2	0	0	0
equal to: 1	9	0	2	3	0	0	0	0	0	0	3	0	0	0	0	1	0	0	0
Cheek																			
Number of genera with cheek equal to: 1	29	0	5	7	0	0	0	3	0	1	3	1	1	1	0	7	0	0	0

Taxon abbreviations are explained in Appendix Table 19.1. The taxa are in the same order as in Figures 5.7 and 12.15. Data for the Lepospondyli (LE), a collection of Carboniferous orphans of uncertain affinities, as well as for genera Incertae Sedis (IS), which includes all other animals of unknown relationships, are listed here but are not treated elsewhere in this paper.

For stapes, otic capsule and cheek, illustration quality scores are given. These subjective assessments of illustration quality may be interpreted as follows: 0, no illustration; 1, one view with no detail; 2, one or two views with some detail; 3, two or more views with highly detailed drawings.

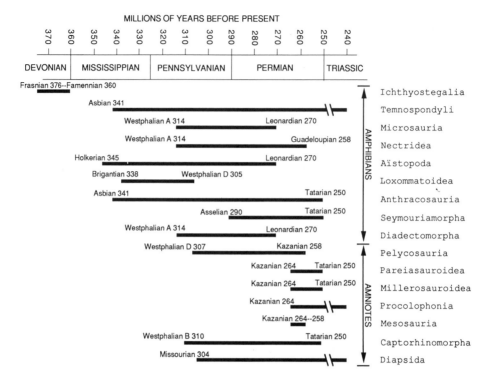

FIGURE 19.5. The major groups of Paleozoic tetrapods considered in this chapter. Contents of the taxa are described in the Appendix. The time range over which each group is presently known to exist is indicated by a black bar. Range is considered to extend through the entire stage (or substage for the Westphalian) in which a representative of the group occurs. Continuity of each group is inferred; a time continuous series of fossils is not available for any of the groups. Range data compiled from Baird and Carroll (1967); Barbarena et al. (1985); Beaumont (1977); Bolt et al. (1988); Campbell and Bell (1977); Carroll and Baird (1972); Carroll and Gaskill (1978); Carroll and Winer (1977); Ivakhnenko (1979); Kitching (1977); Milner (1980a and b); Olson (1962); Panchen and Smithson (1988); Reisz (1986); Ricqlès and Taquet (1982); Smithson (1985); Spjeldnaes (1982); Thommasen and Carroll (1981); Wellstead (1982). Correlations primarily from Haq and Van Eysinga (1987), with substages of the Westphalian from Harland et al. (1982).

Paleozoic tetrapod remains has been fairly steady for 60 years (and more) with World War Two having the expected depressing effect.

2.3 The Major Taxa

Despite the fact that new genera have been discovered steadily for over a hundred years the number of major Paleozoic tetrapod types known (higher taxa) was established fairly early and has not changed dramatically as a consequence of the discovery of new fossils over the last half century. One can think of higher level taxa as "shape categories" that are relatively easy to distinguish. For example, any child with a moment's instruction could successfully sort 99% of the world's living amphibians into the three orders: salamanders, frogs, and caecilians. In Paleozoic tetrapods, excluding certain derived Late Permian groups not considered in this chapter (see Appendix), there are 16 generally recognized "shape categories" and a few unclassifiable orphans. Nine of these higher taxa are recognized as amphibians and seven as amniotes. Not all are of equivalent taxonomic rank: some are orders, some are superfamilies, etc., and these designations change with shifting notions of relationship. All, however, are commonly recognized currency in paleontology.

Some information about these sixteen groups is displayed in Table 19.2 and in Figure 19.5, which shows the time span over which each of the groups is presently known to have existed, the number of genera each contains and some statistics indicating scholarly activity over the past 60 years. Most of the groups arise and go extinct (as far as we know) entirely within the Paleozoic. Three, the temnospondyl amphibians and the procolophonid and diapsid amniotes, continue into the Mesozoic. The absence of organisms from the first 15 million years of Mississippian is glaring. Each of these groups has been studied with approximately equal intensity in proportion to the number of genera encompassed. This does not mean that our knowledge of each is equal (Table 19.2). In general, larger groups such as the temnospondyls and pelycosaurs are better known simply because the sheer volume of material to study is larger.

These 16 taxa are the major types of organisms in which the early experiments in terrestrial hearing are taking place. To understand the origin and early evolution of terrestrial hearing two things must be known about these taxa: The evolutionary relationships among them, and the primitive condition of the ear in each. These topics will be taken up in the next two sections.

3. Evolutionary Relationships

The phylogeny used here is based on cladistics (= phylogenetic systematics), today the dominant philosophy in systematics. Cladistics offers both a theoretically based definition of, and practical procedures for discovering, groups that are comprised of all descendants of a common ancestor — i.e., monophyletic groups. To reconstruct phylogeny, cladistics relies solely on derived characters; primitive characters tell us nothing about relationships. A derived character shared by two or more taxa (whether species, families, or any other rank in the Linnean hierarchy) is a synapomorphy of those taxa. Cladistics uses synapomorphies to discover monophyletic groups, which in genealogical terms are groups whose members are more closely related to one another than they are to members of any other group. Two monophyletic groups of whatever taxonomic rank that are more closely related to each other than either is to any other group, are called sister groups (= sister taxa). In Figure 19.6, for instance, aïstopods and nectrideans are sister taxa. These phylogenetic relationships are diagrammed as cladograms, which show

branching patterns but do not specify ancestor–descendent relationships.

Any practicing systematist knows that application to real organisms of these apparently straightforward procedures is liable to be complicated and difficult. Especially in fossils, it may be hard to identify many — or any — usable derived characters. Also, characters may conflict, in the sense that different sets of characters may diagnose different sister groups, and thus specify different cladograms. Such character conflicts are generally resolved by applying some rule of parsimony. Parsimony in this context means that the preferred cladogram is the one requiring the smallest number of character changes. This is an operating rule which does not imply either that the most parsimonious cladogram is necessarily correct, or that the course of evolution is parsimonious. If few characters and few taxa are involved, it is possible to examine every cladogram and choose the shortest one. As the character x taxon matrix becomes larger, the number of possible cladograms quickly increases beyond the point where exhaustive examination is possible even using computers. A number of computer algorithms have thus been developed that *estimate* the most parsimonious cladogram without examining all possibilities.

There is as yet no generally agreed cladogram of major tetrapod groups, and thus no generally accepted cladistic classification. Indeed, there is no well-researched single published cladogram covering all tetrapod groups that have Paleozoic representatives. The cladogram in Figure 19.6 is thus necessarily a composite. Sources were Smithson (1985) and Panchen and Smithson (1988) for anamniote groups, and Gauthier, Kluge, and Rowe (1988a,b) for amniotes. Group names and contents used here are mostly those given by Carroll (1988), with modification (see Appendix for details).

There is little reason to doubt that most of the groups we use are monophyletic. Pelycosauria and Temnospondyli are exceptions. Given the (traditional) contents of these groups as specified by Carroll, both are paraphyletic. In other words, while they include *only* descendents of a (hypothesized) common ancestor, they do not include *all* of those descendents. So long as this is understood, it presents no serious problem here, because each of these groups generally includes the most primitive

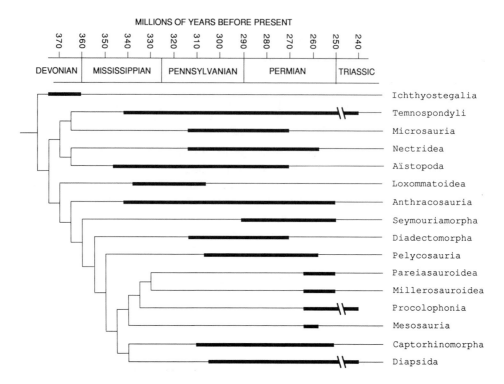

FIGURE 19.6. Cladogram showing stratigraphic ranges and hypothesized relationships of all major groups of tetrapods considered in this chapter that have Paleozoic representatives. Cladogram modified from those presented by Panchen and Smithson (1988) and Gauthier et al. (1988a) (see text). Contents of each terminal group on the cladogram specified in Appendix B. An interrupted range bar, as for Temnospondyli, indicates extension of the group's range into the Mesozoic and possibly beyond. Sources of range data and correlations as for Figure 19.5.

and earliest members of the larger, monophyletic group to which they belong.

Given a robust hypothesis of relationships—i.e., a well-corroborated cladogram—it would be simple to produce an equally robust hypothesis of the morphological evolution of otic components. It is not yet possible to do this, because relationships both between and within our terminal groups are still incompletely analyzed. The interrelationships of amniote groups proposed by Gauthier et al. (1988a) are supported by analysis of 112 characters using a computer program (PAUP) that can reliably estimate the most parsimonious cladogram. Even so, substantial uncertainties remain. For example, in discussing the relationships of mesosaurs, millerettids, pareiasaurs and procolophonians, Gauthier, Kluge and Rowe (1988a) state (p. 145) that ". . . we doubt the accuracy of that part of our cladogram." In contrast, the cladograms proposed by Panchen and Smithson (1988) depend on at most 31 characters by our count. Twelve characters are assigned to the basal node, and the remaining nodes in their cladograms each rest on one to three characters. This would be little cause for concern if each character could be considered reliable. As it is, the authors emphasize the tentative nature of their conclusions, even to presenting two different cladograms for some taxa because they are unable to agree on which characters to use.

Formation of our composite cladogram was not quite as simple as attaching the amniote cladogram to the end of the anamniote one. The two sets of authors differ radically in their placement of Ichthyostegalia, *Crassigyrinus* and the Loxommatoidea. Panchen and Smithson (1988) associate *Crassigyrinus* and the Loxommatoidea with Anthracosauria as first and second outgroups, respectively. We have basically followed their arrangement, by incorporating *Crassigyrinus* into our Anthracosauria, with loxommatoids as outgroup. On the other hand, we have followed Gauthier, Kluge, and Rowe (1988a) and Gauthier et al. (1989) in considering Ichthyostegalia as the sister group of all other tetrapods.

Relationships *within* our terminal groups are generally not well understood. Pelycosaurs and other synapsids have been the subjects of several cladistic studies (e.g., Hopson and Barghusen 1986; Reisz 1986; Gauthier et al. 1988b), and pelycosaur interrelationships are certainly well enough established for present purposes. Interrelationships of temnospondyl groups, however, are very poorly worked out. Cladograms of subsets of Temnospondyli have been published, but there is not a single published well-supported cladogram for the group as a whole.

There is a rough consensus regarding the relationships of extant tetrapod groups to the terminal groups in Figure 19.6. For amniotes, that consensus is reinforced by the conclusions of Gauthier et al. (1988a,b). Thus the sister group of turtles is Captorhinomorpha, and the sister groups of other living reptiles and of birds are found within the Diapsida. The sister group of mammals is most closely related to a subgroup of our Pelycosauria. There is greater diversity of opinion regarding relationships of the living amphibians. The usual working hypothesis, adopted here, is that they comprise a monophyletic Lissamphibia, which is the sister group of dissorophoid temnospondyls—see, for example, Milner (1988) and references therein.

4. Overview of Otic Structure

4.1 The Scope of the Data

Some results from our survey of otic morphology illustrations are summarized in Table 19.2. For the stapes and otic capsule, every "occurrence" for each species is summarized in the table; for the cheek region only the data for the best occurrences are given. An occurrence may include more than one figure: for example, two or three views of the subject, in one plate or scattered through the chapter.

Since 1928, there have been 156 occurrences of stapes illustrations for the taxa considered here. Many of these represent separately published re-illustrations of particularly good specimens so that in actuality the stapes of 85 genera have been illus-

trated. This is 21% of the genera included in this study, an unexpectedly high and encouraging figure. However, most of these are informative only to the degree that they indicate presence of a stapes in the specimen under study. That is, they are either a single, simple outline drawing from one view and showing no detail (28 occurrences where stapes illustration quality score = 3) or are a couple of simple drawings showing only little detail (33 occurrences where stapes score = 2). The former are most often occipital views, the latter occipital plus ventral. Many of the specimens represented by these less informative illustrations hold promise for the future, though, for the author is in fact indicating the presence of a stapes in the matrix of the posterior underside of the skull, and further preparation of the specimen may expose informative morphology. In other cases the stapes may require relatively little work to enable full examination. This is because the stapes is a relatively minor component of the tetrapod skull in the view of many authors. They will expend considerable descriptive effort on the skull roof, cheek, palate and jaws of a good specimen and give only a nod to the stapes. Thus, all but the (possibly excellent) stapes will be well illustrated and described. For these reasons there is, in our view, more promise than presently useful information in the bulk of existing illustrations.

There are 24 genera represented by quite informative illustrations of the stapes, however (stapes score = 1). This represents about 7% of the genera illustrated since 1928 and most of these have been published in the last 20 years (see legend for Figure 19.7). Figure 19.7 illustrates the distribution of these well illustrated stapes by taxon and age. Ten of the sixteen taxa have no well illustrated stapes (and in four of these the stapes is not known at all, Table 19.2). For the remaining seven taxa, the distribution of well known stapes is patchy. Two large monographs account for about a third of the good illustrations: Carroll and Gaskill (1978) for the Microsauria and Romer and Price (1940) for the Pelycosauria. This is encouraging, for it indicates that concentrated effort on one group is likely to yield useful material. Finally, only four genera older than the Permian have well illustrated stapes. This does not make it easy to understand the origin and early evolution of the ear. We are fortunate indeed that the stapes of *Greererpeton*,

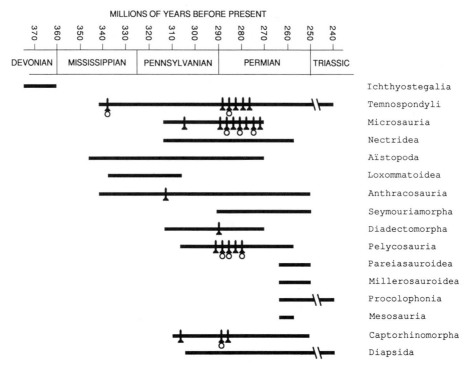

FIGURE 19.7. Distribution by taxon and time of the best illustrated stapes (rod) and otic capsules (circle). Each icon or icon pair indicates a genus for which excellent illustrations exist. Icons side-by-side indicate excellent illustrations for more than one aspect of a genus. For complete data see Table 19.2. Morphological data for stapes from Bolt and Lombard (1985), Boy (1988), Carroll and Baird (1968), Carroll and Gaskill (1978), Clack (1983), Eberth (1985), Godfrey (1989), Heaton (1979), Lombard and Bolt (1988), Olson (1965), Romer (1969), Romer and Price (1940), Romer and Witter (1942), Sawin (1941), Sigogneau-Russell and Russell (1974) and Smithson (1982). Morphological data for otic capsule from Carroll and Baird (1968), Carroll and Gaskill (1978), Eberth (1985), Heaton (1979), Romer (1969), Romer and Price (1940), Sawin (1941), Smithson (1982) and Sigogneau-Russell and Russell (1974).

a mid-Mississippian colosteid temnospondyl, is known from many excellent specimens and has been well illustrated (Godfrey 1989; Lombard and Bolt 1988; Smithson 1982).

Since 1928, there have been 173 occurrences of otic capsule illustrations for the taxa considered here. As with the stapes, many of these represent separately published re-illustrations of particularly good specimens, so that in actuality the otic capsules of 110 genera have been illustrated; that is 31% of the genera included in this study. Again, as with the stapes this is an encouraging figure. However, again, most of the illustrations are not very informative. One hundred and one of the occurrences are for genera where relatively spare outline drawings showing little detail are presented. The 68 least informative (otic capsule score = 3) are usually either occipital or ventral views in which the fenestra ovalis, even if present, would not ordinarily be visible. Thirty-three, where otic capsule score = 2, usually combine both a simple occipital and ventral view. Thus we have available only nine genera with well illustrated otic capsules where several views, each showing some detail, are presented (otic capsule score = 1). The distribution of these nine is shown in Figure 19.7.

Well-preserved and illustrated otic capsules are known for only four of the higher taxa considered here. As with the stapes, most are for Permian genera, only that of *Greererpeton* coming from the first 80 or so million years of tetrapod existence (Smithson 1982). As with the stapes, the monographs of Carroll and Gaskill (1978) and Romer and Price (1940) are major contributors to our

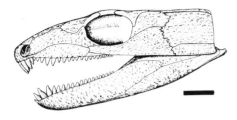

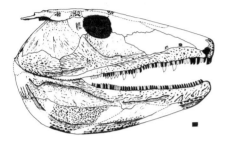

FIGURE 19.8. The two major cheek morphologies in Paleozoic tetrapods as seen in lateral view of restored skulls. Upper: Straight or caudally convex caudal cheek margin. *Eocaptorhinus laticeps*, a captorhinid amniote from the Early Permian of Texas and Oklahoma, USA (from Heaton 1979). Lower: Temporal notch. *Pholiderpeton scutigerum*, an anthracosaur amphibian from the Pennsylvanian of Great Britain (from Clack 1987). The distribution of these cheek types in the higher taxa considered in this chapter may be found in Table 19.3. Scale bar = 1 cm.

TABLE 19.3. Otic Characters of Paleozoic Tetrapod Taxa. Primitive states = 0, derived = 1

Characters	Temporal Notch	Number of Stapedial Heads	Stapes Direction	Fenestra Ovalis
Taxa				
Osteolepiformes	0	0	0	0
Ichthyostegalia	1?	1	1?	?
Temnospondyli	0	0?	1?	1
Microsauria	0	0?	0?	1
Nectridea	0	?	?	?
Aïstopoda	0	1	?	1
Loxommatoidea	1	?	?	1
Anthracosauria	1	1	1?	1
Seymouria-morpha	1	1?	1?	1
Diadectomorpha	0?	0?	0	1
Pelycosauria	0	0	0	1
Pareiasauroidea	0	?	?	1
Millerosauroidea	0	1	?	1
Procolophonia	0	1	?	1
Mesosauria	0	?	?	?
Captorhino-morpha	0	0	0	1
Diapsida	0	0	0	1

knowledge of otic morphology in Paleozoic fossils: in only 9 of 368 genera do we have good information on both the otic capsule and stapes. A reasonable interpretation of otic evolution, in which we can have some confidence, will require far better data.

In the following section we give an overview of the basic types of otic structures represented in the illustrations examined for this study. We limit discussion to those few features which are known for most of the higher taxa of the Paleozoic, and so will deal with very simple aspects of structure.

4.2 Otic Structure

4.2.1 The Cheek

Figure 19.8 illustrates the two types of cheek morphologies found in Paleozoic tetrapods. The illustrations are of restored skulls and thus the overall shape is a best estimate of the original form of the skull. As seen in lateral view, the posterior margin of the cheek may be relatively straight as seen in the captorhinomorph *Eocaptorhinus*. On the other hand the margin can be incised by a temporal notch as in the anthracosaur *Pholiderpeton*. The temporal notch can be rounded and more deeply incised as in many temnospondyls and in extreme cases may become inclosed posteriorly to form a subcircular opening in the skull as in the Permian temnospondyl *Cacops* (Bolt and Lombard 1985).

The primitive condition of the cheek with respect to these two morphologies for the 16 tetrapod taxa considered here is shown in Table 19.3.

4.2.2 Stapes Heads

Figure 19.9 illustrates the two configurations of proximal heads found on the known stapes of Paleozoic tetrapods. The specimens from which the illustrations are made are beautifully preserved and undistorted, and more than one specimen of each is known. Stapes may have one proximal head abutting the otic capsule as in the Mississippian colosteid temnospondyl *Greererpeton*. Commonly the head is unfinished bone as in *Greererpeton*, but may be finished as in the Permian temnospondyl

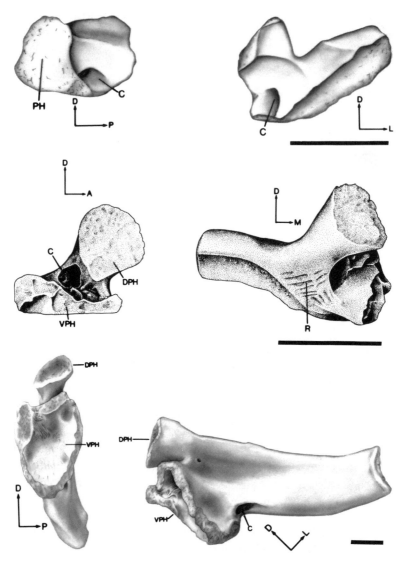

FIGURE 19.9. The three major stapes morphologies known for Paleozoic tetrapods. For each, a medial (left) and posterior (right) view is given. Upper: **single proximal head**. Right stapes (cast, Carnegie Museum of Natural History 11130) of *Greererpeton burkemorani*, a colosteid temnospondyl amphibian from the Mississippian of West Virginia, USA. Middle: **double proximal head, dorsal head in fenestra ovalis**. Left stapes (University of Michigan Museum of Paleontology 11670) of *Trimerorhachis* cf. *T. insignis*, a temnospondyl amphibian from the Early Permian of the four corners region and Texas and Oklahoma, USA. Bottom: **double proximal head, ventral head in fenestra ovalis**. Right stapes (American Museum of Natural History 4155) of *Ophiacodon* sp., a pelycosaur amniote from the Early Permian of Texas and Oklahoma, USA. The distribution of these stapes types in the higher taxa considered in this chapter may be found in Table 19.3. Abbreviations: A, anterior; C, stapedial canal; D, dorsal; DPH, dorsal proximal head of stapes; L, lateral; M, medial; P, posterior; VPH, ventral proximal head of stapes. Scale bar = 1 cm. All from Lombard and Bolt 1988.

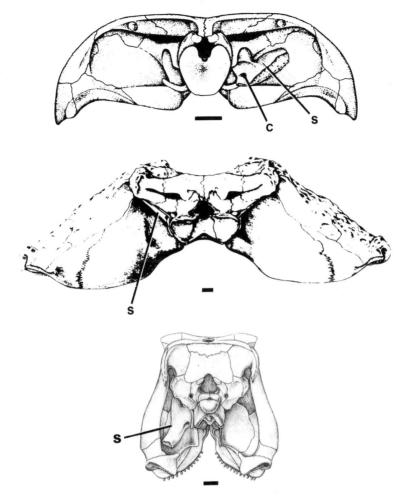

FIGURE 19.10. The two extremes in stapes orientation occurring in Paleozoic tetrapods as seen in occipital view of restored skulls. Upper: **Ventrolateral orientation**. *Ophiacodon uniformis*, an early Permian pelycosaur amniote from Texas, USA. In *Ophiacodon*, the distal end of the stapes is directed to the internal surface of the quadrate bone (from Romer and Price 1940). Middle: **Dorsolateral orientation**. *Greererpeton burkemorani*, a mid-Mississippian colosteid temnospondyl amphibian from West Virginia, USA. In *Greererpeton*, the anterior surface of the stapes abuts against the inside of the epipterygoid bone (from Smithson 1982). Lower: **Dorsolateral orientation**. *Eryops megacephalus*, an Early Permian temnospondyl amphibian from North America and Europe. In *Eryops*, the distal end is directed to an otic notch in the dorsolateral skull roof (reproduced from Sawin 1941). The distribution of these stapes orientations among the higher taxa considered in this chapter may be found in Table 19.3. (c) stapedial canal; (s) stapes. Scale bar = 1 cm.

Doleserpeton (Bolt and Lombard 1985). Alternatively, the stapes may have two heads separated by an incisure. Two examples are shown in Figure 19.9. In that from the temnospondyl, *Trimerorhachis*, it is the dorsal of the two heads which fits into the fenestra ovalis of the otic capsule. The ventral head articulates with the wall of the capsule ventral to the fenestra. In the other example, from the Permian pelycosaur *Ophiacodon*, it is the ventral head which is associated with the fenestra ovalis and the dorsal head forms an articulation with the skull bones dorsal to the otic capsule. We have previously discussed these interesting observations and the implications of variation in fenestral head number (Lombard and Bolt 1988) and will not comment further here. Distribution of stapes with the head number we consider primitive for the higher taxa discussed in this chapter is shown in Table 19.3.

4.2.3 Stapes Orientation

Figure 19.10 contains occipital views of skulls which display the two extremes of stapes orientation found in Paleozoic tetrapods. At one extreme, the distal shaft of the stapes may be oriented dorsolaterally as in *Greererpeton*. In this animal some of the anterior surface of the stapedial shaft abuts against the posterior surface of the epipterygoid bone; there is no temporal notch in the skull. In the ancient Texan *Eryops*, on the other hand, the similarly oriented but more rod-like stapes is directed to a definite temporal notch in the skull. At the other extreme, the stapedial shaft is directed ventrolaterally toward the medial surface of the quadrate bone near the jaw joint as in the pelycosaur *Dimetrodon*. In no animal with this stapes orientation is the skull incised by a temporal notch. Distribution of the condition considered primitive for the higher taxa discussed here is shown in Table 19.3.

4.2.4 Fenestra Ovalis

Figure 19.11 illustrates lateral views of the otic capsule and braincase in three Paleozoic tetrapods, all well-preserved specimens. The stapes is shown in situ for one, *Eocaptorhinus*. In *Greererpeton* and *Eocaptorhinus* the fenestra is reconstructed as being rather large, about one centimeter in diameter. This is a fairly common type of reconstruction in Paleozoic tetrapods where enough of the fenestra is thought to be preserved to enable an estimate. This is relatively huge compared to the fenestra of modern taxa. The morphology of the fenestral rim, although a fenestra is universally presumed to be present, is not known at all for three of our taxa: Ichthyostegalia, Nectridea; and Mesosauria. In most others it is poorly known, either because it is only partly preserved or prepared or because, in many Paleozoic tetrapods, much of the capsule wall was apparently formed in cartilage and hence not preserved at all. In Paleozoic animals the capsule, where well preserved (remember there are relatively few), is formed of the prootic and opisthotic bones with an occasional contribution from the supraoccipital dorsally. Natural internal casts of the capsule are known for *Eryops* (Sawin 1941) and Edops (Romer and Edinger 1942), and a reconstruction from serial ground sections has been made for *Diadectes*, a Permian diadectomorph (Olson 1965). These data have not proven of

much use for reconstructing the morphological or functional evolution of the ear and hearing. We would profit greatly from a concentrated effort to prepare the otic capsules of selected well-preserved skulls – which *are* available.

5. Evolution of the Ear

We are interested here in tracing the evolution of otic components, rather than in reconstructing phylogeny. We therefore take as a given the "consensus" cladogram discussed above, based on the work of others. In the discussion that follows, it will often be necessary to discuss ambiguities and alternatives, and at this stage we must confess that we are still far from being able to resolve many of the most basic issues. Despite these uncertainties, the phylogeny of otic components implied by the "consensus" cladogram is at least indicative of our current understanding, and we believe that it has heuristic value.

Table 19.3 shows the assignment of character states to each of our terminal taxa. The outgroup consisted of "osteolepiforms," which here means *Eusthenopteron*. Outgroup character states were assigned to the outgroup node and scored as 0. The most primitive state of each character found in any representative of a given group, was considered as primitive for that group. This procedure was necessary here because relationships within most groups are uncertain. As it happens, with the exception of temnospondyls assignments made under this rule are consistent with the best available estimates of each group's internal structure. It is ironic that relationships within pelycosaurs are reasonably well understood, because their otic components as we define them are invariant and thus would imply the same primitive states in any case.

We traced character state evolution on our "consensus" cladogram (Fig. 19.6), using the MacClade computer program (version 2.1), written by Wayne and David Maddison, Harvard University. If a character state is unknown for a taxon, MacClade will assign to that taxon the most parsimonious state for a given cladogram. If no single character can be assigned with maximum parsimony to a branch, MacClade will note that assignment is equivocal. For example, in Figure 19.13 the number of stapedial heads in the branch leading to

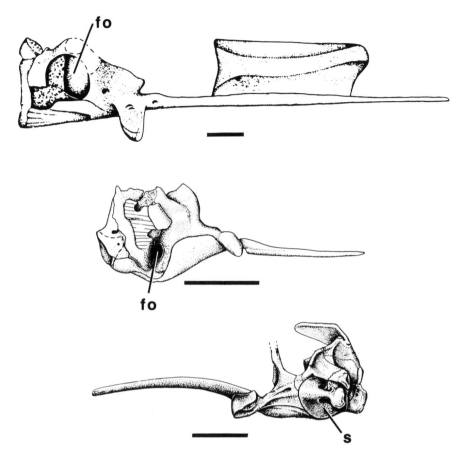

FIGURE 19.11. Lateral views of the braincase and otic capsule as restored in some Paleozoic tetrapods. Upper: *Greererpeton burkemorani*, a colosteid temnospondyl from the mid-Mississippian of West Virginia, USA. Anterior is to the right (from Smithson, 1982). Middle: *Pantylus cordatus*, a microsaur amphibian from the Early Permian of Texas, USA. Anterior is to the right (from Carroll and Baird, 1968). Lower: *Eocaptorhinus laticeps*, a captorhinid amniote from the Early Permian of Texas and Oklahoma, USA. The stapes is in situ covering up the fenestra ovalis. Anterior is to the left (from Heaton, 1979). (fo) fenestra ovalis; (s) stapes. Scale bar = 1 cm.

mesosaurs is equivocal, i.e., the same number of character state transformations is required (1, in this case) whether a single-headed stapes evolved before or after the origin of mesosaurs. In the absence of any reason to think that reversal is impossible, we assume that each of our otic-component characters is reversible. Thus, for example, a single-headed stapes could evolve into a double-headed one and back again, as many times as necessary to achieve maximum parsimony.

We will consider the phylogeny of each otic component in turn from the "outside" in, thus beginning with the cheek. The most discussed aspect of this otic component has been variously referred to in whole or in part as the "otic notch," "temporal notch," "squamosal embayment," or "quadrate angle," and this is not necessarily an exhaustive list. As an otic notch (the most widely used term) this structure was long assumed to carry a tympanum. After the assumption of early and unique evolution of a tympanum came into question, the otic notch was explicitly defined in both morphological and functional terms (Lombard and Bolt 1988). That definition is difficult to apply to some phylogenetically critical taxa because it requires information about stapes orientation which is either missing or equivocal. We therefore adopt the strictly topological definition of Godfrey, Fiorillo, and Carroll (1987, p. 803) for this structure: "a groove, notch, or embayment between the skull

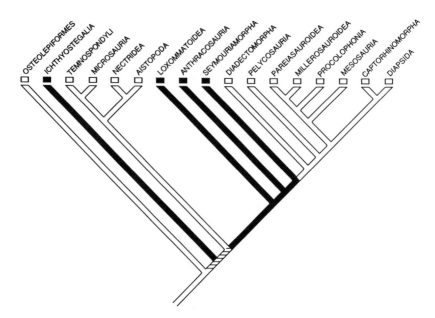

FIGURE 19.12. Evolution of the temporal notch. Cladistic relationships shown here (same as in Fig. 19.6) are treated as a given, as is the distribution of character states shown in Table 19.3. The state in osteolepiforms (temporal notch absent) is considered ancestral for tetrapods. Branches and taxa with this primitive state are white. Presence of the derived state (temporal notch present) is shown by black shading. A rectangle above a branch terminus indicates that we were able to assign a character state (present or absent) to that taxon, and the rectangle is the appropriate shade for that state.

table and cheek in the immediate vicinity of the squamosal, supratemporal, and tabular." We will refer to it as a "temporal notch," following Gauthier, Kluge, and Rowe (1988a), because that term is morphologically descriptive but free of functional implications.

The temporal notch is a simple structure that varies widely among and within our terminal taxa. Morphologists of good will can therefore disagree on its presence or absence, and on which state is primitive for tetrapods. Carroll (in Panchen 1985) has suggested that the temporal notch is homologous with the spiracular opening of osteolepiform fish, coupling this suggestion with a scenario of function. On strictly morphological grounds, we are presently unconvinced that the small spiracular opening, lying in front of the posterior border of the skull and therefore completely surrounded by bone, is the homologue of the tetrapod temporal notch. We therefore assign state 0 (no temporal notch) to osteolepiforms and to the outgroup node. We here consider the temporal notch to be derived for temnospondyls, despite its widespread occur-

rence even in early members of the group. This conclusion follows from the absence of a temporal notch among the Colosteidae. Colosteids (Middle Mississippian-Middle Pennsylvanian) are usually considered to be either very primitive temnospondyls, as we do here, or to be the primitive sister group to temnospondyls or temnospondyls+ microsaurs (see Panchen and Smithson 1988). If colosteids are excluded from temnospondyls, then by our convention absence of the temporal notch in such Triassic groups as the Brachyopidae still forces us to score this condition as primitive even though we strongly suspect that absence of a temporal notch in this family is in fact derived.

The temporal notch has clearly had a complex history (Figure 19.12). It has been lost and redeveloped at least once among our terminal taxa, so that the temporal notch of ichthyostegids is not homologous with that of, say, anthracosaurs. This is so whether or not the notch is primitive for temnospondyls, and even if we follow Gauthier, Kluge, and Rowe (1988a) in considering the temporal notch of osteolepiforms as homologous with that of

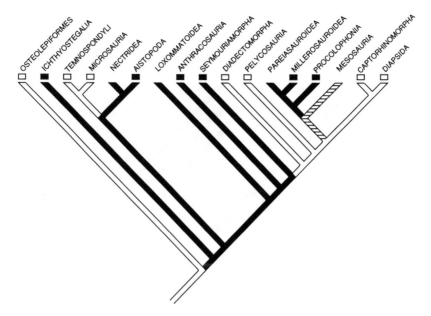

FIGURE 19.13. Evolution of number of stapedial heads. Cladistic relationships shown here (same as in Fig. 19.12) are treated as a given, as is the distribution of character states shown in Table 19.3. The most parsimonious distribution of temporal notch character states is then determined by the MacClade program. The state in osteolepiforms (two stapedial heads) is considered ancestral for tetrapods. Branches and taxa with this primitive state are white. Presence of the derived state (single stapedial head) is shown by black shading. A pattern of horizontal lines indicates branches whose state is equivocal, i.e., either presence or absence would be equally parsimonious. A rectangle above a branch terminus indicates that we were able to assign a character state (present or absent) to that taxon, and the rectangle is the appropriate shade for that state. Absence of a rectangle indicates that we were unable to assign a state to that taxon, in which case MacClade calculates the state. This can be considered a prediction of the state to be expected in this taxon if the given cladogram correctly represents the relationships of the included taxa.

ichthyostegids. In this case its presence would be primitive for tetrapods, and our cladogram would be one step shorter. However, the cladogram still forces the conclusion that the temporal notch has been lost and redeveloped at least once.

The stapes has been, for fossil tetrapods, the most discussed and debated of our otic components (for example, Lombard and Bolt 1988). Much of the recent discussion has concerned the orientation of the stapedial shaft and the number of proximal heads, which we treat as two characters. Most of the discussants agree that the hyomandibula of osteolepiforms, with two proximal heads and the shaft directed ventrolaterally toward the quadrate, represents conditions in the outgroup (scored 0). Derived states are a single proximal head, and lateral or dorsolateral orientation of the shaft (scored 1). The hypotheses of stapedial evolution implied by our character state assignments are shown in Figures 19.13 and 19.14. Two origins of the single-headed stapes are shown, and two reversions to a ventrolateral orientation. The most striking aspect of these cladograms is the implication that the two-headed, ventrolaterally oriented stapes of early reptiles is not primitive, as has long been thought, but instead is derived and thus convergent on the primitive condition. This is particularly interesting in view of Clack's (1989) discovery of a single-headed stapes in the ichthyostegalian *Acanthostega*. Clack suggests that such a stapes was primitive for tetrapods, including reptiles. The majority of workers in this area, including ourselves, consider the stapes of early reptiles to be primitive in orientation and number of heads. Nonetheless, our results if taken at face value do point to the same conclusion reached by Clack.

These hypotheses of stapedial evolution are not supported by a great deal of evidence. In the

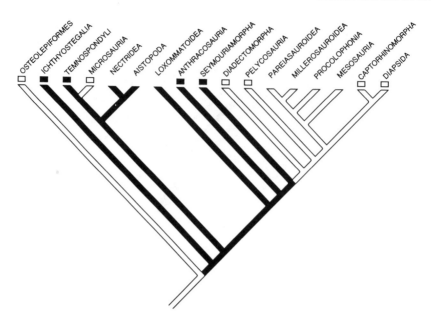

FIGURE 19.14. Evolution of stapedial orientation. Cladistic relationships shown here (same as in Fig. 19.6) are treated as a given, as is the distribution of character states shown in Table 19.3. The most parsimonious distribution of stapedial orientation character states is then determined by the MacClade program. The state in osteolepiforms (hyomandibula/stapes ventrally directed) is considered ancestral for tetrapods. Branches and taxa with this primitive state are white. Presence of the derived state (stapes directed laterally or dorsolaterally) is shown by black shading. A pattern of horizontal lines indicates branches whose state is equivocal, i.e., either presence or absence would be equally parsimonious. A rectangle above a branch terminus indicates that we were able to assign a character state (present or absent) to that taxon, and the rectangle is the appropriate shade for that state. Absence of a rectangle indicates that we were unable to assign a state to that taxon, in which case MacClade calculates the state. This can be considered a prediction of the state to be expected in this taxon if the given cladogram correctly represents the relationships of the included taxa.

branch leading to reptiles, for example, the following weaknesses are apparent on closer examination: (1) The stapes of loxommatids is entirely unknown. (2) The seymouriamorph stapes is known from only a few specimens, where it is apparently incompletely ossified (White 1939; Bystrow 1944; Watson 1954). It is not attached to the margin of the fenestra ovalis in any of these cases, so its orientation is not definitely known. Reconstructions that show the stapes directed into the otic notch are conjectural. Moreover, Smithson (1982) points out, that the anomalous occiput reconstructed by Bystrow (1944) for *Kotlassia* may have been based at least in part on a temnosondyl. (3) Only a single anthracosaur stapes is known (Clack 1983), and that from a member of one of the most derived anthracosaur groups (Clack 1987). (4) In the branch leading to Temnospondyli we are forced, absent a good cladogram of the group, to

consider the two-headed stapes of the Trimerorhachidae (Lombard and Bolt 1988) as primitive for temnospondyls. With only one exception (in a Triassic group), all other temnospondyls, including the Colosteidae, have a single-headed stapes. We are really forced by ignorance rather than any strong positive evidence to accept the trimerorhachid type of stapes as primitive for temnospondyls, and we strongly suspect that it is derived (see discussion in Lombard and Bolt 1988). In view of all these uncertainties, the only reasonable statement possible at this point is that the history of the stapes may just as well have been either more or less complicated than suggested by our figures.

The fenestra ovalis has been reported in every one of our terminal taxa except ichthyostegalians, nectrideans, and mesosaurs. MacClade predicts its presence in the latter two, while the condition in ichthyostegalians is equivocal. Our survey thus

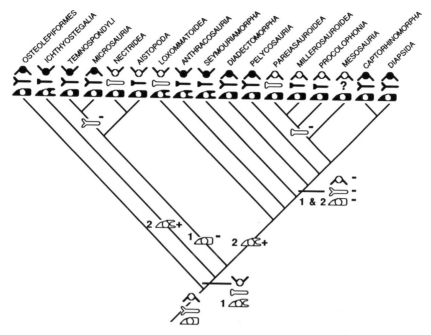

FIGURE 19.15. Most parsimonious hypothesis possible at present for the evolution of the ear region in Paleozoic tetrapods. Cheek morphology, stapes orientation, stapes head configuration and the fenestra ovalis are represented by icons. Icons at the end of a branch indicate what is presently believed to be the primitive condition for the higher taxon named: **solid** where the ear region is known from fossils and **outline** where the condition is predicted (Figs. 19.12–19.14) from presently available data. Icons embedded in the clado- gram indicate where transformations in the states of the features considered here would have to take place in order to minimize reversal and convergence. Reversal is indicated by (−) and convergence by (+). There are two equally parsimonious possibilities for evolu- tion of the posterior margin of the cheek (temporal notch). These possibilities are represented by skull icons numbered (1) and (2). The icon numbered "1 & 2" indicates that the same change in cheek morphology is required in either case.

appears to confirm the widespread belief that a fenestra ovalis is present in all tetrapods, and can in fact be used to diagnose Tetrapoda (Gaffney 1979). However, while we do not wish to claim that the fenestra ovalis is definitely absent in any of our terminal taxa, the fact is that the published evi- dence for its presence is often dubious. The lateral wall of the braincase is often poorly ossified in the vicinity of the fenestra, but the universal expecta- tion of a fenestra ovalis in tetrapods means that any possible segment of its rim will usually be parlayed in reconstructions into an undoubted fenestra. The resulting "reconstructed" fenestra ovalis is often startlingly large by comparison with the fenestra seen in extant tetrapods. That expectation is liable to exceed evidence is further shown by the near- universal failure to describe the surface texture of putative remnants of the fenestral rim. In the majority of cases, it is thus impossible to tell whether or not the preserved portion of the rim is finished bone. In other words, it is usually impossi- ble to determine whether or not any preserved opening was entirely filled by cartilage in life. One of the few exceptions is the Lower Permian aïsto- pod *Phlegethontia*, whose braincase preserves a fenestra ovalis with a completely ossified rim. From McGinnis' (1967) illustrations, the rim appears to be finished bone, as well.

Figure 19.15 summarizes the most parsimonious hypothesis possible at present for evolution of cheek morphology, stapes orientation and number of heads, and the fenestra ovalis.

The gross morphology of our otic components presumably has functional implications for hear- ing. Ideally, these could be discussed as terms in an equation, which would surely include such parameters as areas of tympanum, footplate, and fenestra ovalis, and mass of bony stapes. No practicable equation of this sort has been developed even for living animals. If one did exist, our survey above shows that it would likely be impossible, in the present state of knowledge, to obtain the necessary values in more than a few early tetrapods. Morphologists who claim, for instance, that on the basis of the size of the stapes a given fossil tetrapod did/did not have an "impedance-matching ear" convey by that phrase a spurious impression of quantitative precision. Interpretation of otic function in early tetrapods is necessarily qualitative rather than quantitative, and often must be approached indirectly. We consider here two of the most discussed aspects of otic function in early tetrapods: the existence of a tympanum, and evidence of ability to perceive airborne sound.

There is little direct evidence in early tetrapods for the existence of a tympanum. A change in surface texture of bone in passing from the cheek to the occiput might be so interpreted. Similarly, the occasional presence of a shallow groove on the occipital aspect of the squamosal may mark a tympanic attachment. No lengthy discussion is needed to make the point that such features are open to other plausible interpretations that do not require the presence of a tympanum. On the other hand, we feel that presence of an otic notch *is* evidence for auditory function. We have defined the otic notch as "a concavity in the cheek or lateral part of the dorsal skull roof, toward which the distal end of the stapes is directed" (Lombard and Bolt 1988, p. 42). Lateral or dorsolateral orientation of the stapes is derived for tetrapods, and an auditory function is the most plausible explanation we can think of for this arrangement. A tympanum may or may not have been present even in species with an otic notch, since it is not required for sound perception.

The case for existence of both a tympanum and aerial sound perception is particularly strong in the Paleozoic temnospondyls — dissorophoids — most closely related to living amphibians. We have discussed (Bolt and Lombard 1985) the otic evidence for relationship of dissorophoids and frogs, the only living amphibians with a tympanic ear. Part of this evidence included our identification of the dissorophoid dorsal quadrate process with the anuran tympanic annulus, a cartilaginous ring spanned by the tympanum. The dorsal quadrate process

carries a clearly marked unossified strip that we interpreted as a likely tympanic attachment area. It thus seems very probable that dissorophoids had both a tympanic ear and some aerial hearing ability. It is a logical extension to suggest that other temnospondyls with an otic notch and similar stapedial morphology and orientation had both a tympanum and some ability to hear airborne sounds.

Acceptance of this hypothesis suggests that even such stapes as the 7-cm-long example known in the large Paleozoic temnospondyl *Edops* must have functioned in hearing. The *Edops* stapes is not uniquely large among temnospondyls, but it is far larger than that of any living tetrapod. The stapes of very early tetrapods such as the Upper Devonian *Acanthostega* is indeed "massive" (Clack 1989) compared with that of living tetrapods, but it is small relative to that of *Edops*, *Eryops*, and some other Paleozoic temnospondyls. On the basis of size (= mass) alone, there is no reason whatever to conclude that even the earliest tetrapod stapes known could not have functioned in aerial hearing.

It has long been accepted that the hyomandibula/stapes underwent an evolutionary transformation, perhaps more than once, from a predominantly suspensorial to a predominantly auditory function. The role of the stapes during this transformation has been much debated, and the debate has been renewed by recent discoveries in three early tetrapod species of stapes with virtually identical and unexpected morphology. One occurrence is in the Upper Mississippian temnospondyl *Greererpeton* (Smithson 1982; Godfrey 1989), one in the Middle Pennsylvanian anthracosaur *Pholiderpeton* (Clack 1983), and one in the Upper Devonian ichthyostegalian *Acanthostega* (Clack 1989). In each of these cases the ossified stapes has a single head and a flattened shaft which is directed laterally. *Acanthostega* and *Pholiderpeton* have distinct though not identical temporal notches; *Greererpeton* lacks a temporal notch. That the function of the stapes in these animals is ambiguous is perfectly exemplified by two responses to Clack's announcement of the *Acanthostega* stapes. Both are Letters to Nature, and appear on the same page (1990, p. 116). For Fritzsch, who argues for a tympanum in *Acanthostega*, "The exciting aspects of *Acanthostega* is the unmistakable connection of the stapes with the otic notch and with the ear as a whole, rather than with the jaws." Gottfried and Foreman, on the other hand, feel that "As no auditory function is demonstrated for the element, and as the interpretation presented [by Clack] is that it acted like a fish hyomandibular, we believe that this bone should be called a hyomandibular, not a stapes." We do not know which of these two views more closely approximates the truth. We do suspect, however, that neither view could be so firmly held if the authors reflected on one easily overlooked observation that applies to the stapes of all three species: In each case, the lateral and ventrolateral margins of the stapedial shaft are unfinished bone, and thus evidently continued in cartilage. Conditions in the outgroup, and the comparative anatomy of the otic region in both living and fossil tetrapods, make it plausible that there were cartilaginous continuations of the shaft downward to the quadrate, laterally toward the cheek (and tympanum?), or both or neither. Any reconstruction of these cartilaginous parts must be based almost entirely on hypotheses about the nature of the primitive tetrapod stapes, and about the "appropriate" stapes for a given taxonomic group. If we expected early tetrapods to have a *Greererpeton*-like (or *Acanthostega*-like) stapes, such a morphology could easily be restored in some "good" temnospondyls which current orthodoxy would endow with a "rod-like" stapes. For example, the stapes of a juvenile *Sclerocephalus haeuseri* figured by Boy (1988) is unfinished along both distal and ventral margins of the shaft. The ossified portion is more or less rod-like, but in fact this bone lends itself just as well to reconstruction with a plate-like shaft, as suggested by Boy. As with otic morphology in early tetrapods in general, so here consideration of the quality of the data should compel caution in its interpretation. We are not yet at the point where we can fully reconstruct the morphological, much less functional, history of the tetrapod ear, although recent discoveries have contributed significantly to progress toward that goal.

6. Summary and Conclusions

The nature and quality of morphological data obtainable from fossils are liable to be misunderstood by neontologists, and are sometimes forgotten even by paleontologists. The fossil record of Paleozoic tetrapods is discontinuous in time and space, and represents faunas sampled from a narrow range of environments. Available specimens vary widely in completeness, and even those that preserve some portion of the otic region may or may not provide useful morphological information about it. We surveyed the world literature of Paleozoic tetrapods, 1928-present, for (1) stratigraphic distribution and (2) geographical distribution, of localities producing Paleozoic tetrapod remains, and (3) for original illustrations of otic morphology. We identified 470 publications with at least one original illustration of the stapes, otic capsule, or "cheek," scored each occurrence for illustration quality, and assigned the species illustrated to one of 16 generally recognized higher taxonomic groups. In most of our 16 groups, more than 50% of the genera illustrated since 1928 are unrepresented by a single original illustration of the stapes or otic capsule, and the number of excellent illustrations of each of these otic features was very low or zero within each higher taxon. Using outgroup comparison, we determined primitive and derived states for four otic characters representing the cheek, stapes (two characters), and otic capsule. These four characters were mapped onto a "consensus" cladogram, compiled from several sources, of our 16 Paleozoic tetrapod groups. The computer program MacClade was used to calculate the most parsimonious distribution of missing characters, both among our 16 groups and at each node connecting two groups. This analysis indicated that three of our otic characters have had complex histories, with multiple origins and losses. The exception, presence/absence of a fenestra ovalis, may have had a complex history that is not reflected in published data, due to the nearly universal assumption of its presence. Interpretation of the otic region in Paleozoic tetrapods is frequently dependent on hypotheses of relationship and/or function that may not be explicitly recognized or critically examined, and can lead to dubious or incorrect results. Although these problems should be recognized, our survey suggests that the potential for recovering additional useful information from existing specimens is often good. Careful analysis of the resulting data can contribute to much-improved understanding of the early evolution of the tetrapod ear.

Acknowledgments. We would like to acknowledge the photographic skills and help of Vernon Fergeson in making several of the illustrations. We are especially grateful to Cathryn Easterbrook of the University of Chicago's Crerar Library for her patience, diligence and skill in helping us quarry out some truly obscure publications.

References

Baird D, Carroll RL (1967) *Romeriscus*, the oldest known reptile. Science 157:56–59.

Barbarena MC, Araújo DC, Lavina EL (1985) Late Permian and Triassic tetrapods of southern Brazil. Natl Geogr Res 1(1):5–20.

Beaumont EH (1977) Cranial morphology of the Loxommatidae (Amphibia: Labyrinthodontia). Philos Trans R Soc Lond B 280:29–101.

Bolt JR, Lombard RE (1985) Evolution of the amphibian tympanic ear and the origin of frogs. Biol J Linnean Soc 24:83–99.

Bolt JR, McKay RM, Witzke BJ, McAdams MP (1988) A new Lower Carboniferous tetrapod locality in Iowa. Nature 333:768–770.

Boy JA (1988) Über einige Vertreter der Eryopoidea (Amphibia: Temnospondyli) aus dem europäischen Rotliegend (?höchtes Karbon-Perm) 1. Sclerocephalus. Paläontologisch Zeits 62:107–132.

Boy JA, Bandel K (1973) *Brukterpeton fiebigi* n. gen. n. sp. (Amphibia: Gephyrostegoidea) der erste Tetrapode aus dem rheinischwestfälischen Karbon (Namur B; W-Deutschland) Palaeontographica (A) 145:39–77.

Boyd MJ (1984) The Upper Carboniferous tetrapod assemblage from Newsham, Northumberland. Palaeontology 27:367–392.

Bystrow AP (1944) *Kotlassia prima* Amalitzky. Bull Geol Soc Am 55:379–416.

Campbell KSW, Bell MW (1977) A primitive amphibian from the late Devonian of New South Wales. Alcheringa 1:369–381.

Carroll RL (1984) Problems in the use of terrestrial vertebrates for zoning of the Carboniferous. Ninth International Congress of Carboniferous Stratigraphy and Geology, Compte Rendu, V. 2, Biostratigraphy, pp. 135–147.

Carroll RL (1988) *Vertebrate Paleontology and Evolution.* New York: W. H. Freeman and Co., 698 pp.

Carroll RL, Baird D (1968) The Carboniferous amphibian *Tuditanus* (*Eosauravus*) and the distinction between microsaurs and reptiles. Am Mus Novit 2337:1–50.

Carroll RL, Baird D (1972) Carboniferous stem-reptiles of the Family Romeriidae. Bull Mus Comp Zool 143:321–364.

Carroll RL, Gaskill P (1978) The Order Microsauria. Memoirs of the Am Philos Soc 126:1–211.

Carroll RL, Winer L (1977) Classification of amphibians and list of genera and species known as fossils. Appendix to accompany Ch. 13 in Patterns of Evolution as illustrated by the Fossil Record, A. Hallam, ed.; ch. 13, pp. 403–437. (Self-published and distributed. Compiled 1975–1976.)

Chatterjee S, Hotton N (III) (1986) The Paleoposition of India. J SE Asian Earth Sci 1:145–189.

Clack JA (1983) The stapes of the Coal Measure embolomere *Pholiderpeton scutigerum* Huxley (Amphibia: Anthracosauria) and otic evolution in early tetrapods. Zool J Linnean Soc 79:121–148.

Clack JA (1987) *Pholiderpeton scutigerum* Huxley, an amphibian from the Yorkshire coal measures. Philos Trans R Soc London 318B:1–107.

Clack JA (1989) Discovery of the earliest-known tetrapod stapes. Nature 342:425–427.

Eberth DA (1985) The skull of *Sphenacodon ferocior*, and comparisons with other sphenacodontines (Reptilia: Pelycosauria). New Mexico Bureau of Mines and Mineral Resources Circular 190:1–40.

Fritzsch B (1990) Evolution of tetrapod hearing. Nature 344:116.

Gaffney ES (1979) Tetrapod monophyly: a phylogenetic analysis. Bull Carnegie Mus Nat Hist 13:92–105.

Gauthier JA, Kluge AG, Rowe T (1988a) The early evolution of the Amniota. Syst Assoc Spec Vol 35A:103–155.

Gauthier JA, Kluge AG, Rowe T (1988b) Amniote phylogeny and the importance of fossils. Cladistics 4:105–209.

Gauthier J, Cannatella D, de Queiroz K, Kluge AG, Rowe T (1989) Tetrapod phylogeny. In: Fernholm B, Bremer K, Jörnvall H (eds) The Hierarchy of Life. Chapter 25, pp. 337–353. Elsevier.

Godfrey SJ (1989) Ontogenetic changes in the skull of the Carboniferous tetrapod *Greererpeton burkemorani* Romer, 1969. Philosophical Trans Roy Soc Lond B 323:135–153.

Godfrey SJ, Fiorillo AR, Carroll RL (1987) A newly discovered skull of the temnospondyl amphibian *Dendrerpeton acadianum* Owen. Can J Earth Sci 24:796–805.

Gottfried MD, Foreman B (1990) Nature 344:116.

Haq BU, Van Eysinga FWB (1987) Geological Time Table. Elsevier Science Publishers.

Harland WB, Cox AV, Llewellyn PG, Pickton CAG, et al. (1982) A geologic time scale. Cambridge Earth Science Series, Cambridge, London, New York, New Rochelle, Melbourne, Sydney: Cambridge University Press.

Heaton MJ (1979) Cranial morphology of primitive captorhinid reptiles from the late Pennsylvanian and early Permian, Oklahoma and Texas. Bull Oklahoma Geol Surv 127:1–84.

Hook RW, Baird D (1986) The diamond coal mine of Linton, Ohio, and its Pennsylvanian-age vertebrates. J Vert Paleontol 6:174–190.

Hook RW, Baird D (1988) An overview of the Upper Carboniferous fossil deposit at Linton, Ohio. Ohio J Sci 88:55–60.

Hopson JA, Barghusen HR (1986) An analysis of therapsid relationships. In: Hotton N III, MacLean PD, Roth JJ, Roth EC (eds) The Ecology and Biology of Mammal-like Reptiles. Smithsonian Institution Press, pp. 83–106.

Ivakhnenko MF (1979) (The Permian and Triassic procolophonians from the Russian Platform). In Russian. Trudi, Paleontological Institute, Academy of Sciences of the USSR 164:1–80.

Kitching JW (1977) The distribution of the Karroo vertebrate fauna. University of Witwatersrand, Bernard Price Institute for Palaeontological Research. Memoir 1:1–131.

Lombard RE, Bolt JR (1988) Evolution of the stapes in Paleozoic tetrapods: conservative and radical hypotheses. In: Fritzsch B, Ryan MJ, Wilczynski W, Walkoviak W (eds) The Evolution of the Amphibian Auditory System. pp. 37–67.

McGinnis HJ (1967) The osteology of Phlegethontia, a Carboniferous and Permian aïstopod amphibian. U Calif Pub Geol Sci 71:1–49.

Mapes G, Mapes RH (eds) (1989) Regional Geology and Paleontology of Upper Paleozoic Hamilton Quarry Area in Southeastern Kansas. Kansas Geological Survey Guidebook 6.

Milner AR (1980a) The temnospondyl amphibian Dendrerpeton from the Upper Carboniferous of Ireland. Palaeontology 23:125–141.

Milner AR (1980b) The tetrapod assemblage from Nyrany, Czechoslovakia. In: Panchen AL (ed) The Terrestrial Environment and the Origin of Land Vertebrates, pp. 439–496.

Milner AR (1987) The Westphalian tetrapod fauna; some aspects of its geography and ecology. J Geol Soc Lond 144:495–506.

Milner AR (1988) The relationships and origin of living amphibians. System Assoc Spec Vol 35A:59–102.

Milner AR, Panchen AL (1973) Geographical variation in the tetrapod faunas of the Upper Carboniferous and Lower Permian. In: Tarling DH, Runcorn SK (eds) Implications of Continental Drift to the Earth Sciences, 1:353–368.

Milner AR, Smithson TR, Milner AC, Coates MI, Rolfe WDI (1986) The search for early tetrapods. Mod Geol 10:1–28.

Olson EC (1957) Catalogue of localities of Permian and Triassic terrestrial vertebrates of the territories of the USSR. J Geol 65:196–226.

Olson EC (1962) Late Permian terrestrial vertebrates, U.S.A. and U.S.S.R. Transactions of the American Philosophical Society (n.s.) 52(2):1–224.

Olson EC (1965) Relationships of Seymouria, Diadectes, and Chelonia. Am Zool 5:295–307.

Olson EC (1989) Problems of Permo-Triassic terrestrial vertebrate extinctions. Hist Biol 2:17–35.

Panchen AL (1977) Geographical and ecological distribution of the earliest tetrapods. In: Hecht MK, Goody PC, Hecht BM (eds) Major Patterns in Vertebrate Evolution. pp. 723–238, New York: Plenum Press.

Panchen AL (1985) On the amphibian Crassigyrinus scoticus Watson from the Carboniferous of Scotland. Philos Trans R Soc Lond (B) 309:505–568.

Panchen AL, Smithson TR (1988) The relationships of the earliest tetrapods. System Assoc Spec Vol 35A:1–32.

Reisz RR (1986) Teil 17A. Pelycosauria. Handbuch der Paläoherpetologie Teil 17A:1–102.

Ricqles A de, Taquet P (1982) La faune de vertébrés du Permien supérieur du Niger. I. Le Captorhinomorphe Moradisaurus grandis (Reptilia, Cotylosauria)–Le Crane. Ann Paléontol 68:33–106.

Romer AS (1969) The cranial anatomy of the Permian amphibian Pantylus. Breviora 314:1–37.

Romer AS, Edinger T (1942) Endocranial casts and brains of living and fossil Amphibia. J Comp Neurol 77:355–389.

Romer AS, Price LI (1940) Review of the Pelycosauria. Spec Papers Geol Soc Am 28:1–538.

Romer AS, Witter RV (1942) Edops, a primitive rhachitomous amphibian from the Texas red beds. J Geol 50:925–959.

Sawin HJ (1941) The cranial anatomy of Eryops megalocephalus. Bull Mus Comp Zool 88:407–463.

Sigogneau-Russell D, Russell DE (1974) Etude du premier Caseide (Reptilia, Pelycosauria) d'Europe occidentale. Bulletin du Museum National d'Histoire Naturelle, Paris, Ser. 3, Sciences de la Terre 38:145–215.

Smith AG, Hurley AM, Briden JC (1981) Phanerozoic Paleocontinental World Maps. Cambridge University Press.

Smithson TR (1982) The cranial morphology of Greererpeton burkemorani Romer (Amphibia: Temnospondyli). Zool J Linnean Soc 76:29–90.

Smithson TR (1985) The morphology and relationships of the Carboniferous amphibian Eoherpeton watsoni Panchen. Zool J Linnean Soc 85:317–410.

Spjeldnaes N (1982) Palaeoecology of Ichtyostega [sic] and the origin of terrestrial vertebrates. In: Proceedings of the First International Meeting on "Palaeontology, Essential of Historical Geology," pp. 323–343.

Thommasen H, Carroll RL (1981) Broomia, the oldest known millerettid reptile Palaeontology 24:379–390.

Watson DMS (1954) On *Bolosaurus* and the origin and classification of the reptiles. Bull Mus Comp Zool 111:299–449.

Wellstead CF (1982) A lower Carboniferous aïstopod amphibian from Scotland. Palaeontology 25:193–208.

White TE (1939) Osteology of *Seymouria baylorensis* Broili. Bull Mus Comp Zool 85:325–409.

Appendix

The bulk of the data reported in this chapter result from a literature review. We examined original illustrations published since 1928 from selected taxa of Paleozoic tetrapods. Original illustrations are the most widely accessible data of paleontology and this search was undertaken to find out as precisely as possible what is known about the ear region. Nineteen twenty-eight was chosen as a workable cut-off which would enable us to both complete the project and avoid a number of entangling nomenclatural problems in the older literature. We are continuing into this older literature as part of our ongoing research and thus this is an interim report.

As defined for this work, an original illustration is of a body part or parts (no trackways, coprolites, burrows, etc.) and with no attribution in the figure legend to an earlier publication. Our working source of citations was the *Bibliography of Fossil Vertebrates*, presently published by the Society of Vertebrate Paleontology. The most recent volume is for 1986, so the later publications included come from our own collections and knowledge of the literature. Just over 1300 citations mentioning Paleozoic tetrapods were examined and 470 publications were found to contain original illustrations for members of the higher taxa considered in this chapter.

One could argue that with a cut-off date of 1928, drawings of a large number of taxa would be missing. We are indeed missing some data, but not to the degree that might be surmised; we have probably surveyed 95% of the known genera. Carroll (1988) lists 300 genera, among which are 30 that we have not encountered. Many of these indeed have not been re-illustrated since their original description prior to 1928, some with good reason. Prior to the turn of the century, many genera were named from very scrappy material, material that

TABLE 19.1. Taxa used in literature search.

This Chapter	Carroll, 1988
Ichthyostegalia (IC)	Order Ichthyostegalia
Temnospondyli (TE)	Order Temnospondyli
Microsauria (MI)	Order Microsauria
Nectridea (NE)	Order Nectridea
Aïstopoda (AI)	Order Aïstopoda
Loxommatoidea (LO)	Superfamily Loxommatoidea
Anthracosauria (AN)	Suborder Embolomeri
	Suborder Gephyrostegida
	Family Lanthanosuchidae
	Family Chroniosuchidae
	Family Solenodonsauridae
	Family Tokosauridae
	Family Nycteroleteridae
	Genus *Crassigyrinus*
"protoanthracosaur" (Bolt et al. 1988)	
Seymouriamorpha (SE)	Suborder Seymouriamorpha
Diadectomorpha (DD)	Family Diadectidae
	Family Limnoscelidae
	Family Tseajaiidae
Pelycosauria (PE)	Order Pelycosauria
Pareiasauroidea (PA)	Suborder Pareiasauroidea
Millerosauroidea (ML)	Suborder Millerosauroidea
Procolophonia (PR)	Suborder Procolophonia
Mesosauria (ME)	Order Mesosauria
Captorhinomorpha (CA)	Suborder Captorhinomorpha
	Genus *Eunotosaurus*
Diapsida (DI)	Order Araeoscelida
	Family Mesenosauridae
	Family Coelurosauravidae
	Order Eosuchia
Lepospondyli (LE)	Family Adelogyrinidae
	Family Acherontiscidae
	Order Lysorophia

more recent workers have not deemed worthy of reillustration. Others have been synonymized since Carroll's work. At the moment, our estimate is that there are about 20 genera not represented in our data. For all but a couple of these, ear parts are unknown. On the other hand, we have found illustrations of 360 genera, 60 more than are listed in Carroll. Some of these 60 are synonyms or invalid, but most appear to be perfectly respectable. It is probably safe to say that at this time no one knows exactly how many valid genera of Paleozoic tetrapods there are.

The higher taxa employed here and in the sources of our composite cladogram have long been used in the literature. Use of "traditional" group names ensures some nomenclatural stability, but can be misleading because the taxa included under each name may vary from paper to paper, sometimes radically. To reduce confusion, we have used names and contents from Carroll's (1988) detailed Appendix on vertebrate classification as shown here in Appendix Table 19.1. Our taxa are consistent with our source cladograms and other recent literature, although the allocation of some of the taxa comprising Seymouriamorpha and Diadectomorpha is controversial. For our purposes these problems are unimportant because the morphology of the otic region is unknown for most. For the nonspecialist it is important to realize that the taxa included encompass *all* Paleozoic amphibians and all but the most derived Paleozoic amniotes. The amniote groups not examined include lepidosauromorph diapsids, except those generally thought to be the most primitive, and the Archosauria and Therapsida, both highly derived amniote clades.

For each publication, each *species* illustrated was counted as an occurrence, no matter the number of illustrations of that species. For each occur rence the following data were taken: 1. genus (as given by author); 2. species (as given by author); 3. year species was described; 4. higher taxon (as in Appendix Table 19.1 per Carroll, 1988 even if contra to the author); 5. country of collection; 6. province or state (for large countries only); 7. age (as considered by author). For each species the otic region was scored for illustrations of the 8. stapes, 9. otic capsule and 10. cheek. Each of these three was given a score from 0–3.

0. No illustration
3. One view with no detail (generally an outline drawing)
2. One or two views using simple drawings, but with some detail.
1. Two or more views with good detail.

For each occurrence these ten pieces of information along with appropriate bibliographic information were entered into a dBASE III+ database for study. This database in its ever-evolving form is available on disk to interested colleagues.

20

The Stapes of *Acanthostega gunnari* and the Role of the Stapes in Early Tetrapods

J.A. Clack

1. Introduction

Acanthostega gunnari is a tetrapod from the Upper Devonian of East Greenland, and is therefore one of the earliest tetrapod taxa for which skeletal remains are known. It was exactly contemporary with *Ichthyostega* (Jarvik 1952, 1980), a much more widely known form, but recent discoveries in Greenland (Bendix-Almgreen, Clack, and Olsen 1988) have shown that in many ways *Ichthyostega* and *Acanthostega* are quite different (Clack 1988a,b). Apart from primitive characters, they have little in common, suggesting early diversification among tetrapods.

The stapes of *Ichthyostega* remains unknown, but recently (Clack 1989) I described that of *Acanthostega gunnari*, suggesting that the earliest tetrapods used an open spiracle as part of their fish-like breathing mechanism, and that it was operated by muscles attached to the stapes. Here I shall give further details of the stapes and otic region of *Acanthostega*. I shall put the case that the earliest tetrapods were not fully autostylic, but that the stapes still formed the only link between the palate and the otic region of the braincase. The palate was potentially mobile, and was involved with buccal ventilation. Only when the palatal bones were released from this role by the development of alternative mechanisms, did the stapes specialize as a purely auditory ossicle. The role of the stapes may also have been affected by developments to the vertebral column concerned with terrestrial locomotion and to the skull associated with terrestrial feeding.

2. The Stapes and Otic Region of *Acanthostega gunnari*

The stapes of *Acanthostega* has been discovered in a number of specimens, which also preserve parts of the braincase including otic capsule material. Specimens belong to the Geological Museum, Copenhagen (MGUH), but are as yet uncatalogued. I shall quote their field numbers for identification. Specimen MGUH1227 (Fig. 20.1) shows the skull in dorsal view, with the bones of the skull roof removed and some of the interorbital and cheek bones missing or unprepared. The right stapes is preserved in near enough its original position, and has been prepared out, and the left stapes is also present but covered by crushed braincase material. A second specimen, MGUH 1604 (Fig. 20.2), has the right stapes present, and has been sectioned using a Lastec diamond-wire saw with a 0.12 mm diameter wire, courtesy of Oxford University. A third specimen, MGUH 258, retains the natural mold of the left stapes, visible in lateral view in contact with the otic capsule, and two others, MGUH 1300 (Fig. 20.3) and University Museum of Zoology Cambridge specimen number UMZCT 1300 (Clack 1988a), have displaced stapes. These specimens give an almost complete picture of the bone, and with specimen MGUH 236 (Fig. 20.4), sectioned using a Well diamond-wire saw with a 0.3 mm wire, give some information on the otic capsule and other parts of the braincase.

A

B

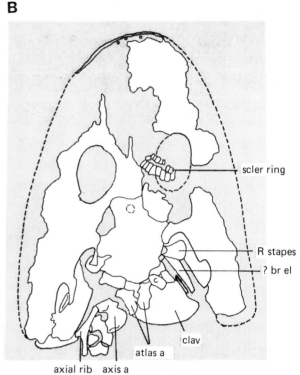

FIGURE 20.1. MGUH field number 1227 (part). Skull of *Acanthostega gunnari* in dorsal view, showing right stapes and other elements. Shaded areas—matrix. (A) photograph; (B) interpretive diagram. Scale bar, 10 mm.

Abbreviations used in figures: artic, articular; atlas a, atlas arch; axis a, axis arch; br el, branchial element; bptpr, basipterygoid process; clav, clavicle; d sk rf, dermal skull roof; ectopt, ectopterygoid; epipter, epipterygoid; int naris, internal naris; L, left; lr jaw, lower jaw; max, maxilla; ot cap, otic capsule; par for, parietal foramen; pal, palatine; pp, postparietal; premax, premaxilla; pter, pterygoid; R, right; scler ring, sclerotic ring; sphet, sphenethmoid; ssc, semicircular canal; st, supratemporal; stap for, stapedial foramen; tab, tabular; tab emb, tabular embayment; vom, vomer.

2.1 Description and Comparative Morphology

The stapes of *Acanthostega* is most closely comparable to those of the colosteid *Greererpeton burkemorani* (Carroll 1980; Smithson 1982; Godfrey 1989), and of the anthracosauroid *Pholiderpeton scutigerum* and its close relative *Palaeoherpeton decorum* (Clack 1983), which until the current discovery were the earliest known tetrapod stapes (Fig. 20.5). As in *Greererpeton* and *Pholiderpeton*, the distal part of the stapes of *Acanthostega* consists of a fan-shaped plate with unfinished lateral and posterior margins. The proximal end is somewhat expanded, and the footplate can be seen in the stapes from MGUH 1300, which has been isolated

from the matrix and prepared in three dimensions. This and the sections from specimen MGUH1604 show that it had only a single head. This is in accord with those of *Greererpeton* and *Pholiderpeton* which also have a single head.

Another similarity with *Greererpeton* is the fact that both furnish us with a number of specimens in which the stapes has remained in close contact with the braincase. This contrasts with the preservation of most other early tetrapods, even those represented by many specimens and/or well-preserved skulls, such as the loxommatid *Megalocephalus* (Beaumont 1977), and the anthracosauroids *Proterogyrinus* (Holmes 1984) and *Archeria* (Clack and Holmes 1988). It suggests that the stapes were firmly attached in life – not necessarily

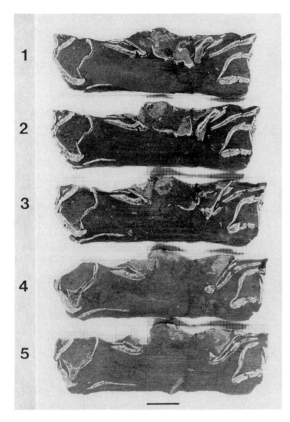
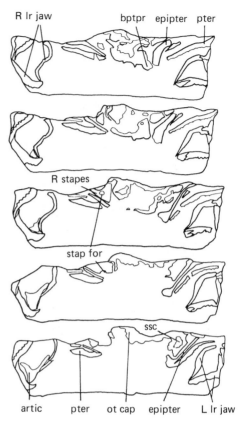

FIGURE 20.2. MGUH field number 1604 (part). Skull of *Acanthostega gunnari*, posterior portion in transverse section. Sections viewed anteriorly, anteriormost section at the top. Right stapes in position, skull roofing bones not present. Scale bar, 10 mm. For abbreviations see Fig. 20.1.

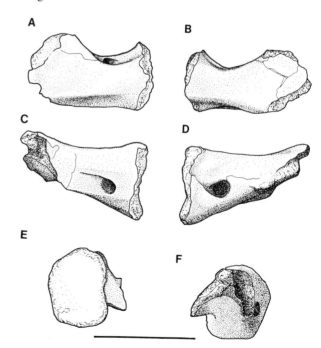

FIGURE 20.3. MGUH field number 1300 (part). Left stapes of *Acanthostega gunnari*. (A) Approximately dorsal view; (B) approximately ventral view; (C) approximately anterior view; (D) approximately posterior view; (E) proximal view; (F) distal view. Scale bar, 10 mm.

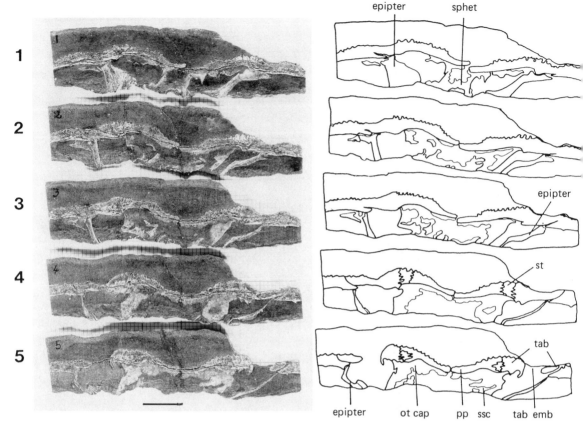

FIGURE 20.4. MGUH field number 236. Skull of *Acanthostega gunnari,* posterior part in transverse section. Sections viewed anteriorly, anteriormost at the top. Ventral parts of the palate and braincase missing, dermal bone sutures omitted from sections 1, 2 and 3. Scale bar, 10 mm. For abbreviations see Fig. 20.1.

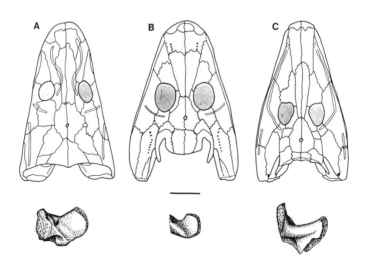

FIGURE 20.5. Skulls and stapes of early tetrapods. (A) *Greererpeton* (snout to quadrate length 131 mm), stapes in dorsal view below; (B) *Acanthostega* (snout to quadrate length 115 mm), restoration of stapes from specimens MGUH 1227 and MGUH 1604 in approximately dorsal view below; (C) *Pholiderpeton* (snout to quadrate length 320 mm), stapes in approximately dorsal view below. Scale bar for stapes, 10 mm. (From Clack 1989. Reprinted by permission from Nature 324:425. Copyright © 1989 *Macmillan Magazines Ltd.*)

sutured, but held with much soft tissue, or embedded deeply into the otic capsule. Smithson (1982) has suggested this for *Greererpeton*, and has described the large opening into which the stapes fitted.

In *Greererpeton*, there is no doubt that the stapes lay in close contact with the palatoquadrate. All the specimens of *Acanthostega*, however, are flattened to an unknown degree. In specimen MGUH1227, the stapes rests on the palatoquadrate, at least distally, but the notch has been widened by compression, and the contact could have been greater in life. The sections of MGUH1604 show the stapes also resting on the palatoquadrate, and in this case, most of the distal plate is involved. While it is not possible to be sure of the degree of contact, given the situation in *Greererpeton*, it seems possible that it was also extensive in *Acanthostega*. The isolated stapes (Fig. 20.3) has been illustrated to conform in approximate orientation to those illustrated by Clack (1983) for *Pholiderpeton* and Smithson (1982) for *Greererpeton*, but it remains uncertain.

2.2 The Temporal Notch of *Acanthostega*

One of the major differences between *Acanthostega* and *Greererpeton* is that while *Greererpeton* has no temporal notch, *Acanthostega* appears to have two (Fig. 20.5). The question arises as to which of the two was homologous with the temporal ('otic') notch of other early tetrapods. (Carrying no functional implications, the term 'temporal notch' does not obscure possible positional homologies). One of these notches lies at the squamosal-tabular suture, the second is bounded entirely by the tabular. It is this which I believe to be the true temporal notch. It has been suggested (Jarvik 1952) that the space beneath the tabular notches is homologous with the fossa bridgei of fishes like *Eusthenopteron*, from which the overlying dermal bones have retreated. If this were so, the opisthotics would be expected to contact the tabulars lateral to the notch, but this is not the case. Large grooves in the braincase, perhaps the nearest equivalent to the post-temporal fossae of early tetrapods like loxommatids (Beaumont 1977), and therefore regarded as homologues of the fish fossa bridgei, lie bounded laterally by downgrowing flanges from the tabulars forming the medial margin of the notch (Fig. 20.4). In addition, the stapes lies with its distal margins enclosed within the tabular notch.

The squamosal-tabular junction is a smooth surface with no interdigitations, and contrasts strongly with other dermal bone sutures in the skull some of which interdigitate in three dimensions (Fig. 20.4). An extension to the tabular appears to have grown around the original notch in relation to the development of the unique skull roof pattern of *Acanthostega*, the elimination of the intertemporal, and the development of its characteristic tabular horn.

2.3 The Otic Capsule of *Acanthostega*

The otic capsules of *Acanthostega* provide another contrast with *Greererpeton* (Smithson 1982), but resemble in some respects those of anthracosauroids such as *Eoherpeton* (Smithson 1985) and the embolomeres *Palaeoherpeton* ('*Palaeogyrinus*,' Panchen 1964), *Pholiderpeton* (Clack 1987) and *Archeria* (Clack and Holmes 1988). They are well-ossified dorsally, presumably from the opisthotics and prootics, though no specimens show sutures. In *Greererpeton* however, the otic capsules were poorly ossified dorsally, with the opisthotics and prootics possibly represented by small endochondral plates (Smithson 1982). In *Acanthostega*, the basioccipital was apparently poorly incorporated into the braincase. The parasphenoid of *Acanthostega* is very short, though it laps under the basisphenoid portion of the otic capsule, and therefore presumably crossed the ventral cranial fissure unlike that of *Ichthyostega* (Jarvik 1980). It probably contacted the basioccipital, but did not suture firmly with it, a situation similar to that in *Crassigyrinus* (Panchen 1985).

The dorsal part of the otic capsule has side-walls that look not unlike those in anthracosauroids. Internally, there are no periosteally lined cavities for semicircular canals; unlined cavities may represent them (Figs. 20.2 and 20.4), as they do in anthracosauroids such as *Eoherpeton* (Smithson 1985), *Palaeoherpeton* (Panchen 1964), and *Pholiderpeton* (Clack 1987). Further anteriorly, there are some periosteally lined nerve foramina, but it is impossible to say yet which nerves might have passed through them. It seems that the otic capsule itself was in open contact with the cranial cavity. This is the situation found in anthracosauroids and loxommatids. However, it is also possible that the capsules were completed in cartilage and were separated from the cranial cavity in life, as for example in *Latimeria*. This would have implications for the possible hearing capabilities of early tetrapods (Fritzsch, Chapter 18).

It is not possible to be sure of the position, if any, of the fenestra ovalis. In many early tetrapods, including *Acanthostega*, *Crassigyrinus* (Panchen 1985) and anthracosauroids (Clack and Holmes 1988), the region of the lateral otic fissure, which separates the otic and occipital components, is frequently poorly ossified. Hence the fenestra ovalis, which lies at the conjunction of these components, is never well represented in these early tetrapods. In some of the sections, the stapes abuts against the side wall of the otic capsule, as it does in MGUH1227 (Fig. 20.1), in an unexpectedly dorsal position. This may not be natural, but be the result of compression. If it were natural, then it would suggest that no fenestra ovalis was present. In specimen MGUH 258 however, the stapes appears to pass into the capsule more ventrally, but the evidence is difficult to interpret.

2.4 Phylogenetic Considerations

I suggested recently (Clack 1989) that the form of the stapes in *Acanthostega*, a clearly notched, but very early tetrapod, provides corroboration for the view that a robust stapes with a single head and a broad fan-shaped distal plate was that which was synapomorphous, though nevertheless primitive for tetrapods. This is based on three lines of evidence. The first is admittedly stratophenetic. The earliest tetrapod stapes known so far are all of this form. We await with anticipation discoveries of stapes from the East Kirkton Viséan locality in Scotland, which has yielded not only the earliest temnospondyls and probably anthracosaurs, but also the earliest reptile (Milner et al. 1986; Smithson 1989), and from new material from the Viséan of Iowa (Bolt et al. 1988).

The second is the occurrence of similar stapes in three skulls of contrasting construction. *Acanthostega* had a uniquely constructed notch, *Greererpeton* had none at all (Carroll 1980), and *Pholiderpeton* had a narrow, wedge-shaped notch (Clack 1987). *Acanthostega* had a flattened skull, with good braincase/skull-roof contact and otic capsules ossified dorsally, *Greererpeton* also had a flattened skull but poorly developed braincase/skull roof

contact, and *Pholiderpeton* had a much deeper skull, but with a similar braincase/skull roof relationship to *Acanthostega*. Therefore, it seems unlikely that the stapes performed an architectural role associated with support of the braincase (Carroll 1980; Smithson 1982) which was similar in each. Their similarity may therefore be a result of their primitive nature.

The third is inferred from the possible relationships of the three genera. *Acanthostega* cannot be clearly associated with any other known group of early tetrapods on shared derived characters, though it shares a number of primitive features with *Ichthyostega*, which have been lost in other tetrapods. Therefore, I regard it as a stem tetrapod. Whether it or *Ichthyostega* will prove to be the more primitive cannot yet be established, but must await cladistic analysis following a detailed description of the rest of its morphology.

The colosteids are widely regarded as either primitive temnospondyls (Smithson 1982; Hook 1983), or as belonging to the primitive sister taxon of temnospondyls plus microsaurs (Milner et al. 1986; Panchen and Smithson 1988). Smithson (1985) and Panchen and Smithson (1988) regard this stock as forming one branch of a dichotomy, the two branches of which gave rise on the one hand to modern amphibia (the batrachomorph clade), and on the other to amniotes (the reptiliomorph clade). They therefore regard *Greererpeton* as a stem amphibian rather than a true temnospondyl, at the base of the batrachomorph clade. The anthracosauroids by contrast, are regarded by many as related to seymouriamorphs and amniotes, that is, according to Panchen and Smithson's scheme, *Pholiderpeton* is a stem reptiliomorph. The anthracosauroids are notable for retaining many characters of the skull which appear to be primitive for tetrapods, and share no derived characters with stem amphibians. Thus *Acanthostega*, *Greererpeton* and *Pholiderpeton* belong to three quite unrelated groups according to this scheme, so that the similarities of the stapes are unlikely to be the result of a shared derived condition.

Temnospondyls and anthracosauroids were for many years placed in the large group known as 'labyrinthodonts' (Romer 1947, 1966). Originally, this classification was based on characters of the teeth and vertebrae. On the basis of these characters, *Acanthostega* would be included as a labyrinthodont, since it has teeth whose bases have folded enamel, and it has vertebrae which are rhachitomous – bipartite centra with large ventral wedge-shaped intercentra and small paired, dorsal pleurocentra. However, these characters are almost certainly primitive for tetrapods, based on comparison with an osteolepiform outgroup.

Recently, Lombard and Bolt (1988) concluded that by comparison with an osteolepiform outgroup, the large, double-headed, ventrally directed stapes of the earliest reptiles may represent the primitive condition for tetrapods (but see Bolt and Lombard, Chapter 19). That being so, the kind of stapes which was found in *Greererpeton* and anthracosauroids must be derived, and they used it as a new synapomorphy uniting temnospondyls and 'batrachosaurs' (i.e., anthracosauroids) (Bolt and Lombard 1985). That group would be almost equivalent to the old group 'labyrinthodonts' and *Acanthostega* would belong there. Taking characters of the stapes alone, this is a logical conclusion, but it needs to be supported by other derived characters linking these three taxa. A double-headed, ventrally directed stapes in a nonamniote tetrapod has not yet been found.

One more point should be made. *Acanthostega*, like *Pholiderpeton* and *Greererpeton* seems to have been a primarily aquatic animal, as we infer from the presence of lepidotrichia in the tail (Clack 1988b) and the form of the forelimb (Coates and Clack 1990). There still remains the possibility that the similarities of the stapes of all three animals relates to their aquatic habits. However, while to have two early tetrapod stapes convergently similar could be regarded as a possibility, to have three looks like too much of a coincidence.

3. The Function of the Stapes and Otic Region in Early Tetrapods

The significance of the plate-like stapes in *Greererpeton* and *Pholiderpeton* has been the subject of much speculation (Carroll 1980; Smithson 1982; Clack 1983; Panchen 1985), as has the possible function of the temporal notch of early tetrapods [Clack 1983; Panchen 1985; Carroll (in Panchen 1985); Clack 1989]. After a brief review of these ideas, I shall present some alternatives.

3.1 Recent Hypotheses

Carroll (1980) suggested that the stapes of *Greererpeton* and possibly other early tetrapods was acting as a brace supporting the braincase. Subsequent to the separation of the head from the shoulder-girdle during the fish tetrapod transition, the connections between the otic region, occipital arch and the skull roof would have been unstable, requiring the additional support of the hyomandibula until such connections were consolidated. Carroll suggested that only when the braincase became securely incorporated into the skull roof by the formation of an integrated occiput, could the stapes lose its bracing role and function as a hearing ossicle. Carroll described this bracing function as being similar to that of the fish hyomandibula, but in fishes, the hyomandibula supports the palate, suspending it from the presumably securely anchored braincase, and providing a pivot about which it can move. Thus the function suggested by Carroll for the stapes of *Greererpeton* is not directly comparable to that of the fish hyomandibula (Smithson 1982).

In *Greererpeton*, there is little bony contact between the braincase and the skull roof. Smithson (1982) proposed that the stapes resisted mechanical stress caused by contraction of the hypaxial muscles, which might otherwise have tended to dislocate the braincase from the skull roof. He saw the stapes acting as a compression member, rather than resisting tensional forces created during jaw closure as does the fish hyomandibula.

The stapes of *Pholiderpeton* (Clack 1983), while broadly similar to that of *Greererpeton*, is unlikely to have been necessary as a brace for the braincase, given the diametrically opposite construction of the skull, and I suggested that it could have had some role in low-frequency or underwater sound conduction. While I saw no possibility of the temporal notch having housed a tympanum, I made no firm suggestions as to alternative functions.

When Panchen (1985) described the aberrant early tetrapod *Crassigyrinus*, he took up a suggestion originally made informally by Carroll, to the effect that the temporal notch housed an open spiracle. The stapes remains unknown in *Crassigyrinus*, but because Panchen regards *Crassigyrinus* as primitive, he suggested that the stapes would have been similar to those of *Pholiderpeton* and *Greererpeton*. He suggested that the spiracular chamber, filled with air, with the stapes embedded in its wall, would have acted as an underwater ear, somewhat as Fritzsch (Chapter 18) is suggesting might have been the case in osteolepiforms and the earliest tetrapods. This, however, would require the otic capsule to have been isolated from the cranial cavity, which is not indicated by the ossified remains of early tetrapod braincases.

In 1989, following Carroll and Panchen, I supported the argument that the notch found so widely in early tetrapods, even in anthracosaurs with a large stapes, was the site of an open spiracle, which could have been used to inhale or exhale air to or from the spiracular pouch. The spiracular pouch would have been permanently filled with air, and may have allowed the animal to breathe with its head under water, or allowed an accessory air supply when the animal was under stress. I extended this idea to suggest that the stapes still had a role to play in the operation of a buccal pump breathing mechanism, and in that of the spiracle, and that its similarity to the fish hyomandibula was greater than even Carroll had supposed.

3.2. Relationships of the Stapes, Palate, Braincase, and Skull Roof in Early Tetrapods—the Fallacy of Autostyly in the Earliest Tetrapods

In osteichthyans (bony fishes, of which tetrapods are a subgroup), the hyomandibula plays a pivotal role in skull function. Primitively, in chondrichthyans, it suspends the lower jaw (Meckel's cartilage) directly. In osteichthyans the connection is less direct, and one or several other bony elements are interposed between hyomandibula and quadrate. In many more recent fishes, the hyomandibula has little direct responsibility for suspension of the lower jaw, whose articulation is incorporated into the dermal skull roof. The hyomandibula retains a role, however, in bony fishes, in supporting the palate and cheek, and in coordinating their movements in ventilation. It provides insertion points for muscles operating the opercular series whose movements are a crucial component of the buccal pump. In the modern but atypical actinopterygian fish *Polypterus*, it also provides origin for muscles operating an openable spiracle (Allis 1922), though it is now clear (Brainerd, Liem, and

Samper 1989) that in the specimens they examined, air does not pass through the spiracle.

In tetrapods, the opercular apparatus covering the gills in fishes is lost, with the exception of small preopercular bones in *Acanthostega* and *Crassigyrinus*, which are incorporated into the suspensorium. Therefore, if the early tetrapods still retained such a fish-like buccal pump breathing mechanism, movements must have been accommodated by flexibility between the braincase and the palate and the palate and the skull roof. It is almost universally assumed, however (for example, Panchen and Smithson 1987) that early tetrapods were fully autostylic.

Autostyly, according to the definition by Huxley (1876), means that the palatoquadrate is united to the braincase directly 'by the part of its own substance which constitutes the suspensorium.' That is to say, the posterior part, which contacts the otic region. True autostyly is found for example in holocephalans, dipnoans, frogs, and urodeles, turtles, crocodiles and mammals. In these animals, the palate is firmly fixed to the braincase, with no possibility of movement. Panchen and Smithson (1987) point out that "autostyly" is not a single condition, and contrast "autodiastyly" in which the basal articulation remains mobile as in early tetrapods, with "autosystyly" in which it is fused as in lungfishes (de Beer 1937). However, a survey of the earliest tetrapods reveals that there is almost no recorded evidence of contact between the palatoquadrate and the otic region of the braincase.

In *Acanthostega*, the epipterygoid sends its vertical ramus dorsally to approach the skull roof in the region of the tabular—squamosal joint, at an unsutured facet formed on the underside of the dermal bones (Fig. 20.4). There is no evidence for an otic process of the epipterygoid. This is true of *Greererpeton* (Smithson 1982), *Palaeoherpeton* (Panchen 1964), and *Pholiderpeton* (Clack 1987) in which the stapes is known, and also in *Proterogyrinus* (Holmes 1984) and *Archeria* (Holmes 1989). In *Crassigyrinus* (Panchen 1985) and *Eoherpeton* (Smithson 1985), this region is too poorly represented for comment.

Only in loxommatids is there any contact described between the palatoquadrate and the otic capsule. Beaumont (1977) reported small facets on the basisphenoid of *Loxomma* which she interpreted as receiving processes from the

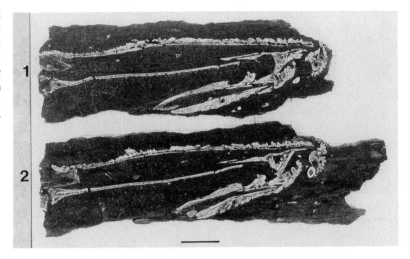

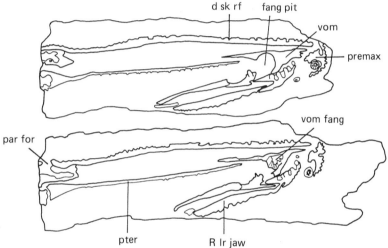

epipterygoid, examples of which were found in *Megalocephalus*. The facets are described as articular.

It is evident therefore that in the four early tetrapods for which the stapes is known, it forms the
only contact between the palatoquadrate and the
otic region of the braincase. By definition, then,
these animals were not autostylic. From the structure of the palate and braincase of other genera,
this condition is inferred to be widespread among
early tetrapods.

Further evidence suggests that the palate in
many early tetrapods was not firmly incorporated
into the rest of the skull. In *Acanthostega*, the epipterygoid contacted the skull roof via a stout buttress; in *Palaeoherpeton* (Panchen 1964) and
Pholiderpeton (Clack 1987) the epipterygoid also
formed a substantial buttress which probably

reached to the skull roof, and in many other early
tetrapods [for example, loxommatids, Beaumont
1977, *Greererpeton* (Smithson 1982), *Proterogyrinus* (Holmes 1984)], the ossified portions of the
columella cranii of the epipterygoid did not reach
the skull roof, though it is assumed that they were
extended dorsally in cartilage. However, there is
no evidence of firm attachment between the epipterygoid and the skull roof in any of these animals.

In *Acanthostega* and other early tetrapods, the
epipterygoid posterior to the columella cranii was
a thin plate, often less than a millimeter thick, as
was the pterygoid itself in many places. Beaumont
(1977) suggested that in loxommatids such bones
might have allowed some flexibility of the palate.

Anteriorly, the contact between the pterygoid
and the vomers is known in only a few early

FIGURE 20.7. MGUH field number 1604 (part). Skull of *Acanthostega gunnari*, in transverse section. Sections viewed posteriorly, anteriormost at the top. Skull roofing bones not present except part of left maxilla. For abbreviations see Fig. 20.1.

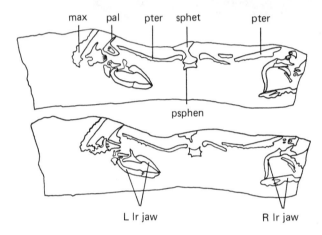

tetrapods. One reason for this is that vomers are frequently missing from even well-preserved specimens, such as *Proterogyrinus* (Holmes 1984) and *Archeria* (Holmes 1989). However, in *Acanthostega*, the contact was between a thin plate extending posteriorly from the vomer which lay dorsal to the pterygoid. The joint shows no interdigitation. By contrast the vomer is sutured to the premaxilla by a very heavily interdigitating suture (Fig. 20.6). This may account for the observation that in many skull specimens of *Acanthostega*, the snout, consisting of the premaxillae and vomers, is missing. In *Pholiderpeton* (Clack 1987), the pterygoid ran between the vomers, and the contact surface is characterized by a tongue and groove arrangement, with some longitudinal striations.

A similar situation occurs at the junctions between the pterygoid and the palatine and ectopterygoid. In *Acanthostega*, narrow, but plain surfaces from the palatines and ectopterygoids overlapped the pterygoids dorsally (Fig. 20.7). This arrangement is found in *Pholiderpeton*, in loxommatids (Beaumont 1977), in *Greererpeton* (Smithson 1982) and in *Archeria* (Holmes 1989). In the latter, the palatine and ectopterygoid are usually missing, and Holmes suggested that they were probably loose, indeed that the whole palate might have been loosely attached to the skull roof. Beaumont (1977) suggested that such simple overlapping sutures may represent areas of elasticity capable of accommodating slight distortion.

The contact between the palatoquadrate and the dermal bones of the suspensorium in early tetrapods is rather variable. Investigations in *Acanthostega* have not yet been completed. In *Pholiderpeton*, the squamosal and palatoquadrate met simply without suturing along the back of the suspensorium, though there was probably contact medially between a dorsal facet of the squamosal and the epipterygoid. The quadrate-quadratojugal suture is characterized by longitudinal grooves. In *Proterogyrinus*, Holmes (1984) describes the contact between the quadrate and quadratojugal as poorly consolidated, though does not describe that between the squamosal and epipterygoid. In loxommatids, Beaumont suggested that the quadrate was movable in relation to the squamosal and quadratojugal, and suggested a degree of kinesis between the palate and cheek. However, in *Greererpeton*, the palatoquadrate was apparently sutured to the back of the cheek (Smithson 1982).

In the earliest tetrapods, for example *Acanthostega*, *Crassigyrinus*, loxommatids, anthracosaurs, and early reptiles (e.g., *Eocaptorhinus*, Heaton 1979), there is a conspicuous articulation between the palate and braincase known as the basal, or basipterygoid articulation. A large process (basipterygoid process) from the braincase fitted into a socket in the palatoquadrate (epipterygoid), and synovial surfaces between the two are evident. This articulation is found likewise in primitive osteichthyans such as the earliest actinopterygians, and sarcopterygians such as osteolepiforms. There has been much discussion about the possible function of this joint and the degree of movement which it would have allowed (Panchen 1964, 1972; Holmes 1984; Clack 1987; Clack and Holmes 1988). In fishes, it presumably allowed the palate to move laterally and ventrally, as the volume of the buccal cavity changed during both ventilation and feeding movements. The shape of the articular facets in the embolomere *Archeria* suggest that similar movements might have occurred (Holmes 1989), though Smithson (1982) has argued that because of the skull structure of *Greererpeton*, no movement was possible at the joint in that animal.

Many of the above observations apply equally to the early reptile *Eocaptorhinus* (Heaton 1979). Not only did it have a synovial basal articulation, but the pterygoid sutures with the palatine were also simple overlaps. There was a muscle scar on the dorsal surface of the pterygoid which Heaton interpreted as left by the levator pterygoideus, by analogy with *Sphenodon* (Ostrom 1962). Likewise, in *Pholiderpeton* there was a massive muscle-scar in a similar position. The levator pterygoideus is a homologue of the levator palatoquadrati, which raises the palate in fishes.

In summary, there is a body of evidence to suggest that the palate was not firmly fixed to the rest of the skull in the earliest tetrapods, certainly it was not firmly fixed to the braincase. They should no longer be thought of as fully autostylic. The palate and braincase contacted at only two points, at the basal articulation and via the stapes, both of which had the potential for mobility. Potential flexibility, rather than a rigid attachment is suggested at other points where the palate and skull roof met. This corresponds more closely to the definition of amphistyly provided by Huxley (1876): "In the amphistylic skull, the palatoquadrate cartilage is quite distinct from the rest of the skull; but it is wholly or almost wholly, suspended by its own ligaments, the hyomandibula being small and contributing little to its support." This definition was based on the condition in the chondrichthyan 'Cestracion' (= *Heterodontus*). It is possible that the stapes was playing a greater role in the support of the palate in early tetrapods than Huxley envisaged. Significantly perhaps, the four earliest tetrapod stapes have cartilage finished distal edges, so not only is it difficult to define the orientation of the distal end, but the full extent of the contact between the palate and stapes cannot be determined. It could have been much greater than that which is preserved. I suggest that the maintenance of this amphistylic condition could, at least initially, have facilitated some movement of the palate associated with the retention of a buccal pump, and also with a persistent spiracle.

3.3 Functional Considerations

If the stapes were originally intimately involved with breathing mechanisms, release of the stapes from this role would have been dependent upon the development of alternatives. Some of these alternatives may be reflected in changes to the skull or postcranial skeletons of early tetrapods. Consideration of this prompts the following speculations.

The earliest group of tetrapods to depart from the pattern of amphistyly and palatal kinesis were the temnospondyls. Based on a cladistic analysis of many characters, it is now widely accepted that temnospondyls are paraphyletic with respect to frogs, or frogs plus urodeles, or to a monophyletic Lissamphibia (Milner 1988), and that all belong to the batrachomorph clade defined by Panchen and Smithson (1988).

Where the earliest tetrapods have closed, bony palates formed mainly by the pterygoids, temno-

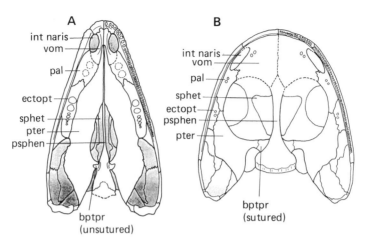

FIGURE 20.8. Palates of Carboniferous tetrapods showing contrast in construction. (A) *Pholiderpeton scutigerum*, an anthracosauroid; (B) *Platyrhinops* *lyelli*, a dissorophoid temnospondyl (original reconstructions). (Not to scale). For abbreviations see Fig. 20.1.

spondyls (excluding *Greererpeton*, which is probably not a true temnospondyl (Panchen and Smithson 1988)), developed large palatal vacuities (Fig. 20.8). From specimens which show denticulated ossicles in the palate (e.g., *Amphibamus*, Carroll 1964), we know that the vacuities were covered with skin in life. However, the remaining portions of the pterygoids were firmly sutured to the rest of the palate, for example as in *Platyrhinops lyelli* which I am currently redescribing with A.R. Milner. Furthermore, in later temnospondyls, the basipterygoid articulation became sutured up to provide a very firm junction, as noted by Watson (1919). Where the braincase is known in detail, for example in *Edops* (Romer and Witter 1942) and *Eryops* (Sawin 1941), there is a stout otic process of the epipterygoid contacting the otic capsule. In other words, temnospondyls became truly autostylic.

It is possible that especially the larger temnospondyls used some form of aspiration breathing, for example by recoil of the body wall as in *Polypterus* (Brainerd, Liem, and Samper 1989), or by movements of internal organs as in crocodiles (Gans 1970). I shall outline below reasons for believing that it was not powered by ribs. It is also parsimonious to suggest that, like frogs and other amphibians they continued to use a buccal pump mechanism in conjunction with cutaneous gas exchange. Arguments against the existence of these mechanisms have been based largely on the idea

that both, by themselves, would have been inadequate to meet the needs of a large animal (Gans 1970). This is perhaps not the place to review these arguments, however, some points at least can be made. Buccal pumping is used successfully by lungfishes, which can grow to a considerable size. Although two of the modern genera have compliant body walls, *Neoceratodus* is scale-covered and many fossil forms had long, substantial ribs almost circling the body, so that it is possible to inflate the lungs by buccal pumping under such circumstances. Cutaneous gas exchange would not have been inhibited by the amour of scales and dermal bones of early tetrapods, since these were often vascularized (Clack 1987, 1988a), and would have had a layer of epidermis externally. We cannot know enough about the physiology of these early tetrapods to make adequate judgements based on modern animals as models. I suggest that many modes of gas exchange and ventilation would have been used in conjunction with one another in the earliest tetrapods.

In temnospondyls the skin covering the palatal vacuities may have provided a more efficient buccal pumping mechanism than the bony palate of early tetrapods. In salamanders (Francis 1934), the levator bulbi muscle is said to act as an accessory respiratory muscle by raising the skin covering the palatal vacuities, thus enlarging the mouth cavity, though this needs to be corroborated using modern techniques. If true it would provide support for the above hypothesis.

In consequence of this possibly more efficient system, the stapes was no longer required to support the palate and operate the ventilation mechanism. The skull became autostylic, losing the connection between the palatoquadrate and the stapes, the latter becoming free to act as a hearing ossicle. The spiracle was replaced by a tympanic membrane and the spiracular pouch by the middle ear cavity. The temnospondyl lineage was the first to evolve an effective terrestrial ear, with a dorsally directed rod-like stapes, associated with an otic region closely comparable with that of frogs, and suggestive of the presence of a tympanum.

In other lineages, loss of the spiracle and/or buccal movements may have accompanied the development of aspiration breathing using ribs, which today is found only among amniotes. Circumstantial evidence suggests that it may have been achieved several times, judging from the diverse mechanisms seen among modern forms (Gans 1970). This suggests that their nearest common ancestor might not have used costal aspiration breathing exclusively. The mammalian pattern is by no means representative of amniotes in general, and seems to have evolved in conjunction with the dorso-ventral flexion of the spinal column during locomotion (Carrier 1987). Turtles cannot use their ribs for aspiration breathing, but have evolved a unique mechanism of their own, and crocodiles have evolved a mechanism functionally similar to a diaphragm (Gans 1970).

No modern amphibian uses costal movements in aspiration, and ribs in temnospondyls are very often rather short. Nevertheless, it is usually suggested that large temnospondyls such as *Eryops* must have used ribs (Romer 1972), since the ribs were sturdy, double headed, and usually bear uncinate processes. However, while recognizing that occurrence of muscle scars is frequently a poor guide to occurrence of muscles in modern animals, Olson's study (1936) reconstructing the musculature of *Eryops*, *Diadectes*, and *Dimetrodon* found evidence of intercostal muscle scars in the latter two (related to or actually amniotes) and none in *Eryops* (a temnospondyl), despite its being a large animal.

Another line of evidence however might suggest absence of costal movements in early tetrapods, and concerns the structure of the vertebrae. In most early tetrapods, the vertebrae were compound structures, with the neural arch in articulation with the centrum, which was bipartite, consisting of an anteroventral element (the intercentrum) and (often paired) posterodorsal elements (the pleurocentra). The neural arch articulated with the pleurocentrum, the capitulum (lower head) of the rib articulated with the intercentrum and the tuberculum (upper head) with the neural arch. In rhachitomous vertebrae, which are probably primitive for tetrapods and are found in *Acanthostega* and temnospondyls, the intercentra were dominant. In gastrocentrous vertebrae which are found in anthracosauroids and their probable relatives the seymouriamorphs and amniotes, the pleurocentra were dominant. In amniotes and seymouriamorphs the pleurocentrum became almost cylindrical, and the intercentrum was very small, becoming sutured to the pleurocentrum, or it disappeared.

Panchen (1967) suggested a possible functional interpretation of the difference between these arrangements. In the rhachitomous forms, emphasis was on locomotion using the myocommatal muscles, which insert on the ribs, a situation analogous to that in fishes. It suggests effective swimming movements. Many early tetrapods, and many temnospondyls, were primarily aquatic. In the gastrocentrous forms, emphasis was given to locomotion using axial muscles, inserting on the neural arches, and in fully gastrocentrous forms, the vertebral column with its firm attachment of the neural arch would have formed an effective girder. This form of vertebral construction is found among the more terrestrially adapted amniotes and seymouriamorphs.

Ribs could perhaps only have functioned efficiently in aspiration breathing in vertebrae which were firmly consolidated. That is to say in forms in which the centrum was a unit and sutured or fused to the neural arch, to give a solid anchorage with which the ribs could articulate. The fully gastrocentrous vertebrae of amniotes satisfy this criterion. There may therefore be a connection between terrestrial locomotion and the development of aspiration ventilation using ribs.

Microsaurs and nectrideans had cylindrical centra whose homologies are often difficult to evaluate (Carroll 1989). However, these two groups and amniotes are all notable for lacking a temporal notch. This could be a coincidental correlation, or it could be explained by the idea that they no longer required the spiracular breathing mechanism suggested for early tetrapods.

If this hypothesis is correct, a prediction can be made from it. It is that an animal which retains multipartite vertebrae should also retain some degree of amphistyly and a primitive i.e., plate-like stapes, unless the animal has palatal vacuities as do temnospondyls. The reverse is not the case, however. I do not suggest that animals with consolidated vertebrae will necessarily have lost the temporal notch, or necessarily have developed a dorsally directed rod-like stapes.

The seymouriamorphs could at first sight provide an exception to prove (i.e., probe) this rule. With fully gastrocentrous vertebrae, unlike amniotes they retain a conspicuous temporal notch (White 1939). However, seymouriamorphs, including *Seymouria* itself and its relatives the discosauriscids, have a specialized otic region, and though poorly ossified, the stapes in *Seymouria* was apparently a dorsally directed rod. Discosauriscids, seymouriamorphs from Czechoslovakia, are currently being described by Klembara. They probably possessed the unique seymouriamorph otic morphology, though the otic capsules were unossified, (Klembara, personal communication) and stapes are preserved in his three-dimensional material (Klembara 1990). We await their description with interest. The evidence suggests that, paralleling temnospondyls, seymouriamorphs developed a terrestrially adapted ear, according to this hypothesis, consequent upon the development of aspiration breathing using ribs. It is therefore significant to note that, while in *Seymouria*, there is no evidence for contact between the epipterygoid and the braincase, and the basal articulation was unsutured, interdigitating sutures united the pterygoids to the palatine and ectopterygoids, and thus to the skull roof, and to each other in the midline (White 1939). Thus seymouriamorphs were effectively autostylic.

Early reptiles, having lost the spiracular notch and developed some form of costal ventilation, nevertheless retained a massive stapes which apparently formed a brace between the braincase and the quadrate. In early reptiles and some modern ones, the basal articulation remained unsutured. As noted above, a levator pterygoidei

scar is present in *Eocaptorhinus*, in which the muscle was apparently well developed. In some individuals of *Sphenodon* (Ostrom 1962), the levator pterygoideus and protactor pterygoideus muscles operate to produce some kinesis of the palate employed during feeding. While there was apparently no protractor pterygoideus scar in *Eocaptorhinus* (Heaton 1979), some mobility in the palate is possibly indicated, as in *Sphenodon* perhaps associated with feeding.

The emergence of the amniotes is associated with full terrestrialization. The earliest amniotes apparently employed a static pressure system of jaw action, presumably in response to feeding on land, which is associated with changes to the proportions of the skull and jaws and the relative importance of the adductor musculature (Olson 1961). No longer required for buccal movements, an essentially primitive palatal construction could have been exploited in this system, in which the stapes maintained its connection with the palate, but in which it functioned as a strut between the palate and the braincase. True autostyly apparently developed independently in several amniote lineages. It occurs in turtles, crocodiles and mammals, the nearest common ancestor of which must have been an early reptile similar to *Eocaptorhinus* in that it was not fully autostylic. It may be no coincidence that these amniote groups have developed true terrestrial hearing also apparently independently. Further investigations of the time and mode of development of autostyly and terrestrial hearing among amniotes could test the hypothesis that these phenomena are connected.

When considering the role of the stapes in early tetrapods, it is instructive to take a holistic view of the animals involved. The problems tetrapods would have faced at the emergence onto land such as terrestrial breathing, locomotion and feeding may all have had their influences on the role of the stapes and the architecture of the skull. The contrasting ways in which different groups solved these problems may all be involved in the evolution of terrestrial hearing.

4. Summary

The stapes of *Acanthostega gunnari* is a stout bone, with a fan-shaped distal plate, resembling those of the colosteid *Greererpeton* and the anthracosauroid

Pholiderpeton. It probably had a single head and articulated with the palatoquadrate. The otic capsule was ossified dorsally, but may have been open to the cranial cavity. The fenestra ovalis is not preserved. The stapes found in *Acanthostega* is probably typical of that to be found in early tetrapods.

Contrary to previous assertions, the early tetrapod skull was not autostylic, and the stapes formed the major point of contact between the otic region of the braincase and the palate, the only other contact being at the basal articulation. Both of these contacts were potentially mobile. Other points of contact between the palate and the skull roof were not firmly consolidated, and could have allowed movement or distortion of the palate. Early tetrapods were thus amphistylic, and the stapes still had a role to play in supporting the palate and coordinating its movements during buccal ventilation. Temnospondyls were the first group to become fully autostylic, thus allowing the development of a terrestrial ear. In other groups, developments to the vertebral column and feeding system may have had indirect effects upon the fate of the stapes.

Acknowledgments. My thanks are due to the Greenland Geological Survey and to Dr. Svend Erik Bendix-Almgreen without whose help this material could not have been collected; the latter also generously allowed me to prepare and study it.

References

Allis AP (1922) The cranial anatomy of *Polypterus*, with special reference to *Polypterus bichir*. J Anat 56: 189–294.

Beaumont E (1977) Cranial morphology of the Loxommatidae (Amphibia: Labyrinthodontia). Phil Trans R Soc Lond B 280:29–101.

Beer GR de (1937) The development of the vertebrate skull. Oxford University Press.

Bendix-Almgreen SE, Clack JA, Olsen H (1988) Upper Devonian and Upper Permian vertebrates collected in 1987 around Keyser Franz Joseph Fjord, central East Greenland. Rapp Grønlands Geol Unders 140:95–102.

Bolt JR, Lombard RE (1985) Evolution of the amphibian tympanic ear and the origin of frogs. Biol J Linn Soc 24:83–99.

Bolt, JR, Lombard RE (1991) Survey of otic morphology in Paleozoic tetrapods. In: Popper A, Webster D, Fay RR (eds) The evolutionary biology of hearing. New York: Springer-Verlag.

Bolt JR, McKay RM, Witzke BJ, McAdams MP (1988) A new Lower Carboniferous tetrapod locality in Iowa. Nature 333:768–770.

Brainerd EL, Liem KF, Samper CT (1989) Air ventilation by recoil aspiration in polypterid fishes. Science 246:1593–1595.

Carrier DR (1987) The evolution of locomotor stamina in tetrapods: circumventing a mechanical constraint. Paleobiology 13:326–341.

Carroll RL (1964) Early evolution of the dissorophid amphibians. Bull Mus Comp Zool Harv 131:1–250.

Carroll RL (1980) The hyomandibular as a supporting element in the skull of primitive tetrapods. In: Panchen AL (ed) The terrestrial environment and the origin of land vertebrates. London: Academic Press.

Carroll RL (1989) Developmental aspects of lepospondyl vertebrae in Paleozoic tetrapods. Historical Biology 3:1–25.

Clack JA (1983) The stapes of the Coal Measures embolomere *Pholiderpeton scutigerum* Huxley (Amphibia: Anthracosauria) and otic evolution in early tetrapods. Zool J Linn Soc 79:121–148.

Clack JA (1987) *Pholiderpeton scutigerum* Huxley, an amphibian from the Yorkshire Coal Measures. Phil Trans R Soc Lond B 318:1–107.

Clack JA (1988a) New material of the early tetrapod *Acanthostega* from the Upper Devonian of East Greenland. Palaeontology 31:699–724.

Clack JA (1988b) Pioneers of the land in East Greenland. Geology Today 4:192–194.

Clack JA (1989) Discovery of the earliest tetrapod stapes. Nature 342:425–430.

Clack JA, Holmes R (1988) The braincase of the anthracosaur *Archeria crassidisca* with comments on the interrelationships of primitive tetrapods. Palaeontology 31:85–107.

Coates MI, Clack JA (1990) Polydactyly in the earliest-known tetrapods limbs. Nature 347:66–69.

Francis ETB (1934) The anatomy of the salamander. Oxford: Clarendon Press.

Fritzsch B (1991) The water-to-land transition: evolution of the tetrapod basilar papilla, middle ear and auditory nuclei. In: Popper A, Webster D, Fay RR (eds) The evolutionary biology of hearing. New York: Springer-Verlag.

Gans C (1970) Respiration in early tetrapods – the frog is a red herring. Evolution 24:723–734.

Godfrey SJ (1989) Ontogenetic changes in the skull of the Carboniferous tetrapod *Greererpeton burkemorani* Romer, 1969. Phil Trans R Soc Lond B 323:135–153.

Heaton MJ (1979) Cranial anatomy of primitive captorhinid reptiles from the late Pennsylvanian and early Permian of Oklahoma and Texas. Bull Okla Geol Surv 127:1–84.

Holmes R (1984) The Carboniferous amphibian *Proterogyrinus scheeli* Romer and the early evolution of tetrapods. Phil Trans R Soc Lond B 306:431–527.

Holmes R (1989) The skull and axial skeleton of the Lower Permian anthracosauroid amphibian *Archeria crassidisca* Cope. Palaeontographica A 207:161–206.

Hook RW (1983) *Colosteus scutellatus* (Newberry), a primitive temnospondyl amphibian from the Middle Pennsylvanian of Linton Ohio. Am Mus Nov 2770: 1–41.

Huxley TH (1876) Contributions to morphology. Ichthyopsida. No. 1. On *Ceratodus forsteri*, with observation on the classification of fishes. Proc Zool Soc Lond 1876:24–59.

Jarvik E (1952) On the fish-like tail in the ichthyostegid stegocephalians. Meddr Grønland 114:1–90.

Jarvik E (1980) Basic structure and evolution of vertebrates, (Vol. 1). London: Academic Press, 575 pp.

Klembara J (1990) Discosauriscids from Boscovice Furrow, Moravia, Czechoslovakia. In: Schweiss D, Heidtke U (eds) New Results on Permocarboniferous fauna. Abstracts. Bad Durkheim: Pollichia-buch.

Lombard RE, Bolt JR (1988) Evolution of the stapes in Paleozoic tetrapods. In: Fritzsch B (ed) The evolution of the amphibian auditory system. New York: John Wiley.

Milner AR (1988) The relationships and origin of living amphibians. In: Benton MJ (ed) The phylogeny and classification of tetrapods, (Vol 1). Oxford: Clarendon Press.

Milner AR, Smithson TR, Milner AC, Coates MI, Rolfe WDI (1986) The search for early tetrapods. Modern Geology 10:1–28.

Olson EC (1936) The dorsal axial musculature of certain primitive Permian tetrapods. J Morphol 59:265–311.

Olson EC (1961) Jaw mechanisms: rhipidistians, amphibians, reptiles. Am Zool 1:205–215.

Ostrom JH (1962) On the constrictor dorsalis muscles of *Sphenodon*. Copeia 4:732–735.

Panchen AL (1964) The cranial anatomy of two Coal Measure anthracosaurs. Phil Trans R Soc Lond B 242:207–281.

Panchen AL (1967) The homologies of the labyrinthdont centrum. Evolution 21:24–33.

Panchen AL (1972) The skull and skeleton of *Eogyrinus attheyi* Watson (Amphibia: Labyrinthodontia). Phil Trans R Soc Lond B 263:279–326.

Panchen AL (1985) On the amphibian *Crassigyrinus scoticus* Watson from the Carboniferous of Scotland. Phil Trans R Soc Lond B 309:461–568.

Panchen AL, Smithson TR (1987) Character diagnosis, fossils and the origin of tetrapods. Biol Rev 62:341–438.

Panchen AL, Smithson TR (1988) The relationships of the earliest tetrapods. In: Benton MJ (ed) The phylogeny and classification of the tetrapods, (Vol. 1). Oxford: Clarendon Press.

Romer AS (1947) Review of the Labyrinthodontia. Bull Mus Comp Zool Harv 99:1–368.

Romer AS (1966) Vertebrate Paleontology 3rd ed. Chicago: Chicago University Press.

Romer AS (1972) Skin breathing – primary or secondary? Resp Physiol 14:183–192.

Romer AS, Witter RV (1942) *Edops*, a primitive rhachitomous amphibian from the Texas red-beds. J Geol 50:925–960.

Sawin HG (1941) The cranial anatomy of *Eryops megacephalus*. Bull Mus Comp Zool Harv 88:405–465.

Smithson TR (1982) The cranial morphology of *Greererpeton burkemorani* Romer (Amphibia: Temnospondyli). Zool J Linn Soc 76:29–90.

Smithson TR (1985) The morphology and relationships of the Carboniferous amphibian *Eoherpeton watsoni* Panchen. Zool J Linn Soc 85:317–410.

Smithson TR (1989) The earliest known reptile. Nature 342:676–678.

Watson DMS (1919) The structure, evolution and origin of the amphibia – the "orders" rachitomi (sic) and stereospondyli. Phil Trans R Soc B 209:1–73.

White TE (1939) Osteology of *Seymouria baylorensis* Broili. Bull Mus Comp Zool Harv 85:325–409.

21
The Effects of Body Size on the Evolution of the Amphibian Middle Ear

Thomas E. Hetherington

1. Introduction

Body size can be an important determinant of many features of animals (Schmidt-Nielsen 1984; Calder 1984). This chapter will examine how body size may guide the evolution of mechanisms of acoustic reception in amphibians. Amphibians are a good group for such an analysis because they possess a diversity of systems responsive to acoustic signals (besides just the standard tympanic middle ear) and many aspects of these systems appear to be size dependent. This chapter will review what is known about acoustic reception in amphibians, will consider size-related aspects of the systems involved, and will discuss how evolutionary changes in body size may have affected the evolution of these systems. Much attention will be focused on the role body size may play in a common evolutionary trend in amphibians, namely the reduction and loss of a tympanic middle ear. A major aim of this chapter is to set forward hypotheses about the relationship between body size and acoustic function; because of a lack of pertinent data in many cases, this is about the best that can be accomplished at this time.

This chapter will focus on anuran amphibians (frogs and toads), simply because most research on amphibian hearing has centered on the elucidating the function of the well-developed tympanic middle ear observed in this group. But anurans also possess the other major middle ear system observed in amphibians, the opercularis system, and provide the best evidence of nontympanic mechanisms of sound reception. Therefore, anurans provide the best opportunity to analyze the functional proper-

ties, and the functional interaction, of these different strategies of acoustic reception. Anurans also come in a wide range of body sizes, occupy a variety of habitats, and display a diversity of lifestyles. This group, therefore, allows a good opportunity to study evolutionary trends and possible selective factors operating within the same lineage.

2. Design and Function of Acoustic Receptive Systems in Amphibians

2.1 The Tympanic Middle Ear

A tympanic middle ear is found only in metamorphosed anurans and is absent in salamanders and gymnophiones (Wever 1985; Wilczynski and Capranica 1984; Jaslow, Hetherington, and Lombard 1988). Most anurans possess a tympanic ear, but it is reduced or absent in some species. The anuran tympanic ear is very similar in design to that of other tetrapod vertebrates. It consists of a tympanum overlying an air-filled middle ear cavity (Fig. 21.1). The tympanum is attached to an extrastapes that passes medially to articulate with an auditory ossicle, the stapes (Fig. 21.1). Many workers use the term columella rather than stapes (Capranica 1976; Wever 1985). I use the latter term because it is very likely that the amphibian stapes is homologous to the primitive vertebrate hyomandibula (Lombard and Bolt 1988). Medially, the stapes has an expanded footplate that fits into the oval window of the otic capsule (Fig. 21.1), thereby relaying motion of the tympanum to the

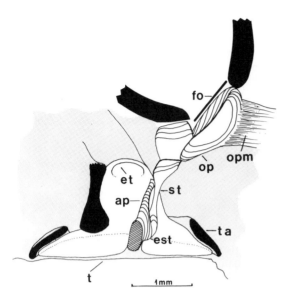

FIGURE 21.1. Semidiagrammatic reconstruction of the right middle ear of the frog *Rana warschewitschii*. This is an example of a well-developed tympanic middle ear. A frontal view looking dorsally. Lateral is down, and anterior to the left. Cut edges of skull elements, otic capsule, and tympanic annulus (ta) are black. Cut edges of middle ear elements are cross-hatched and surrounding soft tissue is stippled. Ventral portions of the tympanic cavity and membrane are not shown. (ap) ascending process of extrastapes; (est) extrastapes; (et) eustachian tube; (fo) oval window; (op) operculum; (opm) opercularis muscle; (st) stapes; (t) tympanum. (From Jaslow et al. 1988. Structure and Function of the Amphibian Middle Ear. Copyright © 1988, John Wiley & Sons. Reprinted by permission.)

inner ear fluids. Typically, there is a cartilaginous tympanic annulus along the margin of the tympanum. The extrastapes is always cartilaginous and has various processes connecting it to the medial surface of the tympanum. There is usually an ascending process of the extrastapes that connects to the dorsal parotic crest of the otic capsule. Medially, the extrastapes has a fibrous connection to the stapes. The latter element is usually ossified. Its lateral portions lie in the middle ear cavity, but medially it passes through connective and muscle tissue to reach the oval window (Wever 1985; Jaslow, Hetherington, and Lombard 1988). This is one difference between the tympanic middle ear of anuran amphibians and that of many other tetrapods; the auditory ossicle(s) of the latter usually traverse only through air-filled spaces of the mid-

dle ear cavity. Medially the stapes expands into a typically cartilaginous footplate that fits into the oval window and has a ventral hinge on the otic capsule. This hinge presumably acts as a fulcrum about which the stapedial footplate rocks in response to tympanic motion. The relatively long in-lever arm (extrastapes and stapes) and short out-lever arm (stapedial footplate) of this lever arrangement probably functions to amplify footplate motion in the oval window. The stapedial footplate sometimes lies in a chamber lateral to the oval window (Wever 1985; Jaslow, Hetherington, and Lombard 1988). In such cases the footplate may be shielded from the oval window by a shelf of bone (see Fig. 21.10). The right and left middle ear cavities typically connect to the pharynx via eustachian tubes, so the two cavities usually are linked together by air-filled channels. This arrangement allows the tympanic middle ear to function as a directionally sensitive pressure-gradient receiver (Feng and Schofner 1981; Pettigrew, Chung, and Anson 1978; Rheinlaender, Walkowiak, and Gerhardt 1981; Eggermont 1988).

2.2 The Opercularis System

A striking feature of the ears of many amphibians is the presence of a middle ear system completely distinct from the tympanic middle ear, the opercularis system. The opercularis system is more widespread than the tympanic middle ear, and is found in all metamorphosed anurans and salamanders (Kingsbury and Reed 1909; Wever 1985; Hetherington, Jaslow, and Lombard 1986). It is absent only in completely aquatic (neotenic) salamanders and in gymnophiones. The opercularis system consists of an opercularis muscle that originates on parts of the pectoral girdle and inserts on an otic operculum, a movable element resting in the oval window (Figs. 21.1, 21.2). The operculum is usually cartilaginous, and is held in the oval window by elastic connective tissue (Hetherington, Jaslow, and Lombard 1986). At some point along its margin it is attached to the otic capsule; this connection appears to act as a hinge about which the operculum can move in and out of the oval window. In some species the operculum has a raised ridge on its posterior surface where the opercularis muscle inserts; this ridge probably functions to increase the moment arm of the muscle and increase force of opercular motion (Hetherington,

Jaslow, and Lombard 1986). The opercularis muscle of different amphibians is not always homologous; in all anurans and many salamanders it is a specialized part of the levator scapulae superior muscle (Monath 1965). In the large family of plethodontid salamanders, and in some hynobiids, it is represented by part of the cucullaris muscle (Monath 1965; Smirnov 1987). Regardless, in all cases it is largely tonic in nature, and is specialized for maintained tension (Becker and Lombard 1977; Hetherington 1987).

Whereas the tympanic middle ear clearly functions in reception of aerial sound, the function of the opercularis system has been a matter of contention. Originally the system was proposed to act in detection of ground vibrations (Kingsbury and Reed 1909), and recent experimental studies have demonstrated that it can indeed function in seismic reception (Hetherington 1985, 1988). The system basically acts to exaggerate inertial differences between the motion of the operculum and inner ear. When the head is held off the substrate (as is typical in frogs and salamanders), the operculum will be linked by the opercularis muscle to a part of the body directly affected by ground motion. The shoulder girdle of frogs and salamanders is extremely responsive to vertical substrate motion acting on the forelimbs and thoracic body regions, and motion of the girdle is often greater than that of the substrate at frequencies below about 100 Hz (Hetherington 1988). The shoulder girdle of many amphibians may be especially responsive to the action of Rayleigh waves, a common type of seismic signal that propagates along the surface of the ground (Lewis and Narins 1985; Narins 1990). Most of the motion associated with these waves is vertically oriented, and much of the energy of the waves is restricted to frequencies below 100 Hz (Lewis and Narins 1985). The tonic opercularis muscle appears to be constantly tensed when amphibians are in terrestrial situations (Hetherington and Lombard 1983; Hetherington 1987), and can therefore effectively transfer motion of the shoulder girdle to the operculum. Movement of the latter relative to the otic capsule will generate fluid motion within the inner ear that can stimulate appropriate receptor areas, such as the saccule (Koyama et al. 1982; Lewis and Lombard 1988).

Other functions, however, have been attributed to the opercularis system. Most significantly,

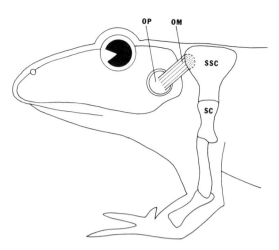

FIGURE 21.2. A diagrammatic representation of the amphibian opercularis system. An opercularis muscle (OM), originating on the shoulder skeleton, inserts on an operculum (OP) that rests in the oval window of the otic capsule. As in salamanders and some anurans, a tympanic middle ear is absent. If present, a stapedial footplate would be positioned in the oval window anterior to the operculum. (SC) scapula; (SSC) suprascapula.

Lombard and Straughan (1974) and Wever (1979; 1985) suggested that the system acts to modulate responsiveness of the tympanic middle ear for purposes of frequency tuning and protection of the inner ear respectively. However, the tonic, slowly contracting nature of the opercularis muscle seems ill suited for rapid modification of stapedial motion (Hetherington, 1987), and the morphology of the "joint" between the stapedial footplate and operculum appears designed to allow independent motion of the two elements in the oval window rather than somehow to link the operculum to the stapes (Hetherington, Jaslow, and Lombard 1986). Also, many amphibians that lack a tympanic middle ear (salamanders and several anurans) still possess an opercularis system, so such an hypothesized function is not generalizable to all amphibians. It therefore seems unlikely that the opercularis system is functionally linked to the tympanic middle ear.

Lombard and Straughan (1974) found that removal of the opercularis muscle of certain anurans decreased midbrain responses to sound below about 1,000 Hz, and suggested that the opercularis system can increase the responsiveness of the tympanic middle ear to low frequency sound. However, their findings may only demonstrate that the opercularis system itself may function in sound reception. Removal of the opercularis muscle can decrease inner ear responses to sound in species that lack a tympanic middle ear, such as salamanders, and this decrease is most pronounced at low frequencies (Hetherington 1989). The large suprascapula forms a broad, thin sheet of cartilage over the shoulder area of many anurans and salamanders, and low-frequency sound may cause a significant motion of this element. For example, in anurans, suprascapular motion may be greater than tympanic motion at frequencies below about 200 to 500 Hz, depending on the species (Hetherington, in preparation). Such motion may be relayed to the operculum and subsequently to the inner ear fluids. The observations of Walkowiak (1980) that midbrain responses in the anuran *Bombina bombina*, a species that lacks a tympanic ear, were highest when a closed sound system delivered stimuli over the shoulder region provides some support for this hypothesis.

Alternatively, the opercularis system could act as an inertial system for reception of aerial sound. Because low-frequency sound appears to effectively penetrate the head tissues of many amphibians (Aertsen et al. 1986), the opercularis muscle may increase the inertial differences in motion between the operculum and inner ear fluids by anchoring the operculum to another body region (Hetherington 1985).

The opercularis system therefore may function in reception of both ground vibrations and aerial sounds. Further study of its possible role in detecting aerial sound is needed, but for the purposes of this chapter the opercularis system will be considered one type of nontympanic pathway of sound reception.

2.3 Other Nontympanic Pathways of Sound Reception

Pathways other than the tympanic ear and opercularis system may be involved in reception and transmission of acoustic stimuli. One general observation is that low-frequency sound is readily absorbed by unspecialized body tissues; for example, Aertsen et al. (1986) found that the head of *Rana temporaria* is nearly transparent to sounds below about 2 kHz. Other researchers have found

similar results, although most have limited such absorption to a lower frequency range, often below about 500 to 1,000 Hz (Feng 1980; Chung, Pettigrew, and Anson 1981). In addition, several studies have minimized the role of the tympanic middle ear in reception of low-frequency sounds. For example, Wilczynski, Resler, and Capranica (1987) found in *Rana pipiens* that inner ear responses to sound below about 500 Hz could be produced as effectively via extratympanic sound transmission as by tympanic sound transmission. Chung, Pettigrew, and Anson (1981) found that the tympanum of *Rana pipiens* was responsive mainly to frequencies above about 800 Hz. There actually has been a great deal of variability in measurements of tympanic responsiveness to sounds of different frequencies. It appears that factors contributing to this variability include extreme sensitivity of the tympanic ear to methods of stimulation, pressure fluctuations in the mouth cavity and lungs during experiments, and even posture of the animals (Moffat and Capranica 1978; Pinder and Palmer 1983; Eggermont 1988). Another factor contributing to such variability is body size (see below). Nonetheless, it appears that generally the anuran tympanic middle ear is most responsive, depending on the species, to frequencies above about 500 to 2,000 Hz, and that reception of low-frequency sound may not require a tympanic middle ear.

What structures and mechanisms are involved in reception of low-frequency sound? One possibility, as discussed in Section 2.2, is the opercularis system. Severing the opercularis muscle decreased midbrain responses to sound below about 1 kHz in some frogs (Lombard and Straughan 1974), and interruption of the opercularis system can produce decreases in inner ear responses to low-frequency sound in both anurans and salamanders (Paton 1971; Hetherington 1989). However, other pathways of sound transfer involving air-filled cavities of the head and body have been demonstrated, and these pathways may act in coordination with the tympanic middle ear or may act independently in stimulation of the inner ear. Low frequency sound apparently can penetrate into the buccal cavity through the floor of the mouth and produce pressure waves in the middle ear cavity that affect tympanic motion (Feng 1980; Vlaming, Aertsen, and Epping 1984). Also, areas of the lateral body wall overlying the lungs may be very responsive to aerial sound (Narins, Ehret, and Tautz 1988; Ehret, Tautz, and Schmitz 1990). Fluctuations in lung pressure generated by body wall motion can pass forward to the middle ear cavities and affect tympanic motion Fig. (21.3), and this route of sound transmission is most effective at relatively low frequencies (Narins, Ehret, and Tautz 1988; Ehret, Tautz, and Schmitz 1990). It also is possible that body wall reception may directly stimulate the inner ear without involving the tympanic middle ear. For example, Narins, Ehret, and Tautz (1988) suggested that pressure fluctuations in the lungs could be transferred to the endolymphatic sac system that extends from the vertebral canal to the inner ear (Fig. 21.3).

These nontympanic pathways may be important for reception of low-frequency sound, but they also probably play a role in improving the directionality of the tympanic middle ear. Vlaming, Aertsen, and Epping (1984) and Eggermont (1988) have demonstrated by modelling studies that an additional pathway for modifying pressure within the middle ear cavity besides the tympanic ear system is required for the tympanic ear to show directional characteristics at relatively low frequencies. Without such an additional pathway the ear would only be directionally sensitive at its natural frequency. One such pathway could be via the body wall and through the lungs to the pharynx and middle ear cavities (Narins, Ehret, and Tautz 1988). Nontympanic pathways of sound transfer may therefore be critical for producing a directionally sensitive tympanic middle ear over a wide frequency range.

2.4 An Integrated View of Acoustic Reception in Amphibians

Several peripheral systems therefore may be involved in acoustic reception in amphibians: the tympanic middle ear, the opercularis, system, and other various nontympanic pathways, including the lateral body wall and lungs. Although much remains to be determined concerning their functional significance, the following generalizations are likely to be correct. The tympanic ear is specialized for reception of aerial sound, and

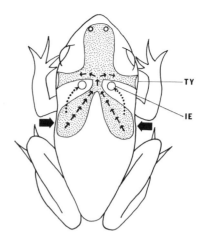

FIGURE 21.3. Schematic diagram of hypothesized pathways of sound reception involving the lateral body wall of frogs. The lungs lie just below the surface of the lateral body wall and external sound pressures (large arrows) can produce motion of the body wall. Body wall motion may be transferred directly (dotted pathways) to the inner ears (IE) or may produce pressure fluctuations inside the lungs that pass forward via the bronchi and glottis to the middle ear cavities (arrow pathways). The latter may modify motion of the tympana (TY). (Modified after Narins et al. 1988.)

certain nontympanic pathways, such as the lateral body wall and lungs, may also function in sound reception. These systems are effective over different frequency ranges; the tympanic ear is more responsive to higher frequencies and the nontympanic pathways are more responsive to lower frequencies. The tympanic middle ear also functions as a pressure-gradient receiver and provides directional information at certain frequencies, but its directional characteristics at low frequencies probably are improved by nontympanic pathways of sound transfer. The opercularis system is functionally independent of the tympanic middle ear and can function in the reception of groundborne vibrations. However, the system also may act as another nontympanic pathway for sound reception of low-frequency sounds. A variety of receptive systems may therefore be operating within the same individual, each involved in the detection of different types of acoustic signals and in the extraction of different types of information.

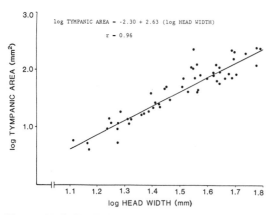

FIGURE 21.4. Graph demonstrating scaling of tympanic area with head width in the American bullfrog, *Rana catesbeiana*. The slope of the allometric curve is 2.63, demonstrating that the tympanum is becoming proportionately larger as size increases.

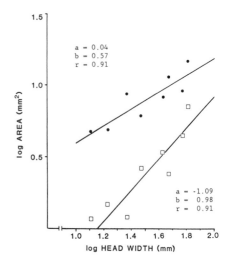

FIGURE 21.5. Graph demonstrating scaling of stapedial footplate area (open squares) and opercular area (solid circles) with head width in the American bullfrog *Rana catesbeiana*. The slopes of the allometric curves are 0.98 and 0.57 for the stapedial footplate and operculum respectively, so both elements are becoming proportionately smaller as size increases. Also, as body size increases the stapedial footplate becomes relatively larger compared to the operculum.

3. Effects of Body Size on Mechanisms of Acoustic Reception

3.1 Scaling of Middle Ear Morphology

No detailed analysis of scaling of the anuran middle ear has been done, but sufficient data are available to suggest some interesting allometric trends. Shofner and Feng (1981) measured ontogenetic changes in size of the tympanum in the bullfrog (*Rana catesbeiana*) and found that the tympanum becomes proportionately larger with increasing body size. Figure 21.4 shows data from the same species that demonstrate the relationship between tympanic area and head width. Tympanic area becomes proportionately greater as size increases; the slope (b) of the allometric curve (2.63) is greater than that expected for geometric scaling (2.00). A similar pattern has been observed in other anuran species, although the tympana of small species tend to scale in a more geometrically similar fashion (personal observation).

Figure 21.5 includes scaling data for stapedial footplate area in the bullfrog. Unlike tympanic area, footplate area becomes proportionately smaller as size increases ($b < 2.00$). These data demonstrate that the ratio of tympanic area to stapedial footplate area, presumably an important

measure of the amplifying characteristics of the middle ear, increases with increasing size. These scaling trends observed in tympanic and stapedial footplate area therefore suggest that sensitivity of the bullfrog tympanic ear increases with size. It must be stressed that scaling data need to be obtained for many other features of the tympanic ear, such as ossicular mass, middle ear stiffness, middle ear cavity volume, etc., before a complete functional picture is clear. However, these preliminary data suggest that morphological and functional characteristics of the anuran tympanic ear are clearly related to body size and that larger frogs possess better-developed and more sensitive tympanic ears.

Scaling of the opercularis system has not been studied before. Figure 21.5 includes scaling data for opercular area in the bullfrog *Rana catesbeiana*. The slope of the allometric curve is 0.57, so the operculum becomes proportionately smaller with increasing size. This finding is probably generalizable to most anurans (personal observation). The decrease in relative size of the opercularis system as body size increases may suggest that the system is more effective for acoustic reception in smaller species. Examination of the allometric curves for opercular and stapedial footplate area in Figure 21.5 shows that the operculum becomes smaller relative to the stapedial footplate as body size increases. One interpretation of these data is that, although the opercularis system may be useful for hearing (as well as vibration reception) in small animals, the tympanic middle ear becomes increasingly more important for sound reception in larger animals. Some data relevant to testing this hypothesis about the relative effectivenesses of the opercularis system and tympanic ear in acoustic reception in small and large anurans are discussed in the following sections.

3.2 Scaling of the Frequency Response of the Tympanic Middle Ear

The tympanic middle ears of vertebrates, depending on specific conditions of mass, stiffness, and friction, show certain patterns of responsiveness to sounds of different frequencies. The frequency response of the anuran tympanic ear varies among species, although there has been no broadly comparative study including species of many different sizes. Several studies have analyzed acoustic responses of the auditory nerve and various brain centers, but their findings may reflect more the frequency tuning of the auditory end organs of the inner ear rather than the frequency response of the middle ear (Shofner and Feng, 1981; Shofner 1988). Indeed, physiological studies on frogs have established that the best excitatory frequency of eighth nerve units is size-dependent, with smaller animals having higher best excitatory frequencies, and such differences have been suggested to be related to inner ear tuning (Narins and Capranica 1976; Wilczynski et al. 1984; Ryan and Wilczynski 1988). However, tympanic motion in response to sound has been directly measured by several workers, and their data suggest a correspondence between body size and the frequency response of the tympanic ear (Moffat and Capranica 1978; Chung, Pettigrew, and Anson 1981; Pinder and Palmer 1983). The ears of smaller frogs appear more responsive to higher frequencies and those of larger frogs appear more responsive to lower

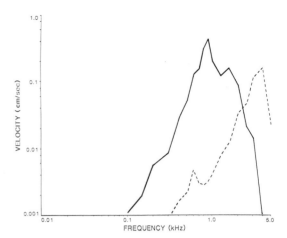

FIGURE 21.6. Comparison of the frequency responses of the tympana of a small toad (a 1.3 g *Bufo quercicus*; dashed line) and a large toad (a 104 g *Bufo americanus*; solid line) as measured with a Polytec laser vibrometer at 90 dB SPL using a laterally positioned sound source in free-field conditions. Tympanic motion is generally greater and shifted toward lower frequencies in the larger species.

frequencies. This matches findings from other terrestrial vertebrates that relatively small tympana and stapedial footplates respond best to higher frequencies (Khanna and Tonndorf 1978; Rosowski and Graybeal 1991). Figure 21.6 compares the frequency response of the tympanum of a very small, 1.3 g bufonid (*Bufo quercicus*) to that of a much larger, 104 g bufonid (*Bufo americanus*). The responsiveness of the tympanum of the small species peaks at around 4 kHz, whereas that of the large species peaks at about 0.9 kHz. The same pattern has been observed in hylid and ranid frogs of different body sizes (Hetherington, in press).

Physiological studies of the midbrain and auditory nerve have demonstrated that anurans are often most sensitive to sounds matching the dominant frequency of behavioral calls (Loftus-Hills 1973; Wilczynski and Capranica 1984). It might be expected that the tympanic ear also is tuned to the same frequency range. This tuning could be achieved through the general correspondence between body size and the frequency response of the tympanic ear noted above. Dominant frequencies of behavioral calls typically show the same size relationship; small frogs have calls of higher frequency than large frogs

(Loftus-Hills 1973; Ryan 1985; Ryan and Wil-czynski 1988). Therefore, frequency matching of tympanic responsiveness and behavioral calls may not have required any special co-evolution of sound-producing and sound-receiving structures; rather, a general matching simply might be retained as evolutionary changes in body size occur. However, there are many exceptions to the general relationship of body size and dominant call frequency. Ryan (1985) demonstrated that frogs of the genus *Physalaemus* have calls of lower frequency than other similarly sized frogs of the same subfamily and suggested that this was caused by intense sexual selection through female preference for lower frequency calls. In fact, most families of anurans will have some species that have a similar body size but have calls with very different dominant frequencies (Loftus-Hills 1973). For example, two species of toads (*Bufo rangeri* and *Bufo carens*) of about the same body size have calls with dominant frequencies of about 2,000 Hz and 200 Hz respectively (Passmore and Carruthers 1979). It is probable that natural selection has molded the tuning characteristics of the tympanic ears of each species to match the call frequencies, but experimental study of such cases is required. As of yet there has been no broad survey to analyze the relationship of tympanic responsiveness and dominant call frequencies. Regardless, it seems safe to assume that although natural selection can complicate the scaling relationship, the tympanic ears of small anurans generally are more responsive to high frequency sound and those of larger anurans are more responsive to lower frequency sound.

The tympana of larger anurans also appear to be generally more sensitive to sound that those of smaller species. Although the responsivenesses of the tympana of large and small anurans peak at different frequencies, the peak velocities observed in large specimens are generally higher than those found in smaller animals (Hetherington, in press). In Figure 21.6, for example, the peak velocity of the tympanum of the larger toad is higher than that of the smaller species. Such higher tympanic responsiveness probably reflects increased sensitivity of the tympanic middle ears of larger animals.

3.3 Scaling of Nontympanic Sound Reception

It was hypothesized above that the tympanic middle ear of anurans is used primarily for reception of high-frequency sound and that nontympanic pathways, including the opercularis system and lateral body wall surface, are used primarily for reception of lower frequency sound. However, as found for the tympanic middle ear, the acoustic responsiveness of these nontympanic systems may be very dependent on body size.

As described in Section 3.1, the opercularis system is proportionally larger in small amphibians, and it can be hypothesized that the opercularis system is more important in aerial sound reception in small species than in large species. Several studies have demonstrated that sound of certain frequencies can effectively penetrate head tissues (Feng 1980; Chung, Pettigrew, and Anson 1981; Aertsen et al. 1986), and the opercularis system could function to maximize inertial differences in motion of the operculum and otic capsule produced by absorbed sound energy. The degree of sound absorption by unspecialized head tissues depends on both the frequency of the sound and the size of the animal. This can be demonstrated by laser vibrometric measurements of motion of the lateral head surface anterior to the tympanum of several species of frogs exposed to aerial sound. In small frogs, less than about 5 g, the lateral head surface is often more responsive than the tympanum to sound up to about 500 to 600 Hz, whereas in larger frogs, over about 100 g, lateral head motion is greater than tympanic motion only at frequencies below about 100 Hz (Hetherington, in press). The same pattern occurs when sound effects on the suprascapular (shoulder) region are measured; suprascapular motion in smaller frogs is greater than tympanic motion up to about 500 Hz, whereas suprascapular motion in larger frogs is greater than tympanic motion only at much lower frequencies of about 50 to 100 Hz (Hetherington, in press). Therefore, if the opercularis system functions in sound reception by utilizing suprascapular motion or by maximizing inertial differences in motion of the operculum and inner ear fluids, its effectiveness should be body-size dependent.

Results from previous studies attempting to determine the role of the opercularis system in sound reception need to be reexamined in light of the influence of body size. Lombard and Straughan (1974) observed that removal of the opercularis muscle of small hylid frogs decreased midbrain responses to sound below about 1,000 Hz. Paton (1971) and Hetherington (1989), however, found that muscle removal in much larger bullfrogs (*Rana catesbeiana*) decreased inner ear responses to sound at only very low frequencies below about 100 Hz. These findings are consistent with the hypothesis that the effectiveness of the opercularis system for sound reception is size-dependent. It may be effective over a larger frequency range in smaller species.

The acoustic responsiveness of the lateral body wall of anurans also is dependent on size (Hetherington, in press). Figure 21.7 shows the acoustic responsiveness of the lateral body walls of the same two animals whose tympanic responses are displayed in Figure 21.6. Motion of the body wall of the small toad is generally greater than tat of the larger toad and peaks at a higher frequency (about 2,000 Hz) than that of the larger toad (about 400 Hz). This pattern, in which the body walls of small species show higher overall motion and higher frequency responses, appears generalizable to most anurans (Hetherington, in press).

3.4 An Overview of the Effects of Body Size on Acoustic Reception

Although more data are needed, it appears that many characteristics of the acoustic systems of amphibians show correlations with body size. The anuran tympanic middle ear is best developed and apparently most sensitive in large animals. The frequency responses of the tympanic ears of large anurans also are shifted to lower frequencies than those of smaller species. Nontympanic systems, in contrast, appear to be most effective in smaller animals. The opercularis system is proportionately largest in small species, and its role in hearing aerial sound may be related to body size. In small species, it may contribute to reception of a relatively wide range of frequencies, whereas in large species it may be involved in reception of only very low-frequency sounds. Another nontympanic

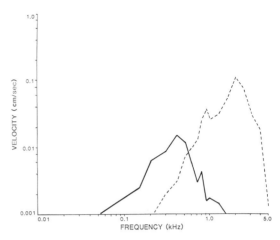

FIGURE 21.7. Graph of the acoustic responsiveness of the lateral body wall of a small toad (*Bufo quercicus*; 1.3 g; dashed line) and a large toad (*Bufo americanus*; 104 g; solid line). Velocity measurements were made with a laser vibrometer in free-field conditions at 90 dB SPL using a laterally positioned sound source. Body wall motion is generally greater and shifted toward higher frequencies in the small species. The frequency responses of the tympana of the same specimens are displayed in Fig. 21.6.

pathway, the lateral body wall and lung, is more responsive to sound in small anurans than in large anurans, and its frequency response is related to body size. The body wall of small species shows a broader frequency response than that of large species; responsiveness in the latter is restricted to lower frequencies. Sound absorption of unspecialized head tissues also is dependent on body size; as size increases, measurable absorption of sound becomes more restricted to low frequencies.

These findings suggest that amphibians may have different options available for sound reception depending on their body size. In small animals, nontympanic systems may be effective for reception of sounds over a wide frequency range. In larger animals, these nontympanic systems may be effective for reception of only very low frequency signals. Larger species, therefore, may require a well-developed tympanic middle ear for effective hearing. Figures 21.8 and 21.9 allow a comparison of the effectiveness of nontympanic sound reception in a small bullfrog (*Rana catesbeiana*) and a large

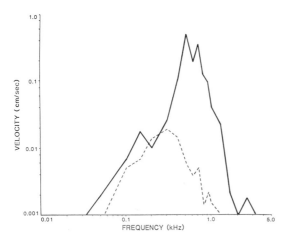

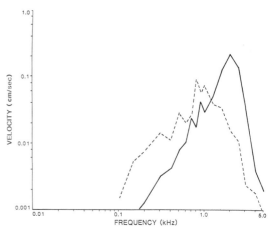

FIGURE 21.8. Comparison of the acoustic responsiveness of the tympanum (solid line) and lateral body wall (dashed line) of a large (386.5 g) bullfrog *Rana catesbeiana*. Velocity measurements were made with a laser vibrometer in free-field conditions at 90 dB SPL using a laterally positioned sound source. Tympanic motion is generally greater than that of the body wall over the entire range of frequencies. Compare with similar plots for a small bullfrog in Fig. 21.9.

FIGURE 21.9. Comparison of the acoustic responsiveness of the tympanum (solid line) and lateral body wall (dashed line) of a small (8.7 g) bullfrog *Rana catesbeiana*. Velocity measurements were made with a laser vibrometer in free-field conditions at 90 dB SPL using a laterally positioned sound source. Tympanic motion is generally greater at higher frequencies and body wall motion is generally greater at lower frequencies. Compare with similar plots for a large bullfrog in Fig. 21.8.

bullfrog. In the large bullfrog (Fig. 21.8), the responsiveness of the tympanum is greater than that of the body wall at almost every frequency examined. In the small bullfrog (Fig. 21.9), the tympanum is more responsive than the body wall only at higher frequencies, above about 1 kHz. This comparison suggests that small anurans can effectively use the body wall (or other nontympanic pathways) for reception of low frequencies and use a tympanic middle ear for reception of higher frequencies. Larger anurans, however, require a tympanic middle ear for appreciable hearing sensitivity at both low and high frequencies.

This hypothesis suggests that a tympanic middle ear becomes less useful as body size decreases. Its sensitivity should diminish and various nontympanic systems should become increasingly effective over a broad frequency range. One expectation of this hypothesis, therefore, is that small animals may display an evolutionary tendency to lose tympanic middle ears. In fact, reduction and loss of the tympanic middle ear is a common evolutionary trend in amphibians, and examination of the correlation of this trend with body size may be used to test the above hypothesis.

4. Loss of the Tympanic Middle Ear—A Major Evolutionary Trend

4.1 Morphological Patterns of Tympanic Middle Ear Reduction

Reduction and loss of the tympanic middle ear is a widespread and striking trend in middle ear evolution in amphibians. First, the tympanic ear is lacking in salamanders and gymnophiones. It is possible that the ancestors of these groups never possessed a tympanic middle ear. However, if extant amphibians are considered a monophyletic group (Duellman and Trueb 1986), it is not unreasonable to assume that during their evolution salamanders and gymnophiones lost a tympanic middle ear. Anurans probably evolved from certain paleozoic temnospondyls that clearly possessed tympanic ears (Lombard and Bolt 1988). If salamanders and gymnophiones also descended from this or a related group, their ancestors may have possessed a tympanic ear that was subsequently lost in these lineages. This is not an unlikely evolutionary occurrence; extant anurans show

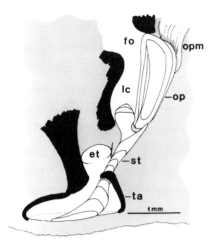

FIGURE 21.10. Semidiagrammatic reconstruction of the right middle ear of the frog *Kaloula pulchra*. This is an example of a reduced tympanic middle ear lacking a specialized tympanum. A frontal view looking dorsally. Lateral is down, and anterior to the left. Cut edges of skull elements, otic capsule, and tympanic annulus (ta) are black and surrounding soft tissue is stippled. Ventral portions of the tympanic cavity and membrane are not shown. (et) eustachian tube; (fo) oval window; (lc) lateral chamber; (op) operculum; (opm) opercularis muscle; (st) stapes. (From Jaslow et al. 1988. Structure and Function of the Amphibian Middle Ear, Jaslow et al., Copyright © 1988. John Wiley & Sons, Inc. Reprinted by permission.)

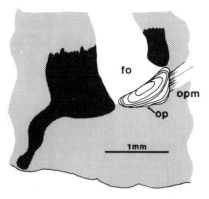

FIGURE 21.11. Semidiagrammatic reconstruction of the right middle ear of *Atelopus chiriquiensis*, a frog completely lacking a tympanic middle ear. Only an opercularis system remains. A frontal view looking dorsally. Lateral is down, and anterior to the left. Cut edges of skull elements and otic capsule are black and surrounding soft tissue is stippled. (fo) oval window; (op) operculum; (opm) opercularis muscle. (From Jaslow et al. 1988. Structure and Function of the Amphibian Middle Ear, Jaslow et al., Copyright © 1988. John Wiley & Sons, Inc. Reprinted by permission.)

many cases of reduction and loss of a tympanic ear (Jaslow, Hetherington, and Lombard 1988).

The simplest and most widespread condition of tympanic ear reduction in anurans involves loss of a well-defined tympanum. In many frogs and toads the tympanum consists of unspecialized skin that is indistinguishable from surrounding skin of the head surface (Fig. 21.10). The skin layer of the tympanic region is usually much thicker than the standard tympanum, and often contains cutaneous glands and other structures typically found in skin on the rest of the body. Although an obvious tympanum may be lacking in many anurans, other components of the tympanic ear, including the tympanic annulus, middle ear cavity, extrastapes, and stapes, may show no significant reduction.

In addition to this minimal state of reduction, other species display a wide spectrum of loss or modification of other components of the tympanic ear (Jaslow, Hetherington, and Lombard 1988). Some anurans, especially largely aquatic and fossorial species, display a reduction or loss of the middle ear cavity and a reduction in the relative size of the stapes and extrastapes. The most extreme case of reduction, of course, involves the complete absence of the tympanic ear. In such cases there is no trace of a stapes, extrastapes, middle ear cavity, tympanum or tympanic annulus (Fig. 21.11). The only middle ear structures remaining are those of the opercularis system. Such complete absence is widely distributed among anuran families and has evolved independently in these lineages (Fig. 21.12).

4.2 Factors Correlated with Reduction and Loss of the Tympanic Ear

There have been several efforts to correlate reduction and loss of the tympanic middle ear with habitat or behavior. Although species of anurans with reduced or absent tympanic ears live in a variety of habitats, several species are fossorial or aquatic. A well-developed tympanic middle ear may be unnecessary underwater or underground where sound or vibrations can effectively penetrate the head tissues. Many fossorial anurans possess well-developed opercularis systems with very

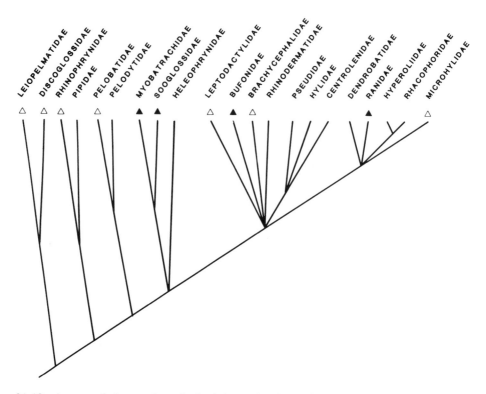

FIGURE 21.12. Anuran phylogeny hypothesized by Duellman and Trueb 1986. Triangles (both solid and hollow) denote families that have at least one species completely lacking a tympanic middle ear. Solid trian- gles denote families with at least one "earless" species with a mating call having a dominant frequency above 1 kHz. (From Jaslow et al. 1988.)

large opercula that may improve sensitivity to vibrations penetrating the head and body by max- imizing inertial properties of the system (Hether- ington, Jaslow, and Lombard 1986). A correspon- dence of tympanic ear reduction with burrowing habits may adequately explain the loss of a tym- panic ear in salamanders and gymnophiones; both groups are generally fossorial and spend little time moving above ground. The opercularis system remains as the only middle ear system in salaman- ders, but it too is lost in gymnophiones. This is not surprising, given that these amphibians lack limbs and girdles, thereby precluding a necessary condi- tion for the existence of an opercularis system. Gymnophiones do possess a relatively large, plate- like stapes resting in the oval window, and it proba- bly acts as an inertially sensitive element respond- ing to vibratory signals passing through the ground and directly penetrating the head (Wever 1985). In a sense, therefore, gymnophiones employ an

inertial strategy of vibratory reception similar to that used by limbed amphibians with an oper- cularis system.

Another obvious correlate of reduction and loss of a tympanic ear might be lack of calling behavior, or at least production of only low-frequency calls. Because nontympanic mechanisms may be more effective in reception of low-frequency sounds than the tympanic middle ear, the latter might be of limited use to species producing such calls. Indeed, several species that lack tympanic ears do not pro- duce calls. Such species are often found in noisy environments, such as near waterfalls or rapidly flowing streams (Jaslow, Hetherington, and Lom- bard 1988). However, many species that do call lack tympanic middle ears as well. Furthermore, among these are species that produce relatively high-frequency calls, above about 1 kHz (Jaslow, Hetherington, and Lombard 1988). Those anuran families including at least one species that both lacks a middle ear and produces calls with dominant frequencies above 1 kHz are displayed in Figure 21.12.

Therefore, there is some correspondence between reduction and loss of the tympanic ear and habitat (noisy environments), habits (fossorial, aquatic), and lack of vocalizations. However, most workers have refrained from declaring any environmental or behavioral factor as the major reason for loss of a tympanic middle ear (Jaslow, Hetherington, and Lombard 1988).

4.3 The Relationship Between Body Size and Loss of the Tympanic Ear

Although various factors may contribute to the reduction and loss of a tympanic ear in specific lineages, the most general factor that may explain this evolutionary phenomenon is body size. For example, the absence of a specialized tympanum is observed in at least some representatives of almost every anuran family, and it appears to be correlated with small body size. Figure 21.13 shows that loss of a conspicuous tympanum in a representative group of treefrogs of the family Hylidae is limited to relatively small species. This same pattern is observed in many other groups of anurans; for example, many small species of hyperoliid frogs lack specialized tympana (Passmore and Carruthers 1979). Also, as described in Section 3.1, the tympanum of frogs becomes proportionately smaller as body size decreases (Fig. 21.4).

Complete absence of a tympanic middle ear also is most evident in relatively small species of anurans. Except for some large species of fossorial anurans, such as *Rhinophrynus dorsalis* and species of *Hemisus*, lack of a tympanic middle ear is generally restricted to small frogs below about 20 to 30 mm in snout-vent length. For example, several small groups of myobatrachid frogs (most species of *Pseudophryne* and some species of *Crinia*) and the small sooglossid frogs lack tympanic ears (Barker and Grigg 1977; Nussbaum, Jaslow, and Watson 1982).

Additional support for the correspondence between body size and loss of the tympanic ear comes from ontogenetic studies (Fig. 21.14). In many anurans, the tympanic ear is quite undeveloped at the end of metamorphosis and the

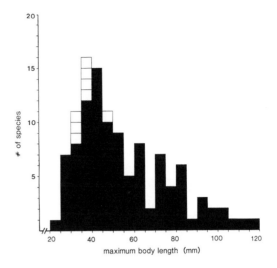

FIGURE 21.13. A histogram showing the maximum body size distribution of a total of 112 species of treefrogs of the family Hylidae found in Mexico and Central America. Solid squares represent species with a distinct tympanum; open squares represent species lacking a specialized tympanum. Note that absence of a specialized tympanum is restricted to smaller species. (Data obtained from Duellman 1970.)

initiation of terrestrial activity (Sedra and Michael 1959; Hetherington 1987). Most of its components develop later after body size has increased and the animal is approaching adult body size.

Several functional reasons can be hypothesized for the association of small size and evolutionary reduction and loss of a tympanic middle ear. As discussed in Section 3.3, small frogs may be able to utilize nontympanic systems, such as the opercularis system and body wall, for sound reception over a wide frequency range. Such pathways would be most effective at low frequencies, but their effectiveness would be dependent on the exact size of the animal and probably on other factors as well. It may be that sensitivity to low-frequency sound is generally more important than sensitivity to higher frequencies for many amphibians. The delayed development of the tympanic ears of many anurans may be especially significant. This suggests that acoustic reception via the tympanic middle ear is most important for detection of mating calls by adults (Hetherington 1987). Mating calls typically consist of relatively high-frequency signals that might best be detected by a tympanic middle ear

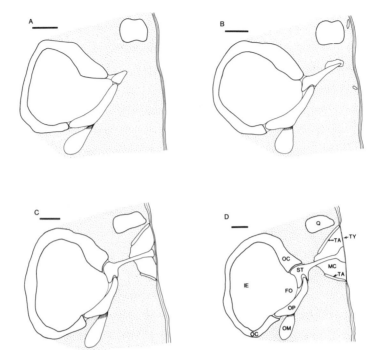

FIGURE 21.14. Diagrammatic representations of the right otic regions of a series of specimens of the frog *Hyla crucifer* demonstrating postmetamorphic development of the tympanic ear. These drawings compress features of the middle ear at different levels, as observed in frontal sections, into one plane to provide a general view of middle ear morphology. Snout-vent lengths (mm) and body weights (g) are provided for every specimen. (A) end of metamorphosis, 10.1 mm, 0.16 g. (B) 20 days postmetamorphic, 12.7 mm, 0.24 g. (C) 40 days post- metamorphic, 14.9 mm, 0.36 g. (D) 60 days postmeta- morphic, 16.1 mm, 0.52 g. The middle ear of (D) is essentially complete, and all of the parts are labelled in this specimen. Note that the tympanic middle ear develops well after metamorphosis and after substantial increase in size. (FO) oval window; (IE) inner ear; (MC) middle ear cavity; (OC) otic capsule; (OM) opercularis muscle; (OP) operculum; (Q) palatoquadrate; (ST) stapes; (TA) tympanic annulus; (TY) tympanum. (From Hetherington 1987.)

rather than by nontympanic systems. The direc- tional characteristics of a tympanic middle ear also may be important for localization of calls. The major selective advantage of a tympanic middle ear may therefore involve reception and localization of relatively high-frequency mating calls. Many frogs, especially small species, may utilize non- tympanic mechanisms for detection of many other types of important, generally low-frequency, acoustic signals. However, some species that lack tympanic ears but still vocalize may use nontym- panic strategies for mating call detection as well.

Tympanic ears clearly can retain functional sig- nificance in even very small anurans, as many of the latter possess well-developed tympanic middle ears. But it appears that the selective advantage of

a tympanic ear decreases with decreasing size. Some species have responded evolutionarily with only a reduction in the size of the tympanic ear, whereas other species display loss of certain com- ponents, such as a specialized tympanum, or loss of the entire tympanic middle ear.

5. Discussion and Summary

In this chapter I have tried to present the hypothe- sis that evolutionary changes in body size have had important effects on the evolution of the amphibian middle ear. More information certainly is needed on the relationship between body size and morpho- logical and functional features of the acoustic systems of amphibians. Nonetheless, many features of these systems, which include the tympanic ear, opercularis system, and additional nontympanic pathways, show correspondences with size. It could be argued that such correspondences are not the result of natural selection but rather are epiphenomena. For example, selection could act to decrease body size for many different reasons, and the modifications observed in the acoustic systems could result simply from a paedomorphic trend. From this perspective, the relationship between small body size and reduction or loss of the tympanic middle ear would inevitably result from a truncation in development of the skull and otic region. Trueb and Alberch (1985) tested a similar hypothesis that many features of miniaturized skulls of anurans were only part of a paedomorphic trend and not functionally significant. They eventually rejected this hypothesis, concluding that functional reasons were important in explaining the morphological features of miniaturized skulls. Similarly, I also reject the hypothesis that the relationship between middle ear morphology and body size is epiphenomenal. For example, paedomorphosis alone cannot explain why only the tympanic middle ear is reduced or lost in many small frogs, whereas the opercularis system is retained and actually proportionally larger in the same animals. Also, many small anurans still retain well-developed tympanic middle ears, showing that reduction and loss of a tympanic ear need not be part of a general paedomorphic trend. Reduction and loss of the tympanic ear probably resulted because of functional reasons, specifically the effectiveness of nontympanic mechanisms in small amphibians. This allowed a relaxation of the selective pressures required to maintain a standard tympanic ear. The consistency in the pattern of differences in the morphological and functional properties of the acoustic systems of small and large amphibians strongly suggests this functional underpinning.

In conclusion, I would suggest that body size has been important in determining what mechanisms of acoustic reception have evolved in amphibians. Amphibians are generally small tetrapods, and perhaps the nontympanic mechanisms observed in them would be infeasible in other larger tetrapod vertebrates. However, it would be interesting to see if other groups of tetrapods include small species that employ strategies of acoustic reception analogous to those of small amphibians. It may be that the standard tympanic ear of tetrapods is an ear designed for relatively large animals; a diversity of alternative possibilities exists for smaller animals.

Acknowledgments. Original research for this chapter was supported by Grant 8-R01-DC00297, National Institutes of Health. Eric Lombard and John Bolt provided additional data and useful suggestions.

References

Aertsen AMHJ, Vlaming MSMG, Eggermont JJ, Johannesma PIM (1986) Directional hearing in the grassfrog (*Rana temporaria* L.): II. Acoustics and modelling of the auditory periphery. Hear Res 21:17–40.

Barker J, Grigg G (1977) A Field Guide to Australian Frogs. Adelaide: Rigby Limited.

Becker RP, Lombard RE (1977) Structural correlates of function in the "opercularis" muscle of amphibians. Cell Tissue Res 175:499–522.

Calder WA (1984) Size, Function, and Life History. Cambridge: Harvard University Press.

Capranica RR (1976) Morphology and physiology of the auditory system. In: Linas R, Precht W (eds) Frog Neurobiology. New York: Springer-Verlag, pp. 443–466.

Chung S-H, Pettigrew AG, Anson M (1981) Hearing in the frog: Dynamics of the middle ear. Proc R Soc Lond B 212:459–485.

Duellman WE (1970) The Hylid Frogs of Middle America. Monograph of the Museum of Natural History, No. 1. The University of Kansas.

Duellman WE, Trueb L (1986) Biology of Amphibians. New York: McGraw-Hill.

Eggermont JJ (1988) Mechanisms of sound localization in anurans. In: Fritzsch B, Ryan M, and others (eds) The Evolution of the Amphibian Auditory System. New York: John Wiley & Sons, pp. 307–336.

Ehret G, Tautz J, Schmitz B (1990) Hearing through the lungs: Lung-eardrum transmission of sound in the frog *Eleutherodactylus coqui*. Naturwissenschaften 77:192–194.

Feng AS (1980) Directional characteristics of the acoustic receiver of the leopard frog (*Rana pipiens*): A study of eighth nerve auditory responses. J Acoust Soc Am 68:1107–1114.

Feng AS, Shofner WP (1981) Peripheral basis of sound localization in anurans, acoustic properties of the frogs ear. Hear Res 5:201–216.

Hetherington TE (1985) The role of the opercularis muscle in seismic sensitivity in the bullfrog *Rana catesbeiana*. J Exp Zool 235:27–34.

Hetherington TE (1987) Physiological features of the opercularis muscle and their effects on vibratory sensitivity in the bullfrog *Rana catesbeiana*. J Exp Biol 131:1–16.

Hetherington TE (1988) Biomechanics of vibration reception in the bullfrog *Rana catesbeiana*. J Comp Physiol 163:43–52.

Hetherington TE (1989) Effect of the amphibian opercularis muscle on auditory responses. Prog Zool 35:356–359.

Hetherington TE The effects of body size on functional properties of middle ear systems of anuran amphibians. In press, Brain, Behavior, and Evolution.

Hetherington TE, Lombard RE (1983) Electromyography of the opercularis muscle of the bullfrog *Rana catesbeiana*: an amphibian tonic muscle. J Morphol 175:17–26.

Hetherington TE, Jaslow AP, Lombard RE (1986) Comparative morphology of the amphibian opercularis muscle. I. General design features and functional interpretation. J Morphol 190:43–61.

Jaslow AP, Hetherington TE, Lombard RE (1988) Structure and function of the amphibian middle ear. In: Fritzsch B, Ryan M, and others (eds) The Evolution of the Amphibian Auditory System. New York: John Wiley & Sons, pp. 69–92.

Khanna SM, Tonndorf J (1978) Physical and physiologic principles controlling auditory sensitivity in primates. In: Noback CR (ed) Neurobioloy of Primates. New York: Plenum Press, pp. 23–52.

Kingsbury BF, Reed HD (1909) The columella auris in Amphibia. J Morphol 20:549–628.

Koyama H, Lewis ER, Leverenz EL, Baird RA (1982) Acute seismic sensitivity in the bullfrog ear. Brain Res 250:168–172.

Lewis ER, Narins PM (1985) Do frogs communicate with seismic signals? Science 227:187–189.

Lewis ER, Lombard RE (1988) The amphibian inner ear. In: Fritzsch B, Ryan M, and others (eds) The Evolution of the Amphibian Auditory System. New York: John Wiley & Sons, pp. 93–124.

Loftus-Hills JJ (1973) Comparative aspects of auditory function in Australian anurans. Aust J Zool 21:353–367.

Lombard RE, Bolt JR (1988) Evolution of the stapes in paleozoic tetrapods: conservative and radical hypotheses. In: Fritzsch B, Ryan M, and others (eds) The Evolution of the Amphibian Auditory System. New York: John Wiley & Sons, pp. 37–68.

Lombard RE, Straughan IR (1974) Functional aspects of anuran middle ear structures. J Exp Biol 61:71–93.

Moffat AJM, Capranica RR (1978) Middle ear sensitivity in anurans and reptiles measured by light scattering spectroscopy. J Comp Physiol 127:97–107.

Monath T (1965) The opercular apparatus of salamanders. J Morphol 116:149–170.

Narins PM (1990) Seismic communication in anuran amphibians. Bioscience 40:268–274.

Narins PM, Capranica RR (1976) Sexual differences in the auditory system of the tree frog *Eleutherodactylus coqui*. Science 192:378–380.

Narins PM, Ehret G, Tautz J (1988) Accessory pathway for sound transfer in a neotropical frog. Proc Natl Acad Sci USA 85:1508–1512.

Nussbaum RA, Jaslow AP, Watson J (1982) Vocalization in frogs of the family Sooglossidae. J Herpetol 16:198–204.

Passmore NI, Carruthers VC (1979) South African Frogs. Johannesburg: Witwatersrand University Press.

Paton JA (1971) Microphonic potentials in the inner ear of the bullfrog. Masters Dissertation, Cornell University.

Pettigrew A, Chung S-H, Anson M (1978) Neurophysiological basis of directional hearing in amphibia. Nature 272:138–142.

Pinder AC, Palmer AR (1983) Mechanical properties of the frog ear: Vibration measurements under free and closed-field acoustic conditions. Proc R Soc Lond B 219:371–396.

Rheinlaender J, Walkowiak W, Gerhardt HC (1981) Directional hearing in the green treefrog: A variable mechanisms. Naturwissenschaften 68:430–421.

Rosowski JJ, Graybeal A (1991) What did *Morganucodon* hear? Zoological Journal of the Linnean Society 101:131–168.

Ryan MJ (1985) The Tungara Frog, A Study in Sexual Selection and Communication. Chicago: University of Chicago Press.

Ryan MJ, Wilczynski W (1988) Coevolution of sender and receiver: effect on local mate preference in cricket frogs. Science 240:1786–1788.

Schmidt-Nielsen K (1984) Scaling: Why is Animal Size So Important? Cambridge: Cambridge University Press.

Sedra SN, Michael MI (1959) The ontogenesis of the sound conducting apparatus of the Egyptian Toad, *Bufo regularis* Reuss, with a review of this apparatus in Salientia. J Morphol 104:359–375.

Shofner WP (1988) Postmetamorphic changes in the auditory system. In: Fritzsch B, Ryan M, Wilczynski W, Hetherington TE, Walkowiak W (eds) The Evolution of the Amphibian Auditory System. New York: John Wiley & Sons.

Shofner WP, Feng AS (1981) Post-metamorphic development of the frequency selectivities and sensitivities of the peripheral auditory system of the bullfrog *Rana catesbeiana*. J Exp Biol 93:181–196.

Smirnov SV (1987) The sound conducting apparatus of urodela: Structure, function and evolutionary pathways. Zool? Zhur? 46:735–745.

Trueb L, Alberch P (1985) Miniaturization and the anuran skull: a case study of heterochrony. Fortsch Zool 30:113–121.

Vlaming MSMG, Aertsen AMHJ, Epping WJM (1984) Directional hearing in the grassfrog (*Rana temporaria* L.). I. Mechanical vibrations of tympanic membrane. Hear Res 14:191–201.

Walkowiak W (1980) The coding of auditory signals in the torus semicircularis of the fire-bellied toad and the grass frog: Responses to simple stimuli and conspecific calls. J Comp Physiol 138:131–148.

Wever EG (1979) The middle ear muscles of the frog. Proc Natl Acad Sci USA 76:3013–3033.

Wever EG (1985) The Amphibian Ear. Princeton: Princeton University Press.

Wilczynski W, Capranica RR (1984) The auditory system of anuran amphibians. Prog Neurobiol 22: 1–38.

Wilczynski W, Zakon HH, Brenowitz EA (1984) Acoustic communication in spring peepers. Call characteristics and neurophysiological aspects. J Comp Physiol 155:577–584.

Wilczynski W, Resler C, Capranica RR (1987) Tympanic and extratympanic sound transmission to the leopard frog. J Comp Physiol 161:659–669.

22
Biological Constraints on Anuran Acoustic Communication: Auditory Capabilities of Naturally Behaving Animals

Peter M. Narins

1. Introduction

As vertebrate animals first adjusted to terrestrial life, strong selection pressure was exerted to develop acoustic communication signals which were at once well-suited for the exchange of biologically significant information between conspecifics, inconspicuous to or poorly localizable by predators, and robust in the face of high-level background noise. In this chapter I attempt to review the literature concerning comparative field studies of hearing in anuran amphibians (frogs and toads) using psychophysical paradigms. I am interested in what we may learn from such investigations and how such studies relate to the standard psychoacoustical experiments carried out in laboratories worldwide. I chose to focus on anuran amphibians as a model taxon for several reasons: these animals are often highly vocal; sounds are known to play a critical role in their courtship and reproductive behavior; their vocal repertoire consists of a small number of species-specific, stereotyped sounds; and these sounds, although often complex, may be readily synthesized and used as stimuli in playback experiments in the field.

Studies designed to determine the absolute capabilities of the auditory system of anurans are typically carried out in anechoic or sound reduced conditions. In natural environments, however, anurans must often communicate in the face of high levels of ambient or background noise. Thus it becomes relevant to determine the behavioral capacity of the anuran auditory system in the animals' natural world. This chapter is concerned with measures of auditory function determined in the frog's natural habitat, with a view toward understanding acoustic behavior of both the individual as well as the group of signaling anurans. Neotropical frogs are ideally suited for such measurements and have produced a great deal of data which provide insights into constraints operating on communication behavior in these animals. These constraints, both intrinsically and extrinsically imposed, are explored in this chapter. Pure-tone detection thresholds as well as intensity, frequency, and temporal discrimination functions for frogs calling in their natural habitat have been measured as well as the deterioration of these functions when masked by broadband noise. Implications of these data will be discussed in an evolutionary framework.

The call parameters of individual frogs in a population exhibit natural variation on which selection may operate. Thus, the call note frequencies, durations, rise-times, amplitude modulation envelope, harmonic structure, and repetition rates are all subject to natural selection. In a real sense, the advertisement calls (those given repeatedly by males throughout the calling period) are the net result of a suite of selective agents acting on the communication system, namely, neighboring males, other interfering sounds generated by conspecific and heterospecific frogs, as well as other animals, and physiological factors such as energetic constraints, limitations of the auditory system, temperature, etc.

Rather than defining the absolute thresholds for a given auditory task, field experiments reveal an animal's performance limits under natural conditions of often remarkably high levels of ambient

439

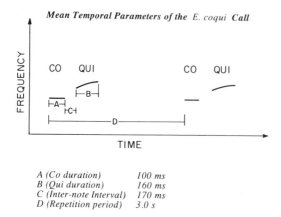

A (Co duration)   100 ms
B (Qui duration)   160 ms
C (Inter-note Interval) 170 ms
D (Repetition period)   3.0 s

FIGURE 22.1. Schematic diagram of two successive advertisement calls from an individual male of the neotropical arboreal frog, *Eleutherodactylus coqui*. The mean temporal parameters from a population of males at an altitude of 900 m are shown.

noise. These studies frequently use standard psychophysical techniques to characterize the auditory system's capability to respond to controlled stimuli under natural conditions.

Since auditory functions obtained in the field are made in the presence of nonstationary background noise, they may not be directly compared, strictly speaking, to measurements made in quiet, anechoic conditions. Moreover, the responses from free-ranging animals are unconditioned, so that we might expect, for example, thresholds determined in this way to be higher than the "absolute" thresholds obtained in sound-quiet conditions, and using conditioning paradigms. Nevertheless, these measures in and of themselves reflect an animal's ability to perform an auditory task in the very environment in which he must carry out all of his life functions (e.g., territorial defense, mate attraction, foraging) and are therefore critical for the understanding of the animal's calling behavior. With these caveats, I shall now discuss various measures of auditory function which have been obtained from playback experiments with male frogs in their natural habitat.

1.1 A "Representative" Neotropical Chorusing Anuran

In the Caribbean National Forest in eastern Puerto Rico, the largest and most well-studied amphibian is the Puerto Rican Coqui, *Eleutherodactylus*

coqui. Males of this arboreal species produce a two-note advertisement call ("Co-Qui"). At the higher elevations of Cerro El Yunque the "Co" note is a constant-frequency tone of about 1.1 kHz lasting nearly 100 ms and functions in male–male territorial interactions. The Co note is followed by a pause of 150 to 200 ms and a second ("Qui") note which sweeps upward in frequency from about 1.8 kHz to 2.1 kHz in 160 ms (Fig. 22.1). The Qui note serves to attract females (Narins and Capranica 1976, 1978). Males give the two-note call every 2 to 4 s from sunset to shortly after midnight throughout 11 months of the year. This species will serve as a model of a chorusing frog communicating in high levels of background noise.

2. Field Psychoacoustics

2.1 Behavioral Response Functions

Behavioral response functions may be obtained from unperturbed, calling males by presenting them with acoustic stimuli which evoke a clear, unambiguous antiphonal response which is either not produced spontaneously or produced at low rates in the absence of a synthetic stimulus. For *E. coqui*, one behavioral assay for stimulus detection is the vocalization change from a two-note call (Co-Qui) to an antiphonal one-note call (Co). By systematically increasing the playback level and arbitrarily defining threshold as a particular one-note call rate, one obtains threshold for a pure-tone stimulus. Using this isoresponse criterion and repeating this measurement over a range of frequencies results in a behavioral response function for the animal under test.

This was done for a sample of males ($n = 16$) of *E. coqui* in the Luquillo Mountains of the Caribbean National Forest in eastern Puerto Rico (Narins and Capranica 1978). Figure 22.2 shows a representative behavioral response function for one male in his natural habitat, and indicates that the frequency at which the threshold response rate could be evoked at minimum intensity closely approximates that of the male's first (Co) call note. Moreover, the threshold intensity increases rapidly for stimulus frequencies above and below the optimal frequency. Thus, the behavioral response function reflects the frequency-dependent thresholds exhibited by the frog in response to a range of acoustic stimuli. This function clearly depends on the stimulus having some significance to the frog, rather than being a measure of absolute threshold or of the limits of hearing for the animal under test.

2.2 Isointensity Response

To determine whether the selectivity of this behavioral response function is level independent, one may present an animal with an acoustic stimulus at a constant level above best threshold, and score the number or percentage of one-note calls evoked over a range of frequencies, as was done for 14 males of *E. coqui* (Fig. 22.2B). While this analysis does not yield absolute behavioral thresholds, it nevertheless demonstrates that the selectivity of the behavioral response function is maintained for suprathreshold stimuli. Moreover, it is apparent from this figure that the best-fit regression line through the thresholds above F_{co} ($y = -0.47 x$) has a greater absolute slope than the line through the threshold points below F_{co} ($y = 0.27 x$). It follows that for equal frequency shifts, it is easier for males of *E. coqui* to synchronize their vocalizations with synthetic call notes lower in frequency than their own than it is for them to synchronize their calls to synthetic calls higher in frequency. This probably reflects the asymmetrical shape of the typical auditory nerve fiber tuning curve, for which the high-frequency slope is steeper than the low-frequency slope (Narins 1983).

2.3 Timing Shifts as a Measure of Threshold

Many species of frogs have been shown to avoid acoustic interference with their neighbors by altering their calling pattern in response to either (1) calls of other frogs or (2) playbacks of natural or artificial stimuli (Foster 1967; Lemon 1971; Loftus-Hills 1971, 1974; Awbrey 1978; Narins and Capranica 1978; Lemon and Struger 1980; Zelick and Narins 1982, 1983, 1985; Wells and Schwartz 1984; Schneider, Joermann, and Hodl 1988; Walkowiak 1988; Brush and Narins 1989). However, the notion of using small but consistent shifts in call timing as a behavioral assay for call detectability by calling frogs is novel and has not been utilized by many workers.

In one study using such a method, thresholds were extracted quite readily from frogs in their

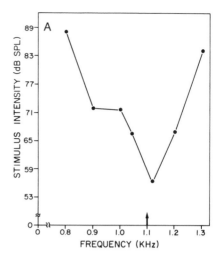

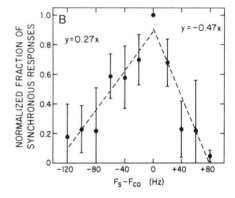

FIGURE 22.2. (A) Behavioral response function for an individual male of *E. coqui* in its natural habitat. The synthetic Co note stimulus used in obtaining this function was a 100-ms tone burst presented every 2 s. Thresholds at each frequency represent the playback level of 50 identical tone bursts which evoked 10 synchronized (one-note, short-latency) calls from the test male. The stimulus frequency which required minimum playback level to evoke the threshold response rate was 1.12 kHz, near the frequency (1.1 kHz) of the frog's natural Co note. (From Narins and Capranica 1978). (B) Frequency-dependence of the synchronous vocal response of *E. coqui* using constant-intensity stimuli. The plot shows the normalized mean number of synchronized responses (± 1 S.D.) as a function of the synthetic Co note stimulus frequency, F_s minus the animal's Co note frequency, F_{co}. For every male tested ($n = 14$), the greatest number of synchronous responses occurred for $F_s = F_{co}$. The function is asymmetrical about its maximum indicating that a stimulus with a given frequency difference from a male's Co note frequency will evoke a higher number of synchronous responses for negative rather than positive deviations. Dashed lines are the best-fit least-squares regression lines through the mean response rates for each frequency tested. (From Narins 1983.)

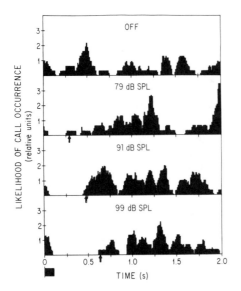

FIGURE 22.3. Call likelihood distributions produced in response to 1.1 kHz, 100-ms tone burst stimuli presented at increasing levels (indicated in each panel) for an individual *E. coqui*. The horizontal bar beneath the distributions represents the stimulus duration. The arrows refer to the end of the poststimulus suppression, which is a function of stimulus intensity. The distribution shown in the top panel was obtained during spontaneous calling. Bin width: 10 ms.

natural habitat by quantifying the shift in call pattern which resulted in response to playbacks of periodic tone bursts at a series of different frequencies (Zelick and Narins 1982). The resulting call pattern may be analyzed for poststimulus suppression, or the cessation of calling immediately following a detectable stimulus. Poststimulus suppression is manifested in the analysis of call timing as an intensity-dependent gap in the frog's call likelihood distribution. In fact, the duration of the gap may be used as a measure for call detectability, and any arbitrarily small gap duration (e.g., 110 ms poststimulus) may serve as a detectability "threshold." A typical result from a male *E. coqui* presented with a series of tone bursts at progressively higher intensities is shown in Figure 22.3. In this case, it is clear that the duration of the gap increases with increasing stimulus intensity and that for the test frequency shown (1.1 kHz), the mean ($n = 4$) detectability threshold is 78 dB SPL. Although this method is reliable and produces unambiguous thresholds, we find a high degree of threshold

variability, probably reflecting the different extents to which individual frogs are vocally interacting with their neighbors just prior to testing.

An underlying assumption for experiments designed to assess the ability of males to adjust their call timing to avoid acoustic interference is that it is beneficial to the calling male to do so (Littlejohn and Martin 1969; Loftus-Hills 1971, 1974; Passmore and Telford 1981; Zelick and Narins 1982; Narins and Zelick 1988). The advantage to the males might come about in at least three ways. First, by avoiding overlap with his nearest neighbors, he ensures that the signal-to-noise ratio at the source (himself) is not sacrificed. Presumably this aids in assessing the number of and distance between neighbors, thus facilitating territorial maintenance (Lemon 1971; Passmore and Telford 1981; Zelick and Narins 1985; Brush and Narins 1989). Second, acoustic overlap results in degradation of the temporal parameters of the call, potentially limiting species discrimination (Littlejohn 1977; Wells and Schwartz 1984). Third, females should be better able to localize unmasked calls than ones overlapped by other calls (Passmore and Telford 1981; Zelick and Narins 1983; Wells and Schwartz 1984). To directly test these hypotheses, Schwartz (1987) carried out a series of elegant acoustic playback experiments with *Hyla versicolor* and *H. microcephala*, species which produce amplitude-modulated (AM) calls, and with *H. crucifer*, which does not. The AM provides temporal fine structure in the advertisement call important for species recognition (Gerhardt 1982; Schwartz 1987). Results of this study indicate that males avoid call or note overlap because it impairs both their ability to assess the intensity of their neighbors' calls and their ability to emit signals attractive to potential mates. Females of *H. versicolor* and *H. microcephala* preferred alternating calls to overlapping ones, but the preference for alternating calls disappeared when the overlapping calls were presented in such a way that the AM peaks were in phase. Unlike the other two species, females of *H. crucifer* did not discriminate between simultaneous calls presented such that temporal information was obscured. These observations suggest that for species which use temporal call information, avoidance of call overlap by males serves to (1) improve the ability of conspecific males to judge the intensity of neighbors and thus effect spacing and other territorial behaviors and (2) enhance female phonotaxis. Call overlap in species which use tonal calls lacking AM (such as *E. coqui*) would thus be predicted to have less of a deleterious effect on call localization by females than in species utilizing AM calls (Narins and Zelick 1988).

It should be added that whereas call alternation seems to be widespread among males in anuran choruses, some species produce overlapping calls. For example, males of *Smilisca sila* appear to suffer less predation from the frog-eating bat, *Trachops cirrhosus*, by calling synchronously rather than asynchronously (Tuttle and Ryan 1982).

2.4 Call Duration Sensitivity

A calling male frog's sensitivity to call-note duration may also be measured in the animal's natural habitat. By systematically changing synthetic call duration and using an appropriate behavioral assay for stimulus detection, one obtains a duration-response function, such as illustrated in Figure 22.4 (Narins 1976). In this case, the number of one-note, short-latency (Co) calls was used as the measure while the stimulus duration was varied around the natural call-note duration (100 ms). This figure shows that the most vigorous response was recorded for tone bursts of 100-ms durations, for both stimulus repetition periods tested. Thus, although the temporal integration time constants for individual auditory nerve fibers in this species are in the range of 200 ms (Dunia and Narins 1989), results of behavioral experiments rather suggest the existence of an optimum signal duration (i.e., 100 ms). The inevitable and interesting interpretation of such results is that males of this species appear to respond more vigorously to tones which approximate their own in duration, rather than those which would be most easily detected. Subsequently, duration sensitivity was behaviorally demonstrated in both *Hyla ebraccata* (Schwartz and Wells 1984) and *H. versicolor* (Wells and Taigen 1986, 1989; Klump and Gerhardt 1987). Moreover, neurons in the anuran midbrain have been identified which respond selectively to tones of particular durations (Potter 1965; Narins and Capranica 1980). It is interesting to note that male preferences for particular stimulus repetition rates have not been demonstrated in frogs to date.

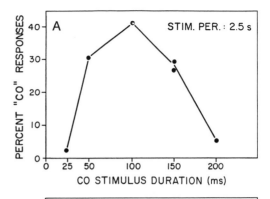

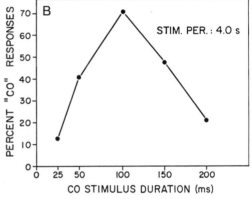

FIGURE 22.4. (A) Percentage of Co responses by a representative male *E. coqui* in his natural habitat as a function of stimulus duration. Stimulus was a series of synthetic Co notes (100-ms tone bursts) whose frequency was adjusted by ear to be that of the calling male under test. The responses to stimulus periods of both (A) 2.5 s and (B) 4.0 s were qualitatively similar, although this particular male responded more vigorously to the slower rate of stimulus presentation.

2.5 Effective Critical Ratio and Effective Critical Band

Critical ratio (Scharf 1970; Fay 1974; Ehret 1975; Long 1977) and critical band (Fletcher 1940; Scharf 1970) are two common behavioral measures of auditory frequency resolution. The former typically refers to an indirectly derived quantity, specifically the ratio between the level of a threshold test tone to the spectrum level of a broadband noise necessary to totally mask the test tone. In contrast, the term critical band is reserved for the quantity obtained by a direct, experimental measurement which requires a manipulation of masker bandwidth (Scharf 1970). Several studies have

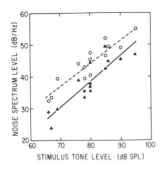

FIGURE 22.5. Stimulus tone level required to evoke the just-suprathreshold response rate vs the spectrum level of a wideband noise masker required to reduce this response by half (*filled symbols* and *solid least-squares regression line*, r² = 0.79) or to reduce this response to zero (*open symbols* and *dashed least-squares regression line*, r² = 0.82). (From Narins 1982b.)

attempted to measure the critical ratio (CR) and critical band (CB) in anurans. Ehret and Gerhardt (1980) used test tones to which female frogs would orient and approach. They found in the green treefrog (*Hyla cinerea*) that the ratio of the level of a 0.9 kHz test tone to that of the broadband noise required to totally mask that tone (CR) was about 22 dB. Using the reflex modification technique, Moss and Simmons (1986) estimated the lowest critical ratios for the same species to be 22 to 24 dB at about 900 Hz and 3,000 Hz, the two dominant frequencies in the male's advertisement call. Similar measures in undisturbed male *E. coqui* in the Luquillo Mountains in eastern Puerto Rico revealed that masked tone thresholds deteriorated rapidly with increasing masking noise level. Figure 22.5 illustrates the relationship between the stimulus tone level necessary to evoke the just-suprathreshold response rate, and the spectrum level of wideband noise necessary to reduce this response rate to one-half the unmasked rate, and to zero. Measurements of this type resulted in an "effective critical ratio" (ECR) of 31 dB (Narins 1982a). Presumably a factor contributing to the difference between the CR and ECR is the high-level, temporally varying ambient noise present in the habitat of *E. coqui*. Thus, the ratio of the ECR to the CR may be interpreted as the deterioration in an animal's signal extraction ability in simultaneous, high-level noise.

How do these measurements relate to the evolution of the anuran's acoustic communication system? Clearly, an important selective pressure on such a system is the effect of potentially interfering vocalizations from neighboring males (both conspecific and heterospecific), since communication systems do not evolve *in vacuo*. That natural noise can have a direct effect on frog calling behavior has also been demonstrated (Schwartz and Wells 1983). Two sympatric species, *Hyla ebraccata* and *H. microcephala*, share an acoustic environment and produce advertisement calls with overlapping spectral compositions. Playback of the chorus sound of *H. microcephala* causes males of *H. ebraccata* to reduce the proportion of aggressive calls given in response to conspecific males. Thus in this natural noise masking situation, *H. ebraccata* is not immune from effects of the chorus, a factor which may be important in the behavioral ecology of other species as well.

There is additional evidence for males gaining a selective advantage by shifting the timing of their vocalizations in response to background noise. In a two-choice experiment, females of *H. ebraccata* preferentially approached a loudspeaker broadcasting clear, unmasked conspecific male advertisement calls over one emitting the same conspecific advertisement calls which overlapped with the natural background sounds of a chorus of the sympatric species, *H. microcephala* (Schwartz and Wells 1983). Female choice experiments with *H. cinerea* using synthetic masking noise yield comparable results (Ehret and Gerhardt 1980; Gerhardt and Klump 1988).

2.6 Intensity Discrimination

2.6.1 Behavioral Responses to Level Shifts

The limits of intensity discrimination in frogs vocalizing in their natural habitat may also be determined. In one case, males of *E. coqui* were presented control-tone stimuli, one second in duration, for which the tone frequency was adjusted by ear to be the same as the Co note of the calling male under test. The tones were presented every 2.5 s, the mean natural calling rate of the population tested, and at a level which was adjusted to be in the range resulting in a clear change in the animal's temporal calling pattern. Immediately following the control tone, a second test tone was presented for the remainder of the stimulus period, namely during a 1.5 s "window." The level of the second tone was varied in discrete steps until the animal no longer preferentially called in the window. Fifty-nine percent of the males tested in this way were able to successfully detect, and avoid acoustic interference with, a window depth of 4 dB (Zelick and Narins 1983). This result clearly demonstrates that male frogs are remarkably adept at detecting small changes in background intensity, and are able to reliably initiate calling immediately following an intensity reduction of as small as 4 dB.

Very recent experiments have attempted to measure the degree of immunity of anuran intensity discrimination to broadband masking noise. Preliminary results of such tests with males of *E. coqui* made in the animal's natural habitat suggest that this ability is rather robust. Spectrum levels of masking noise as high as 47.5 dB/Hz at the frog are necessary to significantly reduce the frog's discrimination performance for pairs of test tones differing in intensity by 2 dB to 20 dB presented at behaviorally relevant sound pressure levels (Narins, unpublished data).

2.6.2 Sensitivity to Gaps in Background Noise

One calling "strategy" adopted by some anuran species to reduce acoustic interference is to vocalize during the "quiet" periods of the noise (Rosen and Lemon 1974; Wells 1977). This is achieved by calling immediately following another frog's call (Littlejohn and Martin 1969) during his ensuing refractory period (Narins 1982a), with the favorable result of lowering the probability of call overlap. Males of *E. coqui* employ this strategy, as do other vertebrates (Drewry and Rand 1983). This behavior benefits a male by ensuring that (1) the local signal-to-noise ratio is high, since the greatest source of interference (his nearest and most intense neighbor) is silent, and (2) the probability of his call being "jammed" by his nearest neighbor is low, since the neighbor would be in his behavioral refractory period (see Section 3.1). To carry out such a strategy, it becomes imperative that a calling male have the ability to both *detect* and

initiate calling during a momentary pause in the ambient noise.

Laboratory psychophysical studies have quantified the abilities of a variety of organisms to detect pauses or gaps in otherwise continuous auditory stimuli. For example, the minimum gap in a monaurally presented 500-ms broadband noise burst detectable by the house finch (*Carpodacus mexicanus*) is 3.8 to 5.0 ms (Dooling, Zoloth, and Baylis 1978). In a similar paradigm, humans exhibit minimum gap-detection thresholds of 2.3 to 4.2 ms (Plomp 1964; Shailer and Moore 1983; Buus and Florentine 1985), whereas for the higher stimulus levels tested in the chinchilla (*Chinchilla lanigera*), the gap duration threshold is 3.0 ms (Giraudi et al. 1980). The close agreement between the minimum-detectable gap for birds and mammals suggests a common mechanism underlying gap detection in these two groups (Klump and Maier 1990). In contrast, Fay (1985) showed that in the goldfish (*Carassius auratus*), minimum-detectable gaps in psychophysical paradigm using stimuli presented at 40 dB SPL, were about 35 ms, about an order of magnitude longer than the durations reported for birds and mammals. These results suggest a fundamental difference between the mechanisms underlying gap detection in fishes and endotherms. To explain these results Fay (1985) proposed that fishes encode gaps at the level of the saccular nerve by the rebound excitation following the gap, whereas mammals (and birds) presumably use a central integrator with an appropriate time constant to process gaps in acoustic stimuli (Zwislocki 1960; Brown and Maloney 1986). Clearly this is an area which warrants additional comparative research.

While the limits of gap detection by amphibians have not yet been reported, shifts in call timing in response to playback of natural or synthetic advertisement calls have been demonstrated in a variety of species (Littlejohn and Martin 1969; Lemon 1971; Rosen and Lemon 1974; Loftus-Hills 1974; Awbrey 1978; Lemon and Struger 1980; Passmore and Telford 1981; Narins 1982a; Zelick and Narins 1982; Schwartz and Wells 1983; Sullivan 1985). A number of these field studies were designed to answer the question: "What is the minimum duration of a gap in a continuous stimulus during which a male frog in his natural habitat will initiate calling?" In practice this takes the form of quantifying

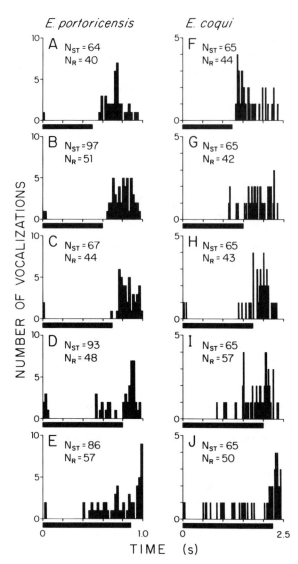

FIGURE 22.6. PST histograms of responses of a representative male of *E. coqui* and *E. portoricensis* to which tone burst stimuli of different durations were presented. Stimulus repetition rate was 2.5 s for *E. coqui* and 1.0 s for *E. portoricensis*. Horizontal bar beneath each histogram indicates stimulus duration. N_{st} is the number of stimulus tone bursts presented; N_r is the number of vocalizations produced by the frog. Bin width: 20 ms for A–E and 10 ms for F–J. Males of both species were able to initiate more calls in the gap between stimuli than would be expected by chance. (From Zelick and Narins 1983.)

the ratio of calls a male emits during the playback of an interfering tone, to the number of calls emitted during the interim (gap) between the interfering tones. This observed ratio is then compared to the expected ratio (based on the stimulus duty cycle) using a goodness of fit test. In a series of playback experiments with males of *Eleutherodactylus portoricensis* and *E. coqui*, two arboreal frogs from Puerto Rico, Zelick and Narins (1983) attempted to evaluate the precision with which a frog is able to adjust the timing of his vocalizations in order to avoid temporal overlap with an interfering acoustic stimulus. Figure 22.6 illustrates typical results of such a study in which one male of each species was presented with a series of pure tones with a frequency equal to that of his first call note. Results show that even when the stimulus tone was lengthened to occupy 90% of the stimulus repetition period, males of both species initiated more calls in the gap than would be expected by chance. This finding also illustrates another feature of anuran acoustic communication: the remarkable capacity of frogs to alter the temporal pattern of their vocalizations in the face of acoustic interference.

Acoustic avoidance of regularly occurring interfering tones could be mediated by periodicity detection and subsequent resetting of the call oscillator to "match" the repetition rate of the interfering signal (Loftus-Hills 1974). If this were the case, aperiodic interfering tones would be expected to confound the male and result in a high percentage of acoustic overlap. In an experiment designed to test this hypothesis, male frogs were presented with both short and long tones interspersed with *pseudo-randomly* occurring, 750-ms gaps. The results of this experiment suggest that the males were capable of nearly complete avoidance of the interfering tones in the absence of periodicity cues (Zelick and Narins 1985). This behavior requires a cycle-by-cycle update of the call oscillator interval, presumably triggered by the sudden stimulus intensity reduction at the gap onset. Thus, the oscillator which drives the spontaneous call mechanism appears to be resettable in midcycle, an extraordinary adaptation for producing vocal signals in the presence of nonpredictable interfering signals.

3. Interindividual Time Constants

Local populations of frogs consist of groups of individual calling males which interact vocally both to advertise their position to conspecifics as well as to attract mates. Chorus behavior may be considered a means by which a formal structure is imposed on the interactions of calling males, but which, by its very nature, is a potential source of acoustic interference which must be taken into account by each individual participating in the chorus. Thus, chorusing simultaneously increases both the order and the complexity of the acoustic environment. The rather formidable problem faced by individual participants in the chorus is that of developing and executing a *strategy* of chorus participation which at once minimizes acoustic interference while maximizing the individual's chance of attracting a mate. Another consideration is the optimization of energy expenditure, since it is known that calling is energetically quite expensive (Bucher, Ryan, and Bartholomew 1982; Wells and Taigen 1984; Taigen and Wells 1985).

How does one go about deciphering the strategy of a single male in the chorus, or even if a male is adopting a strategy at all? The first step might be to obtain the upper limit for antiphonal calling rate for an isolated male interacting with a synthetic acoustic stimulus, which would set an upper bound for the chorus participation rate. In practice, this may be determined by measuring either of two complementary parameters: behavioral refractory period and call synchronization rate.

3.1 Behavioral Refractory Period (BRP)

Analogous to the refractory period of a neuron, this is defined in two parts: the *absolute* behavioral refractory period (BRP) is the time period immediately following a frog's call during which an acoustic stimulus is almost completely ineffective at eliciting synchronized responses from the male (*sensu* Narins 1982a). The *relative* BRP is the period following the absolute BRP during which an acoustic stimulus can evoke synchronized responses with a probability which is an increasing function of the time after stimulus occurrence. The maximum evoked response rate has been shown to

be species-specific for two neotropical frogs (*Hyla ebraccata* and *Eleutherodactylus coqui*) in which the BRPs have been measured. Figure 22.7 shows the BRP measurements for a male of *E. coqui* from eastern Puerto Rico. In this case, the frog's own call was used as the trigger signal for a portable sound synthesizer, in which the delay between the trigger and the resulting synthetic call was continuously adjustable. Using this closed-loop stimulus technique, it was determined that the minimum delay following the frog's call after which the frog would begin to respond to a second stimulus (absolute BRP) for this animal was 1,120 ms and the relative BRP was ca. 500 ms. In contrast, the mean ($n = 6$) absolute BRP for *H. ebraccata* from the Panamanian lowlands was 210 ms. Nevertheless, both the duration of the relative BRP (500 ms) and the functional dependence of the synchronized response rate on the time of stimulus occurrence during the relative BRP were similar for these two species. Thus, the absolute BRP appears to be species-dependent, implying in this case that fewer males of *E. coqui* (longer BRP) would be able to participate in chorus interactions and avoid acoustic overlap than males of *H. ebraccata* (shorter BRP). Obtaining BRP values for individuals from high-density and low-density populations could potentially reveal ecological correlates of chorusing which could help explain vocal interactions among widely diverse species.

3.2 Call Synchronization Rate (CSR)

Male frogs tend to call at higher rates in choruses than when isolated, presumably due to the stimulatory effect of nearby calling individuals. Thus to obtain another behavioral measure of chorus "participation potential" it is reasonable to experimentally determine the upper rate of call synchronization to a playback of a repetitive natural or synthetic stimulus (Loftus-Hills 1974; Awbrey 1978; Lemon and Struger 1980; Zelick and Narins 1985; Moore et al. 1989). To obtain this measure, several series of calls, each with a fixed but different repetition rate, are broadcast to an isolated calling male in its natural habitat. The responses from this male are recorded and subsequently analyzed for their degree of correlation to the stimulus. Loftus-Hills (1974) showed that males

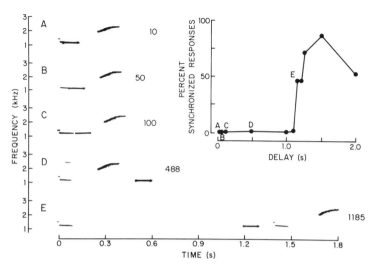

FIGURE 22.7. Determination of the behavioral refractory period (BRP) for an individual calling male of *E. coqui*. Sonagrams show the temporal relationship between two-note ('Co-Qui') call and the single-note ("Co") stimulus triggered by the frog's "Co" note. Delays between trigger and broadcast of "Co" note stimulus are (A) 10 ms; (B) 50 ms; (C) 100 ms; (D) 488 ms; and (E) 1185 ms. Air temperature during recordings: 21 °C. Inset: The percentage of synchronized responses as a function of trigger-stimulus delay, where synchronized responses are those occurring within 300 ms of the stimulus onset. The initial flat portion of the curve corresponds to the absolute BRP and the delayed rising phase is the relative BRP (see text for details).

of Strecker's chorus frog, *Pseudacris streckeri*, are able to entrain their calls to exogenous stimuli over a range which encompassed rates from 18% shorter to 15% longer than their spontaneous call rates. This range of rates corresponded to that found in natural populations. In a study designed in part to determine the range over which "locking" to an acoustic stimulus could occur, Zelick and Narins (1985) presented tone bursts with repetition periods from 1,200 ms to 2,800 ms (tone burst durations from 700 ms to 2,300 ms) to isolated calling males of *E. coqui*. Results of that study show that reducing the acoustic stimulus period caused a concomitant shortening of the frog's response period, to a point where 1:1 following was abruptly replaced by 1:2 following, that is the frog would call once for every two stimuli (Fig. 22.8). The transition point between these two lowest following ratios serves as a clear and unambiguous marker for the upper range of the call oscillator (lower range of call period), and appears to be a function of the animal's spontaneous calling rate. The transition point in this example occurred at about 1,900 ms, which is 30% shorter than the

spontaneous calling period, or nearly twice the shortening exhibited by Strecker's chorus frog.

A similar experiment was carried out with the white-lipped frog, *Leptodactylus albilabris*, a substrate-dweller from Puerto Rico. This burrowing animal has an advertisement call consisting of short (30 to 40 ms) "chirps" repeated 3 to 4 times per second (Lewis and Narins 1985; Lopez et al. 1988). Vocal behavior in this species is unusual in that (a) its repetition period could be decreased by up to 50% below the spontaneous calling period (from ca. 340 ms to 170 ms) in response to an exogenous acoustic stimulus, (b) 1:n following extends to stimulus rates that are six to eight times the natural calling rate of this species, and (c) the calling algorithm for *L. albilabris* may be characterized by a refractory period plus a delayed poststimulus inhibition (Moore et al. 1989).

These studies illustrate the wide variation among different frog species of a single time constant: namely a male's minimum response delay following an external acoustic stimulus. This parameter takes on increasing importance in the development of models and in simulations of intra- and interspecific acoustic interactions among vocalizing organisms.

3.3 Relationship Between BRP and CSR

What is the relationship between the behavioral refractory period and the upper limit of call synchronization rate? BRP is measured using an "echo" stimulus triggered by the frog's own vocalization, and represents the minimum time following a frog's call which results in a subsequent "synchronized call" (see Section 3.1). Since the trigger for the "echo" stimulus is derived from the frog's natural call, the stimulus is clearly *aperiodic*. Moreover, since the frog under test can adjust his calling rate at will, one might expect that the BRP would represent the absolute shortest period between calls. In contrast, the highest rate of call synchronization is obtained in response to a *periodic* stimulus sequence and thus might be expected to yield a longer time constant, since the frog under test would have multiple chances to "adjust" his calling rate over time. In the only species in which both measures have been made (the Puerto Rican coqui, *E. coqui*), this prediction is borne out. The average absolute BRP ($n = 30$) is ca. 1,130 ms and the upper limit on call synchronization rate (lower limit on call period) is ca. 1,900 ms. Direct comparison between these values must be made with caution, however, since these experiments were performed on different populations of frogs, at different ambient temperatures. This is critical, since the temperature dependence of the temporal parameters of anuran vocalizations has been well documented (Zweifel 1968; Nevo and Schneider 1976; Gerhardt 1978; Gerhardt and Mudry 1980; Narins and Smith 1986; Stiebler and Narins 1990).

In theory, a variety of call-timing algorithms can give rise to very similar calling patterns. Thus, the algorithms underlying calling are to some degree protected from the selection pressures acting on the communication system as a whole. Thus, a case may be made that the call timing algorithms used by closely related species may reflect phylogenetic similarities as do more traditional morphological features, e.g., osteology (Moore et al. 1989). Clearly, there is a need to obtain more detailed information on these algorithms from a wide variety of species both to understand more fully the

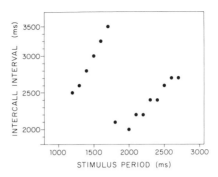

FIGURE 22.8. Change in intercall interval for a single male of *E. coqui* in response to periodic tone bursts of different periods. Between 2,000 and 2,600 ms the frog exhibited 1:1 following with the stimulus, whereas from 1,200 to 1,700 ms 1:2 following occurred. The transition between 1:1 to 1:2 following occurred at a stimulus period of about 1,800 ms. The mean spontaneous calling rate for this animal was 2,700 ms. (From Zelick and Narins 1985.)

dynamics of chorus interactions and perhaps to reexamine phylogenetic relationships of the Anura using a more mechanistic, yet (arguably) conservative criterion.

The assumption is made that the calling behavior of a male is neither energy limited nor energy conservative; that is, the frog calls at a rate somewhere between the maximum and the minimum possible rate (zero calls/minute). Thus, a calling male is thought to produce an intermediate calling rate, but equally important, to generate his calls at *particular* times which are in some way cued to his nearest neighbors' calls. It is the details of this interaction strategy which is of concern in the following section.

4. Collective Acoustic Behavior of Anuran Amphibians

Although the literature is replete with studies on the nature of acoustic behavior in amphibians (see Littlejohn 1977 and Wells 1977 for reviews; Narins and Capranica 1976, 1978; Lemon and Struger 1980; Zelick and Narins 1982, 1983; Arak, 1983; Schwartz and Wells 1983; Wells and Schwartz 1984; Narins and Zelick 1988; Wells 1988; Gerhardt 1988, 1989), the majority of this work has dealt with the responses of either isolated individuals or pairs of frogs to natural or synthetic stimuli. In fact, actual studies of group signaling behavior are rare (see Hardy 1959 and Fox and Wilczynski 1986 for exceptions). This is due in part to the inherent difficulty in monitoring the behavior of more than one or two individuals at a time. Nevertheless, studies of the behavior of individuals within groups must be undertaken if we are to comprehend more fully the dynamics of the vocal interactions between members of such groups. An understanding of the acoustic behavior of an isolated individual coupled with the knowledge of an individual's responses to the vocalizations of its neighbors may lead to elucidation of the underlying mechanisms governing collective calling behavior.

An amphibian chorus may be operationally defined as an assemblage of frogs (conspecific and/or heterospecific males) calling at levels such that each individual in the group is capable of detecting the call of at least one other individual (Narins and Zelick 1988). Arboreal frogs in such an assemblage form a three-dimensional (or two-dimensional in the case of substrate dwellers) spatially distributed array of punctate sound sources, each with unique spectral and temporal characteristics. A human observer placed in such a sound field might report a continuous din of background noise with little or no conspicuous pattern. At the position of an individual frog in the chorus, however, omnidirectional measurements of sound pressure level reveal instead that nearest neighbors' calls are significantly above the background noise and thus should be distinguishable from those of more distant neighbors (Narins and Hurley 1982; Prestwich, Brugger, and Topping 1989). In addition to the sound generated by chorusing frogs, acoustic signals produced by other vertebrates, insects, and sounds of abiotic origin (e.g., wind, rivers, rain, etc.) contribute to the total acoustic environment.

4.1 Dynamic Chorus Structure (DCS)

Critical to the understanding of chorusing behavior in anurans is a quantitative description of the vocal patterns used by the individual frogs comprising the chorus. The approach taken in studying the chorusing behavior of the Puerto Rican coqui, *E. coqui*, has been to develop a model of chorusing

based initially on the spontaneous and evoked calling behavior of individual, isolated males (Brush and Narins 1989). A computer simulation of this model was then extended to groups of calling males which allowed us to generate expected chorus patterns and subsequently to compare these patterns to those actually observed in field experiments. Since it is known that in this species, the properties of the call notes themselves (rather than the internote intervals) are crucial for male aggressive responses and female recognition, an algorithm was chosen which minimizes, from an individual's point of view, call overlap with his nearest neighbors. Simulations of groups of calling males suggest that a given male can maximize effectiveness of each call by actively avoiding acoustic overlap with at most, two of its nearest neighbors.

Field studies of chorus behavior were carried out in Puerto Rico in 1985 and 1987 in order to test the prediction of the model. One microphone was placed next to every frog (up to eight) in a localized chorus of calling males. The timing of each frog call was encoded by a tone generated by a triggerable oscillator with a unique frequency, and the resulting tones were then mixed and recorded on a cassette tape. Subsequent laboratory analysis of the tape consisted of first separating the encoded tones using a bank of filters with center frequencies corresponding to the tone frequencies, then digitizing the filter outputs for entry into a computer. In this way, the complete calling record of an entire chorus of calling males may be stored in the computer, facilitating pair-wise correlation analysis of the calls.

For each chorus of **n** frogs studied, an **n** by **n** correlation matrix was constructed. The off-diagonal (cross) correlograms provide detailed information regarding both the latencies between calls of any pair of frogs in chorus and the significance of all pair-wise vocal interactions. From these correlograms, one may construct a series of arrows between participating frogs such that an arrow from frog **a** to frog **b** indicates that frog **b** was better than any other frog at initiating a call within the inter call interval of frog **a**. When the arrows are superimposed on a representation of the calling site, one immediately appreciates the three-dimensional aspect of the dynamic chorus structure (DCS). An example of such a structure from a representative chorus in Puerto Rico is shown in

FIGURE 22.9. Dynamic chorus structure for the Puerto Rican arboreal frog *E. coqui* superimposed on a drawing depicting a typical scene in the rainforest near El Verde in the Luquillo Mountains of eastern Puerto Rico. Calling males typically assume vertical, head-up postures on tree trunks or branches and time their vocalizations such that they interact with one or two of their nearest neighbors. The principal tree in the montane forest at this alti-tude (ca. 350 m) is the Tabanuco (*Dacryodes excelsa*) shown in the foreground. The sierra palm (*Prestoea montana*) is a subcanopy species which occurs commonly throughout the forest. Other prominent features of the Tabanuco forest include lianas and other climbers (e.g., *Philodendron*), bromeliads, trunk barks mottled with growth and plaster vines, and an open ground story in deep shade. (Drawing by Sarah Landry.)

Figure 22.9. From this type of analysis of calling patterns of natural frog choruses, we have learned that the original hypothesis is strongly supported; namely that the large majority of individual frogs actively avoid call overlap with either one or two neighbors thus maximizing the communicative effectiveness of their calls (Brush and Narins 1989). Future studies shall address the questions of the stability of this structure over time, the effect of removal of an individual (predation) on the DCS, and the effect of a new frog entering the chorus (immigration) on the DCS.

5. Summary and Conclusions

Anuran amphibians (frogs and toads) operate in an environment that is inherently noisy at low frequencies, and in which many other frogs of the same species are calling in competition. This chapter is concerned with measures of auditory function determined in the frog's natural habitat, with a view toward understanding the acoustic behavior of both the individual as well as the group of signaling anurans. Using the neotropical arboreal species *Eleutherodactylus coqui* as a model system, and by utilizing the response behavior to synthetic acoustic stimuli, behavioral response functions, intensity dependence of thresholds, effective critical ratios and effective critical bands, and duration sensitivities may be quantified for undisturbed frogs in nature. Moreover, temporal parameters relevant for inter individual vocal interactions such as behavioral refractory periods, call synchronization rates, and gap sensitivities, as well as intensity discrimination are also amenable to study and have been determined for a variety of new world species.

A significant goal of auditory neuroethologists is to understand the factors affecting the evolution of acoustic communication systems. Clearly, one of these factors is the presence of calling conspecific males in the vicinity of an animal attempting to communicate. One approach to this problem is to fully characterize the acoustic behavior of an isolated individual, and then extrapolate the results to a vocally interacting group. Alternatively, one might model the vocal interactions of a group, predict the "ideal" behavior of an individual in the group from the model, and directly test the predictions by measuring the vocal behavior of a number of individuals in the chorus. Some preliminary success with the second approach is encouraging; future studies using this method might directly assess the effect of factors such as predation, immigration, man-made interference, etc., on both the individual as well as on a group of signaling organisms.

Acknowledgments. Original research for this chapter was supported by NIH Grant no. NS19725 and UCLA Academic Senate Grant no. UR3501. I am grateful to M. Kowalczyk and H. Kabe for their unfailing help with the figure preparations.

References

Arak A (1983) Vocal interactions, call matching, and territoriality in a Sri Lankan treefrog, *Philautus leucorhinus* (Rhacophoridae). Anim Behav 31:292–302.

Awbrey FT (1978) Social interaction among chorusing Pacific tree frogs, *Hyla regilla*. Copeia 1978:208–214.

Brown CH, Maloney CG (1986) Temporal integration in two species of Old World monkeys: Blue monkeys (*Cercopithecus mitis*) and grey-cheeked monkeys (*Cercocebus albigena*). J Acoust Soc Am 79:1058–1064.

Brush JS, Narins PM (1989) Chorus dynamics of a neotropical amphibian assemblage: Comparison of computer simulation and natural behavior. Anim Behav 37:33–44.

Bucher TL, Ryan MJ, Bartholomew GA (1982) Oxygen consumption during resting, calling and nest building in the frog *Physalaemus pustulosus*. Physiol Zool 55:10–22.

Buus S, Florentine M (1985) Gap detection in normal and impaired listeners: The effects of level and frequency. In: Michelsen A (ed) Time Resolution in Auditory Systems. Berlin, Heidelberg, New York, Tokyo: Springer-Verlag.

Dooling RJ, Zoloth SR, Baylis JR (1978) Auditory sensitivity, equal loudness, temporal resolving power, and vocalizations in the house finch (*Carpodacus mexicanus*). J Comp Physiol Psychol 92:867–876.

Drewry GE, Rand AS (1983) Characteristics of an acoustic community: Puerto Rican frogs of the genus *Eleutherodactylus*. Copeia 1983:941–953.

Dunia RD, Narins PM (1989) Temporal integration in an anuran auditory nerve. Hearing Res 39:287–298.

Ehret G (1975) Masked auditory thresholds, critical ratios, and scales of the basilar membrane of the house mouse (*Mus musculus*). J Comp Physiol 103:329–341.

Ehret G, Gerhardt HC (1980) Auditory masking of synthetic mating calls by noise in the green treefrog (*Hyla cinerea*). J Comp Physiol 141:13–18.

Fay RR (1974) Masking of tones by noise for the goldfish (*Carassius auratus*). J Comp Physiol Psychol 87:708–716.

Fay RR (1985) Temporal processing by the auditory system of fishes. In: Michelsen A (ed) Time Resolution in Auditory Systems. Berlin, Heidelberg, New York, Tokyo: Springer-Verlag.

Fletcher H (1940) Auditory patterns. Rev Mod Phys 12:47–65.

Foster WA (1967) Chorus structure and vocal response in the Pacific tree frog *Hyla regilla*. Herpetologica 23:100–104.

Fox JH, Wilczynski W (1986) The augmentation of per-capita active space through chorusing in anurans: a computer model. Soc Neurosci Abstr 12:314.

Gerhardt HC (1978) Mating call recognition in the green treefrog (*Hyla cinerea*): The significance of some fine-temporal properties. J Exp Biol 74:59–73.

Gerhardt HC (1982) Sound pattern recognition in some North American treefrogs (Anura: Hylidae): Implications for mate choice. Amer Zool 22:581–595.

Gerhardt HC (1988) Acoustic properties used in call recognition by frogs and toads. In: Fritzsch B, Ryan M, Wilczynski W, Hetherington T, Walkowiak W (eds) The Evolution of the Amphibian Auditory System. New York: John Wiley and Sons.

Gerhardt HC (1989) Acoustic pattern recognition in anuran amphibians. In: Dooling RJ, Hulse SH (eds) The Comparative Psychology of Audition: Perceiving Complex Sounds. Hillsdale: Lawrence Erlbaum Associates.

Gerhardt HC, Klump GM (1988) Masking of acoustic signals by the chorus background noise in the green tree frog: a limitation on mate choice. Anim Behav 36:1247–1249.

Gerhardt HC, Mudry KM (1980) Temperature effects on frequency preferences and mating call frequencies in the green treefrog, *Hyla cinerea* (Anura: Hylidae). J Comp Physiol 137:1–6.

Giraudi D, Salvi R, Henderson D, Hamernik R (1980) Gap detection by the chinchilla. J Acoust Soc Am 68:802–806.

Hardy DF (1959) Chorus structure in the striped chorus frog, *Pseudacris nigrita*. Herpetologica 15:14–16.

Klump GM, Gerhardt HC (1987) Use of non-arbitrary acoustic criteria in mate choice by female gray tree frogs. Nature 326:286–288.

Klump GM, Maier E (1990) Gap detection in the starling (*Sturnus vulgaris*): I. Psychophysical thresholds. J Comp Physiol 164:531–538.

Lemon RE (1971) Vocal communication by the frog *Eleutherodactylus martinicensis*. Can J Zool 49:211–217.

Lemon RE, Struger J (1980) Acoustic entrainment to randomly generated calls by the frog, *Hyla crucifer*. J Acoust Soc Am 67:2090–2095.

Lewis ER, Narins PM (1985) Do frogs communicate with seismic signals? Science 227:187–189.

Littlejohn MJ (1977) Long-range acoustic communication in anurans: An integrated and evolutionary approach. In: Taylor DH, Guttman SI (eds) The Reproductive Biology of Amphibians. New York: Plenum Press.

Littlejohn MJ, Martin AA (1969) Acoustic interaction between two species of leptodactylid frogs. Anim Behav 17:785–791.

Loftus-Hills JJ (1971) Neural correlates of acoustic behaviour in the Australian bullfrog *Limnodynastes dorsalis* (Anura: Leptodactylidae). Z Vergl Physiol 74:140–152.

Loftus-Hills JJ (1974) Analysis of an acoustic pacemaker in Strecker's chorus frog, *Pseudacris streckeri* (Anura: Hylidae). J Comp Physiol 90:75–87.

Long GR (1977) Masked auditory thresholds from the bat, *Rhinolophus ferrumequinum*. J Comp Physiol 116:247–255.

Lopez PT, Narins PM, Lewis ER, Moore SW (1988) Acoustically-induced call modification in the white-lipped frog, *Leptodactylus albilabris*. Anim Behav 36:1295–1308.

Moore SW, Lewis ER, Narins PM, Lopez PT (1989) The call timing algorithm of the white-lipped frog, *Leptodactylus albilabris*. J Comp Physiol 164:309–319.

Moss CF, Simmons AM (1986) Frequency selectivity of hearing in the green treefrog, *Hyla cinerea*. J Comp Physiol 159:257–266.

Narins PM (1976) Auditory processing of biologically meaningful sounds in the treefrog, *Eleutherodactylus coqui*. PhD Thesis, Cornell University, Ithaca, NY.

Narins PM (1982a) Behavioral refractory period in neotropical treefrogs. J Comp Physiol 148:337–344.

Narins PM (1982b) Effects of masking noise on evoked calling in the Puerto Rican Coqui (Anura: Leptodactylidae). J Comp Physiol 147:438–446.

Narins PM (1983) Synchronous vocal response mediated by the amphibian papilla in a neotropical treefrog: Behavioral evidence. J Exp Biol 105:95–105.

Narins PM, Capranica RR (1976) Sexual differences in the auditory system of the treefrog, *Eleutherodactylus coqui*. Science 192:378–380.

Narins PM, Capranica RR (1978) Communicative significance of the two-note call of the treefrog, *Eleutherodactylus coqui*. J Comp Physiol 127:1–9.

Narins PM, Capranica RR (1980) Neural adaptations for processing the two-note call of the Puerto Rican treefrog, *Eleutherodactylus coqui*. Brain Behav Evol 17:48–66.

Narins PM, Hurley DD (1982) The relationship between call intensity and function in the Puerto Rican Coqui (Anura: Leptodactylidae). Herpetologica 38: 287–295.

Narins PM, Smith SL (1986) Clinal variation in anuran advertisement calls: basis for acoustic isolation? Behav Ecol Sociobiol 19:135–141.

Narins PM, Zelick R (1988) The effects of noise on auditory processing and behavior in amphibians. In: Fritszch B, Ryan M, Wilczynski W, Hetherington T, Walkowiak W (eds) The Evolution of the Amphibian Auditory System. New York: John Wiley and Sons.

Nevo E, Schneider H (1976) Mating call pattern of Green toads in Israel and its ecological correlate. J Zool Lond 178:133–145.

Passmore NI, Telford SR (1981) The effect of chorus organization on mate localization in the painted reed frog (*Hyperolius marmoratus*). Behav Ecol Sociobiol 9:291–293.

Plomp R (1964) Rate of decay of auditory sensation. J Acoust Soc Am 36:277–282.

Potter HD (1965) Patterns of acoustically evoked discharges of neurons in the mesencephalon of the bullfrog. J Neurophysiol 28:1155–1184.

Prestwich KN, Brugger KE, Topping M (1989) Energy and communication in three species of hylid frogs: Power input, power output and efficiency. J Exp Biol 144:53–80.

Rosen M, Lemon RE (1974) The vocal behavior of spring peepers, *Hyla crucifer*. Copeia 1974:940–950.

Scharf B (1970) Critical bands. In: Tobias JV (ed) Foundations of Modern Auditory Theory, Vol. 1. New York and London: Academic Press.

Schneider H, Joermann G, Hodl W (1988) Calling and antiphonal calling in four neotropical anuran species of the family Leptodactylidae. Zool Jb Physiol 92:77–103.

Schwartz JJ (1987) The function of call alternation in anuran amphibians: a test of three hypotheses. Evolution 41:461–471.

Schwartz JJ, Wells KD (1983) The influence of background noise on the behavior of a neotropical treefrog, *Hyla ebraccata*. Herpetologica 39:121–129.

Schwartz JJ, Wells KD (1984) Interspecific acoustic interactions of the neotropical treefrog *Hyla ebraccata*. Behav Ecol Sociobiol 14:211–224.

Shailer MJ, Moore BCJ (1983) Gap detection as a function of frequency, bandwidth, and level. J Acoust Soc Am 74:467–473.

Stiebler I, Narins PM (1990) Temperature-dependence of auditory nerve response properties in the frog. Hearing Res 46:63–82.

Sullivan BK (1985) Male calling behavior in response to playback of conspecific advertisement calls in two bufonids. J Herpetol 19:78–83.

Taigen TL, Wells KD (1985) Energetics of vocalization by an anuran amphibian (*Hyla versicolor*). J Comp Physiol 155 B:163–170.

Tuttle MD, Ryan MJ (1982) The role of synchronized calling, ambient light, and ambient noise, in anti-bat-predator behaviour of a treefrog. Behav Ecol Sociobiol 11:125–131.

Walkowiak W (1988) Two auditory filter systems determine the calling behavior of the fire-bellied toad: a behavioral and neurophysiological characterization. J Comp Physiol 164:31–41.

Wells KD (1977) The social behavior of anuran amphibians. Anim Behav 25:666–693.

Wells KD (1988) The effect of social interactions on anuran vocal behavior. In: Fritszch B, Ryan M, Wilczynski W, Hetherington T, Walkowiak W (eds) The Evolution of the Amphibian Auditory System. New York: John Wiley and Sons.

Wells KD, Schwartz JJ (1984) Vocal communication in a neotropical treefrog, *Hyla ebraccata*: advertisement calls. Anim Behav 32:405–420.

Wells KD, Taigen TL (1984) Reproductive behavior and aerobic capacities of male American toads (*Bufo americanus*): Is behavior constrained by physiology? Herpetologica 40:292–298.

Wells KD, Taigen TL (1986) The effect of social interactions on calling energetics in the gray treefrog (*Hyla versicolor*). Behav Ecol Sociobiol 19: 9–18.

Wells KD, Taigen TL (1989) Calling energetics of a neotropical treefrog, *Hyla microcephala*. Behav Ecol Sociobiol 25:13–22.

Zelick RD, Narins PM (1982) Analysis of acoustically evoked call suppression behaviour in a neotropical treefrog. Anim Behav 30:728–733.

Zelick RD, Narins PM (1983) Intensity discrimination and the precision of call timing in two species of neotropical treefrogs. J Comp Physiol 153: 403–412.

Zelick RD, Narins PM (1985) Characterization of the advertisement call oscillator in the frog *Eleutherodactylus coqui*. J Comp Physiol 156:223–229.

Zweifel RG (1968) Effects of temperature, body size, and hybridization on mating calls of toads, *Bufo a. americanus* and *Bufo woodhousii fowleri*. Copeia 1968:269–285.

Zwislocki J (1960) Theory of temporal auditory summation. J Acoust Soc Am 32:1046–1060.

Abstract E

Some Unique Features of the Ear and the Lateral Line of a Catfish and Their Potential Bearing for Sound Pressure Detection

H. Bleckman, B. Fritzsch*, U. Niemann and H.M. Müller

Catfish are auditory specialists (von Frisch, 1923). While much is known about the organization of their "middle ear" and the Weberian ossicles, little is known about the central anatomy and physiology of the acoustic system of catfish and the possible interactions this system may have with the mechanoreceptive lateral line.

The morphology of the inner ear and adjacent lateral line canals was studied in the bottom dwelling catfish *Ancistrus*. Our anatomical data—derived from 3-D-reconstructions of serial sections—indicate that the sensory epithelia of the sacculus and the lagena may be well suited to perceive sound pressure changes mediated as particle displacement to the inner ear via the Weberian ossicles. In *Ancistrus* a foramen in the otic capsule connects the ear with the middle trunk lateral line canal. This foramen most likely facilitates inner ear particle motions by acting as a pressure release window. Close to the foramen the middle trunk lateral line canal harbors two neuromasts which indirectly may be stimulated by sound pressure waves.

Fibers of the sacculus and the lagena-labeled with HRP or fluorescent dextran amines—selectively terminate in the medial auditory nucleus (MAN). The MAN projects to the midbrain and was assumed to be homologous to the superior olive of tetrapods (Finger and Tong 1984). Our data indicate, however, that the MAN is a primary auditory nucleus of the hindbrain. Based on its connections and its cytoarchitectonic discreteness, the MAN is as distinct as the primary auditory nuclei of tetrapods. We suggest that the MAN is not homologous to the tetrapod auditory nucleus but instead evolved independently.

The central projections of the lateral line neuromasts found close to the lateral foramen of the inner ear (see above) are not strikingly different from those of the anterior and posterior lateral line nerves. However, the fibers which innervate these neuromasts show an unusual course in that they run through the ear but they enter the brain stem together with the posterior lateral line nerve. We suggest that these lateral line afferents are homologous to the middle lateral line nerve of other bony fish (Northcutt 1989).

Many toral units of *Ancistrus* are excited by human voice and noiselike acoustic events. These units also respond to a fast (20 cm/s), but, unlike the mechanoreceptive lateral line, not to a slow (2.3 cm/s) moving object. If monofrequency stimuli are used, midbrain acoustic units of *Ancistrus* show band-pass tuning with low and high-pass slopes of up to -12 and 36 dB/octave, respectively. For the frequencies tested (30 to 3,700 Hz) minimal thresholds were in the 300 to 900 Hz region. Within this region a sound pressure level of -39 dB (rel. 1 microbar) was sufficient to generate a neural response. Supported by DFG-grants (HB 242/3, Fr 572)

References

Finger TE, Song S-L (1984) J Comp Neurol 229:129–151.

Frisch, K von (1923) Biologisches Zentralblatt 43:439–446.

Northcutt G (1989) In: Coombs S, Görner P, Münz H (eds) *The Mechanosensory Lateral Line. Neurobiology and Evolution*. New York: Springer-Verlag, pp. 17–78.

Abstract F

Comparative Analysis of Electrosensory and Auditory Function in a Mormyrid Fish

John D. Crawford

Mormyrid fishes are best known for their highly developed electrosensory system but they also exhibit specialized auditory organs and use sounds in their social behavior. The two extant sensory systems, auditory and electrosensory, are components of the octavolateralis system and likely represent the outcomes of evolutionary processes acting to modify common ancestral mechanosensory systems. In both systems, the peripheral receptors are hair cell like but in one system the receptors are stimulated by electric current and in the other by mechanical movement. The comparison of these systems is useful as a means to further understand the types of modifications evolution has brought about within the octavolateralis system and as a way to identify properties which have been conserved amongst components.

Pollimyrus isidori is a particularly valuable subject for this comparative study because it uses electric and acoustic communication signals and because signals in these two sensory channels interact during courtship. Its electric organ discharge (EOD) is a brief (80 µs), individually stereotyped, triphasic pulse which can be produced in a wide range of inter-EOD time interval patterns (interval sequences). The power spectra of the individual EODs are very broad, with energy from below 100 Hz to well over 50 KHz. In contrast to EODs, interval sequences are highly context dependent. Interval sequences usually have high variance, and include intervals from 8 ms to 250 ms (4 to 125 EODs/sec, intervals > 0.25 s classified as pauses). When a female is ready to spawn, she approaches a territorial male while producing an unusually uniform sequence of inter-EOD intervals (CV ≤ 0.03). While discharging her electric organ in this metronome-like fashion, she is courted by a male with a sonic display consisting of two sounds (*Grunts & Moans*). A third distinct sound is produced by the male (*Growl*) when the female leaves his territory. These sounds have energy between 100 Hz and 1.0 KHz with maxima at about 220 and 440 Hz. They exhibit little frequency modulation and are best described as three distinct acoustic pulse trains, differing in inter-pulse interval (approximately 4 ms, 20 ms, and 40 ms). Playback experiments show that electric signals are potent releasers of sonic courtship. Even though females do have a sex-characteristic EOD waveform, male sonic responses are influenced by interval sequences and not by features of the EOD waveform. Thus, during courtship, the female's display of a highly uniform interval sequence releases male sonic behavior which, in turn, seems to complete a bimodal exchange of information.

The two signal types, acoustic and electric, are physically different. However, several lines of evidence suggest that the salient features of both are temporal patterns that are similar in nature. In both channels the signals are rich in temporal structure and the intervals that are behaviorally significant are of roughly comparable duration. The electrosensory pathway is broadly tuned ($Q_{10dB} \leq 1.0$; Raman 1988) and appears specialized for temporal coding (Amagai et al. 1988; Bell and Grant 1989). Neurophysiological studies of the mormyrid auditory system have also revealed characteristics expected of a good time encoding system. Auditory neurons recorded in the mid-brain Torus (nucleus Medialis Dorsalis) show weak frequency selectivity to tones ($Q_{10dB} \leq 1.2$) with best frequencies in the 100 Hz to 1.0 KHz range. Auditory responses to temporally patterned acoustic pulse trains are currently in progress.

Abstract G
Biophysics of Underwater Hearing in the Clawed Frog, *Xenopus laevis*

J. Christensen-Dalsgaard and A. Elepfandt

Male clawed frogs (*Xenopus laevis*) produce an underwater mating call with frequency peaks at 1.2 and 1.8 kHz, and recent research has shown that receptive females can localize the mating call with considerable accuracy (Schanz and Elepfandt, Verh. Dtsch. Zool. Ges. 81:211 (1988). In order to investigate the biophysics of *Xenopus* underwater hearing, we have studied the vibrations of the eardrum and body wall with laser vibrometry.

Immobilized animals were suspended 15 cm below water surface in a water tank and stimulated with sound from an underwater loudspeaker. Vibration velocities of the eardrum and body wall were measured with a Dantek laser vibrometer. Dental cement casts of the middle ear cavity were made and from these casts the volume of the cavity was calculated. In some experiments, the middle ear cavity was filled with water through the Eustachian tube.

The eardrum vibration velocities were maximal in the frequency range from 1.6 to 2.2 kHz and the peak velocity amplitudes ranged from 0.5 to 2.5 mm/s (corresponding to displacements of approximately 100 nm) at a sound pressure of 10 pa. A small peak was found at lower frequencies (below 1 kHz) and it was shown that this peak corresponded to the peak of the vibrations of the body wall above the lung. Furthermore, the position of this peak could be shifted by inflating the lung. When the middle ear cavity is filled with water, the eardrum vibrations decreased by 10–30 dB. Together with the correspondence between the data and calculations based on a model of fish swimbladder vibrations (Christensen-Dalsgaard, Breithaupt and Elepfandt, Naturwissenschaften 77, *in press* (1990)) this shows that the sound-induced pulsations of the air in the middle ear cavity accounts for most of the observed eardrum vibrations in the frequency range of the mating call.

In a different series of experiments the directionality of the *Xenopus* ear was investigated. The animals were placed in the center of a circular pond (4 m diameter, 25 cm depth) and the eardrum vibration responses to contra- and ipsilateral sound stimulation measured using laser vibrometry. While still preliminary the results showed that there is up to 20 dB difference between ipsi- and contralateral response in the frequency range of the mating call. The reason for this might be that the two ears are connected and the auditory system thus works as an underwater pressure difference receiver.

Section V
Nonmammalian Amniotes

The auditory systems of reptiles and birds have received considerable attention in the past several years: they are accessible for experimental study; and the avian system bears some striking similarities to that of mammals. Moreover, these systems provoke any number of important questions relevant to the evolution of hearing.

The morphology of the auditory portion of the ears of extant reptiles is extensively reviewed by Miller (Chapter 23). He proposes an evolutionary scheme in which the auditory papillae of most lizard families may have been derived from either of two basic types of basilar papillae.

The morphological patterns of the lizard basilar papilla show greater differences both between and within families than those of any other vertebrate groups, according to Miller. This idea is further developed by Köppl and Manley (Chapter 24), who argue that this structural diversity is not accompanied by an equivalent diversity in function. The preservation of similar physiological response patterns in morphologically different ears points to the importance of the coding of certain parameters that allow the basic functions of sound detection and localization, and

in some cases, acoustic communication. This paper strongly supports the arguments made earlier by Ted Lewis (Chapter 11).

The central auditory pathways of reptiles and birds are considered by Carr (Chapter 25). From comparative anatomical studies of central auditory pathways of turtles, lizards, crocodilians, and birds, she suggests a hypothetical morphotype. This morphotype may have existed in the common ancestor of all extant amniotes.

Some comparative aspects of avian hearing are discussed by Dooling in Chapter 26. He demonstrates great similarity in auditory sensitivity among birds that have been tested with standard psychoacoustical techniques. However, he also points out that species differ greatly in higher-order auditory capacities, such as responses to species-specific vocalizations.

Continuing with themes that were enunciated in Chapters 23 and 24 on reptiles, Manley and Gleich (Chapter 27) argue that there are also great similarities in hearing organs between living birds and crocodilians. They present evidence that the similarity of these organs to those of mammals is a case of parallel or convergent evolution.

23
The Evolutionary Implications of the Structural Variations in the Auditory Papilla of Lizards

Malcolm R. Miller

1. Introduction

In 1953 Shute and Bellairs in a light microscopic study suggested that the gross anatomical features of the lizard cochlear duct could have significant bearing on the taxonomic relationships of lizard families. Schmidt (1964) corroborated this assumption. Shortly thereafter I published a gross anatomic study of the cochlear duct of 205 species (131 genera) of 18 lizard families unequivocally demonstrating that cochlear duct anatomy provided important clues to lizard family relationships (Miller 1966a). Irwin Baird (1960) published information concerning lizard cochlear duct anatomy and summarized the status of the reptilian cochlear duct in 1974, as I did somewhat later (Miller 1980a). E.G. Wever (1978) made very extensive studies of the reptilian basilar papilla (186 species, 16 families of lizards) based largely on the differences in tectorial hair cell coverings and hair cell orientation determined by light microscopic serial sections. From these investigations, he proposed taxonomic relationships between lizard families.

In the 1970s and 1980s, SEM studies of a considerable variety of lizard basilar papillae confirmed and extended the taxonomic implications of earlier gross anatomic studies (Baird and Marovitz 1971; Baird 1974; Bagger-Sjöbäck and Wersäll 1973; Mulroy 1974; Miller 1980a, 1981; Köppl 1988, 1989). Likewise, in the 1970s and 1980s, TEM study of both hair cell ultrastructure and innervation, as well as papillar nerve dendrite injection added further detail (Bagger-Sjöback 1976; Nadol et al. 1976; Mulroy 1986; Miller 1985; Miller and Beck 1988, 1990a, 1990b; Köppl and Manley 1990).

Based upon the foregoing, as well as other studies, this chapter will describe the basic gross and microscopic features of the cochlear duct, the basilar papilla, and the innervation of the auditory hair cells in species of lizard families. An attempt will be made to relate the variations in these structural features to taxonomic relationships (closeness or relatedness) in lizard families. Finally, possible sequences of the evolution of lizard auditory papillae will be proposed.

2. Basic Anatomy of the Lizard Cochlear Duct and Basilar Papilla

Figure 23.1 illustrates the lizard membranous labyrinth showing the location of the cochlear duct. Figures 23.2 and 23.3 illustrate the two sensory areas (basilar papilla and macula lagenae) of cochlear ducts, and the specialized limbic tissue supporting the basilar membrane on which the basilar papilla rests. Figure 23.4A is an SEM of the surface of a basilar papilla demonstrating both unidirectional (UHC) and bidirectional hair cells (BHC). Figure 23.4B is a diagrammatic representation of groups of unidirectional and bidirectional hair cells to clarify the meaning of these terms. Various types of tectorial coverings will be illustrated in subsequent figures. It should be pointed out that Wever (1978) differentiates between the tectorial membrane originating from the neural limbus and various types of structures covering the

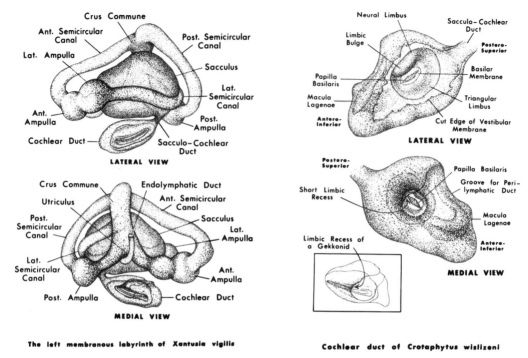

FIGURE 23.1. Lateral and medial views of the left membranous labyrinth of *Xantusia vigilis*. (From Miller 1966a.)

FIGURE 23.2. Lateral and medial views of the left cochlear duct of *Crotaphytus wislizeni*. (From Miller 1966a.)

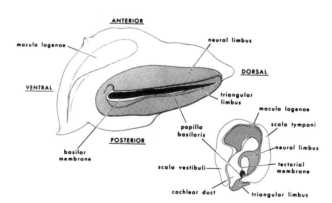

FIGURE 23.3. Upper: Lateral view of the left cochlear duct of *Gekko gecko*. Lower: Cross section of the cochlear duct. (From Miller 1973a.)

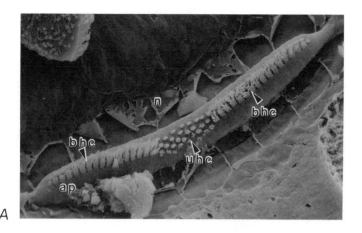

A

NEURAL SIDE OF PAPILLA
Branches of auditory nerve enter this side of the papilla

BIDIRECTIONAL   UNIDIRECTIONAL   BIDIRECTIONAL
HAIR CELLS   HAIR CELLS   HAIR CELLS

ABNEURAL SIDE OF PAPILLA

B   Arrow tips indicate position of the kinocilia

FIGURE 23.4. (A) *Sceloporus magister* (Iguanidae). SEM montage of the left basilar papilla. The central region consists of short-ciliated UHC. The tectorial cap covering these hair cells has been removed. The apical and basal long-ciliated BHC segments are arranged in two oppositely oriented rows and lack any tectorial covering. ×300. (From Miller 1978b.) (B) Diagram illustrating unidirectional and bidirectional hair cell orientation. In the central region of the auditory papilla depicted above, all the hair cells are oriented by their kinocilia in the abneural direction (away from the side of the papilla where the nerve fibers enter). As all the hair cells in this group are oriented in the same direction, they are designated unidirectional hair cells. Unidirectional hair cells are almost always abneurally oriented. The groups of hair cells at each end of the papilla consist of a row of hair cells that are abneurally oriented and an opposing row of hair cells that are neurally oriented. Where groups of hair cells consist of oppositely oriented hair cells, they are designated bidirectional hair cells regions. Note that hair cell orientation in this diagram is the same as that illustrated in the SEM shown in Figure 23.4A. Abbreviations: a, afferent nerve terminal; ab, abneural (direction); ab-b, abneural bidirectional hair cell group; ap, apical (direction); as, apical segment; b or bhc, bidirectional hair cell(s); ba, basal (direction); bhc, bidirectional hair cell; bs, basal segment; c, constriction between segments; cm, culmen; ct, central papillar region; e, efferent nerve terminal; es, efferent synapse; g, gap junction; h, hiatus between neural and abneural hair cell groups; hc, hair cell; n, neural (direction); n-b, neural bidirectional hair cell group; nk, neck region of papilla; sa, sallets; ss, synaptic site; tc, tectorial cap; tm, tectorial membrane; tp, tectorial plate; tr, transitional hair cells (intermediate between uni- and bidirectional hair cells); u or uhc, unidirectional hair cell(s).

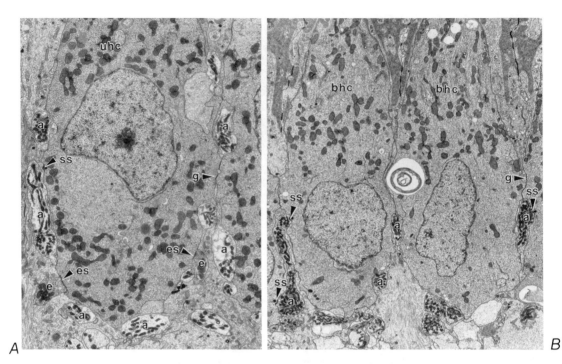

FIGURE 23.5. (A) *Xantusia vigilis*. Portions of two unidirectional hair cells showing a number of afferent nerve terminals, an afferent synaptic site, and two efferent nerve terminals and synapses with the hair cell. The gap junction between these two hair cells is unique to this lizard species. ×3500. (B) *Xantusia vigilis*. Por-tions of three bidirectional hair cells showing numerous afferent nerve terminals, and three afferent synaptic sites. Note the smaller diameter of these hair cells (ca. 10 μm) as compared with UHC (ca. 13 μm) and the absence of efferent nerve terminals. ×3500. For abbreviations, see Figure 23.4.

hair cells. I have followed Wever's convention in this regard. These hair cell coverings may or may not be attached to a tectorial membrane extending from the limbus. Figures 23.5A and 23.5B are TEM sagittal sections of unidirectional and bidirectional hair cells showing some fine structural features and differences between these hair cell types.

3. Comparative Anatomy of the Cochlear Duct and Basilar Papilla of Lizard Families

3.1 Relationship of Lizard Cochlear Duct and Basilar Papilla Anatomy to that of other Reptiles

Figure 23.6 illustrates the basic features of the cochlear ducts and basilar papilla of all the lizard families together with that of *Sphenodon* (the

Tuatara, a so-called living fossil), a snake, a turtle, and an amphisbaenid (Worm Lizard). In 1966 (Miller 1966a, 1966b), I pointed out that the mac-ula lagenae bearing portion of the cochlear duct was clearly separated from the basilar papilla con-taining portion in *Sphenodon*, turtles, and snakes, while in lizards those two regions of the duct were contained in a more unified structure (Fig. 23.6). I also mentioned that while it was generally agreed that turtles and *Sphenodon* were more primitive than any living lizards, it is thought by some inves-tigators that snakes were derived from lizards (Estes et al. 1988). Even though the latter may be the case, the fact that the snake cochlear duct retains more separated lagenar and basilar regions of the duct than does the duct of any living lizard family (Miller 1978a) suggests either a separate or very early common origin of the cochlear duct of snakes. The crocodilian cochlear duct and basilar papilla is very similar to that of birds and is con-siderably different from that of other reptiles

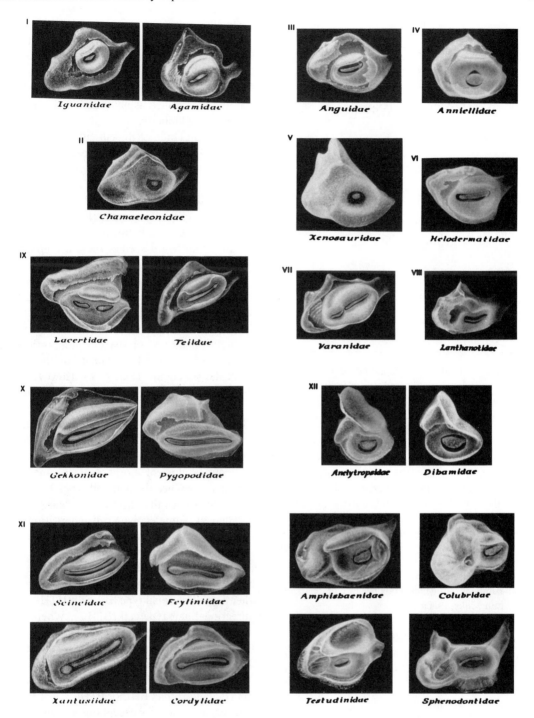

FIGURE 23.6. A summary plate showing the lateral aspects of the left cochlear ducts of representatives of all living lizard families. For comparative purposes, the duct of an amphisbaenid, a terrestrial colubrid snake, a turtle, and *Sphenodon* are also shown. The lizard ducts are arranged in family groups that are thought to be closely related on the basis of cochlear duct and basilar papilla structure. (From Miller 1966c.)

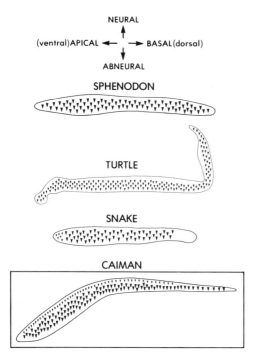

FIGURE 23.7. Graphic representation of the general shape of the auditory papilla and the orientation of the hair cells in *Sphenodon*, a turtle (*Kinosternon* sp.), a snake (*Epicrates cenchris*), and a caiman (*Caiman crocodilus*). The downward pointing arrows indicate that the kinocilium is on the abneural side of the hair cells. All the hair cells in the papillae of these reptiles are unidirectionally oriented. The smaller arrows on the upper edge of the caiman papilla are tall hair cells situated on the limbic tissue. The larger arrows indicate a shorter and wider type of hair cell resting on the basilar membrane. The ducts are not represented to scale, the fixed-dried lengths are as follows: *Sphenodon*, 1.0 mm (estimated); *Kinosternon* sp., 1.5 mm; *Epicrates cenchris*, 0.65 mm; and *Caiman crocodilus*, 4.0 mm. The caiman papilla is outlined in a box because of its greater size. (From Miller 1980a.)

(von Düring et al. 1974). Patterns of orientation of uni-and bidirectional hair cell groups in different reptilian orders are illustrated in Figure 23.7 and in a variety of lizard families in Figure 23.8.

3.2 The Development of Bidirectionally Oriented Hair Cells in Lizards

An important difference between what I consider a "primitive" type auditory papillae and the derived

ones of all extant lizard species is the development in lizards of groups of hair cells that are bidirectionally oriented. In the papilla of a probably primitive type of reptile characterized by *Sphenodon*, all hair cells are abneurally oriented by their kinocilia, and are defined as unidirectionally oriented hair cells. The kinocilia of all hair cells are located on the abneural side of the cuticular plate (top of the hair cell) and the adjoining stereovilli (stereocilia) are situated on the neural side (site of entry of the innervating nerve fibers) of the kinocilium. The hair cells of crocodilians, birds, and mammals are all unidirectionally oriented (the kinocilia of all hair cells are abneurally directed).

In all extant lizard species there are groups of both unidirectionally and bidirectionally oriented hair cell types. In the bidirectionally oriented hair cell regions, the hair cells situated on the approximate neural half of the papilla are oriented abneurally (the kinocila are on the abneural side of the stereovilli), and those on the abneural portion are oriented neurally (the kinocilia are on the neural side of the stereovilli) (Fig. 23.8). However, hair cell orientation in itself is not the most fundamental criterion of hair cell type (Miller and Beck 1988). In lizard species where kinocilial orientation differs from the cytological characteristics (hair cell size and innervation) of the underlying cell body, cytological morphology together with function determines the hair cell type. The essential characteristics that differentiate hair cell type are hair cell diameter, hair cell innervation, tectorial cover, and the range of characteristic frequencies that are related to the hair cell's function. In hair cells where kinocilial orientation differs from cell cytology, I refer to the cell type as either a unidirectional type hair cell (UTHC), or a bidirectional hair type cell (BTHC). Where kinocilial orientation and hair cell morphology are the same, I refer to these simply as unidirectional (UHC) or bidirectional hair cells (BHC).

Both UHC and UTHC are usually larger (ca. 13 μm in diameter) than BHC or BTHC (ca. 10 μm in diameter) and with few exceptions are supplied by a greater number of afferent nerve fibers and afferent synapses. Both UHC and UTHC cells are always covered by a tectorial structure. In species of five families (iguanids, agamids, anguids, xenosaurids, and anniellids) BHC cells lack tectorial cover. UHC and UTHC hair cells are innervated

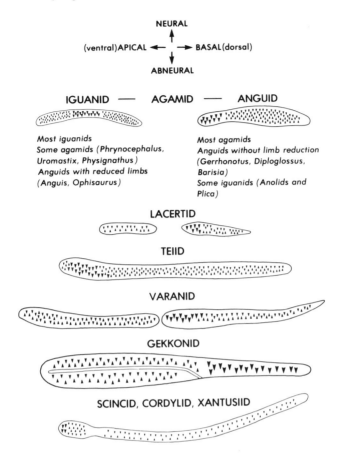

FIGURE 23.8. Graphic representation of the general shape and approximate size of the auditory papillae and the orientation of the hair cells in a variety of lizard families. The larger arrows indicate the location of unidirectional hair cells. The smaller arrows indicate bidirectional hair cells; those arrows pointing downward indicate abneurally oriented hair cells and the upward directed arrows indicate neurally oriented hair cells. In most lizard families, the length of the papilla varies with species size, larger species having larger and longer papillae. The papilla of the teiid in this figure is that of a tupinambid type teiid, not an ameivid type. In the latter there is a pronounced basal constriction (see Fig. 23.6). The unidirectional hair cells are centrally located in *Ameiva*, but more apically in *Tupinambis*. (From Miller 1980a.)

by efferent nerve fibers (with the exception of agamid species) while BHC or BTHC hair cells hair cells lack efferent innervation. Functionally, UHC and UTHC are concerned with low tone reception (frequencies up to ca. 0.9 kHz), and BHC and BTCH with higher frequencies (ca. 1–5 kHz) (Köppl and Manley, Chapter 24). It might be supposed that each oppositely oriented group of hair cells in a bidirectionally oriented papillar segment were sending separate signals to the central nervous system. It should be mentioned that determination of the function of bidirectionality is complicated by the observation that the two oppositely

oriented groups of hair cells comprising the bidirectional hair cell region are innervated in many cases by the same afferent nerve fibers (Miller and Beck 1990a; Köppl and Manley 1990).

As will be presented in the following sections, I propose that species of some lizard families show the development of cytological bidirectionality before kinocilial hair cell orientation is completely bidirectional. This condition I have termed, "hair cell orientation-cytological nonconformity," and these hair cells are referred to as bidirectional type hair cells (BTHC). More often, in regions of cytologically unidirectional hair cells, kinocilial

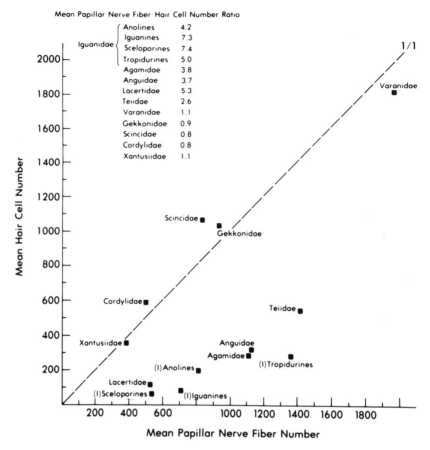

FIGURE 23.9. Graph demonstrating the relationship between papillar nerve fiber number and hair cell number in lizard families. The black squares represent the means of the papillar nerve fibers and the hair cell numbers for each family. The dashed line placed at a 45-degree angle to the x- and y-axes represents points where a ratio of 1/1 would be located. Ratios less than 1/1 would be above this line, and ratios greater than 1/1, below this line. (From Miller 1985.)

orientation may be bidirectional. The hair cells in these instances are termed unidirectional type hair cells (UTHC) and their occurrence will be detailed in the following sections. When hair cell orientation is the same as hair cell cytology, this is "hair cell orientation-cytological conformity" and the hair cells are designated either unidirectional (UHC) or bidirectional hair cells (BHC).

3.3 Gross Anatomical Features of the Cochlear Duct of Species of Different Lizard Families

A primary feature of the lizard cochlear duct and basilar papilla is that the gross morphological features vary sufficiently from family to family so that the duct of each family is different from that of other families (Fig. 23.6). Briefly, the features that vary are: (1) the relative size of the lagenar portion of the duct, (2) the size and shape of the limbus surrounding the basilar membrane, (3) the size and shape of a lateral projection of the limbus (limbic lip) which usually gives rise to the tectorial membrane, (4) size and shape of the basilar papilla, (5) size and shape of the basilar membrane, (6) size and thickness of the papillar bar underlying the papilla basilaris, (7) size and shape of the recesses on the medial limbic wall housing portions of the scala tympani and, (8) the number of papillar hair cells (Miller 1966a, 1985).

It is particularly important that the length (and consequently the number of hair cells) of the basilar papilla varies from family to family, but is relatively similar in each family. Within any one

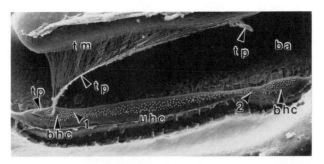

FIGURE 23.10. *Cnemidophorus motaguae* (Teiidae) SEM of part of the cochlear duct and basilar papilla. Attached to the tectorial membrane on its lower side is a relatively thick unspecialized tectorial structure that has been lifted free of all but a few of the apicalmost hair cells. There is a marked constriction between the basal BTHC and the central group of UTHC and a less marked narrowing of the papilla between the apical bidirectional and the central UTHC. The hair cells between arrows 1 and 2 are UTHC. Hair cell orientation is difficult to determine at this low magnification. ×123. (From Miller 1985.) For abbreviations, see Figure 23.4.

family, papillar length is closely related to the adult body length of a particular species within that family; the length of the papilla is proportional to the snout-vent length of the animal (Miller 1966a).

It is also important that although the anatomy of the cochlear duct and basilar papilla anatomy are distinctive for each family, there are similarities between some families indicating taxonomic relationship at least as far as ear structure is concerned. Groupings of lizard families as determined by closeness of ear structure is depicted in Figure 23.6 and will be discussed in greater detail in the following sections.

For the benefit of the reader, although Figure 23.22 primarily illustrates a proposed phylogeny of the lizard papilla, it also clearly shows the disposition of the uni- and bidirectional hair cell groups in the papillae of different lizard families. Table 23.1 summarizes many other morphological characteristics of the papillae of different lizard families. Figure 23.9 shows the relationship between the papillar nerve fiber number and hair cell number in lizard families.

3.4 Groups of Ear-Related Lizard Families Based on Gross and Fine Structural Anatomy

3.4.1 Teiidae (Teiids) and Varanidae (Monitor Lizards) (Figures 23.10–23.14)

The gross structure of the teiid cochlear duct is characterized by a prominent projection from the lateral limbic wall, and the varanid lateral limbus is essentially flat. Still, on the basis of other anatomical features the cochlear ducts and papillae of these two families are similar. Species of these two families are the only lizards in which the hair cells are completely covered by a substantial, relatively unspecialized tectorial structure (Fig. 23.10). Specialized tectorial structures or conditions are perforated caps, sallets, large masses (culmens), or the loss of all tectorial covering. Both the ameivid type teiid and the varanid papillae are constricted at a particular point; in the ameivid type teiid the constriction is near the basal end of the papilla, and in the varanids, near the apical end. In species of both families the papillar length is medium to long depending on species' size.

In three teiid species, *Ameiva ameiva* (commonly known as Ameivas), *Cnemidophorus motaguae*, and *Cnemidophorus tigris* (Whip-tail Lizard), an exceptionally large number of hair cells are oriented abneurally (ca. 90%) (Miller 1973b; 1978b). Since the hair cells in turtles (Miller 1978a; Sneary 1988a) and in *Sphenodon* (Wever 1978) are completely oriented abneurally, it is probable that a high percent of abneurally oriented hair cells is a primitive feature. In the papillae of *Ameiva ameiva* (Fig. 23.11) and *Cnemidophorus tigris*, however, the hair cells in both the unidirectional and bidirectional areas, regardless of kinocilial orientation are in large part, cytologically either uni-or bidirectional types (Fig. 23.12) ("hair cell orientation-cytological nonconformity"). In the papilla of *Ameiva ameiva* there is a group of

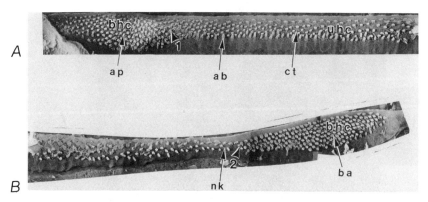

FIGURE 23.11. (A) *Ameiva ameiva* (Teiidae) Montage of the apical half of the papilla. The apical swelling consists of BTHC and the long central region between arrows 1 and 2, UTHC. While it may barely be apparent at this magnification, ca. 90% of the hair cells are abneurally oriented. Most of the hair cells in the most abneural row (ab) are oriented neurally while all the upper rows are oriented abneurally. Figure 23.12 shows that regardless of hair cell orientation, most of the central area is composed of cytologically UTHC, and the apical and basal regions of cytologically BTHC. ×263. (From Miller 1973b.) (B) Basal half of the papilla showing the marked constriction between the central UTHC and the basal BTHC. In the basal bidirectional region only the basal part of the abneuralmost row of hair cells (towards the reader, or the bottom row of hair cells) are neurally oriented (upward). ×263. (From Miller 1973b.) For abbreviations, see Figure 23.4.

hair cells located between the apical BTHC and the central UTHC (Fig. 23.12A) that are transitional in character, intermediate in size between uni- and bidirectional type hair cells and lacking efferent innervation. These possibly are UTHC that are in the process of a cytological transformation from uni- to bidirectional type hair cells. Cytologically, both UHC and UTHC are larger than BHC or BTHC (Fig. 23.12) and, with the exception of agamid species both are innervated by afferent and efferent nerve fibers. BHC or BTHC are smaller and lack efferent innervation. This phenomenon is described and discussed in 3.2 and in Miller and Beck 1988.

Lizard families in which nonconformity of hair cell orientation and cytology has been observed are in species of teiids, lacertids, scincids, and xantusiids. In my opinion, nonconformity of hair cell

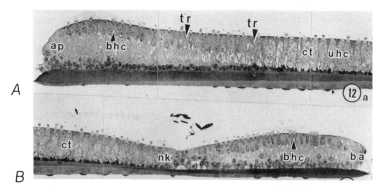

FIGURES 23.12(A) and (B). *Ameiva ameiva* (Teiidae) Light micrograph of a sagittal section of the papilla basilaris. Figure 23.12A is the apical half, and 23.12B the basal half corresponding to Figures 23.11A and 23.11B. The central UTHC are considerably larger than the apical and basal BTHC. Note the transitional hair cells located between the arrows labelled tr that are intermediate in size between the UTHC and the BTHC. The small terminal groups of UTHC are not shown in this micrograph. ×211. (From Miller and Beck 1988.) For abbreviations, see Figure 23.4.

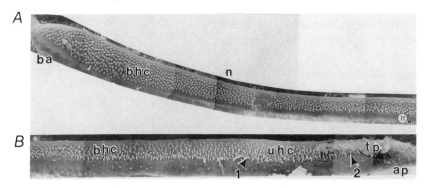

FIGURE 23.13. *Tupinambis teguixin* (Teiidae) Montage of a right papilla. Figure 23.13A is the basal half, and Figure 23.13B, the apical half. The remnants of a thick unspecialized tectorial covering is seen at the apical end of the papilla. The remainder of the covering has been removed to show the hair cell surfaces. Higher magnification views which are not shown here reveal that the UTHC occupy an area near the apical end of the papilla (region between arrows 1 and 2). The most apical hair cells are BTHC. See Figure 23.8 for hair cell orientation. This species lacks the papillar constrictions found in *Ameiva*. ×153. (From Miller 1973b.) For abbreviations, see Figure 23.4.

orientation and hair cell cytology may be a condition of incomplete hair cell differentiation. Primitive papillae probably contained a single type of hair cell that was cytologically unidirectional and abneurally oriented. From some of the UHC in lizards, BHC developed. Ultimately papillae were composed of two hair cell types, each located in specific parts of the papilla, and hair cell orientation corresponding to the cytological characteristics of the cells. The papillae of approximately two-thirds of extant lizard families show conformity between hair cell orientation and cytology and have completed the process of hair cell differentiation into two hair cell types.

Unfortunately, only one other teiid species, *Tupinambis teguixin* (Tegu) (Fig. 23.13), has been studied by SEM. Hair cell orientation is considerably different from the above described species (Miller 1973b). In *Tupinambis* the unidirectional hair cells are more apically located. This possibly represents a stage in the elimination of the apical bidirectional hair cells which would result in an apical location of the unidirectional hair cells, a condition found in most agamids and some iguanids and anguids as well as in scincomorphic species (skinks, xantusiids, and cordylids). Teiid lizards are a remarkably diverse group and further studies of the auditory papillae of species of this family should be of great value.

Regardless of kinocilial hair cell orientation, the papillae of *Ameiva* and *Cnemidophorus* consist for the most part of a large central group of cytologically characteristic unidirectional hair cells and an apical and a basal group of cytologically characteristic bidirectional hair cells (Figs. 23.11, 23.12). In *Ameiva ameiva*, there are also a small group of cytologically unidirectional hair cells appended to the bidirectional hair cells at both ends of the papilla. This unusual configuration together with the group of transitional hair cells described above lends further support to the supposition that the ameivid type teiid papilla is probably similar to an ancestral teiid that was in the process of transforming groups of UHC to BHC. And from this ancestral type, papillae with complete complements of BHC were developed.

In teiids, the papillar nerve fiber number/hair cell number ratio is intermediate, averaging 2.7/1 for the three species studied (Fig. 23.9). In *Ameiva ameiva*, the UTHC are mostly exclusively innervated, but some hair cells are nonexclusively innervated. The BTHC are exclusively innervated (Miller and Beck 1988), mostly by a single afferent nerve fiber.

The papilla of varanids does not exhibit divergence of hair cell orientation from a cytologically characteristic hair cell type (conformity of hair cell orientation and cytological hair cell type) that was observed in the teiids. However, the overall

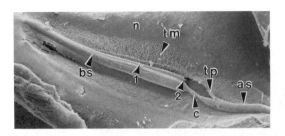

FIGURE 23.14. *Varanus exanthematicus* (Varanidae) SEM view showing that the papilla is divided into a shorter apical and a longer basal segment. The location of the UHC is indicated by the area between the arrows 1 and 2 below the apical quarter of the basal segment. A tectorial membrane extends from the neural limbus to a moderately thick material covering all the hair cells. ×42. (From Miller 1974b.) For abbreviations, see Figure 23.4.

distribution of UHC and BHC is the same as in the teiids. Specifically there is a centrally located group of UHC with BHC located both apically and basally (Fig. 23.14). In some species, as in *Varanus bengalensis*, the apical segment is completely separated from the long basal segment leaving the UHC at the apical end of the basal segment. In this location they are essentially centrally placed in reference to the entire papilla (Miller 1978b). The papillar nerve fiber number/hair cell number ratio is 1.1/1, the number of afferent innervating nerve fibers is relatively small (2 to 3/hair cell), and both UHC and BHC are nonexclusively innervated (Miller and Beck 1990b). It is my supposition that a low papillar nerve fiber number/hair cell number ratio, a small number of afferent nerve fibers innervating each hair cell, and nonexclusive hair cell innervation (afferent nerve fibers innervate more than one hair cell) are primitive features. One other feature observed in *Varanus exanthematicus* as well as in *Ameiva ameiva* is the presence of axo-dendritic junctions (synapses). These neural relationships are not observed in the greater number of lizard species, but are regularly observed in crocodilians, birds, and mammals. Several of the proposed primitive features listed above are also found in turtles (Sneary 1988b). I predict that these conditions will also be found in *Sphenodon*.

3.4.2 Iguanidae-Agamidae-Anguidae (Figures 23.4, 23.15, 23.16)

While the structural features of the cochlear duct and basilar papilla of these three families are remarkably similar, the phylogenetic derivation of the anguids differs from that of the iguanids and agamids (see 4.2.2 and 4.2.5). The cochlear ducts of these families lack prominent lateral limbic projections, and the similarity in basilar papilla surface structure demonstrated by SEM is striking. The papillae of species of these families are relatively short (0.15 to 0.4 mm for most species) and consequently possess fewer hair cells than species with longer papillae. While the UHC may be located either apically or in the midpapillar region, they are covered by a limbic-attached tectorial cap. The BHC cells may be either basal in location or are divided into two segments, one apical, and the other basal to a midpapillar segment of UHC. Both these patterns of UHC and BHC localizations occur in each of these families (Fig. 23.8). The BHC are specialized in that they have no tectorial cover ("free standing hair cells"), a condition found only in the papillae of species of these families and in the regressed papillae of certain anguid related lizards (*Xenosaurus*, and *Anniella*). The kinocilia of the BHC are different from those of all other lizards in that they are shorter than the taller stereovilli and the kinocilial heads are not attached to the tops of the tallest stereovilli. The stereovilli are longest immediately adjacent to the UHC group, and become progressively shorter toward the terminal end of the papilla. In the past, the iguanid-agamid-anguid type papillae have been considered "primitive" (Baird 1974; Wever 1978). In my view, however, they are highly specialized.

Another specialization limited to species of these three families is the nature of the innervation of the hair cells. In species of these families the ratio of the number of innervating nerve fibers to the number of papillar hair cells is higher than in other lizard species with the exception of the lacertids (Fig. 23.9). In 16 iguanid species, the papillar nerve fiber number/hair cell number ratio varies from 3.8/1 to 11.1/1 with an average of 6.2/1, in two species of agamids the average ratio is 3.8/1, and in two species of anguids, 3.6/1 (Miller 1985). It is significant that the pattern of hair cell innervation in iguanids and agamids is largely exclusive (i.e., each afferent nerve fiber innervates only one hair cell), and in anguids innervation may be exclusive or mixed (some afferent nerve fibers innervate one hair cell, while others innervate more than one) (Mulroy 1986; Miller and Beck 1988, 1990b). Another specialization found so far only in agamid

FIGURE 23.15. (A) *Agama agama* (Agamidae). Montage of the right papilla basilaris. The abneural side faces upward and the apical end of the papilla is at the left extremity. Note particularly the small group of long-ciliated BHC apical to the adjacent short-ciliated UHC. The tectorial cap has been removed from the UHC. The great majority of specimens of this species are like the papilla shown in Figure 23.15B where the UHC are all apical to the basal BHC. I consider that the apical BHC shown in this figure are residual in nature and are evidence of incomplete loss of the ancestral apical BHC group. ×287. (From Miller 1978b.) (B)

Agama agama. A typical specimen of this species in which the UHC are apically, and the BHC, basally located. The UHC in this specimen are covered by a limbic attached tectorial cap. The BHC lack any tectorial cover. ×282. (From Miller 1978b.) (C) *Uromastix* sp. (Agamidae) Usually the UHC of agamid species are apically located, but in this genus the pattern of a central group of UHC flanked by BHC is observed. This pattern is observed in most species of iguanids and in burrowing or legless anguids. The tectorial cap covering the UHC has been removed. ×287. (From Miller 1978b.) For abbreviations, see Figure 23.4.

species is a lack of efferent nerve fiber innervation to the unidirectional hair cells (Bagger-Sjöbäck 1976; Miller and Beck 1990b). I believe that relatively high nerve fiber number/hair cell number ratios and exclusive hair cell innervation are specialized features.

3.4.3 Lacertidae (Lacertids) (Figure 23.17)

I group this family separately and view the auditory papilla as possibly intermediate between the teiids and the iguanid-agamid papillae. The anguid papilla, though very similar to the iguanid-agamid papilla in structure, has a different origin (see 4.2.5). In lacertids there is a prominent lateral limbic projection (lip) reminiscent of, but not as pronounced as that of the teiids. The completely divided papilla of the lacertids could have been derived from a type of teiid lizard with a marked

papillar constriction which is possibly evidence of the beginning of separation of one portion of the papilla from the other.

The unspecialized tectorial plate covering the BHC (Miller 1978b) is similar to that of the teiids and varanids. Wever (1978), however, in other lacertid species not studied by myself, describes sallets (specialized tectorial structures) overlying the BHC. This is significant since the BHC of the scincomorphic lizards (skinks, xantusiids, and cordylids) as well as the gekkonids are covered by sallets. The UTHC, on the other hand, are covered by a limbic attached tectorial cap (Fig. 23.17) that is very similar to that of iguanid-agamid-anguid species (Köppl 1989). The UTHC, like those of the ameivid type teiids, although cytologically unidirectional are not all abneurally oriented (see 3.4.1). In the bidirectional regions hair cell orientation as well as cytology are in conformity

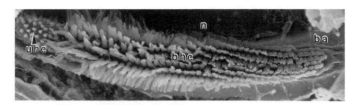

FIGURE 23.16. *Celestus costatus* (Anguidae) An apical group of short-ciliated UHC is observed on the left side and a group of long-ciliated BHC constituting the greater part of the papilla extends basally. The tectorial cap covering the UHC has been removed. The BHC lack tectorial cover. The longest stereovilli of the BHC group are adjacent to the UHC and become progressively shorter in the basal direction. ×374. (From Miller 1985.) For abbreviations, see Figure 23.4.

(i.e., hair cells that are bidirectionally oriented possess the cytological characteristics of bidirectional hair cells). The papillar nerve fiber number/hair cell number ratio is ca. 5/1 which is similar to that of many iguanid species. Unidirectional hair cell innervation is mixed (both exclusive and nonexclusive) and the number of innervating afferent nerve fibers is intermediate (5 or 6/hair cell). Innervation of the BHC is exclusive and the number of afferent nerve fibers few (one to two per hair

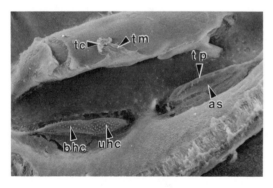

FIGURE 23.17. *Lacerta galloti* (Lacertidae) Entire right cochlear duct showing the completely divided papilla basilaris. The apical segment consists of BHC that are completely covered by a moderately thick tectorial cover. The basal segment consists of a wider apical portion made up of UTHC and a basal part of BHC. These latter, like the apical BHC are covered by an unspecialized structure that has been removed in this specimen. The UTHC were covered by a limbic attached tectorial cap that has been pulled free of the papilla and is now seen still attached to the tectorial membrane above the neural limbus. ×142. (From Miller 1978b.) For abbreviations, see Figure 23.4.

cell), which is similar to that of BTHC innervation in the teiid, *Ameiva ameiva*. Thus the lacertid papilla shares several features in common with both the iguanid-agamid-anguid and the teiid types.

3.4.4 Cordylidae (Girdle-Tailed Lizards), Scincidae (Skinks), Xantusiidae (Night Lizards) (Figures 23.18–23.20)

The cochlear duct of species of these families lack lateral limbic projections. The papillae are intermediate to long in length and consist of a shorter but wider apical group of UTHC and a much longer basal region of BHC. There is often a constricted region between the UHC and the BHC regions, or even separation of these two regions. In skinks, the UTHC are covered by a nonlimbic attached large tectorial mass ("culmen" of Wever) and the basal BHC by nonlimbic attached tectorial modifications known as sallets (Fig. 23.20). In xantusiids, the culmen is limbic attached, but the sallets are not attached. In six skink species the kinocilial heads are unusual in being leaf-shaped rather than oblate or spheroidal as in most lizard species (Miller 1974a, 1974b, 1978b). However, in the bobtail lizard, *Tiliqua rugosa*, the kinocilial heads are not leaf-shaped (Köppl 1988).

The papillar nerve fiber number/hair cell number ratio is among the lowest found in lizards [1.1/1 in *Xantusia vigilis* (Yucca Night Lizard)], and 0.8/1 in *Mabuya multifasciata* (type of Skink) and in *Cordylus vittifer* (type of Girdle-tailed Lizard) (Miller 1985; Miller and Beck 1988). The apical UTHC are all cytologically unidirectional, but hair cell orientation is irregular and divergent (hair cell rows are oriented away from, rather than towards the midpapillar axis) in *Mabuya* (Miller

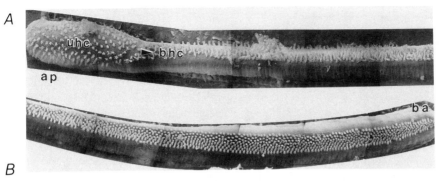

FIGURES 23.18(A) and (B). *Mabuya carinata* (Scincidae) Figure 23.18A is the apical, and 23.18B the basal half, of the papilla. Hair cell orientation of the apical wider part of the papilla is divergent, that is, the hair cell rows are oriented away from the midpapillar axis, rather than towards the midpapillar axis as is usually the case in regions of BHC orientation. The hair cells in this region, however, are cytologically UTHC. The long basal region of the papilla beginning at a point marked by an arrow, is constituted of hair cells that are both bidirectionally oriented and cytologically characteristic BHC. The tectorial sallets normally covering these hair cells have been removed. These are shown in Figure 23.20. Figures 23.18A and 23.18B are each ×143. (From Miller 1974a.) For abbreviations, see Figure 23.4.

and Beck 1988) and in *Tiliqua rugosa* (Köppl 1988). *In Xantusia*, a part of the neural row of hair cells are neurally oriented (Miller and Beck 1990a). In the bidirectional region, hair cell orientation is in conformity with cytology. UTHC are nonexclusively innervated by relatively large numbers of afferent nerve fibers in some species (11–14 in *Mabuya multifasciata*, Miller and Beck 1988). One afferent nerve fiber innervated four to 14 hair cells in *Tiliqua rugosa* (Köppl and Manley 1990)]. By contrast, BHC are innervated by only three afferent nerve fibers in *Mabuya*. In *Xantusia vigilis* innervation is also nonexclusive for both hair cell types, with six or seven afferent fibers innervating the UTHC and an average of 5.4 afferents innervating the BHC (Miller 1985; Miller and Beck 1988, 1990a). The hair cells of *Xantusia*

vigilis are very unusual in that most of the hair cells are interconnected by gap junctions (Miller and Beck 1990a). Hair cell innervation has not been studied in a cordylid.

3.4.5 Gekkonidae (Geckos) and Pygopodidae (Flap-footed Lizards) (Fig. 23.21)

The cochlear duct and basilar papillae of species of these families (Miller 1966a, 1973a) are characterized by a much enlarged lateral limbic projection (limbic lip, Fig. 23.3), and long papillae (some of the longest among the lizards). The UHC are located basally and the apical hair cells consist of two longitudinally arranged strips of BHC hair cells (see Figs. 23.8 and 23.22 for clarification). Wever (1978) describes several species of

FIGURE 23.19. *Scincus scincus* (Scincidae) Low power SEM view of a right cochlear duct. The apical end of the papilla faces right and the UTHC in this region are covered by a thick tectorial mass (culmen). The two arrows indicate the division between the uni- and bidirectional hair cells. The long basal stretch of BHC are covered by tectorial specializations known as sallets. ×107. (From Miller 1974a.)

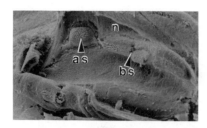

FIGURE 23.20. *Xantusia vigilis* (Xantusiidae) Left coch-lear duct. The papilla consists of an ovoid apical segment of UTHC and a longer basal segment of BHC. As in the skinks, the apical segment is covered by a thick tectorial structure (culmen) and the basal BHC by sallets, both of which have been removed in this specimen. ×62 . (From Miller 1978b.) For abbreviations, see Figure 23.4.

gekkonids in which all the hair cells are bidirec-tionally oriented. In *Lialis burtonis*, a pygopodid, the basal hair cells are also bidirectionally oriented (Wever 1978). However, these bidirectionally oriented hair cells in the basal region, are probably cytologically unidirectional hair cells (UTHC). In *Coleonyx variegatus* (Western Banded Gecko) (Miller and Beck 1988), and in several other spe-cies of gekkonids (Miller, unpublished), the basal region consists of "conforming" UHC. A different type of tectorial structure covers each apical bidirectional hair cell strip and also the basal UHC. The abneuralmost strip of BHC is covered by salletal tectorial modifications somewhat simi-lar to, but with less surface complexity than the sallets covering the BHC of species of cordylids, scincids, and xantusiids. The neural strip of BHC is covered by undifferentiated tectorial material

and the UHC by filamentous strands. It is possible that the basal location of the UHC which is found only in species of these families, arose by the loss of a basal group of BHC from an ancestral form similar to the ameivid type teiid papilla (see 4.2.3). In *Coleonyx variegatus* (Miller and Beck 1988) the papillar nerve fiber number/hair cell number ratio is ca. 0.9/1, and the UHC are exclusively, and the BHC, nonexclusively innervated. The UHC are supplied by 4 or 5 afferent nerve fibers and the BHC by 6 afferent fibers.

3.4.6 Families of Undetermined Relationships

Because of lack of study by either SEM or TEM, in most cases little may be said about the relation-ships of the following families as determined by cochlear duct or papillar morphology.

3.4.6.1 Chamaeleonidae (True Chamaeleons) (Fig. 23.6)

The cochlear duct and basilar papilla of species of this family are much reduced in size and no obvious similarities to other families are observed. Wever (1978) does not state how the hair cells are ori-ented, but his illustrations of chamaeleonid papil-lae indicate that the relatively few hair cells (ca. 50) are mostly unidirectionally oriented with some degree of bidirectionality. Baird (1974) states that in *Chameleo zeglonicus* the hair cells are arranged in oppositely oriented pairs or groups. It is likely that the casque-like skeletal transformations in the temporal region of chamaeleons, together with loss of the external and middle ear, has resulted in

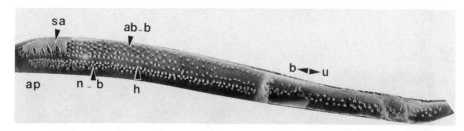

FIGURE 23.21. *Coleonyx variegatus* (Gekkonidae) Mon-tage of a right papilla. The neural side of the papilla faces downward and the apical end faces left. The longer apical portion of the papilla is divided into two groups of BHC by a gap or hiatus separating these two groups one from the other. Except for a few sallets all tectorial coverings have been lost in this specimen. The abneural group of

these BHC are normally covered by tectorial specializa-tions called sallets. A few of these remain at the apical end of the abneural group. A relatively unspecialized material normally covers the neural group of BHC. UHC consti-tute the approximate basal quarter of the papilla and are normally covered by filamentous strands. ×178. (From Miller 1973b.) For abbreviations, see Figure 23.4.

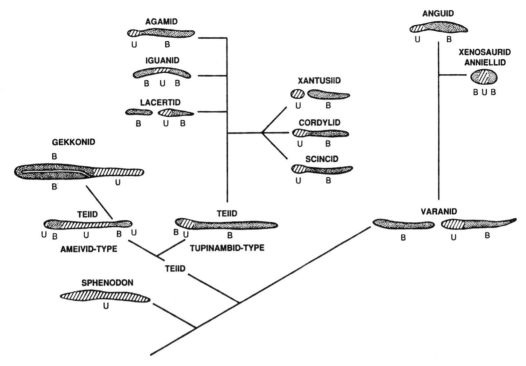

FIGURE 23.22. A diagrammatic representation of the possible phylogenetic relationships of several lizard families based on the structure of the auditory papilla. The papilla of *Sphenodon* is probably close to the ancestral type. The papillae of the teiids and the varanids are considered basic types from which other families could have been derived. This figure primarily shows the lines of derivation of the papillae of other lizard families and particularly the disposition of the unidirectional (U, cross-hatched) and the bidirectional (B, dotted) hair cells. The basic types represented by the teiids and varanids are characterized by more centrally located unidirectional hair cells (U) flanked both apically and basally by bidirectional (B) hair cells. The details of the derivation of the papillae of other lizard families from the base types are described in Section 4. Table 23.1 summarizes the morphological detail in different lizard families.

regression in the size and complexity of the cochlear duct and basilar papilla. A similar regression in size of the cochlear duct and papilla is also observed in burrowing skinks and in *Anniella* (Miller 1966a). Regression in the size of the cochlear duct and the basilar papilla in lizards with loss of the external and middle ear is in marked contrast to the condition in snake species where a fossorial life mode is correlated with elongation and better development of the papilla as well as the cochlear nuclei (Miller 1978a, 1980b).

3.4.6.2 Anniellidae (Legless Lizards) and Xenosauridae (Xenosaurids) (Fig. 23.6)

The basilar papilla of *Anniella*, like that of *Xenosaurus* (Miller 1966a) and *Shinosaurus* (Chinese Earless Monitor, Miller, unpublished), is probably much regressed in size from that of its ancestral condition. This is possibly due to a fossorial life mode leading to reduction of the external and middle ears. The gross structure of the cochlear duct and basilar papilla of *Shinosaurus crocodilurus* is almost identical to that of *Xenosaurus grandis* (Miller, unpublished). Wever (1978) shows that in both *Anniella* and *Xenosaurus*, a tectorial cap partially covers a central region of shorter UHC cells, while longer ciliated, uncovered ("free standing") BHC comprise the remainder of the papilla. In *Anniella* the BHC are irregularly located, but in *Xenosaurus*, the BHC are more regularly arranged. Wever's (1978) observations of the surface structure of the papillae of species of these two families indicates close affinity to the anguid lizards. It is perhaps on this basis that Wever lists these two families as anguoid type lizards.

3.4.6.3 Helodermatidae (Gila Monsters)
(Fig. 23.6)

Both gross anatomic (Miller 1966a) and light microscopic studies (Wever 1978) indicate the close affinity of the two species of this family to the varanid lizards. No SEM or TEM studies of the papillae of these species have been reported.

3.4.6.4 Lanthanotidae (Earless Monitor)
(Fig. 23.6)

Only the gross morphology of the cochlear duct of *Lanthanotus borneensis* has been reported (Miller 1966c). It is probable that this lizard is related to the varanids. SEM and TEM studies need to be undertaken.

3.4.6.5 Anelytropsidae and Dibamidae
(No Common Names, Skink Relatives)
(Fig. 23.6)

Both the cochlear duct and basilar papilla are nearly identical in the species of these two families indicating a very close relationship (Miller 1966c). These structures appear to be reduced or regressed from the ancestral condition. Greer (1985) has proposed single family status for these two families on the basis of a number of other anatomical characters.

4. Phylogeny of Lizard Families Based on Cochlear Duct and Basilar Papilla Structure

4.1 Ancestral Type Auditory Papilla and Its Relationship to the Papillae of the Teiidae and Varanidae

Based on the description of the papilla basilaris of *Sphenodon* (Wever 1978) and the serial TEM studies of the turtle, *Chrysemys scripta elegans* (Sneary 1988a, 1988b), one may postulate that the ancestral lizard basilar papilla was similar to that of *Sphenodon*, turtles, and snakes (Miller 1978a). It is probable then that ancestral lizard papillae were moderate in length (approximating 1 mm), and possessed hair cells with a limbic attached, substantial, unspecialized tectorial covering. There was only one type of hair cell, all

of these abneurally (unidirectionally) oriented (with the exception of occasional aberrantly oriented hair cells) and innervated by both afferent and efferent nerve fibers. The BHC of species of all extant lizard families lack efferent hair cell innervation, while UHC, with the exception of species of the Agamidae, are all innervated by both afferent and efferent nerve fibers. The primitive innervational pattern was probably mixed (that is, some hair cells were exclusively, and other nonexclusively innervated) or may have been completely nonexclusive.

I propose that the cochlear duct and basilar papillae of teiids and varanids are closer to the ancestral type than are those of other lizard families and may be regarded as the basic types from which the papillae of other lizard families may have been derived. The characteristics of these papillae are described in 3.4.1. The papillae of other families could have been derived by (1) loss of either the apical or basal bidirectional region from a teiid or varanid ancestor, (2) loss of all bidirectional hair cell tectorial cover, (3) development of several different kinds of tectorial specializations and (4) modifications of the lateral limbic wall. These innovative developments that characterize the extant families of lizards are described in the following subsections and summarized in Table 23.1.

4.2 Possible Lines of Descent

Refer to Figure 23.22 for illustration of the location of hair cell groups and Table 23.1 for a summary of the morphological features of different lizard families.

4.2.1 Derivation of Lacertidae

If one begins with a teiid type ancestor and the following developments occur in the cochlear duct and basilar papilla a lacertid type papilla would result:

(1) some reduction of the teiid type lateral projection of the limbic lip,
(2) complete separation of the papilla into two separate segments, the apical segment comprised of conforming type BHC and the apical part of the basal segment made up of nonconforming UTHC and the more basal region of conforming BHC,

TABLE 23.1. Morphological characteristics of the auditory papillae of lizard families

Family	Limbic lip[b] (0 to ++++)	Tectorial cover[a] Unspecialized[c] U[j]	B[k]	Specialized[d] U	B	Absent[e] U	B	Papillar nerve fiber no./hair cell no. ratio[f] Hi (4/-11/1)	Med (2/1-4/1)	Low (0.8/1-1.5/1)	Relation of hair cell orientation to hair cell cytology[g] Non-conformity[g] U	B	Conformity[g] U	B	Innervational pattern of hair cells Exclusive[h] U	B	Non-exclusive[h] U	B	Hair cell groups all present or lost[i] Apical BHC	Basal BHC
Ancestral (Based on Sphenodon and Turtle)	0	One HC type	+						+		+		One HC type +				One HC type +		One HC type	
Teiidae	+++	+	+						+		+		+		+		+		all present	
Varanidae	0	+	+							+	+			+			+	+	all present	
Lacertidae	++	+		+				+			+			+	+	+	+		all present	
Iguanidae	+			+			+	+					+	+	+	+			lost in a few sp.	
Agamidae	+			+			+	+					+	+	+	+			lost in most sp.	
Anguidae	+			+			+	+					+	+	+			+	lost in ca 1/2	
Scincidae	0			+	+					+	+			+	+		+	+	lost	
Xantusiidae	0			+	+					+	+			+			+	+	lost	
Gekkonidae	++++			+	+					+			+	+?	+			+		lost

[a]Wever (1978) differentiates between the tectorial membrane extending from the lateral limbic wall and a modified structure covering specific hair cell groups. The modified structure may or may not be attached to a tectorial membrane.

[b]A projection from the lateral limbic wall giving rise to the tectorial membrane.

[c]A plain flat structure of variable thickness covering all or most of the papillar hair cells.

[d]Tectorial coverings in the form of caps, sallets, masses (culmens), or skein-like formations.

[e]The complete absence of any tectorial cover.

[f]The ratio between the number of papillar nerve fibers in the nerve innervating the auditory papilla and the number of papillar hair cells.

[g]In some groups of hair cells, kinocilial orientation may be bidirectional when the underlying hair cell cytology is unidirectional in character. This is a condition of "nonconformity" between hair cell orientation and hair cell cytology. If both hair cell orientation and hair cell cytology is the same, this is "conformity" between hair cell orientation and hair cell cytology.

[h]In exclusive hair cell innervation, each nerve fiber supplies only one hair cell. In nonexclusive innervation, two or more hair cells are supplied by each nerve fiber.

[i]In an ancestral papilla, probably all hair cells were unidirectional in type. In all extant lizard species there are two types of hair cells, namely, uni- and bidirectional. A basic or relatively primitive papilla contains a central group of unidirectional hair cells, flanked on both sides by a group of bidirectional hair cells. In primitive papillae, both hair cell types are covered by an unspecialized tectorial structure. In the development of some species, either an apical or a basal group of hair cells apparently was lost.

[j]Undirectional hair cells.

[k]Bidirectional hair cells.

(3) retention of an unspecialized nonlimbic attached tectorial structure covering the BHC and the development of a specialized limbic attached tectorial cap similar to that of the iguanids, agamids, and anguids covering the UTHC,

(4) development of a moderately high papillar nerve fiber number/hair cell number ratio (ca. 5/1).

4.2.2 Derivation of Iguanidae and Agamidae (see 4.2.5 for Anguidae)

Whether the development of the papillae of species of these families originated from an extinct ancestor or proceeded directly from a tupinamid type teiid ancestor or through a lacertid is not clear, however, from one of these possible sources, the following developments occurred:

(1) further reduction of the limbic lip,
(2) development of a tectorial cap over the UHC (as in the lacertids),
(3) loss of all tectorial cover over the BHC,
(4) retention of the pattern of centrally located UHC and apical and basal groups of BHC in most iguanid species,
(5) loss of the apical bidirectional hair cell group in most agamid species so that the apical region is comprised of UHC and the basal region of BHC. Evidence that the apical BHC are actually lost from the agamid papilla is shown by the observation that in some specimens of *Agama agama*, there is a small group of BHC apical to the UHC that probably represents a condition of incomplete loss of these hair cells (Fig. 23.15A). In most specimens of this same species, *Agama agama*, as in most other agamid species, all apical BHC have been lost (Fig. 23.15B). Further, in one other agamid genus, *Uromastix*, the probable original teiid configuration of a central group of UHC flanked by apical and basal BHC is retained (Fig. 23.15C).

4.2.3 Derivation of Gekkonidae and Pygopodidae

The basilar papillae of species of these families could have been derived from an ameivid type teiid ancestor by the following developments:

(1) further elongation of the papilla,
(2) further lateral extension of the limbic lip,
(3) loss of the basal BHC group leaving the UHC basally located. This development is foreshadowed by the basal constriction of the ameivid type of teiid papilla and occurs only in species of gekkonids and pygopodids. As described in 3.4.5, some species of these lizards have retained only BHC orientation, but it is probable that the basal hair cells are cytologically, unidirectional type hair cells (UTHC),
(4) development of two longitudinal strips of apical BHC each covered by a different type of tectorial structure.

4.2.4 Derivation of Scincidae, Cordylidae, and Xantusiidae

The basilar papillae of species of these families could have been derived from a tupinambid type teiid ancestor by the following developments:

(1) marked reduction of the limbic lip,
(2) loss of the apical BTHC leaving nonconforming UTHC cells apically located,
(3) development of a nonlimbic attached large tectorial structure (culmen) covering the apical UTHC,
(4) development of specialized nonlimbic attached tectorial structures (sallets) covering the BHC,
(5) development of leaf-shaped kinocilial heads in the bidirectional hair cells of several species.

4.2.5 Derivation of Anguidae

In the past I have been particularly uncertain of the phylogenetic relationships between the iguanids and agamids in relation to the anguids on the basis of the structure of the basilar papillae of species of these three families. On a general structural basis, the cochlear duct and papillar structures of species of these families are remarkably similar. In a review paper of the cochlear duct of reptiles (Miller 1980a), I stated (p. 189) "The significance of the structural similarities and differences found in the Iguanidae, Agamidae, and Anguidae is conjectural. Probably species of these families acquired their basic papillar anatomy from a common ancestor. Close relationships of the iguanids and agamids is not surprising, but the relationship of the anguids here is problematical, as consideration of other anatomical systems does not show the anguids to be closely related to the iguanids and agamids. For the present, one must conclude that while the anguids were probably separated early and differ in other anatomical systems from the iguanids and agamids, the ear structure in these three families has remained remarkably similar."

The marked similarity of papilla basilaris structure of the anguid related lizards, *Xenosaurus* and *Anniella*, to the anguid papilla (Wever 1978), together with more recent fine structural studies of the varanid papilla (Miller and Beck 1990b), supports the supposition that the anguid papilla possibly originated from a varanid ancestor and its development was parallel to that of the iguanids and agamids. While the papillae of the iguanids, agamids, and anguids are remarkably similar, their origins are probably different.

Wever (1978) proposed no evolutionary sequence for the development of lizard papillae, but he did state that he considered the iguanid type papilla to be primitive. Baird (1974) was also of this opinion. In my view, however, the loss of all tectorial cover over all the bidirectional hair cells in the papillae of iguanids, agamids, and anguids, as well as the development of exceptionally tall stereovilli, represents a high degree of specialization. In addition, the development of exclusive afferent hair cell innervation is probably a highly derived feature in lizards that has also been developed by the avian tall hair cells and the mammalian inner hair cells for finer frequency analysis. According to Manley et al. (1989), the avian tall hair cells may be the equivalent of the mammalian inner hair cells.

4.2.6 Proposed Scheme of the Developmental Relationships of Lizard Families Based on Cochlear Duct and Papilla Basilaris Structure

Figure 23.22 together with a summary of morphological detail in Table 23.1 presents a possible phylogeny of the auditory papilla of lizards. In brief, from an ancestral form, the still primitive *Sphenodon* was derived. Further along the line two base groups, the teiids and varanids evolved and represent types from whose ancestral forms the auditory papillae of other lizard families could have been derived. The details of how other families may have been derived from the teiids and varanids is presented in Section 4.

5. Possible Taxonomic and Phylogenetic Relationships of Lizard Families Based on Wever's (1978) Reptilian Ear Studies and on that of Other Anatomical Characters

5.1 Wever's Taxonomic Grouping of Lizard Families

Taxonomic groupings based on light microscopic study of a large number of lizard species by serial celloidin embedded sections of the cochlear duct and papilla basilaris by E.G. Wever (1978).

Superfamily	Family
	Iguanidae
Iguanid	Agamidae
	Chamaeleonidae
	Anguidae
Anguoidea	Anniellidae
	Xenosauridae
	Varanidae
Varanoidea	Helodermatidae
	Lanthanotidae
Lacertoidea	Teiidae
	Lacertidae
Gekkota	Gekkonidae
	Pygopodidae
Scincomorpha	Gerrhousauridae
	Xantusiidae
	Feylinidae
	Dibamidae
	Scincidae
	Cordylidae

At present, I am in general agreement with the above taxonomic groupings proposed by Wever. Wever, however, did not propose a phylogenetic sequence.

FIGURE 23.23. Conservative cladogram of squamate relationships derived from character analysis. Metataxa are indicated by an asterisk. (From Estes et al. 1988.)

5.2 Taxonomic Relationships Based on Brain Anatomy

An extensive morphological study of the fore-and midbrain of species of most lizard families by Northcutt (1978) agrees in one important aspect with our ear studies (Miller 1980a, 1985; Miller and Beck 1990b) by indicating a close relationship between the teiids and varanids. However, while these families are closely related on the basis of brain structure, Northcutt does not view them as necessarily basic or stem types. Papillar structure in my view indicates that the teiids and varanids are not far removed from an ancestral type ear. The relationships between other families, when comparing brain and ear structure differ in that Northcutt places the xantusiids close to the Gekkota, while the ear structure of xantusiids is unequivocally scincomorphic. Ear structure also shows the cordylids to be scincomorphic rather than grouped with anguinomorphic types as brain structure indicates.

5.3 Taxonomic Relationships Based on Other Anatomical Characters

An outstanding study based on a large number of anatomical characters (130 osteological and 18 soft

anatomy characters) is that of Estes et al. (1988). Figure 23.23 is a cladogram from this publication. The general taxonomic relationships of lizard families determined from this work are in fairly close agreement with those determined by ear structure. However, my supposition of the ancestor close or basic nature of the papillae of the Teiidae and Varanidae is not indicated in the Estes et al. (1988) cladogram (Fig. 23.23). Also, ear anatomy indicates a scincoid rather than a lacertoid relationship of Xantusiidae. It is probably much simpler to propose a developmental sequence when considering only one anatomical system than many systems. Further, different anatomical systems probably evolve independently of others, and the developmental sequence of one character need not be the same as it is for others.

5.4 Camp's (1923) Classic Proposal of Lizard Family Phylogeny

Camp's scheme indicates greater separation of the iguanid-agamid line from other lines of descent than I have estimated from ear structure. It may be that my proposal of lizard family phylogeny based on ear structure relies too heavily on the significance of the relatively unspecialized tectorial cover and the papillar constructions that are characteristic of the teiids and varanids. However, I believe that the teiids and varanids, with their complete complement and particular distribution of hair cell types, and unspecialized tectorial membranes, exemplify a basic or stem type condition. From this condition I have proposed the derivation of other extant families toward a more specialized condition by the loss of papillar segments and the development of modified tectorial covers. I do not consider it likely that a specialized structure like the iguanid, agamid, scincid, xantusiid, or the gekkonid papilla could have reverted to what I consider a condition capable of giving rise to other types of papillae.

6. Possible Significance of Cochlear Duct and Basilar Papilla Structural Variations

One would expect that the large variations in cochlear duct and basilar papilla structure of lizards would be associated with differences in acoustic performance and with environmental and behavioral adaptations. With the exception of geckos, however, lizards display little vocal activity and the role of sound in the life of lizards is an almost unknown quantity. In all probability sound plays a role in the avoidance of predators and the capture of prey, but little or nothing is known concerning the role of hearing in these activities. As a result, the relationship between the great differences in ear structure and any possible benefit to a species is largely unknown. What little is known has been summarized and reported by Marcellini (1978). Our main recourse to correlating structure with function is neurophysiological study of the lizard ear. The results of such studies are reported in Chapter 24 by Köppl and Manley.

7. Summary and Conclusions

The cochlear duct and basilar papilla of lizards exhibit a considerable range of structural variations. Cochlear duct structure varies particularly in the development of projections from the lateral limbic wall and of enclosed spaces for the scala tympani on the medial limbic wall. The papillae

vary greatly in length and number of hair cells, type of tectorial cover, the specific disposition of unidirectional and bidirectional hair cell groups, the ratio of the papillar nerve fiber number to hair cell number (from 0.8/1 to 12/1), and the pattern of afferent nerve fiber innervation of the hair cells. The structural details of the duct and papilla are sufficient to differentiate one lizard family from another. However, similarities in duct and papilla structure of certain lizard families indicate close taxonomic relationship.

The probable ancestral lizard auditory papilla was intermediate in length (ca. 1 mm in length) and consisted of ca. 800 to 1,000 unidirectionally oriented hair cells covered by a substantial, relatively unspecialized tectorial structure. Each hair cell was innervated by relatively few afferent nerve fibers (two to three) and by efferent nerve fibers, and the pattern of afferent hair cell innervation was mixed in that some hair cells were exclusively innervated (each nerve fiber innervates only one hair cell) and others were nonexclusively innervated (nerve fibers innervate more than one hair cell).

I propose that the auditory papillae of the teiid and varanid lizards, although derived from a simpler ancestral type represent basic stem types from which species of other families with more specialized papillae could have been derived. From a tupinambid type teiid ancestor, I propose two, or possibly three, major lines of development. (1) The scincomorphic (species of the Scincidae, Cordylidae, and Xantusiidae) type of papillae characterized by loss of the apical group of bidirectional hair cells (BHC) from the ancestor so that the apical segment consists of unidirectional type hair cells (UTHC) covered by a heavy tectorial structure (culmen), and a long basal group of BHC covered by tectorial specializations called sallets. (2) The lacertids, characterized by complete separation of the papilla into two separate segments. The apical segment consists of a group of BHC. The basal segment consists of an apical group of UTHC (which are now in an essentially central location) and a longer basal region of BHC. The BHC are covered by an unspecialized structure similar to that of the teiids, and the unidirectional hair cells by a limbic attached cap-like cover similar to that of the iguanids and agamids. (3) Possibly on a separate or third line of development, or from the same line leading to the lacertids, species of the Iguanidae and Agamidae could have been derived from a tupinambic type of teiid. This could have come about either by retention of the teiid pattern of hair cell disposition with a central unidirectional hair cell group covered by a tectorial cap and flanked on both sides by bidirectional hair cells, or with loss of the apical group of bidirectional hair cells so that the unidirectional hair cells become apically located. In both cases the bidirectional hair cells have lost all tectorial cover and are "free standing." (4) On another line of development from an ameivid type of papilla where the papillar constriction is basal in location, the basal bidirectional group of hair cells could have been lost and papillae characteristic of the gekkonids and pygopodids developed. With loss of the BHC cells, the UHC become basal in location, a condition found only in species of these two families of lizards. Additionally, the lateral limbic projection (limbic lip) of the teiid type ancestor has been further extended, and the apical BHC and their tectorial covers have become specialized.

From a varanid type of papilla, the anguid related lizards could have been derived, (1) by either retention of the varanid pattern of centrally located UHC flanked by apical and basal groups of BHC or (2) by loss of the apical group of BHC resulting in the UHC group becoming apical and the basal region, bidirectional, (3) development of a limbic attached tectorial cap similar to that of the iguanids and agamids covering the UHC, and (4) the loss of all tectorial covering of the BHC leaving them "free standing" as in the iguanid and agamid type papillae.

Acknowledgments. The author is most indebted to the following: the United States Public Health Service, division of National Institutes of Health for an almost uninterrupted series of grants over the past 25 years that supported the work reported in this chapter, to three excellent coworkers, Michiko Kasahara, Janet Beck, and Dr. Jeanne Miller who shared completely in this endeavor, to four graduate students, Michael Mulroy, Patricia Leake, Paul Teresi, and Michael Sneary who all added greatly to this work and hopefully will continue to carry these studies onwards, to the Coleman Laboratory of the Department of Otolaryngology of the University of California, San Francisco for use of their TEM facilities and important consultative help. A special debt of gratitude is due Dr. Alan E. Leviton and the Department of Herpetology of the California Academy of Sciences for their generous provision of specimens and an education into the mysteries of herpetological taxonomy. Gratitude is also extended to Professor Geoffrey Manley of the Technical University of Münich (Garching), for many years of encouragement and participation in a process of cross fertilization of ideas concerning ongoing and future studies.

References

Baird IL (1960) A survey of the periotic labyrinth in some recent representative reptiles. Univ Kansas Sci Bull 41:891–981.

Baird IL (1974) Anatomical features of the inner ear in submammalian vertebrates. In: Keidel WD, Neff WD (eds) Handbook of Sensory Physiology, Vol. V, part 1. Berlin-Heidelberg-New York: Springer, pp. 159–212.

Baird IL, Marovitz WF (1971) Some findings of scanning and transmission microscopy of the papilla basilaris of the lizard, *Iguana iguana.* Anat Rec 169:270.

Bagger-Sjöbäck D (1976) The cellular organization and nervous supply of the basilar papilla in the lizard, *Calotes versicolor.* Cell Tiss Res 165:141–156.

Bagger-Sjöbäck D, Wersäll J (1973) The sensory hairs and tectorial membrane of the basilar papilla in the lizard, *Calotes versicolor.* J Neurocytol 2:329–350.

Camp C (1923) Classification of the Lizards. Bull Amer Mus Nat Hist 48:289–481.

Düring M, von Karduck A, Richter H-G (1974) The fine structure of the inner ear in *Caiman crocodilus.* Z Anat Entwickl Gesch 145:41–65.

Estes R, de Queiroz K, Gauthier J (1988) Phylogenetic relationships within squamata. In: Estes R, Pregill G (eds) Phylogenetic relationships of the lizard families. Palo Alto: Stanford University Press.

Greer A (1985) The relationships of the lizard genera *Anelytropsis* and *Dibamus.* J Herpetology 19:116–156.

Köppl C (1988) Morphology of the basilar papilla of the bobtail lizard, *Tiliqua rugosa.* Hearing Res 35:209–228.

Köppl C (1989) Struktur und Funktion des Hörorgans der Echsen: Ein vergleich von Lacertidae und Scincidae. Doctoral dissertation in Zoology, Technical University of Munich.

Köppl C, Manley G (1990) Peripheral auditory processing in the bobtail lizard, *Tiliqua rugosa*: II. Tonotopic organization and innervation pattern of the basilar papilla. J Comp Physiol A 167:101–112.

Manley G, Gleich O, Kaiser A, Brix J (1989) Functional differentiation of sensory cells in the avian auditory periphery. J Comp Physiol A 164:289–296.

Marcellini DL (1978) The acoustic behavior of lizards. In: Greenberg N, MacLean PD (eds) Behaviour and Neurology of lizards. US Dept HEW Natl Inst Mental Health pp. 287–301.

Miller MR (1966a) The cochlear duct of lizards. Proc Calif Acad Sci 33:255–359.

Miller MR (1966b) The cochlear ducts of lizards and snakes. Amer Zool 6:421–429.

Miller MR (1966c) The cochlear ducts of *Lanthanotus* and *Anelytropsis* with remarks on the familial relationship between *Anelytropsis* and *Dibamus*. Occ Papers Calif Acad Sci No 60 pp. 1–15.

Miller MR (1973a) A scanning electron microscope study of the papilla basilaris of *Gekko gecko*. Zeit f Zellforsch 136:307–328.

Miller MR (1973b) Scanning electron microscope studies of some lizard basilar papillae. Am J Anat 138:301–330.

Miller MR (1974a) Scanning electron microscope studies of some skink papillae basilares. Cell Tiss Res 150:125–141.

Miller MR (1974b) Scanning electron microscopy of the lizard papilla basilaris. Brain Behav Evol 10:95–112.

Miller MR (1978a) SEM studies of the papilla basilaris of some turtles and snakes. Am J Anat 151:409–435.

Miller MR (1978b) Further SEM studies of lizard auditory papillae. J Morph 156:381–418.

Miller MR (1980a) The Reptilian Cochlear Duct. In: Popper AN, Fay RR (eds) Comparative Studies of Hearing in Vertebrates. New York: Springer.

Miller MR (1980b) The cochlear nuclei of snakes. J Comp Neurol 192:717–736.

Miller MR (1981) Scanning electron microscope studies of the auditory papillae of some iguanid lizards. Am J Anat 162:55–72.

Miller MR (1985) Quantitative studies of auditory hair cells and nerves in lizards. J Comp Neurol 232:1–24.

Miller MR, Beck J (1988) Auditory hair cell innervational patterns in lizards. J Comp Neurol 271:604–628.

Miller MR, Beck J (1990a) The innervation of a lizard auditory organ having gap junctions between most hair cells: A serial TEM study. J Comp Neurol 293:223–235.

Miller MR, Beck J (1990b) Further serial TEM studies of auditory hair cell innervation in lizards and a snake. Am J Anat 188:175–184.

Mulroy MJ (1974) Cochlear anatomy of the alligator lizard. Brain Behav Evol 10:69–87.

Mulroy MJ (1986) Pattern of afferent synaptic contacts in the alligator lizard's cochlea. J Comp Neurol 248:263–271.

Nadol JB, Mulroy MJ, Goodenough DA, Weiss TF (1976) Tight and gap junctions in a vertebrate inner ear. Am J Anat 147:281–302.

Northcutt RG (1978) Forebrain and midbrain organization in lizards and its phylogenetic significance. In: Greenberg N, MacLean PD (eds) Behavior and Neurology of Lizards. Dept. HEW, Natl. Inst. Mental Health.

Schmidt R (1964) Phylogenetic significance of the lizard cochlea. Copeia (1964) 542–549.

Shute CDD, Bellairs AD'A (1953) The cochlear apparatus of Gekkonidae and Pygopodidae and its bearing on the affinities of these groups of lizards. Proc Zool Soc London 123:695–709.

Sneary MS (1988a) Auditory receptor of the Red-Eared Turtle: I. General ultrastructure. J Comp Neurol 276:573–587.

Sneary MS (1988b) Auditory receptor of the Red-Eared Turtle: II. Afferent and efferent synapses and innervation patterns. J Comp Neurol 276:588–606.

Wever EG (1978) The Reptile Ear. Princeton, NJ: Princeton University Press.

24
Functional Consequences of Morphological Trends in the Evolution of Lizard Hearing Organs

Christine Köppl and Geoffrey A. Manley

1. Introduction

The hearing organ of modern terrestrial amniotic vertebrates shows a fascinating diversity of structure. The morphological patterns of the lizard basilar papilla show greater differences between families and even within families than those in any other vertebrate groups (Chapter 23 by Miller). This, of course, raises the question of the functional consequences of morphological variations and of the clues they might provide to the selection pressures that have led to the diversity seen today. In this chapter, we will try to outline the differences and similarities in the physiology of the diverse hearing organs studied and infer correlations between structure and function which might be related to the selection pressures underlying the evolution of such great diversity.

2. The Hearing of the Red-Eared Turtle (*Pseudemys scripta*) as an Example of the Presumed Ancestral Condition

To understand the evolution of lizard hearing we first have to ask what the hearing of the presumed ancestor of modern lizard families might have been like. The red-eared turtle has been chosen to illustrate this primitive condition for three reasons. (1) It is generally agreed that turtles are the most primitive forms among modern reptiles and thus represent a stage of evolution similar to stem reptiles. (2) Comparative anatomical studies of the lizard basilar papilla (Chapter 23 by Miller), suggest that a papilla similar to the turtles' was the probable ancestral condition. (3) The red-eared turtle is the only primitive reptile whose papillar function has been studied in detail.

2.1 Hearing Range

In 1966, Patterson described a behavioral audiogram for the red-eared turtle. Subsequently, similar data were derived from recordings of cochlear microphonics (Wever 1978) and auditory nerve fibers (Crawford and Fettiplace 1980; Art and Fettiplace 1984). The hearing range of the turtle is low, with the region of best sensitivity (at 35 to 45 dB SPL) between 200 and 600 Hz and lower and upper physiological frequency limits near 10 and 1,000 Hz, respectively (Fig. 24.1).

2.2 Tonotopic Organization of the Turtle's Basilar Papilla

The basilar papilla is tonotopically organized along its length, the lowest frequencies being processed at the apical end of the epithelium (Crawford and Fettiplace 1980; Art, Crawford, and Fettiplace 1986; Art and Fettiplace 1987). The distribution of frequencies along the papilla was found to be roughly exponential, with differing space constants of frequency derived for different recording situations: about 0.1 mm/octave (Crawford and Fettiplace 1980) and about 0.2 mm/octave (Art, Crawford, and Fettiplace 1986).

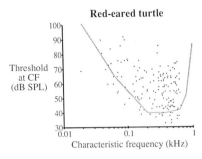

FIGURE 24.1. Scatterplot of thresholds at CF of auditory nerve fibers of the red-eared turtle as a function of CF. (Data replotted from Crawford and Fettiplace 1980.) The gray line joins the threshold values determined by behavioral testing of the same species. (Data from Patterson 1966.)

2.3 Frequency Selectivity

2.3.1 Frequency Selectivity of Hair Cells and Primary Auditory Nerve Fibers

The sound pressure levels (SPL) necessary at different frequencies to elicit a standard receptor potential change in a hair cell or a standard increase in discharge rate in a nerve fiber, respectively, define a frequency tuning curve. Tuning curves of hair cells and of primary fibers were roughly V-shaped with a clear characteristic, or most sensitive, frequency (CF) and a sharp tip. The quality of tuning, expressed as the Q_{10dB}-value (CF/tuning-curve bandwidth, 10 dB above CF threshold), was similar in hair cells and nerve fibers (Crawford and

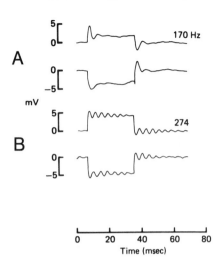

FIGURE 24.3. Examples of oscillations in the membrane potential of turtle hair cells in response to current injection through a recording electrode. The response of two cells (A and B) with different acoustic CF (indicated on the right) to depolarizing (top) and hyperpolarizing (bottom) current injection is shown. Note that the frequency of the induced oscillation depends on the CF. Each trace represents an average of 128 or 256 responses, voltage is given relative to the resting potential. (From Crawford and Fettiplace 1981, used with permission.)

Fettiplace 1980). For nerve fibers, the 10 dB bandwidth averages 125 Hz (Crawford and Fettiplace 1980), irrespective of CF, resulting in a linear increase of Q_{10dB} with increasing CF, from an average of 0.25 at CF 30 Hz to 5 at CF 700 Hz (Fig. 24.2). As Crawford and Fettiplace (1980) pointed out, these Q-values are higher than those of mammalian auditory nerve fibers in the same CF range. Stimulation of efferent fibers both strongly reduced the sensitivity and broadened the tuning of hair cells and fibers (Art and Fettiplace 1984; Art, Fettiplace, and Fuchs 1984; Art et al. 1985).

2.3.2 Electrical Resonance as a Tuning Mechanism

During intracellular recordings, Crawford and Fettiplace (1980) noted that auditory hair cells behaved like resonators. After the end of a puretone stimulus, or in response to a click stimulus, the receptor potential displayed exponentially decaying oscillations at a frequency close to the acoustic CF of the cell. Similar oscillations were

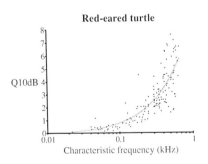

FIGURE 24.2. Scatter plot of tuning-curve sharpness represented as Q_{10dB} for auditory nerve fibers in the red-eared turtle as a function of CF. (Data taken from Crawford and Fettiplace 1980.) The gray line is a linear regression on the data.

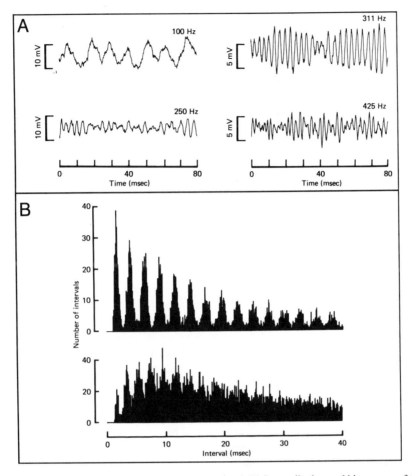

FIGURE 24.4. (A). Spontaneous voltage fluctuations in intracellular recordings from four different turtle hair cells in the absence of acoustic stimulation. The major frequency components in the fluctuations are correlated with the acoustic CF of the cell, given to the right of each trace. (From Crawford and Fettiplace 1980, used with permission.) (B) Interspike-interval histograms of the spontaneous discharge of two auditory nerve fibers in the turtle. Each histogram shows preferred intervals of spontaneous firing, correlated with the acoustic CF of the fiber (380 Hz for the top histogram, 480 Hz for the bottom one. (From Crawford and Fettiplace 1980, used with permission.)

observed in response to extrinsic current injection (Fig. 24.3). The Q_{10dB}-values calculated from the decay times of the oscillations also corresponded closely to those derived from the respective acoustic tuning curves.

Crawford and Fettiplace (1981) successfully modelled the hair cell response to small stimuli as a simple electrical resonator. They concluded that, assuming the normal transducer current is subject to the same filtering operations as injected current, the highly selective tuning curve tip is contributed by an electrical resonance phenomenon within each hair cell. The remaining broad-band part of the tuning curve probably reflects the less selective filter characteristics of the middle ear and basilar membrane. It was subsequently shown that isolated hair cells display all aspects of the resonance phenomenon, thus strengthening the suggestion of an electrical fine-tuning mechanism private to each cell (Art and Fettiplace 1987). In addition, spontaneous membrane-potential fluctuations were observed whose spectrum was concentrated in a frequency band around the CF of the hair cell and which were reflected in preferred intervals in the spontaneous activity of nerve fibers (Fig. 24.4) (Crawford and Fettiplace 1980). Voltage-clamp experiments on solitary hair cells served to illuminate the ionic mechanisms underlying the electrical resonance. A voltage-dependent Ca^{2+}-conductance and a large Ca^{2+}-dependent K^+-conductance are the primary membrane components sufficient to explain the observed hair-cell resonance (Art and Fettiplace 1987). The different resonant frequencies seen in individual hair cells of the turtle papilla are primarily achieved by systematic variations in the kinetics of the K^+-channel and to a lesser extent by variations in the numbers of K^+- and Ca^{2+}-channels (Art and Fettiplace 1987).

2.4 An Active Force-Generating Mechanism in Hair Cells

The membrane potential of turtle hair cells, and therefore also its electrical resonance, is inextricably linked with the micromechanics of the stereovillar bundle (Crawford and Fettiplace 1985; Art, Crawford, and Fettiplace 1986). Upon application of a force step with a flexible glass fiber, bundles displayed damped oscillations of displacement at the CF of the hair cell in synchrony with oscillations of the membrane potential. Similar oscillations of both the bundle and the potential could be elicited by extrinsic current injection, indicating that the movement of the bundle is not simply a passive mechanical resonance (Crawford and Fettiplace 1985). In fact, it was shown that the electrical resonance can be observed if the mechanical oscillation is blocked, but not vice versa: The membrane potential still displayed oscillations in response to a mechanical displacement of the bundle with a rigid glass fiber prohibiting mechanical resonance (Art, Crawford, and Fettiplace 1986). On this and other bases, Crawford and Fettiplace (1985) concluded that the observed mechanical oscillations of the stereovillar bundle require an active contribution from the hair cell. This is further supported by the presence of spontaneous bundle oscillations in unison with the membrane potential (Crawford and Fettiplace 1985). The passive mechanical resonance frequency for single stereovillar bundles was estimated by Crawford and Fettiplace (1985) to be around 50 kHz and thus orders of magnitude away from the actual range of resonant frequencies between 20 and 320 Hz (Crawford and Fettiplace 1985). They suggested

that the passive micromechanical properties of the bundles do not play a major role in determining the resonant frequencies. However, in the intact papilla, the additional mass of a tectorial membrane might well lower the passive resonance frequency into the audio range observed for the electrical resonances (see e.g., Lewis et al. 1985). The presence of a moderate anatomical gradient in the height of the stereovillar bundles along the papilla (Sneary 1988a) is compatible with passive micromechanics influencing the responses.

3. The Hearing of Modern Lizards

The hearing capabilities of lizards were first compared by Wever and his coworkers (summarized in Wever 1978), using audiograms constructed from cochlear-potential recordings in representatives of all major modern lizard families. Since then, a number of lizard species from different families have been studied in more detail, using recordings of single nerve-fiber or hair-cell activity, fiber tracing and mechanical measurement techniques. This section provides a brief summary of the more detailed knowledge in selected lizard species that will be important for the discussion in Sections 4 and 5. The order of presentation was chosen mainly for historical reasons and does not mean to imply any phylogenetic sequence.

3.1 The Alligator Lizard *Gerrhonotus multicarinatus* (Anguidae) and the Granite Spiny Lizard *Sceloporus orcutti* (Iguanidae)

3.11.1 Hearing Range

The alligator lizard and the granite spiny lizard will be treated together, because, although not thought to be closely related, their papillar anatomy (Chapter 23 by Miller) and auditory physiology are extremely similar. The CFs found in auditory nerve fibers ranged from 0.2 kHz up to 4 kHz (at around 22°C body temperature). Both lizards showed a discontinuous distribution of fibers across that range, a lack of fibers around CF 1 kHz defining a low- and a high-CF group. The inferred audiogram is thus two-peaked, with the most sensitive fibers having comparable thresholds in both

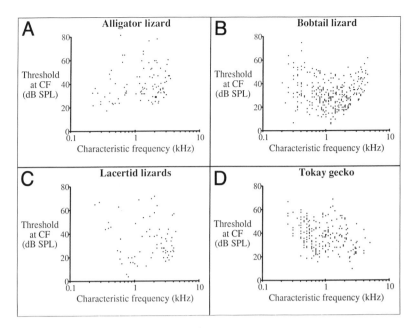

FIGURE 24.5. Threshold for CF tones for single auditory nerve fibers of different species of lizards as a function of CF. (A) the alligator lizard (data taken from Weiss et al. 1976), (B) the bobtail lizard, (C) two lacertid lizard species and (D) the Tokay gecko (data partly from R.A. Eatock, personal communication).

CF-groups at around 20 dB SPL (Fig. 24.5A) (Weiss et al. 1976; Turner 1987).

3.1.2 The Tonotopic Organization of the Basilar Papilla

The tonotopic organization of the basilar papilla (Fig. 24.6A) has been shown indirectly by dye-marking recording sites in the auditory nerve somewhere near the papilla and by stepping the recording electrode along the nerve (Weiss et al. 1976; Turner, Muraski, and Nielsen 1981; Holton and Weiss 1983b; Turner 1987). Low-CF fibers innervate a small, anatomically distinct segment of the papilla located apically in the alligator lizard and centrally in the granite spiny lizard (for a detailed description of the papillar anatomy see Mulroy 1974; Mulroy and Williams 1987; Turner 1987). A tonotopic organization of this low-CF segment was not evident. High-CF fibers ran to the other papillar segment, located basally in the alligator lizard and, curiously, present twice in the granite spiny lizard papilla, both apical and basal of the low-frequency segment. The high CFs were clearly tonotopically organized, the frequency rising exponentially towards the end(s) of the

papilla with small space constants of frequency between about 0.06 and 0.15 mm/octave.

3.1.3 Frequency Selectivity

3.1.3.1 Tuning Curves of Auditory Nerve Fibers

The tuning curves of low- and high-CF nerve fibers differed distinctly in their shapes (Fig. 24.7A). High-CF fiber tuning curves were broad and almost symmetrically V-shaped, whereas low-CF tuning curves are only approximately symmetrical at the tip and thereafter rise very steeply on the high-frequency flank (Weiss et al. 1976; Holton 1980). The Q_{10dB}-values of both nerve-fiber populations were roughly equal (median values near 2; Fig. 24.8A). However, the fibers' Q_{10dB}-values depended on CF threshold, more sensitive tuning-curves having sharper tips (Holton 1980; Holton and Weiss 1983b).

3.1.3.2 Mechanical Resonance of Individual Hair-Cell Bundles as a Tuning Mechanism for the High-Frequency Range

The differences seen between the tuning curves of low- and high-CF fibers and the many anatomical differences seen between the associated papillar

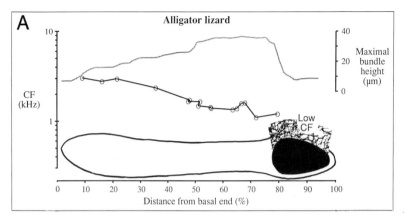

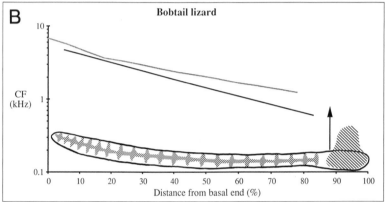

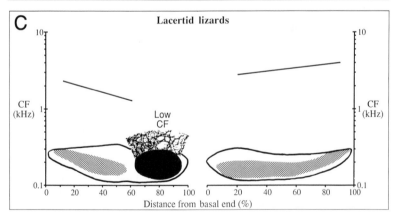

FIGURE 24.6. Schematic illustration of the relationship between structure and tonotopic organization of the basilar papilla in (A) the alligator lizard, (B) the bobtail lizard, and (C) two species of lacertid lizards. In each case, a schematic drawing of the papillar outline with the tectorial structures indicated is included. Basal is to the left, neural to the top in all cases. Tectorial structures not affixed to the limbus are drawn dot- or stripe-shaded. The number of tectorial sallets shown in the bobtail lizard does not represent the true number (about 70 to 90). Tectorial structures attached to the neural limbus are

shown black and the net-like connection with the limbus is indicated. (A) The circles joined by a black line show the tonotopic organization of the basal, high-frequency segment of the papilla as a function of distance of recorded fibers from the basal end (refer to left ordinate; data taken from Holton and Weiss 1983b). The apical segment is marked "low CF," since a tonotopic organization is not known. The gray line is a running average of maximal stereovillar bundle height along the papilla (refer to right ordinate; data taken from Mulroy and Williams 1987). (B) The black line is an exponential regres-

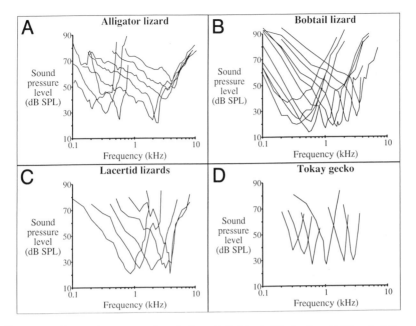

FIGURE 24.7. Representative tuning curves for single auditory nerve fibers in different species of lizards, (A) alligator lizard (data taken from Holton 1980 and Holton and Weiss 1983b), (B) bobtail lizard (all curves from one individual, (C) the lacertid lizards (all curves from one *Podarcis muralis*), and (D) the Tokay gecko (data taken from Eatock et al. 1981).

segments (Mulroy 1974) suggested to Weiss et al. (1976) that the primary determinants of tuning in the alligator lizard are structural. In particular, they implicated differences in the micromechanics of the stereovillar bundles of the high-frequency segment, which lacks a tectorial membrane. Peake and Ling (1980) showed that the basilar-membrane motion is very broadly tuned and does not account for the sharp frequency tuning of hair cells and nerve fibers. Since there is a close correlation between the gradient in maximal height of the stereovillar bundles and the frequency representa-

tion at least for the high-CF papillar segment (Fig. 24.6A) (Turner, Muraski and Nielsen 1981; Holton and Weiss 1983b; Mulroy and Williams 1987; Turner 1987), it was proposed that the additional tuning mechanism was a mechanical resonance of the stereovillar bundles. Based on theoretical studies, Weiss and Leong (1985) and Freeman and Weiss (1988) concluded that the free-standing bundles could resonate passively in the observed audio range. Direct observations of stereovillar-bundle motion in the isolated papilla of the alligator lizard subsequently showed that free-standing bundles

◁──

FIGURE 24.6. *continued*

sion through data derived from single-fiber tracing experiments, representing the tonotopic organization of the basal, high-frequency segment of the papilla. The gray line shows the calculated resonance frequencies of hair cell–tectorial membrane salletal units (see text for details). The arrow over the apical papillar segment indicates that the low-CF tonotopic organization runs across the papilla, from the abneural to the neural side. (C) For the lacertid lizards, the diagram is

split into two, representing the basal (left) and apical (right) subpapillae. The black lines show exponential regressions through data derived from auditory nerve fiber tracing experiments. The slopes and ranges of these tonotopic distributions are different for the basal segment of the basal subpapilla and the apical subpapilla, respectively. The apical segment of the basal subpapilla is marked "low CF," since a tonotopic organization is not known.

respond preferentially to different stimulation frequencies (Frischkopf and DeRosier 1983; Hudspeth and Holton 1986) and the measured resonant frequencies correspond to the known tonotopic organization (Frischkopf and DeRosier 1983).

Little is known about tuning mechanisms in the low-CF range. The tonotopic organization of the low-frequency papillar segment is not yet clear and no attempt has been made so far to predict the micromechanical frequency response from the observed gradients in stereovillar bundle morphology (Mulroy and Williams 1987). Evidence for electrical tuning is lacking; there are, however hardly any data concerning this question. Although "ringing" of the hair-cell receptor potential as seen in the turtle has never been observed (Holton and Weiss 1983a,b), hair-cell recordings are almost exclusively from the high-frequency papillar segment.

3.1.4 Discharge Patterns of Auditory Nerve Fibers

The low- and high-CF populations of auditory nerve fibers also differ in their time-resolution capabilities. Rose and Weiss (1988), found, surprisingly, that low-CF fibers phase-couple to higher stimulus frequencies than high-CF fibers. The corner frequency (-3 dB) of the decline in phase-coupling was on average 500 Hz for low-CF and 370 Hz for high-CF fibers. Rose and Weiss therefore proposed that the two fiber populations fulfill different functions and form the peripheral starting point of separate "timing" and "nontiming" neural pathways.

3.2 The Bobtail Lizard
Tiliqua rugosa (Scincidae)

3.2.1 Hearing Range

The CFs found in the auditory nerve of the bobtail lizard ranged from 0.2 to 4.5 kHz. The audiogram inferred from the neural data has its peak sensitivity near 1.2 kHz at approximately 10 dB SPL (Fig. 24.5B). The best sensitivity declines towards both ends of the CF-range, being about 40 dB SPL at 4 kHz and 25 dB SPL at 0.2 kHz (at 30°C body temperature; Köppl and Manley Abstr 10th Midwinter

Mtg Assoc Res Otolaryngol p. 229, 1987; Manley, Köppl and Johnstone 1990).

3.2.2 Tonotopic Organization of the Basilar Papilla

Both cobalt marking of physiologically characterized nerve fibers and systematic peripheral recordings from the auditory nerve showed that the two anatomically distinct segments of the bobtail lizard papilla process different CF ranges (Köppl and Manley Abstr 10th Midwinter Mtg Assoc Res Otolaryngol p. 229, 1987; Köppl and Manley 1990a; for a detailed description of the anatomy see Köppl 1988). Fibers of low CF (up to 0.8 kHz) innervated hair cells in the smaller apical segment, and fibers of medium to high CF innervated the much longer basal segment. In addition, the tonotopic organization of the two segments differed (Fig. 24.6B). Frequency changed approximately exponentially (frequency space constant 0.56 mm/octave) along the high-CF, basal segment, with the highest CFs located at the basal end. The low-frequency apical segment, in contrast, was organized ACROSS ITS WIDTH, with frequency increasing towards the neural side of the segment.

3.2.3 Frequency Selectivity

3.2.3.1 Tuning Curves of Auditory Nerve Fibers

Based on the shape of the tuning curve (Fig. 24.7B), fibers were divided into a low-CF group (CF up to about 0.8 kHz) and a high-CF group (CF above about 0.8 kHz). Low-CF fiber tuning curves were simply V- or U-shaped. In contrast, high-CF fibers showed a two-component curve with a sharp, sensitive tip (up to 46 dB deep) added to a broader trough-like upper part (Manley, Köppl, and Johnstone 1990). There was a general increase in frequency selectivity with CF; the rate of change of Q_{10dB}, however, was different for the low- and high-CF populations (Fig. 24.8B) (Manley, Köppl, and Johnstone 1990). As the bandwidth 10 dB above CF-threshold was virtually constant at 370 Hz for the low-CF fibers, there was a rapid linear increase of Q_{10dB} from an average of 0.6 at CF 0.2 kHz to 2.3 at CF 0.8 kHz. The subsequent increase in Q_{10dB} with CF within the high-CF fiber population was much slower, resulting in an average of 4.2 at the highest CF.

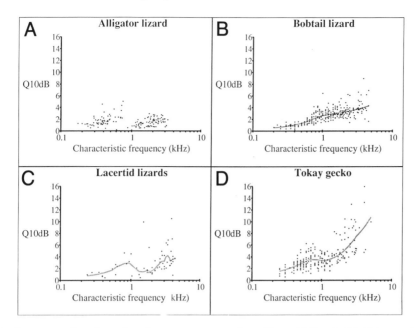

FIGURE 24.8. Sharpness of tuning of auditory nerve fibers in different lizard species, represented as the Q_{10dB} value as a function of CF. (A) alligator lizard (data taken from Weiss et. al. 1976), (B) bobtail lizard, (C) two species of lacertid lizards (D) Tokay gecko (data partly from R.A. E .iock, personal communication). The gray lines in b, c, and d represent locally weighted regressions through the data points.

3.2.3.2 Mechanical Resonance of Groups of Hair Cells and Their Tectorial Structure as a Possible Tuning Mechanism in the High-Frequency Range

Manley, Yates, and Köppl (1988) showed that although the basilar membrane of the bobtail lizard is greatly elongated, its motion does not support a travelling wave and does not contribute any frequency-selective, place-dependent component to the tuning of the auditory nerve fibers. In fact, the average basilar-membrane tuning curve fitted well the broad upper part of high-CF fiber tuning curves. The ratios between the neural and mechanical frequency-response curves resembled a simple high-pass resonant filter characteristic (Manley, Yates, and Köppl 1988; Manley, Köppl, and Yates 1989). The situation therefore resembles that in the alligator lizard; however, the structure of the high-frequency segment in the bobtail lizard is quite different, implying a different basis for any possible mechanical resonance. In particular, in the bobtail lizard a specialized tectorial structure, a chain of sallets, covers the hair cells (Wever 1967; Miller 1974; Köppl 1988). Manley, Köppl, and

Yates (1989) suggested that the prominent subdivision of this salletal chain results in a functional grouping of hair cells. Assuming a mechanical resonance of a simple mass-stiffness type and using quantitative measurements of important morphological parameters (Köppl 1988), the expected passive resonance frequencies of such hair cell-sallet groups were calculated. The result was a gradient of frequencies remarkably similar to the actual CF-gradient (Fig. 24.6B) (Manley, Köppl, and Yates 1989). The structural basis for micromechanical tuning in the high-CF range is thus present.

As in the alligator lizard, our knowledge of possible tuning mechanisms in the low-frequency range is much poorer. Based on anatomical studies (Köppl 1988) it seems very unlikely that micromechanical resonance alone accounts for frequency tuning. In particular, a huge uniform tectorial structure, the culmen, covers the whole low-CF segment. Calculations of the theoretical resonance frequency of all hair cells together with the culmen or parts of the hair-cell population and culmen yielded very unrealistic values around a few Hz. On the other hand, evidence for electrical

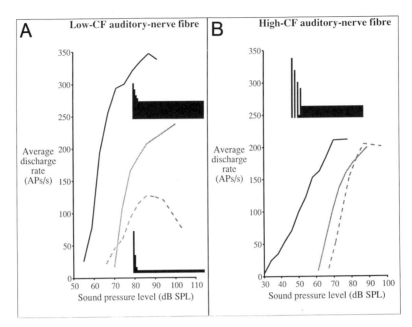

FIGURE 24.9. Typical response patterns of (A) low-CF and (B) high-CF auditory nerve fibers in the bobtail lizard. Each diagram shows three input/output functions for one fiber stimulated at its CF (black line) and at frequencies about one octave below CF (gray solid line) and above CF (gray dashed line), respectively. The insets are schematic representations of typical peristimulus-time histograms at medium to high levels of stimulation with 50 ms long tone bursts. In a, two such histograms are shown, corresponding to the different responses of low-CF fibers at and below CF (upper PSTH) and above CF (lower PSTH), respectively.

tuning is also lacking. Hair-cell recordings, which could produce the most direct evidence, are not available. Preferred intervals in the spontaneous discharge of auditory nerve fibers, as a possible indirect indication of electrical tuning, are not present (Köppl and Manley 1990b). Considering the pronounced convergence of hair-cell input to the nerve fibers in skinks (Miller and Beck 1988; Köppl and Manley 1990a), however, it is highly unlikely that preferred intervals would be observed even if electrical hair-cell tuning were present.

3.2.4 Discharge Patterns of Auditory Nerve Fibers

In the bobtail lizard, fundamentally different stimulus-response patterns were found in low-CF and high-CF fibers (Köppl and Manley 1990b). Low-CF fibers showed a characteristic variation of the discharge pattern with stimulation frequency (Fig. 24.9A). They responded tonically to fre-

quencies below and around CF, changing to an increasingly phasic response pattern towards the high-frequency limit of their response area. Their input/output (I/O) functions consequently flattened considerably with increasing stimulation frequency above CF. In contrast, high-CF fiber I/O functions showed a shallower slope at stimulation frequencies around CF, compared to I/O functions either above or below CF and a chopper-like discharge at all stimulation frequencies (Fig. 24.9B). In the mammalian cochlear nucleus, such chopper response patterns are associated with a poor temporal resolution capability due to pronounced temporal summation resulting from long membrane time constants (Oertel 1985); perhaps the same is true of high-CF fibers in the bobtail lizard. The phase-locking ability of auditory nerve fibers in the bobtail lizard certainly also suggests this (Manley et al. 1990). As in the alligator lizard, low-CF fibers phase lock to higher stimulation frequencies than high-CF fibers.

3.3 The Lacertid Lizards *Podarcis muralis* and *P. sicula* (Lacertidae)

3.3.1 Hearing Range

In the two species of *Podarcis*, CFs of auditory nerve fibers ranged from 0.2 kHz to 5.2 kHz (at 30°C body temperature; Köppl 1989) with two clear sensitivity minima, around 1 kHz at 10 dB SPL and around 3 kHz near 20 dB SPL (Fig. 24.5C). As outlined below, these two peaks correspond to the center frequencies of the two subpapillae (resulting from the complete anatomical division of the basilar papilla, characteristic of Lacertids; see Wever 1978; Miller 1978; Chapter 23 by Miller for descriptions of the anatomy; a more detailed report is in preparation).

3.3.2 Tonotopic Organization of the Basilar Papilla

The tonotopic organization of the two subpapillae was investigated by dye-marking physiologically characterized auditory nerve fibers and tracing them to their innervation sites (Köppl 1989; Köppl and Manley Abstr 12th Midwinter Mtg Assoc Res Otolaryngol p. 177, 1989). The results showed that in these species also, CFs below about 1 kHz are processed in a small, anatomically distinct segment, in this case the apical segment of the basal subpapilla. A tonotopic organization of this segment was not evident. On the remaining segment of the basal subpapilla, the medium CFs between 1 kHz and about 2.5 kHz were represented. Fibers of CF above 2.5 kHz were traced to the apical subpapilla. The medium and high-CF areas were both tonotopically organized, with frequency changing approximately exponentially along the epithelium (Fig. 24.6C). However, the two space constants of frequency were different, being 0.1 mm/octave for the medium-CF range and 0.25 mm/octave for the high-CF range. The curious splitting of the basilar papilla into two completely separate subpapillae is thus correlated with a physical subdivision of the higher-CF range and an extended spatial representation of the upper part of that range.

3.3.3 Frequency Selectivity

3.3.3.1 Tuning Curves of Auditory Nerve Fibers

Tuning curves of *Podarcis* auditory nerve fibers are roughly V-shaped (Köppl 1989) with no particular grouping of fibers according to shape (Fig. 24.7C). The frequency selectivity showed a complex dependency on CF (Fig. 24.8C). The Q_{10dB}-value rose steadily with CF up to a CF of about 1 kHz, dropped again towards CF 2 kHz and finally rose to a plateau value at CF 3 kHz. Thus the transitions from one anatomically distinct segment to the next (at 1 and at 2.5 kHz) are reflected in a change of the frequency selectivity of the innervating nerve fibers.

3.3.3.2 Mechanical Resonance of Groups of Hair Cells and Their Tectorial Structure as a Possible Tuning Mechanism in the High-Frequency Range

Since the two medium- to high-frequency papillar segments show pronounced anatomical gradients along their lengths (Köppl 1989), expected resonance frequencies were calculated analogous to the method applied in the bobtail lizard (Manley, Köppl, and Yates 1989). The tectorial structures in *Podarcis*, however, show no obvious subdivisions in the two papillar segments in question (Köppl 1989), so it was necessary to use arbitrary parameters for the size of a resonant unit. On the assumption that one resonant unit contained two hair cells and a piece of tectorial structure 10 to 20 µm wide, the calculations reproduced well the different frequency ranges of the two segments. The hair cells in the segments in question are relatively widely spaced, with often only two cells of opposite polarity in any one cross-section (Köppl 1989). Thus resonant units of this size are a reasonable assumption.

3.4 The Tokay Gecko *Gekko gecko* (Gekkonidae)

3.4.1 Hearing Range

The CF range of *Gekko* auditory nerve fibers at 24°C body temperature was 0.15 to 5 kHz. The audiogram of the Tokay gecko, derived from these auditory nerve recordings, shows two sensitivity minima, around 0.7 kHz and above 2 kHz, both with thresholds below 20 dB SPL (Fig. 24.5D) (Eatock, Manley, and Pawson 1981).

3.4.2 Frequency Selectivity

3.4.2.1 Tuning Curves of Auditory Nerve Fibers

Tuning curves of Tokay auditory nerve fibers were V-shaped (Fig. 24.7D) and extraordinarily sharply tuned (Eatock, Manley and Pawson 1981). Q_{10dB}-values ranged from about 1 to 14 and were clearly dependent on CF. Subsequent analysis (Köppl 1989) revealed that this increase of Q_{10dB} with CF is not smooth but, as in the other species described, shows an irregularity around CF 1 kHz (Fig. 24.8D). In the Tokay gecko, fibers below and above a CF of about 1 kHz each have an approximately constant but different 10 dB-bandwidth of 200 Hz and 400 Hz, respectively. This produces an increase in Q_{10dB} with CF at different rates over these frequency ranges (Fig. 24.8D).

3.4.2.2 Electrical Resonance as a Possible Tuning Mechanism in the Low-Frequency Range

Manley (1979) and Eatock, Manley, and Pawson (1981) described preferred intervals closely corresponding to the CF-period in the spontaneous discharge of auditory nerve fibers of the Tokay gecko and suggested that they might result from membrane-potential oscillations in the hair cells. All fibers with a CF below 0.4 kHz showed this phenomenon and none above a CF of 0.5 kHz. Based on similar findings in the turtle auditory nerve (Crawford and Fettiplace 1980), these preferred intervals are now generally regarded as indirect evidence for electrical resonance in the cell membrane of the hair cell innervated. It may thus be assumed that at least in the low-frequency range, electrical tuning plays an important role in the Tokay gecko. The sharp upper frequency border at CF 0.5 kHz found for fibers showing preferred intervals may, however, be an artefact of a change in the innervation pattern (Miller and Beck 1988) and/or the phase-locking capabilities of auditory nerve fibers.

3.4.3 Discharge Patterns of Auditory Nerve Fibers

The discharge patterns of *Gekko* nerve fibers in response to pure-tone CF stimuli were classified into three categories, based on the shape of the peristimulus-time-histogram 20 dB above threshold (Eatock, Manley, and Pawson 1981). The three categories, tonic, phasic and intermediate, were found in different fiber CF ranges (up to 0.7, above 0.7 and from 0.7 to 2 kHz, respectively). The characteristics of those different response patterns resemble those described in the bobtail lizard,

suggesting similar and fundamental differences between low- and high-CF fibers, with a poorer time-resolution capability of the high-CF fibers. Indeed, there is additional evidence that fibers with a tonic response phase lock to higher frequencies than fibers with intermediate or phasic responses (Eatock, Manley, and Pawson 1981).

4. How Is the Diverse Anatomy of the Lizard Ear Reflected in its Function?

4.1 The Hearing Range and Papillar Size Do Not Correlate

Audiograms derived from neural recordings are similar between families, in spite of more than an order of magnitude difference in papillar size and hair-cell numbers. The CF range usually extends from about 0.2 kHz to 4 or 5 kHz. The only exception among the species studied neurophysiologically is the Bengal monitor lizard, *Varanus bengalensis*, where CFs from 0.25 to 2.8 kHz were found (Manley 1976, 1977). The Bengal monitor has thus the same lower frequency limit as the other lizard species, but a comparatively low upper limit. The basilar papilla of this lizard is large compared to those of some lizards that have a wider frequency range, but papillae that are about an order of magnitude smaller, such as the alligator lizard or the lacertid lizards. There is thus no correlation between papillar size and the frequency-response range or hearing sensitivity.

The same conclusions on sensitivity were reached by Wever et al. (1965), using audiograms of a standard cochlear-microphonic output to the stimulus fundamental as the threshold criterion. However, the variation in the microphonic audiograms, particularly with regard to sensitivity, was much larger than in the neural data. In our view, caution should be exercised in comparing cochlear-microphonic recordings across lizard species or families, for a number of reasons.

1. In Wever's experiments, body temperature was uniformly held at 24 to 25°C, while the preferred temperatures of lizards vary between about 24° and 40°C. Campbell (1969) has shown that cochlear-potential recordings vary considerably with

body temperature, with peak sensitivity near the preferred temperature of the respective lizard species. Single-fiber neural tuning is also temperature-sensitive (Eatock and Manley 1981), but neural recordings in lizards were generally performed within a few degrees of the species' preferred temperature.

2. As Eatock, Manley, and Pawson (1981) showed for the Tokay gecko, the audiogram derived from microphonic recordings may greatly overestimate thresholds in the high-frequency range, as compared to an audiogram derived from neural single-unit thresholds. This is due to the presence of hair-cell populations with opposite polarity in the high-CF segment of the basilar papilla, whose extracellular potentials would partly cancel and thus reduce the size of the measured cochlear-microphonic fundamental component (Eatock, Manley, and Pawson 1981). Since the relative sizes of such bidirectional hair-cell areas vary characteristically between lizard families (Chapter 23 by Miller), the bias introduced into the cochlear-microphonic sensitivity estimates could vary enormously (Manley 1981).

4.2 There Is a Fundamental Segregation of Low-Frequency and High-Frequency Processing

Despite the bewildering variety in the structure of lizard basilar papillae, one unifying principle has emerged from the functional studies: the processing of low frequencies below about 1 kHz is different from that of the higher frequencies even at the peripheral level of the sensory organ and its afferent nerve fibers (Manley 1990). This conclusion is based on the following general findings (summarized in Table 24.1).

4.2.1 Low and High Frequencies Are Represented on Different, Anatomically Distinct Segments of the Papilla

As summarized by Miller (Chapter 23), all lizards possess two different types of hair cells defined by their different innervation but also distinguishable on other, more variable morphological grounds, e.g., hair-cell orientation pattern or type of tectorial structure. These two types are also always confined to at least two different papillar segments. In the past, different papillar areas were described

TABLE 24.1. List of features that distinguish the two anatomically and physiologically different segments on lizard basilar papillae

Low-frequency segment (CFs below ~1 kHz)	High-frequency segment (CFs above ~1 kHz)
< 100–300 μm long	~ 200–2000 μm long sometimes divided
Unidirectional-type hair cells (afferent and efferent innervation)	Bidirectional-type hair cells (only afferent innervation)
Tectorial covering usually attached to neural limbus	Different specializations of tectorial covering or loss thereof
Anatomical gradients absent or only subtle	Pronounced anatomical gradients
Tonotopic organization across the width?	Tonotopic organization exponential along the length
Electrical fine tuning?	Predominantly micromechanical fine tuning
Good phase-locking of auditory nerve fibers	Inferior phase locking
Frequency-dependent pure-tone response pattern of auditory nerve fibers?	Often chopper-like pure tone response pattern

At present, physiological characteristics have to be generalized from studies in only a few selected lizard species (see text); statements based on information from only one species therefore bear a question mark.

on the basis of the orientation of the hair-cell bundles: In "unidirectional" areas, all hair cells have their bundles oriented in the same direction, whereas in "bidirectional" areas, two populations exist that have their bundles oriented at 180° to each other. The areas defined by hair-cell orientation are often, but not always, identical to the two hair-cell populations (Miller and Beck 1988). To preserve the continuity with the older nomenclature, Miller (Chapter 23) uses the terms "unidirectional type" and "bidirectional type" for the two different hair-cell types. In all lizard species in which the frequency organization of the basilar papilla is known, the low CFs (up to 1 kHz) and the higher CFs are represented separately in anatomically distinct papillar segments corresponding to the two hair-cell types (alligator lizard, Weiss et al. 1976; Holton and Weiss 1983b; granite spiny lizard, Turner, Muraski, and Nielsen 1981; Turner 1987; bobtail lizard, Köppl and Manley 1990a; lacertid lizards, Köppl and Manley 12th Midwinter Mtg Assoc Res Otolaryngol, p. 177, 1989; Bengal monitor lizard, Manley 1977). The low-frequency segment has much in common between the different lizard species studied and can, when compared to the turtle papilla (Sneary 1988a,b), be regarded as the more primitive segment. The hair cells are of the "unidirectional type", are most frequently unidirectionally oriented, and are almost always innervated by both afferent and efferent nerve fibers (Miller and Beck 1988). The tectorial structures overlying the hair cells in the low-frequency segment show some variation, but are generally attached to the neural limbus and show extensive connections to the microvilli of the intervening supporting cells. This segment is always relatively small (<100 to 300 μm long) and anatomical gradients within it are absent or only subtle (Turner, Muraski, and Nielsen 1981; Mulroy and Williams 1987; Köppl 1988). A tonotopic organization of the low-CF segment has only been revealed in the bobtail lizard (Köppl and Manley 1990a), where, curiously, it runs at 90° to that of the high-frequency segment, i.e., across the epithelium. Anatomical characteristics suggest the same for the alligator and granite spiny lizards (Turner, Muraski, and Nielsen 1981; Mulroy and Williams 1987), so this situation may be typical in lizards.

The high-frequency segment varies a lot more between the lizard species studied. Its length varies from about 200 to 2,000 μm. Most of the anatomical peculiarities found only in lizard papillae, such as the division of the epithelium into two completely separate subpapillae, the various types of tectorial structures unattached to the neural limbus or even hair cells without any tectorial covering, occur within this high-frequency segment. Common characteristics of the high-frequency segment are a uniform hair-cell type (the "bidirectional" type, Miller and Beck 1988) receiving only afferent innervation and the presence of pronounced anatomical gradients (Miller 1973; Mulroy 1974; Mulroy and Williams 1987; Turner 1987; Köppl 1988, 1989). Wherever a description of the tonotopic organization is available, it is exponential and oriented along the epithelium; however, the space constants of frequency vary widely with the size of the segment (from about 0.06 to 0.6 mm/octave; Manley 1990).

4.2.2 Are There Different Tuning Mechanisms for the Low- and High-Frequency Range?

4.2.2.1 The High-Frequency Range

To date, the mechanism of fine frequency tuning has only been demonstrated experimentally by direct observation of micromechanical resonance of individual hair-cell stereovillar bundles relative to their cell bodies in the alligator lizard (see Section 3.1.). However, the available evidence suggests similar tuning mechanisms, based on local micromechanical resonance, for the high-frequency range of other lizards as well:

(a) Mechanical measurements of basilar-membrane motion in both the alligator lizard and the bobtail lizard failed to reveal a sharply tuned travelling-wave motion (Peake and Ling 1980; Manley, Yates, and Köppl 1988). Judging from control data on inner-ear sensitivity, artefacts due to physiological insult were highly unlikely in both cases. We thus assume that in lizards generally, local mechanical resonances do not involve the basilar membrane or the whole cochlear partition.

(b) Calculations of theoretically expected resonance frequencies in the bobtail lizard and the lacertid lizards (Sections 3.2.3.2. and 3.3.3.2.), show that the anatomical substrate for a local mechanical resonance is present. Thus morphologically defined units made up of a number of hair-cells with their own portion of the tectorial-structure show calculated resonances in the audio range of the respective species.

(c) A comparison of the auditory nerve fiber Q_{10dB} across and within species reveals characteristic differences correlated with papillar anatomy. A very interesting case in this regard concerns the lacertid lizards, which show an additional morphological subdivision of the high-frequency range into two areas with quite different space constants of CF distribution (Section 3.3.). The general anatomical characteristics of the two papillar segments are the same (Köppl 1989), leaving the available space per octave as the essential difference correlated with a rather sudden increase of the Q_{10dB}-value of auditory nerve fibers over the transitional frequency range. If frequency selectivity is based on a local micromechanical resonance of small hair-cell/tectorial-structure units, lateral mechanical coupling via a continuous tectorial structure as in the lacertids would negatively influence the selectivity of the resonance of closely apposed units. The proposed tuning mechanism is thus consistent with the observed improvement of frequency selectivity based on the extended spatial frequency representation on the highest CF part of the papilla.

Another interesting evolutionary development in this regard is the tectorial sallets of skinks and geckos. The subdivided nature of this tectorial structure suggests a reduced lateral coupling, creating more discrete resonant units. The very high frequency selectivity seen in auditory nerve fibers of the Tokay gecko certainly supports this view. The somewhat lower, but still very good tuning of bobtail lizard nerve fibers correlates with the larger size and less regular arrangement of the sallets in this species compared with *Gekko* (Miller 1973; Köppl 1988). A curious specialization of papillar anatomy with regard to the frequency selectivity of micromechanical resonance is the complete loss of tectorial covering and concomitant great elongation of the hair-cell stereovillar bundles in some families. The average Q_{10dB}-values of alligator lizard and granite spiny lizard auditory nerve fibers innervating such areas are the lowest seen in all lizards in this high-frequency range (Fig. 24.8). Without the additional mass loading of a tectorial membrane, the stereovillar bundles have to be very long and relatively compliant to resonate mechanically in the same frequency range. This in turn makes it more difficult to overcome fluid damping, prohibiting a sharp resonance, and increases the thermal noise (e.g., Lewis et al. 1985).

It is a matter of debate as to whether passive mechanical resonance is possible at all in any of these ears (see e.g., Lewis et al. 1985; Patuzzi and Yates 1987; Freeman and Weiss 1988). In view of the presence of an active force-generating mechanism already present in the primitive turtle papilla, however, it is reasonable to suggest the existence of mechanical feedback in the lizards also, overcoming fluid damping. Active processes, whatever their exact basis may be, seem to be a universal characteristic of vertebrate inner ears, since otoacoustic emissions, often regarded as indirect

evidence of their existence, have been found wherever they have been looked for (review in Zurek 1985; Manley, Schulze, and Oeckinghaus 1987; Köppl and Manley Abstr 14th Midwinter Mtg Assoc Res Otolaryngol p. 83, 1991).

The hypothesis of a mechanical feedback into the movement of the stereovillar bundle is especially attractive in those lizards with a tectorial structure that is not attached to the neural limbus. Such tectorial structures (e.g., sallets) are equally free to move in the abneural and neural direction and any active feedback through the hair-cell bundles would reduce fluid damping and increase the resonance amplitude. The fact that the hair-cell bundles are of opposite polarity would presumably result in a mechanical input to the motion twice per cycle. In this respect, it is interesting to note that in lizard papillar segments that have tectorial structures not attached to the limbus, the hair-cell orientation pattern is always bidirectional. In addition, in such areas the relative numbers of hair cells of both polarities are roughly equal and each polarity is restricted to either the neural or abneural half of the papilla with little mixing around the midline. In other papillar segments where the tectorial structure is attached to the limbus (or is extremely large like the scincid culmen), hair-cell orientation is either unidirectional or irregularly bidirectional.

4.2.2.2 The Low-Frequency Range

Our knowledge of the tuning mechanisms operating in the low-frequency range of lizard basilar papillae are rather rudimentary, the evidence being either negative or indirect. However, there are several reasons to suspect that the mechanism or the relative importance of different mechanisms is different from that in the high-frequency range. Firstly, the absence of pronounced anatomical gradients implies that purely micromechanical mechanisms cannot contribute greatly to the frequency selectivity seen in low-CF auditory nerve fibers. Secondly, the Q_{10dB}-values of auditory nerve fibers always show a change in their dependence on CF at around CF 1 kHz. Thirdly, even though the tonotopic organization within the low-frequency segment is mostly unknown, it is clear that the frequency representation in terms of the available space/octave must always be more compressed than in the high-frequency segment. A change towards very small frequency space constants in the low-frequency range has also been observed in birds and was interpreted as consistent with an electrical tuning mechanism in the low-frequency range (Manley, Brix, and Kaiser 1987; Manley et al. 1988; Gleich 1989). In contrast to almost all mechanically tuned systems, the electrical resonance properties of a single hair cell are independent of those of its neighbors and may therefore vary rapidly along the epithelium without detrimental effects on frequency selectivity. The small space constant of frequency in the turtle basilar papilla is consistent with this view.

The phenomenon of preferred intervals in the spontaneous discharge of auditory nerve fibers also provides evidence for electrical resonance as a tuning mechanism. Although such preferred intervals have so far only been demonstrated in low-CF fibers of one lizard species (the Tokay gecko), systematic studies concerning this phenomenon are mostly lacking. However, the occurrence of preferred intervals would be dependent on the innervation pattern, which differs greatly across lizard families (Miller and Beck 1988). Further investigation is thus necessary to clarify whether electrical resonance is a universal tuning mechanism in the low-frequency range of lizards.

4.2.3 The Stimulus Coding Properties of Low- and High-CF Auditory Nerve Fibers are Different

Based on the poorer phase-locking ability of high-CF fibers compared to low-CF fibers in the alligator lizard, Rose and Weiss (1988) suggested a functional segregation of auditory nerve fibers into a low-frequency "timing" and a high-frequency, "nontiming" pathway. More extensive measurements in the bobtail lizard (Manley et al. 1990) and indirect evidence in the Tokay gecko (Eatock, Manley, and Pawson 1981) showed the same basic difference in the phase-locking behavior of low- and high-CF fibers in these species.

In addition, the temporal response patterns of low- and high-CF fibers to pure tones in the bobtail lizard revealed a fundamental difference; high-CF fibers showed a poorer time-resolution capability (Köppl and Manley 1990b). Although a comparable analysis of discharge patterns to pure-tone stimulation does not exist for any other lizard species, the available information on the Bengal monitor lizard (Manley

1977), the Tokay gecko (Eatock, Manley and Pawson 1981) and the alligator lizard (Turner 1980; Eatock and Weiss Abstr. 9th Midwinter Mtg. Assoc Res Otolaryngol, 1986) is consistent with the observations in the bobtail lizard. This suggests that firstly, a difference between low- and high-CF fibers with regard to the temporal response pattern is typical. Secondly, the characteristics of both low- and high-CF fiber responses are very similar between lizard species; the poorer time-resolution capability of high-CF fibers may in all cases be due to increased temporal summation due to a long membrane time constant in the nerve fiber (see Section 3.2.3).

This fairly uniform behavior of nerve fibers across species is remarkable in view of the widely varying innervation pattern of the different papillae (Miller and Beck 1988). It implies that a segregation into different neuronal pathways encoding different parameters of auditory stimuli is of fundamental importance to sound analysis. A specialization into two such pathways has been demonstrated at different brain levels in the barn owl (Konishi, Sullivan, and Takahashi 1985) and may, indeed, be a common feature of the auditory pathway of amniotes (Carr, Chapter 25). In lizards, however, the specialization is already present at the level of the auditory nerve. As pointed out by Manley (1981), these specializations result in peripheral filtering playing a more important role in the auditory system of reptiles than it does in that of birds and mammals. Peripheral filtering is manifested, for example, in the variety of response patterns found in auditory nerve fibers. The high temporal synchronization in the neuronal "chopper" discharges in high-CF fibers may also play an important role in improving sound localization in these animals that have only small heads, hear only relatively low frequencies, and have no external ear (Eatock, Manley, and Pawson 1981; Manley 1981).

5. Implications for the Evolution of Lizard Hearing

5.1 Differences Between the Hearing of the Turtle and of Lizards

Having described the primitive hearing of the turtle and the hearing of the modern lizards, we finally ask the question: What important changes have taken place in the course of evolution? One obvious change is the significant extension of the upper frequency limit of hearing through the addition of a special group of hair cells analyzing higher frequencies. Whereas the red-eared turtle has its auditory response limit below 1 kHz, the lizards have very sensitive hearing up to 3 to 5 kHz and an upper physiological limit of almost 10 kHz. This difference is far too great to be explained by temperature differences in experiments on different species, although temperature is known to affect the frequency and sensitivity of hearing in nonmammals (Campbell 1969; Eatock and Manley 1981; Smolders and Klinke 1984; Schermuly and Klinke 1985). Interestingly, the only lizard species (the Bengal monitor) with a somewhat lower upper hearing limit than most lizards (see Section 4.1) belongs to a family that is regarded as one of the most primitive among the lizards investigated neurophysiologically (Chapter 23 by Miller). This may be an indication that anatomically more primitive lizard papillae still represent an early stage in the evolutionary trend towards higher frequency hearing within the lizards.

The many differences seen between sensory processing in the low and high frequency ranges in lizards suggest that the expansion of the frequency range was not simply a matter of adapting already existing mechanisms. Almost certainly, new mechanisms were developed or previously unimportant capabilities of hair cells were greatly enhanced to provide the basis for sensitive, frequency-selective responses above 1 kHz. As outlined in Section 4.2, there is evidence that the frequency tuning mechanisms are different in the low- and high-frequency range in lizards, respectively. An electrical resonance, similar to that found in turtle hair cells, probably influences tuning in the low-frequency range. In the high-frequency range, all available information points to a predominantly micromechanically resonant system, involving the stereovillar bundles and, if present, the tectorial structure. The details of exactly how resonant units are created and how a gradient in resonance frequencies is achieved varies with papillar structure across families. We conclude that the bewildering anatomical variety of the high-frequency papillar segment in lizards merely represents variations on a common theme.

5.2 Anatomical Characteristics and the Origin of the High Frequency Papillar Segment in Lizards

As has been outlined, in Section 4.2.1., the low- and high-CF segments of the lizard papilla are anatomically distinct and the structure of the low-frequency segment is more conservative across lizard families than that of the high-frequency segment. In addition, structural characteristics identify the low-frequency segment as more similar to the turtle papilla and therefore more primitive. The papilla of the two most primitive modern lizard families, the Teiidae and Varanidae (Chapter 23 by Miller), has a relatively large low-frequency segment flanked apically and basally by equally sized or smaller high-frequency segments (for the Teiidae, this is inferred from cytological characteristics, since there are no physiological data available). Interestingly, a possible precursor of this condition may already be seen in primitive papillae, such as those of turtles and *Sphenodon*. At the extreme apical and basal end of the turtle basilar papilla, and overlying the limbus, are hair-cell areas which form a second structurally distinct population (Wever 1978; Sneary 1988a,b; Chapter 23 by Miller). The characteristics of these areas distinguishing them from the rest of the papilla are all features considered to be derived (Chapter 23 by Miller), such as a more variable hair-cell orientation pattern, smaller hair-cell size, increased branching of the innervating afferent nerve fibers and longer stereovillar bundles with an increased internal height gradient. Unfortunately, the physiology of these "other" areas of the turtle basilar papilla is unknown, since physiological experiments (Section 2) involved only the hair cells overlying the free basilar membrane. It is tempting to speculate that these areas at the extreme ends of the papilla are the precursors of the high-frequency segments of the lizards and the large central basilar-membrane area is the precursor of the low-frequency segment. Physiological evidence for electrical tuning in the low-CF range of lizards (Section 4.2.2.2) is consistent with this anatomical analogy. If this is true, then the lizards have greatly expanded and refined the terminal regions as high-frequency segments, while also reducing the relative importance of the central low-frequency segment. Miller (Chapter 23) has proposed a detailed scheme by which the different papillae of more advanced modern lizard families may be derived from the primitive arrangement seen in the Varanidae and Teiidae.

5.3 Possible Causes of the Structural Variation in Lizards

Having established that the function and the underlying basic mechanisms differ only to a minor extent between lizard families, one may look for the factors determining the development of so many different structural patterns. One important factor may be their independent evolution along several different lines. Unfortunately, fossil records during the time of emergence of the modern lizard families are virtually lacking, so that very little is known about the specific interrelationships between families (Carroll 1987). It is still a matter of debate whether lizards are mono- or polyphyletic (e.g., Caroll 1987; Estes, de Queiroz, and Gauthier 1988). Phylogenetic trees derived from investigations of various different anatomical parameters (summarized by Miller, Chapter 23) suggest slightly different evolutionary relationships. Studies not considering inner-ear structure generally yield family groups within which considerable variation in papillar structure exists. Miller's scheme based on basilar-papillar structure alone shows of course more consistency of structure along the suggested evolutionary lines. The fact that the papilla of snakes is relatively primitive and their hearing range is low (Wever 1978) would suggest that the high-frequency papillar areas were not strongly differentiated before the lepidosaur radiation when the family groups of lizards diverged from each other. This indicates that some of the diversity of the high-frequency segments is due to independent evolution.

Another approach to explaining structural variation is to look for differences in the auditory performance of different species or in the importance of the acoustic sense for behavior. Such correlations may provide important clues to possible selection pressures leading to the development of different inner-ear structures. Unfortunately, the behavior of lizards indicates a rather inferior importance of the acoustic sense in the daily life of most species. The geckos are a prominent exception, producing vocalizations associated with

mating behavior and territorial defense (Marcellini 1977; Werner, Frankenberg, and Adar 1978). There is also evidence for acoustic prey localization in geckos (Sakaluk and Belwood 1984). It is reasonable to assume that the nocturnal lifestyle of most members of this family enhanced the importance of auditory cues, so that the structure of the basilar papilla of geckos may be regarded as an adaptation to this increased demand. The large gecko papilla shows a very high degree of structural specialization and regularity, particularly in the high-frequency segment (Miller 1973; we infer the frequency organization from cytological characteristics). As outlined in Section 4.2.2.1., this structurally precise arrangement correlates with the exceptionally high frequency selectivity of gecko auditory nerve fibers (Eatock, Manley, and Pawson 1981). Thus in this lizard family, the relative greater importance of the acoustic sense is correlated with a better-than-average auditory performance at the level of the auditory nerve. As far as other lizard families are concerned, however, we know of only one successful psychophysical study, roughly evaluating the hearing range of lacertid lizards using a moderate arousal response (Berger 1924). With the exception of the Tokay gecko, the auditory performance as seen in the frequency selectivity, the sensitivity, or temporal discharge patterns of single auditory nerve fibers varies only mildly between the lizard species investigated and also does not indicate any extraordinary specializations in certain families. One important difference between lizard families is seen in the size of the basilar papilla and thus the number of hair cells and auditory nerve fibers. A larger number of sensory cells and nerve fibers for the same overall hearing range means more channels for parallel processing in the auditory pathway. However, lizards at least partly "destroy" this effect by the increased convergence of hair-cell input onto the nerve fibers and therefore a reduced nerve-fiber/hair-cell ratio in larger papillae as compared to small papillae (Miller 1985; Miller and Beck 1988).

We thus conclude that the acoustic sense in lizards mainly serves general arousal and orientation functions. This implies that there were no strong selection pressures during the evolution of lizard hearing, shaping auditory performance to particular and important tasks. A general simultaneous trend towards expansion of the frequency range to higher frequencies led to the establishment of a variety of slightly different solutions to this problem, whereby in all cases important functional properties were preserved.

6. Summary

1. With the exception of the relatively primitive papilla of the monitor lizard, hearing ranges and sensitivities do not differ significantly between lizard species despite great differences in size and structure.

2. All lizards have a structurally and functionally divided papilla. Low- and high-frequency regions (below and above CF 1 kHz) show a high degree of physiological differentiation involving different frequency-tuning mechanisms and peripheral filtering of important acoustic cues.

3. The low-frequency segment resembles the turtle papilla, shows few anatomical gradients, and probably has electrically tuned hair cells; their afferent fibers respond tonically to pure tones at and below CF and show good stimulus time resolution.

4. The high-frequency segment accounts for most of the characteristic structural variability of the lizard papilla. It always shows pronounced anatomical gradients consistent with primarily micromechanical tuning with or without the aid of a tectorial structure. Active hair-cell motions are probably involved in mechanical resonance. Their fibers respond with a "chopper" pattern and have a lower high-frequency limit to phase locking than low-CF fibers.

5. Compared to the primitive turtle papilla, the lizards have added or extended existing apical and/or basal hair-cell areas to process sound of frequency above 1 kHz. The available evidence indicates that some of the structural diversity of the high-CF region is due to independent evolution. However, the structural diversity is not accompanied by an equivalent functional diversity. Indeed, the preservation of similar physiological response patterns points to the importance of the coding of certain parameters that allow the basic functions of detection and localization of sound stimuli and, in some cases, acoustic communication.

References

Art JJ, Fettiplace R (1984) Efferent desensitization of auditory nerve fibre responses in the cochlea of the turtle *Pseudemys scripta elegans*. J Physiol 356:507–523.

Art JJ, Fettiplace R (1987) Variation in membrane properties in hair cells isolated from the turtle cochlea. J Physiol 385:207–242.

Art JJ, Fettiplace R, Fuchs PA (1984) Synaptic hyperpolarization and inhibition of turtle cochlear hair cells. J Physiol 356:525–550.

Art JJ, Crawford AC, Fettiplace R, Fuchs PA (1985) Efferent modulation of hair cell tuning in the cochlea of the turtle. J Physiol 360:397–421.

Art JJ, Crawford AC, Fettiplace R (1986) Electrical resonance and membrane currents in turtle cochlear hair cells. Hearing Res 22:31–36.

Berger K (1924) Experimentelle Studien über Schallperzeption bei Reptilien. Z vergl Physiol 1:517–540.

Campbell HW (1969) The effects of temperature on the auditory sensitivity of lizards. Physiol Zool 42: 183–210.

Carroll RL (1987) Vertebrate Palaeontology and Evolution. New York: Freeman.

Crawford AC, Fettiplace R (1980) The frequency selectivity of auditory nerve fibres and hair cells in the cochlea of the turtle. J Physiol 306:79–125.

Crawford AC, Fettiplace R (1981) An electrical tuning mechanism in turtle cochlear hair cells. J Physiol 312:377–412.

Crawford AC, Fettiplace R (1985) The mechanical properties of ciliary bundles of turtle cochlear hair cells. J Physiol 364:359–379.

Eatock RA, Manley GA (1981) Auditory nerve fibre activity in the tokay gecko: II: Temperature effect on tuning. J Comp Physiol A 142:219–226.

Eatock RA, Manley GA, Pawson L (1981) Auditory nerve fibre activity in the tokay gecko: I: Implications for cochlear processing. J Comp Physiol A 142: 203–218.

Estes R, de Queiroz K, Gauthier J (1988) Phylogenetic relationships within squamata. In: Estes R, Pregill G (eds) Phylogenetic relationships of the lizard families. Palo Alto: Stanford Univ Press.

Freeman DM, Weiss TF (1988) The role of fluid inertia in mechanical stimulation of hair cells. Hearing Res 35:201–208.

Frischkopf LS, DeRosier DJ (1983) Mechanical tuning of free-standing stereociliary bundles and frequency analysis in the alligator lizard cochlea. Hearing Res 12:393–404.

Gleich O (1989) Auditory primary afferents in the starling: Correlation of function and morphology. Hearing Res 37:255–268.

Holton T (1980) Relations between frequency selectivity and two-tone rate suppression in lizard cochlear-nerve fibres. Hearing Res 2:21:38.

Holton T, Weiss TF (1983a) Receptor potentials of lizard cochlear hair cells with free-standing stereocilia in response to tones. J Physiol 345:205–240.

Holton T, Weiss TF (1983b) Frequency selectivity of hair cells and nerve fibres in the alligator lizard cochlea. J Physiol 345:241–260.

Hudspeth AJ, Holton T (1986) Micromechanical properties of the alligator lizard's basilar papilla contribute to frequency tuning. Hearing Res 22:93.

Konishi M, Sullivan WE, Takahashi T (1985) The owl's cochlear nuclei process different sound localization cues. J Acoust Soc Amer 78:360–364.

Köppl C (1988) Morphology of the basilar papilla of the bobtail lizard Tiliqua rugosa. Hearing Res. 35:209–228.

Köppl C (1989) Struktur und Funktion des Hörorgans der Echsen: Ein Vergleich von Lacertidae und Scincidae. Doctoral dissertation in zoology, Technical University of Munich.

Köppl C, Manley GA (1990a) Peripheral auditory processing in the bobtail lizard Tiliqua rugosa: II. Tonotopic organization and innervation pattern of the basilar papilla. J Comp Physiol A 167:101–112.

Köppl C, Manley GA (1990b) Peripheral auditory processing in the bobtail lizard Tiliqua rugosa: III. Patterns of spontaneous and tone-evoked nerve-fibre activity. J Comp Physiol A 167:113–127.

Lewis ER, Leverenz EL, Bialek WS (1985) The vertebrate inner ear. Boca Raton: CRC Press.

Manley GA (1976) auditory responses from the medulla of the monitor lizard Varanus bengalensis. Brain Res 102:329–334.

Manley GA (1977) Response patterns and peripheral origin of auditory nerve fibres in the monitor lizard, Varanus bengalensis. J Comp Physiol A 118:249–260.

Manley GA (1979) Preferred intervals in the spontaneous activity of primary auditory neurones. Naturwiss 66:582.

Manley GA (1981) A review of the auditory physiology of the reptiles. Prog Sens Physiol 2:49–134.

Manley GA (1990) Peripheral hearing mechanisms in reptiles and birds. Heidelberg, New York: Springer-Verlag.

Manley GA, Brix J, Kaiser A (1987) Developmental stability of the tonotopic organization of the chick's basilar papilla. Science 237:655–656.

Manley GA, Schulze M, Oeckinghaus H (1987) Otoacoustic emissions in a song bird. Hearing Res 26:257–266.

Manley GA, Yates G, Köppl C (1988) Auditory peripheral tuning: evidence for a simple resonance phenomenon in the lizard Tiliqua. Hearing Res 33:181–190.

Manley GA, Brix J, Gleich O, Kaiser A, Köppl C, Yates G (1988) New aspects of comparative peripheral auditory physiology. In: Syka J, RB Masterton (eds) Auditory Pathway— Structure and Function. London, NY: Plenum Press, pp. 3–12.

Manley GA, Köppl C, Yates GK (1989) Micromechanical basis of high-frequency tuning in the bobtail lizard. In: Wilson JP, Kemp DT (eds) Cochlear mechanism. New York: Plenum Press.

Manley GA, Köppl C, Johnstone BM (1990) Peripheral auditory processing in the bobtail lizard Tiliqua rugosa: I. Frequency tuning of auditory-nerve fibres. J Comp Physiol A 167:89–99.

Manley GA, Yates GK, Köppl C, Johnstone BM (1990) Peripheral auditory processing in the bobtail lizard Tiliqua rugosa: IV. Phase-locking of auditory-nerve fibres. J Comp Physiol A 167:129–138.

Marcellini D (1977) Acoustic and visual display behavior of gekkonid lizards. Amer Zool 17:251–260.

Miller MR (1973) A scanning electron microscope study of the papilla basilaris of Gekko gecko. Z Zellforsch 136:307–328.

Miller MR (1974) Scanning electron microscope studies of some skink Papillae basilares. Cell Tiss Res 150:125–141.

Miller MR (1978) Further scanning electron microscope studies of lizard auditory papillae. J Morphol 156:381–418.

Miller MR (1980) The reptilian cochlear duct. In: Popper AN, Fay RR (eds) Comparative studies of hearing in vertebrates. New York, Heidelberg, Berlin: Springer-Verlag, pp. 169–204.

Miller MR (1985) Quantitative studies of auditory hair cells and nerves in lizards. J Comp Neurol 232:1–24.

Miller MR, Beck J (1988) Auditory hair cell innervational patterns in lizards. J Comp Neurol 271:604–628.

Mulroy MJ (1974) Cochlear anatomy of the alligator lizard. Brain Beh Evol 10:69–87.

Mulroy MJ, Williams RS (1987) Auditory stereocilia in the alligator lizard. Hearing Res 25:11–21.

Oertel D (1985) Use of brain slices in the study of the auditory system: spacial and temporal summation of synaptic inputs in cells in the anteroventral cochlear nucleus of the mouse. J Acoust Soc Amer 78:328–333.

Patterson WC (1966) Hearing in the turtle. J Aud Res 6:453–464.

Patuzzi RB, Yates GK (1987) The low-frequency response of inner hair cells in the guinea-pig cochlea: implications for fluid coupling and resonance of the stereocilia. Hearing Res 30:83–98.

Peake WT, Ling A (1980) Basilar-membrane motion in the alligator lizard: its relation to tonotopic organization and frequency selectivity. J Acoust Soc Amer 67:1736–1745.

Rose C, Weiss TF (1988) Frequency dependence of synchronization of cochlear nerve fibres in the alligator lizard: Evidence for a cochlear origin of timing and nontiming neural pathways. Hearing Res 33:151–166.

Sakaluk SK, Belwood JJ (1984) Gecko phonotaxis to cricket calling song: a case of satellite predation. Animal Behaviour 32:659–662.

Schermuly L, Klinke R (1985) Change of characteristic frequencies of pigeon primary auditory afferents with temperature. J Comp Physiol A 156:209–211.

Smolders JWT, Klinke R (1984) Effects of temperature on the properties of primary auditory fibres of the spectacled caiman, *Caiman crocodilus* (L.). J Comp Physiol 155:19–30.

Sneary MG (1988a) Auditory receptor of the red-eared turtle: I. General ultrastructure. J Comp Neurol 276:573–587.

Sneary MG (1988b) auditory receptor of the red-eared turtle: II. Afferent and efferent synapses and innervation pattern. J Comp Neurol 276:588–606.

Turner RG (1980) Physiology and bioacoustics in reptiles. In: Popper AN, Fay RR (eds) Comparative studies of hearing in vertebrates. New York, Heidelberg, Berlin: Springer-Verlag, pp. 205–237.

Turner RG (1987) Neural tuning in the granite spiny lizard. Hearing Res 26:287–299.

Turner RG, Muraski AA, Nielsen DW (1981) Cilium length: Influence on neural tonotopic organization. Science 213:1519–1521.

Weiss TF, Leong R (1985) A model for signal transmission in an ear having hair cells with free-standing stereocilia. III. Micromechanical stage. Hearing Res 20:157–174.

Weiss TF, Mulroy MJ, Turner RG, Pike CL (1976) Tuning of single fibres in the cochlear nerve of the alligator lizard: relation to receptor morphology. Brain Res 115:71–90.

Werner YL, Frankenberg E, Adar O (1978) Further observations on the distinctive vocal repertoire of *Ptyodactylus hasselquistii CF. hasselquistii* (Reptilia: Gekkonidae). Israel J Zool 27:176–188.

Wever EG (1967) The tectorial membrane of the lizard ear: types of structure. J Morphol 122:307–320.

Wever EG (1978) The Reptile Ear. Princeton NJ Princeton Univ Press.

Wever EG, Vernon JA, Crowley DE, Peterson, EA (1965) Electrical output of lizard ear: relation to hair-cell population. Science 150:1172–1174.

Zurek PM (1985) Acoustic emissions from the ear: A summary of results from humans and animals. J Acoust Soc Am 78:340–344.

25
Evolution of the Central Auditory System in Reptiles and Birds

Catherine E. Carr

1. Introduction

Birds and reptiles are considered together in a single chapter because of their close phylogenetic relationship (Fig. 25.1). Despite differences in their auditory periphery, the central auditory pathways in the birds and reptiles are very similar. Furthermore, the different reptile groups (turtles, lizards, and crocodilians) make suitable outgroups for evolutionary comparisons with each other and with birds, enabling us to identify both ancestral and derived features of the auditory system.

Auditory stimuli contain information about the frequency, loudness, and phase or timing of the auditory signal. These variables appear to be encoded in the central nervous system in a similar way in both birds and reptiles. Nevertheless, many apparently homologous nuclei have been given different names in birds and reptiles. In this chapter I shall summarize the homologies suggested in the literature (Table 25.1). We do not yet have sufficient information to establish a unified nomenclature.

In the basic pattern or morphotype, the auditory nerve projects to two cochlear nuclei, the nucleus magnocellularis and the nucleus angularis (Fig. 25.2). These nuclei project to second-order olivary nuclei, to lemniscal nuclei, and to the auditory midbrain (also known as the torus semicircularis, or the nucleus mesencephalicus dorsalis, or the inferior colliculus). The central nucleus of the auditory midbrain projects to a major thalamic target (termed the nucleus medialis or reuniens in reptiles, and the nucleus ovoidalis in birds). The auditory thalamus projects to the striatum and to a specialized auditory region of the forebrain, the medial dorsal ventricular ridge in reptiles (DVR; Field L in birds). There have been many independent developments within this basic pattern, and some nuclei may more correctly be described as field homologues in comparisons between turtles, lizards, crocodilians, and birds. Different groups exhibit different specializations. Most birds are notable auditory specialists, while only some reptiles such as the crocodilians and the geckos are. Examples of specialization for sound localization may be found in nocturnal predators such as the owl, while song birds and geckos have developed specializations for acoustic communication.

2. Phylogenetic Considerations

The phylogeny of the birds and reptiles has been discussed in Chapter 27 by Manley and Gleich, and by Ulinski and Margoliash (1990). Amniotes are thought to be a monophyletic group that evolved from a single stock of primitive tetrapods during the early Carboniferous. They diverged into two major lineages, the Sauropsida and Synapsida. The Sauropsida are a paraphyletic group made up of Chelonia (turtles and tortoises), Lepidosauria (lizards and snakes), Crocodilia (crocodiles and alligators), and Aves (birds). The Synapsida include the modern mammals (see Chapter 28, Allin and Hopson). The reptiles are phylogenetically heterogeneous (Fig. 25.1). For convenience, some gradistic terms will be used throughout this chapter e.g., reptiles, turtles, lizards, birds, etc. The species name of the animal will also be given where possible.

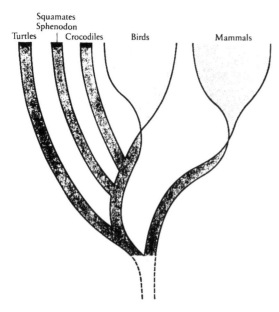

FIGURE 25.1. Simplified phylogeny of amniotes. The class Reptilia includes the three darkly stippled lineages, labeled here as turtles, squamates (including lepidosaurs), and crocodiles. Reptiles share a common ancestry but the Class Reptilia is considered paraphyletic since it does not include all reptilian descendents. Birds and mammals can each be defined on the basis of unique, shared derived characters, leaving reptiles defined as amniotes that lack the specialized characters of birds and mammals. (From Vertebrate Paleontology & Evolution. By R.L. Carroll 1987. Copyright © 1988 by W.H. Freeman and Company. Reprinted with permission.)

2.1 Reptiles

The turtles and tortoises are often considered the most primitive of the modern reptiles, principally because of the lack of temporal openings in their skulls. If the auditory system of turtles is also primitive, comparisons between turtles and other reptiles and birds may be useful in identifying homologous features of the central auditory pathways. The other sauropsids are a large and heterogeneous group: lepidosaurs, crocodiles, and birds. The lepidosaurs are represented by the primitive *Sphenodon* of New Zealand, and by lizards and snakes. The central auditory pathways of the snakes and lizards differ somewhat from those of the turtles, birds and crocodilians. Miller (1975; Chapter 23) has proposed that these differences

represent derived features or independent development (see below). The ancestral lizard morphotype is believed to be similar to that of *Sphenodon* and the turtles.

Crocodilians and birds are more closely related to each other than to the lepidosaurs. During the Triassic, the subclass Archosauria underwent a major radiation [see Hopson (1977) for review]. The basal archosaur group, the order Thecodontia, radiated to produce four orders, the Saurischia and Ornithischia (together known as dinosaurs), the Pterosauria and the Crocodilia. The birds are an offshoot of the dinosaurs, but are placed in their own class, Aves. Crocodilians therefore share a more recent common ancestor with birds than with other living reptiles (Fig. 25.1). Comparisons of the central auditory pathways of modern crocodilians and birds reveal a number of shared characters. It may be reasonably assumed that these characters may also have been found in the dinosaurs (Hopson 1977, 1980).

2.2 Birds

Birds evolved from an archosaur common ancestor in the Mesozoic (for review see Ulinski and Margoliash 1990) and representatives of most modern orders of birds were present at the beginning of the Cenozoic, about 65 M years ago. Bird phylogeny is still somewhat controversial. The modern arrangement of birds reviewed in Carroll (1987) is as follows: There are two superorders, the Plaeognathae or ratites, and the Neognathae. The paleognaths are regarded as primitive. They are generally flightless birds, and include such birds as the ostrich. Little is known about the paleognath auditory system. The neognaths are divided into three groups, the so-called basal landbirds (e.g., chickens, guinea fowls, and pigeons), the so-called higher landbirds (owls, Passeriformes, or songbirds) and the waterbirds (gulls, ducks). Despite the nomenclature, there is little evidence that any one group is more primitive than another. Most modern basal landbird families were present by the end of the Eocene. Most higher land birds arose at the same time as the basal land birds, although the passerine birds only arose in the late Oligocene. Waterbirds arose in the mid Miocene (Carroll 1987).

3. Hearing Range

3.1 Auditory Sensitivity

There are few behavioral studies on the auditory sensitivity of reptiles, but hearing capabilities may be inferred from a number of physiological studies. Behavioral audiograms from the red-eared turtle, *Chrysemys scripta*, showed sensitivity to low-frequency sounds with best sensitivity between 200 and 600 Hz (Patterson 1966). Similar results have been obtained from cochlear microphonics and auditory nerve recordings (Wever 1978; Crawford and Fettiplace 1980; Manley 1990; see Köppl and Manley, Chapter 24).

Wever and his colleagues have carried out extensive cochlear microphonic studies that describe the hearing range of lizards, snakes, and crocodilians (Wever 1978). The primitive relative of the lizards, *Sphenodon*, appears to have similar auditory sensitivity to the red-eared turtle, and a similar cochlea (Gans and Wever 1976; Wever 1978). The sensitivity of the lizard auditory system varies among different groups. The range is from 100 to 5,000 Hz, with the greatest sensitivity between 100 and 3,000 Hz. Among the lizards, geckos have particularly sensitive hearing, and produce species-specific vocalizations. Snakes hear low-frequency sound. Crocodilians produce complex vocalizations (see Hopson 1977), and have large, well-developed ears with best sensitivity between 200 and 2,000 Hz (Wever 1978).

There have been numerous studies of the auditory sensitivity of birds. Audiograms reviewed in Fay (1988) for about 22 birds species show best sensitivities from about −10 to 10 dB SPL in the frequency range between 1 and 3 kHz. The barn owl is an exception to this rule, and displays optimal sensitivity up to 6 to 7 kHz, with a frequency range out to 12 kHz (Konishi 1973b). Studies of hearing thresholds reveal a high-frequency limit of bird hearing of about 10 kHz, with the limit imposed by middle ear function and cochlear mechanisms (Sachs, Sinnott, and Heinz 1978). Fay states that optimal intensity discrimination thresholds for birds are within the range for fishes and mammals, as are frequency discrimination thresholds (Fay 1988; Dooling et al. 1990). Bird intensity-difference limens are worse than those of mammals, although there appear to be few

MORPHOTYPE

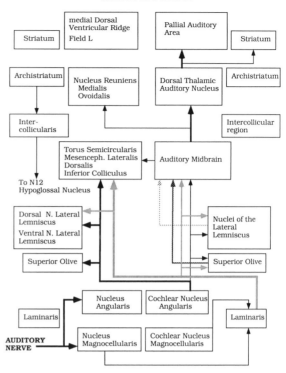

FIGURE 25.2. Proposed plesiomorphic pattern or "morphotype" of the central auditory pathways of reptiles and birds. Both the nucleus angularis and the nucleus laminaris form bilateral ascending projections to the lemniscal nuclei and the central nucleus of the auditory midbrain. This central nucleus projects bilaterally to the auditory thalamus that in turn projects to the auditory region of the dorsal pallium and the striatum. N12, hypoglossal nucleus. Note: In most cases, boxes on the right half give generalized names of auditory structures, while boxes on the left half give specific names or synonyms for these structures. Also note that not all connections have been demonstrated in all species.

differences in peripheral processing between birds and mammals to account for these differences (Sachs et al. 1978).

3.2 Sound Localization

Animals use both interaural time and intensity differences for sound localization. These acoustic cues are frequency dependent. Large interaural intensity differences are most significant at high frequencies because of the scattering properties of the head. When the wavelength of sound becomes longer than the head, then the sound shadow generated by the head decreases. Since most birds and reptiles do not hear well at frequencies above 2 to 5 kHz, interaural intensity differences may not be as useful as the interaural time differences generated when sounds travel over different path lengths to the two tympanic membranes. Pressure gradient mechanisms, however, may be used by some birds to enhance the interaural intensity information generated by low-frequency sounds. Birds appear to use both interaural intensity and interaural time differences for sound localization, while there are little data on sound localization in any reptile.

Crocodilians have the same feeding efficiency in clear and turbid water (Schaller and Crawshaw 1982), and a recent study of *Caiman crocodilus* showed that they do not require vision for underwater prey capture (Fleishman and Rand 1989). In this study, an experiment with a piece of fish showed that motion was required before the food was snapped at, suggesting a role for mechanoreception. It is not clear if crocodilians also use sound localization for prey capture. Wever and Vernon (1957) showed that transmission of sounds through the interaural canal resulted in negligible interaural intensity differences, but suggested that interaural time differences may be useful for sound localization. Because the speed of sound is much greater in water than in air, interaural time differences would be small in water. It would be interesting to find out if crocodilians detect interaural time differences in water or air, and if the auditory system recalibrates as the crocodilian moves from water to air.

In birds, auditory localization is acute, with minimal audible angles of about 2 to 20° (reviewed in Fay 1988). Any degree of acuity is remarkable for two reasons. Birds' heads are small; thus interaural time differences are small, and they do not hear well at high frequencies; thus, interaural intensity differences are also small (Klump, Windt, and Curio 1986; Manley 1990). Birds exhibit a number of adaptations that improve sound localization, however. In many birds, the interaural canal allows for phase interactions that increase the effective separation of the two ears to produce larger interaural time differences than predicted from the head size (Calford and Piddington 1988).

Barn owls are an exception to this rule. They have large heads, and their interaural canals are ineffectual at the high frequencies used for sound localization (Moiseff and Konishi 1981). They have evolved specializations that enable them to localize sound with great accuracy (Payne 1971; Konishi 1973b; see Section 5.4.1).

3.3 Song and Vocalization

Many reptiles produce sounds of some kind (hissing, roaring, etc.), but few make true vocalizations. Some turtles have a vocal repertoire (Gans and Maderson 1973). The geckos appear to be the only lepidosaurs to produce sounds with the consistent form and pattern to convey an intraspecific message and to be described as a vocalization (see review in Marcellini 1978). The most vocal gecko, *Gekko gekko*, produces calls with a complex harmonic structure and a fundamental frequency of about 375 to 450 Hz (Marcellini 1978; Dodd and Capranica 1989). Crocodilians may produce vocalizations, including bellows, pips, and alarm calls (references in Hopson 1977).

Birds are specialized for both the perception and production of sounds. All birds vocalize, producing species-specific calls that are used for both species and individual recognition, and for social communication. The term birdsong is reserved for passerine birds that produce song with complex temporal and spectral patterns (Konishi 1985). The nature of birdsong suggests that songbirds may have particularly acute temporal and spectral discrimination (Marler 1982; Marler and Peters 1981; Dooling et al. 1990). Many psychophysical studies have shown that birds have good sensitivity in frequency and time analysis (reviewed in Fay 1988).

4. The Central Auditory System: Basic Reptilian Plan

4.1 Chelonia (Turtles and Tortoises)

Since the central nervous system of the turtle is considered to be primitive, the organization of the central auditory pathways of turtles is of considerable interest (Fig. 25.3). In the mud turtle, *Kinosternon leucostomum*, the auditory nerve projects to the two cochlear nuclei, a large nucleus magnocellularis and a small, more rostrally located nucleus angularis (Miller and Kasahara 1979). The nucleus magnocellularis contains both large round neurons with few dendrites (bushy cells) and stellate cells (Browner and Marbey 1988). The bushy cells receive endbulbs of Held terminals from the auditory axons and project to the nucleus laminaris in a pattern that is found in most reptiles and in all birds (Browner and Marbey 1988). The nucleus laminaris is not the large, well-developed nucleus found in crocodilians and birds, but is a mass of fusiform cells that lies just ventral and lateral to the nucleus magnocellularis (Miller and Kasahara 1979). Similarly, neither the nucleus angularis nor the nucleus magnocellularis is as well developed in turtles as in the other reptiles. The nucleus angularis is poorly defined in contrast to that of lizards and crocodilians. The nucleus magnocellularis is a single nucleus, and is not subdivided into the medial and lateral cell groups present in lizards (Miller and Kasahara 1979; see below).

The central connections of the cochlear nucleus complex in the turtle, *Pseudemys scripta elegans*, have been described by Künzle (1986). Because injections of tracer were not confined to the cochlear nuclei or nucleus laminaris alone, the projections have been referred to as belonging to the cochlear nucleus complex (Fig. 25.3). There is every reason to suppose that there are separate projections from the different cochlear nuclei, but this hypothesis needs to be tested. The cochlear nucleus complex forms a predominantly contralateral projection to the superior olive, the nuclei of the lateral lemniscus, and the torus semicircularis. The torus is divided into central and laminar nuclei, and both nuclei have a number of distinct cell types (Browner, Kennedy, and Facelle 1981). The major portion of the ascending auditory input is confined to the central nucleus.

Belekhova et al. (1985) have described the central auditory pathways in both the pond turtle, *Emys orbicularis*, and the tortoise, *Testudo horsfieldi*. The cells of the central nucleus of the torus project bilaterally to the nucleus reuniens of the dorsal thalamus (called nucleus medialis in lizards). The nucleus reuniens contains pyriform and multipolar neurons whose dendrites extend into the lateral forebrain bundle. The neurons of the nucleus reuniens project to the ipsilateral medial part of the DVR as well as to the striatum.

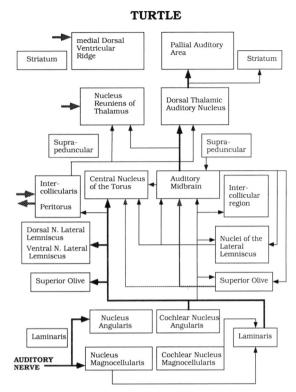

FIGURE 25.3. Generalized plan of the connections of the central auditory system of turtles, compiled from the studies cited in the text. The general pattern of projections is bilateral, although the contralateral projections from the auditory brainstem to the midbrain are heavier than the ipsilateral projections. Partially overlapping somatosensory projections to auditory nuclei are shown as horizontal striped arrows at left.

Although the major projection of the nucleus reuniens is to the DVR, it also projects to other thalamic nuclei. Belekhova et al. (1985) have also shown a partial overlap between the auditory and somatosensory pathways in the mesencephalic and thalamic auditory nuclei. The significance of this observation is not clear, but turtles are sensitive to sound transmitted through their shells (Wever 1978).

4.2 Lepidosauria (Lizards and Snakes)

The brainstem auditory nuclei of lizards and snakes differ from those of all other reptiles and birds; in many cases the basilar papilla projects to a complex of four, rather than two, cochlear nuclei

LIZARD

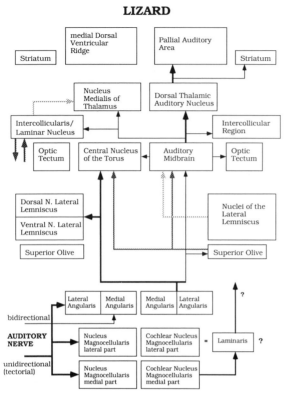

FIGURE 25.4. Generalized plan of the connections of the central auditory system of lizards compiled from the studies cited in the text. Note that since not all lizards have four divisions of the cochlear nucleus, and only some lizards have an obvious nucleus laminaris, a generalized scheme of this sort may not be valid. The (?) indicates uncertainty about the identity and connections of the nucleus laminaris (see text).

(see below) (Fig. 25.4). These differences in the central auditory projections may result from changes in auditory periphery (Miller 1975; see Chapter 23, Miller). Lizards are unique in that they have two populations of hair cells in the basilar papilla, the so-called unidirectional, low center frequency (CF) population and the bidirectional, high CF population (Manley 1990). Each population is cytologically distinct. The unidirectional type hair cells are covered by a tectorial membrane (Weiss et al. 1976; Mulroy 1987; Miller and Beck 1988). The bidirectional population of hair cells may or may not be covered by specialized tectorial structures (see Chapter 23, Miller).

There are significant physiological differences between the responses from the uni- and bidirec-

tional hair cell populations (see Chapter 24, Köppl and Manley; Köppl and Manley 1990a,b; Manley, Köppl, and Johnstone 1990; Manley et al. 1990). In both the alligator lizard, *Gerrhonotus multicarinatus*, and the bobtail lizard, *Tiliqua rugosa*, auditory nerve fibers that receive input from the unidirectional population have low center frequencies (100 to 800 Hz), low spontaneous discharge rates, and sharp, asymmetric tuning curves. These responses resemble those of hair cells in other reptiles, birds, and mammals. Furthermore, these unidirectional type hair cells are contacted by auditory nerve fibers that may correspond to the mammalian type 1 and avian cochlear nerve fibers (Spzir et al. 1990). This arrangement appears to have been present in the stem reptiles from which modern reptiles, birds and mammals evolved (Miller 1980b; Manley 1981; Rose and Weiss 1988). The bidirectional population of hair cells is physiologically and morphologically distinct. In the alligator lizard, fibers from the bidirectional hair population have high center frequencies (900 to 4,000 Hz), high spontaneous rates and broad symmetric tuning curves (for review see Spzir et al. 1990). In the bobtail lizard, these fibers have high CF (900 Hz to 4.5 kHz), high spontaneous rates and V-shaped tuning curves with obvious sharp tips around CF (see Chapter 24, Köppl and Manley; Manley, Köppl, and Johnstone 1990). The relationship of the bidirectional population to hair cells in other species is not clear, and they may be a uniquely derived, autapomorphic feature of lizards (Wever 1978; Miller 1980b; Spzir et al. 1990).

The cochlear nuclei of the lizards show a great deal of interspecific variation (Miller 1975). These differences appear to be related to the changes in the auditory periphery described above, specifically to changes in the population of bidirectional hair cells (see Chapter 23, Miller). In many lizards, Miller recognized four nuclear populations that receive primary auditory inputs: a medial and lateral nucleus magnocellularis and a medial and lateral nucleus angularis. Auditory nerve fibers whose peripheral processes contact unidirectional hair cells project to three of the four divisions of the cochlear nuclei, the medial and lateral nucleus magnocellularis, and the lateral nucleus angularis, forming endbulbs of Held and terminal boutons in the nucleus magnocellularis, and small boutons in the nucleus angularis (Spzir et al. 1990)

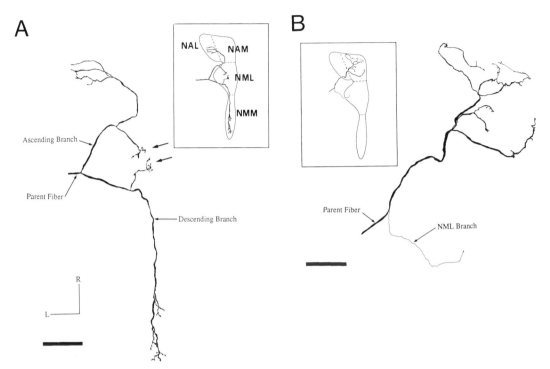

FIGURE 25.5. Projections of auditory nerve fibers in the alligator lizard, showing the terminal arbors of fibers innervating the uni- and bidirectional populations of hair cells. (A) A fiber that receives its input from the unidirectional population follows the pleisiomorphic pattern of termination in the nucleus angularis lateralis (NAL) and in the nucleus magnocellularis (NML and NMM). Arrows mark endbulbs of Held in NML. (B) A fiber from the bidirectional population terminates in a "new" medial region of the nucleus angularis (NAM). (From Spzir et al. 1990.)

(Fig. 25.5A). Neurons that contact the bidirectional population of hair cells project primarily to the medial nucleus angularis (Spzir et al. 1990) (Fig. 25.5B). It can be hypothesized that the unique organization and morphological characteristics of the first-order auditory nuclei in lizards are a reflection of the presence of, and variation in, the bidirectional hair cell population (see Köppl and Manley, Chapter 24). The elaboration of this unique feature in some lizards may have led to the substantial changes in the central auditory system.

Recordings from the lizard auditory nerve reveal different responses from the auditory nerve fibers that innervate uni- and bidirectional hair cell populations described above. Recordings from the cochlear nuclei in *Varanus bergalensis* also reflect this segregation. Manley (1977) recorded low-frequency responses from the lateral region of nucleus angularis and from nucleus magnocellularis, and high-frequency responses only from the nucleus angularis. Tuning curves from the cochlear nuclei also fall into two groups. All units with best frequencies below 1.1 kHz had sharp V-shaped tuning curves, while units with best frequencies from 1.1 to 2.8 kHz had more complex tuning curves (Manley 1977). Despite these differences, both populations gave primary-like responses to the auditory stimulus in both *Varanus* and in the leopard lizard, *Crotophytus wislizenii* (Manley 1974; 1977).

Miller (see Chapter 23) proposed that the wide variation in the lizard papilla has led to variation in the development of the brainstem auditory nuclei. The size of the nucleus magnocellularis as a whole is correlated with the size of the unidirectional hair cell region (Miller and Kasahara 1979). The medial nucleus magnocellularis is thus poorly developed in iguanids, moderate in skinks and anguids, and large in geckos and teiids. The nucleus angularis is fairly constant in size (Miller 1975). The nucleus laminaris is the most variable among medullary auditory nuclei. A morphologically distinct, although not well developed, nucleus laminaris has been found only in those lizards with a primitive basilar papilla, i.e., one with few bidirectional hair cells. Furthermore, in those lizards with a nucleus laminaris, there are differences in the auditory nerve projections. An auditory nerve projection to the "inner cell strand" of nucleus laminaris has been described in the monitor lizard, *Varanus exanthematicus* (Barbas-Henry and Lohman 1986). In the tegu lizard, *Tupinambis nigropunctatus*, however, the auditory nerve projects to the nucleus angularis and the nucleus magnocellularis, but not to the nucleus laminaris (DeFina and Webster 1974). These differences may represent species differences or differences in identification of the nucleus laminaris. Both lizards have a relatively unspecialized basilar papilla (see Chapter 23, Miller). In other lizards with more derived basilar papilla morphology, e.g., in the alligator lizard, a distinct nucleus laminaris cannot be identified at all (Miller 1975).

Conflicting reports exist in the literature because of the variation in the development of the nucleus magnocellularis and laminaris. Only some lizards have a distinct pathway from the nucleus magnocellularis to the nucleus laminaris similar to the basic pattern found in turtles, crocodilians, birds, and mammals. Miller (1975) proposed that the more primitive lizards, e.g., varanids, may have a nucleus laminaris, while it might be lost in other lizards. It is clear that more data are needed on the projections of the nucleus magnocellularis in different species before the relationship between the different cochlear, i.e., first order auditory nuclei, can be clarified. It is reasonable to propose that those cochlear nuclei that project to the auditory midbrain may be homologized to the nucleus angularis and the nucleus laminaris. Those cochlear nucleus neurons that project locally should correspond to the nucleus magnocellularis. Interestingly enough, Foster and Hall (1978) placed injections of HRP into the central nucleus of the torus in the iguana, and found retrogradely labelled cells in only two cochlear nuclei. These cells were mostly in the nucleus angularis (only one division was recognized), but some were found in the "nucleus magnocellularis lateralis." These observations are further supported by the unpublished results of Dodd and Capranica, who found labelled neurons in both the nucleus angularis and

the "nucleus magnocellularis lateralis" after injections of HRP into the torus of the Tokay gecko. Parsimony would suggest that this portion of the nucleus magnocellularis lateralis, which in addition does not receive first-order auditory input, is homologous to the nucleus laminaris described in all other reptiles and birds.

There have been very few physiological investigations of the cochlear nuclei in reptiles, although the auditory periphery has been studied extensively. Manley (1970b) recorded from the cochlear nuclei in the caiman, leopard lizard, and Tokay gecko. He found indications of tonotopic order, and also found similar primary-like responses in both the lizard and caiman, but different results in gecko, where units showed very precise onset responses. Since the Tokay gecko is a nocturnal hunter, Manley speculated that these responses may mediate directional information about interaural time differences. Primary-like responses have also been recorded from the auditory nerve of this gecko (Dodd and Capranica 1989).

The medullary auditory nuclei project to the torus semicicularis and the lemniscal nuclei. The torus and its connections have been described in the gecko (Kennedy and Browner 1981), iguanid (Foster and Hall 1978), and varanid (ten Donkelaar et al. 1987). There are varying reports of these central connections. In the iguanid, *Iguana iguana*, the cochlear nuclear complex forms a mostly contralateral projection to the central nucleus of the torus. The central nucleus of the torus also receives inputs from contralateral superior olive, and from the ipsilateral nucleus of the lateral lemniscus. In the monitor lizard, *Varanus exanthematicus*, ten Donkelaar et al. (1987) described predominantly ipsilateral projections from the nucleus angularis and the nucleus laminaris, and extensive ascending bilateral projections from superior olive. They also found that the nuclei of the lateral lemniscus may be divided into dorsal and ventral divisions. The lemniscal nuclei are made up of very small to medium-sized cells that project to the central nucleus of the torus (ten Donkelaar et al. 1987).

The torus is divided into central, laminar, and superficial nuclei. The central nucleus consists of diffusely arranged small and medium-sized cells. It extends from the lateral part of the laminar nucleus in a mediocaudal direction. The central nucleus receives secondary auditory projections via the lateral lemniscus, and is the primary source of the commissural projection to the contralateral central nucleus (ten Donkelaar et al. 1987). The central nucleus gives rise to ascending projections to nucleus medialis of the dorsal thalamus (Foster and Hall 1978).

Auditory responses have been recorded from the torus in the gecko, probably from the central nucleus (Kennedy 1974; Manley 1981). A Golgi study from Kennedy and Browner (1981) described a variety of complex cell types in the central nucleus of the gecko torus. The laminar nucleus forms part of the compact periventricular cell layer that extends throughout the mesencephalon. As is also the case in the red-eared turtle, this region does not receive ascending auditory projections, but spinal afferents in the iguana (Foster and Hall 1978). Although the inputs to the laminar nucleus in the gecko are unknown, stimulation of this region evokes species-specific vocalizations (Kennedy 1975). Geckos are the most vocal lizards, and are reported to be as sensitive to sound as some birds and mammals (Marcellini 1978). Because of its connections and role in vocalization, the laminar nucleus has been homologized to the intercollicular nuclei of birds (Kennedy and Browner 1981).

The central nucleus of the torus forms an ipsilateral projection to the nucleus medialis of the dorsal thalamus. Nucleus medialis in turn projects ipsilaterally to the medial anterior portion of the DVR in a number of genera including *Tupinambis* (Lohman and van Woerden-Verkley 1978), Gecko, and iguana (Foster and Hall 1978; Bruce and Butler 1984). There is also a sparse projection to the striatum from the nucleus medialis. These projections of the lizard auditory system are comparable to those found in turtles (Hall and Ebner 1970; Parent 1976; Balaban and Ulinski 1981; Belekhova et al. 1985), caiman (Pritz 1974a,b), and pigeon (Karten 1967; 1968). As in turtles (Section 4.1), there may be a partial overlap between the auditory and somatosensory inputs in the nucleus medialis. In *Varanus exanthematicus*, the laminar nucleus of the torus, which receives spinal cord input, projects to the nucleus medialis along with auditory fibers from the central nucleus (Hoogland 1982). As in turtles, the significance of this overlap is unknown.

There are fewer data on the connections of the auditory system of snakes than of lizards. Hartline and Campbell (1969) found auditory responses to tones of 50 to 1,000 Hz in the midbrain of three families of snake. Miller (1980a) found that three burrowing and three nonburrowing species of snake possess both cochlear nuclei and a nucleus laminaris. The nucleus angularis is small, while nucleus magnocellularis has both medial and lateral divisions. There is a nucleus laminaris, and Miller states that the nucleus laminaris in a boid snake was better developed than in any lizard examined. In the burrowing snakes, *Xenopeltis unicolor*, *Cylindrophis rufus* and *Eryx johni*, the cochlear nuclei were larger and better differentiated than in nonburrowing snakes. These findings parallel the observations and predictions of Gans and Wever (1972), who found that members of the burrowing suborder Amphisbaenia responded to auditory cues and showed moderate to good sensitivity to auditory cues between 200 and 2,000 Hz. It is likely that the central auditory pathways of the amphisbaenids are similarly developed. Not all burrowing snakes have good hearing, however. The Typhlopoidea (blindsnakes) have poor auditory sensitivity (Wever 1978).

4.3 Crocodilia (Caiman)

The crocodilian ear is large, with a long basilar membrane and unidirectional population of hair cells covered by a tectorial membrane. Crocodilians share an archosaur common ancestor with the birds, and the central auditory pathways of the two groups are very similar (Fig. 25.6). Manley (1974) recorded from the cochlear nucleus in the caiman, and found primary-like responses that were very similar to those in the leopard lizard. Klinke and his colleagues (Smolders and Klinke 1986) have also recorded from auditory nerve fibers in the caiman. The hearing range of the caiman is between 20 and 2,800 Hz. Auditory nerve units are relatively sharply tuned, and phase lock to frequencies up to 1.5 kHz (Smolders and Klinke 1986).

The central auditory pathways in crocodilians are very similar to those of birds. In the caiman, the auditory nerve projects to the nucleus angularis and the nucleus magnocellularis (Leake 1974). The nucleus angularis is just anterior to the root of the auditory nerve and composed of large and

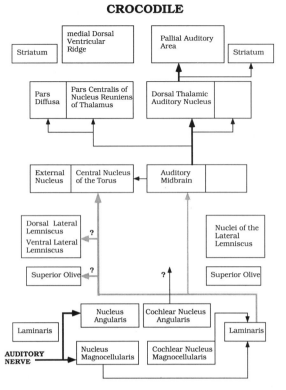

FIGURE 25.6. Known connections of the central auditory system of crocodiles compiled from the studies cited in the text. (?) indicates possible connections of the brainstem auditory nuclei to the central nucleus of the torus.

work on the projections of the brainstem auditory nuclei in the crocodilian, but Brauth (personal communication) has shown that the nucleus angularis and the nucleus laminaris project bilaterally to the torus, as is also the case in other reptiles and birds.

The cytology and connections of the midbrain and thalamic components of the caiman auditory system have been described in a series of papers by Pritz. He showed that the torus is composed of two distinct regions in *Caiman*: an external nucleus continuous with the deep layers of the optic tectum, and a central nucleus. The central nucleus is like that of lizards and birds, while it is not yet clear how the external nucleus is related to the laminar and superficial nuclei of the lizard torus. The external nucleus resembles the superficial nucleus of the inferior colliculus of birds, since it is continuous with the deep layers of the optic tectum. As is also the case in the geckos and birds, the central nucleus of the torus contains a number of different cell types and is tonotopically organized (Pritz 1974b; J. Manley 1971). Characteristic frequencies varied with recording depth in the torus with low frequencies (70 Hz) represented dorsally and high frequencies (1,850 Hz) represented ventrally (J. Manley 1971). Most units recorded in the torus were of the onset type.

The central nucleus of the torus projects bilaterally to the central part of the nucleus reuniens (Pritz 1974b). The central portion of the nucleus reuniens is made up of compact cells with round cell bodies. It is bordered laterally by the pars diffusa of the nucleus reuniens that contains ovoid cells and has been homologized to the nucleus semilunaris in birds (Pritz 1974b). From the central part of the nucleus reuniens, ascending axons enter the dorsal peduncle of the lateral forebrain bundle, and terminate in the caudo-medial region of the ipsilateral DVR (Pritz 1974a). This medial region of the DVR contains neurons that are sensitive to sound stimuli (Weisbach and Schwartzkopff (1966) quoted in Pritz 1974a). There is no overlap between visual and auditory projections in this region (Pritz 1980). This is not surprising because the auditory and visual areas of the DVR are separated by a somatosensory area. Since turtles show an overlap of auditory and somatosensory responses, it would be useful to determine if any overlap existed in the crocodilians.

small ovoid cells. The nucleus magnocellularis contains large round cells and is much larger than the nucleus angularis (Leake 1974). The nucleus magnocellularis has lateral and medial divisions. The caudal part of the medial division was capped by a small-celled component. This is significant because there is a similar small-celled region of the nucleus magnocellularis of the bird that has been reported to receive lagenar input (Boord and Rasmussen 1963). The cochlear nuclei of the caiman are tonotopically organized in a way similar to that of the song bird (Manley 1970a; Konishi 1970).

The crocodilians are the only living reptiles to have a well-developed nucleus laminaris (Fig. 25.7). The nucleus laminaris forms a monolayer sheet of bipolar spindle-shaped cells that is very similar to that seen in the basal land birds (Fig. 25.9). There is unfortunately little published

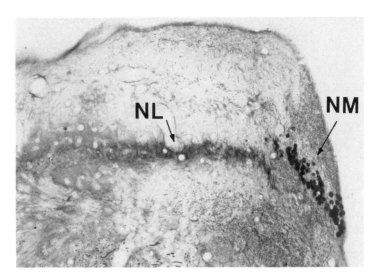

FIGURE 25.7. The nucleus magnocellularis and the nucleus laminaris in crocodilians resemble those of basal land birds. Photomicrograph of a transverse section through the brainstem of *Caiman* labelled with an antibody against calbindin. NM, rostral portion of nucleus magnocellularis, NL, nucleus laminaris. (Courtesy of S. Brauth.)

5. The Central Auditory Pathways: Basic Avian Plan

The three major neognath divisions are made up of the so-called basal landbirds, higher landbirds, and water birds. Basal landbirds include the following orders: Cuculiformes, Falconiformes, Galliformes, Columbiformes, and Psittaciformes. The basic avian plan will be represented by several species of basal landbirds, including the pigeon (Columbiformes), the chicken and guinea fowl (Galliformes). and the budgerigar (Psittaciformes). The so-called higher landbirds include the orders Coraciiformes, Strigiformes, Caprimulgiformes, Apodiformes, Piciformes, and Passeriformes. The auditory system of two higher land bird groups, the owls (Strigiformes) and the song birds (Passeriformes), will be described. The water bird assemblage comprises all other birds. It includes the Gruiformes, Charadriiformes, Anseriformes, Ciconiiformes, Pelecaniformes, Procellariiformes, Gaviiformes, and Sphenisciformes. Little is known about the central auditory pathways of water birds, and the gulls (Charadriiformes) and ducks (Anseriformes) will only be briefly mentioned.

Many birds are auditory specialists, and thus identification of derived features of the auditory system in different groups may provide insight into selective pressures. The basic plan of the avian auditory central nervous system is, however, organized in a pattern common to both birds and reptiles. The avian version of the basic plan or morphotype is as follows: The auditory nerve projects to two cochlear nuclei, the nucleus magnocellularis and the nucleus angularis. These nuclei project to second order olivary nuclei, to the lemniscal nuclei, and to the central nucleus of the nucleus mesencephalicus lateralis dorsalis (MLd, inferior colliculus). The central nucleus of the MLd projects to a major thalamic target, the nucleus ovoidalis. The nucleus ovoidalis projects to a specialized auditory region of the DVR termed Field L. Field L projects to other forebrain nuclei, some of which are involved in the control of song and other vocalizations.

5.1 Basal Landbirds – Pigeon, Chicken, Guinea Fowl, and Budgerigar (Fig. 25.8)

Boord and Rasmussen (1963) showed that the auditory nerve projects to the cochlear nucleus magnocellularis and cochlear nucleus angularis in the pigeon. Each auditory nerve fiber branches to form varicose terminals in the nucleus angularis, and endbulbs of Held on the neurons in the nucleus magnocellularis (Jhaveri and Morest 1982). The

CHICKEN/PIGEON/GUINEA FOWL

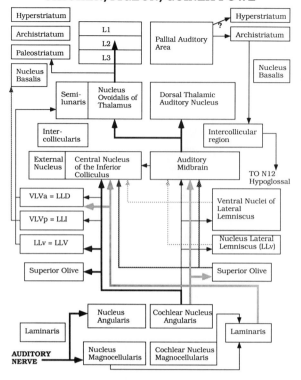

FIGURE 25.8. A composite of known connections of the central auditory system of chickens, pigeons, and guinea fowl compiled from the studies cited in the text. The projections of the nucleus angularis are shown as black arrows, and the projections of the nucleus laminaris as gray arrows.

endbulb synapse preserves the relationship between the discharge of the auditory nerve fibers and the neurons of the nucleus magnocellularis. The auditory nerve of the pigeon shows phase-locked responses to auditory stimulation that are similar to those described in other advanced landbirds (Hill, Stange, and Mo 1989).

In the owl, Sullivan and Konishi (1984) and Takahashi, Moiseff, and Konishi (1984) demonstrated that the nucleus magnocellularis is the origin of a neural pathway that encodes timing information, while a parallel intensity pathway originates with the nucleus angularis. Recent physiological recordings in the chick cochlear nuclei have found a similar segregation of function (Warchol and Dallos 1990). Angularis units are more sensitive to changes in intensity than magnocellular units, and many have nonmonotonic rate-intensity functions.

Magnocellular units phase lock to higher frequencies than angularis units. These similarities between the owl and the chick suggest that the functional separation of time and intensity coding is a common feature of the avian auditory system.

The projection from the nucleus magnocellularis to the nucleus laminaris in the chick has been the subject of numerous studies from Morest, Rubel, and their colleagues (reviewed in Rubel and Parks 1988). In basal land birds the nucleus laminaris is a tonotopically organized monolayer of bipolar cells (Fig. 25.9). These cells receive input from the ipsilateral nucleus magnocellularis onto their dorsal dendrites and input from the contralateral nucleus magnocellularis onto their ventral dendrites. Young and Rubel (1983) found that each magnocellular axon forms an elongated band of endings along the mediolateral length of the nucleus, and proposed that the axons acted as delay lines to form a map of interaural time differences in the nucleus laminaris (Parks and Rubel 1975; Young and Rubel 1983; 1986). The first physiological evidence that magnocellular afferents might be delay lines came from the work of Sullivan and Konishi (1986) in the barn owl, and will be discussed below.

Boord (1968), Leibler (1975), and Conlee and Parks (1986) described the projections of the brainstem auditory nuclei in the pigeon and chicken (Figure 25.10). Given the functional segregation described in the cochlear nuclei by Warchol and Dallos (1990), the finding of Conlee and Parks that the nucleus angularis and the nucleus laminaris projected bilaterally to segregated regions in the central nucleus of the inferior colliculus in the chicken is significant, and suggests that the functional separation of time and intensity information is preserved up to the level of the midbrain. The angularis projection is found rostrally in the central nucleus, and the laminaris projection caudally and medially. Both projections are predominantly contralateral. The separation of time and intensity information is less distinct in the olivary and lemniscal projections of the two systems. The nucleus angularis projects bilaterally to the superior olive, and to the ventral nucleus of the lateral lemniscus (VLV), and to the contralateral nucleus of the lateral lemniscus (LLV). The projections of the nucleus laminaris are generally similar to the projections of the nucleus angularis. Nucleus laminaris projects to the ipsilateral superior olive, and to the contralateral

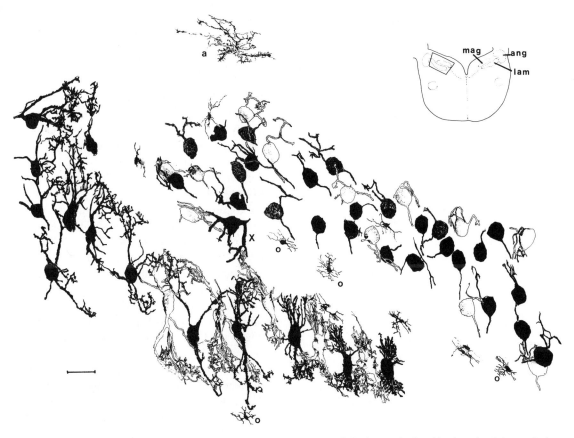

FIGURE 25.9. The cytology of the nucleus magno-cellularis (above) and the nucleus laminaris (below) in the chick. Note the large round cells of the nucleus magnocellularis, and the bipolar dendrites of the cells of the nucleus laminaris. (From Jhaveri and Morest 1982.)

LLV and VLV. Therefore no lemniscal nucleus is innervated purely by nucleus laminaris, as is the case in the barn owl (see below). The ventral nucleus of the lateral lemniscus (VLV) is divided into two divisions in the barn owl (Takahashi and Konishi 1988a; Moiseff and Konishi 1983). These divisions are not as anatomically obvious in the basal land birds, though some studies have provided evidence of their existence (Leibler 1975; Heil and Scheich 1986). It is important to deter-mine in these land birds whether or not the two indistinct divisions of VLV receive nonoverlap-ping projections from the nucleus angularis and the nucleus laminaris.

Coles and Aitkin (1979) investigated the response properties of the neurons of the MLd. Responses were similar to those found in the inferior collicu-lus of the barn owl. Most neurons were excited by inputs from the contralateral ear and inhibited by the ipsilateral ear (EI cells), although units with bilateral excitation (EE) and contralateral excita-tion only (EO) were also found. Both EE and EI cell types have been implicated in detection of interaural time and intensity differences and thus in sound localization. Coles and Aitkin found a population of neurons in the inferior colliculus that were sensitive to changes in interaural intensity and others that responded to interaural time differ-ences, although they caution that these responses were not necessarily within the physiologically useful range for sound localization.

This is an appropriate place for a discussion of nomenclature. It has been stated repeatedly in the literature that the MLd of birds is homologous to the mammalian inferior colliculus (Karten 1967). In 1983, Knudsen introduced a new nomenclature

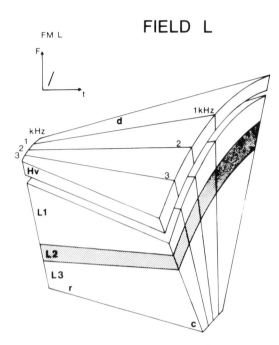

FIELD L

FM L

FIGURE 25.10. The tonotopic and 3-dimensional organization of Field L. The thalamic input layer (L2) is shown stippled. Physiological responses change between the different laminae. HV, hyperstriatum; L1–3, Field L; r, rostral; c, caudal; d, dorsal. (From Scheich 1990.)

in his paper, "Subdivisions of the inferior colliculus in the barn owl." Most individuals not working on owls, however, have retained the use of the MLd nomenclature for the reason that although the MLd as a whole is homologous to the inferior colliculus, the subdivisions may not be strictly homologous in birds and mammals. It would be conservative to describe the mammalian inferior colliculus and MLd as field homologues, and to retain the present terminology simply because it is entrenched in the literature.

In the pigeon, the central nucleus of the MLd projects bilaterally to the nucleus ovoidalis and the nucleus semilunaris in the dorsal thalamus (Karten 1967). The nucleus semilunaris is a small nucleus that surrounds the ventrolateral edge of the nucleus ovoidalis, and the semilunaris–ovoidalis complex has been homologized to the mammalian medial geniculate nucleus (Karten 1967). In both the pigeon and guinea fowl, axons from the nucleus ovoidalis project ipsilaterally to Field L, a caudal and medial region of the DVR, as well as to the striatum (Fig. 25.8) (Karten 1968; Bonke, Bonke,

and Scheich 1979). Similar projections have been described in the budgerigar (Brauth et al. 1987). The projection to Field L in the guinea fowl has been studied by Scheich and his colleagues (Fig. 25.10). Bonke, Bonke, and Scheich (1979) divided Field L into three parallel layers, L1, L2, and L3, and found that the nucleus ovoidalis projects primarily to L2 (Karten 1968). Auditory units in L2 generally have narrow tuning curves with inhibitory sidebands, while the cells of L1 and L3 exhibit more complex responses (Scheich, Langer, and Bonke 1979b).

The tonotopic organization of the auditory system has been described in 2-deoxyglucose studies of several basal land birds (Fig. 25.10; Müller and Scheich 1985; Heil and Scheich 1986). In the MLd, tonotopic organization was in good agreement with that found using electrophysiological techniques (Coles and Aitkin 1979). Although auditory responses were reported from the nucleus ovoidalis and Field L of the ring dove (Biederman-Thorson 1970), electrophysiological demonstration of tonotopy have so far only been obtained from higher land birds. The tonotopic organization of the nucleus ovoidalis has been demonstrated in the starling (Bigalke-Kunz, Rübsamen, and Dörrscheidt 1987), and the tonotopic organization of Field L has been demonstrated in starlings (Leppelsack and Schwartzkopff 1972; Rübsamen and Dörrscheidt 1986) and zebra finches (Zaretsky and Konishi 1976). The projections of Field L have also been investigated in the songbird and will be discussed in detail below.

Basal landbirds may have increased their auditory sensitivity in a novel way. The auditory systems of pigeons, chickens, and guinea fowl contain subsets of neurons that respond to infrasound (below 20 Hz) (Theurich et al. 1984; Warchol and Dallos 1989; Schermuly and Klinke 1990). It is not known whether other birds, or any reptiles and mammals possess similar abilities. Infrasound sensitive units in the auditory nerve have very high levels of "background" activity, and although they do not show reliable increases in discharge rate with infrasound stimuli, their mean discharge rates are modulated by these stimuli (Schermuly and Klinke 1990). Haeseler, Brix, and Manley (1989) have shown that the lagena is not the site of infrasound sensitivity, as had been previously suggested. Warchol and Dallos (1989) suggested that the peripheral site of low-frequency sensitivity was the most distal region of the basilar papilla, an area adjacent to the lagena whose morphology in the chick differs from the rest of the papilla (see Chapter 27, Manley and Gleich). Schermuly and Klinke (submitted) labelled infrasound auditory nerve units with HRP and found terminals in the apical part of the papilla, and not in the lagena.

Responses to infrasound stimuli are found in the cochlear nuclei. The neurons in the chick cochlear nuclei differ from the auditory nerve units in that they show increases in spike rate with infrasound stimuli (Warchol and Dallos 1989). Note, however, that this difference may be due more to technical differences between the sound proof rooms than to neural processing. Responses to infrasound have been found throughout the central auditory system. They were first observed in the MLd of the guinea fowl (Theurich et al. 1984). Field L in chickens and pigeons contains a low-frequency expansion (below 500 Hz) that may be related to low-frequency hearing (Müller and Scheich 1985). Infrasound signals may travel over great distances, and Kreithen and Quine (1979) have suggested that pigeons may use infrasound for orientation. Infrasound sensitivity has as yet only been found in primitive land birds and may not be in other groups (Theurich et al. 1984; Müller and Scheich 1985).

"Nontraditional" auditory pathways have been described between the lemniscal nuclei and various thalamic and forebrain nuclei in a number of birds. The details depend upon a new nomenclature introduced for the lemniscal nuclei by Arends (1981). The lemniscal nuclei in the pigeon were described by Karten and Hodos (1967, atlas) as comprising three components: a nucleus of the lateral lemniscus, pars ventralis (LLv), a nucleus of the lateral lemniscus, pars dorsalis (LLd), and a ventral nucleus of the lateral lemniscus (VLV). LLd is now known not to receive a projection from the lateral lemniscus, and as far as is known, is not an auditory nucleus. It is given no further consideration. Karten and Hodos' (1967) terminology was subsequently accepted by Leibler (1975), Conlee and Parks (1986) and Takahashi and Konishi (1988a). Arends (1981), however, introduced a different terminology for the mallard duck, and this was subsequently used by Arends and Ziegler (1986) and Wild (1987) for the lemniscal nuclei in the pigeon. In this terminology, which corresponds

to that used in mammals, there is a ventral nucleus of the lateral lemniscus (LLV) (=LLv of Karten and Hodos, 1967), and an intermediate nucleus of the lateral lemniscus (LLI) (=VLV of Karten and Hodos, 1967), and a slightly more rostrally situated dorsal nucleus of the lateral lemniscus (LLD) (=VLVa of Leibler 1975). On the basis of retrograde tracing studies in the pigeon, LLI (=VLVp) projects to the basal telencephalon, to or around the nucleus basalis (Arends and Zeigler 1986). This projection may account for the short latency auditory potentials which have repeatedly been recorded in and around the nucleus basalis of many birds, including the pigeon, starling, and duck (Kirsch, Coles and Leppelsack 1980; Maekawa 1987; Arends and Zeigler 1989). Both LLV (=LLv) and LLD (=VLVa) in the pigeon project to the auditory thalamus, primarily to the nucleus semilunaris parovoidalis (Wild 1987) as well as to the MLd. Note that, although Arends' names are the same as those of the lemniscal nuclei in mammals, these nuclei do not appear to be homologous. The lemniscal complexes in reptiles, birds, and mammals should be regarded as field homologues (Takahashi and Konishi 1988a; Carr, Fujita, and Konishi 1989).

5.4 Higher Landbirds

5.4.1 Owls

Barn owls are nocturnal predators that are able to localize objects in space on the basis of auditory cues alone (Payne 1971; Konishi 1973b). Their auditory systems are very well developed (Fig. 25.11). As is also the case in the evolution of the auditory system of lizards and mammals, the evolution of the owl auditory system appears to have been driven in part by a specialization of the auditory periphery. The external ears of some species of owls are oriented asymmetrically in the vertical plane. Because of this asymmetry, interaural intensity differences vary more with the elevation of the sound source than with azimuth. This asymmetry allows owls to use interaural intensity differences to localize sounds in elevation, while they use interaural time differences to determine the azimuthal location of a sound (Norberg 1978; Knudsen and Konishi 1979a,b; Moiseff 1989b; Olsen et al. 1989).

The barn owl has exceptional high frequency hearing for a bird, with characteristic frequencies

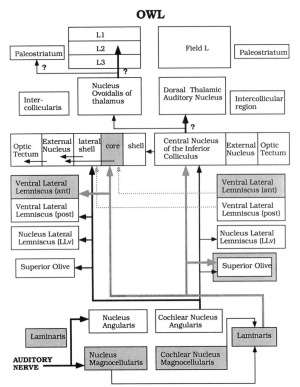

FIGURE 25.11. Known connections of the central auditory system of the barn owl. The nuclei of the time coding system are shown stippled, while the nuclei of the intensity pathway are shown as empty. Time and intensity information is combined in the space-mapped region, or external nucleus of the inferior colliculus.

of up to 9 to 10 kHz in the auditory nerve. The basilar papilla is the longest described in birds, about 11 mm in length (see Chapter 27, Manley and Gleich). Specializations are found in the basal, high-frequency portion of the papilla and its overlying basilar membrane (Fischer, Köppl, and Manley 1988; Smith et al. 1985). The anatomy and physiology of the central auditory pathways below forebrain levels are well understood in the barn owl (for review see Konishi et al. 1988). As in all other birds, the auditory nerve projects to both the nucleus magnocellularis and the nucleus angularis. The auditory nerve, the neurons of the nucleus magnocellularis, and the neurons of the nucleus laminaris phase lock to auditory stimuli, while the cells of the cochlear nucleus angularis do so poorly (Sullivan and Konishi 1984; Carr and Konishi 1990). The terminal morphology of the

two auditory nerve collaterals reflects this functional division into phase and intensity. The endbulbs of Held terminate on the neurons in the nucleus magnocellularis and conserve phase information, while varicose auditory nerve terminals ramify in the nucleus angularis where phase information is not conserved (Konishi, Sullivan, and Takahashi 1985). The endbulb, with its multiple sites of synaptic contact, provides a substrate for the preservation of the relationship between the discharge of the auditory nerve fibers and the neurons of the nucleus magnocellularis. Magnocellular neurons display morphological adaptations for phase locking. They have few medium length dendrites, as is also the case in the chicken (Smith and Rubel 1979; Jhaveri and Morest 1982). The nucleus is tonotopically organized, with low (1 to 4 kHz) frequencies represented caudolaterally and high (5 to 9 kHz) best frequencies represented rostromedially (Konishi 1973a; Rubel and Parks 1975; Takahashi and Konishi 1988b). The cochlear nucleus magnocellularis projects bilaterally to the nucleus laminaris (Takahashi and Konishi 1988b; Carr and Konishi 1988; 1990).

In the barn owl, the nucleus laminaris is vastly expanded from the monolayer structure typical of most basal land birds. The nucleus laminaris is no longer a flat sheet but a 1 mm thick neuropil with the neurons sparsely distributed throughout a plexus of myelinated fibers. It contains four times more neurons than are found in other birds of similar size (barn owl 10,020 vs crow 2,540; Winter and Schwartzkopff 1961).

The projection from the cochlear nucleus magnocellularis to the nucleus laminaris forms a circuit that resembles the model proposed by Jeffress to explain sound localization (Carr and Konishi 1988). In the model, delay lines project to neurons that require simultaneous arrival of spikes from the two sides to elicit a maximal discharge. The magnocellular afferents function as delay lines and the neurons of the nucleus laminaris act as coincidence detectors (Fig. 25.15; Sullivan and Konishi 1986; Carr and Konishi 1988; 1990). Magnocellular inputs reach the laminaris neurons by axonal paths that may be unequal for the two sides. The neurons respond maximally when the disparities in spike conduction time are opposed by an equal interaural time difference (Carr and Konishi 1990). This circuit is therefore responsible for the detection of interaural time differences in the brainstem of the barn owl. The phase differences mapped by the magnocellular axons are also comparable to the range of interaural time differences of about 200 μsec available to the owl's auditory system (Moiseff 1989a).

The neurons of the nucleus laminaris do not code unambiguously for interaural time differences, however. Because they encode interaural phase differences at one stimulus frequency, their responses show phase ambiguity. This ambiguity is resolved in the higher-order projections of the nucleus laminaris. Laminaris neurons project contralaterally to two targets, the anterior division of the ventral nucleus of the lateral lemniscus (VLVa) and to a restricted region of the central nucleus of the inferior colliculus termed the core (Fig. 25.11). Calretinin-like immunoreactivity labels the terminal fields of nucleus laminaris in VLVa and the core (Takahashi et al. 1987). Electrophysiological recordings from the core region of the inferior colliculus reveal a systematic mapping of interaural time differences (Wagner, Takahashi, and Konishi 1987). This region is tonotopically organized, and all neurons in a penetration perpendicular to isofrequency laminae encode one interaural time difference. Thus an interaural time difference is conserved in an array of neurons (Wagner, Takahashi, and Konishi 1987). Each array projects to "space-specific" neurons in the external or space-mapped nucleus of the inferior colliculus, and the collective response of the array may be responsible for the properties of the space-specific neurons (Takahashi, Wagner, and Konishi 1988, also see below).

Less is known about the intensity pathway in the barn owl (see review in Spence and Pearson 1990). The neurons of nucleus angularis respond to changing sound intensity over about a 30 dB range. Each nucleus angularis projects to the contralateral VLVp and to the "shell" of the central nucleus of the inferior colliculus (Fig. 25.11). The cells of VLVp are excited by stimulation of the contralateral ear and inhibited by stimulation of the ipsilateral ear (Moiseff and Konishi 1983; Takahashi and Konishi 1988c; Manley et al. 1988). The source of the inhibition is the contralateral VLVp (Takahashi and Konishi 1988c). Mapping of interaural intensity differences may begin in VLVp. Manley et al. (1988) showed that VLVp neurons are organized topographically, according to their preferred interaural intensity difference. The dorsal region of the nucleus contains neurons that prefer a louder sound in the contralateral ear. This preference changes gradually with depth in the nucleus. VLVp neurons do not encode elevation unambiguously, however, because they are not immune to changes in interaural sound level. Each VLVp projects contralaterally to the lateral shell of the central nucleus of the inferior colliculus. It appears that information about both interaural time and intensity differences are merged in the lateral shell to form "complex field" responses as a prelude to construction of the space map in the external nucleus (Takahashi, Wagner, and Konishi 1988; Knudsen and Konishi 1978).

To produce a map of auditory space, time and intensity information converge on the neurons of the external nucleus, discarding tonotopic information (Knudsen and Konishi 1978; 1979a,b; Konishi 1986). Each space-specific neuron receives inputs from a population of neurons tuned to different frequencies (Takahashi and Konishi 1986). The nonlinear interactions of these different frequency channels act to remove phase ambiguity in the response to interaural time differences (Takahashi and Konishi 1986). The external nucleus contains an ordered representation of auditory space, with most of the region devoted to the contralateral hemifield (Knudsen and Konishi 1978). It projects topographically to the optic tectum (Knudsen and Knudsen 1983). The tectum contains maps of visual and auditory space that are in register (Knudsen and Knudsen 1983). Activity in the tectum directs the rapid head movements made by the owl in response to auditory and visual stimuli (Du Lac and Knudsen 1990). Masino and Knudsen (1990) have recently shown that the sensory map of space in the tectum is transformed into an orthogonal coordinate system representing the direction of head movement.

Very little is presently known about the diencephalic and telencephalic projections of the auditory system in any owl. Karten (1968) noted that the nucleus ovoidalis in the oilbird and the saw-whet owl was larger than in the pigeon. In the barn owl, nucleus ovoidalis contains a number of subdivisions (Takahashi, personal communication). Some neurons in Field L of the barn owl respond to sound location, while other neurons present more conventional responses to sound independent of location (Knudsen, Konishi, and Pettigrew 1977).

5.4.2 Songbirds

The central auditory system will be reviewed in song birds because a great deal is known about those auditory and motor nuclei specialized for the perception and production of song (Nottebohm et al. 1976). The song system has also been a model for studies of development and sexual dimorphism in the brain, but that work will not be reviewed here.

The brainstem and midbrain auditory pathways of song birds are similar to those of other birds, although anatomical studies are needed to confirm this. Auditory nerve responses resemble those reported for other birds and reptiles (Gleich and Narins 1988; Manley et al. 1985; Chapter 27, Manley and Gleich). The cochlear nuclei are tonotopically organized, and the range of best frequencies (1 to 6 kHz) is partly correlated with the vocal range (Konishi 1970). Sachs and Sinnott (1978) recorded from the nucleus magnocellularis and the nucleus angularis in the redwing blackbird, and found very similar results to those obtained later for the barn owl (Sullivan and Konishi 1984) and chick (Warschol and Dallos 1990). The nucleus magnocellularis contained primary-like units, while the nucleus angularis neurons exhibited two response types that differed in their degree of inhibition (Sachs and Sinnott 1978). There is no reason to suppose that the projections of the central auditory system are substantially different in the songbird.

Significant modifications are present in the forebrain nuclei specialized for the perception and production of song. Field L contains auditory responsive units in starlings (Leppelsack and Schwartzkopff 1972; Leppelsack and Vogt 1976) and zebra finches (Zaretsky and Konishi 1976). In starlings, about two-thirds of the neurons in Field L responded to simple sounds, while the others required more complex stimuli (Leppelsack 1978). Auditory information appears to reach the song system via the projections of Field L, although there are no direct projections from Field L to any song system nuclei. Kelley and Nottebohm (1979) first described the projections of Field L. On the basis of anterograde tracking studies, Field L has been suggested to project to regions that are contiguous with the nuclei of the song system, including a distinct region below the nucleus hyperstriatum ventrale (HVc), termed the "shelf" region. A region postero-ventral to Field L also projects to the region around the robust nucleus of the archistriatum (RA).

Note that HVc was named as if it were in the hyperstriatum, but is actually in the neostriatum (Paton et al. 1981). The name HVc has been retained because of familiarity. It is now, in some papers, referred to as HVC, the "higher vocal center" (Simpson and Vicario 1990).

Nottebohm and colleagues (1976) described the neural pathway for the production of song (Fig. 25.12). Auditory information reaches the nucleus HVc via the shelf region. HVc projects to RA, which in turn projects to the motor neurons of the hypoglossal nucleus that control the different muscles of the syrinx (Bottjer and Arnold 1982; Vicario and Nottebohm 1987). RA also projects to the dorsomedial intercollicular nucleus (DM) (Gurney 1981). Neurons in HVc respond to auditory stimuli (Katz and Gurney 1981), and some prefer the bird's own song (Margoliash 1983). Both RA and HVc are apparently directly involved in the control of song because lesions of these nuclei disrupt adult song.

Additional nuclei and connections were described after the original discovery of the song system. HVc receives input from a nucleus uvaeformis (UVA) of the thalamus and from two other forebrain structures, the interfacial nucleus of the striatum (NIF) and the medial portion of the magnocellular nucleus of the neostriatum (mMAN) (Fig. 25.12) (Nottebohm, Kelley, and Paton 1982). Note that NIF is actually in the neostriatum. Very little is known about the role of NIF and UVA except that UVA projects both to NIF and HVc, and the neurons in NIF fire before the neurons of HVc and RA in the singing bird (McCasland 1987). The firing patterns of many of these cells reflect the temporal pattern of the song (for review see Konishi 1989).

HVc is interconnected with the thalamus via a loop of intermediate nuclei that may play a role in song learning (Fig. 25.12). HVc projects to area X in lobus parolfactorius of the forebrain (LPO) (Nottebohm, Stokes, and Paton, 1976; Katz and Gurney 1981), and X in turn projects to the medial portion of the dorsolateral nucleus of the thalamus (DLM) (Bottjer et al. 1989). DLM projects back to the

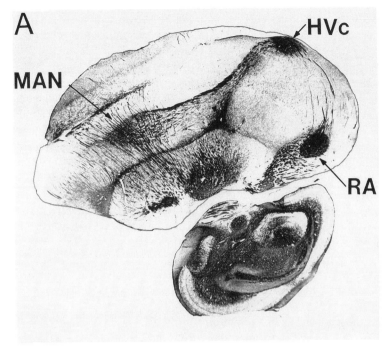

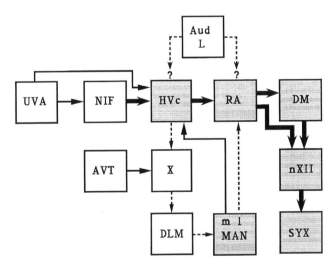

FIGURE 25.12. The song system. (A) A silver stained, parasaggittal section through the brain of a male zebra finch showing the nuclei HVc, RA and MAN, and several fiber tracts. (B) Known connections of the song system. The forebrain nuclei include NIF (interfacial nucleus of the neostriatum), HVc (caudal portion of the ventral nucleus of the hyperstriatum), RA (robust nucleus of the archistriatum), X (Area X) and MAN (magnocellular nucleus of the anterior neostriatum). Thalamic nuclei include DLM (the medial portion of the dorsolateral nucleus of the thalamus) and UVA (uvaeform nucleus). DM is the dorsomedial nucleus of the intercollicular nucleus and ATV is the ventral area of Tsai. nXII is the syringeal portion of the hypoglossal nucleus, and its motor neurons innervate the muscles of the syrinx (SYX). The thick arrows indicate the motor pathway to the syrinx, and the dashed arrows indicate the putative auditory projections within the song system. (From Konishi 1989.)

lateral region of the telencephalic nucleus MAN (lMAN). The medial portion of MAN (mMAN) projects to HVc while lMAN projects to RA. Both MAN and X neurons respond to the bird's own song (Doupe and Konishi 1989), and MAN may be necessary for song learning because lesions in MAN in the juvenile disrupt song learning while lesions in the adult do not affect song (Bottjer and Arnold 1986; Bottjer et al. 1984).

The evolution of the song system is of particular interest because it allows us to ask whether new elements are formed to serve the new behaviors or whether old connections are modified to suit new functions. An important and pertinent observation is whether the song control nuclei are unique to the passerine birds (Nottebohm 1980). It has been hypothesized that HVc, RA, X, and perhaps MAN and NIF are found only in birds capable of vocal imitation (Paton et al. 1981, also see review in Konishi 1985). Thus the song nuclei are present in those birds that develop abnormal vocalizations when deafened (Nottebohm 1980; Kroodsma 1984) and absent in birds that continue to call normally when deafened, like chickens (Konishi 1963). The passerines are not the only birds capable of vocal imitation, however. Budgerigars (Psittaciformes, basal landbirds) also imitate, and have forebrain nuclei similar in appearance and connections to HVc and RA (Paton et al. 1981). Have the Psittaciformes invented a song system independently of the passeriform birds? If the nuclei of the song system developed from preexisting structures (as seems likely, see below) then resulting systems must either be homologous or represent a case of parallel homoplasy (Northcutt 1984). The projections of the vocal control system in the Psittaciformes are different from those of the passeriform system. In canaries, the projections from RA to the intercollicular nucleus (DM) and the hypoglossal nucleus are ipsilateral, while they are bilateral in the budgerigar. Paton et al. (1981) suggest that the bilateral projection may represent the primitive condition. Alternatively, the two song systems may have developed in parallel.

In the guinea fowl, a basal landbird that does not sing, Bonke, Bonke, and Scheich (1979) found projections of Field L to regions of hyperstriatum and archistriatum similar to the projections to the regions around HVc and RA found in the songbird. Similar projections from Field L to RA have been described in the budgerigar (Brauth and McHale

1988). The descending projection from RA to the nucleus intercollicularis conforms to the pattern of projections from the archistriatum in all birds (Bonke, Bonke, and Scheich 1979; Karten and Shimuzu 1989), and may resemble projections from the archistriatum to the intercollicularis found in pigeons (Karten, personal communication).

Stimulation of the archistriatum and intercollicular nucleus results in calls in many species (Potash 1970; Brown 1971; Konishi 1973a). Vocal control in song birds and psittaforms differs from that in other birds and reptiles, however, because the preexisting projection from the archistriatum to the intercollicular region appears to have been modified. In song birds and psittaform birds, RA projects directly to the motor nucleus of the hypoglossal nerve in the medulla, in addition to the nucleus intercollicularis.

5.3 Waterbirds (Duck, Seagull)

There are few references to the central auditory system of waterbirds in the literature. Early studies from Konishi (1973b) showed that the cochlear nuclei of the duck were tonotopically organized, and examination of the brainstem auditory nuclei of a herring gull and a ring billed gull reveals well-developed cochlear nuclei. Furthermore, in the gulls and ducks examined, the nucleus laminaris is not the monolayer typical of the basal land birds, but resembles the hypertrophied nucleus found in the owls (unpublished observation).

The lemniscal nuclei of the duck are also well developed, and appear to have projections similar to those in the chicken and pigeon (Arends 1981; Wild 1987; refer to Section 5.1 for a discussion of lemniscal nuclei nomenclature). Other studies of waterbirds include a comparative study of the organization of Field L of a number of birds, including both ducks and gulls. Field L had a similar tonotopic organization and trilaminar structure in all birds tested (Müller and Scheich 1985).

6. Evolution of Central Auditory Pathways in Birds and Reptiles

6.1 Morphotype

The auditory central nervous system is organized in a pattern common to both birds and reptiles. The morphotype appears to be constructed as follows:

SENSORY SYSTEMS

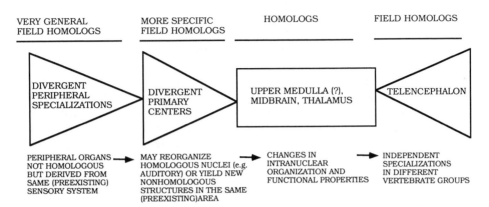

FIGURE 25.13. Model of the general pattern of change in the periphery and CNS resulting from the evolution of specialized sensory systems. Midbrain or thalamic nuclei may not be strict homologs in some systems, but better described as field homologs. (From Wilcynski 1984.)

The auditory nerve projects to two cochlear nuclei, the nucleus magnocellularis and the nucleus angularis. The nucleus magnocellularis projects to the second order nucleus laminaris that in turn projects to the superior olive, to the lemniscal nuclei, and to the central nucleus of the torus (nucleus mesencephalicus lateralis dorsalis, inferior colliculus). The nucleus angularis also projects to the superior olive, to the lemniscal nuclei, and to the central nucleus of the torus. Not enough is known at present to determine the details of these projections. Although brainstem auditory connections are generally bilateral, contralateral projections predominate. The central nucleus of the torus projects bilaterally to a major thalamic target nucleus (reuniens, medialis, ovoidalis). The thalamic nucleus projects to a specialized auditory region of the DVR (medial DVR, Field L). Field L projects to other forebrain nuclei that may be involved in the control of song and other vocalizations. Descending projections from the architriatum to the intercollicular nucleus and to the hypoglossal nucleus appear to mediate vocalization.

Physiological investigations have not been as extensive as the anatomical studies. Where it has been looked for, the neurons of the nucleus magnocellularis have been found to phase lock to the auditory stimulus, while the neurons of the nucleus angularis encode changes in sound level. It is likely that the parallel processing of time and intensity information seen in birds will also be found in turtles, crocodilians and some lizards. Other common features of the auditory system include tonotopic order in many auditory nuclei from the level of the cochlear nuclei to the telencephalon. Neurons exhibit V-shaped tuning curves, often with inhibitory sidebands.

6.2 Specializations in Reptiles

6.2.1 Bidirectional Hair Cells and High-Frequency Hearing in Lizards

Lizards have developed a cytologically distinct population of hair cells, and physiological studies that differentiate between hair cells have shown that this bidirectional population of hair cells is sensitive to higher frequencies than the unidirectional population of hair cells (see section 4.2 Chapters 24, 27 by Miller, and by Köppl and Manley). This evolutionary advance in the periphery appears to have driven changes in the central auditory nuclei in a manner predicted by Wilcynski (1984).

Wilcynski (1984) discussed the effects of changes in the periphery on the central nervous system (Fig. 25.13). He proposed that since new peripheral structures are specializations of preexisting structures, their central systems are also specializations of preexisting areas. The newly specialized areas might be related to the preexisting area either as field homologs or true homologs. A major (genetic) change in the periphery need not require a similar genetic change in the CNS because the "new" inputs could reorganize their targets in a developmental or epigenetic fashion. A major change at one level could therefore cause rearrangements at other levels of a functional system. The development of the bidirectional population of hair cells in the lizard papilla is an example of a major change in the periphery. These new inputs appear to have rearranged the cochlear nuclei, creating a new target zone in the medial nucleus angularis, and also modifying the other cochlear nuclei (Spzir et al. 1990, also see Fig. 25.5). In many cases, the target of the nucleus magnocellularis, the nucleus laminaris, also is reduced in size or disappears. The new nuclei (medial and lateral magnocellularis, medial and lateral angularis) are not strictly homologous to the original structures (i.e., nucleus magnocellularis and angularis of turtles, crocodilians and birds), but should be considered field homologs of these nuclei. It remains to be seen what effect these changes in the brainstem have on the central processing of the auditory stimulus.

6.2.2 Archosaurs (Crocodilians, Dinosaurs, and Birds)

The common features of the archosaur central nervous system should reflect the pattern found in the thecodont ancestor of both the birds and crocodilians. Crocodilians are regarded as conservative descendants of the early thecodont radiation, while birds share a common thecodont ancestor with the dinosaurs (Carroll 1987). The most primitive bird, *Archaeopteryx*, is very similar to the large-brained coelurosaurs (Hopson 1977). The auditory systems of the birds and crocodilians are so similar that it is likely that dinosaurs also had well-developed auditory systems. Hopson (1977) has proposed that dinosaurs used vocalizations for species-specific communication, and heard sounds in the crocodilian range (200 to 2,000 Hz).

6.3 Specializations in Birds

6.3.1 Changes in the Auditory Periphery: Asymmetric Ears

The external ears of barn owls are oriented asymmetrically in the vertical plane. Because of this asymmetry, interaural intensity differences vary more with the elevation of the sound source than

with azimuth (see section 5.4.1). Barn owls are therefore able to use interaural time differences to determine the azimuthal location of a sound, to form a two-dimensional map of auditory space (Norberg 1978; Knudsen 1980; Moiseff 1989a,b; Olsen et al. 1989). Ear asymmetries may take the form of changes in the skull, or changes in the external ears, but all make it possible to localize both the elevation of a sound source. Asymmetric ears are of obvious utility to a nocturnal predator like an owl, and it appears that they have arisen independently at least five times among the owls (Norberg 1978; see review in Volman and Konishi 1990) (Fig. 25.14). The phylogentic relationships of the larger groups of owls are shown in Figure 25.14. Barn owls (Tytonidae) are generally regarded as a sister group to the other owls, the Strigidae. Ear asymmetries are found in all but the stem group.

Volman and Konishi (1989; 1990) investigated the neural basis of sound localization in owls with both symmetrical and asymmetrical ears to determine if ear asymmetry was correlated with responses to changes in the elevation of a sound source. They compared responses from the external nucleus of the inferior colliculus in two species with symmetric ears (great horned owl and burrowing owl) and two species with asymmetric ears (barn owl and long-eared owl). They found that all these owls had spatially restricted receptive fields in the external nucleus, i.e., the nucleus contained a map of auditory space. In owls with symmetric ears, however, the receptive fields were only limited in azimuth, and not in elevation. Thus only azimuth was mapped. These findings contrast with those from the barn owl, whose map of auditory space is two-dimensional and whose receptive fields are restricted in both elevation and azimuth (Knudsen and Konishi 1978; 1979a,b). The barn owl has the most clearly delineated receptive fields, although in the long-eared owl, receptive fields were also somewhat limited in elevation. Similar results have been found in another owl with asymmetric ears, the saw-whet owl (Wise, Frost, and Shaver, submitted).

Volman (1990) hypothesized that sensitivity to interaural intensity differences is a primitive character in the owls. She proposed that ear asymmetries may have evolved from the ancestral condition as follows: owls with symmetric ears use interaural intensity differences as a supplementary cue for determination of azimuth. Since interaural intensity differences become appreciable at higher frequencies when the head acts as a shadow (see below), this creates a selective pressure for perception of higher frequency sound. Any ear asymmetry could then be exploited by owls that were already preadapted for detection of interaural intensity differences. Because these interaural intensity differences were not absolutely required for encoding of azimuth, they could be regarded as a "free" parameter, and converted to the role of encoding elevation. This is different from the case in mammals, where interaural intensity differences are a necessary cue for localization at high frequencies (the duplex theory).

6.3.2 Specializations for Phase Locking

Phase locking to the auditory stimulus is found in all birds and reptiles. In the alligator lizard and bobtail lizard, auditory nerve fibers innervated by the unidirectional (tectorial) hair cell population phase lock better to the auditory stimulus (up to about 600 to 700 Hz) (Rose and Weiss 1988; Manley et al. 1990). Rose and Weiss (1988) suggest that this population subserves auditory tasks that require temporal precision, such as sound localization. The auditory nerve fibers that receive from hair cells from the bidirectional population (free standing) show poorer synchronization to tones, and respond to sound chiefly by an increase in discharge rate. This division into a timing (low CF) and a nontiming pathway has been explored by Köppl and Manley (1990). The degree of phase locking observed in the unidirectional population is still poor with respect to birds and mammals (Sullivan and Konishi 1984; Weiss and Rose 1988; Manley et al. 1990). In mitigation, it should be noted that the highest frequency of phase locking is temperature dependent. Nevertheless, both lizards and birds appear to have independently developed parallel processing of time and intensity components of the auditory stimulus.

The organization of the barn owl's auditory brainstem is similar to that of other reptiles and birds (Rubel and Parks 1988; Takahashi and Konishi 1988a,b). The points where it differs from the basic avian pattern may indicate specializations for the wide frequency range and high temporal resolution of owl's auditory system. The

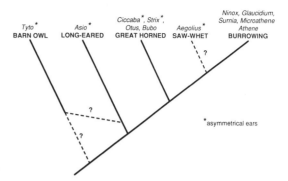

FIGURE 25.14. Phylogenetic relationships among the owls. Note that all except the stem group contain owls with asymmetric ears. Genus names are in *italics*, and those with ear asymmetries are marked with an asterisk. Dotted lines with questions marks indicate relationships that are in doubt. (Courtesy of S. Volman.)

phase-coding portions of the owl's brainstem auditory system are homologous to the mammalian and reptile circuits in which the auditory nerve projects to the bushy cells of the anteroventral cochlear nucleus, and the bushy cells project to the medial superior olive (for review see Harrison and Howe 1974). Specializations for the encoding of temporal information include the robust endbulb synapses from the auditory nerve to the nucleus magnocellularis, the few and/or short dendrites and thick axons of the magnocellular and laminaris neurons. The presence of various calcium-binding proteins in the eighth nerve, cochlear nuclei, and the nucleus laminaris has also been correlated with phase locking (Carr 1986). These shared features appear to be conserved solutions for the preservation and coding of temporally ordered signals in the auditory system.

Phase locking is most conspicuous in the owl. Auditory nerve fibers show phase locking, as measured by vector strength values above 0.2, to acoustic stimuli up to 9 kHz (Sullivan and Konishi 1984), as opposed to about 2 kHz in the pigeon (Hill, Stange, and Mo 1989), about 4 kHz in the starling (Gleich and Narins 1988) 5 to 6 kHz in the blackbird (Woolf and Sachs 1977; Sachs and Sinnott 1978), and about 5 to 6 kHz in the cat (Johnson 1980). Volman and Konishi (1990) provided the following hypothesis to explain the development of high-frequency phase locking in the owl. The barn owl uses a bicoordinate system of sound

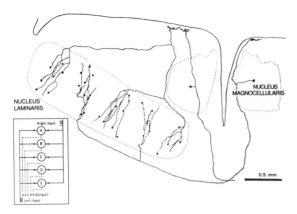

FIGURE 25.15. Mapping of interaural time differences in the nucleus laminaris of the barn owl. Magnocellular afferents from each side act as delay lines and interdigitate in the nucleus laminaris to form many maps of interaural difference oriented in the dorsoventral dimension of the nucleus. Those laminaris neurons close enough to receive synaptic input are shown as filled circles. (From Carr and Konishi 1990.)

Inset: Jeffress (1948) model of a circuit for measuring interaural time differences. The inputs act as delay lines and the postsynaptic cells as coincidence detectors. The pattern of inputs creates conduction delays in the nucleus that are equal but opposite to the interaural time differences. When these time differences are compensated for by the delay line, the postsynaptic cell fires maximally. Thus a map of interaural time difference is created by the array of cells. (From Konishi et al. 1988.)

localization, that is, it uses interaural intensity differences to localize sound in elevation, and interaural time differences to localize sound in azimuth (see above, Konishi, Sullivan, and Takahashi 1985). Thus the barn owl combines time and intensity information to determine the location of a sound source. There are substantial interaural intensity differences at 9 kHz (Coles and Guppy 1988; Moiseff 1989a,b). Therefore the pairing of time and intensity may have provided the selective pressure for the evolution of phase locking to comparatively high frequencies.

6.3.3 Mapping of Interaural Time Differences in the Nucleus Laminaris

The organization of nucleus laminaris in the chicken is representative of the plesiomorphic pattern found in most basal land birds and in the crocodilians, where nucleus laminaris is composed of a monolayer of bipolar neurons which receive input from ipsi- and contralateral cochlear nuclei onto their dorsal and ventral dendrites, respectively (Boord 1968; Jhaveri and Morest 1982; Browner and Marbey 1988) (Fig. 25.9). The magnocellular afferents appear to form a single map of interaural time difference along the mediolateral dimension of the nucleus laminaris in the chick (Parks and Rubel 1975; Young and Rubel 1983, 1986). This pattern is in contrast to the multiple, dorsoventrally directed maps of interaural time difference found in the barn owl (Fig. 25.15). Takahashi and Konishi (1988b) proposed that the morphology of the barn owl's nucleus laminaris is equivalent to the chicken's monolayer map turned by 90° and replicated many times over. There is an obvious adaptive significance in the development of many parallel maps of interaural time difference in the mediolateral dimension of nucleus laminaris, because the iteration of the same calculation would increase the sensitivity of the postsynaptic targets in the central nucleus of the inferior colliculus to interaural time differences. How could such a change come about?

Since the barn owls are a sister group to other owls (see Volman and Konishi 1990), a comparative study of several other species was undertaken to determine if auditory pathways are specialized to the same degree in all owls or if variations are present among various owls and between owls and other birds (Carr 1989). To this end, circuits were compared in the chicken and a number of owls, including the screech owl, *Otus asio*, the great horned owl, *Bubo virginianus*, and the barred owl, *Strix varia* and the burrowing owl, *Athene cunicularia*. The magnocellular-laminaris circuit was similar in all owls, except that there was considerable variation in the dorsoventral dimensions of the nucleus laminaris. The dorsoventral depth of nucleus laminaris is important because if the available interaural time difference is mapped in this dimension in the same way that it is mapped in the barn owl, then the dorsoventral dimension should be proportional to head size. This was not the case. The two most diurnal owls, the great horned and the burrowing owls, have the smallest nucleus laminaris for their size, while nucleus laminaris in the barred, screech, and saw-whet owls resembles that of the barn owl. From these observations, it was predicted that the great horned and burrowing owls represent an intermediate between the plesio-morphic and derived conditions, and should display maps of interaural time difference which change both with depth and mediolateral position in nucleus laminaris (Carr 1989).

It seems likely that independent hypertrophy of nucleus laminaris has occurred many times. This conclusion is supported by the observation of similar development of the nucleus laminaris in a basal land bird, the budgerigar (see Fig. 4 in Brauth 1990), in other advanced landbirds, including the starling and the zebra finch, and in all water birds examined, including several species of gull and ducks (unpublished observations). Cellular mechanisms could include the following: In the chick, nucleus laminaris undergoes 80% cell death during development (Rubel, Smith, and Miller 1975). If cell death were reduced by either decreasing the dependence of laminaris neurons on their target, or by increasing the size of the target, then nucleus laminaris could become larger. As soon as the nucleus laminaris ceased to be a monolayer, time delays could be introduced in the dorsoventral as well as the mediolateral dimension.

6.3.4 Parallel Processing of Time and Intensity Information

In the owl, the nucleus magnocellularis initiates a neural pathway that encodes timing information, while the nucleus angularis encodes intensity information (Sullivan and Konishi 1984). This parallel processing of time and intensity information has also been found in the chick (Warchol and Dallos 1990). The similarity between the owl and chick suggests that functional separation of time and intensity may be a common feature of the bird auditory system. Functional segregation of stimulus variables also appears to have evolved independently in the lizards (see 6.3.2 above).

The time coding pathways of reptiles, birds and mammals appear to be homologous (see above). The intensity-coding pathways of reptiles, birds, and mammals also share similar features, although it is not known whether these reflect homology or convergence. The nucleus angularis of birds and reptiles has many shared features with both the dorsal cochlear nucleus and the posteroventral cochlear nucleus of mammals (Foster and Hall 1978; Sullivan 1985; Takahashi and Konishi 1988b;

Ryugo and Rouiller 1988; Spzir, Sento, and Ryugo 1990; see below).

6.4 Comparisons with the Mammalian Central Auditory System

The evolution of mammals is accompanied by the conversion of jaw elements to middle ear bones. The development of this kind of middle ear made possible improved conduction of high-frequency sounds to the oval window. This major phylogenetic advance may have driven the evolution of the mammalian central auditory system. It seems probable that the development of high-frequency hearing was responsible for the substantial differences between mammals and the reptiles and birds. Table 25.1 outlines suggested homologies between the auditory systems of reptiles, birds and mammals. It also shows the many mammalian brainstem nuclei that do not appear to have homologs in either birds or reptiles.

Some aspects of the auditory system appear to have changed little during phylogeny, however. The best example is the system for encoding and measuring interaural time differences. The mammalian projection from low, best-frequency auditory nerve fibers to spherical bushy cells via an endbulb is homologous to the projection described in birds and reptiles (Sento and Ryugo 1989). Magnocellular neurons in birds and reptiles form a homogeneous population corresponding to the bushy/spherical cells of the mammalian anteroventral cochlear nucleus (Brawer, Morest, and Kane 1974; Jhaveri and Morest 1982; Browner and Pierz 1986). The relationship between endbulb endings and spherical cells in the auditory pathway is homologous in amniotic vertebrates (Boord and Rasmussen 1963; Miller 1975; Rubel and Parks 1975; Miller and Kasahara 1979; Marbey and Browner 1983; Browner and Marbey 1988). There are differences between the groups, however. In the alligator lizard, the barn owl, and the cat, each auditory nerve fiber may form more than one endbulb, while each bushy cell receives 2 to 3 endbulbs (Browner and Marbey 1988; Carr, unpublished observations; Fekete et al. 1982; Rouiller et al. 1986). In the cat, however, each auditory nerve fiber projects to many targets in the cochlear nucleus, indicating that there is more parallel processing of the auditory signal in this mammal

TABLE 25.1. Central auditory nuclei of reptiles and birds

Turtles	Lizards	Crocodiles	Chickens, pigeons	Owls	Mammals
Magnocellularis	Medial magnocellularis LATERAL MAGNOCELLULARIS	Magnocellularis	Magnocellularis	Magnocellularis	Anteroventral cochlear
Angularis	Lateral angularis MEDIAL ANGULARIS	Angularis	Angularis	Angularis	POSTEROVENTRAL COCHLEAR DORSAL COCHLEAR
Laminaris	Laminaris LATERAL MAGNOCELLULARIS	Laminaris	Laminaris	Laminaris	Medial superior olive
Superior olive	Superior olive	Superior olive	Superior olive	Superior olive	LATERAL SUPERIOR OLIVE
Ventral n. lat. lem.	Ventral n. lat. lem.	Ventral n. lat. lem.	N. lateral lemniscus (LLv) INTERMEDIATE N. LAT. LEM.	N. lateral lemniscus (LLv)	VENTRAL N. LAT. LEM
Dorsal n. lat. lem.	Dorsal n. lat. lem.	Dorsal n. lat. lem.	Ventral n. lat. lem. (VLVa) Ventral n. lat. lem. (VLVp)	Ventral n. lat. lem. (VLVa) Ventral n. lat. lem. (VLVp)	INTERMEDIATE N. LAT. LEM. DORSAL N. LAT. LEM.
Torus semicircularis	Torus semicircularis	Torus semicircularis	Mesencephalicus dorsalis	Inferior colliculus	Inferior colliculus
Central nucleus	Central nucleus	Central nucleus	Central nucleus	Central nucleus	Central nucleus
		External	External nucleus Superficial nucleus	External nucleus Superficial nucleus	External nucleus
Peritoral nucleus	Intercollicular/laminar	Intercollicular	Intercollicular	Intercollicular	Intercollicular
Reuniens	Medialis	Reuniens (centralis) Reuniens (diffusa)	Ovoidalis Semilunaris	Ovoidalis Semilunaris	Medial geniculate (ventral) Medial geniculate
Medial DVR	Medial DVR	Medial DVR	Field L2	Field L2	Layer IV auditory cortex

TABLE 25.1. Summary of the nuclei of the central auditory systems of birds and reptiles, with homologies suggested by studies cited in the text. The names of those nuclei that may be field homologs or new nuclei are shown in capital letters.

TABLE 25.2. Future directions

	Anatomy	Physiology
Turtles	Trace projections of individual cochlear nuclei	All central auditory
Lizards	Trace projections of individual cochlear nuclei	All central auditory
	Determine variation related to papilla structure	
Crocodiles	All brainstem auditory connections	Brainstem division into time & intensity
		Sound localization
Paleognaths	All central auditory connections	All central auditory
Primitive land birds	Divisions of MLD/inferior colliculus	Midbrain division into time & intensity
Owls	Ascending connections of inferior colliculus	Thalamic and Field L responses
Songbirds	All brainstem auditory connections	
Water birds	All central auditory connections	All central auditory
All	Interspecific variation	Interspecific variation

TABLE 25.2. Future directions for research include tract tracing studies summarized in the Anatomy column, and electrophysiological and behavioral studies in the Physiology column.

(Rhode, Oertel, and Smith 1983; Rouiller and Ryugo 1984).

Mammalian bushy cells are very like magnocellular neurons in birds and turtles in that they have large, round somata with numerous somatic spines, and thick axons. They have more dendrites than magnocellular neurons, however. Bushy cell physiology also resembles that of magnocellular neurons. Although the bushy cell projection to the medial superior olive has not been described in detail, these cells project bilaterally to the olive in a pattern that appears to be similar to that described in the bird (Stotler 1953; Warr 1966; Young and Rubel 1986; Carr and Konishi 1988; Yin and Chan 1990). It would be important to determine the nature of the magnocellular projection to the nucleus laminaris in such reptiles as crocodilians.

7. Summary

(1) The central auditory pathways are organized in a pattern that is similar in turtles, lizards, crocodilians, and birds, although many homologous nuclei have different names (see Table 25.1).

(2) In the morphotype, the auditory nerve projects to two cochlear nuclei, the nucleus magnocellularis and the nucleus angularis. These nuclei project to second order olivary nuclei, lemniscal nuclei and the midbrain torus (nucleus mesencephalicus lateralis dorsalis; inferior colliculus). The inferior colliculus projects to a major dorsal thalamic target that projects to a specialized auditory region of the dorsal ventricular ridge, and to the striatum.

(3) In the owl and the chicken, physiological techniques have established a functional separation of the processing of phase and intensity information beginning at the level of the cochlear nuclei. This segregation is marked by a number of morphological and physiological specializations that are found in both birds and reptiles. Therefore this functional separation may be found in all bird and reptile auditory systems.

(4) Independent developments in the auditory periphery may drive changes in the central auditory system. Examples are found in both lizards and owls.

(5) Similarities between the living archosaurs (crocodilians) and birds suggest that dinosaurs had similar auditory pathways to the living archosaurs.

(6) Major gaps in our knowledge of the central auditory pathways of birds and reptiles are summarized in Table 25.2.

Acknowledgments. I thank R. Browner, R. Capranica, H. Karten, M. Konishi, C. Köppl, G. Manley, D. Margoliash, R. G. Northcutt, H. Scheich, T. Takahashi, W. Tomlinson, S. Volman, D. Webster and J. M. Wild for helpful discussion. S. Brauth, R. Browner, M. Konishi, K. Morest, H. Scheich, S. Volman, and W. Wilczynski generously allowed me to use unpublished results or figures from their work. C. McCormick provided a most useful, thorough review. Supported by the Sloan Foundation and by NIH DC00436.

References

Arends JJA (1981) Sensory and motor aspects of the trigeminal system in the mallard (*Anas platyrhonchos* L.). State Univ of Leiden, Netherlands.

Arends JJA, Zeigler HP (1986) Anatomical identification of an auditory pathway from a nucleus of the lateral lamniscal system to the frontal telencephalon (nucleus basalis) of the pigeon. Brain Res 398:375–381.

Balaban CD, Ulinski PS (1981) Organization of thalamic afferents to anterior dorsal ventricular ridge in turtles. I. Projections of thalamic nuclei. J Comp Neurol 200:95–129.

Barbas-Henry HA, Lohman AHM (1986) Primary projections and efferent cells of the VIIIth cranial nerve in the monitor lizard, *Varanus exanthematicus*. Brain Res 398:375–381.

Belekhova MG, Zharskaja VD, Khachunys AS, Gaidaenko GV, Tumanova NL (1985) Connections of the mesencephalic, thalamic and telencephalic auditory centers in turtles. Some structural bases for audiosomatic interrelations. J Hirnforsch 26:127–152.

Biederman-Thorson M (1970) Auditory responses of units in the ovoid nucleus and cerebrum (Field L) of the ring dove. Brain Res 24:247–256.

Bigalke-Kunz B, Rübsamen R, Dörrscheidt GJ (1987) Tonotopic organization and functional characterization of the auditory thalamus in a songbird, the European starling. J Comp Physiol 161:255–265.

Bonke BA, Bonke D, Scheich H (1979) Connectivity of the auditory forebrain nuclei in the guinea fowl (*Numida meleagris*). Cell Tissue Res 200:101–121.

Boord RL (1968) Ascending projections of the primary cochlear nuclei and nucleus laminaris in the pigeon. J Comp Neurol 133:523–542.

Boord RL, Rasmussen GL (1963) Projection of the cochlear and lagenar nerves on the cochlear nuclei of the pigeon. J Comp Neurol 120:462–475.

Bottjer SW, Arnold AP (1982) Afferent neurons in the hypoglossal nerve of the zebra finch (*Poephila guttata*): Localization with horseradish peroxidase. J Comp Neurol 210:190–197.

Bottjer SW, Arnold AP (1984) The role of feedback from the vocal organ. I. Maintenance of stereotypical vocalizations by adult zebra finches. J Neurosci 4:2387–2396.

Bottjer SW, Arnold AP (1986) The ontogeny of vocal learning in songbirds. In: Blass EM (ed) Handbook of Behavioral Neurobiology. New York: Plenum Press, pp. 129–161.

Bottjer SW, Halsema KA, Brown SA, Miesner EA (1989) Axonal connections of a forebrain nucleus involved with vocal learning in Zebra Finches. J Comp Neurol 279:312–326.

Brauth SE (1990) Investigation of central auditory nuclei in the budgerigar with cytochrome oxidase histochemistry. Brain Res 508:142–146.

Brauth SE, McHale CM, Brasher CA, Dooling RJ (1987) Auditory pathways in the Budgerigar. Brain Behav Evol 30:174–199.

Brauth SE, McHale CM (1988) Auditory pathways in the budgerigar. II. Intratelencephalic pathways. Brain Behav Evol 32:193–207.

Brawer JR, Morest DK, Kane EC (1974) The neuronal architecture of the cochlear nucleus of the cat. J Comp Neurol 155:251–300.

Brown JL (1971) An exploratory study of vocalization areas in the brain of the red-winged blackbird (*Agelius phoenicerus*). Behavior 24:91–127.

Browner RH, Kennedy MC, Facelle T (1981) The cytoarchitecture of the torus semicircularis in the red-eared turtle. J Morphol 169:207–223.

Browner RH, Pierz DM (1986) Endbulbs of Held in a cochlear nucleus, nucleus magnocellularis in the red-eared turtle, *Chrysemys scripta elegans*. Soc Neurosci Abstr 12:1265.

Browner RH, Marbey D (1988) The nucleus magnocellularis in the red-eared turtle, *Chrysemys scripta elegans*: Eighth nerve endings and neuronal types. Hearing Res 33:257–272.

Bruce LL, Butler AB (1984) Telencephalic connections in lizards. II. Projections to anterior dorsal ventricular ridge. J Comp Neurol 229:602–615.

Calford MB, Piddington RW (1988) Avian interaural canal enhances interaural delay. J Comp Physiol 162:503–510.

Carr CE (1986) Time coding in electric fish and owls. Brain Behav Evol 28:122–133.

Carr CE (1989) Comparative anatomy of the brainstem auditory pathways in owls. In: Georg Thieme Verlag (eds). Neural mechanisms of behavior. Proceedings of the 2nd international congress of neuroethology, Stuttgart: Springer-Verlag, pp. 116.

Carr CE, Konishi M (1988) Axonal delay lines for time measurement in the owl's brainstem. Proc Natl Acad Sci USA 85:8311–8315.

Carr CE, Fujita I, Konishi M (1989) Distribution of GABAergic neurons and terminals in the auditory system of the barn owl. J Comp Neurol 286:190–207.

Carr CE, Konishi M (1990) A circuit for detection of interaural time differences in the brainstem of the barn owl. J Neurosci 10:3227–3246.

Carroll RL (1987) Vertebrate paleontology and evolution. New York: W.H. Freeman & Co.

Coles RB, Aitkin LM (1979) The response properties of auditory neurones in the midbrain of the domestic fowl (*Gallus gallus*) to monaural and binaural stimuli. J Comp Physiol 134:241–251.

Coles RB, Guppy A (1988) Directional hearing in the barn owl (*Tyto alba*). J Comp Physiol 163:117–133.

Conlee JW, Parks TN (1986) Origin of ascending auditory projections to the nucleus mesencephalicus lateralis pars dorsalis in the chicken. Brain Res 367:96–113.

Crawford AC, Fettiplace R (1980) The frequency selectivity of auditory nerve fibers and hair cells in the cochlea of the turtle. J Physiol 306:79–125.

DeFina AV, Webster DB (1974) Projections of the intraotic ganglion to the medullary nuclei in the Tegu lizard, *Tupinambis nigropunctatus*. Brain Behav Evol 10:197–211.

Dodd F, Capranica RR (1989) Frequency and temporal selectivity of single auditory nerve fibers in the tokay gecko. Soc Neurosci Abst 15:348.

Dooling RJ, Brown SD, Park TJ, Okanoya K (1990) Natural perceptual categories for vocal signals in budgerigars (*Melopsittacus undulatus*). In: Stebbins WC, Berkley MA (eds) Comparative perception Vol. II: Complex signals. New York: John Wiley and Sons, pp. 345–374.

Doupe A, Konishi M (1989) Auditory properties of song nuclei of estrildid finches. Soc Neurosci Abstr 15:347.

Du Lac S, Knudsen EI (1990) Neural maps of head movement vector and speed in the optic tectum of the barn owl. J Neurophysiol 63:131–146.

Fay RR (1988) Hearing in vertebrates: A psychophysics databook. Winnetka, Illinois: Hill-Fay Associates.

Fekete DM, Rouiller EM, Liberman MC, Ryugo DK (1982) The central projections of intracellularly labeled auditory nerve fibers in cats. J Comp Neurol 229:432–450.

Fischer FP, Köppl C, Manley GA (1988) The basilar papilla of the barn owl Tyto alba: A quantitative morphological SEM analysis. Hearing Res 34:87–101.

Fleishman JL, Rand AS (1989) *Caiman crocodilus* does not require vision for underwater prey capture. J Herpetol 23:296.

Foster RJ, Hall WJ (1978) The organization of central auditory pathways in a reptile *Iguana iguana*. J Comp Neurol 178:783–832.

Gans C, Maderson PFA (1973) Sound producing mechanisms in Recent Reptiles: Review and comment. Am Zoologist 13:1195–1203.

Gans C, Wever EG (1972) The ear and hearing in Amphisbaenia (Reptilia). J Exp Zool 179:17–34.

Gans C, Wever EG (1976) The ear and hearing in *Sphenodon punctatus*. Proc Natl Acad Sci USA 73:4244–4246.

Gleich O, Narins PM (1988) The phase response of primary auditory afferents in a songbird (*Sturnus vulgaris* L.). Hearing Res 32:81–92.

Gurney M (1981) Hormonal control of cell form and number in the zebra finch song system. J Neurosci 1:658–673.

Hall WC, Ebner FF (1970) Thalamotelencephalic projections in the turtle (*Pseudemys scripta*). J Comp Neurol 140:101–127.

Haeseler C, Brix J, Manley GA (1989) Innervation patterns and spontaneous activity of afferent fibers to the chick's lagenar macula. In: Elsner N, Singer W (eds) Dynamics and Plasticity in Neuronal Systems. Stuttgart: Thieme, p. 282.

Harrison JM, Howe ME (1974) Anatomy of the afferent auditory nervous system in mammals. In: Kreidel WD, Neff WD (eds) Handbook of Sensory Physiology Vol 5/1. Berlin: Springer-Verlag, pp. 283–336.

Hartline PH, Campbell HW (1969) Auditory and vibratory responses in the midbrains of snakes. Science 163:1221–1223.

Heil P, Scheich H (1986) Effects of unilateral and bilateral cochlear removal on 2-deoxyglucose patterns in the chick auditory system. J Comp Neurol 252:279–301.

Hill KG, Stange G, Mo J (1989) Temporal synchronization in the primary auditory response in the pigeon. Hearing Res 39:63–74.

Hoogland PV (1982) Brainstem afferents to the thalamus in a lizard, *Varanus exantematicus*. J Comp Neurol 210:152–162.

Hopson JA (1977) Relative brain size and behavior in archosaurian reptiles. Ann Rev Ecol System 8:429–448.

Hopson JA (1980) Relative brain size in dinosaurs. Implications for dinosaurian endothermy. In: Thomas RDK, Olson EC (eds) A Cold Look at the Warm-Blooded Dinosaurs. AAAS Selected Symposium, pp. 287–310.

Jeffress LA (1984) A place theory of sound localization. J Comp Physiol Psychol 41:35–39.

Jhaverl S, Morest DK (1982) Neuronal architecture in nucleus magnocellularis of the chicken auditory system with observations on nucleus laminaris: A light and electron microscope study. Neuroscience 7:809–836.

Johnson DH (1980) The relationship between spike rate and synchrony in responses of auditory nerve fibers to single tones. J Acoust Soc Am 68:1115–1122.

Karten HJ (1967) The organization of the ascending auditory pathway in the pigeon (*Columba livia*) I. Diencephalic projections of the inferior colliculus (nucleus mesencephali lateralis, pars dorsalis). Brain Res 6:409–427.

Karten HJ (1968) The ascending auditory pathway in the pigeon (*Columba livia*) II. Telencephalic projections of the nucleus ovoidalis thalami. Brain Res 11:134–153.

Karten HJ, Hodos W (1967) A Stereotaxic Atlas of the Brain of the Pigeon (*Columbia livea*). Baltimore: Johns Hopkins Press.

Karten HJ, Shimizu T (1989) The origins of neocortex: connections and lamination as distinct events in evolution. J Cognit Neurosci 1:291–301.

Katz LC, Gurney ME (1981) Auditory responses in the zebra finch's motor system for song. Brain Res 211:192–197.

Kelley DB, Nottebohm F (1979) Projections of a telencephalic auditory nucleus – Field L – in the canary. J Comp Neurol 183:455–470.

Kennedy MC (1974) Auditory multiple-unit activity in the midbrain of the Tokay gecko (*Gekko gekko*, L.). Brain Behav Evol 10:257–264.

Kennedy MC (1975) Vocalization elicited in a lizard by electrical stimulation of the midbrain. Brain Res 91: 321–325.

Kennedy MC, Browner RH (1981) The torus semicircularis in a gekkonid lizard. J Morphol 169:259–274.

Kirsch M, Coles RB, Leppelsack H-J (1980) Unit recordings from a new auditory area in the frontal neostraitum of the awake starling (*Sturnus vulgaris*). Exp Brain Res 38:375–380.

Klump GM, Windt W, Curio E (1986) The great tit's (*Parus major*) auditory resolution in azimuth. J Comp Physiol 158:383–390.

Knudsen EI (1980) Sound localization in birds. In: Popper AN, Fay RR (eds) Comparative studies of hearing in vertebrates. Berlin: Springer Verlag, pp. 287–322.

Knudsen EI, Konishi M, Pettigrew JD (1977) Receptive fields of auditory neurons in the owl. Science 198: 1278–1280.

Knudsen EI (1983) Subdivisions of the inferior colliculus in the barn owl (*Tyto alba*). J Comp Neurol 218:174–186.

Knudsen EI, Knudsen PF (1983) Space-mapped auditory projections from the inferior colliculus to the optic tectum in the barn owl (*Tyto alba*). J Comp Neurol 218:187–196.

Knudsen EI, Konishi M (1978) A neural map of auditory space in the owl. Science 200:795–797.

Knudsen EI, Konishi M (1979a) Mechanisms of sound localization in the barn owl (*Tyto alba*). J Comp Neurol 133:13–21.

Knudsen EI, Konishi M (1979b) Sound localization by the barn owl (*Tyto alba*). J Comp Physiol 133:1–11.

Konishi M (1963) The role of auditory feedback in the vocal behavior of the domestic fowl. J Tierpsychol 20:349–367.

Konishi M (1970) Comparative neurophysiological studies of hearing and vocalization in songbirds. J Comp Physiol 66:257–272.

Konishi M (1973a) Development of auditory neuronal responses in avian embryos. Proc Natl Acad Sci USA 70:1795–1798.

Konishi M (1973b) How the owl tracks its prey. Am Scientist 61:414–424.

Konishi M (1985) Birdsong: From behavior to neuron. Annu Rev Neurosci 8:125–170.

Konishi M (1986) Centrally synthesized maps of sensory space. Trends Neurosci 9:163–168.

Konishi M (1989) Bird song for neurobiologists. Neuron 3:541–549.

Konishi M, Sullivan WE, Takahashi T (1985) The owl's cochlear nuclei process different sound localization cues. J Acoust Soc Am 78:360–364.

Konishi M, Takahashi T, Wagner H, Sullivan WE, Carr CE (1988) Neurophysiological and anatomical substrates of sound localization. In: Edelman GM, Gall WE, Cowan WM (eds) Auditory function: Neurobiological Bases of Hearing. Neuroscience Institute. New York: John Wiley & Sons, pp. 721–745.

Köppl C, Manley GA (1990) Peripheral auditory processing in the bobtail lizard *Tiliqua rugosa*. II. Tonotopic organization and innervation pattern of the basilar papilla. J Comp Physiol 167:101–112.

Köppl C, Manley GA (1990) Peripheral auditory processing in the bobtail lizard *Tiliqua rugosa*. II. Patterns of spontaneous and tone-evoked nerve-fiber activity. J Comp Physiol 167:113–127.

Kreithen ML, Quine DB (1979) Infrasound detection by the homing pigeon: A behavioral audiogram. J Comp Physiol 129:1–4.

Kroodsma DE (1984) Songs of the alder flycatcher (*Empidonax alnorum*) and willow flycatcher (*Empidonax traillii*) are innate. Auk 101:13–24.

Künzle H (1986) Projections from the cochlear nuclear complex to rhombencephalic auditory centers and torus semicircularis in the turtle. Brain Res 379:307–319.

Leake PA (1974) Central projections of the statoacoustic nerve in *Caiman crocodilus*. Brain Behav Evol 10:170–196.

Leibler LM (1975) Monaural and binaural pathways in the ascending auditory system of the pigeon. PhD. Thesis, Massachusetts Institute of Technology.

Leppelsack H-J, Vogt M (1976) Responses of auditory neurons in the forebrain of a songbird to stimulation with species specific sounds. J Comp Physiol 107: 263–274.

Leppelsack H-J (1978) Unit responses to species-specific sounds in the auditory forebrain center of birds. Fed Proc 37:2336–2341.

Leppelsack H-J, Schwartzkopff J (1972) Eigenschaften von aukutishen neuronen im kaudalen Neostriatum von vogeln. J Comp Physiol 80:137–140.

Lohman AHM, van Woerden-Verkley I (1978) Ascending connections to the forebrain in the tegu lizard. J Comp Neurol 182:555–594.

Maekawa M (1987) Auditory responses in the nucleus basalis of the pigeon. Hearing Res 27:231–237.

Manley GA (1970a) Frequency sensitivity of auditory neurons in the *Caiman* cochlear nucleus. Zeit verg Physiol 66:251–256.

Manley GA (1970b) Comparative studies of auditory physiology in reptiles. Zeit verg Physiol 67:363–381.

Manley GA (1974) Activity patterns of neurons in the peripheral auditory system of some reptiles. Brain Behav Evol 10:244–256.

Manley GA (1977) Response patterns and peripheral origin of auditory nerve fibres in the monitor lizard, *Varanus bengalensis*. J Comp Physiol 118:249–260.

Manley GA (1981) A review of the auditory physiology of reptiles. In: Autrum HE, Perl E, Schmidt RF (eds) Progress in Sensory Physiology. Berlin: Springer Verlag, pp. 49–134.

Manley GA, Gleich O, Leppelsack H-J, Oeckinghaus H (1985) Activity patterns of cochlear ganglion neurons in the starling. J Comp Physiol 157:161–181.

Manley GA, Köppl C, Konishi M (1988) A neural map of interaural intensity difference in the brainstem of the barn owl. J Neurosci 8:2665–2677.

Manley GA, Köppl C, Johnstone BM (1990) Peripheral auditory processing in the bobtail lizard *Tiliqua rugosa*. I. Frequency tuning of auditory-nerve fibers. J Comp Physiol 167:88–99.

Manley GA, Yates GK, Köppl C, Johnstone BM (1990) Peripheral auditory processing in the bobtail lizard *Tiliqua rugosa*. IV. Phase locking of auditory nerve fibers. J Comp Physiol 167:129–138.

Manley JA (1971) Single unit studies in the midbrain auditory area of Caiman. Zeit verg Physiol 71:255–261.

Marbey D, Browner RB (1983) A golgi impregnation study of the acoustic tubercle in the red-eared turtle, *Chrysemys scripta elegans*. Soc Neurosci Abstr 9:495.

Marcellini DL (1978) The acoustic behavior of lizards. In: Greenberg N, MacLean PD (eds) Behavior and neurology of lizards. U.S. Department of Health, Education and Welfare, Rockville, MD, pp. 287–300.

Margoliash D (1983) Acoustic parameters underlying the responses of song specific in the white-crowned sparrow. J Neurosci 3:1039–1057.

Marler P (1982) Avian and primate communication: The problem of natural categories. Neurosci Biobehav Rev 6:87–94.

Marler P, Peters S (1981) Birdsong and speech: Evidence for special processing. In: Eimas P, Miller J (eds) Perspectives on the study of speech. Hillsdale, New Jersey: Ehrlbaum, pp. 75–112.

Masino T, Knudsen EI (1990) Horizontal and vertical components of head movement are controlled by distinct neural circuits in the barn owl. Nature 345:434–437.

McCasland JS (1987) Neuronal control of bird song production. J Neurosci 7:23–39.

Miller MR (1975) The cochlear nuclei of lizards. J Comp Neurol 159:375–406.

Miller MR (1980a) The cochlear nuclei of snakes. J Comp Neurol 192:717–736.

Miller MR (1980b) The reptilian cochlear duct. In: Popper AN, Fay RR (eds) Comparative studies of hearing in vertebrates. Berlin: Springer Verlag, pp. 169–204.

Miller MR, Beck J (1988) Auditory hair cell innervational patterns in lizards. J Comp Neurol 271:604–628.

Miller MR, Kasahara M (1979) The cochlear nuclei of some turtles. J Comp Neurol 185:221–236.

Moiseff A (1989a) Bicoordinate sound localization by the barn owl. J Comp Physiol 164:637–644.

Moiseff A (1989b) Binaural disparity cues available to the barn owl for sound localization. J Comp Physiol 164:629–636.

Moiseff A, Konishi M (1981) Neuronal and behavioral sensitivity to binaural time differences in the owl. J Neurosci 1:40–48.

Moiseff A, Konishi M (1983) Binaural characteristics of units in the owl's brainstem auditory pathway: Precursors of restricted spatial receptive fields. J Neurosci 2:2553–2562.

Müller SC, Scheich H (1985) Functional organization of the avian auditory field L. J Comp Physiol 156:1–12.

Mulroy MJ (1987) Auditory stereocilia in the alligator lizard. Hearing Res 25:11–21.

Norberg RA (1978) Skull asymmetry, ear structure and function, and auditory localization in Tengmalm's owl, *Aegolius funerus* (Linne). Philos Trans R Soc Lond 282:325–410.

Northcutt RG (1984) Evolution of the vertebrate central nervous system: patterns and processes. Am Zoologist 24:701–716.

Nottebohm F, Stokes TM, Paton JA (1976) Central control of song in the canary, *Serinus canarius*. J Comp Neurol 207:344–357.

Nottebohm F (1980) Brain pathways for vocal learning in birds: A review of the first 10 years. Prog Pyschobiol Physiol Psychol 9:85–124.

Nottebohm F, Kelley DB, Paton JA (1982) Connections of vocal control nuclei in the canary telencephalon. J Comp Neurol 207:344–357.

Olsen JF, Knudsen EI, Esterly SD (1989) Neural maps of interaural time and intensity differences in the optic tectum of the barn owl. J Neurosci 9:2591–2605.

Okuhata S, Saito N (1987) Synaptic connections of thalamo-cerebral vocal nuclei of the canary. Brain Res Bull 18:35–44.

Parent A (1976) Striatal afferent connections in the turtle (*Chrysemys picta*) as revealed by retrograde axonal transport of horseradish peroxidase. Brain Res 108:25–36.

Parks TN, Rubel EW (1975) Organization and development of brain stem auditory nucleus of the chicken: Organization of projections from N. magnocellularis to N. laminaris. J Comp Neurol 164:435–448.

Paton JA, Manogue KR, Nottebohm F (1981) Bilateral organization of the vocal control pathway in the budgerigar, *Melopsittacus undulatus*. J Neurosci 1:1276–1288.

Patterson WC (1966) Hearing in the turtle. J Audit Res 6:453–464.

Payne RS (1971) Acoustic localization of prey by barn owls (*Tyto alba*). J Exp Biol 54:535–573.

Potash LM (1970) Neuroanatomical regions relevant to production and analysis of vocalization within the avian torus semicircularis. Experientia 26:257–264.

Pritz MB (1974a) Ascending connections of a thalamic auditory area in a crocodile, *Caiman crocodilus*. J Comp Neurol 153:199–214.

Pritz MB (1974b) Ascending connections of a midbrain auditory area in a crocodile, *Caiman crocodilus*. J Comp Neurol 153:179–198.

Pritz MB (1980) Parallels in the organization of auditory and visual systems in crocodiles. In: Ebbesson SOE (ed) Comparative Neurology of the Telencephalon. New York: Plenum Press, pp. 331–342.

Rhode WS, Oertel D, Smith PH (1983) Physiological response properties of cells labeled intracellularly with horseradish peroxidase in cat ventral cochlear nucleus. J Comp Neurol 213:448–463.

Rose C, Weiss TF (1988) Frequency dependence of synchronization of cochlear nerve fibers in the alligator lizard: Evidence for a cochlear origin of timing and non-timing neural pathways. Hearing Res 33:151–166.

Rouiller EM, Ryugo DK (1984) Intracellular marking of physiologically characterized cells in the ventral cochlear nucleus of the cat. J Comp Neurol 225:167–186.

Rouiller EM, Cronin-Schreiber R, Fekete DM, Ryugo DK (1986) The central projections of intracellularly labeled auditory nerve fibers in cats: an analysis of terminal morphology. J Comp Neurol 249:261–278.

Rubel EW, Parks TN (1975) Organization and development of brainstem auditory nuclei of the chicken: Tonotopic organization of N. magnocellularis and N. laminaris. J Comp Neurol 164:411–434.

Rubel EW, Parks TN (1988a) Organization and development of the avian brainstem auditory system. In: Edelman GM, Gall WE, Cowan WM (eds) Auditory function: Neurobiological Bases of Hearing. Neuroscience Institute, New York: John Wiley & Sons, pp. 3–92.

Rübsamen R, Dörrscheidt GJ (1986) Tonotopic organization of the auditory forebrain in a songbird, the European starling. J Comp Physiol 158:639–646.

Rubel EW, Smith ZDJ, Miller LC (1975) Organization and development of brain stem auditory nuclei of the chicken: Ontogeny of nucleus magnocellularis and nucleus laminaris. J Comp Neurol 166:469–490.

Ryugo DK, Rouiller EM (1988) Central projections of intracellularly labeled auditory nerve fibers in cats: morphometric correlations with physiological properties. J Comp Neurol 271:130–142.

Sachs MB, Sinnott JM (1978) Responses to tones of single cells in nucleus magnocellularis and nucleus angularis of the redwing blackbird (*Agelaius phoeniceus*). J Comp Physiol 126:347–361.

Sachs MB, Sinnott JM, Hienz RD (1978) Behavioral and physiological studies of hearing in birds. Fed Proc 37:2329–2335.

Schaller GB, Crawshaw PG (1982) Fishing behavior of Paraguayan caiman (*Caiman crocodilus*). Copeia 82:66–72.

Schermuly L, Klinke R (1990) Infrasound sensitive neurons in the pigeon cochlear ganglion. J Comp Physiol 166:355–363.

Scheich H (1990) Representational geometries of telencephalic auditory maps in birds and mammals. In: Finlay B, Innocenti G (eds) The Neocortex: Ontogeny and Phylogeny. Nato workshop, Plenum Press. (in press).

Scheich H, Bonke BA, Bonke D, Langer G (1979a) Functional organization of some auditory nuclei in the guinea fowl demonstrated by the 2-deoxyglucose technique. Cell Tissue Res 204:17–27.

Scheich H, Langer G, Bonke D (1979b) Responsiveness of units in the auditory neostriatum of the guinea fowl (*Numida meleagris*) to species specific calls and synthetic stimuli. II. Discrimination of iambus-like calls. J Comp Physiol 132:257–276.

Schermuly L, Klinke R (1990) Infrasound sensitive neurones in the pigeon cochlear ganglion. J Comp Physiol 166:355–363.

Sento S, Ryugo DK (1989) Endbulbs of Held and spherical bushy cells in cats: morphological correlates with physiological properties. J Comp Neurol 280:553–562.

Simpson HB, Vicario DS (1990) Brain pathways for learned and unlearned vocalizations differ in zebra finches. J Neurosci 10:1541–1556.

Smith CA, Konishi M, Schuff N (1985) Structure of the barn owl's (*Tyto alba*) inner ear. Hearing Res 17:237–247.

Smith ZDJ, Rubel EW (1979) Organization and Development of brain stem auditory nuclei of the chicken: Dendritic gradients in nucleus laminaris. J Comp Neurol 186:213–239.

Smolders JWT, Klinke R (1986) Synchronized responses of primary auditory fibre-populations in *Caiman crocodilus* (L.) to single tones and clicks. Hearing Res 24:89–103.

Spence CD, Pearson JD (1990) The computation of sound source elevation in the barn owl. In: Advances in neural information processing systems 2. IEEE Conference on neural information processing systems.

Spzir MR, Sento S, Ryugo DK (1990) The central projections of the cochlear nerve fibers in the alligator lizard. J Comp Neurol 295:530–547.

Stotler WA (1953) An experimental study of the cells and connections of the superior olivary complex of the cat. J Comp Neurol 98:401–432.

Sullivan WE (1985) Classification of response patterns in cochlear nucleus of barn owl: Correlation with functional response properties. J Neurophysiol 53:201–216.

Sullivan WE, Konishi M (1984) Segregation of stimulus phase and intensity coding in the cochlear nucleus of the barn owl. J Neurosci 4:1787–1799.

Sullivan WE, Konishi M (1986) Neural map of interaural phase difference in the owl's brainstem. Proc Natl Acad Sci 83:8400–8404.

Takahashi T, Konishi M (1986) Selectivity for interaural time difference in the owl's midbrain. J Neurosci 6:3413–3422.

Takahashi T, Moiseff A, Konishi M (1984) Time and intensity cues are processed independently in the auditory system of the owl. J Neurosci 4:1781–1786.

Takahashi T, Carr CE, Brecha N, Konishi M (1987) Calcium binding protein-like immunoreactivity labels the terminal field of nucleus laminaris of the barn owl. J Neurosci 7:1843–1856.

Takahashi T, Konishi M (1988a) Projections of nucleus angularis and nucleus laminaris to the lateral lemniscal nuclear complex of the barn owl. J Comp Neurol 274:212–238.

Takahashi T, Konishi M (1988b) The projections of the cochlear nuclei and nucleus laminaris to the inferior colliculus of the barn owl. J Comp Neurology 274:190–211.

Takahashi T, Konishi M (1988) Commissural projections mediate inhibition in a lateral lemniscal nucleus of the Barn Owl. Soc Neurosci Abstr 14:323.

Takahashi T, Wagner H, Konishi M (1988) The role of commissural projections in the representation of bilateral space in the Barn Owl's inferior colliculus. J Comp Neurol 281:545–554.

ten Donkelaar HJ, Bangma GC, Barbas-Henry HA, de Boer-van Huizen R, Wolters JG (1987) The brain stem in a lizard, *Varanus exanthematicus*. Adv Anat Embryol Cell Biol 103:56–60.

Theurich M, Langer G, Scheich H (1984) Infrasound responses in the midbrain of the guinea fowl. Neurosci Lett 49:81–86.

Ulinski PS, Margoliash D (1990) Neurobiology of the reptile-bird transition. In: Jones EG, Peters A (eds) Cerebral Cortex: Evolution and comparative anatomy of cerebral cortex. New York: Plenum Press (in press).

Vicario DS, Nottebohm F (1987) Organization of the zebra finch song control system: I. Representation of syringeal muscles in the hypoglossal nucleus. J Comp Neurol 271:346–354.

Volman SF (1990) Neuroethological approaches to the evolution of neural systems. Brain Behav Evol 36:154–165.

Volman SF, Konishi M (1989) Spatial selectivity and binaural responses in the inferior colliculus of the Great Horned Owl. J Neurosci 9:3083–3096.

Volman SF, Konishi M (1990) Comparative physiology of sound localization in four species of owls. Brain Behav Evol (in press).

Wagner H, Takahashi T, Konishi M (1987) Representation of interaural time difference in the central nucleus of the barn owls' inferior colliculus. J Neurosci 7:3105–3116.

Warchol ME, Dallos P (1989) Neural response to very low-frequency sound in the avian cochlear nucleus. J Comp Physiol 166:83–95.

Warchol ME, Dallos P (1990) Neural coding in the chick cochlear nucleus. J Comp Physiol 166:721–734.

Warr WB (1966) Fiber degeneration following lesions in the anterior ventral cochlear nucleus of the cat. Exp Neurol 14:453–474.

Weiss TF, Rose C (1988) A comparison of synchronization filters in different auditory receptor organs. Hearing Res 33:175–180.

Weiss TF, Mulroy MJ, Turner RG, Pike CL (1976) Tuning of single fibers in the cochlear nerve of the alligator lizard: Relation to receptor organ morphology. Brain Res 115:71–90.

Wever EG (1978) The Reptile Ear. Princeton, New Jersey: Princeton University Press.

Wever EG, Vernon JA (1957) Auditory responses in the spectacled caiman. J Cell Comp Physiol 50:333–339.

Wilczynski W (1984) Central neural systems subserving a homoplasous periphery. Am Zoologist 24:755–763.

Wild JM (1987) Nuclei of the lateral lemniscus project directly to the thalamic auditory nuclei in the pigeon. Brain Res 408:303–307.

Winter P, Schwartzkopff J (1961) Form und zellzahl der akustischen nervenzentren in der medulla oblongata von eulen (Striges). Experientia 17:515–516.

Wise LZ, Frost BJ, Shaver SW (1991) The representation of sound location and frequency in the midbrain of the saw-whet owl (*Aegolius acadicus*). J Comp Physiol (submitted).

Woolf NK, Sachs MB (1977) Phase-locking to tones in avian auditory nerve fibers. J Acoust Soc Am 62:46.

Yin TCT, Chan JCK (1990) Interaural time sensitivity in the medial superior olive of the cat. J Neurophysiol 64:465–488.

Young SR, Rubel EW (1983) Frequency-specific projections of individual neurons in chick brainstem auditory nuclei. J Neurosci 7:1373–1378.

Young SR, Rubel EW (1986) Embryogenesis of arborization pattern and topography of individual axons in n. laminaris of the chicken brain stem. J Comp Neurol 254:425–459.

Zaretsky MD, Konishi M (1976) Tonotopic organization in the avian telencephalon. Brain Res 111:167–171.

26
Hearing in Birds

Robert J. Dooling

1. Introduction

Birds are the most completely studied class of vertebrates other than mammals. Recent estimates suggests that fewer than two percent of the living species of birds remain to be described (Carroll 1988; Welty 1982). At present there is general agreement that there are about 8,900 living species of birds grouped into approximately 166 families consisting of about 27 orders (Parkes 1975). While nearly all of the living orders have some fossil record, it has contributed little to the understanding of the interrelation of modern orders. Modern authors maintain that there was a major dichotomy between two large groups of birds – the water bird assemblage and the land bird assemblage and each of these assemblages can be further divided into primitive and advanced forms (Fedducia 1980; Olson 1985).

Given only a sketchy and incomplete evolutionary history, it is impossible to arrive at a clear understanding of the evolution of hearing in birds. But the task ahead is clear. From comparative studies of hearing in extant birds, one would want to correlate species differences in hearing with what is known about evolutionary history. It is already known that on many of the basic measures of hearing, the different species of birds that have been tested are remarkably similar (Dooling 1982). On other measures, however, there is compelling evidence for specialized perceptual adaptations. Because behavioral paradigms require the participation of the whole, conscious organism, the data from behavioral experiments are some ways the most valid – especially when addressing evolutionary questions. Natural selection operates on behavior and it is the behavior of hearing that is intimately related to an animal's survival and adaptation to its natural environment. So, while hearing can be studied from anatomical, physiological, and behavioral perspectives, behavioral studies of hearing in birds are reviewed below.

Hearing usually refers to either differential sensitivity or discrimination (i.e., detecting a change in a stimulus) or absolute sensitivity (i.e., detecting the presence of a stimulus). With recent advances in animal psychoacoustics, it is certainly fair to add the perception and categorization of complex sounds such as vocal signals. While there is a wealth of behavioral data on hearing in birds (e.g., Fay 1988), it is only on a few measures that a sufficient number and diversity of species have been tested so as to provide potentially useful information for understanding the evolution of hearing in birds. Below, data are reviewed from three classes of studies: differential sensitivity, absolute sensitivity, and the perception of complex vocalizations.

2. Sensitivity to Changes in an Acoustic Signal

A number of acoustic discrimination tests have been conducted on the ability of birds to discriminate changes in either the spectral, temporal, intensive, or spatial aspects of acoustic stimuli. Most of these tests involve changes in simple, precisely controlled sounds and the purpose of these tests is to establish basic psychoacoustic capabilities. Frequency and intensity difference limens have been studied in a number of avian species. In general, birds – in their most sensitive region of hearing—approach the levels of human sensitivity on measures of frequency discrimination. Thus, birds are not too different from most other vertebrates (see, for review, Dooling 1982; Fay 1988). Birds appear to be slightly worse than humans on various measures in intensity resolving power (Dooling and Saunders 1975; Dooling and Searcy 1981; Klump and Baur 1990). On various measures of temporal resolving power including duration discrimination (Dooling and Haskell 1978; Maier and Klump 1990), temporal integration (Dooling 1979; Dooling and Searcy 1985; Klump and Maier 1990), and gap detection (Dooling, Zoloth, and Baylis 1978; Klump and Maier 1989; Okanoya and Dooling 1990), birds approach the levels of sensitivity of humans and other vertebrates (Fay 1988).

There is a wealth of masking data (using both tones and noises as maskers) available for birds but only for a few species (Dooling 1982; Okanoya and Dooling 1987; Sinnott, Sachs, and Hienz 1980). The evidence to date indicates that, with the exception of budgerigars (Okanoya and Dooling 1987; Dooling and Saunders 1975a; Saunders, Denny, and Bock 1978), birds are not very different from other vertebrates that have been tested on the same or similar tasks (Fay 1988; Chapter 14).

There is a body of evidence from field and playback studies that birds can and do localize the vocalizations of their own and other species. With the exception of the barn owl, the meager psychophysical data on sound localization in other species of birds are not sufficient to specify the precise cues that are important or the exact mechanisms involved (Dooling 1980; 1982). Some recent progress using modern methods has been made in determining the sound localization and spatial resolution of small birds and the results are interesting but somewhat complicated (Klump, Windt, and Curio, 1986; Park, Okanoya, and Dooling, 1987; Park and Dooling in press).

It appears that if the task is one of absolute localization, which is functionally equivalent to the behavioral head-turning response in the barn owl (Knudsen, Blasdel, and Konishi 1979), then small birds such as the great tit, budgerigar, canary, and zebra finch show poor sound localization ability (MAAs of roughly 20 to 30 degrees). This is generally consistent with what one would expect from an animal with a small head and poor high frequency sensitivity (Klump et al. 1986; Park and Dooling in press). If, on the other hand, the task is one of

relative localization (i.e., detecting a change in the position of a sound source), then these small birds and others show unexpectedly good spatial resolving power for broadband signals (Park, Dooling and Okanoya in preparation; Lewald 1987). Such low thresholds raise the possibility that the small birds may have evolved a mechanism for enhancing the acoustic cues available for sound localization perhaps via the interaural pathway. A final answer to this question in small birds will probably have to await behavioral experiments conducted in a completely anechoic environment or with binaural earphone inserts.

3. Absolute Threshold Sensitivity

The amount of data available on measures of absolute auditory sensitivity is somewhat more promising. Here, behavioral audibility curves are available for 23 species of birds. These species cover 7 (out of 27) Orders (Anseriformes, Columbiformes, Falconiformes, Galliformes, Passeriformes, Psittaciformes, and Strigiformes) and 16 (out of 166) Families.

The comparison among species from different studies is complicated by the fact that all species have not been tested at the same frequencies. Two strategies were used to cope with this problem. First, threshold data from each species were fit with a 4th order polynomial function. New thresholds were taken by interpolation from this function at half-octave steps from 250 Hz to 10.0 kHz. This provided a common basis for species comparison and for measures of variability. The second strategy was to characterize each audiogram in terms of seven arbitrary descriptive parameters. These parameters were: best frequency (kHz), best intensity (dB), bandwidth at 20 dB above best intensity (octaves), low-frequency sensitivity in dB (threshold at 500 Hz), high-frequency cutoff in kHz (i.e., frequency at which threshold reaches 60 dB), low-frequency slope (dB/octave), and high-frequency slope (dB/octave). It is then possible to characterize the differences among groups of birds on these parameters (Dooling 1980). Unless otherwise stated, all of the data presented below are taken from three other comparative reviews of hearing namely those of Dooling (1980; 1982) and Fay (1988).

Figure 26.1 shows the average avian audiogram calculated from the audiograms of 23 species

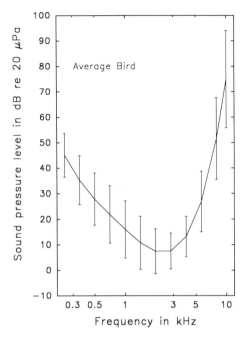

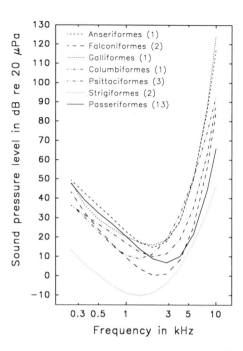

FIGURE 26.1. Average absolute thresholds taken from 23 species of birds. The original threshold data were fit with a 4th order polynomial and thresholds were determined from this function at the same frequency for each bird. Vertical bars represent ± one standard deviation from the mean.

FIGURE 26.2. Average audiograms for the seven different orders of birds calculated from fitted functions.

using thresholds at half-octave steps from 250 Hz to 10 kHz. Vertical bars represent standard deviations. The average avian audiogram—the audiogram of *the* bird—has the following characteristics. It has a best frequency of 2.58 kHz, a best intensity of 4.37 dB, and a hearing bandwidth at 20 dB above best threshold of 3.1 octaves. Threshold at 500 Hz is 27.9 dB and the highest frequency cutoff is 9.0 kHz. Below the most sensitive frequency in the audiogram, threshold declines at the rate of 14.6 dB/octave. Above the most sensitive region of the audiogram, threshold declines at the rate of 46.2 dB/octave.

3.1 Comparison Among the Orders of Birds

The average audiogram for each of the seven orders of birds is shown in Figure 26.2. These are listed in the evolutionarily old to new and there are some differences here which are worth mentioning. Across the entire audible range of frequencies,

Strigiformes (i.e., barn owl, great horned owl) are more sensitive than birds of other orders. Falconiformes as well have unusually good sensitivity in the mid-frequency region. Passeriformes on average show better sensitivity at high frequencies than do nonpasseriformes.

3.2 Comparison of Passerines/ Nonpasserines/Owls

In this sample of 23 audiograms, there are two predatory species whose hearing is central to their predatory lifestyle—both are in the Strigiformes. A logical grouping then is to compare Passeriformes, nonpasseriformes (excluding Strigiformes), and Strigiformes and these audiograms are shown in Figure 26.3. Here there is an interesting pattern. Passeriformes show a best frequency of 2.9 kHz, a best intensity of 5.1 dB, and a hearing bandwidth at 20 dB above best threshold of 2.9 octaves. Threshold at 500 Hz is 32.4 dB and the highest frequency a passerine bird can hear (i.e., threshold

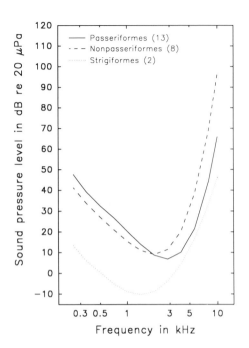

FIGURE 26.3. Average audiograms for Passeriformes, Nonpasseriformes, and Strigiformes.

TABLE 26.1. Comparison of Hearing Characteristics in Three Groups of Birds

	Characteristic						
	1	2	3	4	5	6	7
Passerines (n = 13)							
Mean	32.4	9.7	2.9	5.1	2.9	13.7	33.9
Sdev.	10.2	1.8	.7	8.3	.4	5.4	13.8
Nonpasserines (n = 8)							
Mean	27.0	7.5	2.1	8.5	3.9	12.9	48.4
Sdev.	4.5	1.1	.3	7.3	.4	3.2	8.7
Strigiformes (n = 2)							
Mean	1.0	11.2	2.7	−17.6	3.3	11.1	28.9
Sdev.	2.7	4.9	1.6	3.2	0.3	0.8	17.0

1, Threshold in dB at 500 Hertz; 2, high-frequency cutoff in Kilohertz; 3, best frequency in Kilohertz; 4, best intensity in decibels; 5, bandwidth (kHz) at 20 dB above best intensity; 6, low-frequency slope in dB/octave; 7, high-frequency slope in dB/octave.

of 60 dB) is 9.7 kHz. Below the most sensitive frequency in the audiogram, threshold declines at the rate of 13.7 dB/octave. Over the range of 4 to 8 kHz, threshold sensitivity declines at the rate of 33.9 dB/octave.

Compared to Passeriformes, nonpasseriformes (excluding owls) show a significantly lower best frequency of 2.1 kHz, a best intensity of 8.5 dB, and a significantly wider hearing bandwidth at 20 dB above the best threshold of 3.4 octaves. Threshold at 500 Hz is 27.0 dB and the highest frequency a nonpasserine can hear is 7.5 kHz. Below the most sensitive region of the audiogram threshold declines at the rate of 12.9 dB/octave. Between 4 and 8 kHz, sensitivity declines at the rate of 48.4 dB/octave.

For the two Strigiformes, best frequency is 2.7 kHz, best intensity is significantly lower than that of the other two groups at −17.6 dB, and hearing bandwidth at 20 dB above best intensity is 3.3 octave. Threshold at 500 Hz is 1 dB and high frequency cutoff is 11.2 kHz. For Strigiformes, threshold declines at the rate of 11.1 dB/octave at low frequencies and at high frequencies threshold declines a total of 28.9 dB between 4 and 8 kHz. More data are given in Table 26.1.

For several reasons, extreme caution should be observed in interpreting the difference between Strigiformes and other groups of birds. For one, the audibility curve for Strigiformes represents an average of only two species. Furthermore, only one animal was tested in each of these studies (Trainer 1946; Konishi 1973). Most importantly, the two audibility curves which make up the average function for Strigiformes are markedly different. To reinforce this point even further, the data from the two studies cited above, fit with a 4th order polynomial function are shown in Figure 26.4. Thus, while the unusually low thresholds observed in Strigiformes might invite speculation about auditory specializations and the relation between hearing sensitivity and predatory lifestyle, the data are too limited to provide a very confident picture of auditory sensitivity in the Strigiformes compared to other birds.

3.3 Comparison Among the Families of Passerines

It is generally agreed that passerines are the most recently evolved order of birds. To date, there are behavioral audiograms for 13 species of passerines from six families. Figure 26.5 shows the audiograms of these species by family. The only distinctive patterns here are that the Corvidae (represented here by crows and bluejays — also the largest birds in the sample) show the best absolute sensitivity especially at low- to mid-frequencies and the

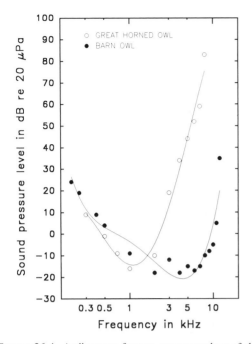

FIGURE 26.4. Audiograms for two representatives of the order Strigiformes, the barn owl and the great horned owl.

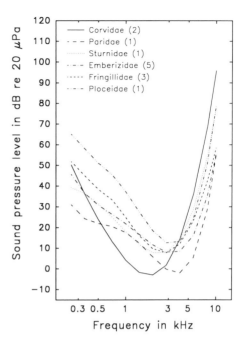

FIGURE 26.5. Average audiograms for six different families of passerines.

poorest at high frequencies. The Emerberizidae (swamp sparrow, song sparrow, blackbird, cowbird, and field sparrow) show generally good sensitivity at high frequencies compared to the other passerines. Otherwise, the passerines are fairly homogeneous in terms of absolute auditory sensitivity with the exception that the Paridae (represented here only by the great tit) show unusually good sensitivity at mid- to high-frequencies.

3.4 Absolute Sensitivity and the Characteristics of Vocal Signals

While there are clearly differences in the audiograms of different orders of birds, one cannot help but be struck by the overall similarities. But, these similarities should probably not be taken to mean that absolute auditory sensitivity in this class of vertebrates is a conservative (i.e., resistant to modification) evolutionary trait. Recent studies on passerines and psittacines that have examined the relation between vocalizations and hearing in fact suggest just the opposite. A number of reports over the years have shown a correlation between the spectral characteristics of species song in passerines and psittacines and the frequency of

best hearing (Dooling, Mulligan, and Miller 1970; Dooling and Saunders 1975; Konishi 1971). Most recently we have worked with two congeneric species of sparrows with distinctly different temporal song organization, the swamp sparrow (Melospiza georgiana) and the song sparrow (Melospiza melodia) (see, for example, Marler and Peters, 1989). These two species also show small differences in absolute auditory sensitivity which parallel small differences in the long term average power spectra of the species songs (Okanoya and Dooling 1988). The following experiment provides an even more intriguing example of a close coupling between hearing and the characteristics of vocal signals in birds.

3.4.2 Hearing and Vocalizations in Canaries

This more dramatic example (suggestive of involving artificial rather than natural selection), comes from a recent series of investigations on hearing and vocalizations in the canary. Over the last few years we have examined absolute auditory sensitivity in different strains of canaries. One strain, the Belgian Waterslager strain has been bred for over two hundred years for loud, low-frequency song (Stresemann 1923). To date, we have now

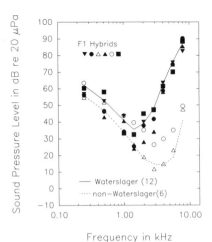

FIGURE 26.6. Audiograms for different strains of canaries: Belgian Waterslager canaries, non-Belgian Waterslager canaries, and six F1 hybrid hybrids from a Waterslager-Roller cross. (From Okanoya, Dooling, and Downing 1990.) Fl hybrids with elevated thresholds are shown with closed symbols, others with open symbols.

obtained behavioral audiograms from a total of 12 Waterslager canaries and 6 non-Waterslager canaries using the same behavioral procedures. These are shown in Figure 26.6. Also shown in this figure are absolute thresholds from F1 hybrid canaries resulting from a cross between a Belgian Waterslager canary and a German Roller canary.

Several things are clear. First, canaries from the Belgian Waterslager strain have extraordinarily poor high-frequency sensitivity for birds in general and for song birds in particular including other strains of canaries. It is extremely unlikely that

these strain differences in auditory sensitivity are due to environmental factors (e.g., loud noise in the aviary, antibiotics in feed, etc.) (Okanoya and Dooling 1987; 1988). Second, the thresholds of F1 hybrids are strongly suggestive of a genetic basis for elevated thresholds in waterslager canaries. What is particularly interesting in the present context is that, on average, the vocal characteristics of these F1 hybrids are intermediate between Waterslager and non-Waterslager canaries (Okanoya, Dooling, and Downing 1990). These results are shown in Figure 26.7.

One possible explanation for these strain differences in hearing and vocalizations is the following. Perhaps in selecting for loud, low pitched vocalizations, breeders in fact selected for poor high-frequency hearing. The idea here is that initial differences in absolute auditory sensitivity (i.e. poor high frequency hearing) led to modifications in the spectral and intensive characteristics of vocal output. Of course, one could argue the reverse case as well. But, regardless of the exact causal sequence, these data and those cited above are clearly suggestive of a close link between the evolution of auditory sensitivity and the evolution of vocalization characteristics in some birds.

In the above examples, the correlation is between fairly crude measures of vocal characteristics (i.e., the overall spectral characteristics of vocal signals) and fairly crude measures of hearing (i.e., the shape of the audibility curve). If there can be parallels at this rather crude level, there must also exist an intimate relation between hearing and vocalizations at other levels as well. Indeed, taking human speech perception as an example, one might expect that using natural vocalizations as test stim-

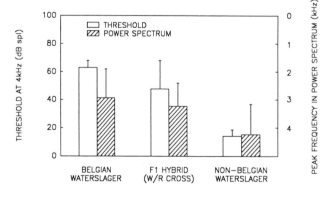

FIGURE 26.7. The relation between high-frequency sensitivity and the peak in the power spectrum of distance calls for Belgian Waterslager, non-Belgian Waterslager, and F1 hybrid canaries. (From Okanoya, Dooling, and Downing 1990.)

uli—in place of pure tones and noises—would reveal a rich domain of special, species-specific auditory processes as the following experiments demonstrate.

4. Vocal Learning and the Perception of Complex Sounds

Vocal learning can be defined as the ability to use auditory information (including auditory feed back) to modify or enhance vocal development (Kroodsma and Miller 1982). Except, of course, for humans, the evidence for vocal learning in mammals is quite sparse. But the phenomenon of vocal learning appears to be pervasive among birds, being especially pronounced among songbirds and parrots. To date, unequivocal evidence for vocal learning occurs in three orders of birds (Psittaciformes, Apodiformes, and Passeriformes) (Kroodsma and Baylis 1982) and it is likely that vocal learning evolved independently in each order (Nottebohm 1972). Both songs and calls can develop through learning but historically it is song, primarily a male endeavor, which has attracted the greatest attention in vocal learning studies.

The phenomena of vocal learning and sexual dimorphism in vocal behavior add interesting dimensions to the study of hearing and auditory perception in birds. That hearing is critically involved in the development of normal vocal behavior has been shown in a number of species by examining the song of birds deafened from birth (see, for example, Konishi and Nottebohm 1969). On a more refined level, some birds reared in isolation also develop abnormal vocalizations (Marler and Mundinger 1971) and some clearly show a selective perceptual bias toward learning only conspecific songs during the sensitive period for song learning (Dooling and Searcy 1980; Marler and Peters 1989).

In the present context of the evolution of the auditory system, one is concerned with evidence of specializations and adaptations in the auditory perceptual system for the processing of species-specific vocalizations. These issues can be addressed in the laboratory using operant conditioning procedures. In the following experiments, birds are trained to discriminate among complex vocal signals in such a way that response latency

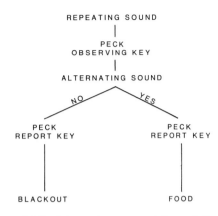

ALTERNATING SOUND TASK

FIGURE 26.8. Schematic representation of the testing procedure for generating response latency data from call discrimination experiments suitable for scaling and clustering.

provides an index of perceptual similarity (Dooling et al. 1987; Okanoya and Dooling 1988). To briefly describe one of these procedures shown schematically in Figure 26.8, birds are trained using operant conditioning to peck microswitches for access to food. In a task referred to as an alternating sound task, the birds are trained to peck a key in response to a change in a repeating background of sound (Okanoya and Dooling 1988). The reinforcement contingencies in this task are structured in such a way that the birds respond more quickly the more discriminable the background and the target sounds are. In other words, response latency provides an index of perceptual similarity between two complex sounds.

In experiments aimed at examining the perception of complex vocalizations, one can then use response latencies as a measure of the "efficiency" with which the auditory system processes complex signals. Furthermore, if a bird is tested on all possible comparisons among a set of complex sounds, it is also possible to arrive at a matrix of response latencies that is equivalent to a "similarity" matrix. With data in this form, there are a variety of statistical procedures such as multidimensional scaling and hierarchical clustering available to assess the perception of complex vocalizations. The following experiments use these procedures to highlight auditory perceptual issues which may be

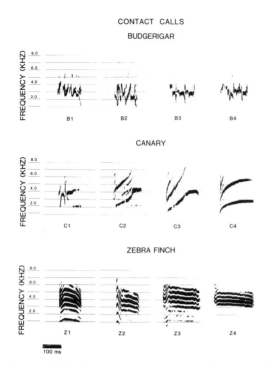

FIGURE 26.9. Sonograms of contact or distance calls from four budgerigars, four canaries, and four zebra finches. (Replotted from Dooling, Park et al. 1987.)

unique to organisms whose vocal repertoire develops through learning.

4.1 Perceptual Categories for Vocal Signals: Species Differences

Many birds produce a short call when separated from conspecifics. This vocalization is usually called a "contact" or "distance" call. Figure 26.9 shows sonograms of four distance calls from each of three species: budgerigars, canaries, and zebra finches. The differences among the calls of the three species are quite evident.

We tested five budgerigars, four canaries, seven zebra finches, and four starlings on these calls. Every call in the set served as both a background and a target stimulus in the procedure described above. At the conclusion of this experiment the response latency matrices of these birds were analyzed by the multidimensional scaling algorithm called SINDSCAL to produce a two-dimensional spatial arrangement of the stimuli for

each species. These spatial maps are shown in Figure 26.10. For budgerigars, canaries, zebra finches, and starlings these spatial maps accounted for 48%, 68%, 49%, and 56% of the variance in the response latencies of each species respectively. For all four species tested with these sounds, the calls of budgerigars, canaries, and zebra finches are arranged in three groups in multidimensional space. In other words, all four species find the acoustic differences between call categories much more discriminable than the acoustic differences among calls of the same category. This provides an operational definition of a perceptual category.

The fact that each species groups these into three categories is perhaps not too surprising since even casual human listeners can readily recognize and classify these sounds as belonging in three groups. The fact that starlings and human listeners, two species without extensive experience with these sounds, still group them into three categories indicates that the constellation of acoustic characteristics defining these categories lead to robust perceptual differences.

While it is of interest to know whether each of these four species can discriminate among the three categories of calls, it is more interesting to know how well each species discriminated among conspecific calls compared to the calls of the other species. The most intuitively simple way of examining these data for evidence of specialized perceptual processes is to compare the average response latencies of the three species when discriminating among conspecific calls versus the calls of the other species. These data are shown in Figure 26.11.

What these results show is that the differences among these natural stimuli are not equivalent for these four species. This finding is interesting for several reasons. First, with the exception of starlings, each of the subjects in these experiments were housed together for periods of several months to several years in the same large aviary consisting of, on average, 50–60 budgerigars, 10–15 canaries, and 10–15 zebra finches. If it were simply the total amount of exposure to certain call types that determined discriminability, each species should find the differences among budgerigar calls most discriminable. This is clearly not the case. Furthermore, starlings provide a control for experience since they had no prior experience with any of

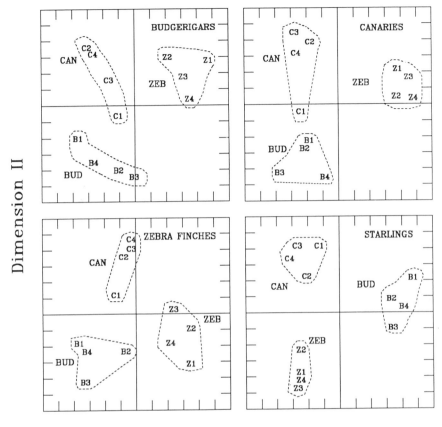

Dimension II

Dimension I

FIGURE 26.10. Two-dimensional spatial arrangements of the distance calls resulting from a SINDSCAL analysis of the response latencies of each of the four species. Dashed lines were placed with the aid of a cluster analysis on the data from each species. (From Dooling, Brown, et al., in press).

these call categories. The results from starlings suggest that canary vocalizations are inherently more discriminable. One can conclude from this pattern of results from these four species that there exist perceptual adaptations for the processing of species-specific vocal signals.

What also supports and refines this conclusion is the fact that these species differences in perception are not easily explained on the basis of species differences in basic hearing capabilities such as absolute thresholds, or various measures of temporal or spectral resolving power – data which are available for all four species. The tentative conclusion, then, is that these perceptual differences reflect more central – rather than peripheral – auditory system adaptations. The degree to which this more effi-

cient processing of species-specific vocal signals is innate or acquired through experience is not clear. But it would not be surprising to find both processes at work in producing the specialized processing for species-specific vocal signals described here. It is also likely that the nature of both prior and on-going experience with species-specific sounds plays a crucial role in the development and maintenance of species-specific perceptual advantages. These experiences would include the interaction between parent and young during development serving to emphasize biologically-relevant stimuli for learning, the interaction among conspecifics during adulthood serving to focus attention on socially meaningful acoustic differences in complex vocalizations, and the constant, unique

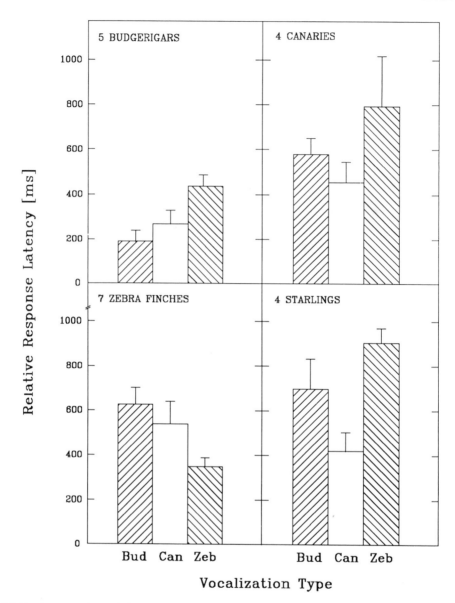

FIGURE 26.11. A comparison of the average response latencies from budgerigars, canaries, zebra finches, and starlings to detect a within-category change among the three species' calls shown in Figure 26.9 (From Dooling, Brown, et al., in press).

stimulation of the auditory system from self-produced vocalizations (Dooling, Brown, Klump, and Okanoya, in press).

4.2 Perceptual Categories for Vocal Signals: Effects of Experience and Sex

The above experiment does not directly address the issue of experience but there is other evidence that perceptual categories for vocal signals can be learned at least in budgerigars. In a recent experiment, three budgerigars that were housed together for several months developed a very similar contact call. When these three cagemates were tested on a set of their own contact calls, they showed evidence of perceptual categories for the calls of each individual. Interestingly, budgerigars housed next to these birds, who were exposed to the same calls but did not learn to produce them, failed to group

FIGURE 26.12. An assortment of calls drawn from the vocal repertoire of the budgerigar. (From Dooling, Park et al. 1987.)

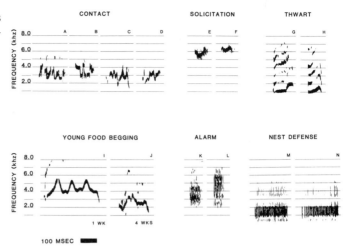

the same calls by individual (Brown, Dooling, and O'Grady, 1988). Such data are consistent with a mechanism of vocal perception which involves a reference to motor production processes.

Experiments showing that some species of birds are selective in what they will learn (see, for example, Marler and Peters, 1989) and other experiments showing birds reared without access to an external auditory model develop abnormal vocalizations raise issues concerning the role that auditory perception might play in guiding the vocal learning process. Evidence of sexual dimorphism in vocal behavior in songbirds also raises questions of whether there are parallel differences in auditory perception (see, for example, Williams and Nottebohm, 1985). Recent studies on the perception of vocal signals in budgerigars is relevant to both these issues in showing that male and female birds differ in their perception of calls drawn from the species vocal repertoire (Dooling, et al, 1990) and that birds reared in acoustic isolation are different from normal birds in their perception of these same vocal signals (Dooling, Park, Brown, and Okanoya, in press) and these experiments are described below.

The budgerigar vocal repertoire contains a number of functionally and acoustically distinct call types. As with many songbirds, it is predominantly the male budgerigar who sings. In the case of budgerigars, there is also a difference between the sexes in the kinds of calls that are produced. Figure 26.12 shows an assortment of budgerigar calls.

Four contact calls and two calls from each of the other major classes of calls in the budgerigar vocal repertoire are shown. Some of these (i.e., nest defense and solicitation calls) are produced only by females, some are produced only by males (i.e., thwart), and some are produced by both sexes (i.e., contact calls, alarm calls).

The question addressed in this experiment is whether male birds and female birds perceive these sounds differently and whether birds reared in isolation perceived these calls differently than birds in large social groups of other budgerigars. To answer this question, six normal budgerigars (3 male, 3 female) and two isolate-reared male birds were tested on these 14 stimuli. The three-dimensional spatial representation generated by SINDSCAL of the perceptual similarity among these calls for the eight budgerigars is shown in Figure 26.13. The distances in this three dimensional solution accounted for 57.7% of the variance in response latencies with the first, second, and third dimensions accounting for 30.3%, 15.7% and 11.7% respectively. The important point to note is that these calls fall out in groups in perceptual space corresponding to functional and acoustic categories. In other words, calls belonging to the same category are perceived as most similar to one another.

The subject weights for this solution are shown in Figure 26.14. Subject weights provide an indication of how much variance in each bird's data is accounted for by the spatial arrangement of stimuli. The subject space for this solution shows

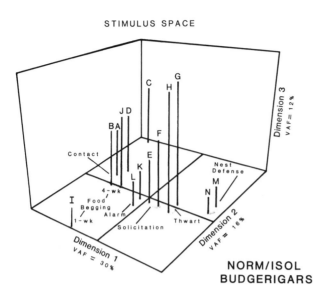

STIMULUS SPACE

FIGURE 26.13. Three-dimensional perceptual space for 8 budgerigars (3 male, 3 female, and 2 isolate-reared males) tested on these 14 calls.

that there are differences between male birds and female birds in the perception of this set of vocalizations. It is also clear that isolate-reared birds are different from socially-reared birds. In other words, these stimuli are not perceived in quite the same way by these three groups of birds. It remains to be seen exactly how these three groups of birds differ in their perception of species-specific sounds and whether these differences might also be reflected in other psychoacoustic measures such as a difference in sensitivity to frequency modulation.

In comparing cluster analyses of the data from normal birds versus birds reared in social and acoustic isolation as well as other response latency measures, it becomes clear that this difference is not large. Birds reared in acoustic isolation — never having heard normal budgerigar vocalizations — have roughly the same perceptual categories for vocal signals as normal birds do. The differences appear to arise in how discriminable the various categories of calls are from one another (Dooling, Park, et al, in press). The fact that isolate-reared birds are not very different from socially-reared birds in the perception of vocal categories is important in that it goes to the issue of whether perceptual categories for vocal signals are learned as a

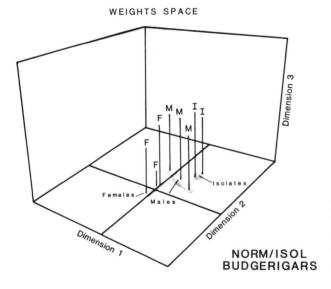

WEIGHTS SPACE

FIGURE 26.14. Subject weight space for 8 budgerigars tested on the assortment of 14 calls from the budgerigar vocal repertoire. Male, female, and isolate-reared birds fall out in three separate clusters.

result of social interaction with conspecifics or whether they are innate: a relatively important issue for an organism whose vocal repertoire develops through learning.

The finding that male and female birds may differ in their perception of species-specific calls is probably not unique to budgerigars. In fact, budgerigars at the moment are probably less interesting for the simple reason that much less is known about the sensory and motor anatomy controlling vocal behavior in this species. By contrast, the zebra finch is a very popular bird for the study of the behavior and neuroanatomy of vocal learning. As a consequence, there have been a tremendous number of song learning experiments on zebra finches (see, for review, Slater, 1988). Sexual dimorphism in vocalizations has been demonstrated at the level of behavior (Immelmann, 1969), neuroanatomy (Gurney, 1981; Nottebohm and Arnold, 1976), and motor neurophysiology (Williams, 1985; Williams and Nottebohm, 1985). Like budgerigars, male and female zebra finches also differ in their perception of species-specific calls (Brown and Dooling, 1991). This finding is consistent with the notion of a "motor theory" of song perception in these birds (Williams and Nottebohm, 1985; Williams, 1985).

5. Summary and Conclusions

A review of the psychophysical data on hearing in birds reveals that birds, in general, approach the levels of sensitivity in discrimination of simple sounds found for humans and other vertebrates. As a rule of thumb, for instance, thresholds for the detection of changes in intensity, frequency, and temporal aspects of a simple signal are roughly what they are for humans. In terms of discriminating spatial aspects of a sound source, there may be significant differences. Birds, with the exception of the barn owl, show very poor absolute localization of sound sources as might be expected from their poor high-frequency hearing and small interaural distance. But these small birds show a much better ability to detect a change in sound source location (relative localization).

There are a sufficient number of species, spread over a number of orders and families that have been tested on detection of pure tones in the quiet (i.e., audiogram) that general trends can be observed

which may have evolutionary significance. There is a tendency for evolutionarily recent birds to have better absolute sensitivity in the middle- and high-frequency regions of the audible range and poorer sensitivity at low frequencies. The Strigiformes—auditory predators—are certainly distinguished in this regard but Passeriformes can be distinguished from nonpasseriformes on roughly the same basis.

Studies of the evolution of hearing invariably focus on the auditory periphery. But there are certainly other examples of hearing behavior in birds, involving learning, attention, and vocal production, that provide a basis for evolution through natural selection to operate. The phenomenon of vocal learning has evolved, probably independently, in at least three orders of birds—Passeriformes, Psittaciformes, and Apodiformes. This behavior brings with it a number of special auditory issues some of which birds may share with humans. One issue is whether or not species show special adaptations for the perception of vocal signals. For budgerigars, canaries, and zebra finches, the answer is unequivocally affirmative. Another issue is whether or not birds can learn perceptual categories. For budgerigars, at least, this is certainly true.

Still another issue is to what extent are perceptual categories for vocal signals innate and to what extent are they determined by social or vocal experience. The issue is whether a young bird—destined to develop a complex vocal repertoire through learning—comes into the world with a reasonably well-specified set of perceptual categories or whether these categories, instead, develop from repeated exposure to normal species-specific vocalizations. The results of experiments on budgerigars reared in acoustic isolation suggest that, while there are subtle but demonstrable effects attributable to isolation, perceptual categories for the major classes of calls in the species repertoire still obtain in the absence of experience with the vocalizations. This is no less an important issue for birds faced with the task of learning a complex vocal repertoire than it is for the human infant embarking on the task of learning language.

In many species of birds, male and females differ dramatically in their vocal behavior. Not surprisingly, psychoacoustic tests with vocal signals reveal evidence of sex differences in perception of vocal signals—paralleling sexual dimorphism in vocal production. This supports not only the notion of a close link between vocal and auditory behavior in birds but also provides support for a "motor" theory of vocal perception—a strategy which may be common to a number of vertebrates and invertebrates.

References

Brown SD, Dooling RJ, O'Grady K (1988) Perceptual organization of acoustic stimuli by budgerigars (*Melopsittacus undulatus*): III Contact calls. J Comp Psychol 102:236–247.

Brown SD, Dooling RJ (1991) Sex differences in perception of distance calls by zebra finches. Paper presented at the Association for Research in Otolaryngology (Feb, 1991).

Carroll RL (1988) Vertebrate Paleontology and Evolution. New York: W.H. Freeman.

Dooling RJ (1979) Temporal summation of pure tones in birds. J Acoust Soc Am 65:1058–1060.

Dooling RJ (1982) Auditory perception in birds. In: Kroodsma D, Miller E (eds) Acoustic Communication in Birds, Vol. 1. New York: Academic Press, pp. 95–130.

Dooling RJ (1980) Behavior and Psychophysics of hearing in birds. In: Popper AN, Fay RR (eds) Comparative Studies of Hearing in Vertebrates. New York: Springer-Verlag, pp. 261–288.

Dooling RJ, Brown SD, Park TJ, Okanoya K (1990) Natural perceptual categories for vocal signals in budgerigars (*Melopsittacus undulatus*). In: Berkley M, Stebbins WC (eds) Comparative Perception, Vol. II: Communication. New York: John Wiley & Sons, pp. 345–374.

Dooling RJ, Haskel RJ (1978) Auditory duration discrimination in the parakeet (*Melopsittacus undulatus*). J Acoust Soc Am 63:1640–1642.

Dooling RJ, Brown SD, Klump GM, and Okanoya K (in press) Auditory perception of conspecific and heterospecific vocalizations in birds: Evidence for special processes. J. Comp Psychol

Dooling RJ, Mulligan JA, Miller JD (1971) Auditory sensitivity and song spectrum of the common canary (*Serinus canarius*). J Acoust Soc Am 50: 700–709.

Dooling RJ, Park TJ, Brown SD, Okanoya K, Soli SD (1987) Perceptual organization of acoustic stimuli by budgerigars (*Melopsittacus undulatus*): II. Vocal signals. J Comp Psyc 101:367–381.

Dooling RJ, Park TJ, Brown SD, Okanoya K. Perception of species-specific vocalizations by isolate-reared budgerigars (*Melopsittacus undulatus*). Int J Comp Psyc (in press).

Dooling RJ, Saunders JC (1975a) Hearing in the parakeet (*Melopsittacus undulatus*): Absolute thresholds, critical ratios, frequency difference limens, and vocalizations. J Comp Physiol Psych 88: 1–20.

Dooling RJ, Saunders JC (1975b) Auditory intensity discrimination in the parakeet (*Melopsittacus undulatus*). J Acoust Soc Am 58:1308–1310.

Dooling RJ, Searcy MH (1980) Forward and backward auditory masking in the Parakeet (*Melopsittacus undulatus*). Hear Res 3:279–284.

Dooling RJ, Searcy MH (1981) Amplitude modulation thresholds for the Parakeet (*Melopsittacus undulatus*). J Comp Physiol 143:383–388.

Dooling RJ, Searcy MH (1985) Temporal integration of acoustic signals by the budgerigar (*Melopsittacus undulatus*). J Acoust Soc Am 77:1917–1920.

Dooling RJ, Zoloth SR, Baylis JR (1978) Auditory sensitivity, equal loudness, temporal resolving power and vocalizations in the house finch (*Carpodacus mexicanus*). J Comp Physiol Psych 92:867–876.

Fay RR (1988) Hearing in Vertebrates: A Psychophysics Databook. Winnetka, IL: Hill-Fay Associates.

Fedducia A (1980) The Age of Birds. Cambridge: Harvard University Press.

Gurney ME (1981) Hormonal control of cell form and number in zebra finch song system. J Neurosci 1:658–673.

Hienz RD, Sinnott JM, Sachs MB (1977) Auditory sensitivity of the red-winged blackbird (*Agelaius phoeniceus*) and brown-headed cowbird (Molothrus ater). J Comp Physiol Psychol 91:1365–1376.

Immelmann K (1969) Song development in zebra finch and other estrildid finches. In: Hinde RA (ed) Bird Vocalizations, Their Relation to Current Problems in Biology and Psychology. London: Cambridge University Press.

Klump GM, Baur A (1990) Intensity discrimination in the European Starling (*Sturnus vulgaris*). Naturwissen 77:545–548.

Klump GM, Maier EH (1989) Gap detection in the starling (*Sturnus vulgaris*). I. Psychophysical thresholds. J Comp Physiol 164:531–538.

Klump GM, Maier EH (1990) Temporal summation in the starling (*Sturnus vulgaris*). J Comp Psych 104:94–100.

Klump GM, Curio E, Windt W (1986) The great tit's (*Parus major*) auditory resolution in azimuth. J Comp Physiol 158:383–390.

Knudsen EI, Blasdel GG, Konishi M (1979) Sound localization by the Barn Owl (*Tyto alba*) measured with the search coil technique. J Comp Physiol 133:1–11.

Konishi M (1971) Comparative neurophysiological studies of hearing and vocalizations in songbirds. Z Vergl Physiol 66:257–272.

Konishi M (1973) How the Barn Owl tracks its prey. Am Sci 61:414–424.

Konishi M, Nottebohm F (1969) Experimental studies in the ontogeny of avian vocalizations. In: Hinde RA (ed) Bird Vocalizations, Their Relation to Current Problems in Biology and Psychology. London: Cambridge University Press, pp. 29–48.

Kroodsma DE, Baylis JR (1982) A world survey of evidence for vocal learning in birds. In: Kroodsma DE, Miller EH (eds) Acoustic Communication in Birds, Vol. 2: Song Learning and its Consequences. New York: Academic Press.

Kroodsma DE, Miller EH (eds) (1982) Acoustic Communication in Birds, Vol. 2: Song Learning and its Consequences. New York: Academic Press.

Lewald J (1987) The acuity of sound localization in the pigeon (*Columbia livia*). Naturwissen 74:296–297.

Maier EH, Klump GM (1990) Auditory duration discrimination in the European starling (*Sturnus vulgaris*). J Acoust Soc Am 88:616–621.

Marler P, Mundinger P (1971) Vocal learning in birds. In: Moltz M (ed) Ontogeny of Vertebrate Behavior. New York: Academic Press, pp. 389–450.

Marler P, Peters S (1989) Species differences in auditory responsiveness in early vocal learning. In: Dooling RJ, Hulse SH (eds) The Comparative Psychology of Audition: Perceiving complex sounds. Hillsdale, NJ: Erlbaum, pp. 243–273.

Nottebohm F (1972) The origins of vocal learning. Am Nat 106:116–140.

Nottebohm F, Arnold A (1976) Sexual dimorphism in vocal control areas of the songbird brain. Science 194: 211–213.

Okanoya K, Dooling RJ (1988) Hearing in the swamp sparrow (*Melospiza georgiana*) and the song sparrow (*Melospiza melodia*). Anim Behav 36:726–732.

Okanoya K, Dooling RJ (1987) Hearing in Passerine and Psittacine birds: A comparative study of masked and absolute auditory thresholds. J Comp Psyc 101:7–15.

Okanoya K, Dooling RJ, Downing J (1990) Hearing and vocalizations in hybrid Waterslager-Roller canaries (*Sernus canaria*). Hear Res 46:271–276.

Okanoya K, Dooling RJ (1990) Detection of gaps in noise by budgerigars (*Melopsittacus undulatus*) and zebra finches (*Poephila guttata*). Hear Res 50:185–192.

Olson SL (1985) The fossil record of birds. In: Farner D, King J, Parkes K (eds) Avian Biology, Vol. 8. Orlando, FL: Academic Press, pp. 79–238.

Park TJ, Okanoya K, Dooling RJ (1987) Sound localization in the budgerigar and the interaural pathways. Paper presented at the 113th Meeting of the Acoustical Society of America. Indianapolis, IN.

Park TJ, Dooling RJ (1991) Sound localization in small birds: Absolute localization in azimuth. J Comp Psychol 105:125–133.

Park TJ, Okanoya K, Dooling RJ (1987) Sound localization in small birds. Paper presented at the 1987 National Meeting of the Animal Behavior Society. Williamstown, MA.

Parkes KC (1975) Special Review Auk 92:818–830.

Saunders JC, Denny RM, Bock GR (1978) Critical bands in the parakeet (*Melopsittacus undulatus*). J Comp Physiol 125:359–365.

Sinnott JM, Sachs M, Hienz R (1980) Aspects of frequency discrimination in passerine birds and pigeons. J Comp Physiol Psych 94:401–415.

Slater PJB, Eales LA, Clayton NS (1988) Song learning in zebra finches (*Taeniopygia guttata*): Progress and prospects. Adv Study Behav 18:1–34.

Stresemann E (1923) Zur Geschichte einiger Kanarianrassen. Ornithol Monatsberichte 31:103–106.

Trainer JE (1946) The auditory acuity of certain birds, PhD Thesis, Cornell University, Ithaca, New York.

Williams H (1985) Sexual dimorphism of auditory activity in the zebra finch song system. Behav Neur Biol 44:470–484.

Williams H, Nottebohm F (1985) Auditory responses in avian vocal motor neurons: A motor theory for song perception in birds. Science 229:279–282.

Welty JC (1982) The Life of Birds. Philadelphia: Saunders.

27
Evolution and Specialization of Function in the Avian Auditory Periphery

Geoffrey A. Manley and Otto Gleich

1. Introduction

Birds, particularly passerines, are generally highly vocal animals. In view of the importance of these vocal communication signals, it would be reasonable to expect that strong selection pressures have influenced the evolution of the inner ear and auditory pathway. In this chapter, we shall discuss the structure and physiology of the hearing system peripheral to and including the auditory nerve. The features of auditory sensitivity, frequency discrimination, and time resolution abilities should be appropriate to the communication tasks at hand, as well as to normal environmental acoustic awareness.

Information concerning the evolution of structure and function can be derived from a number of sources. Data relevant to the present discussion can be obtained from palaeontological, comparative morphological, comparative physiological, and behavioral studies. In practice, however, our main sources of data on the evolution of the soft structures of the hearing organ are comparative anatomical and neurophysiological studies, which will be described below. The discussion is made much more difficult by the fact that birds do not fossilize well, so the course of evolution within the Class Aves is poorly understood (Carroll 1987).

It is of considerable importance to recognize the great similarity of the hearing organs of Aves and Crocodilia (Manley 1990): this fact suggests that the hearing organ has not changed substantially since the divergence of these two groups and that the ancestral avian hearing organ was probably similar in many respects to that of modern birds. One point we shall emphasize is that despite the

structural similarities between the crocodilian-avian hearing organ and that of mammals, the available evidence indicates that the resemblance between these two hearing-organ types is due to the *independent* development of specialized groups of sensory cells (Manley et al. 1989).

2. The Middle Ear and the Hearing Range

Although reptiles and birds show significant variability in the details of their middle-ear structure and in the ossification of columella and extracolumella, all have the single-ossicle, *second-order-lever* middle-ear system (Fig. 27.1; Manley 1981; Manley 1990). Proximally, the columella forms a simple footplate in the oval window and connects distally with the tympanic membrane via a set of radiating, flexible processes of the extracolumella (Fig. 27.1). The long inferior process swings in and out when the tympanic membrane moves; its fulcrum is at the edge of the tympanum and the shaft of the extracolumella is attached somewhere near the middle. This generates a lever system. The transformation from the radial swinging of the inferior process to a piston-like motion takes place within the extracolumella itself, which prohibits a full ossification of the extracolumella. In birds, the middle ear is mechanically more protected than it generally is in reptiles, in being isolated from the buccal cavity and having, on average, a deeper external auditory meatus than the reptiles. Thus, in birds, the middle-ear ossicle is generally more ossified than in reptiles.

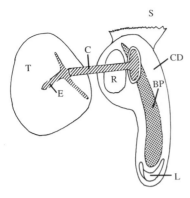

FIGURE 27.1. Schematic diagram of the middle ear and membranous labyrinth of the cochlear duct of a bird. The cochlear duct (CD) is connected to the sacculus (S), not shown here. Within the cochlear duct lies the basilar papilla (BP); the lagena macula (L) is at the apex. The tympanum (T) connects to various processes of the extracolumella (E) and via the columella (C) to the footplate in the oval window adjacent to the round window (R). The basilar papilla is typically 3 to 5 mm long.

All of these nonmammalian middle ears have a "low-pass" characteristic in the displacement transfer function. Although there are no systematic differences in the sensitivity and lever ratio to the mammalian ear, the upper frequency limit of the transfer characteristic is significantly lower than it is in mammals (Manley 1973; Manley 1981) and is strongly influenced by the flexibility of the middle ear (Manley 1972a,b). At high frequencies (>4 kHz), the efficiency of the middle ear deteriorates (Manley 1981), because an increasing amount of the acoustic energy is lost in a flexing motion *within* the inferior process. In addition, however, the inner ear itself ceases to absorb energy and thus dampens high-frequency middle-ear transmission, especially at low levels. Even if we look at the barn owl and at very small passerines, which have the highest high-frequency limits of nonmammals, the absolute limit of avian hearing at physiological sound levels is seen to be near 12 kHz. This value is very low when compared to a "typical" mammal, which has an upper limit near 50 to 70 kHz. Especially in such species as barn owls, oil birds, and cave swiftlets, whose existence depends on the fine analysis of relatively high-frequency sounds, selection pressures must have pushed the inner ear and middle ear to operate at the physiological limit.

In the face of only small differences in the frequency response compared to "normal" birds, however, we conclude that the single-ossicle middle ear really is limited by its basic structure. The inability of the middle ear to transmit high frequencies, even though it does contain a lever system, has strongly limited the evolution of function in the reptiles and birds.

It is highly unlikely that synapsid reptiles at the reptilian-mammalian transition were able to process high-frequency sounds (Manley 1973; Manley 1990). A recent study of such transition mammals confirms this, indicating that they possessed only a very short hearing organ grossly resembling that of modern advanced reptiles (Graybeal et al. 1989). It was thus a fortuitous predisposition of structure in the three-ossicle ear which allowed the modern descendants of the mammals to modify their inner ear for the processing of very high frequencies.

3. Phylogenetic Considerations

3.1 The Phylogenetic Relationships of Extant Reptiles, Birds and Mammals

The vertebrate fossil record is so complete that there remains little doubt concerning the time of origin and the relationships of the major groups that concern us in this chapter. This information is essential for the discussion below and will be outlined briefly. Further detail can be found in textbooks of palaeontology, such as that of Carroll (1987). Although both birds and mammals are derived from reptiles, the times of the divergence of their direct ancestors from the other reptile groups are very different. The *synapsid* mammal-like reptiles are considered to have diverged from the stem reptiles about 300 million years (MY) ago, in the Pennsylvanian period (Fig. 27.2). True mammals originated over 200 MY before the present. The archosaur ancestors of birds and Crocodilia, in contrast, derive much later (and quite independently of mammals) from *diapsid* reptiles of the early triassic period (about 230 MY), with ancient birds emerging at the border of the upper jurassic and early cretaceous periods (140 MY; Carroll 1987; Fig. 27.2).

Thus the nearest common ancestors of birds and mammals are the stem reptiles older than 300 MY,

FIGURE 27.2. Highly simplified "family tree" of recent land vertebrates of interest to the discussion in this chapter, showing the approximate geological time scale of the various adaptive radiations. (After various authors.)

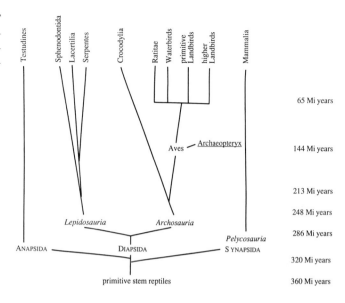

about whose hearing organ we have no direct information. We can, however, propose that modern chelonians, such as turtles, possess an inner ear whose structure has not fundamentally changed over this time period. This notion is supported by its structural similarities to the hearing organ of primitive diapsids such as Sphenodon (see the Chapter 23, Miller). If this is true, then the early mammal-like reptile hearing organ did not show the differentiation seen in the mammalian inner ear today. More specifically, the sensory cells and their innervations were not specialized into two groups placed neurally (that is, on the side where the nerve fibers enter the papilla) and abneurally (that is, on the side opposite the nerve) on the papilla. The similar specialization of the hair-cell populations of birds and Crocodilia is also a later development, with a probable origin soon after their common ancestor diverged from the ancestors of the lepidosaurs such as lizards and snakes (which show a different kind of hair-cell specialization). We thus observe the parallel and convergent acquisition by mammals and birds of auditory papillae having two or more groups of sensory hair cells organized across the papilla and of a specialized innervation pattern. We have discussed new evidence for important functional parallels between these auditory specializations in a recent paper (Manley et al. 1989).

3.2 Relationships Between Extant Families of Birds

Unfortunately, the fossil record of birds is relatively poor, so that there are substantial uncertainties with regard to the details of relationships between different avian groups. As we have no information at all about the hearing of the ratites (flightless birds retaining the ancient palaeognathous palate), we do not need to concern ourselves with the ongoing discussion as to whether this group is monophyletic or not. All other birds are neognathous (have a more modern palate) and can be divided into a water-bird assemblage and a land-bird assemblage (Feduccia 1980; Carroll 1987). The duck and seagull referred to below belong in the water-bird assemblage. The pigeon and chick belong to a primitive subdivision of the land-bird assemblage, whereas the starling and the barn owl belong to a more derived group of land birds (cf. Figs. 27.2 and 27.10).

4. The Starling as a Model of Bird Hearing

As there is more information available on the hearing of starlings than for any other avian species, we shall use its basilar papilla to briefly describe the structural arrangement and the physiology. This is

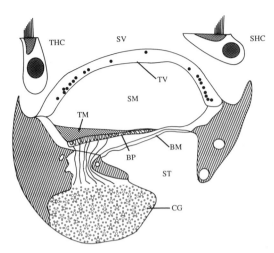

FIGURE 27.3. Schematic drawing of a cross-section through the cochlear duct near the apex of a typical avian basilar papilla. Scala vestibuli (SV) is separated from scala media (SM) by the tegmentum vasculosum (TV) which is equivalent to the mammalian stria vascularis. Above scala tympani (ST) lies the basilar membrane (BM) supporting the abneural part of the basilar papilla (BP), which is covered by the tectorial membrane (dotted pattern, TM). Nerve fibers (very few are illustrated) from the basilar papilla run to the cochlear ganglion (CG). Cut limbic material is shown shaded. The enlarged insets show (top left) a tall hair cell (THC) from the neural side of the papilla and (top right) a short hair cell (SHC) from the abneural side.

not intended to imply that the starling is in any way primitive or that this pattern indicates a kind of ancestral condition for birds.

4.1 The Avian Cochlear Duct

There is a strong similarity to be found in the structural arrangement of the cochlear ducts of birds and Crocodilia, with their lagenar macula and basilar papilla receptor areas (Fig. 27.1). The duct is not coiled, but twisted; this twisting is quite complex in the long ducts of owls (Schwartzkopff and Winter 1960; Fischer, Köppl, and Manley 1988). In general, the average avian auditory sensory epithelium is shorter (mostly less than 4 to 5 mm) and wider than that of a typical mammal. A thick tectorial membrane covers the entire papilla (Fig. 27.3). Almost all authors recognize two to four intergrading hair-cell types, i.e., the tall (THC), intermediate (INHC), short (SHC), and lenticular hair cells (Fig. 27.3; Smith 1985). Not all

types have been recognized in all species and they are, unlike in the papilla of Caiman, frequently difficult to distinguish in a surface view (Fig. 27.4). The hair cells are surrounded by supporting cells. THC are the least specialized and most strongly resemble the typical hair cell of more primitive groups of vertebrates (Takasaka and Smith 1971; Chandler 1984). They are distinguished from SHC by their columnar shape (Fig. 27.3) and different innervation pattern. Except near the apical end, THC are found predominantly supported by the neural limbus and *do not lie over the free basilar membrane*. They can be entirely absent from the basal end. In contrast, the SHC are wider than they are tall (Fig. 27.3). These cells occupy most of the space over the free basilar membrane. INHC are intermediate in both shape and position, but have not been described in all species. A few hair cells at the basal end of the chick papilla and many hair cells of the basal 3 mm of the barn owl papilla have been called lenticular hair cells. They are flattened, with a large apical surface area, only part of which has a cuticular plate (Smith 1985). In the barn owl papilla, short hair cells grade into the lenticular type; at any one location, these cell types can be neighbors (Fischer, Köppl, and Manley 1988). The actual distribution of these hair-cell types is species-specific, the most striking differences being found at the apical end (Smith 1985). In all avian species investigated so far, the orientation of the hair cell bundles changes systematically according to the position across the papilla (Fig. 27.5; Fischer et al., 1988; Gleich and Manley, 1988; Tilney et al., 1987). The innervation of hair-cell types also differs (Takasaka and Smith 1971; Chandler 1984; von Düring, Andres, and Simon 1985; Smith 1985; Singer, Fischer, and Manley 1989).

The total number of sensory hair cells is comparable to that of the mammalian cochlea and ranges from a few thousand in some song birds (e.g., starling; 5,800 hair cells in a 3 mm long papilla) to about 10,000 in the pigeon and chicken papilla (Gleich and Manley 1988). In the papilla of barn owls, which exceed 11 mm in length and is thus more than twice the length of the papillae of starling, chick, or pigeon, there are more than 16,000 hair cells (Schwartzkopff 1968; Fischer, Köppl, and Manley 1988). The avian papilla being short, there can be over 50 hair cells in a single cross section of its widest area (apical), compared to 4 to 6 in mammals.

4.2 Structure of the Starling's Hearing Organ

We (Gleich and Manley 1988) have previously described quantitatively the morphological patterns of the basilar papilla of the starling. It is roughly five times wider at the apical end than at the basal end (190 to 40 μm), reaching its widest point about 80 to 90% of the length from the basal end (cf. Fig. 27.9). The number of hair cells in any one cross section roughly parallels the change in width, rising from 8 basally to 30 apically. The number of stereovilli per hair-cell bundle falls from near 200 at the basal end to near 50 at the apical end, the form of the bundle changing from elongated basally to rounded apically. The height of the tallest stereovilli in the bundles (in the fixed, embedded state) varies from about 2.7 μm basally to 9.4 μm apically, the increase in height being much faster in the apical third. There is no consistent difference between the height of the neural and abneural bundles in each transect, although hair-cell bundles tend to be shortest on centrally lying cells. Hair cells lying at the extreme neural and abneural positions on the papilla have their stereovillar bundles all oriented nearly perpendicularly (±20°) to the edge of the papilla ("abneural"), similar to that seen in mammalian hair cells. Cells in the center of the papilla, however, tend to have their bundles turned towards the apex, the orientation gradually changing in any cross section from either edge towards the middle of the papilla. Although this tendency is hardly noticeable at the base, the orientation angle increases towards the apex to such an extent that the bundles of centrally located apical cells are rotated up to 90° towards the apex (Figs. 27.4 and 27.5).

Relatively little is known about the innervation of the avian auditory papilla. In new studies of the starling and chick papillae using serial ultrastructural sections, Fischer et al. (1991) and Singer et al. (1989) found greater differences between the innervation patterns of THC and SHC than would be predicted from the data of von Düring et al. (1985), especially in the basal half of the papilla. There, whereas THC receive both afferent and efferent innervation, the SHC studied had no afferent synapses at all and thus no afferent connection to the brain. Their synaptic areas were dominated by large efferent endings. These data indicate an unexpectedly

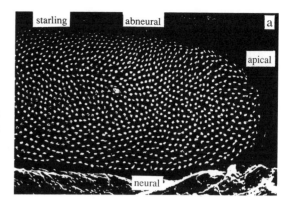

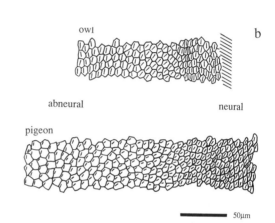

FIGURE 27.4. (a) Scanning electron micrograph of the apical hair-cell mosaic of the starling's basilar papilla. There are no clear rows of hair cells and THC and SHC are not distinguishable in surface view. (b) Schematic drawings of the surfaces of strips of hair cells across near-apical regions of the barn owl's and pigeon's basilar papillae, with the longer axis of the stereovillar bundles represented as bars. Whereas both neural and abneural hair-cell bundles are oriented more-or-less parallel to the neural papillar edge, hair cells lying medially on the papillae have their bundles rotated towards the cochlear apex. In the owl, the transition between differently oriented areas is abrupt. The shaded area of the owl papilla was covered in limbic material and not visible.

high degree of functional separation of THC and SHC in these species.

4.3 Physiology of the Auditory Papilla of the Starling

As there are very few data on the electrical activity of avian basilar-papilla hair cells stimulated by sound, we will briefly describe the activity

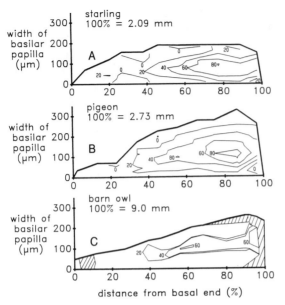

FIGURE 27.5. Highly schematic outlines of the basilar papillae of (A) the starling, (B) the pigeon, and (C) the barn owl, all normalized in length. The lengths given for the three papillae are those measured from SEM micrographs (fixed, dried). Within each papilla are shown a number of iso-orientation contours for the the orientation of the hair-cell stereovillar bundles, in degrees of bundle rotation towards the apex. Shaded areas in the owl could not be analyzed. (From Manley 1990.)

patterns of primary auditory nerve fibers. In this section, we review briefly the starling data as an example of the sensory responses originating in the avian papilla (Manley and Leppelsack 1977; Manley 1979; Manley and Gleich 1984; Manley et al. 1985, 1989; Gleich and Narins 1988; Gleich 1989).

4.3.1 Spontaneous Activity of Single Auditory Nerve Fibers

All primary auditory nerve fibers of the starling are irregularly spontaneously active, with a mean rate of 48 spikes/sec. The distribution of spontaneous rates is unimodal, whereas it is bimodal in mammals (Sachs, Lewis, and Young 1974; Sachs, Woolf, and Sinnott 1980; Manley et al. 1985). However, the overall interval distribution in time-interval histograms (TIH) in the spontaneous activity of nerve fibers in birds generally resembles that in mammals. The pseudo-Poisson interval distribution (Fig. 27.6B) is attributed to stochastic

processes either in the hair cell or in the nerve fiber terminal. At short intervals, these processes are modified by the absolute and relative refractory periods of the fiber and, possibly, limiting factors in the hair cell synapse. The modes (most frequent interval) of the TIH in the starling data are very short, typically 1 to 2 ms (Manley et al. 1985).

The distribution of intervals in spontaneous data is, in many cells, strongly modified by the presence of more or less prominent preferred intervals (Manley 1979; Manley and Gleich 1984; Manley et al. 1985). In such cells, the activity is quasi-periodic, such that certain numerically related intervals occur more often and others less often than expected (Fig. 27.6A). They are not due to inadvertent stimulation or to background noise (Manley et al. 1985; Temchin 1988). The characteristic properties of these preferred intervals are:

1. They are found only in about half of the cells and with a best, or characteristic response frequency (CF) to acoustic stimuli below about 1.7 kHz. The limit in frequency could be related to the fact that the nerve fibers rapidly lose their ability to phase lock and/or the hair cells' AC receptor potentials become very small above 1 kHz (Gleich and Narins 1988).

2. The variations between interpeak intervals within each histogram are almost all <5% (Manley et al. 1985). The mean interpeak intervals are inversely related to the CF of the cell, but the mode of their distribution in the starling is on average 15% longer than the CF-period. This fact also indicates that they do not result from inadvertent noise stimulation (Manley et al. 1985). Even in those low-CF cells that do not show preferred intervals, the mode of the TIH of spontaneous activity is itself correlated with the CF. Here also, the interval of the mode is on average slightly longer than the CF-period (see above); the mode is thus a special case of a preferred interval. Gleich (1987) also found evidence of electrical tuning in the phase-response characteristics of primary auditory fibers of the starling, a tuning whose best frequency was, on average, 20% lower than the acoustic CF of the cell. This percentage difference corresponds extremely well to the discrepancy noted between the basic interval in preferred intervals and the period of the CF.

The most probable explanation of these details is that the spontaneous activity of all low-CF fibers

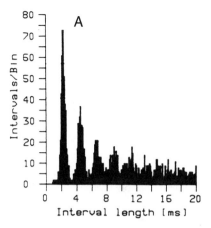
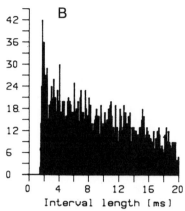
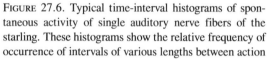

FIGURE 27.6. Typical time-interval histograms of spontaneous activity of single auditory nerve fibers of the starling. These histograms show the relative frequency of occurrence of intervals of various lengths between action potentials. (A) a low-frequency fiber showing very obvious preferred intervals (fiber CF: 0.45 kHz). (B) a low-CF fiber not showing preferred intervals, but whose mode was 1/CF (fiber CF: 0.5 kHz). (From Manley et al. 1985.)

is influenced by rhythmical electrical potentials of the individual hair cells they innervate. In the red-eared turtle, voltage oscillations are observed due to the properties of the ion channels of individual hair cells (Fettiplace 1987). Preferred intervals are thus manifestations of an *electrical tuning mechanism* in hair cells (summary in Manley 1986). Under certain conditions, isolated tall hair cells from the apex of the chick cochlea show electrical resonances to injected current, resembling those previously demonstrated in the turtle basilar papilla and frog sacculus (Fuchs and Mann 1986; Fuchs, Magai, and Evans 1988). The frequencies of these oscillations depended on the original location of the hair cell and were estimated by Fuchs, Nagai, and Evans (1988) to be up to 1 kHz for hair cells from the middle third of the chick papilla when corrected to the temperature of the living animal. Preferred intervals would be most easily seen in nerve fibers that only innervate one single hair cell, as is the case in starling low-CF THC (see above).

4.3.2 Frequency-Response Characteristics of Single Nerve Fibers

In common with all other vertebrate auditory fibers, starling eighth-nerve afferents each have a best, or characteristic frequency (CF) to which they respond at the lowest sound-pressure level

(SPL). Responses to sounds of other frequencies can only be evoked by applying a greater SPL, the tuning curves being highly frequency selective (Fig. 27.7A). This characteristic was certainly inherited by birds from their reptilian ancestors. The CFs range from very low frequencies (below 100 Hz) to an upper limit of about 6 kHz. As noted by Sachs, Woolf, and Sinnott (1980), avian tuning curves are, if anything, more sharply tuned than those of mammals in the equivalent frequency range (Manley et al. 1985), at least when measured as the sharpness of the tip region ($Q_{10\ dB}$). However, starling nerve fiber tuning curves have a different symmetry from those of the mammals (Manley et al. 1985).

4.3.3 Tonotopicity and the Localization of Active Afferents

The various CFs of the nerve or cochlear ganglion of the starling are distributed nonrandomly in space, indicating a tonotopic organization of the papilla. Tonotopicity is also a primitive characteristic of vertebrate hearing organs (Manley 1990). Recently, single-fiber staining techniques permitted tracing the origin of responses in different frequency ranges to specific locations in the papilla of the starling and the chick (Manley, Brix, and Kaiser 1987; Gleich 1989; Manley et al. 1989). Using cobalt stains in the starling and HRP

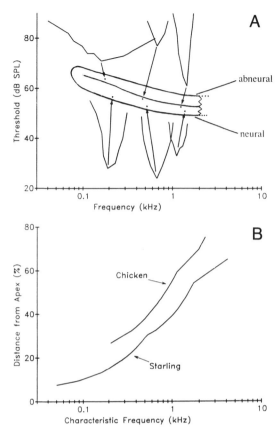

highest. The arrangement of CFs is unequal, the CF distribution in the low-frequency range being about 0.1 mm/octave, whereas at high frequencies it is near 0.6 mm/octave (Manley et al. 1988, 1989; Gleich 1989). This phenomenon is also known, but is not so pronounced, in the cat cochlea (Liberman 1982) and is also typical of lizard papillae (Manley et al. 1988; Manley, Köppl, and Yates 1989; Manley 1990).

Of the traced fibers in the starling, 24 were also successfully localized in transverse sections. This made it possible to describe the relative positions of innervated hair cells in both dimensions of the sensory mosaic. Hair cells were classified as THC (height/width ratio > 1) or SHC (ratio < 1). In virtually all cases, each stained fiber only contacted one single THC (Fig. 27.7A). The locations of these hair cells were described by using their relative position (= rank) across the row of hair cells in the cross sections, calling the neuralmost hair cell number 1. Virtually all hair cells innervated by the stained fibers had a rank of less than 15, even though up to 35 hair cells were found in any one cross-section (Gleich 1989; Manley et al. 1989).

Only two out of 34 stained fibers innervating abneurally lying hair cells were found (Fig. 27.7A); both were in the apical part of the papilla (Gleich 1989; Manley et al. 1989). One of these fibers innervated about 6 hair cells and was the only branched fiber found. In the apical area of the starling papilla, SHC receive very few afferent terminals (Miltz et al. 1990); this might explain why so few stained fibers to SHC were encountered. These two fibers did, however, have unusual response properties–they not only had high thresholds (>70 dB SPL) but also extremely flat, low-frequency tuning curves for which it was hardly possible to define a characteristic frequency (Fig. 27.7A). They resemble the infrasound fibers stained by Schermuly and Klinke (1988) in the apical abneural area of the pigeon's papilla. In the pigeon, these fibers apparently belong to a group of apical fibers not forming part of the large group representing the "normal" frequency map of the avian papilla.

As the THC area of birds is much larger than the equivalent area for mammals, which have only a single row of IHC, we tested the starling for differences in the physiological properties of fibers innervating hair cells at different positions *across*

FIGURE 27.7. Diagram illustrating the systematic distribution of starling nerve-fiber response parameters according to the location of the hair cell they innervate. (A) Frequency tuning curves for six fibers traced to hair cells located at the locations shown on the schematic apical half of the papilla (arrows). The three sensitive tuning curves belonged to fibers that each innervated one neurally lying hair cell. In contrast, the three insensitive fibers innervated more abneurally lying hair cells. The continuous line along the middle of the papilla separates THC from SHC. (B) Best-fit functions illustrating the tonotopic organization of the basilar papillae of the chick and starling (CF of afferent fiber-hair cell connections vs the distance of their innervation site from the apex of the papilla, the papillar lengths being normalized to 100%). For references see text.

in the chick, fibers that had been physiologically characterized were traced to their synaptic contacts. Almost all fibers only contacted THC, and a tonotopic organization was obvious (Fig. 27.7A,B), the apical end of the papilla responding to the lowest frequencies and the basal end to the

the papilla (Gleich 1989; Manley et al. 1989). Neither the sharpness of tuning nor the spontaneous activity of the fibers correlated with position. However, there was a surprisingly strong relationship between the rank of the innervated hair cell and the rate-response threshold of fibers (Fig. 27.7A), such that the most neurally lying cells were more sensitive to sound. According to a linear correlation of the data (n = 12, r = 0.764, P < 0.01), there is a threshold shift of almost 6 dB/hair cell across the papilla (to exclude threshold differences due, for example, to the middle-ear transmission characteristic, we selected the CF range between 0.6 and 1.8 kHz, where the starling audiogram is relatively flat; Kuhn et al. 1982). Considering the morphological variability between different avian groups, it will be necessary to investigate additional species, to see whether this threshold gradient is a general phenomenon in birds. Large threshold differences between neural and medial fibers would explain why it is not unusual in auditory nerve recordings in birds to find threshold ranges for any one frequency region which exceed 50 dB (Manley et al. 1985).

In birds, almost all of the THC are supported by the neural limbus (superior cartilaginous plate) and do not lie over the free basilar membrane. Traditional concepts in auditory physiology suggest that hair cells lying over the free basilar membrane (with the largest displacement amplitudes to sound) should be more sensitive. The highest sensitivity of THC furthest from the basilar membrane is contrary to intuition based on these traditional concepts. However, very little is known about the mechanics of hair-cell stimulation in birds. The pattern of hair-bundle orientation across all avian papillae (see Sections 4.1 and 4.2) suggests a complex pattern of hair-cell stimulation. Assuming a simple radial pattern of stimulation, the change of hair-cell bundle orientation across the starling papilla (in which medial, apical hair cell bundles in the frequency region we analyzed are rotated up to 70° towards the apex, Fig. 27.5) would certainly reduce the effectiveness of stimulation on rotated cells. The maximal threshold effect would, however, be <1 dB per hair cell. Thus the pattern of hair-cell bundle orientation alone cannot explain the range of sensitivity differences we found in the starling. There is evidence in some species that the tectorial membrane

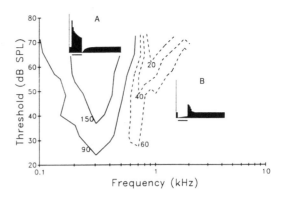

FIGURE 27.8. Tuning curve of a single auditory nerve afferent fiber showing the excitatory area (continuous lines, different response criteria in spikes/sec) and areas of primary suppression (dashed lines). The insets show idealized forms of poststimulus time histograms to short excitatory (A, upper left) and suppressive (B, lower right) pure-tone bursts (bar = stimulus duration).

contributes its mass to a resonance system (bobtail lizard, Manley et al. 1988; Manley, Köppl, and Yates 1989; guinea pig, Zwislocki et al. 1988). If the stimulus to the hair cells depends mainly on a resonance of the tectorial membrane and less or only indirectly on the vibrational amplitude of the basilar membrane, there is no reason why hair cells over the basilar membrane should be more sensitive.

4.3.4 Discharge Patterns of Auditory Nerve Fibers to Pure Tones

4.3.4.1 Discharge Patterns to Single-Tone Stimuli

Avian auditory nerve fibers can respond to a sound stimulus in one of three ways: an increased firing rate above the spontaneous level, a decrease below spontaneous level, or a phase locking with or without a change in discharge rate (Manley 1990). The second of these phenomena, also known as primary or single-tone suppression, has not been observed in mammals and is described in Section 4.3.4.3. The most common response to a tonal stimulus is a tonic increase in discharge rate (Fig. 27.8a; Manley et al. 1985). With increasing sound-pressure level, the response includes an increasingly large phasic component, whose time course may exceed that of the stimulus, that is, a steady state is not reached before the stimulus (50 ms duration) is turned off. Except for near-threshold stimuli, the tonal response is followed at offset by a period of reduced spontaneous activity, the magnitude and duration of the reduction depending on the strength and duration of the stimulus (Fig. 27.8A). The discharge rate increases monotonically with increasing sound pressure, often exceeding 300 spikes/s (averaged over the entire 50 ms stimulus; the instantaneous rates at onset are higher). Such discharge rates are higher than those reported for various mammals and correlate with the higher spontaneous rates in birds (Manley 1983; 1990).

4.3.4.2 Phase Locking
to Tonal Stimuli

In the starling, significant phase locking occurs in most low-CF cells at sound pressures below the mean rate threshold (Gleich and Narins 1988). The difference between phase locking threshold and mean rate threshold decreased with increasing frequency, suggesting that for low-frequency cells, phase locking is more important than a rate increase. No phase locking was observed above a few kHz. In general, the rate of change in the phase of the response towards higher frequencies was faster in lower-CF cells, indicating a greater delay in their responses than those of higher-CF cells. A plot of response phase versus stimulus frequency for single primary nerve fibers of the starling did not, however, always result in a straight line. Below CF, the phase lag was less than expected and was greater than expected for frequencies above CF. The overall phase response could be modelled by the combination of a constant delay plus the phase shift introduced by a standard LRC filter (made up of an inductance, resistance, and capacitance). Gleich (1988) used an iterative procedure to calculate the center frequency and sharpness of the putative LRC filter functions from the curves which resulted from subtracting a straight line response-phase characteristic from the individual phase function. The resonance frequency of the best-fit LRC filter and the fibers' acoustic CFs were correlated. As in the case of the preferred intervals in spontaneous activity (Section 4.3.1), however, individual fibers had a best-filter match in which the center frequency was on average 20% lower than that of the acoustic CF of the fiber. This also indicates

that the tuning of the hair-cells' electrical filters is, in most cases, mismatched to the acoustic CF.

4.3.4.3 Primary and Two-Tone Suppression

Not only can some tones suppress responses to other tones (two-tone suppression or TTRS; a phenomenon well known in the mammalian auditory nerve), but spontaneous activity can often be suppressed by single tones which do not themselves excite the cell (single-tone or primary suppression). Primary suppression has only begun to be studied systematically. Although in the case of very sensitive cells the possibility that the spontaneous activity of some fibers is partly a response to uncontrolled, low-level noise cannot be excluded, many fibers showing this phenomenon are quite insensitive. The observation of such nonclassical responses to sound from avian single fibers has been reported by Gross and Anderson (1976); Temchin (1982); Manley et al. (1985); Temchin (1988); and Hill, Mo, and Stange (1989a,b). Experimental procedures for examining fiber responses to a large matrix of frequencies and SPLs readily reveal the presence of such suppressive side bands on avian tuning curves. In the starling (Fig. 27.8B; Manley et al. 1985), the discharge rate of the cell to single suppressive tones often falls well below the spontaneous rate — sometimes even to zero. Such suppression is often accompanied by an "off" response (Fig. 27.8B). Of course, such effects can only be seen in cells with a significant spontaneous activity.

In both mammals and nonmammals, the phenomenon of two-tone suppression (TTRS) has similar properties to those described here for primary suppression, except of course that the suppressed activity is in response to the first tone (Sachs and Kiang 1968; Manley 1983). The threshold for suppression of a CF tone (10 dB above threshold) by a second tone may be lower for frequencies of the second tone which lie above or below the CF. TTRS areas in the chick and the starling have characteristics similar to those of the primary suppression areas (Manley 1990). That the suppression is not a synaptic (inhibitory) phenomenon is indicated by the fact that it has essentially no latency. In addition, Temchin (1988) observed primary suppression in the pigeon even after severance of the eighth nerve.

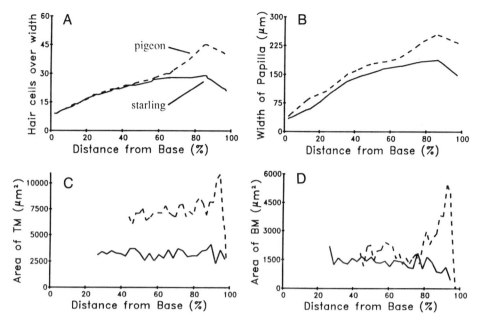

FIGURE 27.9. Four morphological criteria measured along the papillae, to illustrate the apical specialization of the pigeon's papilla compared to that of the starling. (A) number of hair cells across the papilla; (B) papillar width; (C) area of the tectorial membrane in cross section and (D) area of the basilar membrane in cross section. In each case, the pigeon data are drawn as dashed lines.

5. A Comparison of Structural and Functional Data From Other Avian Species

Comparative data on hearing-organ structure are available from the pigeon (Takasaka and Smith 1971; Gleich and Manley 1988), chick (Tanaka and Smith 1978; Tilney and Saunders 1983; Tilney, Tilney, and DeRosier 1987; Manley et al. unpublished data), barn owl (Smith, Konishi, and Schull 1985; Fischer, Köppl, and Manley 1988), duck (Chandler 1984) and seagull (Counter and Tsao 1986). Single-fiber physiological data are available for the pigeon (Sachs, Lewis, and Young 1974; Gross and Anderson 1976; Temchin 1982; Schermuly and Klinke 1985; Schermuly and Klinke 1988; Temchin 1988; Hill, Mo, and Stange 1989a,b) redwing blackbird (Sachs, Woolf, and Sinnott 1980) and chicken (Manley, Brix, and Kaiser 1987 and in preparation; Warchol and Dallos 1989a). There are a large number of structural similarities between the different avian papillae. Here, we shall only refer to data that show

significant deviations from the basic patterns described for the starling.

1. In the pigeon, the basilar papilla does not increase steadily in width from base to apex. The width increases gradually from 40 μm basally to 190 μm at 60% of the length from the basal end and then becomes disproportionately wider to 250 μm at 85%, before tapering somewhat to the apical end (Fig. 27.9B). The apical third of the pigeon papilla has more hair cells than expected even from this disproportionate width increase (Fig. 27.9A). Between 65% and 88% of the length, the number of hair cells in a transect rises from 30 to almost 50. This is accompanied by a dramatic reduction in the surface area of the hair cells of the abneural side of the papilla, which falls from near 120 μm² at 60% of the length from the basal end to <40 μm² at 85% of the length. A smaller size reduction is seen in hair cells in the middle area of the papilla. The patterns in the areas of the stereovillar bundles in the pigeon papilla differ from those seen in the starling, the abneural cells showing more dramatic changes. In the basal region, such abneural hair cells have almost 50% of their surface covered by

the stereovillar bundle. This percentage falls to only about 10% for cells near the middle of the papilla and remains constant to the apical end, the cells and their bundles both getting progressively smaller. This change in the relative surface area of cell and bundle occurs in spite of the fact that the number of stereovilli per cell falls quite steadily from base to apex. In addition, the apical area of the pigeon cochlea shows large increases in the dimensions of the basilar and tectorial membranes. Such dramatic dimensional changes are not seen in the starling (Fig. 27.9C,D). The pigeon specializations are found in the apical area described by Klinke and Schermuly (1986) as giving rise to responses to infrasound stimuli (see Section 6.2).

2. In the chick, the anatomy of the most apical region of the papilla apparently differs from that seen in other species of birds. Lavigne-Rebillard, Cousillas, and Pujol (1985) describe a crescent-shaped apical region they termed "very distal part," in which the hair cells show a greater resemblance to vestibular than to auditory receptors. There is evidence that this area also mediates very low-frequency hearing (see Section 6.2). In a comparative 2-deoxyglucose study of the auditory forebrain of a variety of avian species, Müller and Scheich (1985) found evidence in both the pigeon and the chicken of an area specialized for the processing of infrasound frequencies.

3. A partial tonotopic map is also available for the chick basilar papilla, obtained using horseradish-peroxidase staining of single fibers (21 single afferents or groups of afferents). It differs from that obtained from the starling, in that the curve lies at lower frequencies for equivalent locations on the hearing organ (Fig. 27.7B; Manley, Brix, and Kaiser 1987). This suggests that both the upper and lower frequency limits are lower in the chick.

4. The basilar papilla of the barn own *Tyto alba* is the longest so far described in birds, being over 11 mm in length in the unfixed state and containing over 16,000 hair cells. In the fixed, dried state, it is roughly 9 mm long, being 250 μm wide at the apical end and gradually reducing to 50 μm at the basal end. Whereas the data from the pigeon and the chick are relatively consistent with those from the starling in regard to the height and orientation of the stereovilli, (Tilney and Saunders 1983; Tilney, Tilney, and DeRosier 1987; Gleich and

Manley 1988), the change in hair-cell orientation in the owl papilla is very abrupt. In the apical two-thirds, both the neurally placed and abneurally placed hair cells are oriented parallel to the edge of the papilla (0°). In the center of the papilla, however, is a region where the orientation suddenly changes to at least 50° and up to 90° (apical orientation), and, further over the papilla, just as suddenly back again to 0°. On the neural side of middle and apical areas of the barn-owl papilla, the sudden change in orientation occurs at the place where Smith, Konishi, and Schull (1985) indicate a border between the tall, intermediate, and short hair cells. The sudden change in hair-cell orientation is often accompanied by a sharp rise in the number of stereovilli per hair-cell bundle at exactly the same place, so that a change in orientation angle from 0° to 50° may be correlated with a rise in the number of stereovilli from 120 to 180 per hair cell. Similarly, a return to 0° orientation is accompanied by a fall in the number of stereovilli. In the barn owl, the height of the tallest stereovilli in any one bundle changes from near 5 μm apically to 1 μm basally. Most of the height reduction occurs, however, in the apical half of the papilla, so that by 4.5 mm towards the base, it has already dropped to about 1.5 μm, remaining remarkably constant over the basal half of the papilla. Smith, Konishi, and Schull (1985) found that in the barn owl, this basal end has a marked thickening of the basilar membrane, a feature that has been found in specialized areas of some mammalian (e.g., bat) cochleae. As the basal area also differs in some other respects from that of other birds (Fischer, Köppl, and Manley 1988), it can be regarded as a specialization for high-frequency hearing (Section 6.2).

5. Unlike birds and despite the very similar cochlear anatomy, spontaneous activity rates in Caiman are bimodally distributed, as in mammals (Klinke and Pause 1980). The population of mammal and Caiman units that have spontaneous rates near zero does not exist in birds. Even within the avian data on spontaneous rates of nerve fibers, there are discordances, even for different studies of the same species. Whereas Sachs, Woolf, and Sinnott (1980) report a mean rate of 90 spikes/sec for the pigeon, Temchin (1988) found a mean rate of 78 spikes/sec in the 26 pigeon units he studied and Hill, Mo, and Stange (1989a) report a mean rate of only 35 spikes/sec. However, we will not

discuss this further, as the origin of these discrepancies is not yet understood.

6. Primary suppressive areas in the chick (Manley et al., in preparation) were not, as in the pigeon (Temchin 1988), only found in cells that showed preferred intervals in their spontaneous activity. Although TTRS has also been described for the pigeon by Sachs, Lewis, and Young (1974), it is curious that they report that in the pigeon nerve, "spontaneous activity was never inhibited by acoustic simuli," that is, primary suppression did not occur. This difference may be explained if these authors only refer to single stimuli within the excitatory tuning curve or through the use of threshold-tracking paradigms, which do not detect a reduced discharge rate. In the same species, Hill, Mo, and Stange (1989a) report clear cases of primary suppression.

6. The Evolution of the Avian Hearing Organ

6.1 General Trends in the Early Evolution of the Avian Papilla

Certain features of the avian papilla are regarded as being relatively unchanged from the stem reptiles (i.e., "primitive"; see also Table 27.1). These include the presence and the direction of the tonotopic organization, the presence of a high frequency selectivity (sharp tuning) in the responses of auditory nerve fibers and their general response patterns, including that of the spontaneous activity (irregularity of the spontaneous discharge and the presence of preferred intervals).

As shown in Fig. 27.10, one of the earliest trends in papillar evolution following the divergence of several evolutionary lines from the stem reptiles was an elongation of the papilla. This elongation was accompanied by the extension of the hearing range above the presumed upper limit of the stem reptiles (about 1 kHz). It is not known definitively whether this event occurred independently in the three lines leading to the mammal-like reptiles, to the archosaurs and to the lepidosaurs. At some stage, it is apparent that at the highest frequencies, a mechanical frequency selectivity mechanism became dominant over an electrical mechanism. This change was associated with a change in the

TABLE 27.1. Inherited and derived features of avian basilar papillae

Inherited

- The good frequency selectivity of auditory nerve fibers
- The presence of a tonotopic organization and its direction
- Electrical tuning at low frequencies (e.g., preferred intervals in spontaneous activity)
- A basalward extension of the papilla to higher frequency responses (at least partially mechanical tuning)
- Tonic responses to tones; single- and two-tone suppression

Derived

- Specialization of hair-cell types *across* the papilla
- Specialization of afferent and efferent innervations to different hair-cell populations
- Macromechanics (e.g., nonuniform hair-cell bundle orientation)
- Specific adaptations of some apical (infrasound reception) and basal (extended higher-frequency range) papillar areas

space constant of frequency distribution on the papilla (Manley et al. 1988). Also accompanying the elongation of the papilla in the archosaurs was a differentiation and specialization of the hair cells into recognizable, but intergrading, types and the establishment of clear differences in their innervation patterns. The spontaneous and sound-driven activity patterns of the nerve fibers innervating apical hair cells in the avian papilla probably did not change significantly compared to those inherited from the stem reptiles. The above changes were most likely essentially complete before the ancestors of the Crocodilia and Aves diverged. Later changes in the avian line involved the orientation patterns of the hair-cell bundles, the specialization of the abneural area of the apical papilla for encoding extremely low frequencies, and, at least in owls, changes in the basal papilla to facilitate high-frequency hearing. Some of these later specializations are described in more detail below.

Thus we assess these other features of the avian papilla as being specialized (see also Table 27.1). These include the presence of different types of hair cells across the width of the papilla and their very different afferent and efferent innervation patterns, the elongation of the papilla and probable dominance of mechanical frequency selectivity at higher frequencies, and the changes in hair-cell bundle orientation along and across the papilla, which also implies a new pattern of mechanical stimulation.

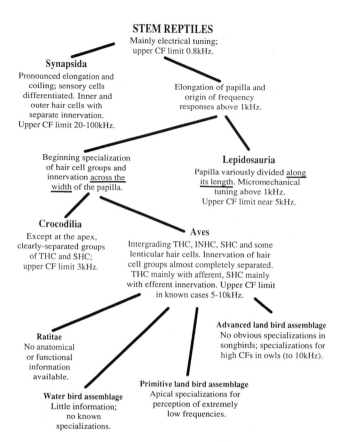

FIGURE 27.10. Schematic diagram to illustrate (top to bottom) our current understanding of the changes in morphology and function of the hearing organ, observed during the evolution of different lines of land vertebrates from stem reptiles.

6.2 Functional Implications of Variations in Avian Papillar Anatomy

Our understanding of the function of the avian basilar papilla is still relatively poor. It is thus difficult to realize the full implications of the anatomical variability discussed above. It should, however, be noted that each of the bird papillae investigated to date is unique. We thus expect that there will be species-specificities in the response patterns of the hair cells and their associated nerve fibers. To date, the following specializations are known:

a: *The mechanics of stimulation of the avian papilla probably differ from those of mammals.* In the chick, the change of hair-cell bundle orientation across the papilla led Tilney, Tilney, and DeRosier (1987) to suggest that there is an unexpected pattern of hair-cell stimulation — unexpected that is, when we consider the essentially radial shear pattern thought to be present in the mammalian cochlea. Since we found similar morphological patterns in the starling, pigeon, and the

barn owl (Fischer, Köppl, and Manley 1988; Gleich and Manley 1988), they can probably be considered as typical of birds. It should also be borne in mind that in all birds, a high proportion of THC do not lie over the free basilar membrane, and thus require their mechanical stimulus to come from the tectorial membrane.

b: *The barn owl (an "advanced land bird") has exceptional high-frequency hearing and a specialized basal papillar area.* Major differences are not seen in the hearing ranges (measured behaviorally or as the range of CFs of the tuning curves) of most avian species studied so far (100 Hz to about 6 to 8 kHz; Konishi 1970). However, Sullivan and Konishi (1984) report recordings in the brainstem of the barn owl which, together with the behavioral audiogram (Konishi 1973) indicate that in this large bird, the CFs in the nerve reach 9 to 10 kHz. The morphological data indicate that, whereas the apical half of the barn owl papilla shows structural patterns that resemble those of the entire papillae of other birds, the basal half is exceptional

(Fischer, Köppl, and Manley 1988). This is presumably a specific adaptation for processing of the high frequencies (5 to 10 kHz) used by the barn owl for sound localization. Also in the barn owl, the abruptness of the change in the orientation of hair-cell bundles across the papilla is striking and appears to correlate in position both with a change in the number of stereovilli on the hair cells and, at least in the neural half of the papilla, with the edge of the neural limbus. That is, hair cells on the two sides of the outer edge of the superior cartilaginous plate have quite different stereovillar-bundle orientation.

c: *The pigeon and the chick ("primitive land birds") have differently specialized apical papillar areas sensitive to very low sound frequencies.* The low-frequency sensitivity of the pigeon disappears upon removal of the cochlear duct (Kreithen and Quine 1979), proving that the receptors lie in the basilar papilla or the lagena macula. Klinke and Schermuly (1986) report finding extremely low-frequency, phase-locked responses in pigeon auditory nerve fibers. When stained, these fibers were found to innervate hair cells in the abneural area of the apical part of the basilar papilla. This area (see Section 5) is specialized in the pigeon, being unusually wide and having an exceptionally large number of sensory cells (Gleich and Manley 1988).

In the chick, Warchol and Dallos (1989a) report finding cochlear nucleus cells that responded to low-frequency sound (10 to 500 Hz). About half of the cells responded with equal sensitivity to frequencies between 10 and 100 Hz, the other half having broad, but more classical, auditory tuning curves with CF near 100 Hz. Many of these cells responded to sound only with a modulation of their spontaneous discharge and had more Gaussian than Poisson distributions of intervals in the spontaneous activity. Warchol and Dallos (1989b) later traced low-CF fibers in the chicken to the apical papillar area, suggesting that the "very distal part" might be specialized in a functionally similar way to the pigeon. As the anatomical patterns are very different, however, it is probable that the pigeon and chick have independently developed a low-frequency specialization. In this respect it is important to note that both the pigeon and the chicken belong to a more primitive group of land birds than the other species described here (Feduccia 1980; Carroll 1987). In most previous

experiments with other species, however, the sound systems were often unable to stimulate adequately below 100 Hz and phase-locking responses were neglected, so it is difficult to estimate the relative occurrence of very low-frequency responses in other avian species.

6.3 Mechanisms of Frequency Selectivity in the Avian Basilar Papilla

It has been suggested for some years now that there is more than one mechanism of frequency analysis in the vertebrate inner ear (Klinke 1979; Manley 1979). More recently, fundamentally different mechanisms of frequency selectivity have been recognized in terrestrial vertebrates, which frequently coexist (Manley 1986, 1990). Although it is not yet possible in any individual case to cleanly separate the different mechanisms, for descriptive purposes they will be treated separately.

There are two fundamental foundations upon which frequency-selectivity mechanisms can operate:

(a) Frequency selectivity resulting from the *electrical* characteristics of the hair-cell membrane, which may operate in addition to:
(b) Frequency selectivity resulting from *mechanical* factors. These factors can be one or a combination of the following:
   b1: mechanical properties of the hair-cell stereovillar bundle if not coupled to a tectorial membrane,
   b2: mechanical interaction between hair-cell bundles and the tectorial membrane, and
   b3: active movement processes in a large number of hair cells, resulting in mechanical interactions between the hair cells and the basilar (and tectorial?) membrane(s). This last mechanism is inextricably mixed with the passive selectivity of the accessory structures (e.g., basilar and tectorial membranes) themselves.

Which of the above selectivity mechanisms have been retained by or have evolved in birds? Electrical tuning is most likely a primordial property of hair cells (Manley 1986, 1990). The presence of both a voltage-sensitive Ca^{2+} conductance and a Ca^{2+}-sensitive K^+ conductance in hair cells can under stimulation produce a resonance whose properties can explain a good deal of the frequency selectivity of certain hair cells. In the frog sacculus and the turtle basilar papilla, the variations in the number and the kinetics of these channels are thought to be responsible for the different CFs of hair cells (Crawford and Fettiplace 1981; Ashmore and Atwell 1985; Hudspeth 1985; Art, Crawford, and Fettiplace 1986; Hudspeth 1986; Fettiplace 1987). Recently, it has also been show that such channels are found in low-CF hair cells of the alligator (Evans and Fuchs 1987) and the chicken (Fuchs, Nagai, and Evans 1988). In the latter, the properties of the channel kinetics also suggest a role of electrical tuning in the frequency selectivity of low-frequency avian THC. Three other factors, the presence of preferred intervals in the spontaneous activity of primary auditory nerve fibers, the finding of specific deviations from expected phase responses and of a temperature sensitivity in the frequency tuning also provide strong evidence that birds have retained electrical tuning in their hair cells, at least at low CF (Klinke 1979; Manley 1979, 1981, 1986; Manley and Gleich 1984; Schermuly and Klinke 1985; Gleich 1987). Thus a major part of the peripheral tuning mechanism in low-CF avian THC resides in the properties of individual cell membranes of the sensory cells. We do not yet know how high in frequency this mechanisms can operate at the high body temperatures of birds. Indeed, one important feature of electrical tuning to be investigated in future research concerns the factors that limit its frequency response. The highest limits found for indicators of electrical tuning in preferred intervals in the spontaneous activity of primary auditory nerve fibers in birds are between 1.5 kHz (Manley et al. 1985) and 2.5 kHz (Temchin 1988). These limits are, however, influenced by the decreasing ability of the nerve fibers or their synapses to follow higher-frequency oscillations of the hair-cell membrane potential. The question of frequency limitations has not been exhaustively analyzed at the hair-cell level. Above the limit of function of electrical tuning, it would be necessary for higher CFs to be analyzed using other, presumably predominantly micromechanical, mechanisms.

The situation with regard to mechanical tuning is much more complex than with electrical tuning, for there is a variety of mechanical structures that can play a role in tuning, the basilar and tectorial membranes and the hair-cell stereovillar bundles being the most obvious candidates. The presence of obvious gradients in the structural parameters of the hair-cell stereovillar bundles in all species examined in this respect has led to the expectation that in all hair cells, the mechanical properties of the stereovillar bundle will play an important role in frequency selectivity. This would be true irrespective of the presence or absence of, for example, a tectorial membrane. Although such gradients certainly influence the tuning, it is at present not possible to quantify the contribution of individual parameters in avian species. In order to gain an impression of the frequency-response parameters of the different cells, it is not enough to know the height of the stereovillar bundles, for this parameter is constant in the basal half of the owl papilla. At low frequencies, even a knowledge of all bundle parameters will not suffice where electrical tuning also plays a role.

In the avian hearing organ, all hair cells are rather firmly connected to the tectorial membrane. This makes it unlikely that the resonance properties of individual stereovillar bundles alone play an important role in frequency selectivity. We would rather expect that larger numbers of hair cells and an area of tectorial membrane would form some sort of resonant unit, as in b2 above.

With regard to possibility b3 above, it now seems virtually certain that in mammals, the macromechanical resonance of the organ of Corti depends at low levels to a very large extent on active mechanical motions of the outer hair cells. These cells respond actively to specific stimuli, which locally greatly increases the motion of the entire cochlear partition. This increased motion influences the inner hair cells, which thus respond to the "net result" of the passive and active motion of the organ of Corti. The active motions of hair cells result under certain conditions in the generation of otoacoustic emissions, sounds that emerge from the cochlea and are often present spontaneously. These emissions, which are currently being studied extensively in mammals, are also present at higher frequencies in the external meatus of the starling (Manley, Schulze, and Oeckinghaus 1987), suggesting that similar patterns of mechanical activity of hair cells are present in birds, at least at higher CFs. The only available evidence on the macromechanics of the avian basilar membrane is from the pigeon (Gummer, Smolders, and Klinke 1986) and suggest that there is a crude equivalent of the travelling wave of mammals. However, there are no equivalent measurements of relatively low-frequency regions in the mammalian cochlea and the difficulties associated with such investigations are very large. In the pigeon, Smolders (personal communication) found a pattern of tonotopic organization very similar to that which we reported for the starling and chick, the frequency map for the basal third strongly resembling that derived from the macromechanical measurements. Our recent experiments marking single avian nerve fibers (Manley, Brix, and Kaiser 1987; Gleich 1989) indicate that it is not unreasonable to speculate that there are substantial functional differences between THC and SHC, which may manifest themselves in a similar functional separation of hair-cell populations in birds as in mammals. There is a large number of hair cells across the papilla in birds, so that if avian hair cells can produce active movement there would be enough cells to drive the papilla. However, the amount of space along the papilla devoted to one octave is less than in mammals and the basilar papilla is thick, both factors that would reduce the selectivity of any active process. In addition, most THC do not sit over the free basilar membrane, so it would be important to take the motion of the tectorial membrane into account. Perhaps an interaction between hair-cell populations through the tectorial membrane would be a more appropriate postulate for the avian situation.

In the hair-cell populations recognized in Crocodilia and in birds, the differences in the position of the cells (neural and abneural part of the papilla, over the limbs or on the free basilar membrane), in their structure (e.g., form of the cell body, form and position of the stereovillar bundle), and in the pattern of the afferent and efferent innervation are substantial and comparable to the differences between inner and outer hair cells of mammals. This raises the important question of the evolutionary origin of separate hair-cell populations. We hypothesize that the specific abilities of primitive vertebrate hair cells (both sensory transduction and active motility of some sort; Crawford and Fettiplace 1985) predispose large arrays of such cells to specialization. It is conceivable that

the selection pressures acting on a large uniform population of hair cells in different vertebrate classes could produce sense organs of similar structure and function based on a "division of labor" between hair-cell groups. Depending on the differences in their stimulus input, hair cells could specialize by emphasizing certain functions above others (e.g., mechanical response to the stimulus rather than accurate transmission through afferent synapses). The development of functionally similar hair-cell populations in birds and mammals is a remarkable case of the independent and convergent evolution of a complex interactive process in a sense organ resulting in specific improvements in the function of their hearing organs.

6.4 Is the Avian Basilar Papilla a Multifunction Sense Organ?

Our finding (Manley, Brix, and Kaiser 1987; Gleich 1989) both in the starling and the young chick, that afferent fibers of the auditory nerve primarily contact tall hair cells indicates that there is a division of labor among the different hair-cell populations of the avian papilla (Manley et al. 1989). That primary afferents primarily contact THC over the neural limbus has recently also been confirmed in the pigeon (Smolders, personal communication). The fibers of the pigeon which responded to infrasound, and which were stained by Klinke and Schermuly (1986), innervate an abneural area of the pigeon papilla which, as the tonotopic organization of the pigeon resembles that of the starling (Smolders, personal communication), lies adjacent to an area of neural cells responding to a few hundred Hz. Fibers innervating abneural hair cells in the starling were also quite different in their properties. It is possible that the fact that THC of birds, at least the lower-frequency THC, show some electrical tuning, allowed the THC of the apical part of the hearing organ to fulfill the function of a frequency-selective organ without necessitating any mechanical interaction between hair-cell populations. In this sense, the apical, abneural hair cells would then be unnecessary for "hearing" of normal acoustical frequencies and could be involved in a different function. The very different structure and innervation of SHC, and their proliferation in the pigeon in an area responding to infrasound suggest this also. One can thus speculate that the avian papilla consists of three functional areas:

a: Apical THC responding selectively and at least partly via electrical tuning to sound up to frequencies of about 1.5 kHz,

b: Apical abneural hair cells (INHC or SHC) encoding very low frequency and infrasound, predominantly via phase locking in their afferents,

c: The basal end of the papilla responding to sound frequencies above 1.5 kHz mainly by mechanical frequency selectivity involving both THC and SHC.

As flying animals, it is certainly important for birds to collect information about their position in space. The infrasound receptors described are in a position to respond to the slow air-pressure changes of winds and of slightly different flight altitudes. Such a function would be best fulfilled by hair cells over the free basilar membrane. Further study of the specializations of hair-cell populations will show whether these speculations have any real substance. Indeed, the investigation of the functional significance of distinct hair-cell populations — those factors accompanying the evolution of complex hearing organs — will be one of the most fruitful future areas of avian auditory research.

7. Summary

Birds have retained the single-ossicle middle ear, which limits their upper frequency of hearing to 12 kHz. In the course of the evolution of the archosaurs from the stem reptiles, the sensory hair cells have become specialized into several intergrading types, the extremes of which have profoundly different structure and innervation patterns. Present evidence suggests a specialization of function during this evolution parallel to that seen in the mammalian organ of Corti. However, the anatomy of the basilar papilla varies between representatives of the different avian groups, suggesting that there may be significant differences in functional patterns. THC are generally not on the free basilar membrane and may be stimulated in a different way than mammalian inner hair cells. Apically, THC also show electrical tuning and analyze low-frequency sound. In some primitive land birds, this abneural, apical end is specialized, transducing

low- and very-low-frequency stimuli which are then coded primarily via phase locking in the afferent fibers. In the basal area, present evidence indicates that high frequencies are analyzed by a mechanical tuning mechanism which may involve interaction between hair-cell types. Barn owls have evolved a specialized basal cochlear area.

Acknowledgments. This work was supported by grants to GAM, mainly from the Deutsche Forschungsgemeinschaft and within the programme of the Sonderforschungsbereich 204.

References

Art JJ, Crawford AC, Fettiplace R (1986) Electrical resonance and membrane currents in turtle cochlear hair cells. Hearing Res 22:31–36.

Ashmore JF, Attwell, D (1985) Models for electrical tuning in hair cells. Proc Roy Soc B 226:325–344.

Carroll RL (1987) Vertebrate Palaeontology and Evolution. New York: Freeman.

Chandler JP (1984) Light and electron microscopic studies of the basilar papilla in the duck, Anas platyrhynchos: I. The hatchling. J Comp Neurol 222:506–522.

Counter SA, Tsao P (1986) Morphology of the seagull's inner ear. Acta Otolaryngol 101:34–42.

Crawford AC, Fettiplace R (1981) An electrical tuning mechanism in turtle cochlear hair cells. J Physiol 312:377–412.

Crawford AC, Fettiplace R (1985) The mechanical properties of ciliary bundles of turtle cochlear hair cells. J Physiol 364:359–379.

Düring M von, Andres KH, Simon K (1985) The comparative anatomy of the basilar papillae in birds. Fortschritte der Zoologie 30:681–685.

Evans MG, Fuchs PA (1987) Tetrodotoxin-sensitive, voltage-dependent sodium currents in hair cells from the alligator cochlea. Biophys J 52:649–652.

Feduccia A (1980) The age of birds. Cambridge: Harvard Univ Press.

Fettiplace R (1987) Electrical tuning of hair cells in the inner ear. Trends in Neurosci 10:421–425.

Fischer FP, Köppl C, Manley GA (1988) The basilar papilla of the barn owl Tyto alba: A quantitative morphological SEM analysis. Hearing Res 34:87–101.

Fischer FP, Brix J, Manley GA (1991) Morphological gradients in the innervation of the chick basilar papilla. Abstr. 14th Mtg Assoc Res Otolaryngol, p. 153.

Fuchs PA, Mann AC (1986) Voltage oscillations and ionic currents in hair cells isolated from the apex of the chick cochlea. J Physiol 371:31P.

Fuchs PA, Nagai T, Evans MG (1988) Electrical tuning in hair cells isolated from the chick cochlea. J Neurosci 8:2460–2467.

Gleich O (1987) Evidence for electrical tuning in the starling inner ear. In: Elsner N, Creutzfeldt O (eds) New Frontiers in Brain Research. Stuttgart, New York: Thieme Verlag, p. 101.

Gleich O (1988) Untersuchungen zur funktionellen Bedeutung der Haarzelltypen und ihrer Innervationsmuster im Hörorgan des Staren. Thesis, Institut für Zoologie, Technische Universität München.

Gleich O (1989) Auditory primary afferents in the starling: Correlation of function and morphology. Hearing Res 37:255–268.

Gleich O, Manley GA (1988) Quantitative morphological analysis of the sensory epithelium of the starling and pigeon basilar papilla. Hearing Res 34:69–86.

Gleich O, Narins PM (1988) The phase response of primary auditory afferents in a songbird (Sturnus vulgaris L.). Hearing Res 32:81–91.

Graybeal A, Rosowski JJ, Ketten DR, Crompton AW (1989) Inner-ear structure in Morganucodon, an early Jurassic mammal. Zool J Linn Soc 96:107–117.

Gross NB, Anderson DJ (1976) Single unit responses recorded from the first order neuron of the pigeon auditory system. Brain Res 101:209–222.

Gummer A, Smolders JWT, Klinke R (1986) The mechanics of the basilar membrane and middle ear in the pigeon. In: Allen JB, Hall JL, Hubbard A, Neely ST, Tubis A (eds) Peripheral auditory mechanisms. Berlin, Heidelberg, New York, Tokyo: Springer-Verlag, pp. 81–88.

Hill KG, Mo J, Stange G (1989a) Excitation and suppression of primary auditory fibres in the pigeon. Hearing Res 39:37–48.

Hill KG, Mo J, Stange G (1989b) Induced suppression in spike responses to tone-on-noise stimuli in the auditory nerve of the pigeon. Hearing Res 39:49–62.

Hudspeth AJ (1985) The cellular basis of hearing: the biophysics of hair cells. Science 230:745–752.

Hudspeth AJ (1986) The ionic channels of a vertebrate hair cell. Hearing Res 22:21–27.

Klinke R (1979) Comparative physiology of primary auditory neurones. In: Hoke M, de Boer E (eds) Models of the auditory system and related signal processing techniques. Scand Audiol Suppl 9:49–61.

Klinke R, Pause M (1980) Discharge properties of primary auditory fibres in Caiman crocodilus: Comparisons and contrasts to the mammalian auditory nerve. Exp Brain Res 38:137–150.

Klinke R, Schermuly L (1986) Inner ear mechanics of the crocodilian and avian basilar papillae in comparison to neuronal data. Hearing Res 22:183–184.

Konishi M (1970) Comparative neurophysiological studies of hearing and vocalizations in songbirds. Z vergl Physiol 66:257–272.

Konishi M (1973) How the owl tracks its prey. Amer Sci 61:414–424.

Kreithen ML, Quine DB (1979) Infrasound detection by the homing pigeon: a behavioural audiogram. J Comp Physiol 129:1–4.

Kuhn A, Müller CM, Leppelsack HJ, Schwartzkopff J (1982) Heart rate conditioning used for determination of auditory threshold in the starling. Naturwiss 69:245–246.

Lavigne-Rebillard M, Cousillas H, Pujol R (1985) The very distal part of the basilar papilla in the chicken: a morphological approach. J Comp Neurol 238:340–347.

Liberman MC (1982) The cochlear frequency map for the cat: Labeling auditory-nerve fibres of known characteristic frequency. J Acoust Soc Amer 72:1441–1449.

Manley GA (1972a) Frequency response of the ear of the tokay gecko. J Exp Zool 181:159–168.

Manley GA (1972b) The middle ear of the tokay gecko. J Comp Physiol 81:239–250.

Manley GA (1973) A review of some current concepts of the functional evolution of the ear in terrestrial vertebrates. Evolution 26:608–621.

Manley GA (1979) Preferred intervals in the spontaneous activity of primary auditory neurones. Naturwiss 66:582.

Manley GA (1981) A review of the auditory physiology of the reptiles. Progr Sens Physiol 2:49–134.

Manley GA (1983) Auditory nerve fibre activity in mammals. In: Lewis B (ed) Bioacoustics. London, New York: Academic Press, pp. 207–232.

Manley GA (1986) The evolution of the mechanisms of frequency selectivity in vertebrates. In: Moore BCJ, Patterson RD (eds) Auditory frequency selectivity. New York, London: Plenum Press, pp. 63–72.

Manley GA (1990) Peripheral hearing mechanisms in reptiles and birds. Berlin, Heidelberg: Springer-Verlag.

Manley GA, Gleich O (1984) Avian primary auditory neurones: the relationship between characteristic frequency and preferred intervals. Naturwiss 71:592–594.

Manley GA, Leppelsack H-J (1977) Preliminary data on activity patterns of cochlear ganglion neurones in the starling. In: Portmann M, Aaron, J-M (eds) Inner ear biology XIVth workshop. INSERM, Paris, pp. 127–136.

Manley GA, Gleich O, Leppelsack H-J, Oeckinghaus H (1985) Activity patterns of cochlear ganglion neurones in the starling. J Comp Physiol A 157:161–181.

Manley GA, Brix J, Kaiser (1987) Developmental stability of the tonotopic organization of the chick's basilar papilla. Science 237:655–656.

Manley GA, Schulze M, Oeckinghaus H (1987) Otoacoustic emissions in a song bird. Hearing Res 26: 257–266.

Manley GA, Yates GK, Köppl C (1988) Auditory peripheral tuning: evidence for a simple resonance phenomenon in the lizard Tiliqua. Hearing Res 33:181–190.

Manley GA, Brix J, Gleich O, Kaiser A, Köppl C, Yates GK (1988) New aspects of comparative peripheral auditory physiology. In: Syka J, Masterton RB (eds) Auditory Pathway—Structure and Function. NY: Plenum Press, London, pp. 3–12.

Manley GA, Köppl C, Yates GK (1989) Micromechanical basis of high-frequency tuning in the bobtail lizard. In: Wilson JP, Kemp D (eds) Cochlear Mechanisms—Structure, Function and Models. NY: Plenum Press, pp. 143–151.

Manley GA, Gleich O, Kaiser A, Brix J (1989) Functional differentiation of sensory cells in the avian auditory periphery. J Comp Physiol A 164:289–296.

Miltz C, Singer I, Fischer FP, Manley GA (1990) Ultrastructure of the hair cells in the basilar papillar of the European starling. In: Elsner N, Roth G (eds) Brain-Perception, Cognition. Stuttgart: Thieme Verlag, p. 135.

Müller SC, Scheich H (1985) Functional organization of the avian auditory field L. J Comp Physiol A 156: 1–12.

Sachs MB, Kiang NY-S (1968) Two-tone inhibition in auditory-nerve fibres. J Acoust Soc Amer 43:1120–1128.

Sachs MB, Lewis RH, Young ED (1974) Discharge patterns of single fibres in the pigeon auditory nerve. Brain Res 70:431–447.

Sachs MB, Woolf NK, Sinnott JM (1980) Response properties of neurons in the avian auditory system: comparisons with mammalian homologues and consideration of the neural encoding of complex stimuli. In: Popper AN, Fay RR (eds) Comparative studies of hearing in vertebrates. New York, Heidelberg, Berlin: Springer-Verlag, pp. 323–353.

Schermuly L, Klinke R (1985) Change of characteristic frequencies of pigeon primary auditory afferents with temperature. J Comp Physiol A 156:209–211.

Schermuly L, Klinke R (1988) Single-fibre staining of infrasound-sensitive neurones in the pigeon inner ear. Pflügers Archiv Suppl 411:R168.

Schwartzkopff J (1968) Structure and function of the ear and of the auditory brain areas in birds. In: deReuck AVS, Knight J (eds) Hearing Mechanisms in vertebrates. Boston: Little, Brown, pp. 41–58.

Schwartzkopff J, Winter P (1960) Zur Anatomie der Vogel-Cochlea unter natürlichen Bedingungen. Biologisches Zentralblatt 79:607–625.

Singer I, Fischer FP, Manley GA (1989) Hair-cell innervation in the basilar papilla of the European starling (Sturnus vulgaris). Abstr. 26th Inner Ear Biology Mtg., Paris, p. 60.

Smith CA (1985) Inner ear. In: King AS, McLeland J (eds) Form and function in birds, Vol 3. London: Academic Press, pp. 273–310.

Smith CA, Konishi M, Schull N (1985) Structure of the barn owl's (Tyto alba) inner ear. Hearing Res 17: 237–247.

Sullivan WE, Konishi M (1984) Segregation of stimulus phase and intensity coding in the cochlear nucleus of the barn owl. J Neurosci 4:1787–1799.

Takasaka T, Smith CA (1971) The structure and innervation of the pigeon's basilar papilla. J Ultrastruct Res 35:20–65.

Tanaka K, Smith CA (1978) Structure of the chicken's inner ear. Am J Anat 153:251–271.

Temchin AN (1982) Acoustical reception in birds. In: Ilyichev VD, Gavrilov VM (eds) Acta XVIII Congressus Internat Ornithologicus, Moscow, August 1982.

Temchin AN (1988) Discharge patterns of single fibres in the pigeon's auditory nerve. J Comp Physiol A 163: 99–115.

Tilney LG, Saunders JC (1983) Actin filaments, stereocilia, and hair cells of the bird cochlea. I. Length, number, width, and distribution of stereocilia of each hair cell are related to the position of the hair cell on the cochlea. J Cell Biol 96:807–821.

Tilney MS, Tilney LG, DeRosier DJ (1987) The distribution of hair cell bundle lengths and orientations suggests an unexpected pattern of hair cell stimulation in the chick cochlea. Hearing Res 25:141–151.

Warchol ME, Dallos P (1989a) Neural response to very low-frequency sound in the avian cochlear nucleus. J Comp Physiol A 166:83–95.

Warchol ME, Dallos P (1989b) Localization of responsiveness to very low frequency sound on the avian basilar papilla. Abstr. 12th Mtg Assoc Res Otolaryngol, p. 125.

Zwislocki JJ, Slepecki NB, Cefaratti LK (1988) Tectorial-membrane stiffness and hair-cell stimulation. Abstr. 11th Mtg Assoc Res Otolaryngol, p. 170.

Abstract H

Tuning in the Turtle: An Evolutionary Perspective

Michael G. Sneary and Edwin R. Lewis

Many investigators believe that tuning in the acoustic sensors of lower vertebrates is accomplished by second order resonances, first described in turtle hair cells (Crawford and Fettiplace 1981) and subsequently found in the hair cells of two amphibian sensors (Pitchford and Ashmore 1987; Lewis and Hudspeth 1987). In contrast, acoustic nerve fiber responses from both amphibian sensors implied that the tuning process employs high-order dynamic structures rather than resonances, with filters exhibiting nearly linear phase vs frequency plots and amplitude curves lacking resonant peaks but exhibiting band edges with sustained, steep slopes (Lewis 1987, 1988). In mammals, similar tuning characteristics are produced by high-order filters in the form of a traveling wave structure. Second-order resonances, acting alone, can provide good spectral resolution or good temporal resolution, but not both. The high-order filters of amphibian and mammalian acoustic sensors, on the other hand, are well suited to both tasks. We have been studying the responses of turtle acoustic nerve fibers to determine whether or not similar filters are present. Previous measurements of phase vs frequency from these fibers implied dynamic orders of at least six and equivalent time delays of as much as 10 ms (Sneary and Lewis 1989). Now we have supplemented those results with impulse responses and tuning curves derived from REVCOR. Axonal time delays of 1 to 3 ms, observed in the impulse responses, imply that the major contribution to the slopes of the phase vs frequency curves was from high-order dynamics in the tuning structure. Furthermore, the shapes of the impulse responses are consistent with high-order dynamics and not second-order resonances. Amplitude tuning curves imply dynamic orders as great as nine. To reconcile the evidence derived from turtle hair cells and nerve fibers we propose an exchange of signal energy between hair cells linked in cascade. In our hypothesis, each nerve fiber is connected to a peripheral tuning structure that comprises not only the electrical and mechanical tuning elements associated directly with the hair cell(s) innervated by that fiber, but also the electrical and mechanical elements of neighboring hair cells. The evolution of acoustic sensors in lower vertebrates and mammals seems to have converged on a common filter design with high-order dynamics. The physical realization of that design evidently employs different biophysical elements in different species. We propose that the adaptive value of this design is its ability to provide good spectral and good temporal resolution.

References

Crawford AC, Fettiplace R (1981) An electrical tuning mechanism in turtle cochlear hair cells. J Physiol 312:377–412.

Lewis ER (1987) Speculations about noise and the evolution of vertebrate hearing. Hear Res 25:83–90.

Lewis ER (1988) Tuning in the bullfrog ear. Biophys J 53:441–447.

Lewis RS, Hudspeth AJ (1983) Frequency tuning and ionic conductances in hair cells of the bullfrog sacculus. In: Klinke R, Hartmann R (eds) *Hearing-Physiological bases and Psychophysics*, Berlin: Springer-Verlag pp. 17–24.

Pitchford S, Ashmore JF (1987) An electrical resonance in the hair cells of the amphibian papilla of the frog Rana temporaria. Hear Res 27:75–84.

Sneary M, Lewis ER (1989) Response properties of turtle auditory afferent nerve fibers: evidence for a high order tuning mechanism. In: Wilson JP, Kemp DT (eds) Cochlear Mechanisms – Structure, Function and Models. NATO ASI SERIES. New York: Plenum Press.

Abstract I

Paratympanic and Spiracular Sense Organs: Phylogenetic Distribution and Theories of Functions, Including Hearing

Christopher S. von Bartheld and Edwin W. Rubel

The paratympanic organ (PTO) and the spiracular sense organ (SSO) are epithelial vesicles or sacs in the dorsal wall of the avian middle ear and in the spiracular cleft of 'primitive' fishes. These organs possess a sensory epithelium with hair cells which are innervated by the facial nerve or the anterior lateral line nerve. Spiracular sense organs (SSOs) are well developed in elasmobranchs and in nonteleost bony fishes; paratympanic organs (PTOs) are well developed in birds. Embryonic and rudimentary PTOs are present in 'primitive' reptiles and mammals (Simonetta 1953).

Various functions have been suggested for the PTO/SSO, from electroreception to hearing. Morphological and physiological evidence is consistent with a proprioceptive function of the SSO (Barry et al. 1988a, b) and a barometric function of the PTO (von Bartheld 1990). Both of these functions appear to be related to the close association of the SSO/PTO with the ligaments connected to the hyomandibula or stapes (columella in birds).

The consistent anatomical and functional relationship of the SSO/PTO with the hyomandibula/stapes suggests that the SSO may have been involved in the evolutionary development of the stapes, possibly by providing a sensory feedback for stapes positioning to optimize transduction of sound to the otic capsule. The SSO thus may have played a role in the development of hearing in amniote vertebrates. After development of an auditory function for the hyomandibula/stapes, the SSO regressed in most tetrapods but not in birds

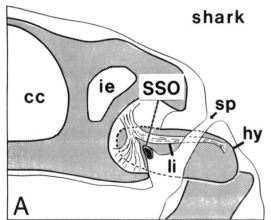
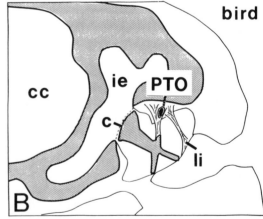

Schemes of cross sections through the ear of sharks (A) and birds (B) show the location of the SSO and the PTO and their relationships with the hyomandibula (hy) or columella (c) and attaching ligaments (A) modified from Barry et al. (1988a). cc, cranial cavity; ie, inner ear; li, ligaments; sp, spiracle.

in which the SSO persisted as a PTO, possibly functioning as a baroreceptor.

(Supported by NIH grants DC 00019 and DC 00395).

References

Barry MA, Hall DH, Bennett MVL (1988a). The elasmobranch spiracular organ. I. Morphological studies. J Comp Physiol A 163:85–92.

Barry MA, White RL, Bennett MVL (1988b). The elasmobranch spiracular organ II. Physiological studies. J Comp Physiol A 163:93–98.

Simonetta A (1953). L'organo di senso dello spiracolo e l'organo paratimpanico nella sistematica dei Vertebrati. Arch ital Anat Embriol 58:266–294.

von Bartheld CS (1990). Development and innervation of the paratympanic organ (Vitali organ) in chick embryos. Brain Behav Evol 35:1–15.

Section VI
Mammals

Although we understand a great deal about mammalian hearing, our knowledge of its evolution is still limited. Many investigators are particularly intrigued by trying to "predict" the hearing capabilities of extinct species (as attempted by Fay in Chapter 14) or in species that cannot be studied for other reasons.

In Chapter 28, Allin and Hopson review the fossil record of the middle ear from early synapsid reptiles to the earliest mammals. In doing so, they both document the dramatic structural changes and interpret them with functional hypotheses.

Rosowski (Chapter 29) reviews the current data on the auditory anatomy of one of the first mammals (*Morganucodon*), and the known relationships between auditory structure and hearing capabilities of current mammals. He concludes that the hearing of the earliest mammals was similar in frequency range to those of many extant small mammals.

Webster and Plassmann (Chapter 30) discuss parally evolved structures in two independently evolved groups of desert rodents and demonstrate both similarities and differences. They show that function is more similar than the detailed structure.

In Chapter 31, Plassmann and Brändle present a mathematical model, based on physical measurements of the outer and middle ears, that allows a prediction of the frequency range of auditory sensitivity in diverse mammals.

Frost and Masterton (Chapter 32) discuss the medial geniculate body, which in the most generalized living mammals, projects substantially to the striatum as well as to the auditory cortex. They point out that in more specialized mammals it projects much less, or not at all, to the striatum, and much more to auditory cortex. From these data, they suggest that the mammalian auditory cortex may have evolved by an unrolling and layering of tissue from the striatum.

Merzenich and Schreiner (Chapter 33) review the single unit physiology of auditory cortex in the few mammalian species in which detailed studies have been done. They show that all these species have at least one tonotopic representation. However, diversity of numbers and organization of tonotopic regions and other auditory specializations vary widely.

Heffner and Heffner (Chapter 34) review the behavioral data on mammalian sound localization with respect to the selective pressures which may have played a role in its evolution. Because of the large diversity of species they have studied, they have been able to demonstrate that auditory localization is almost completely lost in some mammals, and highly developed in others. Environment, among other conditions, appears to be a key factor. They also present interesting data on the importance of monaural cues in high-frequency sound localization, and the relationship between sound localization and vision.

The final two chapters in this section deal with highly specialized mammalian groups. Ketten (Chapter 35) discusses hearing in living and fossil marine mammals. Based upon data from both modern mammals and the fossil record, she makes predictions about the hearing and echolocating abilities of the earliest marine mammals.

Pollak's central theme in Chapter 36, describing the structures and mechanisms of echolocation in the mustache bat, is that the ability to orient by means of sonar did not evolve *de novo*, but rather by an enhancement of abilities to perceive echoes, which are and were common to many mammals.

28
Evolution of the Auditory System in Synapsida ("Mammal-Like Reptiles" and Primitive Mammals) as Seen in the Fossil Record

Edgar F. Allin and James A. Hopson

1. Introduction

Based on evidence from comparative anatomy, embryology, and paleontology, it is well established that the middle ear of existing mammals is morphologically unique, the tympanic bone, malleus, incus, and tensor tympani muscle all being homologous with components of the feeding apparatus of other vertebrates (Fig. 28.1, see below for details). Extant mammals are also unique in having a very elongate cochlea, and sensitivity to a broader range of sound frequencies than other vertebrates, usually extending beyond 10,000 Hz. In addition, they are distinctive in generally having a protruding pinna (auricle) and a long, tubular external auditory meatus. The present chapter examines the extensive fossil evidence concerning the nature of the auditory machinery of early mammals and their antecedents. Both authors of the present account have written on this subject, arriving at different initial interpretations in certain regards (Hopson 1966; Allin 1975). Although our views have since converged substantially (Allin 1986), significant doubts and disagreements remain.

In this chapter, cladistic taxonomic conventions will be followed. Thus, every formal taxon must include a common ancestor plus *all* of its descendants; paraphyletic taxa are not accepted (for example, Reptilia is not an acceptable taxon, cladistically, if it excludes birds and mammals while including their early amniote ancestors). For convenience, paraphyletic "gradistic" terms will sometimes be used, set off in quotation marks (e.g., "reptiles"). Two great radiations of amniotes are recognizable, the Sauropsida and Synapsida.

Sauropsida comprises the birds, all living "reptiles," and various extinct "reptiles." Synapsida, as originally defined, was a paraphyletic assemblage of fossil forms (the "mammal-like reptiles") comprising two paraphyletic groups, "pelycosaurs" (earlier and more primitive) and "therapsids" (later and more advanced). Defined cladistically, Synapsida also includes the mammals, as does the less inclusive taxon Therapsida.

The cladogram of hypothesized relationships of the major groups of synapsids shown in Figure 28.2 will provide the taxonomic framework for the present account. For nontherapsid synapsids ("pelycosaurs") it follows Reisz (1986), and for nonmammalian therapsids it follows Hopson and Barghusen (1986), with modifications. Illustrations of the skull structure of representative genera are given in Figures 28.4 to 28.8. If it is assumed that no features would have been more advanced (apomorphic, modified, derived, or specialized) at any node (branch point) than in the least advanced (most plesiomorphic or primitive) lineage beyond that node, then it is possible to recognize instances of parallel or convergent evolution implicit in a given cladogram. Alternative cladograms might entail different patterns of parallelism or convergence. Ideally, the history of auditory structures would be reconstructed using a cladogram based entirely on nonauditory characters; such is not the case here, but most elements of the diagram are sufficiently robust that this is not a problem.

Reconstruction of soft tissue structures and functional properties based on osteologic evidence can never be certain, but if rigorously done has heuristic value and sometimes merits a high level of

Abbreviations for all figures:

A, articular bone
AF, adductor fossa
An, angular bone
AnK, keel of angular bone
AP, angular process of dentary
B, tympanic bulla formed by alisphenoid
BM, branchiomandibularis muscle sheet
C, coronoid bone
CA, anterior coronoid bone
CE, coronoid eminence
Ch, ceratohyal
Cho, choana
Co, condyle of dentary
CP, coronoid process of dentary
CR, cochleolagenar recess (cochlear canal of mammals)
D, dentary bone
DPr, dorsal process of stapes
E, epipterygoid (alisphenoid) bone
EF, external fossa
Eu, groove for eustachian tube
Eu1, anterior version of eustachian passage
Eu2, posterior version of eustachian passage
FAn, facet for angular bone
FC, facet for coronoid bone
FO, fenestra ovalis (vestibuli)
FPa, facet for prearticular bone
FR, fenestra rotunda (cochleae)
FR-JF, confluent fenestra rotunda and jugular foramen
FSp, facet for splenial bone
G, glenoid fossa of squamosal
H, anterior hyoid cornu *sensu lato*
I, incus
Ih, interhyal
J, jugal bone
JF, jugular (posterior lacerate) foramen
L, ligament
La, larynx
LF, lateral flange of prootic
LH, levator hyoidei muscle
LHA, fossa for probable attachment of levator
   hyoidei muscle
M, malleus
MC, Meckel's cartilage
MD, deep head of masseter muscle

MF, mandibular foramen
MG, groove for Meckel's cartilage
MM, manubrium of malleus
MP, mastoid portion of petrosal
MPtC, medial pterygoid crest
MR, medial ridge of dentary
MS, superficial head of masseter muscle
OC, oral cavity
P, Paaw's cartilage in stapedius tendon
Pa, prearticular bone
PdT, postdentary portion of tympanum
PgP, postglenoid process
PP, paroccipital process
PF, palatopharyngeal fold and muscle
PqT, postquadrate portion of tympanum
Pr, promontorium (bulge over basal portion of cochlea)
PsM, pseudotemporalis muscle
Pt, pterygoid bone
PtF, descending flange of pterygoid bone
PtH, pterygoid hamulus
PtM, primitive ("reptilian") pterygoideus muscle
PtP, process for insertion of pterygoideus muscle
PtR, pterygopalatine ridge
Q, quadrate bone
QA, quadrate attachment site
Qj, quadratojugal bone
QRE, quadrate ramus of epipterygoid
QRPt, quadrate ramus of pterygoid
RL, reflected lamina
RM, recessus mandibularis
RP, retroarticular process
S, stapes
Sa, surangular bone
SaB, surangular boss for articulation with squamosal
   glenoid fossa
Sh, stylohyal
Sq, squamosal bone
SqS, squamosal sulcus (probably for external
   auditory meatus)
St, stapedius muscle
Th, tympanohyal (laterohyal)
TM, temporalis muscle
V, vestibule
Ve, velum (soft palate)

confidence. The more dissimilar the structure of fossil forms to living ones the less confidently can such interpretations be made. Because "pelycosaurs" are osteologically much more like lizards than like mammals, interpretation of their soft tissue anatomy can generally be more confidently based on a "reptilian" model than a "mammalian" model (using for comparison extant "reptiles" or mammals, respectively), whereas for advanced "therapsids" the reverse may be true. However, it would be unwise to assume, ignoring contrary evidence, that "pelycosaurs" had a middle ear similar to that of most living "reptiles," in which the stapes projected directly to a tympanic membrane posterior to the jaw articulation.

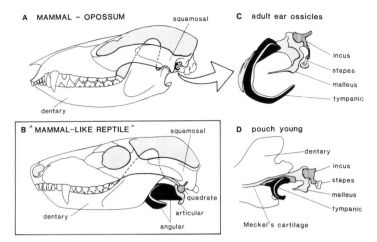

FIGURE 28.1. (A) Skull of the opossum *Didelphis virginiana* showing the dentary-squamosal (temporomandibular) jaw joint characteristic of mammals. (B) Skull of a primitive cynodont, *Thrinaxodon*, showing the primitive (articular-quadrate) jaw joint. The quadratojugal bone is not shown. (C) Middle ear bones of the adult opossum in lateral view. (D) Medial view of the lower jaw of a pouch-young opossum, showing that the middle ear elements are mandibular components in early development (compare with Fig. 28.7G). The same patterning is used in (B), (C), and (D) to indicate homologous entities. The darker stipple in (D) indicates the goniale (prearticular), which forms the anterior process of the adult malleus. (From Hopson 1987 by permission of the National Association of Biology Teachers, Reston, Virginia.)

Advocates of "creation science" are fond of emphasizing the implausibility of nonauditory jaw structures "entering the middle ear" during the evolutionary origin of mammals because, they assert, the transitional forms would have been unable to hear and unable to eat. Actual structurally intermediate specimens are now known in which new and old jaw joints coexist, and accessory jaw bones closely resemble their homologs in extant mammals. Unquestionably, they were capable of feeding; their auditory capabilities will be considered below. Conventional concepts of what

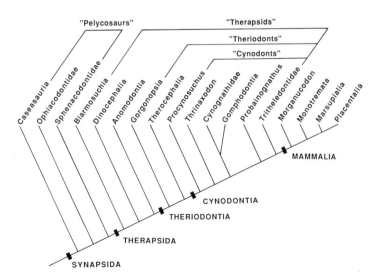

FIGURE 28.2. Cladogram showing a hypothesis of relationships of the principal groups of synapsids. Paraphyletic (noncladistic) groups are in quotation marks.

constitutes a middle ear cannot be applied to pre-mammalian synapsids with security. Posterior portions of the jaw apparatus became associated with an air-filled chamber and reception of airborne sound at some stage in the phylogeny of the Mammalia. As will be shown later, these developments probably antedated the formation of a mammalian jaw joint; however, it remains unclear at what stage they commenced. It will be necessary to consider the structure of the mandible and suspensorium at all relevant nodes in order to elucidate the transformational stages culminating in the definitive mammalian condition of middle ear and mandible.

2. Homologies

During early stages of development the jaws of sauropsids and mammals are quite similar, consisting of cartilages of the first branchial arch. The primary mandible is Meckel's cartilage, which meets another first-arch cartilage at the primary craniomandibular joint. Meckel's cartilage later becomes enveloped by intramembranous ossifications (Fig. 28.1D) and its posterior extremity ossifies to become the articular bone of sauropsids and the main part of the malleus in mammals, while the other cartilage ossifies as the quadrate bone and incus, respectively. Sauropsids retain this pattern throughout life, but mammals transform it dramatically (Gaupp 1913; Goodrich 1930). One of the intramembranous bones, the tooth-bearing dentary, enlarges to become the entire mandible. A process of the dentary grows back to meet the squamosal element of the cranium, forming a secondary craniomandibular joint. Two other intramembranous bones and the incus separate from the mandible as middle ear components. The midportion of Meckel's cartilage degenerates, its sheath becoming the sphenomandibular ligament (pterygomandibular in monotremes) and the anterior ligament of the malleus. One of the intramembranous ossifications, the goniale, becomes the anterior process (processus folii or gracilis) of the malleus, and another becomes the drum-supporting tympanic (ectotympanic) annulus which in some mammals remains loosely suspended and in others fuses to the squamosal and petrosal cranial elements as part of a composite temporal bone. Comparison with sauropsids and advanced nonmammalian cynodonts establishes the following homologies (Fig. 28.1): The mammalian tympanic bone corresponds to the angular element of the multiple-boned mandible, its anterior (morphologically dorsal) limb representing the body of that bone and its posterior (ventral) limb representing the reflected lamina, a process unique to synapsids. The malleus corresponds to fused articular and prearticular mandibular elements, the incus to the quadrate bone of the cranium, and the malleo-incudal joint to the nonmammalian craniomandibular joint between articular and quadrate. The ventrally projecting process of the "cynodont" articular, usually called the retroarticular process but perhaps not homologous with the posteriorly projecting process by that name in sauropsids, is the homolog of the neck of the malleus and the manubrial base (together constituting the pars transversalis of Fleischer 1973, 1978), from which the manubrium is a neomorphic outgrowth. The configuration in which the angular, articular plus prearticular, and quadrate are strictly auditory structures, fully divorced from the feeding apparatus (and renamed tympanic, malleus, and incus), we will refer to as the definitive mammalian middle ear (DMME). The tensor tympani and tensor veli palatini muscles of mammals are homologous with portions of the reptilian pterygoideus muscle (Barghusen 1986), while the stapedius is a levator hyoidei derivative, possibly homologous with the single middle ear muscle of some sauropsids.

In those extant sauropsids that have tympanic ears, the stapes or columella auris consists of a bony proximal part (stapes proper) and a cartilaginous distal part (extrastapes) which has a tympanic process contacting the eardrum. The extrastapes usually has a dorsal process or ligament connecting to the paroccipital process of the otic capsule, and a slender, offset quadrate (internal) process or ligament connecting to the quadrate bone. There may also be a hyoid process, which in early ontogeny (and throughout life in *Sphenodon*) is connected to the anterior hyoid cornu. In mammals the stapes is simpler, having no extrastapes and connecting directly with the incus (= quadrate). During early embryonic development (Fig. 28.3A) the mesenchymal blastema of the second branchial arch skeleton of both sauropsids and mammals has two processes at its upper end, both contacting the otic capsule (de Beer 1937; Presley

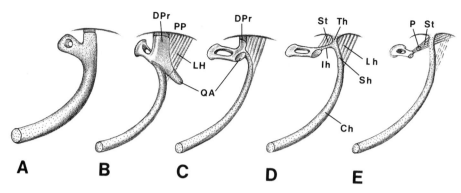

FIGURE 28.3. Diagrammatic representations of the development and evolution of the synapsid hyoid arch skeleton (left side, anterolateral aspect). (A) Mesenchymal primordium in a modern mammalian embryo. Other drawings represent adult synapsids in primitive-to-advanced sequence. (B) "Pelycosaur" condition with a massive stapes, an anteriorly directed foramen for the stapedial artery, a strong dorsal process of the stapes attached to the paroccipital process and situated proximally, a large attachment of the anterior hyoid cornu on the ventral aspect of the stapes, a cartilaginous distal extension of the stapes to articulate with the quadrate, and a large levator hyoidei muscle (should insert on Sh). (C) Primitive cynodont condition with a lighter stapes having a large stapedial foramen directed dorsally, a smaller dorsal process of the stapes situated distally, a smaller (hypothetical) attachment of the anterior hyoid cornu to the stapes, and a large facet on the distal end of the stapes for direct articulation with the quadrate. (D) Hypothetical stage based on ontogeny, with fusion of the dorsal process to the paroccipital process, detachment of the dorsal process and hyoid cornu from the stapes except for a slender ligamentous or cartilaginous connection (interhyal), and beginning differentiation of a stapedius muscle from the levator hyoidei. (E) Modern mammalian condition with a very light stapes, complete detachment of the dorsal process and hyoid cornu from the stapes, reduction of the interhyal to the stapedius tendon and its sesamoid cartilage (of Paaw), complete differentiation of a stapedius and loss of the levator hyoidei as such. Sound transmission via the hyostapedial route is no longer possible.

1984). The medial process is penetrated by the stapedial artery, which may persist or degenerate. This process chondrifies and later ossifies in sauropsids as the stapes proper and in mammals as the entire stapes. The lateral process chondrifies as the dorsal process of the sauropsid extrastapes. In mammals it fuses to the crista parotica of the otic capsule as the tympanohyal (laterohyal); it is continuous inferiorly with the stylohyal (styloid process), which in turn is connected to the anterior hyoid cornu via a stylohyoid ligament or a series of bones. A mesenchymal condensation (interhyal) temporarily connects the embryonic mammalian stapes with the tympanohyal (Fig. 28.3D); its proximal part becomes the tendon of the stapedius muscle, and sometimes a sesamoid cartilage (of Paaw) within this tendon, while its distal part disappears (Fig. 28.3E) except for an occasional derived cartilage (of Spence) alongside the chorda tympani nerve. The cartilages of Paaw and Spence have been interpreted as remnants of a tympanic process, but this is questionable.

In primitive synapsids (Fig. 28.3B) and early sauropsids the mature bony stapes has a dorsal process which articulates with the paroccipital process of the prootic or opisthotic, a ventral or posterior rugosity indicative of a persistent ligamentous or cartilaginous connection to the anterior hyoid cornu, and a direct bony or (by inference) cartilaginous continuation of its shaft toward a depression or facet on the quadrate. There is no convincing evidence of a tympanic process for contact with a tympanic membrane, although small projections (and dorsal or quadrate processes) have been identified as such. Thus, the primitive synapsid stapes appears to include homologs of parts of the sauropsid extrastapes, all but the quadrate process being lost or detached in present-day mammals.

From the work of Gaupp (1913, summarized in Maier 1990) and Presley (1984) it is virtually certain that the mammalian tympanic membrane is not homologous with that of sauropsids. Also, the external auditory meatus of mammals probably does not correspond to that of sauropsids (many of which have none). Finally, the mammalian and sauropsid auditory (eustachian) tube and cavum tympani differ in certain major morphologic relations and appear to be at least partly nonhomologous.

3. Osteologic Features of Fossil Synapsid Groups

3.1 Nontherapsid Synapsids ("Pelycosaurs") (Figs. 28.4A and B, and 28.5A, 28.6A, 28.7A and B)

For all aspects of the structure of early synapsids ("pelycosaurs"), the best source remains Romer and Price (1940). For a brief discussion of early synapsids in general see Kemp (1982).

3.1.1 Mandible (Figs. 28.4A and B, 28.7A)

The compound mandible of early synapsids, like that of other primitive amniotes, consists of the tooth-bearing dentary and several smaller accessory elements: four postdentary bones (articular, prearticular, angular, and surangular), which extend posterior to the dentary; and three paradentary bones (anterior coronoid, posterior coronoid, and splenial), which lie medial to the dentary. The contacts between elements are usually imbricated sutures. The articular is clasped by the other postdentary bones. In life it was continued anteriorly as a substantial meckelian cartilage occupying a large groove and, more anteriorly, a canal bounded by other jaw bones and terminating at the mandibular symphysis. In the most primitive synapsids (Figs. 28.4A and 28.7A) there is no process of the articular projecting posterior to the jaw joint. In sauropsids such a retroarticular process provides leverage for a jaw-opening muscle, the depressor mandibulae, and part of the jaw-closing pterygoideus musculature. However, sphenacodontids (and therapsids) have a ventrally directed "retroarticular" process (Fig. 28.4B) that presumably served for attachment of these muscles. Medial to the base of the retroarticular process is a larger pterygoid process for additional pterygoideus insertion.

Other major sites of muscle insertion included the large adductor fossa on the medial aspect of the jaw and the coronoid eminence (elevated posterodorsal margin of the mandible). The great majority of the jaw-closing musculature attached to the postdentary elements (Barghusen 1968). Except for the external fossa of sphenacodontids (see below), the entire lateral surface of the mandible bears roughened "sculpturing", indicating that it was covered in life by immovably attached scales.

In contrast with early sauropsids, post-caseasaurian synapsids (those beyond the first node in Fig. 28.2) have a mediolaterally compressed mandible with a pronounced ventral keel on the angular. In sphenacodontids the keel is notched posteriorly, exposing a slightly concave surface, the external fossa (Allin 1975, 1986). The posteriorly emarginated keel is called the reflected lamina of the angular; it characterizes sphenacodontids and all therapsids. Allin (1975, 1986) refers to the margin of the notch as the angular gap and the pocket between the planes of the reflected lamina and the external fossa as the angular cleft. Pterygoideus musculature may have passed around the ventral margin of the jaw to occupy the cleft and insert in the external fossa (Barghusen 1968).

3.1.2 Quadrate and Craniomandibular Joint

The quadrate of early synapsids is best known in sphenacodontids, in which it is a deep vertical plate oriented in a posterolateral-to-anteromedial plane. It is barely visible in lateral view, being covered by the squamosal. It has rugose attachments to adjacent bones: laterally the small quadratojugal, posterolaterally the large squamosal, anteromedially the deep quadrate ramus of the pterygoid and the epipterygoid, and posterodorsally the paroccipital process of the opisthotic. It is unlikely that the quadrate could have moved independently of neighboring bones, at the gross level. The quadratojugal of sphenacodontids and "therapsids" is greatly reduced from the primitive synapsid condition and forms a functional unit with the quadrate. Low on the posteromedial surface of the quadrate is a shallow pit which appears to have been an attachment site for the stapes (see below).

The ventral surface of the quadrate has a pair of elongate, anteromedially oriented condyles. The medial condyle extends further ventrally than the

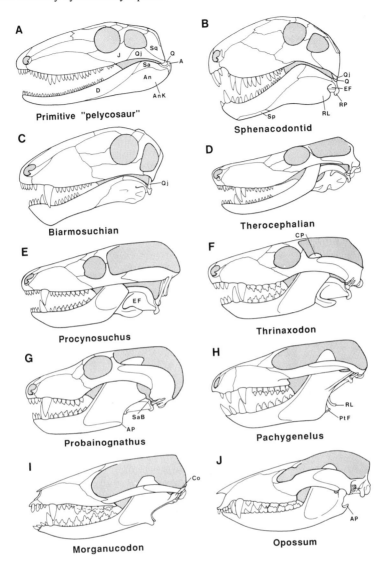

FIGURE 28.4. Lateral views of skulls of synapsids in primitive-to-advanced sequence. Approximate skull lengths in mm are given in brackets. (A) Generalized primitive "pelycosaur" [170]. (B) *Dimetrodon*, a sphenacodontid "pelycosaur" [330] (this and preceding drawing modified after Romer and Price 1940, by permission of the Geological Society of America). (C) *Biarmosuchus*, a very primitive therapsid [170]. (Modified after Sigogneau and Tchudinov 1974.) (D) *Ictidosuch-* *oides*, a primitive therocephalian [113]. (After C. Mendrez-Carroll, unpublished.) (E to H) Nonmammalian cynodonts: (E) *Procynosuchus* [140] (original); (F) *Thrinaxodon* [73] (original); (G) *Probainognathus* [95] (original); (H) *Pachygenelus*, a tritheledontid [75] (original). (I) *Morganucodon*, a very primitive mammal at virtually the same grade as the preceding genus [23]. (Modified after Kermack, et al. 1981.) (I) *Didelphis virginiana*, an opossum [126] (original).

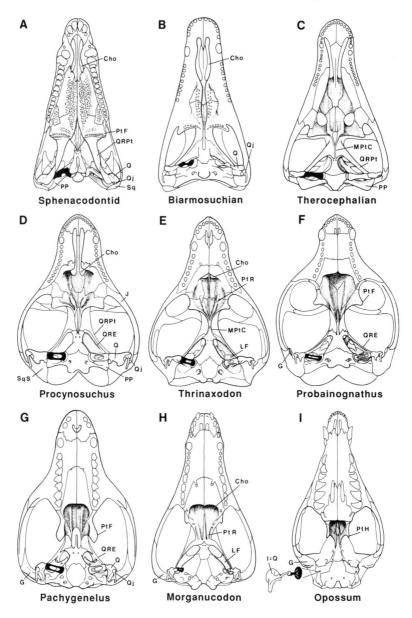

A **Sphenacodontid**
B **Biarmosuchian**
C **Therocephalian**
D **Procynosuchus**
E **Thrinaxodon**
F **Probainognathus**
G **Pachygenelus**
H **Morganucodon**
I **Opossum**

FIGURE 28.5. Ventral views of synapsid crania in primitive-to-advanced sequence. Right stapes is black. Approximate skull lengths in mm are given in brackets. (A) *Dimetrodon*, a sphenacodontid "pelycosaur" [330]. (After Romer and Price 1940.) (B) *Biarmosuchus*, a very primitive therapsid [170] (modified after Sigogneau and Tchudinov 1974, incorporating information from an undescribed biarmosuchian from South Africa). (C) *Ictidosuchoides*, a primitive therocephalian [113]. (After

C. Mendrez-Carroll, unpublished.) (D to G) Nonmammalian cynodonts: (D) *Procynosuchus* [140] (original); (E) *Thrinaxodon* [69]. (Modified after Parrington 1946.) (F) *Probainognathus* [95]. (Modified after Romer 1970 by permission of the Museum of Comparative Zoology, Harvard University.) (G) *Pachygenelus*, a tritheledontid [75] (original). (H) *Morganucodon*, a very primitive mammal [23]. (Modified after Kermack et al. 1981.) (I) *Didelphis virginiana*, an opossum [126] (original).

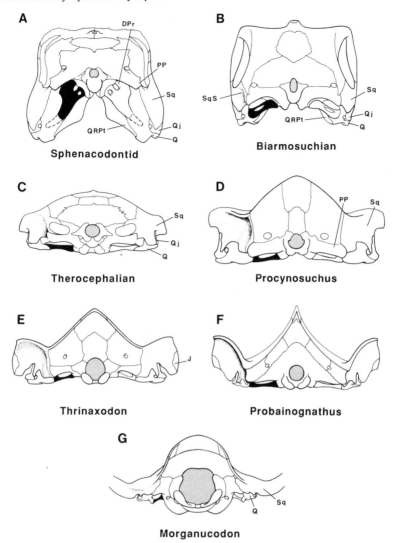

FIGURE 28.6. Posterior views of synapsid crania in primitive-to-advanced sequence. Left stapes is black. Approximate skull widths in mm are given in brackets. (A) *Dimetrodon*, a sphenacodontid "pelycosaur" [180]. (Modified after Romer and Price 1940.) (B) *Biarmosuchus*, a very primitive therapsid [90]. (Modified after Sigogneau and Tchudinov 1974.) (C) *Ictidosuch-* *oides*, a primitive therocephalian [87]. (After C. Mendrez-Carroll, unpublished.) (D to F) Nonmammalian cynodonts: (D) *Procynosuchus* [112] (original); (E) *Thrinaxodon* [48]. (Modified after Parrington 1946.) (F) *Probainognathus* [67]. (Modified after Romer 1970.) (G) *Morganucodon*, a very primitive mammal [15]. (Modified after Kermack et al. 1981.)

lateral. Matching obliquely oriented troughs are present on the articular bone. This peculiar jaw joint appears to have resisted medial displacement of the articular by contraction of the massive pterygoideus musculature or dislocation by external forces such as struggles of large prey (Barghusen 1968).

3.1.3 Stapes (Figs. 28.3B, 28.5A, 28.6A)

The stapes of "pelycosaurs" is remarkably large, both in linear dimensions (up to 75 mm long) and volume. Near its footplate it is penetrated by a canal for the stapedial artery, passing anterodorsad. The long axis of the bone runs posterolaterally

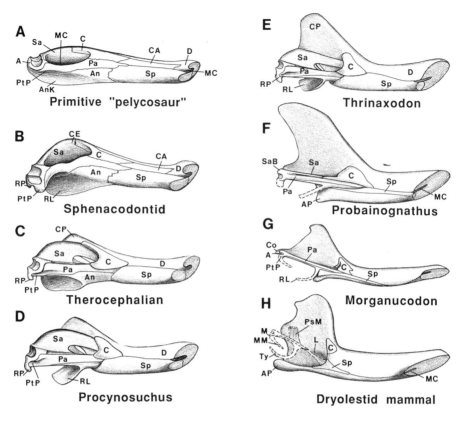

FIGURE 28.7. Diagrammatic medial views of mandibles of synapsids in primitive-to-advanced sequence. (Except as noted, modified after Allin 1975, by permission of Wiley-Liss Division of John Wiley and Sons, Ltd.) Mandibular lengths in mm are given in brackets. (A) Generalized primitive "pelycosaur" [175]. (B) *Dimetrodon*, a sphenacodontid "pelycosaur" [303]. (C) Generalized therocephalian [100]. (D to F) Nonmammalian cynodonts: (D) *Procynosuchus* [112]; (E) *Thrinaxodon* [62]; (F) *Probainognathus* [73]. (G) *Morganucodon*, a very primitive mammal [21]. (Modified after Hopson 1987 and Kermack et al. 1981, by permission of the National Association of Biology Teachers, Reston VA, and the Linnean Society.) (H) An unnamed Upper Jurassic dryolestid mammal [26] (modified after Krebs 1971) with hypothetical middle ear bones not yet fully divorced from the mandible. The ear bones no longer would have moved with complete fidelity to the mandible, being partially stabilized by a slip of pseudotemporalis (or other) muscle.

and, in ophiacodonts and sphenacodontids (in which the quadrate is extremely deep), strongly ventrally. As preserved, the stapes ends a short distance from the posterior pit on the quadrate, into which it is presumed to have sent a cartilaginous extension. On the ventral surface of the stapedial body is an elongate groove interpreted as the hyoid attachment site (Romer and Price 1940). The large footplate is directed medially and its rim broadly overlaps the rim of the smaller fenestra ovalis. This overlapping would have prevented the stapes from being driven into the inner ear; a similar overlapping characterizes the stapes of all non-mammalian synapsids. Lateral to the footplate is a very robust dorsal process which articulates with the paroccipital process. In *Dimetrodon*, this articulation has two parts: laterally, an ovoid condyle and medially an elongate rugose surface which was probably bonded to the paroccipital process by cartilage or connective tissue. The condyle fits into a matching socket on the paroccipital process to form a hinge-like articulation with an anteromedially directed axis, about which some rocking motion appears to have occurred.

3.1.4 Other Cranial Features

Early synapsids lack an otic notch or other indication of an eardrum attachment. Also, they show nothing suggestive of an external acoustic meatus. In the absence of an ossified floor and a clearly defined roof anterior to the narrow paroccipital process, the limits of the cavum tympani (which presumably was present) cannot be determined with certainty. A deep, narrow space lies between the basicranium and the combined quadrate and quadrate ramus of the pterygoid and extends back to include the stapes. Romer and Price (1940) suggest that this space was occupied in life by part of the middle ear cavity and that a eustachian tube broadly open to the pharynx occupied the anterior portion of the space, as in lizards.

A massive tooth-bearing transverse process of the pterygoid braced the medial surface of the mandible; pterygoideus musculature probably took origin from its entire posterior face as well as from the quadrate ramus of the pterygoid. Because no secondary palate is present no tensor veli palatini had differentiated from this musculature, and although the insertion on the pterygoid process of the articular may be homologous with the insertion of the mammalian tensor tympani on the neck of the malleus there is no evidence for differentiation of a similar muscle (originating on the otic capsule).

3.1.5 Auditory Possibilities

In view of the massiveness of the stapes and the absence of any sign of an eardrum attachment it seems very improbable that a refined tympanic ear was present. Low frequency (below 1,000 Hz) airborne sound impinging on acoustically unspecialized superficial tissues overlying the pharyngeal cavity, behind and below the skull, probably actuated the stapes via the hyostapedial route. At least in small individuals, such sounds may have set up audible vibrations of the postdentary portion of the mandible which reached the stapes via the quadrate. However, the relatively inflexible attachments of these bones and their large mass rules out efficient reception by this route.

3.2 Biarmosuchia (Basal Therapsids) (Figs. 28.4C, 28.5B, 28.6B)

The structurally most primitive therapsid known, and one of the earliest, is *Biarmosuchus* (Chudinov

1965; Sigogneau and Tchudinov 1972). The mandible of therapsids is modified from that of sphenacodontids in loss of the anterior coronoid bone and in setting off of the reflected lamina as a more distinct, free-standing structure by anterior extension of the angular gap above it and of the angular cleft deep to it. The lamina becomes a thin sheet with a characteristic therapsid pattern of radiating corrugations. Its lateral surface is no longer sculptured, so tightly adherent thick skin probably no longer covered it.

The quadrate is greatly reduced in height, extending only a short distance below the level of the occiput, and is more transversely oriented. It has an expanded contact with the anterior face of the squamosal and a reduced contact with the slender quadrate ramus of the pterygoid. The quadrate condyles are less prominent. The "screw-shaped" configuration of the condyles is such that as the jaws were opened the articular shifted laterally relative to the quadrate. This is also the case for other primitive therapsids. The stapes is much less massive than that of "pelycosaurs" and extends more directly laterad, contacting a groove on the posterior face of the quadrate. The stapedial foramen is greatly enlarged, passing anteriorly between relatively slender dorsal and ventral crura. The dorsal process is small and lies at the distal end of the stapes. As in all therapsids the distal end of the paroccipital process is broad anteroposteriorly. A possible meatal groove is present on the expanded posterolateral face of the squamosal, terminating medially at a plausible attachment site for a tympanum posterior to the quadrate, at the end of the paroccipital process. Thus, there are indications of at least incipient specializations for aerial sound reception in the postquadrate region. Also, the thinning of the reflected lamina and reduced mass of the postdentary bones and quadrate may have allowed more effective mandibular sound reception and transmission, especially if the widened reflected lamina overlay an air-containing diverticulum of the pharynx (recessus mandibularis).

3.3 Advanced Nontheriodont Therapsids

Dinocephalians and anomodonts are two postbiarmosuchian therapsid clades that diverged from the carnivorous morphology that characterized the "theriodonts" from which mammals arose.

Dinocephalians show few modifications of the mandible and quadrate beyond the biarmosuchian level, but advanced members of the Anomodontia—the dicynodonts—specialize the masticatory and auditory apparatus in ways somewhat similar to what is seen in advanced "theriodonts."

In dicynodonts, the reflected lamina is large but thin and corrugated, the angular gap is long and the angular cleft deep. The quadrate is further reduced in height. It has a simple contact with the squamosal but was probably virtually immobile because of its tight union with the quadratojugal, which in turn was sutured or synostosed to the squamosal. Extensive fore-aft gliding of the mandible occurred at the jaw articulation. The stapes is short and abuts directly against the medial surface of the quadrate. The stapedial foramen is absent in most dicynodonts, having in effect expanded dorsally, eroding away the dorsal crus. Some dicynodonts show evidence that the anterior hyoid cornu articulated directly with the ventromedial aspect of the stapes (Barry 1968).

Both dinocephalians and anomodonts have a groove on the posterior aspect of the squamosal, possibly for an external auditory meatus. Neither group has an otic notch, but some dicynodonts have a process projecting back from the distal end of the paroccipital process which Cox (1959) considers to have supported the upper margin of a postquadrate tympanum.

3.4 Noncynodont Theriodonts (Gorgonopsians and Therocephalians) (Figs. 28.4D, 28.5C, 28.6C, 28.7C)

Characteristic of the theriodonts are the development of a free-standing coronoid process on the dentary and the loss of imbricated contacts between dentary and postdentary elements. The quadrate and quadratojugal are reduced in height and the quadrate lies in a hollow on the anterior surface of the squamosal. In gorgonopsians, the quadrate and quadratojugal have a convex posterodorsal surface which nestles in a deep socket in the squamosal, probably having formed a markedly flexible syndesmosis in life (Kemp 1969, 1982). The gorgonopsian stapes is similar to that of *Biarmosuchus*, but more delicate and probably with a cartilaginous extension into a long groove on the posterior surface of

the quadrate. Possibly, gliding occurred at this junction during movements of the quadrate. As for all theriodonts, there is no direct evidence for a persisting hyostapedial connection, but the persistence of a strong dorsal process on the stapes suggests that it was present. A prominent sulcus on the rear of the flaring squamosal may have been meatal; the ventrolateral margin of the paroccipital process may mark the location of a tympanum.

Therocephalians are the sister taxon of cynodonts, but they possess several unique specializations and retain several primitive features lost in cynodonts. In both groups the postdentary elements form a unit distinct from, and flexibly bonded to, the dentary and paradentary elements (Kemp 1972). The reflected lamina of therocephalians is a large, thin, corrugated plate overlying a deep angular cleft. The angular gap is very long but narrow. Some independent movement of the quadrate-quadratojugal complex appears to have occurred. Kemp (1972) noted that the axes of the two jaw joints of some therocephalians are not in line (each running anteromedially), so the joint would disarticulate as the jaws opened unless mobility occurred at other articulations. This mobility may have occurred at the articulation of the quadrate and quadratojugal with the squamosal (as proposed by Kemp), at the dentary-postdentary interface, at the mandibular symphysis, or some combination of these. The stapes lacks a foramen and dorsal process and articulates with the quadrate much as in gorgonopsians.

Lightening of the stapes is consistent with refinement of a postquadrate tympanic mechanism, yet the stapes remains bulky relative to existing sauropsids with tympanic ears, and its direct attachment to the quadrate indicates that sound transmission from the mandible was important. In therocephalians, especially, a plausible postdentary tympanum could not be solely a membrane spanning the narrow angular gap; it would necessarily include the thin reflected lamina itself. Increased flexibility of the attachments of the postdentary unit and quadrate must have evolved for mechanical reasons related to feeding, but would have enhanced the efficiency of sound reception and transmission.

3.5 Nonmammalian Cynodonts (Figs. 28.4E-H, 28.5D-G, 28.6D-F, 28.7D-F, 28.8A-C)

Cynodonts show many unique modifications of the skull and mandible. Primitive cynodonts (procynosuchids, Figs. 28.4E, 28.5D, 28.6D and 28.7D) possess a shallow fossa on the coronoid process which, with the outward bowing of the zygomatic arch away from the mandible, indicates that part of the external adductor muscle mass had extended its insertion onto the outer surface of the dentary as an incipient masseter (Barghusen 1968, 1972). The reflected lamina is reduced in size so that the angular gap is greatly increased in area. The main mass of the postdentary unit is a deep plate with most of its lateral surface occupied by the smooth external fossa (medial wall of the angular cleft). The quadrate is wide and the craniomandibular joint robust. A poorly defined sulcus on the posterior face of the squamosal ends behind the quadrate at the distal end of the paroccipital process. The stapes is long but light, having a large foramen and slender crura. It has been reoriented so that the foramen passes upward rather than forward and the former dorsal and ventral crura are now posterior and anterior, respectively. A dorsal process extends upward from the posterolateral corner of the stapes toward the paroccipital process. Laterally, the stapes articulates directly with a flat facet on the inner face of the medial quadrate condyle. A new connection is present between the otic capsule and the quadrate rami of the pterygoid and epipterygoid; it is a horizontal process of the prootic, the lateral flange, which roofs the cavum tympani anterior to the paroccipital process. Possibly the mammalian tegmen tympani is homologous, in part, with the lateral flange.

All cynodonts more advanced than procynosuchids show enlargement of the dentary and reduction of the postdentary unit and quadrate. The coronoid process becomes very high and a masseteric fossa covers its entire lateral side, extending onto the body of the mandible. The bulk of the much-expanded jaw-closing musculature now inserted on the dentary. As the dentary extends posteriorly a process develops superior to the postdentary unit (the posterior or condylar process) and a distinct angle forms posteroventrally. We think this is homologous with the

mandibular angle of living mammals but it has been termed a pseudangular process by Jenkins, Crompton and Downs (1983). A lateral ridge strengthens the posterior process, and a medial ridge forms above the postdentary unit to prevent its upward displacement relative to the dentary. The surangular and the body of the angular become reduced in height and a smooth-surfaced trough forms on the medial aspect of the dentary to house the postdentary unit. In very advanced premammalian cynodonts the main mass of the postdentary unit has become a straight, rodlike entity and the adductor fossa is reduced to a narrow space. This reduction in height and mass occurs independently in at least three lineages (Cynognathidae, advanced Gomphodontia, and Probainognathidae plus other advanced groups), if our cladogram is correct (Fig. 28.9). The articular and prearticular fuse together. A groove and canal for Meckel's cartilage persist but are reduced in caliber. In advanced "cynodonts" the reflected lamina is delicate and rarely intact in fossils. Described as short in some genera, it may actually be incomplete. It is long but very slender in *Pachygenelus* (Figs. 28.4H and 28.8C). It may be partially hidden from view by the dentary, in lateral aspect, but in such cases a pocket-like space separates it from the medial surface of the dentary (e.g., *Cynognathus* and *Diademodon*). The downturned retroarticular process of tritylodonts resembles a malleus in having a thin lamina anteriorly, a sharp forwardly pointing tip (incipient manubrium?), and, in one genus (*Kayentatherium*), a rounded thickening of its posteroventral border resembling the orbicular apophysis of some mammals (Sues 1986). The postdentary unit shows conspicuous negative allometry in this genus, being relatively large in small individuals (probably juveniles).

The quadrate becomes considerably reduced in both height and width, its condylar region included. Several genera develop a new component to the craniomandibular joint, lateral to the primary joint, in the form of an articulation between a boss on the posterior end of the surangular and the squamosal (Crompton 1972). This joint would have reduced stresses borne by the small quadrate. In one genus, *Probainognathus*, the surangular boss is convex and meets a concavity on the medial aspect of the squamosal rim. Because the surangular-squamosal joint is not coaxial with

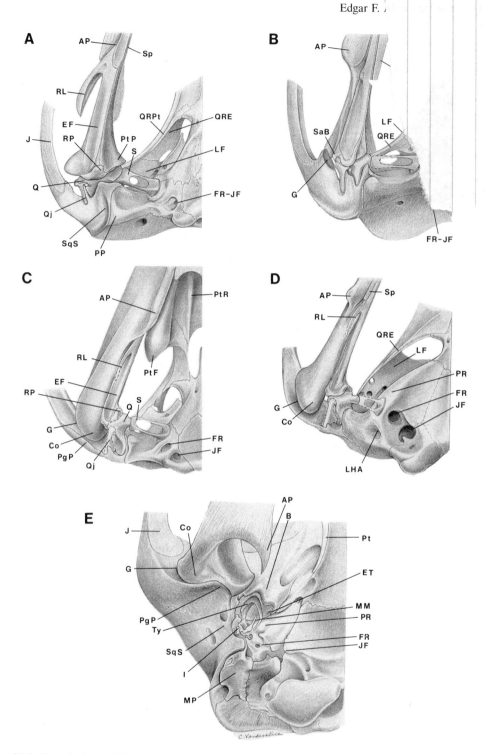

FIGURE 28.8. Ventral views of right ear regions and jaw articulations of Cynodontia in primitive-to-advanced sequence. (A) *Thrinaxodon* (original, based on several specimens). (B) *Probainognathus* (original, largely based on specimen MCZ 4274 from the Museum of Comparative Zoology, Harvard University). (C) *Pachygenelus*, a tritheledontid (original, largely based on specimens SAM K1329 and K1350 from the South African Museum). (D) *Morganucodon*, a very primitive mammal. (Modified after Kermack, et al. 1981.) (E) *Didelphis virginiana*, an opossum, right ventrolateral. (Modified after Wible 1990.)

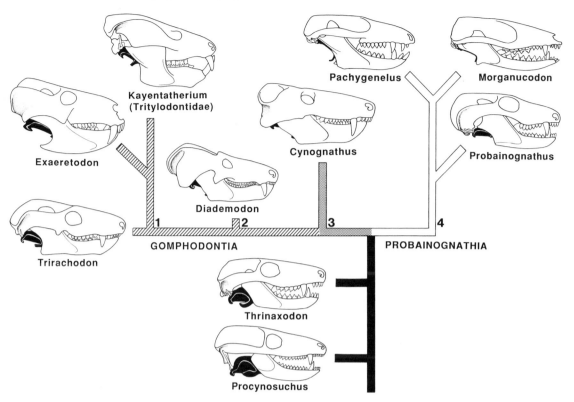

FIGURE 28.9. A "cladogradogram" of several groups of cynodonts, combining cladistic affinities (branching lines) and grades of modification in jaw structure (vertical position corresponds to degree of reduction in relative height of the postdentary unit, as measured in medial view). If this cladogram is correct, and no reversals in this trend occurred, reduction beyond the grade represented by *Trirachodon* must have taken place independently in: 1, advanced Gomphodontia; 2, the primitive gomphodontian *Diademodon*; 3, *Cynognathus*; and 4, Probainognathia.

the articular-quadrate joint, it could only remain engaged during jaw opening if the quadrate rocked anteriorly on the squamosal; the dorsal flange of the quadratojugal in this form is oriented sagittally, as is its notch in the squamosal, permitting direct anteroposterior streptostyly. In other genera the surangular boss is flat or concave, and meets a flat or convex facet on the squamosal. Tritylodonts and tritheledonts (ictidosaurs), the most advanced nonmammalian cynodonts, lack the surangular-squamosal joint yet have very small quadrates. Crompton (1985) has shown that the configuration of jaw muscles in advanced "cynodonts" could have minimized reaction forces at the jaw articulation. Tritheledonts probably had a synovial joint between the posterior end of the dentary and the ventral margin of the squamosal. Tritylodonts do not appear to have had a synovial joint between the dentary and the skull but possibly did have a syndesmodial secondary joint between the posterior end of the lateral ridge of the dentary and the jugal. The axis of the articular-quadrate joint of these forms runs anteromedially, so mobility of the quadrate, the postdentary unit, or the unfused mandibular symphysis must have occurred as the mandible was depressed. The axis has less anteromedial inclination in most earlier cynodonts, including *Cynognathus* in which, however, the anteromedial obliquity of the lateral quadrate condyle necessitated posterolateral movement of the postdentary unit, or anteromedial movement of the quadrate, during jaw depression. In this genus, as in most "cynodonts", the symphysis menti is fused (synostosed) so intramandibular kinesis at the dentary-postdentary interface probably was a significant aspect of jaw mechanics.

In all advanced "cynodonts" for which good specimens are known, the stapes remains similar in form to that of primitive ones, though generally becoming relatively smaller. It continues to articulate directly, and broadly, with the ventromedial end of the quadrate which, in the tritylodont *Kayentatherium*, is an elongate curved process projecting ventromedially (Sues 1985). Whether a dorsal process of the stapes persisted in very advanced "cynodonts" is uncertain. A specimen of *Scalenodon* has a stapes with a finger-like bony process projecting directly laterally from its distal end. This was considered to be a tympanic process by Parrington (1946), but it is probable that the process was an extension of the usual contact with the quadrate.

Most of the later premammalian cynodonts have a conspicuous squamosal sulcus running along the posterior portion of the high zygomatic arch and then downward to terminate behind the quadrate, at the distal end of the paroccipital process. In most Cynognathidae and Gomphodontia this groove is very large and is continuous with a broad surface on the posterior aspect of the squamosal which ends at an abrupt lateral margin. In tritylodonts a groove of similar relations is present, but it is narrow and, in one genus, almost tubular anteriorly (*Oligokyphus*: Kühne 1956), while in others it is fissure-like posteriorly (e.g., *Kayentatherium*, Sues 1985). Most authors since Gregory (1910) have interpreted the squamosal sulcus as having held a tubular meatus leading to a postquadrate eardrum, but Allin (1975) suggested that it may instead have held a depressor mandibulae muscle inserting on the articular and that no postquadrate tympanum existed. Subsequently, Allin (1986) accepted that the sulcus was indeed meatal and led to a tympanum which was at least partly postquadrate, but expressed the view that in primitive cynodonts the meatus was not tubular but rather was broadly open laterally, and that only in advanced forms did it become tubular by fusion of its margins, beginning ventrally (over a portion of the tympanum on the postdentary region of the mandible) and progressing dorsally.

The transverse process of the pterygoid braced the mandible, its lateral face contacting the coronoid bone. It has a long descending flange which is probably the homolog of the mammalian pterygoid hamulus, although there is no groove (pulley) for a tensor veli palatini tendon or other indication of such a muscle, except in *Pachygenelus* (Crompton 1989). "Cynodonts" have a long bony secondary palate, posterior to which the choanae open into an unpaired median groove and paired lateral grooves, separated by prominent pterygopalatine ridges, on what may be termed the bony roof of the nasopharynx. A soft palate, attached to the hard palate rim and the ventral lip of the descending flange, probably overlay these grooves and ridges, converting them into enclosed passages (Fig. 28.10). The median nasopharyngeal passage clearly was for the respiratory airway leading to the larynx. However, the lateral nasopharyngeal passages are more puzzling. Barghusen (1986) argues that they were eustachian (pharyngotympanic), running ventral to the pterygoideus muscle mass to reach the angular cleft laterally and the stapes posteriorly. In favor of this view is the presence in *Pachygenelus* (Figs. 28.4H, 28.5G and 28.8C) of a well-defined continuation of the lateral nasopharyngeal groove along the entire length of the descending pterygoid flange, bounded medially by a lip which becomes very prominent distally. The flange is very long, curves posteriorly, and ends in close proximity to the base of the reflected lamina.

No premammalian cynodont has a fossa muscularis major (tensor tympani origin on the otic capsule anterior to the fenestra ovalis). Several advanced ones have a fossa on the underside of the paroccipital process posterolaterally, very conspicuous in tritylodonts, which may demarcate the origin of a stapedius muscle (making it a fossa muscularis minor) and/or a levator hyoidei.

Several "cynodonts" have a dorsal process of the stapes, so a persisting hyostapedial connection may well have existed (Fig. 28.3C). At the posterolateral end of the paroccipital process of *Probainognathus* is a short ventral projection in some specimens, a circular area of unfinished bone in others; this was considered by Allin (1975) to be evidence that the hyoid cornu had shifted its mooring from stapes to skull (as a styloid process) but it may instead have been a site of muscle attachment similar to the mastoid process of mammals. It probably corresponds to the posterior process of the bifid distal end of the paroccipital process of tritylodonts, which also has a large anterior process on which the quadrate articulates. Posteroventrally, the latter process comes to a sharp point

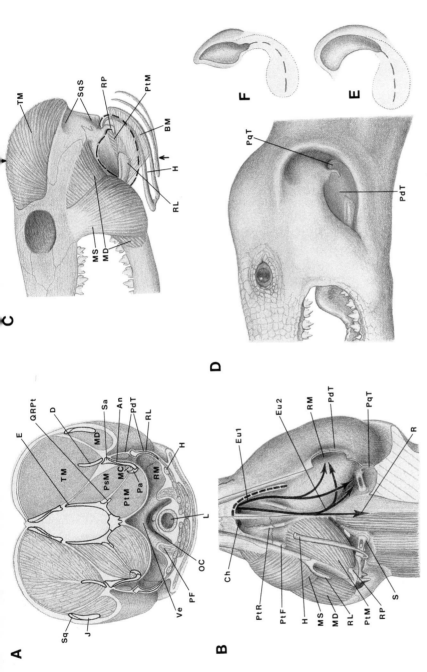

FIGURE 28.10. Reconstructed soft-tissue structures in the early cynodont *Thrinax-odon*. (A) Posterior view of the head, transected in the plane represented by the vertical arrows in (C). Bones seen in section modified from Fourie (1974). (B) Ventral view showing relevant muscles and contour of the roof of the pharyngeal cavity and its extensions. The soft palate is not shown, but its bony attachments are indicated by the dashed line. The respiratory airway is indicated by the arrow labelled R, and two interpretations of the eustachian (pharyngotympanic) passage are indicated by the arrows labelled Eu1 and Eu2. Eu1 follows Barghusen 1986 (and is favored by Allin); Eu2 follows Hopson (1966) but adds an extension into the angular cleft (a recessus mandibularis) following Westoll (1943, 1945). (C) Rele-

vant musculature and hyobranchial apparatus (largely hypothetical). The first cornu is assumed to have connected to the stapes. The dashed line demarcates the limits of the recessus mandibularis and cavum tympani proper. The thin bran-chiomandibularis muscle is shown as transparent to allow the primitive pterygoideus muscle to be seen. (D) Appearance in life. The external auditory meatus is depicted as a depression bounded by the masseter and squamosal anteri-orly and by a raised soft-tissue fold (primitive pinna) posteriorly. The tympanum has postdentary and postquadrate components. (E and F) Stages in fusion of the margins (dashed line) of the ventral parts of the primitive meatus to form a tubular passage, accompanied by development of the definitive pinna.

which is a better candidate for a hyoid attachment (Sues 1986) by analogy with a similar configuration in monotremes. However, it should be noted that if a tympanum was situated between the anterior and posterior processes it would be posterior to the hyoid attachment rather than anterior to it as in mammals.

3.6 Primitive Mammals

The earliest forms generally classed as mammals (Mammaliaformes of Rowe 1988) are mouse-sized creatures of the latest Triassic (Rhaetian) and earliest Jurassic, all of which have a well-developed, unquestionably synovial dentary-squamosal joint with a large rounded condyle. The mandible is otherwise similar to that of advanced nonmammalian cynodonts, the postdentary elements still being attached and the articular-quadrate joint still constituting part of the jaw articulation, although subsidiary to the new joint. There is no surangular-squamosal contact. The best known of very early mammals is *Morganucodon* (Kermack, Mussett, and Rigney 1981), which will now be described (Figs. 28.4I, 28.5H, 28.6G, 28.7G and 28.8D). Only in this genus is the reflected lamina known (but incomplete). It is a very long, flat process partly overlapped by the posterior border of the dentary but separated from it by a pocket-like space. No complete retroarticular process is known. As in tritylodonts, the quadrate is attached to the anterior process of the bifid end of the paroccipital process, with no intervening squamosal lappet (unlike most earlier cynodonts). A long stapedial process of the quadrate corresponds to the crus longum of an incus. The quadratojugal is not known but a notch for its upper end is present on the squamosal; probably it was similar to that of "cynodonts." Only the proximal part of the stapes is known. It is smaller than that of "cynodonts," but similar in having a large foramen. The lateral flange of the prootic is extended anteriorly, roofing the cavum tympani. There is no distinct fossa muscularis major but there is a fossa for a probable stapedius and/or levator hyoidei. An apparent meatal groove is present on the posterior aspect of the squamosal, but it does not continue onto the low zygomatic arch.

Several other genera comparable in age to *Morganucodon* are also known from crania and

mandibles. Informative descriptions have been published for *Sinoconodon* (Crompton and Sun 1985) and *Megazostrodon* (Gow 1986), and other good skulls await description. All are similar to *Morganucodon* in relevant features. Middle and Late Jurassic docodonts are known from good but incompletely described material, and are also similar. *Kuehneotherium*, known only from dentary bones and teeth, is contemporaneous with *Morganucodon* but shows dental characteristics suggestive of affinity with all later therian mammals. Postdentary elements are definitely still mandibular components, because there is a trough for their attachment, overhung by a prominent medial ridge. There is a rounded or transversely expanded dentary condyle. Thus, all adequately known mammals of Late Triassic and Early Jurassic age retained the postdentary unit and quadrate as jaw structures, and at least one group (docodonts) did so into the Late Jurassic. All other mammals for which sufficient evidence is available have at least partial, and usually total, separation of the postdentary bones from the mandible, as shown by loss of the postdentary trough and medial ridge. Most Mesozoic genera are represented only by isolated dentary bones and teeth, so it is unknown whether the postdentary elements and quadrate were retained as auditory structures or lost altogether. In one group of nontherian Mesozoic mammals (multituberculates) they were definitely retained, and in another (triconodontids) there is strong evidence for their retention. Skulls of Late Jurassic and Early Cretaceous Triconodontidae have an apparent attachment site for the short crus of an incus (fossa incudis) on the crista parotica, an epitympanic recess to house the dorsal parts of a malleus and incus (Crompton and Sun 1985), and a well-defined depression anterior to the fenestra ovalis (clearly shown in the figures of Kermack 1963) which is almost certainly a fossa muscularis major. Thus, both incus and malleus appear to have been present in life. Multituberculates, known from numerous skulls, have these same features and in one specimen of *Lambdopsalis* all three ossicles have been preserved in place (Miao and Lillegraven 1986; Miao 1988). The ossicles are rather robust, especially the stapes (which lacks a foramen). Also, the entire middle ear is reoriented so that its morphologic lateral aspect is anterior, as a result of great expansion

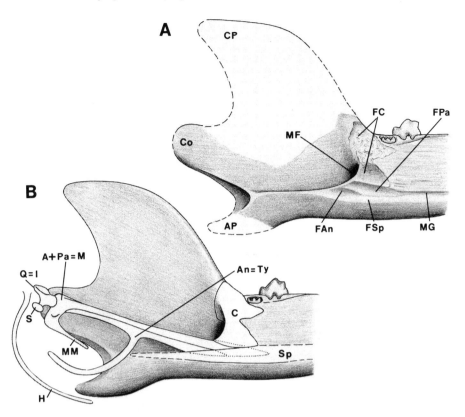

FIGURE 28.11. Mandible of the primitive therian mammal *Amphitherium* in medial aspect, showing (A) apparent attachment sites on the dentary bone for postdentary and paradentary accessory elements (based on specimen I, Oxford University Museum), and (B) a hypothetical reconstruction of these elements. The anterior extremities of the angular, An, and pre-articular, Pa, probably were narrower than shown, not fully occupying their attachment facets. Some gliding motion probably occurred, the postdentary unit moving less than the dentary as the jaw was opened and closed.

of the vestibule of the inner ear in this peculiar multituberculate (see below). Assuming that the tympanic membrane was essentially parallel to the fenestra ovalis, it was probably tilted to face somewhat ventrally in triconodontids and multituberculates, as in almost all existing mammals.

With two apparent exceptions, all therian mammals more advanced than *Kuehneotherium* have entirely lost the postdentary trough and the medial ridge of the dentary. Several retain a slender meckelian groove commencing medial to the mandibular foramen and extending to the mandibular symphysis, as well as a facet for attachment of a small coronoid bone. The exceptions are the Middle Jurassic *Amphitherium* and the Late Jurassic *Peramus*. Both are known from isolated dentary bones that have no sign of a medial ridge or posterior portion of a post-

dentary trough. In their place is a broad fossa (the so-called pterygoid fossa). However, the anterior portion of what appears to be a postdentary trough persists. In *Amphitherium* this is larger than in *Peramus*, and has subsidiary facets (Fig. 28.11). Also, it is shallower and farther anterior, as is the mandibular (inferior alveolar) foramen. Probably the postdentary elements remained attached to the dentary at their anterior extremities in these genera but were free of the dentary posteriorly. The "pterygoid fossa" may have provided access for an external auditory meatus to the postdentary unit, as also may be the case for more derived Dryolestidae (in which the angle of the dentary extends far posteriorly and may have overlapped detached postdentary bones that persisted as part of a definitive mammalian middle ear, as shown in Fig. 28.7H).

The pterygoid transverse processes of *Morganucodon* are similar to those of "cynodonts" and must have prevented any large medial excursions of the mandible. There is no pulley for a tensor veli palatini muscle. The bony nasopharyngeal roof is very similar to that of tritheledonts, having median and lateral troughs separated by very prominent pterygopalatine ridges. This region is unknown or as yet undescribed for other early mammals, with the exception of multituberculates, in some of which a similar system of grooves and ridges is known. Barhusen (1986) has interpreted the lateral nasopharyngeal passages as eustachian.

3.7 Advanced Mammals
(Figs. 28.4J, 28.5I, and 28.8E)

The tympanic bone of extant mammals is a simple incomplete ring (annulus) in the fetus, its posterior (ventral) limb corresponding to the reflected lamina and its anterior (dorsal) limb to the body of the angular. It may retain this form throughout life, remaining very loosely suspended on the cranium, and is generally tightly bonded (often synostosed) to a long anterior process of the malleus. Deformation of the thin stem of the anterior process is then required for vibration of the malleus. Whether the free tympanic is itself a vibratory element, in effect a fourth ossicle, has not been directly tested; this would appear to be unlikely for airborne sound (Fleischer 1978), except in monotremes (Aitkin and Johnstone 1972). However, it is likely that a loose tympanic enhances sensitivity to substrate-borne and bone-conducted sound, by increasing the inertia of the ossicles. Most living mammals have a rigidly moored tympanic which is often fused to the squamosal and petrosal and may contribute to a bony external acoustic meatus by outgrowth laterally, and/or a tympanic bulla by outgrowth medially. A bulla may never form, or may be fashioned from a variety of bony or cartilaginous components (Novacek 1977). A true manubrium is present on the malleus, projecting almost parallel to the anterior process in the less modified forms, but deviated away from it, and from the axis of ossicular vibration, in some derived forms. The malleus may develop a bulbous head. The malleo-incudal joint is no longer a hinge joint, instead having, on each bone, two flat planes which meet at approximately a right angle. The

stapes of most placental mammals is stirrup-shaped, with a large foramen, while that of marsupials is generally more triangular, with the foramen small or absent and with an elongate neck. The monotreme stapes is shaped like a golf tee and has no foramen. Extant mammals do not have pterygopalatine ridges. The auditory (eustachian) tube of monotremes is extremely short and is continuously open, but this may be secondary to extreme posterior extension of the hard palate. In marsupials and placentals the tube is generally long and narrow, and is supported in part by cartilage. In humans, at least, it is opened only during swallowing or yawning, primarily by the tensor veli palatini pulling its membranous lateral wall away from its cartilaginous medial wall. The tensor tympani muscle is always immediately dorsal to the auditory tube and may be elongate or rounded. It may be largely enclosed in a bony housing, as may the stapedius. Sometimes a small levator hyoidei is present in addition to a stapedius. Monotremes have only a levator.

4. Inner Ear

Descriptions of inner ear osteology in fossil synapsids are few. Complete information can only be obtained by serially sectioning (or grinding), which has been done for very few specimens. In primitive synapsids, ossification of the otic capsule is less complete than in advanced ones, hampering interpretation of labyrinthine structure. The osseous internal acoustic meatus opens directly into the vestibule, having no bony terminus. It is very large, often with unfinished surfaces suggesting that substantial amounts of cartilage had been present in life. Similarly, the vestibule is capacious and confluent with the entire medial aspect of the horizontal semicircular canal; probably the inner surface of the vestibule was cartilaginous. There is an indefinite or rounded cochleolagenar recess, and no separation of a fenestra rotunda (cochleae) from the jugular foramen. The fenestra ovalis (vestibuli) is large and is at the end of a pronounced extension from the vestibule. This description applies to *Dimetrodon* (Romer and Price 1940), and to several anomodonts and gorgonopsians (Fig. 28.12A; Cox 1959; Olson 1944; Sigogneau 1974). Early cynodonts are similar except that the bony internal

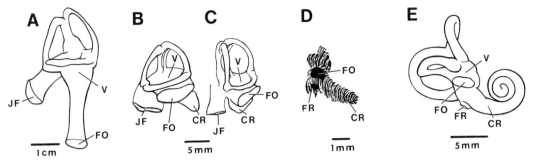

FIGURE 28.12. Endocasts of right osseous labyrinths. (A) Anomodont (dicynodont of uncertain genus, after Cox 1962) in lateral view. (B) The advanced non-mammalian cynodont *Probelesodon* in lateral aspect (after Quiroga 1979). (C) The same in posterolateral aspect. (D) The primitive mammal *Morganucodon* in lateral view. (From Graybeal et al. 1989.) (E) Human in anterolateral aspect.

acoustic meatus is smaller, the semicircular canals are almost completely enclosed in bone, the vestibular extension to the fenestra ovalis is shorter, a distinct round window is seen (but still confluent with the jugular foramen), and the ventrally projecting cochleolagenar recess is slightly more prominent (*Thrinaxodon*: Estes 1961; Fourie 1974) or considerably more prominent (*Diademodon*: Watson 1913; Simpson 1933; *Probelesodon*: Fig. 28.12B & C, Quiroga 1979; *Probainognathus*: Allin 1986). In tritylodonts, tritheledonts, and all mammals the internal auditory meatus is separated from the vestibular cavity by bone and the round window is fully separated from the jugular foramen. In mammals there is no vestibular extension to the oval window, while the cochleolagenar recess is narrower, tubular, longer (except in the earliest mammals), and at its commencement projects anteriorly. Straight or moderately curved in primitive mammals (*Morganucodon*: Fig. 28.12D; Graybeal et al. 1989; triconodontids and multituberculates: Miao 1988; monotremes: Pritchard 1881), the recess becomes a coiled canal in marsupial and placental mammals (termed cochlear rather than cochleolagenar because the lagena has been lost), as shown in Fig. 28.12E. The cochlear duct of monotremes does not conform to its bony enclosure (Z. Luo, pers. comm.) and curves through 180 degrees in *Tachyglossus* and 270 degrees in *Ornithorhynchus* (Kermack and Mussett 1983; Pritchard 1881).

In marsupials and placentals an estimate can be made of the highest frequencies that can be perceived by measuring the thickness and the minimum width (thus stiffness) of the basilar membrane at the proximal end of the cochlea, or approximating these measurements in osteological specimens from the thickness of, and distance between, the free edges of the primary and secondary bony spiral laminae to which it attaches. Fleischer (1976) made such estimates for fossil cetaceans based on basilar membrane widths, but did not consider thicknesses. No estimates of this sort can be made for other groups of mammals or for nonmammalian synapsids because bony laminae are either absent or poorly developed. Elongation of the cochlea is evidence of enhanced discrimination of frequencies but may not demonstrate upward extension of the frequency range because high frequencies are sensed in the proximal (basal), not the distal (apical), region. High-frequency vibrations dissipate rapidly with distance and perhaps, therefore, must be sensed close to the stapedial footplate. Although quantification has not been attempted, it is obvious that the least distance (and amount of liquid) between stapes and round window (or jugular foramen) was vastly greater in primitive synapsids than in mammals; this may imply far less sensitivity to high frequencies.

The vestibule was greatly expanded in certain multituberculates in comparison with existing mammals, indicating enlargement of the sacculus and/or utriculus, possibly to enhance sensitivity to substrate-borne sounds as part of a fossorial adaptation (Miao 1988).

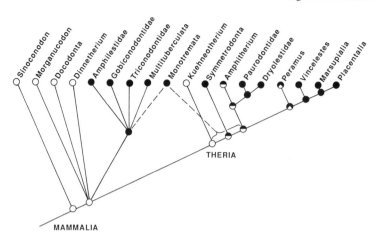

FIGURE 28.13. Cladogram showing relationships of the major groups of Mesozoic mammals (partly after Prothero 1981). The phylogenetic position of monotremes is controversial, dental evidence from Cretaceous and Miocene specimens suggesting a later origin than does evidence from the structure of the braincase and postcranial skeleton. Open circles represent the primitive condition, in which the postdentary bones are entirely mandibular. Partially black circles represent partial detachment of the postdentary elements from the dentary. Completely black circles represent complete separation of the postdentary elements from the mandible. These bones are known to have been retained as middle ear elements in the three groups of living mammals, in Multituberculata, and almost certainly in Triconodontidae and *Vincelestes*. Direct evidence of their retention in other groups is not available.

5. Parallel Originations of the Definitive Mammalian Middle Ear (DMME)

As stated above, the earliest known mammals still retained the postdentary bones and quadrate as components of the mandible and suspensorium. Functionally, they probably already had a mammalian middle ear in the sense that these elements were already involved in aerial sound reception. However, because the postdentary bones moved with the mandible and were subjected to mechanical stresses, they could not have been fully optimized for hearing. Three groups of later mammals (advanced therians, monotremes, and multituberculates) are known to have fully divorced these elements from the mandible as strictly auditory structures, and a fourth group (triconodontids) almost certainly had done so. In addition, several genera of Mesozoic mammals, constituting the Symmetrodonta and Amphilestidae (possibly paraphyletic), also may have achieved the DMME. Intermediate stages in the transference of postdentary elements to the cranium are poorly documented. Indeed, the only fossil evidence on this critical interval is the presence of persistent attachment sites for the anterior end of the postdentary unit in the primitive therians *Amphitherium* and *Peramus*. When all of these observations are plotted on a cladogram of mammals based largely on dental and cranial characters (Fig. 28.13), it will be seen that independent origination of the DMME must be inferred to have taken place more than once. (To infer a single origin of the DMME, one must assume convergent origination of braincase specializations shared by monotremes, multituberculates, and triconodontids, and of dental specializations shared by all therians.) The postcranial skeleton of monotremes is more primitive than that of an unnamed Jurassic therian similar to *Amphitherium*, and multituberculates appear to be closely related to the Rhaetian haramyids (known only from teeth), suggesting early origination of both of these lineages (Krebs 1988).

6. Mode of Transference of Postdentary Elements to the Cranium

Once a dentary-squamosal articulation had been established the postdentary bones and quadrate could be fully relieved of their suspensory role, even while still attached to the mandible. However, as long as the postdentary unit was obliged to follow the dentary in its excursions auditory refinement would still be hampered, an important disadvantage for animals that spent a great deal of time masticating food. Elimination of the medial ridge of the dentary, no longer needed to brace the postdentary unit, would allow the postdentary unit to begin to lag behind as the mandible was depressed. Jaw musculature inserting on the postdentary unit, most likely a portion of the pseudo-temporalis (Fig. 28.7H), may have restrained its excursion. The syndesmotic anterior attachment of the prearticular and angular to the dentary probably became more flexible, allowing these bones to shift somewhat in their facets on the dentary. This may explain why these facets in *Amphitherium* are somewhat greater in height than the expected size of these postdentary elements. Eventually, an elongate ligament connected the postdentary unit to the mandible (medial to the mandibular foramen), and muscular restraint could be replaced by direct bonding to the epipterygoid (alisphenoid) or other cranial bone. Thus, the ligamentous continuation of the anterior end of the postdentary unit became functionally divided into an anterior ligament of the malleus (concerned only with support of the malleus) and a sphenomandibular or pterygomandibular ligament (still participating in jaw suspension). If the condyle of the dentary glided anteriorly during jaw opening in primitive mammals, as it does in some living mammals, stabilization of the postdentary unit would be easier to achieve than if the motion was hinge-like. An additional important factor in mooring the postdentary unit to the cranium during transitional stages must have been attachment of the posterior (ventral) limb of the tympanic annulus to the stylohyal. See Maier (1990) for further discussion.

7. Theories of Auditory Evolution: Comparison of Paradigms

Although there is voluminous osteologic information relating to evolutionary modification of the middle ear in extinct synapsids, attempts to place these observations in a theoretical framework have been hampered by uncertainties regarding pertinent soft-tissue structures. Critical issues are whether premammalian synapsids possessed:

(1) a postquadrate tympanum comparable to that in many sauropsids; (2) a postdentary (mandibular) tympanum held in the angular gap as in living mammals; (3) both of these tympana, either separately or in continuity; or (4) no specialized tympanic membrane. Basic to any conclusions are interpretation of the functional significance of the squamosal sulcus and the angular cleft. The principal alternative interpretations of relevant soft-part anatomy which have been proposed, and their most complete published expositions are the following:

A. Limits of the middle ear cavity in nonmammalian synapsids:
  1. Essentially reptilian, without a submandibular diverticulum (Parrington 1955; Hopson 1966).
  2. With a submandibular extension (recessus mandibularis) into the external fossa of the angular, deep and posterior to the reflected lamina (Westoll 1943, 1945; Watson 1953; Shute 1956; Allin 1975, 1986).
B. Presence or absence of a specialized tympanum in primitive synapsids:
  1. Postquadrate tympanum homologous with that of sauropsids, with a direct stapedial connection, at the end of an external auditory meatus (Westoll 1943, 1945; Parrington 1955; Hopson 1966).
  2. No specialized tympanum (Watson 1953; Tumarkin 1955; Allin 1975, 1986). The latter two authors argue for significant sensitivity to low frequency airborne sounds via unspecialized structures.
C. Presence of a specialized tympanum in nonmammalian Cynodontia:
  1. Postquadrate only, at the end of a long, tubular meatus (Westoll 1943, 1945; Parrington 1955; Hopson 1966).
  2. Postquadrate and postdentary, in continuity (Watson 1953; Allin 1986) or separately (Allin 1986).
  3. Postdentary only (Allin 1975; Kermack, Musset, and Rigney 1981; Kermack 1982; Kermack and Mussett 1983).

We shall now briefly summarize our earlier published views and consider their pros and cons.

Hopson (1966) considered the boundaries of the middle ear cavity of nonmammalian therapsids to have been set by the basicranium medially, the quadrate rami of the pterygoid and epipterygoid laterally, and the paroccipital process (and, in cynodonts, the lateral flange of the prootic) dorsally. This is essentially as in a living lizard. The external fossa of the angular was occupied by insertion of pterygoideus musculature, again by analogy with living sauropsids (Barghusen 1968, 1972). The squamosal sulcus held a long, tubular cartilaginous external acoustic meatus comparable to that of monotremes (Gregory 1910), which led to a tympanum behind the quadrate supported by a semicircular notch in the squamosal at the distal end of the paroccipital process, or by the distal margin of the paroccipital process itself. The posterolateral corner of the stapes would not quite reach an eardrum at this site, so a cartilaginous tympanic process is presumed to have extended from it. The tympanum took on a mammalian aspect only with the establishment of the definitive mammalian jaw joint. Expansion of the dentary and reduction in size of the postdentary elements and quadrate occurred for nonauditory reasons. The diminished articular and quadrate intruded into the middle ear cavity as the tympanum increased in size. Posterior migration of the angular region of the dentary eventually carried the associated reflected lamina into a position medial to the dentary condyle. It could then support the tympanic membrane. The retroarticular process of the vestigial articular provided leverage which enhanced auditory sensitivity, making the tympanic process of the stapes redundant and resulting in its disappearance.

A major difficulty with Hopson's theory is the necessity of a hypothetical extrastapedial connection with the tympanic membrane having been severed in favor of a longer route of transmission via the articular (malleus) and quadrate (incus). More importantly, however, if the postdentary elements became involved in sound transmission only *after* the dentary-squamosal articulation was established why had they become reduced in size and loosely moored to the dentary in Triassic cynodonts? That is, what selective forces caused the postdentary bones and quadrate to take on many features of mammalian auditory ossicles in the absence of an auditory role? Also, it is not obvious why the stapes would not have separated from the quadrate to reduce auditory interference during feeding.

Allin (1975) addressed the latter issues, showing that if the postdentary bones and quadrate became

involved in hearing in primitive therapsids, the observed changes in structure of the jaw make adaptive sense as having taken place, at least partly, to improve auditory performance. He proposed that primitive synapsids lacked a specialized tympanic ear, but that in later therapsids, with the development of a large air-filled diverticulum of the pharynx (recessus mandibularis) extending around the ventral aspect of the pterygoideus muscle mass into the angular cleft, the postdentary portion of the mandible became specialized for aerial sound reception. The outer wall of the recessus, including the reflected lamina as well as a sheet of tissue bridging the angular gap, served as a functional tympanum. Vibrations were conducted via the primitive articular-quadrate-stapes chain. Thinning and expansion of the reflected lamina altered the vibratory properties of this postdentary tympanum, as did later replacement of much of the reflected lamina by broadened membrane in the angular gap. The trend toward loosening and reduction in mass of the postdentary elements and quadrate was given a plausible explanation in terms of selection for a more sensitive sound-conducting system. "Refinement of the cynodont middle ear without fundamental change in mechanism led to the present-day mammalian middle ear" (Allin 1975, p. 420).

Allin's theory had two attractive elements lacking in earlier theories: (1) functional continuity was maintained from an early therapsid stage to modern mammals, with an adaptive explanation for the great reduction in size of the postdentary elements, their progressive loosening from the dentary, and the shifting of muscle insertions onto the dentary; (2) because the reconstructed postdentary tympanum was much larger than the postquadrate drum hypothesized by Hopson (and others) it seems more likely to have provided sensitive impedance matching (see Kermack, Mussett, and Rigney 1981).

To account for the squamosal sulcus of therapsids, which nearly all other authors had interpreted as housing a tubular meatus, Allin proposed that it held a depressor mandibulae muscle, as a similar sulcus does in some turtles. However, he subsequently (1986) modified this view, mainly because in some tritylodonts the anterior part of the sulcus is almost tubular and the posterior part cleft-like, therefore too confining to be a likely muscle origin.

Also, because the lower end of the sulcus always extends, partly at least, to the lateral limit of the paroccipital process posterior to the quadrate he accepted that a tympanum did attach here, either separate from the postdentary tympanum or connected with it and involving transmission via the hyostapedial connection that probably persisted in primitive cynodonts with a stapedial dorsal process. Allin points out that coexistence of dual tympana, postquadrate and postdentary, makes sense only if they received different spectra of sound frequencies. He considers Westoll's (1943, 1945) hypothesis that the recessus mandibularis served initially as a vocal resonator to be plausible and adds that the postdentary apparatus may have served as a sensitive receiver of conspecific vocalizations.

If a tubular meatus homologous with that of mammals ended separately from a tympanic membrane in the angular gap homologous with the pars tensa of the mammalian tympanum, the meatal terminus could not subsequently extend to reach the postdentary drum without interfering with its function. Kemp (1979, p. 115) considers the "cynodont" meatus to have "carried on below and forwards from the ventral termination of the bony meatus", suggesting "a tympanic position in these later cynodonts . . . anteroventral to the quadrate." Kermack (1982, p. 154) supports this view, stating that it "would reconcile the theories of W.K. Gregory and E.F. Allin in a most satisfactory and satisfying way." Crompton (1985) illustrates such a meatus in *Thrinaxodon*. Allin (1986) offers a different interpretation, proposing that the meatus of primitive cynodonts, rather than being tubular, was a laterally open trough extending along the squamosal sulcus and behind the masseter muscle (Fig. 28.10D). A raised soft-tissue margin posterior to this C-shaped depression may have served as a primitive pinna. Fusion of the margins of the meatal groove, commencing ventrally, would convert the groove into a tube ending in relation to both postdentary and postquadrate sound-receptive regions (Fig. 28.10E and F).

Allin's reinterpretation of the squamosal sulcus of nonmammalian therapsids brings his earlier views more in line with those of Hopson (1966), but greatly complicates what was previously a relatively simple, straightforward evolutionary scenario. Allin's (1986) version of the nature of the meatus in early cynodonts seems more acceptable than the closed tube suggested by Kemp (1979), Kermack (1982), and Crompton (1985). In Hopson's current view, the necessity for such a meatus, dictated by the morphology of nonmammalian cynodonts, somewhat weakens the plausibility of the concept of a functional mandibular tympanum. Even so, the attractiveness of Allin's functional explanation for the reduction and loosening of the postdentary elements prior to their departure from the lower jaw makes it the best interpretation offered thus far for how the ear ossicles of mammals evolved from jaw bones. Needless to say, Allin agrees!

8. Summary

As has long been known, mammals have converted primitive jaw elements into a sensitive mechanism for transmitting airborne sound vibrations from tympanum to inner ear. The tympanic ear of mammals evolved independently of that of amphibians and sauropsids (living reptiles and birds). The fossil record of Synapsida (mammals and their "reptilian" predecessors) sets out the sequence of structural modifications of the bones and, less certainly, the muscles and other soft tissues that become components of the mammalian auditory apparatus. In "pelycosaurs," the earliest synapsids, the relevant bones are all large, functional jaw elements (quadrate and articular, forming the craniomandibular joint, and angular plus prearticular mandibular elements). The stapes is massive and its distal end contacts the quadrate. A true tympanum appears to have been absent. In sphenacodontid "pelycosaurs" the angular develops a posteriorly notched ventral keel, the reflected lamina. Low frequency aerial sound may have been transmitted through the lower jaw to the quadrate and thence to the stapes, but sensitivity must have been very low. Another route of transmission was probably through the hyostapedial connection.

In more advanced synapsids, the noncynodont therapsids, the tooth-bearing dentary bone enlarges somewhat, developing a coronoid process in some. The quadrate is reduced in height and may become more flexibly attached to the skull. These changes were presumably part of a reorganization of the feeding machinery. The stapes is also greatly reduced in mass, though retaining its abutment on the quadrate. The reflected lamina becomes a thin, broad plate set off from the body of the angular bone by a large space, the angular cleft (possibly housing an air-filled diverticulum of the middle ear cavity or pharynx). The reflected lamina may have functioned as part of a sound-receiving surface, its vibrations carried to the oval window via the post-dentary bones, quadrate, and stapes. However, a groove on the rear of the squamosal suggests the presence of an external auditory meatus ending at the broadened outer end of the paroccipital process where a postquadrate tympanum may have been attached. If this is so, vibrations may have reached the stapes via a persisting hyostapedial connection (or by a tympanic process for which there is little evidence).

In early cynodonts, the dentary becomes further enlarged and increasingly overlaps the postdentary bones. A masseter muscle develops on its lateral surface. The angular cleft is greatly expanded but the reflected lamina is reduced in size. At this stage, the thin soft tissues overlying the cleft may have served as the major portion of a rather crude postdentary tympanum. The presumed meatal groove on the squamosal becomes prominently developed. In later cynodonts the dentary is pro-gressively enlarged until, in at least one lineage, it forms a new craniomandibular joint with the squamosal. The postdentary elements are reduced to a slender rod flexibly supported in a smooth trough on the dentary, and the reflected lamina becomes a thin prong. The quadrate and stapes are also greatly reduced in size. In the earliest mam-mals and their nearest relatives (tritheledontids), primary and secondary jaw joints are both present, and the mass of the postdentary bones and quad-rate is no greater than that of the malleus and incus in many living mammals. A reduced slip of ptery-goideus musculature inserting on the articular becomes the mammalian tensor tympani, its origin shifting onto the cochlear housing. Another part of this musculature becomes the tensor veli palatini, developing an insertion into the soft palate (a cynodont innovation).

Fossil evidence indicates that reduction in size of the postdentary bones in nonmammalian cyno-donts proceeded independently in several lineages, as did the freeing of these elements from the man-dible in early mammals. Presumably, these modifi-cations were impelled by strong selection for improved auditory capabilities. Accompanying these changes in the middle ear in late "cyno-donts" and early mammals was an increase in length of the cochlear recess, implying enhanced frequency discrimination.

Uncertainty persists regarding the existence and location of a tympanum in nonmammalian synap-sids. If, as seems likely, a tympanic membrane existed in "cynodonts", it may have been situated: (1) in a postquadrate location, at the end of an external auditory meatus within a sulcus in the squamosal (Hopson 1966); (2) in a postdentary location, supported by the reflected lamina (Allin 1975); or (3) in both positions separately (Allin 1986). A compromise is offered here, with (1) and (2) coexisting in continuity, as parts of a single tym-panum. The primary portion, homologous with the mammalian pars tensa, was attached to the postdentary bones. Another portion extended below and behind the articular and quadrate to attach at the end of the squamosal groove. The outer ear would have then consisted of a curved depression, widely open laterally, which may have had a raised posterior rim (primitive pinna). Fusion of the margins of this depression, beginning over the mandibular tympanum, would later form the tubular mammalian meatus.

Acknowledgments. Allin's research was in part supported by the Chicago Community Trust. Hopson's research was supported by National Science Foundation Research Grant BSR-8906619. Except Fig. 28.10 (by E.F.A.), all figures are by Claire Vanderslice, whose skill and effort are greatly appreciated.

References

Aitkin LM, Johnstone BM (1972) Middle ear function in a monotreme: the echidna (*Tachyglossus aculeatus*). J Exp Zool 180:245–250.

Allin EF (1975) Evolution of the mammalian middle ear. J Morph 147:403–438.

Allin EF (1986) The auditory apparatus of advanced mammal-like reptiles and early mammals. In: Hot-ton N III, McLean PD, Roth JJ, Roth EC (eds) The Ecology and Biology of Mammal-like Reptiles. Washington DC: Smithsonian Institution Press, pp. 283–294.

Barghusen HR (1968) The lower jaw of cynodonts (Reptilia, Therapsida) and the evolutionary origin of mammal-like adductor jaw musculature. Postilla 116: 1–49.

Barghusen HR (1972) The origin of the mammalian jaw apparatus. In: Schumacher GH (ed) Morphology of the Maxillo-Mandibular Apparatus. Leipzig: Thieme, pp. 26–32.

Barghusen HR (1986) On the evolutionary origin of the therian tensor veli palatini and tensor tympani muscles. In: Hotton N III et al. (eds) The Ecology and Biology of Mammal-like Reptiles. Washington DC: Smithsonian Institution Press, pp. 253–262.

Barry TH (1968) Sound conduction in the fossil anomodont Lystrosaurus, Ann S Afr Mus 50:275–281.

Chudinov PK (1965) New facts about the upper Permian of the U.S.S.R. J Geol 73:117–130.

Cox CB (1959) On the anatomy of a new dicynodont genus with evidence on the position of the tympanum. Proc Zool Soc Lond 132:697–750.

Cox CB (1962) A natural cast of the inner ear of a dicynodont. Am Mus Novitates 2116:1–6.

Crompton AW (1972) The evolution of the jaw articulation of cynodonts. In: Joysey KA, Kemp TS (eds) Studies in Vertebrate Evolution. Edinburgh: Oliver and Boyd, pp. 231–251.

Crompton AW (1985) Origin of the temporomandibular joint. In: Carlson DS, McNamara JA, Ribbens KA (eds) Developmental Aspects of Temporomandibular Joint Disorders. Monograph 16, Craniofacial Growth Series, Center for Human Growth and Development, U. Mich., Ann Arbor, pp. 1–18.

Crompton AW (1989) The evolution of mammalian mastication. In: Wake DB, Roth G (eds) Complex Organismal Functions: Integration and Evolution in Vertebrates. New York: Wiley and Sons, pp. 23–40.

Crompton AW, Sun A (1985) Cranial structure and relationships of the Liassic mammal Sinoconodon. Zool J Linn Soc 85:99–119.

De Beer GR (1937) The Development of the Vertebrate Skull. Oxford: Clarendon Press.

Estes R (1961) Cranial anatomy of the cynodont reptile Thrinaxodon liorhinus. Bull Mus Comp Zool 6: 165–180.

Fleischer G (1973) Studien am Skelett des Gehörorgans der Säugetiere, einschließlich des Menschen. Säugetierkundl Mitt 40:131–239.

Fleischer G (1976) Hearing in extinct cetaceans as determined by cochlear structure. J Paleont 50:133–152.

Fleischer G (1978) Evolutionary principles of the mammalian middle ear. Adv Anat Embryol Cell Biol 55: 1–70.

Fourie S (1974) The cranial morphology of Thrinaxodon liorhinus Seeley. Ann S Afr Mus 65:337–400.

Gaupp E (1913) Die Reichertsche Theorie. Arch Anat u Physiol Abt Anat Supplement 4:1–417.

Goodrich ES (1930) Studies on the Structure and Development of Vertebrates. London: Macmillan.

Gow CE (1986) A new skull of Megazostrodon (Mammalia, Triconodonta) from the Elliot Formation (Lower Jurassic) of southern Africa. Palaeont Afr 26:13–23.

Graybeal A, Rosowski JJ, Ketten DR, Crompton AW (1989) Inner-ear structure in Morganucodon, an early Jurassic mammal. Zool J Linn Soc 96:107–117.

Gregory WK (1910) The orders of mammals. Bull Am Mus Nat Hist 27:1–524.

Hopson JA (1966) The origin of the mammalian middle ear. Am Zool 6:437–450.

Hopson JA (1987) The mammal-like reptiles: a study of transitional fossils. Am Biol Teacher 49:17–26.

Hopson JA, Barghusen HR (1986) An analysis of therapsid relationships. In: Hotton N III et al. (eds) The Ecology and Biology of Mammal-like Reptiles. Washington, DC: Smithsonian Institution Press, pp. 83–106.

Jenkins FA Jr, Crompton AW, Downs WR (1983) Mesozoic mammals from Arizona: New evidence on mammalian evolution. Science 222:1233–1235.

Kemp TS (1969) On the functional morphology of the gorgonopsid skull. Phil Trans Roy Soc Lond B 256:1–83.

Kemp TS (1972) The jaw articulation and musculature of the whaitsiid Therocephalia. In: Joysey KA, Kemp TS (eds) Studies in Vertebrate Evolution. Edinburgh: Oliver and Boyd, pp. 213–230.

Kemp T (1979) The primitive cynodont Procynosuchus: Functional anatomy of the skull and relationships. Phil Trans Roy Soc Lond B 285:73–122.

Kemp TS (1982) Mammal-like Reptiles and the Origin of Mammals. London: Academic Press.

Kermack KA (1963) The cranial structure of triconodonts. Phil Trans Roy Soc Lond B 246:83–102.

Kermack KA (1982) The ear in the Theropsida. Geobios mem spec 6:151–156.

Kermack KA, Mussett F (1983) The ear in mammal-like reptiles and early mammals. Acta Palaeontologica 28:147–158.

Kermack KA, Mussett F, Rigney HW (1981) The skull of Morganucodon. Zool J Linn Soc 71:1–158.

Krebs B (1971) Evolution of the mandible and lower dentition in dryolestids (Pantotheria, Mammalia). Zool J Lin Soc Suppl 1, 50:89–102.

Krebs B (1988) Mesozoische Säugetiere – Ergebnisse von Ausgrabungen in Portugal. Sber Ges Naturf Freunde Berlin 28:95–107.

Kühne WG (1956) The Liassic therapsid Oligokyphus. London: Brit Mus (Nat Hist).

Maier W (1990) Phylogeny and ontogeny of mammalian middle ear structures. Neth J Zool 40:55–75.

Miao D (1988) Skull morphology of *Lambdopsalis bulla* (Mammalia, Multituberculata) and its implications to mammalian evolution. Contrib Geol Univ Wyoming Spec Pap 4:1–104.

Miao D, Lillegraven JA (1986) Discovery of three ear ossicles in a multituberculate mammal. Nat Geog Res 2:500–507.

Novacek MJ (1977) Aspects of the problem of variation, origin and evolution of the eutherian auditory bulla. Mamm Rev 7:131–149.

Olson EC (1944) Origin of mammals based upon cranial morphology of the therapsid suborders. Geol Soc Amer Spec Pap 55:1–136.

Parrington FR (1946) On the cranial anatomy of some gorgonopsids and the synapsid middle ear. Proc Zool Soc Lond 125:1–40.

Presley R (1984) Lizards, mammals and the primitive tetrapod tympanic membrane. Symp Zool Soc Lond No. 52:127–152.

Pritchard U (1881) The cochlea of the *Ornithorhynchus platypus* compared with that of ordinary mammals and birds. Phil Trans Roy Soc Lond 172:267–282.

Quiroga JC (1979) The inner ear of two cynodonts (Reptilia–Therapsida) and some comments on the evolution of the inner ear from pelycosaurs to mammals. Gegenbaurs Morph Jahrb Leipzig 125:178–190.

Reisz RR (1986) Pelycosauria. Handb Paläoherpetologie 17A:102.

Romer AS (1970) The Chañares (Argentina) Triassic reptile fauna, VI. A chiniquodontid cynodont with an incipient squamosal-dentary jaw articulation. Breviora Mus Comp Zool No 344.

Romer AS, Price LI (1940) Review of the Pelycosauria. Geol Soc Am Spec Pap 28:1–538.

Rowe T (1988) Definition, diagnosis, and origin of Mammalia. J Vert Paleont 8:241–262.

Shute CCD (1956) The evolution of the mammalian eardrum and tympanic cavity. J Anat 90:261–281.

Sigogneau D, Tchudinov PK (1972) Reflections on some Russian eotheriodonts (Reptilia, Synapsida, Therapsida). Palaeovert 5:79–109.

Sigogneau D (1974) The inner ear of *Gorgonops* (Reptilia, Therapsida, Gorgonopsia). Ann S Afr Mus 64:53–69.

Simpson GG (1933) The ear region and the foramina of the cynodont skull. Am J Sci 26:285–294.

Sues H-D (1985) The relationships of the Tritylodontidae (Synapsida). Zool J Linn Soc 85:205–217.

Sues H-D (1986) The skull and dentition of two tritylodontid synapsids from the Lower Jurassic of western North America. Bull Mus Comp Zool 151:217–268.

Tumarkin A (1955) On the evolution of the auditory conducting apparatus: A new theory based on functional considerations. Evol 9:221–242.

Watson DMS (1913) Further notes on the skull, brain, and organs of special sense of *Diademodon*. Ann Mag Nat Hist 12:217–228.

Watson DMS (1953) Evolution of the mammalian ear. Evol 7:159–177.

Westoll TS (1943) The hyomandibular of *Eusthenopteron* and the tetrapod middle ear. Proc Roy Soc Lond B 131:393–414.

Westoll TS (1945) The mammalian middle ear. Nature (London) 155:114–115.

29
Hearing in Transitional Mammals: Predictions from the Middle-Ear Anatomy and Hearing Capabilities of Extant Mammals

John J. Rosowski

1. Introduction

When the ancestors of terrestrial vertebrates moved from a water to a land environment they were confronted with the problem of sensing airborne sounds. One of the evolutionary response to this problem was the development of middle ears, which enabled more efficient collection of acoustic power from the air and transmission of the collected power to the inner ear (Wever and Lawrence 1954; Killion and Dallos 1979; Dallos 1984; Rosowski, Carney, Lynch, and Peake 1986).

One widely held view has been that the evolution of middle ears in different vertebrate groups reflected a continuum of progressive changes from a "primitive" amphibian middle ear with its single ossicle through a more "advanced" single-ossicle reptilian-avian middle ear, culminating in the multiple-ossicle middle ear of mammals — see Henson (1974) for a review of these arguments. This line of reasoning led to suggestions that mammal-like reptiles (the immediate ancestors of mammals) and the earliest mammals must have had hearing capabilities resembling those of extant reptiles (Parrington 1949; 1979) which are sensitive mainly to sounds of relatively low frequency, i.e., below 5 kHz (Turner 1980; Manley 1981). There are two major problems with such a scenario: (1) it requires a rather complicated sequence of events (Hopson 1966; Henson 1974) to explain the interposition of the new "mammalian" ossicles (the malleus and incus, formerly parts of the reptilian jaw articulation) between the functioning tympanic membrane and stapes.[1] (2) It implies that the "reptilian" middle ear was already well

developed 450 million years ago when the break occurred between the ancestral lines leading to modern mammals and reptiles and birds. This implication goes against the existing evidence; the stapes of the earliest mammal-like reptiles were huge, firmly attached to the walls of the skull and unlikely to have been very effective in transmitting airborne sounds (Tumarkin 1955; 1968; Bolt and Lombard Chapter 19).

More recent paleontological evidence suggests that the middle ears of modern amphibians, bird-reptiles and mammals developed independently (Allin 1975; Bolt and Lombard Chapter 19; Clack Chapter 20; Allin and Hopson Chapter 28). In particular, Allin (1975; 1986) has formulated a scheme which suggests that the bones of the "reptilian" jaw joint were interposed between the stapes and the sound-receptive tympanic membrane (or its functional precursor) even in the earliest stages of middle-ear development in mammal-like reptiles. Allin's scheme suggests a double use of the "accessory bones" (Allin 1975) of the jaws of mammal-like reptiles such that they functioned *both* as load-bearing jaw joints and as part of the ossicular chains which coupled acoustic vibrations to the stapes and inner ears of these animals. As evolution proceeded from the most primitive mammal-like reptiles to the mammals, there was a gradual reduction in the size of the jaw-bound

---

[1] In order to reduce confusion produced by class-dependent terms for homologous structures, I will use mammalian terms to refer to homologous structures in nonmammalian vertebrates. Therefore, the reptilian and avian columella will be referred to as the stapes.

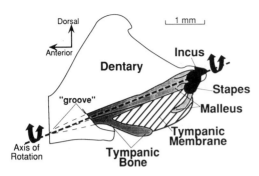

FIGURE 29.1. A view of the medial surface of the right lower jaw of *Morganucodon*. Illustrated are the dentary, the accessory bones of the lower jaw (in light gray) including the tympanic and malleus, the attached incus and stapes (in dark gray), and the axis of ossicular rotation (dashed line) proposed by Allin (1975). The tympanic membrane proposed by Kermack and Musset (1983) is also included. The drawing is a synthesis of several drawings of Kermack and co-workers. (Kermack et al. 1981; Kermack and Kermack 1984.)

"ossicles" and the appearance of new load-bearing joints between the lower jaw and the base of the skull (Crompton and Jenkins 1973; Crompton and Parker 1978; Allin and Hopson Chapter 28). Some of the later mammal-like reptiles and the earliest mammals, e.g., *Morganucodon*, had two jaw joints on each side of the head; a direct articulation between the dentary and the base of the skull as well as the remnants of a "reptilian" jaw joint between the accessory bones of the lower jaw and the homolog of the mammalian incus (Crompton and Jenkins 1979; Jenkins and Crompton 1979). The reduction in the size of the components of the "reptilian" jaw joint and the freeing of the ossicles from a jaw function by the development of a second jaw joint may have been driven, at least in part, by evolutionary pressures to improve hearing sensitivity (Allin 1975).

Allin's suggestion that an auditory function for the three-bone mammalian ossicular chain preceded the complete migration of these bones from the jaw into the middle-ear space has several attractive features: Firstly, it does not require the sudden introduction of the malleus and incus between the tympanic membrane and the stapes. Secondly, it provides a basis for improvements in hearing ability to act as one of the driving forces for the evolution of the mammalian ossicles even while their homologs existed within the jaws of

mammal-like reptiles. Thirdly, it eliminates the argument that the earliest mammals had low-frequency auditory capabilities much like modern reptiles and allows a reexamination of ethological and ecological suggestions (Masterton, Heffner, and Ravizza 1969; Jerison 1973) that "primitive" mammals were sensitive to high-frequency sounds much like extant small mammals.

This chapter approaches the debate about the hearing capabilities in early mammals from a different perspective. The middle-ear structure and auditory capacities of extant mammals both vary widely and there have been attempts to correlate the two (e.g., Fleischer 1978; Plassmann and Brändle, Chapter 31). This chapter will (1) demonstrate that significant relationships exist between middle-ear structure and auditory function, (2) use those relationships to predict the auditory behavior of early mammals for which only structural data are available, and (3) provide some mechanistic arguments for the bases of these relationships.

2. Review of the Structure of the Middle Ear of a Transitional Mammal, *Morganucodon*

The middle ear of *Morganucodon* (Fig. 29.1), one of the earliest known transitional mammals, has been fairly well described (Kermack, Musset, and Rigney 1981; Kermack and Musset 1983; Allin and Hopson Chapter 28). Most of the lower jaw is made up of the dentary bone, but there is a small group of bones which fits into a tapered groove on the inner surface of the dentary (Kermack, Musset, and Rigney 1973). These so-called "accessory bones" include the homologs of the mammalian tympanic bone and the malleus. The malleus articulated with the homolog of the incus with its attached stapes, and Allin (1975; 1986) and Kermack and Musset (1983) have proposed that the tympanic membrane was suspended by the accessory bones.[2] The old "reptilian" jaw joint is between the malleus and the incus which is loosely

---

[2]This membrane is believed by these authors to be analogous to the pars tensa of the mammalian tympanic membrane. Allin (1986) has proposed a second membrane positioned posterior to the quadrate that is analogous to the pars flaccida.

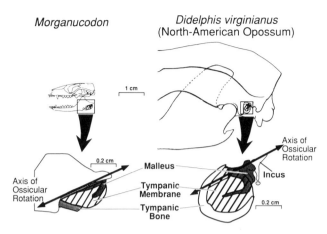

FIGURE 29.2. Comparison of the skulls and middle ears of *Morganucodon* and the North-American Opossum. The upper panel compares the skull of *Morganucodon* with the posterior part of the opossum skull. The bottom panel compares lateral views of the left middle ear in the two animals. The drawing of the skull and lower jaw of *Morganucodon* is after Kermack and Kermack (1984) while the opossum drawings are after those of Crompton and Jenkins (1979). The hypothetical axes of ossicular rotation are those defined for *Morganucodon* by Allin (1975) and for "primitive" and "microtype" ears by Fleischer (1978).

bound to the squamosal bone of the skull (not shown). The condyle of the dentary, partially hidden behind the incus, together with the squamosal forms the new "mammalian" jaw joint. The small size of the middle-ear structures in *Morganucodon* (1 to 2 mm in linear dimensions) is consistent with the small body weight (about 20 gm) estimated for these animals (Jenkins and Crompton 1979; Rosowski and Graybeal 1991).

How could these jaw structures function as a middle ear? According to Allin (1975; 1986) and Kermack and Musset (1983), the coupled accessory bones, tympanic membrane, incus and stapes were able to rotate about an axis which ran along the long axis of the accessory bones, from their most anterior tip in the dentary groove to the most posterior aspect of the incus (the dashed line in Figure 29.1). Rotation about this axis permitted sound-induced motions of the tympanic membrane to be transformed into rocking motions of the stapes footplate. In the reconstruction shown here, the stapes, malleus and tympanic bone and membrane are on the same side of the axis of rotation, such that an inward motion of the tympanic-malleus complex, results in an inward motion of the stapes. The succeeding analysis will assume that the Allin hypothesis is correct, see Parrington (1949; 1979) for an alternative point of view.

3. Comparisons of the Middle Ear of *Morganucodon* with Those of Extant Mammals

3.1 Comparisons of Middle-Ear Type

Despite the fact that key elements of the *Morganucodon* middle ear are obviously still components of the lower jaw there are many similarities between this middle ear and those of some extant mammals. Figure 29.2 compares the skull and middle ear of *Morganucodon* with those of *Didelphis virginianus*, the North American opossum. (The opossum was chosen because its middle ear is similar to that of many small mammals and because it has been classically used in comparative anatomy as an example of a relatively "primitive" mammalian ear.) Direct comparisons of skull and middle-ear sizes show that despite a factor of 5 difference in the length of the two skulls, the tympanic membrane and malleus of the opossum are only twice the size of those of *Morganucodon*. There are also similarities in the middle-ear structures: The tympanic membrane in each animal is partially circumscribed by the tympanic bone and malleus; the two mallei are firmly attached to the tympanic bone; and finally the long axis of the ventral "long arm" of the malleus in both ears roughly parallels the axis of rotation of the ossicles.

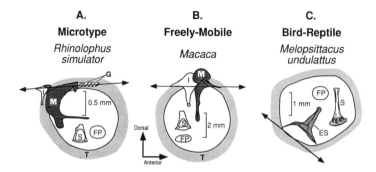

FIGURE 29.3. Medial views of the left middle ear of two extant mammals and a bird which were chosen to illustrate different middle-ear types. The tympanic, malleus, and incus are shown as in their natural position; the stapes has been disconnected from the ossicular chain so that you can see the other ossicles. A rotated view of the stapes is also included to reveal its superstructure. The double-pointed arrows point out the hypothetical axes of rotation of these ears. (Fleischer 1978; Manley 1972a; Saunders 1985.) ES, extra stapes; FP, stapes footplate; G, gonial process of the malleus; I, incus; M, malleus; S, stapes; T tympanic bone. (A) The middle ear of *Rhinolophus simulator*, the lesser horseshoe bat (after Fleischer 1973), an example of a "Microtype" middle ear. The dashed line of the gonial process of the malleus indicates a firm attachment to the tympanic bone. (B) The middle ear of an unspecified macaque (after Fleischer, 1973), an example of a "Freely Mobile" middle ear. The gonial process is reduced to a spicule for the attachment of the anterior ligament of the malleus. (C) The middle ear of *Melopsittacus undulatus*, the parakeet or budgie (after Saunders 1985), an example of a "Bird-Reptile" middle ear. The cartilaginous extra-stapes is shown still embedded in the tympanic membrane, while the boney stapes has been cut and rotated. The stapes is fenestrated near its footplate.

Although the ear of *Morganucodon* is similar in design to the ears of *some* mammals, there is a wide range of middle-ear structure within mammals. Fleischer (1973; 1978) has classified the middle ears of terrestrial mammals into several groups. Figure 29.3A illustrates a medial view of the right middle ear of *Rhinolophus simulator*, a horseshoe bat, which shows all of the characteristics of Fleischer's "microtype" middle ear: (1) The tympanic membrane and ossicles are small relative to those of other mammals. (2) There is a large and firm gonial attachment between the malleus and the tympanic bone. (3) The ventral "long arm" of the malleus parallels the axis of ossicular rotation. (4) The malleus is many times larger than the incus. In his description of microtype ears, Fleischer (1978) suggested that the firm gonial connection between the tympanic and malleus bones in microtype ears greatly stiffens the middle ear and reduces the responsiveness of the ear to low-frequency sound. Both the opossum and *Morganucodon* middle ear (Fig. 29.2) could be classified as "microtype."

At the other extreme of Fleischer's classification is the "freely mobile" middle ear, exemplified by the middle ear of a macaque (Fig. 29.3B). These ears typically; have larger tympanic membranes and ossicles, lack a gonial process so that the malleus swings freely about its ligamentous attachments, contain a long arm of the malleus arm which is oriented perpendicularly to the axis of ossicular rotation, and have inci and mallei of similar size. The lack of the gonial connection between the tympanic bone and malleus is thought to reduce the stiffness of the ear and enable a high degree of ossicular mobility, a condition favoring sensitivity to low-frequency sounds.

Also illustrated for comparative purposes is a medial view of the single-ossicle middle ear of *Melopsittacus undulatus*, the parakeet (Fig. 29.3C). In birds, the tympanic annulus is formed not by the tympanic bone but by the quadrate and the squamosal bones, and there is no firm connection between the ossicle and the annulus. The dimensions of bird-reptile middle ears fall between those of the bat and macaque; while, the lack of a bony connection should enable a high mobility for the tympanic membranes and stapes much like the freely mobile mammalian ear. Such mobility is consistent with the good low-frequency hearing capabilities of birds and reptiles (Dooling 1980; Manley 1981). Comparisons of Figures 29.1, 29.2, and 29.3 demonstrate that the middle ear of *Morganucodon* is clearly different from the ears of modern birds or reptiles and closely resembles the microtype middle ears of some mammals.

3.2 Comparisons of Middle-Ear Size

The dimensions of the middle ear also vary greatly among mammals. Part of this variability can be explained by the scaling of the middle-ear structures with body size (Khanna and Tonndorf 1978; Hunt and Korth 1980), but there also seems to be a relationship between middle-ear dimensions and audible frequency range (Fleischer 1978; Rosowski and Graybeal 1991). Figure 29.4 compares the areas of the *Morganucodon* footplate and tympanic membrane with these areas in extant mammals. Large mammals (e.g., *Loxodonta africana*—the African elephant, and *Equus caballus*—the horse) have large tympanic-membrane and footplate areas, while small mammals (e.g., *Suncus etruscus*—the dwarf shrew, and *Mus musculus*—the mouse) have small middle-ear areas. The *Morganucodon* middle-ear areas, the circled X, fall within ranges circumscribed by the areas of small mammals with good high-frequency hearing—bats, shrews, mouse and rat—but are a factor of 2–10 times smaller than the middle-ear areas of *Meriones* and *Dipodomys*, small rodents with ears which appear to be specialized for superior sensitivity to low-frequency sound (Webster and Plassmann, Chapter 30).

Figure 29.4 also points out that there is a high degree of correlation between the two middle-ear areas. The three dashed lines delineate simple linear relationships between the areas of the footplate and the tympanic membrane. The tympanic membrane is between 10 and 40 times larger than the footplate in all but three of these ears, and 67% of the variance in the data set is accounted for by an area ratio of 20. The tympanic-membrane to footplate area ratio in birds and reptiles is also close to 20 (Rosowski and Graybeal 1991).

4. Dependence of Auditory Function on Middle-Ear Structure

It has been demonstrated that the ear of *Morganucodon* is similar in structure and size to those of

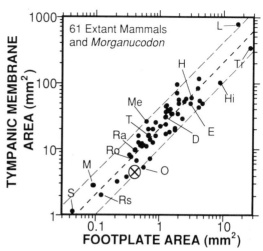

FIGURE 29.4. A comparison of the tympanic-membrane and footplate area of *Morganucodon* with those of 61 extant mammals. Most of the mammals in the sample are terrestrial, but two aquatic animals are also included; *Phoca vitulina*—the harbor seal and *Trichechus*—a manatee. The tympanic-membrane area is defined by the planar area of the pars tensa of the tympanic membrane (Fleischer 1973), which ignores both the curvature of the tympanic membrane and the area of pars flaccida. The circled X is the *Morganucodon* point. The data from extant animals comes from Wever and Lawrence (1954), Kirikae (1960), Fleischer (1973), and Hunt and Korth (1980). The estimates of the *Morganucodon* areas comes from Kermack and Musset (1983) and Graybeal et al. (1989). Several of the extant mammal points are labeled. D, *Dipodomys merriami*—the kangaroo rat; E, *Equus zebra*; H, *Homo*; Hi, *Hippopotamus amphibius*; L, *Loxodonta africana*—the Africana elephant; M, *Mus musculus*—the house mouse; Me, *Meriones unguiculatus*—the Mongolian gerbil; O, *Ornithorhynchus anatinus*—the platypus; Ra, *Rattus norvegicus*—the rat; Ro, *Rousettus aegyptiacus*—a fruit bat; Rs, *Rhinolophus simulator*—a horse-shoe bat; S, *Suncus etruscus*—the dwarf shrew; T, *Tachyglossus aculeatus*—the echidna; Tr, *Trichechus*—a manatee. The dashed lines are iso-ratio contours for tympanic-membrane to footplate area ratios of 10, 20, and 40.

extant small mammals with microtype ears, but I have yet to describe any precise relationships between those physical features and auditory function. This next section will review a common method for quantifying auditory function in animals and then use those measures to investigate relationships between auditory function and middle-ear structure.

4.1 Definition of Limits of Hearing

Various physical and behavioral responses have been used to quantify and compare the auditory capabilities of extant vertebrates, including: (1) physical measurements of middle and inner-ear motion, (Saunders and Johnstone 1972), (2) measurements of electrical responses from the cochlea, auditory nerve or CNS (Wever and Lawrence 1954; Manley 1981), and (3) behavioral audiograms. The two most common techniques used for this purpose have been measures of some behavioral response to sound (e.g., Masterton, Heffner, and Ravizza 1969; Stebbins 1970; Dooling 1980) or the gross cochlear potential (Wever 1978; Dallos 1970; Gates, Saunders, Bock, Aitkin, and Elliott 1974; Peterson, Levison, Lovett, Feng, and Dunn 1974). However, since measurements of cochlear potentials do not correlate well with measures of behavioral auditory performance at the lowest or highest audible frequencies (Dallos 1970; 1973; Lynch et al. 1982) I have chosen to use only behavioral measurements of auditory performance in an attempt to define relationships between middle-ear structure and hearing capabilities.

Previously determined behavioral audiograms for 23 terrestrial mammalian species with known tympanic-membrane and footplate area[3] were gathered from the literature—the audiometric data are all found in Fay (1988), while the primary sources for the anatomic data were Wever and Lawrence (1954), Kirikae (1960), Fleischer (1973), and Hunt and Korth (1980). Four audiometric limits were extracted from each audiogram—see Masterton et al. (1969) or Dooling (1980) for a description of similar procedures. The "best threshold" was defined as the lowest sound pressure (in dB SPL) of a tone of any frequency which produced a positive

---

[3]In three cases the available audiometric and anatomical data were gathered from different species within the same genus. These include, the genus *Rhinolophus*, where the audiometric data are from *R. ferrumequinum* and the anatomical data are from *R. simulator*; the genus *Mustela*, where the selected audiometric data are a mean of data from *M. nivalis* and *M. putorious* and the anatomical data are from *M. erminea*; and the genus *Macaca*, where the audiometric data are a representative audiogram for *M. arctoides*, *M. fascicularis*, *M. mulatta* and *M. nemestrina* and the anatomical data are from an unspecified species.

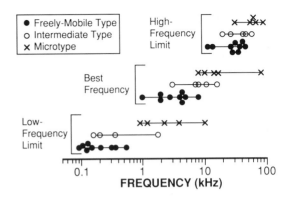

FIGURE 29.5. The relationship between middle-ear type and audiometric limits. Three audiometric frequency limits—Low-Frequency Limit, Best Frequency and High-Frequency Limit—from 19 animals with different middle-ear types (Table 29.1) are scaled together to illustrate type-dependent differences in the range of the limits.

behavioral response. The "best frequency" was defined as the tone frequency at best threshold. The "low-frequency" and "high-frequency" limits of each audiogram were defined by the intersection of the audiogram with an isopressure contour defined by best threshold plus 30 dB. The last three of these four limits can be used to approximate the shape of the audiogram.

4.2 Relationships Between Middle-Ear Type and Hearing Limits

Besides defining distinct types of mammalian middle-ear structures, Fleischer (1973; 1978) also proposed several associations between type and auditory function, including the notion that microtype middle ears were observed in those animals with hearing restricted to high frequencies, and the corollary that freely mobile middle ears were necessary for improved low-frequency hearing. To our knowledge, these hypothetical relationships between middle-ear type and auditory function have never been tested.

Figure 29.5 and Table 29.1 are the results of a compilation of auditory limits for 19 mammalian species, 9 with "freely mobile" middle-ears, 5 "microtype" middle ears, and 5 middle ears classified by Fleischer (1973; 1978) as "intermediate" type. Although there is much overlap between the

TABLE 29.1. Middle-ear Type and audiometric limits.

Species	Common name	Middle-ear Type[a]	Low-frequency limit (kHz)	Best frequency (kHz)	High-frequency limit (kHz)
Cavia procellus	Guinea pig	F	0.55	8	35.0
Chinchilla laniger	Chinchilla	F	0.09	2.8	29.0
Dipodomys merriami	Kangaroo rat	F	0.10	1.0	34.0
Homo sapiens	Human	F	0.15	4.0	14.0
Macaca[b]	Macaques	F	0.13	3.9	27.0
Meriones unguiculatus	Mongolian gerbil	F	0.21	4.0	46.0
Mustela[b]	Weasels	F	0.34	4.9	39.0
Oryctolagus cuniculus	Rabbit	F	0.33	2.0	38.0
Pan troglodytes	Chimpanzee	F	0.13	2.0	11.6
Equus caballus	Horse	I	0.16	2.0	27.0
Felis catus	Cat	I	0.2	8.0	59.0
Galago senegalensis	Bushbaby	I	0.36	8.0	47.0
Phoca vitulina	Harbor seal	I	–	11.0	20.0
Tupaia glis	Tree shrew	I	1.8	16.0	44.0
Didelphis virginianus	Opossum	M	1.2	16.0	64.0
Mus musculus	House mouse	M	3.9	16.0	62.0
Rattus norvegicus	Rat	M	0.9	8.0	55.0
Rhinolophus ferrumequinum	Horseshoe bat	M	10	82.0	90.0
Rousettus aegyptiacus	A fruit bat	M	2.2	10.0	31.0

[a]Fleischer's (1978) middle-ear types: F, Freely mobile; I, Intermediate; M, Microtype.
[b]Represents the average audiometric limits of several members of this genus.

audiometric limits from the three types of ears, there are some clear patterns. (1) The median audiometric limits for each of the three ear types are always arranged in the frequency domain such that the median "freely mobile" limit is of lower frequency than the median "intermediate" limit which is always less than the median "microtype" limit. (2) Five of the six highest values for low-frequency limit are associated with microtype ears, as are five of the highest eight best frequencies and four of the highest five high-frequency limits. An analysis of variance reveals that a significant fraction of the variance in each feature can be attributed to differences in middle-ear type.[4] A similar analysis comparing microtype ears with a grouping of the other two ear types reveals that the probability that the microtype and others are independent samples from the same population is less than 1% for the low- and high-frequency limits and less than 5% for best frequency. Clearly, the audiometric limits vary with middle-ear type and the microtype

middle-ears tend to have audiometric limits which are of higher frequency. A third analysis of variance indicates that the audiometric limits of the freely mobile middle ears are also statistically different from the other two ear types and supports Fleischer's suggestion that these ears are capable of better low-frequency hearing. Indeed, the most obvious differences in the hearing limits of the different middle-ear types is in the low-frequency hearing limit and not in the highs. These larger low-frequency differences are consistent with Fleischer's (1978) views that primitive mammalian middle ears (including microtypes) were primarily sensitive to high-frequency sounds, and that as the ear evolved, sensitivity to low-frequency sounds was added but not at the expense of continued sensitivity to high-frequencies.

4.3 Relationships Between Middle-Ear Areas and Hearing Limits

Considerations of the dependence of the optimal frequencies of acoustic sources and receivers on their size have been used by Khanna and Tonndorf (1978) and others (e.g., Michelsen, Chapter 5) to suggest relationships between ear dimensions and

---

[4]The probability that type does not contribute to the variance in each audiometric limit is less than 5% for the low and high-frequency limits and less than 10% for best frequency.

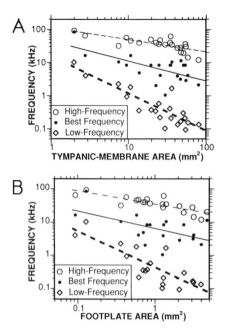

FIGURE 29.6. The dependence of audiometric limits on middle-ear areas. Three audiometric frequency limits plotted vs (A) Tympanic-membrane area, (B) Footplate area. The straight line segments represent power functions fit by least-squares regression analysis (Table 29.2). $n = 19$–22.

audible ranges. Such considerations stem from two different pieces of evidence. (1) The transmission of sound power from a circular source to the surrounding medium (and vice-versa) is best at

frequencies above a critical frequency, where that frequency is dependent on the area of the source, such that the smaller the area, the higher the critical frequency (Beranek, Chapters 5 and 7, 1954). This mechanism predicts that the low-frequency limit of the effectiveness of monopole acoustic receivers should vary as one over the square root (the $-1/2$ power) of the receivers area. (2) Holographic measurements of the motion of the tympanic membrane of cat and man (Khanna and Tonndorf 1972; Tonndorf and Khanna 1972) suggest that there is a size-dependent critical frequency below which the motion of all segments of the tympanic membrane move in phase with the sound stimulus. Above this second critical frequency, parts of the membrane move with different phase with the effect of reducing the overall motion of the membrane. Khanna and Tonndorf (1978) suggest that this second critical frequency can be defined by the frequency where the diameter of the tympanic membrane equals $1/4$ of a wave length of the sound. This second mechanism would lead to a *high*-frequency limit of tympanic-membrane motion which is also inversely proportional to the square root of the area of the tympanic membrane.

A plot of the audiometric limits from our sample population of mammals with known tympanic-membrane areas (Fig. 29.6A) indicates that the simple mechanisms suggested above do not perfectly predict those audiometric limits. The data are consistent with the direction of the dependencies proposed by Khanna and Tonndorf, in that the larger the tympanic-membrane area, the lower the

TABLE 29.2. Power functions fit to audiometric limits and anatomical dimensions.

Limit, y	n	$y = A x^B$ Coefficient, A	Exponent, B	Percent of variance explained	Significance of fit
		x = Area of Pars Tensa of Tympanic Membrane			
Low-frequency limit	21	15	-1.1	43%	$p < 0.1\%$
Best frequency	22	44	-0.61	76%	$p < 0.1\%$
High-frequency limit	20	110	-0.37	56%	$p < 0.1\%$
		x = Footplate Area			
Low-frequency limit	19	0.40	-1.1	67%	$p < 0.1\%$
Best frequency	20	6.2	-0.52	32%	$p < 1\%$
High-frequency limit	20	34	-0.40	68%	$p < 0.1\%$
		x = Basilar-Membrane Length			
High-frequency limit	12	391	-0.85	60%	$p < 1\%$

frequency limits of the audiogram. There are however quantitative differences between the data and the theoretical predictions.

The lines in Figure 29.6 illustrate power functions fit to the data; the coefficients of these functions are listed in Table 29.2. All of the fits are significant at the 0.1% level, and each of the audiometric limits has a different dependence on membrane area.[5] The low-frequency limit appears roughly inversely proportional to tympanic-membrane area (power of −1), the best frequency varies roughly as the −1/2 power of area, and the high-frequency limit varies roughly as the −1/3 power. While the exponent relating high-frequency limit to area is statistically similar to the exponent of −1/2 predicted by Khanna and Tonndorf (1978), the observed inverse proportionality of low-frequency limit and tympanic membrane area is a significantly steeper dependence than the −1/2 power predicted by the simple analysis of sound radiation and reception. As Khanna and Tonndorf note themselves, one likely source of errors in the prediction is the failure to account for the effects of the pinna and external ear on the frequency dependence of sound-power collection (Shaw 1974; Shaw and Stinson 1980; Rosowski, Carney, Lynch, and Peake 1986; Rosowski, Carney, and Peake 1988). Figure 29.6B reveals that the relationships between footplate area and the audiometric limits are similar to those which relate the tympanic-membrane area and hearing.

5. Predicting Hearing Limits from Middle-Ear Dimensions

The qualitative comparisons of Figures 29.2 and 29.3 suggest that the *Morganucodon* middle ear was similar to the microtype middle ears of modern small mammals which are most sensitive to high-frequency sounds (Fig. 29.5). A quantitative method for predicting hearing function from middle-ear anatomy can be fashioned from the additional observations that the limits of hearing are highly correlated with middle-ear area (Fig. 29.6, Table 29.2).

---

[5]The exponents of the fitted power functions are significantly different from each other. The probability that they are the same is less than 5%.

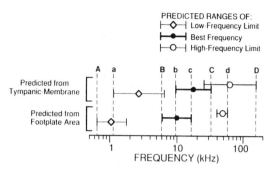

FIGURE 29.7. The 95% confidence limits around predictions of the auditory limits of *Morganucodon* based on the regression analysis of the data in Figure 29.6 and known dimensions of the *Morganucodon* middle ear. The vertical dashed lines labeled with uppercase letters point out the extremes of the limits predicted by the analysis of either middle-ear area, i.e., the lowest Low-Frequency Limit predicted by either analysis (A), the lowest Best Frequency predicted by either (B), the highest Best Frequency predicted by either (C) and the highest High-Frequency Limit predicted by either (D). The vertical lines labeled with lower case letters point out the extremes of the limits which are common to both predictions, i.e. the lowest Low-Frequency Limit predicted by both analyses (a), the lowest Best Frequency predicted by both (b), the highest Best Frequency predicted by both (c), and the highest High-Frequency Limit common to both predictions (d).

5.1 Predictions of the Hearing Limits for *Morganucodon*

The highly significant power-functions fit to the data of Figure 29.6 can be used to predict auditory limits from known middle ear areas. The area of the *Morganucodon* tympanic membrane has been inferred from the dimensions of the malleus and tympanic bone as 4.5 mm[2] (Kermack and Musset 1983), while the area of the *Morganucodon* footplate must fall somewhere in between the 0.28 mm[2] area determined by Kermack et al. (1981) and the 0.5 mm[2] area of the oval window measured by Graybeal et al. (1989). I will assume a footplate area of 0.4 mm[2]. The 95% confidence limits of predictions for the three audiometric limits based on these areas and the regression analyses of Table 29.2 are pictured in Figure 29.7. The confidence limits around the footplate predictions are somewhat narrower than are those around the tympanic-membrane predictions. The footplate

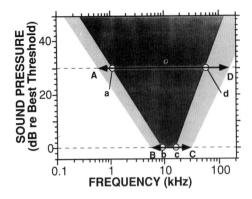

FIGURE 29.8. The likely regions of auditory sensitivity for *Morganucodon* predicted from its middle-ear areas. The broadest region of likely auditory sensitivity (the combination of the light and dark shaded areas) is constructed from the end points of two horizontal lines, one plotted at Best Threshold with end points defined by the lowest and highest Best Frequencies predicted by either middle-ear area (B and C of Figure 7), and the second plotted at 30 dB above Best Threshold with end points defined by the lowest Low-Frequency Limit and highest-High Frequency Limit consistent with either middle-ear area (A and D of Figure 7). The narrower region of likely auditory sensitivity (the darker area) is constructed from similar line segments defined by the lowest and highest audiometric limits that are common to both predictions (a, b, c, and d of Figure 29.7).

predictions are also of somewhat lower frequency, but there is overlap between the two predicted ranges for each limit.

The frequency extremes predicted by the separate regression analyses of tympanic membrane and footplate areas can be used to define likely regions of auditory sensitivity. Two overlapping regions can be constructed; a broader region based on the extremes of the frequency limits predicted by either analysis (the frequencies labeled by upper-case letters in Figure 29.7), and a narrower region, contained within the first, based on the extreme frequency limits that are common to predictions from both the tympanic-membrane and footplate area analysis (the frequencies labeled by lower-case letters in Figure 29.7). The likely auditory regions of *Morganucodon* are illustrated in Figure 29.8, where the details of construction of the regions are included in the figure caption. Since our analyses ignored absolute auditory

threshold, the ordinate of the predicted regions is dB re Best Threshold.

The most likely region for auditory sensitivity in *Morganucodon* suggests a best frequency between 6 and 30 kHz, and low- and high-frequency limits of greater than 1 kHz and less than 100 kHz respectively. Not surprisingly, these limits are much like those of several small mammals used in the analysis which have similar middle-ear areas, and the shapes of the audiograms of two such animals (mouse and rat) fit within the likely auditory region for *Morganucodon* (Figure 29.9A).

5.2 Tests of the Prediction Procedure

Before accepting the likely region of auditory sensitivity predicted for *Morganucodon*, legitimate concerns regarding the power and specificity of the prediction procedure need to be addressed. One method of estimating the strength of the procedure is to test predictions against known facts. Two mammals with known middle-ear dimensions, that were not included in the population used to define the prediction procedures, are the monotremes *Ornithorhynchus anatinus* — the platypus, and *Tachyglossus aculeatus* — the echidna. The presumptively "primitive" middle-ears of these animals (Fleischer 1973; 1978) make them particularly attractive test examples. Although no behavioral audiograms have been reported for these animals, the cochlear potential produced by the platypus ear has been measured by Gates et al. (1974) and Aitkin and Johnstone (1972) have measured the velocity of the echidna stapes in response to sound. Both of these nonbehavioral measures can be used to estimate the audiogram shape for these animals (Dallos 1973).

Comparisons of the shape of the estimated audiograms for echidna and platypus with the likely auditory regions predicted from the areas of their tympanic membrane and footplate (Fig. 9B,C), illustrate that the likely auditory region does not predict the shape of the audiogram, but does a reasonable job of predicting the frequency region in which greatest auditory sensitivity occurs. The estimated audiograms do fit within the likely auditory regions, but in both the echidna and platypus the best frequency is off to one side of the likely auditory region.

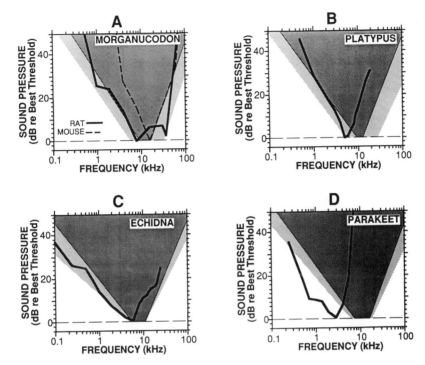

FIGURE 29.9. Regions of likely auditory sensitivity predicted from the dimensions of several middle ears. (A) A comparison of the regions of likely auditory sensitivity predicted for *Morganucodon* and the shapes of the audiograms of *Rattus norvegicus* — the rat (Kelly and Masterton 1980) and *Mus musculus* — the mouse (Heffner and Masterton 1980). (B) A comparison of the region of likely auditory sensitivity predicted for the platypus with an estimate of the shape of its audiogram. The audiogram shape was defined by the cochlear potential measurements of Gates et al. (1974). (C) A comparison of the region of likely auditory sensitivity for the echidna with an estimate of the shape of its audiogram. The audiogram is defined by the inverse of isopressure velocity contours of the echidna stapes footplate. (Aitkin and Johnstone 1972.) (D) A comparison of the region of likely auditory sensitivity predicted for the parakeet (from the mammalian data) with the shape of its audiogram (Dooling and Saunders 1975.)

The question of the specificity of the predicted regions was investigated by comparing the known audiogram of the parakeet (Dooling and Saunders 1975) with the likely auditory region predicted from measurements of parakeet middle-ear dimensions (Saunders 1985) and the relationships of Table 29.2 (Fig. 29.9D). The mismatch between the parakeet audiogram and the predicted likely auditory region is to be expected. Although birds and reptiles have middle-ear dimensions that are similar to those of small mammals, the relationships between audiometric limits and middle ear areas in birds and reptiles are different from those of mammals (Rosowski and Graybeal 1991). In any case, the mismatch in Figure 29.9D is indicative of a specificity in the prediction procedure for the ears of mammals.

5.3 Hearing in *Morganucodon*

Our analysis reveals that the *Morganucodon* middle ear is of a size and structure which are consistent with maximum sensitivity to sounds of moderate to high frequency (6 to 30 kHz). This finding is contrary to hypotheses that these animals had "reptile-like" ears with primarily low-frequency hearing abilities but supports other arguments that suggest a role for high-frequency hearing in small primitive mammals (Masterton et al. 1969; Jerison 1973). It should be remembered that our analysis does not predict the absolute sensitivity of an ear, but only predicts the relative sensitivity of an ear to sounds of different frequencies.

6. Discussion

6.1 The Influence of the Inner Ear

As has been pointed out by Manley (1971; 1972b) and Kermack and Musset (1983) the auditory capabilities of vertebrates are dependent on the combined response of the entire auditory system. A middle ear which is capable of transmitting high-frequency sound would be of no benefit if the inner ear were insensitive to such sounds. Little is known of the structure of the inner ear in *Morganucodon*. The best available information on this issue comes from petrosal bones serially sectioned and reconstructed with the aid of a computer (Graybeal et al. 1989). The reconstructed cochlear fluid spaces are short (≈ 2.5 mm in length), tapered, slightly curved tubes. There are no clearly demarcated scalae in the sectioned material, with only a hint of a spiral lamina and all soft tissue is lost. Could such a short cochlea have been sensitive to high frequency sounds? One thought has been that long cochleas are required for sensitivity to higher frequencies (Kermack and Musset 1983), however this notion is not supported by the evidence. West (1985) and Rosowski and Graybeal (1991) investigated the relationships between cochlear length and audiometric limits within mammals using anatomic and behavioral data available from the literature. One result of these studies was the definition of a highly significant ($p < 1\%$) inverse relationship between basilar-membrane length and high-frequency hearing (Table 29.2) such that the mammals with the shortest basilar membrane were most sensitive to high frequency sounds. Therefore, there is no reason to think that the short length of the *Morganucodon* cochlea placed limits on high-frequency hearing. It is still possible however, that cochlear length is positively correlated with the bandwidth or extent of the audible range.

Since there are no clear indications of the width of the *Morganucodon* basilar membrane, suggested inverse correlations between membrane width and high-frequency hearing (Fleischer 1973; Pye and Hinchcliffe 1976; Ketten 1984, Chapter 35) cannot be utilized to suggest the relative sensitivity of the inner ear to different sound frequencies.[6]

6.2 Absolute Hearing Sensitivity

The analysis presented in this chapter has concentrated on the relative auditory sensitivity for sounds of different frequency and has ignored differences in the absolute threshold of hearing. The conclusion that *Morganucodon* probably was more sensitive to sounds of higher frequency says nothing about the overall sensitivity of it's ear to sounds of varying level. A similar correlation analysis of ear dimensions and auditory behavior (Rosowski and Graybeal 1991) suggested that there was no correlation between auditory structures and sensitivity. On the other hand, measurements of behavioral responses, cochlear potentials, and middle-ear velocities in what might be considered "primitive" mammalian ears such as opossum, and the monotremes (Masterton et al. 1969; Aitkin and Johnstone 1972; Gates et al. 1974) suggest that these animals have higher auditory thresholds than "more advanced" small mammals such as mouse and rat. It is likely then, that *Morganucodon* was also *relatively* insensitive to sound.

6.3 Mechanisms for the Observed Relationships Between Middle-Ear Structure and Hearing Function

Although I have suggested several relationships between middle-ear structure and auditory function, the mechanisms that contribute to those relationships have not been well defined. Potential interactions between the dimensions of the tympanic membrane and sound wave length (Khanna and Tonndorf 1978) are intriguing but need further elucidation before they can be accepted as fact.

---

[6]Another suggestion in the literature is that the high-frequency limit of hearing in vertebrates is dependent on a dimensionless ratio, the "membrane value." Manley (1971, 1972b) describes this membrane value as the product of a shape factor (length/widest width) and a second factor defined by the ratio of the narrowest and widest basilar membrane width. The attractiveness of this scheme was that a single number could be used to correlate basilar-membrane dimensions and high-frequency hearing limit. However, newer data unavailable in 1972 suggest that the hypothesized relationship is much more variable, e.g., humans and mice (Ehret and Frankenreiter 1977) have similar membrane values, but greatly different high-frequency hearing capabilities.

TABLE 29.3. Summary of measurements of the middle-ear stiffness of terrestrial vertebrates.

Species	Common name	Specific acoustic stiffness[a] (Pa-m^{-1})	Area of pars tensa of tympanic membrane (mm^2)	References
Mammalia				
Rhinolophus ferrumequinum	Horse-shoe bat	4.2×10^8	2.0^b	Wilson and Bruns '83; Fleischer '73
Tachyglossus asculeatus	Echidna	2.9×10^8	7.1	Aitkin and Johnstone '72; Fleischer '73
Mus musculus	Mouse	9.5×10^7	2.8	Saunders and Summers '82
Eptesicus pumilis	A bat	6.3×10^7	–	Manley, Irvine and Johnstone '72
Mesocricetus auratus	Golden hamster	5.7×10^7	7.0	Relkin and Saunders '80; Stephens '72
Meriones unguiculatus	Gerbil	2.0×10^{7c}	26	Ravicz '90; Lay '72
Homo sapiens	Human	2.0×10^7	60	Goode et al. '89; Wever and Lawrence '54
Cavia procellus	Guinea pig	1.9×10^7	25	Manley and Johnstone '74; Fleischer '73
Felis catus	Cat	1.3×10^7	36	Buunen and Vlaming '81; Fleischer '73
Chinchilla laniger	Chinchilla	4.3×10^6	71	Ruggero et al. '90; Fleischer '73
Aves				
Melopsittacus undulatus	Parakeet	5.7×10^6	8.7	Saunders '85
Columbia livia	Pigeon	5.4×10^6	18	Gummer et al. '89; Saunders '85
Gallus domesticus	Chicken	3.9×10^{6c}	33	Saunders et al. '86; Saunders, '85
Reptilia				
Gerrhonotus multicarinatus	Southern alligator lizard	5.5×10^6	7.4	Rosowski et al. '85; Ketten, unpublished
Gekko gecko	Tokay gecko	1.8×10^6	48	Manley, '72a; Wever, '78

[a]Stiffness was defined from some low-frequency (less than or equal to 1 kHz) measurements of velocity or displacement of the tympanic membrane.
[b]The tympanic-membrane area of *Rhinolophus simulator*.
[c]For the gerbil and chicken, stiffness is defined by the product of the acoustic stiffness and the tympanic-membrane area.

Furthermore, the importance of the dimensions of the external-ear to sound-power collection (Shaw 1974; Khanna and Tonndorf 1978; Shaw and Stinson 1980; Rosowski et al. 1986; 1988), which was ignored in this chapter, needs to be considered before one can fully understand the relationship between ear structure and hearing capabilities.

A second type of mechanism which is alluded to in the previous sections concerns the effect of the mechanical properties of the middle ear, specifically the stiffness of the ossicular chain, on hearing capabilities. Fleischer (1973; 1978) has suggested both that the gonial-induced increased stiffness of the microtype middle ear acts to limit the ear's sensitivity to low-frequency sounds, and that ears with flaccid or freely mobile ossicular chains are sensitive to low-frequency sounds. The data of Table 29.1 and Figure 29.5 support these ideas. Can a difference in stiffness between the freely mobile middle ears of bird-reptiles and the microtype middle ears of small mammals explain differences in hearing function?

Table 29.3 illustrates the specific acoustic stiffness estimated from low-frequency measurements of ossicular or tympanic-membrane motion in ten mammal, three bird and two reptile species. Five of the six ears with the lowest stiffness are from birds and reptiles; only the chinchilla (a mammal with documented good low-frequency hearing) has a comparable middle-ear stiffness. Other mammalian ears have stiffnesses which are 2.5 to 75 times greater than the stiffest birds-reptile ear measured. These differences are consistent with the idea that the stiffness of the middle ear is one factor which differentiates the middle ears of mammals from those of other amniotes. The four microtype ears in the population (those of the two bats, the echidna and mouse) are at least 10 times more stiff than the bird and reptile middle ears, and there is a suggestion of an inverse relationship between stiffness and tympanic-membrane area. These findings are consistent with Fleischer's suggestion that stiffness and small size are related.

6.4 Effect of Head Size

Several authors (Masterton et al. 1969; Heffner and Masterton 1980; Heffner and Heffner, Chapter 34) have suggested a relationship between high-frequency hearing and head size in mammals. Briefly, the acoustic interaural differences necessary for mammals to localize sources of sound in space are a function of the ratio of head size to sound wave length. Head to wavelength ratios of one or greater are required to produce the interaural intensity differences necessary for sound localization, and the smaller the head, the higher the sound frequency necessary for interaural differences. This dependence of sound localization, a necessary auditory function for many nocturnal mammals, on head size is thought to be responsible for a highly significant relationship observed between interaural distance and high-frequency limit of hearing in extant mammals (Heffner and Heffner Chapter 34). Extant mammals with interaural distances on the order of that of *Morganucodon* (1 to 2 cm) possess high-frequency hearing limits of greater than 80 kHz. These limits are within the likely region of auditory sensitivity that is predicted by our analysis.[7]

---

[7]Differences in the criterion sensation level are a small complication in comparing the high-frequency limits compiled by Heffner and Heffner (Chapter 34) and those used in this analysis. Heffner and Heffner define the high-frequency limit to be the frequency where the audiogram crosses an isopressure contour at a pressure equivalent to 60 dB re Best threshold.

6.5 Speculations About the Niche Filled by *Morganucodon* and Other Early Mammals

The first mammals differentiated from their mammal-like reptile ancestors about 200 million years ago in the late Triassic. The dominant land animals at this time were the archosaurs. It has been speculated (e.g., Jerison 1973; Kermack and Kermack 1984) that intense competition between archosaurs and mammal-like reptiles drove the later into nocturnal niches, which demanded the development of improved smell and hearing senses in order to locate prey and other forms of nourishment in the night. The teeth of *Morganucodon* suggest it was an insectivore, and high-frequency hearing might have been useful in locating stridulating insects. A key question is when did the improvement in hearing occur. Kermack and Musset (1983) have suggested that high-frequency hearing required the development of a coiled cochlea which did not occur until late in the Jurassic. The view of the *Morganucodon* middle ear presented in this paper, and the suggestion that a small, uncoiled cochlea could be sensitive to high-frequency sound suggest that even the earliest of mammals might have had a hearing sense that was broadly "tuned" to sounds of higher frequencies.

7. Summary and Conclusions

The currently available data on the auditory anatomy of transitional mammals and the known relationships between auditory structure and hearing capabilities lead to the conclusion that the earliest mammals had auditory capabilities which were similar in frequency range to those of many modern small mammals, with best sensitivity to high-frequency sounds, although the ear of *Morganucodon* may not have been as sensitive to sound as those of rat and mouse. This conclusion is consistent with both the suggestion of Masterton et al. (1969) that high-frequency hearing is a primitive mammalian trait, and Jerison's (1973) hypothesis that the early mammals existed within nocturnal niches which favor high-frequency hearing abilities. Indeed, since high-frequency hearing and others of the suite of traits which distinguish mammals from reptiles (e.g., homeothermy and the mammalian jaw and palate) are essential characteristics of modern small nocturnal insectivorous mammals, the appearance of all of these traits at the reptile-mammal transition is not surprising.

The conclusions concerning the hearing capabilities of early mammals can be tested by several means, including: (1) Further description of the anatomy of the inner and middle ears of early mammals might enable realistic estimates of the width of the basilar membrane and the stiffness of the ossicular connections that could either support or refute this conclusion. (2) More measurements of the middle-ear and basilar-membrane dimensions of the many small mammals with documented hearing abilities (e.g., Heffner and Heffner, Chapter 34) could be used to further test the relationships used to predict the likely auditory region of *Morganucodon*. (3) A more thorough understanding of the mechanics of the middle ears of modern small mammals is also needed.

Acknowledgments. This work could not have been performed without the aid and support of N.Y.S. Kiang, W.T. Peake and the rest of the staff of the Eaton-Peabody Laboratory especially A. Graybeal. Special thanks go to A.W. Crompton of the Harvard Museum of Comparative Zoology who was the early driving force to the investigation of how early mammals may have heard. This work was supported by grants from the National Institute of Health (NIH-RO1 DC 00194 and NIH PO1 DC00119.)

References

Aitkin LM, Johnstone BM (1972) Middle-ear function in a monotreme: the echidna (*Tachyglossus aculeatus*). J Exp Zool 180:245–250.

Allin EF (1975) Evolution of the mammalian middle ear. Journal of Morphology 147:403–437.

Allin EF (1986) The auditory apparatus of advanced mammal-like reptiles and early mammals. In: Hotton N III, MacLean PDM, Roth JJ, Roth EC (eds) The Ecology and Biology of Mammal-like Reptiles. Washington, DC: Smithsonian Institution Press, pp. 283–294.

Beranek LL (1954) Acoustics. New York: McGraw-Hill.

Buunen TJF, Vlaming MSMG (1981) Laser-Doppler velocity meter applied to tympanic membrane vibrations in cat. J Acoust Soc Am 69:744–750.

Crompton AW, Jenkins Jr FA (1973) Mammals from reptiles: A review of mammalian origins. Annual Rev Earth Planet Sci 1:131–153.

Crompton AW, Jenkins Jr FA (1979) Origin of mammals. In: Lillegraven JA, Kielan-Jaworowska Z, Clemens WA (eds) Mesozoic Mammals: The First Two-Thirds of Mammalian History. Berkeley, CA: University of California Press, pp. 59–73.

Crompton AW, Parker P (1978) Evolution of the mammalian masticatory apparatus. Am Sci 66:192–201.

Dallos P (1970) Low frequency auditory characteristics: species dependence. J Acoust Soc Am 48:489–499.

Dallos P (1973) The Auditory Periphery. New York: Academic Press.

Dallos P (1984) Peripheral mechanisms of hearing. In: Darian-Smith, I (ed) Handbook of Physiology, Section 1: the Nervous System, Volume III Sensory Processes, Part 2. Bethesda, MD: American Physiological Society, pp. 595–638.

Dooling RJ (1980) Behavior and psychophysics of hearing in birds. In: Popper AN, Fay RR (eds) Comparative Studies of Hearing in Vertebrates. New York: Springer-Verlag, pp. 261–288.

Dooling RJ, Saunders JC (1975) Hearing in the parakeet (*Melopsittacus undulatus*): Absolute thresholds, critical ratios, frequency-difference limens and vocalizations. J Comp Physiol Psychol 88:1–20.

Ehret G, Frankenreiter M (1977) Quantitative analysis of cochlear structures in the house mouse in relation to mechanics of acoustical information processing. J Comp Physiol (A) 122:65–85.

Fay RR (1988) Hearing in Vertebrates: a Psychophysics Databook. Winnetka, Illinois: Hill-Fay Associates.

Fleischer G (1973) Studien am Skelett des Gehörorgans der Säugetiere, einschliesslich des Menschen. Säugetierkundl. Mitteilungen (München) 21:131–239.

Fleischer G (1978) Evolutionary principles of the mammalian middle ear. Adv Anat Embryol Cell Biol 55:3–69.

Gates GR, Saunders JC, Bock GR, Aitkin LM, Elliott MA (1974) Peripheral auditory function in the platypus, *Ornithorhynchus anatinus*. J Acoust Soc Am 56:152–156.

Goode RL, Nakamura K, Gyo K, Aritomo H (1989) Comments on "Acoustic transfer characteristics in human middle ears studied by a SQUID magnitometer method." J Acoust Soc Am 86:2446–2449.

Graybeal A, Rosowski JJ, Ketten DR, Crompton AW (1989) Inner-ear structure in *Morganucodon*, an early Jurassic mammal. Zool J Linnean Soc 96:107–117.

Gummer AW, Smolders JWT, Klinke R (1989) Mechanics of a single-ossicle ear: I. The extra-stapedius of the pigeon. Hearing Res 39:1–14.

Heffner HE, Masterton RB (1980) Hearing in Glires: domestic rabbit, cotton rat, feral house mouse and kangaroo rat. J Acoust Soc Am 68:1584–1599.

Henson OW (1974) Comparative anatomy of the middle ear. In: Keidel WD, Neff WD (eds) Handbook of Sensory Physiology: The Auditory System V/1, New York: Springer-Verlag, pp. 39–110.

Hopson JA (1966) The origin of the mammalian middle ear. Am Zoologist 6:437–450.

Hunt RM, Korth WW (1980) The auditory region of Dermoptera: morphology and function relative to other living mammals. J Morphol 164:167–211.

Jenkins Jr FA, Crompton WA (1979) Triconodonta. In: Lillegraven JA, Kielan-Jaworowska Z, Clemens WA (eds) Mesozoic Mammals: The First Two-Thirds of Mammalian History. CA: University of California Press, Berkeley, pp. 74–90.

Jerison HJ (1973) Evolution of Brain and Intelligence. New York: Academic Press.

Kelly JB, Masterton RB (1980) Auditory sensitivity of the albino rat. J Comp Physiol Psychol 91:930–936.

Kermack DM, Kermack KA (1984) The Evolution of Mammalian Characters. London: Croom Helm.

Kermack KA, Musset F (1983) The ear in mammal-like reptiles and early mammals. Acta Palaeontol Polon 28:148–158.

Kermack KA, Musset F, Rigney HW (1973) The lower jaw of *Morganucodon*. Zool J Linnean Soc 53:87–175.

Kermack KA, Musset F, Rigney HW (1981) The skull of *Morganucodon*. Zool J Linnean Soc 71:1–158.

Ketten DR (1984) Correlations of morphology with frequency for Odontocete cochlea: systematics and topology. PhD Thesis, The Johns Hopkins University, Baltimore.

Khanna SM, Tonndorf J (1972) Tympanic membrane vibrations in cats studied by time-average holography. J Acous Soc Am 51:1904–1920.

Khanna SM, Tonndorf J (1978) Physical and physiological principles controlling auditory sensitivity in primates. In: Noback R (ed) Neurobiology of Primates. New York: Plenum Press, pp. 23–52.

Killion MC, Dallos P (1979) Impedance matching by the combined effects of the outer and middle ear. J Acoust Soc Am 66:599–602.

Kirikae I (1960) The Structure and Function of the Middle Ear. University of Tokyo Press.

Lay DM (1972) The anatomy, physiology, functional significance and evolution of specialized hearing organs of Gerbilline rodents. J Morphol 138:41–120.

Lynch TJ III, Nedzelnitsky V, Peake WT (1982) Input impedance of the cochlea in cat. J Acoust Soc Am 72:108–130.

Manley GA (1971) Some aspects of the evolution of hearing in vertebrates. Nature 230:506–509.

Manley GA (1972a) The middle ear of the Tokay Gecko. J Comp Physiol 81:239–250.

Manley GA (1972b) A review of some current concepts of the functional evolution of the ear. Evolution 26:608–621.

Manley GA (1981) A review of the auditory physiology of reptiles. Prog Sens Physiol 2:49–134.

Manley GA, Johnstone BM (1974) Middle-ear function in the guinea pig. J Acoust Soc Am 56:571–576.

Manley GA, Irvine DRF, Johnstone BM (1972) Frequency response of bat tympanic membrane. Nature 237:112–113.

Masterton B, Heffner H, Ravizza R (1969) The evolution of human hearing. J Acoust Soc Am 45:966–985.

Parrington FR (1949) Remarks on a theory of evolution of the tetrapod middle ear. J Laryngol Otol 63:580–595.

Parrington FR (1979) The evolution of the mammalian middle and outer ears: A personal view. Biolog Rev 54:369–387.

Peterson EA, Levison M, Lovett S, Feng A, Dunn SH (1974) The relation between middle ear morphology and peripheral auditory function in rodents, I: Sciuridae. J Audit Res 14:227–242.

Pye A, Hinchcliffe R (1976) The comparative anatomy of the ear. In: Hinchcliffe R, Harrison D (eds) Sci Found Otolaryngol, pp. 184–202.

Ravicz ME (1990) Acoustic impedance of the gerbil ear. MS Thesis, Boston University, Boston MA.

Relkin EM, Saunders JC (1980) Displacement of the malleus in neonatal golden hamsters. Acta Otolaryngol 90:6–15.

Rosowski JJ, Carney LH, Lynch TJ III, Peake WT (1986) The effectiveness of the external and middle ears in coupling acoustic power into the cochlea. In: Allen JB, Hall JL, Hubbard A, Neely ST, Tubis A (eds) Peripheral Auditory Mechanisms, New York: Springer-Verlag, pp. 3–12.

Rosowski JJ, Carney LH, Peake WT (1988) The radiation impedance of the external ear of cat: Measurements and applications. Journal of Acoustical Society of America 84:1695–1708.

Rosowski JJ, Graybeal A (1991) What did *Morganucodon* hear? Zool J Linnean Soc 101:131–168.

Rosowski JJ, Peake WT, Lynch TJ III, Leong R, Weiss TF (1985) A model for signal transmission in an ear having free-standing stereocilia. II. Macromechanical stage. Hearing Research 20:139–155.

Ruggero MA, Rich NC, Robles L, Shivapuja BG (1990) Middle ear response in the chinchilla and its relationship to mechanics at the base of cochlea. J Acoust Soc Am 87:1612–1629.

Saunders JC (1985) Auditory structure and function in the bird middle ear: An evaluation by SEM and capacitive probe. Hearing Res 18:253–268.

Saunders JC, Johnstone BM (1972) A comparative analysis of middle-ear function in nonmammalian vertebrates. Acta Otolaryngol 73:353–361.

Saunders JC, Relkin ER, Rosowski JJ, Bahl C (1986) Changes in middle-ear input admittance during postnatal auditory development in chicks. Hearing Res 24:227–235.

Saunders JC, Summers RM (1982) Auditory structure and function in the mouse middle ear: An evaluation by SEM and capacitive probe. J Comp Physiol A 146: 517–525.

Shaw EAG (1974) The external ear. In: Keidel WD, Neff WD (eds) Handbook of Sensory Physiology, Vol V/1, Auditory System. New York: Springer-Verlag, pp. 455–490.

Shaw EAG, Stinson MR (1980) The human external and middle ear: Models and concepts. In: deBoer E, Viergever MA (eds) Mechanics of Hearing. Delft University Press, pp. 3–10.

Stebbins WC (1970) Animal Psychophysics: The Design and Conduct of Sensory Experiments. New York: Appleton-Century-Crofts.

Stephens CB (1972) Development of the middle and inner ear in the Golden hamster (*Mesocricetus auratus*): A detailed description to establish a norm for physiological study of congenital deafness. Act Otolaryngologica Supplementum Number 296: pp. 1–51.

Tonndorf J, Khanna SM (1972) Tympanic-membrane vibrations in human cadaver ears studied by time-averaged holography. J Acoust Soc Am 52:1221–1233.

Tumarkin A (1955) On the evolution of the auditory conducting apparatus: a new theory based on functional considerations. Evolution 9:221–242.

Tumarkin A (1968) Evolution of the auditory conducting apparatus in terrestrial vertebrates. In: deReuck AVS, Knight J (eds) Ciba Foundation Symposium on Hearing Mechanisms in Vertebrates. London: Churchill, pp. 18–37.

Turner RG (1980) Physiology and Bioacoustics in reptiles. In: Popper AN, Fay RR (eds) Comparative Studies of Hearing in Vertebrates. New York: Springer-Verlag, pp. 205–237.

West CD (1985) The relationship of the spiral turns of the cochlea and the length of the basilar membrane to the range of audible frequencies in ground dwelling mammals. J Acoust Soc Am 77:1091–1101.

Wever EG (1978) The Reptile Ear, Princeton University Press.

Wever EG, Lawrence M (1954) Physiological Acoustics, Princeton University Press.

Wilson JP, Bruns V (1983) Middle-ear mechanics in the CF-bat *Rhinolophus ferrumequinum*. Hearing Research 10:1–13.

30
Parallel Evolution of Low-Frequency Sensitivity in Old World and New World Desert Rodents

Douglas B. Webster and Wolfgang Plassmann

1. Introduction

Parallel evolution occurs when separate groups of organisms, inhabiting similar environments, independently evolve similar adaptations (Simpson 1967). The degree of similarity between these adaptations varies widely, but tends to be greater in extreme environments—such as deserts—where selective pressures are most harsh.

We have studied auditory adaptations in desert rodents for a number of years. The heteromyid rodents (kangaroo rats, *Dipodomys*, and their allies) of the new world deserts, and the gerbilline rodents (gerbils and their allies) of the old world deserts, have independently evolved many similar anatomical, physiological, and behavioral modifications, which have adapted them to survive in their harsh environments.

Deserts are characterized by aridity, sparse food supply, hot daytime temperatures, and little natural cover. In adapting to these conditions, both the new world heteromyids and the old world gerbillines have evolved nocturnal activity: they spend the day in underground burrows and forage and feed at night (Monson and Kessler 1940). By allowing them to avoid heat and sun, this behavior, combined with dietary and some physiological adaptations, enables these animals to survive on metabolic water; they require no free water (Schmidt-Nielsen and Schmidt-Nielsen 1951). Both groups also have evolved primarily bipedal locomotion (Howell 1932).

These groups also exhibit many parallel auditory adaptations. Most striking are the enlarged middle ear cavities, large tympanic membranes, and freely mobile ossicular formations (Howell 1932).

Enlarged middle ears have also evolved in other desert mammals—for instance the springhaas (*Pedetes*) and some elephant shrews (Howell 1932; Segall 1970) and, as we know from the fossil record, in some desert-inhabiting marsupials of South America (Segall 1970).

The heteromyids and gerbillines also have unusual cochlear structures, especially in the apical region (Webster 1961; Lay 1972; Plassmann, Peetz, and Schmidt 1987), and a sensitivity to low frequency sounds which is not often found in small mammals.

The fact that these modifications evolved independently in two different desert rodent groups suggests that they are adaptations to the desert environment. However, other successful desert groups do not have these modifications. For example, desert ground squirrels, grasshopper mice, and deer mice—all of which are sympatric with kangaroo rats and their allies (Oaks 1967)—have adapted to the environment in other ways.

2. Middle Ear Differences Between Heteromyids and Gerbils

Middle ear modifications vary considerably within both the new world heteromyids and the old world gerbillines. In general, the most conservative middle ears are found in the species which live in the least arid environments, and the most extreme or derived middle ears are found in those living in the most extreme desert regions.

Within the heteromyids, the spiny pocket mice (*Heteromys* and *Liomys*) are the most conservative; the pocket mice (*Perognathus*) are intermediate; the kangaroo rats (*Dipodomys*) are more derived; and the kangaroo mice (*Microdipodops*) are the most extreme in their adaptations (Webster and Webster 1975).

Within the much larger and even more diverse gerbilline rodents, middle ear structures also vary from conservative to extremely derived; again, the most conservative forms inhabit the least arid regions and the most derived forms inhabit the most arid regions. For instance, one of the more conservative gerbils is *Tatera*; *Meriones* is somewhat intermediate; *Psammomys* is more derived; *Pachyuromys* is extremely modified (Legouix, Petter, and Wisner 1954; Legouix and Wisner 1955; Oaks 1967; Lay 1972).

To explore the extent of parallelism between these two large groups, we will examine the most derived heteromyid form, the kangaroo mouse (*Microdipodops*), and the most derived gerbilline, the fat-tailed rat (*Pachyuromys*).

Both have extremely large middle ear cavities, with the combined right and left volume exceeding the volume of the cranial cavity. But the enlargement does not occur equally in all elements of the middle ear. In the kangaroo mouse the greatest middle ear inflation is in the epitympanum; next largest is the antrum; smallest is the hypotympanum. In the fat-tailed rat, the greatest inflation is in the hypotympanum; next largest is the antrum; smallest is the epitympanum. Thus, although extremely large middle ear cavities have evolved in both forms, they have done so by different "strategies."

Both *Microdipodops* and *Pachyuromys* have extremely large tympanic membranes for such small rodents. But here, too, there is a dramatic difference between the two genera. In *Microdipodops*, as in all heteromyids, the tympanic membrane is composed entirely of pars tensa and there is essentially no pars flaccida (Webster and Webster 1975). In *Pachyuromys*, although the pars tensa is large, there is also a large and prominent pars flaccida. This is true of many other, but not all, gerbils (Lay 1972).

Although most rodents have a microtype ossicular system (Fleischer 1978), both *Microdipodops* and *Pachyuromys* have a freely mobile ossicular system (Howell 1932; Lay 1972). In this system,

the manubrium of the malleus is roughly parallel to the long process of the incus, and the anterior process of the malleus is not firmly attached to the tympanic annulus. Both *Pachyuromys* and *Microdipodops* also have a large lever ratio, as well as a large areal ratio between the tympanic membrane and the footplate of the stapes. Both these adaptations facilitate a mechanically efficient impedance matching system from the tympanic membrane to the oval window.

The freely mobile ossicular system occurs in all heteromyid rodents, including the neotropical, more conservative genera *Heteromys* and *Liomys* (Webster and Webster 1975). There is greater variation among the gerbilline rodents, where forms such as *Tatera* have a transitional ossicular system (Oaks 1967).

It is therefore evident that these two geographically separated groups have each evolved highly derived large middle ears, large tympanic membranes, and freely mobile ossicular systems with large lever ratios; but that they have done so in quite different ways. This is as we would expect in independently evolving groups which experience similar selective pressures but which would not be expected to have the same mutations in their gene pools.

3. Inner Ear Differences Between Heteromyid and Gerbilline Rodents

Since the middle ear works in concert with the inner ear, it is not surprising that striking cochlear modifications have also evolved in both *Microdipodops* and *Pachyuromys*. Interestingly, there is more similarity between the two genera in their cochlear modifications than in their middle ear modifications. In neither genus is the basilar membrane particularly long: it is 8.7 mm in *Microdipodops* and 11.7 mm in *Pachyuromys* (Webster and Webster 1977; Plassmann, Peetz, and Schmidt 1987). However, in each, it widens rapidly in the first turn from its narrow base and then remains wide throughout the rest of the cochlea.

In each genus there is a prominent concentration of hyaline material ("cottony ground substance") within the zona pectinata of the basilar membrane (Webster and Webster 1977; Lay 1972; Plassmann, Peetz, and Schmidt 1987). This material is sparse in the narrow basal portion of the cochlea, becoming extremely large and prominent throughout the rest of the cochlea where the basilar membrane is wider. Although conspicuous in *Pachyuromys*, it is even more so in *Microdipodops*.

The border cells of Hensen deserve special mention. In *Pachyuromys* they are larger than in other mammals, but do not differ morphologically from "normal" Hensen's cells (Lay 1972; Plassmann, Peetz, and Schmidt 1987). However, in *Microdipodops* they are most unusual, especially in the middle and apical turns. They are flask-shaped, and rest on a cup formed by the innermost Claudius cell. Each cell has a single cytoplasmic process which extends toward the scala media and then expands into an umbrella-shaped canopy, the edges of which form tight junctions with the canopy of adjacent Hensen's cells (Webster and Webster 1977).

Within the gerbillines, there is a general correlation between larger Hensen's cells and more arid environments (Lay 1972; Plassmann, Peetz, and Schmidt 1987). The situation is similar within the heteromyids, where Hensen's cells are unspecialized in the neotropical forms but fully developed and specialized in the desert forms, including *Perognathus* and *Dipodomys* as well as *Microdipodops* (Webster and Webster 1977).

In summary, the cochleae of both groups have similar basilar membrane lengths and widths as well as increased cottony ground substance in the pars pectinata. Both groups have enlarged Hensen's cells, but in *Microdipodops* (as in most heteromyids) the Hensen's cells are also qualitatively different than in other mammals.

4. Low-Frequency Hearing and Its Adaptive Value in Heteromyids and Gerbillines

Both heteromyid and gerbilline rodents have unusually sensitive low-frequency hearing for such small rodents (Webster and Webster 1972; Ryan 1976; Heffner and Masterton 1980), although this is usually in addition to, rather than at the expense of, the high-frequency hearing their size would suggest. Two adaptations facilitate this low-frequency hearing in both groups: large middle ear cavities, which provide large air cushions behind

large tympanic membranes; and freely mobile ossicular systems with large lever and areal ratios and no fusion of the anterior process of the malleus with the annulus. The wide basilar membrane in both genera is also an obvious adaptation for low frequencies.

The possible roles of the modified border cells and of the hyaline material within the basilar membrane are not so obvious. Since the vibratory medium of the inner ear is fluid rather than air, these modifications do not add substantially to the mass of the cochlear partition. They may play a mechanical role in facilitating the movement of the basilar membrane, and/or the shearing action between the tectorial membrane and the stereocilia of the hair cells.

Since it is logical to conclude that the unusual auditory modifications in these two genera promote their unusual low frequency sensitivity, it is also logical to ask: What is the adaptive value of such low frequency sensitivity?

On a theoretical level, one could point to the scarcity of natural cover in their arid environment, their need to travel considerable distances to forage for food, and their nocturnal behavior. In such an environment there would be significant selective value to sensory systems that allow an animal to detect the presence of predators in the dark. There are also direct data to support this. Experiments have shown that at least some gerbils and kangaroo rats can avoid predation by owls and snakes except when their middle ear volumes are reduced — and, in kangaroo rats, when very low light levels deprive them visually as well (Webster and Webster 1971; Lay 1974). Being able to hear the low frequency sounds produced by both snakes and owls just prior to striking (Webster 1962) — and thus to avoid their predation — would seem in these two cases to be the primary selective pressure that favored low-frequency sensitivity.

5. Summary

In both these independently evolving genera, *Microdipodops* and *Pachyuromys*, there has been significant selective pressure for low-frequency hearing. In response, both have evolved greatly enlarged middle ear cavities and tympanic membranes, freely mobile ossicular systems, a wide (and rapidly widening) basilar membrane, a hyaline mass in the zona pectinata, and modified Hensen's cells. The specific evolution of these features is considerably different in the two genera, however. Parallel evolution always occurs through the mutations randomly available within the population, and while it often arrives at the same destination in different groups, it does so by different means.

References

Fleischer G (1978) Evolutionary principles of the mammalian middle ear. Adv Anat Embryol Cell Biol 55:3–70.

Heffner H, Masterton RB (1980) Hearing in glires: Domestic rabbit, cotton rat, feral house mouse, and kangaroo rat. J Acoust Soc Am 68:1584–1599.

Howell AB (1932) The saltatorial rodent *Dipodomys*: The functional and comparative anatomy of its muscular and osseous systems. Proc Am Acad Arts Sci 67:377–536.

Lay DM (1972) The anatomy, physiology, functional significance and evolution of specialized hearing organs of gerbilline rodents. J Morph 138:41–120.

Lay DM (1974) Differential predation on gerbils (*Meriones*) by the little owl (*Athene brahma*). J Mamm 55:608–614.

Legouix JP, Petter F, Wisner A (1954) Étude de l'audition chez des mammières à bulles tympaniques hypertrophiées. Mammalia 18:262–271.

Legouix JP, Wisner A (1955) Rôle functionnel des bulles tympaniques géantes de certains rongeurs (*Meriones*). Acustica 5:209–216.

Monson G, Kessler W (1940) Life history notes on the banner-tailed kangaroo rat, Merriam's kangaroo rat, and the white throated wood rat in Arizona and New Mexico. J Wildlife Management 4:37–43.

Oaks EC (1967) Structure and function of inflated middle ears of rodents. Ph.D. Thesis, Yale University.

Plassmann W, Peetz W, Schmidt M (1987) The cochlea of gerbilline rodents. Brain Behav Evol 30:82–101.

Ryan A (1976) Hearing sensitivity of the Mongolian gerbil, *Meriones unguiculatus*. J Acoust Soc Am 59:1222–1226.

Schmidt-Nielsen B, Schmidt-Nielsen K (1951) A complete account of the water metabolism in kangaroo rats and an experimental verification. J Cell Comp Physiol 38:165–181.

Segall W (1970) Morphological parallelisms of the bulla and auditory ossicles in some insectivores and marsupials. Fieldiana Zool 51:169–205.

Simpson GG (1967) The Meaning of Evolution. New Haven: Yale University Press.

Webster DB (1961) The ear apparatus of the kangaroo rat, *Dipodomys*. Am J Anat 108:123–148.

Webster DB (1962) A function of the enlarged middle-ear cavities of the kangaroo rat, *Dipodomys*. Physiol Zool 35:248–255.

Webster DB, Webster M (1971) Adaptive value of hearing and vision in kangaroo rat predator avoidance. Brain Behav Evol 4:310–322.

Webster DB, Webster M (1972) Kangaroo rat auditory thresholds before and after middle ear reduction. Brain Behav Evol 4:41–53.

Webster DB, Webster M (1975) Auditory systems of Heteromyidae: Functional morphology and evolution of the middle ear. J Morphol 146:343–376.

Webster DB, Webster M (1977) Auditory systems of Heteromyidae: Cochlear diversity. J Morphol 152:153–170.

31
A Functional Model of the Auditory System in Mammals and Its Evolutionary Implications

Wolfgang Plassmann and Kurt Brändle

1. Introductory Deliberations

1.1 Frequency Range and Adaptation

The mammalian peripheral auditory system has predominantly been studied under five main aspects: (1) How does the system achieve impedance match from air to the cochlea as a whole? (2) What is the influence of the morphological ear structures on the different frequency ranges of best sensitivity? (3) What is the relation between auditory structures and behavioral hearing functions? (4) What is (are) the selective pressure(s) that has (have) helped created different frequency ranges reflected in the audiograms of various mammalian species? (5) How has the peripheral auditory system evolved in the various mammalian species?

There is general agreement on the overall functions of the basic acoustic features in the mammalian peripheral auditory system. Impedance match by the outer and middle ear is influenced by properties such as volume stiffness, friction in the mechanical transfer system, area and lever ratios, as well as frequency-dependent vibration of the middle ear structures. The combination of such properties is considered to primarily determine the shapes of the audiograms (e.g., Dallos 1973; Moller 1974; Shaw 1974; Relkin 1988) which can in turn be attributed to major morphological differences in shape, size and arrangement of the structures involved in sound transmission (Fleischer 1978). The ear structures in high frequency-sensitive mammals are, for instance, smaller than those of low frequency-sensitive mammalian species.

The more specific changes in the shapes of middle and inner ear structures are generally interpreted as prime examples for adaptation of peripheral auditory systems to particular needs in special environments. The apical thickening of the basilar membrane in some bats (Neuweiler et al. 1980; see Pollak, Chapter 36) is considered a morphological specialization middle ear volume and the wider and thicker basilar membranes in desert dwelling rodents (Lay 1972; Plassman et al. 1987; Webster and Webster 1980, 1984; see Webster and Plassman, Chapter 30) are viewed as specialized structures for low-frequency perception facilitating predator avoidance in arid environments. Lever arm reduction of middle ear ossicles in subterranean rodents (Rado et al. 1989) is probably an adaptation to low-frequency sound in underground acoustic environments. The existence of ossified tympanic membrane structures in some marine mammals (Fleischer 1973, see Ketten, Chapter 35, this volume) may be considered a specialization for underwater orientation and communication.

In contrast to these examples of widely accepted adaptational features, the presence of adaptation is not as obvious in many other mammals. Several authors (Hopson 1973; Jerison 1973) emphasize a strong selective pressure for high-frequency sensitivity in small mammals for more efficient predator evasion. Hence, modification in the sound-transferring structures of the middle ear are either believed to have resulted in increased sensitivity to higher frequencies for better sound localization (Masterton et al. 1969; Jerison 1973; see Heffner and Heffner, Chapter 34, this volume) or thought to have contributed to efficient intraspecific communication as the frequency of emitted utterances coincides with the receiving system's frequency range of best sensitivity (Manley 1973; Stebbins 1980). Behavioral studies by Heffner and Heffner support the idea of a relation between high-frequency sensitivity and sound localization (Chapter 34). They were able to show that high-frequency limits of hearing are correlated with interaural distances and conclude that at least the upper frequency range of sensitivity appears to be an adaptation to the need of sound localization.

1.2 Adaptation versus Constraints

At this point it appears worthwhile to draw attention to the basic idea formulated by Gould and Lewontnin (1984) with respect to evolutionary interpretation of the shapes of biological structures. Based on their ideas it is conceivable that not all the structural differences between various species that result in different frequency ranges of best sensitivity are adequately described by adaptational concepts. Hence, it appears advisable to always keep in mind that the shapes of auditory structures are also influenced by phylogenetic preconditions that determine and limit the available set of auditory structures, for instance, the presence of ossicles in mammals versus the columnella in nonmammalian terrestrial vertebrates. Similarly, there are internal relations between several features or morphological structures during ontogenetic growth, e.g., the final dimensions of the bulla in rodents appear to match the tympanic membrane dimensions already reached during earlier stages of development (Plassmann, unpublished observation). Finally, there are certain physical constraints, such as material properties inherent in specific auditory structures, e.g., in the tympanic membrane, or in the ossicles. In other words, the actual sizes and shapes of the various peripheral auditory structures are also determined by phylogenetic, developmental, and physical constraints. To make things even more complicated for evolutionary analysis, the interpretation of all structural changes as either adaptation under selective pressure or as the result of various constraints ought to include the results of ontogenetic studies because phylogeny can ultimately be considered as a chain of ontogenetic events.

Some indication of the influence of nonacoustic parameters that may constitute a limitation to

conceivable adaptations in frequency ranges of best sensitivity is provided by Calder (1984). He reported identical correlations between body mass and upper frequency limit, as well as body mass and frequency range of best sensitivity. Since the high-frequency limit of hearing is also correlated with head size (Heffner and Heffner, Chapter 34, this volume), the frequency range of best sensitivity would also have to be correlated with head size.

Based on the above, there are two alternative hypotheses for the explanation of frequency ranges of best sensitivity in mammals: (1) The auditory system has adapted its frequency range of best sensitivity to reach an optimum of directional hearing, to improve intraspecific communication, or to achieve other such goals (adaptation hypothesis). (2) The system's frequency range of best sensitivity is primarily determined by a combination of physical limitations, and phylogenetic and developmental constraints, such as head size, a predefined set of auditory traits, and interdependent and variable growth of head and auditory structures (constraints hypothesis).

Any decision in favor of one or the other hypothesis requires a functional analysis of the peripheral auditory system. Bock (1988) has suggested that the function of any biological system can be theoretically predicted on the basis of the morphological shapes of its component parts. This presupposes that the material properties of the individual structures are known. Hence, the function of the peripheral auditory system can be studied physico-acoustically in a single mammalian species because size, shape and material properties of the structural elements determine the acoustic properties of that system. However, a decision for either the adaptation or the constraints hypothesis demands a comparative and functional study of as many mammalian auditory systems as possible because the functional acoustic properties shared by all the species, as well as the functional differences between these species, can be established only by comparison of many hearing systems of different sizes and frequency ranges of best sensitivity.

1.3 Methodological Deliberations

Our comparative analysis of the peripheral auditory system in a variety of mammalian species is characterized by establishing and testing a functional bauplan pertaining to the frequency range of best sensitivity achieved by that system. We would like to emphasize here that our approach is first of all an attempt to provide an overall functional explanation of the dimensional relations observed to exist between eardrum area and middle ear volume. By introducing our functional bauplan we concentrate on the fundamental frequencies of the structures while neglecting—for the time being—all off frequencies,[1] as well as the acoustic impact of the ossicular chain and the inner ear. For this reason we anticipate later additions to or modifications of our model. In our working hypothesis we maintain that eardrum and middle ear volume have certain acoustic properties in common, which implies that no adaptation is present with respect to a specific hearing range of best sensitivity. This hypothesis is confirmed if we can demonstrate the existence of common functional relations between auditory structures, as well as pertinent acoustic constants that can be combined in a uniform functional bauplan valid for an abundance of mammals with different ear sizes and frequency ranges of best sensitivity. In other words, the applicability of such a bauplan would rule out the presence of specialized systems adapted under specific selective pressure. Failure of this bauplan to cover the auditory system of a specific mammalian species would imply that the hearing organ of this particular species possess specialized auditory structures and may possibly constitute an example of adaptation under selective pressure.

In our analysis we first studied the functional relation between the specific hearing ranges in a variety of mammalian species and the absolute dimensions of their ear structures. On the basis of this functional analysis we attempted to derive a biological evaluation of the different mammalian auditory systems in light of their possible evolutionary history. Based on the results of our own comparative studies of several rodent species and on the abundance of morphological and physiological data and audiograms available in the literature, we were able to specify a functional model of the middle ear which is capable of explaining the different frequency ranges of best sensitivity.

In principle, such a model can be expressed either in words or in mathematical language. We preferred to use a mathematical model for the following three reasons: (1) It facilitates incorporation of the laws of physical acoustics. (2) It allows quantitative precision in the description of interacting components. (3) It finally allows one to establish numerically the physical conditions under which highest sensitivity occurs in different hearing ranges.

Our functional model utilizes morphological parameters that are expected to exert a major influence on the frequency range of best sensitivity. We have incorporated the morphological dimensions of ear canal, eardrum, and middle ear volume of as many mammalian species as we possibly could. Our model does not cover friction by the ossicle chain or inner ear resistance because they only influence both vibratory amplitude and phase behavior of the structures, but appear to have no impact on the resonance frequency of the system (Plassmann and Kadel 1991). On the basis of our functional model we are capable of predicting the theoretical frequency range of best sensitivity from given structural dimensions covered by our model, but for the reasons mentioned above, we are unable to assess the absolute pressure gain achieved by the system. The model specifies the conditions under which theoretically possible hearing ranges of best sensitivity are more or less likely to occur. Verification of whether or not our functional model actually covers the peripheral auditory system of a given mammalian species with known morphometric dimensions has to be proved by independent experimental evidence. Thus, on the one hand, the model constitutes a set of tools for the explanation of structural dimensions and frequency ranges of best sensitivity, and provides on the other hand—in conjunction with independent experimental proof and mathematical testing procedure—first indication of the presence of specializations in the peripheral auditory system that may be interpreted as adaptation under selective pressure during evolution.

In our contribution we will demonstrate that the peripheral auditory system in many mammals, and the middle ear in particular, have evolved independent of a specific frequency range of best sensitivity, while simultaneously striving to obtain a maximum sensitivity. We will further show that the

---

[1] Phase-shift, damping of the oscillators.

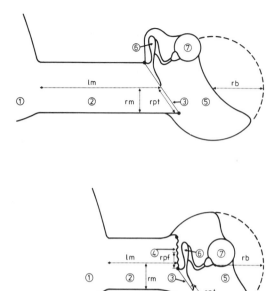

FIGURE 31.1. Schematic drawing of a typical outer and middle ear in mammals (above) and a specialized middle ear in gerbils (below). Shown are structures and dimensions used in the model of the peripheral auditory system. (1) pinna/concha complex, (2) meatus acusticus externus, (3) pars tensa tympani, (4) pars flaccida tympani, (5) middle ear cavity, (6) ossicles, (7) cochlea, (1m) meatus length, (rb) radius of idealized middle ear volume, (rm) radius of meatus area, (rpf) radius of pars flaccida tympani, (rpt) radius of pars tensa tympani.

frequency range of best sensitivity is decisively determined by the dimensions of a few middle ear structures which are, in turn, limited by physical, phylogenetic, and developmental factors. We come to the conclusion that the frequency range of best sensitivity in many mammalian species has most likely not been subject to selective pressure during evolution. This does not rule out, however, that the hearing range may be occasionally expanded under selective pressure toward lower or higher frequencies. For instance, such expansion toward higher frequencies can be brought about by the acoustic properties of the pinna/concha complex which appears to be matched by appropriate changes in the inner ear (Plassmann, unpublished observation). Finally, we derive from our model some criteria that may help identify the presence of adaptation and provide an example of what evolutionary strategy may be used to achieve frequency

adaptation of a particular peripheral auditory system to specific environmental requirements.

1.4 Basic Facts on the Peripheral Auditory System

Mammalian auditory systems basically possess the same structural components. The outer ear, consisting of the pinna/concha complex, opens into the ear canal or meatus acousticus externus. The ear canal frequently has the shape of an oval tube closed off medially by a circular or slightly oval eardrum or tympanic membrane suspended and affixed at its rims to the tympanic ring or anulus tympanicus. Eardrum and tympanic ring are frequently found in a slanted position relative to the longitudinal axis of the ear canal. The middle ear cavity behind the eardrum has bony walls and is of rather irregular and occasionally bean-like shape. In some mammalian species the cavity is divided by bony septi into several chambers of different size. The chambers communicate via openings partly constituted by the vestibular canals. Inside the middle ear cavity the ossicles, a chain consisting of the three bony structures, malleus, incus, and stapes provide a mechanical connection between the eardrum and the inner ear or cochlea. The size of the ossicles, the positioning of the axis of rotation, and the manner in which the malleus is affixed to the tympanic ring via the goniale may vary considerably from species to species (Fleischer 1973, 1978). In small, high-frequency sensitive mammals the rotation axis of the ossicles is usually more parallel to the manubrium of the malleus, and the malleus itself is more firmly affixed to the tympanic ring via the goniale than has been observed in larger species sensitive to lower frequencies. Also, the eardrums and stapes footplates are larger in low-frequency sensitive species than in small species with higher-frequency perception.

Since the impedance of the inner ear as a whole is approximately 1,000 times higher than that of air, the pressure of incoming sound waves has to be amplified by about 60 dB in order to avoid major reflection of sound. The required pressure increase is achieved by the structures of the outer and middle ear. Part of that increase is affected mechanically by the ratio between the larger eardrum and the smaller stapes footplate areas and by the ratio

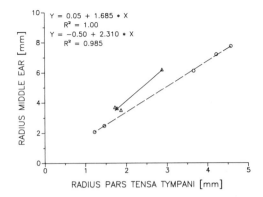

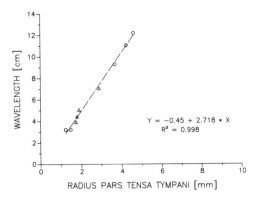

FIGURE 31.2. Allometric relation between radius of pars tensa tympani and radius of middle ear volume for the mammals of group I (circles) and of group II (triangles). Equations: above mammals of group I, below mammals of group II.

FIGURE 31.3. Relation between radius of pars tensa tympani and wavelength of maximum sound pressure increase in the meatus in front of the eardrum. Notice the similar relations for all mammals of group I (circles) and group II (triangles).

between the longer lever arm of the malleus and the shorter one of the incus. This mechanical impedance gain is frequency independent and cannot be used as explanation of why maximum impedance match should occur only in a specific frequency range. Compliance of the middle ear volume and resonance properties of the sound transferring structures, however, have an impact on the frequency range actually transferred and eventually determine the frequency range of best sensitivity (see Dallos 1973; Moller 1974; Relkin 1988).

1.5 Structural Model of the Peripheral Auditory System

Morphometric measurements of ear canal, eardrum, and middle ear volume in conjunction with established concepts of physical acoustics have contributed to formulation of a first, fairly simplified structural model of the peripheral auditory system. It consists of a cylindrical straight tube of circular diameter representing the ear canal, of a circular membrane representing the eardrum, and finally of a cavity lying behind the eardrum. The overall shape of that cavity is idealized to represent a sphere, while the septi and chambers are neglected (Fig. 31.1). The length and cross-sectional area of the meatus, the area of the eardrum, and the volume of the middle ear cavity are parameters that vary from species to species. However,

there are certain defined allometric relations between the dimensions of these parameters within one species.

Regression analyses based on morphometric measurements (Fig. 31.2) in *Homo sapiens*, *Felis catus*, *Chinchilla chinchilla*, *Rattus norvegicus*, and *Acomys cahirinus* reveal that the relation between the radius of the idealized eardrum area and the radius of the idealized middle ear volume is adequately described by equation

$$rv = 0.05 + 1.685 * rt \qquad (1)$$

as shown in (Fig. 31.2). (*rv*) stands for the radius of the middle ear volume and (*rt*) for that of the tympanic membrane. Similarly, the relation between eardrum radius and frequency of maximum sound pressure in the meatus in front of the eardrum—here expressed as wavelength—can be described by equation

$$\lambda = -0.45 + 2.718 * rt \qquad (2)$$

as shown in (Fig. 31.3). (λ) stands for wavelength in air.

1.6 Fundamental Assumptions for the Model

Based on the structural model and on the above relations we formulate the crucial hypothesis of our functional model of the outer and middle ear: The resonance frequency of the eardrum and that of the middle ear volume occur in the same frequency range in which maximum sound pressure amplitudes are found in the meatus. Verification of our functional model is achieved by a mathematical optimization procedure (see Appendix) in which the absolute dimensions of the ear structures in the analyzed species and the respective frequency ranges of maximum sound pressure in the meatus are used as variables. Subsequently, functional constants for mammalian ears are determined by a method of successive approximations, i.e., by an iterative method.

Our approach is based on four fundamental hypotheses: (1) The frequency range of best sensitivity in the audiogram is determined by the properties of the outer and middle ear complex (Dallos 1973). (2) During free-field acoustic stimulation with white noise the frequency of maximum sound pressure amplitude in the ear canal occurs in the same frequency range of best sensitivity as observed in the audiogram (Plassmann 1992a). (3) The best impedance match from air to the cochlea as a whole is found in the frequency range of best sensitivity. (4) Mechanical systems have lowest impedance in the frequency range of their own resonance frequency.

2. Functional Model of Mammalian Outer and Middle Ear

2.1 Assumptions and Acoustic Equations

All hypotheses mentioned above can be explained by assuming: (1) The eardrum has a resonance frequency; (2) the middle-ear volume behaves like a Helmholtz resonator; (3) both structures vibrate with the same resonance frequency, i.e., they are tentatively considered as coupled resonators. Thus,

$$f_d = f_b \qquad (3)$$

in which (f_d) stands for the resonance frequency of the Pars tensa tympani and (f_b) for that of the middle-ear volume.

If the eardrum is considered to constitute a thin and uniform circular membrane fixed at the rim, to be of negligible stiffness, to have perfect elasticity, and to possess even distribution of tension, its resonance frequency can be determined by equation

$$f_d = \frac{H}{2\,\pi\,rt} * M; \quad M = \sqrt{\frac{T}{p}}; \quad rt = \sqrt{\frac{St}{\pi}} \qquad (4)$$

For computation of the membrane's fundamental frequency the value 2.405 is used as a constant (H). The material constant (M) is derived from the quotient of tension per unit length (T) of the membrane and of mass per unit area (p) of the membrane. At present, other possible membrane modes are disregarded for simplicity reasons. (St) stands for the membrane area, (rt) for the radius of membrane (Kinsler et al. 1982).

The resonance frequency of the middle ear cavity (f_b) can be determined by the equation for a standard Helmholtz resonator

$$f_b = \frac{C}{2\,\pi} * \sqrt{\frac{Sm}{L'V}}; \quad L' = 1.5*rm + ln \qquad (5)$$

(Kinsler et al. 1982), provided that the cavity is considered a Helmholtz resonator where the resonator neck opens into the meatus. There is no restriction on resonator shape, as long as the cavity dimensions are shorter as a wave length and the neck is not too wide. This property of a Helmholtz resonator is the reason why we can idealize the shape of the middle ear volume to a sphere. (c) stands for the sound velocity of 352.9 m/s at 36°C (c_{air} = 331.3 + 0.6t m/s; for dry air, (t) for temperature in C°), (L') for effective neck length of the resonator in which the air mass oscillates back and forth, activated by the volume of the middle ear cavity behaving like a spring, (ln) stands for the morphological neck length, which is always shorter than the effective length (L') of the resonator neck (see Eq. 5), (V) for middle ear volume, and (rm) for radius of neck opening area (Sm) of the Helmholtz resonator into the meatus. Equations (4) and (5) are linked via Equation (3) by virtue of identical resonance frequencies in accordance with the crucial hypothesis of our functional model.

2.2 Empirical Basis for Verification of the Functional Model

Our suggested model with its three main assumptions and the two conditional equations for the resonance frequencies can be verified both on the basis of our own measurements of the mammalian species studied, and on the basis of data available in the literature. The mammalian species analyzed are divided in Group I consisting of *Homo sapiens*, *Felis catus*, *Chinchilla chinchilla*, *Rattus norvegicus*, and *Acomys cahirinus*, and in Group II consisting of the low-frequency-sensitive gerbil species *Pachyuromys duprasi*, *Meriones unguiculatus*, *Meriones tristrami*, and *Tatera spec* (Table 31.1). The middle ear volumes of the species in Group II are larger relative to their respective eardrums than those in Group I (Fig. 31.2).

2.3 Mathematical Deliberation

The tentative assumption of coupled resonators is justified only if it can be proved that the measurement values for all the variables in all the species studied fulfill Equations (4) and (5). Measurement values, however, are necessarily erroneous due to lack of absolute precision and on account of varying of dimensions within the individual species.

Moreover, the equations contain unknown parameters. We assumed that the material property of the tympanic membrane (M) is a constant for all mammal species, and we assumed that the neck of a resonator is proportional to the meatus length by a factor.

With respect to all n species analyzed we developed $2*n$ conditional equations derived from Equations (4) and (5) for simultaneous estimation of the unknown parameters and for correction of the measurement values in such a manner that all equations (solved to equal zero) are fulfilled as precisely as possible. Criterion for successful fitting was the demand that the sum of the deviation squares of the corrected variable values becomes as small as possible if compared with the actual measurement values, under the assumption that the errors have normal distribution.

The exact mathematical procedure is contained in the Appendix.

3. Basic Functional Model and Essential Deviations

3.1 Establishment of a Plausible Model Variant

3.1.1 General Description of Procedure

The mathematical procedure described in Section 2.3 and in the Appendix has the advantage that the

TABLE 31.1. Absolute dimensions of ear structures in the mammalian species in Groups I and II analyzed

Species	*lm* (mm)	*pt* (mm²)	*pf* (mm²)	*S* (mm²)	*V* (mm³)	*fm* (kHz)	*fd* (kHz)
Group I							
HS	22, 5	65, 0		44, 0	1,930	2, 9	
CC	20, 0	56, 0		28, 0	1,550	3, 2	
FC	15, 0	42, 0		15, 0	950	3, 8	
RN	5, 7	6, 7		3, 8	63	9, 0	
AC	3, 7	4, 7		3, 1	37	11, 0	
Group II							
PD	3, 0	25, 9	8, 46	13, 6	949	4, 0	2, 0
MT	3, 0	10, 4	3, 0	5, 9	181	7, 0	4, 0
MU	3, 0	9, 9	2, 97	6, 1	200	8, 0	2, 8
TA	3, 0	8, 5	3, 05	5, 1	211	9, 0	3, 0

The values are used in the mathematical computations and constitute the empirical basis for verification of our model. (f_b) resonance frequency of middle ear, (f_m) frequency of maximum amplitude in meatus, which we assume to be equal to the resonance frequency of the eardrum f_d, (*lm*) meatus length, (p_f) area of pars flaccida tympani, (p_t) area of pars tensa tympani, (*S*) meatus area, (*S*) middle ear volume; (HS) *Homo sapiens*, (CC) *Chinchilla chinchilla*, (FC) *Felis catus*, (RN) *Rattus norvegicus*, (AC) *Acomys cahirinus*, (PD) *Pachyuromys duprasi*, (MT) *Meriones tristrami*, (MU) *Meriones unguiculatus*, (TA) *Tatera spec*. The values for meatus and pars tensa in *Homo sapiens*, *Chinchilla* and *Felis* are reported in Shaw (1974). Middle-ear volume in *Homo sapiens* (Teranishi and Shaw 1968). Frequency of maximum amplitude in the meatus of *Homo sapiens* (Teranishi and Shaw 1968), of *Chinchilla* (Bismarck and Pfeiffer 1967) and of *Felis* (Wiener et al. 1966).

empirical data of all the species analyzed under one group can be subjected to simultaneous optimization. This also allows us to determine structural constants and factors that may be characteristic of individual auditory systems. By substituting different measurement values as variables in Equations (4) and (5), as well as by using the material constant (M) and the neck length either as fixed values or as unknown parameters, we are in a position to specify several plausible variants of our basic functional model. As a criterion for deciding which of the many possible variants yields the best approximations to the respective empirical values we used the minimum sum of error squares. However, since the total value of this minimum will increase with increasing interlinkage of variables, a second criterion was required. Therefore, we also compared the discrepancies between empirical and theoretical values for each individual species.

3.1.2 A Few Examples

The material constant (M) in $N/(kg/m)$ in eardrum Equation (4) depends on the empirical tissue properties of that membrane. Since both the membrane's tension and mass have not been measured so far, the computed value for (M) is determined by the constant $H = 2.405$. Thus, the resulting constant value for (M) allows us to compare different variants of the basic model for Groups I and II (see Table 31.1). It also enables us to compare different groups of mammalian species analyzed under only one variant of the model (see Section 3.3). However, the numerical value of (M) in itself has no numerical meaning.

For computation of the opening area of the middle ear resonator neck (Sm) and for the radius of that opening area (rm) in the Helmholtz resonator Equation (5), there were three possibilities. We could either use the tympanic membrane area, or the projection of that area from its inclined location into a vertical position, or the cross-sectional area of the meatus. Best results were obtained when the meatus area was considered as a variable for the opening of the middle ear resonator.

There were also three theoretically equally plausible possibilities for the morphological length (ln) of resonator neck in Equation (5). (1) The first possibility was that the entire meatus length was considered as the morphological length of the resonator neck (constant length $CL = 1$). (2) Or the morphological neck length was considered to be zero $(CL = 0)$. (3) It was also conceivable that the morphological neck length was proportional to the meatus length, which would require introduction of a proportionality constant of the meatus length $(0 < CL < 1)$ into Equation (5).

All these assumptions could be tested by replacing ln in Equation (5):

$$ln = lm * CL \qquad (6)$$

thus the effective neck length (L') in Equation (5) becomes

$$L' = 1.5 * rm + lm * CL$$

with $CL = 1$ (i)

or $CL = 0$ (ii)

or $0 < CL < 1$ (iii)   (7)

where (rm) stands now for meatus radius, and (lm) for the meatus length, and CL as a factor. Depending on the theoretical possibility chosen, different resonance frequencies are obtained for the middle ear resonator, and hence for the unknown material constant (M) of the eardrum, as both conditional equations are interdependent in Equation (3).

3.2 Basic Functional Model and Approximation Procedure

3.2.1 Model for Group I (Variant A + B)

Our functional model for the mammalian outer and middle ear has been developed by means of a best approximation to the empirical data for *Homo sapiens*, *Felis catus*, *Chinchilla chinchilla*, *Rattus norvegicus*, and *Acomys cahirinus* in Group I. The fundamental hypothesis of our model (variant A) presupposes that the resonance frequencies of eardrum and middle ear are identical.

Computation of the material constant (M) in Equation (4) by means of iteration shows that the computed values for the variables of resonance frequency (f_d) and eardrum radius (rt) deviate negligibly from the absolute values measured in the species of Group I (Table 31.2, Variant A). Under the assumption that $H = 2.405$, the material constant (M) for the eardrum assumes a value 34.80. Hence, the resonance frequency of the tympanic membrane is

$$f_d = \frac{2.405}{2\pi * rt} * 34.8 \qquad (8)$$

In Helmholtz resonator Equation (5) the effective length of the resonator neck (L') is unknown. Substitution of a constant morphological neck length $ln = 0$ leads to frequency values exceeding those obtained by measurement. If the total meatus length (lm) is used as a variable the computed frequencies are too low in comparison with the measured values. In sharp contrast to this, however, there are two equally possible variants of our model that both lead to good agreement between computed values and the empirical data. The difference between the two possibilities occurs because the first solution involves use of a constant parameter value for the morphological neck length (ln) of the Helmholtz resonator, while the second solution for (ln) utilizes a proportionality factor

(CL) with respect to the total meatus length (lm). Unfortunately, there is no way we can decide at present which of the two variants is appropriate because our sample size (five species) is too small. The first good solution (Table 31.2, variant A) is obtained when the morphological neck length (ln) has the value of 0.96. In this model variant the material constant (M) equals 35.7. For this variant the resonance frequency of the middle ear is

$$f_b = \frac{C}{2\pi} * \sqrt{\frac{Sm}{L'V}} \qquad (9)$$

where (Sm) stands for the respective meatus area and (V) for the respective middle ear volume. The effective length (L') of the Helmholtz resonator neck is

$$L' = 1.5 * rm + 0.96 \qquad (10)$$

where (rm) stands for radius of the meatus area.

In the second good solution with minimum deviation of the theoretically computed values from the actual measurement values (Variant B, Table 31.2) we used a morphological neck length (ln) which is proportional to the total meatus length (lm) by a factor (CL). Here, the effective length (L') of the Helmholtz resonator neck is

$$L' = 1.5 * rm + 0.084 * lm \qquad (11)$$

3.2.2 Verification of Basic Model

For the verification of our basic functional model for the mammalian species of Group I we stipulate for methodological reasons that the resonance frequency of the eardrum (f_d) is no longer required to equal the resonance frequency of the middle ear (f_b) (Variant C, Table 31.2). Since the assumption

---

(A and B)—The resonance frequencies of tympanic membrane and middle ear volume are interdependent. In variant A the morphological neck length of the middle ear resonator assumes a constant value (ln). In variant B the neck length is proportional to the meatus length. (C)—Both resonance frequencies are now uncoupled. Notice the good agreement between measured values and variables of basic and verified models. $(\Sigma\varepsilon^2)$ Sum of error squares, (M) material constant of tympanic membrane, (CL) proportionality factor for neck length of middle ear resonator, (lm) meatus length, (m) meatus area, (rm) meatus radius, membrane, (p_t) tympanic membrane area, (V) volume of middle ear cavity, (f_b) resonance frequency of middle ear cavity, (f_d) resonance frequency of tympanic membrane.

TABLE 31.2. Comparison of empirical and theoretical values in accordance with basic models (variant A and B) and verification of variant B in variant C for five mammalian species of Group I

Model variant	(A) $f_d = f_b$	(B) $f_d = f_b$	(C) $f_d = f_b$
$\Sigma\varepsilon^2$	1, 19	0, 89	0, 31
M	35, 67	34, 80	34, 91
	ln: 0.96	CL: 0.084	CL: 0.071
Variab Emp	Theor (A)	Theor (B)	Theor (C)
Homo sapiens			
lm 22, 5	–	22, 53	22, 50
m 44, 0	43, 86	43, 88	43, 94
rm 3, 7	4, 60	4, 33	3, 82
p_t 66, 0	65, 95	65, 96	66, 00
V 1,930, 0	1,930, 00	1,930, 00	1,930, 00
f_d 2, 9	2, 98	2, 91	2, 92
f_b –	–	2, 91	3, 13
Chinchilla chinchilla			
lm 20, 0	–	19, 98	20, 00
m 28, 0	28, 08	28, 13	28, 09
rm 3, 0	3, 00	2, 70	2, 96
p_t 56, 0	56, 00	56, 02	56, 00
V 1,550, 0	1,550, 00	1,550, 00	1,550, 00
f_d 3, 2	3, 23	3, 15	3, 17
f_b –	–	3, 15	3, 12
Felis catus			
lm 15, 0	–	14, 97	14, 99
m 15, 0	15, 09	15, 12	15, 03
rm 2, 2	1, 70	1, 61	2, 01
p_t 42, 0	42, 03	42, 04	41, 99
V 950, 0	950, 00	950, 00	950, 00
f_d 3, 8	3, 73	3, 64	3, 66
f_b –	–	3, 64	3, 49
Rattus norvegicus			
lm 5, 7	–	5, 71	5, 71
m 3, 8	3, 90	3, 77	3, 75
rm 1, 1	0, 88	1, 19	1, 25
p_t 6, 7	6, 91	6, 74	6, 77
V 63, 0	62, 99	63, 00	63, 00
f_d 9, 0	9, 21	9, 09	9, 10
f_b –	–	9, 09	9, 04
Acomys cahirinus			
lm 3, 7	–	3, 71	3, 71
m 3, 0	3, 18	3, 02	3, 02
rm 1, 0	0, 83	1, 19	1, 20
p_t 4, 7	4, 81	4, 63	4, 66
V 37, 0	36, 99	37, 01	37, 01
f_d 11, 0	11, 04	10, 98	10, 97
f_b –	–	10, 98	11, 04

of a proportionality factor (*CL*) has provided the best solution when both equations were coupled (Variant B), we used this assumption also for Variant C. If our basic assumption of two resonators with identical resonance frequencies is true, then the two separately computed resonance frequencies (f_d and f_b) ought to resemble each other, and the constant (*M*) and the factor (*CL*) should be fairly similar, even if the resonance frequencies in Equations (4) and (5) are uncoupled. If the resonance frequencies of structures are uncoupled, the material constant (*M*) for the eardrum assumes the value 34.91, and the neck length is proportional to the meatus length by the factor *CL* = 0.074. There is hardly any difference between the resonance frequencies of eardrum and middle ear even though both are computed separately, and what is more important, the resonance frequencies separately obtained are in good agreement with those under the assumption that both resonance frequencies are coupled. If the resonance frequencies are mathematically uncoupled the value $\Sigma \varepsilon^2 = 0.31$ is smaller than that of 0.89, when both frequencies are coupled. As above verification procedure did not give rise to substantial deviation between the variants, where the morphological neck length has a value of 0.96, or is a constant factor of the meatus length (0.084 * *lm*), we are justified to consider both variants of our models as adequate theoretical descriptions of the outer and middle ear complex in the mammalian species of Group I.

Our basic theoretical model applies only to the species analyzed so far. Its validity for other mammalian species remains to be tested. However, since our model is valid for one primate, one carnivore and several rodent species with extreme differences in morphological dimensions and resonance frequencies, and since there is ample independent experimental evidence concerning the existence of allometric relations between eardrum size and middle ear volume in 28 mammalian species (Plassman 1992b), we are fairly confident that our model will cover a wide variety of mammalian species. However, we would like to emphasize at this point that our approach only demonstrates the compatibility of our model with the empirically measured dimensions. We do not claim that the outer and middle ear in fact behave acoustically as described by our model.

3.3 Deviation from the Basic Model in Gerbilline Rodents

The same mathematical procedure described for Group I was applied to the gerbilline rodent species *Tatera spec.*, *Meriones unguiculatus*, *Meriones tristrami*, and *Pachyuromys duprasi* of Group II. Since we know from physiological measurements in gerbilline rodents that the frequency range of maximum amplitudes in the meatus and the range of the middle ear resonance frequency do not coincide (Plassmann and Kadel 1991), we analyzed the species of Group II under the assumption that the resonance frequencies of eardrum and middle ear are not identical.

Computation of the material constant (*M*) in Equation (4) by means of the iterative method shows that the differences in frequency and eardrum radius are negligibly small if compared with their respective measurement values. Under the assumption that *H* = 2.405, (*M*) assumes the value of 36.4. Thus, the tympanic membrane in gerbils possesses a material constant comparable to that of Group I. Hence, the resonance frequency of the tympanic membrane is

$$f_d = \frac{2.405}{2\pi * r_t} * 36.4 \qquad (12)$$

where (r_t) stands for the respective idealized membrane area of the eardrum. Besides the pars tensa tympani, gerbilline rodents also possess a smaller pars flaccida tympani located dorsal of the pars tensa. This flexible membrane has no contact with the ossicles. If the radius of this membrane is used in Equation (4) the material constant (*M*) for the pars flaccida tympani is 8, which is 4.55 times smaller than that of the pars tensa, provided that *H* = 2.405 for both membranes.

For computation of the middle ear resonance frequency in Equation (5) the morphological length (*ln*) of the effective neck length (*L'*) was unknown. For the resonator opening area (*Sm*) we either substitute the eardrum area (*pt*) or the pars flaccida area (*pf*). For (*rb*) in (*L'*) we used the radius of the meatus area (*rm*). Provided that the eardrum area constitutes the resonator opening, that the morphological length of the resonator neck (*ln*) is either zero, or equals the total meatus length, and that the pars flaccida is neglected, the computed values for the middle ear resonance frequency and for the meatus radius deviate substantially from the respective empirical data. By introduction of a proportionality factor (CL), good agreement between computed and empirical data can be obtained. However, in sharp contrast to Group I, the morphological length of the resonator neck in Group II would have to be 4.5 times longer than the total meatus length.

For this reason, we introduced the pars flaccida tympani as the opening area of the resonator neck into Equation (5) for our computations pertaining to Group II. When the neck length (ln) equaled zero, the resulting frequency values were too high in comparison with the measured frequencies. Satisfactory results were either obtained by using a neck length corresponding to the entire meatus length, or to a calculated neck length of 2.60, or by utilizing a constant proportionality factor (CL) of 0.87 (Table 31.3). The similarity of all results is explained by the fact that both results, a constant proportionality factor or a constant neck length, correspond to almost the total meatus length.

Therefore, we propose the following equation for computation of the resonance frequency of the middle ear cavity in the gerbil species under Group II

$$f_b = \frac{C}{2\pi} * \sqrt{\frac{pf}{L'V}} \qquad (13)$$

in which the resonance frequencies of eardrum and middle ear volume are not identical and where (pf) stands for the respective pars flaccida area. The effective neck length of the resonator is either

$$L' = 1.5 * rm + 0.87 * lm \qquad (14)$$

or

$$L' = 1.5 * rm + 2.60 \qquad (15)$$

Both equations revealed similar good approximations. Therefore we cannot decide which of the two solutions is valid. Thus, the outer and middle ear function in gerbilline rodents of Group II appears to deviate substantially from that described in our basic functional model for the species in Group I. While the resonance frequencies of tympanic membrane and middle ear volume are basically identical in the species of Group I, gerbilline rodents must be expected to possess a system of two

TABLE 31.3. Comparison of empirical data and computed values in Group II

Model variant $fd = fb$		
$\Sigma\varepsilon^2$		1, 97
M		36, 40
CL		0, 87
Varia	Emp	Theor
Tatera spec		
lm	3, 0	3, 03
rm	1, 3	1, 32
p_t	9, 3	8, 97
p_f	3, 1	2, 99
V	211, 5	211, 50
f_d	9, 0	8, 24
f_b	3, 0	3, 11
Meriones unguiculatus		
lm	3, 0	3, 06
rm	1, 4	1, 51
p_t	9, 9	9, 82
p_f	3, 0	2, 85
V	200, 0	200, 00
f_d	8, 0	7, 88
f_b	2, 8	3, 02
Meriones tristrami		
lm	3, 0	2, 88
rm	1, 4	1, 16
p_t	11, 0	11, 14
p_f	3, 0	3, 21
V	181, 0	181, 00
f_d	7, 0	7, 40
f_b	4, 0	3, 62
Pachyuromys duprasi		
lm	3, 0	3, 02
rm	2, 1	2, 12
p_t	25, 9	26, 01
p_f	8, 5	8, 44
V	995, 0	995, 00
f_d	5, 0	4, 84
f_b	2, 0	2, 15

The resonance frequencies of the eardrum and middle ear are not assumed to be identical. ($\Sigma\varepsilon^2$) sum of errors squares, (M) material constant of eardrum, (CL) proportionality factor for neck length of middle ear resonator, (lm) meatus length, (rm) radius of meatus, (p_t) area of pars tensa tympani, (p_f) area of pars flaccida tympani, (V) middle ear volume, (f_d) resonance frequency of pars tensa tympani, (f_b) resonance frequency of middle ear cavity.

independent resonators in which the middle ear resonator utilizes the pars flaccida tympani for pressure release. These conclusions are supported by two observations. Failure to consider the pars flaccida in our computations results in middle ear resonance frequencies that significantly exceed the respective empirical values. Furthermore, experimental stiffening of the flexible membrane in *Pachyuromys duprasi* eliminates the low-frequency middle ear resonator and shifts the maximum sound pressure amplitudes to a higher frequency range (Plassmann and Kadel 1991).

4. Basic Functional Model and Implications

4.1 Physical Functioning of the Model

In terms of physical acoustics our basic functional model of the mammalian outer and middle ear can be described as a system of coupled resonators consisting of a resonating membrane and a Helmholtz resonator, represented by eardrum and middle ear volume. The fundamental frequency of the middle ear resonator is tuned to the fundamental frequency of the eardrum. It must remain undecided here whether the quality factors (Qs) of the two resonators are high, low, or medium. However, it is safe to assume that (Q) of the tympanic membrane resonator is fairly low due to the resistance of the inner ear: This assumption is supported by some experimental evidence, for instance, by Moller (1974) in cat and Plassmann and Kadel (1991) in the gerbil species *Pachyuromy duprasi*. If sound waves hit the eardrum, the power at the resonance frequency of that driven oscillator is transferred to the resonating middle ear, and back to the vibrating membrane. Due to the extremely short resonator neck and to the slanted position of the tympanic membrane, it cannot be decided here whether the eardrum lies proximal to the Helmholtz resonator neck which constitutes the point of connection between the lumped mass of that resonator (i.e., the resonator neck) and the lumped spring of the resonator (i.e., the middle ear volume), or whether the tympanic membrane is suspended distal to the resonator neck. Based on different model calculations, however, we assume that the eardrum lies either between the lumped mass and lumped spring of the resonator or even inside the lumped spring of the Helmholtz resonator. Because such models, which included the eardrum area in different tilted positions as neck opening (*St*) of the Helmholtz resonator, did not affect the results.

4.2 Frequency Range of Best Sensitivity

The crucial hypothesis of our functional model demands that the resonance frequencies of eardrum and middle ear cavity are in the same frequency range where maximum sound pressure amplitudes occur in the meatus. It could be argued at this point that it is the ear canal, rather than the coupled resonators of tympanic membrane and middle ear, that brings about the amplitude enhancement observed in the meatus. This concept interprets the ear canal as a resonating pipe, and the middle ear volume as a stiffness component. However, measurement of the transfer characteristics of the outer ear near the eardrum has shown in at least two rodent species that the ear canal is too short in order to convincingly explain the frequency range of sound pressure increase in the meatus (Plassmann 1992a). For instance, a pipe resonating at 9 kHz in *Rattus norvegicus* would require a meatus length of 10 mm, but it is actually only 5.7 mm long. In *Acomys cahirinus*, at 11 kHz, the meatus would have to measure 8 mm, but it reaches only 3.7 mm. For this reason we prefer the concept of coupled resonators over that of a resonating pipe. Furthermore, it cannot be ruled out that additional resonators in the peripheral auditory system may also contribute to broadening of the frequency range of best sensitivity.

Structures such as tympanic membrane and middle ear cavity have lowest impedance at the resonance frequency. Sound energy entering such a system at the resonance frequency is absorbed over that frequency range. The vibratory amplitude of the tympanic membrane is enhanced by acoustic lever action and transferred to the cochlea via the ossicles. As a consequence, the acoustic gain in this frequency range is higher than in any other range of frequencies, the system has a better impedance match from air to the cochlea as a whole, and its sensitivity is highest in that particular frequency range.

4.3 Interdependency of Structures and Resonance Frequencies

Based on our hypothesis that two coupled resonators are active in the peripheral auditory system of mammals we are able to explain the frequency ranges of best sensitivity in many mammalian species on the basis of the absolute dimensions of only a few outer and middle ear structures. Our basic model furthermore allows us to predict the dimensions of individual middle ear structures if the frequency range of best sensitivity in a given mammalian species is known. The prognostic value of our model resides in the fact that outer and middle ear properties are described by means of equations into which the absolute dimensions of the ear structures are incorporated as interdependent variables. Besides its inherent explanatory and prognostic potential, our model constitutes a heuristic means for diagnosis of adaptation in auditory systems during evolution.

In our model the morphological dimensions of middle ear volume, meatus area, and tympanic membrane are adjusted to each other, and their structures are tuned to a specific frequency range. Thus, the outer and middle ears of all mammalian species adequately described by our basic functional model are shaped in accordance with a uniform bauplan, irrespective of individual dimensions and resonance frequencies. Based on a given eardrum radius the most likely frequency range of best sensitivity can be predicted by means of Equation (8). The same equation enables us to predict for two mammalian auditory systems with a one octave difference in sensitivity that their tympanic membrane areas will differ by a factor of four. Based on Equations (9, 10) we expect that the middle ear volumes of these two species will differ by a factor of eight. The accuracy of these predictions becomes apparent in the correlation analysis of tympanic membrane and middle ear (Fig. 31.2 and Equation 1) and in the correlation analysis of pars tensa and wavelength of maximum sound pressure in the meatus (Fig. 31.3 and Equation 2), both in Section 1.2.

4.4 Morphological Constraints for Frequency Shifts

A shift of the frequency range of best sensitivity toward lower frequencies can be achieved by a strategy of increasing the middle ear volume. Under the restriction of a given head size, however, this is possible only in a fairly limited way, because middle ear cavity and tympanic membrane would have to be considerably enlarged. In rodent species with small heads the required dimensional increase would soon have reached its natural limit. Therefore, head size appears to constitute a major constraint for low-frequency sensitivity. If sensitivity in a low-frequency range becomes, nevertheless, vital due to selective pressure, it can also be achieved—at least in the gerbil species of Group II—by abandoning the general functional bauplan. In *Pachyuromys duprasi*, for instance, the strategy of reducing the bulla resonator opening appears to be adopted because this species utilizes the smaller pars flaccida as opening of the bulla resonator instead of the wider meatus area. Both strategies, that of middle ear volume increase and that of neck opening reduction, result in a lower resonance frequency of the middle ear cavity. Both strategies are expressed mathematically in equation (5).

Other mammalian species, such as *Cavia porcellus* or some heteromyids such as *Dipodomys merriami* that do not possess a pars flaccida are apparently also capable of lowering their middle ear resonance frequencies by increase of tympanic membrane area and reduction of their resonator neck opening areas. They achieve this by utilizing an extremely slanted angle for the suspension of their tympanic membranes (Plassmann, unpublished observation). This strategy may help avoid excessive expansion of middle ear volume and provides the first indication of possible low-frequency perception.

Additional sensitivity in the range of higher frequencies can theoretically be achieved by modification of structural dimensions in the pinna/concha complex without abandoning the general bauplan. In rodent species such as *Acomys cahirinus* or *Rattus norvegicus*, these structures are apparently responsible for a sound pressure increase in the frequency range above the resonance frequency of the tympanic membrane and middle ear cavity (Plassmann 1992a). Additional sensitivity in extremely high frequency ranges is achieved in bats by several specializations such as basilar membrane thickening in the cochlea (Neuweiler et al. 1980; Vater et al. 1985; Pollak, Chapter 36).

4.5 Evolutionary Aspects

Ideally speaking, any evolutionary interpretation of the morphological variety encountered in the mammalian peripheral auditory system ought to clearly distinguish between the physical and structural limitations in a given system, the structures' phylogentic preconditions, and the modifications brought about by adaptation. At the same time, the study of such evolutionary development ought to also consider the behavior of species, environmental conditions that may exert selective pressure, as well as the temporal sequence of events. Unfortunately enough, pertinent information is hard to obtain because the auditory systems in recent species have been shaped by at least several of the above factors. This is complicated further by the fact that adaptation can occur in many respects. Selective pressure may have taken place, for example, towards a gain in sensitivity (a central aspect of this contribution), toward improving directional hearing (see Heffner and Heffner, Chapter 34), or toward better frequency analysis by the ear (Fay, Chapter 14). Our bauplan may be irrelevant for the kinds of adaptation in which additional structural changes outside the scope of our bauplan are involved. For instance, directional hearing is predominantly influenced by size, shape, and movability of the pinna (Yost and Gourevitch 1987) and echolocation in bats relies in part on structural changes in the cochlea (Pollak, Chapter 36).

In this context our model provides a heuristic means for initial diagnosis of adaptation with respect to a given frequency range of best sensitivity. By specifying the prime parameters that determine the frequency range of best sensitivity, i.e., cross-sectional area of the ear canal that corresponds to the resonator neck opening, area, and angle of inclination of the tympanic membrane, and the volume of the middle ear, it helps clarify whether or not adaptation has taken place, and if so, how and by means of which specific structural changes it has been achieved, and which overall evolutionary strategies become apparent in those modifications. We hope, in the long run, to be able to discuss adaptation not only on the level of modifications in specific auditory structures, but also in terms of evolutionary strategies adopted by the system. If, for instance, major dimensional deviations from our proposed model and bauplan are encountered in conjunction with a substantial shift of the frequency range of best sensitivity, we are confident to interpret such abandoning of the bauplan as adaptation to selective pressure. Quite a few desert-dwelling rodent species apparently have

adapted toward better low-frequency perception in old-world, as well as in new-world rodent species by abandoning the functional bauplan. In both cases similar strategies appear to have been adopted. For instance, bulla size has substantially increased compared with the area of the tympanic membrane, and hypertrophic hyaline tissue has developed underneath the basilar membrane in the cochlea (see Webster and Plassmann, Chapter 30).

According to our interpretation of the peripheral auditory system, the high-frequency range of best sensitivity for directional hearing in small-headed mammalian species for the purpose of predator evasion cannot be considered a good example of adaptation. We believe that a high-frequency range of best sensitivity in small mammalian species is primarily determined by the structural dimensions required in the bauplan, and these dimensions are in turn correlated with head or body size, as suggested by Fleischer (1973). There is some likelihood, however, that adaptation may have taken place with respect to an expansion of high-frequency perception due to increased dimensions and/or movability of the pinna/concha complex because both factors help improve directional hearing in small mammals (Heffner and Heffner, Chapter 34).

As it appears that our model and bauplan cover different mammalian species of large and small size we conclude that the functioning of the peripheral auditory system has basically remained unchanged during evolution. An abundance of structural modifications is conceivable in the peripheral auditory system that may have occurred at a certain level without abandoning the bauplan. Under the assumption that the growth of auditory structures is closely linked with that of the skull, any phylogenetic change in head size would entail appropriate modifications of the structural dimensions. Such changes that keep within the scope of the general bauplan cannot be considered as genuine adaptation. For example, change of dimensions and rotation axis of the ossicles ought to be considered from the point of view of maintaining the functional bauplan rather than as an adaptation under selective pressure. The question of whether morphological variation in the shape of the ossicles in species from different families has any influence on the systems' acoustic properties must remain undecided here.

Our basic functional model with its structural interdependencies that decisively determine the frequency range of best sensitivity so far applies to the peripheral auditory system from meatus to middle ear. We are confident, however, that this model can be expanded to also cover some relevant inner ear structures. For instance, there are striking allometric relations between the radius of the tympanic membrane and the length of the basilar membrane in the cochlea. This leads us to expect that functional interdependencies can also be formulated for the whole peripheral auditory system in mammals.

5. Summary

Analysis of the peripheral auditory system in a variety of mammalian species revealed allometric relations between eardrum area and middle ear volume. Similarly, a relation was found between the frequency range of best sensitivity and the size of the tympanic membrane. In an attempt to explain the morphological and functional relations we propose a functional model shared by an abundance of species with different ear dimensions. The model shows that eardrum and middle ear volume have identical resonance frequencies in each mammalian species studied. We utilize the physical concepts of a resonating membrane and a Helmholtz resonator. For all the species analyzed we formulate two conditional equations for the resonance frequencies of membrane and middle ear volume by incorporating the following parameters: radius of the pars tensa, radius of the middle ear volume, cross-sectional area and length of the meatus, and frequency of best sensitivity. By means of these equations and by a mathematical optimization procedure we are able to estimate the unknown parameters of a material constant for the eardrum and of the effective length of the Helmholtz resonator neck. On the basis of our functional model, expressed in a system of two equations for each individual species, we cannot only explain the allometric relations between eardrum and middle ear, but also predict the dimensions of these structures in a species of which only the frequency range of best sensitivity is known. We are also in a position to derive the frequency range of best sensitivity from the size of the tympanic membrane.

Our approach, furthermore, reveals that gerbil species have specialized peripheral auditory systems because the resonance frequencies of their eardrums do not coincide with those of their middle ear volumes. This is explained by the fact that the neck opening of the middle ear resonator is no longer the cross-sectional area of the meatus, but the smaller pars flaccida tympani. This configuration, which shifts the resonance frequency of the middle ear to a lower frequency range without excessive expansion of the bulla volume, explains the sensitivity of some gerbil species to low frequencies despite their small head size.

This example illustrates that our model eventually constitutes a heuristic means for initial diagnosis of adaptation with respect to a specific frequency range. It helps clarify whether or not adaptation may be present in a given species, and if so, how and by means of which particular structural changes this has been achieved. It may be possible, in the long run, to discuss adaptation not only at the level of structural modifications, but also in terms of evolutionary strategies adopted by species under selective pressure concerning a specific frequency range of best sensitivity.

Appendix: Mathematical Solution

Based on the basic hypotheses in Section 2.1 there are two conditional equations for the resonance frequency of the eardrum and for the middle ear volume, respectively.

$$\Phi_{dl} = \frac{H\,M}{2\pi*rt_1} - f_{dl} = 0; \quad M = \sqrt{\frac{T}{p}}; \quad rt_1 = \sqrt{\frac{S_1}{\pi}}$$
$$\text{(A1)}$$

$$\Phi_{bl} = \frac{C}{2\pi} * \sqrt{\frac{S_1}{L'\,V_1}} - f_{bl} = 0;$$

$$L' = 1.5\,rm_1 + lm_1 * CL \qquad \text{(A2)}$$

$1 = 1\ldots\ldots n\ (n = \text{number of species})$

$CL = 0 \quad \text{or} \quad CL = 1 \quad \text{or} \quad 0 < CL < 1$

(d) = eardrum; (b) = middle ear cavity; (V_1) (middle ear volume). In both Equations (A1 and A2) we assume that the area of eardrum (St) is identical with the neck opening area of the Helmholtz resonator (Sm), because model calculations including

(Sm) tilted at different degrees of neck opening, did not affect the results. Therefore (St_1) and (Sm_1) are replaced in the equations by (S_1). (rt_1) (radius of the tympanic membrane), (rm_1) (radius of the meatus), (lm_1) (meatus length), (f_{d1}), (f_{b1}) (resonance frequencies of eardrum and middle ear, respectively) are variables (η_i), (M), and (CL) common, but unknown constant parameters (x_j), (H), and (c) physical constants.

Since both equations must be valid for all the mammalian species studied, a system of $n * 2$ equations is formulated for n species. In order to solve this system of equations the unknown parameters (x_j) had to be estimated and the measurement values (y_i) had to be corrected to η_i in such a manner that all $2 * n$ equations were fulfilled.

As conditions for the optimization of the equation system we chose, as a procedure for the fitting of measurement values, the method of least error squares after Gauss under the restriction

$$\Phi_{kl}(x, \eta) = 0; \quad k = d, b; \quad 1 = 1 \ldots n \quad (A3)$$

Since the measurement values in the two conditional equations are interdependent, their most probable values were calculated by Lagrange's method of multipliers (Brandt 1981). Because of the nonlinearity of the conditional equations a Taylor series was established, which is discontinued after the first member. The iterative solution of the Taylor series yields better approximations for the parameters and measurement values:

$$\Phi_{kl}(x, \eta) = \Phi_{kl}(x_0, \eta_0)$$

$$+ \sum \left(\frac{d\Phi_{kl}}{dx_j}\right)_{x_0\eta_0} (x_j - x_{0j})$$

$$+ \sum \left(\frac{d\Phi_{kl}}{d\eta_i}\right)_{x_0\eta_0} (\eta_i - \eta_{0i}) \quad (A4)$$

(j) = 1 . . . . . p (i) = 1 . . . . m
(p) = number of parameters; (m) = number of variables
Let

$$a_{klj} = \left(\frac{d\Phi_{kl}}{dx_j}\right)_{x_0\eta_0}; \quad b_{kli} = \left(\frac{d\Phi_{kl}}{d\eta_i}\right)_{x_0\eta_0} \quad (A5)$$

and

$$c_{kl} = \Phi_{kl}(x_0, \eta_0); \quad \xi_j = x_j - x_{0j};$$

$$\delta_i = \eta_i - \eta_{0i} = -\varepsilon_i \quad (A6)$$

the result in matrix notation is

$$\Phi = c + A\xi + B\delta$$
$$\Rightarrow \Phi = c + A\xi - B\varepsilon \quad (A7)$$

with

$$A = \begin{bmatrix} a_{d11} & a_{d12} & \cdots\cdots & a_{d1p} \\ a_{b11} & a_{b12} & \cdots\cdots & a_{b1p} \\ a_{d21} & a_{d22} & \cdots\cdots & a_{d2p} \\ a_{b21} & a_{b22} & \cdots\cdots & a_{b2p} \\ \cdot & & & \\ \cdot & & & \\ a_{dn1} & a_{dn2} & \cdots\cdots & a_{dnp} \\ a_{bn1} & a_{bn2} & \cdots\cdots & a_{bnp} \end{bmatrix}$$

$$B = \begin{bmatrix} b_{d11} & b_{d12} & \cdots\cdots & b_{d1m} \\ b_{b11} & b_{b12} & \cdots\cdots & b_{b1m} \\ b_{d21} & b_{d22} & \cdots\cdots & b_{d2m} \\ b_{b21} & b_{b22} & \cdots\cdots & b_{b2m} \\ \cdot & & & \\ \cdot & & & \\ b_{dn1} & b_{dn2} & \cdots\cdots & b_{dnm} \\ b_{bn1} & b_{bn2} & \cdots\cdots & b_{bnm} \end{bmatrix}$$

and the column vector

$$c = \begin{bmatrix} \Phi_{d1}(x_0, \eta_0) \\ \Phi_{b1}(x_0, \eta_0) \\ \Phi_{d2}(x_0, \eta_0) \\ \Phi_{b2}(x_0, \eta_0) \\ \cdot \\ \cdot \\ \Phi_{dn}(x_0, \eta_0) \\ \Phi_{bn}(x_0, \eta_0 \end{bmatrix}$$

The requirement, $\Sigma\varepsilon^2 = \min$ with the restriction

$$c + A\xi + B\delta = 0 \text{ (see eq. A7)} \quad (A9)$$

results in the Lagrange function

$$L = \delta^T \delta + 2 \mu^T (c + A\xi + B\delta) \quad (A10)$$

where μ stands for the column vector of the Lagrange multipliers.
$\Phi_{kl}(x, \eta) = 0$ is fulfilled at

$$\frac{dL}{d\delta} = 2\delta + 2B^T \mu = 0 \quad (A11)$$

and

$$\frac{dL}{d\xi} = 2\mu^T A = 0 \quad (A12)$$

Under the assumption that measurement errors follow the normal distribution curve it is

$$y_i = \eta_i + \varepsilon_i \qquad (A13)$$

$$\Rightarrow \quad \varepsilon = y_i - \eta_i = \eta_{oi} - \eta_i = -\delta_i \qquad (A14)$$

Therefore the measurement values (y_i) were used as first approximations for the variables (η_i).

Acknowledgment. We thank Gertraud Plassmann for critical comments and translation of the manuscript.

References

Bismarck GV, Pfeiffer RR (1967) On the sound pressure transformation from free field to eardrum of chinchilla. J Acoust Soc Am (Abstr) 42:1156.

Bock WJ (1988) Explanations in Morphology. Am Zool 28:205–215.

Brandt (1981) Datenanalyse. Mannheim: BI Wiss Verlag.

Calder WA III (1984) Size, Function, and Life History. Cambridge (Mass) and London (Eng): Harvard Univ Press, pp. 236–243.

Dallos P (1973) The Middle Ear. In: Dallos P (ed) The Auditory Periphery. New York: Academic Press, pp. 22–39.

Fleischer G (1973) Studien am Skelett des Gehörorgans der Säugetiere, einschließlich des Menschen. Säugetierk Mitteilg 21(2):131–239.

Fleischer G (1978) Evolutionary Principles of the Mammalian Middle Ear. Berlin, Heidelberg, New York: Springer.

Gould SJ, Lewontnin RC (1984) Sprandels of San Marco and the Panglossian Paradigm. In: Elliott, Sober (eds) Conceptual Issues in Evolutionary Biology. Cambridge (Mass) London (Eng): The MIT-Press.

Hopson JA (1973) Endothermy, small size, and the origin of mammalian reproduction. Am Naturalist 107:446–452.

Jerison HJ (1973) Evolution of the brain and intelligence. New York: Academic Press.

Kinsler LE, Frey AR, Coppens AB, Sanders JV (1982) Fundamentals of Acoustics. New York: John Wiley & Sons.

Lay DM (1972) The Anatomy, Physiology, Functional Significance and Evolution of Specialized Hearing Organs of Gerbilline Rodents. J Morph 138:41–120.

Manley GA (1973) A review of some current concepts of the functional evolution of the ear in terrestrial vertebrates. Evolution 26:608–621.

Masterton B, Heffner HE, Ravizza R (1969) The evolution of human hearing. J Acoust Soc Am 45:996–985.

Møller AR (1974) Function of the Middle Ear. In: Keidel WD, Neff WD (eds) Handbook of Sensory Physiology, Vol V/I, Auditory System. Berlin, Heidelberg, New York: Springer. pp. 491–517.

Neuweiler G, Bruns V, Schuller G (1980) Ears adapted for the detection of motion, or how echolocating bats have exploited the capacities of the mammalian auditory system. J Acoust Soc Am 68(3):741–753.

Plassmann W (1992a) The Peripheral Auditory System in Rodents: I. Transfer Function of Outer and Middle Ear. (accepted)

Plassmann W (1992b) The Peripheral Auditory System in Rodents: II. Functional Bauplan and Frequency range of Best Sensitivity. (accepted)

Plassmann W, Kadel M (1991) Low Frequency Sensitivity of the Gerbilline Rodent Pachyuromys Duprasi: The Tympanic Membrane as a Pressure Gradient Receiver. Brain Behav Evol (in press).

Plassmann W, Peetz W, Schmidt M (1987) The Cochlea in Gerbilline Rodents. Brain Behav Evol 30:82–101.

Rado R, Himelfarb M, Arensburg B, Terkel J, Wollberg Z (1989) Are seismic communication signals transmitted by bone conduction in the blind mole rat? Hearing Research 41, 23–30.

Relkin EM (1988) Introduction to the Analysis of Middle-Ear Function. In: Jahn AF, Santos-Sacchi J (eds) Physiology of the Ear. New York: Raven Press, pp. 103–123.

Shaw EAG (1974) The external ear. In: Keidel WD, Neff WD (eds) Handbook of Sensory Physiology (Vol. V/I), Auditory System. Berlin, Heidelberg, New York: Springer, pp. 455–490.

Stebbins WC (1980) The Evolution of Hearing in the Mammals. In: Popper AN, Fay RR (eds) Comparative Studies of Hearing in Vertebrates. New York, Heidelberg, Berlin: Springer-Verlag, pp. 421–436.

Teranishi R, Shaw EAG (1968) External ear acoustic models with simple geometry. J Acoust Soc Am 44:257–263.

Vater M, Feng AS, Betz M (1985) An HRP-study of the frequency-place map of the horseshoe bat cochlea: Morphological correlates of sharp tuning to a narrow frequency band. J Comp Physiol 157:671–686.

Webster DB, Webster M (1980) Morphological Adaptations of the Ear in the Rodent Family Heteromyidae. Am Zool 20:247–254.

Webster DB, Webster M (1984) The Specialized Auditory System of Kangaroo Rats. Contributions to Sensory Physiol, Vol 8. Acad Press, pp. 161–196.

Wiener FM, Pfeiffer RR, Bachus ASN (1966) On the sound pressure transformation by the head and auditory meatus of the cat. Acta Oto-Laryngol (Stockh) 61:255–269.

Yost WA, Gourevitch G (1987) Directional Hearing. New York, Berlin, Heidelberg: Springer-Verlag.

32
Origin of Auditory Cortex

Shawn B. Frost and R. Bruce Masterton

1. Introduction

All mammals, including the egg-laying mono-tremes, possess an auditory cortex—that is, an auditory area in their cerebral neocortex. In contrast, no reptile or bird possesses even neocortex itself let alone an auditory area within it. Therefore, the question of the evolutionary origin of auditory cortex is entangled in the larger question of the origin of mammalian neocortex. Because there are no animals now extant that serve as an unarguable witness to the transition from the reptilian to the mammalian form of cerebral cortex the question of the origin of mammalian cortex is an old and persistent one. We now have but two snapshots of the forebrain through geological time—a reptilian stage without neocortex and a later mammalian stage with a relatively large amount of neocortex and an auditory area already within it.

1.1 Homology and Homoplasy Among Vertebrate Forebrains

Figure 32.1 summarizes the problem surrounding the evolution of the mammalian forebrain and the similarity in position of reptilian "dorsal cortex" with mammalian "neocortex." Although we know now that the two types of cortex are not homologous in the usual comparative or practical senses of the term—that is, they are sharply different in structural detail—the reasons for the early comparative anatomists such as Elliot Smith (1910) and Herrick (1956) hoping the dorsal cortex and neocortex might be homologous are evident.

Despite the nonhomology, however, the idea of reptilian dorsal cortex together with some other reptilian forebrain structure being joint precursors of mammalian neocortex has persisted and for good reason. For example, Edinger (1908) and Elliot Smith (1910) noted that the thalamus in reptiles projected to the subcortical "epistriatum." Because the projections of the dorsal thalamus are used to define "neocortex" in mammals, this observation suggests that the epistriatum may be a precursor of neocortex (e.g., Rose and Woolsey 1949). Furthermore, the epistriatum is a different type of tissue from the rest of the striatum in reptiles: along with the cerebral cortex of reptiles and mammals, the epistriatum is pallial in embryological origin. Hence, Elliot Smith and Herrick, and certainly Johnston (1915), who later called this noncortical tissue the "dorsal ventricular ridge," foreshadowed some of the discoveries surrounding the thalamotelencephalic projections of the sensory systems in reptiles and mammals made a half-century later when adequate tract-tracing techniques finally became available.

1.2 Contribution of Afferent Projections to the Study of Forebrain Homologies

The new era in tract-tracing, and hence a new era in the pursuit of the reptile–mammal forebrain transition, began in the 1960s and has continued to the present. Walle Nauta and his students and grandstudents, names that now constitute a Who's Who of comparative neuroanatomy, began to use Nauta's sensitive tract-tracing methods to follow the unmyelinated sensory fibers of nonmammals,

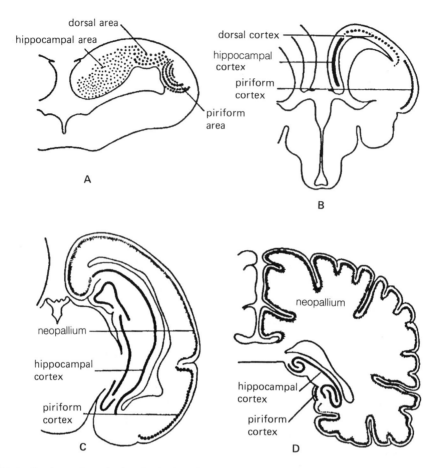

FIGURE 32.1. Sections through the forebrain of (A) amphibian (*Necturus*), (B) tortoise (*Cistudo*), (C) opossum (*Didelphis*), and (D) human, showing the similarity in position of dorsal cortex with neocortex (neopallium). (From Herrick 1956.) Note similar relative position of reptilian dorsal cortex (in B) and mammalian neocortex (in C and D). Also note absence of cerebral cortex of any kind in amphibian (A). Cerebral cortex probably first appeared in reptiles, neocortex in mammals.

synapse by synapse, to the telencephalon (Nauta and Karten 1970).

These new tract-tracing methods when applied to reptiles and birds eventually showed that the dorsal ventricular ridge or DVR (i.e., the "epistriatum" of Edinger 1908, the "hypopallium" of Elliot Smith 1910 and of Kappers, Huber, and Crosby 1936) received separate thalamic projections from the sensory systems in much the same manner as the neocortex receives them in mammals (see Petras and Noback 1969; Ulinski 1986). Therefore, it appeared that the DVR might serve reptiles and birds as the telencephalic head ganglion of the sensory systems in much the same manner as the sensory areas of neocortex serve mammals. Given

this new insight into the reptilian, and presumably the premammalian, organization of the telencephalon, it was only natural to consider whether the dorsal ventricular ridge of reptiles might somehow be a precursor of the sensory areas of neocortex in mammals.

1.3 Relative Locations of Sensory Targets in the Forebrain

The difference between the thalamotelencephalic projections in the reptilian forebrain and the mammalian forebrain are shown in schematic form in Figure 32.2. It can be seen in Figure 32.2A that the reptilian dorsal cortex receives direct thalamic

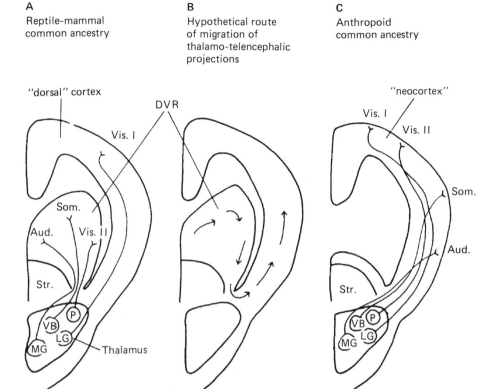

FIGURE 32.2. Schematic drawing of projection from four thalamic sensory nuclei to the telencephalon in (A) reptiles and (C) monkey. (B) Hypothetical route of migration from reptilian dorsal ventricular ridge to mammalian neocortex. See Ulinski (1983) for derivation of A, Karten (1969) for B, Jones and Burton (1976) for C.

Vis. I, first visual field or lateral geniculate (LG) receiving zone; Vis. II, second visual field or puvinar (P) receiving zone; Som., somatosensory field or ventrobasal (VB) receiving zone; Aud., auditory field or medial geniculate (MG) receiving zone; DVR, dorsal ventricular ridge; Str., striatum.

projections from only one sensory system (the retinogeniculate visual system) while the other sensory nuclei of the thalamus project to the dorsal ventricular ridge. The mammalian neocortex, in contrast, receives projections from each of the thalamic sensory nuclei (Fig. 32.2C). It is the possibility of a change in location of the subcortical thalamic targets in reptiles to a location in the neocortex of mammals, as first suggested by Karten (1969), that is the starting point of the present discussion (Fig. 32.2B).

It is now known that the thalamic sensory nuclei in reptiles project to the dorsal ventricular ridge in an orderly fashion (Fig. 32.2A). The second visual system (i.e., the retinotectothalamic system) projects to lateral DVR, the auditory system projects

medially, and the somatosensory system projects in between (e.g., see Ulinski 1983). This ordering of the projections of the sensory systems means that an uncomplicated, order-preserving migration of the thalamic targets from the dorsal ventricular ridge onto the surface of the cerebrum could result in the arrangement of sensory areas seen now on the cortical surface of mammals (Fig. 32.2B and C).

For the present purpose, it can be noted that if the sensory zones in the reptilian DVR were to preserve their relative positions while gradually migrating onto the cortical surface, it would be the auditory area that would be the last to leave a subcortical position and the last to appear on the cortical surface. Indeed, the topographical arrangement of the sensory areas within the neocortex of

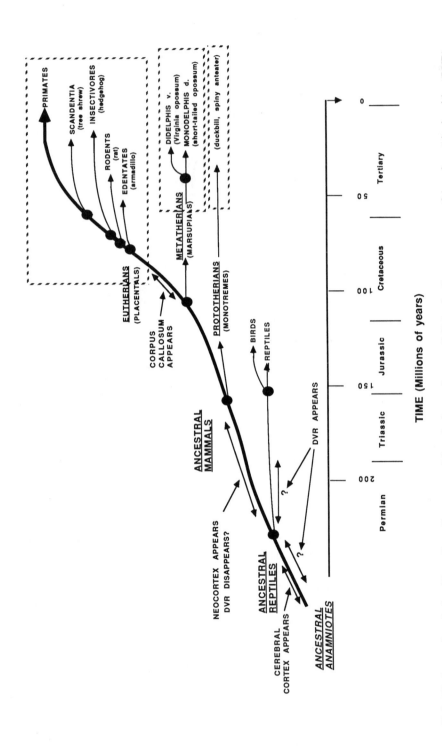

FIGURE 32.3. Phylogenetic relationships of reptiles, marsupial and placental mammals, and the extinct animals in the anthropoid line of descent. (After clado-grams of Clemons 1977 and of Eisenberg 1981.) Queries meant to denote the two possibilities for first appearance of dorsal ventricular ridge (DVR) now seen only in reptiles and birds. If DVR arose before ancestral reptiles, it would have had to disappear between ancestral reptiles and ancestral mammals.

mammals is consistent with this notion. Auditory cortex is found just above the rhinal fissure in mammals – exactly the position one would expect it to occupy if it had been the last to migrate onto the cortical surface by way of that fissure (e.g., Diamond, Chow, and Neff 1958).

1.4 Phyletic Evidence of Nuclear Migration

Adding a cladistic or phylogenetic dimension to this idea, if one hoped to find evidence of a migration or translocation through geological time, it would be in mammals with the most remote common ancestry with placental mammals which might yield evidence of an incomplete migration. Furthermore, if the migratory process remained incomplete in these animals, it would be the auditory tissue which most likely would be caught midroute. That is, if the transition might have been arrested before the entirety of sensory tissue reached the neocortical surface in these animals, then one might expect either an entirely subcortical projection of the auditory system or a dual projection, part to cortex and part to subcortex.

Our idea was that marsupials, with their last common ancestor with placental mammals somewhere along the lineage between ancient reptiles and the last common ancestor of placental mammals, might provide a snapshot of the migratory process arrested at some intermediate stage (Fig. 32.3). This general idea received support from the work of Ebner (1967), who had already shown that in opossum or hedgehog, a lesion in the medial geniculate resulted in degenerating terminals along its projection to cortex including terminals in nearby noncortical telencephalic nuclei.

2. Telencephalic Projections of Medial Geniculate in Marsupials

2.1 Subcortical Projections in *Didelphis virginiana*

It was for this reason that with Motoi Kudo, now of Kanazawa University in Japan, we took up the question in 1983. We began by injecting tritiated leucine into the medial geniculate of one of the most neurologically generalized marsupials available (the Vir-

ginia opossum) in order to trace the auditory fibers to their telencephalic target. We found that, as in any mammal, a large band of fibers leaves the opossum's medial geniculate, streams forward through the thalamus, and then turns sharply laterally, destined for a termination in (auditory) neocortex just above the rhinal fissure (see Fig. 32.4; cf. Ebner 1967, 1969; Diamond and Utley 1963).

However, in addition to the heavy projection to the cortex, many labeled fibers were seen to leave the labeled tract, turn ventrally, and terminate in the lateral amygdala and parts of putamen and caudate (Fig. 32.4; Kudo et al. 1986). This result confirmed Ebner's (1967) results with the Nauta degeneration methods.

2.1.1 Large Subcortical Projection Originates in Caudal Medial Geniculate

The projection to noncortical structures in opossum appeared so heavy that it occurred to us at the time that what seemed to be a medial geniculate projection might not be emanating entirely from medial geniculate alone. That is, the location of the medial geniculate at the caudal end of the diencephalon means that it is surrounded by mesencephalic structures, dorsally, medially, and ventrally and it might be these nonthalamic nuclei which projected to the subcortical structures. Therefore, we sought to label the medial geniculate cells retrogradely from the subcortical target. We injected horseradish peroxidase–wheat germ agglutinin (HRP–WGA) into the subcortical target through the lateral surface of pyriform cortex to avoid breaking and labeling the fibers destined for auditory cortex. We could verify that this requirement was met in each case by the absence of retrogradely labeled corticothalamic cells and by the absence of orthogradely labeled terminals in auditory cortex. Having assured ourselves that we had not interrupted and inadvertently labeled the thalamic projections to the cortex, we could then look for retrogradely labeled cells in the medial geniculate.

As might be expected, some nonthalamic nuclei surrounding the medial geniculate (n. subparafascicularis and n. suprageniculate) were found to contain retrogradely labeled cells. However, a vast number of retrogradely labeled cells were found within the medial geniculate itself. Indeed, the entire caudal one-third of the medial geniculate was

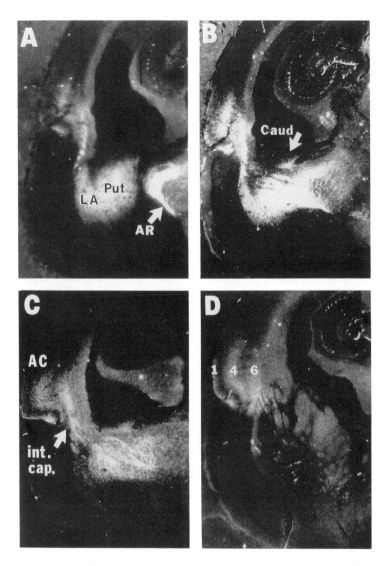

FIGURE 32.4. Darkfield photomicrograph of labeled fibers after tritiated leucine injection in medial geniculate in *Monodelphis domestica*. (A), (B), (C), and (D) are suc-cessively more rostral sections. The fibers are destined for the lateral amygdala and putamen (see B and C) as well as auditory cortex.

filled with heavily labeled cells. Equally important in the present context, no other thalamic sensory nucleus contained labeled cells, only the medial geniculate (Kudo et al. 1986).

Needless to say, this result was far beyond our expectations. When we began this experiment with only the orthograde degeneration studies of Ebner (1967) and the retrograde degeneration studies of Diamond and Utley (1963) as a guide, we hoped to find a remnant of a medial geniculate projection to nonneocortical structures. But what we had found

proved to be a quite substantial projection to sub-cortex paralleling, and almost equalling in size, the projection to auditory cortex.

2.1.2 Two Segregated Geniculotelecephalic Auditory Systems

Before continuing it is of interest to note in the present context that the labeling of the medial genic-ulate projections also showed that the possible migration route of the subcortical target onto the surface of the nearby cortex would have faced no great anatomical or topological barrier. The orthograde tract-tracing results (e.g., Fig. 32.4) show that the MG projections to cortex pass through the subcortical targets, appearing to drop off fibers from the main bundle as they do so.

Although this physical arrangement of the two targets is satisfying for the notion of a translocation, the close proximity of the two targets in the opossum opened the question of whether they might be served by collaterals of one set of MG axons. That is, one and the same set of axons might project both to subcortex and to cortex—implying a somewhat different type of migration than that envisioned.

To explore this question, we began by retrogradely labeling the cortical projection and comparing the location of the contributing MG cells to those retrogradely labeled from the subcortical target. We found that the MG cells retrogradely labeled from the two targets were quite separate. That is, the rostral two-thirds of the opossum's medial geniculate could be almost completely labeled by HRP placed in the cortex alone while the caudal one-third remained entirely unlabeled. The reverse was true for HRP injected into the subcortical target: Here it was the caudal one-third that was almost completely labeled while the rostral two-thirds contained few, if any, retrogradely labeled cells. Once more, no other thalamic nucleus contained labeled cells after either injection.

Therefore, it began to appear that there was not just one projection with collateral axons serving the two targets but instead, two separate projections originating from two separate populations of cells in the opossum's medial geniculate: cells in the caudal one-third projecting subcortically, cells in the rostral two-thirds projecting cortically.

To be sure, we verified the separateness of the two projections using double-labeling methods. We injected True Blue or Nuclear Yellow, both retrograde tracers known to be mutually compatible, one into the cortical target and the other into the subcortical target. Although an injection of a mixture of the two tracers in either target resulted in a very high percentage of double-labeled MG cells, the results obtained with one marker in each target resulted in virtually no double-labeled cells—confirming that the two targets are served by two separate sets of medial geniculate cells (Kudo et al. 1986).

2.1.3 Incomplete Migration of Auditory Targets in the Telencephalon

At this point, it seemed clear that the opossum had not just one but two separate thalamotelencephalic auditory systems—one originating in the rostral two-thirds of MG and ending in cortex just like placental mammals, the other originating in the caudal one-third of MG and ending in subcortical targets—an arrangement reminiscent of nonmammals. The fact that these two systems were not collaterals of each other seemed to suggest that the subcortical projection target of the medial geniculate in premammalian reptiles might indeed have migrated bit-by-bit onto the neocortical surface. If this were the case, what was seen in the opossum could be considered to be a third snapshot, this one with the process arrested relatively late in the migration—after the entire second visual system, the entire somatosensory system, and about two-thirds of the auditory system had reached neocortex.

2.2 Subcortical Projections in *Monodelphis domestica*

At this point the possibility intruded that the Didelphis opossum might be unique in its thalamotelencephalic projections. Therefore, we turned to a different genus of generalized marsupial, the South American gray short-tailed opossum, *Monodelphis domesticus*. Performing the same set of experiments once more, we found the same double projection system as in the Didelphis opossum (Fig. 32.5). In this case, however, the subcortical projection of the medial geniculate proved to be even larger than it is in the Didelphis opossum (Fig. 32.6).

Figure 32.5 shows the orthograde labeling which results from an injection of tritiated leucine into the medial geniculate and also the retrograde labeling of MG cells resulting from an injection of HRP-WGA into the lateral amygdala and putamen in the Monodelphis opossum. Again, after verification that the cortical projection of the medial geniculate was not interrupted or inadvertently labeled by the retrograde marker, a large number of heavily labeled cells was found in the caudal part of the medial geniculate once more. Moreover, in Monodelphis the subcortical system occupies fully 50% of the medial geniculate—measurably more than the 33% of the Didelphis. And once more, injections of the

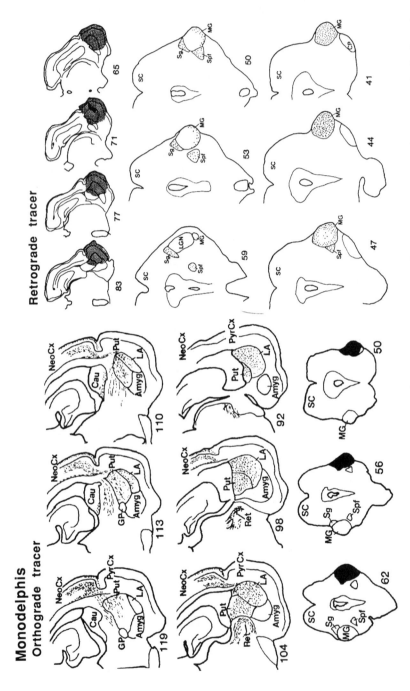

FIGURE 32.5. Left: Distribution of orthogradely labeled fibers and terminals following tritiated leucine injection in the medial geniculate in *Monodelphia domestica*. Note heavy terminal labeling in neocortex, predominantly layers 1, 3, 4, and 6; lateral amygdala; and putamen. Moderate terminal labeling is seen in the caudate nucleus ventrally. Right: Distribution of retrogradely labeled neurons in the medial geniculate following injection of WGA–HRP in the lateral amygdala and putamen. Note heavy retrograde labeling in caudal MG and sparse labeling in rostral MG. If marker is placed in auditory cortex (not shown), retrograde label appears in cells within rostral and not caudal MG. In this and following figures numerals refer to sections numbered from caudal to rostral.

FIGURE 32.6. Darkfield photomicrographs of retrogradely labeled neurons in the medial geniculate of *Monodelphis domestica* following injection of WGA–HRP into the lateral amygdala and putamen (Case 18). (a) Rostral MG, suprageniculate and subparafascicular nuclei (section 50 in Fig. 32.5); (b) caudal MG (section 41 in Fig. 32.5). Virtually every projection neuron in the caudal half of MG projects to subcortical nuclei.

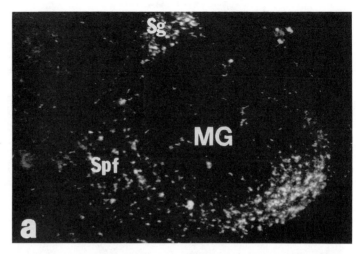

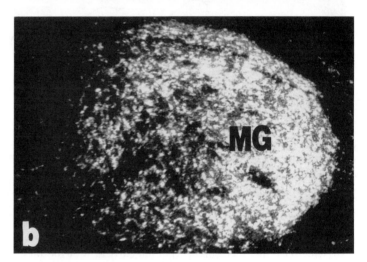

retrograde marker confined to the subcortical target labeled the caudal half with only a few cells in the rostral half of the medial geniculate and labeled cells in no other thalamic sensory nucleus (Fig. 32.6). The reverse was seen again after retrograde labeling from the cortical target.

3. Telencephalic Projections of Medical Geniculate in Placentals

Convinced that Didelphis was not unique in having a substantial subcortical auditory projection and satisfied that the arrangement seen in the two marsupials was consistent with the notion of a migration of the auditory area from a subcortical to a cortical location, we saw two possible ways to proceed (see Fig. 32.3). One was to do the same sort of experimentation in monotremes with their even more remote common ancestry with placental mammals to see if the subcortical projection might be larger and the cortical projection smaller than in the marsupials. This idea was entirely ruled out when the Australian government declared monotremes to be completely protected species. Certainly with this type of experimentation requiring many animals to be properly conducted, experimentation on monotremes was, and still is, out of the question.

The second possibility was to turn in the opposite direction, to placental mammals with more recent common ancestry with primates, to see if a

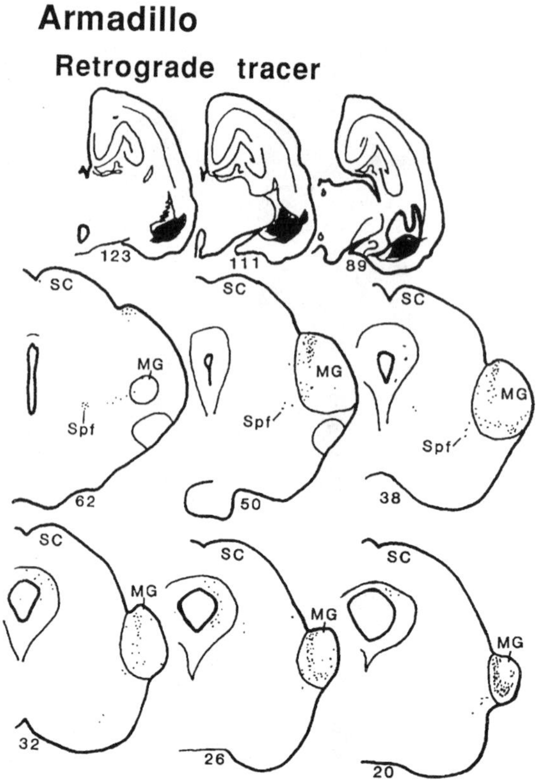

Armadillo
Retrograde tracer

FIGURE 32.7. Distribution of retrogradely labeled neurons in the medial geniculate following injection of WGA–HRP into the lateral amygdala and putamen in armadillo (*Dasypus novemcinctus*). Projection to auditory cortex arises from the blank spaces in MG seen in this figure. In most placental mammals, subcortical projections arise more from medial than from caudal MG.

similar subcortical projection might be present and if so, whether it might be smaller and the cortical projection larger than in the marsupial opossums.

3.1 Subcortical Projections in Placentals

At about this time we learned that no subcortical projection from MG seems to be present in monkey (see Burton and Jones, 1976) and that at least some projections of the medial geniculate body to subcortical regions of the forebrain had been reported to be present in several placental species including the hedgehog (*Paraechinus hypomelas*), tree shrew (*Tupaia glis*), gray squirrel (*Sciurus carolinensis*), cat, and rat (Ebner 1969; Graybiel 1973; Ryugo and Killackey 1974; Ottersen and Ben-Ari 1979; Veening, Cornelissen, and Lieven 1980; Russchen, 1982; Lin, May, and Hall 1984; Maiskii, Gonchar, and Chupega 1984; Takada et al. 1985; LeDoux, Ruggiero, and Reis 1985).

Particularly important to us, the description of the rat's subcortical projection seemed to indicate that it was smaller than we had seen in the opossum (LeDoux, Ruggiero, and Reis 1985). We performed the same experiments in rats and confirmed LeDoux, Ruggiero, and Reis's discovery, noting for our purposes that the rat's subcortical projection, though substantial, was measurably smaller in size than the opossum's— reinforcing the notion of a possible migration once more. Perhaps we now had four independent snapshots of a projection in transition—one at the time of the ancient reptiles, another at the last common ancestor of marsupials and placentals, another at the last common ancestor of rodents and primates, and still another perhaps at the origin of primates. To this we could add the information that the migratory process was probably completed sometime before the origin of Old World monkeys.

3.2 Phyletic Reduction
of Subcortical Projections

Encouraged, we turned to placental mammals with successively more recent common ancestry with primates to fill in still more points along the historical time-line. Using cladograms of Eisenberg (1981) and Clemons (1977) or McKenna (1975) as our guide to the relative timing of the last common ancestry of each animal with primates, we performed the same tract-tracing experiments in armadillo (*Dasypus novemcinctus*), hedgehog (*Paraechinus hypomelas*), and tree shrew (*Tupaia belangeri*).

Figures 32.7, 32.8, and 32.9 show examples of the results of these experiments. Each of these animals proved to possess a subcortical as well as a cortical projection of their medial geniculate. Furthermore, injections of a retrograde marker into the subcortical target resulted in the heavy labeling of a distinct subset of cells in the medial geniculate. Most important for the notion of a relocation, however, the amount of medial geniculate contributing to the subcortical projection was less than in opossums and tended to decrease further as the animal's kinship with primates became closer.

If the several animals are arranged in order of their "recency of common ancestry with primates" and the percentage of medial geniculate projecting to the subcortical target is estimated for each, a gradual loss of the subcortical projection is evident (Fig. 32.10). This loss can be interpreted as showing that among the successive ancestors of primates, the subcortical projection dwindled in size both relatively and absolutely. Among the placentals, the origin of the projection also seemed to shift from the caudal medial geniculate to the dorsal division and medial (or magnocellular) division of the medial geniculate, which may provide a clue to the origins of these divisions.

4. Discussion

In considering the implications of these results for the origin of auditory cortex, it is important to note that the possibility of a migration of auditory tissue from a noncortical to a cortical location need not have started from the dorsal ventricular ridge at least as it is now seen in modern reptiles and birds.

Although we used the orderliness of that projection as an impetus for these experiments, it can be seen that the existence of a DVR in the reptilian-mammalian common ancestry is not a logically necessary condition for the present results to obtain because the subcortical targets of the auditory projection are found in a lateral position even in the marsupials. Therefore, our results yield no information whatever regarding a still more ancient position of the auditory tissue—it might have been more medial such as is seen in the DVR of modern reptiles or it might have been in a more lateral position before the beginning of the migration to cortex. For example, the premammalian reptiles might not have had a DVR at all; that is, as shown in Figure 32.3, modern reptiles may have acquired a DVR only after their divergence from the reptilian lineage leading to mammals (Northcutt 1978, 1981). Evidence for this possibility might be seen in the forebrain of turtles which have the most remote common ancestry with other modern reptiles yet have only a small DVR. If premammalian reptiles had no DVR, the arrangement of thalamic projections to DVR in modern reptiles is beside the point. The present results do not speak to this possibility since a lateral position and a graded loss of subcortical auditory tissue would be an expected result of a migration almost regardless of the starting location of the subcortical auditory tissue in premammalian reptiles. Indeed, Northcutt (1967) has suggested that the premammalian condition of the telencephalon might be better inferred from a still earlier, perhaps anamniotic, ancestry of mammals in which the noncortical thalamic targets might be found more laterally and ventrally than they are found in modern reptiles. This possible location of telencephalic auditory tissue is much closer to that which is now seen as subcortical auditory tissue in opossums; that is, it is much closer to the position of the tissue found just behind and below the opossum's rhinal fissure. Therefore, this alternative to the ancient location of telencephalic auditory tissue is entirely consistent with the present results.

4.1 Translocation of Auditory
Telencephalon

Regardless of the exact location of the telencephalic auditory tissue in premammalian reptiles,

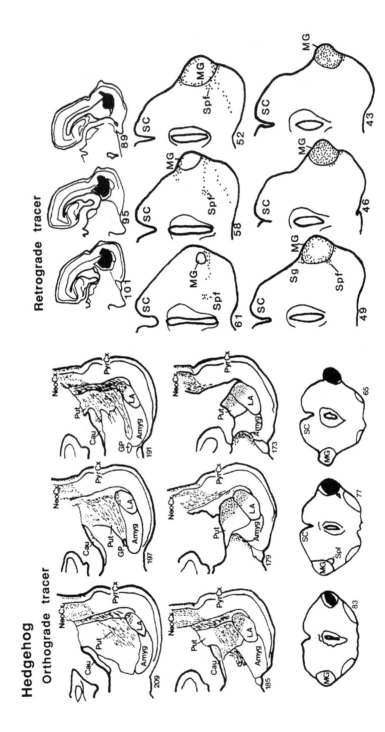

FIGURE 32.8. Left: Distribution of orthogradely labeled fibers and terminals following injection of tritiated leucine in the medial geniculate in hedgehog (*Paraechinus hypomelas*). Right: Distribution of retrogradely labeled neurons in medial geniculate following injection of WGA–HRP in the lateral amygdala and putamen. Subcortical projections of MG in hedgehog arise mostly from caudal MG while cortical projection (not shown) arises from rostral MG.

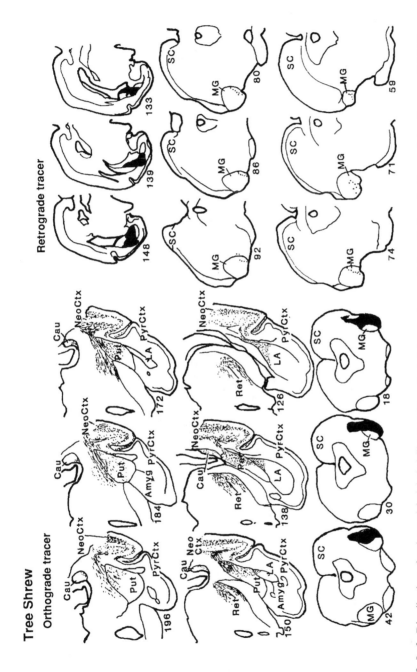

FIGURE 32.9. Left: Distribution of orthogradely labeled fibers and terminals following injection of tritiated leucine in the medial geniculate in tree shrew (*Tupaia belangeri*). Right: Distribution of retrogradely labeled neurons in medial geniculate following injection of WGA-HRP in the lateral amygdala and putamen. Subcortical projections arising from medial MG are present though there are measurably fewer than in opossums, armadillo, or hedgehog. Cortical projections arise from remainder of MG (not shown).

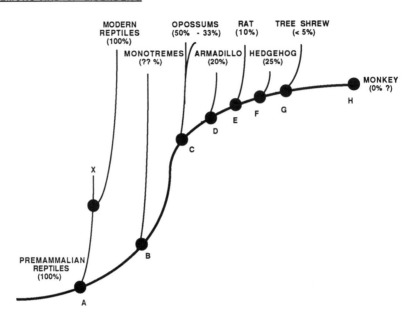

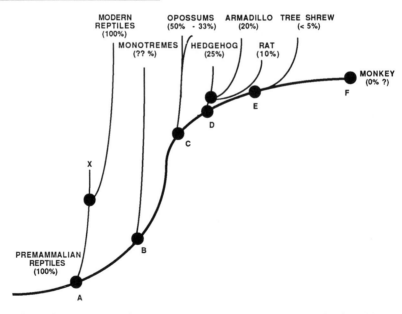

FIGURE 32.10. Estimated percentage of MG cells projecting to subcortical telencephalon in animals aligned on the basis of their recency of common ancestry with anthropoids. See Ulinski (1983) for reptilian estimate, Burton and Jones (1976) for monkey.

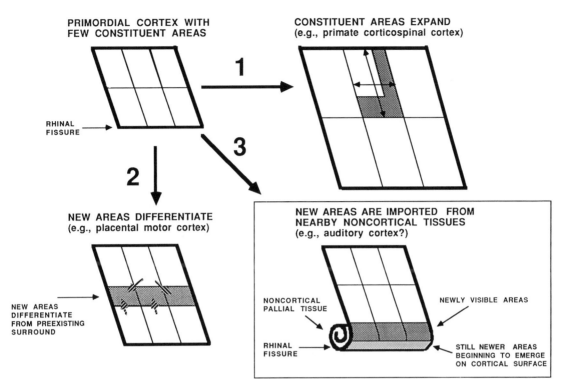

FIGURE 32.11. Hypothetical transformations resulting in an expanded neocortex. 1, Preexisting areas of neocortex expand in size (Lashley and Clark 1946; see Kaas 1987). 2, Differentiation of new areas from preexisting neocortical areas (Brodmann 1909; von Economo 1929; Kaas 1987). 3, Migration, importation, or inclusion of previously nonneocortical areas into neocortex. The present results suggest that auditory cortex arose by means of the third process.

if it is accepted that the common properties of the dual auditory projection among the successive sets of mammals are likely to approximate the trend among their successive common ancestors, the collection of results can be viewed as a series of snapshots of the progress of a transition or relocation of telencephalic auditory tissue: they show a subcortical projection dwindling in size while the cortical projection increases. The auditory system seems to be in the process of shifting its target onto the cortical surface in much the way suggested by Karten (1969). This idea is summarized in the final figure (Fig. 32.11–32.13).

4.2 Expansion of Preexisting Neocortex

Before considering the chief point of Figure 32.11, it should be noted that it has long been suggested that the vast enlargement of the cerebral cortex in

mammals along mankind's ancestral lineage theoretically could have expanded, and probably did expand as a result of at least two types of process (see Kaas 1987, for recent review). In one of these (Fig. 32.11-1), some or all of the preexisting areas of cortex expanded in size, the new total summing to a larger amount of cortex (Lashley and Clark 1946; and see Kaas 1987). Indeed, there is evidence that at least some cortical areas (such as the cortex originating corticospinal fibers in the primate lineage) probably did expand in just this way (Nudo and Masterton 1990a,b).

4.3 Differentiation of Preexisting Neocortex

The second process that would have resulted in the expansion of neocortex is also an old idea: the differentiation of entirely new areas from the old or preexisting areas within the cortex itself (Brodmann 1909; von Economo 1929; Kaas 1987). Once more, there is evidence that this type of contribution to the enlargement of neocortex probably occurred too. For example, "motor cortex" as seen in placental mammals (i.e., Area 4) seems to be entirely absent in the marsupial opossums, and short of springing de novo after the origin of marsupials and before the origin of placental mammals, or of disappearing entirely during the recent history of opossums, it is reasonable to suppose that somehow, it differentiated from the surrounding cortex (e.g., Donoghue and Ebner 1981). The proliferation of subordinate cortical sensory areas (Vis. III and IV or Aud. II and S II) seem to provide other examples (Kaas 1987).

4.4 Importation of Non-Neocortex into Neocortex

Returning to the question of the possible origin of auditory cortex, Figure 32.11-3 shows a third way in which the neocortex might have expanded. It is the notion of the migration (or importation or inclusion) of previously non-neocortical areas into the neocortex. It is pictured here as a figurative unrolling of subcortical tissue onto the cortical surface beginning from a location behind the rhinal fissure. However, the same sort of process might have occurred at the other borders of neocortex too—for example, among the cingulate, subicular, and perhaps entorhinal neighbors of neocortex. It is this third means of gaining more and newer neocortical areas, by importing them from nearby non-neocortical tissues, that we think might have been the origin of auditory cortex.

5. Summary

The question of the evolutionary origin of auditory cortex is entangled in the larger question of the origin of mammalian neocortex. Because of the non-homology of the reptilian dorsal cortex and the mammalian neocortex, the idea that dorsal cortex together with some other reptilian forebrain structure being joint precursors of mammalian neocortex has persisted.

The thalamic sensory nuclei in reptiles project to the "dorsal ventricular ridge" in an orderly fashion, in an arrangement similar to that seen in

the sensory projections to neocortex in mammals. Moreover, if the sensory zones in the reptilian DVR were to preserve their relative positions while migrating onto the cortical surface, it would be the auditory area that would be the last to leave its subcortical position and the last to appear on the neocortical surface. Should this migratory process have remained incomplete in some extant mammals, it would most likely be seen in mammals with the most remote common ancestry with placental mammals: monotremes and marsupials.

Recent tract-tracing experiments have shown that didelphid marsupials have a substantial medial geniculate projection to subcortical areas as well as the usual mammalian projection to neocortex. Comparative examination of several placental mammals with successively more recent common ancestry with primates suggests a gradual loss of the subcortical projection with an increase in the proportion of the neocortical projection.

Insofar as these results can be viewed as a series of snapshots of a transition or relocation of telencephalic auditory tissue through time, the subcortical projection dwindled in size while the cortical projection increased. This shifting of the telencephalic target of the auditory system onto the surface of the cerebrum through the migration, or inclusion, of previously subjacent, non-neocortical tissue may have been the process by which the of auditory cortex, now typical of placental mammals, originally appeared.

References

Brodmann K (1909) Vergleichende Lokalisationslehre der Grosshirnrinde. Leipzig: Verlag Barth.

Burton H, Jones EG (1976) The posterior thalamic region and its cortical projection in new and old world monkeys. J Comp Neurol 168:249–302.

Clemons WA (1977) Phylogeny of the marsupials. In: Stonehouse R, Gilmore D (eds) The Biology of the Marsupials. Baltimore: University Park Press, pp. 51–60.

Diamond IT, Chow KL, Neff WD (1958) Degeneration of caudal medial geniculate body following cortical lesions ventral to auditory area II in the cat. J Comp Neurol 109:349–362.

Diamond IT, Utley JD (1963) Thalamic retrograde degeneration study of sensory cortex in opossum. J Comp Neurol 120:129–160.

Donoghue JP, Ebner FF (1981) The laminar distribution and ultrastructure of fibers projections for three thalamic nuclei to the somatic sensory-motor cortex of the opossum. J Comp Neurol 198:389–420.

Ebner FF (1967) Afferent connections to neocortex in the opossum (*Didelphis virginiana*). J Comp Neurol 129:241–268.

Ebner FF (1969) A comparison of primitive forebrain organization in metatherian and eutherian mammals. In: Petras JM, Noback CR (eds) Comparative and Evolutionary Aspects of the Vertebrate Central Nervous System. Ann NY Acad Sci 167:241–257.

Edinger L (1908) Bau der Nervosen Zentralorgane. Band ii.

Eisenberg JF (1981) The Mammalian Radiations. Chicago: The University of Chicago Press.

Graybiel AN (1973) Some fiber pathways related to the posterior thalamic region in the cat. Brain Behav Evol 6:363–393.

Herrick CJ (1956) The Evolution of Human Nature. Austin: The University of Texas Press.

Johnston JB (1915) The cell masses in the forebrain of the turtle *Cistudo carolina*. J Comp Neurol 25:393–468.

Kaas JH (1987) The organization and evolution of neocortex. In: Wise SP (ed) Higher Brain Functions. New York: John Wiley & Sons, pp. 347–378.

Kappers ACU, Huber GC and Crosby EC (1936) The Comparative Anatomy of the Nervous System of Vertebrates Including Man. New York: MacMillan.

Karten HJ (1969) The organization of the avian telencephalon and some speculations on the phylogeny of the amniote telencephalon. Ann NY Acad Sci 167:164–179.

Kudo M, Glendenning KK, Frost SB, Masterton RB (1986) Origin of mammalian thalamocortical projections. I. Telencephalic projections of the medial geniculate body in the opossum (*Didelphis virginiana*). J Comp Neurol 245:176–197.

Lashley KS, Clark G (1946) The cytoarchitecture of the cerebral cortex of *Ateles*: A critical examination of architectonic studies. J Comp Neurol 85:223–305.

LeDoux JE, Ruggiero DA, Reis DJ (1985) Projections to the subcortical forebrain from anatomically defined regions of the medial geniculate body in the rat. J Comp Neurol 242:182–213.

Lin CS, May PJ, Hall WC (1984) Nonintralaminar thalamostriatal projection in the grey squirrel (*Sciurus carolinensis*) and tree shrew (*Tupaia glis*). J Comp Neurol 230:33–46.

Maiskii VA, Gonchar YA, Chupega EN (1984) Neuronal populations in the posterior group of thalamic nuclei projecting to the amygdala and auditory areas in cats. Neirofiziologiya, 16:213–224.

McKenna MC (1975) Toward a phylogenetic classification of the Mammalia. In: Luckett WP, Szalay FS (eds) Phylogeny of the Primates. New York: Plenum Press, pp. 21–46.

Nauta WJH, Karten HJ (1970) A general profile of the vertebrate brain, with sidelights on the ancestry of cerebral cortex. In: Schmitt FO (ed) The Neuro-

sciences. New York: The Rockefeller University Press, pp. 7–26.

Northcutt RG (1967) Architectonic studies of the telencephalon of *Iguana iguana*. J Comp Neurol 130:109–147.

Northcutt RG (1978) Forebrain and midbrain organization in lizards and its evolutionary significance. In: Greenberg N, MacLean PD (eds) The Behavior and Neurology of Lizards. NIMH, Rockville, Maryland, pp. 11–64.

Northcutt RG (1981) Evolution of the telencephalon in nonmammals. Annu Rev Neurosci 4:301–350.

Nudo RJ, Masterton RB (1990a) Descending pathways to the spinal cord, III: Sites of origin of the corticospinal tract. J Comp Neurol 296:559–583.

Nudo RJ, Masterton RB (1990b) Descending pathways to the spinal cord, IV: Some factors related to the amount of cortex devoted to the corticospinal tract. J Comp Neurol 296:584–597.

Ottersen OP, Ben-Ari Y (1979) Afferent connections to the amygdaloid complex of the rat and cat. I. Projections from thalamus. J Comp Neurol 187:401–424.

Petras JM, Noback CR (eds) (1969) Comparative and Evolutionary Aspects of the Vertebrate Central Nervous System. Ann NY Acad Sci 167.

Rose JE, Woolsey CN (1949) Organization of the mammalian thalamus and its relationships to cerebral cortex. EEG Clin Neurophysiol 1:391–404.

Russchen FT (1982) Amygdalopetal projections in the cat. V. Subcortical afferent connections. A study with retrograde tracing techniques. J Comp Neurol 207:157–176.

Ryugo DK, Killackey HP (1974) Differential telencephalic projections of the ventral and medial divisions of the medial geniculate body of the rat. Brain Res 82:173–177.

Smith GE (1910) Some problems relating to the evolution of the brain. Lancet I:1–6.

Takada M, Itoh K, Yasui Y, Sugimoto T, Mizuno N (1985) Topographical projections from the posterior thalamic regions to the striatum in the cat, with reference to possible tecto-thalamo-striatal connections. Exp Brain Res 60:385–396.

Ulinski PS (1986) Neurobiology of the therapsid-mammal transition. In: Hotton N III, Roth JJ, Roth EC, MacLean PD (eds) The Ecology and Biology of Mammal-Like Reptiles. Washington, D.C.: Smithsonian Institution Press, pp. 149–171.

Ulinski PS (1983) Dorsal Ventricular Ridge; a Treatise on Forebrain Organization. New York: John Wiley & Sons.

Veening JG, Cornelissen FM, Lieven PAJM (1980) The topographical organization of afferents to the caudatoputamen of the rat. A horseradish peroxidase study. Neuroscience 5:1253–1268.

von Economo C (1929) The Cytoarchitectonics of the Human Cortex. Oxford: Oxford University Press.

33
Mammalian Auditory Cortex—
Some Comparative Observations

Michael M. Merzenich and Christoph E. Schreiner

1. Introduction

The organization of auditory cortical fields has been described in a number of marsupial, insectivore, rodent, carnivore, and primate species. These studies have been conducted with limited consideration of the issues of comparative development. Experiments have been piecemeal, and results seldom confirmed. There has been only anecdotal study of the auditory cortex in primitive insectivores, and few or no studies of auditory forebrain anatomy or physiology in several groups of hearing-specialized mammals, including toothed whales, pinnipeds, and burrowing insectivores and rodents. Taken as a whole, these data provide only a limited basis for understanding the phylogenetic development of the auditory cortex in mammals, or for drawing comparative conclusions about cortical field homologies or niche-related specializations.

In this report, studies of the mammalian auditory cortex that provide insights into its common and species-specific organizational features shall be reviewed. They include: (a) Primary auditory cortex has been identified and described by application of cytoarchitectural or myeloarchitectural methods in a number of mammalian species, principally in studies conducted in the first half of this century. (b) Best-frequency electrophysiological mapping experiments have resulted in the functional identification of primary auditory cortex and other auditory cortical fields in a scattered group of mammals. (c) Similarities and differences in the internal organization of the so-called "primary auditory cortical field" or "A1" have been documented, with widely variable precision,

in electrophysiological studies conducted in more than 20 mammals. (d) Anatomical connections of auditory cortical fields, important for inferring cortical field homology, have been defined in piecemeal fashion for a small number of mammals. (e) There is growing evidence for auditory cortical maps that are not strictly best-frequency based—as well as other evidence for auditory forebrain specialization—derived from studies of a small number of hearing-specialized mammals.

After reviewing these studies, we shall briefly discuss them from the perspective of some current views of sensory system development in mammalian evolution.

2. Some Comparative Observations on the Organization of Auditory Cortical Fields

2.1 Cytoarchitectonic and Myeloarchitectonic Studies of Auditory Cortical Fields

The basic cytoarchitecture and the cytoarchitectonic boundaries of cortical field A1 are not as distinctive as are those of primary visual or somatosensory cortical fields (Campbell 1905; Brodmann 1909; von Economo 1927; M. Rose 1935; von Bonin and Bailey 1947; J. Rose 1949). Indeed, the location of an auditory "granular cortex" or "koniocortex" followed myeloarchitectonic studies that revealed an early myelinated cortical region on the superior temporal plane in primates including man, similar in its early myelinization and myeloarchitecture to primary cortical fields in other sensory modalities (see Flechsig 1927). The fact that there are anatomical boundaries *within* A1 in at least primate and carnivore species (von Economo 1927; Beck 1927; von Bonin and Bailey 1947; Sanides 1973; Imig and Brugge 1978; Brugge and Reale 1985) further complicated the identification of A1 field perimeter boundaries, and in fact led von Economo to describe the konio-cortical field in humans as comprised of several narrow bands that are wholly discontinuous from each other. Because the auditory koniocortex is relatively difficult to identify and demarcate on the basis of simple light microscopic anatomical criteria, its identification on strictly cytoarchitectonic terms has often been uncertain, and in fact in several classical studies of this cortical zone, it has been misidentified. These practical difficulties have forestalled any wide light microscopic survey of its location and extent in a larger series of mammals as has been possible in studies of primary visual, somatosensory and motor cortices, and bring historic reconstructions of its location or extent into some question.

It might be noted that early *myelo*architectonic anatomists were able to identify primary auditory cortex, sometimes distinguish subtle distinctions within it, and delimit surrounding presumptively auditory cortical fields in species in which these same distinctions by cytoarchitectonic criteria would be difficult or impossible, for example, in the porpoise *Phocaena* (M. Rose 1935), or in the chimpanzee *Troglodytes* (Beck 1927). However, use of these myelin-staining methodologies in contemporary neuroscience is rare, and there have been no important extensions of these studies since the 1930's.

Given their limitations, these classical anatomical studies coupled with human clinical and animal ablation-behavioral evidence resulted in the early identification of a "koniocortical" or "primary" auditory cortical zone in a number of mammalian species (M. Rose 1935; Brodmann 1909).

2.2 Physiological Identification of the Primary Auditory Cortex

Topographic "maps" of cochlear location and of represented sound frequency approximately

defining the boundaries and internal representational order of a presumptive "primary auditory cortical field" ("A1") were first constructed using surface evoked potential recording techniques and restricted electric and tonal stimulation in the cat and dog, by Walzl (1947) and Tunturi (1950), respectively. They have subsequently been defined using microelectrode mapping techniques and tonal stimuli in a primitive marsupial, the northern native cat, *Dasyurus* (Aitkin et al. 1986); the brush-tailed possum, *Trichosurus* (Gates and Aitkin 1982); the tree shrew, *Tupaia* (Oliver et al. 1976; the mustached bat, *Pteronotus* (Suga 1982, 1984, 1988; Asanuma et al. 1983); the horseshoe bat, *Rhinolophus* (Ostwald 1984); the big brown bat, *Eptesicus* (Jen et al. 1989); the little brown bat, *Myotis* (Wong and Shannon 1988; Berkowitz and Suga 1989); the domestic rabbit, *Lagorhinchus* (McMullen and Glaser 1982); the mouse, *Mus* (Steibler 1987); the white rat, *Rattus* (Sally and Kelly 1988); the Mongolian gerbil, *Meriones* (Steffen et al. 1987); the guinea pig, *Cavia* (Hellweg et al. 1977; Redies et al. 1989); the grey squirrel, *Sciureus* (Merzenich et al. 1976; Luethke et al. 1988); the ferret, *Mustela* (Kelly et al. 1986; Phillips et al. 1988); the cat, *Felis* (Walzl 1947; Merzenich et al. 1975, 1984; Knight 1977; Reale and Imig 1980); the galago, *Galago* (Brugge 1982); the common marmoset, *Callithrix* (Aitkin et al. 1986); the tamarin, *Saguinus* (Luethke et al. 1989); the owl monkey, *Aotus* (Imig et al. 1977); the rhesus and stumptailed monkeys, *Macaca* (Merzenich and Brugge 1973); and the human, *Homo* (Romani et al. 1982).

In several of these studies, the boundaries of cortical field A1 were estimated by using a combination of electrophysiological mapping data and cytoarchitectural analysis. Where there were clear reversals in the representation of sound frequency in these maps, for example, across the anterior or posterior boundaries of field A1 in the cat or macaque, a functional boundary could be relatively sharply defined and clear cytoarchitectonic and/or myeloarchitectonic distinctions drawn between cortical zones away from either side of these functional boundaries (see, for example, Merzenich and Brugge 1973; Merzenich et al. 1975; Oliver et al. 1976; Knight 1977; Reale and Imig 1980; Luethke et al. 1988, 1989). In other instances, both physiologically and anatomically defined boundaries appeared to be graded with no abrupt boundary distinctions, for example as along both the ventral and dorsal boundaries of A1 in the most intensively studied mammal, the cat (see Schreiner and Cynader 1984; Middlebrooks and Zook 1983; Rose 1949; Winer 1984).

2.2.1 Common Features of A1 in Studied Mammals

Of the studies cited above, particularly complete best-frequency maps of cortical field A1 have been derived in the Australian marsupial northern native cat; in the mustached bat, big brown bat and horseshoe bat among the studied chiropterans; in the rat, guinea pig, and grey squirrel among studied rodents; in the cat, ferret, and dog among the carnivores; and in the common marmoset, owl monkey, an macaque monkey among the primates. What common features of the representation of sound frequency or of the cochlear sensory epithelium are recorded in A1 in this widely scattered sample of mammals?

First, there is an orderly representation of sound frequency within A1 (Fig. 33.1). Best frequency or cochlear location is represented (1) systematically and (2) probably in entirety across one cortical field dimension.

Second, in all studied species there is a rerepresentation of sound frequency/cochlear location across the cortical field axis of A1 orthogonal to the frequency axis, its "isofrequency axis." Excepting the specialized echolocating bats in which the frequency organization of A1 is more complicated, any given band of frequency is represented by a cortical slab that extends across A1 from edge to edge (see Fig. 33.1).

Third, in many well-mapped species, again excepting the specialized echolocating bats, there is a disproportionately larger territory of representation of the more basal cochlea, representing higher sound frequencies.

Fourth, although the overall pattern of organization of frequency representation is relatively constant in a given species, there are substantial idiosyncratic differences in the details of best frequency maps derived in different individual animals of the same species. No two maps are exactly alike.

2.2.2 The Orientation of Field A1 Differs in Different Studied Mammals

The tonotopic sequence is oriented in a low-to-high rostral-to-caudal or ventrorostral to dorsocaudal sequence in marsupials and in several lissencephalic rodents (e.g., the grey squirrel). Presumptive cortical field A1 has the reverse (low-to-high, caudal-to-rostral) tonotopic orientation in other studied rodents (the rat and gerbil) and in the insectivore tree shrew. In all studied primates, the low-to-high frequency representational sequence in A1 is oriented rostral-to-caudal or rostrolateral-to-caudomedial on the superior temporal plane and superior temporal gyrus. In carnivores, A1 orientation is reversed from that in primates and some rodents, i.e., with a low-to-high frequency representational sequence oriented caudal-to-rostral (see Fig. 33.2). It is shifted by roughly 90 degrees in orientation in the ferret (Kelly et al. 1986; also see Phillips et al. 1988). These differences in A1 orientation between the ferret and dog and cat, and between these carnivores and rodents and primates have been attributed to differences in sulcation and the formation of a deep sylvian fissure and insula in primates (Woolsey 1960, 1971). However, that cannot explain apparent A1 orientation differences in different studied rodents, and between well-studied rodents and the tree shrew, in which A1 identification is strongly supported by both connectional anatomy and physiology.

2.2.3 Cortical Representational Distortions Do Not Mirror Peripheral Innervation Densities

The peripheral innervation density of the cochlear sensory epithelium varies significantly as a function of cochlear location in any given species, and varies substantially when comparisons are made between studied mammals. These peripheral disproportionalities are especially striking in echolocating bats, in which zones of high innervation density in the inner ear can be roughly centered on organ of Corti zones of representation of prominent echolocation signal frequencies (see Zook and Leake 1989). In other studied mammals, a middle-frequency sector of the inner ear has the highest innervation density (e.g., see Spoendlin 1972; Morrison et al. 1975; Ehret and Frankenreiter 1977; Keithley and Schreiber 1987; Leake and Hradek 1988).

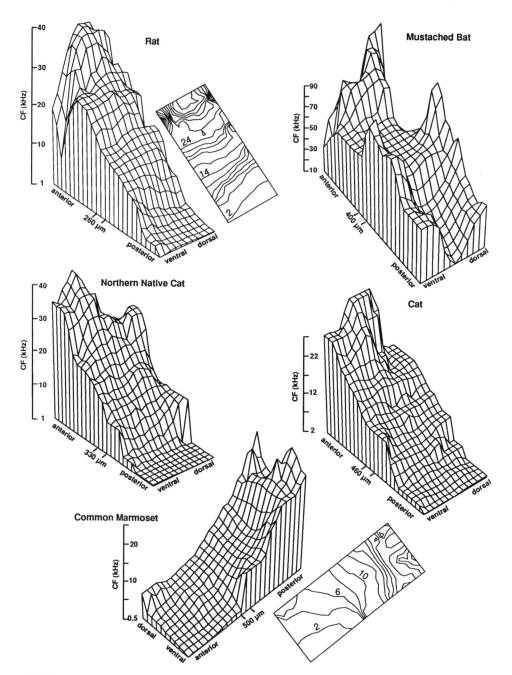

FIGURE 33.1. Frequency organization in primary auditory cortex. The spatial distribution of the characteristic frequency (CF) of locations in the primary auditory cortical fields (left hemisphere) has been reconstructed for five mammalian species: a rodent (rat, adopted from Kelly and Sally 1988); a chiropteran (mustached bat, adopted from Suga 1984); a marsupial (northern native cat, adopted from Aitkin et al. 1986a); a carnivore (cat, adopted from Merzenich et al. 1975); and a primate (common marmoset, adopted from Aitkin et al. 1986b). The CF distribution is shown in a pseudo-three-dimensional projection. Two dimensions correspond to the cortical surface. The elevation (z-axis) of the grid is proportional to the CF. In two cases, contour plots of the CF reconstructions are shown. Points of equal CF—and equal elevation of the grid surface—are represented along the contour lines, the "isofrequency contours" of A1. The frequency interval between contours is 2 kHz.

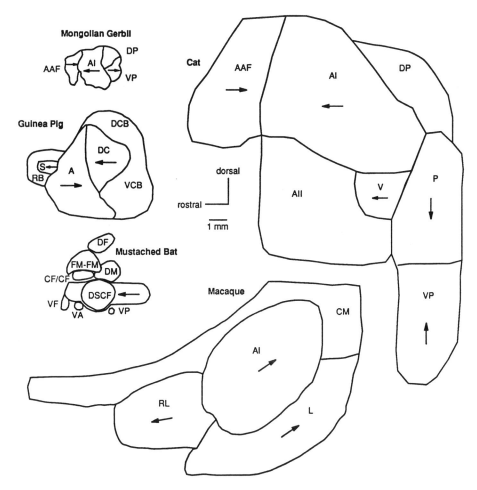

FIGURE 33.2. Multiple cortical auditory fields. The arrangement of multiple auditory cortical fields (left hemisphere) is shown to scale for five species: two rodents (Mongolian gerbil, redrawn from Steffen et al. 1987; guinea pig, redrawn from Redies et al. 1989b); a chiropteran (mustached bat, redrawn from Suga 1984); a carnivore (cat, redrawn from Reale and Imig 1980); a primate (macaque, redrawn from Merzenich and Brugge 1973). The arrows in some of these fields indicate the existence of a reliable tonotopic organization in that cortical subdivision. The directions of the arrows approximate the main tonotopic gradient and point to the high-frequency representation.

As in the visual and somatosensory systems, the proportionalities of auditory cortical maps do not simply reflect peripheral innervation densities, as was once proposed. Although auditory cortical maps reflect this disproportion in studied bats, they do not precisely mirror it. Thus, for example, peripheral innervation densities in cochlear sectors representing the second and third harmonics of the echolocation signal are almost equally magnified in the mustached bat cochlea (see Zook and Leake 1989), but in the cortex, the zone of representation of the second harmonic is differentially enlarged (Suga 1982, 1984; Asanuma et al. 1983; see Fig. 33.1). In another well-studied species, the cat, the cortical representation of middle- to highest-frequency sectors increases progressively (see Merzenich et al. 1975, 1976; 1984), while the inner hair cell innervation density reaches a maximum at a cochlear location approximately 40% of the distance from the cochlear base.

What accounts for the apparently larger cortical representation of highest frequency octaves recorded in *all* studied mammals except the echolocating bats and the gerbil? Keithley and Schreiber (1987) recently mapped the frequency representation of the spiral ganglion of the cat. Lowest frequencies are somewhat compacted in representation in the apical sector of the ganglion. As a consequence, and perhaps serendipitously, the rate of change of best frequency as a function of position along the spiral ganglion (but not along the organ of Corti) closely parallels the cortical function along the frequency representational axis of A1 in this species.

Of course, the distributions of cortical inputs presumably also reflect any unequal subcortical distortions in the neural representation of the cochlea. The pontine brain stem nuclei, for example, receive disproportionate inputs from across the cochlear spiral, with an under-representation of low sound frequencies (the cochlear apex) in the lateral superior olive, and an under-representation of high sound frequencies (the cochlear base) in the medial superior olive (see Guinan et al. 1972). In forward projection, this unequal subcortical treatment of inputs from across the cochlear spiral may contribute to a resultant, disproportionately larger representation of higher frequencies. In this respect, it is interesting that a system of excitatory-inhibitory cortical bands prominent in cortical field A1 in the cat, for example, increase in their dimensions at progressively higher sound frequencies, and are not evident in cortical zones representing frequencies below approximately 2 kHz.

Whatever its origins or functional significance, this high frequency disproportion constitutes a widespread feature of A1 organization in mammals. The few birds whose auditory cortical "field L" input-layer representations have been mapped have similar frequency-isofrequency cortical field topographies (see Bonke et al. 1979; Heil and Scheich 1985). However, the proportionalities of representation of different sound frequency ranges differ significantly from this mammalian pattern.

2.2.4 Rerepresentation of a Narrow Frequency Range Across an Enlarged Sector of the Frequency Representation Dimension

In our own cortical mapping experiments, we often observed restricted cortical map sectors in which a very limited frequency range was represented over a substantial cortical sector. Such extended-representation sectors are frequently located along the extreme high-frequency margins of A1. They are highly idiosyncratic in different studied individuals. Thus, for example, in maps of the marmoset A1, large extended sectors of representations of frequencies near 14 kHz in one animal occupied the entire highest-frequency sector of A1, filling an A1 zone that was more than 2 mm in diameter. In other adult marmosets, the A1 best frequency representation extended continuously up to more than 30 kHz (Aitkin et al. 1988).

These commonly recorded local map distortions are probably attributable to cortical map reorganization following cochlear degeneration and damage, recorded most often in older animals (Robertson and Irvine 1989). It is reasonable to conclude that much of the considerable variability in cortical tonotopic map representations for adults of any given species is attributable to a capacity for plastic modification of representation following peripheral hearing losses, and/or result from differences in predominant acoustic behaviors that generate differences in cortical map detail (see Weinberger and Diamond 1988; Merzenich et al. 1984, 1989; Merzenich 1985, 1987; and discussion below).

2.3 Identification of Other Auditory Cortical Fields

More extensive mapping data directed toward defining the overall organization of the auditory-responsive sector of the neocortex has resulted in an identification of other mappable auditory cortical fields in some of these species, especially the mustached bat (Suga 1982; 1984); the horseshoe bat (Ostwald 1984); the little brown bat (Wong and Shannon 1988); the Mongolian gerbil (Steffen et al. 1987); the guinea pig (Hellweg et al. 1977; Redies et al. 1989); the grey squirrel (Merzenich et al. 1976; Luethke et al. 1988), the cat (Merzenich et al. 1984; Knight 1977; Reale and Imig 1980), the galago (Brugge and Reale 1985); the tamarin (Luethke et al. 1989); the owl monkey (Imig et al. 1977); and the macaque monkey (Merzenich and Brugge 1973) (see Fig. 33.2). In several other studies, the perimeter of the primary field A1 has been mapped in detail, providing some evidence for the existence of "secondary" auditory fields bordering the more-responsive A1 field. In investigations of auditory cortex in the marsupial northern native cat (Aitkin et al. 1986) the big brown bat

(Jen et al. 1989), the rabbit (McMullen and Glaser 1982), the ferret (Phillips et al. 1988), and the marmoset (Aitkin et al. 1986), for example, the perimeter of field A1 was mapped in reasonable detail.

What do these studies indicate about the identities or homologies of cortical fields other than A1? Before initiating that discussion, it should be noted that the completeness of auditory cortical field mapping is always open to some question because nearly all mapping data have been derived in anesthetized animals. Moreover, the failure to identify cortical fields surrounding A1 as auditory may sometimes be attributable to the use of inappropriate (only tonal) sound stimuli in mapping.

With these important reservations, some general conclusions can be drawn from these limited comparative studies.

First, the primary auditory cortical field is at least largely encircled by auditory-responsive cortex and/or by cortical fields with cortico-cortical connections with primary auditory cortex, in well-studied species. In the primitive marsupial native cat and in the brush-tailed opossum, there is no physiological or anatomical evidence for any tonotopically organized fields flanking A1 although recording sites just across A1 borders were sometimes weakly responsive to auditory stimuli. Indeed, in the marsupial native cat, a second, major projection of the auditory thalamus is to the subcortical amygdala (see Kudo et al. 1989), an organizational schema not recorded in any other studied mammal. *In every other well-studied species, at least one additional tonotopically organized cortical field has been described.*

Second, there is a probable increase in the *number* of identifiable auditory cortical fields in more advanced mammals. A single, large identifiable field is recorded in the marsupial native cat and the brush-tailed opossum. It is surrounded by a rim of weakly auditory responsive cortex that does not appear to have response characteristics or thalamocortical, corticothalamic or corticocortical connections consistent with the existence of separate, discrete "secondary" auditory cortical fields like those recorded in other mammals. On the basis of anatomical and physiological criteria, there are three or four or more auditory cortical fields in well-studied insectivores, three, four, five or six auditory cortical fields in studied rodents, and six, seven, eight,

or more auditory cortical fields in the cat, in the tamarin, and in owl and macaque monkeys.

Third, frequency reversals mark the low-frequency boundary of cortical field A1 with a second, mirror-image auditory cortical field in a number of studied species. In some mammals, for example in the macaque, marmoset, tamarin or owl monkey, or in the grey squirrel, this field is the most responsive and prominent of those bordering A1. It is relatively weakly responsive or unresponsive in anesthetized preparations in other mammals (e.g., the cat, guinea pig, rabbit).

Are these fields bordering low-frequency A1 and mirroring its topography homologous in the different mammals in which they appear? Commonalities of cortical cytoarchitecture, connections and functional organization indicate that this field is present and almost certainly homologous in all studied monkeys as well as in the prosimian galago. However, identification of a homologous field in other mammals is uncertain. On the one hand, unlike A1, the frequency representation of this rostral field in studied primates has been found to be incomplete, limited to the representation of lower sound frequencies (see Merzenich and Brugge 1973; Brugge and Reale 1985; Aitkin et al. 1986). The sound frequency representation may also be incomplete in the grey squirrel, and possibly in the cat. On the other hand, the so-called "rostral field" of primates is a cytoarchitecturally distinctive granular cortex. In general, the special cytoarchitectonic features of this field do not obviously apply to the low-frequency reversed field recorded in the same location in nonprimate mammals, e.g., in cats or squirrels. The question of field homology remains unanswered.

Fourth, frequency reversals mark the high-frequency boundary of cortical field A1 with a third, mirror-image auditory cortical field in a number of studied species. This is the most strongly driven of "secondary" auditory cortical fields in several studied mammals (the rabbit, guinea pig, cat, dog), while it is weakly definable or absent in others (e.g., in the grey squirrel, marmoset, macaque). As with the field bordering A1 on its low-frequency representational border, this high-frequency bordering field recorded in different mammals may not be homologous in the species in which it has been identified. Response properties of neurons in this field in cats are highly distinctive (see Knight 1977; Schreiner and Urbas 1986, 1988). While physiological studies in other species are limited, they reveal no clear parallels in the responses of neurons in the positionally corresponding field in other studied mammals. As with the low-frequency bordering field, anatomical data that might apply to this question of homology are too limited to resolve it.

Fifth, a cortical area marked by broadly tuned responses and deriving its input from the caudal part of the medial geniculate body (MGB) of the thalamus has been defined in a similar positional relationship to A1 in carnivores and primates, indicating a common phylogenetic origin in these widely divergent species. While the possible identity of this "AII" field in other mammalian orders is uncertain, a caudal MGB target has been identified in the auditory cortex sector in the tree shrew (Casseday et al. 1976; Oliver et al. 1976) and the rat (Patterson 1976). The primary input source to the caudal MGB is the pericentral nucleus of the inferior colliculus, a structure apparently common to all mammals.

On the basis of such findings, it has been argued that there are two largely anatomically separated auditory forebrain projection systems (see Andersen et al. 1980): (a) A *tonotopic system* feeds projections from the central nucleus of the inferior colliculus via the lateral and dorsal parts of the MGB to tonotopically organized cortical fields. Anatomical connections in this "system" are topographically precise. (b) A *nontonotopic system* feeds projections from the pericentral nucleus of the inferior colliculus via the caudal part of the MGB to cortical fields marked by broad frequency response bandwidths and limited tonotopicity. Anatomical projections in this "system" are highly divergent/convergent, approaching an all-to-all projection scheme.

Given limited anatomical and physiological evidence for its presence in studied insectivores, rodents, carnivores and primates, and given the apparent universality of a pericentral nucleus (and caudal MGB nucleus) in well-studied mammals, this dual-system organization may be general to at least eutherian species. However, further studies shall have to be completed to determine whether that is the case.

Sixth, the forebrain organization of multiple cortical fields in at least one well-studied group of

mammals, the echolocating bats, appears to be truly specialized. In studied species, complex combinations of cue features appear to be represented in different, specific auditory cortical fields. Only one of these fields is strictly tonotopic; orderly representations of specific complex cue continua are arrayed across other studied cortical areas. It should be emphasized that this organization does not of itself infer forebrain specialization because the derivation of the specific responses to these complex echolocation cues is subcortical, and these unique cortical 'maps' may simply reflect this subcortical treatment. On the other hand, limited combined physiological and anatomical experiments conducted in the best neurologically studied echolocating bat, the mustached bat, indicate that (a) different echolocation cue combinations are extracted and represented by different, specific thalamic subdivisions, and (b) thalamic nucleus-specific information is then conveyed forward to specific, specialized cortical areas (see Suga 1982; Suga 1984). Further, (c) the detailed topographic maps of these complex cue features are believed to first emerge at the cortical level (Suga 1982, 1984, 1988).

If more intensive studies confirm these findings, they may manifest a true forebrain specialization in this species. By contrast, the cortical areas in the forebrain of other mammals receive parallel, complexly interwoven – not nucleus-specific – projections from subdivisions of thalamic nuclei (see, for example Patterson 1976; Colwell 1977; Andersen et al. 1980; Merzenich et al. 1982; Morel and Imig 1987). And, by contrast, cortical map topographies generally reflect thalamic topographies.

The forebrain representations in the auditory domain are developed in a special form in at least two other groups of mammals. In toothed whales, the medial geniculate body is massive, and its projection zone encompasses a significant fraction of a large cortical mantle. Although anatomical studies are limited and physiological studies nonexistent, the gross anatomy of the whale brain indicates that roughly 1/4 to 1/3 of the cortical mantle is occupied by auditory-dominated fields (unpublished observations).

Auditory-dominated cortical fields also appear to extend over a differentially great cortical sector in man (see Penfield and Roberts 1959). Further, representations of complex acoustic signals (speech and vocalizations) are lateralized in humans and in the Japanese macaque (see Heffner and Heffner 1986).

2.4 Internal Organization of Auditory Cortical Fields; Topographic Organization Beyond Tonotopicity

In scattered studies in rodents, carnivores and primates, orderly representations of response features other than sound frequency or cochlear position have been recorded within auditory cortical fields. How are these "higher-order" representations alike or different in different well-studied mammals?

2.4.1 Binaural Interaction "Column" Organization in Rats, Cats and Monkeys

Binaural interaction columns were first defined in A1 in the mapping studies of Imig and Adrian (1977). They have subsequently been described in a number of electrophysiological and anatomical studies in cats (see Imig and Brugge 1978; Imig and Reale 1980; Middlebrooks et al. 1980; Middlebrooks and Zook 1983; Schreiner and Cynader 1984; Figure 33.3). In felines, these irregular columns are oriented roughly orthogonal to the isofrequency plane. Columns in which the predominant response is excitatory-inhibitory (i.e., in which neurons are driven by contralateral stimulation with that response inhibited by simultaneous ipsilateral inhibition; the "binaural suppression columns" Figure 33.3) are recorded only over the higher frequency representational sector of A1. Two, three, or four of these suppression bands are usually recorded in any individual cat, separated by interleaved "binaural summation (excitatory-excitatory) columns." These binaural columns have highly idiosyncratic forms in different individual animals (Schreiner and Cynader 1984).

In cats, these "columns" are cytoarchitecturally distinguished by a special population of neurons in layer 3 of summation columns, which convey callosal inputs to summation columns in A1 in the opposite hemisphere. Suppression columns lack this distinctive layer-three pyramidal cell population. It has been argued that suppression and summation columns receive inputs from different thalamic sources (Middlebrooks and Zook 1983; but see Brandner and Redies 1990).

Some primates also appear to have binaural columns. Segregation of excitatory-excitatory and excitatory-inhibitory responses has been noted in A1 of monkeys (Brugge and Merzenich 1973), although it has not been mapped. Bands extending orthogonal to probable isofrequency axes have been described in anatomical studies of A1 in humans (von Economo 1927), and a number of subdivisions can be distinguished *within* cortical field A1 in chimpanzees (Beck 1929). Callosal inputs in the owl monkey A1 appear to be banded, and a distinctive banded pattern of corticocortical projections was recorded in that species following a restricted tracer injection into a cortical area (the "rostral field") anterior to A1 (Fitzpatrick and Imig 1980). By contrast, callosal inputs to A1 from the contralateral A1 cortex do not appear to be banded in either the common marmoset (Aitkin et al. 1988) or tamarin (Luethke et al. 1989). Furthermore, there is no clear evidence of a banded thalamic projection to A1—a striking feature elucidating the thalamocortical origins of binaural bands in the cat (see Merzenich et al. 1982; Middlebrooks and Zook 1983; Morel and Imig 1987)—in any of these several well-studied New World species.

If we consider this scattered evidence collectively, there is some structural evidence and some indirect physiological evidence of a banded organization of A1 in at least one New World monkey and in Old World macaques, apes and man that support the conclusions that: (a) some primates have cytoarchitecturally and myeloarchitecturally distinguishable bands or field subdivisions that may be akin to feline "binaural columns." (b) The functional significance of these presumptive columns in primates is still unknown. Although binaural responses are segregated in at least some primates, this segregation has not yet been directly related to anatomically distinguishable columns. (c) In some primate species, as in the cat, one series of alternating bands appears to be connected callosally; while the interleaved series is not. (d) Again, there is absolutely *no* evidence for binaural columns recorded in investigations on two well-studied primates, the New World tamarin and common marmoset.

Do binaural interaction columns exist in other mammals? Suga and colleagues (Suga 1982, 1984, 1988) have demonstrated a sharp segregation of excitatory-excitatory (summation) and excitatory-inhibitory (suppression) neuron response populations

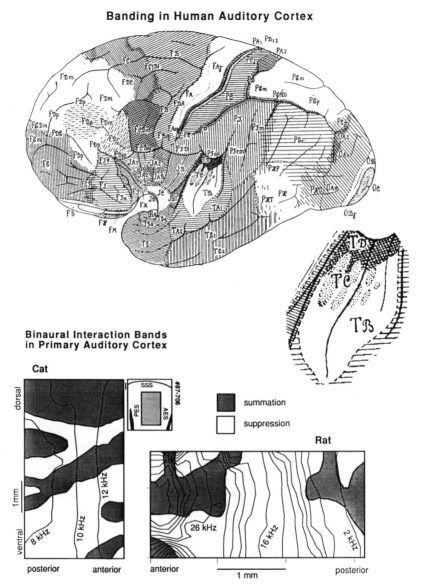

FIGURE 33.3. Subdivisions within auditory cortical fields. Upper panel: Classification of subdivisions in human cortex by von Economo (1927) based on cytoarchitectonic criteria. The temporal fields TB, TC, and TD are reproduced in an enlarged version. Note the banding in those cortical fields as indicated by stippling. Mapping studies in macaques (Merzenich and Brugge, 1973) and humans (Romani et al. 1982) indicates that this banding is probably orthogonal to the isofrequency axis of the A1 field. Lower panel: Binaural interaction classes in primary audi-tory cortex of a cat (Schreiner, unpublished observation) and a rat (reconstructed from Kelly and Sally 1988). Cortical regions that received excitatory input from both ears ("binaural summation") are indicated by dark shading. Locations receiving excitatory input from one ear and inhibitory input from the other ear ("binaural suppression") are unshaded. Locations with response properties that did not match either of these two classes were omitted from the reconstruction. The arrangement of isofrequency contours are super-imposed with intervals of 2 kHz.

in a high-frequency sector of A1 in the mustached bat. A segregation of summation and suppression binaural responses has also been recorded in the rat (Sally and Kelly 1988), although binaural responses are clustered and not clearly banded in that species (see Fig. 33.3). In the grey squirrel, limited callosal projection studies indicate that there is little structure in the contralateral projection pattern consistent with a banding of A1 by callosal projections from summation "columns" or "clusters" (see Luethke et al. 1988), as in the cat and owl monkey.

What can we conclude from these scattered studies? Summation and suppression binaural interaction responses have been found to be segregated in the cortex in the several species in which their distribution has been mapped physiologically. Studies are still piecemeal, so that any statement about possible phylogenetic origins or homologies of this special auditory cortex feature must be reserved. However, it is interesting to note that callosal projections in the grey squirrel, tamarin or marmoset do not form distinct bands (or sharply delimited clusters) as would be recorded with equal-sized, opposite hemisphere injections in the cat or owl monkey. At the same time, evidence for banding orthogonal to the isofrequency axis in A1 in the owl monkey, human and possibly the chimpanzee, as well as in cats, indicates that its emergence in these phylogenetically disparate species may be attributable to a common ancestor or cause. Further study of this very interesting feature of auditory forebrain organization would be of substantial comparative interest.

2.4.2 Representation of Other Response Features Across Auditory Cortical Maps

Representations beyond simple tonotopicity have been described in greatest detail in the elegant studies of echolocating bats conducted by Suga and colleagues (1982, 1984, 1988). These studies have revealed that (1) different cortical regions respond selectively to specific complex echolocation cue features, and (2) beyond field A1, each cortical field represents a particular cue continuum(ua) in an orderly way. Representations differ in different echolocating bat species, reflecting species-specific differences in echolocation vocalization and their behavioral utility (Ostwald 1984; Wong

and Shannon 1988; Jen et al. 1989; Berkowitz and Suga 1989).

How special is this organization? Suga has plausibly argued (1988) that it constitutes a model for the likely cortical bases of representation of features of human speech, as the cue continua for this "FM bat" are similar to those required to make certain distinctions in speech. In any event, the processes by which the cortex *creates* these representational continua are probably general for the neocortex, and presumably apply to all species (Merzenich et al. 1984, 1989).

Schreiner and colleagues have begun to search for higher-order representational continua in the cat model by mapping the representation of basic response features within cortical field A1 (see Schreiner et al. 1988; Mendelson et al. 1988; Schreiner and Langner 1988; Schreiner and Mendelson 1990). They have discovered that there are orderly representations of several response continua along the isofrequency dimension of this field in adult cats, including representations of tuning bandwidths, monotonicity, FM sweep speeds, FM sweep directions, responses to broadband stimuli, and distributions of sideband inhibition (see also Tunturi 1950; Phillips et al. 1985; Pantev et al. 1989; Shamma and Fleshman 1990). Whether these constitute distributed, selective response properties of neurons manifesting complex cue-combination representations akin to those recorded in the echolocating bats remains to be determined.

3. Studies of Phylogenetic Development of Cortical Representations in Other Sensory Systems; Some Relevant Findings and Conclusions

Comparative studies of the mammalian visual cortex indicate some possible parallels between its phylogenetic development and that of auditory cortical fields. There is clear evidence of an increase in the *number* of cortical fields along a number of lines of evolution, with two or possibly three visual areas in the primitive insectivore hedgehog, and progressively larger numbers in animals with progressively more developed neocortex. Cats have 14 or 15 visual cortical areas (see Tusa et al. 1981; Rosenquist 1985); Old World monkeys have more than 20 identified visual cortical areas (see Merzenich and Kaas 1980; Kaas 1987; DeYoe and Van Essen 1989). As in some "secondary" auditory cortical fields, the primary visual field has a complete representation of the peripheral sensory epithelium, while some other fields represent only limited sectors of it.

A possible basis for an increase in cortical field number is suggested by recent ontogenetic development studies of McKinley and colleagues (1987). They have shown that after a neonatal spinal cord transection that deprives a large sector of somatosensory cortical fields of their normal driving inputs, a second mirror-image representation of the forelimb emerges within the somatosensory koniocortex. This suggests that if evolution results in a creation of an enlarged cortical mantle, in-place inputs may be drive to occupy it, with topographically mirrored projections.

Several investigators have suggested that there is a major change between the visual cortical areas with the emergence of primates, expressed by the fact that prosimians as well as New World and Old World monkeys have many clearly homologous cortical fields that have no clear nonprimate equivalents (see Allman 1989; Kaas 1987, 1989). Although less completely described, studied primates including the prosimian galago also appear to have homologous auditory cortical fields, all of which (especially the striking "rostral field" and the commonly occurring "lateral field") do not have clear homologies in mammals in other orders.

There is some indication that an internal map feature of auditory cortex, the physiological/anatomical "columns" orthogonal to the isofrequency axis of A1, is present in advanced primates (Pongids, Hominids) and possibly the New World owl monkey (*Aotus*), in at least one carnivore, the cat (*Felis*), and in at least one insectivore, the echolocating mustached bat (*Pteronotus*). However, studies of the common marmoset and tamarin and of several rodents and the insectivore tree shrew indicate that while there may be some functional segregation of suppression and summation responses in cortical field A1, segregated response zones are not arrayed in elongated strips as in bats, cats and some primates. For the animals in which these elongated "columns" have emerged, they may have evolved in parallel.

Studies of the visual cortex provide a possible precedent for this finding. There, it has been argued that ocular dominance column organization of the primary visual cortical area (area 17) has evolved at least three times (see, for example, Florence et al. 1986; Kaas 1989). Insectivores, rodents, lagomorphs, prosimian primates and some New World monkeys (e.g., the owl and squirrel monkeys) lack ocular dominance columns. Advanced carnivores, some New World monkeys (e.g., the spider and cebus monkeys) and Old World monkeys all have well-developed ocular dominance columns. Perhaps more likely, factors resulting in ocular dominance column formation may be general, but unresolved species differences may result in a wide range of their species-specific expression across mammalian phylogeny (see Allman 1986). In any event, a genetic basis for refinement of binaural column organization, like ocular dominance or orientation columns, may be inherited by virtually all mammals, because this auditory cortex feature emerges in parallel in at least one chiropteran (the mustached bat) as well as in primates and carnivores.

4. Ontogenetic Formation of Cortical "Maps"; Implications for Comparative Studies

One striking aspect of auditory cortex organization is the idiosyncratic differences in the details of cortical representation in different adults of any given species. An examination of the individual maps of A1 or of the fields bordering A1 in different individuals in any careful mammalian study will point out these differences. They are, not uncommonly, as great as the average differences in representation *between* species. What are their origins?

In both the somatosensory and auditory systems, there is growing evidence that the details of cortical representations can be modified by haptic or hearing behaviors (see Weinberger and Diamond 1988; Merzenich et al. 1984, 1989). These use-driven representational changes are clearly a major source of representational variability (see Merzenich 1985; Merzenich et al. 1987). We have earlier argued that in the somatosensory cortex they can dwarf the differences in cortical representation recorded between related primates. Thus, if maps of the hands of monkeys of different monkey species are compared, differences in predominant hand uses related to differences in behavioral niche may largely account for observed map differences; many interspecies map differences cannot be directly attributed to genetic differences, as was earlier believed.

Auditory cortical fields are no less "plastic" than somatosensory fields. There are massive consequences of learning on specific auditory cortex responses (see Weinberger and Diamond 1988 for review) and striking changes in best-frequency topographies can result from a few minutes or tens of minutes of classical conditioning in some cortical fields. Another example of auditory cortical field plasticity has come from studies of the adult mustached bat, in which the cortical representations of echolocation signals have been found to be "personalized"; that is, the *precise* frequencies that dominate any individual bat's cry are represented over a disproportionately larger zone of the bat's cortex (Suga et al. 1987). These personalized frequency representations change progressively as each bat matures. Bat echolocation cue maps are also idiosyncratic in representational detail.

Thus, the details of evident cortical representational topographies are to some extent a function of the earlier predominant behaviors of the animal under study. In open reception (nonecholocating) animals, prior hearing experience is almost always very difficult to document or reconstruct, and significant differences might be expected to emerge for cage-reared versus wild-caught animals. These important nature/nurture issues must be considered in interpretation of detailed electrophysiological data as it relates to these important comparative issues.

As noted earlier, cortical maps are also altered following peripheral hearing losses (Robertson and Irvine 1989). Map variability attributable to cochlear deterioration can be minimized by studying younger animals. Studies of hearing specialization—especially in rodents in which inner ear deterioration is often very rapid in ontogenetic development—must consider possible contributions of inner ear deterioration to physiologically determined cortical maps.

5. Summary; Some General Conclusions

5.1 An apparently homologous primary auditory cortex (A1) has been identified in all studied mammals.

5.2 The intrafield frequency representation of A1 is similar in scattered, studied mammals, including marsupials, insectivores, rodents, carnivores, and primates.

5.3 Binaural summation and suppression responses have been found to be segregated in the few mammalian species in which there have been mapped. Summation and suppression populations form bands that extend across A1 orthogonal to the isofrequency dimension of the field in cats, and probably in New World and Old World monkeys, apes and man. This may represent an example of parallel evolution, as no clear banding is evident in the anatomy and physiology of the tamarin or marmoset, or in studied rodents.

5.4 Excepting marsupials and echolocating bats, all studied mammals have at least one tonotopically organized field bordering A1 that has a intrafield topography roughly mirroring it. In some clear instances the cochlea is only partially represented in these "secondary" cortical fields. Fields with reversed topographies bordering the lowest frequency sector of A1 have been identified in a scattered group of rodents, carnivores, and primates, but it is not clear if these cortical fields are homologous to each other. Similarly, fields with reversed or parallel topographies bordering the highest frequency sector of A1 have been identified in another group of rodents and carnivores, and in primates, respectively. Again, it is not yet clear that these identified cortical entities are homologous to one another.

5.5 At least one cortical field marked by broad frequency response tuning has been identified in all well-studied carnivore and primate species. It derives its inputs via highly divergent/convergent anatomical projections from the caudal MGB, which, in turn, is fed from the pericentral nucleus of the inferior colliculus. This anatomical projection axis has also been identified in insectivores and rodents, and hence this "second" parallel auditory-dominated "system" may be general in eutherian mammals. There is, as yet, no evidence for this second, "nontonotopic" system in the forebrain of marsupials.

5.6 Mammals with highly developed neocortex appear to have a larger *number* of auditory cortical fields than do mammals with less developed neocortex.

5.7 The auditory forebrain in at least three groups of mammals—(a) echolocating bats, (b) toothed whales, and (c) man, and possibly some highly vocal Old World primates including apes — appears to be specialized. The anatomical organizations of MGB-to-cortex projections appear to differ significantly in echolocating bats from that in other studied mammals. The auditory dominated cortices of toothed whales and humans appears to have been enlarged differentially.

5.8 Echolocating bats map complex echolocation cue feature combinations beyond tonotopicity. They create orderly cortical representations of specific complex cue combinations and separately represent them within specific auditory cortical fields. This may reflect a true specialization, or may be a widely occurring feature of cortical maps. Consistent with that possibility, recent experiences in cats have revealed that a number of response properties beyond sound frequency are represented in a systematic way in A1.

5.9 Details of representation of sound inputs can be altered by a mammal's experiences in at least some cortical auditory representations. They can also be altered by deterioration or damage to the inner ear. These effects probably account for most of the striking idiosyncratic variability in representations recorded in different individuals of the same species.

References

Aitkin LM, Irvine DRF, Nelson JE, Merzenich MM, Clarey JC (1986) Frequency representation in the auditory midbrain and forebrain of a marsupial, the northern native cat (*Dasyurus hallucatus*). Brain Behav Evol 29:12–28.

Aitkin LM, Kudo M, Irvine DRF (1988) Connections of the primary auditory cortex of the common marmoset (*Callithrix jaccus*). J Comp Neurol 269: 235–248.

Aitkin LM, Merzenich MM, Irvine DRF, Clarey JC, Nelson JE (1986) Frequency representation in auditory cortex of the common marmoset (*Callithrix jaccus*). J Comp Neurol 252:175–185.

Allman JM (1986) In: Steklis HD, Erwin J (eds) *Comparative Primate Biology*, Volume 4. New York: Alan Liss.

Andersen RA, Knight PL, Merzenich MM (1980) The thalamocortical and corticothalamic connections of A1, AII, and the anterior auditory field (AAF) in the cat: Evidence for two largely segregated systems of connections. J Comp Neurol 194:663–701.

Asanuma A, Wong D, Suga N (1983) Frequency and amplitude representations in anterior primary auditory cortex of the mustached bat. J Neurophysiol 50:1182–1196.

Beck E (1929) Der myeloarchitektonische Bau des in der Sylvischen Furche gelegenen Teiles des Schläfenlappens beim Schimpansen (*Troglodytes niger*). J Psychol Neurol 38:309–420.

Berkowitz A, Suga N (1989) Neural mechanisms of ranging are different in two species of bats. Hearing Res 41:255–264.

Bonin G von, Bailey P (1947) The neocortex of *Macaca mulatta*. In: Allen RB, Kampmeier OF, Schour I, Serles ER (eds) *Illinois Monographs in the Medical Sciences*, Vol. 5. Urbana: University of Illinois Press, pp. 1–163.

Bonke D, Scheich H, Langner G (1979) Responsiveness of units in the auditory neostriatum of the guinea fowl (*Numida meleagris*) to species-specific calls and synthetic stimuli. I. Tonotopy and functional zones of field L. J Comp Physiol 132:243–255.

Brandner S, Redies H (1990) The projection from medial geniculate to field A1 in cat: Organization in the isofrequency dimension. J Neurosci 10:50–61.

Brodmann K (1909) *Vergleichende Lokalizationslehre der Grosshirnrinde in ihren Prinzipien dargestellt auf Grund des Zellenbaues*. Leipzig: Barth.

Brugge JF (1982) Auditory cortical areas in primates. In: Woolsey CN (ed) *Cortical Sensory Organization. Vol. 3. Multiple Auditory Areas*. Clifton, NJ: Humana Press, pp. 59–70.

Brugge JF, Merzenich MM (1973) Responses of neurons in auditory cortex of the macaque monkey to monaural and binaural stimuli. J Neurophysiol 36:1138–1158.

Brugge JF, Reale RA (1985) Auditory Cortex. In: Peters A, Jones EG (eds) *Cerebral Cortex, Vol. 4. Association and Auditory Cortices* New York: Plenum, pp. 229–271.

Campbell AW (1905) *Histological Studies on the Localization of Cerebral Function*. Cambridge: Cambridge U. Press.

Casseday JH, Diamond IT, Harting JK (1976) Auditory pathways to the cortex in *Tupaia glis*. J Comp Neurol 166:303–340.

Colwell SA (1977) *Reciprocal structure in the medial geniculate*. PhD Dissertation, University of California, San Francisco.

DeYoe EA, Van Essen DC (1988) Concurrent processing streams in monkey visual cortex. Trends Neurosci 11:219–226.

Economo C von (1927) *Architecture Cellulaire Normale de l'Écorce Cérébrale*. Paris: Masson.

Ehret G, Frankenreiter M (1977) Quantitative analysis of cochlear structures in the house mouse in relation to mechanisms of acoustical information processing. J Comp Physiol 122:65–85 1977.

Fitzpatrick KA, Imig TJ (1980) Auditory cortico-cortical connections in the owl monkey. J Comp Neurol 192:589–610.

Flechsig PE (1927) *Meine myelinetische Hirnlehre mit biographischer Einleitung*, Berlin: J Springer.

Florence SL, Conley M, Casagrande VA (1986) Ocular dominance columns and retinal projections in new world spider monkeys (*Ateles ater*) J Comp Neurol 243:234–249.

Gates GR, Aitkin LM (1982) Auditory cortex in the marsupial possum, *Trichosurus vuplecula*. Hearing Res 7:1–11.

Guinan JJ, Norris BE, Guinan SS (1972) Single auditory units in the superior olivary complex. II. Locations of unit categories and tonotopic organization. Int J Neurosci 4:147–166.

Heil P, Scheich H (1985) Quantitative analysis and two-dimensional reconstruction of the tonotopic organization of the auditory field L in the chick from 2-deoxy-glucose data. Exp Brain Res 58:532–543.

Heffner HE, Heffner RS (1986) Effect of unilateral and bilateral auditory cortex lesions on the discrimination of vocalizations by Japanese macaques. J Neurophysiol 56:683–701.

Hellweg FC, Koch R, Vollrath M (1977) Representation of the cochlea in the neocortex of guinea pigs. Exp Brain Res 29:467–474.

Imig TJ, Adrian HO (1977) Binaural columns in the primary field (AI) of cat auditory cortex. Brain Res 138:241–257.

Imig TJ, Brugge JF (1978) Sources and terminations of callosal axons related to binaural and frequency maps in primary auditory cortex of the cat. J Comp Neurol 182:637–660.

Imig TJ, Reale RA (1980) Patterns of cortico-cortical connections related to tonotopic maps in cat auditory cortex. J Comp Neurol 192:293–332.

Imig TJ, Ruggero MA, Kitzes LM, Javel E, Brugge JF (1977) Organization of auditory cortex in the owl monkey (*Aotus trivirgatus*). J Comp Neurol 171:111–128.

Jen PH, Sun XD, Lin PJ (1989) Frequency and space representation in the primary auditory cortex of the frequency modulating bat (*Eptesicus fuscus*). J Comp Physiol 165:1–14.

Kaas JH (1987) The organization of neocortex in mammals: Implications for theories of brain function. Ann Rev Psychol 38:129–151.

Kaas JH (1989) The evolution of complex sensory systems in mammals. J Exptl Biol 146:165–176.

Keithley EM, Schreiber RC (1987) Frequency map of the spiral ganglion in the cat. J Acoust Soc Am 81:1036–1042.

Kelly JB, Judge PW, Phillips DP (1986) Representation of the cochlea in primary auditory cortex of the ferret (*Mustela putorius*). Hearing Res 24:111–115.

Kelly JB, Sally SL (1988) Organization of auditory cortex in the albino rat: Binaural response properties. J Neurophysiol 59:1756–1769.

Knight PL (1977) Representation of the cochlea within the anterior auditory field (AAF) of the cat. Brain Res 130:447–467.

Kudo ML, Aitkin M, Nelson JE (1989) Auditory forebrain organization of an Australian marsupial, the northern native cat (*Dasyurus hallucatus*). J Comp Neurol 279:28–42.

Leake PA, Hradek GT (1988) Cochlear pathology of long term neomycin induced deafness in cats. Hearing Res 33:11–34.

Luethke LE, Krubitzer LA, Kaas JH (1988) Cortical connections of electrophysiologically and architectonically defined subdivisions of auditory cortex in squirrels. J Comp Neurol 268:181–203.

Luethke LE, Krubitzer LA, Kaas JH (1989) Connections of primary auditory cortex in the New World monkey, *Saguinus*. J Comp Neurol 285:487–513.

McKinley PA, Jenkins WM, Smith JL, Merzenich MM (1987) Age-dependent capacity for somatosensory cortex organization in chronic spinal cats. Devel Brain Research 31:136–139.

McMullen NT, Glaser EM (1982) Tonotopic organization of rabbit auditory cortex. Exp Neurol 75:208–220.

Mendelson JR, Schreiner CE, Grasse K, Sutter M (1988) Spatial distribution of responses to FM sweeps in cat primary auditory cortex. Assoc Res Otolaryngol Abstr 11:36.

Merzenich MM (1985) Sources of intraspecies and interspecies cortical map variability in mammals: conclusions and hypotheses. In: Cohen M, Strumwasser F (eds) *Comparative Neurobiology: Modes of communication in the nervous system*. New York: John Wiley & Sons, pp. 138–157.

Merzenich MM (1987) Dynamic neocortical processes and the origins of higher brain functions. In: Changeux JP, Konishi M (eds) *The neural and molecular bases of learning*. Chichester: John Wiley & Sons, pp. 337–358.

Merzenich MM, Brugge JF (1973) Representation of the cochlear partition on the superior temporal plane of the macaque monkey. Brain Res 50:275–296.

Merzenich MM, Kaas JH (1980) Principles of organization of sensory-perceptual systems in mammals. Prog Psychobiol Physiol Psychol 9:1–42.

Merzenich MM, Colwell SA, Andersen RA (1982) Auditory forebrain organization. Thalamocortical and corticothalamic connections in the cat. In: Woolsey CN (ed) *Cortical Sensory Organization*, Vol. 3. Clifton, NJ: Humana Press, pp. 43–57.

Merzenich MM, Jenkins WM, Middlebrooks JC (1984) Observations and hypotheses on special organizational features of the central auditory nervous system. In: Edelman GM, Cowan WM, Gall WE (eds) *Dynamic Aspects of Neocortical Function*. New York: John Wiley & Sons, pp. 397–424.

Merzenich MM, Kaas JH, Roth GL (1976) Auditory cortex of the grey squirrel: Tonotopic organization and architectonic fields. J Comp Neurol 166:387–401.

Merzenich MM, Knight PL, Roth GL (1975) Representation of cochlear within primary auditory cortex in the cat. J Neurophysiol 38:231–249.

Merzenich MM, Nelson RJ, Kaas JH, Stryker MP, Jenkins WM, Zook JM, Cynader MS, Schoppmann A (1987) Variability in hand surface representations in areas 3b and 1 in adult owl and squirrel monkeys. J Comp Neurol 258:281–296.

Merzenich MM, Recanzone G, Jenkins WM, Allard TT, Nudo RJ (1989) Cortical representational plasticity. In: Rakic P, Singer W (eds) *Neurobiology of Neocortex*, Chichester: John Wiley & Sons, pp. 41–67.

Middlebrooks JC, Dykes RW, Merzenich MM (1980) Binaural response-specific bands in primary auditory cortex (A1) of the cat: Topographical organization orthogonal to isofrequency contours. Brain Res 181:31–48.

Middlebrooks JC, Zook JM (1983) Intrinsic organization of the cat's medial geniculate body identified by projections to binaural response-specific bands in the primary auditory cortex. J Neurosci 3:203–224.

Morel A, Imig TJ (1987) Thalamic projections to fields A, AI, P, and VP in the cat auditory cortex. J Comp Neurol 265:119–144.

Morrison D, Schindler RA, Wersäll J (1975) A quantitative analysis of the afferent innervation of the organ of Corti in guinea pig. Acta Otolaryngol 79:11–23.

Oliver DL, Merzenich MM, Roth GL, Hall WC, Kaas JH (1976) Tonotopic organization and connections of primary auditory cortex in the tree shrew (*Tupaia glis*). Anat Rec 184:491.

Ostwald J (1984) Tonotopical organization and pure tone response characteristics of single units in the auditory cortex of the Greater Horseshoe Bat (*Rhinolophus ferrumequinum*). J Comp Physiol 155:821–834.

Pantev C, Hoke M, Lütkenhöner B, Lehnertz K (1989) Tonotopic organization of the auditory cortex: Pitch versus frequency representation. Science 246:486–488.

Patterson HA (1976) *An anterograde degeneration and retrograde axonal transport study of the cortical projections of the rat medial geniculate body*. PhD Thesis, Boston University, Boston.

Penfield W, Roberts L (1959) *Speech and Brain-Mechanisms*. Princeton: Princeton University Press.

Phillips DP, Judge PW, Kelly JB (1988) Primary auditory cortex in the ferret (*Mustela putorius*): Neural response properties and topographic organization. Brain Res 443:281–294.

Phillips DP, Orman SS, Musicant AD, Wilson GF (1985) Neurons in the cat's auditory cortex distinguished by their response to tones and wide-spectrum noise. Hearing Res 18:73–86.

Reale RA, Imig TJ (1980) Tonotopic organization in auditory cortex of the cat. J Comp Neurol 192:265–291.

Redies, Brandner CS, Creutzfeldt OD (1989a) Anatomy of the auditory thalamocortical system of the guinea pig. J Comp Neurol 282:489–511.

Redies, Sieben HU, Creutzfeldt OD (1989b) Functional subdivisions in the auditory cortex of the guinea pig. J Comp Neurol 282:473–488.

Robertson D, Irvine DRF (1989) Plasticity of frequency organization in auditory cortex of guinea pigs with partial unilateral deafness. J Comp Neurol 282:456–471.

Romani GL, Williamson SJ, Kaufman L (1982) Tonotopic organization of human auditory cortex. Science 216:1339–1340.

Rose JE (1949) The cellular structure of the auditory region of the cat. J Comp Neurol 91:409–440.

Rose M (1935) Cytoarchitektonik und Myeloarchitektonik der Grosshirnrinde. In: Bumke O, Foerster O (eds) *Handbuch der Neurologie*. Berlin: Springer-Verlag, pp. 588–778.

Rosenquist AC (1985) Connections of visual cortical areas in the cat. In: Peters A, Jones EG (eds) *Cerebral Cortex*. New York: Plenum, pp. 81–117.

Sally SL, Kelly JB (1988) Organization of auditory cortex in the albino rat: Sound frequency. J Neurophysiol 59:1727–1738.

Sanides F (1973) Representation in the cerebral cortex of its areal lamination pattern. In: Bourne GH (ed) *Structure and Function of Nervous Tissue, Volume 5*. New York: Academic Press, pp. 222–262.

Schreiner CE, Cynader MS (1984) Basic functional organization of second auditory field (AII) of the cat. J Neurophysiol 51:1284–1305.

Schreiner CE, Langner G (1988) Coding of temporal patterns in the central auditory nervous system. In: Edelman GM, Gall WE, Cowan WM (eds) *Auditory Function–Neurobiological Bases of Hearing*. New York: John Wiley & Sons, pp. 337–362.

Schreiner CE, Urbas JV (1986) Representation of amplitude modulation in the auditory cortex of cat. I. Anterior auditory field. Hearing Res 21:227–241.

Schreiner CE, Urbas JV (1988) Representation of amplitude modulation in the auditory cortex of cat. II. Comparison between cortical fields. Hearing Res 32:49–64.

Schreiner CE, Mendelson JR (1990) Functional topography of cat primary auditory cortex: Distribution of integrated bandwidth. J Neurophysiol 64:1442–1459.

Schreiner CE, Mendelson JR, Grasse K, Sutter M (1988) Spatial distribution of basic response properties in cat primary auditory cortex. Assoc Res Otolaryngol Abstr 11:36.

Shamma SA, Fleshman JW (1990) Spectral orientation columns in the primary auditory cortex. Assoc for Res Otolaryngol Abstr 13:222.

Spoendlin H (1972) Innervation densities of the cochlea. Acta Otolaryngol 73:235–248.

Steffen H, Simonis C, Thomas H, Tillein J, Scheich H (1987) Auditory cortex: Multiple fields, their architectonics and connections in the Mongolian gerbil. In: Syka J, Masterton RB (eds) Auditory Pathway-Structure and Function. New York: Plenum Press, pp. 223–228.

Steibler I (1987) A distinct ultrasound-processing area in the auditory cortex of the mouse. Naturwissenschaften 74:96–97.

Suga N (1982) Functional organization of the auditory cortex: Representation beyond tonotopy in the bat. In: Woolsey CN (ed) Cortical Sensory Organization, Vol. 3. Multiple Auditory Areas. Clifton, NJ: Humana Press, pp. 157–218.

Suga N (1984) The extent to which biosonar information is represented in the bat auditory cortex. In: Edelman GM, Cowan WM, Gall WE (eds) Dynamic Aspects of Neocortical Function. New York: John Wiley & Sons, pp. 315–373.

Suga N (1988) Auditory neuroethology and speech processing: Complex sound processing by combination-sensitive neurons. In: Edelman GM, Gall WE, Cowan WM (eds) Auditory Function-Neurobiological Bases of Hearing. New York: John Wiley & Sons, pp. 679–720.

Suga N, Niwa H, Taniguchi I, Margoliash D (1987) The personalized auditory cortex of the mustached bat: adaptation for echolocation. J Neurophysiol 58:643–654.

Tunturi AR (1950) Physiological determination of arrangement of the afferent connections to the middle ectosylvian auditory area in the dog. Am J Physiol 162:489–502.

Tusa RJ, Palmer LA, Rosenquist AC (1981) Multiple cortical visual areas: Visual field topography in the cat. In: Woolsey CN (ed) Cortical Sensory Organization. Clifton, NJ: Human Press, pp. 1–31.

Walzl EM (1947) Representation of the cochlea in the cerebral cortex. Laryngoscope 57:778–787.

Weinberger N, Diamond D (1988) Dynamic modulation of the auditory system by associative learning. In: Edelman GM, Gall WE, Cowan WM (eds) Auditory Function-Neurobiological Bases of Hearing. New York: John Wiley and Sons, pp. 485–512.

Winer J (1984) Anatomy of Layer IV in cat primary auditory cortex (AI). J Comp Neurol 224:535–567.

Wong D, Shannon SL (1988) Functional zones in the auditory cortex of the echolocating bat, Myotis lucifugus. Brain Res 453:349–352.

Woolsey CN (1960) Organization of cortical auditory system: A review and a synthesis. In: Rasmussen GL, Windle W (eds) Neural Mechanisms of the Auditory and Vestibular Systems. Springfield, IL: C. Thomas, pp. 165–180.

Woolsey CN (1971) Tonotopic organization of the auditory cortex. In: Sachs MB (ed) Physiology of the Auditory System. Baltimore: National Educational Consultants, pp. 271–282.

Zook JM, Leake PA (1989) Connections and frequency representation in the auditory brainstem of the mustache bat, Pteronotus parnellii. J Comp Neurol 290:243–261.

34
Evolution of Sound Localization in Mammals

Rickye S. Heffner and Henry E. Heffner

1. Introduction

The ability to locate the source of a sound too brief to be either scanned or tracked using head or pinna movements is of obvious advantage to an animal. Since most brief sounds are made by other animals, the ability to localize such sounds enables an animal to approach or avoid other animals in its immediate environment. Moreover, it can be used to direct the eyes, thus bringing another sense to bear upon the source of the sound. Given the value of sound localization to the survival of an animal, it is not surprising that the need to localize sound has been implicated as a primary source of selective pressure in the evolution of mammalian hearing (Masterton et al. 1969; Masterton 1974).

Because of the obvious survival value of sound localization it might seem logical that all animals are under strong selective pressure to localize sound as accurately as possible. However, in the last decade, it has become apparent that this is not true. Not only are there poor localizers whose limited acuity cannot be attributed to a reduction in the availability of locus cues, but there exists at least one species that lacks entirely the ability to localize brief sounds (R. Heffner and Heffner 1990b). This situation indicates that selective pressure for accurate sound localization must vary between different species of mammals.

The purpose of this chapter is to review the behavioral data on mammalian sound localization in a search for the selective pressures that have played a role in its evolution and also to examine how the need to localize sound has exerted selected pressure on mammalian hearing. In doing so, we will address the following points: What are the basic sound-localization cues and how do mammals vary in the use of these? What is the relationship between the use of the binaural cues and the morphology of the superior olivary complex? How has the need to localize sound influenced the evolution of high-frequency hearing? What accounts for the variation in sound-localization acuity?

2. The Cues for Sound Localization

The locus of a sound can be described in terms of its azimuth, elevation, and distance from the observer. At present, discrimination of elevation has been determined for only a few species and almost nothing is known concerning the comparative ability of mammals to discriminate distance. However, there now exists a large body of information on the ability of mammals to discriminate the azimuth of sound sources and it is this aspect that is the primary focus of this chapter.

In comparing the ability of mammals to localize sound, it is helpful to review the cues that animals use to determine the location of a sound source. These cues can be divided into two general categories. The first are the binaural cues in which the azimuth of a sound source is computed by comparing the input from the two ears. The second are the monaural spectral cues which arise from the variation in the spectrum of a sound due to the directionality of the pinnae and the diffraction of sound by the head and torso.

TABLE 34.1. Effect of head size and azimuth on interaural time difference and the frequency at which binaural phase becomes ambiguous

Animal	Radius of head (in mm)	Interaural time difference (in μsec) Azimuth		Frequency of phase ambiguity (in kHz) Azimuth	
		90°	10°	90°	30°
House mouse	4.5	39	7	12.700	25.400
Dog	47.5	415	72	1.210	2.410
Human	90.0	786	136	0.636	1.270
Elephant	366.0	3,200	555	0.156	0.313

Interaural time difference and frequency of phase ambiguity calculated using the formula from Kuhn (1977).

Although most research has centered on the contribution of the binaural cues, monaural spectral cues, especially those generated by the pinnae, provide the primary information needed to determine elevation and to prevent front-back reversals. As we shall see, the need to use monaural spectral cues appears to have played a greater role than previously recognized in the evolution of mammalian hearing.

2.1 Binaural Locus Cues

The two chief binaural locus cues are the difference in the time of arrival of a sound at the two ears, which can be abbreviated "Δt," and the difference in the frequency-intensity spectra of a sound at the two ears, "Δfi." These two cues play a major role in the localization of sound in the horizontal plane. (For a review of the neural encoding of the binaural locus cues, see Phillips and Brugge 1985).

2.1.1 Binaural Time Cue

The difference in the time of arrival of a sound at the two ears, Δt, for a particular angle depends on the size of an animal's head. Animals with large heads have much larger Δt's available to them than do smaller animals. For example, the Δt cue available to the Indian elephant is approximately 80 times greater than that available to a wild house mouse (Table 34.1). Thus the auditory system of a small mammal would have to achieve much greater resolution of binaural time differences than that of a large mammal in order to attain the same degree of sound localization acuity.

It should be noted that a physically large head does not always result in a large Δt. Animals that hear underwater have a smaller *functional* head

size because sound travels faster in water and, in some cases, the sound travels through the head instead of around it (McCormick et al. 1970; Norris and Harvey 1974). Both of these factors act to reduce the difference in the time of arrival of a sound at the two ears.

In order for the auditory system to determine a binaural time difference, it is necessary to compare the time of arrival of a sound at one ear with the arrival of the same portion of the sound at the other ear. In the case of pure tones, this is done by comparing the time of arrival of the same phase of the sine wave at the two ears. This cue is thus referred to as the binaural phase-difference cue or "$\Delta\phi$," a special case of Δt. The existence of a physiological upper limit for the use of the interaural phase-difference cue was suggested by Stevens and Newman (1936), who noted that the auditory nerve is limited in its ability to synchronize with the phase of a stimulus. It has been shown electrophysiologically that phase locking usually begins to decline at about 1 kHz and, in mammals, has not been observed higher than 5 kHz (Rose et al. 1967). Thus, the auditory system is incapable of deriving binaural phase information from high-frequency tones.

Even if the auditory system were capable of phase locking to high frequencies, there is a physical limitation to the upper limit of the usefulness of $\Delta\phi$. Although the auditory system is able to determine the time of arrival of a particular portion of the waveform at the two ears (e.g., the peak of a sine wave) it cannot distinguish a portion of one cycle of a sine wave from the same portion of another cycle. Therefore, in order to use $\Delta\phi$, it is necessary for a particular cycle to reach the far ear before the next cycle reaches the ear nearest the sound source.

The frequency at which the $\Delta\phi$ becomes physically ambiguous can be calculated according to the following formula:

$$F = 1/[6(a/C)\sin\theta]$$

where a is the radius of the head, C is the speed of sound, and θ is the angle of the sound source from the animal's midline.

Basically, $\Delta\phi$ becomes physically ambiguous when the difference in the distance of the two ears from the sound source equals one half of the wavelength of a tone and remains ambiguous for all shorter wavelengths (i.e., higher frequencies). As indicated by this formula, there are two factors which determine the frequency of ambiguity. The first is the size of an animal's head (a): animals with small heads will have a higher frequency of ambiguity than those with large heads due to the smaller time delay between the two ears (Table 34.1). Indeed, $\Delta\phi$ may be physically available to small mammals at such high frequencies that their upper limit for $\Delta\phi$ is determined solely by physiological factors, i.e., their ability to phase lock and to discriminate small binaural time differences. The second factor is the angle of the sound source (θ) from midline: $\Delta\phi$ becomes physically available at progressively higher frequencies as the azimuth of the sound source moves from the side toward the midline owing to the accompanying reduction in effective distance (i.e., time delay) between the two ears (Table 34.1). Thus, the angle of the sound source may determine whether a particular frequency is above or below the frequency of ambiguity.

In contrast to pure tones, mammals are able to extract binaural time information from high-frequency as well as low-frequency complex signals. This is because time information can be obtained from the envelope of high-frequency noise and from its onset, even though the carrier frequency may be above the frequency of ambiguity (e.g., Henning 1974; McFadden and Pasanen 1976). Although the time information extracted from high-frequency signals may not be as reliable as that obtained from low frequencies (e.g., Butler 1986), it nevertheless enables the Δt cue to be used to locate the source of a wide range of sounds.

2.1.2 Binaural Spectral Difference Cue

The heads and pinnae of mammals produce a sound shadow which results in a difference in the fre-

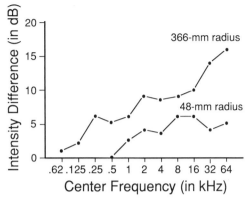

FIGURE 34.1. Sound-shadowing characteristics of two spheres approximating the head size of a dog (48-mm radius) and an elephant (366-mm radius) (cf. Table 34.1). The difference in intensity of a sound at the side nearest and furthest from a loudspeaker was determined for noise bands centered at octave intervals (24 dB/octave rolloff). This comparison supports the idea that small animals must hear higher frequencies than large animals in order for their heads to generate usable binaural intensity differences.

quency–intensity spectra (Δfi) of the sounds at the two ears. The magnitude of the Δfi cue is dependent on the size of the shadowing object so that animals with large heads and pinnae produce a greater sound shadow than small species. This cue is frequency dependent because the attenuation by the head is greater for high frequencies than for low frequencies (Fig. 34.1). The interaction between head size and frequency means that large species not only produce larger overall interaural spectral differences, they also produce differences at lower frequencies than do small species.

The Δfi cue is subject to two limitations. First, low frequencies, whether pure tones or complex sounds, can bend around an animal's head with little or no attenuation. This is a physical limitation in the ability of the head and pinnae to produce a sound shadow. Interestingly, the auditory system is physiologically able to extract spatial information when low-frequency binaural intensity differences are artificially generated by presenting the sounds through headphones even though such intensity differences do not occur in nature (Mills 1972). As illustrated in Figure 34.1, what constitutes a "low" frequency depends upon the size of an animal's head because large heads generate physical intensity differences at lower frequencies than small heads. Thus, unlike the binaural time cue, the binaural spectral-difference cue cannot be obtained from all complex sounds.

The second limitation is the observation that at very high frequencies a sound may be completely shadowed and not detectable at the far ear (e.g., Butler and Flannery 1980). In this case the Δfi cue can only indicate that a sound occurred in the left or right hemifield and may lead to the erroneous perception that the sound source is located in one ear. In such a case it may be advantageous for the auditory system to ignore intensity differences at very high frequencies as they may add no useful information beyond that available from the lower frequencies and could conflict with the information available from the other locus cues.

The Δfi cue is most reliable for complex sounds that contain high frequencies, particularly broadband noise. As the bandwidth of noise is reduced, localization acuity declines (Brown et al. 1980; Butler 1986). In the case of a pure tone, the interaural spectral cue is reduced to a simple interaural intensity difference (ΔI), a subset of Δfi. However, it has been observed that pure tones do not always provide a reliable interaural intensity difference and are subject to left–right reversals with the result that the intensity of a pure tone may in some cases be greater at the far ear than at the near ear (e.g., Harrison and Downey 1970; Martin and Webster 1989). The results of behavioral studies demonstrate a corresponding difficulty in localizing tonal stimuli in both humans and other mammals (Mills 1972; Brown et al. 1978, 1982; Terhune 1985; Martin and Webster 1987). Indeed, some animals appear to take advantage of the fact that pure tones are difficult to localize. For example, the predator warning calls of animals are often more tonal than other calls, thus making the calling individual more difficult to locate (Marler 1955; for reviews see Erulkar 1972; Brown and May 1990). Nevertheless, most natural sounds are complex sounds containing high frequencies which permit the use of the Δfi cue.

2.2 Monaural Spectral Cues

Monaural cues arise from the differences in the spectrum of a sound reaching an ear from different locations. These differences are due to the directionality of the pinna as well as the diffraction of sound around the head and torso (e.g., Butler 1975; Kuhn 1982; Musicant and Butler 1984b). The primary source of these cues is the pinna and the contribution of pinna cues to localization has been investigated in some detail.

The pinna acts as a funnel that selectively admits into the auditory canal, and in some cases amplifies, high frequencies that emanate from sources located along the pinna's directional axis and slightly above the horizon; high frequencies from other directions are attenuated (e.g., Phillips et al. 1982; Calford and Pettigrew 1984; Butler 1987; Carlile and Pettigrew 1987; Humanski and Butler 1988; Middlebrooks 1990). The spectrum of the sound is also affected by the folds of the pinna and distortion of even the diminutive pinna of humans degrades sound location accuracy (for reviews see Butler 1975; Shaw 1974). Low frequencies, on the other hand, are much less strongly affected by the pinna. Thus pinna cues are dependent on high frequencies and in humans, for example, frequencies above 4 kHz must be present in order for pinna cues to be effective (Belendiuk and Butler 1975; Musicant and Butler 1984b). Indeed, pinna cues can provide highly accurate information providing that the pinna is undistorted and the sound is complex and contains high frequencies (Butler 1986).

Monaural spectral cues are most evident in situations where binaural cues are limited or absent. When localizing sound in the horizontal plane, binaural cues are of limited value in preventing front–back reversals and monaural cues are needed to make front–back judgments with reasonable consistency (Flannery and Butler 1981; Musicant and Butler 1984b). Monaural cues also provide necessary information for localizing sounds off to the side near the interaural line where they are more reliable than binaural cues when low frequencies are absent (Butler 1986). With regard to vertical localization, monaural spectral cues are the main, if not the sole source of locus information as binaural differences are slight or nonexistent (e.g., Butler and Belendiuk 1977; Middlebrooks et al. 1989). Furthermore, the signal must contain high frequencies in order for the elevation of the sound source to be accurately determined (Roffler and Butler 1968a). In short, monaural cues play an important role in horizontal localization and an essential role in vertical localization—and they require the presence of high frequencies in order to be effective.

3. Variation in the Use of Binaural Locus Cues Among Mammals

The results of comparative studies from several laboratories have revealed that although most mammals are capable of using both binaural time and intensity differences to localize, there are some species that have reduced or lost their ability to use one or the other of these cues.

3.1 Determining the Use of Binaural Locus Cues

The ability of an animal to use binaural locus cues can be determined with either lateralization tests, in which stimuli are presented through headphones, or localization tests in which stimuli are presented through loudspeakers. In a lateralization test of the ability to use binaural time, Δt, an animal is presented with clicks or noise bursts in which the stimulus to one ear precedes that to the other ear and the animal is trained to indicate which ear received the leading sound (e.g., Masterton and Diamond 1964). An animal that uses Δt should be able to discriminate time differences at least as small as the maximum binaural time difference generated by its head and it is inferred that, like humans, the animal perceives a single sound that is lateralized to the ear receiving the leading stimulus. By substituting pure tones, the ability to use the phase cue, $\Delta \phi$, can be measured and the upper frequency limit of the phase cue can be determined (e.g., Wakeford and Robinson 1974).

The lateralization test is used to determine the ability of an animal to use Δfi in the form of ΔI by presenting identical sounds to the two ears which differ only in level and requiring the animal to indicate which ear received the louder signal (e.g., Masterton and Diamond 1964). Again it is assumed that the animal lateralizes the sound to the ear which receives the louder signal. However, an animal can perform such an intensity lateralization test by listening only to the variation in the signal in one ear. As a result it is necessary to randomize the overall level of the signal in order to prevent the animal from performing the task monaurally (e.g., Yost and Dye 1988).

Because of the technical difficulty of placing headphones on animals, many studies have used free-field tone-localization tests to examine the ability of animals to use the $\Delta \phi$ and ΔI cues. This test is based on the fact that low-frequency tones that bend around the head with little or no intensity difference between the two ears can only be localized using the phase cue, $\Delta \phi$. Tones that lie above the frequency of phase ambiguity, on the other hand, can only be localized using the intensity difference cue, ΔI. Thus, the ability of an animal to localize tones below the frequency of phase ambiguity indicates that it possesses the ability to use $\Delta \phi$, and, presumably Δt, whereas the ability the localize high-frequency tones indicates that it can use ΔI and, presumably Δfi.

Although the ability of an animal to localize pure tones can be determined by obtaining thresholds at various frequencies, a standard procedure is to determine localization performance with loudspeakers placed 30° to the left and right of an animal's midline (Masterton et al. 1975). Because the frequency at which phase becomes ambiguous varies with the angle of the sound source, this test has the advantage of keeping the angle of separation, and thus the frequency of ambiguity, constant (cf. Table 34.1 and Section 2.1.1 of this chapter). An additional important feature of this test is that the tones are presented at a constant level above an animal's threshold (e.g., 40-dB sensation level). Thus, an animal's ability to detect the tones is held constant and any variation in performance across frequency reflects an animal's ability to localize.

3.2 Species Using Binaural Phase and Intensity Difference Cues

Most mammals use both binaural locus cues. The use of both time and intensity differences has been demonstrated in 12 species including insectivores, primates, rodents, cetacea, carnivores, and pinnipeds (Table 34.2). Most of these animals were tested for their ability to localize tones above and below the frequency of phase ambiguity. However, lateralization data are available for man (Mills 1972), squirrel monkey (Don and Starr 1972), macaque (Houben and Gourevitch 1979), and the domestic cat (Wakeford and Robinson 1974) which support the results of the tone-localization tests.

TABLE 34.2. Species of mammals grouped by ability to use the main binaural locus cues

Ability	Species	Source
Use both Δθ and ΔI		
	Tree shrew	Masterton et al. 1975
	Human	Stevens and Newman 1936
	Squirrel monkey	Don and Starr 1972
	Pig-tailed macaque	Brown et al. 1978
	Gerbil	R. Heffner and Heffner 1988c
	Kangaroo rat	Heffner and Masterton 1980
	Norway rat	Masterton et al. 1975
	Bottlenose dolphin	Renaud and Popper 1975
	Red Fox	Isley and Gysel 1975
	Cat	Casseday and Neff 1973
	Least weasel	R. Heffner and Heffner 1987
	Sea lion	Moore and Au 1975
Ability to use Δθ reduced or absent		
	Hedgehog	Masterton et al. 1975
	Spiny mouse	Mooney unpublished Master's thesis
Ability to use ΔI reduced or absent		
	Elephant	R. Heffner and Heffner 1982
	Horse	R. Heffner and Heffner 1986a
	Pig	R. Heffner and Heffner 1989
	Cattle	R. Heffner and Heffner 1986b
	Goat	R. Heffner and Heffner 1986b
Unable to use Δt or Δfi		
	Gopher	R. Heffner and Heffner 1990b

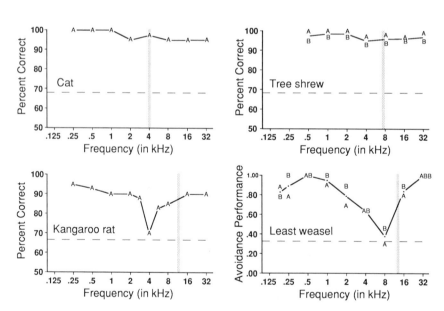

FIGURE 34.2. Sound-localization performance in the horizontal plane as a function of frequency for four species. Tones were presented from two loudspeakers located 30° left and right of midline. Vertical shaded bar indicates the frequency above which binaural phase becomes physically ambiguous. Note that all four species can localize tones above and below the frequency of phase ambiguity indicating that they can use binaural phase and intensity difference cues. Letters indicate individual animals; dashed lines indicate chance level ($p < 0.01$). Note that the tones in this test are presented at a constant level above threshold. (After Masterton et al. 1975; H. Heffner and Masterton 1980; R. Heffner and Heffner 1987.)

Examples of the results of the tone-localization tests are shown in Figure 34.2 for the cat, tree shrew, kangaroo rat, and least weasel. The fact that each of these animals can use both $\Delta\phi$ and ΔI is indicated by their ability to localize tones above and below the frequency at which $\Delta\phi$ becomes physically ambiguous (indicated by the vertical shaded bars).

It can also be seen in Figure 34.2 that some mammals, such as the kangaroo rat and least weasel, show a distinct decrease in performance in the midfrequency range. Such a decrease is usually seen in animals with small heads due to the fact that their ability to use binaural phase declines at frequencies that are too low to generate usable interaural intensity differences. When this occurs, the results can be used to derive an estimate of the upper frequency limit of an animal's ability to use $\Delta\phi$. In using this method, a minimum estimate of an animal's upper limit for $\Delta\phi$ is the frequency just below the point of lowest performance. For the animals in Figure 34.2, this would be 2.8 kHz in the kangaroo rat and 4 kHz in the least weasel. This represents a minimum estimate because it cannot be determined if an animal is using $\Delta\phi$ or ΔI at the point of poorest performance.

However, not all small mammals show a performance decrement in this test. For example, the tree shrew shows no obvious decrement (Fig. 34.2) whereas the Norway rat, which has a head size slightly larger than the tree shrew, has a distinct midfrequency decrement (Masterton et al. 1975; Kelly and Kavanaugh 1986). One interpretation of these results is that the tree shrew's auditory system possesses the ability to use $\Delta\phi$ at higher frequencies than the rat's.

3.3 Species with Limited or Absent Ability to Use Binaural Phase Differences

The idea that all mammals possess the ability to use both binaural cues was first called into question by Masterton and his colleagues (1975). In testing the ability of various mammals to localize pure tones, they discovered that although the hedgehog is able to localize high-frequency tones, it is unable to localize low-frequency tones at frequencies where $\Delta\phi$ is unambiguous (Fig. 34.3).

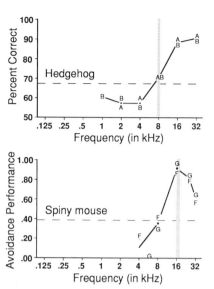

FIGURE 34.3. Sound-localization performance in the horizontal plane as a function of frequency for two species. Tones were presented from two loudspeakers located 30° left and right of midline. Vertical shaded bar indicates the frequency above which binaural phase becomes physically ambiguous. The inability of these two species to localize low-frequency tones suggests that they are unable to use binaural phase cues to localize sound. Letters indicate individual animals; dashed lines indicate chance level ($p < 0.01$). (After Masterton et al. 1975; Mooney unpublished Master's thesis.)

These results indicate that the hedgehog does not use $\Delta\phi$ and, presumably, binaural time (Δt), but is capable of using binaural intensity differences to localize sound.

The possibility exists that other species may also lack the ability to use Δt. In particular, small mammals such as mice which lack good low-frequency hearing may be unable to hear tones low enough for phase locking to occur. One example shown in Figure 34.3 is the spiny mouse whose hearing ranges from 2.2 kHz to 71 kHz (Mooney et al. 1990).

In searching for a reason as to why an animal would relinquish the use of a major binaural locus cue, it might be argued that an animal with a small head and close-set ears may lose the time cue because the available time differences are too small for its nervous system to resolve. However, it should be noted that other small mammals, such as the tree shrew, Norway rat, gerbil, kangaroo rat,

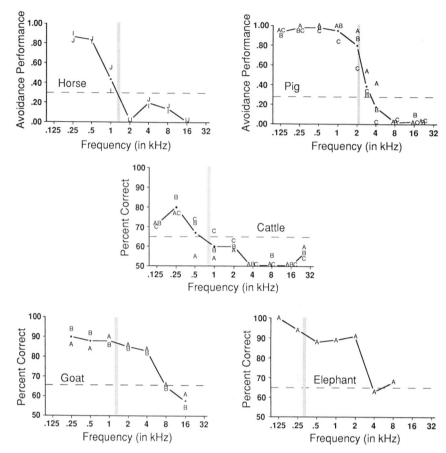

FIGURE 34.4. Sound-localization performance in the horizontal plane as a function of frequency for five species. Tones were presented from two loudspeakers located 30° left and right of midline. Vertical shaded bar indicates the frequency above which binaural phase becomes physically ambiguous. These animals have either partly or completely lost the ability to use binaural intensity cues as indicated by their inability to localize some or all of the tones above the frequency at which binaural phase becomes ambiguous. Letters indicate individual animals; dashed lines indicate chance level ($p < 0.01$). (After R. Heffner and Heffner 1982, 1986a, 1989.)

and least weasel, all have interaural distances smaller than the hedgehog's, and yet are able to use $\Delta\phi$. Thus there is currently no good explanation as to why some animals have relinquished use of $\Delta\phi$.

3.4 Species with Reduced or Absent Ability to Use Binaural Intensity Differences

As a wider variety of species have been examined, it has become apparent that not all mammals use the binaural intensity-difference cue. So far five species of mammals (horse, pig, cattle, goat, and elephant) appear to have lost part or all of the ability to localize pure tones above the frequency of phase ambiguity (Fig. 34.4) (R. Heffner and Heffner 1982, 1986a 1986b, 1989).

Some species, such as goats and elephants, have only partially lost the ability to use ΔI. That is, they are unable to localize pure tones in the upper 2 to 3 octaves of their hearing range, but retain the ability to use ΔI at lower frequencies. Horses, however, are completely unable to localize tones above the frequency of phase ambiguity, suggesting that the inability of this species to use ΔI is total. The results of tests on pigs and cattle suggest that they might retain some residual ability to use ΔI because some individuals were able to localize tones slightly above the frequency of phase ambiguity (R. Heffner and Heffner 1986b, 1989). However, the frequency of phase ambiguity is calculated based on a sphere of the same diameter as the head of an average animal and does not take the actual shape of the head into account. It is thus possible that individual variation in frequency of phase ambiguity may account for the performance of some individuals and that pigs and cattle, like horses, completely lack the ability to use binaural intensity differences.

The ability of horses to perform binaural phase and intensity discriminations was further examined using headphones (R. Heffner and Heffner 1988a). The results of these lateralization tests demonstrated that horses can discriminate binaural phase differences from 250 Hz to 1.2 kHz, but fall to chance at frequencies of 1.3 kHz and above. This determination of the upper frequency limit of binaural phase discrimination is in keeping with the results of the tone localization test (cf. Fig. 34.4). More important, however, was the finding that the animals were unable to discriminate stimuli based on interaural intensity differences when the overall level of the signal was randomized to prevent them from making the discrimination monaurally. Thus, the results of both localization and lateralization tests indicate that horses are unable to use the binaural intensity difference cue, ΔI, and, presumably, Δfi.

In searching for an explanation as to why animals would relinquish ΔI, the possibility arises that it could result from reduced selective pressure to localize sound, perhaps owing to the reduced demands of domestication. However, although some of these species, particularly horses and cattle, are not accurate localizers, others are very accurate. In particular, the 4.5° threshold of the pig makes it more accurate than cats or macaques, while the 1.2° threshold of the elephant makes it as accurate as humans and dolphins. Furthermore, the fact that elephants have not been selectively bred makes it unlikely that the reduction or loss of ΔI is the result of domestication.

Nor have these animals given up the use of ΔI because the cue is unavailable to them. One feature common to these animals is that they are all large mammals. As a result they have relatively large heads and pinnae which generate large binaural intensity differences. Measurements of

the physical cues available to horses, for example, reveal the presence of binaural intensity differences of more than 20 dB (H. Heffner and Heffner 1984). Indeed, we suggest that the reason these animals have reduced or eliminated their use of binaural intensity differences is because the available intensity differences are *too* large.

Given the large heads and pinnae of these animals the situation can arise in which a sound, or the high-frequency component of a sound, is audible in the ear nearest the sound source and *in*audible in the far ear. As previously noted, the spectral difference cue will indicate only the hemifield in which the sound arose and could possibly degrade the locus information derived from the binaural time cue. Thus, if the high-frequency portion of a sound is not audible in both ears, the resulting Δfi cue will add little useful locus information beyond that provided by the other cues.

One way to reduce the occurrence of an unusable Δfi cue is to reduce the ability to hear high frequencies. However, if an animal is under pressure to hear high frequencies for other purposes, such as to prevent front–back reversals, then it might be advantageous for their auditory systems to reduce or eliminate the binaural intensity difference at high frequencies from the computation of locus.

3.5 Species Using Neither Binaural Time Nor Intensity

We now know that there exists at least one species that is incapable of localizing any brief sound—high- or low-frequency, pure tones or complex noise—indicating that it cannot use any of the major locus cues. This animal is the pocket gopher (*Geomys bursarius*), an animal that is specialized for living underground in an essentially one-dimensional world where azimuthal locus may have little significance (R. Heffner and Heffner 1990b). Because gophers can only discriminate the loci of long-duration sounds (i.e., greater than 0.5 sec) and then only at angles greater than 90° separation, they demonstrate little sound localization ability beyond homing or scanning. Although other fossorial species have not been tested systematically, it has been suggested that they also have little response to locus (Burda et al. 1990).

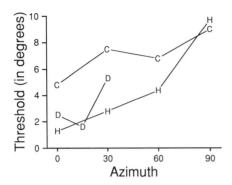

FIGURE 34.5. Sound localization acuity as a function of azimuth for cat (C), human (H), and dolphin (D). Note that cats are as accurate as humans in localizing sound sources centered around 90° azimuth, a front-back discrimination in which monaural spectral cues play a major role. (Data from Renaud and Popper 1975; R. Heffner and Heffner 1988a.)

4. Use of Monaural Spectral Cues

The contributions of monaural spectral cues to sound localization are most evident in preventing front–back reversals and in localizing in the vertical plane—two situations in which binaural cues are relatively ineffective.

4.1 Sound Localization in the Horizontal Plane

Most of what we know about the role of monaural spectral cues in horizontal localization comes from the studies of human sound localization, particularly those of Butler and his colleagues. Monaural locus cues arise primarily from the directionality of the pinna with the diffraction of sound by the head and torso also contributing information (Butler 1975; Kuhn 1982; Musicant and Butler 1984b). Pinna cues require the presence of high frequencies and, in man, frequencies above 4 kHz are necessary in order for the pinna to play an effective role in localization (Belendiuk and Butler 1975; Musicant and Butler 1984b). Furthermore, monaural localization improves as the bandwidth of the stimulus increases (Butler 1986). Thus, the ability to use monaural spectral cues is best demonstrated by tests employing broadband noise containing high frequencies (cf. Brown et al. 1980).

Pinna cues result from the selective amplification and attenuation of different frequencies by the pinna depending on the location of the sound source. This can be demonstrated by measuring the effect of the head and pinna on the sound reaching the auditory meatus (e.g., Shaw 1974). Indeed, the importance of the effect of the pinna on the sound spectrum is demonstrated by the fact that the apparent location of a sound source can be shifted simply by changing the spectral content of the sound with the new perceived location corresponding to the position predicted by the transfer function of the pinna (Butler and Flannery 1980; Flannery and Butler 1981; Musicant and Butler 1984a, 1985; Butler 1987). Furthermore, distorting or filling the convolutions of the pinna dramatically reduce the ability to use monaural localization cues (e.g., Butler 1975).

Although binaural cues play a dominant role in localizing sound near the midline, both in front and in back, monaural spectral cues become increasingly important as the sound source moves off to the side (Musicant and Butler 1984b, 1985; Butler 1986). Furthermore, monaural spectral cues are the primary means for preventing front–back reversals (e.g., Musicant and Butler 1984b; cf. Boring 1942). However, the reliance on monaural spectral cues may be more widespread than is generally realized. As Butler and Flannery (1980) have noted, monaural cues provide the only means of localizing sounds that are inaudible in the ear furthest from the source, a common occurrence with high-frequency transient sounds.

The fact that sound-localization acuity decreases as the sound source is located progressively farther from the midline indicates that monaural cues are generally not as effective as the binaural cues. This decrease in acuity with azimuth has been demonstrated in cats and dolphins, as well as in humans, using complex sounds (Fig. 34.5) and in monkeys using tones (Renaud and Popper 1975; Brown et al. 1982; R. Heffner and Heffner 1988b). However, it is of interest to note that the increase in thresholds in the cat is proportionally less than that for humans. This suggests that the cat's relatively large and mobile pinnae may enhance its ability to use monaural locus cues. Thus, the difference in acuity between monaural and binaural localization in animals with large and mobile pinnae may be significantly less than that found in humans.

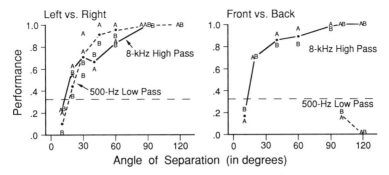

FIGURE 34.6. Localization of continuous (3-sec duration) high- and low-pass noise in the horizontal plane by horses. Either high- or low-pass noise could be localized in the left-right discrimination where the sound sources were centered on midline. However, only the high-pass noise could be localized in the front-back discrimination where the sound sources were centered on 90° azimuth. Horizontal dashed line indicates chance level. (Data from R. Heffner and Heffner 1983.)

A number of studies have examined the directional properties of the pinnae of other mammals. Measurements of the filtering characteristics of the pinnae have been examined in cats (Phillips et al. 1982; Calford and Pettigrew 1984; Irvine 1987), guinea pigs (Carlile and Pettigrew 1987), and bats (Fuzessery and Pollak 1985). These studies have demonstrated that all of these pinnae, despite their different sizes and configurations, act as directional filters in much the same way as the human pinna.

As with humans, other mammals also require the presence of high frequencies in order to make use of monaural spectral cues. This has been demonstrated in the case of the horse (an animal that hears well above 10 kHz and yet does not make use of binaural spectral cues) by assessing sound localization ability for 500-Hz low-pass and 8-kHz high-pass noise. As shown in Figure 34.6, two horses were able to localize either high- or low-frequency noise when the sound sources were centered on the midline (a left–right discrimination) and binaural time cues were available. However, only the 8-kHz high-pass noise could be localized when the sound sources were centered 90° from midline (a front–back discrimination) and monaural spectral differences were the primary cue. This finding demonstrates that horses, which do not use interaural intensity differences, nevertheless require the presence of high frequencies in order to use monaural spectral cues. This in turn suggests that the need to use monaural spectral cues may by itself exert strong selective pressure for good

high-frequency hearing. We will return to this point in Section 6.1.

4.2 Sound Localization in the Vertical Plane

Monaural spectral cues play a primary role in localizing sound in the vertical plane—a situation in which binaural cues are relatively ineffective. As in azimuthal localization, the pinnae selectively amplify and filter different frequencies depending on the elevation of the source (for a review, see Fuzessery 1986). Studies of human localization have demonstrated that vertical localization depends on the same variables as monaural localization in the horizontal plane. Specifically, maximum accuracy is obtained with broad-band complex sound which includes frequencies above 7 kHz and the pinna must be intact (Roffler and Butler 1968a; Butler and Planert 1976). Indeed, all the information necessary for vertical localization in humans appears to be contained in the frequencies above 4 kHz (Butler and Helwig 1983). Furthermore, both ears participate in vertical localization with the ear nearest the sound source playing the major role (Humanski and Butler 1988). Finally, the apparent elevation of a sound varies as a function of frequency (Roffler and Butler 1968b; Butler and Helwig 1983).

In addition to humans, behavioral measures of the role of frequency in vertical localization have been obtained for the cat, macaque, opossum, and porpoise. Most of these animals appear to be highly accurate with thresholds of 4° for the cat (Martin and Webster 1987), 3° for the macaque (Brown et al. 1982), and 1° for the bottlenose dolphin (Renaud and Popper 1975). The opossum, on the other hand, was found to have a threshold of 14° (Ravizza and Masterton 1972). However, this may be an underestimation of the animal's ability because the spectrum of the noise used to test the opossum probably lacked high frequencies as the transducers were hearing-aid receivers which normally roll off at 8 kHz. Indeed, the importance of high frequencies, particularly those above 8 kHz, for accurate vertical localization has been demonstrated in macaques as well as in man (Brown et al. 1982).

4.3 Mobile Pinnae

The fact that most mammals possess mobile pinnae suggests that this character must be of some value. The degree of pinna mobility ranges from the 180° mobility of horse pinnae to the unmoving pinnae of humans. The fact that human pinnae are immobile leaves us unable to conceptualize the advantage of such movement through introspection.

It has often been speculated that pinna mobility confers an advantage in localizing sound (e.g., Jeffress 1975) and to a certain extent this appears to be true. As noted in Section 4.1 of this chapter there is evidence that the mobile pinnae of cats enhance their ability to localize sound in the lateral fields where binaural cues are less effective and it has been shown in the cat that preventing pinna movements by denervating the external ear muscles reduces localization acuity (Siegmund and Santibañez-H 1981). Moreover, some bats appear to require normally mobile pinnae in order to make discriminations in the vertical plane (Mogdans et al. 1988). Thus, pinna mobility may be an effective way of enhancing monaural spectral cues. However, mobile pinnae could potentially interfere with the use of the binaural spectral differences because the magnitude of the cue would depend on the orientation of the pinnae. As a result it would be necessary for the nervous system to take pinna position into account in calculating locus. Although there is some evidence that this does occur in the cat (Stein and Clamann 1981), it would nevertheless be a potential source of error.

It should also be noted that mobile pinnae are not uniquely associated with good sound localization acuity. The best example is the horse, an animal with extremely mobile pinnae and poor sound localization acuity (see Section 7 in this chapter). Although the horse has a large head and correspondingly large binaural cues available to it, it has a midline localization threshold of 22° for noise (H. Heffner and Heffner 1984). Nor is its acuity for localizing off to the side where monaural spectral cues predominate much better even when presented with long-duration sounds that permit scanning (cf. Fig. 34.6). These results suggest that pinna mobility must serve another purpose, at least in the horse. One possibility is that they serve as mobile directional filters, amplifying sounds originating from in front of the opening while attenuating sounds from behind. As such they would enhance an animal's ability to pick out signals embedded in a noisy world without having to move its head.

5. Variation in the Superior Olivary Complex

Numerous studies of the response properties of neurons in the medial and lateral superior olivary nuclei (MSO and LSO, respectively) have been carried out since the early investigations of Galambos and his colleagues (1959) and of Boudreau and Tsuchitani (1968). These have been reviewed in detail elsewhere (e.g., Aitkin et al. 1984; Phillips and Brugge 1985; Irvine 1986) and will not be described here. In general the superior olivary complex has been shown to be the major brainstem structure for neural interaction between input from the two ears. Neurons in the LSO have been shown to be sensitive to intensity differences at the high frequencies to which they are tuned. On the other hand, neurons in the MSO are sensitive to frequencies in the lower portion of an animal's audiogram and have been shown to respond as a function of the interaural time differences. Because mammals differ in their use of binaural time and intensity cues, it is reasonable to expect these differences to be reflected in the morphology of the MSO and LSO.

This may be true for the MSO as it has been demonstrated that the size of this nucleus is directly related to the ability of a species to use the binaural phase cue, $\Delta\phi$ (Masterton et al. 1975). That is, the smaller the MSO, the less an animal is able to use $\Delta\phi$; and an animal that lacks an MSO, i.e., the hedgehog, is completely unable to use this cue. This finding has received recent support from the observation that the spiny mouse, which has almost no MSO, does not appear to use $\Delta\phi$. Thus, there is some truth to the statement by Irving and Harrison (1967) that the MSO is not essential for sound localization per se. However, the presence of an MSO does seem to be necessary for the use of binaural time cues.

For the LSO, however, there does not seem to be any simple correspondence between morphology and ability to use binaural intensity differences. Animals with a well-developed LSO may possess the ability to use ΔI, as in the case of the cat (Masterton et al. 1975), or they may entirely lack the ability to use ΔI as in the case of the pig (R. Heffner and Heffner 1989). Similarly, one can find an undistinguished LSO in animals which possess the ability to use ΔI, such as man (e.g., Mills 1972; Moore 1987), as well as in animals that lack the ability to use ΔI such as horses and cattle (R. Heffner and Heffner 1986a). Therefore, at this time it does not appear that the gross morphology of the LSO simply reflects the degree to which an animal uses the binaural spectral difference cue. (For additional discussions of the relationship between the superior olivary complex and hearing ability see R. Heffner and Heffner 1987, 1990b; R. Heffner and Masterton 1990.)

Apparently the LSO has functions beyond sound localization and these functions may be differentially represented in different species. What these functions might be remains largely unknown, yet at least one can be suggested and others may soon be recognized. In rodents and most bats the LSO has the distinction of being a major source of small olivocochlear efferents (Altschuler et al. 1983; White and Warr 1983; Aschoff and Ostwald 1987). Therefore, it is possible to conceive of a species which retains a nucleus recognizable as an LSO even if that nucleus plays no role in sound localization because its component cells perform other important functions.

6. Evolution of High-Frequency Hearing

The ability of mammals to hear high frequencies is unique among vertebrates as neither fish, amphibians, reptiles, nor birds are able to hear above 10 kHz. However, the ability of mammals to hear high frequencies is not uniform, but varies over a range of nearly 4 octaves from around 10 kHz in elephants and gophers to over 100 kHz in some bats and underwater species (Jacobs and Hall 1972; R. Heffner and Heffner 1982, 1990b; Wenstrup 1984).

There are three commonly suggested explanations for the variation in high-frequency hearing among mammals. Two of these address the question of "why" species differ in their ability to hear high frequencies: One involves the use of high frequencies for sound localization, and the other involves their use in communication. The other explanation addresses the question of "how" animals hear sound; it involves the size of the physical apparatus for transducing sound. Thus, these theories encompass both the functional and the structural levels of biological explanation (see Mayr 1961).

6.1 High-Frequency Hearing and Sound Localization

Over 20 years ago it was proposed that mammals evolved high-frequency hearing for the purpose of localizing sound (Masterton et al. 1969). The basis for this proposal was the observation that the ability of mammals to hear high-frequency sounds is correlated with the size of the head or, more precisely, with the functional distance between the two ears, where functional distance is defined as the time it takes for sound to travel from one ear to the other. Thus, mammals with small heads and close-set ears are better able to hear high-frequency sounds than species with large heads and wide-set ears. Indeed, over the years the correlation between functional interaural distance and high-frequency hearing has remained strong (Fig. 34.7).

At the time, the existence of this relationship was explained in terms of the binaural locus cues. Briefly, the availability of the binaural cues, Δt and Δfi, depend on the functional distance between the two ears and the sound shadow of the head and pinnae. That is, the farther apart the ears, the larger will be the Δt cue for any given direction of a sound source. Similarly, the Δfi cue is greater for animals with wide-set ears both because the sound attenuation is greater over the longer distance between the ears and because animals with wide-set ears usually have large heads or large pinnae which effectively shadow the high-frequency

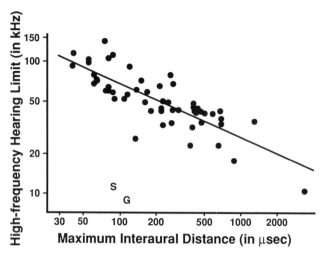

FIGURE 34.7. Relation between maximum interaural distance and the highest frequency that a species can hear at 60 dB SPL ($r = -0.84$, $n = 55$, $p < 0.001$). This relationship is explained by the fact that mammals with small heads need to hear higher frequencies than larger mammals in order to use monaural and binaural spectral cues for sound localization. Not included in the correlation coefficient are two fossorial mammals, the blind mole rat, *Spalax ehrenbergi*, (S) and the pocket gopher, *Geomys bursarius*, (G) —evidence indicates that these animals have lost their high-frequency hearing and their ability to localize sound as part of their adaptation to a subterranean habitat (cf. R. Heffner and Heffner 1990b). (See H. Heffner and Heffner 1985a; R. Heffner and Heffner 1990b,c, for individual points.)

content of sound. While the two binaural locus cues are readily available to animals with large heads, the effectiveness of either cue is diminished in animals with functionally close-set ears. In the case of Δt, the available time difference may be so small that the auditory system can detect only gross changes in sound direction. However, an animal with a small head always has a Δfi cue available, provided that it is able to perceive frequencies that are high enough to be effectively shadowed by its head and pinnae. Therefore, given the ecological importance of sound localization, animals with functionally close-set ears are subjected to more selective pressure to hear high frequencies than animals with more widely set ears.

In recent years evidence has appeared that shows that although this explanation is not incorrect, it is incomplete. The primary source of this evidence comes from the observation that large mammals have reduced, and in some cases abandoned, the use of binaural intensity cues and yet have retained their high-frequency hearing. For example, humans, which hear an octave higher than any nonmammalian vertebrate, rely much more on binaural time than on binaural intensity differ-

ences. Indeed, binaural intensity differences can be eliminated with little effect on localization performance so long as binaural time cues remain (Belendiuk and Butler 1978). A more extreme case is the horse which has completely abandoned the use of binaural intensity differences and yet can hear up to 33.5 kHz. However, both humans and horses require the presence of high frequencies for preventing front–back reversals and for vertical localization.

The above observations suggest that high-frequency hearing evolved in mammals to enable the use of monaural as well as binaural spectral cues for sound localization. Accordingly, the observed correlation between functional head size and high-frequency hearing reflects the need for mammals to hear high frequencies in order to use both types of cues—because the directionality of the pinnae is dependent on the wavelength of sound, animals with small heads and pinnae will need to hear higher frequencies than larger animals in order to obtain usable monaural as well as binaural spectral cues. Indeed, the variation in high-frequency hearing is probably best explained by a combination of factors involving both head and pinna size.

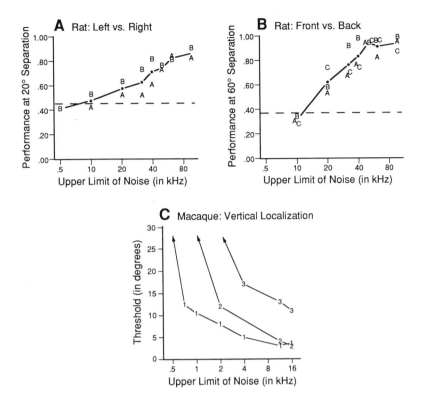

FIGURE 34.8. Effect of removing high frequencies on sound localization. Letters and numbers represent individual animals. Norway rats (A and B) were tested with a train of 100-msec noise bursts; macaques (C) were tested with 300-msec noise bursts. (Data from Brown et al. 1982 and R. Heffner and Heffner 1990a).

The idea that the ability to hear high frequencies provides mammals with monaural and binaural spectral cues is supported by the results of localization experiments in which high frequencies were progressively removed. As shown in Figure 34.8, localization performance declines when high frequencies are removed from broad-band noise signals. This decrease in performance can be seen in a left–right localization task in which Norway rats were required to discriminate two loudspeakers located 20° apart and centered on midline (Fig. 34.8A), a task in which both binaural and monaural cues are available. This result indicates that unlike large mammals, small mammals require the presence of high frequencies for left–right localization (cf. Fig. 34.6). A decrease in performance can also be seen in rats in a front–back localization task in which the loudspeakers were located 60° apart and centered 90° from midline (Fig. 34.8B), a task in which monaural cues predominate. Finally, as Brown and his colleagues have demonstrated,

vertical sound-localization thresholds in monkeys show a progressive increase as high frequencies are removed (Brown et al. 1982; Fig. 34.8C). Thus, it appears that high frequencies are necessary for the optimal performance of small mammals in situations where binaural cues are available and for all mammals in situations requiring the use of monaural spectral cues.

Finally, if high-frequency hearing evolved primarily to support sound localization, then it would be expected that any species which could not localize sound would lose its ability to hear high frequencies. This appears to have happened in fossorial mammals. As previously noted, the pocket gopher has lost virtually all ability to localize sound and there is reason to suspect that the blind mole rat may have done the same (Burda et al. 1990; R. Heffner and Heffner 1990b; see also Section 3.5 of this chapter). As previously shown in Figure 34.7, these two species also have much poorer high-frequency hearing than expected based on their head size. It is also of interest to note that these animals lack significant pinnae which would enhance monaural spectral cues. Thus, while one might speculate on other reasons why fossorial mammals might lose the ability to hear higher than nonmammalian vertebrates, it should be noted that their inability to hear high frequencies is consistent with their relinquishing the ability to localize sound.

6.2 Alternative Explanations of High-Frequency Hearing

6.2.1 Ultrasonic Communication

There has been much interest in the fact that some species of mammals vocalize at ultrasonic frequencies (i.e., frequencies beyond the range of human hearing). In addition to bats, which are known to use ultrasonic vocalizations for echolocation, it has been demonstrated that ultrasonic vocalizations are used for communication by some small rodents and may also be used by other mammals such as insectivores (e.g., Sales and Pye 1974; Geyer and Barfield 1979). Thus, one possibility is that mammals evolved high-frequency hearing in order to use high frequencies in vocal communication.

If mammals evolved high-frequency hearing primarily for communication, then one might expect that those with good high-frequency hearing would either emit high-frequency vocalizations or else use their high-frequency hearing to detect the communications of other species, i.e., to "eavesdrop." Furthermore, the variation in high-frequency hearing should be related to the spectrum of the vocalizations—that is, animals with good high-frequency hearing should vocalize at higher frequencies than those with poorer high-frequency hearing.

Relating the spectrum of the vocalizations to hearing range is not simple. Vocalizations often contain abrupt onsets and offsets generating high frequencies that are "nonfunctional" in the sense that the animals do not use them to identify the sounds and which, in fact, may be beyond their hearing range. Examples of such vocalizations are bird calls which extend beyond 10 kHz (Konishi 1969) and rodent vocalizations which extend beyond 100 kHz (Sales and Pye 1974). However, even if the high-frequency portion of a vocalization

is within an animal's hearing range it may contain no significant information. For example, although human speech contains measurable energy up to 7 kHz, the portion above 3 kHz may be filtered out with no effect on intelligibility (Green 1976). However, functional analyses of animal vocalizations are usually not available.

The main evidence suggesting that high-frequency communication has played a role in the evolution of high-frequency hearing is the observation that some small rodent species emit high-frequency vocalizations (for a review, see Sales and Pye 1974). These vocalizations, which have been observed in approximately twenty species of mice and rats, usually lie in the frequency range above 30 kHz and vocalizations around 50 kHz are not uncommon. Furthermore, behavioral observations indicate that such vocal communications play an important role in reproduction. Thus, it is clear that many small rodents, and perhaps other small mammals as well, use high frequencies for communication.

However, there are other animals with good high-frequency hearing that do not appear to send or receive ultrasonic vocalizations. One example is the tree shrew which is able to hear above 60 kHz (H. Heffner et al. 1969), but which emits vocalizations only within the range from 400 Hz to 15 kHz (Binz and Zimmermann 1989). Other examples of animals whose hearing extends well above the frequency range of their vocalizations are dogs, macaques, pigs, and goats, all of which hear above 40 kHz, as well as horses and cattle which hear above 30 kHz (cf. Kiley 1972; Green 1975; Fox and Cohen 1977; Walser et al. 1981; for hearing ranges, see R. Heffner and Heffner 1990c).

In summary, a comparison of the high-frequency hearing abilities of mammals with the spectra of their vocalizations indicates that many mammals, including humans, hear higher than necessary for perceiving their vocalizations. This suggests that although some mammals use high frequencies for communication, the main source of selective pressure for the development of mammalian high-frequency hearing lies elsewhere.

6.2.2 Size of Auditory Apparatus

Crucial to an understanding of the high-frequency hearing ability of mammals is the analysis of the response properties of the mammalian ear and how

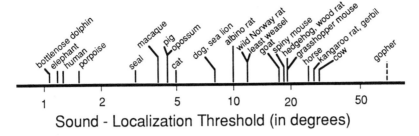

Sound - Localization Threshold (in degrees)

FIGURE 34.9. Variation in acuity for sound localization in the horizontal plane. (The threshold for the Gopher is indicated with a dashed line because it could not be tested using brief stimuli as were all other species.)

these vary between species. One factor believed to be important in the transduction of high-frequency sounds is the size of the various components of the ear. For example, the mass of the middle ear bones, their lever ratios, the volume of the bulla, the area of the tympanic membrane, the length of the basilar membrane, and the number of turns in the cochlea could all selectively enhance or inhibit the reception or transduction of sounds of different frequencies (e.g., Webster 1966; Ehret 1977; Fleischer 1978; Plassmann and Brandle, Chapter 31; Webster and Webster 1984; West 1985). Unlike the previous two factors which ask why mammals hear high frequencies, analysis of the relation between size and frequency response attempts to determine how mammals hear sound.

Because mammals vary in their ability to hear high frequencies, it follows that there must be some anatomical correlates of this functional variation. Because size is an important factor in determining frequency response, it would be expected to be correlated with high-frequency hearing. Furthermore, it is possible that the size of an animal's head may indeed restrict the size of the auditory apparatus thus limiting its hearing range.

However, there are several reasons for believing that the high-frequency hearing of mammals is not simply a function of allometric scaling. First, large animals can always evolve small ears and small animals can, within limits, increase the size of their ears—witness kangaroo rats and gerbils with bullae that can equal the volume of the braincase (e.g., Webster and Webster 1984; Webster and Plassman, Chapter 31). Second, despite the range in size of birds, from small finches to large owls, their high-frequency hearing ability does not vary with the size of the animal (cf. Dooling 1980).

Finally, it seems unlikely that a major sensory attribute such as hearing range is under little or no selective pressure from the environment, but simply varies passively as a function of the physical size of the peripheral auditory apparatus. Instead, it seems more likely that such variation is the result of differences in selective pressure for high-frequency hearing.

7. Evolution of Horizontal Sound-Localization Acuity

Over the years a number of studies have determined the ability of mammals to localize sound. The results of these studies have demonstrated that the sound-localization acuity of mammals as measured by the ability to discriminate sound sources located around the midline in the horizontal plane is far from uniform. As shown in Figure 34.9, midline sound-localization acuity varies from about 1° in humans, elephants, and porpoises to more than 20° in horses and gerbils with at least one species, the pocket gopher, unable to localize brief sounds.

In attempting to account for this variation, a number of ecological and morphological factors have been examined. Among these are the possibility that various lifestyles such as predator vs prey, nocturnal vs diurnal, or surface dwelling vs underground might be associated with variation in localization acuity. However, none of these have proven particularly satisfying; some, such as predator vs prey, are difficult to quantify and all appear to have major exceptions.

Of the many possible explanations, two deserve closer scrutiny. The first is the role of the availability of locus cues in determining sound-localization acuity. The sound is the relation of hearing to vision.

7.1 Availability of Binaural Cues

At one time it was generally believed that all mammals were under strong selective pressure to localize sound as accurately as possible (e.g., Masterton and Diamond 1973). Were this true, then, the main source of variation in sound-localization acuity would be the size of the physical cues available to each animal. Because the magnitude of the binaural locus cues is largely determined by interaural distance, it was expected that any variation in localization acuity would be correlated with interaural distance. Indeed, given the limited data available as recently as 1980, this appeared to be the case: Humans with their large interaural distances were the most accurate localizers, monkeys and cats with intermediate interaural distances were somewhat less accurate, and rats with the smallest interaural distances were least accurate (Fig. 34.10A). The fact that the dolphin with its small functional interaural distance (as a result of underwater hearing and the transmission of sound though its jaw) was also accurate could be attributed to auditory specializations resulting from its use of sonar.

As more species were tested it became apparent that a large interaural distance did not automatically result in good localization acuity. This was particularly true in the case of horses and cattle which, despite their large heads, have poorer acuity than many small rodents (H. Heffner and Heffner 1984; R. Heffner and Heffner 1986b). Furthermore, although a small interaural distance may be a limiting factor in acuity, some small species possess relatively good acuity when compared to many other species with equal or larger interaural distances (e.g., the least weasel and the grasshopper mouse; R. Heffner and Heffner 1987, 1988d).

The relation between head size (functional interaural distance) and sound-localization threshold among all mammals tested to date is illustrated in Figure 34.10B. The correlation coefficient (excluding the echolocating dolphin) is currently -0.57 which, although statistically reliable ($p < 0.01$), accounts for only 32% of the variance. As a result we can no longer accept the notion that all

species are under equal selective pressure to localize accurately and that the variation in sound-localization acuity simply reflects the availability of locus cues.

7.2 The Relation to Vision

In examining the utility of sound localization, it has been observed that one of the most consistent responses to an unexpected sound is the orienting reflex in which the head and eyes are turned toward a sound source (Pumphrey 1950). Indeed, this reflex is faster and more accurate than the visual orienting reflex to a brief flash of light (Whittington et al. 1981). Furthermore, there appears to be some correspondence between brainstem nuclei involved in sound localization and those responsible for eye movement (Irving and Harrison 1967). Given the existence of this close relationship between sound localization and vision, the possibility arises that the degree of sound-localization acuity might be related to some aspect of vision (R. Heffner and Heffner 1985).

In searching for a relation between vision and sound-localization acuity, we have considered several possibilities. Among them are visual acuity and the size of the binocular and panoramic visual fields. However, the one visual parameter that seems to correlate best with sound-localization acuity is the size of the field of best vision.

When visually orienting to a sound, it is not simply the head or the eyes that are directed to the source of the sound, but the part of the visual field with the best visual acuity, that is, the "area centralis." In species such as humans, the area of best vision, i.e., the fovea, subtends an angle of only 1 to 2°. As a result, auditory localization has to be very precise in order to direct the fovea to the sound source. However, most mammals have a broader area of best vision which in some cases covers nearly the entire horizon of the eye and is referred to as a visual streak (Fig. 34.11; for a review of retinal variation, see Hughes 1977). Species with progressively larger fields of best vision, therefore, should require correspondingly less precision in order to orient so that the sound-producing object is placed within their field of best vision. Thus, if the major source of selective pressure for sound localization in mammals is the need to visually locate the source of a sound, then we

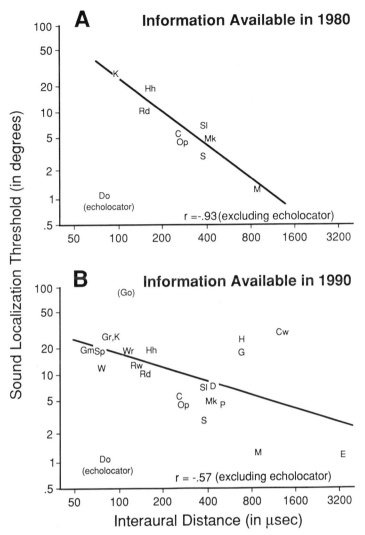

FIGURE 34.10. (A) Relation between functional inter-aural distance and sound-localization acuity among the eight species of mammals for which behavioral thresholds had been published in 1980; note the strong relationship with the exception of the echolocating dolphin. (B) The relationship as it stands with the information available in 1990; with the addition of 12 new species, interaural distance no longer seems to be a useful explanatory factor. C–domestic cat (*Felis catus*) (R. Heffner and Heffner 1988a), Cw–domestic cow (*Bos taurus*) (R. Heffner and Heffner 1986b), D–domestic dog (*Canis familiaris*) (H. Heffner and Heffner 1984), Do–dolphin in water (*Tursiops truncatus*) (Renaud and Popper 1975), E–elephant (*Elephas maximus*) (R. Heffner and Heffner 1982), G–domestic goat (*Capra hircus*) (R. Heffner and Heffner 1986b), Gm–grasshopper mouse (*Onychomys leucogaster*) (R. Heffner and Heffner 1988d), Go–gopher (*Geomys bursarius*) (R. Heffner and Heffner 1990b), Gr–gerbil (*Meriones unguiculatus*) (R. Heffner and Heffner 1988c), H–domestic horse (*Equus caballas*) (H. Heffner and Heffner 1984), Hh–hedgehog (*Paraechinus hypomelas*) (H. Heffner and Heffner 1984), K–kangaroo rat (*Dipodomys meriami*) (H. Heffner and Masterton 1980), M–man (R. Heffner and Heffner 1988a), Mk–macaque (*Macaca nemestrina* and *M. mulatta*) (Brown et al. 1980), Op–opossum (*Didelphis virginiana*) (Ravizza and Masterton 1972), P–domestic pig (*Sus scrofa*) (R. Heffner and Heffner 1989), Rd–domestic rat (*Rattus norvegicus*) (Kelly and Glazier 1978), Rw–wild rat (*Rattus norvegicus*) (H. Heffner and Heffner 1985b), S–harbor seal in air (*Phoca vitulina*) (Terhune 1974), Sl–sea lion (*Zalophus californianus*) (Moore 1975), Sp–spiny mouse (*Acomys cahirinus*) (Mooney, unpublished Master's thesis), W–least weasel (*Mustela nivalis*) (R. Heffner and Heffner 1987), Wr–wood rat (*Neotoma floridana*) (H. Heffner and Heffner 1984). The threshold for the pocket gopher (Go) is for a 1.5-sec noise burst as this animal, unlike the other mammals in this figure, is unable to localize brief sounds.

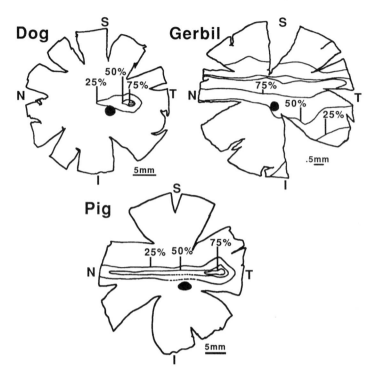

FIGURE 34.11. Ganglion cell density contours in three mammals showing the contour lines encompassing densities 25%, 50%, and 75% of the maximum density for each species, "Area centralis" is operationally defined as the part of the retina containing ganglion cells with packing densities equal to or greater than 75% of maximum. The dog has a small area centralis whereas the gerbil, with its visual streak, has a broad area centralis. The pig has a visual streak combined with a small area of increased density located temporally. I–inferior, N–nasal, S–superior, T–temporal. (Data from R. Heffner and Heffner 1988e.)

would expect sound-localization acuity to be closely correlated with the size of the area centralis.

In order to determine the correlation between sound-localization acuity and size of the field of best vision, we have begun to obtain anatomical measures of the area centralis in animals for which sound-localization thresholds are available. Because the density of retinal ganglion cells is known to correspond to behavioral measures of visual acuity (e.g., Rolls and Cowey 1970), measuring the density of ganglion cells in retinal wholemounts (cf. Stone 1981) can provide an anatomical estimate of the relative acuity of the different parts of the retina. Given a picture of the variation in the density of ganglion cells, it is possible to arrive at a measure of the size of the area centralis.

Using an arbitrary definition of the area centralis as the horizontal width (in degrees) of the region of the retina containing ganglion cell densities equal to or greater than 75% of the maximum density, we have measured the size of the area centralis in 12 species. These measurements have shown that the size of the area centralis varies from approximately 1.5° in man to more than 180° in some species such as gerbils and cattle (cf. fig. 34.11). The relation between the width of the area centralis and sound-localization thresholds is shown in Figure 34.12. As indicated in this figure, the width of the area centralis is positively correlated with sound-localization threshold ($r = 0.91, p < 0.0001$). That is, animals with narrow fields of best vision, such as cat and man, have smaller thresholds (better sound-localization acuity) than those with broader fields of best vision, such as the cow and gerbil.

Although the present sample is small, it includes a broad range of species from five orders of mammals encompassing surface and underground dwellers, nocturnal and diurnal activity patterns, predators and prey, and body sizes ranging through more than five orders of magnitude. Furthermore, included in this sample are animals with good sound-localization acuity as well as those with poor acuity and animals with large as well as those with a small area centralis. Thus, although we are continuing to increase our sample, there is no reason to believe that the high correlation between sound-localization acuity and width of the area centralis is the result of a restricted or unrepresentative sample.

8. Summary: Evolution of Mammalian Sound Localization

The evolution of mammals was accompanied by two events of particular relevance to sound localization. The first was the development of the three-ossicle middle ear which proved able to transduce frequencies beyond the 10-kHz range of nonmammalian vertebrates. The second was the appearance of the pinna which functions as the major source of monaural spectral cues for sound localization—cues that play a crucial role in preventing front–back reversals and in localizing in the vertical plane. The combination of these two anatomical events was fortunate as the pinna can provide reliable monaural spectral cues only if the ear is able to transduce high frequencies. Expanded high-frequency hearing also increased the magnitude of the binaural spectral-difference cue, especially for small mammals. Thus, we suggest that the use of monaural and binaural spectral cues served as major sources of selective pressure for the reception of frequencies above 10 kHz.

Primitive mammals thus had three basic locus cues—binaural time differences, binaural spectral differences, and monaural spectral differences, which they could use to their advantage as they expanded into new niches. In the case of echolocating bats, this expansion was spectacular as their hearing ability allowed them to move into the niche of nocturnal flying predators in which they are unchallenged.

In spite of the obvious survival value of sound localization, not all mammals are able to localize sound accurately. This variation in localization acuity is primarily due *not* to differences in the availability of the physical locus cues, but rather to variation in selective pressure for good acuity. One

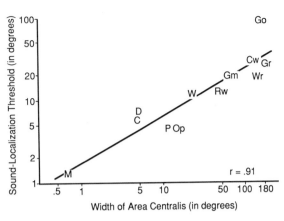

FIGURE 34.12. Relation between the horizontal width of the area centralis (i.e., the field of best vision) and sound localization acuity. Mammals with narrow fields of best vision, such as man, have small localization thresholds, whereas species with wide fields of best vision, such as some rodents and ungulates, have much larger localization thresholds. Letters represent different species (see Fig. 34.10 for key).

of the major uses of sound localization is to enable an animal to direct its gaze to the source of a sound for further scrutiny. Because animals with narrow fields of best vision require more accurate localization to direct their gaze than do animals with broad fields of best vision, sound-localization acuity varies with the size of the field of best vision or area centralis. Apparently it is the *function* of sound localization, i.e., directing the attention of other senses toward the sound-producing object, and not simply the physical cues available to the auditory system, which underlies the variation in mammalian sound-localization acuity.

Moreover, not all mammals have retained full use of the sound-localization cues. There are species that have partially or completely relinquished one or more of the major cues. The ability to use binaural time cues has been lost by some small mammals for reasons that are not yet fully understood. Reliance on the binaural spectral-difference cue has also been reduced or eliminated by some large animals—this may have occurred because the high-frequency components of sounds may not be audible in the ear furthest from the source (with the result that no "difference" cue is available), coupled with the fact that the large binaural time differences available to them provide for sufficient localization acuity. Finally, monaural spectral cues may have been relinquished by aquatic mammals that have lost their pinnae as they adapted to a marine environment. However, all of the locus cues appear to have been relinquished by at least some of the fossorial mammals that have adapted to the one-dimensional world of an underground environment where sound-localization ability may be of little value.

Acknowledgements. Supported by NIH Grant DC 00179.

References

Aitkin LM, Irvine DRF, Webster WR (1984) Central neural mechanisms of hearing. In: Brookhard JM, Mountcastle VB (eds) Handbook of Physiology, Sect. 1: The Nervous System, Vol II, Pt. 2. American Physiological Society, Bethesda, MD, pp. 675–737.

Altschuler RA, Parakkal MH, Fex J (1983) Localization of enkephalin-like immunoreactivity in acetylcholinesterase-positive cells in the guinea-pig lateral superior olivary complex that project to the cochlea. Neuroscience 9:621–630.

Aschoff A, Ostwald J (1987) Different origins of cochlear efferents in some bat species, rats, and guinea pigs. J Comp Neurol 264:56–72.

Belendiuk K, Butler RA (1975) Monaural localization of low-pass noise bands in the horizontal plane. J Acoust Soc Am 58:701–705.

Belendiuk K, Butler RA (1978) Directional hearing under progressive impoverishment of binaural cues. Sensory Proc 2:58–70.

Binz H, Zimmermann E (1989) The vocal repertoire of adult tree shrews (*Tupaia belangeri*). Behaviour 109: 142–162.

Boring EG (1942) Sensation and Perception in the History of Experimental Psychology. New York: Appleton-Century.

Boudreau JC, Tsuchitani C (1968) Binaural interaction in the cat superior olive S segment. J Neurophysiol 31:442–454.

Brown CH, Beecher MD, Moody DB, Stebbins WC (1978) Localization of pure tones by old world monkeys. J Acoust Soc Am 65:1484–1492.

Brown CH, Beecher MD, Moody DB, Stebbins WC (1980) Localization of noise bands by old world monkeys. J Acoust Soc Am 68:127–132.

Brown CH, May BJ (1990) Sound localization and binaural processes. In: Berkley MA, Stebbins WC (eds) Comparative Perception, Vol 1. New York: Wiley, pp. 247–284.

Brown CH, Schessler T, Moody DB, Stebbins W (1982) Vertical and horizontal sound localization in primates. J Acoust Soc Am 72:1804–1811.

Burda H, Bruns V, Müller M (1990) Sensory adaptations in subterranean mammals. In: Nevo E, Reig OA (eds) Evolution of Subterranean Mammals at the Organismal and Molecular Levels. New York: Wiley-Liss, pp. 269–293.

Butler RA (1975) The influence of the external and middle ear on auditory discriminations. In: Keidel WD, Neff WD (eds) Handbook of Sensory Physiology, Vol V/2. New York: Springer-Verlag, pp. 247–260.

Butler RA (1986) The bandwidth effect on monaural and binaural localization. Hearing Res 21:67–73.

Butler RA (1987) An analysis of the monaural displacement of sound in space. Percept Psychophys 41:1–7.

Butler RA, Belendiuk K (1977) Spectral cues utilized in the localization of sound in the median sagittal plane. J Acoust Soc Am 61:1264–1269.

Butler RA, Flannery R (1980) The spatial attributes of stimulus frequency and their role in monaural localization of sound in the horizontal plane. Percept Psychophys 28:449–457.

Butler RA, Helwig CC (1983) The spatial attributes of stimulus frequency in the median sagittal plane and their role in sound localization. American J Otolaryngol 4:165–173.

Butler RA, Planert N (1976) The influence of stimulus bandwidth on localization of sound in space. Percept Psychophys 19:103–108.

Calford MB, Pettigrew JD (1984) Frequency dependence of directional amplification at the cat's pinna. Hearing Res 14:13–19.

Carlile S, Pettigrew AG (1987) Directional properties of the auditory periphery in the guinea pig. Hearing Res 31:111–122.

Casseday JH, Neff WD (1973) Localization of pure tones. J Acoust Soc Am 54:365–372.

Don M, Starr A (1972) Lateralization performance of squirrel monkey (*Samiri sciureus*) to binaural click signals. J Neurophysiol 35:493–500.

Dooling RJ (1980) Behavior and psychophysics of hearing in birds. In: Popper AN, Fay RR (eds) Comparative Studies of Hearing in Vertebrates. New York: Springer-Verlag, pp. 261–288.

Ehret G (1977) Comparative psychoacoustics: Perspective of peripheral sound analysis in mammals. Naturwissenschaften 64:461–470.

Erulkar, SD (1972) Comparative aspects of spatial localization of sound. Physiol Rev 52:237–360.

Flannery R, Butler RA (1981) Spectral cues provided by the pinna for monaural localization in the horizontal plane. Percept Psychophys 29:438–444.

Fleischer G (1978) Evolutionary principles of the mammalian middle ear. Adv Anat Embryol Cell Biol 55:1–70.

Fox MW, Cohen JA (1977) Canid communication. In: Sebeok TA (ed) How Animals Communicate. Bloomington: Indiana University Press, pp. 728–748.

Fuzessery ZM (1986) Speculations on the role of frequency in sound localization. Brain Behav Evol 28:95–108.

Fuzessery ZM, Pollak GD (1985) Determinants of sound location selectivity in bat inferior colliculus: A combined dichotic and free-field stimulation study. J Neurophysiol 54:757–781.

Galambos R, Schwartzkopff J, Rupert A (1959) Macroelectrode studies of superior olivary nuclei. Am J Physiol 197:527–536.

Geyer LA, Barfield RJ (1979) Introduction to the symposium: Ultrasonic communication in rodents. Am Zoologist 19:411.

Green DM (1976) An Introduction to Hearing. Hillsdale, New Jersey, Lawrence Erlbaum.

Green S (1975) Variation of vocal pattern with social situation in the Japanese monkey (*Macaca fuscata*): A field study. In: Rosenblum LA (ed) Primate Behavior, Vol 4. New York: Academic Press, pp. 1–102.

Harrison JM, Downey P (1970) Intensity changes at the ear as a function of the azimuth of a tone source: A comparative study. J Acoust Soc Am 47:1509–1518.

Heffner HE, Heffner RS (1984) Sound localization in large mammals: Localization of complex sounds by horses. Behav Neurosci 98:541–555.

Heffner HE, Heffner RS (1985a) Hearing in two cricetid rodents: Wood rat (*Neotoma floridana*) and grasshopper mouse (*Onychomys leucogaster*). J Comp Psychol 99:275–288.

Heffner HE, Heffner RS (1985b) Sound localization in wild Norway rats (*Rattus norvegicus*). Hearing Res 19:151–155.

Heffner H, Masterton B (1980) Hearing in glires: Domestic rabbit, cotton rat, feral house mouse, and kangaroo rat. J Acoust Soc Am 68:1584–1599.

Heffner HE, Ravizza RJ, Masterton B (1969) Hearing in primitive mammals, III: Tree shrew (*Tupaia glis*). J Audit Res 9:12–18.

Heffner RS, Heffner HE (1982) Hearing in the elephant (*Elephas maximus*): Absolute sensitivity, frequency discrimination, and sound localization. J Comp Physiol Psychol 96:926–944.

Heffner RS, Heffner HE (1983) Sound localization and high-frequency hearing in horses. J Acoust Soc Am 73:S42.

Heffner RS, Heffner HE (1985) Auditory localization and visual fields in mammals. Neurosci Abstr 11:547.

Heffner RS, Heffner HE (1986a) Localization of tones by horses: Use of binaural cues and the role of the superior olivary complex. Behav Neurosci 100:93–103.

Heffner RS, Heffner HE (1986b) Variation in the use of binaural localization cues among mammals. Abstracts of the Ninth Midwinter Research Meeting of the Association for Research in Otolaryngology, 108.

Heffner RS, Heffner HE (1987) Localization of noise, use of binaural cues, and a description of the superior olivary complex in the smallest carnivore, the least weasel (*Mustela nivalis*). Behav Neurosci 101:701–708.

Heffner RS, Heffner HE (1988a) Interaural phase and intensity discrimination in the horse using dichotically presented stimuli. Abstracts of the Eleventh Midwinter Research Meeting of the Association for Research in Otolaryngology, 233.

Heffner RS, Heffner HE (1988b) Sound localization acuity in the cat: Effect of azimuth, signal duration, and test procedure. Hearing Res 36:221–232.

Heffner RS, Heffner HE (1988c) Sound localization and use of binaural cues by the gerbil (*Meriones unguiculatus*). Behav Neurosci 102:422–428.

Heffner RS, Heffner HE (1988d) Sound localization in a predatory rodent, the northern grasshopper mouse (*Onychomys leucogaster*). J Comp Psychol 102:66–71.

Heffner RS, Heffner HE (1988e) The relation between vision and sound localization acuity in mammals. Neurosci Abstr 14:1096.

Heffner RS, Heffner HE (1989) Sound localization, use of binaural cues and the superior olivary complex in pigs. Brain Behav Evol 33:248–258.

Heffner RS, Heffner HE (1990a) The importance of high frequencies for sound localization in mammals. Abstracts of the Thirteenth Midwinter Research Meeting of the Association for Research in Otolaryngology, 110.

Heffner RS, Heffner HE (1990b) Vestigial hearing in a fossorial mammal, the pocket gopher, (*Geomys bursarius*). Hearing Res 46:239–252.

Heffner RS, Heffner HE (1990c) Hearing in domestic pigs (*Sus scrofa*) and goats (*Capra hircus*). Hearing Res 48:231–240.

Heffner RS, Masterton RB (1990) Sound localization in mammals: Brain-stem mechanisms. In: Berkley MA, Stebbins WC (eds) Comparative Perception, Vol I. New York: John Wiley & Sons, pp. 285–314.

Henning GB (1974) Detectability of interaural delay in high-frequency complex waveforms. J Acoust Soc Am 55:84–90.

Houben D, Gourevitch G (1979) Auditory lateralization in monkeys: An examination of two cues serving directional hearing. J Acoust Soc Am 66:1057–1063.

Hughes A (1977) The topography of vision in mammals of contrasting life style: Comparative optics and retinal organisation. In: Crescitelli F (ed) Handbook of Sensory Physiology, Vol VII/5. New York: Springer-Verlag, pp. 613–756.

Humanski RA, Butler RA (1988) The contribution of the near and far ear toward localization of sound in the sagittal plane. J Acoust Soc Am 83:2300–2310.

Irvine DRF (1986) The auditory brainstem. In: Ottoson D (ed) Progress in Sensory Physiology, Vol 7. New York: Springer-Verlag, pp. 1–279.

Irvine DRF (1987) Interaural intensity differences in the cat: Changes in sound pressure level at the two ears associated with azimuthal displacements in the frontal horizontal plane. Hearing Res 26:267–286.

Irving R, Harrison JM (1967) The superior olivary complex and audition: A comparative study. J Comp Neurol 130:77–86.

Isley TE, Gysel LW (1975) Sound-source localization by the red fox. J Mammal 56:397–404.

Jacobs DW, Hall JD (1972) Auditory thresholds of a fresh water dolphin, *Inia geoffrensis* Blainville. J Acoust Soc Am 51:530–533.

Jeffress LA (1975) Localization of sound. In: Keidel WD, Neff WD (eds) Handbook of Sensory Physiology, Vol V/2. New York: Springer-Verlag, pp. 449–459.

Kelly JB, Glazier SJ (1978) Auditory cortex lesions and discrimination of spatial location by the rat. Brain Res 145:315–321.

Kelly JB, Kavanagh GL (1986) Effects of auditory cortical lesions on pure-tone sound localization by the albino rat. Behav Neurosci 100:569–575.

Kiley M (1972) The vocalizations of ungulates, their causation and function. Z Tierpsychol 31:171–222.

Konishi M (1969) Hearing, single-unit analysis, and vocalizations in songbirds. Science 166:1178–1181.

Kuhn GF (1977) Model for the interaural time differences in the azimuthal plane. J Acoust Soc Am 62:157–167.

Kuhn GF (1982) Towards a model for sound localization. In: Gatehouse RW (ed) Localization of Sound: Theory and Applications. Groton, CT: Amphora Press, pp. 51–64.

Marler P (1955) Characteristics of some animal calls. Nature 176:6–8.

Martin RL, Webster WR (1987) The auditory spatial acuity of the domestic cat in the interaural horizontal and median vertical planes. Hearing Res 30:239–252.

Martin RL, Webster WR (1989) Interaural sound pressure level differences associated with sound-source locations in the frontal hemifield of the domestic cat. Hearing Res 38:289–302.

Masterton RB (1974) Adaptation for sound localization in the ear and brainstem of mammals. Fed Proc 33:1904–1910.

Masterton B, Heffner H, Ravizza R (1969) The evolution of human hearing. J Acoust Soc Am 45:966–985.

Masterton RB, Diamond IT (1964) Effects of auditory cortex ablation on discrimination of small binaural time differences. J Neurophysiol 27:15–36.

Masterton B, Diamond IT (1973) Hearing: Central neural mechanisms. In: Carterette EC, Friedman MP (eds) Handbook of Perception, Vol. 3: Biology of Perceptual Systems. Academic Press, New York, pp. 407–448.

Masterton B, Thompson GC, Bechtold JK, RoBards MJ (1975) Neuroanatomical basis of binaural phase-difference analysis for sound localization: A comparative study. J Comp Physiol Psychol 89:379–386.

Mayr E (1961) Cause and effect in biology. Science 134:1501–1506.

McCormick JG, Wever EG, Palin J, Ridgway SH (1970) Sound conduction in the dolphin ear. J Acoust Soc Am 48:1418–1428.

McFadden D, Pasanen EG (1976) Lateralization at high frequencies based on interaural time differences. J Acoust Soc Am 59:634–639.

Middlebrooks JC (1990) Two-dimensional localization of narrowband sound sources. Abstracts of the Thirteenth Midwinter Research Meeting of the Association for Research in Otolaryngology, 109.

Middlebrooks JC, Makous JC, Green DM (1989) Directional sensitivity of sound-pressure levels in the human ear canal. J Acoust Soc Am 86:89–108.

Mills AW (1972) Auditory localization. In: Tobias JV (ed) Foundations of Modern Auditory Theory, Vol II. New York: Academic Press, pp. 303–348.

Mogdans J, Ostwald J, Schnitzler H-U (1988) The role of pinna movement for the localization of vertical and horizontal wire obstacles in the greater horseshoe bat, *Rhinolopus ferrumequinum*. J Acoust Soc Am 84:1676–1679.

Mooney SE, Heffner HE, Heffner RS (1990) Hearing in two species of rodents: Darwin's leaf-eared mouse (*Phyllotis darwini*) and the spiny mouse (*Acomys cahirinus*). Abstracts of the Thirteenth Midwinter Research Meeting of the Association for Research in Otolaryngology, 176.

Moore JK (1987) The human auditory brain stem: A comparative view. Hearing Res 29:1–32.

Moore PWB (1975) Underwater localization of click and pulsed pure-tone signals by the California sea lion (*Zalophus californianus*). J Acoust Soc Am 57:406–410.

Moore PWB, Au WWL (1975) Underwater localization of pulsed pure tones by the California sea lion (*Zalophus californianus*). J Acoust Soc Am 58:721–727.

Musicant AD, Butler RA (1984a) The influence of pinnae-based spectral cues on sound localization. J Acoust Soc Am 75:1195–1200.

Musicant AD, Butler RA (1984b) The psychophysical basis of monaural localization. Hearing Res 14:185–190.

Musicant AD, Butler RA (1985) Influence of monaural spectral cues on binaural localization. J Acoust Soc Am 77:202–208.

Norris KS, Harvey GW (1974) Sound transmission in the porpoise head. J Acoust Soc Am 56:659–664.

Phillips DP, Brugge JF (1985) Progress in neurophysiology of sound localization. Annu Rev Psychol 36:245–274.

Phillips DP, Calford MB, Pettigrew JD, Aitkin LM, Semple MN (1982) Directionality of sound pressure transformation at the cat's pinna. Hearing Res 8:13–28.

Pumphrey RJ (1950) Hearing. Symp Soc Exp Biol 4:1–18.

Ravizza RJ, Masterton B (1972) Contribution of neocortex to sound localization in opossum (*Didelphis virginiana*). J Neurophysiol 35:344–356.

Renaud DL, Popper AN (1975) Sound localization by the bottlenose porpoise *Tursiops truncatus*. J Exp Biol 63:569–585.

Roffler SK, Butler RA (1968a) Factors that influence the localization of sound in the vertical plane. J Acoust Soc Am 43:1255–1259.

Roffler SK, Butler RA (1968b) Localization of tonal stimuli in the vertical plane. J Acoust Soc Am 43:1260–1266.

Rolls ET, Cowey A (1970) Topography of the retina and striate cortex and its relationship to visual acuity in rhesus monkeys and squirrel monkeys. Exp Brain Res 10:298–310.

Rose JE, Brugge JF, Anderson DJ, Hind JE (1967) Phase-locked response to low-frequency tones in single auditory nerve fibers of the squirrel monkey. J Neurophysiol 30:769–793.

Sales GD, Pye JD (1974) Ultrasonic Communication by Animals. New York: John Wiley and Sons.

Shaw EAG (1974) The external ear. In: Keidel WD, Neff WD (eds) Handbook of Sensory Physiology: Auditory System, Vol V/1. New York: Springer-Verlag, pp. 455–490.

Siegmund H, Santibañez-HG (1981) Effects of motor denervation of the external ear muscles on the audio-visual targeting reflex in cats. Acta Neurobiol Exp 41:1–13.

Stein BE, Clamann HP (1981) Control of pinna movements and sensorimotor register in cat superior colliculus. Brain Behav Evol 19:180–192.

Stevens SS, Newman EB (1936) The localization of actual sources of sound. Am J Psychol 48:297–306.

Stone J (1981) The Wholemount Handbook. Sydney: Maitland Publications Pty.

Terhune JM (1974) Directional hearing of a harbor seal in air and water. J Acoust Soc Am 56:1862–1865.

Terhune JM (1985) Localization of pure tones and click trains by untrained humans. Scand Audiol 14:125–131.

Wakeford OS, Robinson DE (1974) Lateralization of tonal stimuli by the cat. J Acoust Soc Am 55:649–652.

Walser ES, Walters E, Hauge P (1981) A statistical analysis of the structure of bleats from sheep of four different breeds. Behaviour 77:67–76.

Webster DB (1966) Ear structure and function in modern mammals. Am Zoologist 6:451–466.

Webster DB, Webster M (1984) The specialized auditory system of kangaroo rats. In: Neff WD (ed) Contributions to Sensory Physiology, Vol 8. New York: Academic Press, pp. 161–196.

Wenstrup JJ (1984) Auditory sensitivity in the fish-catching bat, *Noctilio leporinus*. J Comp Physiol A 155:91–101.

West CD (1985) The relationship of the spiral turns of the cochlea and the length of the basilar membrane to the range of audible frequencies in ground dwelling mammals. J Acoust Soc Am 77:1091–1101.

White JS, Warr WB (1983) The dual origins of the olivocochlear bundle in the albino rat. J Comp Neurol 219:203–214.

Whittington DA, Hepp-Reymond MC, Flood W (1981) Eye and head movements to auditory targets. Experimental Brain Res 41:358–363.

Yost WA, Dye RH Jr (1988) Discrimination of interaural differences of level as a function of frequency. J Acoust Soc Am 83:1846–1851.

35
The Marine Mammal Ear: Specializations for Aquatic Audition and Echolocation

Darlene R. Ketten

1. Introduction

"Marine mammal" is a broad categorization for over 150 species that have one feature in common: the ability to function effectively in an aquatic environment. They have no single common aquatic ancestor and are distributed among four orders (see Appendix 1). Each group arose during the Eocene in either the temperate northern Pacific Ocean or in the Tethys Sea, a paleolithic body of water from which the Mediterranean and middle eastern limnetic basins were formed. Otariids (sea lions), odobenids (walrus), and marine fissipeds (sea otters) developed primarily in the Pacific, while the earliest cetacean (whale), sirenian (manatee and dugong), and phocid (true seal) fossils come from regions bordering Tethys Sea remnants (Kellogg 1936; Domning 1982; Barnes, Domning, and Ray 1985). The level of adaptation to the marine environment varies in marine mammals; many are amphibious and only the Cetacea and Sirenia are fully aquatic, unable to move, reproduce, or feed on land. Structural changes in the ears of marine mammals parallel their degree of aquatic adaptation, ranging from minor in amphibious littoral species, such as otters and sea lions, to extreme in the pelagic great whales.

This chapter focuses on the cetacean ear as the most fully adapted auditory system of marine mammals. It first describes peripheral auditory anatomy in the two extant suborders of Cetacea, the Odontoceti (toothed whales, porpoises, and dolphins) and Mysticeti (baleen or whalebone whales), and then compares these structures with what is known of fossil cetacean ears. A functional analysis is given of generalized cetacean ear anatomy emphasizing how unique structures in cetaceans relate to the ability of a mammalian ear to hear in water. Specific anatomical differences among modern odontocete and mysticete ears are discussed in relation to their role in species-specific frequency ranges, which, in turn, are correlated with differences in habitat and feeding behavior. Lastly, a comparison is made of modern and ancestral cetacean cranial features to allow speculations on the auditory capacity and behavior of extinct species. Since Cetacea evolved from terrestrial species and many specimens represent intermediate stages in the transition to water, this comparison also provides an opportunity to trace the progressive refinement of a mammalian auditory system from terrestrial through amphibious to fully aquatic.

1.1 Adaptive Radiation of Cetacea

Protocetid fossils center on the northern Tethys Sea. It is likely that cetacean radiations are linked to the tectonic uplift and closure of the Tethys, which generated a warm, productive, shallow sea with abundant food supplies (McKenzie 1970; Davis 1972; Lipps and Mitchell 1976). The exploitation of the Tethys shallows 50 to 60 million years ago by an amphibious, mesonychid condylarth, a cat-like, hooved carnivore, led to the development of the Archaeoceti from which the two extant lines of cetaceans are derived (Fig. 35.1) (Kellogg 1936; Barnes and Mitchell 1978; Fordyce 1980; Gingerich et al. 1983). One line, the Odontoceti, has species in virtually every aquatic habitat, from

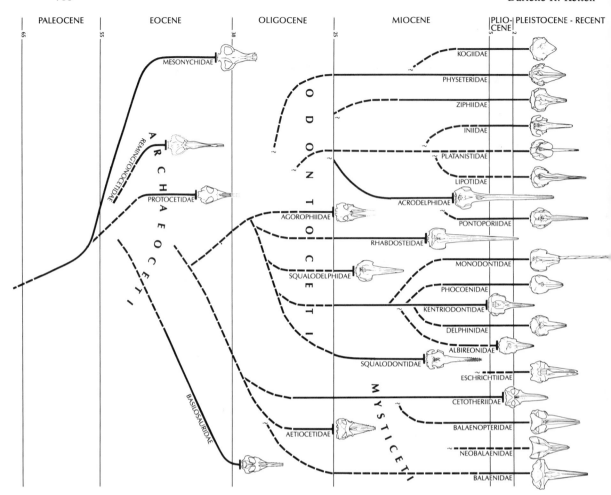

PALEOCENE	EOCENE	OLIGOCENE	MIOCENE	PLIO-CENE	PLEISTOCENE - RECENT

FIGURE 35.1. Cetacean phylogeny. A theoretical phylogenetic tree traces the development of ancestral and modern families of Cetacea. Extinctions are indicated by a cross-bar. Dashed lines indicate estimated links for that family with antecedents. Question marks indicate that links with earlier families cannot be established reliably. This is the case for the freshwater, riverine dolphins which appear abruptly as four distinct lines in the late Miocene and may have evolved in parallel. (Revised version by Barnes and Folkens after Barnes, Domning, and Ray 1985; copyright Pieter A. Folkens.)

estuarine river dolphins to deep-diving, bathypelagic whales. There are over 65 recognized extant odontocete species, of diverse sizes (1 to 30 meters) and shapes, and all are efficient, raptorial carnivores (Leatherwood, Caldwell, and Winn 1976; Leatherwood et al. 1982; Watkins and Wartzok 1985). The second line, the Mysticeti, has 11 species, which are typically large, pelagic planktivores (Ridgway 1972; Gaskin 1976).

Like any mammal, cetaceans are faced with a need for locating food sources, navigating, and finding mates. As Archaeocetes entered the ocean, more of these functions had to be accomplished in water, a dark, dense medium compared to terrestrial environments. The physical demands of water are apparent in virtually every aspect of odontocete and mysticete anatomy. Olfaction and vision in some species are poor compared even to other marine mammals (Dawson 1980; Kastelein, Zweypfenning, and Spekreijse 1990; Kuznetzov 1990; Watkins and Wartzok 1985). It is not surprising, therefore, that sound is believed to be the fundamental sensory and communication channel in Cetacea. All odontocetes tested to date echolocate; i.e., they "image" their environment by analyzing echoes from a self-generated ultrasonic signal of up to 200 kHz (Kellogg 1959; Norris et al. 1961; Pilleri 1983; Kamminga, Engelsma, and Terry 1989). Mysticetes are not believed to echolocate, but they may use infrasonic frequencies[1] (Weston and Black 1965; Watkins et al. 1987; Edds 1988; Clark 1990; Dahlheim and Ljungblad 1990). Cetaceans, as a group, therefore evolved abilities to exploit both ends of the acoustic spectrum and use the broadest range of acoustic channels of any mammalian order.

2. Sound Production Characteristics and Audition

2.1 Audiometric Data

In order to accurately interpret auditory structures of any species, it is necessary to have some measure of its sensitivity. For practical and historical reasons, underwater measures of auditory sensitivity are available for very few marine mammals (Watkins and Wartzok 1985; Thomas, Pawloski, and Au 1990; Awbrey 1990). Consequently, most speculations about cetacean hearing are based on inferences from recordings of emitted sounds or on psychophysical data from experiments on very few odontocete species. The available odontocete data are extensively reviewed in McCormick et al. (1980), Popper (1980), Watkins and Watzok (1985), and Awbrey (1990) and are only briefly summarized here. At present, there are no direct audiometric data for mysticetes.

In odontocetes, electrophysiological and behavioral audiograms indicate best sensitivity (the frequency of a pure tone that can be detected at a lower intensity than all others) varies by species from 12 kHz in *Orcinus orca* (killer whale) (Schevill and Watkins, 1966; Hall and Johnson 1971) to over 100 kHz in *Phocoena phocoena* (harbour porpoise) (Voronov and Stosman 1970; Møhl and Andersen 1973). The majority of species measured are delphinids with best sensitivities in the 40 to 80 kHz range (Johnson 1967; Bullock et al. 1968; Bullock and Ridgway 1972; Ridgway

1980; Ridgway et al. 1981; Thomas, Chun, and Au 1988; Popov and Supin 1990a, 1990b). Interspecies comparisons of audiograms are equivocal since techniques vary widely and reports for even the same species vary by as much as two octaves (see Popper 1980). Critical ratio and critical band measurements indicate odontocetes are generally better than most mammals at detecting signals in noise.[2] Critical ratio functions for dolphins parallel those of humans but the absolute dolphin ratios are narrower and the critical bands are not a constant factor of the ratio over a wide range of frequencies (Johnson 1968; Thomas, Pawloski, and Au 1990). Humans have 24 critical bands which are estimated to be ⅓ of an octave or 2.5 times the critical ratio in the frequency range of speech (Pickles 1982). In *Tursiops truncatus* (bottlenosed dolphin), there are 40 critical bands (Johnson 1968) which vary between 10 times the critical ratio at 30 kHz and 8 times the critical ratio at 120 kHz (Moore and Au 1983).

Au (1990) found that echolocation performance as a function of noise in *Tursiops* is 6 to 8 dB lower than that expected from an ideal receiver. Target detection thresholds as small as 5 cm at 5 meters have been reported, implying a minimal angular resolution of ~0.5°, but the most common range is 1° to 4° for both horizontal and vertical resolution (Bullock and Gurevich 1979; Au 1990). Minimal intensity discrimination is 1 dB (equal to human) and temporal discrimination is approximately 8% of signal duration (superior to human). Frequency discrimination in *Tursiops* (0.3 to 1.5% relative discrimination limens) and *Phocoena* (0.1 to 0.2%) is superior to human and rivals that of microchiropteran bats (Grinnell 1963; Simmons 1973; Sukhoruchenko 1973; Thompson and Herman 1975; Long 1980; Pollak 1980). These data, despite limitations in number or consistency of experiments, suggest odontocetes have no single auditory capacity better than that of some other animal, but their

---

[1] *Infra* (<20 Hz) and *ultra* (>20 kHz) *sonic* are homocentric classifications for sounds beyond the normal human auditory range of 20 to 20,000 Hz (Sales and Pye 1974; Yeowart 1976).

[2] The critical band is a measure of frequency discrimination based on the ability to detect a signal embedded in noise. At some point, as the bandwidth of masking noise is narrowed, the signal becomes far easier to detect; i.e., the detection threshold drops sharply. Noise bandwidth at that point is the critical band. The critical ratio estimates critical bands based on the signal power/noise power ratio.

combination of abilities is an exceptional package geared to frequency and resolution capabilities consistent with aquatic echolocation.

2.2 Cetacean Vocalizations

In contrast to the limited audiometric data, recordings are available of emitted sounds for over 67 species of marine mammals (see Watkins and Wartzok 1985). Although not an optimal measure of sensitivity, spectral and temporal analyses of recorded vocalizations[3] provide indirect estimates of auditory ranges and are currently the most consistent acoustic data base for multispecies comparisons.

There are two functional and three acoustic categories for odontocete signals (Popper 1980):

1. Echolocation signals – broad spectrum clicks with peak energy between 20 and 200 kHz.
2. Communication signals – burst pulse click trains and narrow band, constant frequency (CF) or modulated frequency (FM) whistles ranging from 4 to 12 kHz.

Odontocetes are the only marine mammals known to echolocate (Kellogg 1959; Norris et al. 1961). Individuals can vary pulse repetition rate, interpulse interval, intensity, and spectra of echolocation clicks (Au et al. 1974; Moore 1990), but each species has a characteristic echolocation frequency range (Schevill 1964; Norris 1969; Popper 1980). Based on peak spectra (the frequency of maximum energy) in their typical, broadband echolocation click, odontocete species can be divided into two ultrasonic groups (Ketten 1984): Type I with peak spectra above 100 kHz, and Type II with peak spectra below 80 kHz. These two ultrasonic divisions coincide with differences in habitat and social behavior (Ketten and Wartzok 1990). Type I odontocetes are generally solitary, nonaggregate, inshore or freshwater species whereas Type II species typically form large, offshore groups or pods (Gaskin 1976; Wood and Evans 1980).

Mysticete vocalizations are significantly lower in frequency than those of odontocetes, ranging

from 12 Hz signals in *Balaenoptera musculus* (blue whale) (Cummings and Thompson 1971) to 3 kHz peak spectra calls in *Megaptera novaeangliae* (humpback) (Silber 1986). Most mysticete vocalizations can be categorized as protracted low frequency moans (0.4 to 40 seconds, fundamental frequency <200 Hz); simple (bursts with frequency emphasis <1 kHz) or complex (amplitude and frequency modulated pulses) calls; and "songs," like the now familiar ululations of humpbacks, which have seasonal variations in phrasing and spectra (Thompson, Winn, and Perkins 1979; Watkins 1981; Edds 1982; Payne, Tyack, and Payne 1983; Clark 1990). Infrasonic signals; i.e., below 20 Hz, have been commonly reported in two species of rorquals, *Balaenoptera musculus* (Cummings and Thompson 1971; Edds 1982) and *Balaenoptera physalus* (fin whale) (Watkins 1981; Watkins et al. 1987; Edds 1988). Precise functions for mysticete vocalizations are unclear. Interspecific comparisons of vocalizations are complicated by the diverse categories reported in the literature, and functional analyses currently depend upon field observations of behavior during recordings. Low frequencies have the potential for long distance communication, but this has not been proved.

Clearly, there are significant differences in the frequency ranges of sounds produced by odontocetes and mysticetes. These differences imply different perceptual abilities, which presumably have anatomical correlates in the peripheral and central auditory systems. Fortunately for an evolutionary study, much about auditory capacity in a mammal can be inferred from peripheral auditory structures and from associations of the temporal bone with other skull elements, most of which are preserved in fossil material.

3. Cetacean Cranial Morphology

All odontocetes and mysticetes have extensive modifications of the cranium, nares, sinuses, petrosal bones, and jaws that are linked to feeding, respiration, and the production and reception of sound while submerged. Whatever the driving force for any single modification, evolution in any one cranial component in cetaceans appears to have strongly influenced the anatomy of other typically unrelated structures. Thus, the structure of

---

[3]There is some question about the validity of the term vocalization for Odontoceti considering their nonlaryngeal sound production mechanisms (see Section 3.2); however, the term is correct for the majority of marine mammals and is used here for simplicity.

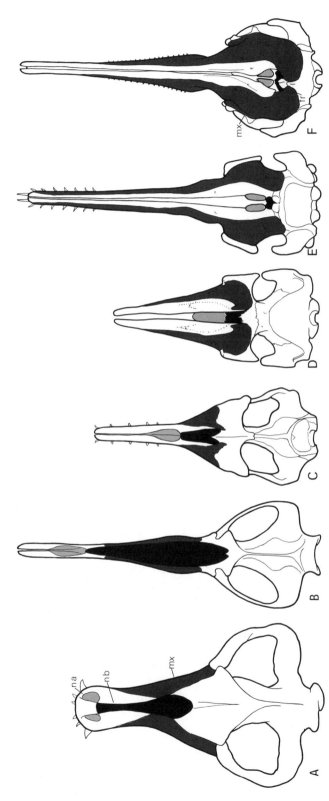

FIGURE 35.2. Telescoping of the cetacean skull. Schematic, dorsal views of skulls of six species illustrate major changes in cranial relationships from Mesonychidae to recent Odontoceti that occur in telescoping. Black areas designate the nasal bones (nb); deep gray, the maxillae (mx); and light gray, the nares (na). The mesonychid condylarth (A–Paleocene) has typical terrestrial cranial relationships. In both *Remingtonocetus harudiensis* (B–early Eocene archaeocete) and *Basilosaurus spp.* (C–early Oligocene archaeocete), the rostrum has narrowed but other relationships are virtually unchanged. The jaw of the primitive odontocetes, *Agorophius pygmaeus* (D–late Oligo-

cene) and *Squalodon spp.* (E–Miocene), closely resembles those of modern odontocetes and the nares are well posterior, implying a fully aquatic existence. The anterior cavity in modern odontocete skulls (*Lipotes vexillifer*–F) accommodates the melon, a spheroid, soft tissue mass implicated in the emission of ultrasonic echolocation signals. The melon is cradled by the latero-posterior expansions of the maxillae that cover the frontal bones (fr) (see also Fig. 35.3). The contiguous soft and bony layers of the rostrum act as a shield acoustically separating the melon from the tympano-periotic complex. (Original artwork and copyright, Pieter A. Folkens.)

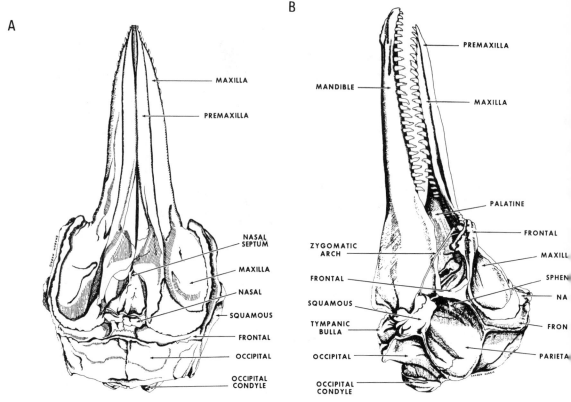

FIGURE 35.3. Modern delphinid skull. The skull of *Tursiops truncatus*, a modern delphinid, shown in (A) dorsal and (B) lateral views. (Reprinted from Ridgway, 1972, courtesy of Charles C. Thomas, Publisher.)

the petrosal bones and inner ear cannot be analyzed in isolation. In odontocetes in particular, it is necessary to examine structures like the jaw and cranial sinuses for atypical acoustic functions, while bearing in mind their conventional roles as well.

3.1 Telescoping

Modern cetaceans have the most derived cranial structure of any marine mammal (Barnes and Mitchell 1978; Barnes, Domning, and Ray 1985). Synapomorphic cranial characters common to all Cetacea include dorso-caudal nares, extensive peribullar and pterygoid sinuses, elongated or extensively reconfigured mandibles and maxillae, petrosal bones detached from the skull, and a foreshortened, concave cranial vault (Norris 1980; Barnes, Domning, and Ray 1985; Gingerich, Smith, and Simons 1990; Oelschläger 1990). The majority of these characters are associated with "telescoping," a term coined by Miller (1923) to

describe the evolutionary revamping of the cranial vault (Fig. 35.2) in which the maxillary bones of the upper jaw expanded back to the vertex of the skull and covered the reduced frontal bones. Concomitantly, the rostrum elongated and the cranial vault foreshortened, pulling the nares and narial passages rearward to a superior position behind the eyes. The product, epitomized by the modern delphinid skull (Fig. 35.3), is a frontally compressed, concave cranium with dorsal nares allowing ventilation with only the most dorsal surface of the head above water. While telescoping is clearly related to changes in the respiratory path it also has significant consequences for channeling sound into and out of the cetacean head.

3.2 Cranial Paths for Emitted Sounds

The mechanisms for sound production and reception in odontocetes have been intensely investigated and vigorously debated for nearly forty years (Evans and Prescott 1962; Fraser and Purves 1954, 1960; Norris 1964, 1968; Purves 1967; McCormick et al. 1970; Pilleri 1983; Goodson and Klinowska 1990). Although exclusively laryngeal mechanisms have been suggested (Purves 1967; Pilleri 1983), the preponderance of data supports a nasal sac theory proposed by Norris and Harvey (Norris 1969; Norris et al. 1972; Norris and Harvey 1974). The controversy is relevant for understanding odontocete hearing since echolocators typically have specialized auditory structures for suppressing reception of their outgoing echolocation signal (Pye 1972). Thus, the form and location of sound generators and exit paths may affect construction of the middle and inner ear.

In odontocetes, telescoping forms a frontal concavity occupied by up to five asymmetrically distributed nasal sacs or diverticulae and the melon, a unique, elliptical, multilayered mass of fibrous tissue and fats. The nasal diverticulae act as pressure driven sound generators that produce clicks when the "pneumatic" lock of the ridged nasal flaps (museau de singe) are forced open by sudden expulsions of air from the sacs (Mackay and Liaw 1981; Amundin and Cranford 1990). Each ventral premaxillary sac is believed to act, in conjunction with the melon, as an acoustic lens to focus and beam anteriorly the outgoing ultrasonic signals (Fig. 35.4) (Norris 1964; Amundin and Cranford 1990). This hypothesis is reinforced anatomically by the extensive innervation of the melon by the trigeminal nerve (V), which rivals the auditory nerve (VIII) for largest cranial nerve fiber count in odontocetes (82,000 fibers in *P. phocoena*) (Jansen and Jansen 1969; Morgane and Jacobs 1972). As an animal ensonifies a target, the melon undulates rapidly. It is likely that the trigeminal, with both sensory and motor roots, controls this motion and may provide the neural mechanism for focusing and monitoring shape of the acoustic lens in echolocation (Ketten unpublished). Anterior reflection of the signal is enhanced by the sandwich of soft and hard tissues of the frontal shield behind the melon (Figs. 35.3, 35.4), which Fleischer (1976) concluded provides a serial impedance mismatch that deflects outgoing pulses generated in the sacs through the melon and away from the tympano-periotic bones. Lastly, the zygomatic arch in odontocetes is exceptionally thin, making it a poor path for bony sound conduction between the rostral and peribullar regions (Fig. 35.3B).

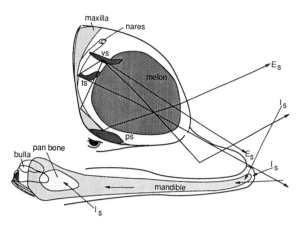

FIGURE 35.4. Sound paths in the Odontocete. Hypothetical sound paths for emission and reception of ultrasonic signals are shown in a schematized dolphin head (revised 1990, after Ketten 1984; copyright DR Ketten). Ultrasonic signals are believed to be generated by the expansion of vestibular (vs) and tubular (ts) nasal sac diverticulae and the subsequent release of air in plosive "gasps." Released air is captured by auxiliary sacs and recycled for subsequent sound production. The signals are reflected off the acoustic shield of the telescoped cranium and the premaxillary sac (ps) and focused by the multilayered fats in the melon into anteriorly directed beams (E_s). Incident sound (I_s) from a target deflecting that beam enters the jaw area where waxy tissues overlain by the mandibular bone channel the sound to the tympano-periotic complex, rather like fiber optic cables channel or conduct light. Ray diagrams of this type are valid only for ultrasonic signals and are not sufficient to explain directed longer wavelength sounds (Mackay 1987). Best reception characteristics from lateral and low frequency signals are found in the area of the pan bone, but the fatty mandibular channels have the lowest acoustic resistance for sounds from an anterior direction (Bullock et al. 1968; Norris and Harvey 1974; Popov and Supin 1990.)

Little is known of the acoustic paths in the mysticete head. Mysticetes do not have a melon and the zygomatic arch is substantial. Both observations are consistent with the assumption that baleen whales do not echolocate. Mysticetes have a larynx but no vocal cords and the cranial sinuses are thought to be involved in phonation (Benham 1901; Hosokawa 1950; Mead 1975; Sukhovskaya and Yablokov 1979; Henry et al. 1983), although no precise mechanism has been demonstrated nor are there comprehensive studies of anatomical correlates of infrasonics in Mysticeti.

B

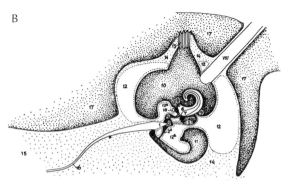

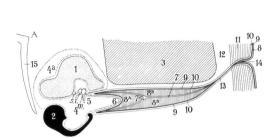

FIGURE 35.5. Cross-sections of the peribullar regions. Schematics demonstrate the relationship of the periotic-tympanic complex to surrounding cranial elements and to the external auditory canal in a generalized mysticete (A) and odontocete (B). The original labelling is retained in these illustrations. Figures are not drawn to a common scale. (Adapted from Reysenbach de Haan 1956.) (A) Frontal section of the tympanic region of a whalebone whale (Mysticeti). (1) periotic; (2) tympanic; (3) squamosal; (4) cavum tympani; (4a) peribullar sinus; (5) tympanic conus; (6) protrusion of tympanic membrane or glove finger into the external auditory canal; (7) interfaces of components of the cerumen plug; (8) stratum corneum; (8a, 8b) cerumen plug; (9) statum germina-

tivum; (10) corium; (11) blubber; (12) connective tissue; (13) blind end of the external auditory canal; (14) external auditory meatus; (15) occipital; (i) incus; (m) malleus; (s) stapes. (B) Schematic dorsoventral section through the auditory apparatus of the Odontoceti. (1) tympanic conus; (2) malleus; (2a) processus gracilis; (3) tensor tympani; (4) incus; (5) stapes in oval window; (6) scala vestibuli; (7) scala media; (8) vestibule; (9) round window; (10) periotic; (11) tympanic; (12) peribullar sinus; (13) one of five ligamentous bands suspending the periotic in the sinus; (14) peribullar plexus; (15) blubber; (16) external auditory meatus; (17) squamosal and basioccipital; (VII) facial nerve; (VIII) acousto-vestibular nerve.

3.3 Cranial Structures for Sound Reception

Whether the external auditory canal is functional in Cetacea is debatable. There are no pinnae, but there is a small external meatus in all species which connects with a relatively narrow auditory canal. In Mysticeti, the canal contains a homogeneous wax and the proximal end flares, covering the "glove finger," a complex, thickened membrane that protrudes laterally from the bulla into the canal and is thought to be derived from the pars flaccida of the tympanic membrane (Figs. 35.5A, 35.7) (Fraser and Purves 1960; Reysenbach de Haan 1956; Lockyer 1974; Van and Utrecht 1981; Ketten in preparation). The glove finger is connected to the tympanic bulla by a fibrous ring, equivalent to the fibrous annulus, but there is no obvious association with any ossicle or with the wax-filled external canal. In odontocetes, the external canal is exceptionally narrow and plugged with cellular debris and dense cerumen (Fig. 35.5B), and the tympanic membrane remains only as a calcified ligament or tympanic conus fused at

its distal and ventral margins with the tympanic bulla (Reysenbach de Haan 1956; Ketten 1984).

Reysenbach de Haan (1956) and Dudok van Heel (1962) were among the first researchers to suggest alternative tissue conduction paths in odontocetes, contradicting the theory of Fraser and Purves (1954, 1960) that the external auditory canal, although occluded with debris, is the principal route to the cochlea. Reysenbach de Haan (1956) reasoned that since the transmission characteristics of blubber and sea water are similar, using a canal occluded with variable substances would be less reliable than tissue or bone conduction. Dudok van Heel (1962) concluded the canal was irrelevant since behavioral measures of the minimum audible angle in *Tursiops* were more consistent with an interbullar critical interaural distance than with intermeatal distances.

A probable alternative path for sound conduction in odontocetes is the lower jaw. The jaw structure of all mysticetes is clearly extensively modified for sieving or gulp feeding and has no evident connection to the temporal bone. The jaw of odontocetes appears to be modified to snap prey, but in actuality, it is a unique composite of fats and bone which serves a second role, to transmit sound to the inner ear. Norris (1968, 1980) observed that the odontocete mandible has two exceptional properties: a concave medial face, which houses a fatty tube that projects from the symphysis back to the temporal bone, and a thin ovoid region, dubbed the "pan bone," near the flared posterior segment of the mandible (Fig. 35.4). The fats in the mandibular channel, like those of the melon, are wax esters with acoustic impedances closer to sea water than any other non-fluid tissues in Cetacea (Varanasi and Malins 1971). Norris (1969) speculated the fat channel acts as an acoustic wave guide and the pan bone, as an acoustic window through which sound is preferentially chan-neled to the petro-tympanic bullae underlying the jaw. Results of several experiments support this hypothesis. Evoked responses (Bullock et al. 1968) and cochlear potentials (McCormick et al. 1970) in *Stenella* and *Tursiops gilli*, the Pacific bottlenosed dolphin, were significantly greater for sound stimuli above 20 kHz placed on or near the mandible. Meas-urements with implanted hydrophones in severed *T. truncatus* heads (Norris and Harvey 1974) found best transmission characteristics for sources directed into the pan bone. In recent behavioral studies, Brill et al. (1988) showed encasing the lower jaw in neoprene significantly impaired performance in echolocation tasks. These results argue strongly that the jaw is an acoustic channel, but they do not preclude alterna-tive paths, including the external auditory canal. Both Popov and Supin (1990b) and Bullock et al. (1968) found minimum thresholds for low frequen-cies were associated with stimuli nearest the external meatus. From the combined results of all these studies, I conclude there may be two parallel systems in odontocetes, one for generation and reception of ultrasonics and one for lower frequency communica-tion signals. No anatomical studies have shown equivalent structural specializations for sound trans-mission in mysticetes, which underscores the poten-tial for the melon and mandible to be efferent and afferent counterparts of odontocete echolocation.

4. The Extant Cetacean Ear

The problems inherent in an aquatic environment for an unmodified terrestrial ear are considerable: increased sound speed, impedance mismatch for an air–fluid refined system, and increased pres-sures. Adaptations that cope with these problems are apparent throughout the middle and inner ear.

4.1 The Tympano-Periotic Complex

The cetacean temporal bone is distinctive and dense, differing from other marine and terrestrial mammalian auditory bullae in appearance, con-struction, cranial associations, and, in some aspects, function (Fig. 35.6). In all modern Ceta-cea, the bulla is separated from the skull and comprised of two components, the periotic and tympanic, both of which are constructed from exceptionally dense, compact bone. This tympano-periotic bullar complex is situated in an extensive peribullar cavity formed by expansions of the mid-dle ear spaces (Figs. 35.5, 35.6). In Mysticeti, a bony flange projects posteromedially from the periotic, dividing the cavity and wedging the bulla tightly between the exoccipital and squamosal (Yamada 1953; Reysenbach de Haan 1956; Kasuya 1973). The peribullar cavity is proportionately larger in odontocetes, completely surrounding the bulla, and, except in physeterids, no bony elements connect the bulla to the skull. The tympano-periotic complex is suspended by five or more sets of ligaments in a peribullar plexus of dense "album-inous foam" which fills the cavity (Fraser and Purves 1954; Jansen and Jansen 1969; Ketten and Wartzok 1990). Fraser and Purves (1960) specu-lated the peribullar spaces were an adaptation in response to the mechanical stress of extreme pres-sures and were correlated with diving ability. They predicted greater development of air spaces in spe-cies from deeper habitats; however, as Oelschläger (1986a) notes, peribullar and pterygoid sinuses are extensively developed in shallow water, river-ine species like *Inia geoffrensis*, an ultrahigh-frequency Amazonian dolphin, and less developed in pelagic mysticetes. These observations imply that peribullar sinus development is related to acoustic isolation rather than mechanical stress. In odontocetes, the composite of bullar structure, vascularized plexus, and ligaments could function as an acoustic isolator for echolocation, analogous to the lamellar construction of bat temporal bones (Simmons 1977).

Size and shape of the tympano-periotic complex are species-specific characteristics, but there are

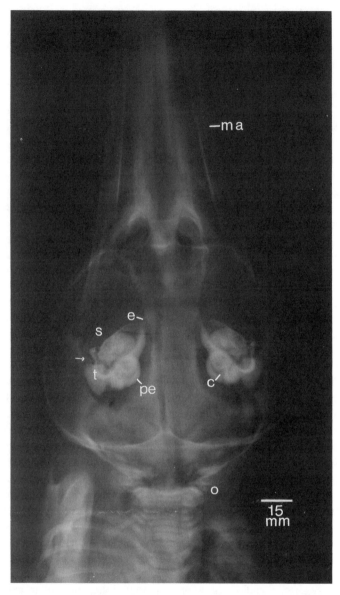

FIGURE 35.6. Radiography of odontocete ears. A single plane X-ray of an odontocete, *Stenella attenuata*, shows the size and density of the bulla in comparison to other skull elements. Radiographs image structures in a grayscale proportional to X-ray attenuation, from white for densest material to black for air. This X-ray image is not enhanced and the whiteness of the bullae in comparison to other skeletal elements is a graphic demonstration of the exceptional mineralization of the odontocete bulla, which has a density (2.7 g/cc) near that of bovine enamel (2.89 g/cc) (Lees, Ahern, and Leonard 1983). The cochlea (C), filled with less dense fluid, is the dark spiral in the periotic (pe), which is medial to the tympanic (t). The eustachian tube (e) is a gray band entering the tympanic anteriorly. Additional structures are the mandible (ma) dotted with dense, white conical teeth; the low density, dark gray region of the foam-filled peribullar sinus completely surrounding each bulla (s); and the occipital condyles (o) anterior to the fused, compressed vertebrae. The frontal ridge is visible only as a white line across the bullae (arrow). In all noncetacean mammals, the tympano-periotic complex is at least partly fused to the brain case.

general characteristics which differentiate mysticete and odontocete bullae at a gross level (Fig.
35.7) (Boenninghaus 1903; Fraser and Purves
1954; Reysenbach de Haan 1956; Kasuya 1973;
Fleischer 1976; Norris and Leatherwood 1981;
Ketten 1984). Mysticete tympanics are hemispher-
ical and nearly twice the volume of the periotic,
which resembles a squat pyramid with the apex
pointed medially. In odontocetes, the periotic and
tympanic are nearly equal in volume, although the
periotic is thicker walled and thus more massive.
The tympanic is conical, tapering anteriorly, while
the periotic is ovoid with a distinct cochlear
promontorium. Species differ in the solidity of the
periotic-tympanic suture, the relative volumes of
tympanic and periotic, the complexity of surface
features, and the degree of attachment to the skull.
Surface measurements scale isometrically with
animal size, and the mass of the bulla can vary a
full magnitude between species (Kasuya 1973;
Ketten 1984; Ketten and Wartzok 1990). Differences in size and shape do not correlate directly
with frequency ranges but they do relate to differences in the anchoring of the tympano-periotic
complex to the skull and thus to its acoustic isolation and reception characteristics.

In vivo, all cetacean bullae are oriented with the
periotic dorsal to the tympanic. The periotic is
relatively uniform in thickness and encloses the
cochlea and vestibular components. The dorsoventral bullar axis is rotated medially 15° to 20° and
the long axis angles ventro-medially, which places
the cochlear apex ventral to the stapes, orthogonal
to conventional terrestrial mammalian formats.
This placement, or displacement, of the cochlea
may result from the spinal flexion and caudal brain
case compression that occurred in Cetacea as they
regressed to a fuselloid shape; its utilitarian effect
is a shorter, less angular pathway for the acousto-
vestibular nerve (VIII), which projects inward
from the dorso-medial edge of the periotic (Figs.
35.5, 35.6, 35.7) and enters a dense, bony canal
leading to the braincase. The VIIIth nerve is thus
"externalized" as it traverses the retroperibullar
space (Ketten and Wartzok 1990). The facial nerve
(VII) does not parallel the VIIIth, as it does in
humans, but remains external to the bulla in many
species. The concha or shell of the tympanic
encloses the ossicular chain (Fig. 35.7) and is lined
with a membranous corpus cavernosum which is a
thin fibrous sheet in mysticetes but is substantially
thicker and highly vascularized in odontocetes. It
has not been determined whether the intratym-
panic space is air-filled in vivo. In all odontocete
bullae examined in situ, a band of fibrous tissue,
analogous to the stylo-hyoid ligaments, joins the
posterolateral edge of the tympanic bulla to the
posterior margin of the mandibular ramus and
stylo-basihyal complex (Fig. 35.4), in effect, align-
ing the bulla with the fatty wave guide of the mandible (Ketten and Wartzok 1990). No such association has been reported in Mysticeti.

4.2 The Middle Ear

Cetacean ossicular anatomy is complex and difficult to interpret in the absence of extensive physiological studies of middle ear function (Belkovich
and Solntseva 1970; McCormick et al. 1970; Fleischer 1976; Solntseva 1987). Anatomical studies
suggest the ossicular chain, like the bullae, has
evolved to accommodate dramatic pressure
changes. In all species, the ossicles are large and
exceptionally dense, and in odontocetes, the structures suggest a compromise between sensitivity
and strength. In bats, high-frequency sensitivity in
the middle ear is achieved by lightening the ossicles and stiffening their attachments (Reysenbach
de Haan 1956; Sales and Pye 1974). Equivalent
structures made of thin bone in an air-filled middle
ear would not withstand the pressure changes in a
dolphin dive. In odontocetes the ossicles are more
massive than in any terrestrial mammal but a bony
ridge, the processus gracilis (Fig. 35.7A), fuses the
malleus to the wall of the tympanic bulla and the
interossicular joints are stiffened with ligaments
and a membranous sheath (Ketten 1984). This
rigid set of attachments is sufficient to transmit
high frequencies (Sales and Pye 1974; McCormick
et al. 1970). In some species, the stapes is fully
ankylosed; in others, it is mobile with a conventional annular ligament (Ketten 1984). Mysticete
ossicles are equally massive, but they are not fused
to the bulla nor are the interossicular joints stiffened with ligaments. Both mysticete and odontocete middle ears contain extensive soft tissue,
but this does not preclude air-filled chambers.
To date, there are only anecdotal reports (McCormick et al. 1970) and no direct evidence of air in
the tympanic cavity.

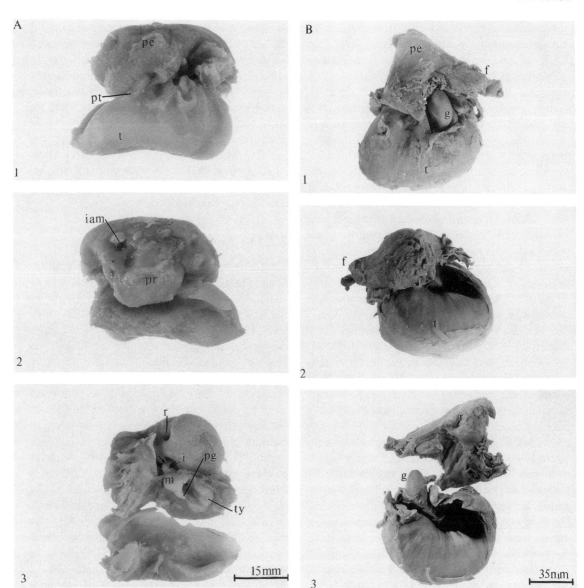

FIGURE 35.7. Cetacean tympano-periotic complex. The left tympano-periotic complex is shown from (A) an odontocete, *Stenella attenuata* (spotted dolphin), and (B) a mysticete, *Eubalaena glacialis* (right whale), as oriented in the cetacean head in (1) lateral and (2) medial views. For (3), the periotic and tympanic were separated in the medial position to show the middle ear space. Characteristic mysticete features include a posterior cranial flange (f); a dense hemispheric tympanic (t); a triangular periotic (pe); and the thick, membranous glove finger (g). The odontocete has a more delicate, conical tympanic and ovoid periotic. The processus tubarius (pt) or accessory ossicle is the outer tympano-periotic connection. The promontorium (Pr) of the periotic houses the cochlea. The corpus cavernosum lining the middle ear in *Stenella* has been removed to reveal the ossicles: (i) incus; (m) malleus; (pg) processus gracilis; (ty) tympanic conus; the stapes crus enters the oval window recess below the round window (r). The size of the VIIIth nerve can be estimated from the diameter of the internal auditory meatus (iam). (Photography by Alison George.)

Based on the principal physiological studies available from delphinids, two possibilities exist for middle ear function: translational bone conduction or conventional ossicular motion. McCormick et al. (1970, 1980) demonstrated in *T. truncatus* and *Lagenorhynchus obliquidens* (pacific white-sided dolphin) that immobilizing the ossicular chain decreased cochlear potentials but disrupting the external canal and tympanic conus had no effect. They concluded sound entering from the mandible by bone conduction produces a "relative motion" between the stapes and the cochlear capsule. Fleischer (1978) disagreed with these findings, suggesting the surgical procedure damaged the precise ossicular mechanism and introduced an artificial conduction pathway. He concluded, from anatomical studies, that the periotic is stable and sound from any path is translated through tympanic vibration to the ossicles which conventionally pulse the oval window. McCormick's theory depends upon tissue conduction and an inertial lag of the cochlear fluids in a vibrating bulla; Fleischer's, on differential resonance of the tympanic and periotic bones. The first theory assumes fixed or fused tympano-periotic joints; the second requires a free moving stapes and flexible tympano-periotic sutures. Neither theory provides a satisfactory or complete general explanation since each is inconsistent with variations in middle ear anatomy in a wide sample of species (Ketten 1984). It should also be considered that the data come from anesthetized vs postmortem specimens. McCormick measured live animals under deep anesthesia after opening the bulla. Fleischer used alcohol-preserved material from previously frozen and thawed *Tursiops* heads in which structural changes, including a loosening of the tympano-periotic sutures, should have occurred. The discrepancies in their conclusions point out a need to consider the complex effects of anesthesia, temperature, and postmortem changes on tissue characteristics (Fitzgerald 1975; Lees, Ahern, and Leonard 1983) as well as the need for replicate and multispecies studies.

To the extent that information can be extrapolated from available anatomical data, the middle ear anatomy of all Cetacea is tailored, in part, to meet environmental pressures and the massiveness and complexity of ossicular structures imply that the middle ear has at least some minimal function.

Mysticetes and odontocetes differ chiefly in the rigidity of the ossicular chain and in the prospect, based on an elaborate tympanic structure, that mysticetes receive auditory stimuli primarily from the ear canal and not from the jaw. If the middle ear space is defined by the volume of the tympanic shell, then the middle ear cavity in Mysticeti is substantially larger than in odontocetes, implying a lower frequency ear (see Webster and Plassman, Chapter 30). In reality, however, these are speculations in search of data and middle ear function remains largely unexplained for any cetacean.

4.3 The Inner Ear

4.3.1 The Vestibular System

The vestibule is large in Cetacea but the semicircular canals are substantially reduced, tapering to fine threads that do not form complete channels, and it is unclear whether all components of the vestibular system are functional (Ketten and Wartzok 1990). Although size is not a criterion for vestibular canal function, cetaceans are exceptional in having semicircular canals that are significantly smaller than their cochlear canal (Jansen and Jansen 1969; Gray 1951). Innervation is proportionately reduced; i.e., only 10% of the cetacean VIIIth nerve is devoted to vestibular fibers, as compared to 40% in most other mammals (Yamada 1953; Jansen and Jansen 1969; Morgane and Jacobs 1972). In the absence of physiological measurements of odontocete vestibular function, we may speculate that the vestibular system of cetaceans acts precisely as van Bergeijk (1967) suggested; i.e., as a "vehicle-oriented accelerometer." If the semicircular canals are vestigial, these animals may obtain only linear acceleration and gravity cues, but no rotational or three-dimensional accelerational inputs. This may be highly adaptive, permitting rapid rotations in a buoyant medium, exemplified by the flying turns of spinner dolphins, without the side-effects of "space-sickness."

4.3.2 The Cochlea

Cetacean cochlea have the prototypic mammalian divisions and relationships. The membranous labyrinth of the scala media (cochlear duct), scala tympani, and scala vestibuli forms an inverted spiral inside the periotic, curving medially and

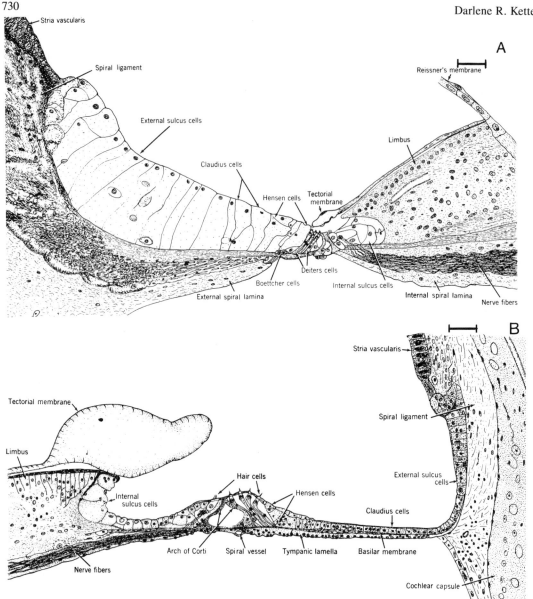

FIGURE 35.8. Cochlear duct cytoarchitecture. Line drawings of two points in the odontocete cochlea (reprinted with permission from Wever et al. 1971a, b) are shown for comparison with light micrographs of 20 µm mid-modiolar sections of the odontocete cochlear duct at equivalent locations (Ketten and Wartzok 1990.) Descriptions in this chapter use conventional neurocentric orientations for the cochlea; i.e., inner or medial are towards the modiolus; outer/lateral refer to the anti-modiolar or abneural side. Although in vivo the cochlear apex points ventrally in dolphins, the images are shown in a standard orientation. Tissues in the photomicro-graphs are from adult animals and represent average odontocete material preserved 5 hours to 4 days post-mortem. They show preservation and processing artifacts similar to those of human temporal bones, including disruption and collapse of Reissner's membrane, absent or necrotic organ of Corti, acidophilic staining of the perilymph, and serous protein deposits in scala media (SM). Each scale bar represents 50 µm. (A) The drawing of the lower basal region of *Tursiops truncatus*, a Type II odontocete, illustrates the classic odontocete features of an osseous outer lamina and heavy cellular buttressing. (B) In the apical region, the osseous

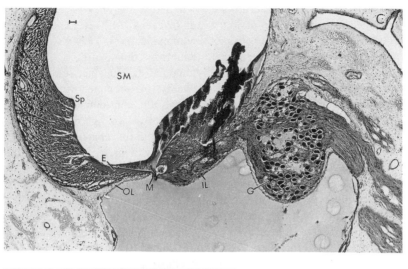

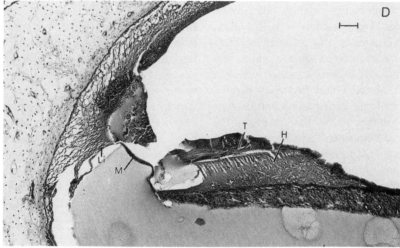

FIGURE 35.8. (*Continued*) outer lamina has disappeared, and the membrane has thinned and broadened. (C) The basilar membrane (M) of *Phocoena phocoena*, a Type I odontocete, in the basal turn (7 mm from the oval window) measures 45 μm × 20 μm and is stretched between inner (IL) and outer (OL) ossified spiral laminae. The outer lamina is 30 to 40 μm thick. The basilar membrane in this species is narrower than that of *Tursiops* throughout the basal turn (see Fig. 35.9). There is heavy staining of the perilymph in scala tympani, but the endolymph of scala media (SM) is not contaminated, indicating the membrane is intact. Blood in scala media is the result of a concussion. A distinctive cellular layer (E) found only in the basal turn in odontocetes lines the lateral basilar membrane recess below the spiral prominence (Sp). Kolmer reportedly dubbed them "ersatzzellen" (Reysenbach de Haan 1956), and although noted by several authors, these cells are unclassified and their function remains unclear. The bulge protruding medially into scala tympani is densely packed with oblate spiral ganglion cells (G) and is characteristic of the spiral ganglia in odontocetes but has not been reported for Mysticeti. (D) In an apical section (4 mm from the helicotrema), the basilar membrane in *Phocoena* is 200 μm wide and 10 μm deep. Only the spiral ligament (Li) supports the lateral edge of the basilar membrane at this point. Huschke's auditory teeth (H) are visible in the spiral limbus immediately below the limbal tectorial membrane (T).

ventrally from the stapes to the helicotrema around a core, the modiolus, containing the auditory branch of the acoustovestibular nerve. Dimensions of the cochlear canal are strongly correlated with animal size, not with frequency. Frequency related variations among species include the number of turns, basilar membrane dimensions, and distributions of membrane support structures (Ketten 1984; Ketten and Wartzok 1990).

Little is known of the comparative structure of the cochlea in mysticetes; i.e., whether there are structural differences related to infrasonic ranges. Present data show few differences from conventional cochlear duct structure in terrestrial mammals (Fraser and Purves 1960; Solntseva 1975; Norris and Leatherwood 1981; Pilleri et al. 1987; Ketten, in preparation). In contrast, odontocete cochlea differ significantly from all other mammalian cochlea. This discussion will outline general characteristics of the cetacean cochlear duct and will discuss in detail two anatomical features of the odontocete inner ear that influence resonance characteristics and frequency perception: basilar membrane construction and osseous spiral laminae configurations. Particularly in the basal turn, these components are exceptional in odontocetes compared to either mysticete or terrestrial ears, and, more important, differences among odontocetes in these two features correlate with the two frequency divisions (Type I and II) of echolocation signals.

A series of papers by Wever et al. (1971a, 1971b, 1971c, 1972) described cellular structure in the cochlear duct for six *T. truncatus* and three *L. obliquidens*. These papers provide the most reliable anatomies of the cochlear duct for these two species available to date and give us a basis for broader discussions of other species from less optimal material. No equivalent data are available for the majority of Cetacea. The cellular descriptions are quite extraordinary compared to other studies (Reysenbach de Haan 1956; Belkovich and Solntseva 1970; Solntseva 1971; Ketten 1984; Ketten and Wartzok 1990) in part because Wever was able to use animals perfused specifically for histology. The papers are a distinctive histologic series which is unlikely to be repeated soon since the animals were obtained in conjunction with electrophysiological studies performed prior to the beginning of current collection restrictions (Ridg-

way and McCormick 1967; Ridgway, McCormick, and Wever 1974; Wever and McCormick, personal communication). They also point out some of the daunting complexities of histology on cetacean temporal bones since only nine bullae out of twenty-five attempted were successfully processed despite the considerable expertise of Dr. Wever's laboratory. Two figures from Wever's papers are included in this chapter (Figs. 35.8A,B) for comparison with micrographs from more conventional odontocete material (Figs. 35.8C,D).

Cellular trends described for *Tursiops truncatus* (Wever et al. 1971a, 1971b, 1971c, 1972) were as follows: a 20-fold reduction in the height of the Claudius cells from base to apex; Boettcher cells distributed throughout the entire length of the cochlear duct with some double rows; substantial cellular buttressing of the basilar membrane in the lower basal turn by Hensen cells; and four rows of outer hair cells in some parts of the apical region. Basilar membrane and tectorial membrane thickness were not discussed, although the basilar membrane was described as a highly differentiated structure with a substantial variation in stiffness. Pilleri (1984) reported deep Azan staining in the basal region of *Monodon monoceros* (narwhal) which he attributed to tonofibrils of the pillar cells. This is consistent with Wever et al.'s observation that the pillar cells are exceptionally thick in the lower basal turn and Reysenbach de Haan's (1956) earlier description of "short and compact" pillar cells in *Phocoena*. Parallel trends were reported in *Lagenorhynchus* (Wever et al. 1972) although absolute numbers and ratios vary: a 15-fold reduction in the Claudius cells; Boettcher cells in single rows with none basally; and irregular hair cell distributions varying from four rows in one specimen to three or even two rows in another. Hair cell distributions are an important finding of these papers since, as Wever et al. (1972) state, previous workers assumed uniform distributions of three rows. Based on Wever's data, odontocetes appear, like microchiropteran bats (Firbas 1972), to have up to four rows of outer hair cells, but the irregular distributions are puzzling. It should be noted, however, that one or more of the Wever animals received aminoglycosidic antibiotics as part of a regular maintenance regimen (Ridgway, McCormick personal communication). When administered to humans, these can have acute toxic effects

TABLE 35.1. Ganglion cell density

Species	Type	Total ganglion cells	Membrane length (mm)	Average density (cells/mm)
Phocoena phocoena	I	66,933	24.31	2,753.3
Lagenorhynchus obliquidens	II	70,000	34.90	2,005.7
Stenella attenuata	II	82,506	37.68	2,189.6
Tursiops truncatus	II	105,043	41.57	2,526.9
Rhinolophus ferrumequinum	–	15,953	16.10	1,000/1,750[a]
Homo sapiens	–	30,500	31.00	983.9

[a] whole cochlea/acoustic fovea region.
Species data compiled from Wever et al. 1972; Bruns and Schmieszek 1980; Schuknecht and Gulya 1986; Ketten and Wartzok 1990.

on the inner ear, including partial or complete loss of hair cells (Schuknecht 1974). Similar ototoxic effects in dolphins may account for the irregular hair cell distributions Wever observed.

Neuronal components of mysticete cochlea have not been carefully described, but it is clear that they do not have densities equivalent to those found in odontocetes. The diameter of the auditory nerve, the volume of cells in Rosenthal's canal, and the number of habenular fibers are all disproportionately large in odontocetes (Fig. 35.8), consistent with an hypertrophy of the entire odontocete auditory system (Wever et al. 1971c, 1972; Morgane and Jacobs 1972; Fleischer 1976; Ketten and Wartzok 1990). Ganglion cell to hair cell ratios appear to be proportional to peak frequency in both bats and odontocetes and it is likely that high afferent ratios in odontocetes are directly related to the complexity of information extracted from echolocation signals. Total ganglion cell counts and ganglion cell densities of *Phocoena*, *Tursiops*, *Stenella*, and *Lagenorhynchus* are compared with bat and human data in Table 35.1. Ganglion cell densities in odontocetes are higher than in any other mammal and range from 2000 cells/mm in *Lagenorhynchus* to 2,700 cells/mm in *Phocoena*, (Wever et al. 1971c, 1972; Ketten and Wartzok 1990). Using a mammalian average of 100 inner hair cells/mm (Kiang personal communication) and four rows of outer hair cells/inner hair cell, these data imply a ganglion to hair cell ratio of nearly 6:1 for *Phocoena phocoena*, 5:1 for *Tursiops truncatus*, 4.4:1 for *Stenella attenuata*, and 4:1 for *Lagenorhynchus obliquidens*. The human ratio is 2.4:1; cats, 3:1; and bats average 4:1 (Firbas 1972; Bruns and Schmieszek 1980). Since 90 to 95% of all afferent spiral ganglion cells innervate inner hair cells, the average ganglion cell to inner hair cell ratio is 24:1 for odontocetes, or more than twice the average ratio in bats and three times that of humans (Firbas 1972). Wever et al. (1971c) speculated that additional innervation is required primarily in the basal region to relay greater detail about ultrasonic signals to the CNS in echolocation analyses. Electrophysiological results are consistent with this speculation. CNS recordings in both porpoises and bats imply increased ganglion cells may correspond to multiple response sets that are parallel processed at the central level. Bullock et al. (1968) found three distinct categories of response units in the inferior colliculus of dolphins; i.e., those that were signal duration specific, those that responded to changes in signal rise time, and those that were specialized to short latencies with no frequency specificity. This division of signal properties among populations of neurons is consistent with, although not identical to, observations in bats of multiple categories of facilitation and analysis neurons (Schnitzler 1983; Suga 1983). The sum of the data implies extensive monitoring of signal characteristics other than frequency occurs from inputs in the basilar membrane region that encodes ultrasonic echolocation signals.

4.3.3 Frequency and Shape

Basilar membrane dimensions are thought to be an important component of the resonance characteristics of the cochlea (von Békésy 1960; Iurato 1962; Zwislocki 1981). In mammalian cochlea, thickness and width vary inversely from base to apex, with highest frequencies encoded in the thicker, narrow, basal region and progressively lower frequencies encoded towards the apex as the

TABLE 35.2. Basilar membrane dimensions

Acoustic Group and Species	Membrane length (mm)	Outer osseous lamina length (mm)	Basal/apical width (μm)	Basal/apical thickness (μm)	Peak vocalization frequency (kHz)
Type I					
Phocoena phocoena	25.93	17.6	30/290	25/5	120
Type II					
Grampus griseus	40.5	–	40/420	20/5	–
Lagenorhynchus albirostris	34.9	8.5	30/360	20/5	40
Stenella attenuata	36.9	8.35	40/400	20/5	60
Tursiops truncatus	40.65	10.3	30/380	25/5	60–70
Mysticete					
Balaena mysticetus	61.3	<10	120/1,670	7.5/2.5	<0.20
Balaenoptera acutorostrata	–	–	100/1,500	–	–
Balaenoptera physalus	–	–	100/2,200	–	.02
Eubalaena glacialis	55.6	–	100/1,400	–	<0.20

Species averages compiled from Fleischer 1976; Norris and Leatherwood 1981; Ketten 1984.

membrane broadens and thins. In bats it has been shown that frequency varies inversely with basal turn membrane widths (Hinchcliffe and Pye 1968; Brown and Pye 1975). Wever's data (1971b, 1972) imply a similar relationship for dolphins; i.e., minimum basal width averaged 30 μm for *Tursiops* and 35 μm for *Lagenorhynchus*; apical widths averaged 350 μm. The peak frequency of echolocation signals for these two species are 60 and 40 kHz respectively (Diercks et al. 1971). Recent studies (Ketten 1984; Ketten and Wartzok 1990) found similar membrane widths in most odontocete species (Table 35.2). Thickness decreased 5-fold from 25 μm to 5 μm base to apex, while widths increased 9- to 14-fold. Therefore, the generic odontocete basilar membrane has a nearly square basal cross-section that thins and broadens apically to a 5 μm strip 300 to 400 μm in width (Fig. 35.8). Based upon dimensions alone, odontocete basilar membranes are highly differentiated, anisotropic structures capable of an exceptionally wide frequency response. By contrast, mysticete basilar membranes are consistently thinner and wider; in *Balaena mysticetus* (bowhead whale) the membrane is 7.5 μm thick, 120 μm wide at the base, and 2.5 μm thick, 1,600 μm wide at the apex.

Multiple species comparisons (Ketten and Wartzok 1990) demonstrated thickness-to-width ratios were a more significant correlate of frequency than width alone. Comparing bat, odontocete, and mysticete basilar membrane ratios (Fig. 35.9), the odontocete ratios are 2 to 3 times greater than that

of the bat in the most basal, ultrasonic regions, and all three echolocating species have significantly higher basal ratios than the mysticete. The maximum ratio occurs in *Phocoena*, a Type I odontocete. Bats and odontocetes have similar apical ratios, but the mysticete value is significantly lower, which is consistent with a broad, floppy membrane for encoding low frequencies. The basal ratio of the mysticete is equivalent to membrane ratios for the lower apical region of odontocetes. Were all other cochlear duct components equal among these four species, the differences in the basilar membrane dimensions alone could be a significant determinant of the different auditory capacity of each species.

Fleischer (1976) categorized all Cetacea into high (odontocete) vs low (mysticete) frequency users based on basilar membrane width estimates from dehydrated and fossil bullae. Although his absolute values are larger than those in other studies, his curves of basilar membrane base to apex widths have an average slope of 0.3 for odontocetes and 0.8 for mysticetes, consistent with other rates of membrane width changes (Table 35.2) (Wever 1971b, 1972; Norris and Leatherwood 1981; Ketten and Wartzok, 1990). Fleischer also observed that basilar membrane support in the basal turn is stronger in Odontoceti than in Mysticeti and that mysticetes have a greater cochlear height-to-diameter ratio than odontocetes. He concluded perception of high frequencies was "favored" by small ratios and states that common models of the cochlea oversimplify cochlear relationships, adding that "... the cochlear parameter ... least understood is the mode of coiling."

To date, there is no practical means of estimating frequency ranges for Cetacea based on cochlear dimensions. Basilar membrane lengths in Cetacea, like those of terrestrial mammals, scale isomorphically with body size (Ketten 1984; West 1985). Greenwood (1961, 1962) used membrane lengths as a major variable to predict critical bands and to estimate maximal perceived frequencies for terrestrial species. A major assumption in the equations is that critical bands are equidistant in all mammals (see Fay, Chapter 14). Practically, this means coefficients in the elasticity function are scaled based on human to animal membrane length ratios. Since membrane lengths are proportional to animal size, as the animal gets larger, calculated frequency maxima get lower. For nonspecialized terrestrial mammals, this relationship is correct; i.e., larger animals have lower frequency ranges (Heffner and Heffner 1980; West 1985). Norris and Leatherwood (1981), using Greenwood's equations, estimated a maximal frequency capacity for *Balaena mysticetus* of 12 kHz, similar to the value Greenwood calculated for the elephant, but it is unclear whether these equations apply to aquatic mammals. Odontocetes do not have the same distribution of critical bands as humans (see Section 2.1) which violates Greenwood's primary assumption, and the maximal frequencies the equations predict for any odontocete from the equations; e.g., 15 kHz for *Tursiops truncatus*, are well below any estimations of their auditory capacity based upon sound production and echolocation ability. Similar anomalies need to be considered before using any method to predict hearing in fossil species, particularly for those which may have special or atypical ears.

In terrestrial species, an outer ossified lamina in lieu of a spiral ligament supporting the basilar membrane implies a high frequency ear (Hinchcliffe and Pye 1969; Reysenbach de Haan 1956; Sales and Pye 1974). Inner and outer ossified spiral laminae are present throughout most of the basal turn in all odontocetes (Table 35.2), and the extent and development of these laminae are among the most striking features of the odontocete cochlea (Reysenbach de Haan 1956; Wever et al. 1971a, 1971b, 1971c, 1972; Ketten 1984). Detailed

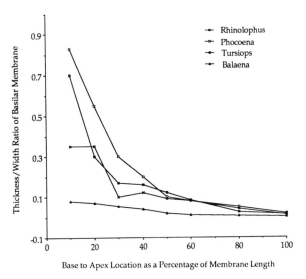

FIGURE 35.9. Basilar membrane ratios. Average thickness to width basilar membrane ratios for the horseshoe bat (*Rhinolophus ferrumequinum*), harbour porpoise (Type I) (*Phocoena phocoena*), bottlenosed dolphin (Type II) (*Tursiops truncatus*), and bowhead whale (Type M) (*Balaena mysticetus*) are plotted as a percentage of cochlear length. (Data from Bruns 1976; Norris and Leatherwood 1981; and Ketten and Wartzok 1990.) High ratios for the bat, porpoise and dolphin reflect a thicker, stiffer membrane which responds to ultrasonic frequencies. Differences in the basal basilar membrane ratios for the three echolocators are consistent with the peak frequency in each species (*Phocoena*, 130 kHz; *Tursiops*, 70 kHz; *Rhinolophus* 40 kHz). Basal ratios in the mysticete cochlea are equivalent to apical ratios for the other three species. The low apical ratio in the bowhead is consistent with a broad, flaccid membrane that encodes extremely low frequencies.

comparisons of the laminae show that they are a major correlate of odontocete ultrasonic ranges (Ketten and Wartzok 1990). The internal osseous spiral laminae, tunneled by the foramina nervosa or nerve fiber tracts, form a bilayered wedge which supports the medial margin (pars arcuata) of the basilar membrane (Fig. 35.8). The thickness of the laminar wedge varies inversely with distance from the stapes. In the lower basal turn, the paired laminae average 50 μm from tympanal to medial lip. In the middle to upper basal turn, the tympanal layer disappears and the medial edge thins to 5 μm, forming a single shelf supporting the spiral limbus. The outer lamina in the lower basal turn in all odontocetes is 30 to 40 μm thick, is heavily

Table 35.3. Cochlear Canal Spiral Parameter

Species	Cochlear type	Turns	Scalae length (mm)	Basal diameter (mm)	Axial height (mm)	Axial pitch[a] (mm)	Basal ratio[b]	Pulse peak frequency (kHz)
Recent Odontoceti								
Inia geoffrensis	I	1.5	38.2	8.5	–	–	–	200
Phocoena phocoena	I	1.5	25.93	5.25	1.47	0.982	0.280	130
Grampus griseus	II	2.5	40.5	8.73	5.35	2.14	0.614	–
Lagenorhynchus albirostris	II	2.5	34.9	8.74	5.28	2.11	0.604	40
Stenella attenuata	II	2.5	36.9	8.61	4.36	1.75	0.507	60
Tursiops truncatus	II	2.25	40.65	9.45	5.03	2.24	0.532	60–70
Physeter catodon	I,II	1.75	72.21	14.3	3.12	1.78	0.218	–
Recent Mysticeti								
Balaenoptera acutorostrata	M	2.2	–	17	7.5	3.41	0.441	–
Eubalaena glacialis	M	2.5	55.6	9.67	6.7	2.68	0.57	<0.20
Extinct Cetacea								
Dorudon osiris	I,II,M	2.5	–	8.2	7	2.8	0.854	–
Parietobalaena palmeri	M	2.3	–	13.5	6.6	2.87	0.489	–
Rhabdosteus spp.	I,II	1.5	–	9.5	3.4	2.27	0.358	–
Squalodon spp.	I,II	1.6	–	10.5	–	–	–	–
Zygorhiza kochii	I,II,M	2.0	–	10.5	6.75	3.38	0.643	–

[a] $\dfrac{\text{axial height}}{\text{turns}}$ [b] $\dfrac{\text{axial height}}{\text{basal turn diameter}}$

Species averages from Kellogg 1936; Fleischer 1976; Norris and Leatherwood 1981; Ketten and Wartzok 1990.

calcified, and functions as a lateral attachment for the basilar membrane and as a housing for the spiral ligament. Thus, in the extreme basal end, the thick basilar membrane is firmly anchored at both margins to a bony shelf. Differences in the length of outer laminae between Type I and Type II species are consistent with the two acoustic divisions of echolocation signals. In delphinids, all of which are Type II echolocators with typical peak frequency ranges of 40 to 80 kHz, this bony anchor is present for only 25% of the cochlear duct (Table 35.2). In phocoenids, Type I echolocators with peak frequencies above 100 kHz, an outer lamina is present for over 60% of the cochlear duct. The basilar membrane therefore has substantial buttressing at both edges over twice as much of its length, proportionally, in Type I than in Type II odontocetes. Type I species use, and presumably hear, higher ultrasonic signals. A longer outer lamina in Type I cochlea increases membrane stiffness, which increases the resonant frequency of that portion of the membrane compared to an equivalent membrane that lacks bony support in Type II cochlea. When combined with the differences observed in membrane ratios, differences in the extent or proportion of outer bony laminae provide a simple but important mecha-

nistic link for species differences in ultrasonic ranges in Odontoceti.

Outer laminae are found in mysticetes as well. Measurements are not available for most Mysticeti, but even qualitative descriptions make it apparent the laminae are not functional equivalents of those in odontocetes (Fleischer 1976; Norris and Leatherwood 1981). The bone is characterized as spongy and meshlike and the outer lamina disappears within the first half turn. These descriptions of mysticete laminae suggest a basilar membrane system opposite that of odontocetes; i.e., a broad thin membrane with insubstantial support. It is likely that the presence of an outer lamina in mysticetes is a residual ancestral condition rather than a derived structure related to mysticete frequency ranges.

Three-dimensional cochlear spiral measurements and reconstructions show striking differences in the construction of the cochlea between odontocetes and mysticetes and between Type I and Type II odontocete species (Ketten 1984; Ketten and Wartzok 1990). Data for eight representative odontocetes are compared with data for two mysticetes and five extinct cetaceans in Table 35.3. There is a strong negative correlation

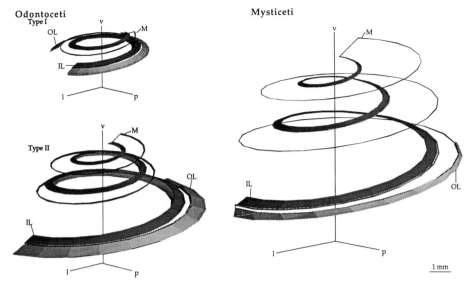

FIGURE 35.10. Basilar membrane and spiral laminae distributions in cetacea. Three-dimensional reconstructions, generated from standardized species data, schematically represent major cochlear duct structural components (IL) inner osseous spiral lamina; (M) lateral edge of basilar membrane attached to the spiral ligament; (OL) outer osseous spiral lamina of Type I and Type II odontocetes and a generalized mysticete. The composites were produced by combining spiral model parameters with cochlear canal data (Tables 35.2, 35.3). Basilar membrane width is equivalent to the distance between the inner osseous lamina and the outer lamina or the spiral ligament, represented here by a thin black line after the outer osseous laminae end. The cochlea are shown inverted from in vivo orientations (l lateral; p posterior; v ventral). Differences in membrane buttressing among the cochlea are clear. The Type I cochlea has proportionately twice as much membrane supported by bony laminae as Type II, and the mysticete laminae is neither strong nor extensive. The basal region of the mysticete membrane is three times as wide and one-third as thick as that of the odontocetes; at the apex it is four times the width and half the thickness of the odontocete membranes. The Type II membrane is broader than the Type I at the apex, suggesting Type II species may resolve lower frequencies than Type I. Differences in laminar support imply Type I cochlea have a higher ultrasonic range.

$(-0.968 < r < -0.791)$ for characteristic frequency with all spiral variables except scalae length and basal diameter. These two variables are positively correlated with animal length $(0.84 < r < 0.92)$. The data indicate three spiral configurations in extant Cetacea differentiated by turns, height, pitch, slope, and basal ratios. Two formats occur exclusively in odontocetes and coincide with Type I and II acoustic divisions. Type I cochlea are shallow, < 2 turn spirals, and Type II are steeper with > 2 turns. Type I is nearly Archimedian; i.e., it is a constant interturn radius curve, like that formed by a tightly coiled rope. Type II is equiangular, like a nautilus shell, with logarithmically increasing interturn radii. This is also the presumed configuration for most mammalian cochlea. The available mysticete data do not allow a full three-dimensional analysis of spiral shape, but the basic measurements indicate a broad spiral with a steeper pitch than that of odontocetes. The spiral data for modern cetaceans, in combination with basilar membrane and outer osseous laminae data, produce three prototypic cochlear shapes, each of which characterizes a major acoustic division. Schematic three-dimensional reconstructions illustrate the major features of each cochlear format (Fig. 35.10).

If auditory capacity is correlated with habitat and behavior, differences in spiral formats should correlate with specific environments and lifestyles as well. In modern Cetacea, Type I spirals have been found only in inshore phocoenids and riverine dolphins (Table 35.3, App. 1), which are the highest frequency group of aquatic mammals, with echolocation signals that reach 200 kHz (Purves and Pilleri 1983; Ketten 1984; Feng et al. 1990).

TABLE 35.4. Bullar-cranial associations

Group	Cochlear type	Nonsynostotic joints with the skull	Synostotic joints with the skull	Pedicles (connections)
Protocetidae	–	X	X	4 (t–b)
Dorudontinae	I,II,M	X	(in some species)	1.5 (p–t)
Squalodontidae	I,II	X	–	2 (p–t)
Delphinidae	II	(in some species)	–	2 (p–t)
Phocoenidae	I	–	–	2 (p–t)
Platanistoidea	I	(in one species)	–	2 (p–t)
Physeteridae	I,M	(in some species)	–	2 (p–t)
Mysticeti	M	X	X (p–b)	1 (p–t)

(t), tympanic; (b), basioccipital region; (p), periotic.
Data compiled from Kellogg 1936; Kasuya 1973; Gingerich et al. 1983; Ketten 1984.

These species live in turbid waters where ultrahigh frequency, short wavelength signals would be advantageous for distinguishing fine detail. Type II formats are common to delphinids which are off-shore and pelagic species with lower frequency echolocation signals. The mysticete format is known only in low-frequency pelagic planktivores. Both groups of echolocators are active predators; the mysticetes are opportunistic feeders. With these characteristics in mind, it is possible to speculate on the functional implications of the cochlear structure of extinct species.

5. The Extinct Cetacean Ear

Fossil evidence indicates that the ability to utilize high frequencies may have originated early in cetacean history, but we are not able to link specific auditory structures with entry into the water. Osteological remnants of the earliest Eocene cetaceans, the protocetid Archaeoceti (Fig. 35.1, App. 1), show relatively few changes from the typical terrestrial mammalian skull (Fig. 35.2), although small accessory air sinuses and a separate periotic are already present. These may reflect "preadaptive" features in Mesonychidae since they are also found in ungulates and are not found in other noncetacean mammals (Barnes and Mitchell 1978; Barnes et al. 1985; Oelschläger 1986a, 1986b, 1990). Karyotypic and seral studies (Boyden and Gemeroy 1950; Ishihara et al. 1958; Lowenstein 1987) indicate close relationships for Cetacea with ungulates, particularly suids, lending credibility to preadaptation theories. In the most ancient Archaeocetes, *Pakicetus inachus*, *Pappocetus lugardi*,

and *Protocetus atavus* (Protocetidae), little is known of the postcranial skeleton, making it difficult to judge their level of adaptation to water, but teeth and sinus patterns suggest they were predatory echolocators (Fordyce 1980; Gingerich and Russell 1981; Gingerich et al. 1983). The protocetids have the cetacean cranial characters of a thin zygomatic arch, a large concave mandible, and a well-defined periotic, although long anterior and posterior processes wedge the periotic between the squamosal and mastoid bones, making separation of the tympanic and periotic bones from the skull incomplete (Gingerich and Russell 1981; Gingerich et al. 1983; Oelschläger 1986a, 1986b). The tympanic in *Pakicetus* has four nonsynostotic articulations with the surrounding skull elements (Table 35.4) (Gingerich and Russell 1981). Gingerich et al. (1983) suggest that the malleus was fused to the tympanic bulla, as in modern odontocetes, and that this permitted *Pakicetus* to hear while submerged; however, they also noted that the air sacs were small and that the tensor tympani fossa was exceptionally large, implying a functional tympanic membrane which is difficult to link theoretically with a fused malleus. They concluded that protocetids, which are common in fluvial sediments, were probably amphibious, freshwater carnivores that were not fully adapted to water.

Basilosauridae exhibit a mixture of plesiomorphic (primitive) and apomorphic (derived) characters which makes them a reasonable, although unestablished stem point for the separation of mysticete and odontocete lineages (Fig. 35.1). They are found only in marine sediments and are pivotal in the development of modern Cetacea (Fig. 35.2C) (Fordyce 1980; Gingerich et al. 1983). One

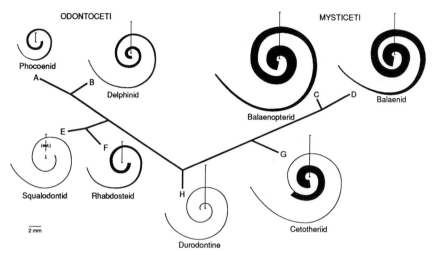

FIGURE 35.11. Two-Dimensional basilar membrane reconstructions. Two-dimensional reconstructions illustrate differences in the shape and dimensions of the cochlear spiral and basilar membrane in four extinct and four modern species. (Data compiled from Kellogg 1936; Fleischer 1976; Norris and Leatherwood 1981; Ketten and Wartzok 1990.) Basilar membrane widths are represented by filled areas. The vertical bar extending from the origin of each spiral represents the axial height. In some species, complete data are not available, and only the path of the cochlear canal is shown (see Table 35.3). In Cetacea, frequency ranges are inversely correlated with basilar membrane width, axial height, and number of turns. Based on these criteria, the spirals shown can be divided into 5 broad categories: mixed high and low frequency ancestral (H *Zygorhiza kochii*; 2 turns; tall); high frequency stem odontocete (E *Squalodon*; F *Rhabdosteus*; ≤ 2 turns, narrow membrane, average height); low ultrasonic Type II odontocete (B *Tursiops truncatus*; > 2 turns, narrow membrane, average height); high ultrasonic Type I odontocete (A *Phocoena phocoena*; < 2 turns, narrow membranes, low basal ratios); and low frequency stem or modern mysticete (C *Balaenoptera acutorostrata*; D *Eubalaena glacialis*; G *Parietobalaena*; >2 turns, wide membranes, tall). Differences may exist among fossil and recent mysticetes for infrasonic perception which are not reflected in these data.

basilosaurine, *Basilosaurus isis*, an Archaeocete originally misclassified and infamously misnamed as a reptile (Kellogg 1936), has recently been shown to be an important intermediate form with hindlimbs that are completely formed but too insubstantial for terrestrial locomotion (Gingerich, Smith, and Simons 1990). All basilosaurids show some cranial modification consistent with a modern cetacean cranial format. Smaller dorudontine basilosaurids (e.g., *Dorudon osiris*, *Dorudon intermedius*, *Zygorhiza* spp.) have the most extensive changes, including elongated maxillae, inflated bullae, large mandibular channels, higher occipital shields, posterior migration of the nares, enlarged sinuses, massive ossicles, and a periotic decoupled from the mastoid (Kellogg 1936; Barnes and Mitchell 1978; Oelschläger 1986a, 1990). Their most primitive characteristic is that the periotic-tympanic complex remains affixed to the

braincase with anterior and posterior flanges wedged between the squamosal and occipital (Table 35.4). A review of Kellogg's descriptions (1936) in the context of modern spiral data shows the dorudontines have a bullar structure and cochlear spirals that are composites of Type I, Type II, and mysticete parameters (Table 35.3, Fig. 35.11). Like mysticetes, they have an inflated, bulbous tympanic which is closed anteriorly but, like odontocetes, they have inner and outer pedicles and prominent posterior tympanic processes (Table 35.4). The dorudontine periotic is distinctly ovoid and strongly resembles those of modern physeterids. *Zygorhiza* has two turns but a steep spiral with an axial pitch and basal ratio that mix characteristic modern mysticete and odontocete values. A few basilosaurid species like *Kekenodon* are sufficiently modern, with full inner and outer bullar pedicles, medial and lateral prominences, and

extensive air sacs, that their status as late Archaeocetes has been questioned (Kellogg 1936; Kasuya
1973; Barnes, Domning, and Ray 1985). Despite
the absence of full telescoping in the anterior cranial
bones in some specimens, dorudontines, at least in
terms of acoustic structures, are the probable ancestral link to squalodonts which superseded the
Archaeocetes and are the link to modern cetaceans.

Within 10 million years after the extinction of
Zygorhiza, Agorophiidae (Fig. 35.2D), an Oligocene squalodontoid family, displays most overt
odontocete skull traits (Barnes, Domning and Ray
1985): (1) telescoping of the skull; (2) nares at the
vertex of the skull; (3) hollow mandible; and (4)
tympano-periotic isolation in an extensive cranial
sinus. In modern Cetacea, these features are associated with significant soft tissue developments
which are principally related to underwater
echolocation; e.g., the presence of a melon or sper-
maceti organ that channels sounds outward, the
lining of the mandibular concavity with fat that
acts as an acoustic wave guide, and filling of the
peribullar sinus with vascularized foam that acts as
an acoustic insulator (Reysenbach de Haan 1956;
Fraser and Purves 1960; Norris 1968; McCormick
1972; McCormick et al. 1975; Ketten 1984). The
presence and extent of related osteological changes
(telescoping, concave mandible, separate bullae,
and enlarged peribullar spaces) in agorophiids are
sufficient for them to be considered intermediate
to the squalodonts from which modern odontocetes
are derived (Kellogg 1936; Barnes, Domning, and
Ray 1985; Oelschläger 1986a; Pilleri, Gihr, and
Kraus 1986).

Spiral parameters in later Squalodontoidae; e.g.,
Rhabdosteus and *Squalodon* (Fig. 35.2E), have
characteristics of both phocoenid (Type I) and delphinid (Type II) cochlea (Table 35.3, Fig. 35.11).
Squalodontid bullae, similarly, show mixed, overt
characteristics of modern platanistid, physeterid,
and ziphiid forms (Kellogg 1936). In most respects,
therefore, these earliest odontocetes are considered to already have the functional acoustic
properties of modern odontocetes; it is likely they
were carnivorous echolocators. From this point
forward, cetacean development follows family
lines which are still fully represented today (Fig.
35.1). Kasuya (1973), in a comprehensive analysis
of dried odontocete bullae, devised a series of
phyletic subdivisions based on surface morphom-

etry and patterns of tympano-periotic fusion. He
concluded the physeterids and platanistids follow a
primitive pattern while the delphinids have the
most recent structure and the phocoenids fall
between primitive and recent forms. Phocoenids
and delphinids both appear to be descended from
kentriodonts and have developed similar skull
asymmetries associated with complex dorsal air
sinuses. It is unclear whether river dolphins arose
from a common ancestor or, alternatively that they
developed as four separate, parallel lineages. The
mysticetes are generally considered modern but
they have some osteological features in common
with more primitive species; e.g., there is only one
inner pedicle, the periotic retains long anterior and
posterior processes which wedge firmly against the
skull, and all species except some Balaenidae have
a distinctive hemispheric bullae (Table 35.4) (Kellogg 1936; Kasuya 1973). This construction, which
is consistent with a low frequency, nonecholocating ear, is not evident prior to *Parietobalaena* (Fig.
35.11) in the middle Miocene.

A controversy arises in the literature at this point
concerning both terminology and function of the
bony bullar structures (Yamada 1953; Reysenbach de Haan 1956; Kasuya 1973; Oelschläger
1986a, 1986b, 1990). All authors agree that a
major structural development tied to a fully aquatic existence for Cetacea was the disassociation of
the tympanic and periotic bones from the skull. To
produce modern odontocete ears from Archaeoceti, it was necessary to isolate the periotic by
replacing anterior and posterior skull processes
with bony, synostotic tympano-periotic connections. During this transition, the tympanic lost its
associations with the squamosal and pterygoid and
formed pedicular or pillar-like attachments to the
periotic; i.e., the nonsynostotic articulations of the
bullar flanges with the squamosal, exoccipital, and
basioccipital found in Protocetidae are gradually
replaced with two synostotic tympano-periotic
pedicles (the posterior petrotympanic process and
the processus tubarius) in modern odontocetes
(Fig. 35.7A) (Oelschläger 1986a, 1986b; Pilleri,
Gihr, and Kraus 1987). Modern mysticete bullae
characteristically have one lateral pedicle and a
distinct mastoid flange (Fig. 35.7B), which connects either or both bullar components to the skull
and may function in transmitting low frequencies
(Yamada 1953; Reysenbach de Haan 1956; Kasuya

1973); i.e., to produce a mysticete ear, it was necessary to retain and fuse one posterior process to the
skull, forming only one auxiliary pedicle (Table
35.4). Authors disagree about the role of the mastoid in this progression, particularly since its actual
location and whether it fuses with the tympanic and
periotic are not clear in present Cetacea (Kasuya
1973, Oelschläger 1986b). Recent work by Oelschläger 1986b, 1990) provides evidence that the
mastoid does not form a fused structure with either
the periotic or tympanic, as traditionally suggested,
but is retained in its relative position to the
squamosal and exoccipital, resulting in a reduction
of the periotic processes and rotation of the
tympano-periotic complex to the present position.

The presence of a cranial process is coincident
with a lower frequency ear, while its absence may
be an indicator of echolocation abilities. Consequently, the form and number of attachments of the
periotic with the tympanic and with the skull are
important diagnostic characteristics distinguishing
mysticetes from odontocetes, and thus can be used
to classify intermediate forms (Kasuya 1973). Different stages in this tympano-periotic metamorphosis are found in dorudonts, squalodonts, and
kentriodonts. Shared characters of Mysticeti and
Odontoceti; e.g., horizontal flukes, median dorsal
fins, extensions of the middle ear sinus, vertebral
ankylosis, and cranial distortions, imply a fairly
early common ancestor, probably a dorudontine
Archaeocete. Odontocetes antedate mysticetes,
therefore, the earliest cetacean was probably a
high, but not ultrahigh, frequency user, and the low
frequency characteristics of mysticete ears are a
relatively recent development.

6. Cetacean Auditory Adaptations

6.1 Comparative Speculations

Piecing together the fossil, cochlear, and acoustic
data, we expect the terrestrial ancestor of Cetacea
to be a small, high-frequency carnivorous mammal, possibly with some ultrasonic capacity, that
exploited the niche vacated by the icthyosaurs.
On entering the water, it faced substantial competition and predation from resilient, ancient, well-
adapted species like sharks. Perhaps the likeliest
animal to succeed would have been a nocturnal

carnivore, a predator preadapted to a dark environment. It would be interesting, therefore, to examine
mesonychid fossil assemblages for evidence of
ultrasonic cochlear adaptations. It must be noted,
however, that even if evidence of a high frequency
mesonychid were found, it is fairly certain the early
Archaeocetes were not *aquatic* echolocators, since
they have no melon. Echolocation is a two-way
function. For an ancestral cetacean to qualify as an
effective echolocator, there must be a coordinated
means of generating a highly directional signal and
receiving its altered echo. Modern odontocetes are
true echolocators, not simply ultrasonic receptors,
and it is important to determine at what point the
ultrasonic source (nasal diverticuli; rostral concavity) and receiver (isolated tympano-periotic and
narrow basilar membrane) coexist in Cetacea.

Cochlear data may also be used to speculate on
the development of the Mysticeti. Their appearance occurs within a reasonable geologic time
scale of the breakup of Gondwanaland (Fordyce
1977). This dispersal of the continents resulted in
the opening of new oceanic regions in the southern
oceans and the creation of the circumpolar Antarctic currents. While these changes produced terrifi-
cally productive waters which are still major
repositories of marine biomass today, they also
brought about substantial reductions in surface
temperatures in the higher latitudes. Cetacea
inhabiting those regions would be faced with an
abundance of food and less pressure to compete,
but also with an even greater risk of hypothermia
than was faced in warmer, northern waters. In
colder southern waters, increased size offers a substantial metabolic advantage. Since surface area
increases more slowly than volume as a structure
expands, increasing size may help retain body
heat; i.e., a larger whale is a warmer whale. We
also know that odontocete cochlea scale isometrically with animal size. It is likely that mysticete
cochlea scale in a similar if not identical way. If
basilar membranes broadened and lengthened
without thickening as cetaceans increased in size, a
lower frequency encoding cochlea would result as
a consequence of the greater mass in mysticetes.
With less pressure to echolocate as a foraging
strategy in more productive waters, a decrease in
the audibility of higher frequencies may not have
been a significant disadvantage. We might hypothesize that colder, richer feeding grounds provided

the appropriate pressure for development of larger, low frequency baleen whales. If correct, there should be a predictable temporal and latitudinal distribution of fossils with odontocetes dominating earlier northern faunas and larger mysticetes increasingly common in more recent southern fossil records. One difficulty in obtaining such data may be that it resides literally at the ocean bottom, since the majority of baleen deaths may deposit carcasses offshore in major pelagic regions that have not shifted significantly despite climatic changes.

Lastly, little is known of the primitive amphibious mammalian ear. Origins of two orders, the extinct Desmostylidae ("sea horses") and Sirenia ("sea cows," manatees, and dugongs), are poorly understood in comparison to Cetacea. Little acoustic or anatomical data are available for the manatee and dugong, although compared to Cetacea, Sirenia are more accessible, and they offer substantial promise of useful data related to low-frequency aquatic adaptations. Both Desmostylia and Sirenia show clear anatomical affinities with two other low frequency mammals, elephants and hippopotami (Barnes, Domning, and Ray 1985; Domning 1982). Recent studies show exceptional low-frequency capability in the elephant (Heffner and Heffner 1980; Payne, Langbauer, and Thomas 1986); but the hippopotamus is virtually unknown acoustically. It is reputed to produce audible clicks underwater and to swim by flicking its rear legs in tandem rather than trotting in the shallows (K. Norris personal communication). Since its adaptation to water is less complete than that of most marine mammals, studies on hippopotamus may provide intriguing insights into the behavioral adaptations of early, amphibious species.

6.2 Future Directions and Open Questions

A great many functional aspects of the cetacean ear are not fully understood. There is no satisfactory model of hearing in Cetacea, yet it is clear they have a substantially different cochlear and middle ear construction from terrestrial mammals. A pressing problem related to developing such a model is that cetacean cochlear anatomy differs significantly among families and variations are not fully described. The second difficulty is that the anatomy that has been well described is perplex-

ing. Facetiously, the odontocete ear could be described as a jaw with windows overlying a rock with strings attached. Although purposely trivial, this description conveys the difficulty of interpreting the elegant and subtle relationships of these complex structures without more extensive studies both in basic anatomy and audiometry, particularly for the mysticetes. Specific answers are needed for: (1) What are the transmission paths for sound to the ear in Mysticeti? (2) Do the sinuses play a significant role in directional hearing in all Cetacea or is their function primarily in echolocation? (3) Does the extreme density of the bulla resist extreme pressure or is it a necessary component for signal detection? (4) Are the ossicles functional and how do they receive sound? A more global question is whether there are, at least in the odontocetes, perhaps two parallel systems for processing sound: one for echolocation and one for lower frequencies. Is the odontocete cochlea a shared property, with two anatomical inputs from two acoustic channels? It is a proposition that fits well with the conflicting evidence for two potential sites of sound production and of many potential paths for sound reception, particularly in light of the two types of sounds odontocetes produce. To my knowledge, this is a novel alternative that has yet to be researched.

One of the most difficult tasks in producing this chapter was to determine a means of characterizing the auditory capability of marine mammals. Aside from the diversity of species, and range of adaptations that implies, comparatively little data exist that match the fairly rigorous and conventionally accepted means of classifying hearing in a mammal. As Watkins and Wartzok (1985) pointed out, information and research in marine mammals ranges ". . . from intensive to eclectic." Much of the available data are difficult to synthesize into a coherent analysis since techniques vary widely and sample sizes are often small. This is not a reflection of poor science in the field. Indeed many experiments verge on the ingenious and heroic considering the environmental, practical, and legal complications implicit in marine mammal research. It is apparent, however, that attention to two areas of experimental design are needed. First, it is imperative that the data base be expanded. Even within a relatively homogeneous group like Odontoceti, one species cannot be used to reliably characterize the entire suborder. Cochlear variations coupled with species differences in echolocation pulse intervals and frequency imply that more than one echolocation model exists, and no ultrasonic model is likely to explain infrasonic mechanisms in mysticetes. Secondly, if experiments are not carried out on live animals under their normal environmental conditions, they must be carefully interpreted with full assessments of the response changes an abnormal, nonaquatic environment may induce. Similar caveats hold for any work with postmortem tissue. These are not trivial analyses to make, but they are sorely needed and an expansion of our knowledge of these animals is imperative for any realistic understanding of their adaptations and abilities.

7. Summary

By the late Miocene, four major cranial trends associated with the environmental pressures of an aquatic habitat and audition were established in both the Odontoceti and Mysticeti: (1) telescoping of the skull; (2) dorso-caudal nares; (3) enlarged peribullar sinuses; and (4) a tympano-periotic bullar complex partly or wholly disassociated from the skull. Environmental influences are equally evident in the gross anatomy of the cetacean auditory system. There are no pinnae and no major pneumatized areas analogous to the mastoid cavities. Cetacean periotics, tympanics, and ossicles are all similarly constructed of massive, porcelaneous bone. The odontocete tympano-periotic complex is completely detached and acoustically isolated from the skull; mysticetes retain a medioposterior skull connection. In Odontoceti, telescoping produced a frontal concavity that accommodates the melon and nasal sacs which function in production of ultrasonic echolocation signals. The position, construction, and ligamentous associations of odontocete bulla support the "pan bone" theory of jaw conduction in which ultrasonic echoes are received by a fatty acoustic wave guide in the mandible. The path of sound reception in mysticetes is unknown, but they retain a highly derived tympanic membrane analogue and the external auditory canal may be functional.

Modern cetaceans divide into three acoustic groups: mysticetes (potentially infrasonic); Type I odontocetes (upper range ultrasonics); and Type II odontocetes (lower range ultrasonics), which parallel habitat and behavioral divisions. The acoustic divisions coincide with three cochlear formats that differ principally in the construction and support of the basilar membrane. The odontocete cochlea is clearly adapted for ultrasonic perception, with an exceptionally narrow basilar membrane, high spiral ganglion cell densities, and extensive bony outer spiral lamina. Basilar membrane cross-sectional dimensions interact with its composition and support to determine resonance characteristics. Membrane thickness-to-width ratios are higher for the basal turn of odontocetes than for any other mammal investigated to date. Mysticete basilar membranes are exceptionally wide and thin, implying that they are specialized for encoding extremely low frequencies, but there is insufficient data to determine whether low frequency specializations are present in mysticete cochlear ducts.

Major indicators for assessing the auditory capacity and level of aquatic adaptation of extinct species include the presence or absence of skull attachments, the number of tympano-periotic pedicles, and cochlear spiral morphometry. Protocetid cranial structure implies the earliest Archaeocetes were amphibious predators. Later dorudontine Archaeoceti were fully aquatic and had the cetacean characters of enlarged air sinuses and reduced attachments of the auditory bulla to the skull. Squalodonts from the late Oligocene have a nearly fully telescoped skull, a well-isolated tympano-periotic complex, and a cochlear spiral with both Type I and Type II characteristics, implying they were at least protoaquatic echolocators. Oligocene paleobalaenids clearly fit low frequency cochlear formats. It is likely that, auditorially, Cetacea are derived from a high-frequency form of mesonychid, but there is little evidence for echolocation in the Archaeoceti. It is suggested that scaling of cochlear structures may place mechanical limitations on the resonance characteristics of the basilar membrane and that size of the great whales limits their auditory capacity to lower frequencies.

Acknowledgments. Original research for this chapter was supported by the ARCS Foundation and NSF grant BNS8118072. Key specimens were obtained and processed through the efforts of

Barbara Burgess, Diane DeLeo Jones, Gregory Early, Joseph Geraci, James Gilpatrick, Richard Lammertson, Daniel Odell, William Perrin, James Mead, and Charles Potter. Radiographic studies were provided by Frank Starr, III. James Anderson, Arthur Rosenbaum, and Alan Walker provided knowledge about structure and opportunities for research. Douglas Whittington gave encouraging support and helpful criticisms at each stage. Extensive and insightful reviews of the manuscript were provided by Peggy Edds, Nelson Kiang, James Mead, Douglas Wartzok, and Alexander Werth, who also gave invaluable advice on paleontological literature and nomenclature. Reconstructions were produced with the cooperation of the Cochlear Implant Research Laboratory and the Eaton-Peabody Laboratory for Auditory Physiology. Lastly, I am particularly grateful to the organizers of this symposium, Arthur Popper, Douglas Webster, and Richard Fay, for providing the impetus and opportunity for many stimulating discussion with participants.

References

Amundin M, Cranford T (1990) Forehead anatomy of *Phocoena phocoena* and *Cephalorhynchus commersonii*: 3-dimensional computer reconstructions with emphasis on the nasal diverticula. In: Thomas JA, Kastelein RA (eds), Sensory Abilities of Cetaceans: Laboratory and Field Evidence. New York: Plenum Press, pp. 1–18.

Au WWL (1990) Target detection in noise by echolocating dolphins. In: Thomas JA, Kastelein RA (eds) Sensory Abilities of Cetaceans: Laboratory and Field Evidence. New York: Plenum Press, pp. 203–216.

Au WWL, Floyd RW, Penner RH, Murchison AE (1974) Measurement of echolocation signals of the Atlantic bottle-nosed dolphin *Tursiops truncatus* Montagu in open waters. J Acoust Soc Am 56:1280–1290.

Awbrey FT (1990) Concluding comments on cetacean hearing and echolocation. In: Thomas JA, Kastelein RA (eds) Sensory Abilities of Cetaceans: Laboratory and Field Evidence. New York: Plenum Press, pp. 427–433.

Barnes LG and Mitchell E (1978) Cetacea. In: Maglio VJ, Cooke HBS (eds) Evolution of African Mammals. Cambridge: Harvard Univ Press, pp. 582–602.

Barnes LG, Domning DP, Ray CE (1985) Status of studies on fossil marine mammals. Mar Mamm Sci 1:15–53.

von Békésy G (1960) Experiments in Hearing. EG Wever (trans). New York: McGraw-Hill.

Belkovich VM, Solntseva GN (1970) Anatomy and function of the ear in dolphins. US Gov Res Develop Rep 70(11):275–282 (read as eng summ).

Benham WB (1901) On the larynx of certain whales (*Cogia* (sic), *Balaenoptera* and *Ziphius*). Proc Zool Soc London 1:278–300.

van Bergeijk WA (1967) The evolution of vertebrate hearing. In: Neff WD (ed) Contributions to Sensory Physiology, Vol. 1. New York: Academic Press, pp. 1–41.

Boenninghaus G (1903) Das Ohr des Zahnwales zyugleich ein Beitrag zur Theorie der Schalleitung. Zool Gahrb (Anatomie) 17:189–360 (not read in original).

Boyden A, Gemeroy D (1950) The relative position of the Cetacea among the orders of Mammalia as indicated by precipitation tests. Zoologica 35:145–151.

Brill RL, Sevenich ML, Sullivan TJ, Sustman JD, Witt RE (1988) Behavioral evidence for hearing through the lower jaw by an echolocating dolphin, *Tursiops truncatus*. Mar Mamm Sci 4(3):223–230.

Brown AM, Pye JD (1975) Auditory sensitivity at high frequencies in mammals. Adv Comp Physiol Biochem 6:1–73.

Bruns V (1976) Peripheral Auditory Tuning for Fine Frequency Analysis by the CF-FM Bat, *Rhinolophus ferrumequinum*: 1. Mechanical Specializations of the Cochlea. J Comp Physiol 106:77–86.

Bruns V, Schmieszek ET (1980) Cochlear innervation in the greater horseshoe bat: Demonstration of an acoustic fovea. Hearing Res 3:27–43.

Bullock TH, Gurevich VS (1979) Soviet literature on the nervous system and psychobiology of cetaceans. Int Rev Neurol 21:47–127.

Bullock T, Ridgway S (1972) Evoked potentials in the central auditory system of alert porpoises to their own and artificial sounds. J Neurobiol 3:79–99.

Bullock TH, Grinnell AD, Ikezono E, Kameda K, Katsuki Y, Nomoto M, Sato O, Suga N, Yanagisawa K (1968) Electrophysiological studies of central auditory mechanisms in cetaceans. Z vergl Physiol 59:117–156.

Clark CW (1990) Acoustic behavior of mysticete whales. In: Thomas JA, Kastelein RA (eds) Sensory Abilities of Cetaceans: Laboratory and Field Evidence. New York: Plenum Press, pp. 571–584.

Cummings WC, Thompson PO (1971) Underwater sounds from the blue whale, *Balaenoptera musculus*. J Acoust Soc Am 50:1193–1198.

Dahlheim M, Ljungblad DK (1990) Preliminary hearing study on gray whales (*Eschrichtius robustus*) in the field. In: Thomas JA, Kastelein RA (eds) Sensory Abilities of Cetaceans: Laboratory and Field Evidence. New York: Plenum Press, pp. 335–346.

Davis RA, Jr (1972) Principles of Oceanography. Menlo Park: Addison-Wesley.

Dawson WW (1980) The Cetacean Eye. In: Herman LM (ed) Cetacean Behavior: Mechanisms and Functions. New York: John Wiley and Sons.

Diercks KJ, Trochta RT, Greenlaw RL, Evans WE (1971) Recording and analysis of dolphin echolocation signals. J Acoust Soc Am 49:1729–1732.

Domning DP (1982) Evolution of manatees: a speculative history. J Paleontol 56:599–619.

Dudok van Heel WH (1962) Sound and Cetacea. Neth J Sea Res 1:407–507.

Edds PL (1982) Vocalizations of the blue whale, *Balaenoptera musculus*, in the St Lawrence Rivers. J Mamm 63(2):345–347.

Edds PL (1988) Characteristics of finback, *Balaenoptera physalus*, vocalizations in the St Lawrence Estuary. Bioacoustics 1:131–149.

Evans WE, Prescott JH (1962) Observations of the sound production capabilities of the bottlenose porpoise: A study of whistles and clicks. Zoologica 47:121–128.

Feng W, Liang C, Wang J, Wang X (1990) Morphometric and Stereoscopic Studies on the Spiral and Vestibular Ganglia of *Lipotes vexillifer* (in press).

Firbas W (1972) Über anatomische Anpassungen des Hörorgans an die Aufnahme hoher Frequenzen. Monatsschr Ohr Laryn-Rhinol 106:105–156.

Fitzgerald E (1975) Dynamic mechanical measurements during the life to death transition in animal tissues. Biorheology 12:397–408.

Fleischer G (1976) Hearing in extinct cetaceans as determined by cochlear structure. J Paleontol 50:133–152.

Fleischer G (1978) Evolutionary principles of the mammalian middle ear. Adv Anat Embryol Cell Biol 55:1–70.

Fordyce RE (1977) The development of the Circum Antarctic Current and the evolution of the Mysticeti (Mammalia:Cetacea). Palaeogeog Palaeoclim Palaeoecol 21:265–271.

Fordyce RE (1980) Whale evolution and Oligocene southern ocean environments. Palaeogeog Palaeoclim Palaeoecol 31:319–336.

Fraser F, Purves P (1954) Hearing in cetaceans. Bull Br Mus Nat Hist 2:103–116.

Fraser F, Purves P (1960) Hearing in cetaceans: Evolution of the accessory air sacs in the structure and function of the outer and middle ear in recent cetaceans. Bull Br Mus Nat Hist 7:1–140.

Gaskin DE (1976) The Evolution, Zoogeography, and Ecology of Cetacea. Ocean Mar Biol Annu Rev 14:247–346.

Gingerich PD, Russell DE (1981) *Pakicetus inachus*, A new Archaeocete (Mammalia Cetacea) from the early-

middle Eocene Kuldana formation of Kohat (Pakistan). Cont Mus Paleont Univ of Mich 25:235–246.

Gingerich PD, Wells NA, Russell DE, Shah SM (1983) Origin of Whales in Epicontinental remnant seas: New Evidence from the Early Eocene of Pakistan. Science 220:403–406.

Gingerich PD, Smith BH, Simons EL (1990) Hind limbs of Eocene *Basilosaurus*: Evidence of feet in whales. Science 249:154–156.

Goodson AD, Klinowska M (1990) A proposed echolocation receptor for the bottlenose dolphin, (*Tursiops truncatus*): Modelling the receive directivity from tooth and lower jaw geometry. In: Thomas JA, Kastelein RA (eds) Sensory Abilities of Cetaceans: New York: Plenum Press, pp. 255–268.

Gray O (1951) An introduction to the study of the comparative anatomy of the labyrinth. J Laryng Otol 65:681–703.

Greenwood DG (1961) Critical bandwidth and the frequency coordinates of the basilar membrane. J Acoust Soc Am 33:1344–1356.

Greenwood DG (1962) Approximate calculation of the dimensions of traveling-wave envelopes in four species. J Acoust Soc Am 34:1364–1384.

Grinnell AD (1963) The neurophysiology of audition in bats: Intensity and frequency parameters. J Physiol 167:38–66.

Hall J, Johnson CS (1971) Auditory thresholds of a killer whale, *Orcinus orca* Linnaeus. J Acoust Soc Am 51:515–517.

Heffner R, Heffner H (1980) Hearing in the Elephant (*Elephas maximus*). Science 208:518–520.

Henry RW, Haldiman JT, Albert TF, Henk WG, Abdelbaki YZ, Duffield DW (1983) Gross anatomy of the respiratory system of the bowhead whale, *Balaena mysticetus*. Anat Rec 207:435–449.

Hinchcliffe R, Pye A (1968) The cochlea in Chiroptera: A quantitative approach. Int Audiol 7:259–266.

Hinchcliffe R, Pye A (1969) Variations in the middle ear of the Mammalia. J Zool 157:277–288.

Hosokawa H (1950) On the cetacean larynx with special remarks on the laryngeal sac of the sei whale and the aryteno-epiglottideal tube of the sperm whale. Sci Rep Whales Res Inst 3:23–62.

Ishihara Y, Saito T, Ito Y, Fujino M (1958) Structure of sperm and sei whale insulins and their breakdown by whale pepsin. Nature 181:1468–1469.

Iurato S (1962) Functional implications of the nature and submicroscopic structure of tectorial and basilar membranes. J Acoust Soc Am 34:1368–1395.

Jansen J, Jansen JKS (1969) The nervous system of Cetacea. In: Anderson HT (ed) The Biology of Marine Mammals, New York: Academic Press, pp. 175–252.

Johnson CS (1967) Sound detection thresholds in marine mammals. In: Tavolga WN (ed) Marine Bioacoustics. New York: Pergamon Press, 2:247–260.

Johnson CS (1968) Masked tonal thresholds in the bottlenosed porpoise. J Acoust Soc Am 44:965–967.

Kamminga CF, Engelsma FJ, Terry RP (1989) Acoustic observations and comparison on wild captive and open water *Sotalia* and *Inia*. Eighth Bienn Conf Biol Mar Mamm 33.

Kastelein RA, Zweypfenning RCVJ, Spekreijse H (1990) Anatomical and histological characteristics of the eyes of a month-old and an adult harbor porpoise (*Phocoena*). In: Thomas JA, Kastelein RA (eds) Sensory Abilities of Cetaceans: Laboratory and Field Evidence. New York: Plenum Press, pp. 463–480.

Kasuya T (1973) Systematic consideration of recent toothed whales based on the morphology of tympanoperiotic bone. Sci Rep Whales Res Inst 25:1–103.

Kellogg AR (1936) A Review of the Archeoceti. Carnegie Inst Wash Publ 482:1–366.

Kellogg WN (1959) Auditory perception of submerged objects by porpoises. J Acoust Soc Am 31:1–6.

Ketten DR (1984) Correlations of morphology with frequency for Odontocete cochlea: Systematics and Topology. PhD thesis, The Johns Hopkins University, Baltimore.

Ketten DR, Wartzok D (1990) Three-dimensional reconstruction of the dolphin cochlea. In: Thomas JA, Kastelein RA (eds) Sensory Abilities of Cetaceans: Laboratory and Field Evidence. New York: Plenum Press, pp. 81–106.

Kuzentzov VB (1990) Chemical sense of dolphins: quasiolfaction. In: Thomas JA, Kastelein RA (eds) Sensory Abilities of Cetaceans: Laboratory and Field Evidence. New York: Plenum Press, pp. 481–504.

Leatherwood S, Caldwell DK, Winn H (1976) Whales, Dolphins, and Porpoises of the Western North Atlantic: A Guide to Their Identification. NOAA Tech Rpt NMFS Circ 396, US Dept of Comm NOAA NMFS Seattle, Wash.

Leatherwood S, Reeves RR, Perrin WF, Evans WE (1982) Whales, Dolphins, and Porpoises of the Eastern North Pacific and Adjacent Arctic Waters: A Guide to Their Identification. NOAA Tech Rpt NMFS Circ 444 US Dept of Comm NOAA NMFS Seattle, Wash.

Lees S, Ahern JM, Leonard M (1983) Parameters influencing the sonic velocity in compact calcified tissues of various species. J Acoust Soc Am 74:28–33.

Lipps JH, Mitchell ED (1976) Trophic model for the adaptive radiations and extinctions of pelagic marine mammals. Paleobiology 2:147–155.

Lockyer C (1974) Investigation of the ear plug of the southern sei whale, *Balaenoptera borealis*, as a valid means of determining age. J Cons Int Explor Mer 36(1):71–81.

Long GR (1980) Some psychophysical measurements of frequency in the greater horseshoe bat. In: van den Brink G, Bilsen F (eds) Psychophysical, Psychological and Behavioural Studies in Hearing. Delft: Delft University Press.

Lowenstein JM (1987) Marine mammal evolution: The Molecular evidence. Sixth Bienn Conf Biol Mar Mamm 7:192.

Mackay RS (1987) Whale heads and ray diagrams. Mar Mamm Sci 3(3):283–285.

Mackay RS, Liaw HM (1981) Dolphin vocalization mechanisms. Science 212:676–678.

McCormick JG (1972) The physiology of hearing in cetaceans. In: Ridgway SH (ed) Mammals of the Sea: Biology and Medicine. Springfield: Charles C Thomas, pp. 731–747.

McCormick JG, Weaver EG, Palin G, Ridgway SH (1970) Sound conduction in the dolphin ear. J Acoust Soc Am 48:1418–1428.

McCormick JG, Weaver EG, Harrill JA, Miller HE (1975) Anatomical and physiological adaptations of marine mammals for the prevention of diving induced middle ear barotrauma and round window fistula. J Acoust Soc Am 58 Suppl 1 p S88.

McCormick JG, Wever EG, Ridgway SH, Palin J (1980) Sound reception in the porpoise as it relates to echolocation. In: Busnel R-G, Fish JF (eds) Animal Sonar Systems. New York: Plenum Press, pp. 449–467.

McKenzie DP (1970) Plate Tectonics and Continental Drift. Endeavour 29:39–44.

Mead JG (1975) Anatomy of the external nasal passages and facial complex in the Delphinidae (Mammalia:Cetacea). Smiths Contrib Zool 207:1–71.

Miller GS (1923) The telescoping of the cetacean skull. Smithsonian Misc Coll 76:1–67.

Mitchell ED (1989) A New cetacean from the late Eocene La Meseta Formation, Seymour Island, Antarctic Peninsula. Can J Fish Aquat Sci 46:2219–2235.

Møhl B, Andersen S (1973) Echolocation: High-frequency component in the click of the harbor porpoise (*Phocoena phocoena* L). J Acoust Soc Am 57: 1368–1372.

Moore PWB, Pawloski DA (1990) Investigations on the control of echolocation pulses in the dolphin (*Tursiops truncatus*) In: Thomas JA, Kastelein RA (eds) Sensory Abilities of Cetaceans: Laboratory and Field Evidence. New York: Plenum Press, pp. 305–316.

Moore PWB, Au WWL (1983) Critical ratio and bandwidth of the Atlantic bottlenose dolphin (*Tursiops truncatus*) J Acoust Soc Am Suppl 1:74.

Morgane PJ, Jacobs MS (1972) The comparative anatomy of the cetacean nervous system. In: Harrison RJ (ed) Functional Anatomy of Marine Mammals. New York: Academic Press, pp. 109–239.

Norris J, Leatherwood K (1981) Hearing in the Bowhead Whale, *Balaena mysticetus*, as estimated by cochlear morphology. Hubbs Sea World Rsch Inst Tech Rpt no 81-132:151–1549.

Norris KS (1964) Some problems of echolocation in cetaceans. In: Tavolga WN (ed) Marine Bio-Acoustics. New York: Pergamon Press, pp. 317–336.

Norris KS (1968) The evolution of acoustic mechanisms in odontocete cetaceans, In: Drake ET (ed) Evolution and Environment, pp. 297–324.

Norris KS (1969) The echolocation of marine mammals. In: Andersen HJ (ed) The Biology of Marine Mammals. New York: Academic Press.

Norris KS (1980) Peripheral sound processing in odontocetes. In: Busnel R-G, Fish JF (eds) Animal Sonar Systems. New York: Plenum Press, pp. 495–509.

Norris KS, Harvey GW (1974) Sound transmission in the porpoise head. J Acoust Soc Am 56:659–664.

Norris KS, Prescott JH, Asa-Dorian PV, Perkins P (1961) An experimental demonstration of echolocation behavior in the porpoise, *Tursiops truncatus* Montagu. Biol Bull 120:163–176.

Norris KS, Harvey GW, Burzell LA, Krishna Kartha DK (1972) Sound production in the freshwater porpoise, *Sotalia* cf *fluviatilis* Gervais and Deville and *Inia geoffrensis* Blainville, in the Rio Negro. Brazil Invest Cetacea 4:251–262.

Oelschläger HA (1986a) Comparative morphology and evolution of the otic region in toothed whales, Cetacea, Mammalia. Am J Anat 177(3):353–368.

Oelschläger HA (1986b) Tympanohyal bone in toothed whales and the formation of the tympano-periotic complex (Mammalia: Cetacea). J Morphol 188:157–165.

Oelschläger HA (1990) Evolutionary morphology and acoustics in the dolphin skull. In: Thomas JA, Kastelein RA (eds) Sensory Abilities of Cetaceans: Laboratory and Field Evidence. New York: Plenum Press, pp. 137–162.

Payne KB, Langbauer WJ, Jr, Thomas EM (1986) Infrasonic cells of the Asian elephant (*Elephas maximus*). Behav Ecol Soc Biol 18:297–301.

Payne KB, Tyack P, Payne RS (1983) Progressive changes in the songs of humpback whales (*Megaptera novaeangliae*). In: Payne RS (ed) Communication and Behavior of Whales. AAAS Selected Symposium Series. Boulder: Westview Press, pp. 9–57.

Pickles JO (1982) An Introduction to the Physiology of Hearing. London: Academic Press.

Pilleri G (1983) The sonar system of the dolphins. Endeavour New Series 7(2):59–64.

Pilleri G (1984) Concerning the ear of the narwhal, *Monodon monoceros*. Invest Cetacea 15:175–184.

Pilleri G, Gihr M, Kraus C (1986) Evolution of the echolocation system in cetaceans, a contribution to paleoacoustics. Invest Cetacea 18:13–130.

Pilleri G, Gihr M, Kraus C (1987) The organ of hearing in cetaceans 1: recent species. Invest Cetacea 20: 43–177.

Pollak GD (1980) Organizational and encoding features of single neurons in the inferior colliculus of bats. In: Busnel R-G, Fish JF (eds) Animal Sonar Systems. New York: Plenum Press.

Popov V, Supin A (1990a) Electrophysiological studies on hearing in some cetaceans and a manatee. In: Thomas JA, Kastelein RA (eds) Sensory Abilities of Cetaceans: Laboratory and Field Evidence. New York: Plenum Press, pp. 405–416.

Popov V, Supin A (1990b) Localization of the acoustic window at the dolphin's head. In: Thomas JA, Kastelein RA (eds) Sensory Abilities of Cetaceans: Laboratory and Field Evidence. New York: Plenum Press, pp. 417–426.

Popper AN (1980) Sound emission and detection by delphinids. In: Herman LM (ed) Cetacean Behavior: Mechanisms and Functions. New York: John Wiley and Sons.

Purves PE (1967) Anatomical and experimental observations on the cetacean sonar system. In: Busnel RG (ed) Whales, Dolphins, Animal Sonar Systems: Biology and Bionics. Laboratoire de Physiologie Acoustique pp. 197–270.

Purves PE, Pilleri GE (1983) Echolocation in Whales and Dolphins. London: Academic Press.

Pye A (1972) Variations in the structure of the ear in different mammalian species. Sound 6:14–18.

Reysenbach de Haan FW (1956) Hearing in whales. Acta Otolaryngol Suppl 134:1–114.

Ridgway SH (1972) Mammals of the Sea: Biology and Medicine. Springfield: Charles C Thomas.

Ridgway SH (1980) Electrophysiological experiments on hearing in odontocetes. In: Busnel R-G, Fish JF (eds) Animal Sonar Systems. New York: Plenum Press.

Ridgway SH, McCormick JG (1967) Anesthetization of porpoises for major surgery. Science 158:510–512.

Ridgway SH, McCormick JG, Wever EG (1974) Surgical approach to the dolphin's ear. J Exp Zool 188: 265–276.

Ridgway SH, Bullock TH, Carder DA, Seeley RL, Woods D, Galambos R (1981) Auditory brainstem response in dolphins. Proc Natl Acad Sci USA 78(3): 1943–1947.

Sales G, Pye D (1974) Ultrasonic Communication by Animals. New York: John Wiley and Sons.

Schevill WE, Watkins WA (1966) Sound structure and directionality in *Orcinus* (killer whale). Zoologica 51:71–76.

Schnitzler HU (1983) Fluttering target detection in horseshoe bats. J Acoust Soc Am 74:Suppl 1 S31–S32.

Schuknecht HF (1974) Pathology of the Ear. Cambridge: Harvard University Press.

Schuknecht HF, Gulya AJ (1986) Anatomy of the Temporal Bone with Surgical Implications. Philadelphia: Lea and Feibiger.

Silber GK (1986) The relationships of social vocalizations to surface behavior and aggression in the Hawaiian humpback whale (*Megaptera novaeangliae*). Can J Zool 64:2075–2080.

Simmons JA (1973) The Resolution of target range by echolocating bats. J Acoust Soc Am 54:157–173.

Solntseva GN (1971) Comparative anatomical and histological characteristics of the structure of the external and inner ear of some dolphins. Tr Atl Nauchno Issled Inst Rybn Khoz Okeanogr (read as eng summ).

Solntseva GN (1975) Morphofunctional aspects of the hearing organ in terrestrial semi-aquatic and aquatic mammals. Zool Zh 54(10):1529–1539 (read as eng summ).

Solntseva GN (1987) Direction of the evolutionary transformations of the peripheral portion of the acoustic analyzer in mammals from different habitats. Zh Obshch Biol 48(3):403–410 (read as eng summ).

Suga N (1983) Neural representation of bisonar (sic) information in the auditory cortex of the mustached bat. J Acoust Soc Am 74(S1):31.

Sukhoruchenko MN (1973) Frequency discrimination of dolphin (*Phocoena phocoena*). Fiziol Ah SSSR im IM Sechenova. 59:1205 (read as eng summ).

Sukhovskaya LI, Yablokov AV (1979) Morphofunctional characteristics of the larynx in balaenopteridae. Invest Cetacea 10:205–214.

Thomas J, Chun N, Au W (1988) Underwater audiogram of a false killer whale (*Pseudorca crassidens*). J Acoust Soc Am 84:936–940.

Thomas JA, Pawloski JL, Au WWL (1990) Masked hearing abilities in a false killer whale (*Pseudorca crassidens*). In: Thomas JA, Kastelein RA (eds) Sensory Abilities of Cetaceans: Laboratory and Field Evidence. New York: Plenum Press, pp. 395–404.

Thompson RKR, Herman LM (1975) Underwater frequency discrimination in the bottlenose dolphin (1-140 kHz) and in human (1-8 kHz). J Acoust Soc Am 57:943.

Thompson TJ, Winn HE, Perkins PJ (1979) Mysticete Sounds, In: Winn HE, Olla BL (eds) Behavior of Marine Animals, Current Perspectives in Research Volume 3: Cetaceans. New York: Plenum Press, pp. 403–431.

Van W, Utrecht L (1981) Comparison of accumulation patterns in layered dentinal tissue of some Odontoceti and corresponding patterns in baleen plates and ear plugs of balaenopteridae. Beaufortia 31(6):111–122.

Varnassi U, Malins DG (1971) Unique lipids of the porpoise (*Tursiops gilli*): Differences in triacyl glycerols and wax esters of acoustic (mandibular canal and melon) and blubber tissues. Biochem Biophys Acta 231:415.

Voronov VA, Stosman IM (1977) Frequency-threshold characteristics of subcortical elements of the auditory analyzer of the *Phocoena phocoena* porpoise. Zh Evol Biokh I Fiziol 6:719.

Watkins WA (1981) The activities and underwater sounds of fin whales. Sci Rep Whales Res Inst 33: 83–117.

Watkins WA, Wartzok D (1985) Sensory biophysics of marine mammals. Mar Mamm Sci 3:219–260.

Watkins WA, Tyack P, Moore KE, Bird JE (1987) The 20 Hz signals of finback whales (*Balaenoptera physalus*). J Acoust Soc Am 82:1901–1912.

West CD (1985) The relationship of the spiral turns of the cochlea and the length of the basilar membrane to the range of audible frequencies in ground dwelling mammals. J Acoust Soc Am 77(3):1091–1101.

Weston DE, Black RI (1965) Some unusual low-frequency biological noises underwater. Deep Sea Res 12:295–298.

Wever EG, McCormick JG, Palin J, Ridgway SH (1971a) The cochlea of the dolphin, *Tursiops truncatus*: General Morphology. Proc Natl Acad Sci USA 68(10):2381–2385.

Wever EG, McCormick JG, Palin J, Ridgway SH (1971b) The cochlea of the dolphin *Tursiops truncatus*: The basilar membrane. Proc Natl Acad Sci USA 68(11):2708–2711.

Wever EG, McCormick JG, Palin J, Ridgway SH (1971c) The cochlea of the dolphin *Tursios truncatus*: Hair cells and ganglion cells. Proc Natl Acad Sci USA 68(12):2908–2912.

Wever EG, McCormick JG, Palin J, Ridgway SH (1972) Cochlear structure in the dolphin *Lagenorhynchus obliquidens*. Proc Natl Acad Sci USA 69: 657–661.

Wood FG, Evans WE (1980) Adaptiveness and ecology of echolocation in toothed whales. In: Busnel R-G, Fish JF (eds) Animal Sonar Systems. New York: Plenum Press.

Yamada M (1953) Contribution to the anatomy of the organ of hearing of whales. Sci Rep Whales Res Inst 8:1–79.

Yeowart NS (1976) Thresholds of Hearing and Loudness for very low frequencies. In: Tempest W (ed) Infrasound and Low Frequency Vibration. London: Academic Press, pp. 37–64.

Zwislocki J (1981) Sound analyses in the ear: A history of discoveries. Am Sci 69:184–192.

Appendix I—Marine Mammal Divisions

This listing is provided as a general reference, since many species and groups mentioned in the text may be unfamiliar to most readers. Common names are listed for representative recent species; extinct divisions are designated by †. Geologic periods indicate the point in the fossil record at which the taxa were first clearly represented. Although conventional scientific names are used and ordered in an apparent hierarchy, the subdivisions are only relative and no attempt is made at formal classification rankings. Classification of marine mammals is undergoing continual revision, and as recently as 1989 a new family of Archaeoceti was proposed as were major changes in the distribution of families in the suborders Mysticeti and Odontoceti. For an accurate classification, the reader is referred to Barnes, Domning, and Ray (1985) and Mitchell (1989).

Throughout this chapter, the conventional terms dolphin, whale, and porpoise are used sparingly since they represent largely false distinctions. Whale actually relates to size and is correctly applied to both odontocetes and mysticetes. The term dolphin is used often to designate smaller beaked delphinids, and porpoise, nonbeaked phocoenids, but the distinctions are blurred and all of these animals porpoise. Consequently, the terms odontocete (toothed whales/dolphin/porpoise), mysticete (baleen/whalebone whales), and cetacean (all whales), although somewhat formal, are preferred.

Cetacea—Whales, dolphins, and porpoises
  Archaeoceti†—Early Eocene
    Protocetidae—Eocene (*Pakicetus*)
    Remingtonocetidae—early Eocene
    Basilosauridae—late Eocene
      Basilosaurinae (*Basilosaurus* = *Zeuglodon*)
      Dorudontinae (*Dorudon, Zygorhiza*)
      Kekenodontinae
  Odontoceti—Early Oligocene
    Squalodontoidea—shark-toothed stem odontocetes
      Agorophiidae†—late Oligocene
      Squalodontidae†—Miocene (*Squalodon*)
      Rhabdosteidae†—Miocene (*Rhabdosteus*)
      Squalodelphidae†—late Oligocene (*Squalodelphis*)

Delphinoidea—dolphins and small toothed whales
  Kentriodontidae†—early Miocene (*Kentriodon*)
  Albireonidae†—late Miocene
  Monodontidae—white whales

Monodon monoceros	Narwhal

  Phocoenidae—porpoises

Phocoena phocoena	Harbour porpoise

  Delphinidae—dolphins, coastal toothed whales, orcans

Delphinus delphis	Common dolphin
Feresa attenuata	Pygmy killer whale
Globicephala macrorhynchus	Short-finned pilot whale
Grampus griseus	Risso's dolphin
Lagenorhynchus albirostris	White-beaked dolphin
Stenella attenuata	Spotted dolphin
Stenella coeruleoalba	Striped dolphin
Stenella longirostris	Long-beaked spinner
Tursiops truncatus	Bottlenosed dolphin

Platanistoidea—river dolphins
  Acrodelphidae†—Miocene
  Iniidae

Inia geoffrensis	Amazonian boutu

  Lipotidae—beiji
  Platanistidae—Asian river dolphins
  Pontoporiidae—franciscana
Ziphioidea
  Ziphiidae—beaked whales
Physeteroidea
  Kogiidae—pygmy sperm whale
  Physeteridae—sperm whale, cachalot

Physeter catodon	sperm whale

Mysticeti—Miocene
  Aetiocetidae†—late Oligocene (*Aetiocetus*)
  Cetotheriidae†—Pliocene (*Paleocetus*)
  Balaenidae—right whales

Eubalaena glacialis	Northern right whale
Eubalaena australis	Southern right whale
Balaena mysticetus	Bowhead

  Neobalaenidae

Caperea marginata	Pygmy right whale

  Eschrichtiidae

Eschrichtius robustus	Grey

  Balaenopteridae—rorquals

Balaenoptera acutorostrata	Minke
Balaenoptera borealis	Sei
Balaenoptera edeni	Bryde's
Balaenoptera musculus	Blue
Balaenoptera physalus	Fin
Megaptera novaeangliae	Humpback

Desmostylia† — Miocene — sea horses
    Desmostylidae
    Paleoparadoxidae

Carnivora
  Fissipedia — Pleistocene — sea otters and sea minks†
    Mustelidae
  Pinnipedia — Oligocene — seals, sea lions, walrus
    Phocidae — true seals
    Otarioidea

    Otariidae
      Enaliarctinae†
      Otariinae — sea lions
    Odobenidae — walruses
Sirenia — Eocene — sea cows
    Prorastomidae†
    Protosirenidae†
    Trichechidae — manatees
    Dugongidae — dugongs, Steller's sea cow

36
Adaptations of Basic Structures and Mechanisms in the Cochlea and Central Auditory Pathway of the Mustache Bat

George D. Pollak

1. Introduction

The ability of bats to orient and successfully avoid obstacles in total darkness has been of interest to scientists for more than two centuries. Although audition has always been strongly associated with this ability, general acceptance of orientation by sound came only around 1940 with the elegant studies of Griffin and his colleagues (Griffin and Galambos 1941; Galambos and Griffin 1942; an excellent summary of this work is provided in Griffin 1958). They showed that bats are not only able to navigate through complex environments but they also can detect, identify and locate prey in the night sky by emitting loud ultrasonic calls and listening to the echoes that are reflected from nearby insects. Griffin (1944) coined the term echolocation to describe this form of biological sonar.

This chapter is concerned with cochlear and central auditory adaptations which have evolved specifically for echolocation in the mustache bat, one of the most specialized of all echolocating bats. My central theme, however, is based on the hypothesis that the ability to orient by means of sonar did not evolve de novo, but rather by enhancing a preexisting ability to perceive echoes, which is common to all higher vertebrates. I suggest this hypothesis for two related reasons. The first is that the ability to interpret echoes with a precision sufficient for acoustic orientation appeared at least four times in vertebrate evolution; once in certain birds, such as oil birds and cave swiftlets (Griffin 1958); a second time in the microchiropteran bats (Griffin 1958), a third time in megachiropteran bats of the genus Rousettus (Möhres and Kulzer 1956; Novick

1958), and a fourth time in porpoises (Kellogg 1961). The second reason is that the perception of echoes can be enhanced to a substantial degree with practice and learning in organisms that normally do not use sonar for orientation. This is well illustrated by humans who can be trained to detect and recognize objects from echoes which they generate by tapping a cane or using a clicker (Supra et al. 1944; Cotzin and Dallenbach 1950; Rice 1967).

A corollary to the above hypothesis is that the structural and functional underpinnings of echolocation also evolved from modifications of features that were invented before microchiropteran bats took to the night skies. To be sure, specialized features are present in the auditory systems of bats, some of which are the topic of this chapter. It is noteworthy, however, that only one specialized feature is universally present in all echolocating animals. That feature is an enhanced ability to respond to each of two sounds presented in rapid succession, simulating the reception of an emitted call and an echo shortly thereafter (Grinnell 1963, 1970, 1973; Grinnell and Hagiwara 1972a,b; Bullock et al. 1968). Except for a fast recovery of responsiveness, I know of no structural or mechanistic feature that is unique to all echolocating animals, and which, therefore, could be considered a necessary addition for the perception of echoes. On the contrary, the overwhelming impression of the auditory systems of all bats that have been studied is one of structural and functional communality with other mammals (Pollak and Casseday 1989).

In the sections below I will begin with a brief description of the orientation signals emitted by mustache bats, how the bat manipulates the

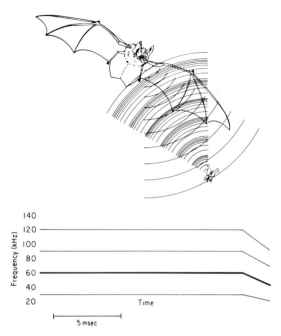

FIGURE 36.1. At the top is a drawing of an echolocating mustache bat and the frequency modulations in the echo due to the wing movements of a nearby moth. The emitted constant frequency component is depicted as a series of regularly spaced waves whereas the echo reflected from the moth is shown as a series of irregularly spaced waves (i.e., the frequency modulations) that are repeated periodically. Below is a time-frequency analysis (sonogram) of one echolocation call. Each call has a relatively long constant frequency component that can be as long as 30 msec and is terminated with a brief frequency-modulated portion. Four harmonics are usually present but the 60 kHz second harmonic is always dominant, as indicated by the thicker line.

frequency of its calls and the significance of these behavioral manipulations for the extraction of certain types of information. I will then describe some cochlear and neural specializations of the mustache bat and discuss how special adaptations enhance the processing of certain features of acoustic signals and what competitive advantages accrue to such processing. Throughout I will also present reasons for concluding that the adaptations evolved from a basic mammalian auditory system, and why the specialized aspects of the auditory system represent exaggerated features of a basic design present in all mammals.

2. The Doppler Based Biosonar System of the Mustache Bat

Mustache bats belong to a nontaxonomic group known as the long constant frequency bats. Other members of this group include the horseshoe bats. The name derives from the type of echolocation calls they emit. The calls of the mustache bat are characterized by an initial long constant frequency (CF) component, effectively a tone burst, that has a duration up to 30 msec, and each call is terminated by a 2 to 4 msec, downward sweeping frequency modulated (FM) component (Novick and Vaisnys 1964; Henson et al. 1980; Suga 1984). The CF component of the mustache bat's orientation call is emitted with four harmonics, at 30, 60, 90, and 120 kHz (Fig. 36.1). Most energy is in the 60 kHz second harmonic. The FM of the second harmonic sweeps from about 60 kHz down to about 45 kHz, and is used for target ranging, and probably for detecting other target attributes (Simmons et al. 1975). The key feature of this biosonar system, however, is the long 60 kHz CF component because it is through this frequency that this animal "sees" much of its world.

As bats pursue their targets, they emit a steady stream of orientation calls that the animal modifies in various ways at different stages of the pursuit. The most remarkable modification is that mustache bats, and other long CF bats such as horseshoe bats, adjust the frequency of the CF component of their pulses to compensate for Doppler shifts in the echoes that reach their ears (Schnitzler 1967; Schnitzler 1970; Schuller et al. 1974; Simmons 1974; Henson et al. 1980; Henson et al. 1982) (Fig. 36.2). This behavior, called Doppler-shift compensation, was discovered by Hans-Ulrich Schnitzler (1967) and is the expression of an extreme sensitivity for motion.

The main advantage of using a "long" CF signal and compensating for Doppler shifts is that it enhances the ability to hunt in acoustically cluttered environments. The reasons are that the bat's ability to distinguish a fluttering insect from background objects is improved as is its ability to recognize and distinguish among the various insects which exist in its hunting territory (Neuweiler 1983, 1984, 1990). All long CF/FM bats that have

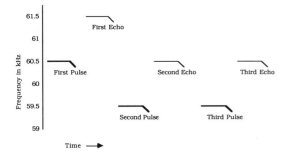

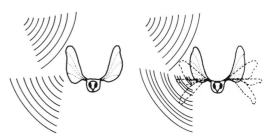

FIGURE 36.2. Schematic illustration of Doppler-shift compensation in a flying mustache bat. The first pulse is emitted at about 60.5 kHz, and the Doppler shifted echo returns with a higher frequency, at about 61.5 kHz. The mustache bat detects the difference between the pulse and echo and lowers the frequency of its subsequent emitted pulses by an amount almost equal to the Doppler-shift. Thus the frequencies of the subsequent echoes are held constant and return at a frequency very close to that of the first emitted pulse.

FIGURE 36.3. Drawings to illustrate the echoes reflected from an insect in which the wings are stationary (left) and from an insect whose wings are in motion (right). Note that the echo returning from the nonflying moth is simply an unmodulated tone, whereas the echo reflected from a fluttering moth has pronounced frequency modulations due to the Doppler shifts induced by the velocity of the moving wings.

been studied hunt for flying insects in dense foliage beneath the forest canopy (Bateman and Vaughn 1974; Neuweiler 1983, 1984, 1990). As they fly in this environment, the echo CF component from stationary background objects, such as trees, leaves, bushes etc., is Doppler shifted upward due to the relative motion of the bat towards stationary background (Trappe and Schnitzler 1982). Long CF bats compensate for the Doppler shifts in the echoes by lowering the frequency of subsequent emitted pulses by an amount nearly equal to the upward frequency shift in the echo. Consequently, these bats clamp the echo CF component and hold it within a narrow frequency band that varies only slightly from pulse to pulse. As discussed in subsequent sections, the mustache bat devotes a large portion of its auditory system to processing the frequency at which it clamps the echo CF component.

However, flight speed differences between the bat and stationary objects are not the only source of Doppler shifts that would return to the ear of a bat hunting in the night sky. A small fluttering insect that happens to cross the path of mustache bat will generate echoes substantially different from those of trees and other background objects. As the emitted CF component strikes and is reflected from the insect, the motion of the wings creates Doppler shifts which vary with the velocity, and hence the movements of the wings (Schnitzler et al. 1983: Schuller 1984). These Doppler shifts create periodic frequency modulations that are superimposed on the CF component of the echoes (Figs. 36.1 and 36.3). In addition, the wing motion, or flutter, presents a reflective surface area of alternating size, thereby also creating periodic amplitude modulations in the echo CF component. Thus the echoes from fluttering insects return with a center frequency close to the frequency of echoes from stationary background objects, but are distinguished from the echoes of background objects by the rich pattern of frequency and amplitude modulations. There is now substantial evidence that the periodic frequency and amplitude changes allow the bat to distinguish fluttering insects from background and provide the information for recognizing different insect types (Goldman and Henson 1977; Schnitzler et al. 1983; Schnitzler and Fleiger 1983: Link et al. 1986; von der Emde 1988). By compensating for Doppler shifts, then, the bat ensures itself that modulated echoes will always have a center frequency to which a large number of neural elements are tuned. The large neural representation of this frequency in combination with the cochlear specializations described next, provide the neural substrate for the fine resolution of target features.

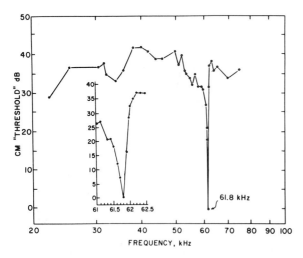

FIGURE 36.4. Cochlear microphonic audiogram of mustache bat. Each point represents the "threshold" response of cochlear microphonic potential for that frequency. The very sharp tuning of the audiogram is apparent. Insert shows expanded view of threshold changes around 61.8 kHz.

3. Adaptations of the Peripheral Auditory System for Processing 60 kHz

Because mustache bats "see" much of their world through a narrow band of frequencies around 60 kHz, many of the specializations in their auditory system are designed to process the 60 kHz CF component of the echo (e.g., Suga 1978; Pollak and Casseday 1989). The special adaptations are particularly evident in the cochlea. The first evidence that the mustache bat's cochlea is adapted for processing 60 kHz came from monitoring cochlear microphonic responses (Pollak et al. 1972). This study showed that the cochlear microphonic audiogram has a region of high sensitivity around 60 kHz where the tuning of the audiogram is remarkably sharp (Fig. 36.4). The sound pressure required to elicit a response at the tuned frequency is up to 40 dB lower than the sound pressures needed to elicit a response of the same amplitude with sounds just a few hundred Hertz higher or lower in frequency. The audiogram is of particular interest because the tuned frequency of the audiogram and the frequency at which mustache bats clamp their echo CF component when compensat-

ing for Doppler shifts are nearly identical (Henson et al. 1980, 1982). The functional consequences of this coincidence will be discussed in later sections.

In subsequent studies, Suga and his colleagues (Suga et al. 1975; Suga and Jen 1977) showed that the sharp tuning of the cochlear microphonic audiogram results from a pronounced cochlear resonance. The resonance is readily seen in cochlear microphonic responses to brief tone bursts at or around 60 kHz which exhibit rise and decay times much longer than those of the acoustic signal. For example, presentation of a brief tone burst at a frequency a few hundred Hertz below the resonance frequency evokes a cochlear microphonic response with a frequency equal to the frequency of the tone, but the frequency of the prolonged decay is always at the frequency of the sharply tuned notch of the audiogram. In short, frequencies within a few kilohertz of the resonant frequency also create motion at the place of resonant frequency, which continues after the termination of the sound. These sustained oscillations, or ringing, are reflected as the prolonged decay of the evoked cochlear microphonic potentials.

Recent studies suggest that active processes are at least in part responsible for the resonance. The evidence is from studies of Kössl and Vater (1985a) who showed that otoacoustic emissions can be recorded from the external ear canal of mustache bats during continuous and after transient acoustic stimulation. The frequency of the emissions, moreover, is always very close to the resonant frequency of the bat's cochlea and can have amplitudes as great as 70 dB SPL. Otoacoustic emissions at other frequencies are either nonexistent or are too small to measure with the equipment used. Of interest is that manipulations such as cooling the ear or anesthetizing the animal have pronounced effects on both the frequency of the emission and its amplitude. This suggests that both the production of the resonance and the exact frequency it assumes are metabolically dependent, and thus the result of some active process.

It is important to stress that the mechanisms that produce the cochlear resonance, and hence the sharp tuning of the cochlear microphonic audiogram and the otoacoustic emissions, are poorly understood. The main reason is that there is little available information about cochlear functioning in the mustache bat. There have, for example, been

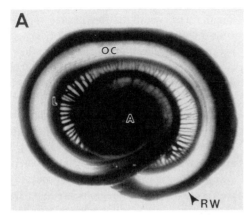
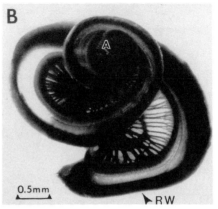
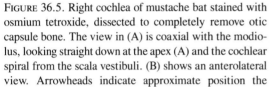

FIGURE 36.5. Right cochlea of mustache bat stained with osmium tetroxide, dissected to completely remove otic capsule bone. The view in (A) is coaxial with the modiolus, looking straight down at the apex (A) and the cochlear spiral from the scala vestibuli. (B) shows an anterolateral view. Arrowheads indicate approximate position the round window (RW), which is under the basilar partition in the intact specimen. OC, organ of Corti; L, spiral ligament. Note that the ligament is greatly thickened in the upper basal turn (in region where label is positioned) as compared to the thinner structure seen basal to that region. (From Zook and Leake 1989.)

no measurements of membrane stiffness, nor have there been studies of the mechanics or cellular physiology of the mustache bat's cochlea. Investigators have focused, instead, on anatomical features. They have revealed morphological specializations in the basal region where 60 kHz is represented, which in turn have led to speculations about the structural basis of the sharp tuning.

3.1 Anatomical Features of the Basal Region of the Mustache Bat's Cochlea

Even before the resonant properties of the cochlea were discovered, Henson (1967, 1970) noted the remarkable size of the mustache bat's inner ear. The bony otic capsule, which unlike many other mammals is not embedded in the petrosal bone, lies in the cranial cavity loosely connected with skeletal elements and occupies roughly a third of the volume of the brain case. The cochlea is coiled about three turns and has an average length of about 14 mm (Henson 1978; Kössl and Vater 1985b; Zook and Leake 1989). The major portion of the cochlea is devoted to the basal turn, which together with hook region (the most basal portion of the basilar membrane) encompasses about 10 mm, or more than two-thirds, of the total cochlear length.

Numerous features of the basal turn are noteworthy. The most apparent are the exceptional dimensions of the round window and cochlear aqueduct, both of which are proportionately much larger than in other mammals (Henson 1970; Henson and Pollak 1973). The curvature of the basal cochlear also has an unusual feature (Henson 1978; Kössl and Vater 1985b; Henson and Henson 1988; Zook and Leake 1989). About midway along the basal turn, the basilar membrane is almost straight, in contrast to the degree of curvature that characterizes the remainder of its length (Fig. 36.5). Along the straight region the basilar membrane is nearly flat, whereas adjacent sectors are steeply tilted. Coincident with the straight region is a pronounced enlargement of the stria vascularis and spiral ligament (Fig. 36.7). The region of their maximal size is centered in the straight region, which lead the Hensons (Henson and Henson 1988) to suggest that the straight region may in fact be created by the enlarged spiral ligament that distorts the normal curvature of the basilar membrane. The spiral ligament is also of interest due to the presence of contractile proteins in its anchoring cells (Henson et al. 1984, 1985). Thus, they suggest that the greater number of tension generating elements in the enlarged spiral ligament produce an increased stiffness of the basilar membrane in the straight region (Henson et al. 1984, 1985; Henson and Henson 1988). I shall return to this point below.

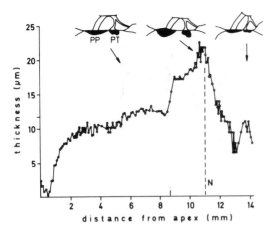

FIGURE 36.6. Change in basilar membrane thickness along cochlear length in the mustache bat. Upper part of figure shows schematic drawings of radial sections through organ of Corti in the hook region (right), the region slightly apical to the notch (indicated by N and is the narrowest point of scala vestibuli) which is in the middle drawing, and the second half turn (left). The two regions of thickened basilar membrane, the pars pectinata (PP) and pars tecta (PT), are blackened. Membrane thickness was measured from pars pectinata and plotted along the length of the cochlea. Note the minimum thickness in the region at about 13 mm from apex, and the maximum thickness at 10.8 mm near the notch (N). Also noteworthy is the abrupt decrease in membrane thickness at about 8.5 mm from the apex. (Adapted from Kössl and Vater 1985b.)

The scala vestibuli is also exceptionally large and exhibits a number of changes in dimensions within the basal turn. In about the middle of the basal turn, just basal to the straight region, the scala bends sharply and narrows substantially at that point (Kössl and Vater 1985b; Henson and Henson 1988). Kössl and Vater (1985b) refer to the point of narrowing as the "notch." Further apically the scala vestibuli broadens again, and beginning in the second turn it suddenly narrows and then remains uniformly small to the end of the apical turn. In contrast to the scala vestibuli, the scala tympani has a large volume in the most basal portion of the cochlea and then diminishes progressively from base to apex (Henson and Henson 1988).

Perhaps the most important cochlear feature is the variation in basilar membrane dimensions along the basal turn. There are marked differences in the changes of width and thickness along its length. The width of the basilar membrane is almost constant from base to apex (Henson 1978), a trend that differs considerably from the progressive increase in width seen in most other mammals. In contrast to the uniformity of width, Kössl and Vater (1985b) found substantial changes in basilar membrane thickness along its length, which are shown in Figure 36.6. The mammalian basilar membrane is subdivided into an inner pars tecta and an outer pars pectinata. Variations in the thickenings of the basilar membrane, especially of the pars pectinata in apical regions, are found in rodents and are associated with regions of enhanced sensitivity for low frequencies (Webster 1961; Pye 1965; Ehret and Frankenreiter 1977). In mustache bats, there are sudden changes in the thickness of the pars pectinata in the basal turn, as illustrated in Figure 36.6. Taking the hook region as a starting point, the basilar membrane is relatively thin and becomes progressively thicker apically. The thickness reaches a maximum at a point close to the notch, i.e., the place where the scala vestibuli suddenly narrows. The membrane remains thick for about 2 mm apically, which encompasses the straight region, and then exhibits a sudden reduction in thickness. Farther apically there is a gradual reduction in thickness.

3.2 Two Regions of the Basal Cochlea Are Densely Innervated

Before discussing some of the functional consequences of the features mentioned above, I shall describe the cochlear innervation and then how frequency is mapped on the cochlear partition. These features are both dramatic and provide reference points for associating regional variations in cochlear morphology with the frequency represented in those regions.

The hair cells in the mustache bat's organ of Corti are organized as one row of inner and three rows of outer hair cells, as they are in other mammals (Henson 1978; Kössl and Vater 1985b; Zook and Leake 1989). Neither the inner nor outer hair cells exhibit variations in density along the length of the cochlear partition (Fig. 36.7), and thus both cell types are uniformly distributed along the organ of Corti (Zook and Leake 1989).

Both the outer and inner hair cells receive afferent and efferent innervation (Bishop and Henson

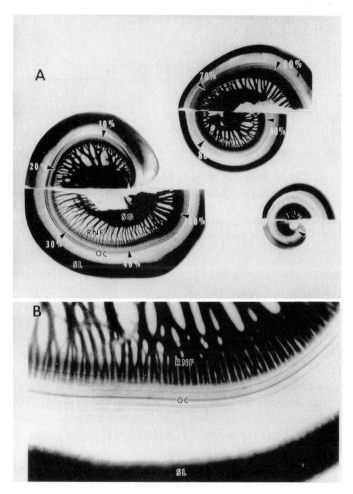

FIGURE 36.7. (A) Surface preparation of a right cochlea to demonstrate neuronal density along cochlear spiral. Basilar membrane length was measured at 10% intervals from base to apex. Note the striking increase in density of radial nerve fibers (RNF) in the region approximately 45 to 5% from the basal extreme of the cochlea. Also note the straight region of the organ of Corti (OC) just basal to this area of increased density, at a point about 40 to 45% from the base. SL, spiral ligament and stria vascularis; SG, spiral ganglion. (B) Higher magnification of the region 40 to 50% from the base where the transition from low innervation density (at left of micrograph) to maximum innervation density (at right) occurs. Note also that the innervation density increases just apical to the straight region, which is apparent although not labeled. (From Zook and Leake 1989.)

1988). Turning first to the outer hair cells, each cell is innervated by one efferent fiber which makes a characteristic synaptic connection on the center of the base of the cell (Bishop and Henson 1988). The efferent fiber is surrounded by a ring of six or seven afferent fibers. This arrangement is uniform in all three rows of outer hair cells, from base to apex. The innervation of the outer hair cells is unique, in that the outer hair cells of most other mammals are contacted by many efferent fibers

with an innervation density that varies along the length of the cochlea and in different rows.

The innervation of the inner hair cells in the basal turn is very different from the innervation of the outer hair cells (Henson 1978; Kössl and Vater 1985b; Zook and Leake 1989). Two regions of dense innervation are apparent (Fig. 36.8). The first is at the beginning of the basal turn, near the round window, and extends about 2.5 mm. The second is at the apical end of the basal turn and also

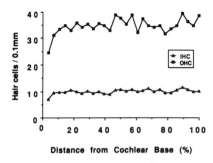

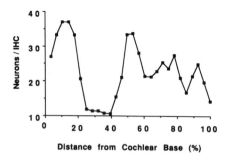

FIGURE 36.8. Left: Cytocochleogram shows the small variation in the numbers of inner (IHC) and outer hair cells (OHC) as a function of cochlear region. The number of inner hair cells per 0.1-mm segment is represented by triangular data points and the outer hair cells counts are indicated by square symbols. Right: The number of neurons per IHC is shown as a function of cochlear position. Note that in the sparsely innervated region at 25 to 40%, each IHC receives only 10 to 12 fibers, whereas in the regions of maximum innervation density there are more than 30 neurons per hair cell. (From Zook and Leake 1989.)

extends for about 2.5 mm. An area of sparse innervation separates these two densely innervated regions. Since there is a constant number of inner hair cells along the cochlea, the large number of fibers in the two densely innervated regions, and the smaller number of fibers in the sparsely innervated region, must both innervate the same number of inner hair cells per unit length of membrane. Thus, each hair cell in the densely innervated regions is contacted by 30 to 37 afferent fibers whereas each hair cell in the intervening sparsely innervated region is contacted by only 10 to 12 afferent fibers (Fig. 36.8) (Zook and Leake 1989).

3.3 The Frequency-Place Map of the Mustache Bat's Cochlea

Frequency mapping studies (Kössl and Vater 1985b; Zook and Leake 1989) have shown that there is an orderly representation of frequency along the basilar membrane: high frequencies are represented basally and lower frequencies are represented in progressively more apical regions, an arrangement similar in principle to that observed in other mammals. There is, however, a distortion of frequency representation in the basal cochlea associated with the pattern of innervation densities (Fig. 36.9). The first, most basal, region of dense innervation represents frequencies from about 110 kHz to about 70 kHz. The highest frequencies are represented most basally, in the hook, whereas the lowest frequency (70 kHz) is represented apically, close to the beginning of sparse innervation. Thus

the major portion of the first densely innervated region largely represents the third harmonic components (92 to 70 kHz) of the biosonar signals. The sparsely innervated region, which comprises about 20% of the basilar membrane, represents frequencies from about 70 kHz to 64 kHz.

The most important feature is that the second, more apical region of dense innervation represents the resonant frequency of the cochlea at 60 kHz (Fig. 36.9). Since the second region of dense innervation extends for about 2 mm, the resonant frequency is represented by about 15% of the total length of the cochlear partition.

Another notable feature is the representation of frequencies from about 60 to 45 kHz. These frequencies comprise the terminal FM component of the second harmonic and are mapped on a 2 to 3 mm length of basilar membrane just apical to the representation of the resonance frequency (Kössl and Vater 1985b). Thus, frequencies ranging from about 62 to 45 kHz, which comprise the CF and FM components of the second harmonic, are represented by approximately 5 mm, or about 35% of the total length of the basilar membrane.

3.4 Speculations on the Functional Consequences of the Anatomical Specializations

As mentioned above, there is no satisfactory explanation of how the anatomical features described above translate into functional properties, although a number of hypotheses have been proposed. Kössl and Vater (1985a,b, 1990) suggest that a number of features basal to the resonant segment may be important for the production of the sharp tuning and the otoacoustic emissions. One of the features they emphasize is the discontinuities in the scala vestibuli and basilar membrane in the region of the notch. Their idea is that the abrupt changes of anatomical features at the notch create a change of acoustic impedance, and thus a reflection point for wave motion. As a traveling wave reaches the notch, part of its energy is transferred to more apical regions but a portion is reflected back through the oval window. Reflections of this sort could then propagate back through the oval window and emerge as otoacoustic emissions. Unexplained in this hypothesis is why the resonant frequency is seen in the otoacoustic emissions and not the higher frequencies which are represented in the region of the notch, since the notch is the putative locus of the reflections.

The feature that has the clearest significance is the sudden change in basilar membrane thickness at the point where the resonant segment begins basally. In the straight region, just basal to the resonant segment, the basilar membrane should be exceptionally stiff. The increased stiffness is thought to be due to both the increase in membrane thickness and the enlargment of the spiral ligament in this region. The enlarged spiral ligament has a greater number of contractile elements which should exert a larger radial force on the basilar membrane in the straight region than in other regions where the spiral ligament is smaller, and has fewer contractile elements. Thus, Kössl and Vater (1985b) suggest that as a traveling wave propagates apically, the amplitude of displacement should increase in the resonant segment. The reason is that small movements and large forces in the basal, stiff, straight segment should be transformed into large movements and small forces in the more compliant resonant segment, apically.

Perhaps the most perplexing question concerns the mechanism(s) of the active process(es) that underlie the resonance at 60 kHz. For reasons given above, an active process of some type most likely contributes to the resonance, but an active process that can function at 60 kHz is difficult to visualize. Below I suggest two possibilities and point out why current thinking finds both unlikely.

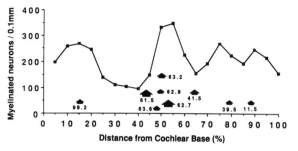

FIGURE 36.9. The cochlear location and distribution of spiral ganglion cells labeled following HPR deposits in various frequency regions of the cochlear nucleus in nine mustache bats. The best frequency at the deposit site is indicated by the number adjacent to each arrow. Note that in five cases in which the best frequency in the cochlear nucleus was in the narrow band of frequencies between 61 and 63 kHz (the filter frequency of the particular bat) the labeled ganglion cells all fell within the region of increased innervation density located 45 to 55% from the base. (From Zook and Leake 1989.)

The first relates to the regional amplification of basilar membrane motion discussed above and the recent discovery of the ability of outer hair cells to contract rapidly. It seems reasonable to suppose that larger excursions of the membrane generate proportionately larger changes in the membrane potentials in outer hair cells. Outer hair cells have two forms of motility that are influenced by membrane potential: a slow motility which is driven by contractile proteins, and thus is ATP dependent, and a fast motility which does not require contractile proteins and is not ATP dependent (Ashmore 1987). If one assumes that the mustache bat's outer hair cells have a fast motility at 60 kHz, then the amplified motion of the basilar membrane, produced by the compliance change at the junction of the straight region and resonant segment, would generate an enhanced depolarization of the outer hair cells in the resonant segment of the cochlea. The enhanced depolarization, in turn, could increase the contraction amplitude of the outer hair cells, which, in turn, could further augment the motion of the basilar membrane. It should be noted, however, that an ability of outer hair cells to contract at a rate of 60,000 times per second has yet to be demonstrated. Moreover, an enhancement of membrane motion by rapid contractions of outer hair cells has been questioned by some investigators (Roberts et al. 1988).

The second possibility is that the hair cells in the resonant region are electrically tuned to 60 kHz. Electrical tuning was invented early in vertebrate evolution and is retained by a wide variety of animals. The frequency selectivity of certain types of electroreceptors of fishes (Zakon 1986), and the hair cells of some amphibians (Lewis and Hudspeth 1983; Pitchford and Ashmore 1987), reptiles (Crawford and Fettiplace 1981) and birds (Fuchs et al. 1988) is achieved largely through the cell's electrical properties. Electrical tuning in both electroreceptors and hair cells is a consequence of the interplay between voltage-sensitive K^+ channels, Ca^{2+} activated K^+ channels and voltage sensitive Ca^{2+} channels (Lewis and Hudspeth 1983; Art and Fettiplace 1987; Hudspeth and Lewis 1988a,b; Roberts et al. 1988). The kinetics of the channels and the number of channels per cell are two of the major factors that determine the cell's resonant frequency. By adjustments of these features, different animals presumably can fine tune the resonance frequencies and sharpness of tuning of their receptor cells to closely match the frequencies that are of importance for survival. In birds and turtles, the tuned frequency of each hair cell is matched to the location of the cell along the cochlea (Fuchs et al. 1988) or basilar papilla (Art and Fettiplace 1987), whereas in fishes, the tuning of the electroreceptors is closely matched to the frequency of the fish's own electric organ discharge.

Given these features, it seems reasonable to at least suggest that the hair cells in the overrepresented region of the mustache bat's cochlea may be electrically tuned. Electrical filters of the sort seen in the hair cells of other animals would sharpen the tuning at 60 kHz and produce exactly the sort of ringing observed in the cochlear microphonic responses evoked by tone bursts around 60 kHz.

The electrical tuning hypothesis for mustache bats, however, suffers from a major weakness. The weakness is that the electrical tuning of the hair cells that have been studied to date, as well as the electroreceptors that have been studied, is only to low frequencies, i.e., less than a few kilohertz. Additionally, theoretical arguments, based on maximum channel activation times and maximum channel densities, suggest that electrical tuning above about 5 kHz is unlikely (Roberts et al. 1988).

In summary, the mustache bat's cochlea exhibits dramatic features both physiologically and anatom-

ically. The pronounced anatomical features must surely contribute to the sharp tuning and to the production of the resonance, but exactly how the anatomical features affect the mechanical and electrical actions of the cochlea has not been shown experimentally and thus is not well understood. Moreover, although seemingly compelling arguments can be advanced to show why fast motility and electrical tuning of hair cells are unlikely explanations for the cochlear resonance, it seems equally likely that some form of active process must exist and must function at what appears to be the impossibly high rate of 60 kHz. But whatever those processes might turn out to be, it is noteworthy the most of the features of the mustache bat's cochlea, whether structural or functional, are also seen in other mammals, but in less pronounced forms (Henson et al. 1984, 1985; Kössl and Vater 1985a,b; Zook and Leake 1989; Neuweiler 1990). Elucidating how the mustache bat's cochlea works, then, is an important challenge for auditory physiology and should reveal much about the cochlear mechanisms of all mammals.

4. Response Properties of Auditory Nerve Fibers Reflect Cochlear Features

The cochlear resonance and the large number of afferent fibers that innervate the resonant region of the organ of Corti create exceptionally narrow tuning curves in a large population of auditory nerve fibers (Suga et al. 1975; Suga and Jen 1977; Kössl and Vater 1990). Both features are illustrated quantitatively in Figure 36.10, which shows the distribution of Q_{10db} values among the population of auditory nerve fibers. Fibers whose best frequencies correspond to the resonant frequency of the cochlea have Q_{10db} values that range from 80 to 500, and are often referred to as "filter units" because of the limited range of frequencies to which they respond (Neuweiler and Vater 1977; Pollak and Casseday 1989). In contrast, units having best frequencies above or below the resonant frequency commonly have small Q_{10kdb} values that range from 3 to 18. The complete spectrum of frequencies to which the cochlea responds is, of course, rerepresented in the auditory nerve, but

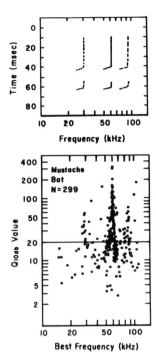

FIGURE 36.10. Sonograms (upper figures) and distributions of Q_{10dB} values of neurons (lower graphs) in the auditory nerve of the mustache bat. Note the mustache bat (C) has a pronounced over-representation of neurons having large Q_{10dB} values at 60 kHz, a frequency corresponding to the dominant CF component of this species. A smaller proportion of neurons at 30 and 90 kHz also have high Q_{10dB} values. (Adapted from Suga and Jen 1977.)

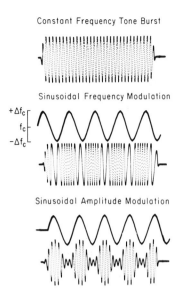

FIGURE 36.11. Illustration of fine structure of a tone burst (upper record), a sinusoidally frequency modulated (SFM) burst (middle record), and a sinusoidally amplitude modulated (SAM) burst (lower record). The tone burst in upper record is a shaped sine wave, where the frequency, or fine structure, of the signal remains constant. The fine structure of the SFM burst changes sinusoidally in frequency. The modulation waveform is shown above the record of the signal's fine structure. In the SAM burst, the fine structure of the signal is a constant frequency, but the amplitude varies with the sinusoidal modulating waveform.

the representation is greatly weighted in favor of the filter units.

4.1 Sharply Tuned Filter Units Are Sensitive to Amplitude and Frequency Modulations

The sharp tuning of the 60 kHz neurons appears to be a specialization for the fine analysis of the frequency modulation patterns imposed on the echo CF component from a fluttering insect (Suga et al. 1975; Suga and Jen 1977; Schuller and Pollak 1979; Schuller 1979, 1984; Pollak and Schuller 1981; Bodenhamer and Pollak 1983). Suga and Jen (1977) were the first to describe the coding of such modulation patterns in the auditory nerve of the mustache bat. They mimicked insect echoes by modulating the frequency or amplitude of a 60 kHz

carrier tone with low frequency sinusoids of 50 to 100 Hz (Fig. 36.11). The low frequency sinusoid is the modulating waveform, and is distinguished from the 60 kHz carrier frequency. This arrangement created either sinusoidally amplitude modulated (SAM) or sinusoidally frequency modulated (SFM) signals (Fig. 36.5). With SFM signals the depth of modulation, the amount by which the frequency varies around the carrier, mimics the degree of Doppler shift created by the motion of the wings. The other modulating parameter, the modulation rate, simulates the insect's wingbeat frequency.

Suga and Jen (1977) observed that 60 kHz filter neurons routinely discharge in register with the phase of the modulating waveform of an SFM signal, and some do so even when the modulation depth varied by as little as ± 10 Hz around the 60 kHz carrier frequency. Such sensitivity for

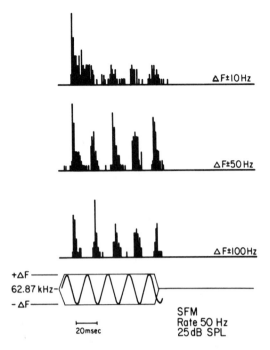

FIGURE 36.12. Peri-stimulus time histograms of a
sharply tuned 60 kHz neuron in the mustache bat's
inferior colliculus that phase-locked to SFM signals.
Signal envelope and SFM waveform are shown below.
This was an unusually sensitive neuron and phase-
locked when the frequency swings (Δf) were as small
as ± 10 Hz around the 62.87 kHz center frequency.
Scale bar is 20 msec.

60 kHz neurons, and their tuning curves are correspondingly wider.

It should now be apparent why Doppler-shift
compensation provides an advantage for hunting
insects in acoustically cluttered environments.
Doppler compensation is the auditory equivalent
of foveation in vision, where moving the eyes
keeps images of interest fixated on the fovea
(Schuller and Pollak 1979; Pollak et al. 1986; Pollak and Casseday 1989). Long CF/FM bats move
the frequency of their voices in accord with the
Doppler shifts in the echoes they receive. By so
doing, the animal ensures itself that the echo from
background objects will return at a frequency that
will stimulate its "acoustic fovea," the resonant segment of the cochlea that is richly innervated, and
thus will excite a large population of filter neurons.
When a fluttering insect comes into the bat's acoustic space, the echoes from the insect can easily be
distinguished from echoes generated by background objects. Echoes from insects will have
complex acoustic signatures due to the modulations imposed upon the emitted CF component by
the motion of their fluttering wings. The modulations will evoke phase-locked discharges from the
population of filter neurons and thereby provide
the neural substrate for distinguishing fluttering
insects from stationary background and for the
identification of preferred prey.

frequency modulation is truly remarkable. Although-
primary auditory neurons in other animals also
display phase locking to SFM signals, such high
sensitivity occurs only in filter neurons of mustache and horseshoe bats.

The sharply tuned neurons at all levels of the
mustache bat's auditory system are extremely
sensitive to the periodic frequency modulations
created by insect wingbeats. When presented with
sinusoidal frequency modulations that sweep as
little as ± 10 Hz around a 60 kHz carrier frequency, sharply tuned neurons in the mustache
bat's inferior colliculus also respond with discharges that are phase locked to the periodicity of
the modulation waveform (Bodenhamer and Pollak
1983) (Fig. 36.12). Neurons having best frequencies above or below the resonant frequency of the
cochlea are one or two orders of magnitude less
sensitive to sinusoidal modulations than are the

5. Peripheral Adaptations are Conserved in the Central Auditory Pathway

The primary auditory pathway consists of a series
of ascending parallel pathways in which the cochlear surface is remapped upon each succeeding
auditory region. The parallel pathways originate as
each auditory nerve fiber enters the brain, where it
divides into an ascending branch that innervates
the anteroventral cochlear nucleus and a descending branch that innervates the posteroventral and
dorsal cochlear nucleus. The orderly representation of frequency established in the cochlea is
preserved in each division of the cochlear nucleus
as three separate tonotopic maps. The principal
auditory pathways originate from these three
major divisions of the cochlear nucleus (Fig.

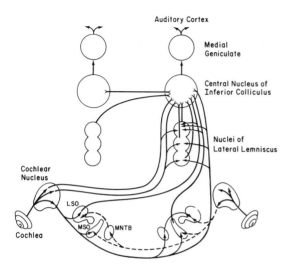

FIGURE 36.13. Schematic diagram of the principal connections of the brainstem auditory system. Note the parallel pathways that originate from the cochlear nucleus. Some pathways converge on nuclei that receive innervation from the two ears, such as the MSO and LSO, while others project directly to the inferior colliculus. MNTB, medial nucleus of the trapezoid body; MSO, medial superior olivary nucleus; LSO, lateral superior olivary nucleus.

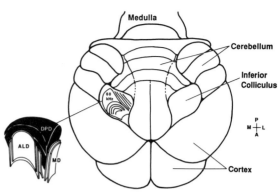

FIGURE 36.14. Schematic drawing of a dorsal view of the mustache bat's brain to show the location of the 60 kHz region in the inferior colliculus. The hypertrophied inferior colliculi protrude between the cerebral cortex and cerebellum. In the left colliculus are shown the isofrequency contours, as determined in anatomical and physiological studies. A three-dimensional representation of the laminar arrangement in the inferior colliculus is shown on far left. Low-frequency contours, representing an orderly progression of frequencies from about 59 kHz to about 10 kHz, fill the anterolateral division (ALD). High frequencies, from about 64 kHz to over 120 kHz, occupy the medial division (MD). The 60 kHz isofrequency contour is the dorsoposterior division (DPD), and is the sole representation of the sharply tuned neurons in the bat's colliculus. (Data from Zook et al. 1985.)

36.13). Each pathway ascends in parallel with the others, and has a unique pattern of connectivity with the subdivisions of the superior olivary complex and the nuclei of the lateral lemniscus. Many of these centers, such as the superior olivary nuclei, are the initial sites of convergence of information from the two ears, and thus play an important role in sound localization, while others, such as the cochlear nuclei, process information from only one ear (e.g., Goldberg 1975; Rothe et al. 1978; Brunso-Bechtold et al. 1981; Aitkin 1985; Irvine 1986; Zook and Casseday 1982, 1985, 1987; Ross et al. 1988). All of these pathways ultimately terminate in an orderly fashion in the central nucleus of the inferior colliculus (ICc), where the multiple tonotopic maps of the lower nuclei are reconstituted into a single tonotopic arrangement.

One of the distinguishing features of the mustache bat's auditory system is that the overrepresentation and sharp tuning of 60 kHz neurons described above are conserved throughout the auditory system, including the ICc (Suga et al. 1975, 1976; Suga 1984; Kössl and Vater 1985; Neuweiler 1980; Pollak 1980; Pollak and Bodenhamer 1981;

Pollak and Casseday 1989; Zook et al. 1985). Indeed, Zook and Leake (1990) maintain that the overrepresentation of 60 kHz is further expanded in the central nervous system by a divergence of fibers emanating from the cochlear nucleus.

The 60 kHz overrepresentation is most dramatically expressed in the tonotopy of the central nucleus of the inferior colliculus (Fig. 36.14). The tonotopy of the ICc in less specialized mammals is manifest as an orderly stacking of sheets of neurons (Rockel and Jones 1973; FitzPatrick 1975; Oliver and Morest 1984; Aitkin 1985), which imparts a laminated appearance to the ICc in Golgi impregnated material. The neuronal population of any one sheet, or lamina, is most sensitive to a particular frequency, and each lamina, therefore, represents a segment of the cochlear surface (Merzenich and Reid 1974; FitzPatrick 1975; Semple and Aitkin 1979; Serviere et al. 1984). The orderly arrangement of isofrequency laminae recreates the cochlear frequency-place map along one axis of the

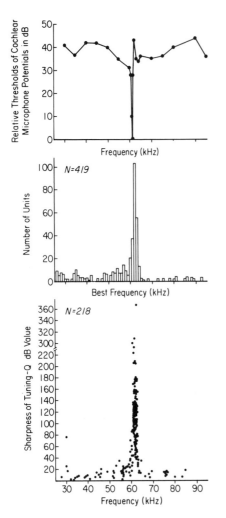

The representation of frequency in the mustache bat's ICc is split into three divisions: the anterolateral division, the medial division and the dorsoposterior division (Fig. 36.14). Each division represents a different region of the cochlear surface. The sheet-like isofrequency laminae in the anterolateral division provide an orderly representation of only the lower frequencies, from about 10 kHz to about 59 kHz; the laminae of the medial division contain an orderly representation only of the high frequencies, ranging from about 64 kHz to over 120 kHz; and the dorsoposterior division (DPD), which occupies about a third of the ICc volume, contains exclusively the 60 kHz representation. The isofrequency composition of the dorsoposterior division is seen in the similarity of the best frequencies of its sharply tuned neurons, which differ by only ± 300 Hz among the population (Pollak and Bodenhamer 1981; Zook et al. 1985; Wenstrup et al. 1986a; Ross et al. 1988). Additionally, the best frequencies coincide closely with the resonant frequency of the individual bat's cochlea (Fig. 36.15). The dorsoposterior division, then, is the midbrain representation of the "resonant" segment of the bat's cochlear partition, and corresponds to the frequency at which the bat clamps the echo CF component when Doppler compensating.

6. Collicular Features Derived from Processing in the Central Auditory Pathway: Monaural and Binaural Representation

A number of important features of the ICc are derived from processing that occurs within lower central auditory structures. Among these derived characteristics are binaural response types. Unlike the visual system, which first combines information from the two eyes in the cortex, binaural processing occurs early in the auditory pathway, initially in the superior olivary nuclei in the brainstem (Boudreau and Tsuchitani 1968; Goldberg 1975; Irvine 1986). Since these, as well as monaural regions such as the cochlear nucleus, send projections to the ICc, neurons having various monaural or binaural response properties are common in the inferior colliculus (Roth et al. 1978; Semple and

FIGURE 36.15. (A) Cochlear microphonic audiogram recorded from one mustache bat showing the sharp sensitive region around 60 kHz. (B) The distribution of best frequencies recorded from single units in the mustache bat's inferior colliculus. There is a pronounced overrepresentation of neurons with best frequencies around 60 kHz, and these frequencies correspond closely to the most sensitive frequency of the cochlear microphonic audiogram. (C) Histogram showing the Q_{10dB} values (tuning sharpness) of single units recorded from the mustache bat's inferior colliculus. The overrepresented 60-kHz neurons have much sharper tuning curves than do units with higher or lower best frequencies.

ICc. In mustache bats, however, the orderly tonotopic sequence of isofrequency laminae is distorted due to the overrepresentation of 60 kHz (Pollak et al. 1983; Zook et al. 1985; Ross et al. 1988).

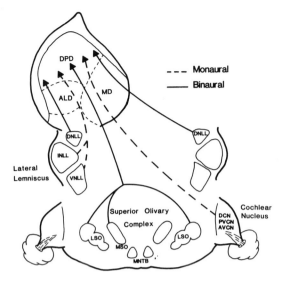

FIGURE 36.16. Ascending projections to the dorso-posterior division (DPD) of the mustache bat's inferior colliculus. Projections from binaural auditory nuclei that receive innervation from the two ears are shown as solid lines whereas projections from monaural auditory nuclei that receive innervation from only one ear are shown as dashed lines. Note that the DPD receives projections from all lower auditory nuclei, and thus receives the same types of inputs as the entire inferior colliculus. Although not shown, the neurons that project to the DPD in each of the lower auditory nuclei are restricted to parts that presumably represent 60 kHz. Abbreviations: ALD, anterolateral division of inferior colliculus; AVCN, anteroventral cochlear nucleus; DCN, dorsal cochlear nucleus; DNLL, dorsal nucleus of the lateral lemniscus; INLL, intermediate nucleus of the lateral lemniscus; LSO, lateral superior olive; MD, medial division of inferior colliculus; MNTB, medial nucleus of the trapezoid body; MSO, medial superior olive; PVCN, posteroventral cochlear nucleus; VNLL, ventral nucleus of the lateral lemniscus. (Drawing based on Ross et al. 1988.)

Aitkin 1979; Irvine 1986; Aitkin 1985). Monaural ICc neurons almost always receive excitatory input from the contralateral ear, and are designated as E–O neurons, the E referring to contralateral excitation and the O referring to the absence of ipsilateral influence. There are two major types of binaural cells: (1) E–E neurons receive excitatory inputs from both ears; and (2) E–I neurons receive excitation from

the contralateral ear and inhibition from the ipsilateral ear.

The enlarged 60 kHz region of the mustache bat's inferior colliculus is an excellent model for the analysis of convergence and separation of inputs within one isofrequency contour of the inferior colliculus. Injections of HRP confined to this isofrequency contour reveal that the dorsoposterior division receives projections from the same set of lower auditory nuclei that project to the entire central nucleus of the inferior colliculus (Ross et al. 1988) (Fig. 36.16). These include: (1) contralateral projections from the cochlear nucleus and inferior colliculus; (2) ipsilateral projections from the medial superior olive, and from the ventral and intermediate nuclei of the lateral lemniscus; and (3) bilateral projections from the lateral superior olive and dorsal nucleus of the lateral lemniscus. The projections arise from discrete segments of the various projecting nuclei, each of which presumably represents 60 kHz.

Since the dorsoposterior division receives projections from several lower nuclei that are monaural (e.g., cochlear nucleus and ventral nucleus of the lateral lemniscus) as well as several that are binaural (e.g., medial and lateral superior olives), it is not surprising that monaural neurons and the two major binaural types are common in the dorsoposterior division, as they are in other isofrequency contours of the mustache bat's inferior colliculus (Fuzessery and Pollak 1985; Wenstrup et al. 1986a). However, neurons having common monaural or binaural properties are topographically organized within the dorsoposterior division (Wenstrup et al. 1986a), and probably in the other isofrequency contours of the mustache bat's inferior colliculus as well (Wenstrup et al. 1986b). There are several zones in the dorsoposterior division, each having a predominant monaural or binaural type (Fig. 36.17). Monaural (E–O) neurons are located along the dorsal and lateral parts of the dorsoposterior division. E–E cells occur in two regions, one in the ventrolateral and the other in the dorsomedial region of the dorsoposterior division. E–I neurons also have two zones: the main population is in the ventromedial region, and a second population occurs along the very dorsolateral margin of the dorsoposterior division, perhaps extending into the external nucleus of the inferior colliculus.

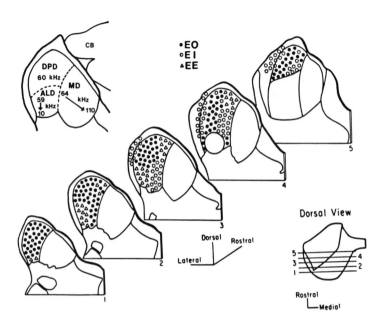

FIGURE 36.17. Schematic illustration of the segregation of binaural response properties in the dorsoposterior division (DPD) of the mustache bat's inferior colliculus. Transverse sections of the inferior colliculus are arranged in caudal to rostral manner. Each section is separated by 180 μm, and the rostrocaudal position of each section is illustrated on the dorsal view of the inferior colliculus (upper left). The medial division, where high frequencies are represented, is the large area to the right of the DPD in each section. The anterolateral division, representing frequencies below 60 kHz, is shown below the DPD in sections 4 and 5. Symbols (EO neurons, o; EE Δ, EI neurons-o) indicate binaural properties of unit clusters recorded in DPD and adjacent external nucleus of the inferior colliculus. (Adapted from Wenstrup et al. 1986.)

7. The Source of Ascending Projections to Each of the Aural Regions of the Dorsoposterior Division of the Mustache Bat's Inferior Colliculus

The topographic arrangement of neurons with particular binaural and monaural response properties in the dorsoposterior division suggests that the aural properties in each region are a consequence of the projections that terminate in each region. Linda Ross and I exploited the greatly enlarged 60 kHz region to determine the source of ascending projections to the four different aural regions of the dorsoposterior division (Ross and Pollak 1989). Small iontophoretic deposits of HRP were made within each of the physiologically defined aural regions, and the locations and numbers of retrogradely labeled cells in the auditory brainstem nuclei were determined. With this method, we demonstrated that each monaural and binaural region within the dorsoposterior division receives its chief inputs from a different subset of nuclei in the lower brainstem. For most regions, the response properties reflect the subset of inputs.

Below I describe the nuclei that project to three binaural areas of the dorsoposterior division: the projections to the monaural regions are not considered here. Each binaural area has a distinctive pattern of inputs from monaural and binaural nuclei. Considered first are the projections to the ventromedial part of the 60 kHz contour, an area in which the neurons are excited by sound to the contralateral ear and inhibited by sound to the ipsilateral ear (E–I cells). Figure 36.18 shows that the input to this area comes largely from binaural nuclei, especially the dorsal nucleus of the lateral lemniscus and less so from the lateral superior olive. A major input also originates in the intermediate nucleus of the lateral lemniscus (not shown in Fig. 36.18), which is a monaural nucleus. The robust inputs from the dorsal nucleus of the lateral lemniscus and lateral superior olive distinguishes the EI region from the other aural regions of the 60 kHz contour.

The other two binaural areas contain neurons that are excited by sound at either ear (E–E cells). One E–E area is situated ventrolaterally and the other dorsomedially in the dorsoposterior division as shown in Figures 36.17 and 36.18. Figure 36.18 shows that a major input to the ventrolateral E–E is from medial superior olive and another is from ventral nucleus of the lateral lemniscus. There is little or no input from lateral superior olive. This pattern is in marked contrast to projections to the E–I area, which arise from the lateral superior olive and from the dorsal and intermediate nuclei of the lateral lemniscus.

Finally, the projections to the dorsomedial E–E area (Fig. 36.18) differ substantially from both the E–I region and ventrolateral E–E region. The dominant input arises from the contralateral inferior colliculus. Substantial input also arises from both the ventral and intermediate nuclei of the lateral lemniscus, but there is almost no input from the binaural centers below the tectum.

These projection patterns suggest that the binaural properties of the medial E–E area arise from interaction via the two colliculi, whereas the binaural properties in the dorsolateral E–E area arise from the binaural centers in the lower brainstem, especially MSO (Fig. 36.18). In contrast, the binaural properties of the E–I area are shaped largely by the binaural nucleus of the lateral lemniscus, the DNLL, and to a lesser extent by the LSO (Fig. 36.18).

These studies show that each aural region of the dorsoposterior division is distinguished both by its neural response properties and by the unique pattern of ascending projections it receives. The connectional differences among the various aural regions are also reflected in functional differences. Some functional distinctions are apparent, such as between monaural compared to binaural neurons and between E–I and E–E neurons. Below the different ways in which E–E and E–I cells code for sound location are considered in detail. However, our understanding of how the response properties of individual neurons are shaped by the particular set of connections they receive does not reach much beyond the more obvious differences. For example

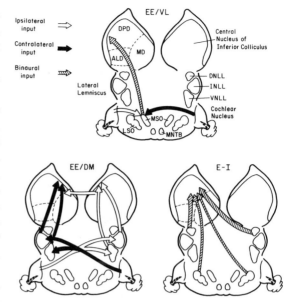

FIGURE 36.18. Schematic diagram showing some of the major projections to the ventrolateral EE region (top panel), and dorsolateral EE region (lower left panel) to illustrate two ways that E–E response properties are initially formed. Inputs from the ipsilateral ear are shown in white, inputs from the contralateral ear are shown in black and projections from binaural nuclei are stripped. Lower right panel shows projections from binaural nuclei to ventromedial EI region. (Adapted from Ross and Pollak 1989.)

it is clear just from the connectional patterns that there must be at least two major subtypes of E–E neurons, but we have little insight into exactly how the connectional differences are expressed physiologically. Moreover, we presently can only speculate about how the convergence of inputs from several lower nuclei could shape the response characteristics of a collicular neuron expressing a particular aural type. Clarifying these issues represents a major challenge for the future.

8. The Population of E–I Neurons have Different Sensitivities for Interaural Intensity Disparities

The population of E–I neurons is of particular interest because they differ in their sensitivities to interaural intensity disparities (IIDs) (Wenstrup

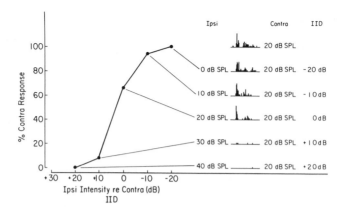

FIGURE 36.19. Responses of a single E–I neuron as a function of interaural intensity disparity (IID). On the right are shown the peri-stimulus time histograms of discharges evoked by the contralateral stimulus alone (top histogram), and the inhibition produced as the intensity of the ipsilateral (inhibitory) signal was increased. The best frequency of this neuron was 62 kHz and its threshold was 10 dB SPL. The inhibitory threshold, defined as the IID value at which the discharge rate declined by 50%, was +4 dB (at an ipsilateral intensity 4 dB higher than the contralateral intensity).

et al. 1985, 1986, 1988a,b). These E–I neurons compare the sound intensity at one ear with the intensity at the other by subtracting the activity generated in one ear from that in the other. Suprathreshold sounds delivered to the excitatory (contralateral) ear evoke a certain discharge rate that is unaffected by low intensity sounds presented simultaneously to the inhibitory (ipsilateral) ear. However, when the ipsilateral intensity reaches a certain level, and thus generates a particular IID, the discharge rate declines sharply, and even small increases in ipsilateral intensity will, in most cases, completely inhibit the cell (Fig. 36.19). Thus each E–I neuron has a steep IID function and reaches a criterion inhibition at a specified IID that remains relatively constant over a wide range of intensities (Wenstrup et al. 1986, 1988a,b). The criterion we adopted is the IID that produces a 50% reduction in the discharge rate evoked by the excitatory stimulus presented alone. This IID is called the neuron's inhibitory threshold. An inhibitory threshold has a value of 0 dB if equally intense signals in the two ears elicit the criterion inhibition. An inhibitory threshold is assigned a positive value if the inhibition occurs when the ipsilateral signal is louder than the contralateral signal, and is assigned a negative value if the intensity at the ipsilateral ear is lower than the contralateral ear when the discharge rate is reduced by 50%. The inhibitory thresholds of E–I neurons in the dorsoposterior

division vary from +30 dB to −20 dB (Fuzessery and Pollak 1984, 1985; Wenstrup et al. 1988a), encompassing much of the range of IIDs that the bat would experience.

8.1 Ultrasonic Frequencies Generate Large Interaural Intensity Disparities

Frequencies having wavelengths shorter than the animal's head generate substantial IIDs due to acoustic shadowing and the directional properties of the ear. Moreover, the value of an IID varies with the location of the sound along the azimuth (Fuzessery and Pollak 1984, 1985; Wenstrup et al. 1989b), as shown in Figure 36.20 for 60 kHz in the mustache bat. When the sound emanates from directly in front of the bat, at 0° elevation and 0° azimuth, equal sound intensities reach both ears, and an interaural intensity disparity of 0 dB is generated. The largest interaural intensity disparities originate at about 40° azimuth and 0° elevation, where the sounds are about 30 dB louder in one ear than in the other. Within the azimuthal sound field from roughly 40° on either side of the midline, the range of interaural intensity disparities created by the head and ears at 60 kHz is about 60 dB (+30 dB to −30 dB). Thus the interaural intensity disparities change on the average by about 0.75 dB/degree. If recordings were made on the left side of the brain and the sound source was located in the left hemifield at 40° azimuth and 0° elevation, the terminology would refer to this as a +30 dB interaural intensity disparity. Figure 36.20 (right panel) shows a schematic representation of the interaural intensity disparities generated by 60 kHz sounds in both azimuth and elevation. Notice that an interaural intensity disparity is not uniquely associated with one position in space, but rather a given intensity disparity can be generated by 60 kHz from several spatial locations, a feature that I shall address in a later section. Since the intensity disparity at the ears varies with the position of a sound source in space, a neuron's sensitivity to interaural intensity disparities is suggestive of how it will respond to sounds emanating from various spatial locations.

8.2 The Spatial Selectivities of 60 kHz E–I Neurons Are Determined by the Disparities Generated by the Ears and the Neuron's Inhibitory Threshold

To determine more precisely how binaural properties shape a neuron's receptive field, Fuzessery and his colleagues (Fuzessery and Pollak 1984, 1985; Fuzessery et al. 1985; Fuzessery 1986; Wenstrup et al. 1988b) first determined how interaural intensity disparities of 60 kHz sounds vary around the bat's hemifield. They then evaluated the binaural properties of collicular neurons with loudspeakers inserted into the ear canals, and subsequently determined the spatial properties of the same neurons with free-field stimulation delivered from loudspeakers around the bat's hemifield. With this battery of information, the quantitative aspects of binaural properties could be directly associated with the neuron's spatially selective properties.

The binaural properties of three 60 kHz E–I units and their receptive fields are shown in Figure 36.21. The first noteworthy point is that the spatial position at which each unit had its lowest threshold was the same among all 60 kHz E–I units, at about −40° azimuth and 0° elevation. This location corresponds to the position in space where the largest interaural intensity disparities are generated, i.e., the position in space at which the sound is always most intense in the excitatory ear and least intense in the inhibitory ear. The thresholds

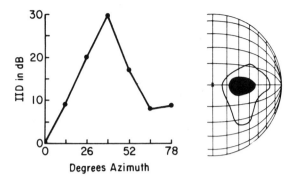

FIGURE 36.20. Left: Interaural intensity disparities (IID) of 60 kHz sounds for azimuthal locations of 0 degrees elevation. Right: Schematic illustration of IIDs of 60 kHz sounds at different azimuths and elevations. Blackened area is region of space where all IIDs are 20 dB or greater. The area enclosed by the solid line is region of space from which all IIDs are 10 dB or greater. (Data from Fuzessery and Pollak 1985.)

increase almost as circular rings away from this area of maximal sensitivity. The highest thresholds are in the ipsilateral sound field, and some units are totally unresponsive to sound emanating from these regions of space.

The second noteworthy point is in that for each 60 kHz E–I unit there is a position along the azimuth that demarcates the spatial region where sounds can evoke discharges from spatial regions where sounds are ineffective in evoking discharges. That azimuthal position of demarcation, and the interaural intensity disparity associated with it, is different for each E–I cell and correlates closely with the neuron's inhibitory threshold. In 60 kHz units having low inhibitory thresholds, those demarcating loci occur along the midline or even in the contralateral sound field, and sounds presented ipsilateral to those loci are incapable of eliciting discharges, even with intensities as high as 110 dB SPL (Fig. 36.21B and C). Units with higher inhibitory thresholds require a more intense stimulation of the inhibitory ear for complete inhibition, and therefore the demarcating loci of these units are in the ipsilateral sound field. Some units with high inhibitory thresholds (Fig. 36.21A) could never be completely inhibited with dichotic stimuli. When tested with free-field stimulation these units display high thresholds in the ipsilateral acoustic field.

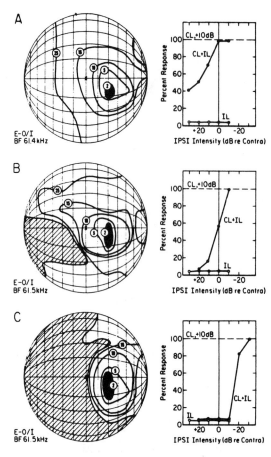

FIGURE 36.21. Spatial selectivity and IID functions for
three 60 kHz E–I units in the mustache bat's inferior colliculus. All lines on globes connect isothreshold points.
Isothreshold contours are drawn in increments of 5 dB.
Blackened area on right is region of space from which
lowest thresholds were elicited. All thresholds in this
region differed by at most 2 dB. The uppermost unit (A)
had a high inhibitory threshold, and could not be completely inhibited regardless of the intensity of the inhibitory sound. The inhibitory threshold of the unit in (B)
was lower, and the neuron was completely inhibited
when the ipsilateral sound was about 10 dB more intense
than the contralateral sound. Its spatial selectivity, on
the left, shows that sounds in portions of the ipsilateral
acoustic field, indicated by the stripped area, were
ineffective in firing the unit even with intensities of over
100 dB SPL. Similar arguments apply to the unit in (C).
The inhibitory threshold of this unit was even lower than
for the unit in B. Correspondingly, the region of space
from which sounds could not evoke discharges in this
unit was also expanded in both the ipsilateral and into
the contralateral acoustic field. (From Fuzessery and
Pollak 1985.)

8.3 Inhibitory Thresholds of 60 kHz E–I Neurons are Topographically Organized Within the Dorsoposterior Division and Create a Representation of Acoustic Space

A finding of particular importance is that the
inhibitory thresholds are topographically arranged
within the ventromedial E–I region of the DPD
(Wenstrup et al. 1985, 1986) (Fig. 36.22). E–I neurons with high, positive inhibitory thresholds (i.e.,
neurons requiring a louder ipsilateral stimulus than
contralateral stimulus to produce inhibition) are
located in the dorsal E–I region, and ventrally E–I
neurons display a progressive shift to lower inhibitory thresholds. The most ventral E–I neurons have
the lowest inhibitory thresholds; they are suppressed by ipsilateral sounds equal to or less
intense than the contralateral sounds.

The topographic representation of IID sensitivities has implications for how the azimuthal, i.e.,
horizontal, position of a sound is represented in
the mammalian inferior colliculus. The fact that
intensity disparities of 60 kHz tones change systematically with azimuth suggests that the value of an
IID can be represented within the dorsoposterior
division as a "border" separating a region of discharging cells from a region of inhibited cells (Wenstrup et al. 1986a, 1989b; Pollak et al. 1986). Consider, for instance, the pattern of activity in the
dorsoposterior division on one side generated by a
60 kHz sound that is 15 dB louder in the ipsilateral
ear than in the contralateral ear (Fig. 36.23). The
IID in this case is +15 dB. Since neurons with low
inhibitory thresholds are situated ventrally, the high
relative intensity in the ipsilateral (inhibitory) ear
will inhibit all the E–I neurons in ventral portions of
the dorsoposterior division. The same sound, however, will not be sufficiently loud to inhibit the E–I
neurons in the more dorsal dorsoposterior division,
where neurons require a relatively more intense
ipsilateral stimulus for inhibition. The topology of
inhibitory thresholds and the steep IID functions of
E–I neurons, then, can create a border between
excited and inhibited neurons within the ventromedial dorsoposterior division. The locus of the
border, in turn, should shift with changing IID, and
therefore should shift correspondingly with changing sound location, as shown in Figure 36.23.

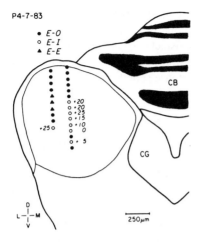

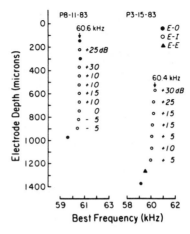

FIGURE 36.22. Left: Systematic shift in the inhibitory thresholds of E–I unit clusters shown in transverse section through the mustache bat's inferior colliculus. Numbers indicate inhibitory thresholds. All clusters were sharply tuned to frequencies between 63.1 and 64.0 kHz. Right: Systematic decrease in inhibitory thresholds of E–I neurons with depth in the dorsoposterior division of the inferior colliculus recorded in two dorsoventral penetrations from different bats. (From Wenstrup et al. 1985.)

9. One Group of E–E Neurons Code for Elevation Along the Midline

A population of E–E units has been found that are most sensitive to sounds presented close to or along the vertical midline. However, the elevation at which these cells are most sensitive is determined by the directional properties of the ears for the frequency to which the neuron is tuned (Fig. 36.24 bottom panel) (Fuzessery and Pollak 1985; Fuzessery 1986; Fuzessery et al. 1990). Their azimuthal selectivities are shaped by their binaural properties that exhibit either a summation or a facilitation of discharges with binaural stimulation. The contralateral ear is always dominant, having the lowest threshold and evoking the greatest discharge rate. For 60 kHz, the ear is most sensitive, and generates the greatest interaural intensity disparities at 0° elevation and about −40° along the azimuth. The position along the azimuth where these cells are maximally sensitive, unlike E–I cells, is not so much a function of ear directionality, but rather is a direct consequence of the interplay between the potency of the excitatory binaural inputs. A sound, for example, presented

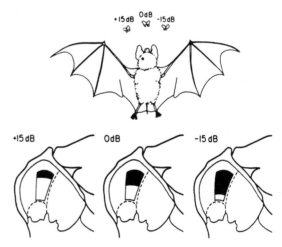

FIGURE 36.23. Schematic illustration of the relationship between the value of the interaural intensity disparity produced by a sound source at a given location (moths at top of figure) and the pattern of activity in the ventromedial E–I region of the left dorsoposterior division, where IID sensitivities are topographically organized. The activity in this region, indicated by the blackened area, spreads ventrally as a sound source moves from the ipsilateral to the contralateral sound field. (From Wenstrup et al. 1986.)

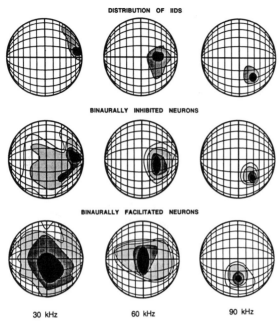

DISTRIBUTION OF IIDS

BINAURALLY INHIBITED NEURONS

BINAURALLY FACILITATED NEURONS

30 kHz   60 kHz   90 kHz

FIGURE 36.24. Interaural intensity disparities (IIDs) generated by 30, 60, and 90 kHz sounds are shown in top row. The blackened areas in each panel indicate the spatial locations where IIDs are 20 dB or greater, and the gray areas indicate those locations where IIDs are at least 10 dB. The panels in the middle row show the spatial selectivity of three E–I units, one tuned to 30 kHz, one to 60 kHz and one to 90 kHz. The blackened areas indicate the spatial locations were the lowest thresholds were obtained. Isothreshold contours are draw for threshold increments of 5 dB. The panels in the bottom row show the spatial selectivities of three E–E units, tuned to each of the three harmonics. (Adapted from Fuzessery 1986.)

from 40° contralateral will create the greatest intensity at the contralateral ear, due to the directional properties of the ear for 60 kHz. However, as the sound is moved towards the midline, the intensity at the contralateral ear diminishes, but simultaneously, the intensity at the ipsilateral ear increases. Since excitation of the ipsilateral ear facilitates the response of the neuron, the net result is that the response is stronger, and more sensitive, at azimuthal positions closer to, or at the midline then are responses evoked by sounds from the more contralateral positions. In short, many E–E neurons can be thought of as midline units because they are maximally sensitive to positions around 0° azimuth. However, the elevation to which they

are most sensitive is determined by the directional properties of the ear. Thus 60 kHz E–E units in the mustache bat are always most sensitive at about 0° elevation, because the pinna generates the most intense sounds at that elevation.

10. Spatial Properties of Neurons Tuned to Other Frequencies

The chief difference among neurons tuned to other harmonics of the mustache bat's echolocation calls, at 30 and 90 kHz, is that their spatial properties are expressed in regions of space that differ from 60 kHz neurons, a consequence of the directional properties of the ear for those frequencies (Fig. 36.24 top panel) (Fuzessery and Pollak 1985; Fuzessery 1986). The binaural processing of neurons tuned to those frequencies, however, are essentially the same as those described for 60 kHz neurons. Moreover, there is even evidence for an orderly representation of interaural intensity disparities in the 90 kHz isofrequency laminae, further supporting the generality of a topology of inhibitory thresholds among E–I cells across isofrequency contours (Wenstrup et al. 1986b). In short, the sort of binaural processing found in the 60 kHz lamina appears to be representative of binaural processing within other isofrequency contours. This feature is most readily appreciated by considering the spatial properties of E–I units tuned to 90 kHz, the third harmonic of the mustache bat's orientation calls (Fig. 36.24, middle panel on right). The maximal interaural intensity disparities generated by 90 kHz occur at roughly 40° along the azimuth and −40° in elevation (Fig. 36.24 top panel on right). The 90 kHz E–I neurons, like those tuned to 60 kHz, are most sensitive to sounds presented from the same spatial location at which the maximal interaural intensity disparities are generated. Additionally, the inhibitory thresholds of these neurons determine the azimuthal border defining the region in space from which 90 kHz sounds can evoke discharges from the region where sounds are incapable of evoking discharges (Fuzessery and Pollak 1985; Fuzessery et al. 1985). The population of 90 kHz E–I neurons have a variety of inhibitory thresholds that appear to be topographically arranged within that contour. Therefore a particular interaural intensity disparity will be encoded by a population of 90 kHz E–I cells, having a border separating the inhibited from the excited neurons, in a fashion similar to that shown for 60 kHz lamina. The same argument can be applied to the 30 kHz cells, but in this case the maximal interaural intensity disparity is generated from the very far lateral regions of space (Fig. 36.20, top panel on left).

The spatial behavior of E–E cells tuned to 30 and 90 kHz are likewise similar to those tuned to 60 kHz. The distinction is only in their elevational selectivity, since their azimuthal sensitivities are for sounds around the midline, at 0° azimuth (Fig. 36.24 bottom panel).

10.1 The Representation of Auditory Space in the Mustache Bat's Inferior Colliculus

We can now begin to see how the cues for azimuth and elevation are derived. The directional properties of the ears generate different interaural intensity disparities among frequencies at a particular location. The neural consequence is a specific pattern of activity across each isofrequency contour in the bat's midbrain. Figure 36.25 shows a stylized illustration of the interaural intensity disparities generated by 30, 60, and 90 kHz within the bat's hemifield, and below is shown the loci of borders that would be generated by a biosonar signal containing the three harmonics emanating from different regions of space. Consider first a sound emanating from 40° along the azimuth and 0° elevation. This position creates a maximal interaural intensity disparity at 60 kHz, a lesser interaural intensity disparity at 30 kHz and an interaural intensity disparity close to 0 dB at 90 kHz. The borders created within each of the isofrequency contours by these interaural intensity disparities are shown in Fig. 36.25 (bottom panel). Next consider the interaural intensity disparities created by the same sound but from a slightly different position in space, at about 45° azimuth and −20° elevation. In this case there is a decline in the 60 kHz interaural intensity disparity, an increase in the 90 kHz interaural intensity disparity, but the 30 kHz interaural intensity disparity will be the same as it was when the sound emanated from the previous position. The constant interaural intensity disparity at 30 kHz is a crucial point, and it occurs

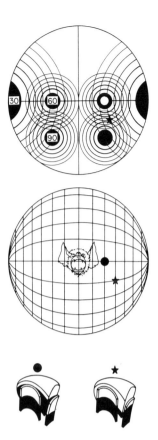

FIGURE 36.25. Loci of borders in 30, 60, and 90 kHz E–I regions generated by biosonar signals emanating from two regions of space. The upper panel is a schematic representation of IIDs of 30, 60, and 90 kHz that occur in the mustache bat's acoustic field. The blackened areas indicate the regions in space where the maximum IID is generated for each harmonic. The middle panel depicts the bat's head and spatial positions of two sounds. The lower panel shows the borders separating the regions of excited from regions of inhibited neurons in the 30, 60, and 90 kHz isofrequency contours.

because for a given frequency an interaural intensity disparity is not uniquely associated with one position in space, but rather can be generated from a variety of positions. It is for this reason that the accuracy with which a sound can be localized with one frequency is ambiguous (Blauert 1969/70; Butler 1974; Musicant and Butler 1985; Fuzessery 1986). However, spatial location, in both azimuth and elevation, is rendered unambiguous by the simultaneous comparison of three interaural intensity disparities, because their values in combination are uniquely associated with a spatial location.

This representation becomes ineffective along the vertical midline, at 0° azimuth, where the interaural intensity disparities will be 0 dB at all frequencies and all elevations. The borders among the E–I populations will thus not change with elevation because the interaural intensity disparities remain constant. The population of E–E neurons tuned to different frequencies may be important in this regard (Fuzessery and Pollak 1984, 1985; Fuzessery 1986; Fuzessery et al. 1990). E–E units tuned to different frequencies exhibit selectivities for different elevations, and thus as the elevation along the vertical meridian shifts, the strength of responding also shifts for E–E units as a function of their frequency tuning. In combination then, the two major binaural types provide a neural representation of sound located anywhere within the bat's acoustic hemifield.

A striking feature of the above scenario is its similarity, in principle, to the ideas about sound localization proposed previously by Pumphrey (1948) and Grinnell and Grinnell (1965). What these recent studies provide are the details of how interaural disparities are encoded and how they are topologically represented in the acoustic midbrain.

11. Conclusions

The overrepresentation and sharp tuning of 60 kHz neurons are the most prominent adaptations of the mustache bat's auditory system. It is noteworthy that these adaptations are functionally similar to aspects of the visual and somatosensory systems. In both of these sensory systems the portions of the sensory surface subserving fine discrimination are densely innervated with neurons having small receptive fields and are greatly magnified in the central nervous system (Hubel and Wiesel 1977; Sur et al. 1980). The analogy with vision is even clearer when the integrated actions of both sensory and motor systems are considered. In vision, changing the direction of gaze keeps images of interest fixated on the fovea. Similarly, Doppler compensation ensures that acoustic signals of interest are processed by a region of the sensory surface innervated by a large number of primary fibers having exceptionally narrow tuning curves. The resonant segment of the bat's cochlea, then, is essentially an "acoustic fovea," and the central

representations of that region, such as the dorsoposterior division, can be thought of as the foveal regions of the auditory system (Schuller and Pollak 1979; Pollak et al. 1986; Pollak and Casseday 1989).

The acoustic features requiring fine resolution are for prey recognition and identification. With a constant frequency sonar system, a fluttering insect's "acoustic signature" is conveyed by the particular modulation patterns in the echo CF component, and is displayed prominently in the inferior colliculus by a unique spatiotemporal pattern of activity across the dorsoposterior division. Although the chief function of the expanded frequency representation in long CF/FM bats is to represent an insect's acoustic profile, it is not the only attribute represented in the dorsoposterior division. The reception of the echo CF component at the two ears creates an interaural intensity disparity whose value is a function of the spatial position of the target. The coding of sound location is represented, at least in part, by the locus of the border separating populations of E–I neurons discharging in registry with the modulation waveforms from inhibited populations of E–I neurons in the ventromedial dorsoposterior division.

These experiments have shown how features of biologically important acoustic stimuli are encoded by the nervous system. The dorsoposterior division has been of value in these studies because its specialized features provide an enlarged "picture window" through which we can obtain a magnified view of how multiple attributes of the external world are represented in the mammalian auditory midbrain.

References

Art JJ, Fettiplace R (1987) Variation of membrane properties in hair cells isolated from the turtle cochlea. J Physiol 385:207–242.

Ashmore JF (1987) A fast motile response in guinea pig outer hair cells: the cellular basis of the cochlear amplifier. J Physiol 388:323–347.

Aitkin LM (1985) The Auditory Midbrain: Structure and Function in the Central Auditory Pathway. Clifton, New Jersey: Humana Press.

Bateman GC, Vaughan TA (1974) Nightly activities of mormoopid bats. J Mammal 55:45–65.

Bishop AL, Henson OW Jr (1988) The efferent auditory system in Doppler compensating bats. In: Nachtigall PE, Moore PWB (eds) Animal Sonar: Processes and Performance. New York: Plenum Press, pp. 307–310.

Blauert J (1969/1970) Sound localization in the median plane. Acoustica 22:205–213.

Bodenhamer RD, Pollak GD (1983) Response characteristics of single units in the inferior colliculus of mustache bats to sinusoidally frequency modulated signals. J Comp Physiol 153:67–79.

Boudreau JC, Tsuchitani C (1968) Binaural interaction in the cat superior olive S segment. J Neurophysiol 31:445–454.

Brunso-Bechtold JK, Thompson GC, Masterton RB (1981) HRP study of the organization of auditory afferents ascending to the central nucleus of the inferior colliculus in the cat. J Comp Neurol 97:705–722.

Bullock TH, Grinnell AD, Ikezono E, Kamuda K, Katsuki Y, Sato O, Suga N, Yanagisawa R (1968) Electrophysiological studies of central auditory mechanisms in cetaceans. Z Verg Physiol 59:117–156.

Butler RA (1974) Does tonotopy subserve the perceived elevation of a sound? Fed Proc 33:1920–1923.

Cotzin M, Dallenbach KM (1950) Facial vision: the role of pitch and loudness in the perception of obstacles by the blind. Am J Psychol 63:485–515.

Crawford AC, Fettiplace R (1981) An electrical tuning mechanism in turtle cochlear hair cells. J Physiol (London) 312:377–412.

Ehert G, Frankenreiter M (1977) Quantitative analysis of cochlear structures in the housemouse in relation to mechanisms of acoustical information processing. J Comp Physiol 122:65–85.

FitzPatrick KA (1975) Cellular architecture and topographic organization of the inferior colliculus of the squirrel monkey. J Comp Neurol 164:185–208.

Fuchs PA, Nagai T, Evans MG (1988) Electrical tuning in hair cells isolated from chick cochlea. J Neurosci 8:2460–2467.

Fuzessery ZM (1986) Speculations on the role of frequency in sound localization. Brain Behav Evol 28:95–108.

Fuzessery ZM, Pollak GD (1984) Neural mechanisms of sound localization in an echolocating bat. Science 225:725–728.

Fuzessery ZM, Pollak GD (1985) Determinants of sound location selectivity in bat inferior colliculus: A combined dichotic and free-field stimulation study. J Neurophysiol 54:757–781.

Fuzessery ZM, Wenstrup JJ, Pollak GD (1985) A representation of horizontal sound location in the inferior colliculus of the mustache bat (Pteronotus p. parnellii). Hearing Res 20:85–89.

Fuzessery ZM, Wenstrup JJ, Pollak GD (1990) Determinants of horizontal sound location selectivity of binaurally excited neurons in the inferior colliculus of an isofrequency region of the mustache bat inferior colliculus. J Neurophysiol 63:1128–1147.

Galambos R, Griffin DR (1942) Obstacle avoidance by flying bats. J Exp Zool 89:475–490.

Goldberg JM (1975) Physiological studies of the auditory nuclei of the pons. In: Keidel WD, Neff WD (eds) Handbook of Sensory Physiology, Vol. V. Auditory System, Part 2. New York: Springer-Verlag, p. 109.

Goldman LJ, Henson OW Jr (1977) Prey recognition and selection by the constant frequency bat, Pteronotus p. parnellii. Behav Ecol Sociobiol 2:411–419.

Griffin DR (1944) Echolocation by blind men and bats. Science 100:589–590.

Griffin DR (1958) Listening in the Dark. Yale University Press, New Haven, Conn.

Griffin DR, Galambos R (1941) The sensory basis of obstacle avoidance by flying bats. J Exp Zool 86:481–506.

Grinnell AD (1963a) The neurophysiology of audition in bats: Intensity and frequency parameters. J Physiol 167:38–66.

Grinnell AD (1963b) The neurophysiology of audition in bats: Temporal parameters. J Physiol 167:67–96.

Grinnell AD (1963c) The neurophysiology of audition in bats: Directional localization and binaural interactions. J Physiol 167:97–113.

Grinnell AD (1963d) The neurophysiology of audition in bats: Resistance to interference. J Physiol 167:114–127.

Grinnell, AD (1967) Mechanisms of overcoming interference in echolocating animals. In: Busnel R-G (ed) Animal Sonar Systems, Vol. I. Jouy-en-Josas 78, France: Laboratoire de Physiologie Acoustique, p. 451.

Grinnell AD (1970) Comparative neurophysiology of neotropical bats employing different echolocation signals. Z Vergl Physiol 68:117–153.

Grinnell AD (1973) Neural processing mechanisms in echolocating bats correlated with differences in emitted sounds. J Acoust Soc Am 54:147–156.

Grinnell AD, Grinnell VS (1965) Neural correlates of vertical localization by echolocating bats. J Physiol 181:830–851.

Grinnell AD, Hagiwara S (1972a) Adaptations of the auditory nervous system for echolocation: Studies of New Guinea bats. Z Vergl Physiol 76:41–81.

Grinnell AD, Hagiwara S (1972b) Studies of auditory neurophysiology in non-echolocating bats and adaptations for echolocation in one genus of Rousettus. Z Vergl Physiol 76:82–96.

Harnischfeger G, Neuweiler G, Schlegel P (1985) Interaural time and intensity coding in the superior olivary complex and inferior colliculus of the echolocating bat, Molossus ater. J Neurophysiol 53:89–109.

Henson MM (1978) The basilar membrane of the bat, Pteronotus parnellii. Anat Rec 153:143–158.

Henson MM, Henson OW Jr, Jenkins DB (1984) The attachment of the spiral ligament to the cochlear wall: Anchoring cells and the creation of tension. Hearing Res 16:231–242.

Henson MM, Burridge K, Fitzpatrick DC, Jenkins DB, Pillsbury HC, Henson OW Jr (1985) Immunocytochemical localization of contractile and contraction associated proteins in the spiral ligament of the cochlea. Hearing Res 20:207–214.

Henson OW Jr (1967) The perception and analysis of biosonar signals by bats. In: Busnel R-G (ed) Animal Sonar Systems. Vol. II. France: Lab Physiol Acoust, Jouy-en-Josas 78, p. 949.

Henson OW Jr (1970) The ear and audition. In: Wimsatt WA (ed) The Biology of Bats, Vol. 2. New York: Academic Press, pp. 181–263.

Henson OW Jr, Pollak GD (1973) A technique for the chronic implantation of electrodes in the cochlea of bats. Physiol Behav 8:1185–1187.

Henson OW Jr, Henson MM, Kobler JB, Pollak GD (1980) The constant frequency component of the biosonar signals of the bat, Pteronotus p. parnellii. In: Busnel R-G, Fish JF (eds) Animal Sonar Systems. New York: Plenum Press, p. 913.

Henson OW Jr, Pollak GD, Kobler JB, Henson MM, Goldman LJ (1982) Cochlear microphonics elicited by biosonar signals in flying bats. Pteronotus p. parnellii. Hearing Res 7:127–147.

Henson OW Jr, Schuller G, Vater M (1985) A comparative study of the physiological properties of the inner ear in Doppler shift compensating bats (Rhinolophus rouxi and Pteronotus parnellii). J Comp Physiol 157: 587–607.

Henson OW Jr, Henson MM (1988) Morphometric analysis of cochlear structures in the mustached bat, Pteronotus parnellii. In: Nachtigall PE, Moore PWB (eds) Animal Sonar: Processes and Performance. New York: Plenum Press, pp. 301–306.

Hubel DH, Wiesel TN (1977) Functional architecture of macaque monkey visual cortex. Ferrier Lecture Proc R Soc Lond 198:1–59.

Hudspeth AJ, Lewis RS (1988a) Kinetic analysis of voltage-and ion-dependent conductances in saccular hair cells of the bull frog, Rana catesbeiana. J Physiol 400:237–274.

Hudspeth AJ, Lewis RS (1988b) A model for electrical resonance and frequency tuning in saccular hair cells of the bull frog, Rana catesbeiana. J Physiol 400:275–297.

Irvine DRF (1986) The Auditory Brainstem. Progress in Sensory Physiology 7, Autrum H, Ottoson D (eds) Berlin-Heidelberg: Springer-Verlag.

Kellogg W (1961) Porpoises and Sonar. Chicago: University of Chicago Press.

Kössl M, Vater M (1985a) Evoked acoustic emissions and cochlear microphonics in the mustache bat, teronotus parnellii. Hearing Res 19:157–170.

Kössl M, Vater M (1985b) The frequency place map of the bat, Pteronotus parnellii. J Comp Physiol 157: 687–697.

Kössl M, Vater M (1990) Resonance phenomena in the cochlea of the mustache bat and their contribution to neuronal response characteristics in the cochlear nucleus. J Comp Physiol (in press).

Lewis R, Hudspeth AJ (1983) Voltage and ion-dependent conductances in solitary vertebrate hair cells. Nature 304:538–541.

Link A, Marimuthu G, Neuweiler G (1986) Movement as a specific stimulus for prey catching behavior in rhinolophid and hipposiderid bats. J Comp Physiol 159:403–413.

Merzenich MM, Reid MD (1974) Representation of the cochlea within the inferior colliculus of the cat. Brain Res 77:397–415.

Möhres FP, Kulzer E (1956) Uber die Orientatierung der Flughunde (Chiroperta, Pteropodidae). Z Vergl Physiol 38:1–29.

Musicant AD, Butler RA (1985) Influence of monaural spectral cues on binaural localization. J Acoust Soc Am 77:202–208.

Neuweiler G, Vater M (1977) Response patterns to pure tones of cochlear nucleus units in the CF-FM bat, Rhinolophus ferrumequinum. J Comp Physiol 115: 119–133.

Neuweiler G (1980) Auditory processing of echoes: Peripheral processing. In: Busnel R-G, Fish JF (eds) Animal Sonar Systems. New York: Plenum Press, p. 519.

Neuweiler G 1983) Echolocation and adaptivity to ecological constraints. In: Huber F, Markl H (eds) Neuroethology and Behavioral Physiology: Roots and Growing Pains. Berlin Heidelberg New York Tokyo: Springer-Verlag, p. 280.

Neuweiler G (1984a) Auditory basis of echolocation in bats. In: Bolis L, Keynes RD, Maddrell SHP (eds) Comparative Physiology of Sensory Systems. Cambridge: Cambridge University Press, p. 115.

Neuweiler G (1984b) Foraging, echolocation and audition in bats. Naturwissenschaften 71:46–455.

Neuweiler G (1990) Auditory adaptations for prey capture in echolocating bats. Physiol Rev (in press).

Novick A (1958) Orientation in paleotropical bats: II. Megachiroptera, J Exp Zool 137:443–462.

Novick A, Vaisnys JR (1964) Echolocation of flying insects by the bat, Chilonycteris parnellii. Biol Bull 127:478–488.

Oliver DL, Morest DK (1984) The central nucleus of the inferior colliculus in the cat. J Comp Neurol 222: 237–264.

Pitchford S, Ashmore JF (1987) An electrical resonance in hair cells of the amphibian papilla of frog, Rana temporaria. Hearing Res 27:75–83.

Pollak GD (1980) Organizational and encoding features of single neurons in the inferior colliculus of bats. In: Busnel R-G, Fish JF (eds) Animal Sonar Systems. New York: Plenum Press, p. 549.

Pollak GD, Henson OW Jr, Novick A (1972) Cochlear microphonic audiograms in the pure tone bat, Chilonycteris parnellii parnellii. Science 176:66–68.

Pollak GD, Henson OW Jr, Johnson R (1979) Multiple specializations in the peripheral auditory system of the CF-FM bat, Pteronotus parnellii. J Comp Physiol 131:255–266.

Pollak GD, Schuller G (1981) Tonotopic organization and encoding features of single units in the inferior colliculus of horseshoe bats: Functional implications for prey identification. J Neurophysiol 45:208–226.

Pollak GD, Bodenhamer RD (1981) Specialized characteristics of single units in inferior colliculus of mustache bat: frequency representation, tuning, and discharge patterns. J Neurophysiol 46:605–619.

Pollak GD, Bodenhamer RD, Zook JM (1983) Cochleotopic organization of the mustache bat's inferior colliculus. In: Ewert J-P, Capranica RR, Ingle DJ (eds) Advances in Vertebrate Neuroethology. New York: Plenum Press, pp. 925–935.

Pollak GD, Wenstrup JJ, Fuzessery ZM (1986) Auditory processing in the mustache bat's inferior colliculus. Trends in Neurosci 9:556–561.

Pollak GD, Casseday JH (1989) The Neural Basis of Echolocation in Bats. Heidelberg: Springer-Verlag.

Pumphrey RJ (1948) The sense organs of birds. Ibis 90:171–199.

Pye A (1965) The auditory apparatus of the Heteromyidae (Rodentia, Sciuromorpha). J Anat 99:161–174.

Rice CE (1967) Human echo perception. Science 155:656–664.

Roberts WM, Howard J, Hudspeth AJ (1988) Hair cells: Transduction, tuning and transmission in the inner ear. Ann Rev Cell Biol 4:63–92.

Rockel AS, Jones EG (1973) The neuronal organization of the inferior colliculus of the adult cat. I. The central nucleus. J Comp Neurol 147:11–60.

Ross LS, Pollak GD, Zook JM (1988) Origin of ascending projections to an isofrequency region of the mustache bat's inferior colliculus. J Comp Neurol 270:488–505.

Ross LS, Pollak GD (1989) Differential projections to aural regions in the 60 kHz isofrequency contour of the mustache bat's inferior colliculus. J Neurosci 9:2819–2834.

Roth GL, Aitkin LM, Andersen RA, Merzenich MM (1978) Some features of the spatial organization of the central nucleus of the inferior colliculus of the cat. J Comp Neurol 182:661–680.

Schuller G (1979a) Coding of small sinusoidal frequency and amplitude modulations in the inferior colliculus of the CF-FM bat, Rhinolophus ferrumequinum. Exp Brain Res 34:117–132.

Schuller G (1979b) Vocalization influences auditory processing in collicular neurons of the CF-FM bat, Rhinolophus ferrumequinum. J Comp Physiol 132:39–46.

Schuller G (1984) Natural ultrasonic echoes form wing beating insects are coded by collicular neurons in the CF-FM bat, Rhinolophus ferrumequinum. J Comp Physiol 155:121–128.

Schuller G, Beuter K, Schnitzler H-U (1974) Response to frequency shifted artificial echoes in the bat, Rhinolophus ferrumequinum. J Comp Physiol 89:275–286.

Schuller G, Pollak GD (1979) Disproportionate frequency representation in the inferior colliculus of horseshoe bats: Evidence for an "acoustic fovea." J Comp Physiol 132:47–54.

Schnitzler HU (1967) Discrimination of thin wires by flying horseshoe bats (Rhinolophidae). In: Busnel, R-G (ed) Animal Sonar Systems, Vol. I. Jouy-en-Josas 78, France: Laboratoire de Physiologie Acoustique, p. 69.

Schnitzler H-U (1970) Comparison of echolocation behavior in Rhinolophus ferrumequinum and Chilonycteris rubiginosa. Bijdr Dierkd 40:77–80.

Schnitzler H-U, Flieger E (1983) Detection of oscillating target movements by echolocation in the greater horseshoe bat. J Comp Physiol 153:385–391.

Schnitzler H-U, Ostwald J (1983) Adaptations for the detection of fluttering insects by echolocation in horseshoe bats. In: Ewert J-P, Capranica RR, Ingle DJ (eds) Advances in Vertebrate Neuroethology. New York: Plenum Press, p. 801.

Schnitzler H-U, Menne D, Kober R, Heblich K (1983) The acoustical image of fluttering insects in echolocating bats. In: Huber F, Markl H (eds) Neuroethology and Behavioral Physiology: Roots and Growing Pains. Berlin Heidelberg New York Tokyo: Springer-Verlag, p. 235.

Semple MN, Aitkin LM (1979) Representation of sound frequency and laterality by units in the central nucleus of the cat's inferior colliculus. J Neurophysiol 42:1626–1639.

Serviere J, Webster WR, Calford MB (1984) Iso-frequency labelling revealed by a combined [14C]-2-deoxyglucose, electrophysiological and horseradish peroxidase study of the inferior colliculus of the cat. J Comp Neurol 228:463–477.

Simmons JA (1971) The sonar receiver of the bat. Ann NY Acad Sci 188:161–184.

Simmons JA (1973) The resolution of target range by echolocating bats. J Acoust Soc Amer 54:157–173.

Simmons JA (1974) Response of the Doppler echolocation system in the bat, Rhinolophus ferrumequinum. J Acoust Soc Amer 56:672–682.

Simmons JA, Howell DJ, Suga N (1975) Information content of bat sonar echoes. Amer Sci 63:204–215.

Suga N (1964a) Recovery cycles and responses to frequency modulated tone pulses in auditory neurons of echolocating bats. J Physiol 175:50–80.

Suga N (1964b) Single unit activity in the cochlear nucleus and inferior colliculus of echolocating bats. J Physiol 172:449–474.

Suga N (1978) Specialization of the auditory system for reception and processing of species-specific sounds. Fed Proc 37:2342–2354.

Suga N (1984) The extent to which biosonar information is represented in the bat auditory cortex. In: Edelman GM, Gall WE, Cowan WM (eds) Dynamic Aspects of Neocortical Function. New York: John Wiley & Sons, pp. 315.

Suga N, Simmons JA, Jen PHS (1975) Peripheral specializations for fine frequency analysis of Doppler-shifted echoes in the CF-FM bat, Pteronotus parnellii. J Exp Biol 63:161–192.

Suga N, Neuweiler G, Moller J (1976) Peripheral auditory tuning for fine frequency analysis by the CF-FM bat, Rhinolophus ferrumequinum. IV. Properties of peripheral auditory neurons. J Comp Physiol 106:111–125.

Suga N, Jen PHS (1977) Further studies on the peripheral auditory system of "CF-FM" bats specialized for the fine frequency analysis of Doppler-shifted echoes. J Exp Biol 69:207–232.

Supa M, Cotzin M, Dallenbach KM (1944) "Facial vision." The perception of obstacles by the blind. Am J Psychol 57:133–183.

Sur M, Merzenich MM, Kass JH (1980) Magnification, receptive field area and "hypercolumn" size in areas 3b and 1 of somatosensory cortex in owl monkeys. J Neurophysiol 44:295–311.

Trappe M, Schnitzler H-U (1982) Doppler-shift compensation in insect-catching horseshoe bats. Naturwissenschaften 69:193–194.

Vater M (1987) Narrow-band frequency analysis in bats. In: Fenton MB, Racey P, Rayner JMV (eds) Recent Advances in the Study of Bats. Cambridge: Cambridge University Press, p. 200.

von der Emme G (1988) Greater horseshoe bats learn to discriminate simulated echoes of insects fluttering with different wingbeat rates. In: Nachtigall PE, Moore PWB (eds) Animal Sonar: Processes and Performance. New York: Plenum Press, pp. 495–500.

Webster DB (1961) The ear apparatus of the kangaroo rat, Dipodomys. Amer J Anat 108:123–148.

Wenstrup JJ, Ross LS, Pollak GD (1985) A functional organization of binaural responses in the inferior colliculus. Hearing Res 17:191–195.

Wenstrup JJ, Fuzessery ZM, Pollak GD (1986a) Binaural response organization within a frequency-band representation of the inferior colliculus: Implications for sound localization. J Neurosci 6:692–973.

Wenstrup JJ, Ross LS, Pollak GD (1986b) Organization of IID sensitivity in isofrequency representations of the mustache bat's inferior colliculus. IUPS Symposium on Hearing, University of California, San Francisco, CA, Abstract 415.

Wenstrup JJ, Fuzessery ZM, Pollak GD (1988a) Binaural neurons in the mustache bat's inferior colliculus: I. Responses of 60 kHz EI units to dichotic sound stimulation. J Neurophysiol 60:1369–1383.

Wenstrup JJ, Fuzessery ZM, Pollak GD (1988b) Binaural neurons in the mustache bat's inferior colliculus: II. Determinants of spatial responses among 60 kHz EI units. J Neurophysiol 60:1384–1404.

Zakon H (1986) The electroreceptive periphery. In: Bullock TH, Heiligenberg W (eds) Electroreception. New York: John Wiley & Sons, pp. 103–156.

Zook JM, Casseday JH (1982) Origin of ascending projections to inferior colliculus in the mustache bat, Pteronotus parnellii. J Comp Neurol 207:14–28.

Zook JM, Winer JA, Pollak GD, Bodenhamer RD (1985) Topology of the central nucleus of the mustache bat's inferior colliculus: Correlation of single unit properties and neuronal architecture. J Comp Neurol 231:530–546.

Zook JM, Casseday JH (1985) Projections from the cochlear nuclei in the mustache bat, Pteronotus parnellii. J Comp Neurol 237:307–324.

Zook JM, Casseday JH (1987) Convergence of ascending pathways at the inferior colliculus of the mustache bat, Pteronotus parnellii. J Comp Neurol 261:347–361.

Zook JM, Leake PA (1989) Correlation of cochlear morphology specializations with frequency representation in the cochlar nucleus and superior olive of the mustache bat, Pteronotus parnellii. J Comp Neurol 290:243–261.

Abstract J

The Story of the Evolution of Hearing: Identifying the Sources of Sound

William A. Yost

Humans have the remarkable ability to determine the sources of sounds, and in general, sound source identification appears to be the most general function of hearing and the auditory system. A century of human psychoacoustical research has revealed a wealth of information about processing the basic properties of sounds. However, far less attention has been paid to how the auditory system uses this information to determine or identify sound sources. Psychoacoustical data, models, and theories suggest that the human's ability to resolve spectral components of complex sounds and to localize sounds could be responsible for sound source identification. But sound sources provide a number of other characteristics that could be used by an auditory system to aid it in sound source identification. A complex auditory system should be able to use these characteristics to help form auditory images of sound sources, and these auditory images would provide the basis for sound source identification. One such characteristic is the slow temporal modulation of sound generated by almost all sound sources. Consider for instance, the slow frequency vibrato and amplitude jitter present in all voiced utterances. We argue that the human auditory system uses such slow temporal modulations to identify sound sources. We show that imparting a slow rate (20Hz) of amplitude modulation to a subset of four spectral components from a larger set of 32 components, allows the subset to be heard as distinct from the rest of the 32-component stimulus. This demonstration suggests that amplitude modulation can be used as a basis for forming auditory images. Processing these modulations has certain consequences for auditory perception of complex sounds in that the auditory system appears to group together tonal components that share a common pattern of temporal modulation even when the tonal components are widely spaced in frequency. Our data indicate that listeners have great difficulty in processing the temporal modulation information of any one spectral component when all the spectral components of a complex sound are amplitude-modulated at the same rate. We call this form of interference, Modulation Detection Interference (MDI, see Yost and Sheft, *J. Acoust. Soc. Amer.* 85, 1989). We show that MDI exists for two tones modulated at the same rate, even when the tones are separated by as many as five octaves. It is argued that MDI occurs because the two tones are modulated at the same rate, and, therefore, are fused into one auditory image. However, when each tone of the two-tone complex is modulated at a different rate, there is a significant reduction in MDI. This reduction in MDI may be because the two tones are modulated at different rates, and, therefore, are no longer fused into one auditory image. These data seem consistent with the assumption that slow temporal modulation may be one of the many cues used by the auditory system to form auditory images. The next step in understanding auditory image formation is to understand the mechanisms which use cues such as slow temporal modulation to fuse spectral components into a single perceptual unit. We argue that these mechanisms are unlikely to be found at the auditory periphery where the neural code for sound is established, but rather they will be located in more central auditory pathways. [Work supported by grants from the National Institute on Deafness and Other Communication Disorders and the Air Force Office of Scientific Research.]

Abstract K
Evolution of Ultrasonic and Supersonic Hearing in Man

Martin L. Lenhardt and Alex M. Clarke

The average mammalian upper frequency limit of hearing is approximately 55 kHz. Within a species, the upper limit is inversely related to the distance between the ears, as Masterton et al. (JASA 45:966, 1969) have shown that all mammalian species tested to date, with the singular exception of man, fit this relation in a well-correlated fashion. For the relation to also hold in man, a 30 to 40 kHz upper frequency limit should exist, as opposed to the limit normally measured at approximately an octave lower. Mammals with very high-frequency hearing appear to have modifications of the basilar membrane, which likely account for the observed high-frequency hearing ability. In contrast, man has a relatively broad, radially undifferentiated basilar membrane, consistent with the measured lower-frequency sensitivity. In both our laboratory and others (Corso, JASA 35:1738, 1963), the upper limit of perception of auditory-like stimuli is found to be in the range of 60 kHz to 90 kHz, provided that it is delivered via bone conduction. Use of these data would bring man more into line with those of other mammalian species. Since man does not have a high-frequency specialized cochlea, we hypothesize that the otolith organs as well respond to bone conduction stimuli above 20 kHz. Data from presbychotic subjects reveal no loss of sensitivity above 20 kHz accompanying the normal loss in the audiometric frequencies. This finding argues against the cochlea as the site of supersonic transduction. Individuals with severe hearing loss perceive supersonic stimulation only if their vestibular system is intact. This, in context with the animal results of Casals, et al. (Science 210:83, 1980) suggest the saccule is an organ of supersonic hearing in man. The question remains, however, as to why man would evolve supersonic hearing in man. The question remains, however, as to why man would evolve supersonic hearing abilities. Since the only auditory-like input to the saccule is via bone conduction, man cannot use this organ for perception of spatial orientation or localization. Berlin et al. (Trans Am. Acad. Ophth. Otolar: 86, 111, 1978) have noted that the deaf who exhibit good speech production have normal or near normal ultrasonic hearing abilities. Our vibrational analysis of vowel vocalization reveals the presence of ultrahigh-frequency energy up to and including some of the supersonic range, when measured via our bone conduction detection techniques. Thus it appears that the saccule can play a speech monitoring role in a mammal having an extensive vocal repertoire as man.

Abstract L

Broad Frequency Selectivity at High Sound Pressure Levels is Important for Speech Coding in the Cochlear Nucleus

S. Greenberg and W.S. Rhode

Speech is remarkably resistant to acoustic interference. Background noise, reverberation and "cocktail party" conditions rarely disrupt the speech decoding process despite the dramatic spectral alterations imposed. We propose that this resistance to masking reflects two principal properties of acoustic transduction in the auditory periphery and cochlear nucleus.

The first involves the broad frequency selectivity observed in the auditory nerve and cochlear nucleus for sound pressure levels typical for conversational speech (70–80 dB). At such intensities, the information-laden, low-frequency (<2 kHz) components of the speech signal dominate the response of many auditory nerve fibers and cochlear nucleus cells whose most sensitive (characteristic) frequencies lie in the mid- and high-frequency portion of the spectrum.

However, broad frequency selectivity must work in concert with neural phase locking to provide a representation of the speech spectrum distributed among parallel channels. Phase locking provides the means by which the activity of tonotopically distributed cells can be grouped as an entity and traced to its driving stimulus, a feature of particular importance for encoding spectrally complex signals.

Distributed representation of key spectral features of the speech signal, such as the first formant and fundamental frequency, maximizes the probability that such information is accurately encoded under conditions of variable and unpredictable acoustic interference by transmission over multiple frequency channels. Viewed in this light, parallel encoding of spectral information may be a primary selection factor steering the evolutionary development of the mammalian auditory system, and is a potentially significant force in shaping the spectral characteristics of vocal communication systems.

(Supported by NIH-NINCDS grants NS-26274 and NS-17590.)

Abstract M

Direction-Dependent Acoustical Transformation in the External Ear of the Cat: Effects of Pinna Movement

A.D. Musicant, J.C.K. Chan, and J.E. Hind

Free field-to-eardrum transfer functions were determined in anesthetized cats inside an anechoic chamber by recording the response to a click stimulus with a tiny probe tube microphone surgically implanted near the tympanic membrane. Complexities of these transfer functions, which appear to result primarily from the acoustical properties of the pinna, include the presence of prominent spectral notches in the 8 to 18 kHz frequency range. The frequency of these notches varies systematically with both the azimuth and elevation of the sound source. These spectral features are comparable, although at higher frequencies, to those reported by Shaw (in "Localization of Sound: Theory and Applications," ed. R. Gatehouse, 1982) for human subjects. In contrast to humans, most mammals can move their pinnae. We studied the spectral changes that occur when the pinna of the cat is rotated.

Moving the loudspeaker through the same angles as the pinna was rotated results in very similar, though not identical, spectral patterns for a sound source ipsilateral to the rotated pinna. For a contralateral source the patterns are dissimilar for sources within about 45 degrees of the horizon. Within this region we could not find a loudspeaker location that has a similar interaural spectrum after rotation of the pinna. Although we have not studied the effects of rotation both pinnae at the same time, our data suggest that even if both ears moved it would not be possible (except for high elevations) to match the interaural spectrum. Either the brain must recalculate the relationship between the binaural spectral information and the source location (which seems improbable), or else the information from the near ear (including pinna position) must play a dominant role in sound localization.

[This study was supported by NIH grants NS12732 and NS07026.]

Abstract N

Toward Understanding Mammalian Hearing Tractability: Preliminary Underwater Acoustical Perception Thresholds in the West Indian Manatee, *Trichechus manatus*

Geoffrey W. Patton and Edmund Gerstein

Sirenians (manatees and dugongs) share common ancestry with the Proboscideans (elephants). As in elephants, manatees may exhibit low-frequency hearing capabilities. Such an ability could be as extraordinary as odontocete (toothed whales, porpoises, and dolphins) ultrasonic echolocation and communication. Knowing the thresholds for these species will aid management strategies for preserving this endangered species.

A simple, two-choice paradigm determined the ability of an adult manatee to detect a range of sound frequencies projected underwater. Trained on command to discriminate between two submerged paddles, the animal pushed one if a tone was detected, and the other if no tone were detected. Electrical switches on each paddle ensured the reliable recording of paddle selections. Double-blind and randomized tone presentations served as precautions against experimental and "Clever Hans" biases.

A signal generator delivered single tones through an underwater speaker, while a hydrophone monitored background and acoustical signal strength. After establishing a reliable behavioral baseline for paddle selection at 80% accuracy, 16 test trials were run for a total of 240 discrimination trials. A nominal frequency range of hearing, 0.15 to 15 kHz, was demonstrated ($p < .05$). Results show significant hearing beyond the frequency limits previously determined through behavioral and physiological evoked brain potential studies.

The discrimination and task-learning ability of this subject has encouraged further study to quantify the quiet thresholds, masked thresholds, and directional sensitivity of this species by testing this and other wild and captive animals.

Abstract O

The Acoustic Spatial Environment of the Mustache Bat within the Context of Echolocation

Z.M. Fuzessery

Echolocating animals are, to varying degrees, able to create and control their own acoustic environment. It is reasonable to assume that the emitter and receiver components of echolocation have evolved in harmony to optimize reception and processing of information. This study focuses on the acoustic spatial environment of the mustache bat when both the radiation pattern of the echolocation pulse and the directional sensitivity of the ears are considered. It describes the directionality of the echolocation system. The radiation patterns of the pulse are derived from a study by Hartley and Suthers (J Acoust Soc Amer 1990 87:2756–2772), and the ear directionality and neuronal spatial tuning are derived from our previous studies (Fuzessery and Pollack 1985 J Neurophysiol 54:757–781; Wenstrup et al. 1988 J Neurophysiol 60:1384–1404; Fuzessery et al. 1990 J Neurophysiol 63:1128–1147).

The directionality of the echolocation system at 60 kHz, a dominant frequency of the emitted pulse, was modeled by adding, at each point in space, the sound attenuation due to the echolocation pulse radiation pattern, and the attenuation due to external ear directionality. These two factors determine the spatial distribution of acoustic energy during echolocation. The emission pattern has a single sound intensity peak at the center of the sound field, while the combined directionality of the two ears shows two peaks of maximum sensitivity. The combined values reveal a plateau of maximum energy extending 40° along the azimuth to either side of the vertical midline. Lateral to each of these peaks, the available energy decreases rapidly by 40 dB. The result is a relatively constant intensity level at the center of the sound field where the bat focuses its attention. One interpretation, in terms of spatial processing, is that this provides a mechanism for keeping intensity level constant along the horizontal axis, allowing the bat to resolve interaural intensity differences without compensation for intensity level. This interpretation is attractive because the mustache bat is known to "intensity compensate," changing emission intensity to maintain a constant intensity (Kobler et al. 1985 Hear Res 20:99–108). Considered in concert, these results suggest a two-stage process in which the bat clamps echo intensity at an absolute level, and stabilizes this level across the center of the sound field. The interaction of the emission radiation pattern and ear directionality may be one bat's solution to the ubiquitous problem of resolving stimulus quantity and quality.

Section VII
Epilogue

37
Epilogue to the Conference on The Evolutionary Biology of Hearing

Douglas B. Webster

1. Definition of Hearing

Dr. Douglas Webster opened the final half-day discussion by suggesting that we focus on the Evolutionary Biology of *Hearing*, rather than of the ear. To do this it was first necessary to reach consensus on the definition of "hearing." During the course of the conference it had become obvious that "hearing" means different things to different investigators in different contexts. For instance, is hearing the detection of far-field (pressure) waves only, or does it also include detection of near-field particle and bulk movement? Does it include vibration detection? Where does detection of "vibration" end and "hearing" begin? Does hearing require an ear, or simply a vibration-sensitive organ?

It also became clear that these questions have importance beyond mere semantics if we are to understand and meaningfully discuss the evolutionary phenomena having to do with hearing.

The following definitions of hearing and sound were proposed and largely written by Dr. Ted Bullock, with input from other participants in the conference:

"*Hearing* is the sensory modality concerned with the perception of sound acting upon the ears of vertebrates or the special acoustic sense organs of invertebrates, at low and moderate intensities, as distinct from the perceptions of touch, vibration, and pain.

"*Sound* is the physical stimulus causing the sensation of hearing. Historically, the word 'sound' gets its definition by extension from human hearing. It is that part of the mechanical disturbance in the medium in which we live that includes the pressure wave arriving from, commonly, distant sources. By extension, it is the adequate stimulus for the auditory portions of the ear of verte-brates, and of other sense organs specialized for acoustic reception in other animals, such as insects and animals living in water. Therefore it is not confined to pressure waves but includes the small particle displacement component of mechanical disturbance from sources that are commonly distant. Sound includes frequencies from ca. 0.1 Hz (called infrasonic), to ca. 200 kHz (called ultrasonic).

"*Hearing*, as usually understood, is distinguished from the modalities that cause sensations of vibration, pain, proprioception, and touch, each of which has its own receptors and pathways. In some animals substrate vibration can normally operate to activate acoustic receptors. Hearing is also distinguished from rheoreception or the reception of fluid currents, turbulence, and bulk water movement such as is mediated in aquatic vertebrates by the mechanoreceptive lateral line sense organs, or equivalent receptors in other species.

"This definition leaves a gray zone where sources are not distant and air or water disturbances generate both particle and bulk movement as well as pressure waves. Such cases, insofar as they are not the artificial result of nearby stimulation, which would be resolvable by increasing the distance, should be decided by the degree of similarity to either ear or lateral line reception, for example by estimating the relative importance of particle displacement vs bulk motion. Intermediate cases are to be expected even after attempting this criterion.

"NOTE: This definition is not anthropocentric but reminds us that the dictionary meanings of the words come from human experience and have been then extended by usage to other species. We are not at liberty, a priori, to call the lateral line modality or the distinct sensation of vibration 'hearing,' any more than we can call the reception of infrared 'vision,' without making a good case for it. It reminds us that the term 'vibration' will generally be understood to mean distinct sensations usually caused by motions in a solid medium. (Bone conduction of such stimuli can, of course, cause sensations of hearing.) It reminds us that, as in other sensory modalities, other forms of stimuli than the usually adequate stimulus can in special cases activate acoustic receptors. It reminds us that the vertebrate ear normally houses several modalities, including gravity, slow acceleration and vibration, in addition to acoustic reception, and that each of these occurs also in other organs, especially in invertebrates. It reminds us that other organs mediate closely related, even overlapping modalities, such as rheoreception (bulk water movement reception) and electroreception. It reminds us that sound cannot be defined as pressure waves only. It underlines that a principal problem boundary is that between acoustic and lateral lines or similar receptors of bulk movement. It avoids the words 'near-field' and 'far-field' but invokes the distinction between sources of disturbance close by or far away in terms of significant or insignificant bulk movement of the medium, and recognizes intermediate cases."

Such long and cumbersome definitions seem unavoidable if one is to both be precise and inclusive. Consensus came after much discussion, as did awareness of the need for a more succinct definition of hearing. Dr. Arthur Popper reminded us that in 1974 Dr. Glen Wever, to whom this conference is dedicated, published an elegantly succinct definition of hearing in his seminal paper on the evolution of vertebrate hearing:

> "Hearing is the response of an animal to sound vibrations by means of a special organ for which such vibrations are the most effective stimulus."

Wever stressed that the term "most effective" is a critical part of the definition, since almost any sense organ will respond to a sound if it is sufficiently intense.

Wever's definition has much appeal, so long as we remember "sound vibrations" is an imprecise term meaning different things to different people.

2. How Often and Where Has Hearing Evolved?

2.1 Invertebrates

In response to a question as to which groups of invertebrates can hear, Dr. Ulli Budelman said that no specialized sense organs for sound reception are known among nonarthropod invertebrates, and therefore, apparently, they do not hear, as defined by either Bullock or Wever. Some noninsect arthropods, however, such as spiders, have hearing organs. Among insects, hearing organs and the ability to hear have evolved at least ten times, according to Dr. Michelsen. Truly, insects are the great hearers of the invertebrate world.

2.2 Fishes

None of today's agnathans (jawless fishes) have a sense organ whose most effective stimulus has been shown to be sound. However, Dr. Sheryl Coombs pointed out that both lampreys and hagfish have mass-loaded hair cell patches in their inner ears and therefore might detect sound vibrations in the same manner as many teleosts. Both Dr. Coombs and Dr. Catherine McCormick agreed that living agnatha probably cannot detect pressure waves but may detect bulk-movement vibrations.

In the chondrichthyeans (sharks and rays), Dr. Jeffrey Corwin has identified the macula neglecta as a sense organ whose most effective stimulus is probably near-field sound, but not pressure waves. If it is true that the macula neglecta is more sensitive to sound than to any other stimulus, it would by definition be a hearing organ. Dr. Coombs noted that it is possible that other sensory epithelia in the inner ear of chondrichthyeans may be sensitive to bulk-movement vibrations.

Among the actinopterygians, Dr. Popper and Dr. Coombs agreed, different organs of hearing (i.e., utricle, saccule, lagena) have evolved adaptations for pressure wave detection in different groups of teleost fishes.

In sarcopterygians (excluding tetrapods, as is commonly done), we know of no sense organs for which sound is the most effective stimulus. However, Dr. Bernd Fritzsch pointed out that in some fossil sarcopterygians, as in modern bottom-dwelling catfish, the swim bladder is totally covered with bone, which, as in the catfish, may have been an adaptation for sound reception. Dr. Michelsen thought it was more likely an adaptation for buoyancy.

2.3 Transition from Water to Land

Dr. Ted Lewis noted that in order to understand the evolution of airborne hearing sensitivity among different groups of tetrapods one must understand the physics of impedance matching.

"It is often said that the middle ear 'matches the acoustical impedance of air to that of water.' This definitely is not correct. The impedance presented to the stapes at the oval window is not the acoustic impedance of water. The acoustic impedance of air or liquid for compressional waves is determined by the density and the adiabatic bulk modulus of the medium (being proportional to the square root of the product of those two parameters). According to all the best cochlear theories, it is the compliance of the basilar membrane (rather than that of water) that supports the propagated wave in the cochlea. Therefore the impedance presented to the stapes at the oval window is independent of the adiabatic bulk modulus of water, and depends instead on the properties of the basilar membrane. In other words, that impedance is not the acoustic impedance of water. In nonmammalian terrestrial vertebrates, one expects similar physics. At the highest audible frequencies for such animals (approximately 10 kHz), the wavelength of a compressional wave in water is approximately 150 cm, much longer than any inner ear structure. Therefore, the impedances presented at the oval windows of those inner ears will be determined by the stiffnesses of the membranes bounding the water rather than by the stiffness (adiabatic bulk modulus) of the water itself. The water will move, effectively, as an incompressible slug."

Did the sarcopterygians that gave rise to terrestrial vertebrates during the transition from water to land have a tympanic ear that was impedance-matched to air borne sound? During this conference Dr. Fritzsch (Chapter 18) argued that there may have been a membrane sealing the spiracular opening in sarcopterygian fishes that could have served as a receptor for airborne sound. Dr. Jennifer Clack (Chapter 20) from her studies of *Acanthostega*, argued that the stem tetrapods had a spiracle but not a tympanic ear. Dr. Clack suggested that the earliest vertebrate ear receptive to airborne sound was in the Temnospondyls. Dr. John Bolt pointed out that the presence of large otic or temporal notches in many of the early tetrapods does not necessarily prove the presence of functional tympanic ears.

Since synapsids separated from other reptiles so early in tetrapod evolution, it is appropriate to ask whether their common ancestor had a tympanic ear sensitive to airborne sounds. Dr. James Hopson pointed out that a paleontologist may be able to say whether an animal had a tympanic ear, but cannot say whether it heard. He added that the presence of a tympanic ear shows that there was a mechanism for impedance matching, assuming that there was also a fenestra ovalis. In many fossil forms, however, it is not known whether there was a fenestra ovalis. Moreover, many early reptiles and amphibians had such massive stapes that, even with a tympanic ear and fenestra ovalis, hearing must have been extremely limited.

He suggested that even a conservative interpretation of the data indicates several evolutions of sensitive airborne hearing during tetrapod evolution: first in the anuran amphibians, again in two or three groups of living reptiles, and again in the synapsid line leading to the evolution of mammals.

Dr. Webster pointed out that in the synapsids the stapes was particularly massive among the pelycosaurs, which undoubtedly meant that they had poor hearing for airborne sounds. Even as the synapsids became smaller and more mammal-like, hearing probably remained poor since the quadrate and articular were part of the jaw mechanism as well as being attached to the stapes. This suggests the interesting possibility that sensitive hearing of airborne sounds arose "suddenly," when the posterior jaw elements were freed to form a stiff, microtype middle ear. Because of its small size and its stiffness, such an ear would have given the capacity to hear frequencies above 12,000 Hz, provided the inner ear could cope with these higher frequencies. Among the vertebrates, this hearing range is unique to mammals.

3. Is the Evolution of Hearing More Complex Than the Evolution of Other Sensory Systems?

Dr. Bullock suggested that since the evolution of vertebrate vision occurred so early, and only once, it is better understood than the evolution of hearing. Drs. Hopson and Gans, however, pointed out that the fossil record tells us very little about vision or other senses, while the bony portions of the ear cause its evolutionary development to be better preserved. Dr. Webster suggested that more people study vision than hearing in living forms because of the challenge: the morphology is more complex, and the problems of stimulus control are greater.

Dr. Fritzsch suggested that because the evolution of vision occurred only once in vertebrates, it is inherently less interesting than the multiple appearances of vertebrate hearing. Dr. Webster pointed out that the relatively late evolution of hearing in different groups of vertebrates has produced a better fossil record and more diversity among living forms, which adds to our understanding of the mechanisms involved.

4. Central Auditory System of Vertebrates

Dr. Catherine Carr pointed out that a similar central auditory pattern occurs in all hearing vertebrates, as Dr. McCormick (Chapter 17) has described in her paper. Dr. McCormick added that there is greater variation in the peripheral sense organs, including the otolithic organs, than in the central auditory pathway. Dr. R. Bruce Masterton stated that in general there probably is more intense selective pressure on organs and cells directly facing the environment than on those facing only neurotransmitters—if only because there is more variation in their microenvironment. However, as he has documented (Chapter 32), dramatic changes have taken place in the evolution of forebrain structures for hearing.

5. Hair Cells

Tom Lewis suggested that hair cells are the key to understanding the evolution of hearing, because in vertebrates they have limited variability and are always the transducer cells. Dr. Popper pointed out that there are two basic types of hair cells for hearing in fishes. Drs. Manley and Köppl added that in birds there are three types: short, tall, and intermediate. Dr. Malcolm Miller (Chapter 23) distinguishes between two different types of lizard hair cells, unidirectional and bidirectional. In mammals there are distinct differences between inner and outer hair cells, as well as between the two distinct types of ganglion cells that innervate them, the type I and type II spiral ganglion cells. Vertebrate auditory hair cells therefore exhibit considerable diversity. Selective pressures have resulted

in different hair cell types among disparate groups of vertebrates.

6. Cochlear Emissions

Considerable attention has recently been given to cochlear emissions in mammals. It was asked whether they are unique to mammals or are found in other vertebrates with sensitive hearing. Dr. Narins pointed out that frogs produce multiple spontaneous emissions from the basilar papilla, which is an auditory sense organ without a basilar membrane. One can also record evoked emissions from frogs.

Dr. Manley related that in the bobtail lizard, which does not have a tuned basilar membrane, there are also many spontaneous emissions. Manley's group has not found spontaneous cochlear emissions in birds. Dr. Köppl pointed out that in lizards there are no efferents on the hair cells in the frequency range of the spontaneous emissions. Dr. Ted Lewis said this is also true in frogs, which amply demonstrates that cochlear emissions do not require efferent innervation to the hair cells. It is not known, however, if the emissions are altered or eliminated when these hair cells are damaged.

Dr. Ted Lewis pointed out that, thermodynamically, all that the presence of emissions need imply is the existence of reverse transduction—from the electrical side of the hair cell to the mechanical side. The active (energy amplifying) processes required for spontaneous emissions already are present in the forward transduction.

7. Selective Pressure for Hearing

The group next addressed the question of what selective pressures have influenced the evolution of hearing, hearing sensitivity, and the evolution of high-frequency hearing in mammals.

Dr. Rickye Heffner gave a solid argument that high-frequency hearing would be selected for in mammals because it enhances auditory localization, particularly for small mammals like the earliest mammals to evolve. Dr. Stebbins pointed out that in addition to being small, the earliest mammals were nocturnal; limited visual ability in a nocturnal environment would increase the selective pressure for hearing. Dr. Edgar Allin pointed out that the dual needs to find prey and to avoid being preyed upon would cause heavy selective pressure for the development of auditory localization. Dr. Yost added that identifying a sound is sometimes as important as locating it: was it something to be ignored, approached, or avoided? Dr. Masterton suggested that this is probably the reason for the parallel pathways in the auditory brainstem. Octopus cells detect sound—any sound. The superior olivary complex via the AVCN locates the source. Systems bypassing SOC are likely "identifying" it by spectral templates. Dr. Gans, citing the single ear of the preying mantis, reiterated that it is not necessary to definitely locate the source of a sound in order to avoid predation. Another example is seen in kangaroo rats and other desert rodents, which make a large but erratic leap just after hearing the final cue for a predator's strike: again, exact auditory localization is not required, just rapid movement out of the way. Yet another example is rabbits, that simply freeze.

The body wall ear of amphibians is another indication of the strength of selective pressure for hearing, said Dr. Narins. Dr. Allin, citing the work of Peter Hartline, added that the same thing is found in snakes. Dr. Michelsen, again recalling Dr. Gans' paper (Chapter 1), commented that hearing need not be perfect, only adequate, to be selected for. With body wall hearing there is certainly distortion due to breathing; this is also true in the hearing organs of many insects.

In general, is the selective pressure greater for hearing than for other senses? Dr. Webster recalled Helen Keller's remark that deafness was more devastating than blindness because it isolated her from people and communication, while blindness isolated her from things.

Recalling the classical Reichert homologies, Dr. Webster suggested that the evolution of the first two branchial arches—from respiratory to feeding to hearing organs—provides ample evidence that the selective pressure for hearing is enormous.

Dr. Bullock asked if there are subsystems of hearing, as there are of vision, and what the adaptive value of these might be; he pointed out that dolphins have distinct systems for echolocation and social communication, as shown by both their anatomy and physiology. Dr. Webster suggested that the vestibulo-ocular reflex, the stapedius reflex, and the relationship of hearing and vision—as Drs. R. and H. Heffner pointed out in their paper (Chapter 34)—all support the idea of parallel systems in hearing.

Finally, Drs. Ted Lewis, Manley, Michelsen, and others discussed how strong the selective pressure would have been for the evolution not just of hearing but of the ability to separate signal from noise. Dr. Manley, asking why vertebrates have the ears they do, pointed out that the auditory inner ears of lizards are extremely diverse in structure, although neither functional differences nor their adaptive values have been demonstrated. What then is the adaptive value of this diversity? If selective pressure is simply acting to preserve a minimal amount of hearing, then much of the diversity among lizards could be due to genetic drift rather than to specific selective pressures in unique habitats.

It is difficult to ask a single question about the very diverse specializations of the auditory system that would apply to all animals. However, as Dr. Masterton noted, specializations can only be recognized if one first defines commonalities of structure, function, etc., which lead to an understanding of common ancestry. Once a specialized character is identified, the question to be asked is: what selective pressure(s) favored its development. Selective pressures vary widely in different animals because they result from the interaction of animal, environment, and behavior.

Dr. Stebbins suggested that selective pressures lead to increasingly finer adaptations which diversify from general to special, and include localization, nocturnal habits, social communication, and extraction of signal from noise. Dr. Yost suggested that, if a major function of hearing is to determine sound *sources*, extracting signals (wanted sources) from noise (unwanted sources) does not appear to be an auditory function, but rather one of selective attention. Rather, the auditory system, not knowing what is wanted or unwanted, would evolve to maximize the organism's ability to preserve as much information about the sound source in its environment as possible. He felt this is an important point, since he does not believe auditory systems "evolved" to extract "signals" from noise.

Dr. Michelsen encouraged the idea of looking at the individual animal's total behavior when trying to understand the evolution of hearing in the organism. He also commented that many of our behavioral and physiological tests are overly simplistic and do not account for what is happening in the real world.

8. Topics Not Adequately Covered in the Conference

Although great depth and breadth of evolutionary aspects of hearing were presented throughout the conference, some major aspects were not adequately, or only cursorily, covered.

8.1 Hearing Ability and Mechanisms in Nonteleost Fishes

Earlier in this chapter the hearing ability of nonteleost fishes is briefly discussed, but no chapter deals with it in detail. This fascinating topic was reviewed from an evolutionary perspective by van Bergeijk (1967). A more thorough and complete analysis of the diversity of fish hearing is found in Tavolga, Popper, and Fay (1981).

8.2 Evolutionary Biology of Middle Ear in Extant Mammals

The tremendous diversity of middle ear structures in living mammals and their adaptive value is not covered in this volume, with the exception of the extreme modifications found in desert rodents (Chapter 30). However, this topic is exhaustively covered by Fleischer (1973, 1978). In these works, he describes the microtype versus the freely mobile type in mammals, as well as transitional types. His analysis is based on a large range of extant mammals from generalized to specialized, and covers both functional and adaptive aspects of ossicular organization. He describes the skeletal components of the mammalian middle ear cavity and its diversity—from the open cavity of most marsupials and insectivores to the enclosed bony cavities of most living mammals.

8.3 Evolutionary Biology of Cochlea in Extant Mammals

Three chapters (Webster and Plassmann, Chapter 30; Ketten, Chapter 35; Pollak, Chapter 36)

include cochlear structure in very specialized mammals. One (Plassmann and Brändle, Chapter 31) describes a mathematical model of the mammalian cochlea. However, the organization of the cochlea in generalized mammals is not included in this volume. Although the number of turns of the cochlea varies greatly among mammals from less than one to over four (West 1985), the cellular organization of the cochlea varies little among living mammals. A single row of inner hair cells appears universal, with the exception of monotremes which may have two rows (Smith and Takasaka 1971). There is also a general pattern of three rows of outer hair calls. However, in many mammals there are regions with four, or even five, rows of outer hair cells, including humans (Smith and Takasaka 1971).

8.4 Evolutionary Biology of Auditory Brainstem in Extant Mammals

The only specific mammalian central auditory chapters in this volume deal with the geniculo-cortical portion of the system (Chapter 32, Frost and Masterton; Chapter 33, Merzenich and Schreiner). In Chapter 25, on the central auditory system in reptiles and birds, Carr includes the basic organization of the central auditory system of mammals, but only in reference to that found in birds and reptiles. This general organization, from eighth nerve through inferior colliculus, is found in all living mammals. However, there is considerable diversity in the relative sizes and nuclear organization at each level of the brainstem. Perhaps the most striking diversity is found in the organization of the superior olivary complex, as analyzed by Irving and Harrison (1967), and by Moore and Moore (1971).

9. Summary

It is difficult to summarize a five-day, 36-paper, 15-poster conference. The excitement for those attending and, one hopes, now for those reading this volume, lay in the impressive diversity of disciplines and perspectives that focused on the subject of hearing.

But diversity needs a rationale—a framework. That was provided by our common theme of evolution. For in hearing science, as in all other biological sciences, evolution is the intellectual framework that brings excitement, awe, joy, and understanding. In this case, it allowed an overview of hearing which would not have been possible otherwise.

With this overview both the things we know and those we do not become more compelling. As Darwin said in the final paragraph of *On The Origin of Species*, "There is grandeur in this view of life."

References

Darwin C (1859) On the Origin of Species by Means of Natural Selection. London: John Murray, Albemarle Street.

Fleischer G von (1973) Studien am Skelett des Gehörorgans der Säugetiere, einschliesslich des Menschen. Säugetierkundliche Mitteilungen 21:131–239.

Fleischer G (1978) Evolutionary principles of the mammalian middle ear. Adv Anat Embryol Cell Biol 55:3–70.

Irving R, Harrison JM (1967) The superior olivary complex and audition: A comparative study. J Comp Neurol 130:77–86.

Moore JK, Moore RY (1971) A comparative study of the superior olivary complex in the primate brain. Folia Primatol 16:35–51.

Smith CA, Takasaka T (1971) Auditory receptor organs of reptiles, birds, and mammals. In: Neff WD (ed) Contributions to Sensory Physiology, Vol. 5. New York: Academic Press, pp. 129–178.

Tavolga WN, Popper AN, Fay RR (1981) Hearing and Sound Communication in Fishes. New York: Springer-Verlag.

van Bergeijk WA (1967) The evolution of vertebrate hearing. In: Neff WD (ed) Contributions to Sensory Physiology, Vol. 2. New York: Academic Press, pp. 1–49.

Wever EG (1974) The evolution of vertebrate hearing. In: Keidel WD, Neff WD (eds) Handbook of Sensory Physiology, Vol. 1. New York: Springer-Verlag, pp. 423–454.

West CD (1985) The relationship of the spiral turns of the cochlea and the length of the basilar membrane to the range of audible frequencies in ground dwelling mammals. J Acoust Soc Am 77:1091–1101.

Author Index

Animal Index

This index includes both the genus names and common names of the animals discussed in this volume. For higher taxonomic groups, refer to the subject index and individual chapters.

Subject Index

Specific genera are not included in as main headings in the subject index but can be found in the species index.

A1, see auditory cortex
Absolute threshold, see audiogram
Acetylcholine, as efferent transmitter, 186, 193, 196
Acoustic behavior, also see communications
  amphibians, 439ff, 449–451
  fish, 302–303, 457
Acoustic cues for communication, 90
Acoustic environment, 295–299, 297
Acoustic images, 164
Acoustic isolation, bird, 557
Acoustic laws, 296–297
Acoustic senses, definition, 163ff
Acoustic startle response, see startle response
Acoustic tubercle, 50, 54
  (*see also* auditory pathways)
Acousticolateral area, 50
  (*see also* auditory pathways)
Acousticolateralis hypothesis, 50–51, 53, 54
Acridadae, see locust, grasshoppers
Actinopterygians, auditory pathways, 330
Adaptation, 16, 17
  mammalian ear, 637ff
Adaptive radiation, 5
Affine behavior, 165
Agnatha
  ear, 53, 56, 317
  ear structure and function, 323–325
  hindbrain organization, 323–325
  lagena, 357
  semicircular canals, 52
Alar plate, 369–372
Amphibian papilla, 51, 52, 163ff, 168, 337ff
  bullfrog, 173ff
  efferent effects, frogs, 188
  evolution of, 368–369

  evolutionary changes, 353
  innervation, 354
  relationship to macula neglecta, 368
  temperature effects, 178
Amphibians
  auditory pathways, 369ff, 370–372
  basilar papilla, 356–357
  critical bands, 443–444
  directional hearing, 459
  ear, evolution, 51, 53
  electroreceptive pathways, 369–370
  hearing capabilities, 440ff
  hearing, 51, 439ff
  inner ear projections, 338ff
  inner ear structure and function, 337ff, (*see also* amphibian papilla)
  lateral line system, 269–270, 369–370
  lateral line pathways, 369–370
  medulla, 338ff
  metamorphosis and CNS pathways, 369–370
  middle ear, evolution, 421ff, 430ff
  nontympanic pathways, 424–425
  nontympanic middle ear, scaling, 428–429
  phylogeny, 432
  saccule, 177ff
  tympanum, size, 426ff
  underwater hearing, 459
  vocalizations, 439–440
Ampullary
  electroreceptors, development, 29
  electroreceptive system, pathways, 369–370
  electroreceptive organs, loss in tetrapods, 369–372
Ancestral tetrapods, hearing in, 369
Anguidae, inner ear, 474–475
Angular acceleration, (*see also* semicircular canals)
  detection, 148
  receptors, crustaceans, 134–135
Anleytropsidae, inner ear, 480